Classics in Mathematics

Tosio Kato Perturbation Theory for Linear Operators

Tosio Kato

Perturbation Theory for Linear Operators

Reprint of the 1980 Edition

Springer

Tosio Kato
Department of Mathematics, University of California
Berkeley, CA 94720-3840
USA

Originally published as Vol. 132 of the
Grundlehren der mathematischen Wissenschaften

Mathematics Subject Classification (1991): 46BXX, 46CXX, 47AXX, 47BXX, 47D03, 47E05, 47F05, 81Q10, 81Q15, 81UXX

ISBN 3-540-58661-X Springer-Verlag Berlin Heidelberg New York

CIP data applied for

SPIN 11403258 41/3111 – 5 4 3 – Printed on acid-free paper

Tosio Kato

Perturbation Theory for Linear Operators

Corrected Printing of the Second Edition

Springer-Verlag
Berlin Heidelberg New York 1980

Dr. Tosio Kato
Professor of Mathematics, University of California, Berkeley

AMS Subject Classification (1970): 46Bxx, 46Cxx, 47Axx, 47Bxx, 47D05, 47Exx, 47Fxx, 81A09, 81A10, 81A45

ISBN 3-540-07558-5 2. Auflage Springer-Verlag Berlin Heidelberg New York
ISBN 0-387-07558-5 2nd edition Springer-Verlag New York Heidelberg Berlin

ISBN 3-540-03526-5 1. Auflage Berlin Heidelberg New York
ISBN 0-387-03526-5 1st edition New York Heidelberg Berlin

Library of Congress Cataloging in Publication Data. Kato, Tosio, 1917-Perturbation theory for linear operators. (Grundlehren der mathematischen Wissenschaften; 132). Bibliography: p. Includes indexes. 1. Linear operators. 2. Perturbation (Mathematics). I. Title. II. Series: Die Grundlehren der mathematischen Wissenschaften in Einzeldarstellungen; Bd. 132.
QA329.2.K37 1976. 515'.72. 76-4553.

Printed in Germany.

Typesetting, printing and binding: Brühlsche Universitätsdruckerei Gießen
Springer-Verlag Berlin Heidelberg New York
a member of BertelsmannSpringer
Science+Business Media GmbH

To the memory
of my parents

Preface to the Second Edition

In view of recent development in perturbation theory, supplementary notes and a supplementary bibliography are added at the end of the new edition. Little change has been made in the text except that the paragraphs V-§ 4.5, VI-§ 4.3, and VIII-§ 1.4 have been completely rewritten, and a number of minor errors, mostly typographical, have been corrected. The author would like to thank many readers who brought the errors to his attention.

Due to these changes, some theorems, lemmas, and formulas of the first edition are missing from the new edition while new ones are added. The new ones have numbers different from those attached to the old ones which they may have replaced.

Despite considerable expansion, the bibliography is not intended to be complete.

Berkeley, April 1976 TOSIO KATO

Preface to the First Edition

This book is intended to give a systematic presentation of perturbation theory for linear operators. It is hoped that the book will be useful to students as well as to mature scientists, both in mathematics and in the physical sciences.

Perturbation theory for linear operators is a collection of diversified results in the spectral theory of linear operators, unified more or less loosely by their common concern with the behavior of spectral properties when the operators undergo a small change. Since its creation by RAYLEIGH and SCHRÖDINGER, the theory has occupied an important place in applied mathematics; during the last decades, it has grown into a mathematical discipline with its own interest. The book aims at a mathematical treatment of the subject, with due consideration of applications.

The mathematical foundations of the theory belong to functional analysis. But since the book is partly intended for physical scientists, who might lack training in functional analysis, not even the elements of that subject are presupposed. The reader is assumed to have only a basic knowledge of linear algebra and real and complex analysis. The necessary tools in functional analysis, which are restricted to the most elementary part of the subject, are developed in the text as the need for them arises (Chapters I, III and parts of Chapters V, VI).

An introduction, containing a brief historical account of the theory, precedes the main exposition. There are ten chapters, each prefaced by a summary. Chapters are divided into sections, and sections into paragraphs. I-§ 2.3, for example, means paragraph three of section two of chapter one; it is simply written § 2.3 when referred to within the same chapter and par. 3 when referred to within the same section. Theorems, Corollaries, Lemmas, Remarks, Problems, and Examples are numbered in one list within each section: Theorem 2.1, Corollary 2.2, Lemma 2.3, etc. Lemma I-2.3 means Lemma 2.3 of chapter one, and it is referred to simply as Lemma 2.3 within the same chapter. Formulas are numbered consecutively within each section; I-(2.3) means the third formula of section two of chapter one, and it is referred to as (2.3) within the same chapter. Some of the problems are disguised theorems, and are quoted in later parts of the book.

Numbers in [ ] refer to the first part of the bibliography containing articles, and those in ⟦ ⟧ to the second part containing books and monographs.

There are a subject index, an author index and a notation index at the end of the book.

The book was begun when I was at the University of Tokyo and completed at the University of California. The preparation of the book has been facilitated by various financial aids which enabled me to pursue research at home and other institutions. For these aids I am grateful to the following agencies: the Ministry of Education, Japan; Commissariat Général du Plan, France; National Science Foundation, Atomic Energy Commission, Army Office of Ordnance Research, Office of Naval Research, and Air Force Office of Scientific Research, U.S.A.

I am indebted to a great many friends for their suggestions during the long period of writing the book. In particular I express my hearty thanks to Professors C. Clark, K. O. Friedrichs, H. Fujita, S. Goldberg, E. Hille, T. Ikebe, S. Kakutani, S. T. Kuroda, G. Neubauer, R. S. Phillips, J. and O. Todd, F. Wolf, and K. Yosida. I am especially obliged to Professor R. C. Riddell, who took the pains of going through the whole manuscript and correcting innumerable errors, mathematical as well as linguistic. I am indebted to Dr. J. Howland, Dr. F. McGrath, Dr. A. McIntosh, and Mr. S.-C. Lin for helping me in proofreading various parts of the book. I wish to thank Professor F. K. Schmidt who suggested that I write the book and whose constant encouragement brought about the completion of the book. Last but not least my gratitudes go to my wife, Mizue, for the tedious work of typewriting the manuscript.

Berkeley
August, 1966

Tosio Kato

Contents

Chapter Three

Introduction to the theory of operators in Banach spaces

Chapter Four

Stability theorems

Chapter Five
Operators in Hilbert spaces

Chapter Six
Sesquilinear forms in Hilbert spaces and associated operators

Chapter Seven

Analytic perturbation theory

Chapter Ten

Perturbation of continuous spectra and unitary equivalence

Supplementary Notes

Introduction

Throughout this book, "perturbation theory" means "perturbation theory for linear operators". There are other disciplines in mathematics called perturbation theory, such as the ones in analytical dynamics (celestial mechanics) and in nonlinear oscillation theory. All of them are based on the idea of studying a system deviating slightly from a simple ideal system for which the complete solution of the problem under consideration is known; but the problems they treat and the tools they use are quite different. The theory for linear operators as developed below is essentially independent of other perturbation theories.

Perturbation theory was created by RAYLEIGH and SCHRÖDINGER (cf. SZ.-NAGY [1]). RAYLEIGH gave a formula for computing the natural frequencies and modes of a vibrating system deviating slightly from a simpler system which admits a complete determination of the frequencies and modes (see RAYLEIGH [1], §§ 90, 91). Mathematically speaking, the method is equivalent to an approximate solution of the eigenvalue problem for a linear operator slightly different from a simpler operator for which the problem is completely solved. SCHRÖDINGER developed a similar method, with more generality and systematization, for the eigenvalue problems that appear in quantum mechanics (see SCHRÖDINGER [1], [1]).

These pioneering works were, however, quite formal and mathematically incomplete. It was tacitly assumed that the eigenvalues and eigenvectors (or eigenfunctions) admit series expansions in the small parameter that measures the deviation of the "perturbed" operator from the "unperturbed" one; no attempts were made to prove that the series converge.

It was in a series of papers by RELLICH that the question of convergence was finally settled (see RELLICH [1]–[5]; there were some attempts at the convergence proof prior to RELLICH, but they were not conclusive; see e. g. WILSON [1]). The basic results of RELLICH, which will be described in greater detail in Chapters II and VII, may be stated in the following way. Let $T(\varkappa)$ be a bounded selfadjoint operator in a Hilbert space H, depending on a real parameter $\varkappa$ as a convergent power series

$$T(\varkappa) = T + \varkappa T^{(1)} + \varkappa^2 T^{(2)} + \cdots . \tag{1}$$

Suppose that the unperturbed operator $T = T(0)$ has an isolated eigenvalue λ (isolated from the rest of the spectrum) with a finite multiplicity m. Then $T(\varkappa)$ has exactly m eigenvalues $\mu_j(\varkappa)$, $j = 1, \ldots, m$

(multiple eigenvalues counted repeatedly) in the neighborhood of λ for sufficiently small $|\varkappa|$, and these eigenvalues can be expanded into convergent series

$$(2) \qquad \mu_j(\varkappa) = \lambda + \varkappa\, \mu_j^{(1)} + \varkappa^2\, \mu_j^{(2)} + \cdots, \quad j = 1, \ldots, m .$$

The associated eigenvectors $\varphi_j(\varkappa)$ of $T(\varkappa)$ can also be chosen as convergent series

$$(3) \qquad \varphi_j(\varkappa) = \varphi_j + \varkappa\, \varphi_j^{(1)} + \varkappa^2\, \varphi_j^{(2)} + \cdots, \quad j = 1, \ldots, m ,$$

satisfying the orthonormality conditions

$$(4) \qquad (\varphi_j(\varkappa), \varphi_k(\varkappa)) = \delta_{jk} ,$$

where the φ_j form an orthonormal family of eigenvectors of T for the eigenvalue λ.

These results are exactly what were anticipated by RAYLEIGH, SCHRÖDINGER and other authors, but to prove them is by no means simple. Even in the case in which H is finite-dimensional, so that the eigenvalue problem can be dealt with algebraically, the proof is not at all trivial. In this case it is obvious that the $\mu_j(\varkappa)$ are branches of algebroidal functions of $\varkappa$, but the possibility that they have a branch point at $\varkappa = 0$ can be eliminated only by using the selfadjointness of $T(\varkappa)$. In fact, the eigenvalues of a selfadjoint operator are real, but a function which is a power series in some fractional power $\varkappa^{1/p}$ of $\varkappa$ cannot be real for both positive and negative values of $\varkappa$, unless the series reduces to a power series in $\varkappa$. To prove the existence of eigenvectors satisfying (3) and (4) is much less simple and requires a deeper analysis.

Actually RELLICH considered a more general case in which $T(\varkappa)$ is an unbounded operator; then the series (1) requires new interpretations, which form a substantial part of the theory. Many other problems related to the one above were investigated by RELLICH, such as estimates for the convergence radii, error estimates, simultaneous consideration of all the eigenvalues and eigenvectors and the ensuing question of uniformity, and non-analytic perturbations.

Rellich's fundamental work stimulated further studies on similar and related problems in the theory of linear operators. One new development was the creation by FRIEDRICHS of the perturbation theory of continuous spectra (see FRIEDRICHS [2]), which proved extremely important in scattering theory and in quantum field theory. Here an entirely new method had to be developed, for the continuous spectrum is quite different in character from the discrete spectrum. The main problem dealt with in Friedrichs's theory is the similarity of $T(\varkappa)$ to T, that is, the existence of a non-singular operator $W(\varkappa)$ such that $T(\varkappa) = W(\varkappa)\, T\, W(\varkappa)^{-1}$.

The original results of RELLICH on the perturbation of isolated eigenvalues were also generalized. It was found that the analytic theory gains in generality as well as in simplicity by allowing the parameter $\varkappa$ to be complex, a natural idea when analyticity is involved. However, one must then abandon the assumption that $T(\varkappa)$ is selfadjoint for all $\varkappa$, for an operator $T(\varkappa)$ depending on $\varkappa$ analytically cannot in general be selfadjoint for all $\varkappa$ of a complex domain, though it may be selfadjoint for all real $\varkappa$, say. This leads to the formulation of results for non-selfadjoint operators and for operators in Banach spaces, in which the use of complex function theory prevails (SZ.-NAGY [2], WOLF [1], T. KATO [6]). It turns out that the basic results of RELLICH for selfadjoint operators follow from the general theory in a simple way.

On the other hand, it was recognized (TITCHMARSH [1], [2], T. KATO [1]) that there are cases in which the formal power series like (2) or (3) diverge or even have only a finite number of significant terms, and yet approximate the quantities $\mu_j(\varkappa)$ or $\varphi_j(\varkappa)$ in the sense of asymptotic expansion. Many examples, previously intractable, were found to lie within the sway of the resulting asymptotic theory, which is closely related to the singular perturbation theory in differential equations.

Other non-analytic developments led to the perturbation theory of spectra in general and to stability theorems for various spectral properties of operators, one of the culminating results being the index theorem (see GOHBERG and KREIN [1]).

Meanwhile, perturbation theory for one-parameter semigroups of operators was developed by HILLE and PHILLIPS (see PHILLIPS [1], HILLE and PHILLIPS [1]). It is a generalization of, as well as a mathematical foundation for, the so-called time-dependent perturbation theory familiar in quantum mechanics. It is also related to time-dependent scattering theory, which is in turn closely connected with the perturbation of continuous spectra. Scattering theory is one of the subjects in perturbation theory most actively studied at present.

It is evident from this brief review that perturbation theory is not a sharply-defined discipline. While it incorporates a good deal of the spectral theory of operators, it is a body of knowledge unified more by its method of approach than by any clear-cut demarcation of its province. The underpinnings of the theory lie in linear functional analysis, and an appreciable part of the volume is devoted to supplying them. The subjects mentioned above, together with some others, occupy the remainder.

Chapter One

Operator theory in finite-dimensional vector spaces

This chapter is preliminary to the following one where perturbation theory for linear operators in a finite-dimensional space is presented. We assume that the reader is more or less familiar with elementary notions of linear algebra. In the beginning sections we collect fundamental results on linear algebra, mostly without proof, for the convenience of later reference. The notions related to normed vector spaces and analysis with vectors and operators (convergence of vectors and operators, vector-valued and operator-valued functions, etc.) are discussed in somewhat more detail. The eigenvalue problem is dealt with more completely, since this will be one of the main subjects in perturbation theory. The approach to the eigenvalue problem is analytic rather than algebraic, depending on function-theoretical treatment of the resolvents. It is believed that this is a most natural approach in view of the intended extension of the method to the infinite-dimensional case in later chapters.

Although the material as well as the method of this chapter is quite elementary, there are some results which do not seem to have been formally published elsewhere (an example is the results on pairs of projections given in §§ 4.6 and 6.8).

§ 1. Vector spaces and normed vector spaces

1. Basic notions

We collect here basic facts on finite-dimensional vector spaces, mostly without proof[1]. A *vector space* X is an aggregate of elements, called *vectors*, $u, v, \ldots$, for which *linear operations* (addition $u + v$ of two vectors u, v and multiplication αu of a vector u by a *scalar* α) are defined and obey the usual rules of such operations. Throughout the book, the scalars are assumed to be *complex numbers* unless otherwise stated (complex vector space). αu is also written as $u \alpha$ whenever convenient, and $\alpha^{-1} u$ is often written as u/α. The *zero vector* is denoted by 0 and will not be distinguished in symbol from the scalar zero.

Vectors $u_1, \ldots, u_n$ are said to be *linearly independent* if their *linear combination* $\alpha_1 u_1 + \cdots + \alpha_n u_n$ is equal to zero only if $\alpha_1 = \cdots = \alpha_n = 0$; otherwise they are *linearly dependent*. The *dimension* of X, denoted by $\dim \mathsf{X}$, is the largest number of linearly independent vectors that exist in X. If there is no such finite number, we set $\dim \mathsf{X} = \infty$. In the present chapter, all vector spaces are assumed to be finite-dimensional ($0 \leq \dim \mathsf{X} < \infty$) unless otherwise stated.

---

[1] See, e. g., GELFAND [1], HALMOS [2], HOFFMAN and KUNZE [1].

A subset M of X is a *linear manifold* or a *subspace* if M is itself a vector space under the same linear operations as in X. The dimension of M does not exceed that of X. For any subset S of X, the set M of all possible linear combinations constructed from the vectors of S is a linear manifold; M is called the linear manifold determined or spanned by S or simply the *(linear) span* of S. According to a basic theorem on vector spaces, the span M of a set of n vectors $u_1, \ldots, u_n$ is at most n-dimensional; it is exactly n-dimensional if and only if $u_1, \ldots, u_n$ are linearly independent.

There is only one 0-dimensional linear manifold of X, which consists of the vector 0 alone and which we shall denote simply by 0.

Example 1.1. The set $X = C^N$ of all ordered N-tuples $u = (\xi_j) = (\xi_1, \ldots, \xi_N)$ of complex numbers is an N-dimensional vector space (the complex euclidean space) with the usual definition of the basic operations $\alpha u + \beta v$. Such a vector u is called a *numerical vector*, and is written in the form of a *column vector* (in vertical arrangement of the components ξ_j) or a *row vector* (in horizontal arrangement) according to convenience.

Example 1.2. The set of all complex-valued continuous functions $u: x \to u(x)$ defined on an interval I of a real variable x is an infinite-dimensional vector space, with the obvious definitions of the basic operations $\alpha u + \beta v$. The same is true when, for example, the u are restricted to be functions with continuous derivatives up to a fixed order n. Also the interval I may be replaced by a region[1] in the m-dimensional real euclidean space R^m.

Example 1.3. The set of all solutions of a linear homogeneous differential equation

$$u^{(n)} + a_1(x)\, u^{(n-1)} + \cdots + a_n(x)\, u = 0$$

with continuous coefficients $a_j(x)$ is an n-dimensional vector space, for any solution of this equation is expressed as a linear combination of n *fundamental solutions*, which are linearly independent.

2. Bases

Let X be an N-dimensional vector space and let $x_1, \ldots, x_N$ be a family[2] of N linearly independent vectors. Then their span coincides with X, and each $u \in X$ can be *expanded* in the form

$$u = \sum_{j=1}^{N} \xi_j x_j \tag{1.1}$$

in a unique way. In this sense the family $\{x_j\}$ is called a *basis*[3] of X, and the scalars ξ_j are called the *coefficients* (or *coordinates*) of u with respect to this basis. The correspondence $u \to (\xi_j)$ is an *isomorphism*

[1] By a region in R^m we mean either an open set in R^m or the union of an open set and all or a part of its boundary.

[2] We use the term "family" to denote a set of elements depending on a parameter.

[3] This is an *ordered basis* (cf. HOFFMAN and KUNZE [1], p. 47).

of X onto C^N (the set of numerical vectors, see Example 1.1) in the sense that it is one to one and preserves the linear operations, that is, $u \to (\xi_j)$ and $v \to (\eta_j)$ imply $\alpha u + \beta v \to (\alpha \xi_j + \beta \eta_j)$.

As is well known, any family $x_1, \ldots, x_p$ of linearly independent vectors can be enlarged to a basis $x_1, \ldots, x_p, x_{p+1}, \ldots, x_N$ by adding suitable vectors $x_{p+1}, \ldots, x_N$.

Example 1.4. In C^N the N vectors $x_j = (\ldots, 0, 1, 0, \ldots)$ with 1 in the j-th place, $j = 1, \ldots, N$, form a basis (the *canonical basis*). The coefficients of $u = (\xi_j)$ with respect to the canonical basis are the ξ_j themselves.

Any two bases $\{x_j\}$ and $\{x'_j\}$ of X are connected by a system of linear relations

$$x_k = \sum_j \gamma_{jk} x'_j, \quad k = 1, \ldots, N. \tag{1.2}$$

The coefficients ξ_j and ξ'_j of one and the same vector u with respect to the bases $\{x_j\}$ and $\{x'_j\}$ respectively are then related to each other by

$$\xi'_j = \sum_k \gamma_{jk} \xi_k, \quad j = 1, \ldots, N. \tag{1.3}$$

The inverse transformations to (1.2) and (1.3) are

$$x'_j = \sum_k \hat{\gamma}_{kj} x_k, \quad \xi_k = \sum_j \hat{\gamma}_{kj} \xi'_j, \tag{1.4}$$

where $(\hat{\gamma}_{jk})$ is the inverse of the matrix (γ_{jk}):

$$\sum_i \hat{\gamma}_{ji} \gamma_{ik} = \sum_i \gamma_{ji} \hat{\gamma}_{ik} = \delta_{jk} = \begin{cases} 1 & (j = k) \\ 0 & (j \neq k), \end{cases} \tag{1.5}$$

$$\det(\gamma_{jk}) \det(\hat{\gamma}_{jk}) = 1. \tag{1.6}$$

Here $\det(\gamma_{jk})$ denotes the determinant of the matrix (γ_{jk}).

The systems of linear equations (1.3) and (1.4) are conveniently expressed by the matrix notation

$$(u)' = (C)(u), \quad (u) = (C)^{-1}(u)', \tag{1.7}$$

where (C) is the matrix (γ_{jk}), $(C)^{-1}$ is its inverse and (u) and $(u)'$ stand for the *column vectors* with components ξ_j and ξ'_j respectively. It should be noticed that (u) or $(u)'$ is conceptually different from the "abstract" vector u which it represents in a particular choice of the basis.

3. Linear manifolds

For any subset S and S' of X, the symbol S + S' is used to denote the *(linear) sum* of S and S', that is, the set of all vectors of the form $u + u'$ with $u \in S$ and $u' \in S'$[1]. If S consists of a single vector u, S + S'

[1] S + S' should be distinguished from the *union* of S and S', denoted by $S \cup S'$. The *intersection* of S and S' is denoted by $S \cap S'$.

is simply written $u + S'$. If M is a linear manifold, $u + M$ is called the *inhomogeneous linear manifold* (or linear variety) through u parallel to M. The totality of the inhomogeneous linear manifolds $u + M$ with a fixed M becomes a vector space under the linear operation

$$\alpha(u + M) + \beta(v + M) = (\alpha u + \beta v) + M . \tag{1.8}$$

This vector space is called the *quotient space* of X by M and is denoted by X/M. The elements of X/M are also called the *cosets* of M. The zero vector of X/M is the set M, and we have $u + M = v + M$ if and only if $u - v \in M$. The dimension of X/M is called the *codimension* or *deficiency* of M (with respect to X) and is denoted by codim M. We have

$$\dim M + \operatorname{codim} M = \dim X . \tag{1.9}$$

If M_1 and M_2 are linear manifolds, $M_1 + M_2$ and $M_1 \cap M_2$ are again linear manifolds, and

$$\dim(M_1 + M_2) + \dim(M_1 \cap M_2) = \dim M_1 + \dim M_2 . \tag{1.10}$$

The operation $M_1 + M_2$ for linear manifolds (or for any subsets of X) is associative in the sense that $(M_1 + M_2) + M_3 = M_1 + (M_2 + M_3)$, which is simply written $M_1 + M_2 + M_3$. Similarly we can define $M_1 + M_2 + \cdots + M_s$ for s linear manifolds M_j.

X is the *direct sum* of the linear manifolds $M_1, \ldots, M_s$ if $X = M_1 + \cdots + M_s$ and $\sum u_j = 0$ $(u_j \in M_j)$ implies that all the $u_j = 0$. Then we write

$$X = M_1 \oplus \cdots \oplus M_s . \tag{1.11}$$

In this case each $u \in X$ has a unique expression of the form

$$u = \sum_j u_j , \quad u_j \in M_j , \quad j = 1, \ldots, s . \tag{1.12}$$

Also we have

$$\dim X = \sum_j \dim M_j . \tag{1.13}$$

Problem 1.5. If $X = M_1 \oplus M_2$, then $\dim M_2 = \operatorname{codim} M_1$.

4. Convergence and norms

Let $\{x_j\}$ be a basis in a finite-dimensional vector space X. Let $\{u_n\}$, $n = 1, 2, \ldots$, be a sequence of vectors of X, with the coefficients ξ_{nj} with respect to the basis $\{x_j\}$. The sequence $\{u_n\}$ is said to *converge* to 0 or have *limit* 0, and we write $u_n \to 0$, $n \to \infty$, or $\lim_{n\to\infty} u_n = 0$, if

$$\lim_{n\to\infty} \xi_{nj} = 0 , \quad j = 1, \ldots, N . \tag{1.14}$$

If $u_n - u \to 0$ for some u, $\{u_n\}$ is said to *converge* to u (or have limit u), in symbol $u_n \to u$ or $\lim u_n = u$. The limit is unique when it exists.

This definition of convergence is independent of the basis $\{x_j\}$ employed. In fact, the formula (1.3) for the coordinate transformation shows that (1.14) implies $\lim \xi'_{nj} = 0$, where the ξ'_{nj} are the coefficients of u_n with respect to a new basis $\{x'_j\}$.

The linear operations in X are *continuous* with respect to this notion of convergence, in the sense that $\alpha_n \to \alpha$, $\beta_n \to \beta$, $u_n \to u$ and $v_n \to v$ imply $\alpha_n u_n + \beta_n u_n \to \alpha u + \beta v$.

For various purposes it is convenient to express the convergence of vectors by means of a *norm*. For example, for a fixed basis $\{x_j\}$ of X, set

$$\|u\| = \max_j |\xi_j| , \tag{1.15}$$

where the ξ_j are the coefficients of u with respect to $\{x_j\}$. Then (1.14) shows that $u_n \to u$ is equivalent to $\|u_n - u\| \to 0$. $\|u\|$ is called the norm of u.

(1.15) is not the only possible definition of a norm. We could as well choose

$$\|u\| = \sum_j |\xi_j| \tag{1.16}$$

or

$$\|u\| = \Big(\sum_j |\xi_j|^2\Big)^{1/2} . \tag{1.17}$$

In each case the following conditions are satisfied:

$$\begin{aligned} &\|u\| \geqq 0 ; \quad \|u\| = 0 \quad \text{if and only if} \quad u = 0 . \\ &\|\alpha u\| = |\alpha| \, \|u\| \quad \text{(homogeneity)} . \\ &\|u + v\| \leqq \|u\| + \|v\| \quad \text{(the triangle inequality)} . \end{aligned} \tag{1.18}$$

Any function $\|u\|$ defined for all $u \in \mathsf{X}$ and satisfying these conditions is called a *norm*. Note that the last inequality of (1.18) implies

$$\big| \|u\| - \|v\| \big| \leqq \|u - v\| \tag{1.19}$$

as is seen by replacing u by $u - v$.

A vector u with $\|u\| = 1$ is said to be *normalized*. For any $u \neq 0$, the vector $u_0 = \|u\|^{-1} u$ is normalized; u_0 is said to result from u by *normalization*.

When a norm $\| \ \|$ is given, the convergence $u_n \to u$ can be defined in a natural way by $\|u_n - u\| \to 0$. This definition of convergence is actually independent of the norm employed and, therefore, coincides with the earlier definition. This follows from the fact that any two norms $\| \ \|$

and $\| \ \|'$ in the same space X are *equivalent* in the sense that

$$\alpha' \|u\| \leqq \|u\|' \leqq \beta' \|u\| , \quad u \in \mathsf{X} , \tag{1.20}$$

where α', β' are positive constants independent of u.

We note incidentally that, for *any* norm $\| \ \|$ and *any* basis $\{x_j\}$, the coefficients ξ_j of a vector u satisfy the inequalities

$$|\xi_j| \leqq \gamma \|u\| , \quad j = 1, \ldots, N , \tag{1.21}$$

$$\|u\| \leqq \gamma' \max_j |\xi_j| , \tag{1.22}$$

where γ, γ' are positive constants depending only on the norm $\| \ \|$ and the basis $\{x_j\}$. These inequalities follow from (1.20) by identifying the norm $\| \ \|'$ with the special one (1.15).

A norm $\|u\|$ *is a continuous function of* u. This means that $u_n \to u$ implies $\|u_n\| \to \|u\|$, and follows directly from (1.19). It follows from the same inequality that $u_n \to u$ implies that $\{u_n\}$ is a *Cauchy sequence*, that is, the *Cauchy condition*

$$\|u_n - u_m\| \to 0 , \quad m, n \to \infty , \tag{1.23}$$

is satisfied. Conversely, it is easy to see that the Cauchy condition is sufficient for the existence of $\lim u_n$.

The introduction of a norm is not indispensable for the definition of the notion of convergence of vectors, but it is a very convenient means for it. For applications it is important to choose a norm most suitable to the purpose. A vector space in which a norm is defined is called a *normed (vector) space*. Any finite-dimensional vector space can be made into a normed space. The same vector space gives rise to different normed spaces by different choices of the norm. In what follows we shall often regard a given vector space as a normed space by introducing an appropriate norm. The notion of a finite-dimensional normed space considered here is a model for (and a special case of) the notion of a *Banach space* to be introduced in later chapters.

5. Topological notions in a normed space

In this paragraph a brief review will be given on the topological notions associated with a normed space[1]. Since we are here concerned primarily with a finite-dimensional space, there is no essential difference from the case of a real euclidean space. The modification needed in the infinite-dimensional spaces will be indicated later.

---

[1] We shall need only elementary notions in the topology of metric spaces. As a handy textbook, we refer e. g. to ROYDEN [1].

A normed space X is a special case of a *metric space* in which the distance between any two points is defined. In X the distance between two points (vectors) u, v is defined by $\|u - v\|$. An (open) *ball* of X is the set of points $u \in \mathsf{X}$ such that $\|u - u_0\| < r$, where u_0 is the *center* and $r > 0$ is the *radius* of the ball. The set of u with $\|u - u_0\| \leqq r$ is a *closed ball*. We speak of the *unit ball* when $u_0 = 0$ and $r = 1$. Given a $u \in \mathsf{X}$, any subset of X containing a ball with center u is called a *neighborhood* of u. A subset of X is said to be *bounded* if it is contained in a ball. X itself is not bounded unless $\dim \mathsf{X} = 0$.

For any subset S of X, u is an *interior point* of S if S is a neighborhood of u. u is an *exterior point* of S if u is an interior point of the complement S' of S (with respect to X). u is a boundary point of S if it is neither an interior nor an exterior point of S. The set $\partial \mathsf{S}$ of all boundary points of S is the *boundary* of S. The union $\bar{\mathsf{S}}$ of S and its boundary is the *closure* of S. S is *open* if it consists only of interior points. S is *closed* if S' is open, or, equivalently, if $\mathsf{S} = \bar{\mathsf{S}}$. The closure of any subset S is closed: $\bar{\bar{\mathsf{S}}} = \bar{\mathsf{S}}$. Every linear manifold of X is closed (X being finite-dimensional).

These notions can also be defined by using convergent sequences. For example, $\bar{\mathsf{S}}$ is the set of all $u \in \mathsf{X}$ such that there is a sequence $u_n \in \mathsf{S}$ with $u_n \to u$. S is closed if and only if $u_n \in \mathsf{S}$ and $u_n \to u$ imply $u \in \mathsf{S}$.

We denote by $\operatorname{dist}(u, \mathsf{S})$ the distance of u from a subset S:

$$\operatorname{dist}(u, \mathsf{S}) = \inf_{v \in \mathsf{S}} \|u - v\| . \tag{1.24}$$

If S is closed and $u \notin \mathsf{S}$, then $\operatorname{dist}(u, \mathsf{S}) > 0$.

An important property of a finite-dimensional normed space X is that the theorem of BOLZANO-WEIERSTRASS holds true. From each bounded sequence $\{u_n\}$ of vectors of X, it is possible to extract a subsequence $\{v_n\}$ that converges to some $v \in \mathsf{X}$. This property is expressed by saying that X is *locally compact*[1]. A subset $\mathsf{S} \subset \mathsf{X}$ is *compact* if any sequence of elements of S has a subsequence converging to an element of S.

6. Infinite series of vectors

The convergence of an infinite series

$$\sum_{n=1}^{\infty} u_n \tag{1.25}$$

of vectors $u_n \in \mathsf{X}$ is defined as in the case of numerical series. (1.25) is said to converge to v (or have the sum v) if the sequence $\{v_n\}$ consisting of the partial sums $v_n = \sum_{k=1}^{n} u_k$ converges (to v). The sum v is usually denoted by the same expression (1.25) as the series itself.

[1] The proof of (1.20) depends essentially on the local compactness of X.

A sufficient condition for the convergence of (1.25) is

$$\sum_n \|u_n\| < \infty . \tag{1.26}$$

If this is true for some norm, it is true for any norm in virtue of (1.20). In this case the series (1.25) is said to *converge absolutely*. We have

$$\|\sum_n u_n\| \leqq \sum_n \|u_n\| . \tag{1.27}$$

Problem 1.6. If u_n and v have respectively the coefficients ξ_{nj} and η_j with respect to a basis $\{x_j\}$, (1.25) converges to v if and only if $\sum\limits_n \xi_{nj} = \eta_j$, $j = 1, \ldots, N$. (1.25) converges absolutely if and only if the N numerical series $\sum\limits_n \xi_{nj}$, $j = 1, \ldots, N$, converge absolutely.

In an absolutely convergent series of vectors, the order of the terms may be changed arbitrarily without affecting the sum. This is obvious if we consider the coefficients with respect to a basis (see Problem 1.6). For later reference, however, we shall sketch a more direct proof without using the coefficients. Let $\sum u_n'$ be a series obtained from (1.25) by changing the order of terms. It is obvious that $\sum \|u_n'\| = \sum \|u_n\| < \infty$. For any $\varepsilon > 0$, there is an integer m such that $\sum\limits_{n=m+1}^{\infty} \|u_n\| < \varepsilon$. Let p be so large that $u_1, \ldots, u_m$ are contained in $u_1', \ldots, u_p'$. For any $n > m$ and $q > p$, we have then $\left\|\sum\limits_{j=1}^{q} u_j' - \sum\limits_{k=1}^{n} u_k\right\| \leqq \sum\limits_{k=m+1}^{\infty} \|u_k\| < \varepsilon$, and going to the limit $n \to \infty$ we obtain $\left\|\sum\limits_{j=1}^{q} u_j' - \sum\limits_{k=1}^{\infty} u_k\right\| \leqq \varepsilon$ for $q > p$. This proves that $\sum u_n' = \sum u_n$.

This is an example showing how various results on numerical series can be taken over to series of vectors. In a similar way it can be proved, for example, that an absolutely convergent double series of vectors may be summed in an arbitrary order, by rows or by columns or by transformation into a simple series.

7. Vector-valued functions

Instead of a sequence $\{u_n\}$ of vectors, which may be regarded as a function from the set $\{n\}$ of integers into X, we may consider a function $u_t = u(t)$ defined for a real or complex variable t and taking values in X. The relation $\lim\limits_{t \to a} u(t) = v$ is defined by $\|u(t) - v\| \to 0$ for $t \to a$ (with the usual understanding that $t \neq a$) with the aid of *any* norm. $u(t)$ is *continuous* at $t = a$ if $\lim\limits_{t \to a} u(t) = u(a)$, and $u(t)$ is continuous in a region E of t if it is continuous at every point of E.

The derivative of $u(t)$ is given by

$$u'(t) = \frac{du(t)}{dt} = \lim_{h\to 0} h^{-1}(u(t+h) - u(t)) \tag{1.28}$$

whenever this limit exists. The formulas

$$\begin{aligned} &\frac{d}{dt}(u(t) + v(t)) = u'(t) + v'(t), \\ &\frac{d}{dt}\phi(t)\, u(t) = \phi(t)\, u'(t) + \phi'(t)\, u(t) \end{aligned} \tag{1.29}$$

are valid exactly as for numerical functions, where $\phi(t)$ denotes a complex-valued function.

The integral of a vector-valued function $u(t)$ can also be defined as for numerical functions. For example, suppose that $u(t)$ is a continuous function of a real variable t, $a \leqq t \leqq b$. The Riemann integral $\int_a^b u(t)\, dt$ is defined as an appropriate limit of the sums $\sum (t_j - t_{j-1})\, u(t_j)$ constructed for the partitions $a = t_0 < t_1 < \cdots < t_n = b$ of the interval $[a, b]$. Similarly an integral $\int_C u(t)\, dt$ can be defined for a continuous function $u(t)$ of a complex variable t and for a rectifiable curve C. The proof of the existence of such an integral is quite the same as for numerical functions; in most cases it is sufficient to replace the absolute value of a complex number by the norm of a vector. For these integrals we have the formulas

$$\begin{aligned} &\int (\alpha\, u(t) + \beta\, v(t))\, dt = \alpha \int u(t)\, dt + \beta \int v(t)\, dt, \\ &\left\| \int u(t)\, dt \right\| \leqq \int \|u(t)\|\, |dt|. \end{aligned} \tag{1.30}$$

There is no difficulty in extending these definitions to improper integrals. We shall make free use of the formulas of differential and integral calculus for vector-valued functions without any further comments.

Although there is no difference in the formal definition of the derivative of a vector-valued function $u(t)$ whether the variable t is real or complex, there is an essential difference between these two cases just as with numerical functions. When $u(t)$ is defined and differentiable everywhere in a domain D of the complex plane, $u(t)$ is said to be *regular (analytic)* or *holomorphic* in D. Most of the results of complex function theory are applicable to such vector-valued, holomorphic functions[1].

---

[1] Throughout this book we shall make much use of complex function theory, but it will be limited to elementary results given in standard textbooks such as Knopp [1, 2]. Actually we shall apply these results to vector- or operator-valued functions as well as to complex-valued functions, but such a generalization usually offers no difficulty and we shall make it without particular comments. For the theorems used we shall refer to Knopp whenever necessary.

Thus we have Cauchy's integral theorem, Taylor's and Laurent's expansions, Liouville's theorem, and so on. For example, if $t = 0$ is an isolated singularity of a holomorphic function $u(t)$, we have

$$(1.31) \qquad u(t) = \sum_{n=-\infty}^{+\infty} t^n a_n, \quad a_n = \frac{1}{2\pi i} \int_C t^{-n-1} u(t)\, dt,$$

where C is a closed curve, say a circle, enclosing $t = 0$ in the positive direction. $t = 0$ is a regular point (removable singularity) if $a_n = 0$ for $n < 0$, a *pole* of order $k > 0$ if $a_{-k} \neq 0$ whereas $a_n = 0$ for $n < -k$, and an *essential singularity* otherwise[1].

Problem 1.7. If $t = 0$ is a pole of order k, then $\|u(t)\| = O(|t|^{-k})$ for $t \to 0$.

Problem 1.8. Let $\xi_j(t)$ be the coefficients of $u(t)$ with respect to a basis of X. $u(t)$ is continuous (differentiable) if and only if all the $\xi_j(t)$ are continuous (differentiable). $u'(t)$ has the coefficients $\xi_j'(t)$ for the same basis. Similarly, $\int u(t)\, dt$ has the coefficients $\int \xi_j(t)\, dt$.

§ 2. Linear forms and the adjoint space

1. Linear forms

Let X be a vector space. A complex-valued function $f[u]$ defined for $u \in \mathsf{X}$ is called a *linear form* or a *linear functional* if

$$(2.1) \qquad f[\alpha u + \beta v] = \alpha f[u] + \beta f[v]$$

for all u, v of X and all scalars α, β.

Example 2.1. If $\mathsf{X} = \mathsf{C}^N$ (the space of N-dimensional numerical vectors), a linear form on X can be expressed in the form

$$(2.2) \qquad f[u] = \sum_{j=1}^{N} \alpha_j \xi_j \quad \text{for} \quad u = (\xi_j).$$

It is usual to represent f as a *row vector* with the components α_j, when u is represented as a *column vector* with the components ξ_j. (2.2) is the matrix product of these two vectors.

Example 2.2. Let X be the space of continuous functions $u = u(x)$ considered in Example 1.2. The following are examples of linear forms on X:

$$(2.3) \qquad f[u] = u(x_0), \quad x_0 \text{ being fixed.}$$

$$(2.4) \qquad f[u] = \int_a^b \phi(x)\, u(x)\, dx, \quad \phi(x) \text{ being a given function.}$$

Let $\{x_j\}$ be a basis of X ($\dim \mathsf{X} = N < \infty$). If $u = \sum \xi_j x_j$ is the expansion of u, we have by (2.1)

$$(2.5) \qquad f[u] = \sum \alpha_j \xi_j$$

where $\alpha_j = f[x_j]$. Each linear form is therefore represented by a numerical vector (α_j) with respect to the basis and, conversely, each numerical

[1] See KNOPP [1], p. 117.

vector (α_j) determines a linear form f by (2.5). (2.5) corresponds exactly to (2.2) for a linear form on C^N.

The same linear form f is represented by a different numerical vector (α'_j) for a different basis $\{x'_j\}$. If the new basis is connected with the old one through the transformation (1.2) or (1.4), the relation between these representations is given by

$$\alpha'_j = f[x'_j] = \sum_k \hat{\gamma}_{kj} f[x_k] = \sum_k \hat{\gamma}_{kj} \alpha_k ,\qquad (2.6)$$
$$\alpha_k = \sum_j \gamma_{jk} \alpha'_j .$$

In the matrix notation, these may be written

$$(f)' = (f)(C)^{-1}, \quad (f) = (f)'(C) , \qquad (2.7)$$

where (C) is the matrix (γ_{jk}) [see (1.7)] and where (f) and $(f)'$ stand for the *row vectors* with components (α_j) and (α'_j) respectively.

2. The adjoint space

A complex-valued function $f[u]$ defined on X is called a *semilinear* (or *conjugate-linear* or *anti-linear*) form if

$$f[\alpha u + \beta v] = \bar{\alpha} f[u] + \bar{\beta} f[v] , \qquad (2.8)$$

where $\bar{\alpha}$ denotes the complex conjugate of α. It is obvious that $f[u]$ is a semilinear form if and only if $\overline{f[u]}$ is a linear form. For the sake of a certain formal convenience, we shall hereafter be concerned with semilinear rather than with linear forms.

Example 2.3. A semilinear form on C^N is given by (2.2) with the ξ_j on the right replaced by the $\bar{\xi}_j$, where $u = (\xi_j)$.

Example 2.4. Let X be as in Example 2.2. The following are examples of semilinear forms on X:

$$f[u] = \overline{u(x_0)} , \qquad (2.9)$$

$$f[u] = \int_a^b \phi(x) \overline{u(x)}\, dx . \qquad (2.10)$$

The linear combination $\alpha f + \beta g$ of two semilinear forms f, g defined by

$$(\alpha f + \beta g) [u] = \alpha f[u] + \beta g[u] \qquad (2.11)$$

is obviously a semilinear form. Thus the set of all semilinear forms on X becomes a vector space, called the *adjoint* (or *conjugate*) *space* of X and denoted by X*. The zero vector of X*, which is again denoted by 0, is the zero form that sends every vector u of X into the complex number zero.

It is convenient to treat X^* on the same level as X. To this end we write

$$f[u] = (f, u) \tag{2.12}$$

and call (f, u) the *scalar product* of $f \in X^*$ and $u \in X$. It follows from the definition that (f, u) is linear in f and semilinear in u:

$$\begin{aligned} (\alpha f + \beta g, u) &= \alpha (f, u) + \beta (g, u) , \\ (f, \alpha u + \beta v) &= \bar{\alpha} (f, u) + \bar{\beta} (f, v) . \end{aligned} \tag{2.13}$$

Example 2.5. For $X = C^N$, X^* may be regarded as the set of all row vectors $f = (\alpha_j)$ whereas X is the set of all column vectors $u = (\xi_j)$. Their scalar product is given by

$$(f, u) = \sum \alpha_j \bar{\xi}_j . \tag{2.14}$$

Remark 2.6. In the algebraic theory of vector spaces, the *dual space* of a vector space X is defined to be the set of all *linear* forms on X. Our definition of the adjoint space is chosen in such a way that the adjoint space of a unitary space (see § 6) X can be identified with X itself[1].

3. The adjoint basis

Let $\{x_j\}$ be a basis of X. As in the case of linear forms, for each numerical vector (α_k) there is an $f \in X^*$ such that $(f, x_k) = \alpha_k$. In particular, it follows that for each j, there exists a unique $e_j \in X^*$ such that

$$(e_j, x_k) = \delta_{jk} , \quad j, k = 1, \ldots, N . \tag{2.15}$$

It is easy to see that the e_j are linearly independent. Each $f \in X^*$ can be expressed in a unique way as a linear combination of the e_j, according to

$$f = \sum_j \alpha_j e_j \quad \text{where} \quad \alpha_j = (f, x_j) . \tag{2.16}$$

In fact, the difference of the two members of (2.16) has scalar product zero with all the x_k and therefore with all $u \in X$; thus it must be equal to the zero form.

Thus the N vectors e_j form a basis of X^*, called the basis *adjoint* to the basis $\{x_j\}$ of X. Since the basis $\{e_j\}$ consists of N elements, we have

$$\dim X^* = \dim X = N . \tag{2.17}$$

For each $u \in X$ we have

$$u = \sum_j \xi_j x_j \quad \text{where} \quad \xi_j = \overline{(e_j, u)} . \tag{2.18}$$

---

[1] See e. g. HALMOS [2]. Sometimes one defines X^* as the set of all linear forms f on X but defines αf by $(\alpha f) [u] = \bar{\alpha} f [u]$, so that $f[u]$ is linear in u and semilinear in f (see e. g. LORCH [1]). Our definition of X^* is the same as in RIESZ and SZ.-NAGY [1] in this respect.

It follows from (2.16) and (2.18) that

$$(f, u) = \sum \alpha_j \xi_j = \sum (f, x_j)(e_j, u) . \tag{2.19}$$

Let $\{x_j\}$ and $\{x'\}$ be two bases of X related to each other by (1.2). Then the corresponding adjoint bases $\{e_j\}$ and $\{e_j'\}$ of X^* are related to each other by the formulas

$$e_j' = \sum_k \bar{\gamma}_{jk} e_k , \quad e_k = \sum_j \bar{\bar{\gamma}}_{kj} e_j' . \tag{2.20}$$

Furthermore we have

$$\bar{\gamma}_{jk} = (e_j', x_k) , \quad \bar{\bar{\gamma}}_{kj} = (e_k, x_j') . \tag{2.21}$$

4. The adjoint space of a normed space

Since X^* is an N-dimensional vector space with X, the notion of convergence of a sequence of vectors of X^* is defined as in § 1.4. For the same reason a norm could be introduced into X^*. Usually the norm in X^* is not defined independently but is correlated with the norm of X.

When a norm $\|u\|$ in X is given so that X is a normed space, X^* is by definition a normed space with the norm $\|f\|$ defined by[1]

$$\|f\| = \sup_{0 \neq u \in \mathsf{X}} \frac{|(f, u)|}{\|u\|} = \sup_{\|u\| = 1} |(f, u)| . \tag{2.22}$$

That $\|f\|$ is finite follows from the fact that the continuous function $|(f, u)|$ of u attains a maximum for $\|u\| = 1$ (because X is locally compact). It is easily verified that the norm $\|f\|$ thus defined satisfies the conditions (1.18) of a norm. There is no fear of confusion in using the same symbol $\| \;\|$ for the two norms.

Example 2.7. Suppose that the norm in X is given by (1.15) for a fixed basis $\{x_j\}$. If $\{e_j\}$ is the adjoint basis in X^*, we have $|(f, u)| \leqq (\sum |\alpha_j|) \|u\|$ by (2.19). But the equality holds if u is such that $|\xi_1| = |\xi_2| = \cdots = |\xi_N|$ and all $\alpha_j \bar{\xi}_j$ are real and nonnegative. This shows that

$$\|f\| = \sum |\alpha_j| . \tag{2.23}$$

Similarly it can be shown that, when the norm in X is given by (1.16), the norm in X^* is given by

$$\|f\| = \max |\alpha_j| . \tag{2.24}$$

Thus we may say that the norms (1.15) and (1.16) are adjoint to each other.

(2.22) shows that

$$|(f, u)| \leqq \|f\| \, \|u\| , \quad f \in \mathsf{X}^* , \quad u \in \mathsf{X} . \tag{2.25}$$

This is called the *Schwarz inequality* in the generalized sense. As we have deduced it, it is simply the definition of $\|f\|$ and has an essential meaning only when we give $\|f\|$ some independent characterization (as, for example, in the case of a unitary space; see § 6).

[1] Here we assume $\dim \mathsf{X} > 0$; the case $\dim \mathsf{X} = 0$ is trivial.

(2.25) implies that $\|u\| \geqq |(f, u)|/\|f\|$. Actually the following stronger relation is true[1]:

$$\|u\| = \sup_{0 \neq f \in X^*} \frac{|(f, u)|}{\|f\|} = \sup_{\|f\|=1} |(f, u)| . \tag{2.26}$$

This follows from the fact that, for any $u_0 \in X$, there is an $f \in X^*$ such that

$$(f, u_0) = \|u_0\| , \quad \|f\| = 1 . \tag{2.27}$$

The proof of (2.27) requires a deeper knowledge of the nature of a norm and will be given in the following paragraph.

Problem 2.8. $(f, u) = 0$ for all $u \in X$ implies $f = 0$. $(f, u) = 0$ for all $f \in X^*$ implies $u = 0$.

A simple consequence of the Schwarz inequality is the fact that *the scalar product (f, u) is a continuous function of f and u*. In fact[2],

$$\begin{aligned} |(f', u') - (f, u)| &= |(f' - f, u) + (f, u' - u) + (f' - f, u' - u)| \\ &\leqq \|f' - f\| \, \|u\| + \|f\| \, \|u' - u\| + \|f' - f\| \, \|u' - u\| . \end{aligned} \tag{2.28}$$

In particular, $u_n \to u$ implies $(f, u_n) \to (f, u)$ for every $f \in X^*$ and $f_n \to f$ implies $(f_n, u) \to (f, u)$ for every $u \in X$. Similarly, the convergence of a series $\sum u_n = u$ implies the convergence $\sum (f, u_n) = (f, u)$ for every $f \in X^*$ (that is, term by term multiplication is permitted for the scalar product). Conversely, $(f, u_n) \to (f, u)$ for all $f \in X^*$ implies $u_n \to u$; this can be seen by expanding u_n and u by a fixed basis of X.

5. The convexity of balls

Let S be an open ball of X. S is a *convex* set: for any two points (vectors) u, v of S, the segment joining u and v belongs to S. In other words,

$$\lambda u + (1 - \lambda) v \in S \quad \text{if} \quad u, v \in S \quad \text{and} \quad 0 \leqq \lambda \leqq 1 . \tag{2.29}$$

In fact, denoting by u_0 the center and by r the radius of S, we have $\|\lambda u + (1 - \lambda) v - u_0\| = \|\lambda (u - u_0) + (1 - \lambda)(v - u_0)\| < \lambda r + (1 - \lambda) r = r$, which proves the assertion. In what follows we assume S to be the unit ball ($u_0 = 0$, $r = 1$).

Since X is isomorphic with the N-dimensional complex euclidean space C^N, X is isomorphic with the $2N$-dimensional real euclidean space R^{2N} as a *real vector space* (that is, when only real numbers are regarded as scalars). Thus S may be regarded as a convex set in R^{2N}. It follows from a well-known theorem[3] on convex sets in R^n that, for

[1] Again $\dim X^* = \dim X > 0$ is assumed.

[2] The continuity of (f, u) follows immediately from (2.19). But the proof in (2.28) has the advantage that it is valid in the ∞-dimensional case.

[3] See, e. g., EGGLESTON [1].

each vector u_0 lying on the boundary of S (that is, $\|u_0\| = 1$), there is a *support hyperplane* of S through u_0. This implies that there exists a *real-linear* form $g[u]$ on X such that

$$g[u_0] = 1 \quad \text{whereas} \quad g[u] < 1 \quad \text{for} \quad u \in S\,. \tag{2.30}$$

That g is real-linear means that $g[u]$ is real-valued and $g[\alpha u + \beta v] = \alpha g[u] + \beta g[v]$ for all real numbers α, β and $u, v \in X$.

g is neither a linear nor a semilinear form on the *complex* vector space X. But there is an $f \in X^*$ related to g according to[1]

$$(f, u) = f[u] = g[u] + i\, g[i\, u]\,. \tag{2.31}$$

To see that this f is in fact a semilinear form on X, it suffices to verify that $f[(\alpha + i\beta)u] = (\alpha - i\beta) f[u]$ for real α, β, for it is obvious that $f[u+v] = f[u] + f[v]$. This is seen as follows:

$$\begin{aligned} f[(\alpha + i\beta)u] &= g[\alpha u + i\beta u] + i\, g[i\alpha u - \beta u] \\ &= \alpha g[u] + \beta g[i u] + i\alpha g[i u] - i\beta g[u] \\ &= (\alpha - i\beta)(g[u] + i\, g[i u]) = (\alpha - i\beta) f[u]\,. \end{aligned}$$

Now this f has the following properties:

$$(f, u_0) = 1\,, \quad \|f\| = 1\,. \tag{2.32}$$

To see this, set $(f, u) = Re^{i\theta}$, θ real and $R \geqq 0$. It follows from what was just proved that $(f, e^{i\theta} u) = e^{-i\theta}(f, u) = R$ and hence that $|(f, u)| = R = \operatorname{Re}(f, e^{i\theta} u) = g[e^{i\theta} u] < 1$ if $\|e^{i\theta} u\| = \|u\| < 1$. This shows that $\|f\| \leqq 1$. In particular we have $|(f, u_0)| \leqq 1$. But since $\operatorname{Re}(f, u_0) = g[u_0] = 1$, we must have $(f, u_0) = 1$. This implies also $\|f\| = 1$.

Note that (2.32) is equivalent to (2.27) in virtue of the homogeneity of the norm.

6. The second adjoint space

The adjoint space X^{**} to X^* is the aggregate of semilinear forms on X^*. An example of such a semilinear form F is given by $F[f] = \overline{(f, u)}$ where $u \in X$ is fixed. With each $u \in X$ is thus associated an element F of X^{**}. This correspondence of X with X^{**} is *linear* in the sense that $\alpha u + \beta v$ corresponds to $\alpha F + \beta G$ when u, v correspond to F, G, respectively. The fact that $\dim X^{**} = \dim X^* = \dim X$ shows that the whole space X^{**} is exhausted in this way; in other words, to each $F \in X^{**}$ corresponds a $u \in X$. Furthermore when X and therefore X^*, X^{**} are normed spaces, the norm in X^{**} is identical with the norm in X: $\|F\| = \|u\|$, as is seen from (2.26). In this way we see that X^{**} can be *identified* with X, not only as a vector space but as a normed space.

[1] i is the imaginary unit.

In this sense we may write $F[f]$ as $u[f] = (u, f)$, so that

$$(u, f) = \overline{(f, u)} . \tag{2.33}$$

It should be noted that these results are essentially based on the assumption that $\dim X$ is finite.

Problem 2.9. If $\{e_j\}$ is the basis of X^* adjoint to the basis $\{x_j\}$ of X, $\{x_j\}$ is the basis of $X^{**} = X$ adjoint to $\{e_j\}$.

We write $f \perp u$ or $u \perp f$ when $(f, u) = 0$. When $f \perp u$ for all u of a subset S of X, we write $f \perp S$. Similarly we introduce the notation $u \perp S'$ for $u \in X$ and $S' \subset X^*$. The set of all $f \in X^*$ such that $f \perp S$ is called the *annihilator* of S and is denoted by $S^\perp$. Similarly the annihilator $S'^\perp$ of a subset S' of X^* is the set of all $u \in X$ such that $u \perp S'$.

For any $S \subset X$, $S^\perp$ is a linear manifold. *The annihilator* $S^{\perp\perp}$ *of* $S^\perp$ *is identical with the linear span* M *of* S. In particular we have $M^{\perp\perp} = M$ for any linear manifold M of X.

Problem 2.10. $\operatorname{codim} M = \dim M^\perp$.

§ 3. Linear operators

1. Definitions. Matrix representations

Let X, Y be two vector spaces. A function T that sends every vector u of X into a vector $v = Tu$ of Y is called a *linear transformation* or a *linear operator* on X to Y if T preserves linear relations, that is, if

$$T(\alpha_1 u_1 + \alpha_2 u_2) = \alpha_1 T u_1 + \alpha_2 T u_2 \tag{3.1}$$

for all u_1, u_2 of X and all scalars α_1, α_2. X is the *domain space* and Y is the *range space* of T. If $Y = X$ we say simply that T is a linear operator *in* X. In this book an operator means a linear operator unless otherwise stated.

For any subset S of X, the set of all vectors of the form Tu with $u \in S$ is called the *image* under T of S and is denoted by TS; it is a subset of Y. If M is a linear manifold of X, TM is a linear manifold of Y. In particular, the linear manifold TX of Y is called the *range* of T and is denoted by $R(T)$. The dimension of $R(T)$ is called the *rank* of T; we denote it by $\operatorname{rank} T$. The deficiency (codimension) of $R(T)$ with respect to Y is called the *deficiency* of T and is denoted by $\operatorname{def} T$. Thus

$$\operatorname{rank} T + \operatorname{def} T = \dim Y . \tag{3.2}$$

For any subset S' of Y, the set of all vectors $u \in X$ such that $Tu \in S'$ is called the *inverse image* of S' and is denoted by $T^{-1} S'$. The inverse image of $0 \subset Y$ is a linear manifold of X; it is called the *kernel* or *null space* of T and is denoted by $\mathsf{N}(T)$. The dimension of $\mathsf{N}(T)$ is called the *nullity* of T, which we shall denote by nul T. We have

$$\text{rank}\, T + \text{nul}\, T = \dim \mathsf{X}\,. \tag{3.3}$$

To see this it suffices to note that T maps the quotient space $\mathsf{X}/\mathsf{N}(T)$ (which has dimension $\dim \mathsf{X} - \text{nul}\, T$) onto $\mathsf{R}(T)$ in a one-to-one fashion.

If both nul T and def T are zero, then T maps X onto Y one to one. In this case the *inverse operator* T^{-1} is defined; T^{-1} is the operator on Y to X that sends Tu into u. Obviously we have $(T^{-1})^{-1} = T$. T is said to be *nonsingular* if T^{-1} exists and *singular* otherwise. For T to be nonsingular it is necessary that $\dim \mathsf{X} = \dim \mathsf{Y}$. If $\dim \mathsf{X} = \dim \mathsf{Y}$, each of $\text{nul}\, T = 0$ and $\text{def}\, T = 0$ implies the other and therefore the nonsingularity of T.

Let $\{x_k\}$ be a basis of X. Each $u \in \mathsf{X}$ has the expansion (1.1), so that

$$T u = \sum_{k=1}^{N} \xi_k\, T x_k\,, \quad N = \dim \mathsf{X}\,. \tag{3.4}$$

Thus an operator T on X to Y is determined by giving the values of $T x_k$, $k = 1, \ldots, N$. Furthermore, these values can be prescribed arbitrarily in Y; then it suffices to define T by (3.4) to make T linear.

If $\{y_j\}$ is a basis of Y, each $T x_k$ has the expansion

$$T x_k = \sum_{j=1}^{M} \tau_{jk}\, y_j\,, \quad M = \dim \mathsf{Y}\,. \tag{3.5}$$

Substituting (3.5) into (3.4), we see that the coefficients η_j of $v = Tu$ with respect to the basis $\{y_j\}$ are given by

$$\eta_j = \sum_k \tau_{jk}\, \xi_k\,, \quad j = 1, \ldots, M\,. \tag{3.6}$$

In this way an operator T on X to Y is represented by an $M \times N$ matrix (τ_{jk}) with respect to the bases $\{x_k\}$, $\{y_j\}$ of X, Y, respectively. Conversely, to each $M \times N$ matrix (τ_{jk}) there is an operator T on X to Y represented by it with respect to the given bases.

Let (τ'_{jk}) be the matrix representing the same operator T with respect to a new pair of bases $\{x'_k\}$, $\{y'_j\}$. The relationship between the matrices (τ'_{jk}) and (τ_{jk}) is obtained by combining (3.5) and a similar expression for $T x'_k$ in terms of $\{y'_j\}$ with the formulas (1.2), (1.4) of the coordinate transformation and the corresponding formulas in Y. The result is

$$\tau'_{jk} = \sum_{i,h} \gamma'_{ji}\, \tau_{ih}\, \hat{\gamma}_{hk}\,. \tag{3.7}$$

Thus the matrix (τ'_{jk}) is the product of three matrices (γ'_{jk}), (τ_{jk}) and $(\hat{\gamma}_{jk})$.

If T is an operator on X to itself, it is usual to set $y_j = x_j$ and $y_j' = x_j'$; we have then

$$(\tau_{jk}') = (\gamma_{jk})\,(\tau_{jk})\,(\hat{\gamma}_{jk})\,. \tag{3.8}$$

It follows by (1.6) that

$$\det(\tau_{jk}') = \det(\tau_{jk})\,. \tag{3.9}$$

Thus $\det(\tau_{jk})$ is determined by the operator T itself and does not depend on the basis employed. It is called the *determinant* of T and is denoted by $\det T$. Similarly, the trace $\sum \tau_{jj}$ of the matrix (τ_{jk}) does not depend on the basis; it is called the *trace* of T and is denoted by $\operatorname{tr} T$.

Problem 3.1. If $\{f_j\}$ is the basis of Y^* adjoint to $\{y_j\}$, then

$$\tau_{jk} = (T x_k, f_j)\,. \tag{3.10}$$

Problem 3.2. Let $\{x_j\}$ and $\{e_j\}$ be the bases of X and X^*, respectively, which are adjoint to each other. If T is an operator on X to itself, we have

$$\operatorname{tr} T = \sum_j (T x_j, e_j)\,. \tag{3.11}$$

2. Linear operations on operators

If T and S are two linear operators on X to Y, their linear combination $\alpha S + \beta T$ is defined by

$$(\alpha S + \beta T)\,u = \alpha(S u) + \beta(T u) \tag{3.12}$$

for all $u \in \mathsf{X}$, and is again a linear operator on X to Y. Let us denote by $\mathscr{B}(\mathsf{X}, \mathsf{Y})$ the set of all operators on X to Y; $\mathscr{B}(\mathsf{X}, \mathsf{Y})$ is a vector space with the linear operations defined as above. The zero vector of this vector space is the zero operator 0 defined by $0\,u = 0$ for all $u \in \mathsf{X}$.

Problem 3.3. $\operatorname{rank}(S + T) \leqq \operatorname{rank} S + \operatorname{rank} T$.

The dimension of the vector space $\mathscr{B}(\mathsf{X}, \mathsf{Y})$ is equal to $N M$, where $N = \dim \mathsf{X}$ and $M = \dim \mathsf{Y}$. To see this, let $\{x_k\}$ and $\{y_j\}$ be bases of X and Y, respectively, and let P_{jk} be the operator on X to Y such that

$$P_{jk}\, x_h = \delta_{kh}\, y_j\,, \quad k, h = 1, \ldots, N\,; \quad j = 1, \ldots, M\,. \tag{3.13}$$

These $M N$ operators P_{jk} are linearly independent elements of $\mathscr{B}(\mathsf{X}, \mathsf{Y})$, and we have from (3.5)

$$T = \sum \tau_{jk}\, P_{jk}\,. \tag{3.14}$$

Thus $\{P_{jk}\}$ is a basis of $\mathscr{B}(\mathsf{X}, \mathsf{Y})$, which proves the assertion. $\{P_{jk}\}$ will be called the basis of $\mathscr{B}(\mathsf{X}, \mathsf{Y})$ associated with the bases $\{x_k\}$ and $\{y_j\}$ of X and Y, respectively. (3.14) shows that the matrix elements τ_{jk} are the coefficients of the "vector" T with respect to the basis $\{P_{jk}\}$, and (3.7) or (3.8) is the formula for coordinate transformation in $\mathscr{B}(\mathsf{X}, \mathsf{Y})$.

The product TS of two linear operators T, S is defined by

$$(TS)\, u = T(Su) \tag{3.15}$$

for all $u \in \mathsf{X}$, where X is the domain space of S, provided the domain space of T is identical with the range space Y of S. The following relations hold for these operations on linear operators:

$$\begin{aligned} &(TS)\, R = T(SR), \text{ which is denoted by } TSR\,, \\ &(\alpha T)\, S = T(\alpha S) = \alpha(TS), \text{ denoted by } \alpha TS\,, \\ &(T_1 + T_2)\, S = T_1 S + T_2 S\,, \\ &T(S_1 + S_2) = TS_1 + TS_2\,. \end{aligned} \tag{3.16}$$

Problem 3.4. $\operatorname{rank}(TS) \leqq \max(\operatorname{rank} T, \operatorname{rank} S)$.

Problem 3.5. If S, T have the matrices (σ_{jk}), (τ_{jk}) with respect to some fixed bases, $S + T$ and TS have the matrices $(\sigma_{jk}) + (\tau_{jk})$, $(\tau_{jk})\,(\sigma_{jk})$ respectively (whenever meaningful). If T^{-1} exists, its matrix is the inverse matrix of (τ_{jk}).

3. The algebra of linear operators

If S and T are operators on X to itself, their product TS is defined and is again an operator on X to itself. Thus the set $\mathscr{B}(\mathsf{X}) = \mathscr{B}(\mathsf{X}, \mathsf{X})$ of all linear operators in X is not only a vector space but an *algebra.* $\mathscr{B}(\mathsf{X})$ is not *commutative* for $\dim \mathsf{X} \geqq 2$ since $TS = ST$ is in general not true. When $TS = ST$, T and S are said to *commute* (with each other). We have $T0 = 0T = 0$ and $T1 = 1T = T$ for every $T \in \mathscr{B}(\mathsf{X})$, where 1 denotes the *identity operator* (defined by $1u = u$ for every $u \in \mathsf{X}$). Thus 1 is the *unit element* of $\mathscr{B}(\mathsf{X})$[1]. The operators of the form $\alpha 1$ are called *scalar operators*[2] and in symbol will not be distinguished from the scalars α. A scalar operator commutes with every operator of $\mathscr{B}(\mathsf{X})$.

We write $TT = T^2$, $TTT = T^3$ and so on, and set $T^0 = 1$ by definition. We have

$$T^m T^n = T^{m+n}\,, \quad (T^m)^n = T^{mn}\,, \quad m, n = 0, 1, 2, \ldots \tag{3.17}$$

For any polynomial $p(z) = \alpha_0 + \alpha_1 z + \cdots + \alpha_n z^n$ in the indeterminate z, we define the operator

$$p(T) = \alpha_0 + \alpha_1 T + \cdots + \alpha_n T^n\,. \tag{3.18}$$

The mapping $p(z) \to p(T)$ is a *homomorphism* of the algebra of polynomials to $\mathscr{B}(\mathsf{X})$; this means that $p(z) + q(z) = r(z)$ or $p(z)\, q(z) = r(z)$

[1] Note that $1 \neq 0$ if (and only if) $\dim \mathsf{X} \geqq 1$.

[2] This should not be confused with the notion of scalar operators in the theory of spectral operators due to DUNFORD (see DUNFORD [1]).

inplies $p(T) + q(T) = r(T)$ or $p(T)\,q(T) = r(T)$ respectively. In particular, it follows that $p(T)$ and $q(T)$ commute.

Problem 3.6. The operators $P_{jk} \in \mathscr{B}(\mathsf{X})$ given by (3.13) with $\mathsf{Y} = \mathsf{X}$, $y_j = x_j$ satisfy the relations

$$P_{jk} P_{ih} = \delta_{ki} P_{jh}, \quad j, k, i, h = 1, \ldots, N. \tag{3.19}$$

Problem 3.7. Set $\mathsf{R}_n = \mathsf{R}(T^n)$ and $\mathsf{N}_n = \mathsf{N}(T^n)$, $n = 0, 1, 2, \ldots$. The sequence $\{\mathsf{R}_n\}$ is nonincreasing and $\{\mathsf{N}_n\}$ is nondecreasing. There is a nonnegative integer $m \leq \dim \mathsf{X}$ such that $\mathsf{R}_n \neq \mathsf{R}_{n+1}$ for $n < m$ and $\mathsf{R}_n = \mathsf{R}_{n+1}$ for $n \geq m$.

If $T \in \mathscr{B}(\mathsf{X})$ is nonsingular, the inverse T^{-1} exists and belongs to $\mathscr{B}(\mathsf{X})$; we have

$$T^{-1} T = T T^{-1} = 1. \tag{3.20}$$

If T has a *left inverse* T' (that is, a $T' \in \mathscr{B}(\mathsf{X})$ such that $T' T = 1$), T has nullity zero, for $Tu = 0$ implies $u = T' T u = 0$. If T has a *right inverse* T'' (that is, $T T'' = 1$), T has deficiency zero because every $u \in \mathsf{X}$ lies in $\mathsf{R}(T)$ by $u = T T'' u$. If $\dim \mathsf{X}$ is finite, either of these facts implies that T is nonsingular and that $T' = T^{-1}$ or $T'' = T^{-1}$, respectively.

If S and T are nonsingular, so is TS and

$$(TS)^{-1} = S^{-1} T^{-1}. \tag{3.21}$$

For a nonsingular T, the negative powers T^{-n}, $n = 1, 2, \ldots$, can be defined by $T^{-n} = (T^{-1})^n$. In this case (3.17) is true for any integers m, n.

The following relations on determinants and traces follow directly from Problem 3.5:

$$\begin{aligned} &\det TS = (\det T)(\det S), \\ &\operatorname{tr}(\alpha S + \beta T) = \alpha \operatorname{tr} S + \beta \operatorname{tr} T, \\ &\operatorname{tr} ST = \operatorname{tr} TS. \end{aligned} \tag{3.22}$$

Problem 3.8. The last formula of (3.22) is true even when $S \in \mathscr{B}(\mathsf{X}, \mathsf{Y})$ and $T \in \mathscr{B}(\mathsf{Y}, \mathsf{X})$ so that $ST \in \mathscr{B}(\mathsf{Y})$ and $TS \in \mathscr{B}(\mathsf{X})$.

4. Projections. Nilpotents

Let M, N be two *complementary* linear manifolds of X; by this we mean that

$$\mathsf{X} = \mathsf{M} \oplus \mathsf{N}; \tag{3.23}$$

see § 1.3. Thus each $u \in \mathsf{X}$ can be uniquely expressed in the form $u = u' + u''$ with $u' \in \mathsf{M}$ and $u'' \in \mathsf{N}$. u' is called the *projection of u on* M *along* N. If $v = v' + v''$ in the same sense, $\alpha u + \beta v$ has the projection $\alpha u' + \beta v'$ on M along N. If we set $u' = Pu$, it follows that P is a linear operator on X to itself. P is called the *projection operator* (or simply the *projection*) *on* M *along* N. $1 - P$ is the projection on N along M. We have

$Pu = u$ if and only if $u \in \mathsf{M}$, and $Pu = 0$ if and only if $u \in \mathsf{N}$. The range of P is M and the null space of P is N. For convenience we often write $\dim P$ for $\dim \mathsf{M} = \dim \mathsf{R}(P)$. Since $Pu \in \mathsf{M}$ for every $u \in \mathsf{X}$, we have $PPu = Pu$, that is, P is *idempotent*:

$$P^2 = P. \tag{3.24}$$

Conversely, any idempotent operator P is a projection. In fact, set $\mathsf{M} = \mathsf{R}(P)$ and $\mathsf{N} = \mathsf{R}(1 - P)$. $u' \in \mathsf{M}$ implies that $u' = Pu$ for some u and therefore $Pu' = P^2u = Pu = u'$. Similarly $u'' \in \mathsf{N}$ implies $Pu'' = 0$. Hence $u \in \mathsf{M} \cap \mathsf{N}$ implies that $u = Pu = 0$, so that $\mathsf{M} \cap \mathsf{N} = 0$. Each $u \in \mathsf{X}$ has the expression $u = u' + u''$ with $u' = Pu \in \mathsf{M}$ and $u'' = (1 - P)u \in \mathsf{N}$. This shows that P is the projection on M along N.

Problem 3.9. If P is a projection, we have

$$\operatorname{tr} P = \dim P. \tag{3.25}$$

The above results can be extended to the case in which there are several linear manifolds $\mathsf{M}_1, \ldots, \mathsf{M}_s$ such that

$$\mathsf{X} = \mathsf{M}_1 \oplus \cdots \oplus \mathsf{M}_s. \tag{3.26}$$

Each $u \in \mathsf{X}$ is then expressed in the form $u = u_1 + \cdots + u_s$, $u_j \in \mathsf{M}_j$, $j = 1, \ldots, s$, in a unique way. The operator P_j defined by $P_j u = u_j$ is the projection on M_j along $\mathsf{N}_j = \mathsf{M}_1 \oplus \cdots \oplus \mathsf{M}_{j-1} \oplus \mathsf{M}_{j+1} \oplus \cdots \oplus \mathsf{M}_s$. Furthermore, we have

$$\sum P_j = 1, \tag{3.27}$$

$$P_k P_j = \delta_{jk} P_j. \tag{3.28}$$

Conversely, let $P_1, \ldots, P_s$ be operators satisfying the conditions (3.27) and (3.28)[1]. If we write $\mathsf{M}_j = \mathsf{R}(P_j)$, it is easily seen that (3.26) is satisfied and the P_j are the projections defined as above. In particular consider the case $s = 3$ and set $P = P_1 + P_2$. Then $P_1 = P_1 P = P P_1 = P P_1 P$; P_1 is a projection commuting with P and with $\mathsf{R}(P_1) \subset \mathsf{R}(P)$. Such a P_1 will be called a *subprojection of* P (a proper subprojection if $P_1 \neq P$ in addition), in symbol $P_1 \leq P$.

A basis $\{x_j\}$ of X is said to be *adapted* to the decomposition (3.26) if the first several elements of $\{x_j\}$ belong to M_1, the following several ones belong to M_2, and so on. With respect to such a basis $\{x_j\}$, each P_j is represented by a diagonal matrix with diagonal elements equal to 0 or 1, the number of 1's being equal to $\dim \mathsf{M}_j$. Conversely, such a matrix always represents a projection.

[1] Such a family is sometimes called a complete orthogonal family of projections. We do not use this term to avoid a possible confusion with the notion of an orthogonal projection to be introduced in a unitary or Hilbert space.

For each linear manifold M of X, there is a complementary manifold N [such that (3.23) is true]. Thus every linear manifold has a projection on it. Such a projection is not unique, however.

A linear operator $T \in \mathscr{B}(\mathsf{X})$ is called a ***nilpotent (operator)*** if $T^n = 0$ for some positive integer n. A nilpotent is necessarily singular.

Let us consider the structure of a nilpotent T in more detail. Let n be such that $T^n = 0$ but $T^{n-1} \neq 0$ (we assume $\dim \mathsf{X} = N > 0$). Then $\mathsf{R}(T^{n-1}) \neq 0$; let $\{x_1^1, \ldots, x_{p_1}^1\}$ be a basis of $\mathsf{R}(T^{n-1})$. Each x_i^1 has the form $x_i^1 = T^{n-1} x_i^n$ for some $x_i^n \in \mathsf{X}$, $i = 1, \ldots, p_1$. If $n > 1$, set $T^{n-2} x_i^n = x_i^2$ so that $T x_i^2 = x_i^1$. The vectors x_i^k, $k = 1, 2$, $i = 1, \ldots, p_1$, belong to $\mathsf{R}(T^{n-2})$ and are linearly independent; in fact $\sum \alpha_i x_i^2 + \sum \beta_i x_i^1 = 0$ implies $\sum \alpha_i x_i^1 = 0$ on application of T and hence $\alpha_i = 0$ for all i, hence $\sum \beta_i x_i^1 = 0$ and $\beta_i = 0$ for all i. Let us enlarge the family $\{x_i^k\}$ to a basis of $\mathsf{R}(T^{n-2})$ by adding, if necessary, new vectors $x_{p_1+1}^2, \ldots, x_{p_2}^2$; here we can arrange that $T x_i^2 = 0$ for $i > p_1$.

If $n > 2$ we can proceed in the same way. Finally we arrive at a basis $\{x_j^k\}$ of X with the following properties: $k = 1, \ldots, n$, $j = 1, \ldots, p_k$, $p_1 \leqq p_2 \leqq \cdots \leqq p_n$,

$$T x_j^k = \begin{cases} x_j^{k-1}, & 1 \leqq j \leqq p_{k-1}, \\ 0, & p_{k-1} + 1 \leqq j \leqq p_k, \end{cases} \tag{3.29}$$

where we set $p_0 = 0$.

If we arrange the basis $\{x_j^k\}$ in the order $\{x_1^1, \ldots, x_1^n, x_2^1, \ldots, x_2^n, \ldots\}$, the matrix of T with respect to this basis takes the form

$$\left(\begin{array}{ccccc:cccc:c} 0 & 1 & & & & & & & & \\ & 0 & 1 & & & & & & & \\ & & \ddots & \ddots & & & & & & \\ & & & 0 & 1 & & & & & \\ & & & & 0 & & & & & \\ \hdashline & & & & & 0 & 1 & & & \\ & & & & & & 0 & 1 & & \\ & & & & & & & \cdot & 1 & \\ & & & & & & & & 0 & \\ \hdashline & & & & & & & & & \ddots \end{array}\right) \tag{3.30}$$

(all unspecified elements are zero)

Problem 3.10. If T is nilpotent, then $T^N = 0$ for $N = \dim \mathsf{X}$.

Problem 3.11. If T is nilpotent, then $\operatorname{tr} T = 0$ and $\det(1 + T) = 1$.

5. Invariance. Decomposition

A linear manifold M is said to be ***invariant*** under an operator $T \in \mathscr{B}(\mathsf{X})$ if $T\mathsf{M} \subset \mathsf{M}$. In this case T ***induces*** a linear operator T_M on M to M, defined by $T_\mathsf{M} u = T u$ for $u \in \mathsf{M}$. T_M is called the ***part of T in*** M.

Problem 3.12. $\mathsf{R}_n = \mathsf{R}(T^n)$, $n = 0, 1, 2, \ldots$, are invariant under T. If m is defined as in Problem 3.7, the part of T in R_n is singular if $n < m$ and nonsingular if $n \geqq m$.

Problem 3.13. If M is invariant under T, M is also invariant under $p(T)$ for any polynomial $p(z)$, and $p(T)_{\mathsf{M}} = p(T_{\mathsf{M}})$.

If there are two invariant linear manifolds M, N for T such that $\mathsf{X} = \mathsf{M} \oplus \mathsf{N}$, T is said to be *decomposed* (or *reduced*) by the pair M, N. More generally, T is said to be decomposed by the set of linear manifolds $\mathsf{M}_1, \ldots, \mathsf{M}_s$ if (3.26) is satisfied and all the M_j are invariant under T [or we say that T is decomposed according to the decomposition (3.26)]. In this case T is completely described by its parts T_{M_j}, $j = 1, \ldots, s$. T is called the *direct sum* of the T_{M_j}. If $\{P_j\}$ is the set of projections corresponding to (3.26), T commutes with each P_j. In fact we have, successively, $P_j u \in \mathsf{M}_j$, $T P_j u \in \mathsf{M}_j$, $P_k T P_j u = \delta_{jk} T P_j u$, and the addition of the last equalities for $j = 1, \ldots, s$ gives $P_k T u = T P_k u$ or $P_k T = T P_k$. Conversely, it is easy to see that T is decomposed by $\mathsf{M}_1, \ldots, \mathsf{M}_s$ if T commutes with all the P_j.

If we choose a basis $\{x_j\}$ adapted to the decomposition (3.26), T is represented by a matrix which has non-zero elements only in s smaller submatrices along the diagonal (which are the matrices of the T_{M_j}). Thus the matrix of T is the direct sum of the matrices of the T_{M_j}.

Problem 3.14. With the above notations, we have

$$\det T = \prod_j \det T_{\mathsf{M}_j}, \quad \operatorname{tr} T = \sum_j \operatorname{tr} T_{\mathsf{M}_j}. \tag{3.31}$$

Remark 3.15. The operator $P_j T = T P_j = P_j T P_j$ coincides with T and also with T_{M_j} when applied to a $u \in \mathsf{M}_j$; it is sometimes identified with T_{M_j} when there is no possibility of misunderstanding.

6. The adjoint operator

Let $T \in \mathscr{B}(\mathsf{X}, \mathsf{Y})$. For each $g \in \mathsf{Y}^*$ and $u \in \mathsf{X}$, the scalar product (g, Tu) is defined and is a semilinear form in u. Therefore, it can be written as $f[u] = (f, u)$ with an $f \in \mathsf{X}^*$. Since f is determined by g, a function T^* on Y^* to X^* is defined by setting $f = T^* g$. Thus the defining equation of T^* is

$$(T^* g, u) = (g, Tu), \quad g \in \mathsf{Y}^*, \quad u \in \mathsf{X}. \tag{3.32}$$

T^* is a linear operator on Y^* to X^*, that is, $T^* \in \mathscr{B}(\mathsf{Y}^*, \mathsf{X}^*)$. In fact, we have $(T^*(\alpha_1 g_1 + \alpha_2 g_2), u) = (\alpha_1 g_1 + \alpha_2 g_2, Tu) = \alpha_1 (g_1, Tu) + \alpha_2 (g_2, Tu) = \alpha_1 (T^* g_1, u) + \alpha_2 (T^* g_2, u) = (\alpha_1 T^* g_1 + \alpha_2 T^* g_2, u)$ so that $T^*(\alpha_1 g_1 + \alpha_2 g_2) = \alpha_1 T^* g_1 + \alpha_2 T^* g_2$. T^* is called the *adjoint (operator)* of T.

The operation $*$ has the following properties:

$$(\alpha S + \beta T)^* = \bar{\alpha} S^* + \bar{\beta} T^*, \quad (TS)^* = S^* T^*. \tag{3.33}$$

In the second formula it is assumed that $T \in \mathscr{B}(\mathsf{Y}, \mathsf{Z})$ and $S \in \mathscr{B}(\mathsf{X}, \mathsf{Y})$ so that TS is defined and belongs to $\mathscr{B}(\mathsf{X}, \mathsf{Z})$; note that $S^* \in \mathscr{B}(\mathsf{Y}^*, \mathsf{X}^*)$ and $T^* \in \mathscr{B}(\mathsf{Z}^*, \mathsf{Y}^*)$ so that $S^* T^* \in \mathscr{B}(\mathsf{Z}^*, \mathsf{X}^*)$. The proof of (3.33) is simple; for example, the second formula follows from $((TS)^* h, u) = (h, TSu) = (T^* h, Su) = (S^* T^* h, u)$ which is valid for all $h \in \mathsf{Z}^*$ and $u \in \mathsf{X}$.

Problem 3.16. $0^* = 0$, $1^* = 1$ (the 0 on the left is the zero of $\mathscr{B}(\mathsf{X}, \mathsf{Y})$ while the 0 on the right is the zero of $\mathscr{B}(\mathsf{Y}^*, \mathsf{X}^*)$; similarly for the second equality, in which we must set $\mathsf{Y} = \mathsf{X}$).

If $T \in \mathscr{B}(\mathsf{X}, \mathsf{Y})$, we have $T^* \in \mathscr{B}(\mathsf{Y}^*, \mathsf{X}^*)$ and $T^{**} \in \mathscr{B}(\mathsf{X}^{**}, \mathsf{Y}^{**})$. If we identify X^{**} and Y^{**} with X and Y respectively (see § 2.6), it follows from (3.32) that

$$T^{**} = T . \tag{3.34}$$

If we take bases $\{x_k\}$ and $\{y_j\}$ in X and Y respectively, an operator $T \in \mathscr{B}(\mathsf{X}, \mathsf{Y})$ is represented by a matrix (τ_{jk}) according to (3.5) or (3.6). If $\{e_k\}$ and $\{f_j\}$ are the adjoint bases of X^* and Y^*, respectively, the operator $T^* \in \mathscr{B}(\mathsf{Y}^*, \mathsf{X}^*)$ can similarly be represented by a matrix (τ^*_{kj}). These matrix elements are given by $\tau_{jk} = (T x_k, f_j)$ and $\tau^*_{kj} = (T^* f_j, x_k) = (f_j, T x_k)$ in virtue of (3.10). Thus

$$\tau^*_{kj} = \overline{\tau_{jk}} , \quad \begin{array}{l} k = 1, \ldots, N = \dim \mathsf{X} , \\ j = 1, \ldots, M = \dim \mathsf{Y} , \end{array} \tag{3.35}$$

and T and T^* are represented by mutually *adjoint* (Hermitian conjugate) matrices with respect to the bases which are adjoint to each other.

Problem 3.17. If $T \in \mathscr{B}(\mathsf{X})$, we have

$$\det T^* = \overline{\det T} , \quad \operatorname{tr} T^* = \overline{\operatorname{tr} T} . \tag{3.36}$$

Let $T \in \mathscr{B}(\mathsf{X}, \mathsf{Y})$. A $g \in \mathsf{Y}^*$ belongs to the annihilator of $\mathsf{R}(T)$ if and only if $(g, Tu) = 0$ for all $u \in \mathsf{X}$. (3.32) shows that this is equivalent to $T^* g = 0$. Thus *the annihilator of the range of T is identical with the null space of T^**. In view of (3.34), the same is true when T and T^* are exchanged. In symbol, we have

$$\mathsf{N}(T^*) = \mathsf{R}(T)^\perp , \quad \mathsf{N}(T) = \mathsf{R}(T^*)^\perp . \tag{3.37}$$

It follows that [see (3.2), (3.3) and (2.17)]

$$\operatorname{nul} T^* = \operatorname{def} T , \quad \operatorname{nul} T = \operatorname{def} T^* , \quad \operatorname{rank} T^* = \operatorname{rank} T . \tag{3.38}$$

If in particular $\mathsf{Y} = \mathsf{X}$, (3.38) shows that T^* is nonsingular if and only if T is; in this case we have

$$(T^*)^{-1} = (T^{-1})^* . \tag{3.39}$$

For the proof it suffices to note that $T^* (T^{-1})^* = (T^{-1} T)^* = 1^* = 1$.

Problem 3.18. If $T \in \mathscr{B}(X)$, we have

$$\text{nul}\, T^* = \text{nul}\, T\,, \quad \text{def}\, T^* = \text{def}\, T\,. \tag{3.40}$$

If $P \in \mathscr{B}(X)$ is a projection, the adjoint $P^* \in \mathscr{B}(X^*)$ is likewise a projection, for $P^2 = P$ implies $P^{*2} = P^*$. The decompositions of the spaces X and X*

$$X = M \oplus N\,, \qquad M = R(P)\,, \qquad N = R(1 - P)\,, \tag{3.41}$$

$$X^* = M^* \oplus N^*\,,\ M^* = R(P^*)\,, \quad N^* = R(1 - P^*)\,, \tag{3.42}$$

are related to each other through the following equalities:

$$N^* = M^\perp\,, \quad M^* = N^\perp\,, \quad \dim M^* = \dim M\,, \quad \dim N^* = \dim N\,, \tag{3.43}$$

as is seen from (3.37) and (3.40).

Similar results hold when there are several projections. If $\{P_j\}$ is a set of projections in X satisfying (3.27–3.28), $\{P_j^*\}$ is a similar set of projections in X*. The ranges $M_j = R(P_j)$, $M_j^* = R(P_j^*)$ are related by

$$\dim M_j^* = \dim M_j\,, \quad j = 1, 2, \ldots, \tag{3.44}$$

$$M_1^* = (M_2 \oplus \cdots)^\perp\,, \quad M_1^\perp = (M_2^* \oplus \cdots)^\perp\,, \quad \text{etc.} \tag{3.45}$$

Problem 3.19. Let $\{x_j\}$ be a basis of X adapted to the decomposition $X = M_1 \oplus \oplus \cdots \oplus M_s$, and let $\{e_j\}$ be the adjoint basis of X*. Then $\{e_j\}$ is adapted to the decomposition $X^* = M_1^* \oplus \cdots \oplus M_s^*$. For any $u \in X$ we have

$$P_j\, u = \sum_{r=1}^{m_j} (u, e_{jr})\, x_{jr}\,, \tag{3.46}$$

where $\{x_{j1}, \ldots, x_{jm_j}\}$ is the part of $\{x_j\}$ belonging to M_j and $m_j = \dim M_j$.

§ 4. Analysis with operators

1. Convergence and norms for operators

Since the set $\mathscr{B}(X, Y)$ of all linear operators on X to Y is an MN-dimensional vector space, where $N = \dim X < \infty$ and $M = \dim Y < \infty$, the notion of convergence of a sequence $\{T_n\}$ of operators of $\mathscr{B}(X, Y)$ is meaningful as in the case of a sequence of vectors of X. If we introduce the matrix representation (τ_{njk}) of T_n with respect to fixed bases $\{x_k\}$, $\{y_j\}$ of X, Y, respectively, $T_n \to T$ is equivalent to $\tau_{njk} \to \tau_{jk}$ for each j, k, for the τ_{njk} are the coefficients of T_n for the basis $\{P_{jk}\}$ of $\mathscr{B}(X, Y)$ (see § 3.2). But the τ_{njk} are the coefficients of $T x_k$ with respect to the basis $\{y_j\}$; hence $T_n \to T$ is equivalent to $T_n x_k \to T x_k$ for all k and therefore to $T_n u \to T u$ for all $u \in X$. This could have been used as the definition of $T_n \to T$.

As it was convenient to express the convergence of vectors by means of a norm, so it is with operators. But an operator-norm is usually introduced only in correlation with the vector-norms. More precisely, when X and Y are normed spaces $\mathscr{B}(\mathsf{X},\mathsf{Y})$ is defined to be a normed space with the norm given by

$$(4.1)\qquad \|T\| = \sup_{0\neq u\in\mathsf{X}} \frac{\|Tu\|}{\|u\|} = \sup_{\|u\|=1}\|Tu\| = \sup_{\|u\|\leq 1}\|Tu\|\,,\quad T\in\mathscr{B}(\mathsf{X},\mathsf{Y})\,.$$

The equality of the various expressions in (4.1) is easily verified[1]. We can replace "sup" by "max" in (4.1) because the set of u with $\|u\| = 1$ or $\|u\| \leq 1$ is compact (see a similar remark for the norm of an $f\in\mathsf{X}^*$ in § 2.4); this shows that $\|T\|$ is finite. It is easy to verify that $\|T\|$ defined on $\mathscr{B}(\mathsf{X},\mathsf{Y})$ by (4.1) satisfies the conditions (1.18) of a norm. Hence it follows that $T_n\to T$ is equivalent to $\|T_n - T\|\to 0$. A necessary and sufficient condition that $\{T_n\}$ converge to some T is given by the Cauchy condition $\|T_n - T_m\|\to 0$, $n, m\to\infty$.

Another convenient expression for $\|T\|$ is[2]

$$(4.2)\qquad \|T\| = \sup_{\substack{0\neq u\in\mathsf{X}\\ 0\neq f\in\mathsf{Y}^*}} \frac{|(Tu,f)|}{\|f\|\,\|u\|} = \sup_{\substack{\|u\|=1\\ \|f\|=1}} |(Tu,f)|\,.$$

The equivalence of (4.2) with (4.1) follows from (2.26).

If we introduce different norms in the given vector spaces X and Y, $\mathscr{B}(\mathsf{X},\mathsf{Y})$ acquires different norms accordingly. As in the case of norms in X, however, all these norms in $\mathscr{B}(\mathsf{X},\mathsf{Y})$ are equivalent, in the sense that for any two norms $\|\ \|$ and $\|\ \|'$, there are positive constants α', β' such that

$$(4.3)\qquad \alpha'\|T\| \leq \|T\|' \leq \beta'\|T\|\,.$$

This is a special case of (1.20) applied to $\mathscr{B}(\mathsf{X},\mathsf{Y})$ regarded as a normed space. Similarly, the inequalities (1.21) and (1.22) give the following inequalities:

$$(4.4)\qquad |\tau_{jk}| \leq \gamma\|T\|\,,\quad j = 1,\ldots, M;\ k = 1,\ldots, N\,,$$

$$(4.5)\qquad \|T\| \leq \gamma'\max|\tau_{jk}|\,,$$

where (τ_{jk}) is the matrix of T with respect to the bases of X and Y. The constants γ and γ' depend on these bases and the norm employed, but are independent of the operator T.

As in the case of vectors, $\alpha S + \beta T$ is a continuous function of the scalars α, β and the operators $S, T\in\mathscr{B}(\mathsf{X},\mathsf{Y})$, and $\|T\|$ is a continuous function of T. As a new feature of the norm of operators, we should note the inequality

$$(4.6)\qquad \|TS\| \leq \|T\|\,\|S\| \quad\text{for}\quad T\in\mathscr{B}(\mathsf{Y},\mathsf{Z}) \quad\text{and}\quad S\in\mathscr{B}(\mathsf{X},\mathsf{Y})\,.$$

[1] The second and third members of (4.1) do not make sense if $\dim\mathsf{X} = 0$; in this case we have simply $\|T\| = 0$.

[2] Here we assume $\dim\mathsf{X}\geq 1$, $\dim\mathsf{Y}\geq 1$.

This follows from $\|TSu\| \leqq \|T\| \|Su\| \leqq \|T\| \|S\| \|u\|$; note also that (4.6) would not be valid if we chose arbitrary norms in $\mathscr{B}(X, Y)$ etc. regarded simply as vector spaces.

Problem 4.1. $\|1\| = 1$ ($1 \in \mathscr{B}(X)$ is the identity operator, $\dim X > 0$). If $P \in \mathscr{B}(X)$ is a projection and $P \neq 0$, then $\|P\| \geqq 1$.

TS is a continuous function of S and T. In other words, $T_n \to T$ and $S_n \to S$ imply $T_n S_n \to TS$. The proof is similar to (2.28); it suffices to use (4.6). In the same way, it can be shown that Tu is a continuous function of T and u. In particular $u_n \to u$ implies $Tu_n \to Tu$. In this sense a linear operator T is a continuous function. It is permitted, for example, to operate with T term by term on a convergent series of vectors:

$$T\Big(\sum_n u_n\Big) = \sum_n T u_n . \tag{4.7}$$

If X is a normed space, $\mathscr{B}(X) = \mathscr{B}(X, X)$ is a *normed algebra* (or *ring*) with the norm given by (4.2). In particular, (4.6) is true for $T, S \in \mathscr{B}(X)$.

If $T \in \mathscr{B}(X, Y)$, then $T^* = \mathscr{B}(Y^*, X^*)$ and

$$\|T^*\| = \|T\| . \tag{4.8}$$

This follows from (4.2), according to which $\|T^*\| = \sup |(T^* f, u)| = \sup |(f, Tu)| = \|T\|$ where $u \in X^{**} = X$, $\|u\| = 1$ and $f \in X^*$, $\|f\| = 1$.

2. The norm of T^n

As an example of the use of the norm and also with a view to later applications, we consider the norm $\|T^n\|$ for $T \in \mathscr{B}(X)$. It follows from (4.6) that

$$\|T^{m+n}\| \leqq \|T^m\| \|T^n\| , \quad \|T^n\| \leqq \|T\|^n , \quad m, n = 0, 1, 2, \ldots . \tag{4.9}$$

We shall show that $\lim_{n \to \infty} \|T^n\|^{1/n}$ *exists and is equal to* $\inf_{n = 1, 2, \ldots} \|T^n\|^{1/n}$. This limit is called the *spectral radius* of T and will be denoted by $\operatorname{spr} T$. As is seen later, $\operatorname{spr} T$ is independent of the norm employed in its definition.

Set $a_n = \log \|T^n\|$. What is to be proved is that

$$a_n/n \to b \equiv \inf_{n = 1, 2, \ldots} a_n/n . \tag{4.10}$$

The inequality (4.9) gives

$$a_{m+n} \leqq a_m + a_n . \tag{4.11}$$

(Such a sequence $\{a_n\}$ is said to be *subadditive*.) For a fixed positive integer m, set $n = mq + r$, where q, r are nonnegative integers with $0 \leqq r < m$. Then (4.11) gives $a_n \leqq q\, a_m + a_r$ and

$$\frac{a_n}{n} \leqq \frac{q}{n} a_m + \frac{1}{n} a_r .$$

If $n \to \infty$ for a fixed m, $q/n \to 1/m$ whereas r is restricted to one of the numbers 0, 1, ..., $m-1$. Hence $\limsup_{n\to\infty} a_n/n \leqq a_m/m$. Since m is arbitrary, we have $\limsup a_n/n \leqq b$. On the other hand, we have $a_n/n \geqq b$ and so $\liminf a_n/n \geqq b$. This proves (4.10)[1].

Remark 4.2. The above result may lead one to conjecture that $\|T^n\|^{1/n}$ is monotone nonincreasing. This is not true, however, as is seen from the following example. Let $\mathsf{X} = C^2$ with the norm given by (1.17) (X is a two-dimensional unitary space, see § 6). Let T be given by the matrix

$$T = \begin{pmatrix} 0 & a^2 \\ b^2 & 0 \end{pmatrix}, \quad a > b > 0 .$$

It is easily seen that (1 is the unit matrix)

$$T^{2n} = a^{2n} b^{2n} 1 , \quad T^{2n+1} = a^{2n} b^{2n} T , \quad \|T^{2n}\|^{1/2n} = a\, b ,$$
$$\|T\| = a^2 , \quad \|T^{2n+1}\|^{1/(2n+1)} = a\, b\, (a/b)^{1/(2n+1)} > a\, b .$$

Next let us consider $\|T^{-1}\|$ and deduce an inequality estimating $\|T^{-1}\|$ in terms of $\|T\|$ and $\det T$, assuming that T is nonsingular. The relation $Tu = v$ is expressed by the linear equations (3.6). Solving these equations for ξ_k, we obtain a fractional expression of which the denominator is $\det T$ and the numerator is a linear combination of the η_k with coefficients that are equal to minors of the matrix (τ_{jk}). These minors are polynomials in the τ_{jk} of degree $N-1$ where $N = \dim \mathsf{X}$. In virtue of the inequalities $\|u\| \leqq \gamma' \max|\xi_j|$ [see (1.22)], $|\tau_{jk}| \leqq \gamma'' \|T\|$ [see (4.4)] and $|\eta_j| \leqq \gamma''' \|v\|$ [see (1.21)], it follows that there is a constant γ such that $\|u\| \leqq \gamma \|v\| \, \|T\|^{N-1}/|\det T|$ or

$$\|T^{-1}\| \leqq \gamma \frac{\|T\|^{N-1}}{|\det T|} . \tag{4.12}$$

The constant γ is independent of T, depending only on the norm employed[2].

3. Examples of norms

Since the norm $\|T\|$ of an operator T is determined in correlation with the norms adopted in the domain and the range spaces X, Y, there is not so much freedom in the choice of $\|T\|$ as in the choice of the norms for vectors. For the same reason it is not always easy to compute the exact value of $\|T\|$. It is often required, however, to estimate an upper bound of $\|T\|$. We shall illustrate by examples how such an estimate is obtained.

Most commonly used among the norms of vectors is the *p-norm* defined by

$$\|u\| = \|u\|_p = \Big(\sum_j |\xi_j|^p\Big)^{1/p} \tag{4.13}$$

with a fixed $p \geqq 1$, where the ξ_j are the coefficients of u with respect to a fixed basis $\{x_j\}$ (which will be called the *canonical basis*). The conditions (1.18) for a

---

[1] See PÓLYA and SZEGÖ [1], p. 17. Cf. also HILLE and PHILLIPS [1], pp. 124, 244.

[2] We can set $\gamma = 1$ if X is a unitary space; see T. KATO [13].

norm are satisfied by this p-norm [the third condition of (1.18) is known as the Minkowski inequality[1]]. The special cases $p = 1$ and 2 were mentioned before; see (1.16) and (1.17). The norm $\|u\| = \max|\xi_j|$ given by (1.15) can be regarded as the limiting case of (4.13) for $p = \infty$.

Suppose now that the p-norm is given in X and Y with the same p. We shall estimate the corresponding norm $\|T\|$ of an operator T on X to Y in terms of the matrix (τ_{jk}) of T with respect to the canonical bases of X and Y. If $v = Tu$, the coefficients ξ_k and η_j of u and v, respectively, are related by the equations (3.6). Let

$$\tau'_j = \sum_k |\tau_{jk}|, \quad \tau''_k = \sum_j |\tau_{jk}|, \tag{4.14}$$

be the row sums and the column sums of the matrix $(|\tau_{jk}|)$. (3.6) then gives

$$\frac{|\eta_j|}{\tau'_j} \leqq \sum_k \frac{|\tau_{jk}|}{\tau'_j} |\xi_k| .$$

Since the nonnegative numbers $|\tau_{jk}|/\tau'_j$ for $k = 1, \ldots, N$ with a fixed j have the sum 1, the right member is a weighted average of the $|\xi_k|$. Since λ^p is a convex function of $\lambda \geqq 0$, the p-th power of the right member does not exceed the weighted average of the $|\xi_k|^p$ with the same weights[2]. Thus

$$\left(\frac{|\eta_j|}{\tau'_j}\right)^p \leqq \sum_k \frac{|\tau_{jk}|}{\tau'_j} |\xi_k|^p ,$$

and we have successively

$$|\eta_j|^p \leqq \tau'^{\,p-1}_j \sum_k |\tau_{jk}|\, |\xi_k|^p \leqq \left(\max_j \tau'_j\right)^{p-1} \sum_k |\tau_{jk}|\, |\xi_k|^p ,$$

$$\|v\|^p = \sum_j |\eta_j|^p \leqq \left(\max_j \tau'_j\right)^{p-1} \sum_k \tau''_k\, |\xi_k|^p \leqq \left(\max_j \tau'_j\right)^{p-1} \left(\max_k \tau''_k\right) \|u\|^p ,$$

hence

$$\|Tu\| = \|v\| \leqq \left(\max_j \tau'_j\right)^{1-\frac{1}{p}} \left(\max_k \tau''_k\right)^{\frac{1}{p}} \|u\| . \tag{4.15}$$

This shows that[3]

$$\|T\| \leqq \left(\max_j \tau'_j\right)^{1-\frac{1}{p}} \left(\max_k \tau''_k\right)^{\frac{1}{p}} . \tag{4.16}$$

If $p = 1$, the first factor on the right of (4.16) is equal to 1 and does not depend on the τ'_j. On letting $p \to \infty$, it is seen that (4.16) is true also for $p = \infty$; then the second factor on the right is 1 and does not depend on the τ''_k.

Problem 4.3. If (τ_{jk}) is a diagonal matrix ($\tau_{jk} = 0$ for $j \neq k$), we have for any p

$$\|T\| \leqq \max_j |\tau_{jj}| . \tag{4.17}$$

4. Infinite series of operators

The convergence of an infinite series of operators $\sum T_n$ can be defined as for infinite series of vectors and need not be repeated here. Similarly, the *absolute convergence* of such a series means that the series

---

[1] The proof may be found in any textbook on real analysis. See e. g. HARDY, LITTLEWOOD and PÓLYA [1], p. 31; ROYDEN [1], p. 97.

[2] For convex functions, see e. g. HARDY, LITTLEWOOD and PÓLYA [1], p. 70.

[3] Actually this is a simple consequence of the convexity theorem of M. RIESZ (see HARDY, LITTLEWOOD and PÓLYA [1], p. 203).

$\sum \|T_n\|$ is convergent for some (and hence for any) norm $\|\ \|$. In this case $\sum T_n$ is convergent with $\|\sum T_n\| \leqq \sum \|T_n\|$.

Owing to the possibility of multiplication of operators, there are certain formulas for series of operators that do not exist for vectors. For example, we have

$$S(\sum T_n) = \sum S T_n\,, \quad (\sum T_n)\, S = \sum T_n S\,, \tag{4.18}$$

whenever $\sum T_n$ is convergent and the products are meaningful. This follows from the continuity of $S T$ as function of S and T. Two absolutely convergent series can be multiplied term by term, that is

$$(\sum S_m)\,(\sum T_n) = \sum S_m T_n \tag{4.19}$$

if the products are meaningful. Here the order of the terms on the right is arbitrary (or it may be regarded as a double series). The proof is not essentially different from the case of numerical series and may be omitted.

Example 4.4 (Exponential function)

$$e^{tT} = \exp(t\,T) = \sum_{n=0}^{\infty} \frac{1}{n!}\, t^n\, T^n\,, \quad T \in \mathscr{B}(\mathsf{X})\,. \tag{4.20}$$

This series is absolutely convergent for every complex number t, for the n-th term is majorized by $|t|^n \|T\|^n/n!$ in norm. We have

$$\|e^{tT}\| \leqq e^{|t|\,\|T\|}\,. \tag{4.21}$$

Example 4.5 (Neumann series)

$$(1 - T)^{-1} = \sum_{n=0}^{\infty} T^n\,, \ \|(1 - T)^{-1}\| \leqq (1 - \|T\|)^{-1}\,, \ T \in \mathscr{B}(\mathsf{X}). \tag{4.22}$$

This series is absolutely convergent for $\|T\| < 1$ in virtue of $\|T^n\| \leqq \|T\|^n$. Denoting the sum by S, we have $T S = S T = S - 1$ by term by term multiplication. Hence $(1 - T)\, S = S(1 - T) = 1$ and $S = (1 - T)^{-1}$. It follows that *an operator* $R \in \mathscr{B}(\mathsf{X})$ *is nonsingular if* $\|1 - R\| < 1$.

It should be noted that whether or not $\|T\| < 1$ (or $\|1 - R\| < 1$) may depend on the norm employed in X; it may well happen that $\|T\| < 1$ holds for some norm but not for another.

Problem 4.6. The series (4.22) is absolutely convergent if $\|T^m\| < 1$ for some positive integer m or, equivalently, if $\operatorname{spr} T < 1$ (for $\operatorname{spr} T$ see § 4.2), and the sum is again equal to $(1 - T)^{-1}$.

In the so-called *iteration method* in solving the linear equation $(1 - T)\, u = v$ for u, the partial sums $S_n = \sum_{k=0}^{n} T^k$ are taken as approximations for $S = (1 - T)^{-1}$ and $u_n = S_n v$ as approximations for the solution $u = S v$. The errors incurred in such approximations can be estimated by

$$\|S - S_n\| = \left\| \sum_{k=n+1}^{\infty} T^k \right\| \leq \sum_{k=n+1}^{\infty} \|T\|^k = \frac{\|T\|^{n+1}}{1 - \|T\|} . \tag{4.23}$$

For $n = 0$ (4.23) gives $\|(1 - T)^{-1} - 1\| \leq \|T\| (1 - \|T\|)^{-1}$. With $R = 1 - T$, this shows that $R \to 1$ implies $R^{-1} \to 1$. In other words, R^{-1} is a continuous function of R at $R = 1$. This is a special case of the theorem that *T^{-1} is a continuous function of T*. More precisely, if $T \in \mathscr{B}(X, Y)$ is nonsingular, any $S \in \mathscr{B}(X, Y)$ with sufficiently small $\|S - T\|$ is also nonsingular, and $\|S^{-1} - T^{-1}\| \to 0$ for $\|T - S\| \to 0$. In particular, *the set of all nonsingular elements of $\mathscr{B}(X, Y)$ is open*. [Of course X and Y must have the same dimension if there exist nonsingular elements of $\mathscr{B}(X, Y)$.]

To see this we set $A = S - T$ and assume that $\|A\| < 1/\|T^{-1}\|$. Then $\|A T^{-1}\| \leq \|A\| \|T^{-1}\| < 1$, and so $1 + A T^{-1}$ is a nonsingular operator of $\mathscr{B}(Y)$ by the above result. Since $S = T + A = (1 + A T^{-1}) T$, S is also nonsingular with $S^{-1} = T^{-1} (1 + A T^{-1})^{-1}$.

Using the estimates for $\|(1 + A T^{-1})^{-1}\|$ and $\|(1 + A T^{-1})^{-1} - 1\|$ given by (4.22) and (4.23), we obtain the following estimates for $\|S^{-1}\|$ and $\|S^{-1} - T^{-1}\|$:

$$\|S^{-1}\| \leq \frac{\|T^{-1}\|}{1 - \|A\| \|T^{-1}\|}, \quad \|S^{-1} - T^{-1}\| \leq \frac{\|A\| \|T^{-1}\|^2}{1 - \|A\| \|T^{-1}\|} \tag{4.24}$$
$$\text{for } S = T + A, \quad \|A\| < 1/\|T^{-1}\| .$$

Remark 4.7. We assumed above that $\|A\| < \|T^{-1}\|$ to show the existence of S^{-1}. This condition can be weakened if $X = Y$ and $TS = ST$. In this case A commutes with T and hence with T^{-1}. Therefore

$$\begin{aligned} \operatorname{spr} A T^{-1} &= \lim \|(A T^{-1})^n\|^{1/n} = \lim \|A^n T^{-n}\|^{1/n} \leq \\ &\leq [\lim \|A^n\|^{1/n}] [\lim \|T^{-n}\|^{1/n}] = (\operatorname{spr} A) (\operatorname{spr} T^{-1}) . \end{aligned} \tag{4.25}$$

It follows that $S^{-1} = T^{-1}(1 + A T^{-1})^{-1}$ exists if

$$\operatorname{spr} A < (\operatorname{spr} T^{-1})^{-1} . \tag{4.26}$$

5. Operator-valued functions

Operator-valued functions $T_t = T(t)$ defined for a real or complex variable t and taking values in $\mathscr{B}(X, Y)$ can be defined and treated just as vector-valued functions $u(t)$ were in § 1.7. A new feature for $T(t)$ appears again since the products $T(t) u(t)$ and $S(t) T(t)$ are defined. Thus we have, for example, the formulas

$$\begin{aligned} \frac{d}{dt} T(t) u(t) &= T'(t) u(t) + T(t) u'(t) , \\ \frac{d}{dt} T(t) S(t) &= T'(t) S(t) + T(t) S'(t) , \end{aligned} \tag{4.27}$$

whenever the products are meaningful and the derivatives on the right exist. Also we have

$$\frac{d}{dt} T(t)^{-1} = - T(t)^{-1} T'(t) T(t)^{-1} \tag{4.28}$$

whenever $T(t)^{-1}$ and $T'(t)$ exist. This follows from the identity

$$S^{-1} - T^{-1} = - S^{-1}(S - T) T^{-1} \tag{4.29}$$

and the continuity of T^{-1} as function of T proved in par. 4.

For the integrals of operator-valued functions, we have formulas similar to (1.30). In addition, we have

$$\begin{aligned} &\int S u(t)\, dt = S \int u(t)\, dt\,, \quad \int T(t)\, u\, dt = \left(\int T(t)\, dt\right) u\,, \\ &\int S T(t)\, dt = S \int T(t)\, dt\,, \quad \int T(t)\, S\, dt = \left(\int T(t)\, dt\right) S\,. \end{aligned} \tag{4.30}$$

Of particular importance again are *holomorphic* functions $T(t)$ of a complex variable t; here the same remarks apply as those given for vector-valued functions (see § 1.7). It should be added that $S(t)\, T(t)$ and $T(t)\, u(t)$ are holomorphic if all factors are holomorphic, and that $T(t)^{-1}$ is holomorphic whenever $T(t)$ is holomorphic and $T(t)^{-1}$ exists [the latter follows from (4.28)].

Example 4.8. The exponential function e^{tT} defined by (4.20) is an *entire function* of t (holomorphic in the whole complex plane), with

$$\frac{d}{dt} e^{tT} = T e^{tT} = e^{tT} T\,. \tag{4.31}$$

Example 4.9. Consider the Neumann series

$$S(t) = (1 - tT)^{-1} = \sum_{n=0}^{\infty} t^n T^n \tag{4.32}$$

with a complex parameter t. By Problem 4.6 this series is absolutely convergent for $|t| < 1/\operatorname{spr} T$. Actually, *the convergence radius r of* (4.32) *is exactly equal to* $1/\operatorname{spr} T$. Since $S(t)$ is holomorphic for $|t| < r$, the Cauchy inequality gives $\|T^n\| \leqq M_{r'} r'^{-n}$ for all n and $r' < r$ as in the case of numerical power series[1] ($M_{r'}$ is independent of n). Hence $\operatorname{spr} T = \lim \|T^n\|^{1/n} \leqq r'^{-1}$ and, going to the limit $r' \to r$, we have $\operatorname{spr} T \leqq r^{-1}$ or $r \leqq 1/\operatorname{spr} T$. Since the opposite inequality was proved above, this gives the proof of the required result. Incidentally, it follows that $\operatorname{spr} T$ is independent of the norm used in its definition.

6. Pairs of projections

As an application of analysis with operators and also with a view to later applications, we shall prove some theorems concerning a pair of

[1] We have $T^n = (2\pi i)^{-1} \int_{|t|=r'} t^{-n-1} S(t)\, dt$ and so $\|T^n\| \leqq (2\pi)^{-1} \int_{|t|=r'} r'^{-n-1} \cdot \|S(t)\|\, |dt| \leqq r'^{-n} M_{r'}$, where $M_{r'} = \max_{|t|=r'} \|S(t)\| < \infty$.

projections (idempotents)[1]. As defined in § 3.4, a projection P is an operator of $\mathscr{B}(\mathsf{X})$ such that $P^2 = P$. $1 - P$ is a projection with P.

Let $P, Q \in \mathscr{B}(\mathsf{X})$ be two projections. Then

$$R = (P - Q)^2 = P + Q - PQ - QP \tag{4.33}$$

commutes with P and Q; this is seen by noting that $PR = P - PQP = RP$ and similarly for Q. For the same reason $(1 - P - Q)^2$ commutes with P and Q because $1 - P$ is a projection. Actually we have the identity

$$(P - Q)^2 + (1 - P - Q)^2 = 1 \tag{4.34}$$

as is verified by direct computation. Another useful identity is

$$(PQ - QP)^2 = (P - Q)^4 - (P - Q)^2 = R^2 - R\,, \tag{4.35}$$

the proof of which is again straightforward and will be left to the reader.

Set

$$U' = QP + (1 - Q)(1 - P)\,, \quad V' = PQ + (1 - P)(1 - Q)\,. \tag{4.36}$$

U' maps $\mathsf{R}(P) = P\mathsf{X}$ into $Q\mathsf{X}$ and $(1 - P)\mathsf{X}$ into $(1 - Q)\mathsf{X}$, whereas V' maps $Q\mathsf{X}$ into $P\mathsf{X}$ and $(1 - Q)\mathsf{X}$ into $(1 - P)\mathsf{X}$. But these mappings are not inverse to each other; in fact it is easily seen that

$$V'U' = U'V' = 1 - R\,. \tag{4.37}$$

A pair of mutually inverse operators U, V with the mapping properties stated above can be constructed easily, however, since R commutes with P, Q and therefore with U', V' too. It suffices to set

$$\begin{aligned} U &= U'(1 - R)^{-1/2} = (1 - R)^{-1/2}\, U'\,, \\ V &= V'(1 - R)^{-1/2} = (1 - R)^{-1/2}\, V'\,, \end{aligned} \tag{4.38}$$

provided the inverse square root $(1 - R)^{-1/2}$ of $1 - R$ exists. A natural definition of this operator is given by the binomial series

$$(1 - R)^{-1/2} = \sum_{n=0}^{\infty} \binom{-1/2}{n} (-R)^n\,. \tag{4.39}$$

This series is absolutely convergent if $\|R\| < 1$ or, more generally, if

$$\operatorname{spr} R < 1\,, \tag{4.40}$$

[1] The following results, which are taken from T. Kato [9], are true even when X is an ∞-dimensional Banach space. For the special case of projections in a unitary (Hilbert) space, see § 6.8. For related results cf. Akhiezer and Glazman [1], Sz.-Nagy [1], [2], Wolf [1].

and the sum T of this series satisfies the relation $T^2 = (1 - R)^{-1}$ just as in the numerical binomial series. Thus[1]

$$VU = UV = 1, \quad V = U^{-1}, \quad U = V^{-1}. \tag{4.41}$$

Since $U'P = QP = QU'$ and $PV' = PQ = V'Q$ as is seen from (4.36), we have $UP = QU$ and $PV = VQ$ by the commutativity of R with all other operators here considered. Thus we have

$$Q = UPU^{-1}, \quad P = U^{-1}QU. \tag{4.42}$$

Thus P and Q are *similar* to each other (see § 5.7). They are isomorphic to each other in the sense that any linear relationship such as $v = Pu$ goes over to $v' = Qu'$ by the one-to-one linear mapping $U: u' = Uu$, $v' = Uv$. In particular their ranges $P\mathsf{X}$, $Q\mathsf{X}$ are *isomorphic*, being mapped onto each other by U and U^{-1}. Thus

$$\dim P = \dim Q, \quad \dim(1 - P) = \dim(1 - Q). \tag{4.43}$$

An immediate consequence of this result is

Lemma 4.10. *Let $P(t)$ be a projection depending continuously on a parameter t varying in a (connected) region of real or complex numbers. Then the ranges $P(t)\mathsf{X}$ for different t are isomorphic to one another. In particular* $\dim P(t)\mathsf{X}$ *is constant.*

To see this, it suffices to note that $\|P(t') - P(t'')\| < 1$ for sufficiently small $|t' - t''|$ so that the above result applies to the pair $P(t'), P(t'')$.

Problem 4.11. Under the assumption (4.40), we have $PQ\mathsf{X} = P\mathsf{X}$, $QP\mathsf{X} = Q\mathsf{X}$ [hint: $PQ = PV' = PU^{-1}(1 - R)^{1/2}$].

Problem 4.12. For any two projections P, Q, we have

$$(1 - P + QP)(1 - Q + PQ) = 1 - R \quad [R \text{ is given by (4.33)}], \tag{4.44}$$

$$(1 - P + QP)^{-1} = (1 - R)^{-1}(1 - Q + PQ) \quad \text{if } \operatorname{spr} R < 1. \tag{4.45}$$

If $\operatorname{spr} R < 1$, $W = 1 - P + QP$ maps $P\mathsf{X}$ onto $Q\mathsf{X}$ and $Wu = u$ for $u \in (1 - P)\mathsf{X}$, while W^{-1} maps $Q\mathsf{X}$ onto $P\mathsf{X}$ and $W^{-1}u = u$ for $u \in (1 - P)\mathsf{X}$, and we have $\mathsf{X} = Q\mathsf{X} \oplus (1 - P)\mathsf{X}$.

Problem 4.13. For any two projections P, Q such that $\operatorname{spr}(P - Q)^2 < 1$, there is a family $P(t)$, $0 \leqq t \leqq 1$, of projections depending holomorphically on t such that $P(0) = P$, $P(1) = Q$. [hint: set $2P(t) = 1 + (2P - 1 + 2t(Q - P)) \cdot (1 - 4t(1 - t)R)^{-1/2}$.]

§ 5. The eigenvalue problem

1. Definitions

In this section X denotes a given vector space with $0 < \dim \mathsf{X} = N < \infty$, but we shall consider X a normed space whenever convenient by introducing an appropriate norm.

---

[1] As is shown later (§ 6.7), U and V are unitary if X is a unitary space and P, Q are orthogonal projections. The same U appears also in Sz.-Nagy [1] and Wolf [1] with a different expression; its identity with the above U was shown in T. Kato [9].

Let $T \in \mathscr{B}(\mathsf{X})$. A complex number λ is called an *eigenvalue (proper value, characteristic value)* of T if there is a non-zero vector $u \in \mathsf{X}$ such that

(5.1) $$Tu = \lambda u .$$

u is called an *eigenvector (proper vector, characteristic vector)* of T belonging to (associated with, etc.) the eigenvalue λ. The set N_λ of all $u \in \mathsf{X}$ such that $Tu = \lambda u$ is a linear manifold of X; it is called the *(geometric) eigenspace* of T for the eigenvalue λ, and dim N_λ is called the *(geometric) multiplicity* of λ. N_λ is defined even when λ is not an eigenvalue; then we have $\mathsf{N}_\lambda = 0$. In this case it is often convenient to say that N_λ is the eigenspace for the eigenvalue λ with multiplicity zero, though this is not in strict accordance with the definition of an eigenvalue[1].

Problem 5.1. λ is an eigenvalue of T if and only if $\lambda - \zeta$ is an eigenvalue of $T - \zeta$. N_λ is the null space of $T - \lambda$, and the geometric multiplicity of the eigenvalue λ of T is the nullity of $T - \lambda$. $T - \lambda$ is singular if and only if λ is an eigenvalue of T.

It can easily be proved that *eigenvectors of T belonging to different eigenvalues are linearly independent*. It follows that *there are at most N eigenvalues of T*. The set of all eigenvalues of T is called the *spectrum* of T; we denote it by $\Sigma(T)$. Thus $\Sigma(T)$ is a finite set with not more than N points.

The *eigenvalue problem* consists primarily in finding all eigenvalues and eigenvectors (or eigenspaces) of a given operator T. A vector $u \neq 0$ is an eigenvector of T if and only if the one-dimensional linear manifold $[u]$ spanned by u is invariant under T (see § 3.5). Thus the eigenvalue problem is a special case of the problem of determining all invariant linear manifolds for T (a generalized form of the eigenvalue problem).

If M is an invariant subspace of T, the part T_M of T in M is defined. As is easily seen, any eigenvalue [eigenvector] of T_M is an eigenvalue [eigenvector] for T. For convenience an eigenvalue of T_M is called an *eigenvalue of T in M*.

If there is a projection P that commutes with T, T is decomposed according to the decomposition $\mathsf{X} = \mathsf{M} \oplus \mathsf{N}$, $\mathsf{M} = P\mathsf{X}$, $\mathsf{N} = (1 - P)\,\mathsf{X}$ (see § 3.5). To solve the eigenvalue problem for T, it is then sufficient to consider the eigenvalue problem for the parts of T in M and in N[2].

[1] There are generalized eigenvalue problems (sometimes called *nonlinear eigenvalue problems*) in which one seeks solutions of an equation $T(\lambda)\, u = 0$, where $T(\lambda)$ is a linear operator depending on a parameter λ; for example, $T(\lambda) = T_0 + \lambda T_1 + \cdots + \lambda^n T_n$. In general a solution $u \neq 0$ will exist only for some particular values of λ (eigenvalues). A special problem of this kind will be considered later (VII, § 1.3).

[2] If $Tu = \lambda u$, then $TPu = PTu = \lambda Pu$ so that $Pu \in \mathsf{M}$ is, if not 0, an eigenvector of T (and of T_M) for the eigenvalue λ, and similarly for $(1 - P)\, u$. Thus any eigenvalue of T must be an eigenvalue of at least one of T_M and T_N, and any eigenvector of T is the sum of eigenvectors of T_M and T_N for the same eigenvalue. The eigenspace of T for λ is the direct sum of the eigenspaces of T_M and T_N for λ.

The part T_{M} of T in M may be identified with the operator $PT = TP = PTP$ in the sense stated in Remark 3.15. It should be noticed, however, that TP has an eigenvalue zero with the eigenspace N and therefore with multiplicity $N - m$ (where $m = \dim P\mathsf{X}$) in addition to the eigenvalues of T_{M}[1].

Problem 5.2. No eigenvalue of T exceeds $\|T\|$ in absolute value, where $\| \ \|$ is the norm of $\mathscr{B}(\mathsf{X})$ associated with any norm of X.

Problem 5.3. If T has a diagonal matrix with respect to a basis, the eigenvalues of T coincide with (the different ones among) all its diagonal elements.

2. The resolvent

Let $T \in \mathscr{B}(\mathsf{X})$ and consider the inhomogeneous linear equation

$$(T - \zeta)\, u = v\,, \tag{5.2}$$

where ζ is a given complex number, $v \in \mathsf{X}$ is given and $u \in \mathsf{X}$ is to be found. In order that this equation have a solution u for every v, it is necessary and sufficient that $T - \zeta$ be nonsingular, that is, ζ be different from any eigenvalue λ_h of T. Then the inverse $(T - \zeta)^{-1}$ exists and the solution u is given by

$$u = (T - \zeta)^{-1}\, v\,. \tag{5.3}$$

The operator-valued function

$$R(\zeta) = R(\zeta, T) = (T - \zeta)^{-1} \tag{5.4}$$

is called the ***resolvent***[2] of T. The complementary set of the spectrum $\Sigma(T)$ (that is, the set of all complex numbers different from any of the eigenvalues of T) is called the ***resolvent set*** of T and will be denoted by $\mathrm{P}(T)$. The resolvent $R(\zeta)$ is thus defined for $\zeta \in \mathrm{P}(T)$.

Problem 5.4. $R(\zeta)$ commutes with T. $R(\zeta)$ has exactly the eigenvalues $(\lambda_h - \zeta)^{-1}$.

An important property of the resolvent is that it satisfies the (first) *resolvent equation*

$$R(\zeta_1) - R(\zeta_2) = (\zeta_1 - \zeta_2)\, R(\zeta_1)\, R(\zeta_2)\,. \tag{5.5}$$

This is seen easily if one notes that the left member is equal to $R(\zeta_1) \cdot (T - \zeta_2)\, R(\zeta_2) - R(\zeta_1)\,(T - \zeta_1)\, R(\zeta_2)$. In particular, (5.5) implies that *$R(\zeta_1)$ and $R(\zeta_2)$ commute*. Also we have $R(\zeta_1) = [1 - (\zeta_2 - \zeta_1)\, R(\zeta_1)] \cdot$

---

[1] Strictly speaking, this is true only when T_{M} has no eigenvalue 0. If T_{M} has the eigenvalue 0 with the eigenspace L, the eigenspace for the eigenvalue 0 of TP is $\mathsf{N} \oplus \mathsf{L}$.

[2] The resolvent is the operator-valued function $\zeta \to R(\zeta)$. It appears, however, that also the value $R(\zeta)$ of this function at a particular ζ is customarily called the resolvent. Sometimes $(\zeta - T)^{-1}$ instead of $(T - \zeta)^{-1}$ is called the resolvent. In this book we follow the definition of STONE [1].

$\cdot R(\zeta_2)$. According to the results on the Neumann series (see Example 4.5), this leads to the expansion

$$(5.6)\quad R(\zeta) = [1 - (\zeta - \zeta_0)\, R(\zeta_0)]^{-1} R(\zeta_0) = \sum_{n=0}^{\infty} (\zeta - \zeta_0)^n R(\zeta_0)^{n+1},$$

the series being absolutely convergent at least if

$$(5.7)\quad |\zeta - \zeta_0| < \|R(\zeta_0)\|^{-1}$$

for *some* norm. We shall refer to (5.6) as the *first Neumann series for the resolvent*.

(5.6) shows that $R(\zeta)$ is holomorphic in ζ with the Taylor series[1] shown on the right; hence

$$(5.8)\quad \left(\frac{d}{d\zeta}\right)^n R(\zeta) = n!\, R(\zeta)^{n+1}, \quad n = 1, 2, 3, \ldots.$$

According to Example 4.9, the convergence radius of the series of (5.6) is equal to $1/\operatorname{spr} R(\zeta_0)$. Hence this series is convergent if and only if

$$(5.9)\quad |\zeta - \zeta_0| < 1/\operatorname{spr} R(\zeta_0) = (\lim \|R(\zeta_0)^n\|^{1/n})^{-1}.$$

For large $|\zeta|$, $R(\zeta)$ has the expansion

$$(5.10)\quad R(\zeta) = -\zeta^{-1}(1 - \zeta^{-1} T)^{-1} = -\sum_{n=0}^{\infty} \zeta^{-n-1} T^n,$$

which is convergent if and only if $|\zeta| > \operatorname{spr} T$; thus $R(\zeta)$ is holomorphic at infinity.

Problem 5.5. We have

$$(5.11)\quad \|R(\zeta)\| \leqq (|\zeta| - \|T\|)^{-1}, \quad \|R(\zeta) + \zeta^{-1}\| \leqq |\zeta|^{-1}(|\zeta| - \|T\|)^{-1}\|T\| \quad \text{for} \quad |\zeta| > \|T\|.$$

The spectrum $\Sigma(T)$ *is never empty*; T *has at least one eigenvalue*. Otherwise $R(\zeta)$ would be an entire function such that $R(\zeta) \to 0$ for $\zeta \to \infty$ [see (5.11)]; then we must have $R(\zeta) = 0$ by Liouville's theorem[2]. But this gives the contradiction $1 = (T - \zeta)\, R(\zeta) = 0$[3].

It is easily seen that each eigenvalue of T is a singularity[4] of the analytic function $R(\zeta)$. Since there is at least one singularity of $R(\zeta)$ on the convergence circle[5] $|\zeta| = \operatorname{spr} T$ of (5.10), $\operatorname{spr} T$ coincides with the

---

[1] This is an example of the use of function theory for operator-valued functions; see KNOPP [1], p. 79.

[2] See KNOPP [1], p. 113. Liouville's theorem implies that $R(\zeta)$ is constant; since $R(\zeta) \to 0$ for $\zeta \to \infty$, this constant is 0.

[3] Note that we assume $\dim \mathsf{X} > 0$.

[4] Suppose λ is a regular point (removable singularity) of the analytic function $R(\zeta)$. Then $\lim_{\zeta \to \lambda} R(\zeta) = R$ exists and so $(T - \lambda) R = \lim_{\zeta \to \lambda} (T - \zeta)\, R(\zeta) = 1$. Thus $(T - \lambda)^{-1} = R$ exists and λ is not an eigenvalue.

[5] See KNOPP [1], p. 101.

largest (in absolute value) eigenvalue of T:

(5.12) $$\operatorname{spr} T = \max_h |\lambda_h| .$$

This shows again that $\operatorname{spr} T$ is independent of the norm used in its definition.

Problem 5.6. $\operatorname{spr} T = 0$ if and only if T is nilpotent. [hint for "only if" part: If $\operatorname{spr} T = 0$, we have also $\operatorname{spr} T_M = 0$ for any part T_M of T in an invariant subspace M, so that T_M is singular. Thus the part of T in each of the invariant subspaces $T^n X$ (see Problem 3.12) is singular. Hence $X \supset TX \supset T^2X \supset \cdots$ with all inclusions $\supset$ being proper until 0 is reached.]

3. Singularities of the resolvent

The singularities of $R(\zeta)$ are exactly the eigenvalues λ_h, $h = 1, \ldots, s$, of T. Let us consider the Laurent series[1] of $R(\zeta)$ at $\zeta = \lambda_h$. For simplicity we may assume for the moment that $\lambda_h = 0$ and write

(5.13) $$R(\zeta) = \sum_{n=-\infty}^{\infty} \zeta^n A_n .$$

The coefficients A_n are given by

(5.14) $$A_n = \frac{1}{2\pi i} \int_\Gamma \zeta^{-n-1} R(\zeta)\, d\zeta ,$$

where Γ is a positively-oriented small circle enclosing $\zeta = 0$ but excluding other eigenvalues of T. Since Γ may be expanded to a slightly larger circle Γ' without changing (5.14), we have

$$\begin{aligned} A_n A_m &= \left(\frac{1}{2\pi i}\right)^2 \int_{\Gamma'}\int_{\Gamma} \zeta^{-n-1} \zeta'^{-m-1} R(\zeta) R(\zeta')\, d\zeta\, d\zeta' \\ &= \left(\frac{1}{2\pi i}\right)^2 \int_{\Gamma'}\int_{\Gamma} \zeta^{-n-1} \zeta'^{-m-1} (\zeta' - \zeta)^{-1} [R(\zeta') - R(\zeta)]\, d\zeta\, d\zeta' , \end{aligned}$$

where the resolvent equation (5.5) has been used. The double integral on the right may be computed in any order. Considering that Γ' lies outside Γ, we have

(5.15) $$\begin{aligned} \frac{1}{2\pi i} \int_\Gamma \zeta^{-n-1} (\zeta' - \zeta)^{-1}\, d\zeta &= \eta_n \zeta'^{-n-1} \\ \frac{1}{2\pi i} \int_{\Gamma'} \zeta'^{-m-1} (\zeta' - \zeta)^{-1}\, d\zeta' &= (1 - \eta_m)\, \zeta^{-m-1} \end{aligned}$$

where the symbol η_n is defined by

(5.16) $$\eta_n = 1 \quad \text{for} \quad n \geqq 0 \quad \text{and} \quad \eta_n = 0 \quad \text{for} \quad n < 0 .$$

---

[1] See KNOPP [1], p. 117.

Thus

$$(5.17)\quad A_n A_m = \frac{\eta_n + \eta_m - 1}{2\pi i} \int_\Gamma \zeta^{-n-m-2} R(\zeta)\, d\zeta = (\eta_n + \eta_m - 1)\, A_{n+m+1}.$$

For $n = m = -1$ this gives $A_{-1}^2 = -A_{-1}$. Thus $-A_{-1}$ is a projection, which we shall denote by P. For $n, m < 0$, (5.17) gives $A_{-2}^2 = -A_{-3}$, $A_{-2} A_{-3} = -A_{-4}$, ... On setting $-A_{-2} = D$, we thus obtain $A_{-k} = -D^{k-1}$ for $k \geqq 2$. Similarly we obtain $A_n = S^{n+1}$ for $n \geqq 0$ with $S = A_0$.

Returning to the general case in which $\zeta = \lambda_h$ is the singularity instead of $\zeta = 0$, we see that the Laurent series takes the form

$$(5.18)\quad R(\zeta) = -(\zeta - \lambda_h)^{-1} P_h - \sum_{n=1}^{\infty} (\zeta - \lambda_h)^{-n-1} D_h^n + \sum_{n=0}^{\infty} (\zeta - \lambda_h)^n S_h^{n+1}.$$

Setting in (5.17) $n = -1, m = -2$ and then $n = -1, m = 0$, we see that

$$(5.19)\quad P_h D_h = D_h P_h = D_h, \quad P_h S_h = S_h P_h = 0.$$

Thus the two lines on the right of (5.18) represent a decomposition of the operator $R(\zeta)$ according to the decomposition $\mathsf{X} = \mathsf{M}_h \oplus \mathsf{M}_h'$, where $\mathsf{M}_h = P_h \mathsf{X}$ and $\mathsf{M}_h' = (1 - P_h)\mathsf{X}$. As the principal part of a Laurent series at an isolated singularity, the first line of (5.18) is convergent for $\zeta - \lambda_h \neq 0$, so that the part of $R(\zeta)$ in M_h has only the one singularity $\zeta = \lambda_h$, and the spectral radius of D_h must be zero. According to Problem 5.6, it follows that D_h is nilpotent and therefore (see Problem 3.10)

$$(5.20)\quad D_h^{m_h} = 0, \quad m_h = \dim \mathsf{M}_h = \dim P_h.$$

Thus the principal part in the Laurent expansion (5.18) of $R(\zeta)$ is finite, $\zeta = \lambda_h$ being a *pole* of order not exceeding m_h. Since the same is true of all singularities λ_h of $R(\zeta)$, $R(\zeta)$ is a *meromorphic function*[1].

The P_h for different h satisfy the following relations:

$$(5.21)\quad P_h P_k = \delta_{hk} P_h, \quad \sum_{h=1}^{s} P_h = 1, \quad P_h T = T P_h.$$

The first relation can be proved in the same way as we proved $P_h^2 = P_h$ above [which is a special case of (5.17)]; it suffices to notice that

$$(5.22)\quad P_h = -\frac{1}{2\pi i} \int_{\Gamma_h} R(\zeta)\, d\zeta$$

where the circles Γ_h for different h lie outside each other. The second equality of (5.21) is obtained by integrating $R(\zeta)$ along a large circle enclosing all the eigenvalues of T and noting the expansion (5.10) of $R(\zeta)$ at infinity. The commutativity of P_h with T follows immediately from (5.22).

[1] See KNOPP [2], p. 34.

Since $R(\zeta)$ is meromorphic and regular at infinity, the decomposition of $R(\zeta)$ into ***partial fractions***[1] takes the form

$$R(\zeta) = -\sum_{h=1}^{s}\Big[(\zeta-\lambda_h)^{-1}P_h + \sum_{n=1}^{m_h-1}(\zeta-\lambda_h)^{-n-1}D_h^n\Big]. \tag{5.23}$$

Problem 5.7. $\operatorname{spr} R(\zeta) = \big[\min_h |\zeta - \lambda_h|\big]^{-1} = [\operatorname{dist}(\zeta, \Sigma(T))]^{-1}$.

Problem 5.8. We have

$$P_h D_k = D_k P_h = \delta_{hk} D_h;\quad D_h D_k = 0\,,\quad h \neq k\,. \tag{5.24}$$

Problem 5.9. For any simple closed (rectifiable) curve Γ with positive direction and not passing through any eigenvalue λ_h, we have

$$\frac{1}{2\pi i}\int_\Gamma R(\zeta)\,d\zeta = -\sum P_h\,, \tag{5.25}$$

where the sum is taken for those h for which λ_h is inside Γ.

Multiplying (5.14) by T from the left or from the right and noting that $TR(\zeta) = R(\zeta)T = 1 + \zeta R(\zeta)$, we obtain $TA_n = A_nT = \delta_{n0} + A_{n-1}$. If the singularity is at $\zeta = \lambda_h$ instead of $\zeta = 0$, this gives for $n = 0$ and $n = -1$

$$\begin{aligned}(T-\lambda_h)S_h &= S_h(T-\lambda_h) = 1 - P_h\,,\\ P_h(T-\lambda_h) &= (T-\lambda_h)P_h = D_h\,.\end{aligned} \tag{5.26}$$

For each $h = 1, \ldots, s$, the holomorphic part in the Laurent expansion (5.18) will be called the ***reduced resolvent*** of T with respect to the eigenvalue λ_h; we denote it by $S_h(\zeta)$:

$$S_h(\zeta) = \sum_{n=0}^{\infty}(\zeta-\lambda_h)^n S_h^{n+1}\,. \tag{5.27}$$

It follows from (5.19) and (5.26) that

$$S_h = S_h(\lambda_h)\,,\quad S_h(\zeta)P_h = P_h S_h(\zeta) = 0\,, \tag{5.28}$$

$$(T-\zeta)S_h(\zeta) = S_h(\zeta)(T-\zeta) = 1 - P_h\,. \tag{5.29}$$

The last equalities show that the parts of $T - \zeta$ and of $S_h(\zeta)$ in the invariant subspace $\mathsf{M}_h' = (1 - P_h)\mathsf{X}$ are inverse to each other.

Problem 5.10. We have

$$(T-\lambda_h)D_h^n = D_h^{n+1}\,,\quad n = 1, 2, \ldots\,. \tag{5.30}$$

$$(T-\lambda_h)^{m_h}P_h = 0\,. \tag{5.31}$$

$$S_h(\zeta) = -\sum_{k\neq h}\Big[(\zeta-\lambda_k)^{-1}P_k + \sum_{n=1}^{m_k-1}(\zeta-\lambda_k)^{-n-1}D_k^n\Big]. \tag{5.32}$$

Problem 5.11. Each $S_h(\zeta)$ satisfies the resolvent equation [see (5.5)] and

$$\left(\frac{d}{d\zeta}\right)^n S_h(\zeta) = n!\,S_h(\zeta)^{n+1}\,,\quad n = 1, 2, \ldots\,. \tag{5.33}$$

4. The canonical form of an operator

The result of the preceding paragraph leads to the *canonical form* of the operator T. Denoting as above by M_h the range of the projection

[1] See KNOPP [2], p. 34.

$P_h, h = 1, \ldots, s$, we have

$$\mathsf{X} = \mathsf{M}_1 \oplus \cdots \oplus \mathsf{M}_s \,. \tag{5.34}$$

Since the P_h commute with T and with one another, the M_h are invariant subspaces for T and T is decomposed according to the decomposition (5.34) (see § 3.5). M_h is called the *algebraic eigenspace* (or *principal subspace*) for the eigenvalue λ_h of T, and $m_h = \dim \mathsf{M}_h$ is the *algebraic multiplicity* of λ_h. In what follows P_h will be called the *eigenprojection* and D_h the *eigennilpotent* for the eigenvalue λ_h of T. Any vector $u \neq 0$ of M_h is called a *generalized eigenvector* (or *principal vector*) for the eigenvalue λ_h of T.

It follows from (5.26) that

$$T P_h = P_h T = P_h T P_h = \lambda_h P_h + D_h \,, \quad h = 1, \ldots, s \,. \tag{5.35}$$

Thus the part T_{M_h} of T in the invariant subspace M_h is the sum of the scalar operator λ_h and a nilpotent D_{h,M_h}, the part of D_h in M_h. As is easily seen, T_{M_h} has one and only one eigenvalue λ_h.

Addition of the s equations (5.35) gives by (5.21)

$$T = S + D \tag{5.36}$$

where

$$S = \sum_h \lambda_h P_h \tag{5.37}$$

$$D = \sum_h D_h \,. \tag{5.38}$$

An operator S of the form (5.37) where $\lambda_h \neq \lambda_k$ for $h \neq k$ and the P_h satisfy (3.27–3.28) is said to be *diagonalizable* (or *diagonable* or *semisimple*). S is the direct sum (see § 3.5) of scalar operators. D is nilpotent, for it is the direct sum of nilpotents D_h and so $D^n = \sum D_h^n = 0$ for $n \geqq \max m_h$. It follows from (5.24) that D commutes with S.

(5.36) shows that *every operator* $T \in \mathscr{B}(\mathsf{X})$ *can be expressed as the sum of a diagonable operator* S *and a nilpotent* D *that commutes with* S.

The eigenvalue λ_h of T will also be said to be *semisimple* if the associated eigennilpotent D_h is zero, and *simple*[1] if $m_h = 1$ (note that $m_h = 1$ implies $D_h = 0$). T is diagonable if and only if all its eigenvalues are semisimple. T is said to be *simple* if all the eigenvalues λ_h are simple; in this case T has N eigenvalues.

(5.36) is called the *spectral representation* of T. The spectral representation is unique in the following sense: if T is the sum of a diagonable operator S and a nilpotent D that commutes with S, then S and D must be given by (5.37) and (5.38) respectively. To show this, we first note that any operator R that commutes with a diagonable operator S of the form (5.37) commutes with every P_h so that M_h is invariant under R.

[1] An eigenvalue which is not simple is said to be *degenerate*.

In fact, multiplying $RS = SR$ from the left by P_h and from the right by P_k, we have by $P_h P_k = \delta_{hk}$

$$\lambda_k P_h R P_k = \lambda_h P_h R P_k \quad \text{or} \quad P_h R P_k = 0 \quad \text{for} \quad h \neq k .$$

Addition of the results for all $k \neq h$ for fixed h gives $P_h R(1 - P_h) = 0$ or $P_h R = P_h R P_h$. Similarly we obtain $R P_h = P_h R P_h$ and therefore $R P_h = P_h R$.

Suppose now that

$$T = S' + D' , \quad S' = \sum_{h=1}^{s'} \lambda'_h P'_h , \quad D'S' = S'D' ,$$

is a second expression of T in the form (5.36). By what was just proved, we have $D' = \sum D'_h$ with $D'_h = P'_h D' = D' P'_h$. Hence $T - \zeta = \sum [(\lambda'_h - \zeta) P'_h + D'_h]$ and therefore

$$(5.39) \qquad (T - \zeta)^{-1} = - \sum_{h=1}^{s'} [(\zeta - \lambda'_h)^{-1} P'_h + (\zeta - \lambda'_h)^{-2} D'_h + \\ + \cdots + (\zeta - \lambda'_h)^{-N} D'^{N-1}_h] ;$$

this is easily verified by multiplying (5.39) from the left or the right by the expression for $T - \zeta$ given above (note that $D'^N_h = D'^N = 0$ since D' is nilpotent). Since the decomposition of an operator-valued meromorphic function into the sum of partial fractions is unique (just as for numerical functions), comparison of (5.39) with (5.23) shows that s', λ'_h, P'_h and D'_h must coincide with s, λ_h, P_h and D_h respectively. This completes the proof of the uniqueness of the spectral representation.

The spectral representation (5.36) leads to the *Jordan canonical form* of T. For this it is only necessary to recall the structure of the nilpotents D_h (see § 3.4). If we introduce a suitable basis in each M_h, the part D_{h, M_h} of D_h in M_h is represented by a matrix of the form (3.30). Thus T_{M_h} is represented by a triangular matrix $(\tau^{(h)}_{j k})$ of the form (3.30) with all the diagonal elements replaced by λ_h. If we collect all the basis elements of $\mathsf{M}_1, \ldots, \mathsf{M}_s$ to form a basis of X, the matrix of T is the direct sum of submatrices $(\tau^{(h)}_{j k})$. In particular, it follows that (see Problem 3.14)

$$(5.40) \qquad \det(T - \zeta) = \prod_{h=1}^{s} (\lambda_h - \zeta)^{m_h} , \quad \operatorname{tr} T = \sum_{h=1}^{s} m_h \lambda_h .$$

It is often convenient to count each eigenvalue λ_h of T repeatedly m_h times (m_h is the *algebraic* multiplicity of λ_h) and thus to denote the eigenvalues by $\mu_1, \mu_2, \ldots, \mu_N$ (for example $\mu_1 = \cdots = \mu_{m_1} = \lambda_1$, $\mu_{m_1+1} = \cdots = \mu_{m_1+m_2} = \lambda_2, \ldots$). For convenience we shall call $\mu_1, \ldots, \mu_N$ the *repeated eigenvalues of* T. Thus every operator $T \in \mathscr{B}(\mathsf{X})$ has exactly N repeated eigenvalues. (5.40) can be written

$$(5.41) \qquad \det(T - \zeta) = \prod_{k=1}^{N} (\mu_k - \zeta) , \quad \operatorname{tr} T = \sum_{k=1}^{N} \mu_k .$$

Problem 5.12. The geometric eigenspace N_h of T for the eigenvalue λ_h is a subset of the algebraic eigenspace M_h. $\mathsf{N}_h = \mathsf{M}_h$ holds if and only if λ_h is semisimple.

Problem 5.13. λ_h is semisimple if and only if $\zeta = \lambda_h$ is a simple pole (pole of order 1) of $R(\zeta)$.

Problem 5.14. If n is sufficiently large ($n \geqq m$, say), rank T^n is equal to $N - m$, where m is the algebraic multiplicity of the eigenvalue 0 of T (it being agreed to set $m = 0$ if 0 is not an eigenvalue of T, see § 1.1).

Problem 5.15. We have

$$|\text{tr}\, T| \leqq (\text{rank}\, T)\, \|T\| \leqq N\, \|T\| \,. \tag{5.42}$$

Problem 5.16. The eigenvalues λ_h of T are identical with the roots of the algebraic equation of degree N *(characterictic equation)*

$$\det(T - \zeta) = 0 \,. \tag{5.43}$$

The multiplicity of λ_h as the root of this equation is equal to the algebraic multiplicity of the eigenvalue λ_h.

Let (τ_{jk}) be an $N \times N$ matrix. It can be regarded as the representation of a linear operator T in a vector space X, say the space C^N of N-dimensional numerical vectors. This means that the matrix representing T with respect to the canonical basis $\{x_j\}$ of X coincides with (τ_{jk}). Let (5.36) be the spectral representation of T, and let $\{x'_j\}$ be the basis of X used in the preceding paragraph to bring the associated matrix (τ'_{jk}) of T into the canonical form.

The relationship between the bases $\{x_j\}$ and $\{x'_j\}$ is given by (1.2) and (1.4). Since the x'_k are eigenvectors of T in the generalized sense, the numerical vectors $(\hat{\gamma}_{1k}, \ldots, \hat{\gamma}_{Nk})$, $k = 1, \ldots, N$, are generalized eigenvectors of the matrix (τ_{jk}). The relationship between the matrices (τ_{jk}) and (τ'_{jk}) is given by (3.8). Thus the transformation of (τ_{jk}) into the simpler form (τ'_{jk}) is effected by the matrix $(\hat{\gamma}_{jk})$, constructed from the generalized eigenvectors of (τ_{jk}), according to the formula $(\tau'_{jk}) = (\hat{\gamma}_{jk})^{-1} (\tau_{jk}) (\hat{\gamma}_{jk})$.

If the eigenvalue λ_h is semisimple, the h-th submatrix of (τ'_{jk}) is a diagonal matrix with all diagonal elements equal to λ_h. If T is diagonable, (τ'_{jk}) is itself a diagonal matrix with the diagonal elements $\lambda_1, \ldots, \lambda_s$, where λ_h is repeated m_h times (that is, with the diagonal elements $\mu_1, \ldots, \mu_N$, where the μ_j are the repeated eigenvalues). In this case $(\hat{\gamma}_{1k}, \ldots, \hat{\gamma}_{Nk})$, $k = 1, \ldots, N$, are eigenvectors of (τ_{jk}) in the proper sense.

Problem 5.17. $(\gamma_{j1}, \ldots, \gamma_{jN})$, $j = 1, \ldots, N$, are generalized eigenvectors of the transposed matrix of (τ_{jk}).

5. The adjoint problem

If $T \in \mathscr{B}(\mathsf{X})$, then $T^* \in \mathscr{B}(\mathsf{X}^*)$. There is a simple relationship between the spectral representations of the two operators T, T^*. If (5.36) is the spectral representation of T, then that of T^* is given by

$$T^* = S^* + D^* = \sum_{h=1}^{s} (\bar{\lambda}_h P_h^* + D_h^*) \,, \tag{5.44}$$

for the P_h^* as well as the P_h satisfy the relations (3.27–3.28) and the D_h^* are nilpotents commuting with the P_h^* and with one another (note the uniqueness of the spectral representation, par. 4). In particular, it follows that *the eigenvalues of T^* are the complex conjugates of the eigenvalues of T, with the same algebraic multiplicities.* The corresponding geometric multiplicities are also equal; this is a consequence of (3.40).

Problem 5.18. $R^*(\zeta) = R(\zeta, T^*) = (T^* - \zeta)^{-1}$ has the following properties:

$$R(\zeta)^* = R^*(\bar{\zeta}), \tag{5.45}$$

$$R^*(\zeta) = -\sum_{h=1}^{s} \left[(\zeta - \bar{\lambda}_h)^{-1} P_h^* + \sum_{n=1}^{m_h - 1} (\zeta - \bar{\lambda}_h)^{-n-1} D_h^{*n} \right]. \tag{5.46}$$

6. Functions of an operator

If $p(\zeta)$ is a polynomial in ζ, the operator $p(T)$ is defined for any $T \in \mathscr{B}(X)$ (see § 3.3). Making use of the resolvent $R(\zeta) = (T - \zeta)^{-1}$, we can now define functions $\phi(T)$ of T for a more general class of functions $\phi(\zeta)$. It should be noted that $R(\zeta_0)$ is itself equal to $\phi(T)$ for $\phi(\zeta) = (\zeta - \zeta_0)^{-1}$.

Suppose that $\phi(\zeta)$ is holomorphic in a domain Δ of the complex plane containing all the eigenvalues λ_h of T, and let $\Gamma \subset \Delta$ be a simple closed smooth curve with positive direction enclosing all the λ_h in its interior. Then $\phi(T)$ is defined by the *Dunford-Taylor integral*

$$\phi(T) = -\frac{1}{2\pi i} \int_\Gamma \phi(\zeta)\, R(\zeta)\, d\zeta = \frac{1}{2\pi i} \int_\Gamma \phi(\zeta)\, (\zeta - T)^{-1}\, d\zeta. \tag{5.47}$$

This is an analogue of the Cauchy integral formula[1] in function theory. More generally, Γ may consist of several simple closed curves Γ_k with interiors Δ_k' such that the union of the Δ_k' contains all the eigenvalues of T. Note that (5.47) does not depend on Γ as long as Γ satisfies these conditions.

It is easily verified that (5.47) coincides with (3.18) when $\phi(\zeta)$ is a polynomial. It suffices to verify this for monomials $\phi(\zeta) = \zeta^n$, $n = 0, 1, 2, \ldots$. The proof for the case $n = 0$ is contained in (5.25). For $n \geqq 1$, write $\zeta^n = (\zeta - T + T)^n = (\zeta - T)^n + \cdots + T^n$ and substitute it into (5.47); all terms except the last vanish on integration by Cauchy's theorem, while the last term gives T^n.

Problem 5.19. If $\phi(\zeta) = \sum \alpha_n \zeta^n$ is an entire function, then $\phi(T) = \sum \alpha_n T^n$.

The correspondence $\phi(\zeta) \to \phi(T)$ is a homomorphism of the algebra of holomorphic functions on Δ into the algebra $\mathscr{B}(X)$:

$$\phi(\zeta) = \alpha_1 \phi_1(\zeta) + \alpha_2 \phi_2(\zeta) \text{ implies } \phi(T) = \alpha_1 \phi_1(T) + \alpha_2 \phi_2(T), \tag{5.48}$$

$$\phi(\zeta) = \phi_1(\zeta)\, \phi_2(\zeta) \text{ implies } \phi(T) = \phi_1(T)\, \phi_2(T). \tag{5.49}$$

[1] See KNOPP [1], p. 61.

This justifies the notation $\phi(T)$ for the operator defined by (5.47). The proof of (5.48) is obvious. For (5.49), it suffices to note that the proof is exactly the same as the proof of (5.17) for $n, m < 0$ (so that $\eta_n + \eta_m - 1 = -1$). In fact (5.17) is a special case of (5.49) for $\phi_1(\zeta) = \zeta^{-n-1}$ and $\phi_2(\zeta) = \zeta^{-m-1}$.

The spectral representation of $\phi(T)$ is obtained by substitution of (5.23) into (5.47). If $T = \sum (\lambda_h P_h + D_h)$ is the spectral representation of T, the result is

$$\phi(T) = \sum_{h=1}^{s} [\phi(\lambda_h) P_h + D'_h] \tag{5.50}$$

where[1]

$$D'_h = \phi'(\lambda_h) D_h + \cdots + \frac{\phi^{(m_h-1)}(\lambda_h)}{(m_h-1)!} D_h^{m_h-1}. \tag{5.51}$$

Since the D'_h are nilpotents commuting with each other and with the P_k, it follows from the uniqueness of the spectral representation that (5.50) is the spectral representation of $\phi(T)$.

Thus $\phi(T)$ has the same eigenprojections P_h as T, and the eigenvalues of $\phi(T)$ are $\phi(\lambda_h)$ with multiplicities m_h. Strictly speaking, these statements are true only if the $\phi(\lambda_h)$ are different from one another. If $\phi(\lambda_1) = \phi(\lambda_2)$, for example, we must say that $\phi(\lambda_1)$ is an eigenvalue of $\phi(T)$ with the eigenprojection $P_1 + P_2$ and multiplicity $m_1 + m_2$.

That the $\phi(\lambda_h)$ are exactly the eigenvalues of $\phi(T)$ is a special case of the so-called *spectral mapping theorem*.

(5.48) and (5.49) are to be supplemented by another functional relation

$$\phi(\zeta) = \phi_1(\phi_2(\zeta)) \quad \text{implies} \quad \phi(T) = \phi_1(\phi_2(T)) . \tag{5.52}$$

Here it is assumed that ϕ_2 is holomorphic in a domain Δ_2 of the sort stated before, whereas ϕ_1 is holomorphic in a domain Δ_1 containing all the eigenvalues $\phi_2(\lambda_h)$ of $S = \phi_2(T)$. This ensures that $\phi_1(S)$ can be constructed by

$$\phi_1(S) = \frac{1}{2\pi i} \int_{\Gamma_1} \phi_1(z) (z - S)^{-1} dz \tag{5.53}$$

where Γ_1 is a curve (or a union of curves) enclosing the eigenvalues $\phi_2(\lambda_h)$ of S. But we have

$$(z - S)^{-1} = (z - \phi_2(T))^{-1} = \frac{1}{2\pi i} \int_{\Gamma_2} (z - \phi_2(\zeta))^{-1} (\zeta - T)^{-1} d\zeta \tag{5.54}$$

for $z \in \Gamma_1$; this follows from (5.49) and the fact that both $z - \phi_2(\zeta)$ and $(z - \phi_2(\zeta))^{-1}$ are holomorphic in ζ in an appropriate subdomain of Δ_2 containing all the λ_h (the curve Γ_2 should be taken in such a subdomain

[1] Note that $(2\pi i)^{-1} \int_\Gamma (\zeta - \lambda)^{-n-1} \phi(\zeta) d\zeta = \phi^{(n)}(\lambda)/n!$ if λ is inside Γ.

so that the image of Γ_2 under ϕ_2 lies inside Γ_1). Substitution of (5.54) into (5.53) gives

$$\phi_1(S) = \left(\frac{1}{2\pi i}\right)^2 \int_{\Gamma_1}\int_{\Gamma_2} \phi_1(z)\,(z - \phi_2(\zeta))^{-1}(\zeta - T)^{-1}\,d\zeta\,dz \tag{5.55}$$
$$= \frac{1}{2\pi i}\int_{\Gamma_2} \phi_1(\phi_2(\zeta))\,(\zeta - T)^{-1}\,d\zeta = \phi(T)$$

as we wished to show.

Example 5.20. As an application of the Dunford integral, let us define the logarithm

$$S = \log T \tag{5.56}$$

of an operator $T \in \mathscr{B}(\mathsf{X})$, assuming that T is nonsingular. We can take a simply connected domain Δ in the complex plane containing all the eigenvalues λ_h of T but not containing 0. Let Γ be a simple closed curve in Δ enclosing all the λ_h. Since $\phi(\zeta) = \log\zeta$ can be defined as a holomorphic function on Δ, the application of (5.47) defines a function

$$S = \log T = \phi(T) = -\frac{1}{2\pi i}\int_\Gamma \log\zeta\, R(\zeta)\,d\zeta\,. \tag{5.57}$$

Since $\exp(\log\zeta) = \zeta$, it follows from (5.52) that

$$\exp(\log T) = T\,. \tag{5.58}$$

It should be noted that the choice of the domain Δ and the function $\phi(\zeta) = \log\zeta$ is not unique. Hence there are different operators $\log T$ with the above porperties. In particular, each of the eigenvalues of $\log T$ with one choice may differ by an integral multiple of $2\pi i$ from those with a different choice. If m_h is the algebraic multiplicity of λ_h, it follows that

$$\operatorname{tr}(\log T) = \sum m_h \log\lambda_h + 2n\pi i\,, \quad n = \text{integer}\,, \tag{5.59}$$

$$\exp(\operatorname{tr}(\log T)) = \prod \lambda_h^{m_h} = \det T\,. \tag{5.60}$$

7. Similarity transformations

Let U be an operator from a vector space X to another one X' such that $U^{-1} \in \mathscr{B}(\mathsf{X}', \mathsf{X})$ exists (this implies that $\dim\mathsf{X} = \dim\mathsf{X}'$). For an operator $T \in \mathscr{B}(\mathsf{X})$, the operator $T' \in \mathscr{B}(\mathsf{X}')$ defined by

$$T' = U T U^{-1} \tag{5.61}$$

is said to result from T by *transformation* by U. T' is said to be *similar*[1] to T. T' has the same internal structure as T, for the one-to-one correspondence $u \to u' = Uu$ is invariant in the sense that $Tu \to T'u' = UTu$. If we choose bases $\{x_j\}$ and $\{x_j'\}$ of X and X', respectively, in such a way that $x_j' = U x_j$, then T and T' are represented by the same matrix.

[1] We have considered in § 4.6 an example of pairs of mutually similar operators.

If

$$T = \sum (\lambda_h P_h + D_h) \tag{5.62}$$

is the spectral representation of T, then

$$T' = \sum (\lambda_h P'_h + D'_h), \quad P'_h = U P_h U^{-1}, \quad D'_h = U D_h U^{-1}, \tag{5.63}$$

is the spectral representation of T'.

§ 6. Operators in unitary spaces

1. Unitary spaces

So far we have been dealing with general operators without introducing any special assumptions. In applications, however, we are often concerned with Hermitian or normal operators. Such notions are defined only in a special kind of normed space H, called a *unitary space*, in which is defined an *inner product* (u, v) for any two vectors u, v. In this section we shall see what these notions add to our general results, especially in eigenvalue problems. We assume that $0 < \dim \mathsf{H} < \infty$.

The inner product (u, v) is complex-valued, (Hermitian) symmetric:

$$\overline{(u, v)} = (v, u) \tag{6.1}$$

and *sesquilinear*, that is, linear in u and semilinear in v. (6.1) implies that (u, u) is always real; it is further assumed to be positive-definite:

$$(u, u) > 0 \quad \text{for} \quad u \neq 0\text{[1]}. \tag{6.2}$$

We shall see in a moment that the inner product (u, v) may be regarded as a special case of the scalar product defined between an element of a vector space and an element of its adjoint space. This justifies the use of the same symbol (u, v) for the two quantities.

Since a unitary space H is a vector space, there could be defined different norms in H. There exists, however, a distinguished norm in H *(unitary norm)* defined in terms of the inner product, and it is this one which is meant whenever we speak of the norm in a unitary space. [Of course any other possible norms are equivalent to this particular norm, see (1.20).] The unitary norm is given by

$$\|u\| = (u, u)^{1/2}. \tag{6.3}$$

Obviously the first two of the conditions (1.18) for a norm are satisfied. Before verifying the third condition (triangle inequality), we note the *Schwarz inequality*

$$|(u, v)| \leq \|u\| \, \|v\|, \tag{6.4}$$

[1] In any finite-dimensional vector space X, one can introduce a positive-definite sesquilinear form and make X into a unitary space.

where equality holds if and only if u and v are linearly dependent. (6.4) follows, for example, from the identity

$$\| \|v\|^2 u - (u, v) v\|^2 = (\|u\|^2 \|v\|^2 - |(u, v)|^2) \|v\|^2 . \tag{6.5}$$

The triangle inequality is a consequence of (6.4): $\|u + v\|^2 = (u + v, u + v) = \|u\|^2 + 2 \operatorname{Re}(u, v) + \|v\|^2 \leqq \|u\|^2 + 2\|u\| \|v\| + \|v\|^2 = (\|u\| + \|v\|)^2$.

Example 6.1. For numerical vectors $u = (\xi_1, \ldots, \xi_N)$ and $v = (\eta_1, \ldots, \eta_N)$ set

$$(u, v) = \sum \xi_j \bar{\eta}_j , \quad \|u\|^2 = \sum |\xi_j|^2 . \tag{6.6}$$

With this inner product the space C^N of N-dimensional numerical vectors becomes a unitary space.

Problem 6.2. The unitary norm has the characteristic property

$$\|u + v\|^2 + \|u - v\|^2 = 2\|u\|^2 + 2\|v\|^2 . \tag{6.7}$$

Problem 6.3. The inner product (u, v) can be expressed in terms of the norm by

$$(u, v) = \frac{1}{4} (\|u + v\|^2 - \|u - v\|^2 + i\|u + i v\|^2 - i\|u - i v\|^2) . \tag{6.8}$$

Problem 6.4. For any pair (ξ_j), (η_j) of numerical vectors, we have

$$\begin{gathered} |\sum \xi_j \eta_j| \leqq (\sum |\xi_j|^2)^{\frac{1}{2}} (\sum |\eta_j|^2)^{\frac{1}{2}} , \\ (\sum |\xi_j + \eta_j|^2)^{\frac{1}{2}} \leqq (\sum |\xi_j|^2)^{\frac{1}{2}} + (\sum |\eta_j|^2)^{\frac{1}{2}} . \end{gathered} \tag{6.9}$$

2. The adjoint space

A characteristic property of a unitary space H is that the adjoint space H* can be identified with H itself: H* = H.

For any $u \in H$ set $f_u[v] = (u, v)$. f_u is a semilinear form on H so that $f_u \in H^*$. The map $u \to f_u$ is linear, for if $u = \alpha_1 u_1 + \alpha_2 u_2$ then $f_u[v] = (u, v) = \alpha_1 (u_1, v) + \alpha_2 (u_2, v) = \alpha_1 f_{u_1}[v] + \alpha_2 f_{u_2}[v]$ so that $f_u = \alpha_1 f_{u_1} + \alpha_2 f_{u_2}$. Thus $u \to f_u = T u$ defines a linear operator T on H to H*. T is isometric: $\|T u\| = \|u\|$. In fact $\|T u\| = \|f_u\| = \sup\limits_v |(u, v)|/\|v\| = \|u\|$ by (2.22) [note the Schwarz inequality and $(u, u) = \|u\|^2$]. In particular T is one to one. Since $\dim H^* = \dim H$, it follows that T maps H onto the whole of H* isometrically. This means that every $f \in H^*$ has the form f_u with a uniquely determined $u \in H$ such that $\|f\| = \|u\|$. It is natural to identify f with this u. It is in this sense that we identify H* with H.

Since $(u, v) = f[v] = (f, v)$, the inner product (u, v) is seen to coincide with the scalar product (f, v).

We can now take over various notions defined for the scalar product to the inner product (see § 2.2). If $(u, v) = 0$ we write $u \perp v$ and say that u, v are mutually *orthogonal* (or *perpendicular*). u is orthogonal to a subset S of H, in symbol $u \perp S$, if $u \perp v$ for all $v \in S$. The set of all $u \in H$ such that $u \perp S$ is the annihilator of S and is denoted by $S^\perp$. Two subsets S, S' of H are *orthogonal*, in symbol $S \perp S'$, if every $u \in S$ and every $v \in S'$ are orthogonal.

Problem 6.5. (u, v) is a continuous function of u, v.
Problem 6.6. The Pythagorean theorem:

$$\|u + v\|^2 = \|u\|^2 + \|v\|^2 \quad \text{if} \quad u \perp v . \tag{6.10}$$

Problem 6.7. $u \perp \mathsf{S}$ implies $u \perp \mathsf{M}$ where M is the span of S. $\mathsf{S}^\perp$ is a linear manifold and $\mathsf{S}^\perp = \mathsf{M}^\perp$.

Consider two unitary spaces H and H'. A complex-valued function $\mathfrak{t}[u, u']$ defined for $u \in \mathsf{H}$ and $u' \in \mathsf{H}'$ is called a *sesquilinear form* on $\mathsf{H} \times \mathsf{H}'$ if it is linear in u and semilinear in u'. If in particular $\mathsf{H}' = \mathsf{H}$, we speak of a sesquilinear form on H. The inner product of H is a special case of a sesquilinear form on H. For a general sesquilinear form $\mathfrak{t}[u, v]$ on H, there is no relation between $\mathfrak{t}[u, v]$ and $\mathfrak{t}[v, u]$ so that the *quadratic form* $\mathfrak{t}[u] \equiv \mathfrak{t}[u, u]$ need not be real-valued. In any case, however, we have the relation *(polarization principle)*

$$\mathfrak{t}[u, v] = \frac{1}{4} (\mathfrak{t}[u + v] - \mathfrak{t}[u - v] + i\, \mathfrak{t}[u + i v] - i\, \mathfrak{t}[u - i v]) \tag{6.11}$$

similar to (6.8). Thus the sesquilinear form $\mathfrak{t}[u, v]$ is determined by the associated quadratic form $\mathfrak{t}[u]$. In particular $\mathfrak{t}[u, v] = 0$ identically if $\mathfrak{t}[u] = 0$ identically. $\mathfrak{t}[u, v]$ is called the *polar form* of $\mathfrak{t}[u]$.

Problem 6.8. If $|\mathfrak{t}[u]| \leqq M\|u\|^2$ for all $u \in \mathsf{H}$, then $|\mathfrak{t}[u, v]| \leqq 2M\|u\| \|v\|$.
Problem 6.9. If T is a linear operator on H to H', $\|T u\|^2$ is a quadratic form on H with the polar form $(T u, T v)$.

Remark 6.10. The validity of (6.11) is closely related to the existence of the scalar i. The quadratic form $\mathfrak{t}[u]$ does *not* determine $\mathfrak{t}[u, v]$ in a *real* vector space.

3. Orthonormal families

A family of vectors $x_1, \ldots, x_n \in \mathsf{H}$ is called an *orthogonal family* if any two elements of this family are orthogonal. It is said to be *orthonormal* if, in addition, each vector is normalized:

$$(x_j, x_k) = \delta_{jk} . \tag{6.12}$$

As is easily seen, the vectors of an orthonormal family are linearly independent. An orthonormal family $\{x_1, \ldots, x_n\}$ is *complete* if $n = N = \dim \mathsf{H}$. Thus it is a basis of H, called an *orthonormal basis*.

Let M be the span of an orthonormal family $\{x_1, \ldots, x_n\}$. For any $u \in \mathsf{H}$, the vector

$$u' = \sum_{j=1}^{n} (u, x_j)\, x_j \tag{6.13}$$

has the property that $u' \in \mathsf{M}$, $u - u' \in \mathsf{M}^\perp$. u' is called the *orthogonal projection* (or simply the projection) of u on M. The Pythagorean theorem

(6.10) gives

$$\|u\|^2 = \|u'\|^2 + \|u - u'\|^2 = \sum_{j=1}^{n} |(u, x_j)|^2 + \|u - u'\|^2 , \tag{6.14}$$

and hence

$$\sum_{j=1}^{n} |(u, x_j)|^2 \leqq \|u\|^2 \quad \textit{(Bessel's inequality)} . \tag{6.15}$$

For any linearly independent vectors $u_1, \ldots, u_n$, it is possible to construct an orthonormal family $x_1, \ldots, x_n$ such that for each $k = 1, \ldots, n$, the k vectors $x_1, \ldots, x_k$ span the same linear manifold as the k vectors $u_1, \ldots, u_k$. This is proved by induction on k. Suppose that $x_1, \ldots, x_{k-1}$ have been constructed. If M_{k-1} denotes the span of $x_1, \ldots, x_{k-1}$, then by hypothesis M_{k-1} is identical with the span of $u_1, \ldots, u_{k-1}$. Set $u_k'' = u_k - u_k'$ where u_k' is the projection of u_k on M_{k-1} (if $k = 1$ set $u_1'' = u_1$). The linear independence of the u_j implies that $u_k'' \neq 0$. Set $x_k = \|u_k''\|^{-1} u_k''$. Then the vectors $x_1, \ldots, x_k$ satisfy the required conditions. The construction described here is called the *Schmidt orthogonalization process*.

Since every linear manifold M of H has a basis $\{u_j\}$, it follows that M has an orthonormal basis $\{x_j\}$. In particular there exists a complete orthonormal family in H. An arbitrary vector x_1 with $\|x_1\| = 1$ can be the first element of an orthonormal basis of H.

It follows also that any $u \in \mathsf{H}$ has an orthogonal projection u' on a given linear manifold M. u' is determined uniquely by the property that $u' \in \mathsf{M}$ and $u'' = u - u' \in \mathsf{M}^\perp$. Thus H is the direct sum of M and $\mathsf{M}^\perp$:

$$\mathsf{H} = \mathsf{M} \oplus \mathsf{M}^\perp . \tag{6.16}$$

In this sense $\mathsf{M}^\perp$ is called the *orthogonal complement* of M. We have

$$\mathsf{M}^{\perp\perp} = \mathsf{M} , \quad \dim \mathsf{M}^\perp = N - \dim \mathsf{M} . \tag{6.17}$$

When $\mathsf{N} \subset \mathsf{M}$, $\mathsf{M} \cap \mathsf{N}^\perp$ is also denoted by $\mathsf{M} \ominus \mathsf{N}$; it is the set of all $u \in \mathsf{M}$ such that $u \perp \mathsf{N}$.

In the particular case $\mathsf{M} = \mathsf{H}$, we have $\mathsf{M}^\perp = 0$ and so $u'' = 0$. Thus (6.13) gives

$$u = \sum_{j=1}^{N} (u, x_j)\, x_j . \tag{6.18}$$

This is the *expansion* of u in the orthonormal basis $\{x_j\}$. Multiplication of (6.18) from the right by v gives

$$(u, v) = \sum_j (u, x_j)\,(x_j, v) = \sum_j (u, x_j)\,\overline{(v, x_j)} . \tag{6.19}$$

In particular

$$\|u\|^2 = \sum_j |(u, x_j)|^2 \quad \textit{(Parseval's equality)} . \tag{6.20}$$

The following lemma will be required later.

Lemma 6.11. *Let* M, M′ *be two linear manifolds of* H *with dimensions* n, n' *respectively. Then* $\dim(\mathsf{M}' \cap \mathsf{M}^\perp) \geqq n' - n$.

This follows from (1.10) in view of the relations $\dim \mathsf{M}' = n'$, $\dim \mathsf{M}^\perp = N - n$, $\dim(\mathsf{M}' + \mathsf{M}^\perp) \leqq N$.

Let $\{x_j\}$ be a (not necessarily orthonormal) basis of H. The adjoint basis $\{e_j\}$ is a basis of $\mathsf{H}^* = \mathsf{H}$ satisfying the relations

$$(e_j, x_k) = \delta_{jk} \tag{6.21}$$

(see § 2.3). $\{x_j\}$ and $\{e_j\}$ are also said to form a *biorthogonal family* of elements of H. The basis $\{x_j\}$ is *selfadjoint* if $e_j = x_j$, $j = 1, \ldots, N$. Thus *a basis of* H *is selfadjoint if and only if it is orthonormal.*

4. Linear operators

Consider a linear operator T on a unitary space H to another unitary space H′. We recall that the norm of T is defined by $\|T\| = \sup \|Tu\|/\|u\| = \sup |(Tu, u')|/\|u\| \, \|u'\|$ [see (4.1), (4.2)].

The function

$$\mathfrak{t}[u, u'] = (Tu, u') \tag{6.22}$$

is a sesquilinear form on $\mathsf{H} \times \mathsf{H}'$. Conversely, an arbitrary sesquilinear form $\mathfrak{t}[u, u']$ on $\mathsf{H} \times \mathsf{H}'$ can be expressed in this form by a suitable choice of an operator T on H to H′. Since $\mathfrak{t}[u, u']$ is a semilinear form on H′ for a fixed u, there exists a unique $w' \in \mathsf{H}'$ such that $\mathfrak{t}[u, u'] = (w', u')$ for all $u' \in \mathsf{H}'$. Since w' is determined by u, we can define a function T by setting $w' = Tu$. It can be easily seen that T is a linear operator on H to H′. If in particular $\mathfrak{t}$ is a sesquilinear form on $\mathsf{H}\,(\mathsf{H}' = \mathsf{H})$, then T is a linear operator on H to itself.

In the same way, $\mathfrak{t}[u, u']$ can also be expressed in the form

$$\mathfrak{t}[u, u'] = (u, T^* u'), \tag{6.23}$$

where T^* is a linear operator on H′ to H, called the *adjoint (operator)* of T. T^* coincides with the adjoint operator defined in § 3.6 by the identification of H^*, H'^* with H, H′ respectively.

T^*T is a linear operator on H to itself. The relation

$$(u, T^*Tv) = (T^*Tu, v) = (Tu, Tv) \tag{6.24}$$

shows that T^*T is the operator associated with the sesquilinear form (Tu, Tv) on H. Note that the first two members of (6.24) are the inner product in H while the last is that in H′. It follows from (6.24) and (4.2) that $\|T^* T\| = \sup |(Tu, Tv)|/\|u\| \, \|v\| \geqq \sup \|Tu\|^2/\|u\|^2 = \|T\|^2$. Since, on the other hand, $\|T^*T\| \leqq \|T^*\| \|T\| = \|T\|^2$, we have

$$\|T^*T\| = \|T\|^2 . \tag{6.25}$$

In particular, $T^*T = 0$ implies $T = 0$.

Problem 6.12. $\mathsf{N}(T^*T) = \mathsf{N}(T)$.

Problem 6.13. We have the polarization principle for an operator T on H to H

$$(Tu, v) = \frac{1}{4}[(T(u+v), u+v) - (T(u-v), u-v) + \\ + i(T(u+iv), u+iv) - i(T(u-iv), u-iv)] . \tag{6.26}$$

Problem 6.14. If T is an operator on H to itself, $(Tu, u) = 0$ for all u implies $T = 0$.

The matrix representation of an operator is related to a pair of adjoint bases in the domain and range spaces [see (3.10)]. This suggests that the choice of selfadjoint (orthonormal) bases is convenient for the matrix representation of operators between unitary spaces.

Let T be an operator from a unitary space H to another one H' and let $\{x_k\}$, $\{x_j'\}$ be orthonormal bases of H, H', respectively. The matrix elements of T for these bases are then

$$\tau_{jk} = (T x_k, x_j') = (x_k, T^* x_j') , \tag{6.27}$$

as is seen from (3.10) by setting $f_j = x_j'$. More directly, (6.27) follows from the expansion

$$T x_k = \sum_j (T x_k, x_j') x_j' . \tag{6.28}$$

Recall that if $\mathsf{H}' = \mathsf{H}$ it is the convention to take $x_j' = x_j$.

Problem 6.15. If $\{x_k\}$ is an orthonormal basis of H and if T is an operator on H to itself,

$$\operatorname{tr} T = \sum_k (T x_k, x_k) . \tag{6.29}$$

The matrix of T^* with respect to the same pair of bases $\{x_j'\}$, $\{x_k\}$ is given by $\tau^*_{kj} = (T^* x_j', x_k)$. Comparison with (6.27) gives

$$\tau^*_{kj} = \overline{\tau_{jk}} . \tag{6.30}$$

Thus the matrices of T and T^* (for the same pair of orthogonal bases) are Hermitian conjugate to each other.

5. Symmetric forms and symmetric operators

A sesquilinear form $\mathfrak{t}[u, v]$ on a unitary space H is said to be *symmetric* if

$$\mathfrak{t}[v, u] = \overline{\mathfrak{t}[u, v]} \quad \text{for all} \quad u, v \in \mathsf{H} . \tag{6.31}$$

If $\mathfrak{t}[u, v]$ is symmetric, the associated quadratic form $\mathfrak{t}[u]$ is real-valued. The converse is also true, as is seen from (6.11).

A symmetric sesquilinear form (or the associated quadratic form) $\mathfrak{t}$ is *nonnegative* (in symbol $\mathfrak{t} \geqq 0$) if $\mathfrak{t}[u] \geqq 0$ for all u, and *positive* if $\mathfrak{t}[u] > 0$ for $u \neq 0$. The Schwarz and triangle inequalities are true for any nonnegative form as for the inner product (which is a special positive sesquilinear form):

$$(6.32)\qquad \begin{aligned} |\mathfrak{t}[u, v]| &\leqq \mathfrak{t}[u]^{1/2}\, \mathfrak{t}[v]^{1/2} \leqq \frac{1}{2}(\mathfrak{t}[u] + \mathfrak{t}[v])\,, \\ \mathfrak{t}[u+v]^{1/2} &\leqq \mathfrak{t}[u]^{1/2} + \mathfrak{t}[v]^{1/2}\,, \\ \mathfrak{t}[u+v] &\leqq 2\mathfrak{t}[u] + 2\mathfrak{t}[v]\,. \end{aligned}$$

Note that strict positivity was not used in the proof of similar inequalities for the inner product.

The *lower bound* γ of a symmetric form $\mathfrak{t}$ is defined as the largest real number such that $\mathfrak{t}[u] \geqq \gamma \|u\|^2$. The upper bound γ' is defined similarly. We have

$$(6.33)\qquad |\mathfrak{t}[u, v]| \leqq M\, \|u\|\, \|v\|\,, \quad M = \max(|\gamma|, |\gamma'|)\,.$$

To see this, we note that the value $\mathfrak{t}[u, v]$ under consideration may be assumed to be real, for (6.33) is unchanged by multiplying u with a scalar of absolute value one. Since $\mathfrak{t}[u]$ is real-valued, we see from (6.11) that $\mathfrak{t}[u, v] = 4^{-1}(\mathfrak{t}[u+v] - \mathfrak{t}[u-v])$. Since $|\mathfrak{t}[u]| \leqq M\|u\|^2$ for all u, it follows that $|\mathfrak{t}[u, v]| \leqq 4^{-1} M(\|u+v\|^2 + \|u-v\|^2) = 2^{-1} M(\|u\|^2 + \|v\|^2)$. Replacement of u, v respectively by $\alpha u, v/\alpha$, with $\alpha^2 = \|v\|/\|u\|$, yields (6.33).

The operator T associated with a symmetric form $\mathfrak{t}[u, v]$ according to (6.22) has the property that

$$(6.34)\qquad T^* = T$$

by (6.22), (6.23) and (6.31). An operator T on H to itself satisfying (6.34) is said to be *(Hermitian) symmetric* or *selfadjoint*. Conversely, a symmetric operator T determines a symmetric form $\mathfrak{t}[u, v] = (Tu, v)$ on H. Thus *(Tu, u) is real for all $u \in \mathsf{H}$ if and only if T is symmetric*. A symmetric operator T is *nonnegative (positive)* if the associated form is nonnegative (positive). For a nonnegative symmetric operator T we have the following inequalities corresponding to (6.32):

$$(6.35)\qquad \begin{aligned} |(Tu, v)| &\leqq (Tu, u)^{1/2} (Tv, v)^{1/2}\,, \\ (T(u+v), u+v)^{1/2} &\leqq (Tu, u)^{1/2} + (Tv, v)^{1/2}\,. \end{aligned}$$

We write $T \geqq 0$ to denote that T is nonnegative symmetric. More generally, we write

$$(6.36)\qquad T \geqq S \quad \text{or} \quad S \leqq T$$

if S, T are symmetric operators such that $T - S \geqq 0$. The upper and lower bounds of the quadratic form (Tu, u) are called the upper and lower bounds of the symmetric operator T.

Problem 6.16. If T is symmetric, $\alpha T + \beta$ is symmetric for real α, β. More generally, $p(T)$ is symmetric for any polynomial p with real coefficients.

Problem 6.17. For any linear operator T on H to H' (H, H' being unitary spaces), $T^* T$ and $T T^*$ are nonnegative symmetric operators in H and H', respectively.

Problem 6.18. If T is symmetric, then $T^2 \geqq 0$; $T^2 = 0$ if and only if $T = 0$. If T is symmetric and $T^n = 0$ for some positive integer n, then $T = 0$.

Problem 6.19. $R \leqq S$ and $S \leqq T$ imply $R \leqq T$. $S \leqq T$ and $S \geqq T$ imply $S = T$.

6. Unitary, isometric and normal operators

Let H and H' be unitary spaces. An operator T on H to H' is said to be *isometric*[1] if

$$\|Tu\| = \|u\| \quad \text{for every} \quad u \in \mathsf{H}\,. \tag{6.37}$$

This is equivalent to $((T^*T - 1)\,u, u) = 0$ and therefore (see Problem 6.14)

$$T^*T = 1\,. \tag{6.38}$$

This implies that

$$(Tu, Tv) = (u, v) \quad \text{for every} \quad u, v \in \mathsf{H}\,. \tag{6.39}$$

An isometric operator T is said to be ***unitary*** if the range of T is the whole space H'. Since (6.37) implies that the mapping by T is one to one, it is necessary for the existence of a unitary operator on H to H' that $\dim \mathsf{H}' = \dim \mathsf{H}$. Conversely, if $\dim \mathsf{H}' = \dim \mathsf{H} < \infty$ any isometric operator on H to H' is unitary. As we shall see later, this is not true for infinite-dimensional spaces.

Problem 6.20. A $T \in \mathscr{B}(\mathsf{H}, \mathsf{H}')$ is unitary if and only if $T^{-1} \in \mathscr{B}(\mathsf{H}', \mathsf{H})$ exists and

$$T^{-1} = T^*\,. \tag{6.40}$$

Problem 6.21. T is unitary if and only if T^* is.

Problem 6.22. If $T \in \mathscr{B}(\mathsf{H}', \mathsf{H}'')$ and $S \in \mathscr{B}(\mathsf{H}, \mathsf{H}')$ are isometric, $TS \in \mathscr{B}(\mathsf{H}, \mathsf{H}'')$ is isometric. The same is true if "isometric" is replaced by "unitary".

Symmetric operators and unitary operators on a unitary space into itself are special cases of ***normal operators***. $T \in \mathscr{B}(\mathsf{H})$ is said to be normal if T and T^* commute:

$$T^*T = T\,T^*\,. \tag{6.41}$$

This is equivalent to (again note Problem 6.14)

$$\|T^*u\| = \|Tu\| \quad \text{for all} \quad u \in \mathsf{H}\,. \tag{6.42}$$

---

[1] Isometric operators can be defined more generally between any two normed spaces, but we shall have no occasion to consider general isometric operators.

An important property of a normal operator T is that

(6.43) $$\|T^n\| = \|T\|^n, \quad n = 1, 2, \ldots .$$

This implies in particular that (spr denotes the spectral radius, see § 4.2)

(6.44) $$\operatorname{spr} T = \|T\| .$$

To prove (6.43), we begin with the special case in which T is symmetric. We have then $\|T^2\| = \|T\|^2$ by (6.25). Since T^2 is symmetric, we have similarly $\|T^4\| = \|T^2\|^2 = \|T\|^4$. Proceeding in the same way, we see that (6.43) holds for $n = 2^m$, $m = 1, 2, \ldots$. Suppose now that T is normal but not necessarily symmetric. Again by (6.25) we have $\|T^n\|^2 = \|T^{n*} T^n\|$. But since $T^{n*} T^n = (T^* T)^n$ by (6.41) and $T^* T$ is symmetric, we have $\|T^n\|^2 = \|(T^* T)^n\| = \|T^* T\|^n = \|T\|^{2n}$ for $n = 2^m$. This proves (6.43) for $n = 2^m$. For general n, we take an m such that $2^m - n = r \geqq 0$. Since (6.43) has been proved for n replaced by $n + r = 2^m$, we have $\|T\|^{n+r} = \|T^{n+r}\| \leqq \|T^n\| \, \|T^r\| \leqq \|T^n\| \, \|T\|^r$ or $\|T\|^n \leqq \|T^n\|$. But since the opposite inequality is obvious, (6.43) follows.

Problem 6.23. If T is normal, $p(T)$ is normal for any polynomial p.

Problem 6.24. T^{-1} is normal if T is normal and nonsingular.

Problem 6.25. If T is normal, $T^n = 0$ for some integer n implies $T = 0$. In other words, a normal operator T is nilpotent if and only if $T = 0$.

7. Projections

An important example of a symmetric operator is an *orthogonal projection*. Consider a subspace M of H and the decomposition $\mathsf{H} = \mathsf{M} \oplus \mathsf{M}^\perp$ [see (6.16)]. The projection operator $P = P_{\mathsf{M}}$ on M along $\mathsf{M}^\perp$ is called the orthogonal projection on M. P is *symmetric and nonnegative*, for (with the notation of par. 3)

(6.45) $$(Pu, u) = (u', u' + u'') = (u', u') \geqq 0$$

in virtue of $u' \perp u''$. Thus

(6.46) $$P^* = P, \quad P \geqq 0, \quad P^2 = P .$$

Conversely, it is easy to see that a symmetric, idempotent operator $P \in \mathscr{B}(\mathsf{H})$ is an orthogonal projection on $\mathsf{M} = \mathsf{R}(P)$.

Problem 6.26. $1 - P$ is an orthogonal projection with P. If P is an orthogonal projection, we have

(6.47) $$0 \leqq P \leqq 1,^{1} \quad \|P\| = 1 \text{ if } P \neq 0 .$$

Problem 6.27. $\|(1 - P_{\mathsf{M}}) u\| = \operatorname{dist}(u, \mathsf{M})$, $u \in \mathsf{H}$.

---

[1] The notation $0 \leqq P \leqq 1$, which is used to denote the order relation defined for symmetric operators [see (6.36)], does not conflict with the notation introduced earlier for projections (see § 3.4).

Problem 6.28. $\mathsf{M} \perp \mathsf{N}$ is equivalent to $P_\mathsf{M} P_\mathsf{N} = 0$. The following three conditions are equivalent: $\mathsf{M} \supset \mathsf{N}$, $P_\mathsf{M} \geqq P_\mathsf{N}$, $P_\mathsf{M} P_\mathsf{N} = P_\mathsf{N}$.

Let $P_1, \ldots, P_n$ be orthogonal projections such that

$$P_j P_k = \delta_{jk} P_j . \tag{6.48}$$

Then their sum

$$P = \sum_{j=1}^{n} P_j \tag{6.49}$$

is also an orthogonal projection, of which the range is the direct sum of the ranges of the P_j.

Orthogonal projections are special kinds of projections. We can of course consider more general "oblique" projections in a unitary space H. Let $\mathsf{H} = \mathsf{M} \oplus \mathsf{N}$ with M, N not necessarily orthogonal, and let P be the projection on M along N. Then P^* is the projection on $\mathsf{N}^\perp$ along $\mathsf{M}^\perp$ [see (3.43)].

Problem 6.29. $\|P\| \geqq 1$ for a projection $P \neq 0$; $\|P\| = 1$ holds if and only if P is an orthogonal projection. [hint for "only if" part: Let $u \in \mathsf{N}^\perp$. Then $u \perp (1 - P) u$. Apply the Pythagorean theorem to obtain $\|Pu\|^2 = \|u\|^2 + \|(1 - P) u\|^2$. From this and $\|P\| = 1$ deduce $\mathsf{N}^\perp \subset \mathsf{M}$. Consider P^* to deduce the opposite inclusion.]

Problem 6.30. A normal projection is orthogonal.

Problem 6.31. If P is a projection in a unitary space with $0 \neq P \neq 1$, then $\|P\| = \|1 - P\|$[1].

8. Pairs of projections

Let us now consider a pair P, Q of projections in H and recall the results of § 4.6 that $\mathsf{R}(P)$ and $\mathsf{R}(Q)$ are isomorphic if $P - Q$ is sufficiently small. A new result here is that *the operator U given by* (4.38) *is unitary if P, Q are orthogonal projections*. This is seen by noting that $U'^* = V'$ and $R^* = R$ [see (4.36) and (4.33)], which imply $U^* = V = U^{-1}$. Thus

Theorem 6.32. *Two orthogonal projections P, Q such that $\|P - Q\| < 1$ are unitarily equivalent, that is, there is a unitary operator U with the property $Q = U P U^{-1}$.*

Problem 6.33. $\|P - Q\| \leqq 1$ for any pair P, Q of orthogonal projections [see (4.34)].

A similar but somewhat deeper result is given by the following theorem.

Theorem 6.34.[2] *Let P, Q be two orthogonal projections with $\mathsf{M} = \mathsf{R}(P)$, $\mathsf{N} = \mathsf{R}(Q)$ such that*

$$\|(1 - Q) P\| = \delta < 1 . \tag{6.50}$$

---

[1] See T. Kato [13].

[2] See T. Kato [12], Lemma 221. This theorem is true even for $\dim \mathsf{H} = \infty$. A similar but slightly weaker result was given earlier by Akhiezer and Glazman [1], § 34.

Then there are the following alternatives. Either

i) *Q maps M onto N one to one and bicontinuously, and*

$$\|P - Q\| = \|(1 - P)\, Q\| = \|(1 - Q)\, P\| = \delta\,; \quad \textit{or} \tag{6.51}$$

ii) *Q maps M onto a proper subspace N_0 of N one to one and bicontinuously and, if Q_0 is the orthogonal projection on N_0,*

$$\begin{aligned} \|P - Q_0\| = \|(1 - P)\, Q_0\| = \|(1 - Q_0)\, P\| = \|(1 - Q)\, P\| = \delta\,, \\ \|P - Q\| = \|(1 - P)\, Q\| = 1\,. \end{aligned} \tag{6.52}$$

Proof. For any $u \in \mathsf{M}$, we have $\|u - Qu\| = \|(1 - Q)\, Pu\| \leqq \delta \|u\|$ so that $\|Qu\| \geqq (1 - \delta)\, \|u\|$. Thus the mapping $u \to Qu$ of M into N is one to one and bicontinuous (continuous with its inverse mapping). Therefore, the image $Q\mathsf{M} = \mathsf{N}_0$ of this mapping is a closed subspace of N. Let Q_0 be the orthogonal projection on N_0.

For any $w \in \mathsf{H}$, $Q_0 w$ is in N_0 and hence $Q_0 w = Qu$ for some $u \in \mathsf{M}$. If $Q_0 w \neq 0$ then $u \neq 0$ and, since $(1 - P)\, u = 0$,

$$\begin{aligned} \|(1 - P)\, Q_0 w\| &= \|(1 - P)\, Qu\| = \|(1 - P)\, (Qu - \|Qu\|^2 \|u\|^{-2} u)\| \\ &\leqq \|(Qu - \|Qu\|^2 \|u\|^{-2} u)\|\,. \end{aligned} \tag{6.53}$$

Hence

$$\begin{aligned} \|(1 - P)\, Q_0 w\|^2 &\leqq \|Qu\|^2 - \|Qu\|^4 \|u\|^{-2} = \\ &= \|Qu\|^2 \|u\|^{-2} (\|u\|^2 - \|Qu\|^2) = \|u\|^{-2} \|Q_0 w\|^2 \|(1 - Q)\, u\|^2 \leqq \\ &\leqq \|u\|^{-2} \|w\|^2 \|(1 - Q)\, Pu\|^2 \leqq \delta^2 \|w\|^2\,. \end{aligned} \tag{6.54}$$

This inequality is true even when $Q_0 w = 0$. Hence

$$\|(1 - P)\, Q_0\| \leqq \delta = \|(1 - Q)\, P\|\,. \tag{6.55}$$

For any $w \in \mathsf{H}$, we have now

$$\begin{aligned} \|(P - Q_0)\, w\|^2 &= \|(1 - Q_0)\, Pw - Q_0 (1 - P)\, w\|^2 \\ &= \|(1 - Q_0)\, Pw\|^2 + \|Q_0 (1 - P)\, w\|^2 \end{aligned} \tag{6.56}$$

since the ranges of $1 - Q_0$ and Q_0 are orthogonal. Noting that $P = P^2$ and $1 - P = (1 - P)^2$, we see from (6.56) that

$$\|(P - Q_0)\, w\|^2 \leqq \|(1 - Q_0)\, P\|^2 \|Pw\|^2 + \|Q_0 (1 - P)\|^2 \|(1 - P)\, w\|^2. \tag{6.57}$$

Since $Q_0 P = Q_0 Q P = QP$ by the definition of Q_0, we have $\|(1 - Q_0)\, P\| = \|(1 - Q)\, P\| = \delta$, and $\|Q_0 (1 - P)\| = \|(Q_0 (1 - P))^*\| = \|(1 - P)\, Q_0\| \leqq \delta$ by (6.55). Hence

$$\|(P - Q_0)\, w\|^2 \leqq \delta^2 (\|Pw\|^2 + \|(1 - P)\, w\|^2) = \delta^2 \|w\|^2\,. \tag{6.58}$$

This gives $\|P - Q_0\| \leqq \delta$. Actually we have equality here, for

$$\delta = \|(1 - Q)\, P\| = \|(1 - Q_0)\, P\| = \|(P - Q_0)\, P\| \leqq \|P - Q_0\| \leqq \delta\,. \tag{6.59}$$

The fact that $\|P - Q_0\| = \delta < 1$ implies that P maps $\mathsf{N}_0 = \mathsf{R}(Q_0)$ *onto* M one to one (see Problem 4.11). Applying the above result (6.55) to the pair P, Q replaced by Q_0, P, we thus obtain $\|(1 - Q_0) P\| \leqq \|(1 - P)Q_0\|$. Comparison with (6.59) then shows that we have equality also in (6.55).

If $\mathsf{N}_0 = \mathsf{N}$, this completes the proof of i). If $\mathsf{N}_0 \neq \mathsf{N}$, it remains to prove the last equality of (6.52). Let v be an element of N not belonging to N_0. Since $P\mathsf{N}_0 = \mathsf{M}$ as noted above, there is a $v_0 \in \mathsf{N}_0$ such that $Pv_0 = Pv$. Thus $w = v - v_0 \in \mathsf{N}$, $w \neq 0$ and $Pw = 0$, so that $(P - Q)\, w = -w$ and $Q(1 - P)\, w = Qw = w$. Thus $\|P - Q\| \geqq 1$ and $\|(1 - P)\, Q\| = \|Q(1 - P)\| \geqq 1$. Since we have the opposite inequalities (see Problem 6.33), this gives the desired result.

As an application of Theorem 6.34, we deduce an inequality concerning pairs of oblique and orthogonal projections.

Theorem 6.35. *Let P', Q' be two oblique projections in H and let $\mathsf{M} = \mathsf{R}(P')$, $\mathsf{N} = \mathsf{R}(Q')$. Let P, Q be the orthogonal projections on M, N respectively. Then*

$$\|P - Q\| \leqq \|P' - Q'\| . \tag{6.60}$$

Proof. Since $\|P - Q\|$ never exceeds 1 (see Problem 6.33), it suffices to consider the case $\|P' - Q'\| = \delta' < 1$. For any $u \in \mathsf{H}$, we have (see Problem 6.27) $\|(1 - Q)\, Pu\| = \operatorname{dist}(Pu, \mathsf{N}) \leqq \|Pu - Q'\, Pu\| = \|(P' - Q')\, Pu\| \leqq \delta' \|Pu\| \leqq \delta' \|u\|$. Hence $\|(1 - Q)\, P\| \leqq \delta' < 1$. Similarly we have $\|(1 - P)\, Q\| \leqq \delta' < 1$. Thus Theorem 6.34 is applicable, where the case ii) is excluded, so that $\|P - Q\| = \|(1 - P)\, Q\| \leqq \delta' = \|P' - Q'\|$ by (6.51).

Problem 6.36. If P', Q' are oblique projections with $\|P' - Q'\| < 1$, then there is a unitary operator U such that $U\mathsf{M} = \mathsf{N}$, $U^{-1}\mathsf{N} = \mathsf{M}$, where $\mathsf{M} = \mathsf{R}(P')$, $\mathsf{N} = \mathsf{R}(Q')$. (This proposition can be extended directly to the infinite-dimensional case, in which it is not trivial.)

Problem 6.37. Let $P'(\varkappa)$ be an oblique projection depending continuously on $\varkappa$ for $0 \leqq \varkappa \leqq 1$, and let $P(\varkappa)$ be the orthogonal projection on $\mathsf{M}(\varkappa) = \mathsf{R}(P'(\varkappa))$. Then $P(\varkappa)$ is also continuous in $\varkappa$, and there is a family of unitary operators $U(\varkappa)$, depending continuously on $\varkappa$, such that $U(\varkappa)\, \mathsf{M}(0) = \mathsf{M}(\varkappa)$, $U(\varkappa)^{-1}\, \mathsf{M}(\varkappa) = \mathsf{M}(0)$.

9. The eigenvalue problem

We now consider the eigenvalue problem for an operator in a unitary space H. For a general operator T, there is not much simplification to be gained by the fact that the underlying space is unitary; this is clear if one notes that any vector space may be made into a unitary space by introducing an inner product, whereas the eigenvalue problem can be formulated without reference to any inner product, even to any norm. The advantage of considering a unitary space appears when the operator T has some special property peculiar to the context of a unitary space, such as being symmetric, unitary or normal.

Theorem 6.38. *A normal operator is diagonable, and its eigenprojections are orthogonal projections.*

Proof. Let T be normal. Since T and T^* commute, we have

$$(T - \zeta)(T^* - \zeta') = (T^* - \zeta')(T - \zeta) . \tag{6.61}$$

If ζ is not an eigenvalue of T and ζ' is not an eigenvalue of T^*, the inverse of (6.61) exists and we have

$$R^*(\zeta') R(\zeta) = R(\zeta) R^*(\zeta') \tag{6.62}$$

where $R(\zeta)$ and $R^*(\zeta)$ are the resolvents of T and T^* respectively. (6.62) shows that these resolvents commute. In view of the expression (5.22) for the eigenprojection P_h associated with the eigenvalue λ_h of T, a double integration of (6.62) along appropriate paths in the ζ and ζ' planes yields the relation

$$P_k^* P_h = P_h P_k^* , \quad h, k = 1, \ldots, s ; \tag{6.63}$$

recall that T^* has the eigenvalues $\bar{\lambda}_h$ and the associated eigenprojections P_h^* (§ 5.5). In particular P_h and P_h^* commute, which means that the P_h are normal. This implies that the P_h are orthogonal projections (see Problem 6.30):

$$P_h^* = P_h , \quad h = 1, \ldots, s \tag{6.64}$$

The eigennilpotents are given by $D_h = (T - \lambda_h) P_h$ and $D_h^* = (T^* - \bar{\lambda}_h) P_h^* = (T^* - \bar{\lambda}_h) P_h$ by (5.26). Since P_h and T commute, $P_h = P_h^*$ commutes with T^* and T^* commutes with T, it follows that D_h commutes with D_h^*, that is, D_h is normal. As a normal nilpotent D_h must be zero (Problem 6.25). Thus T is diagonable.

The spectral representation of a normal operator T thus takes the form

$$\begin{gathered} T = \sum_{h=1}^{s} \lambda_h P_h , \quad T^* = \sum_{h=1}^{s} \bar{\lambda}_h P_h , \\ P_h^* = P_h , \quad P_h P_k = \delta_{hk} P_h , \quad \sum_{h=1}^{s} P_h = 1 . \end{gathered} \tag{6.65}$$

T and T^* have the same set of eigenspaces, which are at the same time algebraic and geometric eigenspaces and which are orthogonal to one another. It follows further from (6.65) that

$$T^* T = T T^* = \sum_{h=1}^{s} |\lambda_h|^2 P_h . \tag{6.66}$$

Hence

$$\begin{aligned} \|Tu\|^2 = (T^* T u, u) = \sum_h |\lambda_h|^2 (P_h u, u) \leq (\max |\lambda_h|^2) \sum (P_h u, u) = \\ = (\max |\lambda_h|^2) \|u\|^2 , \end{aligned}$$

which shows the $\|T\| \leqq \max|\lambda_h|$. On the other hand $\|Tu\| = |\lambda_h| \|u\|$ for $u \in \mathsf{M}_h = \mathsf{R}(P_h)$. Thus we obtain

(6.67) $$\|T\| = \max_h |\lambda_h|$$

for a normal operator T.

If we choose an orthonormal basis in each M_h, the elements of these bases taken together constitute an orthonormal basis of H. In other words, there is an orthonormal basis $\{\varphi_n\}$ of H such that

(6.68) $$T\varphi_n = \mu_n \varphi_n, \quad n = 1, \ldots, N,$$

in which μ_n, $n = 1, \ldots, N$, are the repeated eigenvalues of T. The matrix of T for this basis has the elements

(6.69) $$\tau_{jk} = (T\varphi_k, \varphi_j) = \mu_k \delta_{jk};$$

this is a diagonal matrix with the diagonal elements μ_j.

Problem 6.39. An operator with the spectral representation (6.65) is normal.

Problem 6.40. A symmetric operator has only real eigenvalues. A normal operator with only real eigenvalues is symmetric.

Problem 6.41. Each eigenvalue of a unitary operator has absolute value one. A normal operator with this property is unitary.

Problem 6.42. A symmetric operator is nonnegative (positive) if and only if its eigenvalues are all nonnegative (positive). The upper (lower) bound of a symmetric operator is the largest (smallest) of its eigenvalues.

Problem 6.43. If T is normal, then

(6.70) $$\|R(\zeta)\| = 1/\min_h |\zeta - \lambda_h| = 1/\mathrm{dist}(\zeta, \Sigma(T)),$$
$$\|S_h\| = 1/\min_{k \neq h} |\lambda_h - \lambda_k|,$$

where $R(\zeta)$ is the resolvent of T and S_h is as in (5.18).

10. The minimax principle

Let T be a symmetric operator in H. T is diagonable and has only real eigenvalues (Problem 6.40). Let

(6.71) $$\mu_1 \leqq \mu_2 \leqq \cdots \leqq \mu_N$$

be the repeated eigenvalues of T arranged in the ascending order. For each subspace M of H set

(6.72) $$\mu[\mathsf{M}] = \mu[T, \mathsf{M}] = \min_{\substack{u \in M \\ \|u\| = 1}} (Tu, u) = \min_{0 \neq u \in M} \frac{(Tu, u)}{\|u\|^2}.$$

The *minimax* (or rather *maximin*) *principle* asserts that

(6.73) $$\mu_n = \max_{\mathrm{codim}\,\mathsf{M} = n-1} \mu[\mathsf{M}] = \max_{\mathrm{codim}\,\mathsf{M} \leqq n-1} \mu[\mathsf{M}],$$

where the max is to be taken over all subspaces M with the indicated property. (6.73) is equivalent to the following two propositions:

(6.74) $$\mu_n \geqq \mu[\mathsf{M}] \quad \text{for any} \quad \mathsf{M} \quad \text{with} \quad \operatorname{codim} \mathsf{M} \leqq n-1\,;$$

(6.75) $$\mu_n \leqq \mu[\mathsf{M}_0] \quad \text{for some} \quad \mathsf{M}_0 \quad \text{with} \quad \operatorname{codim} \mathsf{M}_0 = n-1\,.$$

Let us prove these separately.

Let $\{\varphi_n\}$ be an orthonormal basis with the property (6.68). Each $u \in \mathsf{H}$ has the expansion

(6.76) $$u = \sum \xi_n \varphi_n\,, \quad \xi_n = (u, \varphi_n)\,, \quad \|u\|^2 = \sum |\xi_n|^2\,,$$

in this basis. Then

(6.77) $$T u = \sum \xi_n T \varphi_n = \sum \mu_n \xi_n \varphi_n\,, \ (T u, u) = \sum \mu_n |\xi_n|^2\,.$$

Let M be any subspace with $\operatorname{codim} \mathsf{M} \leqq n-1$. The n-dimensional subspace M′ spanned by $\varphi_1, \ldots, \varphi_n$ contains a nonzero vector u in common with M (this is a consequence of Lemma 6.11, where M is to be replaced by $\mathsf{M}^\perp$). This u has the coefficients $\xi_{n+1}, \xi_{n+2}, \ldots$ equal to zero, so that $(T u, u) = \sum \mu_k |\xi_k|^2 \leqq \mu_n \sum |\xi_k|^2 = \mu_n \|u\|^2$. This proves (6.74).

Let M_0 be the subspace consisting of all vectors orthogonal to $\varphi_1, \ldots, \varphi_{n-1}$, so that $\operatorname{codim} \mathsf{M}_0 = n-1$. Each $u \in \mathsf{M}_0$ has the coefficients $\xi_1, \ldots, \xi_{n-1}$ zero. Hence $(T u, u) \geqq \mu_n \sum |\xi_k|^2 = \mu_n \|u\|^2$, which implies (6.75).

The minimax principle is a convenient means for characterizing the eigenvalues μ_n without any reference to the eigenvectors. As an application of this principle, we shall prove the *monotonicity principles*.

Theorem 6.44. *If S, T are symmetric operators such that $S \leqq T$, then the eigenvalues of S are not larger than the corresponding eigenvalues of T, that is,*

(6.78) $$\mu_n[S] \leqq \mu_n[T]\,, \quad n = 1, \ldots, N\,.$$

Here $\mu_n[T]$ denotes the n-th eigenvalue of T in the ascending order as in (6.71).

The proof follows immediately from the minimax principle, for $S \leqq T$ implies $(S u, u) \leqq (T u, u)$ and therefore $\mu[S, \mathsf{M}] \leqq \mu(T, \mathsf{M})$ for any subspace M.

Problem 6.45. For every pair of symmetric operators S, T,

(6.79) $$\mu_1[S] + \mu_1[T] \leqq \mu_1[S + T] \leqq \mu_1[S] + \mu_N[T]\,.$$

Let M be a subspace of H with the orthogonal projection P. For any operator T in H, $S = P T P$ is called the *orthogonal projection of T on* M. M is invariant under S, so that we can speak of the eigenvalues of S in M (that is, of the part S_M of S in M). Note that S_M has $N - r$ repeated eigenvalues, where $r = \operatorname{codim} \mathsf{M}$.

Theorem 6.46. *Let T, S, S_{M} be as above. If T is symmetric, S and S_{M} are also symmetric, and*

$$\mu_n[T] \leqq \mu_n[S_{\mathsf{M}}] \leqq \mu_{n+r}[T], \quad n = 1, \ldots, N - r. \tag{6.80}$$

Proof. The symmetry of S and of S_{M} is obvious. $\mu_n[S_{\mathsf{M}}]$ is equal to $\max \mu[S_{\mathsf{M}}, \mathsf{M}']$ taken over all $\mathsf{M}' \subset \mathsf{M}$ such that the codimension of M' relative to M is equal to $n - 1$ ($\dim \mathsf{M}/\mathsf{M}' = n - 1$). But we have $\mu[S_{\mathsf{M}}, \mathsf{M}'] = \mu[S, \mathsf{M}'] = \mu[T, \mathsf{M}']$ because $(S_{\mathsf{M}} u, u) = (Su, u) = (Tu, u)$ for any $u \in \mathsf{M}$, and $\dim \mathsf{M}/\mathsf{M}' = n - 1$ implies $\operatorname{codim} \mathsf{M}' = \dim \mathsf{H}/\mathsf{M}' = n + r - 1$. Therefore $\mu_n[S_{\mathsf{M}}]$ does not exceed $\mu_{n+r}[T] = \max \mu[T, \mathsf{M}']$ taken over all $\mathsf{M}' \subset \mathsf{H}$ with $\operatorname{codim} \mathsf{M}' = n + r - 1$. This proves the second inequality of (6.80).

On the other hand we have $\mu_n[T] = \mu[T, \mathsf{M}_0]$ where M_0 is the same as in (6.75). Hence $\mu_n[T] \leqq \mu[T, \mathsf{M}_0 \cap \mathsf{M}] = \mu[S_{\mathsf{M}}, \mathsf{M}_0 \cap \mathsf{M}]$. But $\mathsf{M}_0 \cap \mathsf{M}$ has codimension not larger than $n - 1$ relative to M because $\operatorname{codim} \mathsf{M}_0 = n - 1$. Thus $\mu[S_{\mathsf{M}}, \mathsf{M}_0 \cap \mathsf{M}] \leqq \mu_n[S_{\mathsf{M}}]$ by (6.74). This proves the first inequality of (6.80).

Chapter Two

Perturbation theory in a finite-dimensional space

In this chapter we consider perturbation theory for linear operators in a finite-dimensional space. The main question is how the eigenvalues and eigenvectors (or eigenprojections) change with the operator, in particular when the operator depends on a parameter analytically. This is a special case of a more general and more interesting problem in which the operator acts in an infinite-dimensional space.

The reason for discussing the finite-dimensional case separately is threefold. In the first place, it is not trivial. Second, it essentially embodies certain features of perturbation theory in the general case, especially those related to isolated eigenvalues. It is convenient to treat them in this simplified situation without being bothered by complications arising from the infinite dimensionality of the underlying space. The modifications required when going to the infinite-dimensional case will be introduced as supplements in later chapters, together with those features of perturbation theory which are peculiar to the infinite-dimensional case. Third, the finite-dimensional theory has its own interest, for example, in connection with the numerical analysis of matrices. The reader interested only in finite-dimensional problems can find what he wants in this chapter, without having to disentangle it from the general theory.

As mentioned above, the problem is by no means trivial, and many different methods of solving it have been introduced. The method used here is based on a function-theoretic study of the resolvent, in particular on the expression of eigenprojections as contour integrals of the resolvent. This is the quickest way to obtain general results as well as to deduce various estimates on the convergence rates of the perturbation series. In a certain sense the use of function theory for operator-valued functions is not altogether elementary, but since students of applied mathematics are as a rule well-acquainted with function theory, the author hopes that its presence in this form will not hinder those who might use the book for applications.

§ 1. Analytic perturbation of eigenvalues

1. The problem

We now go into one of our proper subjects, the perturbation theory for the eigenvalue problem in a finite-dimensional vector space X[1]. A typical problem of this theory is to investigate how the eigenvalues and eigenvectors (or eigenspaces) of a linear operator T change when T is subjected to a small perturbation[2]. In dealing with such a problem, it is often convenient to consider a family of operators of the form

$$T(\varkappa) = T + \varkappa T' \tag{1.1}$$

where $\varkappa$ is a scalar parameter supposed to be small. $T(0) = T$ is called the *unperturbed operator* and $\varkappa T'$ the *perturbation*. A question arises whether the eigenvalues and the eigenvectors of $T(\varkappa)$ can be expressed as power series in $\varkappa$, that is, whether they are holomorphic functions of $\varkappa$ in the neighborhood of $\varkappa = 0$. If this is the case, the change of the eigenvalues and eigenvectors will be of the same order of magnitude as the perturbation $\varkappa T'$ itself for small $|\varkappa|$. As we shall see below, however, this is not always the case.

(1.1) can be generalized to

$$T(\varkappa) = T + \varkappa T^{(1)} + \varkappa^2 T^{(2)} + \cdots . \tag{1.2}$$

More generally, we may suppose that an operator-valued function $T(\varkappa)$ is given, which is holomorphic in a given domain D_0 of the complex $\varkappa$-plane[3].

The eigenvalues of $T(\varkappa)$ satisfy the characteristic equation (see Problem I-5.16)

$$\det(T(\varkappa) - \zeta) = 0 . \tag{1.3}$$

This is an algebraic equation in ζ of degree $N = \dim \mathsf{X}$, with coefficients which are holomorphic in $\varkappa$; this is seen by writing (1.3) in terms of the matrix of $T(\varkappa)$ with respect to a basis $\{x_j\}$ of X, for each element of this matrix is a holomorphic function of $\varkappa$ [see I-(3.10)]. It follows from a well-

---

[1] In this section we assume that $0 < \dim \mathsf{X} = N < \infty$. Whenever convenient, X will be considered a normed space with an appropriately chosen norm.

[2] There are very few papers that deal specifically with perturbation theory in a finite-dimensional space; see parts of RELLICH [1] and [8], DAVIS [1], B. L. LIVŠIC [1], VIŠIK and LYUSTERNIK [1]. Reference should be made to papers dealing with analytic perturbation theory in Banach spaces. Basic papers in this direction are: RELLICH [1]–[5], SZ.-NAGY [1], [2], WOLF [1], T. KATO [1], [3], [6], DUNFORD-SCHWARTZ [1], RIESZ and SZ.-NAGY [1]. See also BAUMGÄRTEL [1], PORATH [1], [2], RELLICH [6], ROSENBLOOM [1], SCHÄFKE [3]–[5], SCHRÖDER [1]–[3], ŠMUL'YAN [1].

[3] One can restrict $\varkappa$ to real values, but since (1.2) given for real $\varkappa$ can always be extended to complex $\varkappa$, there is no loss of generality in considering complex $\varkappa$.

known result[1] in function theory that the roots of (1.3) are (branches of) *analytic functions of $\varkappa$ with only algebraic singularities*. More precisely, the roots of (1.3) for $\varkappa \in D_0$ constitute one or several branches of one or several analytic functions that have only algebraic singularities in D_0.

It follows immediately that the number of eigenvalues of $T(\varkappa)$ is a constant s independent of $\varkappa$, with the exception of some special values of $\varkappa$. There are only a finite number of such *exceptional points* $\varkappa$ in each compact subset of D_0. This number s is equal to N if these analytic functions (if there are more than one) are all distinct; in this case $T(\varkappa)$ is simple and therefore diagonable for all non-exceptional $\varkappa$. If, on the other hand, there happen to be identical ones among these analytic functions, then we have $s < N$; in this case $T(\varkappa)$ is said to be *permanently degenerate*.

Example 1.1. Here we collect the simplest examples illustrating the various possibilities stated above. These examples are concerned with a family $T(\varkappa)$ of the form (1.1) in a two-dimensional space ($N = 2$). For simplicity we identify $T(\varkappa)$ with its matrix representation with respect to a basis.

a) $$T(\varkappa) = \begin{pmatrix} 1 & \varkappa \\ \varkappa & -1 \end{pmatrix}.$$

The eigenvalues of $T(\varkappa)$ are

(1.4) $$\lambda_\pm(\varkappa) = \pm (1 + \varkappa^2)^{1/2}$$

and are branches of one double-valued analytic function $(1 + \varkappa^2)^{1/2}$. Thus $s = N = 2$ and the exceptional points are $\varkappa = \pm i$, $T(\pm i)$ having only the eigenvalue 0.

b) $$T(\varkappa) = \begin{pmatrix} 0 & \varkappa \\ \varkappa & 0 \end{pmatrix}, \quad s = N = 2.$$

The eigenvalues are $\pm\varkappa$; these are two distinct entire functions of $\varkappa$ (the characteristic equation is $\zeta^2 - \varkappa^2 = 0$ and is reducible). There is one exceptional point $\varkappa = 0$, for which $T(\varkappa)$ has only one eigenvalue 0.

c) $$T(\varkappa) = \begin{pmatrix} 0 & \varkappa \\ 0 & 0 \end{pmatrix}, \quad s = 1.$$

$T(\varkappa)$ is permanently degenerate, the only eigenvalue being 0 for all $\varkappa$; we have two identical analytic functions zero. There are no exceptional points.

d) $$T(\varkappa) = \begin{pmatrix} 0 & 1 \\ \varkappa & 0 \end{pmatrix}, \quad s = 2.$$

The eigenvalues are $\pm\varkappa^{1/2}$, constituting one double-valued function $\varkappa^{1/2}$. There is one exceptional point $\varkappa = 0$.

e) $$T(\varkappa) = \begin{pmatrix} 1 & \varkappa \\ 0 & 0 \end{pmatrix}, \quad s = 2.$$

The eigenvalues are 0 and 1. There are no exceptional points.

f) $$T(\varkappa) = \begin{pmatrix} \varkappa & 1 \\ 0 & 0 \end{pmatrix}, \quad s = 2.$$

The eigenvalues are 0 and $\varkappa$, which are two distinct entire functions. There is one exceptional point $\varkappa = 0$.

[1] See KNOPP [2], p. 119, where algebraic functions are considered. Actually (1.3) determines ζ as *algebroidal* (not necessarily algebraic) functions, which are, however, locally similar to algebraic functions. For detailed function-theoretic treatment of (1.3), see BAUMGÄRTEL [1].

2. Singularities of the eigenvalues

We now consider the eigenvalues of $T(\varkappa)$ in more detail. Since these are in general multiple-valued analytic functions of $\varkappa$, some care is needed in their notation. If $\varkappa$ is restricted to a simply-connected[1] subdomain D of the fundamental domain D_0 containing no exceptional point (for brevity such a subdomain will be called a *simple subdomain*), the eigenvalues of $T(\varkappa)$ can be written

(1.5) $$\lambda_1(\varkappa), \lambda_2(\varkappa), \ldots, \lambda_s(\varkappa),$$

all s functions $\lambda_h(\varkappa)$, $h = 1, \ldots, s$, being holomorphic in D and $\lambda_h(\varkappa) \neq \lambda_k(\varkappa)$ for $h \neq k$.

We next consider the behavior of the eigenvalues in the neighborhood of one of the exceptional points, which we may take as $\varkappa = 0$ without loss of generality. Let D be a small disk near $\varkappa = 0$ but excluding $\varkappa = 0$. The eigenvalues of $T(\varkappa)$ for $\varkappa \in D$ can be expressed by s holomorphic functions of the form (1.5). If D is moved continuously around $\varkappa = 0$, these s functions can be continued analytically. When D has been brought to its initial position after one revolution around $\varkappa = 0$, the s functions (1.5) will have undergone a permutation among themselves. These functions may therefore be grouped in the manner

(1.6) $$\{\lambda_1(\varkappa), \ldots, \lambda_p(\varkappa)\}, \{\lambda_{p+1}(\varkappa), \ldots, \lambda_{p+q}(\varkappa)\}, \ldots,$$

in such a way that each group undergoes a cyclic permutation by a revolution of D of the kind described. For brevity each group will be called a *cycle* at the exceptional point $\varkappa = 0$, and the number of elements of a cycle will be called its *period*.

It is obvious that the elements of a cycle of period p constitute a branch of an analytic function (defined near $\varkappa = 0$) with a branch point (if $p \geqq 2$) at $\varkappa = 0$, and we have *Puiseux series* such as[2]

(1.7) $$\lambda_h(\varkappa) = \lambda + \alpha_1 \omega^h \varkappa^{1/p} + \alpha_2 \omega^{2h} \varkappa^{2/p} + \cdots, \quad h = 0, 1, \ldots, p-1,$$

where $\omega = \exp(2\pi i/p)$. It should be noticed that here no negative powers of $\varkappa^{1/p}$ appear, for the coefficient of the highest power ζ^N in (1.3) is $(-1)^N$ so that the $\lambda_h(\varkappa)$ are continuous at $\varkappa = 0$[3]. $\lambda = \lambda_h(0)$ will be called the *center* of the cycle under consideration.

(1.7) shows that $|\lambda_h(\varkappa) - \lambda|$ is in general of the order $|\varkappa|^{1/p}$ for small $|\varkappa|$ for $h = 1, \ldots, p$. If $p \geqq 2$, therefore, the rate of change at an exceptional point of the eigenvalues of a cycle of period p is infinitely large compared with the change of $T(\varkappa)$ itself[4].

[1] See KNOPP [1], p. 19.
[2] See KNOPP [2], p. 130.
[3] See KNOPP [2], p. 122.
[4] This fact is of some importance in the numerical analysis of eigenvalues of matrices.

Problem 1.2. The sum of the $\lambda_h(\varkappa)$ belonging to a cycle is holomorphic at the exceptional point in question.

In general there are several cycles with the same center λ. All the eigenvalues (1.7) belonging to cycles with center λ are said to depart from the unperturbed eigenvalue λ by *splitting* at $\varkappa = 0$. The set of these eigenvalues will be called the *λ-group*, since they cluster around λ for small $|\varkappa|$.

Remark 1.3. An exceptional point need not be a branch point of an analytic function representing some of the eigenvalues. In other words, it is possible that all cycles at an exceptional point $\varkappa = \varkappa_0$ are of period 1. In any case, however, some two different eigenvalues for $\varkappa \neq \varkappa_0$ must coincide at $\varkappa = \varkappa_0$ (definition of an exceptional point). Thus there is always splitting at (and only at) an exceptional point.

Example 1.4. Consider the examples listed in Example 1.1. We have a cycle of period 2 at the exceptional points $\varkappa = \pm i$ in a) and also at $\varkappa = 0$ in d). There are two cycles of period 1 at $\varkappa = 0$ in b) and f). There are no exceptional points in c) and e).

3. Perturbation of the resolvent

The resolvent

$$R(\zeta, \varkappa) = (T(\varkappa) - \zeta)^{-1} \tag{1.8}$$

of $T(\varkappa)$ is defined for all ζ not equal to any of the eigenvalues of $T(\varkappa)$ and is a meromorphic function of ζ for each fixed $\varkappa \in D_0$. Actually we have

Theorem 1.5. *$R(\zeta, \varkappa)$ is holomorphic in the two variables ζ, $\varkappa$ in each domain in which ζ is not equal to any of the eigenvalues of $T(\varkappa)$.*

Proof. Let $\zeta = \zeta_0$, $\varkappa = \varkappa_0$ belong to such a domain; we may assume $\varkappa_0 = 0$ without loss of generality. Thus ζ_0 is not equal to any eigenvalue of $T(0) = T$, and

$$\begin{aligned} T(\varkappa) - \zeta &= T - \zeta_0 - (\zeta - \zeta_0) + A(\varkappa) \\ &= [1 - (\zeta - \zeta_0 - A(\varkappa))\, R(\zeta_0)]\, (T - \zeta_0)\,, \end{aligned} \tag{1.9}$$

$$A(\varkappa) = T(\varkappa) - T = \sum_{n=1}^{\infty} \varkappa^n T^{(n)}\,, \tag{1.10}$$

where $R(\zeta) = R(\zeta, 0) = (T - \zeta)^{-1}$ and we assumed the Taylor expansion of $T(\varkappa)$ at $\varkappa = 0$ in the form (1.2). Hence

$$R(\zeta, \varkappa) = R(\zeta_0)\, [1 - (\zeta - \zeta_0 - A(\varkappa))\, R(\zeta_0)]^{-1}\,, \tag{1.11}$$

exists if the factor $[\;]^{-1}$ can be defined by a convergent Neumann series (see Example I-4.5), which is the case if, for example,

$$|\zeta - \zeta_0| + \sum_{n=1}^{\infty} |\varkappa|^n \, \|T^{(n)}\| < \|R(\zeta_0)\|^{-1}\,, \tag{1.12}$$

since $|\zeta - \zeta_0| + \|A(\varkappa)\|$ is not greater than the left member of (1.12). This inequality is certainly satisfied for sufficiently small $|\zeta - \zeta_0|$ and $|\varkappa|$, and then the right member of (1.11) can be written as a double power series in $\zeta - \zeta_0$ and $\varkappa$. This shows that $R(\zeta, \varkappa)$ is holomorphic in ζ and $\varkappa$ in a neighborhood of $\zeta = \zeta_0$, $\varkappa = 0$.

For later use it is more convenient to write $R(\zeta, \varkappa)$ as a power series in $\varkappa$ with coefficients depending on ζ. On setting $\zeta_0 = \zeta$ in (1.11), we obtain

$$(1.13)\quad R(\zeta, \varkappa) = R(\zeta)\,[1 + A(\varkappa)\,R(\zeta)]^{-1} = R(\zeta) \sum_{p=0}^{\infty} [-A(\varkappa)\,R(\zeta)]^p = R(\zeta) + \sum_{n=1}^{\infty} \varkappa^n R^{(n)}(\zeta)\,,$$

where

$$(1.14)\quad R^{(n)}(\zeta) = \sum_{\substack{\nu_1 + \cdots + \nu_p = n \\ \nu_j \geq 1}} (-1)^p\, R(\zeta)\, T^{(\nu_1)} R(\zeta)\, T^{(\nu_2)} \ldots T^{(\nu_p)} R(\zeta)\,,$$

the sum being taken for all combinations of positive integers p and $\nu_1, \ldots, \nu_p$ such that $1 \leq p \leq n$, $\nu_1 + \cdots + \nu_p = n$.

(1.13) will be called the *second Neumann series* for the resolvent. It is uniformly convergent for sufficiently small $\varkappa$ and $\zeta \in \Gamma$ if Γ is a compact subset of the resolvent set $\mathrm{P}(T)$ of $T = T(0)$; this is seen from (1.12) with $\zeta_0 = \zeta$, where $\|R(\zeta)\|^{-1}$ has a positive minimum for $\zeta \in \Gamma$.

Example 1.6. The resolvent for the $T(\varkappa)$ of Example 1.1, a) is given by

$$(1.15)\quad R(\zeta, \varkappa) = (\zeta^2 - 1 - \varkappa^2)^{-1} \begin{pmatrix} -1-\zeta & -\varkappa \\ -\varkappa & 1-\zeta \end{pmatrix}.$$

Problem 1.7. Find the resolvents of the $T(\varkappa)$ of b) to f) in Example 1.1.

4. Perturbation of the eigenprojections

Let λ be one of the eigenvalues of $T = T(0)$, with multiplicity[1] m. Let Γ be a closed positively-oriented curve, say a circle, in the resolvent set $\mathrm{P}(T)$ enclosing λ but no other eigenvalues of T. As noted above, the second Neumann series (1.13) is then convergent for sufficiently small $|\varkappa|$ uniformly for $\zeta \in \Gamma$. The existence of the resolvent $R(\zeta, \varkappa)$ of $T(\varkappa)$ for $\zeta \in \Gamma$ implies that there are no eigenvalues of $T(\varkappa)$ on Γ.

The operator

$$(1.16)\quad P(\varkappa) = -\frac{1}{2\pi i} \int_\Gamma R(\zeta, \varkappa)\, d\zeta \text{ [2]}$$

is a projection and is equal to the sum of the eigenprojections for all the eigenvalues of $T(\varkappa)$ lying inside Γ (see Problem I-5.9). In particular

[1] By "multiplicity" we mean the *algebraic* multiplicity unless otherwise stated.

[2] This integral formula is basic throughout the present book. In perturbation theory it was first used by Sz.-Nagy [1] and T. Kato [1], greatly simplifying the earlier method of Rellich [1]—[5].

$P(0) = P$ coincides with the eigenprojection for the eigenvalue λ of T. Integrating (1.13) term by term, we have

$$P(\varkappa) = P + \sum_{n=1}^{\infty} \varkappa^n P^{(n)} \tag{1.17}$$

with

$$P^{(n)} = -\frac{1}{2\pi i} \int_\Gamma R^{(n)}(\zeta)\, d\zeta\,. \tag{1.18}$$

The series (1.17) is convergent for small $|\varkappa|$ so that $P(\varkappa)$ is holomorphic near $\varkappa = 0$. It follows from Lemma I-4.10 that the range $\mathsf{M}(\varkappa)$ of $P(\varkappa)$ is isomorphic with the (algebraic) eigenspace $\mathsf{M} = \mathsf{M}(0) = P\mathsf{X}$ of T for the eigenvalue λ. In particular we have

$$\dim P(\varkappa) = \dim P = m\,. \tag{1.19}$$

Since (1.19) is true for all sufficiently small $|\varkappa|$, it follows that the eigenvalues of $T(\varkappa)$ lying inside Γ form exactly the λ-group. For brevity we call $P(\varkappa)$ the *total projection*, and $\mathsf{M}(\varkappa)$ the *total eigenspace*, for the λ-group.

If $\varkappa = 0$ is not an exceptional point, there is no splitting at $\varkappa = 0$ of the eigenvalue λ in question. In this case there is exactly one eigenvalue $\lambda(\varkappa)$ of $T(\varkappa)$ in the neighborhood of λ, and $P(\varkappa)$ is itself the eigenprojection for this eigenvalue $\lambda(\varkappa)$. (1.19) shows that the multiplicity of $\lambda(\varkappa)$ is equal to m. Similar results hold when $\varkappa = 0$ is replaced by any other non-exceptional point $\varkappa = \varkappa_0$.

Now consider a simple subdomain D of the $\varkappa$-plane and the set (1.5) of the eigenvalues of $T(\varkappa)$ for $\varkappa \in$ D, and let $P_h(\varkappa)$ be the eigenprojection for the eigenvalue $\lambda_h(\varkappa)$, $h = 1, \ldots, s$. The result just proved shows that each $P_h(\varkappa)$ is holomorphic in D and that each $\lambda_h(\varkappa)$ has constant multiplicity m_h. Here it is essential that D is simple (contains no exceptional point); in fact, $P_1(\varkappa_0)$ is not even defined if, for example, $\lambda_1(\varkappa_0) = \lambda_2(\varkappa_0)$ which may happen if $\varkappa_0$ is exceptional.

Let $\mathsf{M}_h(\varkappa) = P_h(\varkappa)\,\mathsf{X}$ be the (algebraic) eigenspace of $T(\varkappa)$ for the eigenvalue $\lambda_h(\varkappa)$. We have [see I-(5.34)]

$$\mathsf{X} = \mathsf{M}_1(\varkappa) \oplus \cdots \oplus \mathsf{M}_s(\varkappa)\,, \tag{1.20}$$

$$\dim \mathsf{M}_h(\varkappa) = m_h\,, \quad \sum_{j=1}^{s} m_h = N\,, \quad \varkappa \in \mathrm{D}\,.$$

The eigennilpotent $D_h(\varkappa)$ for the eigenvalue $\lambda_h(\varkappa)$ is also holomorphic for $\varkappa \in$ D, for

$$D_h(\varkappa) = (T(\varkappa) - \lambda_h(\varkappa))\, P_h(\varkappa) \tag{1.21}$$

by I-(5.26).

5. Singularities of the eigenprojections

Let us now consider the behavior of the eigenprojections $P_h(\varkappa)$ near an exceptional point, which we may again assume to be $\varkappa = 0$. As was shown above, each eigenvalue λ of T splits in general into several eigenvalues of $T(\varkappa)$ for $\varkappa \neq 0$, but the corresponding total projection is holomorphic at $\varkappa = 0$ [see (1.17)]. Take again the small disk D near $\varkappa = 0$ considered in par. 2; the eigenvalues $\lambda_h(\varkappa)$, the eigenprojections $P_h(\varkappa)$ and the eigennilpotents $D_h(\varkappa)$ are defined and holomorphic for $\varkappa \in \mathrm{D}$ as shown above. When D is moved around $\varkappa = 0$ and brought to the initial position in the manner described in par. 2, each of the families $\{\lambda_h(\varkappa)\}$, $\{P_h(\varkappa)\}$ and $\{D_h(\varkappa)\}$ is subjected to a permutation by the analytic continuation. This permutation must be identical for the three families, as is seen from the following consideration.

The resolvent $R(\zeta, \varkappa)$ of $T(\varkappa)$ has the partial-fraction expression

$$R(\zeta, \varkappa) = -\sum_{h=1}^{s} \left[\frac{P_h(\varkappa)}{\zeta - \lambda_h(\varkappa)} + \frac{D_h(\varkappa)}{(\zeta - \lambda_h(\varkappa))^2} + \cdots + \frac{D_h(\varkappa)^{m_h-1}}{(\zeta - \lambda_h(\varkappa))^{m_h}}\right] \tag{1.22}$$

[see I-(5.23)], where ζ is assumed to be somewhere distant from the spectrum of T so that $\zeta \in \mathrm{P}(T(\varkappa))$ for all $\varkappa$ considered. If $\lambda_1(\varkappa), \ldots, \lambda_p(\varkappa)$ constitute a cycle (see par. 2) of eigenvalues, the permutation mentioned above takes $\lambda_h(\varkappa)$ into $\lambda_{h+1}(\varkappa)$ for $1 \leq h \leq p-1$ and $\lambda_p(\varkappa)$ into $\lambda_1(\varkappa)$. But as $R(\zeta, \varkappa)$ should be unchanged by the analytic continuation under consideration, the permutation must take $P_h(\varkappa)$ into $P_{h+1}(\varkappa)$ for $1 \leq h \leq p-1$ and $P_p(\varkappa)$ into $P_1(\varkappa)$ [1]; the possibility that $P_h(\varkappa) = P_k(\varkappa)$ for some $h \neq k$ is excluded by the property $P_h(\varkappa)\, P_k(\varkappa) = \delta_{hk}\, P_h(\varkappa)$. Similar results hold for the eigennilpotents $D_h(\varkappa)$ by (1.21), except that some pair of the $D_h(\varkappa)$ may coincide [in fact all $D_h(\varkappa)$ can be zero).

We shall now show that $P_h(\varkappa)$ and $D_h(\varkappa)$ have at most algebraic singularities. Since $D_h(\varkappa)$ is given by (1.21), it suffices to prove this for $P_h(\varkappa)$. To this end we first note that

$$\|P_h(\varkappa)\| = \left\|\frac{1}{2\pi i}\int_{\Gamma_h(\varkappa)} R(\zeta, \varkappa)\, d\zeta\right\| \leq \varrho_h(\varkappa) \max_{\zeta \in \Gamma_h(\varkappa)} \|R(\zeta, \varkappa)\| \tag{1.23}$$

where $\Gamma_h(\varkappa)$ is a circle enclosing $\lambda_h(\varkappa)$ but excluding all other $\lambda_k(\varkappa)$ and where $\varrho_h(\varkappa)$ denotes the radius of $\Gamma_h(\varkappa)$. On the other hand, we see from I-(4.12) that

$$\begin{aligned} \|R(\zeta, \varkappa)\| &= \|(T(\varkappa) - \zeta)^{-1}\| \leq \\ &\leq \gamma \|T(\varkappa) - \zeta\|^{N-1} / |\det(T(\varkappa) - \zeta)| \leq \\ &\leq \gamma (\|T(\varkappa)\| + |\zeta|)^{N-1} \Big/ \prod_{k=1}^{s} |\zeta - \lambda_k(\varkappa)|^{m_k}, \end{aligned} \tag{1.24}$$

---

[1] This is due to the uniqueness of the partial-fraction representation of $R(\zeta, \varkappa)$ as a function of ζ. A similar argument was used in I-§ 5.4.

where γ is a constant depending only on the norm employed. Hence

$$(1.25)\qquad \|P_h(\varkappa)\| \leqq \gamma\, \varrho_h(\varkappa) \max_{\zeta\in\Gamma_h(\varkappa)} (\|T(\varkappa)\| + |\zeta|)^{N-1} \Big/ \prod_{k=1}^{s} |\zeta - \lambda_k(\varkappa)|^{m_k}.$$

Suppose that $\varkappa \to 0$, assuming again that $\varkappa = 0$ is an exceptional point. Then we have to choose the circle $\Gamma_h(\varkappa)$ smaller for smaller $|\varkappa|$ in order to ensure that it encloses $\lambda_h(\varkappa)$ but no other $\lambda_k(\varkappa)$, for the $\lambda_k(\varkappa)$ of the λ-group will approach $\lambda_h(\varkappa)$ indefinitely. But we know that the distances $|\lambda_h(\varkappa) - \lambda_k(\varkappa)|$ between these eigenvalues tend to zero for $\varkappa \to 0$ at most with some definite fractional order of $|\varkappa|$ because all $\lambda_k(\varkappa)$ have at most algebraic singularities at $\varkappa = 0$ [see (1.7)]. By choosing $\varrho_h(\varkappa) = |\varkappa|^\alpha$ with an appropriate $\alpha > 0$, we can therefore ensure that $\prod |\zeta - \lambda_k(\varkappa)|^{m_k} \geqq \gamma' |\varkappa|^{\alpha N}$ for $\zeta \in \Gamma_h(\varkappa)$ with some constant $\gamma' > 0$. Then we have

$$(1.26)\qquad \|P_h(\varkappa)\| \leqq \text{const.}\ |\varkappa|^{-(N-1)\alpha}\,.$$

This shows that, when $P_h(\varkappa)$ is represented by a Laurent series in $\varkappa^{1/p}$, the principal part is finite.

These results may be summarized in

Theorem 1.8. *The eigenvalues $\lambda_h(\varkappa)$, the eigenprojections $P_h(\varkappa)$ and the eigennilpotents $D_h(\varkappa)$ of $T(\varkappa)$ are (branches of) analytic functions for $\varkappa \in D_0$ with only algebraic singularities at some (but not necessarily all) exceptional points. $\lambda_h(\varkappa)$ and $P_h(\varkappa)$ have all branch points in common (including the order of the branch points), which may or may not be branch points for $D_h(\varkappa)$. If in particular $\lambda_h(\varkappa)$ is single-valued near an exceptional point $\varkappa = \varkappa_0$ (cycle of period 1), then $P_h(\varkappa)$ and $D_h(\varkappa)$ are also single-valued there.*

6. Remarks and examples

Although the $P_h(\varkappa)$ and $D_h(\varkappa)$ have algebraic singularities as well as the $\lambda_h(\varkappa)$, there are some important differences in their behavior at the singular points. Roughly speaking, $P_h(\varkappa)$ and $D_h(\varkappa)$ have stronger singularities than $\lambda_h(\varkappa)$.

We recall that these singular points are exceptional points, though the converse is not true. As we have already noted, the $\lambda_h(\varkappa)$ are continuous even at exceptional points and, therefore, have no poles. But $P_h(\varkappa)$ and $D_h(\varkappa)$ are in general undefined at exceptional points. In particular they may be single-valued and yet have a pole at an exceptional point (see Example 1.12 below).

Even more remarkable is the following theorem[1].

Theorem 1.9. *If $\varkappa = \varkappa_0$ is a branch point of $\lambda_h(\varkappa)$ (and therefore also of $P_h(\varkappa)$) of order $p - 1 \geqq 1$, then $P_h(\varkappa)$ has a pole there; that is, the*

[1] This theorem is due to Butler [1].

Laurent expansion of $P_h(\varkappa)$ in powers of $(\varkappa - \varkappa_0)^{1/p}$ necessarily contains negative powers. In particular $\|P_h(\varkappa)\| \to \infty$ for $\varkappa \to \varkappa_0$.

Proof. Suppose that this were not the case and let

$$P_h(\varkappa) = P_h + \varkappa^{1/p} P'_h + \cdots, \quad h = 1, \ldots, p,$$

be the Laurent expansions of the $P_h(\varkappa)$ belonging to the cycle under consideration. Here we again assume for simplicity that $\varkappa_0 = 0$. When $\varkappa$ is subjected to a revolution around $\varkappa = 0$, $P_h(\varkappa)$ is changed into $P_{h+1}(\varkappa)$ for $1 \leqq h \leqq p - 1$ and $P_p(\varkappa)$ into $P_1(\varkappa)$. Hence we must have $P_{h+1} = P_h$ for $1 \leqq h \leqq p - 1$. On the other hand, the relation $P_h(\varkappa) P_{h+1}(\varkappa) = 0$ for $\varkappa \to 0$ gives $P_h P_{h+1} = 0$, and the idempotent character of $P_h(\varkappa)$ gives $P_h^2 = P_h$. Hence $P_h = P_h^2 = P_h P_{h+1} = 0$. But this contradicts the fact that $\dim P_h(\varkappa) \mathsf{X} = m_h > 0$, which implies that $\|P_h(\varkappa)\| \geqq 1$ (see Problem I-4.1).

As regards the order $p - 1$ of the branch point $\varkappa = \varkappa_0$ for $\lambda_h(\varkappa)$ or, equivalently, the period p of the cycle $\{\lambda_1(\varkappa), \ldots, \lambda_p(\varkappa)\}$, we have the following result. *An eigenvalue λ of T with multiplicity m does not give rise to a branch point of order larger than $m - 1$.* This is an obvious consequence of the fact that such an eigenvalue can never split into more than m eigenvalues [see (1.19)].

Theorem 1.10. *Let X be a unitary space. Let $\varkappa_0 \in \mathrm{D}_0$ (possibly an exceptional point) and let there exist a sequence $\{\varkappa_n\}$ converging to $\varkappa_0$ such that $T(\varkappa_n)$ is normal for $n = 1, 2, \ldots$. Then all the $\lambda_h(\varkappa)$ and $P_h(\varkappa)$ are holomorphic at $\varkappa = \varkappa_0$, and the $D_h(\varkappa) = 0$ identically.*

Proof. We have $\|P_h(\varkappa_n)\| = 1$ since $T(\varkappa_n)$ is normal [see I-(6.64)]. Thus $\varkappa = \varkappa_0$ is not a branch point for any $\lambda_h(\varkappa)$ by Theorem 1.9. Consequently the $\lambda_h(\varkappa)$ are holomorphic at $\varkappa = \varkappa_0$. Then the $P_h(\varkappa)$ are single-valued there and, since they cannot have a pole for the same reason as above, they must be holomorphic. Then the $D_h(\varkappa)$ vanish identically, since the holomorphic functions $D_h(\varkappa) = (T(\varkappa) - \lambda_h(\varkappa)) P_h(\varkappa)$ vanish at $\varkappa = \varkappa_n \to \varkappa_0$.

Remark 1.11. In general the $P_h(\varkappa)$ and $D_h(\varkappa)$ are not defined at an exceptional point $\varkappa_0$. But they can have a removable singularity at $\varkappa_0$ as in Theorem 1.10. In such a case $P_h(\varkappa_0)$ and $D_h(\varkappa_0)$ are well-defined, but they need not be the eigenprojection and eigennilpotent for the eigenvalue $\lambda_h(\varkappa_0)$ of $T(\varkappa_0)$. If, for example, $\lambda_1(\varkappa_0) = \lambda_2(\varkappa_0) \neq \lambda_k(\varkappa_0)$, $k \geqq 3$, then $P_1(\varkappa_0) + P_2(\varkappa_0)$ (and not $P_1(\varkappa_0)$) is the eigenprojection for $\lambda_1(\varkappa_0)$. Again, the eigennilpotent for $\lambda_h(\varkappa_0)$ need not vanish even if $D_h(\varkappa) \equiv 0$, as is seen from Example 1.12 a), d), f) below.

Example 1.12. Consider the eigenprojections and eigennilpotents of $T(\varkappa)$ for the operators of Example 1.1.

a) The resolvent $R(\zeta, \varkappa)$ is given by (1.15), integration of which along small circles around $\lambda_\pm(\varkappa)$ gives by (1.16)

$$P_\pm(\varkappa) = \pm \frac{1}{2(1+\varkappa^2)^{1/2}} \begin{pmatrix} 1 \pm (1+\varkappa^2)^{1/2} & \varkappa \\ \varkappa & -1 \pm (1+\varkappa^2)^{1/2} \end{pmatrix}. \tag{1.27}$$

The reader is advised to verify the relations $P_\pm(\varkappa)^2 = P_\pm(\varkappa)$ and $P_+(\varkappa)\, P_-(\varkappa) = P_-(\varkappa)\, P_+(\varkappa) = 0$. The eigenprojections $P_\pm(\varkappa)$ are branches of a double-valued algebraic function with branch points $\varkappa = \pm i$. Since $s = N = 2$, $T(\varkappa)$ is simple and the eigennilpotents $D_\pm(\varkappa)$ are zero for $\varkappa \neq \pm i$. At the exceptional points $\varkappa = \pm i$, we have quite a different spectral representation of $T(\varkappa)$; there is a double eigenvalue 0, and the spectral representation of $T(\pm i)$ is

$$T(\pm i) = 0 + D_\pm, \tag{1.28}$$

that is, $T(\pm i)$ is itself the eigennilpotent.

b) Integration of the resolvent as in a) leads to the eigenprojections

$$P_1(\varkappa) = \frac{1}{2}\begin{pmatrix} 1 & 1 \\ 1 & 1 \end{pmatrix}, \quad P_2(\varkappa) = \frac{1}{2}\begin{pmatrix} 1 & -1 \\ -1 & 1 \end{pmatrix}. \tag{1.29}$$

for the eigenvalues $\lambda_1(\varkappa) = \varkappa$ and $\lambda_2(\varkappa) = -\varkappa$. Again we have $D_1(\varkappa) = D_2(\varkappa) = 0$ for $\varkappa \neq 0$. The exceptional point $\varkappa = 0$ is not a singular point for any of $\lambda_h(\varkappa)$, $P_h(\varkappa)$ or $D_h(\varkappa)$.

c) The eigenprojection and eigennilpotent for the unique eigenvalue $\lambda(\varkappa) = 0$ of $T(\varkappa)$ are given by $P(\varkappa) = 1$, $D(\varkappa) = T(\varkappa)$.

d) We have

$$P_\pm(\varkappa) = \frac{1}{2}\begin{pmatrix} 1 & \pm\varkappa^{-1/2} \\ \pm\varkappa^{1/2} & 1 \end{pmatrix}, \quad D_\pm(\varkappa) = 0, \quad \varkappa \neq 0, \tag{1.30}$$

for $\lambda_\pm(\varkappa) = \pm\varkappa^{1/2}$. The exceptional point $\varkappa = 0$ is a branch point for the eigenvalues and the eigenprojections. For $\varkappa = 0$, the eigenvalue is zero and the spectral representation is $T(0) = 0 + D$ with $D = T = T(0)$. The operator of this example resembles that of a), with the difference that there is only one exceptional point here.

e) We have

$$P_1(\varkappa) = \begin{pmatrix} 1 & \varkappa \\ 0 & 0 \end{pmatrix}, \quad P_2(\varkappa) = \begin{pmatrix} 0 & -\varkappa \\ 0 & 1 \end{pmatrix}, \quad D_h(\varkappa) = 0, \tag{1.31}$$

for $\lambda_1(\varkappa) = 1$ and $\lambda_2(\varkappa) = 0$. Everything is holomorphic for finite $\varkappa$ since there are no exceptional points. Note that the $P_h(\varkappa)$ are not holomorphic at $\varkappa = \infty$ whereas the $\lambda_h(\varkappa)$ are. This is a situation in a sense opposite to that of the following example.

f) The eigenprojections are

$$P_1(\varkappa) = \begin{pmatrix} 1 & \varkappa^{-1} \\ 0 & 0 \end{pmatrix}, \quad P_2(\varkappa) = \begin{pmatrix} 0 & -\varkappa^{-1} \\ 0 & 1 \end{pmatrix}, \quad \varkappa \neq 0, \tag{1.32}$$

for $\lambda_1(\varkappa) = \varkappa$ and $\lambda_2(\varkappa) = 0$. Note that the $P_h(\varkappa)$ have a pole at the exceptional point $\varkappa = 0$ notwithstanding that the $\lambda_h(\varkappa)$ are holomorphic there. The situation is reversed for $\varkappa = \infty$. At $\varkappa = 0$ the spectral representation is the same as in d).

7. The case of $T(\varkappa)$ linear in $\varkappa$

The foregoing general results are somewhat simplified in the case (1.1) in which $T(\varkappa)$ is linear in $\varkappa$. Then $T(\varkappa)$ is defined in the whole complex plane, which will be taken as the domain D_0. The coefficients of the characteristic equation (1.3) are polynomials in $\varkappa$ of degree not exceeding

N. Hence the eigenvalues $\lambda_h(\varkappa)$ are branches of algebraic functions of $\varkappa$. If the algebraic equation (1.3) is *irreducible*, there is only one N-valued algebraic function so that we have $s = N$. If (1.3) is *reducible*, the eigenvalues $\lambda_h(\varkappa)$ can be classified into several groups, each group corresponding to an algebraic function. If there happen to be identical ones among these algebraic functions, we have $s < N$ (permanent degeneracy)[1].

The algebraic functions $\lambda_h(\varkappa)$ have no pole at a finite value of $\varkappa$. At $\varkappa = \infty$ they have at most a pole of order 1; this is seen by writing (1.1) in the form

$$T(\varkappa) = \varkappa(T' + \varkappa^{-1} T), \tag{1.33}$$

for the eigenvalues of $T' + \varkappa^{-1} T$ are continuous for $\varkappa^{-1} \to 0$. More precisely, these eigenvalues have the expansion $\mu_h + \beta_h(\varkappa^{-1})^{1/p} + \cdots$ (Puiseux series in $\varkappa^{-1}$), so that the eigenvalues of $T(\varkappa)$ have the form

$$\lambda_h(\varkappa) = \mu_h \varkappa + \beta_h \varkappa^{1-\frac{1}{p}} + \cdots, \quad \varkappa \to \infty. \tag{1.34}$$

Note that $P_h(\varkappa)$ or $D_h(\varkappa)$ may be holomorphic at $\varkappa = \infty$ even when $\lambda_h(\varkappa)$ is not [see Example 1.12, f)].

8. Summary

For convenience the main results obtained in the preceding paragraphs will be summarized here[2].

Let $T(\varkappa) \in \mathscr{B}(X)$ be a family holomorphic in a domain D_0 of the complex $\varkappa$-plane. The number s of eigenvalues of $T(\varkappa)$ is constant if $\varkappa$ is not one of the exceptional points, of which there are only a finite number in each compact subset of D_0. In each simple subdomain (simply connected subdomain containing no exceptional point) D of D_0, the eigenvalues of $T(\varkappa)$ can be expressed as s holomorphic functions $\lambda_h(\varkappa)$, $h = 1, \ldots, s$, the eigenvalue $\lambda_h(\varkappa)$ having constant multiplicity m_h. The $\lambda_h(\varkappa)$ are branches of one or several analytic functions on D_0, which have only algebraic singularities and which are everywhere continuous in D_0. [For simplicity these analytic functions will also be denoted by $\lambda_h(\varkappa)$.] An exceptional point $\varkappa_0$ is either a branch point of some of the $\lambda_h(\varkappa)$ or a regular point for all of them; in the latter case the values of some of the different $\lambda_h(\varkappa)$ coincide at $\varkappa = \varkappa_0$.

The eigenprojections $P_h(\varkappa)$ and the eigennilpotents $D_h(\varkappa)$ for the eigenvalues $\lambda_h(\varkappa)$ of $T(\varkappa)$ are also holomorphic in each simple subdomain D, being branches of one or several analytic functions [again denoted by $P_h(\varkappa)$ and $D_h(\varkappa)$] with only algebraic singularities. The analytic functions $P_h(\varkappa)$ and $\lambda_h(\varkappa)$ have common branch points of the

[1] The results stated here are also true if $T(\varkappa)$ is a polynomial in $\varkappa$ of any degree.

[2] For more detailed and precise statement see BAUMGÄRTEL [1].

same order, but $P_h(\varkappa)$ always has a pole at a branch point while $\lambda_h(\varkappa)$ is continuous there. $P_h(\varkappa)$ and $D_h(\varkappa)$ may have poles even at an exceptional point where $\lambda_h(\varkappa)$ is holomorphic.

If $\lambda_1(\varkappa), \ldots, \lambda_r(\varkappa)$ are the λ-group eigenvalues [the totality of the eigenvalues of $T(\varkappa)$ generated by splitting from a common eigenvalue λ of the unperturbed operator $T = T(0)$, $\varkappa = 0$ being assumed to be an exceptional point] and if $P_1(\varkappa), \ldots, P_r(\varkappa)$ are the associated eigenprojections, the total projection $P(\varkappa) = P_1(\varkappa) + \cdots + P_r(\varkappa)$ for this λ-group is holomorphic at $\varkappa = 0$. The total multiplicity $m_1 + \cdots + m_r$ for these eigenvalues is equal to the multiplicity m of the eigenvalue λ of T. The λ-group is further divided into several cycles $\{\lambda_1(\varkappa), \ldots, \lambda_p(\varkappa)\}$, $\{\lambda_{p+1}(\varkappa), \ldots\}, \ldots, \{\ldots\}$ and correspondingly for the eigenprojections. The elements of each cycle are permuted cyclically among themselves after analytic continuation when $\varkappa$ describes a small circle around $\varkappa = 0$. The sum of the eigenprojections in each cycle [for example $P_1(\varkappa) + \cdots + P_p(\varkappa)$] is single-valued at $\varkappa = 0$ but need not be holomorphic (it may have a pole).

§ 2. Perturbation series

1. The total projection for the λ-group

In the preceding section we were concerned with the general properties of the functions $\lambda_h(\varkappa)$, $P_h(\varkappa)$ and $D_h(\varkappa)$ representing respectively the eigenvalues, eigenprojections and eigennilpotents of an operator $T(\varkappa) \in \mathscr{B}(\mathsf{X})$ depending holomorphically on a complex parameter $\varkappa$. In the present section we shall construct explicitly the Taylor series (if they exist) for these functions at a given point $\varkappa$ which we may assume to be $\varkappa = 0$. Since the general case is too complicated to be dealt with completely, we shall be content with carrying out this program under certain simplifying assumptions. Furthermore, we shall give only formal series here; the convergence radii of the series and the error estimates will be considered in later sections[1].

We start from the given power series for $T(\varkappa)$:

$$T(\varkappa) = T + \varkappa T^{(1)} + \varkappa^2 T^{(2)} + \cdots . \tag{2.1}$$

Let λ be one of the eigenvalues of the unperturbed operator $T = T(0)$ with (algebraic) multiplicity m, and let P and D be the associated eigen-

[1] The perturbation series have been studied extensively in quantum mechanics, starting with Schrödinger [1]. Any textbook on quantum mechanics has a chapter dealing with them (see e. g. Kemble [1], Chapter 11 or Schiff [1] Chapter 7). In most cases, however, the discussion is limited to selfadjoint (symmetric) operators $T(\varkappa)$ depending on a real parameter $\varkappa$. In this section we shall consider general nonsymmetric operators, assuming $0 < \dim \mathsf{X} = N < \infty$ as before.

projection and eigennilpotent. Thus (see I-§ 5.4)

(2.2) $TP = PT = PTP = \lambda P + D,\ \dim P = m,\ D^m = 0,\ PD = DP = D.$

The eigenvalue λ will in general split into several eigenvalues of $T(\varkappa)$ for small $\varkappa \neq 0$ (the λ-group), see § 1.8. The total projection $P(\varkappa)$ for this λ-group is holomorphic at $\varkappa = 0$ [see (1.17)]

$$P(\varkappa) = \sum_{n=0}^{\infty} \varkappa^n P^{(n)}, \quad P^{(0)} = P, \tag{2.3}$$

with $P^{(n)}$ given by (1.18). The subspace $\mathsf{M}(\varkappa) = P(\varkappa)\,\mathsf{X}$ is m-dimensional [see (1.19)] and invariant under $T(\varkappa)$. The λ-group eigenvalues of $T(\varkappa)$ are identical with all the eigenvalues of $T(\varkappa)$ in $\mathsf{M}(\varkappa)$ [that is, of the part of $T(\varkappa)$ in $\mathsf{M}(\varkappa)$]. In order to determine the λ-group eigenvalues, therefore, we have only to solve an eigenvalue problem in the subspace $\mathsf{M}(\varkappa)$, which is in general smaller than the whole space X.

The eigenvalue problem for $T(\varkappa)$ in $\mathsf{M}(\varkappa)$ is equivalent to the eigenvalue problem for the operator

$$T_r(\varkappa) = T(\varkappa)\,P(\varkappa) = P(\varkappa)\,T(\varkappa) = P(\varkappa)\,T(\varkappa)\,P(\varkappa), \tag{2.4}$$

see I-§ 5.1. Thus the λ-group eigenvalues of $T(\varkappa)$ are exactly those eigenvalues of $T_r(\varkappa)$ which are different from zero, provided that $|\lambda|$ is large enough to ensure that these eigenvalues do not vanish for the small $|\varkappa|$ under consideration[1]. The last condition does not restrict generality, for T could be replaced by $T + \alpha$ with a scalar α without changing the nature of the problem.

In any case it follows that

$$\hat{\lambda}(\varkappa) = \frac{1}{m}\operatorname{tr}(T(\varkappa)\,P(\varkappa)) = \lambda + \frac{1}{m}\operatorname{tr}\big((T(\varkappa) - \lambda)\,P(\varkappa)\big) \tag{2.5}$$

is equal to the *weighted mean* of the λ-group eigenvalues of $T(\varkappa)$, where the weight is the multiplicity of each eigenvalue [see I-(5.40) and I-(3.25)]. If there is no splitting of λ so that the λ-group consists of a single eigenvalue $\lambda(\varkappa)$ with multiplicity m, we have

$$\hat{\lambda}(\varkappa) = \lambda(\varkappa); \tag{2.6}$$

in particular this is always true if $m = 1$. In such a case the eigenprojection associated with $\lambda(\varkappa)$ is exactly the total projection (2.3) and the eigennilpotent is given by [see I-(5.26)]

$$D(\varkappa) = (T(\varkappa) - \lambda(\varkappa))\,P(\varkappa). \tag{2.7}$$

These series give a complete solution to the eigenvalue problem for the λ-group in the case of no splitting, $\lambda(\varkappa)$, $P(\varkappa)$ and $D(\varkappa)$ being all holomorphic at $\varkappa = 0$.

[1] Note that $T_r(\varkappa)$ has the eigenvalue 0 with multiplicity $N - m$, with the eigenprojection $1 - P(\varkappa)$. Cf. also footnote [1] on p. 36.

Let us now consider the explicit form of the series (2.3) and (2.5) in terms of the coefficients $T^{(n)}$ of (2.1). It should be remarked at this point that we could use as well the coefficients of the series (2.4) instead of $T^{(n)}$, for the eigenvalues and eigenprojections are the same for $T(\varkappa)$ and $T_r(\varkappa)$ so far as concerns the λ-group[1].

The coefficients $P^{(n)}$ of (2.3) are given by (1.14) and (1.18). Thus

$$(2.8)\quad P^{(n)} = -\frac{1}{2\pi i} \sum_{\substack{\nu_1+\cdots+\nu_p=n\\ \nu_j\geqq 1}} (-1)^p \int_\Gamma R(\zeta)\, T^{(\nu_1)} R(\zeta)\, T^{(\nu_2)} \ldots T^{(\nu_p)} R(\zeta)\, d\zeta,$$

where Γ is a small, positively-oriented circle around λ. To evaluate the integral (2.8), we substitute for $R(\zeta)$ its Laurent expansion I-(5.18) at $\zeta = \lambda$, which we write for convenience in the form

$$(2.9)\qquad R(\zeta) = \sum_{n=-m}^{\infty} (\zeta - \lambda)^n S^{(n+1)}$$

with

$$(2.10)\qquad S^{(0)} = -P\,,\quad S^{(n)} = S^n\,,\quad S^{(-n)} = -D^n\,,\quad n \geqq 1\,.$$

Here $S = S(\lambda)$ is the value at $\zeta = \lambda$ of the reduced resolvent of T (see loc. cit.); thus we have by I-(5.19) and (5.26)

$$(2.11)\qquad SP = PS = 0\,,\quad (T-\lambda)\,S = S(T-\lambda) = 1 - P\,.$$

Substitution of (2.9) into the integrand of (2.8) gives a Laurent series in $\zeta - \lambda$, of which only the term with the power $(\zeta-\lambda)^{-1}$ contributes to the integral. The result is given by the *finite* sum

$$(2.12)\quad P^{(n)} = -\sum_{p=1}^{n} (-1)^p \sum_{\substack{\nu_1+\cdots+\nu_p=n\\ k_1+\cdots+k_{p+1}=p\\ \nu_j\geqq 1,\ k_j\geqq -m+1}} S^{(k_1)}\, T^{(\nu_1)}\, S^{(k_2)} \ldots S^{(k_p)}\, T^{(\nu_p)}\, S^{(k_{p+1})}$$

for $n \geqq 1$. For example

$$\begin{aligned}(2.13)\quad P^{(1)} &= \sum_{k_1+k_2=1} S^{(k_1)}\, T^{(1)}\, S^{(k_2)}\\ &= -D^{m-1}\, T^{(1)}\, S^m - \cdots - D\,T^{(1)}\, S^2 - P\,T^{(1)}\, S - S\,T^{(1)}\, P\\ &\quad - S^2\, T^{(1)}\, D - \cdots - S^m\, T^{(1)}\, D^{m-1}\,,\end{aligned}$$

$$P^{(2)} = \sum_{k_1+k_2=1} S^{(k_1)}\, T^{(2)}\, S^{(k_2)} - \sum_{k_1+k_2+k_3=2} S^{(k_1)}\, T^{(1)}\, S^{(k_2)}\, T^{(1)}\, S^{(k_3)}.$$

If in particular λ is a semisimple eigenvalue of T (see I-§ 5.4), we have $D = 0$ and only nonnegative values of k_j contribute to the sum

[1] This remark will be useful later when we consider eigenvalue problems for unbounded operators in an infinite-dimensional space; it is then possible that the series (2.1) does not exist but (2.4) has a series expansion in $\varkappa$. See VII-§ 1.5.

(2.12). Thus we have, for example,

$$(2.14)\quad P^{(1)} = -PT^{(1)}S - ST^{(1)}P\,,$$

$$\begin{aligned}P^{(2)} = &-PT^{(2)}S - ST^{(2)}P + PT^{(1)}ST^{(1)}S + ST^{(1)}PT^{(1)}S +\\ &+ ST^{(1)}ST^{(1)}P - PT^{(1)}PT^{(1)}S^2 - PT^{(1)}S^2T^{(1)}P - S^2T^{(1)}PT^{(1)}P\,,\end{aligned}$$

$$\begin{aligned}P^{(3)} = &-PT^{(3)}S - ST^{(3)}P + PT^{(1)}ST^{(2)}S + PT^{(2)}ST^{(1)}S +\\ &+ ST^{(1)}PT^{(2)}S + ST^{(2)}PT^{(1)}S + ST^{(1)}ST^{(2)}P + ST^{(2)}ST^{(1)}P -\\ &- PT^{(1)}PT^{(2)}S^2 - PT^{(2)}PT^{(1)}S^2 - PT^{(1)}S^2T^{(2)}P - PT^{(2)}S^2T^{(1)}P -\\ &- S^2T^{(1)}PT^{(2)}P - S^2T^{(2)}PT^{(1)}P - PT^{(1)}ST^{(1)}ST^{(1)}S -\\ &- ST^{(1)}PT^{(1)}ST^{(1)}S - ST^{(1)}ST^{(1)}PT^{(1)}S - ST^{(1)}ST^{(1)}ST^{(1)}P +\\ &+ PT^{(1)}PT^{(1)}ST^{(1)}S^2 + PT^{(1)}PT^{(1)}S^2T^{(1)}S + PT^{(1)}ST^{(1)}PT^{(1)}S^2 +\\ &+ PT^{(1)}S^2T^{(1)}PT^{(1)}S + PT^{(1)}ST^{(1)}S^2T^{(1)}P + PT^{(1)}S^2T^{(1)}ST^{(1)}P +\\ &+ ST^{(1)}PT^{(1)}S^2T^{(1)}P + S^2T^{(1)}PT^{(1)}ST^{(1)}P + ST^{(1)}PT^{(1)}PT^{(1)}S^2 +\\ &+ S^2T^{(1)}PT^{(1)}PT^{(1)}S + ST^{(1)}S^2T^{(1)}PT^{(1)}P + S^2T^{(1)}ST^{(1)}PT^{(1)}P -\\ &- PT^{(1)}PT^{(1)}PT^{(1)}S^3 - PT^{(1)}PT^{(1)}S^3T^{(1)}P -\\ &- PT^{(1)}S^3T^{(1)}PT^{(1)}P - S^3T^{(1)}PT^{(1)}PT^{(1)}P\,.\end{aligned}$$

2. The weighted mean of eigenvalues

We next consider the series (2.4) for $T_r(\varkappa) = T(\varkappa)\,P(\varkappa)$. For computation it is more convenient to consider the operator $(T(\varkappa) - \lambda)\,P(\varkappa)$ instead of $T_r(\varkappa)$ itself. We have from (1.16)

$$(2.15)\qquad (T(\varkappa) - \lambda)\,P(\varkappa) = -\frac{1}{2\pi i}\int_\Gamma (\zeta - \lambda)\,R(\zeta, \varkappa)\,d\zeta$$

since $(T(\varkappa) - \lambda)\,R(\zeta, \varkappa) = 1 + (\zeta - \lambda)\,R(\zeta, \varkappa)$ and the integral of 1 along Γ vanishes. Noting that $(T - \lambda)\,P = D$ by (2.2), we have

$$(2.16)\qquad (T(\varkappa) - \lambda)\,P(\varkappa) = D + \sum_{n=1}^{\infty} \varkappa^n\,\tilde{T}^{(n)}$$

with

$$(2.17)\quad \tilde{T}^{(n)} = -\frac{1}{2\pi i}\sum_{\substack{\nu_1 + \cdots + \nu_p = n\\ \nu_j \geqq 1}} (-1)^p \int_\Gamma R(\zeta)\,T^{(\nu_1)}\ldots T^{(\nu_p)}\,R(\zeta)\,(\zeta - \lambda)\,d\zeta$$

for $n \geqq 1$; this differs from (2.8) only by the factor $\zeta - \lambda$ in the integrand. Hence it follows that

$$(2.18)\qquad \tilde{T}^{(n)} = -\sum_{p=1}^{\infty} (-1)^p \sum_{\substack{\nu_1 + \cdots + \nu_p = n\\ k_1 + \cdots + k_{p+1} = p-1\\ \nu_j \geqq 1,\ k_j \geqq -m+1}}$$

with the same summand as in (2.12). For example

$$\begin{aligned}(2.19)\quad \tilde{T}^{(1)} = &-D^{m-1}\,T^{(1)}\,S^{m-1} - \cdots - D\,T^{(1)}\,S + P\,T^{(1)}\,P - S\,T^{(1)}\,D\\ &- \cdots - S^{m-1}\,T^{(1)}\,D^{m-1}\,.\end{aligned}$$

Again these expressions are simplified when λ is a semisimple eigenvalue of $T(D = 0)$; for example

(2.20) $\tilde{T}^{(1)} = P T^{(1)} P$,

$\tilde{T}^{(2)} = P T^{(2)} P - P T^{(1)} P T^{(1)} S - P T^{(1)} S T^{(1)} P - S T^{(1)} P T^{(1)} P$,

$\tilde{T}^{(3)} = P T^{(3)} P - P T^{(1)} P T^{(2)} S - P T^{(2)} P T^{(1)} S -$
$- P T^{(1)} S T^{(2)} P - P T^{(2)} S T^{(1)} P - S T^{(1)} P T^{(2)} P - S T^{(2)} P T^{(1)} P +$
$+ P T^{(1)} P T^{(1)} S T^{(1)} S + P T^{(1)} S T^{(1)} P T^{(1)} S +$
$+ P T^{(1)} S T^{(1)} S T^{(1)} P + S T^{(1)} P T^{(1)} P T^{(1)} S + S T^{(1)} P T^{(1)} S T^{(1)} P +$
$+ S T^{(1)} S T^{(1)} P T^{(1)} P - P T^{(1)} P T^{(1)} P T^{(1)} S^2 - P T^{(1)} P T^{(1)} S^2 T^{(1)} P -$
$- P T^{(1)} S^2 T^{(1)} P T^{(1)} P - S^2 T^{(1)} P T^{(1)} P T^{(1)} P$.

The series for the weighted mean $\hat{\lambda}(\varkappa)$ of the λ-group eigenvalues is obtained from (2.5) and (2.16):

$$\hat{\lambda}(\varkappa) = \lambda + \sum_{n=1}^{\infty} \varkappa^n \hat{\lambda}^{(n)} \tag{2.21}$$

where

$$\hat{\lambda}^{(n)} = \frac{1}{m} \operatorname{tr} \tilde{T}^{(n)}, \quad n \geqq 1 . \tag{2.22}$$

The substitution of (2.18) for $\tilde{T}^{(n)}$ will thus give the coefficients $\hat{\lambda}^{(n)}$.

But there is another expression for $\hat{\lambda}(\varkappa)$ which is more convenient for calculation, namely

$$\hat{\lambda}(\varkappa) - \lambda = - \frac{1}{2\pi i m} \operatorname{tr} \int_\Gamma \log\left[1 + \left(\sum_{n=1}^{\infty} \varkappa^n T^{(n)}\right) R(\zeta)\right] d\zeta . \tag{2.23}$$

Here the logarithmic function $\log(1 + A)$ is defined by

$$\log(1 + A) = \sum_{p=1}^{\infty} \frac{(-1)^{p-1}}{p} A^p \tag{2.24}$$

which is valid for $\|A\| < 1$. Note that (2.24) coincides with I-(5.57) for a special choice of the domain Δ (take as Δ a neighborhood of $\zeta = 1$ containing the eigenvalues of $1 + A$).

To prove (2.23), we start from (2.5) and (2.15), obtaining

$$\hat{\lambda}(\varkappa) - \lambda = - \frac{1}{2\pi i m} \operatorname{tr} \int_\Gamma (\zeta - \lambda) R(\zeta, \varkappa) d\zeta . \tag{2.25}$$

Substitution for $R(\zeta, \varkappa)$ from (1.13) gives

$$\hat{\lambda}(\varkappa) - \lambda = - \frac{1}{2\pi i m} \operatorname{tr} \int_\Gamma \sum_{p=1}^{\infty} (\zeta - \lambda) R(\zeta) (-A(\varkappa) R(\zeta))^p d\zeta ; \tag{2.26}$$

note that the term for $p = 0$ in (2.26) vanishes because $\operatorname{tr} D = 0$ (see Problem I-3.11).

Now we have, in virtue of the relation $dR(\zeta)/d\zeta = R(\zeta)^2$ [see I-(5.8)],

$$\begin{aligned} \frac{d}{d\zeta} (A(\varkappa) R(\zeta))^p &= \frac{d}{d\zeta} [A(\varkappa) R(\zeta) \ldots A(\varkappa) R(\zeta)] \\ &= A(\varkappa) R(\zeta) \ldots A(\varkappa) R(\zeta)^2 + \cdots + A(\varkappa) R(\zeta)^2 \ldots A(\varkappa) R(\zeta) . \end{aligned} \tag{2.27}$$

Application of the identity $\operatorname{tr} AB = \operatorname{tr} BA$ thus gives

$$(2.28) \qquad \operatorname{tr} \frac{d}{d\zeta} (A(\varkappa) R(\zeta))^p = p \operatorname{tr} R(\zeta) (A(\varkappa) R(\zeta))^p ,$$

and (2.26) becomes[1]

$$(2.29) \quad \hat{\lambda}(\varkappa) - \lambda = -\frac{1}{2\pi i m} \operatorname{tr} \int_\Gamma \sum_{p=1}^{\infty} \frac{1}{p} (\zeta - \lambda) \frac{d}{d\zeta} (-A(\varkappa) R(\zeta))^p \, d\zeta$$

$$= \frac{1}{2\pi i m} \operatorname{tr} \int_\Gamma \sum_{p=1}^{\infty} \frac{1}{p} (-A(\varkappa) R(\zeta))^p \, d\zeta \quad \text{(integration by parts)}$$

which is identical with (2.23) [recall the definition (1.10) of $A(\varkappa)$].

If the logarithmic function in (2.23) is expanded according to (2.24) and the result is arranged in powers of $\varkappa$, the coefficients in the series for $\hat{\lambda}(\varkappa)$ are seen to be given by

$$(2.30) \quad \hat{\lambda}^{(n)} = \frac{1}{2\pi i m} \operatorname{tr} \sum_{\nu_1 + \cdots + \nu_p = n} \frac{(-1)^p}{p} \int_\Gamma T^{(\nu_1)} R(\zeta) \ldots R(\zeta) T^{(\nu_p)} R(\zeta) \, d\zeta ; \; n \geqq 1 .$$

This can be treated as (2.8) and (2.17); the result is

$$(2.31) \quad \hat{\lambda}^{(n)} = \frac{1}{m} \sum_{p=1}^{n} \frac{(-1)^p}{p} \sum_{\substack{\nu_1 + \cdots + \nu_p = n \\ k_1 + \cdots + k_p = p-1}} \operatorname{tr} T^{(\nu_1)} S^{(k_1)} \ldots T^{(\nu_p)} S^{(k_p)} .$$

This formula is more convenient than (2.22), for the summation involved in (2.31) is simpler than in (2.18). For example,

$$(2.32) \quad \hat{\lambda}^{(1)} = \frac{1}{m} \operatorname{tr} T^{(1)} P ,$$

$$\hat{\lambda}^{(2)} = \frac{1}{m} \Big[\operatorname{tr} T^{(2)} P + \frac{1}{2} \sum_{k_1 + k_2 = 1} \operatorname{tr} T^{(1)} S^{(k_1)} T^{(1)} S^{(k_2)} \Big]$$

$$= \frac{1}{m} [\operatorname{tr} T^{(2)} P - \operatorname{tr} (T^{(1)} S^m T^{(1)} D^{m-1} + \cdots + T^{(1)} S T^{(1)} P)] ,$$

where we have again used the identity $\operatorname{tr} AB = \operatorname{tr} BA$ [2].

These formulas are simplified when the eigenvalue λ is semisimple. Again making use of the identity mentioned, we thus obtain

$$(2.33) \quad \hat{\lambda}^{(1)} = \frac{1}{m} \operatorname{tr} T^{(1)} P ,$$

$$\hat{\lambda}^{(2)} = \frac{1}{m} \operatorname{tr} [T^{(2)} P - T^{(1)} S T^{(1)} P]$$

$$\hat{\lambda}^{(3)} = \frac{1}{m} \operatorname{tr} [T^{(3)} P - T^{(1)} S T^{(2)} P - T^{(2)} S T^{(1)} P +$$

$$+ T^{(1)} S T^{(1)} S T^{(1)} P - T^{(1)} S^2 T^{(1)} P T^{(1)} P] ,$$

---

[1] The trace operation and the integration commute. The proof is similar to that of I-(4.30) and depends on the fact that tr is a linear functional on $\mathscr{B}(X)$.

[2] For example $(1/2) (\operatorname{tr} T^{(1)} S T^{(1)} P + \operatorname{tr} T^{(1)} P T^{(1)} S) = \operatorname{tr} T^{(1)} S T^{(1)} P$. Similar computations are made in the formulas (2.33).

$$\lambda^{(4)} = \frac{1}{m} \operatorname{tr}[T^{(4)} P - T^{(1)} S T^{(3)} P - T^{(2)} S T^{(2)} P - T^{(3)} S T^{(1)} P$$
$$+ T^{(1)} S T^{(1)} S T^{(2)} P + T^{(1)} S T^{(2)} S T^{(1)} P + T^{(2)} S T^{(1)} S T^{(1)} P -$$
$$- T^{(1)} S^2 T^{(1)} P T^{(2)} P - T^{(1)} S^2 T^{(2)} P T^{(1)} P - T^{(2)} S^2 T^{(1)} P T^{(1)} P -$$
$$- T^{(1)} S T^{(1)} S T^{(1)} S T^{(1)} P + T^{(1)} S^2 T^{(1)} S T^{(1)} P T^{(1)} P +$$
$$+ T^{(1)} S T^{(1)} S^2 T^{(1)} P T^{(1)} P + T^{(1)} S^2 T^{(1)} P T^{(1)} S T^{(1)} P -$$
$$- T^{(1)} S^3 T^{(1)} P T^{(1)} P T^{(1)} P] .$$

Problem 2.1. If $T(\varkappa)$ is linear in $\varkappa$ ($T^{(n)} = 0$ for $n \geqq 2$), we have

$$\hat{\lambda}^{(n)} = \frac{1}{mn} \operatorname{tr}(T^{(1)} P^{(n-1)}), \quad n = 1, 2, 3, \ldots . \tag{2.34}$$

[hint: Compare (2.8) and (2.30).]

Remark 2.2. These expressions for the $\hat{\lambda}^{(n)}$ take more familiar (though more complicated) form if they are expressed in terms of bases chosen appropriately. For simplicity assume that the unperturbed operator T is diagonable. Let λ_h, P_h, $h = 1, 2, \ldots$, be the eigenvalues and eigenprojections of T different from the ones λ, P under consideration. Let $\{x_1, \ldots, x_m\}$ be a basis of $\mathsf{M} = \mathsf{R}(P)$ and let $\{x_{h1}, \ldots, x_{hm_h}\}$ be a basis of $\mathsf{M}_h = \mathsf{R}(P_h)$ for each h. The union of all these vectors x_j and x_{hj} forms a basis of X consisting of eigenvectors of T and adapted to the decomposition $\mathsf{X} = \mathsf{M} \oplus \mathsf{M}_1 \oplus \cdots$ of X. The adjoint basis of X^* is adapted to the corresponding decomposition $\mathsf{X}^* = \mathsf{M}^* \oplus \mathsf{M}_1^* \oplus \cdots$, where $\mathsf{M}^* = \mathsf{R}(P^*)$, $\mathsf{M}_1^* = \mathsf{R}(P_1^*)$, $\ldots$; it consists of a basis $\{e_1, \ldots, e_m\}$ of M^*, $\{e_{11}, \ldots, e_{1m_1}\}$ of M_1^*, etc. (see Problem I-3.19).

Now we have, for any $u \in \mathsf{X}$,

$$P u = \sum_{j=1}^{m} (u, e_j)\, x_j, \quad P_h u = \sum_{j=1}^{m_h} (u, e_{hj})\, x_{hj}, \quad h = 1, 2, \ldots,$$

and for any operator $A \in \mathscr{B}(\mathsf{X})$,

$$\operatorname{tr} A P = \sum_{j=1}^{m} (A x_j, e_j), \quad \operatorname{tr} A P_h = \sum_{j=1}^{m_h} (A x_{hj}, e_{hj}), \quad h = 1, 2, \ldots .$$

The operator S is given by I-(5.32) where the subscript h should be omitted. Hence

$$S u = \sum_h (\lambda_h - \lambda)^{-1} P_h u = \sum_{h,j} (\lambda_h - \lambda)^{-1} (u, e_{hj})\, x_{hj} .$$

Thus we obtain from (2.33) the following expressions for the $\hat{\lambda}^{(n)}$:

$$\hat{\lambda}^{(1)} = \frac{1}{m} \sum_j (T^{(1)} x_j, e_j), \tag{2.35}$$

$$\hat{\lambda}^{(2)} = \frac{1}{m} \sum_j (T^{(2)} x_j, e_j) - \frac{1}{m} \sum_{i,j,h} (\lambda_h - \lambda)^{-1} (T^{(1)} x_i, e_{hj}) (T^{(1)} x_{hj}, e_i) .$$

Suppose, in particular, that the eigenvalue λ of T is simple: $m = 1$. Let φ be an eigenvector of T for this eigenvalue; then we can take $x_1 = \varphi$. Then $e_1 = \psi$ is an eigenvector of T^* for the eigenvalue $\bar{\lambda}$. We shall renumber the other eigenvectors x_{hj} of T in a simple sequence $\varphi_1, \varphi_2, \ldots$, with the corresponding eigenvalues $\mu_1, \mu_2, \ldots$ which are different from λ but not necessarily different from one another (repeated eigenvalues). Correspondingly, we write the e_{hj} in a simple sequence ψ_j so that $\{\psi, \psi_1, \psi_2, \ldots\}$ is the basis of X^* adjoint to the basis $\{\varphi, \varphi_1, \varphi_2, \ldots\}$ of X.

Then the above formulas can be written (note that $\hat{\lambda}^{(n)} = \lambda^{(n)}$ since there is no splitting)

$$\begin{aligned} \lambda^{(1)} &= (T^{(1)} \varphi, \psi), \\ \lambda^{(2)} &= (T^{(2)} \varphi, \psi) - \sum_j (\mu_j - \lambda)^{-1} (T^{(1)} \varphi, \psi_j) (T^{(1)} \varphi_j, \psi) . \end{aligned} \tag{2.36}$$

These are formulas familiar in the textbooks on quantum mechanics[1], except that here neither T nor the $T^{(n)}$ are assumed to be symmetric (Hermitian) and, therefore, we have a biorthogonal family of eigenvectors rather than an orthonormal family. (In the symmetric case we have $\psi = \varphi$, $\psi_j = \varphi_j$.)

3. The reduction process

If λ is a semisimple eigenvalue of T, $D = 0$ and (2.16) gives

$$\tilde{T}^{(1)}(\varkappa) \equiv \frac{1}{\varkappa} (T(\varkappa) - \lambda) P(\varkappa) = \sum_{n=0}^{\infty} \varkappa^n \tilde{T}^{(n+1)} . \tag{2.37}$$

Since $\mathsf{M}(\varkappa) = \mathsf{R}(P(\varkappa))$ is invariant under $T(\varkappa)$, there is an obvious relationship between the parts of $T(\varkappa)$ and $\tilde{T}^{(1)}(\varkappa)$ in $\mathsf{M}(\varkappa)$. Thus the solution of the eigenvalue problem for $T(\varkappa)$ in $\mathsf{M}(\varkappa)$ reduces to the same problem for $\tilde{T}^{(1)}(\varkappa)$. Now (2.37) shows that $\tilde{T}^{(1)}(\varkappa)$ is holomorphic at $\varkappa = 0$, so that we can apply to it what has so far been proved for $T(\varkappa)$. This process of reducing the problem for $T(\varkappa)$ to the one for $\tilde{T}^{(1)}(\varkappa)$ will be called the *reduction process*. The "unperturbed operator" for this family $\tilde{T}^{(1)}(\varkappa)$ is [see (2.20)]

$$\tilde{T}^{(1)}(0) = \tilde{T}^{(1)} = P T^{(1)} P . \tag{2.38}$$

It follows that each eigenvalue of $\tilde{T}^{(1)}$ splits into several eigenvalues of $\tilde{T}^{(1)}(\varkappa)$ for small $|\varkappa|$. Let the eigenvalues of $\tilde{T}^{(1)}$ in the invariant subspace $\mathsf{M} = \mathsf{M}(0) = \mathsf{R}(P)$ be denoted by $\lambda_j^{(1)}$, $j = 1, 2, \ldots$ [the eigenvalue zero of $\tilde{T}^{(1)}$ in the complementary subspace $\mathsf{R}(1 - P)$ does not interest us]. The spectral representation of $\tilde{T}^{(1)}$ in M takes the form

$$\begin{aligned} \tilde{T}^{(1)} &= P T^{(1)} P = \sum_j (\lambda_j^{(1)} P_j^{(1)} + D_j^{(1)}) , \\ P &= \sum_j P_j^{(1)} , \quad P_j^{(1)} P_k^{(1)} = \delta_{jk} P_j^{(1)} . \end{aligned} \tag{2.39}$$

Suppose for the moment that all the $\lambda_j^{(1)}$ are different from zero. By perturbation each $\lambda_j^{(1)}$ will split into several eigenvalues (the $\lambda_j^{(1)}$-group) of $\tilde{T}^{(1)}(\varkappa)$, which are power series in $\varkappa^{1/p_j}$ with some $p_j \geq 1$.[2] The corresponding eigenvalues of $T(\varkappa)$ have the form

$$\lambda + \varkappa \lambda_j^{(1)} + \varkappa^{1+\frac{1}{p_j}} \alpha_{jk} + \cdots , \quad k = 1, 2, \ldots . \tag{2.40}$$

[1] See Kemble [1] or Schiff [1], loc. cit.

[2] In general there are several cycles in the $\lambda_j^{(1)}$-group, but all eigenvalues of this group can formally be expressed as power series in $\varkappa^{1/p_j}$ for an appropriate common integer p_j.

If some $\lambda_j^{(1)}$ is zero, the associated eigenspace of $\tilde{T}^{(1)}$ includes the subspace $\mathsf{R}(1 - P)$. But this inconvenience may be avoided by adding to $T(\varkappa)$ a term of the form $\alpha\varkappa$, which amounts to adding to $\tilde{T}^{(1)}(\varkappa)$ a term $\alpha P(\varkappa)$. This has only the effect of shifting the eigenvalues of $T^{(1)}(\varkappa)$ in $\mathsf{M}(\varkappa)$ [but not those in the complementary subspace $\mathsf{R}(1 - P(\varkappa))$] by the amount α, leaving the eigenprojections and eigennilpotents unchanged. By choosing α appropriately the modified $\lambda_j^{(1)}$ can be made different from zero. Thus the assumption that $\lambda_j^{(1)} \neq 0$ does not affect the generality, and we shall assume this in the following whenever convenient.

The eigenvalues (2.40) of $T(\varkappa)$ for fixed λ and $\lambda_j^{(1)}$ will be said to form the $\lambda + \varkappa\lambda_j^{(1)}$-group. From (2.40) we see immediately that the following theorem holds.

Theorem 2.3. *If λ is a semisimple eigenvalue of the unperturbed operator T, each of the λ-group eigenvalues of $T(\varkappa)$ has the form* (2.40) *so that it belongs to some $\lambda + \varkappa\lambda_j^{(1)}$-group. These eigenvalues are continuously differentiable near $\varkappa = 0$ (even when $\varkappa = 0$ is a branch point). The total projection $P_j^{(1)}(\varkappa)$ (the sum of eigenprojections) for the $\lambda + \varkappa\lambda_j^{(1)}$-group and the weighted mean of this group are holomorphic at $\varkappa = 0$.*

The last statement of the theorem follows from the fact that $P_j^{(1)}(\varkappa)$ is the total projection for the $\lambda_j^{(1)}$-group of the operator $\tilde{T}^{(1)}(\varkappa)$. The same is true for the weighted mean $\hat{\lambda}_j^{(1)}(\varkappa)$ of this $\lambda_j^{(1)}$-group.

The reduction process described above can further be applied to the eigenvalue $\lambda_j^{(1)}$ of $\tilde{T}^{(1)}$ if it is semisimple, with the result that the $\lambda_j^{(1)}$-group eigenvalues of $\tilde{T}^{(1)}(\varkappa)$ have the form $\lambda_j^{(1)} + \varkappa\lambda_{jk}^{(2)} + o(\varkappa)$. The corresponding eigenvalues of $T(\varkappa)$ have the form

$$\lambda + \varkappa\lambda_j^{(1)} + \varkappa^2\lambda_{jk}^{(2)} + o(\varkappa^2)\,. \tag{2.41}$$

These eigenvalues with fixed j, k form the $\lambda + \varkappa\lambda_j^{(1)} + \varkappa^2\lambda_{jk}^{(2)}$-group of $T(\varkappa)$. In this way we see that the reduction process can be continued, and the eigenvalues and eigenprojections of $T(\varkappa)$ can be expanded into formal power series in $\varkappa$, as long as the unperturbed eigenvalue is semisimple at each stage of the reduction process.

But it is not necessary to continue the reduction process indefinitely, even when this is possible. Since the splitting must end after a finite number, say n, of steps, the total projection and the weighted mean of the eigenvalues at the n-th stage will give the full expansion of the eigenprojection and the eigenvalue themselves, respectively.

Remark 2.4. But how can one know that there will be no splitting after the n-th stage? This is obvious if the total projection at that stage has dimension one. Otherwise there is no *general* criterion for it. In most applications, however, this problem can be solved by the following reducibility argument.

Suppose there is a set $\{A\}$ of operators such that $AT(\varkappa) = T(\varkappa)A$ for all $\varkappa$. Then A commutes with $R(\zeta, \varkappa)$ and hence with any eigenprojection of $T(\varkappa)$ [see I-(5.22)]. If there is any splitting of a semisimple eigenvalue λ of T, then, each $P_j^{(1)}(\varkappa)$ in Theorem 2.3 commutes with A and so does $P_j^{(1)} = P_j^{(1)}(0)$. Since $P_j^{(1)}$ is a proper subprojection (see I-§ 3.4) of P, we have the following result: *If λ is semisimple and P is irreducible in the sense that there is no subprojection of P which commutes with all A, then there is no splitting of λ at the first stage.* If it is known that the unperturbed eigenvalue in question is semisimple at every stage of the reduction process, then the irreducibility of P means that there is no splitting at all. Similarly, if the unperturbed eigenprojection becomes irreducible at some stage, there will be no further splitting.

4. Formulas for higher approximations

The series $P_j^{(1)}(\varkappa)$ for the total projection of the $\lambda + \varkappa\,\lambda_j^{(1)}$-group of $T(\varkappa)$ can be determined from the series (2.37) for $\tilde{T}^{(1)}(\varkappa)$ just as $P(\varkappa)$ was determined from $T(\varkappa)$. To this end we need the reduced resolvent for $\tilde{T}^{(1)}$, which corresponds to the reduced resolvent S of T used in the first stage. This operator will have the form

$$S_j^{(1)} - \frac{1}{\lambda_j^{(1)}}(1 - P) \tag{2.42}$$

where the second term comes from the part of $\tilde{T}^{(1)}$ in the subspace $(1 - P)\,\mathsf{X}$ in which $\tilde{T}^{(1)}$ is identically zero, and where

$$S_j^{(1)} = -\sum_{k \neq j}\left[\frac{P_k^{(1)}}{\lambda_j^{(1)} - \lambda_k^{(1)}} + \frac{D_k^{(1)}}{(\lambda_j^{(1)} - \lambda_k^{(1)})^2} + \cdots\right] \tag{2.43}$$

comes from the part of $\tilde{T}^{(1)}$ in $\mathsf{M} = P\mathsf{X}$ [see I-(5.32)]. We note that

$$S_j^{(1)}P = PS_j^{(1)} = S_j^{(1)}, \quad S_j^{(1)}P_j^{(1)} = P_j^{(1)}S_j^{(1)} = 0\,. \tag{2.44}$$

Application of the results of § 1.1 now gives

$$P_j^{(1)}(\varkappa) = P_j^{(1)} + \varkappa\,P_j^{(11)} + \varkappa^2\,P_j^{(12)} + \cdots \tag{2.45}$$

where the coefficients are calculated by (2.12), in which the $T^{(\nu)}$ are to be replaced by $\tilde{T}^{(\nu+1)}$ given by (2.18) and $S^{(k)}$ by $(S_j^{(1)} - (1 - P)/\lambda_j^{(1)})^k$ for $k \geqq 1$, by $-P_j^{(1)}$ for $k = 0$ and by $-(D_j^{(1)})^{-k}$ for $k \leqq -1$. If, for example, $\lambda_j^{(1)}$ is semisimple ($D_j^{(1)} = 0$), we have by (2.14)

$$P_j^{(11)} = -P_j^{(1)}\,\tilde{T}^{(2)}\left(S_j^{(1)} - \frac{1}{\lambda_j^{(1)}}(1 - P)\right) + (\text{inv}) \tag{2.46}$$

where (inv) means an expression obtained from the foregoing one by inverting the order of the factors in each term. Substitution of $\tilde{T}^{(2)}$

from (2.20) gives

$$P_j^{(11)} = -P_j^{(1)} T^{(2)} S_j^{(1)} - \frac{1}{\lambda_j^{(1)}} P_j^{(1)} T^{(1)} P T^{(1)} S + P_j^{(1)} T^{(1)} S T^{(1)} S_j^{(1)} + (\text{inv}).$$

But we have $P_j^{(1)} T^{(1)} P = P_j^{(1)} T^{(1)} P_j^{(1)} = \lambda_j^{(1)} P_j^{(1)}$ because $\lambda_j^{(1)}$ is a semisimple eigenvalue of $P T^{(1)} P$. Hence

$$(2.47)\quad P_j^{(11)} = -P_j^{(1)} T^{(2)} S_j^{(1)} - P_j^{(1)} T^{(1)} S + P_j^{(1)} T^{(1)} S T^{(1)} S_j^{(1)} + (\text{inv}).$$

Note that the final result does not contain $\lambda_j^{(1)}$ explicitly. Similarly $P_j^{(12)}$ can be calculated, though the expression will be rather complicated.

The weighted mean $\hat{\lambda}_j^{(1)}(\varkappa)$ for the $\lambda_j^{(1)}$-group eigenvalues of $\tilde{T}^{(1)}(\varkappa)$ is given by

$$(2.48)\quad \hat{\lambda}_j^{(1)}(\varkappa) = \lambda_j^{(1)} + \varkappa\, \hat{\lambda}_j^{(12)} + \varkappa^2\, \hat{\lambda}_j^{(13)} + \cdots,$$

where the coefficients $\hat{\lambda}_j^{(1n)}$ are obtained from (2.31) by replacing m, P, S, $T^{(\nu)}$ by $m_j^{(1)} = \dim P_j^{(1)} \mathsf{X}$, $P_j^{(1)}$, (2.42) and $\tilde{T}^{(\nu+1)}$ respectively. For example, assuming that $\lambda_j^{(1)}$ is semisimple,

$$\begin{aligned}
(2.49)\quad \hat{\lambda}_j^{(12)} &= \frac{1}{m_j^{(1)}} \operatorname{tr} \tilde{T}^{(2)} P_j^{(1)} \\
&= \frac{1}{m_j^{(1)}} \operatorname{tr} [T^{(2)} P_j^{(1)} - T^{(1)} S T^{(1)} P_j^{(1)}], \\
\hat{\lambda}_j^{(13)} &= \frac{1}{m_j^{(1)}} \operatorname{tr} \left[\tilde{T}^{(3)} P_j^{(1)} - \tilde{T}^{(2)} \left(S_j^{(1)} - \frac{1}{\lambda_j^{(1)}} (1 - P)\right) \tilde{T}^{(2)} P_j^{(1)}\right] \\
&= \frac{1}{m_j^{(1)}} \operatorname{tr} [T^{(3)} P_j^{(1)} - T^{(1)} S T^{(2)} P_j^{(1)} - T^{(2)} S T^{(1)} P_j^{(1)} + \\
&\quad + T^{(1)} S T^{(1)} S T^{(1)} P_j^{(1)} - \lambda_j^{(1)} T^{(1)} S^2 T^{(1)} P_j^{(1)} - \\
&\quad - T^{(2)} S_j^{(1)} T^{(2)} P_j^{(1)} + T^{(1)} S T^{(1)} S_j^{(1)} T^{(2)} P_j^{(1)} + \\
&\quad + T^{(2)} S_j^{(1)} T^{(1)} S T^{(1)} P_j^{(1)} - T^{(1)} S T^{(1)} S_j^{(1)} T^{(1)} S T^{(1)} P_j^{(1)}].
\end{aligned}$$

Here we have again used the identity $\operatorname{tr} A B = \operatorname{tr} B A$ and the relations (2.44) and (2.11). The weighted mean of the $\lambda + \varkappa \lambda_j^{(1)}$-group eigenvalues of $T(\varkappa)$ is given by

$$\begin{aligned}
(2.50)\quad \hat{\lambda}_j(\varkappa) &= \lambda + \varkappa\, \hat{\lambda}_j^{(1)}(\varkappa) \\
&= \lambda + \varkappa\, \lambda_j^{(1)} + \varkappa^2\, \hat{\lambda}_j^{(12)} + \varkappa^3\, \hat{\lambda}_j^{(13)} + \cdots.
\end{aligned}$$

If there is no splitting in the $\lambda + \varkappa \lambda_j^{(1)}$-group (that is, if this group consists of a single eigenvalue), (2.50) is exactly this eigenvalue. In particular this is the case if $m_j^{(1)} = 1$.

Remark 2.5. At first sight it might appear strange that the *third* order coefficient $\hat{\lambda}_j^{(13)}$ of (2.49) contains a term such as $-(1/m_j^{(1)}) \operatorname{tr} T^{(2)} S_j^{(1)} T^{(2)} P_j^{(1)}$ which is *quadratic in* $T^{(2)}$. But this does not involve any contradiction, as may be attested by the following example. Let $N = 2$ and let

$$T = 0\,, \quad T^{(1)} = \begin{pmatrix} 1 & 0 \\ 0 & -1 \end{pmatrix}, \quad T^{(2)} = \begin{pmatrix} 0 & \alpha \\ \alpha & 0 \end{pmatrix}, \quad T^{(n)} = 0 \quad \text{for } n \geq 3\,.$$

The eigenvalues of $T(\varkappa)$ are

$$\pm \varkappa (1 + \alpha^2 \varkappa^2)^{1/2} = \pm \left(\varkappa + \frac{1}{2} \alpha^2 \varkappa^3 + \cdots\right),$$

in which the coefficient of the third order term is $\alpha^2/2$ and is indeed quadratic in α (that is, in $T^{(2)}$).

5. A theorem of Motzkin-Taussky

As an application of Theorem 2.3, we shall prove some theorems due to Motzkin and Taussky [1], [2].

Theorem 2.6. *Let the operator $T(\varkappa) = T + \varkappa T'$ be diagonable for every complex number $\varkappa$. Then all eigenvalues of $T(\varkappa)$ are linear in $\varkappa$ (that is, of the form $\lambda_h + \varkappa\, \alpha_h$), and the associated eigenprojections are entire functions of $\varkappa$.*

Proof. The eigenvalues $\lambda_h(\varkappa)$ of $T(\varkappa)$ are branches of algebraic functions of $\varkappa$ (see § 1.7). According to Theorem 2.3, on the other hand, the $\lambda_h(\varkappa)$ are continuously differentiable in $\varkappa$ at every (finite) value of $\varkappa$. Furthermore, we see from (1.34) that the $d\lambda_h(\varkappa)/d\varkappa$ are bounded at $\varkappa = \infty$. It follows that these derivatives must be constant (this is a simple consequence of the maximum principle for analytic functions[1]). This proves that the $\lambda_h(\varkappa)$ are linear in $\varkappa$.

Since $\lambda_h(\varkappa)$ has the form $\lambda_h + \varkappa\,\alpha_h$, the single eigenvalue $\lambda_h(\varkappa)$ constitutes the $\lambda_h + \varkappa\,\alpha_h$-group of $T(\varkappa)$. Thus the eigenprojection $P_h(\varkappa)$ associated with $\lambda_h(\varkappa)$ coincides with the total projection of this group and is therefore holomorphic at $\varkappa = 0$ (see Theorem 2.3). The same is true at every $\varkappa$ since $T(\varkappa)$ and $\lambda_h(\varkappa)$ are linear in $\varkappa$. Thus $P_h(\varkappa)$ is an entire function.

$P_h(\varkappa)$ may have a pole at $\varkappa = \infty$ [see Example 1.12, e)]. But if T' is also diagonable, $P_h(\varkappa)$ must be holomorphic even at $\varkappa = \infty$ because the eigenprojections of $T(\varkappa) = \varkappa(T' + \varkappa^{-1} T)$ coincide with those of $T' + \varkappa^{-1} T$, to which the above results apply at $\varkappa = \infty$. Hence each $P_h(\varkappa)$ is holomorphic everywhere including $\varkappa = \infty$ and so must be a constant by Liouville's theorem[2]. It follows that T and T' have common eigenprojections [namely $P_h(0) = P_h(\infty)$] and, since both are diagonable, they must commute. This gives

Theorem 2.7. *If T' is also diagonable in* Theorem 2.6, *then T and T' commute.*

---

[1] Since $\mu(\varkappa) = d\lambda_h(\varkappa)/d\varkappa$ is continuous everywhere (including $\varkappa = \infty$), $|\mu(\varkappa)|$ must take a maximum at some $\varkappa = \varkappa_0$ (possibly $\varkappa_0 = \infty$). Hence $\mu(\varkappa)$ must be constant by the maximum principle; see Knopp [1], p. 84. [If $\varkappa_0$ is a branch point of order $p - 1$, apply the principle after the substitution $(\varkappa - \varkappa_0)^{1/p} = \varkappa'$; if $\varkappa_0 = \infty$, apply it after the substitution $\varkappa^{-1} = \varkappa'$.]

[2] See Knopp [1], p. 112.

These theorems can be given a homogeneous form as follows.

Theorem 2.8. *Let A, $B \in \mathscr{B}(\mathsf{X})$ be such that their linear combination $\alpha A + \beta B$ is diagonable for all ratios $\alpha:\beta$ (including ∞) with possibly a single exception. Then all the eigenvalues of $\alpha A + \beta B$ have the form $\alpha\lambda_h + \beta\mu_h$ with λ_h, μ_h independent of α, β. If the said exception is excluded, then A and B commute.*

Remark 2.9. The above theorems are *global* in character in the sense that the diagonability of $T(\varkappa)$ $(\alpha A + \beta B)$ for *all* finite $\varkappa$ (*all* ratios $\alpha:\beta$ with a single exception) is essential. The fact that $T(\varkappa)$ is diagonable merely for all $\varkappa$ in some domain D of the $\varkappa$-plane does not even imply that the eigenvalues of $T(\varkappa)$ are holomorphic in D, as is seen from the following example.

Example 2.10. Let $N = 3$ and

$$T(\varkappa) = \begin{pmatrix} 0 & \varkappa & 0 \\ 0 & 0 & \varkappa \\ \varkappa & 0 & 1 \end{pmatrix}. \tag{2.51}$$

It is easy to see that $T(\varkappa)$ is diagonable for all $\varkappa$ with the exception of three values satisfying the equation $\varkappa^3 = -4/27$[1]. Thus $T(\varkappa)$ is diagonable for all $\varkappa$ in a certain neighborhood of $\varkappa = 0$. But the Puiseux series for the three eigenvalues of $T(\varkappa)$ for small $\varkappa$ have the forms

$$\pm\varkappa^{3/2} + \cdots, \quad 1 + \varkappa^3 + \cdots, \tag{2.52}$$

two of which are not holomorphic at $\varkappa = 0$.

Remark 2.11. Theorem 2.6 is not true in the case of infinite-dimensional spaces without further restrictions. Consider the differential operator

$$T(\varkappa) = -\frac{d^2}{dx^2} + x^2 + 2\varkappa x$$

regarded as a linear operator in the Hilbert space $\mathsf{L}^2(-\infty, +\infty)$ (such differential operators will be dealt with in detail in later chapters). $T(\varkappa)$ has the set of eigenvalues $\lambda_n(\varkappa)$ and the associated eigenfunctions $\varphi_n(x, \varkappa)$ given by

$$\lambda_n(\varkappa) = 2n - \varkappa^2, \qquad n = 0, 1, 2, \ldots,$$

$$\varphi_n(x, \varkappa) = \exp\left(-\frac{1}{2}x^2 - \varkappa x\right) H_n(x + \varkappa),$$

where the $H_n(x)$ are Hermite polynomials. The eigenfunctions φ_n form a complete set in the sense that every function of L^2 can be approximated with arbitrary precision by a linear combination of the φ_n. This is seen, for example, by noting that the set of functions of the form $\exp(-(x + \mathrm{Re}\,\varkappa)^2/2) \times$ (polynomial in x) is complete and that multiplication of a function by $\exp(-i\,x\,\mathrm{Im}\,\varkappa)$ is a unitary operator. Therefore, $T(\varkappa)$ may be regarded as *diagonable* for every finite $\varkappa$. Nevertheless, the $\lambda_n(\varkappa)$ are not linear in $\varkappa$.

6. The ranks of the coefficients of the perturbation series

The coefficients $P^{(n)}$ and $\tilde{T}^{(n)}$ of the series (2.3) and (2.16) have characteristic properties with respect to their ranks. Namely

---

[1] The characteristic equation for $T(\varkappa)$ is $\zeta^3 - \zeta^2 - \varkappa^3 = 0$. This cubic equation has 3 distinct roots so that $T(\varkappa)$ is diagonable, except when $\varkappa = 0$ or $\varkappa^3 = -4/27$. But $T(0)$ is obviously diagonable (it has already a diagonal matrix).

$$\text{(2.53)} \qquad \operatorname{rank} P^{(n)} \leqq (n+1)\, m\,, \quad \operatorname{rank} \tilde{T}^{(n)} \leqq (n+1)\, m\,, \qquad n = 1, 2, \ldots .$$

This follows directly from the following lemma.

Lemma 2.12. *Let* $P(\varkappa) \in \mathscr{B}(X)$ *and* $A(\varkappa) \in \mathscr{B}(X)$ *depend on* $\varkappa$ *holomorphically near* $\varkappa = 0$ *and let* $P(\varkappa)$ *be a projection for all* $\varkappa$. *Let* $A(\varkappa)\, P(\varkappa)$ *have the expansion*

$$\text{(2.54)} \qquad A(\varkappa)\, P(\varkappa) = \sum_{n=0}^{\infty} \varkappa^n B_n\,.$$

Then we have

$$\text{(2.55)} \qquad \operatorname{rank} B_n \leqq (n+1)\, m\,, \quad n = 0, 1, 2, \ldots,$$

where $m = \dim P(0)$. *Similar results hold when the left member of* (2.54) *is replaced by* $P(\varkappa)\, A(\varkappa)$.

Proof. Let

$$\text{(2.56)} \qquad P(\varkappa) = \sum_{n=0}^{\infty} \varkappa^n P_n\,.$$

The coefficients P_n satisfy recurrence formulas of the form

$$\text{(2.57)} \quad P_n = P_0 P_n + Q_{n1} P_0 P_{n-1} + \cdots + Q_{nn} P_0\,, \quad n = 0, 1, 2, \ldots,$$

where Q_{nk} is a certain polynomial in $P_0, P_1, \ldots, P_n$. (2.57) is proved by induction. The identity $P(\varkappa)^2 = P(\varkappa)$ implies that

$$\text{(2.58)} \qquad P_n = P_0 P_n + P_1 P_{n-1} + \cdots + P_n P_0\,,$$

which proves (2.57) for $n = 1$. If (2.57) is assumed to hold with n replaced by $1, 2, \ldots, n-1$, we have from (2.58)

$$\begin{aligned} P_n &= P_0 P_n + P_1 (P_0 P_{n-1} + Q_{n-1,1} P_0 P_{n-2} + \cdots + Q_{n-1,n-1} P_0) + \\ &\quad + P_2 (P_0 P_{n-2} + Q_{n-2,1} P_0 P_{n-3} + \cdots + Q_{n-2,n-2} P_0) + \cdots + P_n P_0 \\ &= P_0 P_n + P_1 P_0 P_{n-1} + (P_1 Q_{n-1,1} + P_2) P_0 P_{n-2} + \cdots + \\ &\quad + (P_1 Q_{n-1,n-1} + P_2 Q_{n-2,n-2} + \cdots + P_n) P_0\,, \end{aligned}$$

which is of the form (2.57). This completes the induction.

Now if $A(\varkappa) = \sum \varkappa^n A_n$ is the expansion of $A(\varkappa)$, we have from (2.54), (2.56) and (2.57)

$$\begin{aligned} \text{(2.59)} \quad B_n &= A_0 P_n + A_1 P_{n-1} + \cdots + A_n P_0 \\ &= A_0 P_0 P_n + (A_0 Q_{n1} + A_1) P_0 P_{n-1} + (A_0 Q_{n2} + A_1 Q_{n-1,1} + \\ &\quad + A_2) P_0 P_{n-2} + \cdots + (A_0 Q_{nn} + A_1 Q_{n-1,n-1} + \cdots + A_n) P_0\,. \end{aligned}$$

Thus B_n is the sum of $n+1$ terms, each of which contains the factor P_0 and therefore has rank not exceeding rank P_0 (see Problem I-3.4). This proves the required inequality (2.55). It is obvious how the above argument should be modified to prove the same results for $P(\varkappa)\, A(\varkappa)$ instead of $A(\varkappa)\, P(\varkappa)$.

§ 3. Convergence radii and error estimates

1. Simple estimates[1]

In the preceding sections we considered various power series in $\varkappa$ without giving explicit conditions for their convergence. In the present section we shall discuss such conditions.

We start from the expression of $R(\zeta, \varkappa)$ given by (1.13). This series is convergent for

$$\|A(\varkappa) R(\zeta)\| = \left\| \left(\sum_{n=1}^{\infty} \varkappa^n T^{(n)} \right) R(\zeta) \right\| < 1 , \tag{3.1}$$

which is satisfied if

$$\sum_{n=1}^{\infty} |\varkappa|^n \|T^{(n)} R(\zeta)\| < 1 . \tag{3.2}$$

Let $r(\zeta)$ be the value of $|\varkappa|$ such that the left member of (3.2) is equal to 1. Then (3.2) is satisfied for $|\varkappa| < r(\zeta)$.

Let the curve Γ be as in § 1.4. It is easily seen that (1.13) is uniformly convergent for $\zeta \in \Gamma$ if

$$|\varkappa| < r_0 = \min_{\zeta \in \Gamma} r(\zeta) , \tag{3.3}$$

so that the series (1.17) or (2.3) for the total projection $P(\varkappa)$ is convergent under the same condition (3.3). Thus r_0 is a lower bound for the convergence radius of the series for $P(\varkappa)$. Obviously r_0 is also a lower bound for the convergence radii of the series (2.21) for $\hat{\lambda}(\varkappa)$ and (2.37) for $\tilde{T}^{(1)}(\varkappa)$. Γ may be any simple, closed, rectifiable curve enclosing $\zeta = \lambda$ but excluding other eigenvalues of T, but we shall now assume that Γ is *convex*. It is convenient to choose Γ in such a way that r_0 turns out as large as possible.

To estimate the coefficients $\hat{\lambda}^{(n)}$ of (2.21), we use the fact that the λ-group eigenvalues of $T(\varkappa)$, and therefore also their weighted mean $\hat{\lambda}(\varkappa)$, lie inside Γ as long as (3.3) is satisfied[2]. On setting

$$\varrho = \max_{\zeta \in \Gamma} |\zeta - \lambda| , \tag{3.4}$$

we see that the function $\hat{\lambda}(\varkappa) - \lambda$ is holomorphic and bounded by ϱ for (3.3). It follows from Cauchy's inequality[3] for the Taylor coefficients that

$$|\hat{\lambda}^{(n)}| \leqq \varrho\, r_0^{-n} , \quad n = 1, 2, \ldots . \tag{3.5}$$

---

[1] The following method, based on elementary results on function theory, is used by Sz.-Nagy [1], [2], T. Kato [1], [3], [6], Schäfke [3], [4], [5].

[2] The convexity of Γ is used here.

[3] See Knopp [1], p. 77.

Such estimates are useful in estimating the error incurred when the power series (2.21) is stopped after finitely many terms. Namely

$$\left|\lambda(\varkappa) - \lambda - \sum_{p=1}^{n} \varkappa^p \lambda^{(p)}\right| \leq \sum_{p=n+1}^{\infty} |\varkappa|^p |\lambda^{(p)}| \leq \frac{\varrho |\varkappa|^{n+1}}{r_0^n (r_0 - |\varkappa|)} . \tag{3.6}$$

Example 3.1. Assume that

$$\|T^{(n)}\| \leq a\, c^{n-1}, \quad n = 1, 2, \ldots \tag{3.7}$$

for some nonnegative constants a, c. Such constants always exist since (1.2) is assumed convergent. Now (3.2) is satisfied if $a|\varkappa| (1 - c|\varkappa|)^{-1} \|R(\zeta)\| < 1$, that is, if $|\varkappa| < (a\|R(\zeta)\| + c)^{-1}$. Thus we have the following lower bound for the convergence radius

$$r_0 = \min_{\zeta \in \Gamma} (a\|R(\zeta)\| + c)^{-1} . \tag{3.8}$$

2. The method of majorizing series

Another method for estimating the coefficients and the convergence radii makes systematic use of majorizing series[1]. We introduce a *majorizing function (series)* $\Phi(\zeta - \lambda, \varkappa)$ for the function $A(\varkappa) R(\zeta)$ that appears in (1.13) and (2.29):

$$\begin{aligned} A(\varkappa) R(\zeta) &= \sum_{n=1}^{\infty} \varkappa^n T^{(n)} R(\zeta) \ll \Phi(\zeta - \lambda, \varkappa) \\ &\equiv \sum_{k=-\infty}^{\infty} \sum_{n=1}^{\infty} c_{kn} (\zeta - \lambda)^k \varkappa^n . \end{aligned} \tag{3.9}$$

By this we mean that each coefficient c_{kn} on the right is not smaller than the *norm* of the corresponding coefficient in the expansion of the left member in the double series in $\zeta - \lambda$ and $\varkappa$. Since $R(\zeta)$ has the Laurent expansion I-(5.18) with λ_h, P_h, D_h replaced by λ, P, D, respectively, this means that

$$\begin{aligned} &\|T^{(n)} D^k\| \leq c_{-k-1,n}, \quad \|T^{(n)} P\| \leq c_{-1,n}, \\ &\|T^{(n)} S^k\| \leq c_{k-1,n}, \quad k > 0 . \end{aligned} \tag{3.10}$$

We assume that $c_{kn} = 0$ for $k < -m$ so that $\Phi(z, \varkappa)$ has only a pole at $z = 0$; this is allowed since $D^m = 0$.

In particular (3.9) implies

$$\left\| \sum_{n=1}^{\infty} \varkappa^n T^{(n)} R(\zeta) \right\| \leq \Phi(|\zeta - \lambda|, |\varkappa|) . \tag{3.11}$$

Thus the series in (1.13) is convergent if $\Phi(|\zeta - \lambda|, |\varkappa|) < 1$. If we choose as Γ the circle $|\zeta - \lambda| = \varrho$, it follows that a lower bound r for the con-

[1] The use of majorizing series was begun by Rellich [4], and was further developed by Schröder [1]–[3]. Their methods are based on recurrence equations for the coefficients of the series, and differ from the function-theoretic approach used below.

vergence radius of the series for $P(\varkappa)$ as well as for $\hat{\lambda}(\varkappa)$ is given by the smallest positive root of the equation

$$(3.12)\qquad \Phi(\varrho, r) = 1 .$$

It is convenient to choose ϱ so as to make this root r as large as possible.

Also Φ can be used to construct a majorizing series for $\hat{\lambda}(\varkappa) - \lambda$. Such a series is given by

$$(3.13)\qquad \Psi(\varkappa) = \frac{1}{2\pi i} \int_{|\zeta-\lambda|=\varrho} -\log(1 - \Phi(\zeta - \lambda, \varkappa))\, d\zeta .$$

To see this we first note that, when the integrand in (2.29) is expanded into a power-Laurent series in $\varkappa$ and $\zeta - \lambda$ using (2.9), only those terms which contain at least one factor P or D contribute to the integral. Such terms are necessarily of rank $\leqq m$ and, therefore, their traces do not exceed m times their norms [see I-(5.42)]. Thus a majorizing series for $\hat{\lambda}(\varkappa) - \lambda$ is obtained if we drop the factor $1/m$ and the trace sign on the right of (2.29) and replace the coefficients of the expansion of the integrand by their norms. Since this majorizing function is in turn majorized by (3.13), the latter is seen to majorize $\hat{\lambda}(\varkappa) - \lambda$.

As is well known in function theory, (3.13) is equal to the sum of the zeros minus the sum of the poles of the function $1 - \Phi(z, \varkappa)$ contained in the interior of the circle $|z| = \varrho$[1]. But the only pole of this function is $z = 0$ and does not contribute to the sum mentioned. In this way we have obtained

Theorem 3.2. *A majorizing series for $\hat{\lambda}(\varkappa) - \lambda$ is given by (the Taylor series representing) the sum of the zeros of the function $1 - \Phi(z, \varkappa)$ (as a function of z) in the neighborhood of $z = 0$ when $\varkappa \to 0$, multiple zeros being counted repeatedly. This majorizing series, and a fortiori the series for $P(\varkappa)$ and $\hat{\lambda}(\varkappa)$, converge for $|\varkappa| < r$, where r is the smallest positive root of (3.12); here ϱ is arbitrary as long as the circle $|\zeta - \lambda| = \varrho$ encloses no eigenvalues of T other than λ.*

Example 3.3. Consider the special case in which $T^{(n)} = 0$ for $n \geqq 2$ and λ is a semisimple eigenvalue of $T (D = 0)$. From (3.10) we see that we may take $c_{-1,1} = \|T^{(1)} P\|$, $c_{k1} \geqq \|T^{(1)} S^{k+1}\|$, $k \geqq 0$, all other c_{kn} being zero. For the choice of c_{k1}, we note that $S^{k+1} = S(S - \alpha P)^k$ for any α because $SP = PS = 0$. Thus we can take $c_{k1} = \|T^{(1)} S\| \, \|S - \alpha P\|^k$ and obtain

$$(3.14)\qquad \Phi(z, \varkappa) = \varkappa \left(\frac{p}{z} + \frac{q}{1 - s z}\right)$$

as a majorizing series, where

$$(3.15)\qquad p = \|T^{(1)} P\| , \quad q = \|T^{(1)} S\| , \quad s = \|S - \alpha P\| \quad \text{for any } \alpha .$$

[1] See Knopp [1], p. 134; note that $\int \log f(z)\, dz = - \int f'(z) f(z)^{-1} z\, dz$ by integration by parts.

For small $|\varkappa|$ there is a unique zero $z = \Psi(\varkappa)$ of $1 - \Phi(z, \varkappa)$ in the neighborhood of $z = 0$. This $\Psi(\varkappa)$ is a majorizing series for $\hat{\lambda}(\varkappa) - \lambda$ by Theorem 3.2. A simple calculation gives

$$\begin{aligned}\Psi(\varkappa) &= p\varkappa + \frac{1}{2s}\,[1 - (ps + q)\,\varkappa - \Omega(\varkappa)] \\ &= p\varkappa + 2pq\varkappa^2[1 - (ps + q)\,\varkappa + \Omega(\varkappa)]^{-1} \\ &= p\varkappa + pq\varkappa^2 + \frac{2pq(ps+q)\,\varkappa^3 + 2p^2q^2s\varkappa^4}{1 - (ps+q)\,\varkappa - 2pqs\varkappa^2 + \Omega(\varkappa)}\,,\end{aligned} \tag{3.16}$$

where

$$\Omega(\varkappa) = \{[1 - (ps + q)\,\varkappa]^2 - 4pqs\varkappa^2\}^{1/2}. \tag{3.17}$$

Each coefficient of the power series of $\Psi(\varkappa)$ gives an upper bound for the corresponding coefficient of the series for $\hat{\lambda}(\varkappa) - \lambda$. Hence the remainder of this series after the n-th order term majorizes the corresponding remainder of $\hat{\lambda}(\varkappa)$. In this way we obtain from the second and third expressions of (3.16)

$$|\hat{\lambda}(\varkappa) - \lambda - \varkappa\,\hat{\lambda}^{(1)}| \leqq 2pq|\varkappa|^2/[1 - (ps + q)\,|\varkappa| + \Omega(|\varkappa|)]\,, \tag{3.18}$$

$$|\hat{\lambda}(\varkappa) - \lambda - \varkappa\,\hat{\lambda}^{(1)} - \varkappa^2\,\hat{\lambda}^{(2)}| \leqq \frac{2pq\,|\varkappa|^3(ps + q + pqs\,|\varkappa|)}{1 - (ps+q)\,|\varkappa| - 2pqs\,|\varkappa|^2 + \Omega(|\varkappa|)}\,. \tag{3.19}$$

Substitution of (3.14) into (3.12) gives a lower bound r for the convergence radii for the series of $P(\varkappa)$ and $\hat{\lambda}(\varkappa)$. The choice of

$$\varrho = p^{1/2}\,s^{-1/2}\,[(ps)^{1/2} + q^{1/2}]^{-1} \tag{3.20}$$

gives the best value[1]

$$r = [(ps)^{1/2} + q^{1/2}]^{-2}\,. \tag{3.21}$$

Note that the choice of (3.20) is permitted because $\varrho \leqq s^{-1} = \|S - \alpha P\|^{-1} \leqq d$, where d is the *isolation distance* of the eigenvalue λ of T (the distance of λ from other eigenvalues λ_k of T). In fact, I-(5.32) implies $(S - \alpha P)\,u = -(\lambda - \lambda_k)^{-1}\,u$ if $u = P_k u$ (note that $P_j P_k = 0$, $j \neq k$, $PP_k = 0$). Hence $\|S - \alpha P\| \geqq |\lambda - \lambda_k|^{-1}$ for all k.

Problem 3.4. In Example 3.3 we have

$$|\hat{\lambda}^{(1)}| \leqq p\,, \quad |\hat{\lambda}^{(2)}| \leqq pq\,, \quad |\hat{\lambda}^{(3)}| \leqq pq(ps + q)\,, \ldots\,. \tag{3.22}$$

Remark 3.5. The series for $P(\varkappa)$ can also be estimated by using the majorizing function Φ. In virtue of (1.16), (1.13) and (3.9), we have

$$P(\varkappa) \ll \frac{1}{2\pi i}\int\limits_{|\zeta - \lambda| = \varrho} \Phi_1(\zeta - \lambda)\,(1 - \Phi(\zeta - \lambda, \varkappa))^{-1}\,d\zeta \tag{3.23}$$

where $\Phi_1(\zeta - \lambda)$ is a majorizing series for $R(\zeta)$. The right member can be calculated by the method of residues if Φ and Φ_1 are given explicitly as in Example 3.3.

3. Estimates on eigenvectors

It is often required to calculate eigenvectors rather than eigenprojections. Since the eigenvectors are not uniquely determined, however, there are no definite formulas for the eigenvectors of $T(\varkappa)$ as

[1] This is seen also from (3.16), which has an expansion convergent for $|\varkappa| < r$ with r given by (3.21).

functions of $\varkappa$. If we assume for simplicity that $m = 1$ (so that $D = 0$), a convenient form of the eigenvector $\varphi(\varkappa)$ of $T(\varkappa)$ corresponding to the eigenvalue $\lambda(\varkappa)$ is given by

$$\varphi(\varkappa) = (P(\varkappa)\,\varphi, \psi)^{-1}\, P(\varkappa)\,\varphi\,, \tag{3.24}$$

where φ is an unperturbed eigenvector of T for the eigenvalue λ and ψ is an eigenvector of T^* for the eigenvalue $\bar{\lambda}$ normalized by $(\varphi, \psi) = 1$. Thus

$$P\varphi = \varphi\,,\quad P^*\psi = \psi\,,\quad (\varphi, \psi) = 1\,. \tag{3.25}$$

That (3.24) is an eigenvector of $T(\varkappa)$ is obvious since $P(\varkappa)\,\mathsf{X}$ is one-dimensional. The choice of the factor in (3.24) is equivalent to each of the following normalization conditions:

$$(\varphi(\varkappa), \psi) = 1\,,\quad (\varphi(\varkappa) - \varphi, \psi) = 0\,,\quad P(\varphi(\varkappa) - \varphi) = 0\,. \tag{3.26}$$

The relation $(T(\varkappa) - \lambda(\varkappa))\,\varphi(\varkappa) = 0$ can be written

$$(T - \lambda)\,(\varphi(\varkappa) - \varphi) + (A(\varkappa) - \lambda(\varkappa) + \lambda)\,\varphi(\varkappa) = 0\,, \tag{3.27}$$

where $A(\varkappa) = T(\varkappa) - T$ [note that $(T - \lambda)\,\varphi = 0$]. Multiplying (3.27) from the left by S and using $S(T - \lambda) = 1 - P$ and (3.26), we have

$$\varphi(\varkappa) - \varphi + S\,[A(\varkappa) - \lambda(\varkappa) + \lambda]\,\varphi(\varkappa) = 0\,. \tag{3.28}$$

Noting further that $S\varphi = 0$, we obtain [write $\varphi(\varkappa) = \varphi(\varkappa) - \varphi + \varphi$ in the last term of (3.28)]

$$\begin{aligned}\varphi(\varkappa) - \varphi &= -\,(1 + S\,[A(\varkappa) - \lambda(\varkappa) + \lambda])^{-1}\, S A(\varkappa)\,\varphi \\ &= -\,S[1 + A(\varkappa)\,S - (\lambda(\varkappa) - \lambda)\,S_\alpha]^{-1}\, A(\varkappa)\,\varphi\end{aligned} \tag{3.29}$$

if $\varkappa$ is sufficiently small, where $S_\alpha = S - \alpha P$ and α is arbitrary. This is a convenient formula for calculating an eigenvector.

In particular (3.29) gives the following majorizing series for $\varphi(\varkappa) - \varphi$:

$$\varphi(\varkappa) - \varphi \prec \|S\|\,(1 - \Phi_2(\varkappa) - \|S_\alpha\|\,\Psi(\varkappa))^{-1}\,\Phi_3(\varkappa)\,, \tag{3.30}$$

where $\Phi_2(\varkappa)$ and $\Phi_3(\varkappa)$ are majorizing series for $A(\varkappa)\,S$ and $A(\varkappa)\,\varphi$[1] respectively [note that $\Psi(\varkappa)$ is a majorizing series for $\lambda(\varkappa) - \lambda$]. The eigenvector $\varphi(\varkappa)$ is useful if $|\varkappa|$ is so small that the right member of (3.30) is smaller than $\|\varphi\|$, for then $\varphi(\varkappa)$ is certainly not zero.

Multiplication of (3.29) from the left by $T - \lambda$ gives

$$(T - \lambda)\,\varphi(\varkappa) = -\,(1 - P)\,[1 + A(\varkappa)\,S - (\lambda(\varkappa) - \lambda)\,S_\alpha]^{-1}\,A(\varkappa)\,\varphi \tag{3.31}$$

and hence

$$(T - \lambda)\,\varphi(\varkappa) \prec (1 - \Phi_2(\varkappa) - \|S_\alpha\|\,\Psi(\varkappa))^{-1}\,\Phi_3(\varkappa)\,. \tag{3.32}$$

---

[1] A majorizing series (function) for a vector-valued function can be defined in the same way as for an operator-valued function.

Example 3.6. If $T^{(n)} = 0$ for $n \geqq 2$, we have $A(\varkappa) = \varkappa T^{(1)}$ so that we may take

$$\Phi_2(\varkappa) = \varkappa \| T^{(1)} S \| = \varkappa q, \quad \Phi_3(\varkappa) = \varkappa \| T^{(1)} \varphi \| . \tag{3.33}$$

Thus (3.30) and (3.32), after substitution of (3.16), give ($s_0 = \|S\|$)

$$\varphi(\varkappa) - \varphi \ll \frac{2\varkappa s_0 \| T^{(1)} \varphi \|}{1 - (p s + q)\varkappa + \{(1 - (p s + q)\varkappa)^2 - 4 p q s \varkappa^2\}^{1/2}}, \tag{3.34}$$

$$(T - \lambda)\varphi(\varkappa) \ll \text{[the right member of (3.34) with the factor } s_0 \text{ omitted]} . \tag{3.35}$$

Problem 3.7. Under the assumptions of Example 3.6, we have for $|\varkappa| < r$ [r is given by (3.21)]

$$\varphi(\varkappa) = \varphi - \varkappa S T^{(1)} \varphi + \varkappa^2 S (T^{(1)} - \lambda^{(1)}) S T^{(1)} \varphi - \cdots, \tag{3.36}$$

$$\varphi(\varkappa) - \varphi \ll s_0 \varkappa (1 + (p s + q)\varkappa + \cdots) \| T^{(1)} \varphi \| , \tag{3.37}$$

$$\| \varphi(\varkappa) - \varphi \| \leqq |\varkappa| \frac{s_0}{(p q s)^{1/2}} ((p s)^{1/2} + q^{1/2})^2 \| T^{(1)} \varphi \| , \tag{3.38}$$

$$\| \varphi(\varkappa) - \varphi + \varkappa S T^{(1)} \varphi \| \leqq |\varkappa|^2 \frac{s_0}{(p q s)^{1/2}} ((p s)^{1/2} + q^{1/2})^2 (p s + q + (p q s)^{1/2}), \tag{3.39}$$

$$\| (T - \lambda)(\varphi(\varkappa) - \varphi) \| \leqq s_0^{-1} \text{[right member of (3.38)]} , \tag{3.40}$$

$$\| (T - \lambda)(\varphi(\varkappa) - \varphi + \varkappa S T^{(1)} \varphi) \| \leqq s_0^{-1} \text{[right member of (3.39)]} . \tag{3.41}$$

[hint for (3.38) and (3.39): Set $\varkappa = r$ after taking out the factor $|\varkappa|$ and $|\varkappa|^2$, respectively, from the majorizing series.]

4. Further error estimates

In view of the practical importance of the estimates for the remainder when the series of $\hat{\lambda}(\varkappa)$ is stopped after finitely many terms, we shall give other estimates of the coefficients $\hat{\lambda}^{(n)}$ than those given by (3.5) or by the majorizing series (3.13).

We write the integral expression (2.30) of $\hat{\lambda}^{(n)}$ in the following form:

$$\hat{\lambda}^{(n)} = \frac{1}{2\pi i m} \sum_{\nu_1 + \cdots + \nu_p = n} \frac{(-1)^p}{p} \int_\Gamma \operatorname{tr}[T^{(\nu_1)} R(\zeta) \ldots T^{(\nu_p)} R(\zeta) - \\ - T^{(\nu_1)} S(\zeta) \ldots T^{(\nu_p)} S(\zeta)]\, d\zeta . \tag{3.42}$$

Here $S(\zeta)$ is the reduced resolvent of T with respect to the eigenvalue λ (see I-§ 5.3), that is,

$$R(\zeta) = R_0(\zeta) + S(\zeta), \quad R_0(\zeta) = P R(\zeta) = R(\zeta) P , \tag{3.43}$$

is the decomposition of $R(\zeta)$ into the principal part and the holomorphic part at the pole $\zeta = \lambda$. Note that the second term in the [ ] of (3.42) is holomorphic and does not contribute to the integral. Now this expression in [ ] is equal to

$$T^{(\nu_1)} R_0(\zeta) T^{(\nu_2)} R(\zeta) \ldots T^{(\nu_p)} R(\zeta) + T^{(\nu_1)} S(\zeta) T^{(\nu_2)} R_0(\zeta) \ldots T^{(\nu_p)} R(\zeta) + \\ + \cdots + T^{(\nu_1)} S(\zeta) \ldots T^{(\nu_{p-1})} S(\zeta) T^{(\nu_p)} R_0(\zeta) , \tag{3.44}$$

each term of which contains one factor $R_0(\zeta) = P R(\zeta)$. Since this factor has rank $\leqq m$, the same is true of each term of (3.44) and, consequently, the rank of (3.44) does not exceed $\min(p\,m, N)$, where $N = \dim \mathsf{X}$. Thus the trace in (3.42) is majorized in absolute value by $\min(p\,m, N)$ times the norm of this expression [see I-(5.42)]. This leads to the estimate

$$(3.45)\quad |\hat{\lambda}^{(n)}| \leqq \frac{1}{2\pi} \sum_{\nu_1+\cdots+\nu_p=n} \min\left(1, \frac{N}{p\,m}\right) \int_\Gamma \| T^{(\nu_1)} R(\zeta) \ldots T^{(\nu_p)} R(\zeta) - - T^{(\nu_1)} S(\zeta) \ldots T^{(\nu_p)} S(\zeta)\| \, |d\zeta| .$$

A somewhat different estimate can be obtained in the special case in which $T^{(n)} = 0$ for $n \geqq 2$. In this case we have (2.34), where $P^{(n-1)}$ has rank $\leqq \min(n\,m, N)$ by (2.53). Hence we have, again estimating the trace by $\min(n\,m, N)$ times the norm,

$$(3.46)\qquad |\hat{\lambda}^{(n)}| \leqq \min\left(1, \frac{N}{n\,m}\right) \|T^{(1)} P^{(n-1)}\| , \quad n = 1, 2, \ldots .$$

On the other hand (2.8) gives

$$(3.47)\quad \|T^{(1)} P^{(n-1)}\| \leqq \frac{1}{2\pi} \int_\Gamma \|(T^{(1)} R(\zeta))^n\| \, |d\zeta| \leqq \frac{1}{2\pi} \int_\Gamma \|T^{(1)} R(\zeta)\|^n \, |d\zeta| \leqq \frac{1}{2\pi} \|T^{(1)}\|^n \int_\Gamma \|R(\zeta)\|^n \, |d\zeta| .$$

Substitution of (3.47) into (3.46) gives an estimate of $\hat{\lambda}^{(n)}$.

Remark 3.8. Under the last assumptions, the $T^{(1)}$ in (3.46–47) for $n \geqq 2$ may be replaced by $T^{(1)} - \alpha$ for any scalar α. This follows from the fact that the replacement of $T^{(1)}$ by $T^{(1)} - \alpha$ changes $T(\varkappa)$ only by the additive term $-\alpha\varkappa$ and does not affect $\hat{\lambda}^{(n)}$ for $n \geqq 2$. In particular, the $\|T^{(1)}\|$ in the last member of (3.47) may be replaced by

$$(3.48)\qquad a_0 = \min_\alpha \|T^{(1)} - \alpha\| .$$

5. The special case of a normal unperturbed operator

The foregoing results on the convergence radii and error estimates are much simplified in the special case in which X is a unitary space and T is normal. Then we have by I-(6.70)

$$(3.49)\qquad \|R(\zeta)\| = 1/\mathrm{dist}(\zeta, \Sigma(T))$$

for every $\zeta \in \mathrm{P}(T)$.

If we further assume that the $T^{(n)}$ satisfy the inequalities (3.7), then (3.8) gives

$$(3.50)\qquad r_0 = \min_{\zeta\in\Gamma} \left(\frac{a}{\mathrm{dist}(\zeta, \Sigma(T))} + c\right)^{-1}$$

as a lower bound for the convergence radii for $P(\varkappa)$ and $\hat{\lambda}(\varkappa)$. If we choose as Γ the circle $|\zeta - \lambda| = d/2$ where d is the isolation distance of the eigenvalue λ of T (see par. 2), we obtain

$$r_0 = \left(\frac{2a}{d} + c\right)^{-1}. \tag{3.51}$$

In the remainder of this paragraph we shall assume that $T^{(n)} = 0$ for $n \geqq 2$. Then we can take $c = 0$ and $a = \|T^{(1)}\|$, and (3.51) becomes

$$r_0 = d/2a = d/2\,\|T^{(1)}\|\,. \tag{3.52}$$

In other words, we have

Theorem 3.9.[1] *Let* X *be a unitary space, let* $T(\varkappa) = T + \varkappa T^{(1)}$ *and let* T *be normal. Then the power series for* $P(\varkappa)$ *and* $\lambda(\varkappa)$ *are convergent if the "magnitude of the perturbation"* $\|\varkappa T^{(1)}\|$ *is smaller than half the isolation distance of the eigenvalue* λ *of* T.

Here the factor 1/2 is the best possible, as is seen from

Example 3.10. Consider Example 1.1, a) and introduce in X the unitary norm I-(6.6) with respect to the canonical basis for which $T(\varkappa)$ has the matrix a). It is then easy to verify that T is normal (even symmetric), $\|T^{(1)}\| = 1$ and $d = 2$ for each of the two eigenvalues ± 1 of T. But the convergence radii of the series (1.4) are exactly equal to $r_0 = 1$ given by (3.52).

Remark 3.11. $a = \|T^{(1)}\|$ in (3.52) can be replaced by a_0 given by (3.48) for the same reason as in Remark 3.8.

For the coefficients $\lambda^{(n)}$, the formula (3.5) gives

$$|\lambda^{(1)}| \leqq a\,, \quad |\lambda^{(n)}| \leqq a_0^n \left(\frac{2}{d}\right)^{n-1}, \quad n \geqq 2\,, \tag{3.53}$$

for we have $\varrho = d/2$ for the Γ under consideration (see Remark 3.11). The formulas (3.46–47) lead to the same results (3.53) if the same Γ is used.

But (3.46) is able to give sharper estimates than (3.53) in some special cases. This happens, for example, when T is *symmetric* so that the eigenvalues of T are real. In this case, considering the eigenvalue λ of T, we can take as Γ the pair Γ_1, Γ_2 of straight lines perpendicular to the real axis passing through $(\lambda + \lambda_1)/2$ and $(\lambda + \lambda_2)/2$, where λ_1 and λ_2 denote respectively the largest eigenvalue of T below λ and the smallest one above. On setting

$$d_1 = \lambda - \lambda_1\,, \quad d_2 = \lambda_2 - \lambda\,, \tag{3.54}$$

we have

$$\|R(\zeta)\| = \left(\frac{d_j^2}{4} + \eta^2\right)^{-1/2}, \quad \zeta \in \Gamma_j\,, \quad j = 1, 2\,, \quad \eta = \operatorname{Im}\zeta\,. \tag{3.55}$$

[1] See T. Kato [1], Schäfke [4].

Hence (3.46–47) with Remark 3.8 give for $n \geqq 2$

(3.56)
$$|\hat{\lambda}^{(n)}| \leqq \frac{1}{2\pi} \min\left(1, \frac{N}{nm}\right) a_0^n \left[\int_{-\infty}^{\infty} \left(\frac{d_1^2}{4} + \eta^2\right)^{-\frac{n}{2}} d\eta + \int_{-\infty}^{\infty} \left(\frac{d_2^2}{4} + \eta^2\right)^{-\frac{n}{2}} d\eta \right]$$
$$= \min\left(1, \frac{N}{nm}\right) \frac{\Gamma\left(\frac{n-1}{2}\right)}{\sqrt{\pi}\,\Gamma\left(\frac{n}{2}\right)} a_0^n \cdot \frac{1}{2} \left[\left(\frac{2}{d_1}\right)^{n-1} + \left(\frac{2}{d_2}\right)^{n-1}\right],$$

in which Γ denotes the gamma function. It should be noted that if λ is the smallest or the largest eigenvalue of T, we can set $d_1 = \infty$ or $d_2 = \infty$, respectively, thereby improving the result.

It is interesting to observe that (3.56) is "the best possible", as is seen from the following example.

Example 3.12. Again take Example 3.10. We have $N = 2$, $\lambda = 1$, $m = 1$, $d_1 = 2$, $d_2 = \infty$ and $a_0 = \|T^{(1)}\| = 1$ and (3.56) gives

(3.57)
$$|\lambda^{(n)}| = |\hat{\lambda}^{(n)}| \leqq \frac{\Gamma\left(\frac{n-1}{2}\right)}{\sqrt{\pi}\, n\, \Gamma\left(\frac{n}{2}\right)}.$$

The correct eigenvalue $\lambda(\varkappa)$ is

(3.58)
$$\lambda(\varkappa) = (1 + \varkappa^2)^{1/2} = \sum_{p=0}^{\infty} \binom{1/2}{p} \varkappa^{2p}.$$

The coefficient $\lambda^{(n)}$ of $\varkappa^n$ in this series is $\binom{1/2}{n/2}$ for even n and is exactly equal to the right member of (3.57).

The factor $\alpha_n = \Gamma\left(\frac{n-1}{2}\right) \Big/ \sqrt{\pi}\,\Gamma\left(\frac{n}{2}\right)$ in (3.56) has the following values for smaller n:

n	α_n
2	$1 = 1.0000$
3	$2/\pi = 0.6366$
4	$1/2 = 0.5000$
5	$4/3\pi = 0.4244$
6	$3/8 = 0.3750$

α_n has the asymptotic value $n^{-1/2}$ for $n \to \infty$[1]. Thus (3.56) shows that $\hat{\lambda}^{(n)}$ is at most of the order

(3.59)
$$\text{const} \left(\frac{2}{d}\right)^{n-1} n^{-3/2}, \quad d = \min(d_1, d_2).$$

But $n^{-3/2}$ must be replaced by $n^{-1/2}$ if $N = \infty$, and this should be done for practical purposes even for finite N if it is large.

---

[1] The Γ-function has the asymptotic formula $\Gamma(x + 1) = (2\pi)^{1/2} x^{x+1/2} e^{-x}(1 + O(x^{-1}))$.

Problem 3.13. (3.56) is sharper than (3.53).

Problem 3.14. Why do the equalities in (3.57) hold for even n in Example 3.10?

Let us now compare these results with the estimates given by the method of majorizing series. We first note that

$$D = 0\,, \quad \|P\| = 1\,, \quad \|S\| = 1/d\,, \tag{3.60}$$

since T is assumed to be normal. If we replace (3.15) by $p = \|T^{(1)}\| = a$, $q = \|T^{(1)}\|\,\|S\| = a/d$ and $s = \|S\| = 1/d$, (3.21) gives $r = d/4a$ as a lower bound for the convergence radii, a value just one-half of the value (3.52). On the other hand, the majorizing series (3.16) for $\hat{\lambda}(\varkappa) - \lambda$ becomes after these substitions

$$\Psi(\varkappa) = a\varkappa + \frac{1}{d}\cdot\frac{2a^2\varkappa^2}{1 - \frac{2a\varkappa}{d} + \left(1 - \frac{4a\varkappa}{d}\right)^{1/2}} = \sum_{n=1}^{\infty} (-1)^{n-1}\binom{1/2}{n} 2^{2n-1}\frac{a^n}{d^{n-1}}\varkappa^n\,. \tag{3.61}$$

The estimates for $\hat{\lambda}^{(n)}$ obtained from this majorizing series are

$$|\hat{\lambda}^{(1)}| \leqq a\,, \quad |\hat{\lambda}^{(2)}| \leqq a_0^2/d\,, \quad |\hat{\lambda}^{(3)}| \leqq 2a_0^3/d^2\,, \quad |\hat{\lambda}^{(4)}| \leqq 5a_0^4/d^3, \ldots \tag{3.62}$$

(replacement of a by a_0 is justified by Remark 3.8). (3.62) is sharper than (3.53) for $n \leqq 5$ but not for $n \geqq 6$. It is also sharper than (3.56) (with d_1 and d_2 replaced by d) for $n \leqq 3$ but not for $n \geqq 4$. In any case the majorizing series gives rather sharp estimates for the first several coefficients but not for later ones. In the same way the majorizing series (3.34) for the eigenfunction becomes (in the case $m = 1$)

$$\varphi(\varkappa) - \varphi \ll \frac{2\varkappa\,\|T^{(1)}\varphi\|/d}{1 - \frac{2a\varkappa}{d} + \left(1 - \frac{4a\varkappa}{d}\right)^{1/2}} = \frac{\|T^{(1)}\varphi\|}{d}\sum_{n=1}^{\infty}(-1)^n\binom{1/2}{n+1}2^{2n+1}\left(\frac{a}{d}\right)^{n-1}\varkappa^n\,. \tag{3.63}$$

For the first several coefficients of the expansion $\varphi(\varkappa) - \varphi = \sum \varkappa^n\varphi^{(n)}$, this gives (replacing a by a_0 as above)

$$\begin{aligned} &\|\varphi^{(1)}\| \leqq \|T^{(1)}\varphi\|/d\,, \quad \|\varphi^{(2)}\| \leqq 2\|T^{(1)}\varphi\|\,a_0/d^2\,, \\ &\|\varphi^{(3)}\| \leqq 5\|T^{(1)}\varphi\|\,a_0^2/d^3\,, \quad \|\varphi^{(4)}\| \leqq 14\|T^{(1)}\varphi\|\,a_0^3/d^4, \ldots \end{aligned} \tag{3.64}$$

Here $\|T^{(1)}\varphi\|$ may also be replaced by $\min_\alpha\|(T^{(1)} - \alpha)\varphi\|$ for the same reason as above[1].

6. The enumerative method

An estimate of $\hat{\lambda}^{(n)}$ can also be obtained by *computing directly the number of terms* in the explicit formula (2.31)[2]. To illustrate the method, we assume for simplicity that X is a unitary space, $T^{(n)} = 0$ for $n \geqq 2$ and that T is normal. Recalling that $S^{(k)} = S^k$, $S^{(0)} = -P$ and $S^{(-k)} = D^k = 0$, $k > 0$, and noting (3.60), we obtain

$$|\hat{\lambda}^{(n)}| \leqq \frac{1}{nm}\sum_{k_1+\cdots+k_n=n-1}|\operatorname{tr} T^{(1)}S^{(k_1)}\ldots T^{(1)}S^{(k_n)}|\,. \tag{3.65}$$

---

[1] For related results see Rellich [4] and Schröder [1]–[3].

[2] Cf. Bloch [1].

Now the expression after the tr sign in (3.65) contains at least one factor $S^{(0)} = -P$, so that this expression is of rank $\leqq m$. Hence this trace can be majorized by m times the norm of this operator as we have done before, giving

$$(3.66) \qquad |\hat{\lambda}^{(n)}| \leqq \frac{(2n-2)!}{n!(n-1)!} \frac{a_0^n}{d^{n-1}} = \beta_n a_0^n \left(\frac{2}{d}\right)^{n-1}, \quad n \geqq 2 .$$

Here we have replaced $\|T^{(1)}\|$ by a_0 for the same reason as above. The numerical factor $2^{n-1} n \beta_n = (2n-2)!/((n-1)!)^2$ is the number of solutions of $k_1 + \cdots + k_n = n - 1$. For smaller values of n we have

n	β_n
2	1 $= 1.0000$
3	1/2 $= 0.5000$
4	5/8 $= 0.6250$
5	7/8 $= 0.8750$
6	21/16 $= 1.3125$

This shows that the estimate (3.66) is sharper than the simple estimate (3.53) only for $n \leqq 5$. In fact β_n has the asymptotic value $\pi^{-1/2} n^{-3/2} 2^{n-1}$ for $n \to \infty$, which is very large compared with unity.

Actually (3.66) is only a special case of the result obtained by the majorizing series: (3.66) is exactly the n-th coefficient of (3.61) for $n \geqq 2$ (again with a replaced by a_0).

Thus the enumerative method does not give any new result. Furthermore, it is more limited in scope than the method of majorizing series, for it is not easy to estimate effectively by enumeration the coefficients $\hat{\lambda}^{(n)}$ in more general cases.

Summing up, it may be concluded that the method of majorizing series gives in general rather sharp estimates in a closed form, especially for the first several terms of the series. In this method, however, it is difficult to take into account special properties (such as normality) of the operator. In such a special case the simpler method of contour integrals appears to be more effective. The estimates (3.50), (3.51) and (3.52) have so far been deduced only by this method.

§ 4. Similarity transformations of the eigenspaces and eigenvectors

1. Eigenvectors

In the preceding sections on the perturbation theory of eigenvalue problems, we have considered eigenprojections rather than eigenvectors (except in § 3.3) because the latter are not uniquely determined. In some cases, however, it is required to have an expression for eigenvectors $\varphi_h(\varkappa)$ of the perturbed operator $T(\varkappa)$ for the eigenvalue $\lambda_h(\varkappa)$. We shall deduce such formulas in the present section, but for simplicity we shall be content with considering the *generalized eigenvectors*; by this we mean any non-zero vector belonging to the algebraic eigenspace $\mathsf{M}_h(\varkappa) = P_h(\varkappa)\mathsf{X}$ for the eigenvalue $\lambda_h(\varkappa)$ (see I-§ 5.4). Of course a generalized eigenvector is an eigenvector in the proper sense if $\lambda_h(\varkappa)$ is semisimple.

These eigenvectors can be obtained simply by setting

$$\varphi_{hk}(\varkappa) = P_h(\varkappa)\,\varphi_k\,, \tag{4.1}$$

where the φ_k are fixed, linearly independent vectors of X. For each h and k, $\varphi_{hk}(\varkappa)$ is an analytic function of $\varkappa$ representing a generalized eigenvector of $T(\varkappa)$ *as long as it does not vanish*. This way of constructing generalized eigenvectors, however, has the following inconveniences, apart from the fact that it is rather artificial. First, $\varphi_{hk}(\varkappa)$ may become zero for some $\varkappa$ which is not a singular point of $\lambda_h(\varkappa)$ or of $P_h(\varkappa)$. Second, the $\varphi_{hk}(\varkappa)$ for different k need not be linearly independent; in fact there do not exist more than m_j linearly independent eigenvectors for a $\lambda_h(\varkappa)$ with multiplicity m_j.

These inconveniences may be avoided to some extent by taking the vectors φ_k from the subspace $\mathsf{M}_h(\varkappa_0)$ where $\varkappa_0$ is a fixed, non-exceptional point. Thus we can take exactly m_j linearly independent vectors $\varphi_k \in \mathsf{M}_h(\varkappa_0)$, and the resulting m_j vectors (4.1) for fixed h are linearly independent for sufficiently small $|\varkappa - \varkappa_0|$ since $P_h(\varkappa)$ is holomorphic at $\varkappa = \varkappa_0$. Since $\dim \mathsf{M}_h(\varkappa) = m_h$, the m_h vectors $\varphi_{hk}(\varkappa)$ form a basis of $\mathsf{M}_h(\varkappa)$. In this way we have obtained a basis of $\mathsf{M}_h(\varkappa)$ depending holomorphically on $\varkappa$.

But this is still not satisfactory, for the $\varphi_{hk}(\varkappa)$ may not be linearly independent (and some of them may vanish) for some $\varkappa$ which is not exceptional. In the following paragraphs we shall present a different procedure which is free from such inconveniences.

2. Transformation functions[1]

Our problem can be set in the following general form. Suppose that a projection $P(\varkappa)$ in X is given, which is holomorphic in $\varkappa$ in a domain D of the $\varkappa$-plane. Then $\dim P(\varkappa)\,\mathsf{X} = m$ is constant by Lemma I-4.10. It is required to find m vectors $\varphi_k(\varkappa)$, $k = 1, \ldots, m$, which are holomorphic in $\varkappa$ and which form a basis of $\mathsf{M}(\varkappa) = P(\varkappa)\,\mathsf{X}$ for *all* $\varkappa \in$ D.

We may assume without loss of generality that $\varkappa = 0$ belongs to D. Our problem will be solved if we construct an operator-valued function $U(\varkappa)$ [hereafter called a *transformation function* for $P(\varkappa)$] with the following properties:

(1) The inverse $U(\varkappa)^{-1}$ exists and both $U(\varkappa)$ and $U(\varkappa)^{-1}$ are holomorphic for $\varkappa \in$ D;

(2) $U(\varkappa)\,P(0)\,U(\varkappa)^{-1} = P(\varkappa)$.

[1] The results of this and the following paragraphs were given by T. Kato [2] in connection with the adiabatic theorem in quantum mechanics.

The second property implies that $U(\varkappa)$ maps $\mathsf{M}(0)$ onto $\mathsf{M}(\varkappa)$ one to one (see I-§ 5.7). If $\{\varphi_k\}$, $k = 1, \ldots, m$, is a basis of $\mathsf{M}(0)$, it follows that the vectors

$$\varphi_k(\varkappa) = U(\varkappa)\,\varphi_k\,, \quad k = 1, \ldots, m\,, \tag{4.2}$$

form a basis of $\mathsf{M}(\varkappa)$, which solves our problem.

We shall now construct a $U(\varkappa)$ with the above properties under the assumption that D is *simply connected*. We have

$$P(\varkappa)^2 = P(\varkappa) \tag{4.3}$$

and by differentiation

$$P(\varkappa)\,P'(\varkappa) + P'(\varkappa)\,P(\varkappa) = P'(\varkappa)\,, \tag{4.4}$$

where we use $'$ to denote the differentiation $d/d\varkappa$. Multiplying (4.4) by $P(\varkappa)$ from the left and from the right and noting (4.3), we obtain [we write P in place of $P(\varkappa)$, etc., for simplicity]

$$P P' P = 0\,. \tag{4.5}$$

We now introduce the *commutator* Q of P' and P:

$$Q(\varkappa) = [P'(\varkappa), P(\varkappa)] = P'(\varkappa)\,P(\varkappa) - P(\varkappa)\,P'(\varkappa)\,. \tag{4.6}$$

Obviously P' and Q are holomorphic for $\varkappa \in$ D. It follows from (4.3), (4.5) and (4.6) that

$$PQ = -PP'\,, \quad QP = P'P\,. \tag{4.7}$$

Hence (4.4) gives

$$P' = [Q, P]\,. \tag{4.8}$$

Let us now consider the differential equation

$$X' = Q(\varkappa)\,X \tag{4.9}$$

for the unknown $X = X(\varkappa)$. Since this is a *linear* differential equation, it has a unique solution holomorphic for $\varkappa \in$ D when the initial value $X(0)$ is specified. This can be proved, for example, by the method of successive approximation in the same way as for a linear system of ordinary differential equations[1].

[1] In fact (4.9) is equivalent to a system of ordinary differential equations in a matrix representation. But it is more convenient to treat (4.9) as an operator differential equation without introducing matrices, in particular when $\dim \mathsf{X} = \infty$ (note that all the results of this paragraph apply to the infinite-dimensional case without modification). The standard successive approximation, starting from the zeroth approximation $X_0(\varkappa) = X(0)$, say, and proceeding by $X_n(\varkappa) = X(0) + \int_0^\varkappa Q(\varkappa)\,X_{n-1}(\varkappa)\,d\varkappa$, gives a sequence $X_n(\varkappa)$ of holomorphic operator-valued functions; it is essential here that D is simply-connected. It is easy to show that $X_n(\varkappa)$ converges to an $X(\varkappa)$ uniformly in each compact subset of D and that $X(\varkappa)$ is the unique holomorphic solution of (4.9) with the given initial value $X(0)$. Here it is essential that the operation $X \to Q(\varkappa)\,X$ is a *linear* operator acting in $\mathscr{B}(\mathsf{X})$.

Let $X(\varkappa) = U(\varkappa)$ be the solution of (4.9) for the initial condition $X(0) = 1$. The general solution of (4.9) can then be written in the form

$$X(\varkappa) = U(\varkappa)\, X(0)\,. \tag{4.10}$$

In fact (4.10) satisfies (4.9) and the initial condition; in view of the uniqueness of the solution it must be the required solution.

In quite the same way, the differential equation

$$Y' = -Y Q(\varkappa) \tag{4.11}$$

has a unique solution for a given initial value $Y(0)$. Let $Y = V(\varkappa)$ be the solution for $Y(0) = 1$. We shall show that $U(\varkappa)$ and $V(\varkappa)$ are inverse to each other. The differential equations satisfied by these functions give $(VU)' = V'U + VU' = -VQU + VQU = 0$. Hence VU is a constant and

$$V(\varkappa)\, U(\varkappa) = V(0)\, U(0) = 1\,. \tag{4.12}$$

This proves that $V = U^{-1}$ and, therefore, we have also

$$U(\varkappa)\, V(\varkappa) = 1\,. \tag{4.13}$$

We shall give an independent proof of (4.13), for later reference, for (4.13) is not implied by (4.12) if the underlying space is of infinite dimension. We have as above

$$(UV)' = QUV - UVQ = [Q, UV]\,. \tag{4.14}$$

This time it is not obvious that the right member of (4.14) is zero. But (4.14) is also a linear differential equation for $Z = UV$, and the uniqueness of its solution can be proved in the same way as for (4.9) and (4.11). Since $Z(\varkappa) = 1$ satisfies (4.14) as well as the initial condition $Z(0) = 1 = U(0)\, V(0)$, UV must coincide with Z. This proves (4.13)[1].

We now show that $U(\varkappa)$ satisfies the conditions (1), (2) required above. (1) follows from $U(\varkappa)^{-1} = V(\varkappa)$ implied by (4.12) and (4.13). To prove (2), we consider the function $P(\varkappa)\, U(\varkappa)$. We have

$$(PU)' = P'U + PU' = (P' + PQ)\, U = QPU \tag{4.15}$$

by (4.9) and (4.8). Thus $X = PU$ is a solution of (4.9) with the initial value $X(0) = P(0)$ and must coincide with $U(\varkappa)\, X(0) = U(\varkappa)\, P(0)$ by (4.10). This is equivalent to (2).

Remark 4.1. In virtue of (4.7), the Q in the last member of (4.15) can be replaced by P'. Thus the function $W(\varkappa) = U(\varkappa)\, P(0) = P(\varkappa)\, U(\varkappa)$ satisfies the differential equation

$$W' = P'(\varkappa)\, W\,, \tag{4.16}$$

[1] (4.13) can also be deduced from (4.12) by a more general argument based on the *stability of the index*; cf. X-§ 5.5.

which is somewhat simpler than (4.9). Similarly, it is seen that $Z(\varkappa) = P(0)\,U(\varkappa)^{-1}$ satisfies

$$Z' = Z\,P'(\varkappa)\,. \tag{4.17}$$

Remark 4.2. $U(\varkappa)$ and $U(\varkappa)^{-1}$ can be continued analytically as long as this is possible for $P(\varkappa)$. But it may happen that they are not single-valued even when $P(\varkappa)$ is, if the domain of $\varkappa$ is not simply connected.

Remark 4.3. The construction of $U(\varkappa)$ can be carried out even if $\varkappa$ is a real variable. In this case $P(\varkappa)$ need not be holomorphic; it suffices that $P'(\varkappa)$ exists and is continuous (or piecewise continuous). Then $U(\varkappa)$ has a continuous (piecewise continuous) derivative and satisfies (1), (2) except that it is not necessarily holomorphic in $\varkappa$.

Remark 4.4. The transformation function $U(\varkappa)$ is not unique for a given $P(\varkappa)$. Another $U(\varkappa)$ can be obtained from the result of I-§ 4.6, at least for sufficiently small $|\varkappa|$. Substituting $P(0)$, $P(\varkappa)$ for the P, Q of I-§ 4.6, we see from I-(4.38) that[1]

$$U(\varkappa) = [1 - (P(\varkappa) - P(0))^2]^{-1/2}[P(\varkappa)\,P(0) + (1 - P(\varkappa))\,(1 - P(0))] \tag{4.18}$$

is a transformation function if $|\varkappa|$ is so small that $\|P(\varkappa) - P(0)\| < 1$. (4.18) is simpler than the $U(\varkappa)$ constructed above in that it is an algebraic expression in $P(\varkappa)$ and $P(0)$ while the other one was defined as the solution of a differential equation. But (4.18) has the inconvenience that it may not be defined for all $\varkappa \in \mathrm{D}$.

3. Solution of the differential equation

Since we are primarily interested in the mapping of $\mathsf{M}(0)$ onto $\mathsf{M}(\varkappa)$ by the transformation function $U(\varkappa)$, it suffices to consider $W(\varkappa) = U(\varkappa)\,P(0)$ instead of $U(\varkappa)$. To determine W it suffices to solve the differential equation (4.16) for the initial condition $W(0) = P(0)$.

Let us solve this equation in the case where $P(\varkappa)$ is the total projection for the λ-group eigenvalues of $T(\varkappa)$. $P(\varkappa)$ has the form (2.3), so that

$$P'(\varkappa) = \sum_{n=0}^{\infty} (n+1)\,\varkappa^n\,P^{(n+1)}\,. \tag{4.19}$$

Since $W(0) = P(0) = P$, we can write

$$W(\varkappa) = P + \sum_{n=1}^{\infty} \varkappa^n\,W^{(n)}\,. \tag{4.20}$$

Substitution of (4.19) and (4.20) into (4.16) gives the following recurrence formulas for $W^{(n)}$:

$$n\,W^{(n)} = n\,P^{(n)}\,P + (n-1)\,P^{(n-1)}\,W^{(1)} + \cdots + P^{(1)}\,W^{(n-1)}\,, \quad n = 1, 2, \ldots\,. \tag{4.21}$$

---

[1] (4.18) was given by Sz.-Nagy [1] in an apparently different form.

The $W^{(n)}$ can be determined successively from (4.21) by making use of the expressions (2.12) for $P^{(n)}$. In this way we obtain

$$W^{(1)} = P^{(1)} P , \tag{4.22}$$
$$W^{(2)} = P^{(2)} P + \frac{1}{2} [P^{(1)}]^2 P ,$$
$$W^{(3)} = P^{(3)} P + \frac{2}{3} P^{(2)} P^{(1)} P + \frac{1}{3} P^{(1)} P^{(2)} P + \frac{1}{6} [P^{(1)}]^3 P .$$

If the eigenvalue λ of T is semisimple, we have by (2.14)

$$W^{(1)} = -S T^{(1)} P , \tag{4.23}$$
$$W^{(2)} = -S T^{(2)} P + S T^{(1)} S T^{(1)} P - S^2 T^{(1)} P T^{(1)} P - -\frac{1}{2} P T^{(1)} S^2 T^{(1)} P .$$

In case the λ-group consists of a single eigenvalue (no splitting), $\mathsf{M}(\varkappa) = P(\varkappa) \mathsf{X}$ is itself the algebraic eigenspace of $T(\varkappa)$ for this eigenvalue, and we have a set of generalized eigenvectors

$$\varphi_k(\varkappa) = W(\varkappa) \varphi_k , \quad k = 1, \ldots, m , \tag{4.24}$$

where $\{\varphi_k\}$, $k = 1, \ldots, m$, is a basis of $\mathsf{M}(0)$. According to the properties of $W(\varkappa)$, the m vectors (4.24) form a basis of $\mathsf{M}(\varkappa)$.

The function $Z(\varkappa) = P(0) U(\varkappa)^{-1}$ can be determined in the same way. Actually we need not solve the differential equation (4.17) independently. (4.17) differs from (4.16) only in the order of the multiplication of the unknown and the coefficient $P'(\varkappa)$. Thus the series $\sum \varkappa^n Z^{(n)}$ for $Z(\varkappa)$ is obtained from that of $W(\varkappa)$ by inverting the order of the factors in each term. This is true not only for the expression of $Z^{(n)}$ in terms of the $P^{(n')}$ but also in terms of P, S, $T^{(1)}$, $T^{(2)}$, ... as in (4.23). This is due to the fact that the expressions (2.12) for the $P^{(n)}$ are invariant under the inversion of the type described. This remark gives, under the same assumption that λ is semisimple,

$$Z^{(1)} = -P T^{(1)} S , \tag{4.25}$$
$$Z^{(2)} = -P T^{(2)} S + P T^{(1)} S T^{(1)} S - P T^{(1)} P T^{(1)} S^2 - -\frac{1}{2} P T^{(1)} S^2 T^{(1)} P .$$

Remark 4.5. The other transformation function $U(\varkappa)$ given by (4.18) can also be written as a power series in $\varkappa$ under the same assumptions. As is easily seen, the expansion of $U(\varkappa) P$ coincides with that of $W(\varkappa)$ deduced above up to the order $\varkappa^2$ inclusive.

4. The transformation function and the reduction process

The transformation function $U(\varkappa)$ for the total projection $P(\varkappa)$ for the λ-group constructed above can be applied to the reduction process described in § 2.3. Since the λ-group eigenvalues are the eigenvalues of $T(\varkappa)$ in the invariant subspace $\mathsf{M}(\varkappa) = \mathsf{R}(P(\varkappa))$ and since $U(\varkappa)$ has the property $P(\varkappa) = U(\varkappa)\, P U(\varkappa)^{-1}$, the eigenvalue problem for $T(\varkappa)$ in $\mathsf{M}(\varkappa)$ is equivalent to the eigenvalue problem for the operator

$$U(\varkappa)^{-1}\, T(\varkappa)\, U(\varkappa) \tag{4.26}$$

considered in the subspace $\mathsf{M} = \mathsf{M}(0) = \mathsf{R}(P)$ (which is invariant under this operator). In fact, (4.26) has the same set of eigenvalues as $T(\varkappa)$, whereas the associated eigenprojections and eigennilpotents of (4.26) are related to those of $T(\varkappa)$ by the similarity transformation with $U(\varkappa)^{-1}$. Since we are interested in the λ-group only, it suffices to consider the operator

$$P U(\varkappa)^{-1}\, T(\varkappa)\, U(\varkappa)\, P = Z(\varkappa)\, T(\varkappa)\, W(\varkappa) \tag{4.27}$$

with the Z and W introduced in the preceding paragraphs.

(4.27) is holomorphic in $\varkappa$ and in this sense is of the same form as the given operator $T(\varkappa)$. Thus the problem for the λ-group has been reduced to a problem within a fixed subspace M of X. This reduction of the original problem to a problem in a smaller subspace M has the advantage that M is independent of $\varkappa$, whereas in the reduction process considered in § 2.3 the subspace $\mathsf{M}(\varkappa)$ depends on $\varkappa$. For this reason the reduction to (4.27) is more complete at least theoretically, though it has the practical inconvenience that the construction of $U(\varkappa)$ is not simple.

In particular it follows that the weighted mean $\hat{\lambda}(\varkappa)$ of the λ-group eigenvalues is equal to m^{-1} times the trace of (4.27):

$$\begin{aligned} \hat{\lambda}(\varkappa) &= m^{-1} \operatorname{tr} Z(\varkappa)\, T(\varkappa)\, W(\varkappa) \\ &= \lambda + m^{-1} \operatorname{tr} Z(\varkappa)\, (T(\varkappa) - \lambda)\, W(\varkappa)\,. \end{aligned} \tag{4.28}$$

Substitution of (4.23) and (4.25) for the coefficients of $W(\varkappa)$ and $Z(\varkappa)$ leads to the same results as in (2.33).

Problem 4.6. Verify the last statement.

5. Simultaneous transformation for several projections

The $U(\varkappa)$ considered in par. 2 serves only for a single projection $P(\varkappa)$. We shall now consider several projections $P_h(\varkappa)$, $h = 1, \ldots, s$, satisfying the conditions

$$P_h(\varkappa)\, P_k(\varkappa) = \delta_{hk}\, P_h(\varkappa)\,, \tag{4.29}$$

and construct a transformation function $U(\varkappa)$ such that

$$U(\varkappa)\, P_h(0)\, U(\varkappa)^{-1} = P_h(\varkappa)\,, \quad h = 1\,, \ldots, s\,. \tag{4.30}$$

As a consequence, we can find a basis $\{\varphi_{h1}(\varkappa), \ldots, \varphi_{hm_h}(\varkappa)\}$ of each subspace $M_h(\varkappa) = R(P_h(\varkappa))$ by setting

(4.31) $$\varphi_{hj}(\varkappa) = U(\varkappa)\, \varphi_{hj},$$

where $\{\varphi_{h1}, \ldots, \varphi_{hm_h}\}$ is a basis of $M_h = M_h(0)$.

As before we assume that the $P_h(\varkappa)$ are either holomorphic in a simply-connected domain D of the complex plane or continuously differentiable in an interval of the real line. We may assume that the set $\{P_h(\varkappa)\}$ is *complete* in the sense that

(4.32) $$\sum_{h=1}^{s} P_h(\varkappa) = 1 .$$

Otherwise we can introduce the projection $P_0(\varkappa) = 1 - \sum_{h=1}^{s} P_h(\varkappa)$; the new set $\{P_h(\varkappa)\}$, $h = 0, 1, \ldots, s$, will satisfy (4.29) as well as the completeness condition, and $U(\varkappa)$ satisfies (4.30) for the old set if and only if it does the same for the new set.

The construction of $U(\varkappa)$ is similar to that for a single $P(\varkappa)$. We define $U(\varkappa)$ as the solution of the differential equation (4.9) for the initial value $X(0) = 1$, in which $Q(\varkappa)$ is to be given now by

(4.33) $$Q(\varkappa) = \frac{1}{2} \sum_{h=1}^{s} [P_h'(\varkappa), P_h(\varkappa)] = \sum_{h=1}^{s} P_h'(\varkappa)\, P_h(\varkappa)$$
$$= - \sum_{h=1}^{s} P_h(\varkappa)\, P_h'(\varkappa) .$$

The equality of the three members of (4.33) follows from (4.32), which implies that $\Sigma P_h' P_h + \Sigma P_h P_h' = \Sigma (P_h^2)' = \Sigma P_h' = 0$. Note also that this $Q(\varkappa)$ coincides with (4.6) in the case of a single $P(\varkappa)$; the apparent difference due to the presence in (4.33) of the factor 1/2 arises from the fact that in (4.33) we have enlarged the single $P(\varkappa)$ to the pair $\{P(\varkappa), 1 - P(\varkappa)\}$ so as to satisfy the completeness condition (4.32).

The argument of par. 2 [in particular (4.15)] shows that (4.30) is proved if we can show that

(4.34) $$P_h'(\varkappa) = [Q(\varkappa), P_h(\varkappa)] , \quad h = 1, \ldots, s .$$

To prove this we differentiate (4.29), obtaining

(4.35) $$P_h' P_k + P_h P_k' = \delta_{hk} P_h' .$$

Multiplication from the left by P_h gives $P_h P_h' P_k + P_h P_k' = \delta_{hk} P_h P_h' = P_k P_h P_h'$, which may be written

(4.36) $$- [P_h P_h', P_k] = P_h P_k' .$$

Summation over $h = 1, \ldots, s$ gives (4.34) in virtue of (4.33).

Obviously a simultaneous transformation function $U(\varkappa)$ can be constructed in this way for the set of all eigenprojections $P_h(\varkappa)$ of $T(\varkappa)$ in any simply connected domain D of the $\varkappa$-plane in which the $P_h(\varkappa)$ are holomorphic[1].

6. Diagonalization of a holomorphic matrix function

Let $(\tau_{jk}(\varkappa))$ be an $N \times N$ matrix whose elements are holomorphic functions of a complex variable $\varkappa$. Under certain conditions such a matrix function can be diagonalized, that is, there is a matrix $(\gamma_{jk}(\varkappa))$ with elements holomorphic in $\varkappa$ such that

$$(\tau'_{jk}(\varkappa)) = (\gamma_{jk}(\varkappa))^{-1} (\tau_{jk}(\varkappa)) \, (\gamma_{jk}(\varkappa)) \tag{4.37}$$

is a diagonal matrix for every $\varkappa$ considered.

This problem can be reduced to the one considered in this section. It suffices to regard the given matrix as an operator $T(\varkappa)$ acting in the space $\mathsf{X} = \mathrm{C}^N$ of numerical vectors and apply the foregoing results. If the value of $\varkappa$ is restricted to a simply-connected domain D containing no exceptional points, we can construct the $U(\varkappa)$ and therefore a basis (4.31) consisting of vector functions holomorphic for $\varkappa \in$ D and adapted to the set $\{P_h(\varkappa)\}$ of eigenprojections of $T(\varkappa)$. With respect to such a basis, the matrix representation of $T(\varkappa)$ takes the simple form described in I-§ 5.4. In particular it is a diagonal matrix with diagonal elements $\lambda_h(\varkappa)$ if the $D_h(\varkappa)$ are all identically zero [which happens, for example, if all $m_h = 1$ or $T(\varkappa)$ is normal for real $\varkappa$, say]. As is seen from I-§ 5.4, this is equivalent to the existence of a matrix function $(\gamma_{jk}(\varkappa))$ with the required property (4.37). Note that the column vectors $(\gamma_{1k}(\varkappa), \ldots, \gamma_{Nk}(\varkappa))$ are eigenvectors of the given matrix $(\tau_{jk}(\varkappa))$.

§ 5. Non-analytic perturbations

1. Continuity of the eigenvalues and the total projection

In the preceding sections we considered the eigenvalue problem for an operator $T(\varkappa) \in \mathscr{B}(\mathsf{X})$ holomorphic in $\varkappa$ and showed that its eigenvalues and eigenprojections are analytic functions of $\varkappa$. We now ask what conclusions can be drawn if we consider a more general type of dependence of $T(\varkappa)$ on $\varkappa$[2].

First we consider the case in which $T(\varkappa)$ is only assumed to be *continuous* in $\varkappa$. $\varkappa$ may vary in a domain D_0 of the complex plane or in

[1] The transformation function $U(\varkappa)$ has important applications in the adiabatic theorem in quantum mechanics, for which we refer to T. Kato [2], Garrido [1], Garrido and Sancho [1].

[2] This question was discussed by Rellich [1], [2], [8] (in greater detail in [8]) for symmetric operators.

an interval I of the real line. Even under this general assumption, some of the results of the foregoing sections remain essentially unchanged.

The resolvent $R(\zeta, \varkappa) = (T(\varkappa) - \zeta)^{-1}$ is now continuous in ζ and $\varkappa$ jointly in each domain for which ζ is different from any eigenvalue of $T(\varkappa)$. This is easily seen by modifying slightly the argument given in § 1.3; we need only to note that the $A(\varkappa) = T(\varkappa) - T$ in (1.11) is no longer holomorphic but tends to zero for $\varkappa \to 0$ $(T = T(0))$.

It follows that $R(\zeta, \varkappa)$ exists when ζ is in the resolvent set $\mathrm{P}(T)$ of T provided that $|\varkappa|$ is small enough to ensure that

$$\|T(\varkappa) - T\| < \|R(\zeta)\|^{-1} \quad (R(\zeta) = R(\zeta, 0)); \tag{5.1}$$

see (1.12). Furthermore, $R(\zeta, \varkappa) \to R(\zeta)$ for $\varkappa \to 0$ uniformly for ζ belonging to a compact subset of $\mathrm{P}(T)$.

Let λ be one of the eigenvalues of T, say with (algebraic) multiplicity m. Let Γ be a closed curve enclosing λ but no other eigenvalues of T. $\|R(\zeta)\|^{-1}$ has a positive minimum δ for $\zeta \in \Gamma$, and $R(\zeta, \varkappa)$ exists for all $\zeta \in \Gamma$ if $\|T(\varkappa) - T\| < \delta$. Consequently the operator $P(\varkappa)$ is again defined by (1.16) and is continuous in $\varkappa$ near $\varkappa = 0$. As in the analytic case, $P(\varkappa)$ is the total projection for the eigenvalues of $T(\varkappa)$ lying inside Γ. The continuity of $P(\varkappa)$ again implies that

$$\dim \mathsf{M}(\varkappa) = \dim \mathsf{M} = m\,, \quad \mathsf{M}(\varkappa) = P(\varkappa)\,\mathsf{X}\,, \quad \mathsf{M} = \mathsf{M}(0) = P\mathsf{X}\,, \tag{5.2}$$

where $P = P(0)$ is the eigenprojection of T for the eigenvalue λ. (5.2) implies that the sum of the multiplicities of the eigenvalues of $T(\varkappa)$ lying inside Γ is equal to m. These eigenvalues are again said to form the λ-group.

The same results are true for each eigenvalue λ_h of T. In any neighborhood of λ_h, there are eigenvalues of $T(\varkappa)$ with total multiplicity equal to the multiplicity m_h of λ_h provided that $|\varkappa|$ is sufficiently small. Since the sum of the m_h is N, there are no other eigenvalues of $T(\varkappa)$. This proves (and gives a precise meaning to) the proposition that *the eigenvalues of* $T(\varkappa)$ *are continuous in* $\varkappa$.

We assumed above that $T(\varkappa)$ is continuous in a domain of $\varkappa$. But the same argument shows that the eigenvalues of $T(\varkappa)$ and the total projection $P(\varkappa)$ are continuous at $\varkappa = 0$ if $T(\varkappa)$ is continuous at $\varkappa = 0$. To see this it suffices to notice that $R(\zeta, \varkappa) \to R(\zeta)$, $\varkappa \to 0$, uniformly for $\zeta \in \Gamma$. We may even replace $T(\varkappa)$ by a sequence $\{T_n\}$ such that $T_n \to T$, $n \to \infty$. Then it follows that the eigenvalues as well as the total projections of T_n tend to the corresponding ones of T for $n \to \infty$.

Summing up, we have

Theorem 5.1. *Let* $T(\varkappa)$ *be continuous at* $\varkappa = 0$. *Then the eigenvalues of* $T(\varkappa)$ *are continuous at* $\varkappa = 0$. *If* λ *is an eigenvalue of* $T = T(0)$, *the*

λ-group is well-defined for sufficiently small $|\varkappa|$ and the total projection $P(\varkappa)$ for the λ-group is continuous at $\varkappa = 0$. If $T(\varkappa)$ is continuous in a domain of the $\varkappa$-plane or in an interval of the real line, the resolvent $R(\zeta, \varkappa)$ is continuous in ζ and $\varkappa$ jointly in the sense stated above.

2. The numbering of the eigenvalues

The fact proved above that the eigenvalues of $T(\varkappa)$ change continuously with $\varkappa$ when $T(\varkappa)$ is continuous in $\varkappa$ is not altogether simple since the number of eigenvalues of $T(\varkappa)$ is not necessarily constant. It is true that the same circumstance exists even in the analytic case, but there the number s of (different) eigenvalues is constant for non-exceptional $\varkappa$. In the general case now under consideration, this number may change with $\varkappa$ quite irregularly; the splitting and coalescence of eigenvalues may take place in a very complicated manner.

To avoid this inconvenience, it is usual to count the eigenvalues repeatedly according to their (algebraic) multiplicities as described in I-§ 5.4 (repeated eigenvalues). The repeated eigenvalues of an operator form an ***unordered*** N-tuple of complex numbers. Two such N-tuples $\mathfrak{S} = (\mu_1, \ldots, \mu_N)$ and $\mathfrak{S}' = (\mu_1', \ldots, \mu_N')$ may be considered close to each other if, *for suitable numbering of their elements*, the $|\mu_n - \mu_n'|$ are small for all $n = 1, \ldots, N$. We can even define the *distance* between such two N-tuples by

$$\operatorname{dist}(\mathfrak{S}, \mathfrak{S}') = \min \max_n |\mu_n - \mu_n'| \tag{5.3}$$

where the min is taken over all possible renumberings of the elements of one of the N-tuples. For example, the distance between the triples (0, 0, 1) and (0, 1, 1) is equal to 1, though the set {0, 1} of their elements is the same for the two triples. It is easy to verify that the distance thus defined satisfies the axioms of a distance function.

The continuity of the eigenvalues of $T(\varkappa)$ given by Theorem 5.1 can now be expressed by saying that *the N-tuple $\mathfrak{S}(\varkappa)$ consisting of the repeated eigenvalues of $T(\varkappa)$ changes with $\varkappa$ continuously*. This means that the distance of $\mathfrak{S}(\varkappa)$ from $\mathfrak{S}(\varkappa_0)$ tends to zero for $\varkappa \to \varkappa_0$ for each fixed $\varkappa_0$.

The continuity thus formulated is the continuity of the repeated eigenvalues *as a whole*. It is a different question whether it is possible to define N single-valued, continuous functions $\mu_n(\varkappa)$, $n = 1, \ldots, N$, which for each $\varkappa$ represent the repeated eigenvalues of $T(\varkappa)$. Such a parametrization is in general impossible. This will be seen from Example 1.1, d), in which the two eigenvalues are $\pm\varkappa^{1/2}$; here it is impossible to define two single-valued continuous functions representing the two eigenvalues in a domain of the complex plane containing the branch point $\varkappa = 0$.

A parametrization is possible if either i) the parameter $\varkappa$ changes over an interval of the real line or ii) the eigenvalues are always real. In case ii) it suffices to number the eigenvalues $\mu_n(\varkappa)$ in ascending (or descending) order:

$$\mu_1(\varkappa) \leqq \mu_2(\varkappa) \leqq \cdots \leqq \mu_N(\varkappa) . \tag{5.4}$$

It should be noted, however, that this way of numbering is not always convenient, for it can destroy the differentiability of the functions which may exist in a different arrangement.

The possibility of a parametrization in case i) is not altogether obvious. This is contained in the following theorem.

Theorem 5.2. *Let $\mathfrak{S}(\varkappa)$ be an unordered N-tuple of complex numbers, depending continuously on a real variable $\varkappa$ in a (closed or open) interval* I. *Then there exist N single-valued, continuous functions $\mu_n(\varkappa)$, $n = 1, \ldots, N$, the values of which constitute the N-tuple $\mathfrak{S}(\varkappa)$ for each $\varkappa \in$ I. [$(\mu_n(\varkappa))$ is called a representation of $\mathfrak{S}(\varkappa)$.]*

Proof. For convenience we shall say that a subinterval I_0 of I has *property* (A) if there exist N functions defined on I_0 with the properties stated in the theorem. What is required is to prove that I itself has property (A). We first show that, whenever two subintervals I_1, I_2 with property (A) have a common point, then their union $I_0 = I_1 \cup I_2$ also has the same property. Let $(\mu_n^{(1)}(\varkappa))$ and $(\mu_n^{(2)}(\varkappa))$ be representations of $\mathfrak{S}(\varkappa)$ in I_1 and I_2, respectively, by continuous functions. We may assume that neither I_1 nor I_2 contains the other (otherwise the proof is trivial) and that I_1 lies to the left of I_2. For a fixed $\varkappa_0$ lying in the intersection of I_1 and I_2, we have $\mu_n^{(1)}(\varkappa_0) = \mu_n^{(2)}(\varkappa_0)$, $n = 1, \ldots, N$, after a suitable renumbering of $(\mu_n^{(2)})$, for both $(\mu_n^{(1)}(\varkappa_0))$ and $(\mu_n^{(2)}(\varkappa_0))$ represent the same $\mathfrak{S}(\varkappa_0)$. Then the functions $\mu_n^{(0)}(\varkappa)$ defined on I_0 by

$$\mu_n^{(0)}(\varkappa) = \begin{cases} \mu_n^{(1)}(\varkappa) , & \varkappa \leqq \varkappa_0 , \\ \mu_n^{(2)}(\varkappa) , & \varkappa \geqq \varkappa_0 , \end{cases} \tag{5.5}$$

are continuous and represent $\mathfrak{S}(\varkappa)$ on I_0.

It follows that, whenever a subinterval I′ has the property that each point of I′ has a neighborhood with property (A), then I′ itself has property (A).

With these preliminaries, we now prove Theorem 5.2 by induction. The theorem is obviously true for $N = 1$. Suppose that it has been proved for N replaced by smaller numbers and for any interval I. Let Γ be the set of all $\varkappa \in$ I for which the N elements of $\mathfrak{S}(\varkappa)$ are identical, and let Δ be the complement of Γ in I. Γ is closed and Δ is open relative to I. Let us now show that each point of Δ has a neighborhood having property (A). Let $\varkappa_0 \in \Delta$. Since the N elements of $\mathfrak{S}(\varkappa_0)$ are not all identical, they can be divided into two separate groups with N_1 and N_2 elements, where $N_1 + N_2 = N$. In other words, $\mathfrak{S}(\varkappa_0)$ is composed of an N_1-tuple and an N_2-tuple with separate elements ("separate" means that there is no element of one group equal to an element of the other). The continuity of $\mathfrak{S}(\varkappa)$ implies that for sufficiently small $|\varkappa - \varkappa_0|$, $\mathfrak{S}(\varkappa)$ consists likewise of an N_1-tuple and an N_2-tuple each of which is continuous in $\varkappa$. According to the induction hypothesis, these N_1 and N_2-tuples can be represented in a neighborhood Δ' of $\varkappa_0$ by families of continuous functions $(\mu_1(\varkappa), \ldots, \mu_{N_1}(\varkappa))$ and $(\mu_{N_1+1}(\varkappa), \ldots, \mu_N(\varkappa))$, respectively. These N functions taken together then represent $\mathfrak{S}(\varkappa)$ in Δ'. In other words, Δ' has property (A).

Since Δ is open in I, it consists of at most countably many subintervals $\mathrm{I}_1, \mathrm{I}_2, \ldots$. Since each point of Δ has a neighborhood with property (A), it follows from the remark above that each component interval I_p has property (A). We denote by $\mu_n^{(p)}(\varkappa)$, $n = 1, \ldots, N$, the N functions representing $\mathfrak{S}(\varkappa)$ in I_p. For $\varkappa \in \Gamma$, on the other hand, $\mathfrak{S}(\varkappa)$ consists of N identical elements $\mu(\varkappa)$. We now define N functions $\mu_n(\varkappa)$, $n = 1, \ldots, N$, on I by

$$\mu_n(\varkappa) = \begin{cases} \mu_n^{(p)}(\varkappa), & \varkappa \in \mathrm{I}_p, \quad p = 1, 2, \ldots, \\ \mu(\varkappa), & \varkappa \in \Gamma. \end{cases} \tag{5.6}$$

These N functions represent $\mathfrak{S}(\varkappa)$ on the whole interval I. It is easy to verify that each $\mu_n(\varkappa)$ is continuous on I. This completes the induction and the theorem is proved.

3. Continuity of the eigenspaces and eigenvectors

Even when $T(\varkappa)$ is continuous in $\varkappa$, the eigenvectors or eigenspaces are not necessarily continuous. We have shown above that the total projection $P(\varkappa)$ for the λ-group is continuous, but $P(\varkappa)$ is defined only for small $|\varkappa|$ for which the λ-group eigenvalues are not too far from λ.

If $T(\varkappa)$ has N distinct eigenvalues $\lambda_h(\varkappa)$, $h = 1, \ldots, N$, for all $\varkappa$ in a simply connected domain of the complex plane or in an interval of the real line, we can define the associated eigenprojections $P_h(\varkappa)$ each of which is one-dimensional. Each $P_h(\varkappa)$ is continuous since it is identical with the total projection for the eigenvalue $\lambda_h(\varkappa)$. But $P_h(\varkappa)$ cannot in general be continued beyond a value of $\varkappa$ for which $\lambda_h(\varkappa)$ coincides with some other $\lambda_k(\varkappa)$. In this sense the eigenprojections behave more singularly than the eigenvalues. It should be recalled that even in the analytic case $P_h(\varkappa)$ may not exist at an exceptional point where $\lambda_h(\varkappa)$ is holomorphic [Example 1.12, f)]; but $P_h(\varkappa)$ has at most a pole in that case (see § 1.8). In the general case under consideration, the situation is much worse. That the eigenspaces can behave quite singularly even for a very smooth function $T(\varkappa)$ is seen from the following example due to RELLICH[1].

Example 5.3. Let $N = 2$ and

$$(5.7) \qquad T(\varkappa) = e^{-\frac{1}{\varkappa^2}} \begin{pmatrix} \cos\frac{2}{\varkappa} & \sin\frac{2}{\varkappa} \\ \sin\frac{2}{\varkappa} & -\cos\frac{2}{\varkappa} \end{pmatrix}, \quad T(0) = 0 .$$

$T(\varkappa)$ is not only continuous but is infinitely differentiable for all *real* values of $\varkappa$, and the same is true of the eigenvalues of $T(\varkappa)$, which are $\pm e^{-1/\varkappa^2}$ for $\varkappa \neq 0$ and zero for $\varkappa = 0$. But the associated eigenprojections are given for $\varkappa \neq 0$ by

$$(5.8) \qquad \begin{pmatrix} \cos^2\frac{1}{\varkappa} & \cos\frac{1}{\varkappa}\sin\frac{1}{\varkappa} \\ \cos\frac{1}{\varkappa}\sin\frac{1}{\varkappa} & \sin^2\frac{1}{\varkappa} \end{pmatrix}, \quad \begin{pmatrix} \sin^2\frac{1}{\varkappa} & -\cos\frac{1}{\varkappa}\sin\frac{1}{\varkappa} \\ -\cos\frac{1}{\varkappa}\sin\frac{1}{\varkappa} & \cos^2\frac{1}{\varkappa} \end{pmatrix}.$$

These matrix functions are continuous (even infinitely differentiable) in any interval not containing $\varkappa = 0$, but they cannot be continued to $\varkappa = 0$ as continuous functions. Furthermore, it is easily seen that there does not exist any eigenvector of $T(\varkappa)$ that is continuous in the neighborhood of $\varkappa = 0$ and that does not vanish at $\varkappa = 0$.

It should be remarked that (5.7) is a symmetric operator for each real $\varkappa$ (acting in C^2 considered a unitary space). In particular it is normal and, therefore, the eigenprojections would be *holomorphic* at $\varkappa = 0$ if $T(\varkappa)$ were *holomorphic* (Theorem 1.10). The example is interesting since it shows that this smoothness of the eigenprojections can be lost completely if the holomorphy of $T(\varkappa)$ is replaced by infinite differentiability.

4. Differentiability at a point

Let us now assume that $T(\varkappa)$ is not only continuous but differentiable. This does not in general imply that the eigenvalues of $T(\varkappa)$ are differentiable; in fact they need not be differentiable even if $T(\varkappa)$ is holomorphic [Example 1.1, d)]. However, we have

Theorem 5.4. *Let $T(\varkappa)$ be differentiable at $\varkappa = 0$. Then the total projection $P(\varkappa)$ for the λ-group is differentiable at $\varkappa = 0$:*

$$(5.9) \qquad P(\varkappa) = P + \varkappa P^{(1)} + o(\varkappa),$$

where[2] $P^{(1)} = -PT'(0)S - ST'(0)P$ *and S is the reduced resolvent of T for λ* (see I-§ 5.3). *If λ is a semisimple eigenvalue of T, the λ-group eigenvalues of $T(\varkappa)$ are differentiable at $\varkappa = 0$:*

$$(5.10) \qquad \mu_j(\varkappa) = \lambda + \varkappa \mu_j^{(1)} + o(\varkappa), \quad j = 1, \ldots, m,$$

where the $\mu_j(\varkappa)$ are the repeated eigenvalues of the λ-group and the $\mu_j^{(1)}$ are the repeated eigenvalues of $PT'(0)P$ in the subspace $\mathsf{M} = P\mathsf{X}$ (P is

[1] See RELLICH [1]; for convenience we modified the original example slightly.

[2] Here $o(\varkappa)$ denotes an operator-valued function $F(\varkappa)$ such that $\|F(\varkappa)\| = o(\varkappa)$ in the ordinary sense.

the eigenprojection of T for λ). If T is diagonable, all the eigenvalues are differentiable at $\varkappa = 0$.

Remark 5.5. The above theorem needs some comments. That the eigenvalues of $T(\varkappa)$ are differentiable means that the N-tuple $\mathfrak{S}(\varkappa)$ consisting of the repeated eigenvalues of $T(\varkappa)$ is differentiable, and similarly for the differentiability of the λ-group eigenvalues. That an (unordered) N-tuple $\mathfrak{S}(\varkappa)$ is differentiable at $\varkappa = 0$ means that $\mathfrak{S}(\varkappa)$ can be represented in a neighborhood of $\varkappa = 0$ by N functions $\mu_n(\varkappa)$, $n = 1, \ldots, N$, which are differentiable at $\varkappa = 0$. The N-tuple $\mathfrak{S}'(0)$ consisting of the $\mu_n'(0)$ is called the derivative of $\mathfrak{S}(\varkappa)$ at $\varkappa = 0$. It can be easily proved (by induction on N, say) that $\mathfrak{S}'(0)$ is thereby defined independently of the particular representation $(\mu_n(\varkappa))$ of $\mathfrak{S}(\varkappa)$. If $\mathfrak{S}(\varkappa)$ is differentiable at each $\varkappa$ and $\mathfrak{S}'(\varkappa)$ is continuous, $\mathfrak{S}(\varkappa)$ is said to be continuously differentiable.

Note that $\mathfrak{S}(0)$ and $\mathfrak{S}'(0)$ together need not determine the behavior of $\mathfrak{S}(\varkappa)$ even in the immediate neighborhood of $\varkappa = 0$. For example, $\mathfrak{S}_1(\varkappa) = (\varkappa, 1 - \varkappa)$ and $\mathfrak{S}_2(\varkappa) = (-\varkappa, 1 + \varkappa)$ have the common value $(0, 1)$ and the common derivative $(1, -1)$ at $\varkappa = 0$.

Proof of Theorem 5.4. We first note that the resolvent $R(\zeta, \varkappa)$ is differentiable at $\varkappa = 0$, for it follows from I-(4.28) that

$$\left[\frac{\partial}{\partial \varkappa} R(\zeta, \varkappa)\right]_{\varkappa = 0} = -R(\zeta)\, T'(0)\, R(\zeta)\,. \tag{5.11}$$

Here it should be noted that the derivative (5.11) exists *uniformly* for ζ belonging to a compact subset of $\mathrm{P}(T)$, for $R(\zeta, \varkappa) \to R(\zeta)$, $\varkappa \to 0$, holds uniformly (see par. 1). Considering the expression (1.16) for $P(\varkappa)$, it follows that $P(\varkappa)$ is differentiable at $\varkappa = 0$, with

$$\begin{aligned} P'(0) &= -\frac{1}{2\pi i} \int_\Gamma \left[\frac{\partial}{\partial \varkappa} R(\zeta, \varkappa)\right]_{\varkappa = 0} d\zeta = \frac{1}{2\pi i} \int_\Gamma R(\zeta)\, T'(0)\, R(\zeta)\, d\zeta \\ &= -P\, T'(0)\, S - S\, T'(0)\, P = P^{(1)} \quad [\text{cf. } (2.14)]\,. \end{aligned} \tag{5.12}$$

This proves (5.9).

As in the analytic case, if λ is semisimple the λ-group eigenvalues of $T(\varkappa)$ are of the form

$$\mu_j(\varkappa) = \lambda + \varkappa\, \mu_j^{(1)}(\varkappa)\,, \qquad j = 1, \ldots, m\,, \tag{5.13}$$

where the $\mu_j^{(1)}(\varkappa)$ are the (repeated) eigenvalues of the operator

$$\tilde{T}^{(1)}(\varkappa) = \varkappa^{-1}(T(\varkappa) - \lambda)\, P(\varkappa) = -\frac{\varkappa^{-1}}{2\pi i} \int_\Gamma (\zeta - \lambda)\, R(\zeta, \varkappa)\, d\zeta\,, \tag{5.14}$$

in the subspace $\mathsf{M}(\varkappa) = P(\varkappa)\, \mathsf{X}$ [see (2.37)]. Since $T(\varkappa)$ and $P(\varkappa)$ are differentiable at $\varkappa = 0$ and $(T - \lambda)\, P = 0$ if λ is semisimple, $\tilde{T}^{(1)}(\varkappa)$ is continuous at $\varkappa = 0$ if we set

$$\tilde{T}^{(1)}(0) = T'(0)\, P + (T - \lambda)\, P'(0) = P\, T'(0)\, P \tag{5.15}$$

where we have used (5.12) and $(T - \lambda)\, S = 1 - P$ [see (2.11)]. Hence the eigenvalues of $\tilde{T}^{(1)}(\varkappa)$ are continuous at $\varkappa = 0$. In particular, the eigenvalues $\mu_j^{(1)}(\varkappa)$ of $\tilde{T}^{(1)}(\varkappa)$ in the invariant subspace $\mathsf{M}(\varkappa)$ are continuous at $\varkappa = 0$ (see Theorem 5.1), though they need not be continuous for $\varkappa \neq 0$. In view of (5.13), this proves (5.10).

5. Differentiability in an interval

So far we have been considering the differentiability of the eigenvalues and eigenprojections at a single point $\varkappa = 0$. Let us now consider the differentiability in a domain of $\varkappa$, assuming that $T(\varkappa)$ is differentiable in this domain. If this is a domain of the complex plane, $T(\varkappa)$ is necessarily holomorphic; since this case has been discussed in detail, we shall henceforth assume that $T(\varkappa)$ is defined and differentiable in an interval I of the *real line*[1].

According to Theorem 5.4, the N-tuple $\mathfrak{S}(\varkappa)$ of the repeated eigenvalues of $T(\varkappa)$ is differentiable for $\varkappa \in$ I *provided that* $T(\varkappa)$ *is diagonable for* $\varkappa \in$ I. However, it is by no means obvious that there exist N functions $\mu_n(\varkappa)$, $n = 1, \ldots, N$, representing the repeated eigenvalues of $T(\varkappa)$, which are single-valued and differentiable on the whole of I Actually this is true, as is seen from the following general theorem.

Theorem 5.6. *Let* $\mathfrak{S}(\varkappa)$ *be an unordered N-tuple of complex numbers depending on a real variable* $\varkappa$ *in an interval* I, *and let* $\mathfrak{S}(\varkappa)$ *be differentiable at each* $\varkappa \in$ I *(in the sense of* Remark 5.5*). Then there exist N complex-valued functions* $\mu_n(\varkappa)$, $n = 1, \ldots, N$, *representing the N-tuple* $\mathfrak{S}(\varkappa)$ *for* $\varkappa \in$ I, *each of which is differentiable for* $\varkappa \in$ I.

Proof. A subinterval of I for which there exist N functions of the kind described will be said to have property (B); it is required to prove that I itself has property (B). It can be shown, as in the proof of Theorem 5.2, that for any overlapping[2] subintervals I_1, I_2 with property (B), their union $\mathrm{I}_1 \cup \mathrm{I}_2$ has also property (B). The only point to be noted is that care must be taken in the renumbering of the $\mu_n^{(2)}(\varkappa)$ to make sure that the continuation (5.5) preserves the differentiability of the functions at $\varkappa = \varkappa_0$; this is possible owing to the assumption that $\mathfrak{S}(\varkappa)$ is differentiable at $\varkappa = \varkappa_0$.

---

[1] The differentiability of eigenvalues was investigated in detail in Rellich [8] in the case when $T(\varkappa)$ is symmetric for each real $\varkappa$. It should be noted that the problem is far from trivial even in that special case.

[2] Here "overlapping" means that the two intervals have common interior points.

With this observation, the proof is carried out in the same way as in Theorem 5.2. A slight modification is necessary at the final stage, for the functions defined by (5.6) may have discontinuities of derivatives at an *isolated point* of Γ. To avoid this, we proceed as follows. An isolated point $\varkappa_0$ of Γ is either a boundary point of I or a common boundary of an I_p and an I_q. In the first case we have nothing to do. In the second case, it is easy to "connect" the two families $(\mu_n^{(p)}(\varkappa))$ and $(\mu_n^{(q)}(\varkappa))$ "smoothly" by a suitable renumbering of the latter, for these two families of differentiable functions represent $\mathfrak{S}(\varkappa)$ at the different sides of $\varkappa = \varkappa_0$ and $\mathfrak{S}(\varkappa)$ is differentiable at $\varkappa = \varkappa_0$. It follows that the interval consisting of I_p, I_q and $\varkappa_0$ has property (B).

Let Γ' be the set of isolated points of Γ. $\varDelta \cup \Gamma'$ is relatively open in I and consists of (at most) countably many subintervals I_k'. Each I_k' consists in turn of countably many subintervals of the form I_p joined with one another at a point of Γ' in the manner stated above. By repeated applications of the connection process just described, the functions representing $\mathfrak{S}(\varkappa)$ in these I_p can be connected to form a family of N functions differentiable on I_k'. This shows that each I_k' has property (B).

The construction of N differentiable functions $\mu_n(\varkappa)$ representing $\mathfrak{S}(\varkappa)$ on the whole interval I can now be carried out by the method (5.6), in which the I_p and Γ should be replaced by the I_k' and Γ'' respectively, Γ'' being the complement of Γ' in Γ. The differentiability at a point $\varkappa_0$ of Γ'' of the $\mu_n(\varkappa)$ thus defined follows simply from the fact that the derivative $\mathfrak{S}'(\varkappa_0)$ consists of N identical elements, just as does $\mathfrak{S}(\varkappa_0)$. This completes the proof of Theorem 5.6.

Theorem 5.7. *If in* Theorem 5.6 *the derivative* $\mathfrak{S}'(\varkappa)$ *is continuous, the* N *functions* $\mu_n(\varkappa)$ *are continuously differentiable on* I.

Proof. Suppose that the real part of $\mu_n'(\varkappa)$ is discontinuous at $\varkappa = \varkappa_0$. According to a well-known result in differential calculus, the values taken by $\mathrm{Re}\,\mu_n'(\varkappa)$ in any neighborhood of $\varkappa_0$ cover an interval of length larger than a fixed positive number[1]. But this is impossible if $\mathfrak{S}'(\varkappa)$ is continuous, for any value of $\mu_n'(\varkappa)$ is among the N elements of $\mathfrak{S}'(\varkappa)$. For the same reason $\mathrm{Im}\,\mu_n'(\varkappa)$ cannot be discontinuous at any point.

Remark 5.8. We have seen (Theorem 5.4) that the eigenvalues of $T(\varkappa)$ are differentiable on I if $T(\varkappa)$ is differentiable and diagonable for $\varkappa \in$ I.

[1] Let a real-valued function $f(t)$ of a real variable t be differentiable for $a \leqq t \leqq b$; then $f'(t)$ takes in this interval all values between $\alpha = f'(a)$ and $\beta = f'(b)$. To prove this, we may assume $\alpha < \beta$. For any $\gamma \in (\alpha, \beta)$ set $g(t) = f(t) - \gamma t$. Then $g'(a) = \alpha - \gamma < 0$, $g'(b) = \beta - \gamma > 0$, so that the continuous function $g(t)$ takes a minimum at a point $t = c \in (a, b)$. Hence $g'(c) = 0$ or $f'(c) = \gamma$. If $f'(t)$ is discontinuous at $t = t_0$, then there is $\varepsilon > 0$ such that, in any neighborhood of t_0, there exist t_1, t_2 with $|f'(t_1) - f'(t_2)| > \varepsilon$. It follows from the above result that $f'(t)$ takes all values between $f'(t_1)$ and $f'(t_2)$. Hence the values of $f'(t)$ in any neighborhood of t_0 covers an interval of length larger than ε.

It is quite natural, then, to conjecture that the eigenvalues are continuously differentiable if $T(\varkappa)$ is continuously differentiable and diagonable. *But this is not true,* as is seen from the following example[1].

Example 5.9. Let $N = 2$ and

$$T(\varkappa) = \begin{pmatrix} |\varkappa|^\alpha & |\varkappa|^\alpha - |\varkappa|^\beta \left(2 + \sin \frac{1}{|\varkappa|}\right) \\ -|\varkappa|^\alpha & -|\varkappa|^\alpha \end{pmatrix}, \quad \varkappa \neq 0; \quad T(0) = 0. \tag{5.16}$$

$T(\varkappa)$ is continuously differentiable for all real $\varkappa$ if $\alpha > 1$ and $\beta > 2$. The two eigenvalues of $T(\varkappa)$ are

$$\mu_\pm(\varkappa) = \pm |\varkappa|^{\frac{\alpha+\beta}{2}} \left(2 + \sin \frac{1}{|\varkappa|}\right)^{\frac{1}{2}}, \quad \varkappa \neq 0, \quad \mu_\pm(0) = 0. \tag{5.17}$$

Since the $\mu_\pm(\varkappa)$ are different from each other for $\varkappa \neq 0$, $T(\varkappa)$ is diagonable, and $T(0) = 0$ is obviously diagonable. The $\mu_\pm(\varkappa)$ are differentiable everywhere, as the general theory requires. But their derivatives are discontinuous at $\varkappa = 0$ if $\alpha + \beta \leqq 4$. This simple example shows again that a non-analytic perturbation can be rather pathological in behavior.

Remark 5.10. If $T(\varkappa)$ is continuously differentiable in a neighborhood of $\varkappa = 0$ and λ is a semisimple eigenvalue of $T(0)$, the total projection $P(\varkappa)$ for the λ-group is continuously differentiable and $\tilde{T}^{(1)}(\varkappa)$ is continuous in a neighborhood of $\varkappa = 0$, as is seen from (5.12) and (5.14).

6. Asymptotic expansion of the eigenvalues and eigenvectors

The differentiability of the eigenvalues considered in the preceding paragraphs can be studied from a somewhat different point of view. Theorem 5.4 may be regarded as giving an *asymptotic* expansion of the eigenvalues $\mu_j(\varkappa)$ up to the first order of $\varkappa$ when $T(\varkappa)$ has the asymptotic form $T(\varkappa) = T + \varkappa T' + o(\varkappa)$, where $T' = T'(0)$. Going to the second order, we can similarly inquire into the asymptotic behavior of the eigenvalues when $T(\varkappa)$ has the asymptotic form $T + \varkappa T^{(1)} + \varkappa^2 T^{(2)} + o(\varkappa^2)$. In this direction, the extension of Theorem 5.4 is rather straightforward.

Theorem 5.11. *Let* $T(\varkappa) = T + \varkappa T^{(1)} + \varkappa^2 T^{(2)} + o(\varkappa^2)$ *for* $\varkappa \to 0$. *Let* λ *be an eigenvalue of* T *with the eigenprojection* P. *Then the total projection* $P(\varkappa)$ *for the* λ*-group eigenvalues of* $T(\varkappa)$ *has the form*

$$P(\varkappa) = P + \varkappa P^{(1)} + \varkappa^2 P^{(2)} + o(\varkappa^2) \tag{5.18}$$

where P, $P^{(1)}$ *and* $P^{(2)}$ *are given by* (2.13). *If the eigenvalue* λ *of* T *is semisimple and if* $\lambda_j^{(1)}$ *is an eigenvalue of* $P T^{(1)} P$ *in* PX *with the eigen-*

[1] The operator of this example is not symmetric. For symmetric operators better behavior is expected; see Theorem 6.8.

projection $P_j^{(1)}$, *then* $T(\varkappa)$ *has exactly* $m_j^{(1)} = \dim P_j^{(1)}$ *repeated eigenvalues (the* $\lambda + \varkappa\,\lambda_j^{(1)}$*-group) of the form* $\lambda + \varkappa\,\lambda_j^{(1)} + o(\varkappa)$. *The total eigenprojection* $P_j^{(1)}(\varkappa)$ *for this group has the form*

$$P_j^{(1)}(\varkappa) = P_j^{(1)} + \varkappa\,P_j^{(11)} + o(\varkappa)\,. \tag{5.19}$$

If, in addition, the eigenvalue $\lambda_j^{(1)}$ *of* $P\,T^{(1)}\,P$ *is semisimple, then* $P_j^{(11)}$ *is given by* (2.47) *and the* $m_j^{(1)}$ *repeated eigenvalues of the* $\lambda + \varkappa\,\lambda_j^{(1)}$*-group have the form*

$$\mu_{jk}(\varkappa) = \lambda + \varkappa\,\lambda_j^{(1)} + \varkappa^2\,\mu_{jk}^{(2)} + o(\varkappa^2)\,,\quad k = 1, \ldots, m_j^{(1)}\,, \tag{5.20}$$

where $\mu_{jk}^{(2)}$, $k = 1, \ldots, m_j^{(1)}$, *are the repeated eigenvalues of* $P_j^{(1)}\,\tilde{T}^{(2)}\,P_j^{(1)} = P_j^{(1)}\,T^{(2)}\,P_j^{(1)} - P_j^{(1)}\,T^{(1)}\,S\,T^{(1)}\,P_j^{(1)}$ *in the subspace* $P_j^{(1)}\mathsf{X}$.

Proof. The possibility of the asymptotic expansion of $T(\varkappa)$ up to the second order implies the same for the resolvent:

$$\begin{aligned} R(\zeta,\varkappa) &= R(\zeta) - R(\zeta)\,(T(\varkappa) - T)\,R(\zeta) + R(\zeta)\,(T(\varkappa) - T)\,\times \\ &\quad\times R(\zeta)\,(T(\varkappa) - T)\,R(\zeta) + \cdots + \\ &= R(\zeta) - \varkappa\,R(\zeta)\,T^{(1)}\,R(\zeta) + \varkappa^2[-R(\zeta)\,T^{(2)}\,R(\zeta) + \\ &\quad + R(\zeta)\,T^{(1)}\,R(\zeta)\,T^{(1)}\,R(\zeta)] + o(\varkappa^2)\,. \end{aligned} \tag{5.21}$$

Here $o(\varkappa^2)$ is uniform in ζ in each compact subset of $\mathrm{P}(T)$. Substitution of (5.21) into (1.16) yields (5.18), just as in the analytic case.

It follows that if λ is semisimple, the $\tilde{T}^{(1)}(\varkappa)$ of (5.14) has the form

$$\tilde{T}^{(1)}(\varkappa) = P\,T^{(1)}\,P + \varkappa\,\tilde{T}^{(2)} + o(\varkappa) \tag{5.22}$$

where $\tilde{T}^{(2)}$ is given by (2.20). The application of Theorem 5.4 to $\tilde{T}^{(1)}(\varkappa)$ then leads to the result of the theorem. Again the calculation of the terms up to the order $\varkappa^2$ is the same as in the analytic case.

7. Operators depending on several parameters

So far we have been concerned with an operator $T(\varkappa)$ depending on a single parameter $\varkappa$. Let us now consider an operator $T(\varkappa_1, \varkappa_2)$ depending on two variables $\varkappa_1, \varkappa_2$ which may be complex or real.

There is nothing new in regard to the continuity of the eigenvalues. The eigenvalues are continuous in $\varkappa_1, \varkappa_2$ (in the sense explained in par. 1,2) if $T(\varkappa_1, \varkappa_2)$ is continuous. Again, the same is true for partial differentiability (when the variables $\varkappa_1$, $\varkappa_2$ are real). But something singular appears concerning *total differentiability*. The total differentiability of $T(\varkappa_1, \varkappa_2)$ does not necessarily imply the same for the eigenvalues even if $T(\varkappa_1, \varkappa_2)$ is diagonable (cf. Theorem 5.4).

Example 5.12.[1] Let $N = 2$ and

$$T(\varkappa_1, \varkappa_2) = \begin{pmatrix} \varkappa_1 & \varkappa_2 \\ \varkappa_2 & -\varkappa_1 \end{pmatrix}\,. \tag{5.23}$$

[1] See Rellich [1].

$T(\varkappa_1, \varkappa_2)$ is totally differentiable in $\varkappa_1, \varkappa_2$ and diagonable for all real values of $\varkappa_1, \varkappa_2$. But its eigenvalues

$$\lambda_\pm(\varkappa_1, \varkappa_2) = \pm(\varkappa_1^2 + \varkappa_2^2)^{1/2} \tag{5.24}$$

are not totally differentiable at $\varkappa_1 = \varkappa_2 = 0$.

We could also consider the case in which $T(\varkappa_1, \varkappa_2)$ is holomorphic in the two variables. But the eigenvalues of $T(\varkappa_1, \varkappa_2)$ might have rather complicated singularities, as is seen from the above example[1].

Remark 5.13. (5.23) is symmetric for real $\varkappa_1, \varkappa_2$ if the usual inner product is introduced into $\mathsf{X} = \mathbf{C}^2$. Thus the appearence of a singularity of the kind (5.24) shows that the situation is different from the case of a single variable, where the eigenvalue is holomorphic at $\varkappa = 0$ if $T(\varkappa)$ is normal for real $\varkappa$ (see Theorem 1.10).

Similar remarks apply to the case in which there are more than two variables[2].

8. The eigenvalues as functions of the operator

In perturbation theory the introduction of the parameter $\varkappa$ is sometimes rather artificial, although it sometimes corresponds to the real situation. We could rather consider the change of the eigenvalues of an operator T when T is changed by a small amount, without introducing any parameter $\varkappa$ or parameters $\varkappa_1, \varkappa_2, \ldots$. From this broader point of view, the eigenvalues of T should be regarded as functions of T itself. Some care is necessary in this interpretation, however, since the eigenvalues are not fixed in number. Again it is convenient to consider the unordered N-tuple $\mathfrak{S}[T]$, consisting of the N repeated eigenvalues of T, as a function of T. This is equivalent to regarding $\mathfrak{S}[T]$ as a function of the N^2 elements of the matrix representing T with respect to a fixed basis of X.

---

[1] But simple eigenvalues and the associated eigenprojections are again holomorphic in $\varkappa_1, \varkappa_2$; this follows from Theorem 5.16.

[2] In connection with a family $T(\varkappa_1, \varkappa_2)$ depending on two parameters, we can ask when $T(\varkappa_1, \varkappa_2)$ has a non-zero null space. This is a generalization of the perturbation theory of eigenvalues we have considered so far, which is equivalent to the condition that $T(\varkappa) - \lambda$, a special operator depending on two parameters, have a non-zero null space. The general case $T(\varkappa_1, \varkappa_2)$ gives rise to the perturbation problem for "nonlinear" eigenvalue problems of the type noted in footnote [1], p. 35. We shall not consider such a general perturbation theory. We note, however, that in some special cases the nonlinear problem can be reduced to the ordinary problem discussed above. Suppose we have a family of the form $T(\lambda) - \varkappa$. The ordinary theory will give the "eigenvalues" $\varkappa$ as analytic functions of the "parameter" λ. If we find the inverse function of these analytic functions, we obtain the "nonlinear" eigenvalues λ as analytic functions of the parameter $\varkappa$. There are other devices for dealing with "nonlinear" problems. See e. g., CLOIZEAUX [1].

Theorem 5.14. *$\mathfrak{S}[T]$ is a continuous function of T. By this it is meant that, for any fixed operator T, the distance between $\mathfrak{S}[T+A]$ and $\mathfrak{S}[T]$ tends to zero for $\|A\| \to 0$.*

The proof of this theorem is contained in the result of par. 1,2 where the continuity of $\mathfrak{S}(\varkappa)$ as a function of $\varkappa$ is proved. An examination of the arguments given there will show that the use of the parameter $\varkappa$ is not essential.

This continuity of $\mathfrak{S}[T]$ is naturally *uniform* on any bounded region of the variable T (that is, a region in which $\|T\|$ is bounded), for the variable T is equivalent to N^2 complex variables as noted above. But the degree of continuity may be very weak at some special T (non-diagonable T), as is seen from the fact that the Puiseux series for the eigenvalues of $T + \varkappa T^{(1)} + \cdots$ can have the form $\lambda + \alpha \varkappa^{1/p} + \cdots$ [see (1.7) and Example 1.1, d)].

Let us now consider the differentiability of $\mathfrak{S}[T]$. As we have seen, the eigenvalues are not always differentiable even in the analytic case $T(\varkappa)$. If T is diagonable, on the other hand, the eigenvalues of $T + \varkappa T^{(1)}$ are differentiable at $\varkappa = 0$ for any $T^{(1)}$ (in the sense of par. 4), and the diagonability of T is necessary in order that this be true for every $T^{(1)}$. This proves

Theorem 5.15. *$\mathfrak{S}[T]$ is partially differentiable at $T = T_0$ if and only if T_0 is diagonable.*

Here "partially differentiable" means that $\mathfrak{S}[T + \varkappa T^{(1)}]$ is differentiable at $\varkappa = 0$ for any fixed $T^{(1)}$, and it implies the partial differentiability of $\mathfrak{S}[T]$ in each of the N^2 variables when it is regarded as a function of the N^2 matrix elements.

Theorem 5.15 is not true if "partially" is replaced by "totally". This is seen from Example 5.12, which shows that $\mathfrak{S}[T]$ need not be totally differentiable even when the change of T is restricted to a two-dimensional subspace of $\mathscr{B}(\mathsf{X})$. In general a complex-valued function $\mu[T]$ of $T \in \mathscr{B}(\mathsf{X})$ is said to be totally differentiable at $T = T_0$ if there is a function $\nu_{T_0}[A]$, *linear* in $A \in \mathscr{B}(\mathsf{X})$, such that

$$\|A\|^{-1} |\mu[T_0 + A] - \mu[T_0] - \nu_{T_0}[A]| \to 0 \quad \text{for} \quad \|A\| \to 0 . \tag{5.25}$$

This definition does not depend on the particular norm used, for all norms are equivalent. $\nu_{T_0}[A]$ is the *total differential* of $\mu[T]$ at $T = T_0$. It is easily seen that $\mu[T]$ is totally differentiable if and only if it is totally differentiable as a function of the N^2 matrix elements of T.

In reality we are here not considering a single complex-valued function $\mu[T]$ but an unordered N-tuple $\mathfrak{S}[T]$ as a function of T. If $\mathfrak{S}[T]$ were an *ordered* N-tuple, the above definition could be extended immediately to $\mathfrak{S}[T]$. But as $\mathfrak{S}[T]$ is unordered, this is not an easy matter and we shall not pursue it in this much generality. We shall rather restrict ourselves to the case in which T_0 is not only diagonable but simple (has N distinct eigenvalues). Then the same is true of $T = T_0 + A$ for sufficiently small $\|A\|$ in virtue of the continuity of $\mathfrak{S}[T]$, and the eigenvalues of T can be expressed in a neighborhood of T_0 by N single-valued, continuous functions $\lambda_h[T]$, $h = 1, \ldots, N$. We shall now prove

Theorem 5.16. *The functions $\lambda_h[T]$ are not only totally differentiable but holomorphic in a neighborhood of $T = T_0$.*

Remark 5.17. A complex-valued function $\mu[T]$ of T is said to be holomorphic at $T = T_0$ if it can be expanded into an absolutely convergent *power series* (Taylor series) in $A = T - T_0$:

$$\mu[T_0 + A] = \mu[T_0] + \mu^{(1)}[T_0, A] + \mu^{(2)}[T_0, A] + \cdots \tag{5.26}$$

in which $\mu^{(n)}[T_0, A]$ is a form of degree n in A, that is,

$$\mu^{(n)}[T_0, A] = \mu^{(n)}[T_0; A, \ldots, A] \tag{5.27}$$

where $\mu^{(n)}[T_0; A_1, \ldots, A_n]$ is a symmetric n-linear form[1] in n operators $A_1, \ldots, A_n$. As is easily seen, $\mu[T]$ is holomorphic at $T = T_0$ if and only if $\mu[T_0 + A]$ can be expressed as a convergent power series in the N^2 matrix elements of A. In the same way holomorphic dependence of an operator-valued function $R[T] \in \mathscr{B}(X)$ on $T \in \mathscr{B}(X)$ can be defined.

Proof of Theorem 5.16. First we show that the one-dimensional eigenprojection $P_h[T]$ for the eigenvalue $\lambda_h[T]$ is holomorphic in T. We have, as in (1.17),

$$P_h[T_0 + A] = -\frac{1}{2\pi i} \int_{\Gamma_h} \sum_{n=0}^{\infty} R_0(\zeta) (-A R_0(\zeta))^n d\zeta, \tag{5.28}$$

where $R_0(\zeta) = (T_0 - \zeta)^{-1}$ and Γ_h is a small circle around $\lambda_h[T_0]$. The series in the integrand of (5.28) is uniformly convergent for $\zeta \in \Gamma_h$ for $\|A\| < \delta_h$, where δ_h is the minimum of $\|R_0(\zeta)\|^{-1}$ for $\zeta \in \Gamma_h$. Since the right member turns out to be a power series in A, we see that $P_h[T]$ is holomorphic at $T = T_0$.

Since $P_h[T]$ is one-dimensional, we have

$$\lambda_h[T_0 + A] = \operatorname{tr}\{(T_0 + A) P_h[T_0 + A]\}. \tag{5.29}$$

Substitution of the power series (5.28) shows that $\lambda_h[T_0 + A]$ is also a power series in A, as we wished to show.

[1] A function $f(A_1, \ldots, A_n)$ is *symmetric* if its value is unchanged under any permutation of $A_1, \ldots, A_n$. It is n-linear if it is linear in each variable A_k.

§ 6. Perturbation of symmetric operators

1. Analytic perturbation of symmetric operators

Many theorems of the preceding sections can be simplified or strengthened if X is a unitary space H. For the reason stated at the beginning of I-§ 6., we shall mainly be concerned with the perturbation of *symmetric* operators.

Suppose we are given an operator $T(\varkappa)$ of the form (1.2) in which $T, T^{(1)}, T^{(2)}, \ldots$ are all symmetric. Then the sum $T(\varkappa)$ is also symmetric for *real* $\varkappa$. Naturally it cannot be expected that $T(\varkappa)$ be symmetric for all $\varkappa$ of a domain of the complex plane.

More generally, let us assume that we are given an operator-valued function $T(\varkappa) \in \mathscr{B}(\mathsf{H})$ which is holomorphic in a domain D_0 of the $\varkappa$-plane intersecting with the real axis and which is symmetric for real $\varkappa$:

$$T(\varkappa)^* = T(\varkappa) \quad \text{for real } \varkappa\,. \tag{6.1}$$

For brevity the family $\{T(\varkappa)\}$ will then be said to be *symmetric*. Also we shall speak of a symmetric perturbation when we consider $T(\varkappa)$ as a perturbed operator. $T(\bar\varkappa)^*$ is holomorphic for $\varkappa \in \bar D_0$ (the mirror image of D_0 with respect to the real axis) and coincides with $T(\varkappa)$ for real $\varkappa$. Hence $T(\bar\varkappa)^* = T(\varkappa)$ for $\varkappa \in D_0 \cap \bar D_0$ by the unique continuation property of holomorphic functions. Thus

$$T(\varkappa)^* = T(\bar\varkappa) \tag{6.2}$$

as long as both $\varkappa$ and $\bar\varkappa$ belong to D_0. This can be used to continue $T(\varkappa)$ analytically to any $\varkappa$ for which one of $\varkappa$ and $\bar\varkappa$ belongs to D_0 but the other does not. Thus we may assume without loss of generality that D_0 is symmetric with respect to the real axis.

Since a symmetric operator is normal, the following theorem results directly from Theorem 1.10.

Theorem 6.1. *If the holomorphic family $T(\varkappa)$ is symmetric, the eigenvalues $\lambda_h(\varkappa)$ and the eigenprojections $P_h(\varkappa)$ are holomorphic on the real axis, whereas the eigennilpotents $D_h(\varkappa)$ vanish identically*[1].

Problem 6.2. If $T(\varkappa) = T + \varkappa T^{(1)}$ with T and $T^{(1)}$ symmetric, the smallest eigenvalue of $T(\varkappa)$ for real $\varkappa$ is a piecewise holomorphic, concave function of $\varkappa$. [hint: Apply I-(6.79)].

Remark 6.3. Theorem 6.1 cannot be extended to the case of two or more variables. The eigenvalues of a function $T(\varkappa_1, \varkappa_2)$ holomorphic in $\varkappa_1, \varkappa_2$ and symmetric for real $\varkappa_1, \varkappa_2$ need not be holomorphic for real $\varkappa_1, \varkappa_2$, as is seen from Example 5.12.

Remark 6.4. A theorem similar to Theorem 6.1 holds if $T(\varkappa)$ is normal for real $\varkappa$ or, more generally, for all $\varkappa$ on a curve in D_0. But such a

[1] See Remark 1.11, however.

theorem is of little practical use, since it is not easy to express the condition in terms of the coefficients $T^{(n)}$ of (1.2).

The calculation of the perturbation series given in § 2 is also simplified in the case of a symmetric perturbation. Since the unperturbed operator T is symmetric, any eigenvalue λ of T is semisimple ($D = 0$) and the reduction process described in § 2.3 is effective. The operator function $\tilde{T}^{(1)}(\varkappa)$ given by (2.37) is again symmetric, for $P(\varkappa)$ is symmetric and commutes with $T(\varkappa)$. *The reduction process preserves symmetry*. Therefore, the reduction can be continued indefinitely. The splitting of the eigenvalues must come to an end after finitely many steps, however, and the eigenvalues and the eigenprojections are finally given explicitly by the formulas corresponding to (2.5) and (2.3) in the final stage of the reduction process where the splitting has ceased to occur. In this way *the reduction process gives a complete recipe for calculating explicitly the eigenvalues and eigenprojections in the case of a symmetric perturbation.*

Remark 6.5. Again there is no general criterion for deciding whether there is no further splitting of the eigenvalue at a given stage. But the reducibility principle given in Remark 2.4 is useful, especially for symmetric perturbations. Since the unperturbed eigenvalue at each stage is automatically semisimple, there can be no further splitting if the unperturbed eigenprojection at that stage is irreducible with respect to a set $\{A\}$ of operators.

In applications such a set $\{A\}$ is often given as a *unitary group* under which $T(\varkappa)$ is invariant. Since the eigenprojection under consideration is an orthogonal projection, it is irreducible under $\{A\}$ if and only if there is no proper subspace of the eigenspace which is invariant under all the unitary operators A.

Remark 6.6. The general theory is simplified to some extent even if only the unperturbed operator T is symmetric or even normal. For example, all the eigenvalues $\lambda_h(\varkappa)$ are then continuously differentiable at $\varkappa = 0$ in virtue of the diagonability of T (Theorem 2.3). The estimates for the convergence radii and the error estimates are also simplified if the unperturbed operator T is symmetric or normal, as has been shown in § 3.5.

Remark 6.7. The estimate (3.52) is the best possible of its kind even in the special case of a symmetric perturbation, for Example 3.10 belongs to this case.

2. Orthonormal families of eigenvectors

Consider a holomorphic, symmetric family $T(\varkappa)$. For each real $\varkappa$ there exists an orthonormal basis $\{\varphi_n(\varkappa)\}$ of H consisting of eigenvectors of $T(\varkappa)$ [see I-(6.68)]. The question arises *whether these orthonormal eigenvectors $\varphi_n(\varkappa)$ can be chosen as holomorphic functions of $\varkappa$*. The answer is yes for real $\varkappa$.

Since the eigenvalues $\lambda_h(\varkappa)$ and the eigenprojections $P_h(\varkappa)$ are holomorphic on the real axis (Theorem 6.1), the method of § 4.5 can be applied to construct a holomorphic transformation function $U(\varkappa)$ satisfying (4.30). Furthermore, *$U(\varkappa)$ is unitary for real $\varkappa$*. To see this we recall that $U(\varkappa)$ was constructed as the solution of the differential equation $U' = Q(\varkappa)\, U$ with the initial condition $U(0) = 1$, where $Q(\varkappa)$ is given by (4.33). Since the $P_h(\varkappa)$ are symmetric, we have $P_h(\varkappa)^* = P_h(\bar{\varkappa})$ as in (6.2) and so the same is true of $P_h'(\varkappa)$. Hence $Q(\varkappa)$ is skew-symmetric: $Q(\bar{\varkappa})^* = -Q(\varkappa)$ and $U(\bar{\varkappa})^*$ satisfies the differential equation

$$\frac{d}{d\varkappa} U(\bar{\varkappa})^* = -U(\bar{\varkappa})^*\, Q(\varkappa)\,. \tag{6.3}$$

On the other hand, $V(\varkappa) = U(\varkappa)^{-1}$ satisfies the differential equation $V' = -V Q(\varkappa)$ with the initial condition $V(0) = 1$. In view of the uniqueness of the solution we must have

$$U(\bar{\varkappa})^* = U(\varkappa)^{-1}\,. \tag{6.4}$$

This shows that $U(\varkappa)$ is unitary for real $\varkappa$.

It follows that the basis $\varphi_{hk}(\varkappa) = U(\varkappa)\, \varphi_{hk}$ as given by (4.31) is orthonormal for real $\varkappa$ if the φ_{hk} form an orthonormal basis (which is possible since T is symmetric). It should be noted that the $\varphi_{hk}(\varkappa)$ are (not only generalized but proper) eigenvectors of $T(\varkappa)$ because $T(\varkappa)$ is diagonable. The existence of such an orthonormal basis depending smoothly on $\varkappa$ is one of the most remarkable results of the analytic perturbation theory for symmetric operators. That the analyticity is essential here will be seen below.

3. Continuity and differentiability

Let us now consider non-analytic perturbations of operators in H. Let $T(\varkappa) \in \mathscr{B}(\mathsf{H})$ depend continuously on the parameter $\varkappa$, which will now be assumed to be real. The eigenvalues of $T(\varkappa)$ then depend on $\varkappa$ continuously, and it is possible to construct N continuous functions $\mu_n(\varkappa)$, $n = 1, \ldots, N$, representing the repeated eigenvalues of $T(\varkappa)$ (see § 5.2). In this respect there is nothing new in the special case where $T(\varkappa)$ is symmetric, except that all $\mu_n(\varkappa)$ are real-valued and so a simple numbering such as (5.4) could also be used.

A new result is obtained for the differentiability of eigenvalues.

Theorem 6.8.[1] *Assume that $T(\varkappa)$ is symmetric and continuously differentiable in an interval* I *of $\varkappa$. Then there exist N continuously differentiable functions $\mu_n(\varkappa)$ on* I *that represent the repeated eigenvalues of $T(\varkappa)$.*

---

[1] This theorem is due to Rellich [8]. Recall that the result of this theorem is not necessarily true for a general (non-symmetric) perturbation (see Remark 5.8).

Proof. The proof of this theorem is rather complicated[1]. Consider a fixed value of $\varkappa$; we may set $\varkappa = 0$ without loss of generality. Let λ be one of the eigenvalues of $T = T(0)$, m its multiplicity, and P the associated eigenprojection. Since λ is semisimple, the derivatives at $\varkappa = 0$ of the repeated eigenvalues of $T(\varkappa)$ belonging to the λ-group are given by the m repeated eigenvalues of $PT'(0)P$ in the subspace $\mathsf{M} = P\mathsf{X}$ (Theorem 5.4). Let $\lambda'_1, \ldots, \lambda'_p$ be the *distinct* eigenvalues of $PT'(0)P$ in M and let $P_1, \ldots, P_p$ be the associated eigenprojections. The $\mathsf{M}_j = P_j\mathsf{H}$ are subspaces of M. It follows from the above remark that the λ-group (repeated) eigenvalues of $T(\varkappa)$ for small $|\varkappa| \neq 0$ are divided into p subgroups, namely the $\lambda + \varkappa\lambda'_j$-groups, $j = 1, \ldots, p$. Since each of these subgroups is separated from other eigenvalues, the total projections $P_j(\varkappa)$ for them are defined. $P_j(\varkappa)$ is at the same time the total projection for the λ'_j-group of the operator $\tilde{T}^{(1)}(\varkappa)$ given by (5.14). But $\tilde{T}^{(1)}(\varkappa)$ is continuous in a neighborhood of $\varkappa = 0$ as was shown there (the continuity for $\varkappa \neq 0$ is obvious). Hence $P_j(\varkappa)$ is continuous in a neighborhood of $\varkappa = 0$ by Theorem 5.1, and the same is true of

$$T_j(\varkappa) = P_j(\varkappa)\, T'(\varkappa)\, P_j(\varkappa) \tag{6.5}$$

because $T'(\varkappa) = dT(\varkappa)/d\varkappa$ is continuous by hypothesis.

The $\lambda + \varkappa\lambda'_j$-group of $T(\varkappa)$ consists in general of several distinct eigenvalues, the number of which may change discontinuously with $\varkappa$ in any neighborhood of $\varkappa = 0$. Let $\lambda(\varkappa_0)$ be one of them for $\varkappa = \varkappa_0 \neq 0$ and let $Q(\varkappa_0)$ be the associated eigenprojection. This $\lambda(\varkappa_0)$ may further split for small $|\varkappa - \varkappa_0| \neq 0$, but the derivative of any of the resulting eigenvalues must be an eigenvalue of $Q(\varkappa_0)\, T'(\varkappa_0)\, Q(\varkappa_0)$ in the subspace $Q(\varkappa_0)\mathsf{H}$ (again by Theorem 5.4). But we have $Q(\varkappa_0)\mathsf{H} \subset P_j(\varkappa_0)\mathsf{H}$ because $\lambda(\varkappa_0)$ belongs to the $\lambda + \varkappa\lambda'_j$-group, so that $Q(\varkappa_0)\, T'(\varkappa_0)\, Q(\varkappa_0) = Q(\varkappa_0)\, T_j(\varkappa_0)\, Q(\varkappa_0)$. Therefore, the derivatives under consideration are eigenvalues of the orthogonal projection (in the sense of I-§ 6.10) of the operator $T_j(\varkappa_0)$ on a certain subspace of $\mathsf{M}_j(\varkappa_0) = P_j(\varkappa_0)\mathsf{H}$. Since $T_j(\varkappa_0)$ is symmetric, it follows from Theorem I-6.46 that these eigenvalues lie between the largest and the smallest eigenvalues of $T_j(\varkappa_0)$ in the subspace $\mathsf{M}_j(\varkappa_0)$. But as $T_j(\varkappa)$ is continuous in $\varkappa$ as shown above, the eigenvalues of $T_j(\varkappa_0)$ in $\mathsf{M}_j(\varkappa_0)$ tend for $\varkappa_0 \to 0$ to the eigenvalues of $P_j T'(0) P_j$ in M_j, which are all equal to λ'_j. It follows that the derivatives of the $\lambda + \varkappa\lambda'_j$-group eigenvalues of $T(\varkappa)$ must also tend to λ'_j for $\varkappa \to 0$. This proves the required continuity of the derivatives of the eigenvalues $\mu_n(\varkappa)$ constructed by Theorem 5.6. [In the above proof, essential use has been made of the symmetry of $T(\varkappa)$ in the application of Theorem I-6.46. This explains why the same result does not necessarily hold in the general case].

[1] The original proof due to RELLICH is even longer.

Remark 6.9. As in the general case, the eigenprojections or eigenvectors have much less continuity than the eigenvalues even in the case of a symmetric perturbation, once the assumption of analyticity is removed. Example 5.3 is sufficient to illustrate this; here the function $T(\varkappa)$ is infinitely differentiable in $\varkappa$ and symmetric (by the usual inner product in $\mathbf{C}^2$), but it is impossible to find eigenvectors of $T(\varkappa)$ that are continuous at $\varkappa = 0$ and do not vanish there.

4. The eigenvalues as functions of the symmetric operator

As in § 5.8, we can regard the eigenvalues of a symmetric operator T as functions of T itself. As before, the eigenvalues are continuous functions of T in the sense explained there. The situation is however much simpler now because we can, if we desire, regard the repeated eigenvalues as forming an *ordered* N-tuple by arranging them in the ascending order

$$\mu_1[T] \leqq \mu_2[T] \leqq \cdots \leqq \mu_N[T]\,. \tag{6.6}$$

This defines N real-valued functions of T, T varying over all symmetric operators in H. The continuity of the eigenvalues is expressed by the continuity of these functions ($T' \to T$ implies $\mu_n[T'] \to \mu_n[T]$).

The numbering (6.6) of the eigenvalues is very simple but is not always convenient, for the $\mu_n[T]$ are not necessarily even partially differentiable. This is seen, for example, by considering the $\mu_n[T + \varkappa T']$ as functions of $\varkappa$, where T, T' are symmetric and $\varkappa$ is real. The eigenvalues of $T + \varkappa T'$ can be represented as holomorphic functions of $\varkappa$ (Theorem 6.1). The graphs of these functions may cross each other at some values of $\varkappa$ (exceptional points). If such a *crossing* takes place, the graph of $\mu_n[T + \varkappa T']$ jumps from one smooth curve to another, making a corner at the crossing point. In other words, the $\mu_n[T + \varkappa T']$ are continuous but not necessarily differentiable. In any case, they are *piecewise holomorphic*, since there are only a finite number of crossing points (exceptional points) in any finite interval of $\varkappa$.

Thus it is sometimes more convenient to return to the old point of view of regarding the repeated eigenvalues of T as elements of an unordered N-tuple $\mathfrak{S}[T]$. Then it follows from the result of the preceding paragraph that $\mathfrak{S}[T]$ *is partially continuously differentiable*. But $\mathfrak{S}[T]$ is totally differentiable only at those T with N distinct eigenvalues (again Example 5.12). In the neighborhood of such a T, however, the functions (6.6) are not only differentiable but holomorphic in T.

5. Applications. A theorem of Lidskii

Perturbation theory is primarily interested in small changes of the various quantities involved. Here we shall consider some problems related to the change of the eigenvalues when the operator is subjected to a *finite* change[1]. More specifically, we consider the problem of estimating the relation between the eigenvalues of two symmetric operators A, B in terms of their difference $C = B - A$.

Let us denote respectively by α_n, β_n, γ_n, $n = 1, \ldots, N$, the repeated eigenvalues of A, B, C in the ascending order as in (6.6). Let

$$T(\varkappa) = A + \varkappa C, \quad 0 \leqq \varkappa \leqq 1, \tag{6.7}$$

so that $T(0) = A$ and $T(1) = B$, and denote by $\mu_n(\varkappa)$ the repeated eigenvalues of $T(\varkappa)$ in the ascending order. As shown in the preceding paragraph, the $\mu_n(\varkappa)$ are continuous and piecewise holomorphic, with $\mu_n(0) = \alpha_n$, $\mu_n(1) = \beta_n$. In the interval $0 \leqq \varkappa \leqq 1$ there are only a finite number of exceptional points where the derivatives of the $\mu_n(\varkappa)$ may be discontinuous.

According to par. 2, we can choose for each $\varkappa$ a complete orthonormal family $\{\varphi_n(\varkappa)\}$ consisting of eigenvectors of $T(\varkappa)$:

$$(T(\varkappa) - \mu_n(\varkappa))\, \varphi_n(\varkappa) = 0, \quad n = 1, \ldots, N, \tag{6.8}$$

in such a way that the $\varphi_n(\varkappa)$ are piecewise holomorphic. In general they are discontinuous at the exceptional points; this is due to the rather unnatural numbering of the eigenvalues $\mu_n(\varkappa)$. If $\varkappa$ is not an exceptional point, differentiation of (6.8) gives

$$(C - \mu_n'(\varkappa))\, \varphi_n(\varkappa) + (T(\varkappa) - \mu_n(\varkappa))\, \varphi_n'(\varkappa) = 0. \tag{6.9}$$

Taking the inner product of (6.9) with $\varphi_n(\varkappa)$ and making use of the symmetry of $T(\varkappa)$, (6.8) and the normalization $\|\varphi_n(\varkappa)\| = 1$, we obtain

$$\mu_n'(\varkappa) = (C\, \varphi_n(\varkappa), \varphi_n(\varkappa)). \tag{6.10}$$

Since the $\mu_n(\varkappa)$ are continuous and the $\varphi_n(\varkappa)$ are piecewise continuous, integration of (6.10) yields

$$\beta_n - \alpha_n = \mu_n(1) - \mu_n(0) = \int_0^1 (C\, \varphi_n(\varkappa), \varphi_n(\varkappa))\, d\varkappa. \tag{6.11}$$

Let $\{x_j\}$ be an orthonormal basis consisting of the eigenvectors of C:

$$C x_n = \gamma_n x_n, \quad n = 1, \ldots, N. \tag{6.12}$$

We have

$$(C\, \varphi_n(\varkappa), \varphi_n(\varkappa)) = \sum_j (C\, \varphi_n(\varkappa), x_j)\, (x_j, \varphi_n(\varkappa)) = \sum_j \gamma_j\, |(\varphi_n(\varkappa), x_j)|^2$$

and (6.11) becomes

$$\beta_n - \alpha_n = \sum \sigma_{nj} \gamma_j, \tag{6.13}$$

$$\sigma_{nj} = \int_0^1 |(\varphi_n(\varkappa), x_j)|^2\, d\varkappa. \tag{6.14}$$

---

[1] For a more general study on finite changes of eigenvalues and eigenvectors, see DAVIS [1].

The orthonormality of $\{\varphi_n(\varkappa)\}$ and $\{x_j\}$ implies that

$$\sum_j \sigma_{nj} = 1\,, \quad \sum_n \sigma_{nj} = 1\,, \quad \sigma_{nj} \geqq 0\,. \tag{6.15}$$

Now it is well known in matrix theory that a square matrix (σ_{nj}) with the properties (6.15) lies in the convex hull of the set of all permutation matrices[1]. Thus (6.13) leads to the following theorem due to LIDSKII [1].

Theorem 6.10. *Let A, B, C, α_n, β_n, γ_n be as above. The N-dimensional numerical vector $(\beta_1 - \alpha_1, \ldots, \beta_N - \alpha_N)$ lies in the convex hull of the vectors obtained from $(\gamma_1, \ldots, \gamma_N)$ by all possible permutations of its elements.*

Another consequence of (6.13) is

Theorem 6.11. *For any convex function $\Phi(t)$ of a real variable t, the following inequality holds*:

$$\sum_n \Phi(\beta_n - \alpha_n) \leqq \sum_n \Phi(\gamma_n)\,. \tag{6.16}$$

The proof follows easily from (6.13), (6.15) and the convexity of Φ, for

$$\Phi(\beta_n - \alpha_n) = \Phi\Big(\sum_j \sigma_{nj}\, \gamma_j\Big) \leqq \sum_j \sigma_{nj}\, \Phi(\gamma_j)\,.$$

Example 6.12. Let $\Phi(t) = |t|^p$ with $p \geqq 1$. Then (6.16) gives[2]

$$\sum_n |\beta_n - \alpha_n|^p \leqq \sum_n |\gamma_n|^p\,, \quad p \geqq 1\,. \tag{6.17}$$

Chapter Three

Introduction to the theory of operators in Banach spaces

This chapter is again preliminary; we present an outline of those parts of operator theory in Banach spaces which are needed in the perturbation theory developed in later chapters. The material is quite elementary, but the presentation is fairly complete, reference being made occasionally to the first chapter, so that this chapter can be read without previous knowledge of Banach space theory. It is also intended to be useful as an introduction to operator theory. To keep the chapter within reasonable length, however, some basic theorems (for example,

---

[1] See BIRKHOFF [1]. A permutation matrix (σ_{jk}) is associated with a permutation $j \to \pi(j)$ of $\{1, 2, \ldots, n\}$ by the relation $\sigma_{jk} = 1$ if $k = \pi(j)$ and $\sigma_{jk} = 0$ otherwise.

[2] It was shown by HOFFMAN and WIELANDT [1] that (6.17) is true for $p = 2$ in a more general case in which A, B are only assumed to be normal, if the right member is replaced by $\mathrm{tr}\, C^* C$ and if a suitable numbering of $\{\alpha_n\}$ is chosen. Note that $C = B - A$ need not be normal for normal A, B.

the Baire category theorem and the Hahn-Banach extension theorem) are stated without proof[1].

Again emphasis is laid on the spectral theory of operators, where the resolvent theory is the central subject. The results related specifically to Hilbert spaces are not included, being reserved for Chapters V and VI for detailed treatment.

§ 1. Banach spaces

1. Normed spaces

From now on we shall be concerned mainly with an infinite-dimensional vector space X. Since there does not exist a finite basis in X, it is impossible to introduce the notion of convergence of a sequence of vectors of X in such a simple fashion as in a finite-dimensional space. For our purposes, it is convenient to consider a *normed (vector) space* from the outset.

A normed space is a vector space X in which a function $\| \ \|$ is defined and satisfies the conditions of a norm I-(1.18). In X the convergence of a sequence of vectors $\{u_n\}$ to a $u \in X$ can be defined by $\|u_n - u\| \to 0$. It is easily seen that the limit u, whenever it exists, is uniquely determined by $\{u_n\}$. As we have shown in I-§ 1.4, every finite-dimensional vector space can be made into a normed space.

In what follows we shall consider exclusively normed spaces X. Finite-dimensional spaces are not excluded, but we shall always assume that $\dim X > 0$.

Example 1.1. Let X be the set of all numerical vectors $u = (\xi_k)$ with a countably infinite number of complex components ξ_k, $k = 1, 2, \ldots$. X is an infinite-dimensional vector space with the customary definition of linear operations. Let m be the subset of X consisting of all $u = (\xi_k)$ such that the sequence $\{\xi_k\}$ is bounded. m is a linear manifold of X and it is itself a vector space. For each $u \in$ m define its norm by

$$\|u\| = \|u\|_{\mathrm{m}} = \|u\|_\infty = \sup_k |\xi_k| . \tag{1.1}$$

It is easy to see that this norm satisfies I-(1.18), so that m is a normed space. (1.1) is a generalization of I-(1.15). Let l be the set of all $u = (\xi_k) \in X$ such that

$$\|u\| = \|u\|_l = \|u\|_1 = \sum_k |\xi_k| \tag{1.2}$$

is finite. l is a normed space if the norm is defined by (1.2). More generally we can define the normed space l^p by introducing the norm[2]

---

[1] Standard textbooks on operator theory in Banach spaces are BANACH [1], DIEUDONNÉ [1], DUNFORD and SCHWARTZ [1], GOLDBERG [1], HILLE and PHILLIPS [1], LORCH [1], LYUSTERNIK and SOBOLEV [1], RIESZ and SZ.-NAGY [1], SOBOLEV [1], TAYLOR [1], YOSIDA [1], ZAANEN [1].

[2] The triangle inequality for the norm (1.3) is known as the Minkowski inequality. The proof, together with the proof of the Hölder inequalities, may be found in any textbook on real analysis; see e. g. ROYDEN [1] or HARDY, LITTLEWOOD and PÓLYA [1].

(1.3) $$\|u\| = \|u\|_{l^p} = \|u\|_p = \Big(\sum_k |\xi_k|^p\Big)^{1/p},$$

where p is a fixed number with $p \geqq 1$. m can be regarded as the limiting case of l^p for $p \to \infty$, and is also denoted by l^∞. l^p is a proper subset of l^q if $p < q$.

Example 1.2. The most important examples of normed spaces are *function spaces*. The simplest example of a function space is the set $C[a, b]$ of all complex-valued, continuous functions $u = u(x)$ on a finite closed interval $[a, b]$[1] of a real variable (see Example I-1.2). $C[a, b]$ is a normed space if the norm is defined by

(1.4) $$\|u\| = \|u\|_{C[a,b]} = \|u\|_\infty = \max_{a \leqq x \leqq b} |u(x)|.$$

More generally, the set $C(E)$ of all continuous functions $u = u(x) = u(x_1, \ldots, x_m)$ defined on a compact region E in the m-dimensional space R^m (or, more generally, on any compact topological space E) is a normed space if the norm is defined by

(1.5) $$\|u\| = \|u\|_{C(E)} = \|u\|_\infty = \max_{x \in E} |u(x)|.$$

Example 1.3. We could introduce other norms in $C[a, b]$, for example

(1.6) $$\|u\| = \|u\|_{L^p} = \|u\|_p = \Big(\int |u(x)|^p\, dx\Big)^{1/p}, \quad p \geqq 1.$$

This would make $C[a, b]$ into a different normed space. Actually a wider class of functions can be admitted when the norm (1.6) is used. We denote by $L^p(a, b)$ the set of all complex-valued, Lebesgue-measurable functions $u = u(x)$ on a finite or infinite interval (a, b) for which the integral (1.6) is finite. It can be shown that $L^p(a, b)$ is a normed space by the customary definition of linear operations and by the norm (1.6). It should be remarked that in $L^p(a, b)$, any two functions u, v are identified whenever they are ***equivalent*** in the sense that $u(x) = v(x)$ almost everywhere in (a, b); this convention makes the first condition of I-(1.18) satisfied[2].

More generally, for any measurable subset E of R^m the set $L^p(E)$ of all Lebesgue-measurable functions $u = u(x) = u(x_1, \ldots, x_m)$ on E such that (1.6) is finite is a normed space if the norm is defined by (1.6). Again two equivalent functions should be identified. In the limiting case $p \to \infty$, $L^p(E)$ becomes the space $L^\infty(E) = M(E)$ consisting of all ***essentially bounded*** functions on E with the norm

(1.7) $$\|u\| = \|u\|_{M(E)} = \|u\|_\infty = \operatorname*{ess\,sup}_{x \in E} |u(x)|.$$

In other words, $\|u\|_\infty$ is the smallest number M such that $|u(x)| \leqq M$ almost everywhere in E.

In the same way, we can define the space $L^p(E, d\mu)$ over any set E on which a measure $d\mu$ is defined[3], the norm being given by $\Big(\int_E |u(x)|^p\, d\mu(x)\Big)^{1/p}$.

Example 1.4. Let $C'[a, b]$ be the set of all continuously differentiable functions on a finite closed interval $[a, b]$. $C'[a, b]$ is a normed space if the norm is defined by

(1.8) $$\|u\| = \|u\|_\infty + \|u'\|_\infty,$$

where $\|u\|_\infty$ is given by (1.4) and $u' = du/dx$.

[1] We denote by $[a, b]$ a closed interval, by (a, b) an open interval, by $[a, b)$ a semi-closed interval, etc.

[2] In this sense L^p is a set of equivalence classes of functions rather than a set of functions. But it is customary to represent an element of L^p by a function, with the identification stated in the text.

[3] When we consider function spaces like L^p, we have to assume that the reader is familiar with the basic facts on real analysis, including Lebesgue integration, for which we may refer to standard textbooks (e. g., Royden [1]). In most cases, however, it is restricted to examples, and the main text will be readable without those prerequisites.

Remark 1.5. The following are known as the *Hölder inequalities*[1]. If $p \geqq 1$ and $q \geqq 1$ are related by $p^{-1} + q^{-1} = 1$ ($p = \infty$ or $q = \infty$ is permitted), then

$$\left|\sum_k \xi_k \eta_k\right| \leqq \|u\|_p \|v\|_q, \quad u = (\xi_k), \quad v = (\eta_k), \tag{1.9}$$

where $\|u\|_p$, $\|v\|_q$ are given by (1.3), and

$$\left|\int u(x)\, v(x)\, dx\right| \leqq \|u\|_p \|v\|_q \tag{1.10}$$

where $\|u\|_p$, $\|v\|_q$ are given by (1.6).

Problem 1.6. For $\| \;\|_p$ defined by (1.6) we have

$$\|u\, v\|_s \leqq \|u\|_p \|v\|_q \quad \text{for } s^{-1} = p^{-1} + q^{-1}, \tag{1.11}$$

$$\|u\, v\, w\|_s \leqq \|u\|_p \|v\|_q \|w\|_r \quad \text{for } s^{-1} = p^{-1} + q^{-1} + r^{-1}, \tag{1.12}$$

etc., where $u\, v = u(x)\, v(x)$, etc. Similar inequalities exist for $\| \;\|_p$ given by (1.3). [hint: Apply (1.10) to the function $|u\, v|^s$ etc.]

2. Banach spaces

In a normed space X the convergence $u_n \to u$ was defined by $\|u_n - u\| \to 0$. As in the finite-dimensional case, this implies the Cauchy condition $\|u_n - u_m\| \to 0$ [see I-(1.23)]. In the infinite-dimensional case, however, a *Cauchy sequence* $\{u_n\}$ (a sequence that satisfies the Cauchy condition) need not have a limit $u \in \mathsf{X}$. A normed space in which every Cauchy sequence has a limit is said to be *complete.* A complete normed space is called a *Banach space.* The notion of Banach space is very useful since, on the one hand, the completeness is indispensable for the further development of the theory of normed spaces and, on the other, most of normed spaces that appear in applications are complete[2]. Recall that a finite-dimensional normed space is complete (I-§ 1.4).

Example 1.7. In the space $\mathsf{C}(E)$ of continuous functions (Example 1.2), $u_n \to u$ means the *uniform convergence* of $u_n(x)$ to $u(x)$ on E. The Cauchy condition I-(1.23) means that $|u_n(x) - u_m(x)| \to 0$ uniformly. It is well known[3] that this implies the uniform convergence of $u_n(x)$ to a continuous function $u(x)$. Hence $\mathsf{C}(E)$ is complete.

Example 1.8. The spaces l^p, $1 \leqq p \leqq \infty$, (Example 1.1) are complete. The function spaces $\mathsf{L}^p(E)$ (Example 1.3) are complete. We shall not give the proof here[4].

Most of the topological notions introduced in I-§ 1.5 for a finite-dimensional normed space can be taken over to a Banach space. Here we shall mention only a few additions and modifications required.

---

[1] For the proof see e. g. Royden [1], p. 97.

[2] Besides, any normed space X can be *completed.* This means that X is *identified* with a linear manifold in a complete normed space $\tilde{\mathsf{X}}$. Furthermore, $\tilde{\mathsf{X}}$ can be so chosen that X is dense in $\tilde{\mathsf{X}}$. $\tilde{\mathsf{X}}$ is constructed as the set of all equivalence classes of Cauchy sequences $\{u_n\}$ in X; two Cauchy sequences $\{u_n\}$ and $\{v_n\}$ are equivalent by definition if $\lim (u_n - v_n) = 0$. For details see e. g. Yosida [1].

[3] See e. g. Knopp [1], p. 71.

[4] See any textbook on real analysis, e. g. Royden [1].

A linear manifold M of a Banach space X need not be closed. A closed linear manifold M of X is itself a Banach space. In this sense it is a *subspace* of X. The closure of a linear manifold is a closed linear manifold.

Lemma 1.9. *If* M *is a closed linear manifold, the linear manifold* M' *spanned by* M *and a finite number of vectors* $u_1, \ldots, u_m$ *is closed.*

Proof. It suffices to consider the case $m = 1$; the general case is obtained by successive application of this special case. If $u_1 \in M$, we have $M' = M$ so that M' is closed. If $u_1 \notin M$, we have $\operatorname{dist}(u_1, M) = d > 0$, because M is closed. M' is the set of all u' of the form $u' = \xi u_1 + v$, $v \in M$. We have

$$|\xi| \leqq \|u'\|/d . \tag{1.13}$$

In fact, $\|\xi^{-1} u'\| = \|u_1 + \xi^{-1} v\| \geqq d$ if $\xi \neq 0$ while (1.13) is trivial if $\xi = 0$.

Suppose now that $u'_n \in M'$, $u'_n \to u'$, $n \to \infty$; we have to show that $u' \in M'$. Let $u'_n = \xi_n u_1 + v_n$, $v_n \in M$. Application of (1.13) to $u' = u'_n - u'_m$ gives $|\xi_n - \xi_m| \leqq \|u'_n - u'_m\|/d \to 0$. Hence $\xi_n \to \xi$ for some ξ, and $v_n = u'_n - \xi_n u_1 \to u' - \xi u_1$. Since M is closed and $v_n \in M$, $v = u' - \xi u_1 \in M$. Hence $u' = \xi u_1 + v \in M'$ as required.

For any subset S of X, there is a *smallest closed linear manifold* containing S (that is, a closed linear manifold M such that any closed linear manifold $M' \supset S$ contains M). It is equal to the closure of the (linear) span of S, and is called the *closed linear manifold spanned by* S or simply the *closed span* of S.

Example 1.10. In l^p the set of all $u = (\xi_k)$ with $\xi_1 = 0$ is a closed linear manifold. In l^∞ the set c of all $u = (\xi_k)$ such that $\lim \xi_k = \xi$ exists is a closed linear manifold. The subset c_0 of c consisting of all u such that $\xi = 0$ is a closed linear manifold of l^∞ and of c. In $C[a, b]$ the set of all functions $u(x)$ such that $u(a) = 0$ is a closed linear manifold. The same is true of the set of u such that $u(a) = u(b) = 0$. $C[a, b]$ itself may be considered a closed linear manifold of $L^\infty(a, b)$.

A subset S of X is *(everywhere) dense* if the closure $\bar{S}$ coincides with X. In this case each $u \in X$ can be *approximated* by an element of S, in the sense that for any $\varepsilon > 0$ there is a $v \in S$ such that $\|u - v\| < \varepsilon$. More generally, for two subsets S_1, S_2 of X, S_1 is *dense with respect to* S_2 if $\bar{S}_1 \supset S_2$ (or dense *in* S_2 if $S_1 \subset S_2$ in addition). Then each $u \in S_2$ can be approximated by an element of S_1. If S_1 is dense with respect to S_2 and S_2 is dense with respect to S_3, then S_1 is dense with respect to S_3.

Example 1.11. In $C[a, b]$ (finite interval) the set of all polynomials is everywhere dense (theorem of WEIERSTRASS)[1]. The same is true for $L^p(a, b)$ for $1 \leqq p < \infty$. In $L^p(a, b)$ (not necessarily finite interval) the set $C_0^\infty(a, b)$ of all infinitely differentiable functions with compact support[2] is everywhere dense. If E is an open set of

[1] See e. g. ROYDEN [1], p. 150.

[2] A function has *compact support* if it vanishes outside a compact set. Thus $u \in C_0^\infty(a, b)$ if all the $d^n u/dx^n$ exist and $u(x) = 0$ except for $a < a' \leqq x \leqq b' < b$ (where a', b' depend on u).

R^m, the set of all infinitely differentiable functions with compact support in E is dense in $L^p(E)$, $1 \leqq p < \infty$[1].

Unlike a finite-dimensional normed space, an infinite-dimensional Banach space X is not locally compact (see I-§ 1.5). Thus X contains a bounded sequence $\{u_n\}$ that contains no convergent subsequence. $\{u_n\}$ is such a sequence if

$$\|u_n\| = 1\,, \quad \|u_n - u_m\| \geqq 1 \quad \text{for} \quad n \neq m\,. \tag{1.14}$$

$\{u_n\}$ can be constructed by induction. Suppose that $u_1, \ldots, u_n$ have been constructed. Let M_n be their span. Then there is a $u \in X$ such that $\|u\| = 1$ and $\operatorname{dist}(u, M_n) = 1$, as is seen from Lemma 1.12 proved below. It suffices to set $u_{n+1} = u$.

Lemma 1.12. *For any closed linear manifold* $M \neq X$ *of* X *and any* $\varepsilon > 0$, *there is a* $u \in X$ *such that* $\|u\| = 1$ *and* $\operatorname{dist}(u, M) > 1 - \varepsilon$. *We can even achieve* $\operatorname{dist}(u, M) = 1$ *if* $\dim M < \infty$.

Proof. There is a $u_0 \in X$ not belonging to M, so that $\operatorname{dist}(u_0, M) = d > 0$. Hence there is a $v_0 \in M$ such that $\|u_0 - v_0\| < d/(1 - \varepsilon)$. Set $u_1 = u_0 - v_0$. Then $\|u_1\| < d/(1-\varepsilon)$ and $\operatorname{dist}(u_1, M) = \inf_{v \in M} \|u_1 - v\| = \inf_{v \in M} \|u_0 - v\| = \operatorname{dist}(u_0, M) = d > (1 - \varepsilon)\|u_1\|$. The required u is obtained by normalizing u_1: $u = u_1/\|u_1\|$. If $\dim M < \infty$, let u_0 be as above and let X_0 be the linear manifold spanned by M and u_0. We can apply the result just obtained to the subspace M of X_0. Thus there exists a $u_n \in X_0$ such that $\|u_n\| = 1$ and $\operatorname{dist}(u_n, M) > 1 - n^{-1}$. Since $\dim X_0 < \infty$, X_0 is locally compact and there is a convergent subsequence of $\{u_n\}$. It is easy to see that its limit u satisfies $\|u\| = 1$ and $\operatorname{dist}(u, M) = 1$.

A subset $S \subset X$ is said to be *fundamental* if the closed span of S is X (in other words, if the span of S is everywhere dense). S is *separable* if S contains a countable subset which is dense in S. For the separability of X, it is sufficient that X contain a countable subset S' which is fundamental; for the set of all linear combinations of elements of S' with *rational* coefficients is a countable set dense in X. A subset of a separable set is separable[2].

Example 1.13. $C[a, b]$ is separable; this is a consequence of the theorem of WEIERSTRASS, for the set of monomials $u_n(x) = x^n$, $n = 0, 1, \ldots$, is fundamental in $C[a, b]$ (see Example 1.11). l^p is separable if $1 \leqq p < \infty$. The canonical basis consisting of $u_n = (\delta_{nk})$, $n = 1, 2, \ldots$, is fundamental in l^p. $L^p(a, b)$ is also separable if $1 \leqq p < \infty$. The set of functions $u_{(a', b')}(x)$, which is equal to 1 on (a', b') and zero otherwise, is fundamental when a' and b' vary over all rational numbers in (a, b)[3]. Similarly $L^p(R^m)$ is separable for $1 \leqq p < \infty$. It then follows that $L^p(E)$ is also separable if E is any measurable subset of R^m, for $L^p(E)$ may be regarded as the subspace of $L^p(R^m)$ consisting of all functions vanishing outside E.

---

[1] For the proof see e. g. SOBOLEV [1], p. 13.

[2] See DUNFORD and SCHWARTZ [1], p. 21.

[3] This follows from the properties of the Lebesgue integral; see ROYDEN [1].

An important consequence of the completeness of a Banach space is the *category theorem of* BAIRE[1]:

Theorem 1.14. *If* X *is the union of a countable number of closed subsets* S_n, $n = 1, 2, \ldots$, *at least one of the* S_n *contains interior points (that is, contains a ball).*

In spite of some essential differences in topological structure between a general Banach space X and a finite-dimensional space, most of the results regarding sequences $\{u_n\}$, infinite series Σu_n and vector-valued functions $u(t)$ stated in I-§ 1.6–7 remain valid in X. Such results will be used freely without comment in the sequel. It should be noted that the completeness of X is essential here. For example, the existence of the sum of an absolutely convergent series Σu_n depends on it, and similarly for the existence of the integral $\int u(t)\, dt$ of a continuous function $u(t)$. Also we note that the analyticity of a vector-valued function $u(t)$ is defined and the results of complex function theory are applicable to such a function (see loc. cit.). On the other hand, results of I-§ 1 based on the explicit use of a basis are in general not valid in the general case.

3. Linear forms

A linear form $f[u]$ on a Banach space X can be defined as in I-§ 2.1. But we here consider also linear forms $f[u]$ that are defined only on a certain linear manifold D of X. Such a form f will be called a linear form *in* X, and $\mathsf{D} = \mathsf{D}(f)$ will be called the *domain* of f. f is an *extension* of g (and g is a *restriction* of f) if $\mathsf{D}(f) \supset \mathsf{D}(g)$ and $f[u] = g[u]$ for $u \in \mathsf{D}(g)$; we write $f \supset g$ or $g \subset f$.

$f[u]$ is continuous at $u = u_0 \in \mathsf{D}$ if $\|u_n - u_0\| \to 0$, $u_n \in \mathsf{D}$, implies $f[u_n] \to f[u_0]$. Since $f[u_n] - f[u_0] = f[u_n - u_0]$, it follows that $f[u]$ *is continuous everywhere in* D *if and only if it is continuous at* $u = 0$. Such an f is simply said to be *continuous*.

In a finite-dimensional space every linear form is continuous. This is not true in the general case, though it is not easy to give an example of a discontinuous linear form defined everywhere on a Banach space X[2].

If a linear form f is continuous, there is a $\delta > 0$ such that $\|u\| < \delta$ implies $|f[u]| \leq 1$. By homogeneity, it follows that

$$|f[u]| \leq M\|u\| \quad \text{for every} \quad u \in \mathsf{D}(f) \tag{1.15}$$

where $M = 1/\delta$. A linear form f with the property (1.15) is said to be *bounded*. The smallest number M with this property is called the *bound* of f and is denoted by $\|f\|$. It is easy to see that, conversely, (1.15) implies that f is continuous. Thus *a linear form is continuous if and only if it is bounded.*

---

[1] See e. g. ROYDEN [1], p. 121, or any textbook on functional analysis.

[2] See footnote [3] of p. 133.

Lemma 1.15. *A bounded linear form f is determined if its values are given in a subset D' dense in $D(f)$.*

Proof. For each $u \in D(f)$, there is a sequence $u_n \in D'$ such that $u_n \to u$. Thus $f[u] = \lim f[u_n]$ by continuity.

Theorem 1.16 *(Extension principle). A bounded linear form with domain D can be extended to a bounded linear form with domain $\bar{D}$ (the closure of D). This extension is unique and the bound is preserved in the extension.*

Proof. The uniqueness of the extension follows from Lemma 1.15. To construct such an extension, let $u \in \bar{D}$ and let $u_n \in D$ be such that $u_n \to u$. The $f[u_n]$ form a Cauchy sequence since $|f[u_n] - f[u_m]| = |f[u_n - u_m]| \leq \|f\| \|u_n - u_m\| \to 0$, $n, m \to \infty$. Let the limit of $f[u_n]$ be denoted by $f'[u]$. $f'[u]$ is determined by u independently of the choice of the sequence $\{u_n\}$, for $u_n \to u$ and $v_n \to u$ imply $u_n - v_n \to 0$ and hence $f[u_n] - f[v_n] = f[u_n - v_n] \to 0$. It is now easy to show that $f'[u]$ is a linear form with domain $\bar{D}$. That $\|f'\| = \|f\|$ follows from the inequality $|f'[u]| \leq \|f\| \|u\|$, which is the limit of $|f[u_n]| \leq \|f\| \|u_n\|$.

Example 1.17. In l^p, $1 \leq p \leq \infty$, consider a linear form f given by

$$f[u] = \sum_k \alpha_k \xi_k \quad \text{for} \quad u = (\xi_k) . \tag{1.16}$$

If we take as $D(f)$ the set of all $u \in l^p$ such that only a finite number of the ξ_k are not zero, the coefficients α_k are arbitrary. But such an f is in general not bounded. f is bounded if

$$M = (\sum |\alpha_k|^q)^{1/q} < \infty \quad \text{where} \quad p^{-1} + q^{-1} = 1 . \tag{1.17}$$

Then we have $|f[u]| \leq M\|u\|$ by the Hölder inequality (1.9). In this case we can take $D(f) = l^p$; then it can easily be shown that $\|f\| = M$. It is further known that if $p < \infty$ any bounded linear form on l^p can be expressed in the form (1.16) with the α_k satisfying (1.17)[1].

Example 1.18. Let $X = C[a, b]$; set $f[u] = u(x_0)$ for every $u \in X$, with a fixed x_0, $a \leq x_0 \leq b$ [see I-(2.3)]. f is a bounded linear form with domain X. More generally, let $f(x)$ be a complex-valued function of bounded variation over $[a, b]$. Then the Stieltjes integral

$$f[u] = \int_a^b u(x)\, df(x) \tag{1.18}$$

defines a linear form on $C[a, b]$. f is bounded, for $|f[u]| \leq M \|u\|$ where M is the *total variation* of f. Hence $\|f\| \leq M$. Actually it is known that $\|f\| = M$, and that any bounded linear form on $C[a, b]$ can be expressed by a function f of bounded variation in the form (1.18)[2].

Example 1.19. For each $u \in C'[a, b]$ (see Example 1.4) set $f[u] = u'(x_0)$. f is a bounded linear form on $C'[a, b]$. f can also be regarded as a linear form in $X = C[a, b]$ with $D(f) = C'[a, b] \subset X$. In this interpretation f is not bounded, for $|u'(x_0)|$ may be arbitrarily large for $\|u\| = \max|u(x)| = 1$[3].

[1] See e. g. ROYDEN [1], p. 103, TAYLOR [1], p. 193, YOSIDA [1], p. 117.

[2] See e. g. TAYLOR [1], p.382.

[3] This f is a simple example of an unbounded linear form, but $D(f)$ is not the whole space $X = C[a, b]$. An unbounded linear form with domain X could be obtained by extending f, but it would require the use of the axiom of choice.

Example 1.20. For $u \in \mathsf{L}^p(\mathsf{E})$ and $f \in \mathsf{L}^q(\mathsf{E})$, $p^{-1} + q^{-1} = 1$, set

$$f[u] = \int_{\mathsf{E}} f(x)\, u(x)\, dx\,. \tag{1.19}$$

For a fixed f, $f[u]$ is a bounded linear form on $\mathsf{L}^p(\mathsf{E})$ with $\|f\| \leqq \|f\|_q$ by (1.10). It is known that $\|f\| = \|f\|_q$ and that any bounded linear form on $\mathsf{L}^p(\mathsf{E})$, $p < \infty$, can be expressed in the form (1.19) in terms of an $f \in \mathsf{L}^q(\mathsf{E})$[1].

4. The adjoint space

Semilinear forms and bounded semilinear forms in a Banach space X can be defined in the same way as above (cf. I-§ 2.2). The *adjoint space* X^* of X is defined as the set of all *bounded* semilinear forms on X, and X^* is a normed vector space if the norm of $f \in \mathsf{X}^*$ is defined as the bound $\|f\|$ of f. As before we introduce the scalar product $(f, u) = f[u]$ for every $f \in \mathsf{X}^*$ and $u \in \mathsf{X}$. The generalized Schwarz inequality I-(2.25) is again a consequence of the definition of $\|f\|$. Other notions and obvious results of the finite-dimensional case will be taken over without further comment.

X^* *is a Banach space.* To prove the completeness of X^*, consider a Cauchy sequence $\{f_n\}$ in X^*. Then

$$|(f_n - f_m, u)| \leqq \|f_n - f_m\|\, \|u\| \to 0\,, \quad n, m \to \infty\,, \tag{1.20}$$

for every $u \in \mathsf{X}$ so that $\lim (f_n, u) = f[u]$ exists. It is easily seen that $f[u]$ is a semilinear form in u. On letting $n \to \infty$ in $|(f_n, u)| \leqq \|f_n\|\, \|u\|$ we have $|f[u]| \leqq M \|u\|$, where $M = \lim \|f_n\| < \infty$ since the $\|f_n\|$ form a Cauchy sequence of positive numbers. Thus f is bounded with $\|f\| \leqq M$. Now (1.20) shows for $m \to \infty$ that $|(f_n - f, u)| \leqq \lim\limits_{m\to\infty} \|f_n - f_m\|\, \|u\|$ or $\|f_n - f\| \leqq \lim\limits_{m\to\infty} \|f_n - f_m\|$. Hence $\lim\limits_{n\to\infty} \|f_n - f\| \leqq \lim\limits_{n,m\to\infty} \|f_n - f_m\| = 0$. Thus $f_n \to f$ and X^* is complete.

All these considerations would be of little use if there were no bounded semilinear forms on X except the trivial form 0. For this reason the *Hahn-Banach theorem,* which assures the existence of "sufficiently many" bounded linear (or equivalently, semilinear) forms on X, is basic in Banach space theory. We shall state this theorem in a rather restricted form convenient for our use.

Theorem 1.21. *Any bounded linear form in a Banach space* X *(with domain* $\mathsf{D} \subset \mathsf{X}$*) can be extended to a bounded linear form on the whole of* X *without increasing the bound.*

We do not give the proof of this theorem[2], but we add the following comments. If this theorem has been proved for real Banach spaces, the complex case can be dealt with by using the method of I-§ 2.5. In the case of a real space, the theorem admits the following geometric interpretation. Let S be the open unit ball of X. The intersection S_0 of S

[1] See e. g. Royden [1], p. 103, Taylor [1], p. 382, Yosida [1], p. 115.
[2] See e. g. Royden [1], p. 162, or any book on functional analysis.

with D is the unit ball of D. If there is in D a support hyperplane M_0 to S_0 (see I-§ 2.5), then there is in X a support hyperplane M to S containing M_0 [1].

A consequence of Theorem 1.21 is

Theorem 1.22. *Let* M *be a closed linear manifold of* X *and let* $u_0 \in X$ *not belong to* M. *Then there is an* $f \in X^*$ *such that* $(f, u_0) = 1$, $(f, u) = 0$ *for* $u \in M$ *and* $\|f\| = 1/\text{dist}(u_0, M)$.

Proof. Let M' be the span of M and u_0. As in the proof of Lemma 1.9, each $u \in M'$ has the form $u = \xi u_0 + v$, $v \in M$. ξ is determined by u, so that we can define a function $f[u] = \bar{\xi}$ on M'. f is obviously semilinear and bounded by (1.13), with $\|f\| \leq 1/d$ where $d = \text{dist}(u_0, M)$. Actually we have $\|f\| = 1/d$, for there is a $u \in M'$ for which $\|u\| = 1$ and $\text{dist}(u, M) > 1 - \varepsilon$ (apply Lemma 1.12 with X replaced by M'); for this u we have $1 - \varepsilon < \text{dist}(u, M) = \text{dist}(\xi u_0 + v, M) = |\xi| \, \text{dist}(u_0, M) = |\xi| \, d$ or $|f[u]| > (1 - \varepsilon) \|u\|/d$. This f can now be extended to X preserving the bound. Denoting again by f the extended form, we see easily that the assertions of the theorem are satisfied.

Corollary 1.23. *For any two vectors* $u \neq v$ *of* X, *there is an* $f \in X^*$ *such that* $(f, u) \neq (f, v)$. *Thus* X^* *contains sufficiently many elements to distinguish between elements of* X.

Corollary 1.24. *For any* $u_0 \in X$, *there is an* $f \in X^*$ *such that* $(f, u_0) = \|u_0\|$, $\|f\| = 1$ [see I-(2.27)].

A consequence of Corollary 1.24 is that I-(2.26) is valid in a general Banach space:

$$\|u\| = \sup_{0 \neq f \in X^*} \frac{|(f, u)|}{\|f\|} = \sup_{\|f\| \leq 1} |(f, u)| = \sup_{\|f\| = 1} |(f, u)| . \tag{1.21}$$

Example 1.25. Let $u = (\xi_k) \in l^p$ and $f = (\alpha_k) \in l^q$ with $p^{-1} + q^{-1} = 1$, $1 \leq p < \infty$. For a fixed f, $f[u] = \sum \alpha_k \bar{\xi}_k$ is a bounded semilinear form on X, and any bounded semilinear form on X is expressed in this form by an $f \in l^q$ (Example 1.17). For this reason the adjoint space of l^p is *identified* with l^q, and we write $(f, u) = \sum \alpha_k \bar{\xi}_k$. Similarly, the adjoint space of $L^p(E)$, $1 \leq p < \infty$, is identified with $L^q(E)$ where $p^{-1} + q^{-1} = 1$. We write

$$(f, u) = \int_E f(x) \overline{u(x)} \, dx \,, \quad u \in L^p(E) \,, \quad f \in L^q(E) \,. \tag{1.22}$$

The adjoint space of $C[a, b]$ is the space $BV[a, b]$ of all functions $f(x)$ of bounded variation properly normalized, with the norm $\|f\|$ equal to the total variation of f. The scalar product is

$$(f, u) = \int_a^b \overline{u(x)} \, df(x) \,, \quad u \in C[a, b] \,, \quad f \in BV[a, b] \,. \tag{1.23}$$

Problem 1.26. Each finite-dimensional linear manifold M of X has a complementary subspace N: $X = M \oplus N$. [hint: It suffices to consider the case $\dim M = 1$. Let $0 \neq u \in M$ and let $f \in X^*$, $(f, u) = 1$. Let N be the set of all $v \in X$ such that $(f, v) = 0$.]

---

[1] The same is true if the open unit ball is replaced by any open convex subset of X.

The adjoint space X^{**} of X^* is again a Banach space. As in the finite-dimensional case, each $u \in X$ may be regarded as an element of X^{**} (see I-§ 2.6). In this sense we may again write $(u, f) = \overline{(f, u)}$ for $u \in X$ and $f \in X^*$. This does not imply, however, that X^{**} can be identified with X as in the finite-dimensional case, for there may be semilinear forms $F[f]$ on X^* that cannot be expressed in the form $\overline{(f, u)}$ with a $u \in X$. If there are no such forms F, X is said to be *reflexive* and X is identified with X^{**}. In general X is identified with a certain subspace of X^{**}.

The results of I-§ 2.6 on annihilators should be modified and supplemented accordingly. For any subset S of X, the annihilator $S^\perp$ is a closed linear manifold of X^* (since the scalar product is a continuous function of its arguments). The annihilator $S^{\perp\perp}$ of $S^\perp$ is a closed linear manifold of X^{**} but need not be a subset of X (under the identification stated above). In any case we have

$$S^{\perp\perp} \cap X = [S] \tag{1.24}$$

where $[S]$ is the closed span of S. Since $S^{\perp\perp} \supset S$ is clear, we have $S^{\perp\perp} \supset [S]$. To prove (1.24), it is therefore sufficient to show that any $u \in X$ not belonging to $[S]$ does not satisfy $(f, u) = 0$ for all $f \in S^\perp = [S]^\perp$. But this is implied by Theorem 1.21.

5. The principle of uniform boundedness

The following are among the basic theorems in Banach space theory. In what follows X denotes a Banach space.

Theorem 1.27. *Let $\{u_n\}$ be a sequence of vectors of X such that the numerical sequence $\{(f, u_n)\}$ is bounded for each fixed $f \in X^*$. Then $\{u_n\}$ is bounded: $\|u_n\| \leq M$.*

Theorem 1.28. *Let $\{f_n\}$ be a sequence of vectors of X^* such that the numerical sequence $\{(f_n, u)\}$ is bounded for each fixed $u \in X$. Then $\{f_n\}$ is bounded: $\|f_n\| \leq M$.*

These are special cases of the following theorem.

Theorem 1.29. *Let $\{p_\lambda[u]\}$ be a family of nonnegative continuous functions defined for all $u \in X$ such that*

$$p_\lambda[u' + u''] \leq p_\lambda[u'] + p_\lambda[u''] . \tag{1.25}$$

If $\{p_\lambda[u]\}$ is bounded for each fixed u, then it is uniformly bounded for $\|u\| \leq 1$.

Proof. Let S_n be the set of all $u \in X$ such that $p_\lambda[u] \leq n$ and $p_\lambda[-u] \leq n$ for all λ. The S_n are closed since the $p_\lambda[u]$ are continuous in u. The assumption implies that for each $u \in X$ there is an n such that $p_\lambda[u] \leq n$ and $p_\lambda[-u] \leq n$ for all λ. Hence $u \in S_n$, and X is the union of the S_n, $n = 1, 2, \ldots$. It follows from the category theorem (Theorem 1.14) that at least one S_n contains a ball K, say with center u_0 and radius r.

Any $u \in X$ with $\|u\| \leqq 2r$ can be written in the form $u = u' - u''$ with $u', u'' \in K$; it suffices to set $u' = u_0 + u/2$, $u'' = u_0 - u/2$. Hence $p_\lambda[u] \leqq p_\lambda[u'] + p_\lambda[-u''] \leqq n + n = 2n$ by (1.25). Thus $\{p_\lambda[u]\}$ is uniformly bounded for $\|u\| \leqq 2r$. If $2r \geqq 1$, this proves the assertion. If $2r < 1$, let m be an integer larger than $1/2r$. Then $\|u\| \leqq 1$ implies $\|u/m\| \leqq 2r$ so that $p_\lambda[u/m] \leqq 2n$. Repeated application of (1.25) then gives $p_\lambda[u] \leqq 2m\, n$ for $\|u\| \leqq 1$.

To deduce Theorem 1.28 from Theorem 1.29, it suffices to take $p_n[u] = |(f_n, u)|$. To deduce Theorem 1.27, we replace the X of Theorem 1.29 by X^* and set $p_n[f] = |(f, u_n)|$ [note (1.21)].

Problem 1.30. Let $f_n \in X^*$ be such that $\lim_{n\to\infty} (f_n, u) = f[u]$ exists for all $u \in X$. Then $f \in X^*$ and $\|f\| \leqq \liminf \|f_n\| < \infty$.

6. Weak convergence

A sequence $u_n \in X$ is said to *converge weakly* if (u_n, f) converges for every $f \in X^*$. If this limit is equal to (u, f) for some $u \in X$ for every f, then $\{u_n\}$ is said to *converge weakly to u* or have *weak limit u*. We denote this by the symbol $u_n \underset{w}{\to} u$ or $u = \text{w-lim}\, u_n$. It is easily seen that a sequence can have at most one weak limit[1]. To distinguish it from weak convergence, the convergence $u_n \to u$ defined earlier by $\|u_n - u\| \to 0$ is called *strong* convergence. To stress strong convergence, we sometimes write $u_n \underset{s}{\to} u$ or $u = \text{s-lim}\, u_n$. But in this book convergence will mean strong convergence unless otherwise stated.

It is obvious that strong convergence implies weak convergence. The converse is in general not true unless X is finite-dimensional. Furthermore, a weakly convergent sequence need not have a weak limit. If every weakly convergent sequence has a weak limit, X is said to be *weakly complete.*

A weakly convergent sequence is bounded. This is an immediate consequence of Theorem 1.27. Also we note that

$$(1.26) \qquad \|u\| \leqq \liminf \|u_n\| \quad \text{for} \quad u = \text{w-lim}\, u_n\,.$$

This follows from $(u, f) = \lim (u_n, f)$ where $f \in X^*$ is such that $\|f\| = 1$, $(u, f) = \|u\|$ (see Corollary 1.24).

Lemma 1.31. *Let $u_n \in X$ be a bounded sequence. In order that u_n converge weakly (to u), it suffices that (u_n, f) converge (to (u, f)) for all f of a fundamental subset S^* of X^*.*

[1] Weak convergence is related to the *weak topology* of X, as strong convergence is related to the norm topology. In this book we do not need the deeper results on weak topology; the use of the simple notion of weak convergence is sufficient for our purposes.

Proof. Let D^* be the span of S^*; D^* is dense in X^*. Obviously (u_n, f) converges [to (u, f)] for all $f \in \mathsf{D}^*$. Let $g \in \mathsf{X}^*$ and $\varepsilon > 0$. Since D^* is dense in X^*, there is an $f \in \mathsf{D}^*$ such that $\|g - f\| < \varepsilon$. Since (u_n, f) converges, there is an N such that $|(u_n - u_m, f)| < \varepsilon$ for $n, m > N$. Thus $|(u_n, g) - (u_m, g)| \leqq |(u_n, g - f)| + |(u_n - u_m, f)| + |(u_m, f - g)| \leqq (2M + 1)\,\varepsilon$ for $n, m > N$, where $M = \sup \|u_n\|$. This shows that (u_n, g) converges for all $g \in \mathsf{X}^*$. When $(u_n, f) \to (u, f)$ for $f \in \mathsf{D}^*$, we can apply the same argument with $u_n - u_m$ replaced by $u_n - u$ to conclude that $(u_n, g) \to (u, g)$ for all $g \in \mathsf{X}^*$.

The relationship between strong and weak convergence is given by

Theorem 1.32. *A sequence $u_n \in \mathsf{X}$ converges strongly if and only if $\{(u_n, f)\}$ converges uniformly for $\|f\| \leqq 1$, $f \in \mathsf{X}^*$.*

Proof. The "only if" part follows directly from $|(u_n, f) - (u_m, f)| \leqq$ $\leqq \|u_n - u_m\|\,\|f\| \leqq \|u_n - u_m\|$. To prove the "if" part, suppose that (u_n, f) converges uniformly for $\|f\| \leqq 1$. This implies that for any $\varepsilon > 0$, there exists an N such that $|(u_n - u_m, f)| \leqq \varepsilon$ if $n, m > N$ and $\|f\| \leqq 1$. Hence $\|u_n - u_m\| = \sup_{\|f\| \leqq 1} |(u_n - u_m, f)| \leqq \varepsilon$ for $n, m > N$ by (1.21).

Remark 1.33. In Theorem 1.32, it suffices to assume the uniform convergence of (u_n, f) for all f of a set dense in the unit ball of X^*, for $|(u_n - u_m, f)| \leqq \varepsilon$ for such f implies the same for all f of the unit ball.

Problem 1.34. Let M be a closed linear manifold of X. Then $u_n \in \mathsf{M}$ and $u_n \underset{w}{\to} u$ imply $u \in \mathsf{M}$. [hint: Theorem 1.22.]

Let us now consider vector-valued functions $u(t)$ of a real or complex variable t. We have already noticed that the notions of strong continuity, strong differentiability, and strong analyticity of such functions, and the integrals $\int u(t)\, dt$ of continuous functions $u(t)$, can be defined as in the finite-dimensional case (see par. 2).

$u(t)$ is *weakly continuous* in t if $(u(t), f)$ is continuous for each $f \in \mathsf{X}^*$. Obviously strong continuity implies weak continuity. $u(t)$ is *weakly differentiable* if $(u(t), f)$ is differentiable for each $f \in \mathsf{X}^*$. If the derivative of $(u(t), f)$ has the form $(v(t), f)$ for each f, $v(t)$ is the *weak derivative* of $u(t)$.

If $u(t)$ is weakly continuous at $t = t_0$, $\|u(t)\|$ is bounded near $t = t_0$; this follows from Theorem 1.27. If $u(t)$ is weakly continuous on a compact set of t, then $\|u(t)\|$ is bounded there.

Lemma 1.35. *If $u(t)$ is weakly differentiable for $a < t < b$ with weak derivative identically zero, then $u(t)$ is constant.*

Proof. The assumption implies that $(u(t), f)$ has derivative zero, so that $(u(t'), f) = (u(t''), f)$ for any t', t'' and $f \in \mathsf{X}^*$. Hence $u(t') = u(t'')$ by Corollary 1.23.

Lemma 1.35 implies that $u(t)$ is constant if $u(t)$ is strongly differentiable with $du(t)/dt = 0$. It is desirable, however, to have a direct proof of this fact, for the proof of Lemma 1.35 is not altogether elementary since it is based on the Hahn-Banach theorem. We shall give a direct proof in a slightly generalized case.

Lemma 1.36. *If $u(t)$ is strongly continuous for $a \leqq t < b$ and has the strong right derivative $D^+ u(t) = 0$, then $u(t)$ is constant.*

Proof. $D^+ u(t)$ is defined as the strong limit of $h^{-1}[u(t+h) - u(t)]$ for $h \searrow 0$[1]. We may assume without loss of generality that $a = 0$ and $u(0) = 0$. We shall prove that $\|u(t)\| \leqq \varepsilon t$ for any $\varepsilon > 0$; then $u(t) = 0$ follows on letting $\varepsilon \to 0$. Suppose ε is given. Since $D^+ u(0) = 0$, we have $\|u(t)\| \leqq \varepsilon t$ for sufficiently small t. Let $[0, c)$ be the maximal subinterval of $[0, b)$ in which $\|u(t)\| \leqq \varepsilon t$ is true; we shall show that $c = b$. If $c < b$, we have $\|u(c)\| \leqq \varepsilon c$ by continuity. Then it follows from $D^+ u(c) = 0$ that $\|u(c+h)\| = \|u(c)\| + o(h) \leqq \varepsilon c + o(h) \leqq \varepsilon(c+h)$ for sufficiently small $h > 0$. But this contradicts the definition of c.

We can also define a weakly holomorphic function $u(t)$ of a complex variable t. But such a function is necessarily strongly holomorphic. Namely, we have

Theorem 1.37. *Let $u(\zeta) \in X$ be defined in a domain Δ of the complex plane and let $(u(\zeta), f)$ be holomorphic in Δ for each $f \in X^*$. Then $u(\zeta)$ is holomorphic in the strong sense (strongly differentiable in ζ).*

Proof. Let Γ be a positively-oriented circle in Δ. We have the Cauchy integral formula for the holomorphic function $(u(\zeta), f)$:

$$(u(\zeta), f) = \frac{1}{2\pi i} \int_\Gamma \frac{(u(\zeta'), f)}{\zeta' - \zeta} d\zeta'$$

for ζ inside Γ. It follows that

$$\begin{aligned}(1.27) \qquad & \frac{1}{\eta}(u(\zeta+\eta) - u(\zeta), f) - \frac{d}{d\zeta}(u(\zeta), f) \\ & = \frac{\eta}{2\pi i} \int_\Gamma \frac{(u(\zeta'), f)}{(\zeta' - \zeta - \eta)(\zeta' - \zeta)^2} d\zeta' .\end{aligned}$$

But $u(\zeta)$ is bounded on Γ because it is weakly continuous, so that $|(u(\zeta'), f)| \leqq M\|f\|$ for $\zeta' \in \Gamma$. Hence (1.27) is majorized by a number of the form $|\eta| M' \|f\|$ for small $|\eta|$. This shows that $\eta^{-1}(u(\zeta+\eta) - u(\zeta), f)$ converges as $\eta \to 0$ to $d(u(\zeta), f)/d\zeta$ uniformly for $\|f\| \leqq 1$. It follows from Theorem 1.32 that $\eta^{-1}(u(\zeta+\eta) - u(\eta))$ converges strongly. This proves that $u(\zeta)$ is strongly differentiable and hence holomorphic.

Remark 1.38. If $\|u(\zeta)\|$ is assumed to be locally bounded, it suffices in Theorem 1.37 to assume that $(u(\zeta), f)$ is holomorphic for all f of a fundamental subset of X^* (note Remark 1.33).

---

[1] $h \searrow 0$ means $h > 0$ and $h \to 0$.

7. Weak* convergence

In the adjoint space X^* there is another notion of convergence called weak* convergence[1]. A sequence $f_n \in X^*$ is said to converge to f weak* if $(u, f_n) \to (u, f)$ for each $u \in X$. Weak* convergence is weaker than weak convergence considered in the Banach space X^*, for the latter requires the convergence of (F, f_n) for all $F \in X^{**}$. We use the notations $f_n \underset{w^*}{\to} f$ or $\text{w}^*\text{-}\lim f_n = f$ for weak* convergence.

A weak convergent sequence is bounded.* This follows immediately from Theorem 1.28.

It should be noted that if (u, f_n) converges for each $u \in X$, then $\text{w}^*\text{-}\lim f_n = f \in X^*$ exists (see Problem 1.30). In this sense X^* is weak* complete.

The following results can be proved in the same way as for weak convergence.

$\{f_n\}$ converges strongly if and only if $\{(u, f_n)\}$ converges uniformly for $\|u\| \leqq 1$, $u \in X$.

If $\{f_n\}$ is bounded, it suffices for the weak* convergence of $\{f_n\}$ that $\{(u, f_n)\}$ converge for all u of a fundamental subset of X.

If $f(\zeta) \in X^*$ is holomorphic in a domain Δ of the complex plane in the weak* sense [that is, if $(f(\zeta), u)$ is holomorphic in Δ for each $u \in X$], then $f(\zeta)$ is holomorphic in the strong sense.

Problem 1.39. Let $u_n, u \in X$, $f_n, f \in X^*$. Then $(u_n, f_n) \to (u, f)$ if (i) $u_n \underset{s}{\to} u$ and $f_n \underset{w^*}{\to} f$ or (ii) $u_n \underset{w}{\to} u$ and $f_n \underset{s}{\to} f$.

8. The quotient space

If M is a linear manifold of a vector space X, the quotient space $\tilde{X} = X/M$ is defined as the set of all cosets $\tilde{u} = u + M$ modulo M (or all inhomogeneous linear manifolds parallel to M) with the linear operations defined by I-(1.8). If X is a normed space and M is closed, $\tilde{X}$ becomes a normed space by the introduction of the norm

$$\|\tilde{u}\| = \inf_{v \in \tilde{u}} \|v\| = \inf_{z \in M} \|u - z\| = \text{dist}(u, M) . \tag{1.28}$$

It is easy to verify that (1.28) satisfies the conditions I-(1.18) of a norm. Recall that $\tilde{u} = \tilde{u}'$ if and only if $u - u' \in M$.

$\tilde{X}$ is a Banach space if X is a Banach space. To prove this, let $\{\tilde{u}_n\}$ be a Cauchy sequence in $\tilde{X}$. Let $n(k)$ be an integer such that $\|\tilde{u}_n - \tilde{u}_m\| \leqq 2^{-k}$ for $n, m \geqq n(k)$. We may further assume that $n(1) \leqq n(2) \leqq \cdots$. Set

$$\tilde{v}_k = \tilde{u}_{n(k+1)} - \tilde{u}_{n(k)}, \; k = 1, 2, \ldots .$$

Then $\|\tilde{v}_k\| \leqq 2^{-k}$, and we can choose a $v_k \in \tilde{v}_k$ for each k in such a way that $\|v_k\| \leqq \|\tilde{v}_k\| + 2^{-k} \leqq 2^{1-k}$. Set $u = u_{n(1)} + \sum_{k=1}^{\infty} v_k$; this series

[1] Again we shall not consider the deeper notion of weak* topology.

converges absolutely and defines a vector of X since X is complete. Denoting by w_k the partial sums of this series, we have $\tilde{w}_k = \tilde{u}_{n(k+1)}$. Since $\|\tilde{w}_k - \tilde{u}\| \leqq \|w_k - u\| \to 0$ as $k \to \infty$, we have $\|\tilde{u}_{n(k)} - \tilde{u}\| \to 0$. Choose k so large that $\|\tilde{u}_{n(k)} - \tilde{u}\| < \varepsilon$ as well as $2^{-k} < \varepsilon$; then $\|\tilde{u}_n - \tilde{u}\| \leqq \leqq \|\tilde{u}_n - \tilde{u}_{n(k)}\| + \|\tilde{u}_{n(k)} - \tilde{u}\| < 2\varepsilon$ for $n \geqq n(k)$. This shows that $\{\tilde{u}_n\}$ has the limit $\tilde{u} \in \tilde{X}$ and completes the proof of the completeness of $\tilde{X}$.

The *codimension* or *deficiency* of a linear manifold M of X is defined by $\operatorname{codim} M = \dim X/M$ as before (I-§ 1.3).

Lemma 1.40. *If* M *is closed, then* $\operatorname{codim} M = \dim M^\perp$ *and* $\operatorname{codim} M^\perp = \dim M$.

Proof. Suppose that $\operatorname{codim} M = m < \infty$. Then there is a finite basis $\{\tilde{x}_j\}$, $j = 1, \ldots, m$, of $\tilde{X} = X/M$. Let $x_j \in \tilde{x}_j$. For any $u \in X$, $\tilde{u}$ can be expressed uniquely in the form $\tilde{u} = \xi_1 \tilde{x}_1 + \cdots + \xi_m \tilde{x}_m$. Hence u admits a unique expression

$$u = \xi_1 x_1 + \cdots + \xi_m x_m + v\,, \quad v \in M\,. \tag{1.29}$$

Let M_j be the span of M and $x_1, \ldots, x_{j-1}, x_{j+1}, \ldots, x_m$. M_j is closed by Lemma 1.9, so that there exists an $f_j \in X^*$ such that $f_j \in M_j^\perp$ and $(f_j, x_j) = 1$ (Theorem 1.22). In other words $f_j \in M^\perp$ and $(f_j, x_k) = \delta_{jk}$. It is easily seen that the f_j are linearly independent.

Let $f \in M^\perp$ and $\alpha_j = (f, x_j)$. Then $f - \alpha_1 f_1 - \cdots - \alpha_m f_m$ has scalar product 0 with all the x_k and $v \in M$, hence with all $u \in X$ by (1.29). Hence it is 0 and $f = \alpha_1 f_1 + \cdots + \alpha_m f_m$. Thus $M^\perp$ is spanned by $f_1, \ldots, f_m$: $\dim M^\perp = m$.

If $\operatorname{codim} M = \infty$, then there exists an infinite sequence M_n of closed linear manifolds such that $M \subset M_1 \subset M_2 \subset \cdots$, all inclusions being proper. Thus $M^\perp \supset M_1^\perp \supset M_2^\perp \supset \cdots$ with all inclusions proper [cf. (1.24)]. Thus $\dim M^\perp = \infty$.

If $\dim M = m < \infty$, let $\{x_1, \ldots, x_m\}$ be a basis of M. As above we can construct $f_j \in X^*$, $j = 1, \ldots, m$, with $(f_j, x_k) = \delta_{jk}$. Each $f \in X^*$ can be expressed in the form $f = \sum_{k=1}^{m} (f, x_k) f_k + f'$ with $(f', x_j) = 0, j = 1, \ldots, m$, that is, $f' \in M^\perp$. Hence $\tilde{f}$ is a linear combination of the $\tilde{f}_k$, where $\tilde{f}$ and the $\tilde{f}_k$ are elements of $X^*/M^\perp$. Since the $\tilde{f}_k$ are linearly independent, this proves that $\operatorname{codim} M^\perp = m$.

If $\dim M = \infty$, there is an infinite sequence M_n of finite-dimensional linear manifolds such that $M_1 \subset M_2 \subset \cdots \subset M$, all inclusions being proper. Thus $M_1^\perp \supset M_2^\perp \supset \cdots \supset M^\perp$ with all inclusions proper. This proves that $\operatorname{codim} M^\perp = \infty$.

Corollary 1.41. *If* M *is a finite-dimensional linear manifold, then* $M^{\perp\perp} = M$.

Problem 1.42. If $\operatorname{codim} M < \dim N$ (which implies that $\operatorname{codim} M = m < \infty$), then $\dim (M \cap N) > 0$. [hint: $u \in M$ is expressed by the m conditions $(u, f_j) = 0$, where the f_j are as in the first part of the proof of Lemma 1.40.]

§ 2. Linear operators in Banach spaces

1. Linear operators. The domain and range

In what follows $\mathsf{X}, \mathsf{Y}, \mathsf{Z}, \ldots$ will denote Banach spaces unless otherwise stated. The definition of a *linear operator* (or simply an *operator*) is similar to that given in the finite-dimensional case (I-§ 3.1). But there are some generalizations of definition that we have to make for later applications.

We find it necessary to consider operators T not necessarily defined for all vectors of the domain space. Thus we define an *operator T from X to Y* as a function which sends every vector u in a certain linear manifold D of X to a vector $v = Tu \in \mathsf{Y}$ and which satisfies the linearity condition I-(3.1) for all $u_1, u_2 \in \mathsf{D}$. D is called the *domain of definition*, or simply the *domain*, of T and is denoted by $\mathsf{D}(T)$. The *range* $\mathsf{R}(T)$ of T is defined as the set of all vectors of the form Tu with $u \in \mathsf{D}(T)$. X and Y are respectively called the *domain* and *range spaces*[1]. If $\mathsf{D}(T)$ is dense in X, T is said to be *densely defined*. If $\mathsf{D}(T) = \mathsf{X}$, T is said to be defined *on* X. If $\mathsf{Y} = \mathsf{X}$, we shall say that T is an operator *in* X. The *null space* $\mathsf{N}(T)$ of T is the set of all $u \in \mathsf{D}(T)$ such that $Tu = 0$.

The consideration of operators not necessarily defined on the whole domain space leads to the notion of *extension* and *restriction* of operators, as in the case of linear forms. If S and T are two operators from X to Y such that $\mathsf{D}(S) \subset \mathsf{D}(T)$ and $Su = Tu$ for all $u \in \mathsf{D}(S)$, T is called an extension of S and S a restriction of T; in symbol

$$T \supset S, \quad S \subset T . \tag{2.1}$$

T is called a *finite extension* of S and S a *finite restriction* of T if $T \supset S$ and $[T/S] \equiv \dim \mathsf{D}(T)/\mathsf{D}(S) = m < \infty$. m is the *order* of the extension or restriction. If $m = 1$, we shall say that T is a *direct extension* of S and S is a *direct restriction* of T.

For any subset S of the domain space X of T, we denote by $T\mathsf{S}$ the *image* of $\mathsf{S} \cap \mathsf{D}(T)$, that is, the set of all $v = Tu$ with $u \in \mathsf{S} \cap \mathsf{D}(T)$; $T\mathsf{S}$ is a subset of the range space Y. For any subset S' of Y, the *inverse image* $T^{-1}\mathsf{S}'$ is the set of all $u \in \mathsf{D}(T)$ such that $Tu \in \mathsf{S}'$[2].

[1] It might appear that one need not complicate the matter by introducing operators not defined everywhere in the domain space, for T could be regarded as an operator on $\mathsf{D}(T)$ to Y or to $\mathsf{R}(T)$. We do not take this point of view, however, for $\mathsf{D}(T)$ is in general not closed in X and hence is not a Banach space (with the norm of X). In particular when $\mathsf{X} = \mathsf{Y}$, it is often convenient and even necessary to regard T as an operator in X rather than as an operator between different spaces $\mathsf{D}(T)$ and X.

[2] The inverse image $T^{-1}\mathsf{S}'$ is defined even if the inverse T^{-1} (see below) does not exist. When T^{-1} exists, $T^{-1}\mathsf{S}'$ coincides with the image of S' under T^{-1}.

The *inverse* T^{-1} of an operator T from X to Y is defined if and only if the map T is one to one, which is the case if and only if $Tu = 0$ implies $u = 0$. T^{-1} is by definition the operator from Y to X that sends Tu into u. Thus

$$\text{(2.2)} \qquad \mathsf{D}(T^{-1}) = \mathsf{R}(T)\,, \quad \mathsf{R}(T^{-1}) = \mathsf{D}(T)\,.$$

$$\text{(2.3)} \qquad T^{-1}(Tu) = u\,, \quad u \in \mathsf{D}(T)\,; \quad T(T^{-1}v) = v\,, \quad v \in \mathsf{R}(T)\,.$$

T is said to be *invertible* if T^{-1} exists. Any restriction of an invertible operator is invertible.

Example 2.1. A linear form in X is an operator from X to C (the one-dimensional space of complex numbers).

Example 2.2. When the domain and range spaces X, Y are function spaces such as C(E), L^p(E), an operator T defined by multiplication by a fixed function is called a *multiplication operator*. For example, let X = L^p(E), Y = L^q(E) and define T by $Tu(x) = f(x)\,u(x)$, where $f(x)$ is a fixed complex-valued measurable function defined on E. D(T) must be such that $u \in$ D(T) implies $fu \in L^q$. If D(T) is maximal with this property [that is, D(T) consists of all $u \in L^p$(E) such that $fu \in L^q$(E)], T is called the *maximal* multiplication operator by $f(x)$. T is invertible if and only if $f(x) \neq 0$ almost everywhere in E. If in particular $p = q$, the maximal multiplication operator T is defined on the whole of L^p(E) if and only if $f(x)$ is essentially bounded on E.

Example 2.3. An infinite matrix $(\tau_{jk})_{j,k=1,2,\ldots}$, can be used to define an operator T from X to Y, where X, Y may be any of the sequence spaces c and l^p, $1 \leq p \leq \infty$. Formally T is given by $Tu = v$, where $u = (\xi_j)$ and $v = (\eta_j)$ are related by

$$\text{(2.4)} \qquad \eta_j = \sum_{k=1}^{\infty} \tau_{jk}\,\xi_k\,, \quad j = 1, 2, \ldots\,.$$

D(T) ⊂ X must be such that, for any $u \in$ D(T), the series (2.4) is convergent for every j and the resulting vector v belongs to Y. T is called the *maximal* operator defined by (τ_{jk}) and X, Y if D(T) consists exactly of all such u. How large D(T) is depends on the property of the given matrix.

Assume for instance that there are finite constants M', M'' such that

$$\text{(2.5)} \qquad \tau'_j = \sum_{k=1}^{\infty} |\tau_{jk}| \leq M'\,,\ j = 1, 2, \ldots;\ \tau''_k = \sum_{j=1}^{\infty} |\tau_{jk}| \leq M''\,,\ k = 1, 2, \ldots\,.$$

If we choose X = Y = l^p, $1 \leq p \leq \infty$, the same calculation as in I-(4.15) shows that Tu exists for any $u \in$ X and that

$$\text{(2.6)} \qquad \|Tu\| = \|v\| \leq M'^{1-\frac{1}{p}}\, M''^{\frac{1}{p}}\, \|u\| \leq \max(M', M'')\, \|u\|\,.$$

Thus the maximal operator T is defined on the whole of l^p. In particular this is the case if (τ_{jk}) is a diagonal matrix with bounded diagonal elements.

Example 2.4. An operator of the form[1]

$$\text{(2.7)} \qquad Tu(y) = v(y) = \int_E t(y, x)\, u(x)\, dx\,, \quad y \in F\,,$$

---

[1] For simplicity we write $Tu(y)$ in place of $(Tu)(y)$. There is no possibility of confusion, for $T[u(y)]$ does not make sense [$u(y)$ is a complex number, T is an operator].

is called an *integral operator* with the *kernel* $t(y, x)$. $t(y, x)$ is a complex-valued measurable function defined for $x \in \mathsf{E}$, $y \in \mathsf{F}$ (E, F being, for example, subsets of euclidean spaces, not necessarily of the same dimension). If X and Y are function spaces defined over E and F respectively, say $\mathsf{X} = \mathsf{L}^p(\mathsf{E})$ and $\mathsf{Y} = \mathsf{L}^q(\mathsf{F})$, then (2.7) defines an operator T from X to Y by an appropriate specification of $\mathsf{D}(T)$. $\mathsf{D}(T)$ must be such that for $u \in \mathsf{D}(T)$, the integral on the right of (2.7) exists (in a suitable sense) for almost every y and the resulting $v(y)$ belongs to y. If $\mathsf{D}(T)$ consists exactly of all such u, T is called the *maximal* integral operator defined by the given kernel and X, Y.

For example, assume that there exist finite constants M', M'' such that

$$\int_{\mathsf{E}} |t(y, x)|\, dx \leq M', \quad y \in \mathsf{F}; \quad \int_{\mathsf{F}} |t(y, x)|\, dy \leq M'', \quad x \in \mathsf{E}. \tag{2.8}$$

If we choose $\mathsf{X} = \mathsf{L}^p(\mathsf{E})$ and $\mathsf{Y} = \mathsf{L}^p(\mathsf{F})$, the maximal operator T is defined on the whole of X, and we have the inequality

$$\|Tu\| = \|v\| \leq M'^{1-\frac{1}{p}} M''^{\frac{1}{p}} \|u\| \leq \max(M', M'')\, \|u\|. \tag{2.9}$$

This can be proved as in I-(4.15) by using the inequalities for integrals instead of those for series. One has to invoke Fubini's theorem[1], however, to prove rigorously the existence of the integral of (2.7) for almost all y.

If, in particular, E, F are compact regions and $t(y, x)$ is continuous in x, y, (2.8) is obviously satisfied. In this case T can also be regarded as an operator from $\mathsf{X} = \mathsf{C}(\mathsf{E})$ or any $\mathsf{L}^p(\mathsf{E})$ to $\mathsf{Y} = \mathsf{C}(\mathsf{F})$ defined on X, for $Tu(y)$ is continuous for all integrable $u(x)$.

Problem 2.5. Let $\int_{\mathsf{F}} \int_{\mathsf{E}} |t(y, x)|^2\, dx\, dy = M^2 < \infty$. Then

$$Tu(y) = v(y) = \int_{\mathsf{E}} t(y, x)\, u(x)\, dx$$

defines an operator T from $\mathsf{X} = \mathsf{L}^2(\mathsf{E})$ to $\mathsf{Y} = \mathsf{L}^2(\mathsf{F})$ with domain X, and $\|Tu\| \leq M\|u\|$. [hint: $|v(y)|^2 \leq \|u\|^2 \int_{\mathsf{E}} |t(y, x)|^2\, dx$ by the Schwarz inequality.]

Example 2.6. One is led to the notion of extension and restriction for operators in a natural way by considering *differential operators*. The simplest example of a differential operator T is

$$Tu(x) = u'(x) = \frac{du(x)}{dx}. \tag{2.10}$$

To be more precise, set $\mathsf{X} = \mathsf{C}[a, b]$ for a finite interval and regard T as an operator in X. Then $\mathsf{D}(T)$ must consist only of continuously differentiable functions. If $\mathsf{D}(T)$ comprises all such functions, T is the *maximal* operator given by (2.10) in X. $\mathsf{D}(T)$ is a proper subset of $\mathsf{X} = \mathsf{C}[a, b]$. Let D_1 be the subset of $\mathsf{D}(T)$ consisting of all $u \in \mathsf{D}(T)$ satisfying the *boundary condition* $u(a) = 0$. Then the operator T_1 in X defined by $\mathsf{D}(T_1) = \mathsf{D}_1$ and $T_1 u = u'$ is a direct restriction of T. Similarly, a direct restriction T_2 of T is defined by the boundary condition $u(b) = 0$. Another possible boundary condition is $u(b) = k\, u(a)$ with a constant k; the resulting direct restriction of T will be denoted by T_3, with domain D_3. Furthermore, the boundary condition $u(a) = u(b) = 0$ gives another restriction T_0 (with domain D_0) of T of order 2. T_0 is a direct restriction of each of T_1, T_2, T_3. All these operators are simple examples of differential operators. The maximal differential operator T is not invertible, for $Tu = 0$ if $u(x) = \text{const}$. T_1, T_2, T_0 are invertible. $\mathsf{D}(T_1^{-1}) = \mathsf{R}(T_1)$ is the whole space $\mathsf{X} = \mathsf{C}[a, b]$ and $T_1^{-1} v(x) = \int_a^x v(t)\, dt$ for every $v \in \mathsf{X}$; thus T_1^{-1}

---

[1] See, e. g., ROYDEN [1], p. 233.

is an integral operator with domain X. T_0 is invertible as a restriction of T_1 or T_2, but the domain of T_0^{-1} is not the whole space X; it is the set of all $v \in \mathsf{X}$ such that $\int_a^b v(x)\,dx = 0$. T_3 is invertible if and only if $k \neq 1$; in this case T_3^{-1} has domain X and

$$T_3^{-1} v(x) = \frac{1}{k-1} \left(k \int_a^x v(t)\,dt + \int_x^b v(t)\,dt \right). \tag{2.11}$$

It should be noted that T is densely defined but the T_n, $n = 0, 1, 2, 3$, are not.

Example 2.7. The differential operator (2.10) can be considered in other function spaces, $\mathsf{X} = \mathsf{L}^p(a, b)$, say. Then $Tu(x) = u'(x)$ need not be continuous, and it is convenient to interpret the differentiation in a slightly generalized sense: $u'(x)$ is supposed to exist if $u(x)$ is *absolutely continuous* [that is, $u(x)$ is an indefinite integral of a locally Lebesgue-integrable function $v(x)$; then $u' = v$ by definition]. Thus the maximal differential operator defined from (2.10) in $\mathsf{X} = \mathsf{L}^p(a, b)$ has the domain $\mathsf{D}(T)$ consisting of all absolutely continuous functions $u(x) \in \mathsf{L}^p(a, b)$ such that $u'(x) \in \mathsf{L}^p(a, b)$. The various boundary conditions considered in the preceding example can also be introduced here, leading to the restrictions $T_0, \ldots, T_3$ of the maximal operator T defined in a similar way. All the operators $T, T_0, \ldots, T_3$ are *densely defined* if $1 \leq p < \infty$. As in Example 2.6, the inverses of T_1, T_2, T_3 exist and are integral operators.

So far we have assumed that (a, b) is a finite interval. The operator (2.10) can also be considered on the whole real line $(-\infty, \infty)$ or on a semi-infinite interval such as $(0, \infty)$, with some modifications. The maximal operator T can be defined in exactly the same way as above. It is convenient to define also the *minimal operator* $\dot{T}$ as the restriction of T with $\mathsf{D}(\dot{T}) = \mathsf{C}_0^\infty(a, b) \subset \mathsf{X}$ (see Example 1.11). [$\dot{T}$ can also be defined when (a, b) is finite.] In the case of a semi-infinite interval $(0, \infty)$, we can again define $T_1 \subset T$ with the boundary condition $u(0) = 0$.

When we consider boundary conditions such as $u(a) = 0$ for T defined in $\mathsf{L}^p(a, b)$, it is important to see what $u(a)$ means. $u(a)$ has no meaning for a general $u \in \mathsf{L}^p(a, b)$, for equivalent functions are identified in $\mathsf{L}^p(a, b)$ (see Example 1.3). But $u(a)$ can be given a meaning if u is equivalent to a continuous function, as the value at $x = a$ of this (necessarily unique) continuous function. Actually each $u \in \mathsf{D}(T)$ is equivalent to a function continuous on $[a, b)$ if $a > -\infty$, as is easily seen from the condition $u' \in \mathsf{L}^p(a, b)$.

2. Continuity and boundedness

An operator T from X to Y is *continuous* at $u = u_0 \in \mathsf{D}(T)$ if $\|u_n - u_0\| \to 0$, $u_n \in \mathsf{D}(T)$, implies $\|Tu_n - Tu_0\| \to 0$. As in the case of a linear form (§ 1.3), T is continuous everywhere in its domain if it is continuous at $u = 0$. Again, T is continuous if and only if T is *bounded*: $\|Tu\| \leq M\|u\|$, $u \in \mathsf{D}(T)$. The smallest number M with this property is called the *bound* of T and is denoted by $\|T\|$. An unbounded operator is sometimes said to have bound ∞.

The *extension principle* proved for a linear form (Theorem 1.16) can be taken over to a bounded operator from X to Y. The only point to be noted in the proof is that the completeness of Y is to be used in showing that $\{Tu_n\}$ has a limit $v \in \mathsf{Y}$ if u_n is a convergent sequence in X.

Problem 2.8. An operator with a finite-dimensional domain is bounded.

Problem 2.9. Let T be bounded with $\mathsf{R}(T)$ dense in Y. If $\mathsf{D}' \subset \mathsf{D}(T)$ is dense in $\mathsf{D}(T)$, $T\mathsf{D}'$ is dense in Y.

Problem 2.10. T^{-1} exists and is bounded if and only if there is an $m > 0$ such that

$$\|Tu\| \geqq m\|u\|, \quad u \in \mathsf{D}(T). \tag{2.12}$$

Example 2.11. The maximal multiplication operator T of Example 2.2 for $q = p$ is bounded if and only if $f(x)$ is essentially bounded on E; we have $\|T\| = \|f\|_\infty$. The operator T of Example 2.3 defined from a matrix (τ_{jk}) is bounded if (2.5) is assumed; we have $\|T\| \leqq M'^{1-1/p} M''^{1/p}$ by (2.6). The integral operator T of Example 2.4 is bounded under the condition (2.8); we have $\|T\| \leqq M'^{1-1/p} M''^{1/p}$ by (2.9). The integral operator T of Problem 2.5 is also bounded with $\|T\| \leqq M$.

Example 2.12. The differential operators considered in Examples 2.6–2.7 are all unbounded, for $\|u'\|$ can be arbitrarily large for $\|u\| = 1$; this is true whether $\mathsf{X} = \mathsf{C}[a, b]$ or $\mathsf{L}^p(a, b)$ and whether boundary conditions are imposed or not[1]. But the inverses T_k^{-1} are bounded; this follows, for example, from the preceding example since these inverses are simple integral operators.

3. Ordinary differential operators of second order

We considered very simple differential operators in Examples 2.6–2.7. Let us consider here in some detail second order ordinary differential operators and their inverses[2]. Let

$$Lu = p_0(x)\, u'' + p_1(x)\, u' + p_2(x)\, u \tag{2.13}$$

be a *formal differential operator* defined on a finite interval $a \leqq x \leqq b$. We assume that p_0, p_1 and p_2 are real-valued, p_0'', p_1', p_2 are continuous on $[a, b]$ and $p_0(x) < 0$. From the formal operator L various operators in (or between) different function spaces can be constructed.

First take the space $\mathsf{X} = \mathsf{C}[a, b]$. Let D be the set of all functions u with u'' continuous on $[a, b]$. By

$$Tu = Lu, \quad u \in \mathsf{D}, \tag{2.14}$$

we define an operator T in X with $\mathsf{D}(T) = \mathsf{D}$. Restriction of the domain of T by means of boundary conditions gives rise to different operators. We shall not consider all possible boundary conditions but restrict ourselves to several typical ones, namely,

$$T_1, \quad \mathsf{D}(T_1) = \mathsf{D}_1: \quad u(a) = 0, \quad u(b) = 0 \quad \text{(zero boundary conditions)}. \tag{2.15}$$

$$T_2, \quad \mathsf{D}(T_2) = \mathsf{D}_2: \quad u'(a) - h_a u(a) = 0, \quad u'(b) + h_b u(b) = 0 \quad \text{(elastic boundary conditions)}. \tag{2.16}$$

$$T_3, \quad \mathsf{D}(T_3) = \mathsf{D}_3: \quad u(a) = 0, \quad u'(a) = 0 \quad \text{(zero initial conditions)}. \tag{2.17}$$

$$T_0, \quad \mathsf{D}(T_0) = \mathsf{D}_0: \quad u(a) = u'(a) = u(b) = u'(b) = 0. \tag{2.18}$$

T is the *maximal operator* in X constructed from the formal operator L. T_1, T_2, T_3 are restrictions of T of order 2 and also extensions of T_0 of order 2.

---

[1] But $T = d/dx$ is bounded if it is regarded as an operator on $\mathsf{C}'[a, b]$ to $\mathsf{C}[a, b]$.

[2] For more details on ordinary differential operators, see CODDINGTON and LEVINSON [1], GOLDBERG [1], NAIMARK [1], STONE [1].

T is not invertible, for $Tu = 0$ has two linearly independent solutions u_1, u_2 belonging to X. The other operators T_k, $k = 0, 1, 2, 3$, are invertible, possibly under certain additional conditions. T_3^{-1} always exists and has domain X, since an initial-value problem is always solvable in a unique way; it is an integral operator of the Volterra type[1], given by

$$(2.19)\qquad T_3^{-1} v(y) = \int_a^y [u_1(y)\, u_2(x) - u_2(y)\, u_1(x)] \frac{v(x)\, dx}{-W(x)\, p_0(x)},$$

where u_1, u_2 are any two linearly independent solutions of $Tu = 0$ and $W(x)$ is the *Wronskian*

$$(2.20)\qquad W(x) = u_1(x)\, u_2'(x) - u_2(x)\, u_1'(x) = \text{const.} \exp\left(-\int^x \frac{p_1}{p_0}\, dx\right).$$

T_0 is also invertible since it is a restriction of T_3.

T_1^{-1} is also an integral operator

$$(2.21)\qquad T_1^{-1} v(y) = \int_a^b g(y, x)\, v(x)\, dx,$$

where $g(y, x)$, the *Green function*[2] for the zero boundary conditions, is given by

$$(2.22)\qquad g(y, x) = \frac{u_1(y)\, u_2(x)}{-p_0(x)\, W(x)},\ y \leqq x; \qquad = \frac{u_2(y)\, u_1(x)}{-p_0(x)\, W(x)},\ y \geqq x.$$

Here u_1, u_2 are the nontrivial solutions of $Lu = 0$ such that $u_1(a) = 0$, $u_2(b) = 0$ and W is their Wronskian given by (2.20). g is well-defined if $u_1(b) \neq 0$, for then $W(b) \neq 0$ and hence $W(x)$ is nowhere zero. This is the case if, for example, $p_2 > 0$ on $[a, b]$. In fact if $u_1(b) = u_1(a) = 0$, $u_1(x)$ takes either a positive maximum or a negative minimum[3]. If $u_1(x_0)$ is a positive maximum, $x_0 \in (a, b)$ and we must have $u_1'(x_0) = 0$, $u_1''(x_0) \leqq 0$, which is a contradiction since $Lu_1(x_0) = 0$, $p_0(x_0) < 0$, $p_2(x_0) > 0$. A negative minimum is excluded in the same way. Thus T_1 is invertible if $p_2 > 0$ on $[a, b]$, and T_1^{-1} is the integral operator (2.21) defined on the whole of X [note that $g(y, x)$ is continuous in x and y].

Similarly it can be shown that T_2^{-1} exists, has domain X and is an integral operator of the form (2.21) with an appropriate Green function g under some additional conditions (for example $p_2 > 0$ on $[a, b]$ and $h_a, h_b \geqq 0$).

Problem 2.13. The domain of T_0^{-1} is not the whole of X; it is the set of all $v \in$ X satisfying the conditions

$$(2.23)\qquad \int_a^b r(x)\, u_k(x)\, v(x)\, dx = 0, \quad k = 1, 2,$$

$$r(x) = \frac{1}{-p_0(x)} \exp\left(\int^x \frac{p_1}{p_0}\, dx\right).$$

Let us estimate $\|T_1^{-1}\|$ explicitly, assuming $p_2 > 0$. We note that $g(y, x) \geqq 0$ and

$$(2.24)\qquad \int_a^b g(y, x)\, dx \leqq c^{-1}, \quad c = \min_x p_2(x) > 0.$$

---

[1] A kernel $t(y, x)$ (or the associated integral operator) is of Volterra type if $t(y, x) = 0$ identically for $y < x$ (or for $y > x$).

[2] See, e. g., Coddington and Levinson [1] for the Green functions and other elementary results used below.

[3] Since L has real coefficients, $u_1(x)$ may be assumed to be real without loss of generality.

$g \geqq 0$ can be seen from (2.22) where we may take $u_1, u_2 \geqq 0$ for the reason stated above [or, more directly, from the fact that $g(y, x)$ cannot have a negative minimum in y for a fixed x since it satisfies $L_y g = 0$ for $y \neq x$ and has a characteristic singularity[1] at $y = x$]. To prove (2.24) denote by $u_0(y)$ the left member. $u_0(y)$ satisfies the differential equation $L u_0 = 1$ with the boundary conditions $u_0(a) = u_0(b) = 0$. Let $u_0(x_0)$ be the maximum of $u_0(x)$. We have then $u_0'(x_0) = 0$, $u_0''(x_0) \leqq 0$ so that $L u_0(x_0) = 1$ gives $p_2(x_0)\, u_0(x_0) \leqq 1$, which implies (2.24).

Now it follows from (2.21) that for $u = T_1^{-1} v$,

$$\|u\| = \max_y |u(y)| \leqq \max_x |v(x)| \max_y \int_a^b g(y, x)\, dx \leqq \|v\|\, c^{-1}.$$

Hence

$$\|T_1^{-1}\| \leqq c^{-1}. \tag{2.25}$$

Problem 2.14. (2.25) is true for T_1 replaced by T_2 if $h_a, h_b \geqq 0$.

Let us now consider the differential operator L in a different Banach space. This time we shall take $\mathsf{X} = \mathsf{L}^p(a, b)$, $1 \leqq p \leqq \infty$. Since $\mathsf{C}[a, b]$ is a subset of $\mathsf{L}^p(a, b)$, the operators T and T_n considered above may as well be regarded as operators in $\mathsf{X} = \mathsf{L}^p(a, b)$. But the above definition of $\mathsf{D}(T)$ is rather arbitrary since now u'' need not be continuous in order that Lu belong to X. It is more natural to assume only that u' is absolutely continuous on (a, b) and $u'' \in \mathsf{X} = \mathsf{L}^p$[2]. Let D be the set of all u with these properties and define T by $Tu = Lu$ with $\mathsf{D}(T) = \mathsf{D}$. The restrictions T_0 to T_3 of T can be defined exactly as above by restricting the domain by the boundary conditions (2.15) to (2.18). [Note that u' is continuous on $[a, b]$ by $u'' \in \mathsf{L}^p$ so that the boundary values $u(a)$, $u'(a)$, etc. are well defined, cf. Example 2.7.]

The results concerning the inverses of T and T_n in the space $\mathsf{C}[a, b]$ stated above are now valid in the new space $\mathsf{X} = \mathsf{L}^p(a, b)$. It suffices to add the following remarks. The Green functions are determined by the formal operator L without reference to the Banach space X. That the T_n^{-1} are bounded and defined on the whole of X follows from the fact that, for example, the function (2.21) belongs to D_1 if and only if $v \in \mathsf{X}$, which is in turn a simple consequence of the property of the Green functions.

The estimate (2.25) for T_1^{-1} is not valid, however. To obtain a similar estimate in the present case we use the inequality

$$\int_a^b g(y, x)\, dy \leqq c'^{-1}, \quad c' = \min_x (p_2 - p_1' + p_0''), \tag{2.26}$$

where we assume that $c' > 0$. This follows from (2.24) by noting that $g(y, x)$ is, when the arguments x, y are exchanged, the Green function of the *adjoint equation* $Mv = 0$ to $Lu = 0$, where

$$Mv = (p_0 v)'' - (p_1 v)' + p_2 v. \tag{2.27}$$

Thus we obtain by Example 2.11

$$\|T_1^{-1}\| \leqq 1/c^{1-1/p} c'^{1/p} \leqq 1/\min(c, c'). \tag{2.28}$$

Problem 2.15. The functions $r u_k$, $k = 1, 2$, in (2.23) are solutions of the adjoint equation $Mv = 0$.

---

[1] $\partial g(y, x)/\partial y$ is discontinuous in y at $y = x$ with a jump equal to $1/p_0(x)$.

[2] We need not even assume $u'' \in \mathsf{X}$ in order that $Tu = Lu$ define an operator in X; it suffices to assume only that $u \in \mathsf{X}$ and $Lu \in \mathsf{X}$. But this seemingly broader definition leads to the same T as above; see Remark 2.16 below.

Remark 2.16. We assumed above that (a, b) is a finite interval, p_0'', p_1', p_2 are continuous on the *closed* interval $[a, b]$ and that $p_0 < 0$. In such a case L is said to be *regular*. Let us now consider *singular* cases in which these conditions are not necessarily satisfied; we assume only that p_0'', p_1' and p_2 are continuous in the open interval (a, b) and $p_0 \neq 0$. The p_k may even be complex-valued, and the interval (a, b) may be finite or infinite.

In a singular case we cannot define all the operators T and T_k given above. But we can define at least the maximal operator T and the minimal operator $\dot{T}$ in $X = L^p(a, b)$, $1 \leqq p < \infty$. T is defined by $Tu = Lu$ where $D(T)$ is the set of all $u \in X$ such that $u(x)$ is continuously differentiable in (a, b), $u'(x)$ is absolutely continuous in (a, b) [so that $u''(x)$ is defined almost everywhere] and $Lu \in X$. $\dot{T}$ is the restriction of T with $D(\dot{T}) = C_0^\infty(a, b)$ (see Example 1.11). T and $\dot{T}$ are densely defined.

In a singular case it is in general not easy to define restrictions of T with "good" boundary conditions such as the T_k considered in the regular case. But T is often a "good" operator itself, in a sense to be explained later.

It should also be remarked that, when applied to the regular case, the maximal operator T just defined is apparently different from the T defined earlier, for the earlier definition requires $u'' \in X$ whereas the new one only $Lu \in X$. Actually, however, these two definitions are equivalent, for $Lu \in X$ implies $u'' \in X$ in the regular case. This is due to the basic property of a solution of the ordinary differential equation $Lu = f \in X$, namely, that u is continuous with u' on $[a, b]$ in the regular case.

§ 3. Bounded operators

1. The space of bounded operators

We denote by $\mathscr{B}(X, Y)$ the set of all bounded operators *on* X to Y. This corresponds to the set of *all* operators on X to Y, denoted by the same symbol, in the finite-dimensional case (see I-§ 3.2). We write $\mathscr{B}(X)$ for $\mathscr{B}(X, X)$.

Since every operator belonging to $\mathscr{B}(X, Y)$ has domain X and range in Y, the meaning of the linear combination $\alpha S + \beta T$ of $S, T \in \mathscr{B}(X, Y)$ is clear (I-§ 3.2). The resulting operator is again linear and bounded. Thus $\mathscr{B}(X, Y)$ is a normed space with the norm $\|T\|$ defined as the bound of T (see I-§ 4.1). Similarly, the product TS is defined for $T \in \mathscr{B}(Y, Z)$, $S \in \mathscr{B}(X, Y)$ by I-(3.15) and belongs to $\mathscr{B}(X, Z)$.

Example 3.1. Consider the operator T of Example 2.3 defined from a matrix. As a maximal operator from $X = l^p$ to itself, T is defined everywhere on X and is bounded if (2.5) is assumed (see Example 2.11). Thus $T \in \mathscr{B}(X)$. The set of all operators of this type is a linear manifold of $\mathscr{B}(X)$. If T, S are operators of this type defined by the matrices (τ_{jk}), (σ_{jk}) respectively, $TS \in \mathscr{B}(X)$ is again of the same type and is defined by the matrix which is equal to the product of the two matrices.

Example 3.2. Consider an integral operator of the form (2.7). Regarded as a maximal integral operator from $X = L^p(E)$ to $Y = L^p(F)$, T is defined on X and is bounded with $\|T\| \leqq M'^{1-1/p} M''^{1/p}$ if the condition (2.8) is satisfied. Thus T belongs to $\mathscr{B}(X, Y)$. The set of all integral operators of this type is a linear manifold of $\mathscr{B}(X, Y)$.

Consider another integral operator $S \in \mathscr{B}(\mathsf{Y}, \mathsf{Z})$ of the same type with a kernel $s(z, y)$ where $y \in \mathsf{F}$ and $z \in \mathsf{G}$ and where $\mathsf{Z} = \mathsf{L}^p(\mathsf{G})$. Then the product ST is defined and belongs to $\mathscr{B}(\mathsf{X}, \mathsf{Z})$. ST is an integral operator with the kernel

$$r(z, x) = \int_{\mathsf{F}} s(z, y)\, t(y, x)\, dy . \tag{3.1}$$

This follows from the expression of $(ST)u = S(Tu)$ by a change of order of integration, which is justified by Fubini's theorem since all the integrals involved are absolutely convergent. It is easily seen that the kernel r given by (3.1) satisfies (2.8) if both s and t do.

Problem 3.3. The null space $\mathsf{N}(T)$ of an operator $T \in \mathscr{B}(\mathsf{X}, \mathsf{Y})$ is a closed linear manifold of X.

$\mathscr{B}(\mathsf{X}, \mathsf{Y})$ *is a Banach space.* To prove the completeness of $\mathscr{B}(\mathsf{X}, \mathsf{Y})$, let $\{T_n\}$ be a Cauchy sequence of elements of $\mathscr{B}(\mathsf{X}, \mathsf{Y})$. Then $\{T_n u\}$ is a Cauchy sequence in Y for each fixed $u \in \mathsf{X}$, for $\|T_n u - T_m u\| \leqq \|T_n - T_m\|\, \|u\| \to 0$. Since Y is complete, there is a $v \in \mathsf{Y}$ such that $T_n u \to v$. We define an operator T by setting $v = Tu$. By an argument similar to that used to prove the completeness of X^* in § 1.4, it is seen that T is linear and bounded so that $T \in \mathscr{B}(\mathsf{X}, \mathsf{Y})$ and that $\|T_n - T\| \to 0$.

Most of the results on $\mathscr{B}(\mathsf{X}, \mathsf{Y})$ deduced in the finite-dimensional case can be taken over to the present case (except those based on the explicit use of a basis). In particular we note the expression I-(4.2) for $\|T\|$, the inequality I-(4.6) for $\|TS\|$ and various results on infinite series and operator-valued functions given in I-§ 4.

Contrary to the finite-dimensional case, however, different kinds of convergence can be introduced into $\mathscr{B}(\mathsf{X}, \mathsf{Y})$. Let T, $T_n \in \mathscr{B}(\mathsf{X}, \mathsf{Y})$, $n = 1, 2, \ldots$. The convergence of $\{T_n\}$ to T in the sense of $\|T_n - T\| \to 0$ [convergence in the normed space $\mathscr{B}(\mathsf{X}, \mathsf{Y})$] is called *uniform convergence* or *convergence in norm*. $\{T_n\}$ is said to converge *strongly* to T if $T_n u \to Tu$ for each $u \in \mathsf{X}$. $\{T_n\}$ converges in norm if and only if $\{T_n u\}$ converges uniformly for $\|u\| \leqq 1$. $\{T_n\}$ is said to converge *weakly* if $\{T_n u\}$ converges weakly for each $u \in \mathsf{X}$, that is, if $(T_n u, g)$ converges for each $u \in \mathsf{X}$ and $g \in \mathsf{Y}^*$. If $\{T_n u\}$ has a weak limit Tu for each $u \in \mathsf{X}$, $\{T_n\}$ has the *weak limit* T. (T is uniquely determined by $\{T_n\}$.) $\{T_n\}$ converges in norm if and only if $(T_n u, g)$ converges uniformly for $\|u\| \leqq 1$ and $\|g\| \leqq 1$ (see Theorem 1.32). A weakly convergent sequence has a weak limit if Y is weakly complete (see § 1.6). Convergence in norm implies strong convergence, and strong convergence implies weak convergence. We use the notations $T = \text{u-lim}\, T_n$, $T_n \xrightarrow[\text{u}]{} T$ for convergence in norm, $T = \text{s-lim}\, T_n$, $T_n \xrightarrow[\text{s}]{} T$ for strong convergence and $T = \text{w-lim}\, T_n$, $T_n \xrightarrow[\text{w}]{} T$ for weak convergence.

Problem 3.4. If $\{T_n u\}$ converges strongly for each $u \in \mathsf{X}$, then $\{T_n\}$ converges strongly to some $T \in \mathscr{B}(\mathsf{X}, \mathsf{Y})$.

If $\{T_n\}$ is weakly convergent, it is uniformly bounded, that is, $\{\|T_n\|\}$ is bounded. To see this we recall that $\{\|T_n u\|\}$ is bounded for each

$u \in X$ because $\{T_n u\}$ is weakly convergent (§ 1.6). The assertion then follows by another application of the principle of uniform boundedness (apply Theorem 1.29 with $p_\lambda[u] = \|T_n u\|$). We note also that

$$(3.2) \qquad \|T\| \leq \liminf \|T_n\| \quad \text{for} \quad T = \text{w-lim}\, T_n ,$$

as is seen from (1.26).

The following lemmas will be used frequently in this treatise. Here all operators belong to $\mathscr{B}(X, Y)$ unless otherwise stated.

Lemma 3.5. *Let $\{T_n\}$ be uniformly bounded. Then $\{T_n\}$ converges strongly (to T) if $\{T_n u\}$ converges strongly (to Tu) for all u of a fundamental subset of X.*

Lemma 3.6. *Let $\{T_n\}$ be uniformly bounded. Then $\{T_n\}$ converges weakly (to T) if $\{(T_n u, g)\}$ converges (to (Tu, g)) for all u of a fundamental subset of X and for all g of a fundamental subset of Y^*.*

The proof of these two lemmas is similar to that of Lemma 1.31 and may be omitted.

Lemma 3.7. *If $T_n \underset{s}{\to} T$ then $T_n u \underset{s}{\to} Tu$ uniformly for all u of a compact subset S of X.*

Proof. We may assume that $T = 0$; otherwise we have only to consider $T_n - T$ instead of T_n. As is every compact subset of a metric space, S is totally bounded, that is, for any $\varepsilon > 0$ there are a finite number of elements $u_k \in S$ such that each $u \in S$ lies within distance ε of some u_k[1]. Since $T_n u_k \to 0$, $n \to \infty$, there are positive numbers n_k such that $\|T_n u_k\| < \varepsilon$ for $n > n_k$. For any $u \in S$, we have then $\|T_n u\| \leq \|T_n(u - u_k)\| + \|T_n u_k\| \leq (M + 1)\varepsilon$ if $n > \max n_k$, where u_k is such that $\|u - u_k\| < \varepsilon$ and M is an upper bound of $\|T_n\|$ (M is finite by the above remark).

Lemma 3.8. *If $T_n \underset{s}{\to} T$ in $\mathscr{B}(Y, Z)$ and $S_n \underset{s}{\to} S$ in $\mathscr{B}(X, Y)$, then $T_n S_n \underset{s}{\to} TS$ in $\mathscr{B}(X, Z)$.*

Proof. $T_n S_n u - TSu = T_n(S_n - S)u + (T_n - T)Su \to 0$ for each $u \in X$. Note that $\{T_n\}$ is uniformly bounded so that $\|T_n(S_n - S)u\| \leq M\|(S_n - S)u\| \to 0$.

Lemma 3.9. *If $T_n \underset{w}{\to} T$ in $\mathscr{B}(Y, Z)$ and $S_n \underset{s}{\to} S$ in $\mathscr{B}(X, Y)$, then $T_n S_n \underset{w}{\to} TS$ in $\mathscr{B}(X, Z)$.*

Proof. Similar to the proof of Lemma 3.8; note that $|(T_n(S_n - S)u, g)| \leq M\|g\| \|(S_n - S)u\| \to 0$ for each $u \in X$ and $g \in Z^*$.

Problem 3.10. If $u_n \underset{s}{\to} u$ and $T_n \underset{s}{\to} T$, then $T_n u_n \underset{s}{\to} Tu$. If $u_n \underset{s}{\to} u$ and $T_n \underset{w}{\to} T$, then $T_n u_n \underset{w}{\to} Tu$.

---

[1] If this were not true for some $\varepsilon > 0$, there would exist an infinite sequence $u_n \in S$ such that $\|u_n - u_m\| \geq \varepsilon$, $n \neq m$, and $\{u_n\}$ would have no convergent subsequence.

For an operator-valued function $t \to T(t) \in \mathscr{B}(X, Y)$ of a real or complex variable t, we can introduce different kinds of continuity according to the different kinds of convergence considered above. $T(t)$ is continuous *in norm* if $\|T(t+h) - T(t)\| \to 0$ for $h \to 0$. $T(t)$ is *strongly continuous* if $T(t)\,u$ is strongly continuous for each $u \in X$. $T(t)$ is *weakly continuous* if $T(t)\,u$ is weakly continuous for each $u \in X$, that is, if $(T(t)\,u, g)$ is continuous for each $u \in X$ and $g \in Y^*$.

$\|T(t)\|$ is continuous if $T(t)$ is continuous in norm. $\|T(t)\|$ is not necessarily continuous if $T(t)$ is only strongly continuous, but $\|T(t)\|$ *is locally bounded and lower semicontinuous if* $T(t)$ *is weakly continuous.* The local boundedness follows from the fact that $\{\|T(t_n)\|\}$ is bounded for $t_n \to t$, and the lower semicontinuity follows from (3.2).

Similarly, different kinds of differentiability can be introduced. $T(t)$ is *differentiable in norm* if the difference coefficient $h^{-1}[T(t+h) - T(t)]$ has a limit in norm, which is the derivative in norm of $T(t)$. The strong derivative $T'(t) = dT(t)/dt$ is defined by $T'(t)\,u = \lim h^{-1}[T(t+h)\,u - T(t)\,u]$, and similarly for the weak derivative.

Problem 3.11. If $u(t) \in X$ and $T(t) \in \mathscr{B}(X, Y)$ are strongly differentiable, then $T(t)\,u(t) \in Y$ is strongly differentiable and $(d/dt)\,T(t)\,u(t) = T'(t)\,u(t) + T(t)\,u'(t)$.

We can also define different kinds of integrals $\int T(t)\,dt$ for an operator-valued function $T(t)$ of a real variable t. If $T(t)$ is continuous in norm, the integral $\int T(t)\,dt$ can be defined as in the finite-dimensional case (or as for numerical-valued functions). If $T(t)$ is only strongly continuous, we can define the integral $v = \int T(t)\,u\,dt$ for each $u \in X$. Then $\|v\| \leq \int \|T(t)\,u\|\,dt \leq \|u\| \int \|T(t)\|\,dt$; note that $\|T(t)\|$ need not be continuous but is bounded and lower semicontinuous and hence integrable[1]. Thus the mapping $u \to v = Su$ defines an operator $S \in \mathscr{B}(X, Y)$ with bound $\leq \int \|T(t)\|\,dt$. If we write $S = \int T(t)\,dt$, we have defined a "strong" integral $\int T(t)\,dt$ for a strongly continuous function $T(t)$, with the properties

$$(3.3) \qquad \left(\int T(t)\,dt\right) u = \int T(t)\,u\,dt\,, \quad \left\|\int T(t)\,dt\right\| \leq \int \|T(t)\|\,dt\,,$$

$$(3.4) \qquad \frac{d}{dt}\int^{t} T(s)\,ds = T(t) \quad \text{(strong derivative!)}\,.$$

Similarly, the integral of a function $T(\zeta)$ of a complex variable ζ along a curve can be defined.

When we consider holomorphic operator-valued functions, there is no distinction among uniform, strong and weak holomorphy, as in vector-valued functions. More precisely, we have

Theorem 3.12. *Let* $T(\zeta) \in \mathscr{B}(X, Y)$ *be defined on a domain* Δ *of the complex plane and let* $(T(\zeta)\,u, g)$ *be holomorphic in* $\zeta \in \Delta$ *for each* $u \in X$

[1] If a real-valued function f is lower-semicontinuous, the set of all t such that $f(t) > \alpha$ is open for any α. Hence f is Lebesgue-measurable.

and $g \in \mathsf{Y}^*$. Then $T(\zeta)$ is holomorphic in Δ in the sense of norm (differentiable in norm for $\zeta \in \Delta$).

The proof is similar to that of Theorem 1.37 and may be omitted. Again it should be remarked that if $\|T(\zeta)\|$ is assumed to be locally bounded, it suffices to assume that $(T(\zeta)\,u, g)$ be holomorphic for all u of a fundamental subset of X and for all g of a fundamental subset of Y^*.

Problem 3.13. Let $T_n \in \mathscr{B}(\mathsf{X}, \mathsf{Y})$ and let $\{(T_n u, g)\}$ be bounded for each $u \in \mathsf{X}$ and $g \in \mathsf{Y}^*$. Then $\{\|T_n\|\}$ is bounded.

2. The operator algebra $\mathscr{B}(\mathsf{X})$

$\mathscr{B}(\mathsf{X}) = \mathscr{B}(\mathsf{X}, \mathsf{X})$ is the set of all bounded operators on X to itself. In $\mathscr{B}(\mathsf{X})$ not only the linear combination of two operators S, T but also their product $S\,T$ is defined and belongs to $\mathscr{B}(\mathsf{X})$. Thus $\mathscr{B}(\mathsf{X})$ is a *complete normed algebra* (Banach algebra) (see I-§ 4.1). Again, most of the results of the finite-dimensional case can be taken over except those which depend on the explicit use of a basis. It should be noted that the completeness of $\mathscr{B}(\mathsf{X})$ is essential here; for example, the existence of the sum of an absolutely convergent series of operators depends on completeness [cf. the Neumann series I-(4.22) and the exponential function I-(4.20)].

$T \in \mathscr{B}(\mathsf{X})$ is said to be *nonsingular* if T^{-1} exists and belongs to $\mathscr{B}(\mathsf{X})$. Actually it suffices to know that T^{-1} has domain X; then it follows that T^{-1} is bounded by the closed graph theorem (see Problem 5.21 below).

$1 - T$ is nonsingular if $\|T\| < 1$; this is proved by means of the Neumann series as in I-§ 4.4. It follows that T^{-1} is a continuous function of T [on the set of all nonsingular operators, which is open in $\mathscr{B}(\mathsf{X})$]. In the same way we see that if $T(\zeta) \in \mathscr{B}(\mathsf{X})$ is holomorphic in ζ and nonsingular, then $T(\zeta)^{-1}$ is also holomorphic (see I-§ 4.5)[1].

The *spectral radius* $\operatorname{spr} T = \lim \|T^n\|^{1/n}$ can also be defined for every $T \in \mathscr{B}(\mathsf{X})$ as in I-§ 4.2. T is said to be *quasi-nilpotent* if $\operatorname{spr} T = 0$.

The trace and determinant of $T \in \mathscr{B}(\mathsf{X})$ are in general not defined (these are examples of notions defined by explicit use of a basis in the finite-dimensional case). But we shall later define the trace and determinant for certain special classes of operators of $\mathscr{B}(\mathsf{X})$.

Example 3.14. The maximal integral operator T of Example 2.4 belongs to $\mathscr{B}(\mathsf{X})$ if $\mathsf{F} = \mathsf{E}$ and $\mathsf{Y} = \mathsf{X} = \mathsf{L}^p(\mathsf{E})$ and if (2.8) is satisfied. Similarly for the maximal matrix operator of Example 2.3. The inverses of the differential operators T_1, T_2, T_3 of Examples 2.6—2.7 belong to $\mathscr{B}(\mathsf{X})$, but T_0^{-1} does not (it is bounded but not defined everywhere on X). Similarly, the inverses of the differential operators T_1, T_2, T_3 of § 2.3 belong to $\mathscr{B}(\mathsf{X})$ (under appropriate conditions stated there) but T_0^{-1} does not.

---

[1] The same is true when $T(\zeta) \in \mathscr{B}(\mathsf{X}, \mathsf{Y})$ is holomorphic in ζ and $T(\zeta)^{-1} \in \mathscr{B}(\mathsf{Y}, \mathsf{X})$.

Example 3.15. An integral operator of Volterra type is usually (but not always) quasi-nilpotent. For example consider a kernel $t(y, x)$ which is continuous on the triangle $a \leqq x \leqq y \leqq b$ and which vanishes for $a \leqq y < x \leqq b$. The associated integral operator T in $\mathsf{C}[a, b]$ or in $\mathsf{L}^p(a, b)$ is quasi-nilpotent. To see this, let $t_n(y, x)$ be the kernel of T^n, $n = 1, 2, \ldots$. It is easy to see that the t_n are also of Volterra type ($t_n(y, x) = 0$ for $y < x$). We have for $y > x$

$$(3.5) \qquad |t_n(y, x)| \leqq \frac{M^n(y-x)^{n-1}}{(n-1)!} \leqq \frac{M^n(b-a)^{n-1}}{(n-1)!}, \quad n = 1, 2, \ldots,$$

where $M = \max|t(y, x)|$, as is seen by induction based on the composition formula (3.1) corresponding to $T^n = T^{n-1} T$. (3.5) gives $\|T^n\| \leqq M^n(b-a)^n/(n-1)!$ (see Example 2.11) so that $\operatorname{spr} T = 0$.

Example 3.16. Let $\mathsf{X} = l^p$, $1 \leqq p \leqq \infty$, and let T be the maximal operator in X defined from a matrix (τ_{jk}) with all elements equal to zero except possibly $\tau_{j,j+1} = \tau_j$, $j = 1, 2, \ldots$. Such a T will be called a *left shift operator* and $\{\tau_j\}$ the *defining sequence* for T. It is easily seen that $T \in \mathscr{B}(\mathsf{X})$ and

$$(3.6) \qquad T x_1 = 0, \quad T x_2 = \tau_1 x_1, \quad T x_3 = \tau_2 x_2, \ldots,$$

where $\{x_j\}$ is the canonical basis of X. Also we have

$$(3.7) \qquad \|T\| = \sup_j |\tau_j|.$$

The iterates T^m of T are represented by matrices $(\tau_{jk}^{(m)})$ such that $\tau_{j,j+m}^{(m)} = \tau_j \tau_{j+1} \ldots \tau_{j+m-1}$, all other elements being zero. Thus

$$(3.8) \qquad \|T^m\| = \sup_j |\tau_j \tau_{j+1} \ldots \tau_{j+m-1}|.$$

It follows, for example, that

$$(3.9) \qquad \operatorname{spr} T = \tau \quad \text{if} \quad |\tau_j| \to \tau, \quad j \to \infty.$$

Similarly a *right shift operator* T' is defined by a matrix (τ'_{jk}) with $\tau'_{j+1,j} = \tau'_j$ and all other elements equal to 0. We have $T' \in \mathscr{B}(\mathsf{X})$ and

$$(3.10) \qquad T' x_1 = \tau'_1 x_2, \quad T' x_2 = \tau'_2 x_3, \ldots.$$

Also we have formulas similar to (3.7) to (3.9).

3. The adjoint operator

For each $T \in \mathscr{B}(\mathsf{X}, \mathsf{Y})$, the adjoint T^* is defined and belongs to $\mathscr{B}(\mathsf{Y}^*, \mathsf{X}^*)$ as in the finite-dimensional case (see I-§§ 3.6, 4.1). For each $g \in \mathsf{Y}^*$, $u \to (g, Tu)$ is a *bounded* semilinear form on X by virtue of $|(g, Tu)| \leqq \|g\| \|Tu\| \leqq \|T\| \|g\| \|u\|$, so that it can be written (f, u) with an $f \in \mathsf{X}^*$; T^* is defined by $T^* g = f$. As before $\|T^* g\| = \|f\| = \sup_{\|u\| \leqq 1} |(f, u)| \leqq \|T\| \|g\|$ gives $\|T^*\| \leqq \|T\|$. To deduce the more precise result $\|T^*\| = \|T\|$, we need a slight modification of the argument of I-§ 4.1 since $\mathsf{X} = \mathsf{X}^{**}$ is no longer true in general. Applying the above result to T replaced by T^*, we have $\|T^{**}\| \leqq \|T^*\| \leqq \|T\|$. But $T^{**} \supset T$ if X is identified with a subspace of X^{**}, for $(T^{**} u, g) = (u, T^* g) = \overline{(T^* g, u)} = \overline{(g, Tu)}$ shows that the semilinear form $T^{**} u$ on Y^* is represented by $Tu \in \mathsf{Y}$ and therefore $T^{**} u = Tu$ by identification. Since $T^{**} \supset T$ implies $\|T^{**}\| \geqq \|T\|$, we obtain the required result $\|T^*\| = \|T\|$.

Again, most of the results of the finite-dimensional case remain valid. Among the exceptions, it should be mentioned that the second relation of I-(3.37) is to be replaced by

$$\mathsf{N}(T) = \mathsf{R}(T^*)^\perp \cap \mathsf{X}\,. \tag{3.11}$$

The relations between the nullities, deficiencies and ranks of T and T^* such as I-(3.38) must be reconsidered accordingly. This problem will be discussed in more detail in Chapter IV.

The relation between the inverses T^{-1} and T^{*-1} given before should be altered in the following way. If $T \in \mathscr{B}(\mathsf{X}, \mathsf{Y})$ has inverse $T^{-1} \in \mathscr{B}(\mathsf{Y}, \mathsf{X})$, then T^* has inverse $T^{*-1} = (T^{-1})^* \in \mathscr{B}(\mathsf{X}^*, \mathsf{Y}^*)$. This follows simply by taking the adjoints of the relations $T^{-1}T = 1_\mathsf{X}$, $T T^{-1} = 1_\mathsf{Y}$ (1_X is the identity operator on X). Conversely, the existence of $T^{*-1} \in \mathscr{B}(\mathsf{X}^*, \mathsf{Y}^*)$ implies that of $T^{-1} \in \mathscr{B}(\mathsf{Y}, \mathsf{X})$; but the proof is not trivial and will be given later in the more general case of an unbounded operator (see Theorem 5.30).

Example 3.17. Consider the maximal integral operator T of Example 2.4 for $\mathsf{X} = \mathsf{L}^p(E)$, $\mathsf{Y} = \mathsf{L}^p(F)$, so that $T \in \mathscr{B}(\mathsf{X}, \mathsf{Y})$ by Example 3.2. Thus $T^* \in \mathscr{B}(\mathsf{Y}^*, \mathsf{X}^*)$ where $\mathsf{X}^* = \mathsf{L}^q(E)$, $\mathsf{Y}^* = \mathsf{L}^q(F)$, $p^{-1} + q^{-1} = 1$ (see Example 1.25; here we assume $p < \infty$). Actually T^* is itself an integral operator of the same type with kernel the Hermitian conjugate to that of T:

$$t^*(x, y) = \overline{t(y, x)}\,. \tag{3.12}$$

This follows from the identity

$$(T^* g, u) = (g, Tu) = \int_F g(y)\, dy \int_E \overline{t(y, x)\, u(x)\, dx} = \int_E \overline{u(x)}\, dx \int_F \overline{t(y, x)}\, g(y)\, dy$$

valid for $u \in \mathsf{X}$ and $g \in \mathsf{Y}^*$; the change of the order of integration is justified under the assumption (2.8) since the double integral is absolutely convergent (Fubini's theorem).

Example 3.18. Let T be the maximal operator T of Example 2.3 defined from a matrix (τ_{jk}) for $\mathsf{X} = \mathsf{Y} = l^p$, $1 \leq p < \infty$. We have $T \in \mathscr{B}(\mathsf{X})$ if (2.5) is satisfied (Example 3.1). Thus $T^* \in \mathscr{B}(\mathsf{X}^*)$ where $\mathsf{X}^* = l^q$ with $p^{-1} + q^{-1} = 1$ (Example 1.25). T^* is also defined from a matrix of the same type

$$\tau^*_{jk} = \overline{\tau_{kj}}\,. \tag{3.13}$$

The proof is similar to that in the preceding Example.

Problem 3.19. The operators T and T' in Example 3.16 are the adjoints of one another if T, T' are defined in l^p, l^q, respectively, where $p^{-1} + q^{-1} = 1$, $1 < p < \infty$, and $\tau'_j = \overline{\tau}_j$.

4. Projections

An idempotent operator $P \in \mathscr{B}(\mathsf{X})$ ($P^2 = P$) is called a *projection*. We have the decomposition

$$\mathsf{X} = \mathsf{M} \oplus \mathsf{N} \tag{3.14}$$

where $\mathsf{M} = P\mathsf{X}$ and $\mathsf{N} = (1 - P)\,\mathsf{X}$, see I-§ 3.4. It should be added that M, N are *closed* linear manifolds of X. This follows from the fact that M and N are exactly the null spaces of $1 - P$ and P, respectively (see Problem 3.3).

Conversely, a decomposition (3.14) of a Banach space into the direct sum of two closed linear manifolds defines a projection P (on M along N) (see loc. cit.). It is easily seen that P is a linear operator on X to itself, but the proof of the boundedness of P is not simple. This will be given later as an application of the closed graph theorem (Theorem 5.20).

For a given closed linear manifold M of X, it is not always possible to find a complementary subspace N such that (3.14) is true[1]. In other words, M need not have a projection on it. On the other hand, M may have more than one projections.

Problem 3.20. Let $v \in \mathsf{X}$ and $f \in \mathsf{X}^*$ be given. The operator P defined by $Pu = (u, f)\, v$ for all $u \in \mathsf{X}$ is a projection if and only if $(v, f) = 1$. In this case $P\mathsf{X}$ is the one-dimensional manifold $[v]$ spanned by v, and $\mathsf{N}(P)$ is the closed linear manifold of X consisting of all u with $(u, f) = 0$ $(\mathsf{N}(P) = [f]^{\perp} \cap \mathsf{X})$. We have $\|P\| \leqq \|f\| \|v\|$.

Problem 3.21. The results of I-§ 4.6 on pairs of projections are valid for projections P, Q in a Banach space X. In particular $P\mathsf{X}$ and $Q\mathsf{X}$ are isomorphic[2] if $\|P - Q\| < 1$.

Also the results of I-§ 3.4 on a family $P_1, \ldots, P_s$ of projections in X satisfying $P_j P_k = \delta_{jk} P_j$ can be extended to a Banach space without modification.

If P is a projection in X, P^* is a projection in X^*. Again we have $\mathsf{M}^* = P^* \mathsf{X}^* = \mathsf{N}(1 - P^*) = \mathsf{R}(1 - P)^{\perp} = \mathsf{N}^{\perp}$ and similarly $\mathsf{N}^* = (1 - P^*)\, \mathsf{X}^* = \mathsf{M}^{\perp}$.

Example 3.22. Let $\mathsf{X} = C[-a, a]$, $a > 0$. Let M, N be the subsets of X consisting of all even and odd functions, respectively. M, N are complementary closed linear manifolds of X. The projection P on M along N is given by

$$Pu(x) = \frac{1}{2}\,(u(x) + u(-x))\,. \tag{3.15}$$

It follows easily that $\|P\| = \|1 - P\| = 1$. M can further be decomposed into the direct sum of two closed linear manifolds M_0, M_1 in such a way that, considered on the subinterval $[0, a]$, each $u \in \mathsf{M}_0$ is even and each $u \in \mathsf{M}_1$ is odd with respect to the center $x = a/2$. N is likewise decomposed as $\mathsf{N} = \mathsf{N}_0 \oplus \mathsf{N}_1$, so that $\mathsf{X} = \mathsf{M}_0 \oplus \mathsf{M}_1 \oplus \mathsf{N}_0 \oplus \mathsf{N}_1$. Each of the four projections associated with this decomposition of X again has norm one. It is easy to see that the same results hold for $\mathsf{X} = L^p(-a, a)$.

Example 3.23. Let $\mathsf{X} = C[-\pi, \pi]$ and

$$P_n u(x) = \frac{1}{\pi}\left(\int_{-\pi}^{\pi} \cos n x\, u(x)\, dx\right) \cos n x\,, \qquad n = 1, 2, \ldots\,. \tag{3.16}$$

The P_n are projecitons (see Problem 3.20) and satisfy $P_m P_n = \delta_{mn} P_n$, with

$$\|P_n\| \leqq \frac{1}{\pi} \|\cos n x\|_1 \|\cos n x\|_\infty = \frac{4}{\pi}\,. \tag{3.17}$$

[1] See DUNFORD and SCHWARTZ [1], p. 553.

[2] Two Banach spaces X, Y are said to be isomorphic (or equivalent) if there is a $U \in \mathscr{B}(\mathsf{X}, \mathsf{Y})$ with $U^{-1} \in \mathscr{B}(\mathsf{Y}, \mathsf{X})$.

The same is true if $X = C[-\pi, \pi]$ is replaced by $L^p(-\pi, \pi)$, with the exception that the $\| \|_1$ and $\| \|_\infty$ should be replaced by $\| \|_q$ and $\| \|_p$ with $p^{-1} + q^{-1} = 1$. In particular we have $\|P_n\| = 1$ for $p = 2$.

§ 4. Compact operators

1. Definition

There is a class of bounded operators, called compact (or completely continuous) operators, which are in many respects analogous to operators in finite-dimensional spaces. An operator $T \in \mathscr{B}(X, Y)$ is *compact* if the image $\{T u_n\}$ of any bounded sequence $\{u_n\}$ of X contains a Cauchy subsequence.

Example 4.1. The integral operator T of Example 2.4 is compact if E, F are compact sets and $t(y, x)$ is *continuous* in x, y and if T is regarded as an operator from $X = L^1(E)$ to $Y = C(F)$. To see this, we note that

$$(4.1)\qquad \begin{aligned} |Tu(y') - Tu(y'')| &\leq \int_E |t(y', x) - t(y'', x)|\, |u(x)|\, dx \\ &\leq \|u\| \max_x |t(y', x) - t(y'', x)| . \end{aligned}$$

Since $t(y, x)$ is continuous, it is uniformly continuous and $|Tu(y') - Tu(y'')|$ can be made arbitrarily small by taking $|y' - y''|$ small. How small $|y' - y''|$ should be depends only on $\|u\|$. In other words, the Tu are *equicontinuous* for a bounded set of u. Since $\{Tu\}$ is *equibounded* for a bounded set $\{u\}$ for a similar reason, we conclude from the theorem of Ascoli[1] that $\{Tu_n\}$ contains a uniformly convergent subsequence if $\{u_n\}$ is bounded. Since this means that $\{Tu_n\}$ contains a Cauchy subsequence in $Y = C(F)$, T is compact.

The same is true when T is regarded as an operator from X to Y, if X is any one of $C(E)$ and $L^p(E)$ and Y is any one of $C(F)$ and $L^q(F)$, $1 \leq p, q \leq \infty$. This is due to the facts that a bounded sequence in X is a fortiori bounded in the norm $\|u\|_1$ and that a Cauchy sequence in $C(F)$ is a fortiori a Cauchy sequence in Y[2].

Example 4.2. Let $X = C'[a, b]$ and $Y = C[a, b]$ with the norm defined as in (1.8) and (1.4). Since X is a subset of Y, the operator T that sends every $u \in X$ into the same function $u \in Y$ is an operator on X to Y. T is compact. This is again a consequence of the Ascoli theorem, for $\{Tu\}$ is equibounded and equicontinuous if $\{\|u\|\}$ is bounded[3].

Problem 4.3. Every operator $T \in \mathscr{B}(X, Y)$ is compact if at least one of Y, X is finite-dimensional.

Problem 4.4. The identity operator 1_X in a Banach space X is compact if and only if X is finite-dimensional. This is another expression of the proposition that X is locally compact if and only if X is finite-dimensional (see § 1.2).

Problem 4.5. A projection $P \in \mathscr{B}(X)$ is compact if and only if the range of P is finite-dimensional.

Problem 4.6. The inverses of the differential operators T_k, $k = 1, 2, 3$, of Examples 2.6–2.7 are compact. The same is true for the second-order differential operators considered in § 2.3.

---

[1] See, e. g. Royden [1], p. 155.

[2] Many other elementary examples of compact operators are given in Lyusternik and Sobolev [1].

[3] Note that $|u(t) - u(s)| = \left| \int_s^t u'(x)\, dx \right| \leq |t - s|\, \|u\|_X$.

2. The space of compact operators

We denote by $\mathscr{B}_0(X, Y)$ the set of all compact operators of $\mathscr{B}(X, Y)$.

Theorem 4.7. *$\mathscr{B}_0(X, Y)$ is a closed linear manifold of the Banach space $\mathscr{B}(X, Y)$. Thus $\mathscr{B}_0(X, Y)$ is itself a Banach space with the same norm as in $\mathscr{B}(X, Y)$.*

Proof. $\mathscr{B}_0(X, Y)$ is a linear manifold. Since it is obvious that αT is compact with T, it suffices to show that $T' + T''$ is compact whenever T', T'' are. Let $\{u_n\}$ be a bounded sequence in X. Take a subsequence $\{u'_n\}$ of $\{u_n\}$ such that $\{T' u'_n\}$ is a Cauchy sequence in Y, and then take a subsequence $\{u''_n\}$ of $\{u'_n\}$ such that $\{T'' u''_n\}$ is Cauchy. Then $\{(T' + T'') u''_n\}$ is a Cauchy sequence. Hence $T' + T''$ is compact.

To prove that $\mathscr{B}_0(X, Y)$ is closed, let $\{T_k\}$ be a sequence of compact operators such that $\|T_k - T\| \to 0$, $k \to \infty$, for some $T \in \mathscr{B}(X, Y)$; we have to show that T is also compact. Let $\{u_n\}$ be a bounded sequence in X. As above, take a subsequence $\{u_n^{(1)}\}$ of $\{u_n\}$ such that $\{T_1 u_n^{(1)}\}$ is Cauchy, then a subsequence $\{u_n^{(2)}\}$ of $\{u_n^{(1)}\}$ such that $\{T_2 u_n^{(2)}\}$ is Cauchy, and so on. Then the diagonal sequence $\{u_n^{(n)} = v_n\}$ has the property that $\{T v_n\}$ is Cauchy. In fact, since $\{v_n\}$ is a subsequence of every sequence $\{u_n^{(k)}\}$, each $\{T_k v_n\}$ is Cauchy for fixed k. For any $\varepsilon > 0$, take a k so large that $\|T_k - T\| < \varepsilon$ and then take N so large that $\|T_k v_n - T_k v_{n+p}\| < \varepsilon$ for $n > N$, $p > 0$. Then

$$\text{(4.2)} \qquad \|T v_n - T v_{n+p}\| \leqq \|(T - T_k)(v_n - v_{n+p})\| + \|T_k(v_n - v_{n+p})\| \leqq \leqq (2M + 1)\,\varepsilon\,,$$

where $M = \sup\|u_n\| < \infty$. Since (4.2) is true whenever $n > N$, $\{T v_n\}$ is Cauchy. This proves that T is compact.

Theorem 4.8. *The product of a compact operator with a bounded operator is compact. More precisely, let $T \in \mathscr{B}_0(X, Y)$ and $A \in \mathscr{B}(Y, Z)$, $B \in \mathscr{B}(W, X)$, W, X, Y, Z being Banach spaces. Then $A T \in \mathscr{B}_0(X, Z)$ and $T B \in \mathscr{B}_0(W, Y)$.*

Proof. Let $\{u_n\}$ be a bounded sequence in X. Take a subsequence $\{u'_n\}$ of $\{u_n\}$ such that $\{T u'_n\}$ is a Cauchy sequence in Y. Then $\{A T u'_n\}$ is a Cauchy sequence in Z; this shows that $A T$ is compact. Again, let $\{v_n\}$ be a bounded sequence in W. Then $B v_n$ is bounded in X and therefore contains a subsequence $\{B v'_n\}$ such that $\{T B v'_n\}$ is Cauchy. This shows that $T B$ is compact.

If in particular $Y = X$, it follows from Theorem 4.8 that the product (in either order) of operator of $\mathscr{B}_0(X) = \mathscr{B}_0(X, X)$ with any operator of $\mathscr{B}(X)$ is again compact. This is expressed by saying that *$\mathscr{B}_0(X)$ is a closed two-sided ideal of the Banach algebra $\mathscr{B}(X)$.*

Problem 4.9. If $\dim X = \infty$, every $T \in \mathscr{B}_0(X)$ is singular [$T^{-1} \in \mathscr{B}(X)$ is impossible].

Theorem 4.10. *The adjoint of a compact operator is compact, that is,* $T \in \mathscr{B}_0(\mathsf{X}, \mathsf{Y})$ *implies* $T^* \in \mathscr{B}_0(\mathsf{Y}^*, \mathsf{X}^*)$.

Proof. We begin by proving that *the range of a compact operator T is separable.* To this end it suffices to show that the image $T\mathsf{S}$ of the unit ball S of X is separable, for $\mathsf{R}(T)$ is the union of $T\mathsf{S}$, $2T\mathsf{S}$, $3T\mathsf{S}$, ... each of which is similar to $T\mathsf{S}$. For any positive integer n, there are a finite number p_n of elements of $T\mathsf{S}$ such that any point of $T\mathsf{S}$ is within the distance $1/n$ from some of these p_n points. Otherwise there would exist an infinite number of points Tu_n ($u_n \in \mathsf{S}$) separated from one another with distance larger than $1/n$; these points would form a sequence containing no Cauchy sequence, contrary to the assumption that T is compact and S is bounded. Now the assembly of these p_n points for all $n = 1, 2, 3, \ldots$ is countable and dense in $T\mathsf{S}$, which shows that $T\mathsf{S}$ is separable.

Returning to the proof of the theorem, let $T \in \mathscr{B}_0(\mathsf{X}, \mathsf{Y})$. Then $T^* \in \mathscr{B}(\mathsf{Y}^*, \mathsf{X}^*)$. We have to show that, from any bounded sequence $\{g_n\}$ in Y^*, we can extract a subsequence $\{g_n'\}$ such that $\{T^* g_n'\}$ is a Cauchy sequence in X^*. Let $\{v_k\}$ be a sequence dense in $\mathsf{R}(T) \subset \mathsf{Y}$; the existence of such a sequence was just proved. Since every numerical sequence $\{(g_n, v_k)\}$ with fixed k is bounded, the "diagonal method" can be applied to extract a subsequence $\{f_n\}$ of $\{g_n\}$ such that $\{(f_n, v_k)\}$ is a Cauchy sequence for every fixed k. Since $\{v_k\}$ is dense in $\mathsf{R}(T)$, it follows that $\{(f_n, v)\}$ is Cauchy for every $v \in \mathsf{R}(T)$. (The arguments are similar to those used in the proof of Theorem 4.7.)

Set $\lim (f_n, v) = f[v]$. f is a semilinear form on $\mathsf{R}(T)$ and is bounded because $\{f_n\}$ is bounded. f can be extended to a bounded semilinear form on Y (the Hahn-Banach theorem), which will again be denoted by f. Thus we have $(f_n, v) \to (f, v)$, $f \in \mathsf{Y}^*$, for every $v \in \mathsf{R}(T)$, that is,

$$(f_n, Tu) \to (f, Tu), \quad n \to \infty, \quad \text{for all} \quad u \in \mathsf{X}. \tag{4.3}$$

We shall now show that $T^* f_n \underset{s}{\to} T^* f$. Since $\{f_n\}$ is a subsequence of $\{g_n\}$, this will complete the proof. Set $f_n - f = h_n$; we have to show that $T^* h_n \underset{s}{\to} 0$. If this were not true, there would exist a $\delta > 0$ such that $\|T^* h_n\| \geqq \delta$ for an infinite number of subscripts n. We may assume that it is true for all n, for otherwise we need only to replace $\{h_n\}$ by a suitable subsequence. Since $\|T^* h_n\| = \sup |(T^* h_n, u)|$ for $\|u\| = 1$, there is for each n a $u_n \in \mathsf{X}$ such that

$$|(h_n, Tu_n)| = |(T^* h_n, u_n)| \geqq \delta/2, \quad \|u_n\| = 1. \tag{4.4}$$

The compactness of T implies that $\{Tu_n\}$ has a Cauchy subsequence. Again replacing $\{Tu_n\}$ by a suitable subsequence if necessary, we may assume that $\{Tu_n\}$ is itself Cauchy. For any $\varepsilon > 0$, there is thus an N

such that $m, n > N$ implies $\|Tu_n - Tu_m\| < \varepsilon$. Then it follows from (4.4) that

$$\delta/2 \leq |(h_n, Tu_n)| \leq |(h_n, Tu_n - Tu_m)| + |(h_n, Tu_m)|$$
$$\leq M\,\varepsilon + |(h_n, Tu_m)|\,, \quad M = \sup\|h_n\|\,.$$

Going to the limit $n \to \infty$ for a fixed m gives $\delta/2 \leq M\,\varepsilon$, for (4.3) implies $(h_n, Tu) \to 0$. But since $\varepsilon > 0$ was arbitrary, this contradicts $\delta > 0$.

3. Degenerate operators. The trace and determinant

An operator $T \in \mathscr{B}(\mathsf{X}, \mathsf{Y})$ is said to be *degenerate* if rank T is finite, that is, if $\mathsf{R}(T)$ is finite-dimensional. This range is necessarily closed. Since a finite-dimensional space is locally compact (I-§ 1.5), a *degenerate operator is compact*. It is easy to see that the set of all degenerate operators is a linear manifold in $\mathscr{B}(\mathsf{X}, \mathsf{Y})$, but it is in general not closed.

Also it is easily verified that the product of a degenerate operator T with a bounded operator is degenerate, with its rank not exceeding rank T (a more precise formulation should be given in a form similar to Theorem 4.8). In particular, the set of all degenerate operators of $\mathscr{B}(\mathsf{X})$ is a (not necessarily closed) two-sided ideal of the algebra $\mathscr{B}(\mathsf{X})$.

Problem 4.11. $T \in \mathscr{B}(\mathsf{X}, \mathsf{Y})$ is degenerate if and only if the codimension of the null space $\mathsf{N}(T)$ is finite ($\dim \mathsf{X}/\mathsf{N}(T) < \infty$).

Problem 4.12. Let T_n be degenerate and let $\|T_n - T\| \to 0$. Then T is compact, though not necessarily degenerate.

A degenerate operator $T \in \mathscr{B}(\mathsf{X}, \mathsf{Y})$ can be described conveniently by making use of a basis $y_1, \ldots, y_m$ of $\mathsf{R}(T)$, where $m = \operatorname{rank} T$. Since $Tu \in \mathsf{R}(T)$ for every $u \in \mathsf{X}$, we can write

$$Tu = \sum_{j=1}^{m} \eta_j\, y_j\,. \tag{4.5}$$

The coefficients η_j are uniquely determined by u and are obviously linear in u. Furthermore, these linear forms are bounded, for $|\eta_j| \leq \leq \gamma\|Tu\| \leq \gamma\|T\|\,\|u\|$ by I-(1.21). Hence we can write $\eta_j = \overline{(e_j, u)} = (u, e_j)$ with some $e_j \in \mathsf{X}^*$, and (4.5) becomes

$$Tu = \sum_{j=1}^{m} (u, e_j)\, y_j\,. \tag{4.6}$$

It is convenient to write, in place of (4.6),

$$T = \sum (\ \ , e_j)\, y_j\,. \tag{4.7}$$

For any $g \in \mathsf{Y}^*$ we have

$$(T^* g, u) = (g, Tu) = \sum_j (e_j, u)\,(g, y_j) = \Big(\sum_j (g, y_j)\, e_j, u\Big)\,. \tag{4.8}$$

Since this is true for all $u \in X$, we have

$$(4.9)\qquad T^* g = \sum_{j=1}^{m} (g, y_j)\, e_j \quad \text{or} \quad T^* = \sum (\ , y_j)\, e_j\, .$$

This shows that $R(T^*)$ is spanned by m vectors $e_1, \ldots, e_m$. Thus T^* is also degenerate with rank $T^* \leq$ rank T. Application of this result to T replaced by T^* shows that T^{**} is also degenerate with rank $T^{**} \leq$ $\leq$ rank T. Since, however, T may be regarded as a restriction of T^{**} (see § 3.3), we have also rank $T \leq$ rank T^{**}. Thus we have proved

Theorem 4.13. *T is degenerate if and only if T^* is. If T is degenerate, we have*

$$(4.10)\qquad \operatorname{rank} T^* = \operatorname{rank} T\, .$$

(4.10) *is true for all $T \in \mathscr{B}(X, Y)$ if ∞ is allowed as a value of* rank T.

Problem 4.14. An integral operator with a kernel of the form $\sum_{j=1}^{m} \overline{f_j(x)}\, g_j(y)$ is degenerate ($f_j \in X^*$, $g_j \in Y$).

Problem 4.15. In (4.9) the e_j form a basis of $R(T^*)$, which is correlated to the basis $\{y_j\}$ of $R(T)$. If we take an arbitrary basis $\{e_j\}$ of $R(T^*)$, then we have the expressions $T = \sum \alpha_{jk}(\ , e_j)\, y_k$, $T^* = \sum \overline{\alpha_{jk}}(\ , y_k)\, e_j$.

An important property of degenerate operators $T \in \mathscr{B}(X)$ is that the determinant can be defined for $1 + T$ and the trace can be defined for T[1]. $R(T)$ is finite-dimensional and invariant under T. Let T_R be the part of T in $R = R(T)$. We define the *determinant* of $1 + T$ by

$$(4.11)\qquad \det(1 + T) = \det(1_R + T_R)$$

where 1_R is the identity operator in R. [For $\det(1_R + T_R)$ see I-§ 3.1.]

Any subspace M of X containing R is likewise invariant under T. If M is finite-dimensional, $\det(1_M + T_M)$ is defined. We have

$$(4.12)\qquad \det(1_M + T_M) = \det(1 + T)\, .$$

This can proved by introducing into R a basis and extending it to a basis of M; the associated matrix for T_M has the property that all rows corresponding to the basis elements not in R contain only zero elements. Thus the determinant of this matrix is equal to $\det(1_R + T_R)$, which is equal to $\det(1 + T)$ by definition.

If T has the form (4.7) where the $y_j = x_j$ form a basis of R, the matrix elements of T_R for this basis are given by (x_k, e_j). Hence

$$(4.13)\qquad \det(1 + T) = \det(\delta_{jk} + (x_j, e_k))\, .$$

[1] There is a wider class of operators than that of degenerate operators for which the trace or the determinant can be defined; see Ruston [1], Grothendieck [1] (for operators in Hilbert space see X-§ 1.4).

Note that this is true even if T is defined by (4.7) with $y_j = x_j \in \mathsf{X}$ which are not necessarily linearly independent. This can be proved by an appropriate limiting procedure.

The determinant thus defined for degenerate operators has the property

$$\det((1 + S)(1 + T)) = \det(1 + S)\det(1 + T)\,. \tag{4.14}$$

Note that $(1 + S)(1 + T) = 1 + R$ where $R = S + T + ST$ is degenerate if S, T are. To prove (4.14), let M be a finite-dimensional subspace containing both $\mathsf{R}(S)$ and $\mathsf{R}(T)$. Then $\mathsf{R}(R) \subset \mathsf{M}$ and $\det((1 + S)\cdot$ $\cdot(1 + T)) = \det(1 + R) = \det(1_{\mathsf{M}} + R_{\mathsf{M}}) = \det(1_{\mathsf{M}} + S_{\mathsf{M}} + T_{\mathsf{M}} + S_{\mathsf{M}}T_{\mathsf{M}})$ $= \det((1_{\mathsf{M}} + S_{\mathsf{M}})(1_{\mathsf{M}} + T_{\mathsf{M}})) = \det(1_{\mathsf{M}} + S_{\mathsf{M}})\det(1_{\mathsf{M}} + T_{\mathsf{M}}) = \det(1 + S)\cdot$ $\cdot\det(1 + T)$ by (4.12).

It follows from (4.13) that $\det(1 + \varkappa T)$ is a polynomial in $\varkappa$ with degree not exceeding $\operatorname{rank} T = \dim \mathsf{R}$. The coefficient of $\varkappa$ in this polynomial is by definition the *trace* of T. Thus we have

$$\operatorname{tr} T = \operatorname{tr} T_{\mathsf{R}} = \operatorname{tr} T_{\mathsf{M}} \tag{4.15}$$

with the same notations as above. If T is given by (4.7) with $x = y$, we have from (4.13)

$$\operatorname{tr} T = \sum_{j=1}^{m} (x_j, e_j)\,. \tag{4.16}$$

The trace has the property

$$\operatorname{tr} TA = \operatorname{tr} AT \tag{4.17}$$

where T is a degenerate operator of $\mathscr{B}(\mathsf{X}, \mathsf{Y})$ and A is any operator of $\mathscr{B}(\mathsf{Y}, \mathsf{X})$. Note that TA is a degenerate operator in Y and AT is a degenerate operator in X. To prove (4.17), let $\mathsf{R} = \mathsf{R}(T) \subset \mathsf{Y}$ and $\mathsf{S} = A\mathsf{R} \subset \mathsf{X}$. R and S are finite-dimensional and $\mathsf{R}(AT) = A\mathsf{R}(T) = \mathsf{S}$ while $\mathsf{R}(TA) \subset \mathsf{R}(T) = \mathsf{R}$. Therefore, it suffices by (4.15) to show that $\operatorname{tr}(TA)_{\mathsf{R}} = \operatorname{tr}(AT)_{\mathsf{S}}$. But it is obvious that $(AT)_{\mathsf{S}} = A'T'$ where T' is the operator on S to R induced by T (that is, $T'u = Tu$ for $u \in \mathsf{S}$) and A' is the operator on R to S induced by A, and similarly $(TA)_{\mathsf{R}}$ $= T'A'$. Thus what we have to show is reduced to $\operatorname{tr} T'A' = \operatorname{tr} A'T'$, a familiar result in the finite-dimensional case (Problem I-3.8)

Problem 4.16. $\det(1 + T^*) = \overline{\det(1 + T)}$, $\operatorname{tr} T^* = \overline{\operatorname{tr} T}$.

Problem 4.17. $\det(1 + TS) = \det(1 + ST)$ [where one of $T \in \mathscr{B}(\mathsf{X},\mathsf{Y})$ and $S \in \mathscr{B}(\mathsf{Y},\mathsf{X})$ is degenerate].

Problem 4.18. Generalize I-(3.25) and I-(5.42) to the case of a degenerate operator.

Problem 4.19. If $T \in \mathscr{B}(\mathsf{X})$ is degenerate and nilpotent, then $\operatorname{tr} T = 0$, $\det(1 + T)$ $= 1$.

§ 5. Closed operators

1. Remarks on unbounded operators

We have seen in the preceding sections that most of the important results of the operator theory developed for finite-dimensional spaces apply without essential modification to operators of $\mathscr{B}(\mathsf{X}, \mathsf{Y})$ [or $\mathscr{B}(\mathsf{X})$]. For unbounded operators with domain not identical with the whole domain space, the situation is quite different and we encounter various difficulties.

For such operators, the construction of linear combinations and products stated in § 3.1 needs some modifications. The linear combination $\alpha S + \beta T$ of two operators S, T from X to Y is again defined by I-(3.12), but the domain of this operator is by definition the intersection of the domains of S and of T:

$$\mathsf{D}(\alpha S + \beta T) = \mathsf{D}(S) \cap \mathsf{D}(T) . \tag{5.1}$$

In fact, $\alpha S u + \beta T u$ would be meaningless if u did not belong to this intersection. It may happen that (5.1) consists of a single element $u = 0$; in such a case $\alpha S + \beta T$ is a trivial operator whose domain consists of $0 \in \mathsf{X}$ alone (its range also consists of $0 \in \mathsf{Y}$ alone).

Problem 5.1. $0 T \subset 0$, $0 + T = T + 0 = T$ for any operator T.

Problem 5.2. For any three operators R, S, T from X to Y, we have $(R + S) + T = R + (S + T)$, which will be written $R + S + T$, and also $S + T = T + S$, but we have only $(S + T) - T \subset S$ (the inclusion cannot in general be replaced by equality).

The product TS of an operator T from Y to Z and an operator S from X to Y is again defined by I-(3.15), but the domain of TS is by definition the set of all $u \in \mathsf{D}(S)$ such that $Su \in \mathsf{D}(T)$; otherwise the right member of this defining equation is meaningless. Thus

$$\mathsf{D}(TS) = S^{-1}\, \mathsf{D}(T) . \tag{5.2}$$

Again it is possible that $\mathsf{D}(TS)$ consist of the single element 0.

Problem 5.3. Let S, T be as above and let R be a third operator with range space X. Then we have $(TS) R = T(SR)$, which will be written TSR. Also we have $(\alpha T) S = T(\alpha S) = \alpha(TS)$ for a scalar $\alpha \neq 0$, but this is not true for $\alpha = 0$; it should be replaced by $(0 T) S = 0(TS) \subset T(0 S)$, for the domains of $(0 T) S$ and $0(TS)$ are the same as (5.2) while that of $T(0S)$ is equal to $\mathsf{D}(S)$. Again, we have $(T_1 + T_2) S = T_1 S + T_2 S$ but $T(S_1 + S_2) \supset TS_1 + TS_2$, where $\supset$ cannot in general be replaced by $=$. If 1_X denotes the identity operator in X, we have $1_\mathsf{Y} T = T 1_\mathsf{X} = T$. Compare these results with I-(3.16).

Problem 5.4. If T is an invertible operator from X to Y, then $T^{-1} T \subset 1_\mathsf{X}$, $T T^{-1} \subset 1_\mathsf{Y}$.

Remark 5.5. These results show that one must be careful in the formal manipulation of operators with restricted domains. For such operators it is often more convenient to work with vectors rather than with operators themselves. For example, we write $T^{-1}(Tu) = u$, $u \in \mathsf{D}(T)$, instead of $T^{-1}T \subset 1_{\mathsf{X}}$. As a rule, we shall make free use of the various operations on operators themselves only for operators of the class $\mathscr{B}(\mathsf{X}, \mathsf{Y})$

2. Closed operators

Among unbounded operators there are certain ones, called closed operators, which admit rather detailed treatment and which are also important for application.

Let T be an operator from X to Y. A sequence $u_n \in \mathsf{D}(T)$ will be said to be *T-convergent* (to $u \in \mathsf{X}$) if both $\{u_n\}$ and $\{Tu_n\}$ are Cauchy sequences (and $u_n \to u$). We shall write $u_n \underset{T}{\to} u$ to denote that $\{u_n\}$ is T-convergent to u. T is said to be *closed* if $u_n \underset{T}{\to} u$ implies $u \in \mathsf{D}(T)$ and $Tu = \lim Tu_n$; in other words if, for any sequence $u_n \in \mathsf{D}(T)$ such that $u_n \to u$ and $Tu_n \to$ $\to v$, u belongs to $\mathsf{D}(T)$ and $Tu = v$. In appearance closedness resembles continuity, but in reality the two notions are quite different.

The set of all closed operators from X to Y will be denoted by $\mathscr{C}(\mathsf{X}, \mathsf{Y})$[1]. Also we write $\mathscr{C}(\mathsf{X}, \mathsf{X}) = \mathscr{C}(\mathsf{X})$.

A bounded operator T is closed if and only if $\mathsf{D}(T)$ is closed. In fact, $u_n \to u$ with $u_n \in \mathsf{D}(T)$ implies automatically the existence of $\lim Tu_n = v$. Therefore, the closedness of T is equivalent to the condition that $u_n \to u$, $u_n \in \mathsf{D}(T)$, implies $u \in \mathsf{D}(T)$.

In particular, every $T \in \mathscr{B}(\mathsf{X}, \mathsf{Y})$ is closed: $\mathscr{B}(\mathsf{X}, \mathsf{Y}) \subset \mathscr{C}(\mathsf{X}, \mathsf{Y})$.

Problem 5.6. $T + A$ is closed if T is closed and A is bounded with $\mathsf{D}(A) \supset \mathsf{D}(T)$.

Problem 5.7. If $T \in \mathscr{C}(\mathsf{Y}, \mathsf{Z})$, $S \in \mathscr{C}(\mathsf{X}, \mathsf{Y})$ and $T^{-1} \in \mathscr{B}(\mathsf{Z}, \mathsf{Y})$, then $TS \in \mathscr{C}(\mathsf{X}, \mathsf{Z})$.

In dealing with closed operators, it is convenient to consider the *graph* of the operator. Consider the *product space* $\mathsf{X} \times \mathsf{Y}$ consisting of all (ordered) pairs $\{u, v\}$ of elements $u \in \mathsf{X}$ and $v \in \mathsf{Y}$. $\mathsf{X} \times \mathsf{Y}$ is a vector space if the linear operation is defined by

$$\alpha_1\{u_1, v_1\} + \alpha_2\{u_2, v_2\} = \{\alpha_1 u_1 + \alpha_2 u_2, \alpha_1 v_1 + \alpha_2 v_2\}. \tag{5.3}$$

Furthermore, $\mathsf{X} \times \mathsf{Y}$ becomes a normed space if the norm is defined by[2]

$$\|\{u, v\}\| = (\|u\|^2 + \|v\|^2)^{1/2}. \tag{5.4}$$

[1] In IV-§ 2 we shall introduce a metric in $\mathscr{C}(\mathsf{X}, \mathsf{Y})$ so as to make it into a metric space.

[2] Other choices of the norm are possible; for example, $\|\{u, v\}\| = \|u\| + \|v\|$ or $\max(\|u\|, \|v\|)$. We employ (5.4) mainly because it ensures that $(\mathsf{X} \times \mathsf{Y})^* = \mathsf{X}^* \times \mathsf{Y}^*$ (as Banach spaces), whereas this is not true for other norms unless different choices are made for $\mathsf{X} \times \mathsf{Y}$ and $\mathsf{X}^* \times \mathsf{Y}^*$. $(\mathsf{X} \times \mathsf{Y})^* = \mathsf{X}^* \times \mathsf{Y}^*$ means the following: (i) each element $\{f, g\} \in \mathsf{X}^* \times \mathsf{Y}^*$ defines an element $F \in (\mathsf{X} \times \mathsf{Y})^*$ by $(\{u, v\}, F) = (u, f) + (v, g)$ and, conversely, each $F \in (\mathsf{X} \times \mathsf{Y})^*$ is expressed in this form by a unique $\{f, g\} \in \mathsf{X}^* \times \mathsf{Y}^*$; (ii) the norm of the above $F \in (\mathsf{X} \times \mathsf{Y})^*$ is exactly equal to $\|\{f, g\}\| = (\|f\|^2 + \|g\|^2)^{1/2}$. It is easy to see that (i) is true. To prove (ii), it suffices to note that $|(\{u, v\}, \{f, g\})| \leqq |(u, f)| + |(v, g)| \leqq \|u\| \|f\| + \|v\| \|g\| \leqq (\|u\|^2 + \|v\|^2)^{1/2} (\|f\|^2 + \|g\|^2)^{1/2} = \|\{u, v\}\| \|\{f, g\}\|$ and that, for fixed $\{f, g\}$ and any $\varepsilon > 0$, there is a $\{u, v\}$ such that $\|u\| = \|f\|$, $\|v\| = \|g\|$, $(u, f) \geqq (1 - \varepsilon) \|f\|^2$, $(v, g) \geqq (1 - \varepsilon) \|g\|^2$ so that $|(\{u, v\}, \{f, g\})| \geqq (1 - \varepsilon)(\|f\|^2 + \|g\|^2)$.

It is easily seen that $X \times Y$ is complete and hence is a Banach space.

The graph $G(T)$ of an operator T from X to Y is by definition the subset of $X \times Y$ consisting of all elements of the form $\{u, Tu\}$ with $u \in D(T)$. $G(T)$ is a linear manifold of $X \times Y$. Now it is clear that a sequence $\{u_n\}$ of vectors of X is T-convergent if and only if $\{u_n, Tu_n\}$ is a Cauchy sequence in $X \times Y$. It follows that *T is closed if and only if $G(T)$ is a closed linear manifold of* $X \times Y$.

Problem 5.8. $S \subset T$ is equivalent to $G(S) \subset G(T)$.

Problem 5.9. If $T \in \mathscr{C}(X, Y)$, the null space $N(T)$ of T is a closed linear manifold of X.

Problem 5.10. In order that a linear manifold M of $X \times Y$ be the graph of an operator from X to Y, it is necessary and sufficient that no element of the form $\{0, v\}$ with $v \neq 0$ belong to M. Hence a linear submanifold of a graph is a graph.

Problem 5.11. A finite extension (see § 2.1) of a closed operator is closed. [hint: Lemma 1.9.]

Problem 5.12. Let $T \in \mathscr{C}(X, Y)$. If $u_n \in D(T)$, $u_n \underset{w}{\to} u \in X$ and $Tu_n \underset{w}{\to} v \in Y$, then $u \in D(T)$ and $Tu = v$. [hint: Apply Problem 1.34 to $G(T)$].

If S is an operator from Y to X, the graph $G(S)$ is a subset of $Y \times X$. Sometimes it is convenient to regard it as a subset of $X \times Y$. More precisely, let $G'(S)$ be the linear manifold of $X \times Y$ consisting of all pairs of the form $\{Sv, v\}$ with $v \in D(S)$. We shall call $G'(S)$ the *inverse graph* of S. As in the case of the graph, $G'(S)$ is a closed linear manifold if and only if S is closed.

If an operator T from X to Y is invertible, then clearly

$$G(T) = G'(T^{-1}) . \tag{5.5}$$

Thus *T^{-1} is closed if and only if T is.*

Problem 5.13. A linear manifold M of $X \times Y$ is an inverse graph if and only if M contains no element of the form $\{u, 0\}$ with $u \neq 0$.

Example 5.14. The differential operators considered in Examples 2.6–2.7 are all closed. In fact, T_1 is closed because $T_1^{-1} \in \mathscr{B}(X)$ is closed. The same is true of T_2 and T_3. T_0 is closed since it is the largest common restriction of T_1 and T_2 [in other words, $G(T_0) = G(T_1) \cap G(T_2)$]. In the same way, we see that all the differential operators of § 2.3 are closed.

Problem 5.15. T is closed if $R(T)$ is closed and there is an $m > 0$ such that $\|Tu\| \geq m\|u\|$ for all $u \in D(T)$.

3. Closable operators

An operator T from X to Y is said to be *closable* if T has a closed extension. It is equivalent to the condition that the graph $G(T)$ is a submanifold of a closed linear manifold which is at the same time a graph. It follows that *T is closable if and only* if the closure $\overline{G(T)}$ of $G(T)$ is a graph (note Problem 5.10). We are thus led to the criterion: T is closable if and only if no element of the form $\{0, v\}$, $v \neq 0$, is the limit of elements of the form $\{u, Tu\}$. In other words, *T is closable if and only if*

$$u_n \in D(T),\ u_n \to 0 \quad \text{and} \quad Tu_n \to v \quad \text{imply} \quad v = 0 . \tag{5.6}$$

When T is closable, there is a closed operator $\tilde{T}$ with $\mathsf{G}(\tilde{T}) = \overline{\mathsf{G}(T)}$. $\tilde{T}$ is called the *closure* of T. It follows immediately that $\tilde{T}$ is the *smallest closed extension* of T, in the sense that any closed extension of T is also an extension of $\tilde{T}$. Since $u \in \mathsf{D}(\tilde{T})$ is equivalent to $\{u, \tilde{T}u\} \in \overline{\mathsf{G}(T)}$, *$u \in \mathsf{X}$ belongs to $\mathsf{D}(\tilde{T})$ if and only if there exists a sequence $\{u_n\}$ that is T-convergent to u.* In this case we have $\tilde{T}u = \lim T u_n$.

Let T be a closed operator. For any closable operator S such that $\tilde{S} = T$, its domain $\mathsf{D}(S)$ will be called a *core* of T. In other words, a linear submanifold D of $\mathsf{D}(T)$ is a core of T if the set of elements $\{u, Tu\}$ with $u \in \mathsf{D}$ is dense in $\mathsf{G}(T)$. For this it is necessary (but not sufficient in general) that D be dense in $\mathsf{D}(T)$.

Problem 5.16. If T is bounded and closed, any linear submanifold D of $\mathsf{D}(T)$ dense in $\mathsf{D}(T)$ is a core of T.

Problem 5.17. Every bounded operator is closable (the extension principle, see § 2.2).

Problem 5.18. Every closable operator with finite rank is bounded. (Thus an unbounded linear form is never closable.)

Problem 5.19. Let T be an operator from X to Y with $T^{-1} \in \mathscr{B}(\mathsf{Y}, \mathsf{X})$. Then $\mathsf{D}' \subset \mathsf{D}(T)$ is a core of T if and only if $T\mathsf{D}'$ is dense in Y.

4. The closed graph theorem

We have seen above that a bounded operator with domain X is closed. We now prove a converse of this proposition.

Theorem 5.20. *A closed operator T from X to Y with domain X is bounded. In other words, $T \in \mathscr{C}(\mathsf{X}, \mathsf{Y})$ and $\mathsf{D}(T) = \mathsf{X}$ imply $T \in \mathscr{B}(\mathsf{X}, \mathsf{Y})$.*

Proof. Let S be the inverse image under T of the unit open ball of Y (it is not yet known whether S is open or not). Since $\mathsf{D}(T) = \mathsf{X}$, X is the union of $\mathsf{S}, 2\mathsf{S}, 3\mathsf{S}, \ldots$. It follows by an argument used in the proof of Theorem 1.29 that the closure $\bar{\mathsf{S}}$ of S contains a ball K, say with center u_0 and radius r.

Any $u \in \mathsf{X}$ with $\|u\| < 2r$ can be written in the form $u = u' - u''$ with $u', u'' \in \mathsf{K}$ (loc. cit.). Since $\mathsf{K} \subset \bar{\mathsf{S}}$, there are sequences $u_n', u_n'' \in \mathsf{S}$ such that $u_n' \to u'$, $u_n'' \to u''$. $\|T(u_n' - u_n'')\| \leq \|Tu_n'\| + \|Tu_n''\| < 2$ shows that $u_n' - u_n'' \in 2\mathsf{S}$. Thus $u = \lim(u_n' - u_n'') \in \overline{2\mathsf{S}}$. It follows by homogeneity that for any $\lambda > 0$, the ball $\|u\| < \lambda r$ of X is a subset of $\overline{\lambda \mathsf{S}}$.

Now take an arbitrary $u \in \mathsf{X}$ with $\|u\| < r$ and an arbitrary ε with $0 < \varepsilon < 1$. Since $u \in \bar{\mathsf{S}}$ as remarked above, there is a $u_1 \in \mathsf{S}$ within distance εr of u, that is, $\|u - u_1\| < \varepsilon r$ and $\|Tu_1\| < 1$. Hence $u - u_1 \in \overline{\varepsilon \mathsf{S}}$ by the same remark and, therefore, there is a $u_2 \in \varepsilon\,\mathsf{S}$ within distance $\varepsilon^2 r$ of $u - u_1$, that is, $\|u - u_1 - u_2\| < \varepsilon^2 r$, $\|Tu_2\| < \varepsilon$. Proceeding in this way, we can construct a sequence $\{u_n\}$ with the properties

$$\|u - u_1 - \cdots - u_n\| < \varepsilon^n r, \ \|Tu_n\| < \varepsilon^{n-1}, \ n = 1, 2, \ldots .$$

If we set $w_n = u_1 + \cdots + u_n$, we have $\|u - w_n\| < \varepsilon^n r \to 0$, $n \to \infty$, and

$$\|T w_n - T w_{n+p}\| \leqq \sum_{k=n+1}^{n+p} \|T u_k\| < \varepsilon^n + \varepsilon^{n+1} + \cdots \leqq (1-\varepsilon)^{-1} \varepsilon^n \to 0 .$$

This implies that $w_n \underset{T}{\to} u$. Since T is closed, we must have $Tu = \lim T w_n$. But since $\|T w_n\| < 1 + \varepsilon + \varepsilon^2 + \cdots = (1-\varepsilon)^{-1}$, we conclude that $\|Tu\| \leqq (1-\varepsilon)^{-1}$. Since this is true for every $u \in \mathsf{X}$ with $\|u\| < r$, T is bounded with $\|T\| \leqq (1-\varepsilon)^{-1} r^{-1}$. Since ε was arbitrary, it follows that $\|T\| \leqq 1/r$.

As an application of Theorem 5.20, we shall prove the boundedness of the projection P on M along N defined by (3.14). It suffices to show that P is closed, for P is defined everywhere on X and linear. Let $\{u_n\}$ be a P-convergent sequence: $u_n \to u$, $P u_n \to v$. Since $P u_n \in \mathsf{M}$ and M is closed, we have $v \in \mathsf{M}$. Since $(1-P)\, u_n \in \mathsf{N}$ and N is closed, $u - v = \lim (u_n - P u_n) \in \mathsf{N}$. Thus $Pu = v$ by definition and P is closed.

Problem 5.21. Let $T \in \mathscr{C}(\mathsf{X}, \mathsf{Y})$ with $\mathsf{R}(T) = \mathsf{Y}$. If T is invertible, then $T^{-1} \in \mathscr{B}(\mathsf{Y}, \mathsf{X})$.

Problem 5.22. Let $T \in \mathscr{B}(\mathsf{X}, \mathsf{Y})$ and let S be a closable operator from Y to Z with $\mathsf{D}(S) \supset \mathsf{R}(T)$. Then $ST \in \mathscr{B}(\mathsf{X}, \mathsf{Z})$. [hint: ST is closable with domain X, hence closed.]

5. The adjoint operator

Consider an operator T from X to Y and an operator S from Y^* to X^*. T and S are said to be *adjoint* to each other if

$$(g, Tu) = (Sg, u), \quad u \in \mathsf{D}(T), \quad g \in \mathsf{D}(S) . \tag{5.7}$$

For each operator T from X to Y, there are in general many operators from Y^* to X^* that are adjoint to T. If T is *densely defined*, however, there is a unique maximal operator T^* adjoint to T. This means that T^* is adjoint to T while any other S adjoint to T is a restriction of T^*. T^* is called *the adjoint (operator)* of T.

T^* is constructed in the following way. $\mathsf{D}(T^*)$ consists of all $g \in \mathsf{Y}^*$ such that there exists an $f \in \mathsf{X}^*$ with the property

$$(g, Tu) = (f, u) \quad \text{for all} \quad u \in \mathsf{D}(T) . \tag{5.8}$$

The $f \in \mathsf{X}^*$ is determined uniquely by g, for $(f, u) = (f', u)$ for all $u \in \mathsf{D}(T)$ implies $f = f'$ because $\mathsf{D}(T)$ is dense in X by assumption. Therefore, an operator T^* from Y^* to X^* is defined by setting $T^* g = f$. Obviously T^* is a linear operator, and comparison of (5.7) with (5.8) shows that $S \subset T^*$ holds for any S adjoint to T while T^* itself is adjoint to T.

The adjointness relation (5.7) admits a simple interpretation in terms of the graphs. Consider the product Banach space $\mathsf{X} \times \mathsf{Y}$ introduced in par. 2. Now (5.7) can be written $(-Sg, u) + (g, Tu) = 0$, which implies that $\{u, Tu\} \in \mathsf{X} \times \mathsf{Y}$ is annihilated by $\{-Sg, g\} \in \mathsf{X}^* \times \mathsf{Y}^* = (\mathsf{X} \times \mathsf{Y})^*$[1]. In other words, *$T$ and S are adjoint to each other if and only if the graph of T and the inverse graph of $-S$ annihilate each other*: $\mathsf{G}(T) \perp \mathsf{G}'(-S)$.

Similarly (5.8) shows that *the inverse graph of $-T^*$ is the annihilator of the graph of T*:

$$\mathsf{G}'(-T^*) = \mathsf{G}(T)^\perp . \tag{5.9}$$

The assumption that T is densely defined guarantees that $\mathsf{G}(T)^\perp$ is indeed an inverse graph. Since an annihilator is closed (see § 1.4), it follows that *T^* is a closed operator*. Note that this is true even if T is not closed or closable, but it may happen that T^* is trivial (has domain 0).

Problem 5.23. If $T \in \mathscr{B}(\mathsf{X}, \mathsf{Y})$, the above definition of T^* coincides with that of § 3.3.

Problem 5.24. If T and S are adjoint to each other and T is closable, then $\tilde{T}$ and S are also adjoint. In particular we have $T^* = (\tilde{T})^*$.

Problem 5.25. $T \subset T'$ implies $T^* \supset T'^*$ (if T is densely defined).

Problem 5.26. If T is from Y to Z, S is from X to Y and if TS is densely defined in X, then $(TS)^* \supset S^* T^*$. Here $\supset$ may be replaced by $=$ if $T \in \mathscr{B}(\mathsf{Y}, \mathsf{Z})$.

Problem 5.27. For any densely defined T, we have

$$\mathsf{N}(T^*) = \mathsf{R}(T)^\perp . \tag{5.10}$$

The notion of adjointness gives a very convenient criterion for closability, namely

Theorem 5.28. *Let T from X to Y and S from Y^* to X^* be adjoint to each other. If one of T, S is densely defined, the other is closable.*

Proof. If T is densely defined, T^* exists, is closed and $T^* \supset S$. Hence S is closable. If S is densely defined, $\mathsf{G}'(-S)^\perp$ is a graph in $\mathsf{X}^{**} \times \mathsf{Y}^{**}$ [just as $\mathsf{G}(T)^\perp$ is an inverse graph in $\mathsf{X}^* \times \mathsf{Y}^*$ if T is densely defined]. Since $\mathsf{G}(T)$ annihilates $\mathsf{G}'(-S)$, it is a subset of $\mathsf{G}'(-S)^\perp$ and the same is true of its closure $\overline{\mathsf{G}(T)}$ (regarded as a subset of $\mathsf{X}^{**} \times \mathsf{Y}^{**}$). Hence $\overline{\mathsf{G}(T)}$ is a graph and T is closable.

Theorem 5.29. *Let X, Y be reflexive. If an operator T from X to Y is densely defined and closable, then T^* is closed and densely defined and $T^{**} = \tilde{T}$.*

Proof. Since X, Y are reflexive, we have $\mathsf{G}(T)^{\perp\perp} = \overline{\mathsf{G}(T)} = \mathsf{G}(\tilde{T})$ by (1.24) (we idenitfy X^{**}, Y^{**} with X, Y respectively). Hence $\mathsf{G}(\tilde{T}) = \mathsf{G}'(-T^*)^\perp$, which implies that T^* is densely defined; otherwise there would exist a $v \in \mathsf{Y}$ such that $0 \neq v \perp \mathsf{D}(T^*)$, hence $\{0, v\} \in \mathsf{G}'(-T^*)^\perp = \mathsf{G}(\tilde{T})$, contradicting the fact that $\mathsf{G}(\tilde{T})$ is a graph. Thus T^{**} is defined as an operator from $\mathsf{X}^{**} = \mathsf{X}$ to $\mathsf{Y}^{**} = \mathsf{Y}$, and $\mathsf{G}(T^{**}) = \mathsf{G}'(-T^*)^\perp = \mathsf{G}(\tilde{T})$, which implies $T^{**} = \tilde{T}$.

---

[1] See footnote [1] of p. 164.

Theorem 5.30. *Let $T \in \mathscr{C}(X, Y)$ be densely defined. If T^{-1} exists and belongs to $\mathscr{B}(Y, X)$, then T^{*-1} exists and belongs to $\mathscr{B}(X^*, Y^*)$, with*

$$T^{*-1} = (T^{-1})^* . \tag{5.11}$$

Conversely, if T^{*-1} exists and belongs to $\mathscr{B}(X^*, Y^*)$, then T^{-1} exists and belongs to $\mathscr{B}(Y, X)$ and (5.11) holds.

Proof. Assume $T^{-1} \in \mathscr{B}(Y, X)$. Then $(T^{-1})^* \in \mathscr{B}(X^*, Y^*)$. For each $g \in D(T^*) \subset Y^*$ and $v \in Y$, we have $((T^{-1})^* T^* g, v) = (T^* g, T^{-1} v) = (g, T T^{-1} v) = (g, v)$; hence $(T^{-1})^* T^* g = g$. On the other hand, for each $f \in X^*$ and $u \in D(T) \subset X$ we have $((T^{-1})^* f, Tu) = (f, T^{-1} Tu) = (f, u)$; hence $(T^{-1})^* f \in D(T^*)$ and $T^* (T^{-1})^* f = f$ by the definition of T^* given above. The two relations thus proved show that T^{*-1} exists and equals $(T^{-1})^*$.

Conversely, assume $T^{*-1} \in \mathscr{B}(X^*, Y^*)$. For each $f \in X^*$ and $u \in D(T)$ we have $(T^{*-1} f, Tu) = (T^* T^{*-1} f, u) = (f, u)$. For any $u \in X$, however, there exists an $f \in X^*$ such that $\|f\| = 1$ and $(f, u) = \|u\|$ (see Corollary 1.24). For this f we have $\|u\| = (T^{*-1} f, Tu) \leqq \|T^{*-1}\| \, \|Tu\|$. This implies that T is invertible with $\|T^{-1}\| \leqq \|T^{*-1}\|$. Since T^{-1} is thus bounded and closed, $R(T)$ is closed. It remains to be shown that this range is the whole space Y. For this it suffices to show that no $g \neq 0$ of Y^* annihilates $R(T)$. But this is obvious from (5.10) since $T^* g = 0$ implies $g = 0$ by the invertibility of T^*.

Example 5.31. Let us determine the adjoints of the T, T_n, $n = 0, 1, 2, 3$, of Example 2.7 defined from the formal differential operator $L = d/dx$ in $X = L^p(a, b)$. We denote by S, S_n the same operators defined in $X^* = L^q(a, b)$, $p^{-1} + q^{-1} = 1$, assuming $1 \leqq p < \infty$. It is easily seen that T and $-S_0$ are adjoint to each other. We shall show that $T^* = -S_0$. To this end let $g \in D(T^*)$, $f = T^* g$; then

$$\int_a^b f \bar{u} \, dx = (f, u) = (T^* g, u) = (g, Tu) = \int_a^b g \bar{u}' \, dx \tag{5.12}$$

for every $u \in D(T)$. Set $h(x) = \int_a^x f \, dx$. Then $f = h'$ and $h(a) = 0$ so that (5.12) gives, after integration by parts,

$$\int_a^b (g + h) \bar{u}' \, dx - h(b) \overline{u(b)} = 0 . \tag{5.13}$$

For any $v \in X$, there is a $u \in X$ such that $u' = v$ and $u(b) = 0$. Hence $g + h \in X^*$ annihilates all $v \in X$ and so $g + h = 0$. Then (5.13) gives $h(b) \overline{u(b)} = 0$. But since there are $u \in D(T)$ with $u(b) \neq 0$, we have $h(b) = 0$. Hence $g = -h$ is absolutely continuous with $g' = -h' = -f \in X^*$ and $g(a) = g(b) = 0$, so that $g \in D(S_0)$ with $T^* g = f = -S_0 g$. This proves the desired result $T^* = -S_0$, since $T^* \supset -S_0$ as noted above.

In the same way it can be proved that $T_0^* = -S$, $T_1^* = -S_2$, $T_2^* = -S_1$, $T_3(k)^* = -S_3(1/\bar{k})$. Similarly, we have $\dot{T}^* = -S$ in the general case where (a, b) need not be finite ($\dot{T}$ is the minimal operator, see Example 2.7).

Example 5.32. Consider the operators of § 2.3 constructed from the formal second-order differential operator L [(2.13)]. First we consider the general (singular) case (Remark 2.16) and define the maximal and minimal operators T, $\dot{T}$ in $\mathsf{X} = \mathsf{L}^p(a, b)$, $1 \leqq p < \infty$. Similarly we define the maximal and minimal operators S, $\dot{S}$ from the formal adjoint M to L [see (2.27)]. A basic result in this connection is

$$\dot{T}^* = S . \tag{5.14}$$

The fact that $\dot{T}$ and S are adjoint to each other is easily seen from the Lagrange identity

$$\int_\alpha^\beta (\bar{v}\, L u - u\, \overline{M v})\, d x = [p_0\, u' \bar{v} - u (p_0 \bar{v})' + p_1\, u\, \bar{v}]_\alpha^\beta \tag{5.15}$$

where $u \in \mathsf{D}(\dot{T}) = \mathsf{C}_0^\infty(a, b)$, $v \in \mathsf{D}(S)$ and $u(x) = 0$ outside of the finite interval (α, β).

To prove the stronger result (5.14), we introduce an integral operator $K = K_\varepsilon$ with the kernel

$$k(y, x) = |y - x|\, \eta(y - x) , \tag{5.16}$$

where $\eta(t)$ is an infinitely differentiable function of the real variable t such that $\eta(t) = 1$ identically for $-\varepsilon/2 \leqq t \leqq \varepsilon/2$ and $\eta(t) = 0$ identically for $|t| \geqq \varepsilon$, ε being a sufficiently small positive constant. $k(y, x)$ is infinitely differentiable except for $y = x$ and $k(y, x) = 0$ for $|y - x| \geqq \varepsilon$.

Let $w(x)$ be an infinitely differentiable function vanishing identically outside the interval $(a + 2\varepsilon, b - 2\varepsilon)$ and let $u = K w$. Since the kernel $k(y, x)$ vanishes for $|y - x| \geqq \varepsilon$, $u(x)$ vanishes outside the interval $(a + \varepsilon, b - \varepsilon)$. Since $k(y, x)$ is continuous at $y = x$, it follows that $u' = K' w$ where K' is the integral operator with the kernel $k'(y, x) = \partial k(y, x)/\partial y$. Since $k'(y, x)$ has a discontinuity at $y = x$ with a jump of 2, the second derivative u'' cannot be obtained simply by differentiation under the integral sign; the correct expression for u'' is

$$u'' = 2 w + K'' w \tag{5.17}$$

where K'' is the integral operator with the kernel $k''(y, x) = \partial^2 k(y, x)/\partial y^2$. Note that k'' is infinitely differentiable *everywhere*. Thus u is infinitely differentiable and vanishes identically outside $(a + \varepsilon, b - \varepsilon)$, so that $u \in \mathsf{D}(\dot{T})$ and

$$\dot{T} u = L u = 2 p_0\, w + p_0\, K'' w + p_1\, K' w + p_2\, K w . \tag{5.18}$$

Now let $g \in \mathsf{D}(\dot{T}^*)$ and $f = \dot{T}^* g$. We have $(g, \dot{T} u) = (f, u)$ for all u of the above form, so that

$$(2 p_0\, g + K''^* p_0\, g + K'^* p_1\, g + K^* p_2\, g, w) = (K^* f, w) . \tag{5.19}$$

Note that K, K', K'' are bounded operators on X to X (since the kernels k, k', k'' are bounded functions; see Examples 2.4, 2.11), and their adjoints are again integral operators with the Hermitian conjugate kernels (Example 3.17). (5.19) is true for every infinitely differentiable function w vanishing outside $(a + 2\varepsilon, b - 2\varepsilon)$. Since such functions form a dense set in $\mathsf{L}^p(a + 2\varepsilon, b - 2\varepsilon)$, it follows that

$$g(x) = \frac{1}{2 p_0(x)} [K^* f(x) - K^* p_2\, g(x) - K'^* p_1\, g(x) - K''^* p_0\, g(x)] \tag{5.20}$$

for almost all $x \in (a + 2\varepsilon, b - 2\varepsilon)$.

Since it is known that $f, g \in \mathsf{X}^* = \mathsf{L}^q$, the right member of (5.20) is a continuous function of x (note that the kernels k, k', k'' are smooth or piecewise smooth functions). Hence $g(x)$ is continuous for $x \in (a + 2\varepsilon, b - 2\varepsilon)$. But since $\varepsilon > 0$ is arbitrary, $g(x)$ is continuous on (a, b). Returning to (5.20) we then see that

$g(x)$ is continuously differentiable on $(a + 2\varepsilon, b - 2\varepsilon)$ and hence on (a, b). This argument can be continued to the point that g' is absolutely continuous and g'' belongs to L^p locally [in any proper subinterval of (a, b)], for (5.20) shows that $g'' - f/p_0$ is continuous.

Once this local property of g is known, an integration by parts gives $(f, u) = (g, \dot{T} u) = (Mg, u)$ for every $u \in D(\dot{T})$ so that $Mg = f \in X^* = L^p(a, b)$. Thus $g \in D(S)$ and $Sg = Mg = f = \dot{T}^* g$. Since $S \subset \dot{T}^*$, this completes the proof of (5.14).

Let us now assume that L is regular (see Remark 2.16) and determine T^* and the T_n^*. Since $T_1 \supset \dot{T}$, we have $T_1^* \subset \dot{T}^* = S$. Thus the Lagrange identity gives

$$(5.21) \qquad [p_0 \bar{u}' g - \bar{u}(p_0 g)' + p_1 \bar{u} g]_a^b = (g, T u) - (Sg, u) \\ = (g, T_1 u) - (T_1^* g, u) = 0$$

for every $u \in D(T_1)$ and $g \in D(T_1^*) \subset D(S)$. Since $u'(a)$ and $u'(b)$ can take any values while $u(a) = u(b) = 0$, (5.21) implies that $g(a) = g(b) = 0$. Thus g satisfies the boundary conditions for S_1 and hence $T_1^* = S_1$. (The S_n are defined in X^* from M in the same way as the T_n are defined in X from L.)

Similarly it can be shown that $T^* = S_0$, $T_2^* = S_2$ (the constants h_a, h_b in the boundary conditions for T_2 and S_2 should be correlated appropriately), $T_3^* = S_4$, $T_4^* = S_3$ (the subscript 4 refers to the case 3 with a and b exchanged) and $T_0^* = S$.

These results show again that S (in the general case) and the S_n (in the regular case) are closed, since they are the adjoints of certain operators. Since the relationship between L and M is symmetric, the T and T_n are also closed (at least for $1 < p < \infty$).

Problem 5.33. If $u''(x)$ is continuous and $u(x) = 0$ outside a closed subinterval $[a', b']$ of (a, b), then $u \in D(\tilde{\dot{T}})$.

6. Commutativity and decomposition

Two operators $S, T \in \mathscr{B}(X)$ are said to commute if $ST = TS$, just as in the finite-dimensional case. It is not easy to extend this definition to unbounded operators in X because of the difficulty related to the domains. Usually this extension is done partly, namely to the case in which one of the operators belongs to $\mathscr{B}(X)$. An operator T in X is said to commute with an $A \in \mathscr{B}(X)$ (or A commute with T) if

$$(5.22) \qquad AT \subset TA .$$

It means that whenever $u \in D(T)$, Au also belongs to $D(T)$ and $TAu = ATu$.

Problem 5.34. (5.22) is equivalent to the old definition $AT = TA$ if $T \in \mathscr{B}(X)$.

Problem 5.35. Every operator T in X commutes with every scalar operator $\alpha 1$ (1 is the identity operator in X).

Problem 5.36. If $T \in \mathscr{C}(X)$ commutes with $A_n \in \mathscr{B}(X)$ and if $A_n \xrightarrow[w]{} A \in \mathscr{B}(X)$, then T commutes with A. [hint: Problem 5.12.]

Problem 5.37. If an invertible operator T in X commutes with $A \in \mathscr{B}(X)$, then T^{-1} commutes with A.

For an operator $T \in \mathscr{B}(X)$, the notion of a subspace M of X being *invariant* under T can be defined as in the finite-dimensional case by the condition $T\mathsf{M} \subset \mathsf{M}$. It is difficult to extend this notion to unbounded operators in X, for $T\mathsf{M} \subset \mathsf{M}$ would be satisfied whenever M has only 0 in common with $\mathsf{D}(T)$ (see § 2.1 for the notation $T\mathsf{M}$).

However, the notion of the *decomposition* of T by a pair M, N of complementary subspaces [see (3.14)] can be extended. T is said to be decomposed according to $\mathsf{X} = \mathsf{M} \oplus \mathsf{N}$ if

$$P\mathsf{D}(T) \subset \mathsf{D}(T), \quad T\mathsf{M} \subset \mathsf{M}, \quad T\mathsf{N} \subset \mathsf{N}, \tag{5.23}$$

where P is the projection on M along N (see § 3.4). Note that the first condition excludes the singular case mentioned above.

(5.23) is equivalent to the condition that T commutes with P:

$$TP \supset PT. \tag{5.24}$$

In fact, (5.23) implies that for any $u \in \mathsf{D}(T)$, $Pu \in \mathsf{D}(T)$ and $TPu \in \mathsf{M}$, $T(1-P)u \in \mathsf{N}$. Hence $(1-P)TPu = 0$ and $PT(1-P)u = 0$ so that $TPu = PTPu = PTu$, which implies (5.24). Similarly, it is easy to verify that (5.24) implies (5.23).

When T is decomposed as above, the *parts* T_M, T_N of T in M, N, respectively, can be defined. T_M is an operator in the Banach space M with $\mathsf{D}(T_\mathsf{M}) = \mathsf{D}(T) \cap \mathsf{M}$ such that $T_\mathsf{M} u = Tu \in \mathsf{M}$. T_N is defined similarly. If T is closed, the same is true of T_M and T_N, for $\mathsf{G}(T_\mathsf{M})$ is the intersection of $\mathsf{G}(T)$ with the closed set $\mathsf{M} \times \mathsf{M}$ (regarded as a subset of $\mathsf{X} \times \mathsf{X}$)

The notion of decomposition can be extended to the case in which there are several projections $P_1, \ldots, P_s$ satisfying $P_h P_k = \delta_{hk} P_h$, with the associated decomposition of the space $\mathsf{X} = \mathsf{M}_1 \oplus \cdots \oplus \mathsf{M}_s$ where $\mathsf{M}_h = P_h \mathsf{X}$. T is decomposed according to this decomposition of X if T commutes with all the P_h. The part T_{M_h} of T in M_h can be defined as above.

Problem 5.38. Let T be densely defined in X. If T is decomposed as described above, the part T_{M_h} is densely defined in M_h.

§ 6. Resolvents and spectra

1. Definitions

The eigenvalue problem considered in I-§ 5 in the finite-dimensional case requires an essential modification when we deal with operators T in a Banach space X[1]. Again, an eigenvalue of T is defined as a complex number λ such that there exists a nonzero $u \in \mathsf{D}(T) \subset \mathsf{X}$, called an eigenvector, such that $Tu = \lambda u$. In other words, λ is an eigenvalue if

[1] Recall that we always assume $\dim \mathsf{X} > 0$.

the null space $\mathsf{N}(T-\lambda)$ is not 0; this null space is the geometric eigenspace for λ and its dimension is the geometric multiplicity of the eigenvalue λ.

These definitions are often vacuous, however, since it may happen that T has no eigenvalue at all or, even if T has, there are not "sufficiently many" eigenvectors.

To generalize at least partly the results of the finite-dimensional case, it is most convenient to start from the notion of the resolvent. *In what follows T is assumed to be a closed operator in* X. Then the same is true of $T-\zeta$ for any complex number ζ. If $T-\zeta$ is invertible with

$$R(\zeta) = R(\zeta, T) = (T-\zeta)^{-1} \in \mathscr{B}(\mathsf{X}), \tag{6.1}$$

ζ is said to belong to the *resolvent set* of T. The operator-valued function $R(\zeta)$ thus defined on the resolvent set $\mathrm{P}(T)$ is called the *resolvent* of T. Thus $R(\zeta)$ has domain X and range $\mathsf{D}(T)$ for any $\zeta \in \mathrm{P}(T)$. This definition of the resolvent is in accordance with that given in the finite-dimensional case (I-§ 5.2)[1].

Problem 6.1. $\zeta \in \mathrm{P}(T)$ if and only if $T-\zeta$ has inverse with domain X (see Theorem 5.20).

Problem 6.2. If $\zeta \in \mathrm{P}(T)$, we have

$$R(\zeta)\, T \subset T R(\zeta) = 1 + \zeta R(\zeta) \in \mathscr{B}(\mathsf{X}). \tag{6.2}$$

Thus T commutes with $R(\zeta)$ (see § 5.6).

Problem 6.3. If $\mathrm{P}(T)$ is not empty, $\mathsf{D}' \subset \mathsf{D}(T)$ is a core of T if and only if $(T-\zeta)\,\mathsf{D}'$ is dense in X for some (or all) $\zeta \in \mathrm{P}(T)$ (see Problem 5.19).

Problem 6.4. If A is a closable operator from X to Y with $\mathsf{D}(A) \supset \mathsf{D}(T)$, then $A R(\zeta, T) \in \mathscr{B}(\mathsf{X}, \mathsf{Y})$ for every $\zeta \in \mathrm{P}(T)$. [hint: Problem 5.22.]

Theorem 6.5. *Assume that* $\mathrm{P}(T)$ *is not empty. In order that T commute with* $A \in \mathscr{B}(\mathsf{X})$, *it is necessary that*

$$R(\zeta)\, A = A\, R(\zeta) \tag{6.3}$$

for every $\zeta \in \mathrm{P}(T)$, *and it is sufficient that this hold for some* $\zeta \in \mathrm{P}(T)$.

Proof. This follows directly from Problem 5.37.

Problem 6.6. The $R(\zeta)$ for different ζ commute with each other.

The resolvent $R(\zeta)$ satisfies the *resolvent equation* I-(5.5) for every $\zeta_1, \zeta_2 \in \mathrm{P}(T)$. The proof is the same as before; it should only be observed that $T R(\zeta)$ is defined everywhere on X [see (6.2)]. From this it follows again that the Neumann series I-(5.6) for the resolvent is valid, but the proof is not trivial. Denote for the moment by $R'(\zeta)$ the right member

[1] We have defined $\mathrm{P}(T)$ and $\Sigma(T)$ only for closed operators in X. They can be defined for more general linear operators T in X. If T is closable, we set $\mathrm{P}(T) = P(\tilde{T})$, $\Sigma(T) = \Sigma(\tilde{T})$. If T is not closable, $\mathrm{P}(T)$ is empty and $\Sigma(T)$ is the whole plane.

of I-(5.6), which exists for small $|\zeta - \zeta_0|$. Then we have $R'(\zeta)(T-\zeta)u = u$ for every $u \in D(T)$, since $R(\zeta_0)(T-\zeta)u = u - (\zeta-\zeta_0)R(\zeta_0)u$. Similarly, we have *formally* $(T-\zeta)R'(\zeta)v = v$ for every $v \in X$. The assumed closedness of T then ensures that actually $R'(\zeta)v \in D(T)$ and the result is correct. This shows that $\zeta \in P(T)$ and $R'(\zeta) = R(\zeta)$ for any ζ for which the series converges. We have thus proved

Theorem 6.7. $P(T)$ *is an open set in the complex plane, and* $R(\zeta)$ *is (piecewise) holomorphic for* $\zeta \in P(T)$. *("Piecewise" takes into account that* $P(T)$ *need not be connected.) Each component of* $P(T)$ *is the natural domain of* $R(\zeta)$ *(*$R(\zeta)$ *cannot be continued analytically beyond the boundary of* $P(T)$*).*

The complementary set $\Sigma(T)$ (in the complex plane) of $P(T)$ is called the *spectrum* of T. Thus $\zeta \in \Sigma(T)$ if either $T - \zeta$ is not invertible or it is invertible but has range smaller than X. In the finite-dimensional case $\Sigma(T)$ consisted of a finite number of points (eigenvalues of T), but the situation is much more complicated now. It is possible for $\Sigma(T)$ to be empty or to cover the whole plane. Naturally we are interested in the moderate case nearer to the situation in the finite-dimensional case, but it happens frequently that the spectrum is an uncountable set.

Example 6.8. Consider the differential operators T and T_n of Example 2.6. $\Sigma(T)$ is the whole plane. In fact, the equation $(T-\zeta)u = u' - \zeta u = 0$ has always a nontrivial solution $u(x) = e^{\zeta x}$, which belongs to X. The restriction T_1 of T with the boundary condition $u(a) = 0$, on the other hand, has an empty spectrum. In fact the resolvent $R_1(\zeta) = R(\zeta, T_1)$ exists for every ζ and is given by

$$R_1(\zeta)v(y) = e^{\zeta y}\int_a^y e^{-\zeta x} v(x)\,dx\,. \tag{6.4}$$

Similarly $\Sigma(T_2)$ is empty. The spectrum of T_3 consists of a countable number of isolated points λ_n (which are exactly the eigenvalues of T_3) given by

$$\lambda_n = \frac{1}{b-a}(\log k + 2n\pi i)\,, \quad n = 0, \pm 1, \pm 2, \ldots\,. \tag{6.5}$$

If ζ is different from any of the λ_n, the resolvent $R_3(\zeta) = R(\zeta, T_3)$ exists and is given by the integral operator [cf. (2.11)]

$$R_3(\zeta)v(y) = \frac{e^{\zeta y}}{k - e^{(b-a)\zeta}}\left[k\int_a^y e^{-\zeta x}v(x)\,dx + e^{(b-a)\zeta}\int_y^b e^{-\zeta x}v(x)\,dx\right]. \tag{6.6}$$

Finally, $\Sigma(T_0)$ is again the whole plane. It is true that $(T_0 - \zeta)^{-1}$ exists and is bounded for every ζ but its domain is not the whole space X. In fact, each $v \in D((T_0-\zeta)^{-1}) = R(T_0 - \zeta)$ has the form $v = u' - \zeta u$ with u satisfying the boundary conditions $u(a) = u(b) = 0$, so that v is subjected to the condition

$$\int_a^b e^{-\zeta x}v(x)\,dx = 0\,. \tag{6.7}$$

These results remain true when the differential operator d/dx is considered in $X = L^p(a, b)$ for a finite (a, b) (Example 2.7).

In these examples, $R(\zeta, T)$ does not exist since the domain of T is too large while $R(\zeta, T_0)$ does not exist since $D(T_0)$ is too small. In a certain sense T_1, T_2, T_3 are "reasonable" operators.

Problem 6.9. Consider d/dx in $\mathsf{X} = \mathsf{L}^p(0, \infty)$ and define T and T_1 as in Example 2.7. Then $\mathrm{P}(T)$ is the right half-plane $\mathrm{Re}\,\zeta > 0$ and $\mathrm{P}(T_1)$ is the left half-plane, with

$$(6.8)\qquad \begin{aligned} R(\zeta, T)\, v(y) &= -\int_y^\infty e^{-\zeta(x-y)}\, v(x)\, dx\,, \quad \mathrm{Re}\,\zeta > 0\,, \\ R(\zeta, T_1)\, v(y) &= \int_0^y e^{\zeta(y-x)}\, v(x)\, dx\,, \quad \mathrm{Re}\,\zeta < 0\,. \end{aligned}$$

Problem 6.10. Consider d/dx on $(-\infty, \infty)$ and construct the maximal operator T in $\mathsf{X} = \mathsf{L}^p(-\infty, \infty)$. Then the two half-planes $\mathrm{Re}\,\zeta \gtrless 0$ belong to $\mathrm{P}(T)$ and

$$(6.9)\qquad R(\zeta)\, v(y) = \begin{cases} -\displaystyle\int_y^\infty e^{-\zeta(x-y)}\, v(x)\, dx\,, & \mathrm{Re}\,\zeta > 0\,, \\ \displaystyle\int_{-\infty}^y e^{\zeta(y-x)}\, v(x)\, dx\,, & \mathrm{Re}\,\zeta < 0\,. \end{cases}$$

Example 6.11. Consider the differential operators T and T_n of § 2.3 in $\mathsf{X} = \mathsf{C}[a, b]$. $\mathrm{P}(T)$ is empty, for $(T - \zeta)\, u = 0$ has always two linearly independent solutions in X. On the other hand, $\Sigma(T_3)$ is empty since $R_3(\zeta) = R(\zeta, T_3)$ exists for every ζ and is an integral operator analogous to (2.19).

$\Sigma(T_1)$ is neither empty nor the whole plane. The solution of $(T_1 - \zeta)\, u = v$ is given by an integral operator of the form (2.21) with the kernel $g(y, x)$ replaced by

$$(6.10)\qquad g(y, x; \zeta) = \frac{u_1(y;\zeta)\, u_2(x;\zeta)}{-p_0(x)\, W(x;\zeta)}\,, \quad y \le x\,; \qquad = \frac{u_2(y;\zeta)\, u_1(x;\zeta)}{-p_0(x)\, W(x;\zeta)}\,, \quad y \ge x\,.$$

Here u_1, u_2 are the solutions of $(L - \zeta)\, u = 0$ for the initial conditions $u_1(a;\zeta) = 0$, $u_1'(a;\zeta) = 1$ and $u_2(b;\zeta) = 0$, $u_2'(b;\zeta) = 1$. $W(x;\zeta)$ is the Wronskian of these two solutions and

$$(6.11)\qquad W(x;\zeta) = W_0(\zeta) \exp\left(-\int_a^x \frac{p_1}{p_0}\, dx\right), \quad W_0(\zeta) = -u_2(a;\zeta)\,.$$

$R_1(\zeta) = R(\zeta, T_1)$ exists if and only if $W_0(\zeta) \neq 0$. Since $W_0(\zeta)$ is an entire function of ζ, its zeros form a countable set $\{\lambda_n\}$ (the λ_n are eigenvalues of T_1). Thus $\Sigma(T_1)$ is a countable set consisting of eigenvalues. That $\Sigma(T_1)$ is not empty can be seen, for example, by observing that the eigenvalue problem for T_1 can be converted into a selfadjoint form by a simple transformation[1]. This remark shows also that the λ_n are all real.

Let us further recall that the Green function (2.22) exists if $\min p_2(x) = c > 0$ and that the estimate (2.25) holds. Applying this result to the operator L replaced by $L - \zeta$, we see that $R_1(\zeta)$ exists and

$$(6.12)\qquad \|R_1(\zeta)\| \le \frac{1}{c - \mathrm{Re}\,\zeta} \quad \text{if} \quad \mathrm{Re}\,\zeta < c = \min_x p_2(x)\,,$$

at least if ζ is real. We shall show that (6.12) is true also for complex ζ. If $\mathrm{Re}\,\zeta < c$, we have $|\mu + \zeta| < \mu + c$ for sufficiently large real μ. Since this implies that $-\mu < c$, $R_1(-\mu)$ exists by what was just proved and $\|R_1(-\mu)\| \le 1/(\mu + c)$. Then it follows from I-(5.7) that $R_1(\zeta)$ exists, and (6.12) follows from I-(5.6) (set $\zeta_0 = -\mu$) when $\mu \to \infty$.

Similarly $\Sigma(T_2)$ consists of a countable number of eigenvalues. The half-plane $\mathrm{Re}\,\zeta < c$ again belongs to the resolvent set and the estimate (6.12) holds, provided that $h_a, h_b \ge 0$.

[1] The differential equation $L u = \lambda u$ can be transformed into $(p_0 v')' + q v = \lambda v$, where $v = (-p_0)^{-1/2} \exp\left[(1/2) \int^x (p_1/p_0)\, dx\right] u$ and $q = p_2 - (p_0' - p_1)^2/4p_0 + (p_0'' - p_1')/2$. This is a *selfadjoint* eigenvalue problem, and there exists a countable set $\{\lambda_n\}$ of real (and only real) eigenvalues (see V-§ 3.6).

These results are also true if we consider the operators in $X = L^p(a, b)$; the only modification is to replace the constant c by $\min(c, c')$ where c' is given by $\min(p_2 - p_1' + p_0'')$ [see (2.26)].

2. The spectra of bounded operators

Consider now an operator $T \in \mathscr{B}(X)$. Then *neither* $P(T)$ *nor* $\Sigma(T)$ *is empty*. More precisely, $P(T)$ contains the exterior of the circle

$$|\zeta| = \operatorname{spr} T = \lim_{n\to\infty} \|T^n\|^{1/n} = \inf_{n\geq 1} \|T^n\|^{1/n} \tag{6.13}$$

(which reduces to the single point $\zeta = 0$ if and only if $\operatorname{spr} T = 0$, that is, T is quasi-nilpotent), whereas there is at least one point of $\Sigma(T)$ on this circle[1]. In particular $\Sigma(T)$ is a subset of the closed disk $|\zeta| \leqq \|T\|$. We note also that

$$\|\zeta R(\zeta) + 1\| \to 0, \quad \zeta \to \infty. \tag{6.14}$$

These are known in the finite-dimensional case (see I-§ 5.2), and the proof in the general case is not essentially different. We see that the Neumann series on the right of I-(5.10) converges for ζ outside the circle (6.13). That the sum of this series is equal to the resolvent $R(\zeta)$ can be seen as in the proof of Theorem 6.7. Since the convergence domain of this series is $|\zeta| > \operatorname{spr} T$, it follows that there is at least one point of $\Sigma(T)$ on (6.13) provided that $\operatorname{spr} T > 0$. If $\operatorname{spr} T = 0$, $\zeta = 0$ belongs to $\Sigma(T)$ because otherwise $R(\zeta)$ would be an entire function, contradicting (6.14) and Liouville's theorem.

Problem 6.12. Consider the shift operator $T \in \mathscr{B}(X)$, $X = l^p$, such that $T x_1 = 0$, $T x_n = x_{n-1}$ ($n \geqq 2$). $\Sigma(T)$ is the unit disk.

3. The point at infinity

The partition given above of the complex plane into the resolvent set and the spectrum of an operator does not refer to the point at infinity. For some purposes it is useful to consider this point in the partition. Before doing so we prove

Theorem 6.13. *Let* $T \in \mathscr{C}(X)$ *and let* $P(T)$ *contain the exterior of a circle. Then we have the alternatives:*

i) $T \in \mathscr{B}(X)$; $R(\zeta)$ *is holomorphic at* $\zeta = \infty$ *and* $R(\infty) = 0$.

ii) $R(\zeta)$ *has an essential singularity at* $\zeta = \infty$.

Proof. Suppose that $\zeta = \infty$ is not an essential singularity of $R(\zeta)$. Since $R(\zeta)$ is not identically zero, we have the expansion

$$R(\zeta) = \zeta^k A + \zeta^{k-1} B + \cdots, \quad A, B, \ldots \in \mathscr{B}(X), \quad A \neq 0, \tag{6.15}$$

for large $|\zeta|$, where k is an integer. Then

$$T R(\zeta) = 1 + \zeta R(\zeta) = 1 + \zeta^{k+1} A + \zeta^k B + \cdots. \tag{6.16}$$

[1] Hence $\operatorname{spr} T = \sup_{\lambda \in \Sigma(T)} |\lambda|$.

First we show that $k \leqq -1$. If $k \geqq 0$, we should have $\zeta^{-k-1} R(\zeta) \to 0$, $T \zeta^{-k-1} R(\zeta) \to A$ for $\zeta \to \infty$, which implies $A = 0$ by the closedness of T, contradicting the assumption $A \neq 0$. Hence $k \leqq -1$ and so $R(\zeta) \to 0$, $T R(\zeta) \to 1 + (\lim \zeta^{k+1}) A$ for $\zeta \to \infty$. Again the closedness of T requires that $1 + (\lim \zeta^{k+1}) A = 0$, which is possible only if $k = -1$ and $A = -1$. Thus $\zeta R(\zeta) u \to -u$ and $T \zeta R(\zeta) u \to B u$ for every $u \in \mathsf{X}$. The closedness of T again implies that $u \in \mathsf{D}(T)$, $Tu = -Bu$. In other words, $T = -B \in \mathscr{B}(\mathsf{X})$.

In view of Theorem 6.13, it is natural to include $\zeta = \infty$ in the resolvent set of T if $T \in \mathscr{B}(\mathsf{X})$ and in the spectrum of T otherwise. When it is desirable to distinguish these extended notions of the resolvent set and the spectrum (as subsets of the extended complex plane) from the proper ones defined before, we shall speak of the ***extended resolvent set*** and the ***extended spectrum*** and use the notations $\widetilde{\mathrm{P}}(T)$, $\widetilde{\Sigma}(T)$. Thus $\zeta = \infty \in \widetilde{\mathrm{P}}(T)$ if and only if $T \in \mathscr{B}(\mathsf{X})$. An unbounded operator always has $\zeta = \infty$ in its extended spectrum $\widetilde{\Sigma}(T)$; if it is an isolated point of $\widetilde{\Sigma}(T)$, it is an essential singularity of $R(\zeta)$.

Problem 6.14. $\widetilde{\Sigma}(T)$ is never empty. (Recall that $\dim \mathsf{X} > 0$.)

Theorem 6.15. *Let T be a closed invertible operator in X. $\widetilde{\Sigma}(T)$ and $\widetilde{\Sigma}(T^{-1})$ are mapped onto each other by the mapping $\zeta \to \zeta^{-1}$ of the extended complex plane*[1].

Proof. Let $0 \neq \zeta \in \mathrm{P}(T)$ so that $R(\zeta)$ exists. Set $S(\zeta) = T R(\zeta) = 1 + \zeta R(\zeta) \in \mathscr{B}(\mathsf{X})$. For every $u \in \mathsf{X}$ we have $S(\zeta) u = T R(\zeta) u$ and $T^{-1} S(\zeta) = R(\zeta) u = \zeta^{-1}(S(\zeta) - 1) u$. Hence

$$-\zeta (T^{-1} - \zeta^{-1}) S(\zeta) u = u \,. \tag{6.17}$$

This shows that $T^{-1} - \zeta^{-1}$ has range X. Moreover, this operator is invertible, for $(T^{-1} - \zeta^{-1}) v = 0$ implies $v = \zeta T^{-1} v$, $Tv = \zeta v$, $v = 0$. Thus it follows from (6.17) that $(T^{-1} - \zeta^{-1})^{-1} = -\zeta S(\zeta) \in \mathscr{B}(\mathsf{X})$, and $\zeta^{-1} \in \mathrm{P}(T^{-1})$.

If $\zeta = 0$ belongs to $\mathrm{P}(T)$, $T^{-1} \in \mathscr{B}(\mathsf{X})$ so that $0^{-1} = \infty \in \widetilde{\mathrm{P}}(T^{-1})$ by definition. If $\zeta = \infty$ belongs to $\widetilde{\mathrm{P}}(T)$, $T \in \mathscr{B}(\mathsf{X})$ and therefore $0 = \infty^{-1} \in \mathrm{P}(T^{-1})$. Thus $\widetilde{\mathrm{P}}(T)$ is mapped by $\zeta \to \zeta^{-1}$ onto $\widetilde{\mathrm{P}}(T^{-1})$. The same is true for the complementary sets $\widetilde{\Sigma}(T)$ and $\widetilde{\Sigma}(T^{-1})$.

Problem 6.16. The spectrum of $R(\zeta_0)$ is the bounded set obtained from $\widetilde{\Sigma}(T)$ by the transformation $\zeta \to \zeta' = (\zeta - \zeta_0)^{-1}$, and

$$R((\zeta - \zeta_0)^{-1}, R(\zeta_0)) = -(\zeta - \zeta_0) - (\zeta - \zeta_0)^2 R(\zeta) \,. \tag{6.18}$$

Furthermore, $\operatorname{spr} R(\zeta_0) = 1/\operatorname{dist}(\zeta_0, \Sigma(T))$.

---

[1] Theorem 6.15 is a special case of the *spectral mapping theorem*, which asserts that the spectrum of a "function" $\phi(T)$ of T is the image under ϕ of $\Sigma(T)$. $\phi(T)$ is defined by the Dunford-Taylor integral as in I-(5.47). We shall not consider this theorem in the general form (see Dunford and Schwartz [1]).

4. Separation of the spectrum

Sometimes it happens that the spectrum $\Sigma(T)$ of a closed operator T contains a bounded part Σ' separated from the rest Σ'' in such a way that a rectifiable, simple closed curve Γ (or, more generally, a finite number of such curves) can be drawn so as to enclose an open set containing Σ' in its interior and Σ'' in its exterior. (For most applications we consider in the following, the part Σ' will consist of a finite number of points.) Under such a circumstance, we have the following *decomposition theorem*.

Theorem 6.17. *Let $\Sigma(T)$ be separated into two parts Σ', Σ'' in the way described above. Then we have a decomposition of T according to a decomposition $\mathsf{X} = \mathsf{M}' \oplus \mathsf{M}''$ of the space (in the sense of § 5.6) in such a way that the spectra of the parts $T_{\mathsf{M}'}$, $T_{\mathsf{M}''}$ coincide with Σ', Σ'' respectively and $T_{\mathsf{M}'} \in \mathscr{B}(\mathsf{M}')$. Thus $\tilde{\Sigma}(T_{\mathsf{M}'}) = \Sigma'$ whereas $\tilde{\Sigma}(T_{\mathsf{M}''})$ may contain $\zeta = \infty$.*

Proof. Set

$$P = -\frac{1}{2\pi i} \int_\Gamma R(\zeta)\, d\zeta \in \mathscr{B}(\mathsf{X}) . \tag{6.19}$$

A calculation analogous to that used to deduce I-(5.17) shows that $P^2 = P$. Thus P is a projection on $\mathsf{M}' = P\mathsf{X}$ along $\mathsf{M}'' = (1 - P)\,\mathsf{X}$. Furthermore

$$PR(\zeta) = R(\zeta)\, P\,, \quad \zeta \in \mathrm{P}(T)\,, \tag{6.20}$$

so that P commutes with T (Theorem 6.5), which means that T is decomposed according to $\mathsf{X} = \mathsf{M}' \oplus \mathsf{M}''$ and the parts $T_{\mathsf{M}'}$, $T_{\mathsf{M}''}$ are defined.

It is readily seen that the parts of $R(\zeta)$ in M', M'', which we denote by $R_{\mathsf{M}'}(\zeta)$, $R_{\mathsf{M}''}(\zeta)$, are the inverses of $T_{\mathsf{M}'} - \zeta$, $T_{\mathsf{M}''} - \zeta$, respectively. This shows that both $\mathrm{P}(T_{\mathsf{M}'})$ and $\mathrm{P}(T_{\mathsf{M}''})$ contain $\mathrm{P}(T)$. Actually, however, $\mathrm{P}(T_{\mathsf{M}'})$ also contains Σ''. To see this we first note that $R_{\mathsf{M}'}(\zeta)\, u = R(\zeta)\, u = R(\zeta)\, Pu$ for $u \in \mathsf{M}'$, $\zeta \in \mathrm{P}(T)$. But for any $\zeta \in \mathrm{P}(T)$ not on Γ, we have

$$R(\zeta)\, P = -\frac{1}{2\pi i} \int_\Gamma R(\zeta)\, R(\zeta')\, d\zeta' = -\frac{1}{2\pi i} \int_\Gamma (R(\zeta) - R(\zeta'))\, \frac{d\zeta'}{\zeta - \zeta'} \tag{6.21}$$

by (6.19) and the resolvent equation I-(5.5). If ζ is outside Γ, this gives

$$R(\zeta)\, P = \frac{1}{2\pi i} \int_\Gamma R(\zeta')\, \frac{d\zeta'}{\zeta - \zeta'}\,. \tag{6.22}$$

Since the right member of (6.22) is holomorphic outside Γ, it follows that $R(\zeta)\, P$, and hence $R_{\mathsf{M}'}(\zeta)$ also, has an analytic continuation holomorphic outside Γ. That this continuation of $R_{\mathsf{M}'}(\zeta)$ is the resolvent of $T_{\mathsf{M}'}$ can be seen from Theorem 6.7. Thus $\mathrm{P}(T_{\mathsf{M}'})$ contains the exterior of Γ and therefore $\Sigma(T_{\mathsf{M}'}) \subset \Sigma'$.

Similarly, it follows from (6.21) that

$$R(\zeta)\, P = R(\zeta) + \frac{1}{2\pi i} \int_\Gamma R(\zeta') \frac{d\zeta'}{\zeta - \zeta'} \tag{6.23}$$

if ζ is inside Γ. This shows that $R(\zeta)\,(1 - P)$ has an analytic continuation holomorphic inside Γ. As above, this leads to the conclusion that $\Sigma(T_{\mathsf{M}''}) \subset \Sigma''$.

On the other hand, a point $\zeta \in \Sigma$ cannot belong to both $\mathrm{P}(T_{\mathsf{M}'})$ and $\mathrm{P}(T_{\mathsf{M}''})$; otherwise it would belong to $\mathrm{P}(T)$ because $R_{\mathsf{M}'}(\zeta)\, P + R_{\mathsf{M}''}(\zeta)\,(1 - P)$ would be equal to the inverse of $T - \zeta$. This shows that we have $\Sigma(T_{\mathsf{M}'}) = \Sigma'$, $\Sigma(T_{\mathsf{M}''}) = \Sigma''$.

Finally we shall show that

$$P\,T \subset T\,P = -\frac{1}{2\pi i} \int_\Gamma T\,R(\zeta)\, d\zeta = -\frac{1}{2\pi i} \int_\Gamma \zeta\, R(\zeta)\, d\zeta \in \mathscr{B}(\mathsf{X})\,. \tag{6.24}$$

$P\,T \subset T\,P$ expresses the known fact that T commutes with P. The second equality of (6.24) is obvious since $T\,R(\zeta) = 1 + \zeta\, R(\zeta)$. The first equality is obtained by a formal multiplication of (6.19) from the left by T. This multiplication is justified by the closedness of T [approximate (6.19) by a finite sum and use the boundedness of $T\,R(\zeta) = 1 + \zeta\, R(\zeta)$].

(6.24) implies that $T_{\mathsf{M}'} \in \mathscr{B}(\mathsf{M}')$. This completes the proof.

We note the following facts proved above. $R(\zeta)$ can be written in the form

$$\begin{aligned} &R(\zeta) = R'(\zeta) + R''(\zeta)\,, \\ &R'(\zeta) = R(\zeta)\, P\,, \quad R''(\zeta) = R(\zeta)\,(1 - P)\,. \end{aligned} \tag{6.25}$$

$R'(\zeta)$ is holomorphic outside Σ' and coincides with $R_{\mathsf{M}'}(\zeta)$ when restricted to M' while it vanishes on M''; similarly $R''(\zeta)$ is holomorphic outside Σ'' and coincides with $R_{\mathsf{M}''}(\zeta)$ on M'' while vanishing on M'.

Theorem 6.17 can easily be extended to the case in which $\Sigma(T)$ is separated into several parts $\Sigma_1, \ldots, \Sigma_s$ and Σ_0, where each Σ_h with $1 \leqq h \leqq s$ is bounded and is enclosed in a closed curve (or a system of closed curves) Γ_h running in $\mathrm{P}(T)$ and lying outside one another, whereas Σ_0 may be unbounded and is excluded by the Γ_h. Then the operators P_h defined by (6.19) with $\Gamma = \Gamma_h$ satisfy $P_h P_k = \delta_{hk} P_h$, $h, k = 1, \ldots, s$. T commutes with every P_h so that T is decomposed according to the decomposition $\mathsf{X} = \mathsf{M}_1 \oplus \cdots \oplus \mathsf{M}_s \oplus \mathsf{M}_0$, $\mathsf{M}_h = P_h \mathsf{X}$, where $P_0 = 1 - P_1 - \cdots - P_s$. The part T_{M_h} of T in M_h has spectrum Σ_h, and $T_{\mathsf{M}_h} \in \mathscr{B}(\mathsf{M}_h)$ for $h \geqq 1$.

5. Isolated eigenvalues[1]

Suppose that the spectrum $\Sigma(T)$ of $T \in \mathscr{C}(X)$ has an *isolated point* λ. Obviously $\Sigma(T)$ is divided into two separate parts Σ', Σ'' in the sense of the preceding paragraph, where Σ' consists of the single point λ; any closed curve enclosing λ but no other point of $\Sigma(T)$ may be chosen as Γ. The operator $T_{M'}$ described in Theorem 6.17 has spectrum consisting of the single point λ. Therefore, $T_{M'} - \lambda$ is quasi-nilpotent (see par. 2). The Neumann series I-(5.10) applied to $T_{M'} - \lambda$ beomes

$$R_{M'}(\zeta) = - \sum_{n=0}^{\infty} (\zeta - \lambda)^{-n-1} (T_{M'} - \lambda)^n \tag{6.26}$$

and converges except for $\zeta = \lambda$. (6.26) is equivalent to

$$R'(\zeta) = R(\zeta)\, P = - \frac{P}{\zeta - \lambda} - \sum_{n=1}^{\infty} \frac{D^n}{(\zeta - \lambda)^{n+1}} \tag{6.27}$$

where

$$D = (T - \lambda)\, P = - \frac{1}{2\pi i} \int_\Gamma (\zeta - \lambda)\, R(\zeta)\, d\zeta \in \mathscr{B}(X) \tag{6.28}$$

is likewise quasi-nilpotent and

$$D = D P = P D\,. \tag{6.29}$$

On the other hand $R_{M''}(\zeta)$ is holomorphic at $\zeta = \lambda$ and admits the Taylor expansion I-(5.6) with $\zeta_0 = \lambda$. This is equivalent to

$$R''(\zeta) = R(\zeta)\,(1 - P) = \sum_{n=0}^{\infty} (\zeta - \lambda)^n\, S^{n+1} \tag{6.30}$$

where

$$S = R_{M''}(\lambda)\,(1 - P) = \lim_{\zeta \to \lambda} R(\zeta)\,(1 - P)\,. \tag{6.31}$$

(Note that $R(\lambda)$ does not exist.) $R''(\zeta)$ will be called the *reduced resolvent* of T for the eigenvalue λ.

It follows from (6.27) and (6.30) that

$$R(\zeta) = - \frac{P}{\zeta - \lambda} - \sum_{n=1}^{\infty} \frac{D^n}{(\zeta - \lambda)^{n+1}} + \sum_{n=0}^{\infty} (\zeta - \lambda)^n\, S^{n+1}\,. \tag{6.32}$$

This is the Laurent series for $R(\zeta)$ at the isolated singularity $\zeta = \lambda$.

S has properties similar to those of the operator S_h introduced in the finite-dimensional case (I-§ 5.3), namely

$$S = \frac{1}{2\pi i} \int_\Gamma R(\zeta)\, \frac{d\zeta}{\zeta - \lambda} \in \mathscr{B}(X)\,, \tag{6.33}$$

$$S T \subset T S \in \mathscr{B}(X)\,, \quad (T - \lambda)\, S = 1 - P\,, \quad S P = P S = 0\,. \tag{6.34}$$

[1] The expression "an isolated eigenvalue" is somewhat ambiguous. We mean by it an eigenvalue which is an isolated point of the spectrum (not just an isolated point in the set of eigenvalues).

The Laurent expansion (6.32) is similar to I-(5.18) in the finite-dimensional case, with the sole difference that the principal part (with negative powers of $\zeta - \lambda$) may be an infinite series. The principal part is finite, however, if M′ is finite-dimensional, for $D_{M'} = T_{M'} - \lambda$ is then nilpotent (see Problem I-5.6) and the same is true of D. In this case λ is an *eigenvalue* of T. In fact, since λ belongs to the spectrum of the finite-dimensional operator $T_{M'}$, it must be an eigenvalue of $T_{M'}$ and hence of T. In this case dim M′ is again called the *(algebraic) multiplicity* of the eigenvalue λ of T, and P and D are the *eigenprojection* and the *eigennilpotent* associated with λ. λ may or may not be an eigenvalue of T if $\dim M' = \infty$.

These results can be extended to the case in which we consider several isolated points $\lambda_1, \ldots, \lambda_s$ of $\Sigma(T)$. The remark at the end of par. 4 leads immediately to

$$R(\zeta) = -\sum_{h=1}^{s}\left[\frac{P_h}{\zeta-\lambda_h} + \sum_{n=1}^{\infty}\frac{D_h^n}{(\zeta-\lambda_h)^{n+1}}\right] + R_0(\zeta)\,. \tag{6.35}$$

Here the P_h are projections and the D_h are quasi-nilpotents such that

$$P_h P_k = \delta_{hk} P_h\,,\quad P_h D_h = D_h P_h = D_h\,,\quad (T-\lambda_h) P_h = D_h\,. \tag{6.36}$$

$R_0(\zeta)$ is holomorphic at $\zeta = \lambda_h$, $h = 1, \ldots, s$, and

$$R_0(\zeta) = R(\zeta) P_0\,,\quad P_0 = 1 - (P_1 + \cdots + P_s)\,. \tag{6.37}$$

Again λ_h is an eigenvalue of T if $M_h = P_h X$ is finite-dimensional, and P_h and D_h are respectively the associated eigenprojection and eigen-nilpotent. We have further

$$T P = \sum_{h=1}^{s} (\lambda_h P_h + D_h)\,,\quad P = P_1 + \cdots + P_s\,. \tag{6.38}$$

Here we have a *spectral representation* of T in a restricted sense. This is not so complete as the one in the finite-dimensional case (I-§ 5.4), since the isolated points do not in general exhaust the spectrum $\Sigma(T)$ and, even if this is the case, there are in general an infinite number of points λ_h. Nevertheless, it gives a fairly complete description of the operator T if one is interested only in a limited portion of the complex plane where there are only a finite number of points of $\Sigma(T)$ which are eigenvalues with finite multiplicities. For brevity a finite collection $\lambda_1, \ldots, \lambda_s$ of such eigenvalues will be called a *finite system of eigenvalues*. For a finite system of eigenvalues, the situation is much the same as in the finite-dimensional case discussed in detail in I-§ 5. Most of the results deduced there can now be taken over, and this will be done in the sequel without further comment whenever there is no particular difficulty.

Problem 6.18. Suppose that $\dim M' = m < \infty$ in Theorem 6.17. Then Σ' consists of a finite system of eigenvalues with the total multiplicity m.

Example 6.19. Consider the differential operator T_3 of Example 2.6 (or 2.7). The spectrum of T_3 consists of the isolated points λ_n given by (6.5). Let us find the associated eigenprojections P_n. Integrating the resolvent $R_3(\zeta)$ given by (6.6) along a small circle around $\zeta = \lambda_n$ (which is a zero of $k - e^{(b-a)\zeta}$), we have by (6.19)

$$(6.39)\qquad P_n v(y) = -\frac{1}{2\pi i}\int R_3(\zeta)\, v(y)\, d\zeta = \frac{e^{\lambda_n y}}{b-a}\int_a^b e^{-\lambda_n x} v(x)\, dx\,,$$

where a simple calculation of the residue has been made[1]. P_n is a degenerate integral operator of rank one with the kernel

$$(6.40)\qquad p_n(y, x) = \frac{1}{b-a}\, e^{\lambda_n (y-x)}\,.$$

Each λ_n is an isolated eigenvalue of T_3 with multiplicity one (simple eigenvalue), and the associated eigennilpotent is zero.

Example 6.20. Consider the differential operator T_1 of § 2.3. $\Sigma(T_1)$ consists of isolated points λ_n which are the zeros of the entire function $W_0(\zeta)$, see Example 6.11. The eigenprojections P_n can be calculated in the same way as above: the resolvent $R_1(\zeta)$ is an integral operator with the kernel $g(y, x; \zeta)$ given by (6.10), and P_n is obtained by (6.19) as in (6.39). Since there is a constant k such that $u_2(x, \lambda_n) = k\, u_1(x, \lambda_n)$ in virtue of the vanishing of the Wronskian at $\zeta = \lambda_n$, a simple calculation of the residue gives

$$(6.41)\qquad P_n v(y) = \frac{k\,\varphi_n(y)}{W_0'(\lambda_n)}\int_a^b \frac{\varphi_n(x)\, v(x)}{p_0(x)}\, e^{\int_a^x \frac{p_1}{p_0} dx}\, dx\,,$$

$$\varphi_n(x) = u_1(x, \lambda_n)\,.$$

$\varphi_n(x)$ is an eigenfunction of T_1 for the eigenvalue λ_n. P_n is a degenerate integral operator with rank one. Incidentally, we note that $P_n^2 = P_n$ is equivalent to

$$(6.42)\qquad \frac{1}{k}\, W_0'(\lambda_n) = \int_a^b \frac{\varphi_n(x)^2}{p_0(x)}\, e^{\int_a^x \frac{p_1}{p_0} dx}\, dx\,,$$

which can also be verified directly from the differential equation satisfied by $\varphi_n(x)$. (6.42) implies that $W_0'(\lambda_n) \neq 0$, for $\varphi_n(x)$ is real because λ_n is real[2] (see Example 6.11).

---

[1] Strictly speaking, the validity of relations such as (6.39) or (6.41) needs a proof, for the integral in (6.19) is an integral of an operator-valued function $R(\zeta)$ whereas (6.39) or (6.41) is concerned with the values of functions. For $X = C[a, b]$ this proof is trivial since $u(y)$ for each fixed y is a bounded linear form in $u \in X$. It is not so simple for $X = L^p(a, b)$. In this case we first note, for example, that (6.19) implies $(Pv, f) = -(2\pi i)^{-1}\int (R(\zeta)\, v, f)\, d\zeta$ for $v \in X$ and $f \in X^* = L^q(a, b)$, $p^{-1} + q^{-1} = 1$. On calculating the right member and noting that $f \in X^*$ is arbitrary, we see that (6.39) or (6.41) holds for almost all y.

[2] $W_0'(\lambda_n)$ may be zero if some of p_0, p_1, p_2 are not real or if we consider non-real boundary conditions. If $W_0'(\lambda_n) = 0$, P_n is no longer of rank 1 and $R(\zeta)$ may have a pole of order higher than 1.

Example 6.21. As a more specific example, consider the operator

$$Tu = -u'', \quad 0 \leq x \leq \pi, \quad \text{with boundary condition} \quad u(0) = u(\pi) = 0. \tag{6.43}$$

We shall regard T as an operator in $\mathsf{X} = \mathsf{C}[0, \pi]$ so that it is a special case of the T_1 of the preceding example, with $a = 0$, $b = \pi$, $p_0 = -1$, $p_1 = 0$, $p_2 = 0$. The eigenvalues and normalized eigenvectors are

$$\lambda_n = n^2, \quad \varphi_n(x) = \sin n x, \quad n = 1, 2, 3, \ldots. \tag{6.44}$$

The resolvent $R(\zeta) = R(\zeta, T)$ is an integral operator with the kernel equal to the Green function of the equation $u'' + \zeta u = 0$, namely

$$g(y, x; \zeta) = \frac{\sin\sqrt{\zeta}\, y \sin\sqrt{\zeta}\,(\pi - x)}{\sqrt{\zeta}\sin\pi\sqrt{\zeta}}, \quad y \leq x, \tag{6.45}$$

and x, y exchanged for $x \leq y$. The poles of $g(y, x; \zeta)$ as a function of ζ are exactly the eigenvalues $\lambda_n = n^2$.

The Laurent expansion of $g(y, x; \zeta)$ in powers of $\zeta - n^2$ corresponds to (6.32). This remark yields expressions for P and S for $\lambda = \lambda_n = n^2$: these are integral operators with the kernels p and s, respectively, given by

$$\begin{aligned} p(y, x) &= \frac{2}{\pi}\sin n y \sin n x, \\ s(y, x) &= \frac{2}{\pi}\Big[-\frac{y}{2n}\cos n y \sin n x + \frac{\pi - x}{2n}\sin n y \cos n x + \\ &\qquad + \frac{1}{4n^2}\sin n y \sin n x\Big], \quad y \leq x, \end{aligned} \tag{6.46}$$

and with x, y exchanged for $x \leq y$. Note that $D = 0$.

For later reference we deduce some estimates for $R(\zeta)$. According to Example 2.4 (final remark), $\|R(\zeta)\|$ is not larger than $\max_y \int |g(y, x; \zeta)|\, dx$. Since $|\sin z| \leq \cosh(\operatorname{Im} z)$ for any complex number z, a simple calculation yields

$$\|R(\zeta)\| \leq \frac{\sinh\pi\beta}{\beta|\zeta|^{1/2}(\sin^2\pi\alpha + \sinh^2\pi\beta)^{1/2}}, \quad \alpha = \operatorname{Re}\sqrt{\zeta}, \quad \beta = \operatorname{Im}\sqrt{\zeta}. \tag{6.47}$$

This further leads to the estimates

$$\begin{aligned} \|R(\zeta)\| &\leq \frac{1}{|\beta|\,|\zeta|^{1/2}} \leq \begin{cases} \dfrac{1}{|\alpha\beta|} = \dfrac{2}{|\operatorname{Im}\zeta|}, \\[2mm] \dfrac{1}{|\beta|^2}, \end{cases} \\ \|R(\zeta)\| &\leq \frac{\pi}{|\zeta|^{1/2}\,|\sin\pi\alpha|} \leq \frac{\pi}{|\alpha\sin\pi\alpha|}. \end{aligned} \tag{6.48}$$

The curve in the ζ-plane defined by $\operatorname{Re}\sqrt{\zeta} = \alpha = \text{const.}$ is a parabola

$$\xi = \alpha^2 - \frac{\eta^2}{4\alpha^2} \quad \text{where} \quad \xi = \operatorname{Re}\zeta, \quad \eta = \operatorname{Im}\zeta. \tag{6.49}$$

It follows from (6.48) that $\|R(\zeta)\| \leq \pi/|\alpha\sin\pi\alpha|$ along such a parabola.

6. The resolvent of the adjoint

There is a simple relation between the resolvent of a closed operator T in X and that of the adjoint T^* (assuming that T is densely defined so that T^* exists). The following theorem is a direct consequence of Theorem 5.30.

Theorem 6.22. $\mathrm{P}(T^*)$ *and* $\Sigma(T^*)$ *are respectively the mirror images of* $\mathrm{P}(T)$ *and* $\Sigma(T)$ *with respect to the real axis, and*

$$R(\zeta, T^*) = R(\bar{\zeta}, T)^*, \quad \bar{\zeta} \in \mathrm{P}(T). \tag{6.50}$$

According to this theorem, any result on the spectrum of T has its counterpart for T^*. For example, if $\Sigma(T)$ is separated into two parts Σ', Σ'' by a curve Γ as in par. 4, then $\Sigma(T^*)$ is separated by $\bar{\Gamma}$ into two parts $\bar{\Sigma}'$, $\bar{\Sigma}''$ ($\bar{\Gamma}$ etc. being the mirror images of Γ etc.). The resulting decompositions of the spaces $\mathsf{X} = \mathsf{M}' \oplus \mathsf{M}''$, $\mathsf{X}^* = \mathsf{M}'^* \oplus \mathsf{M}''^*$ described in Theorem 6.17 have the projection P and its adjoint P^*:

$$\mathsf{M}' = P\mathsf{X},\ \mathsf{M}'' = (1-P)\,\mathsf{X},\ \mathsf{M}'^* = P^*\,\mathsf{X}^*,\ \mathsf{M}''^* = (1-P^*)\,\mathsf{X}^*. \tag{6.51}$$

This follows from the expressions

$$P = -\frac{1}{2\pi i}\int_{\Gamma} R(\zeta, T)\,d\zeta\,, \quad P^* = -\frac{1}{2\pi i}\int_{\bar{\Gamma}} R(\zeta, T^*)\,d\zeta\,, \tag{6.52}$$

by noting (6.50) and that the two integrals of (6.52) are taken along Γ and $\bar{\Gamma}$ in the *positive* direction. (6.51) implies

$$\dim \mathsf{M}' = \dim \mathsf{M}'^*\,, \quad \dim \mathsf{M}'' = \dim \mathsf{M}''^*\,, \tag{6.53}$$

see (4.10) and the remark thereafter.

In particular suppose that $\Sigma(T)$ contains several isolated points $\lambda_1, \ldots, \lambda_s$, so that $R(\zeta) = R(\zeta, T)$ has the form (6.35). The corresponding expression for $R^*(\zeta) = R(\zeta, T^*)$ is

$$R^*(\zeta) = -\sum_{h=1}^{s}\left[\frac{P_h^*}{\zeta - \bar{\lambda}_h} + \sum_{n=1}^{\infty}\frac{D_h^{*n}}{(\zeta - \bar{\lambda}_h)^{n+1}}\right] + R_0^*(\zeta) \tag{6.54}$$

where the P_h^* are projections satisfying $P_h^* P_k^* = \delta_{hk} P_h^*$ and where $R_0^*(\zeta) = R_0(\bar{\zeta})^*$ is holomorphic at $\zeta = \bar{\lambda}_h$, $h = 1, \ldots, s$. If $\mathsf{M}_h = P_h\mathsf{X}$ is finite-dimensional, the same is true of $\mathsf{M}_h^* = P_h^*\mathsf{X}^*$ and $\dim \mathsf{M}_h^* = \dim \mathsf{M}_h$, and $\bar{\lambda}_h$ is an eigenvalue of T^* with (algebraic) multiplicity equal to that of λ_h for T.

Remark 6.23. An isolated eigenvalue λ of T with finite multiplicity m (such as λ_h considered above) has properties quite similar to an eigenvalue of a finite-dimensional operator. For instance, not only is $\bar{\lambda}$ an eigenvalue of T^* with (algebraic) multiplicity m, but the *geometric* multiplicity of $\bar{\lambda}$ for T^* is equal to that of λ for T. Again, the linear equation $(T - \lambda)\,u = v$ is solvable if and only if $v \perp \mathsf{N}(T^* - \bar{\lambda})$ whereas $(T^* - \bar{\lambda})\,g = f$ is solvable if and only if $f \perp \mathsf{N}(T - \lambda)$. These results follow immediately if one notes that the problem is reduced to the finite-dimensional problem for the parts $T_{\mathsf{M}'}$ and $T^*_{\mathsf{M}'^*}$, each of which may be considered the adjoint of the other.

Remark 6.24. If $\dim \mathsf{M}_h = \infty$, it is possible that λ_h is an eigenvalue of T but $\bar{\lambda}_h$ is not an eigenvalue of T^* or vice versa.

Example 6.25. We continue to consider Example 6.20. P_n as given by (6.41) is an integral operator with the kernel $p_n(y, x) = \varphi_n(y)\, \psi_n(x)$ where $\psi_n(x) = k\, \varphi_n(x) \exp\left(\int^x_a \frac{p_1}{p_0}\, dx\right) \Big/ W_0'(\lambda_n)\, p_0(x)$. $\bar{\lambda}_n$ should be a simple eigenvalue of T_1^* with the associated eigenprojection P_n^*, which is an integral operator with the kernel $p_n^*(y, x) = \psi_n(y)\, \varphi_n(x)$. Thus $\psi_n(x)$ is the eigenfunction of T_1^* for the eigenvalue λ_n (note that $\lambda_n, \varphi_n, \psi_n$ are all real). Here we consider T_1 as an operator in $\mathsf{X} = \mathsf{L}^p(a, b)$ rather than in $\mathsf{C}[a, b]$ since T_1 is not densely defined in the latter case.

7. The spectra of compact operators

The spectrum of a compact operator T in X has a simple structure analogous to that of an operator in a finite-dimensional space.

Theorem 6.26. *Let $T \in \mathscr{B}(\mathsf{X})$ be compact. $\Sigma(T)$ is a countable set with no accumulation point different from zero. Each nonzero $\lambda \in \Sigma(T)$ is an eigenvalue of T with finite multiplicity, and $\bar{\lambda}$ is an eigenvalue of T^* with the same multiplicity.*

Proof. We give the proof in several steps.

I. First we prove that *the eigenvalues of T do not accumulate at a point $\lambda \neq 0$.* Otherwise there would be a sequence $\{\lambda_n\}$ of distinct eigenvalues of T with eigenvectors u_n such that $0 \neq \lambda_n \to \lambda \neq 0$. Let M_n be the subspace spanned by the n vectors $u_1, \ldots, u_n$. M_n is invariant under T. Since $u_1, u_2, \ldots$ are linearly independent, M_{n-1} is a proper subspace of M_n and there is a $v_n \in \mathsf{M}_n$ such that $\|v_n\| = 1$ and $\operatorname{dist}(v_n, \mathsf{M}_{n-1}) = 1$ (see Lemma 1.12). With the sequence $\{v_n\}$ thus defined, we shall show that $\{\lambda_n^{-1} T v_n\}$ contains no Cauchy subsequence, contradicting the assumption that T is compact (note that $\{\lambda_n^{-1} v_n\}$ is bounded). Now we have for $m < n$

$$\lambda_n^{-1} T v_n - \lambda_m^{-1} T v_m = v_n - (\lambda_m^{-1} T v_m - \lambda_n^{-1}(T - \lambda_n)\, v_n)$$

where the second term on the right belongs to M_{n-1} because $v_m \in \mathsf{M}_{n-1}$, M_{n-1} is invariant under T and $(T - \lambda_n)\, v_n \in \mathsf{M}_{n-1}$. Since $\operatorname{dist}(v_n, \mathsf{M}_{n-1}) = 1$, it follows that each element of the sequence $\{\lambda_n^{-1} T v_n\}$ has distance $\geqq 1$ from any other one, showing that no subsequence of this sequence can be convergent.

II. Next we prove that $\mathsf{R}(T - \zeta)$ *is closed if $\zeta \neq 0$ and ζ is not an eigenvalue of T.* Suppose that $(T - \zeta)\, u_n \to v$; we have to show that $v \in \mathsf{R}(T - \zeta)$. If $\{u_n\}$ is bounded, $\{T u_n\}$ contains a Cauchy sequence; by replacing $\{u_n\}$ by a subsequence, we may assume that $\{T u_n\}$ itself is Cauchy. Let $T u_n \to w$. Then $\zeta\, u_n = T u_n - (T - \zeta)\, u_n \to w - v$. Application of T gives $\zeta\, T u_n \to T(w - v)$. Thus $\zeta\, w = T w - T v$ or $v = \zeta^{-1}(T - \zeta)\,(w - v) \in \mathsf{R}(T - \zeta)$. It remains to show that $\{u_n\}$ is bounded. Otherwise we may assume that $\|u_n\| \to \infty$, replacing $\{u_n\}$

by a subsequence if necessary. Set $u'_n = u_n/\|u_n\|$. Then $\{u'_n\}$ is a bounded sequence and $(T - \zeta)\, u'_n \to 0$. The same argument as above then leads to the results $T u'_n \to w$, $(T - \zeta)\, w = 0$ and $\zeta\, u'_n \to w$. Thus $\|w\| = \lim \|\zeta\, u'_n\| = |\zeta| > 0$ and w must be an eigenvector of T for the eigenvalue ζ, contrary to the assumption.

III. For the moment a complex number ζ will be said to be *exceptional* if either ζ is an eigenvalue of T or $\bar{\zeta}$ is an eigenvalue of T^*. Since T^* is compact with T (Theorem 4.10), it follows from the result proved above that the set of exceptional numbers is countable and has no accumulation point different from zero. Every nonexceptional point $\zeta \neq 0$ belongs to $\mathrm{P}(T)$. In fact, since we have just shown that $\mathsf{R}(T - \zeta)$ is closed, it suffices to note that $\mathsf{R}(T - \zeta)^\perp = \mathsf{N}(T^* - \bar{\zeta}) = 0$. On the other hand, an exceptional point obviously belongs to $\Sigma(T)$ (note Theorem 6.22). Thus $\Sigma(T)$ is exactly the set of exceptional points. In view of the results of the preceding paragraphs, the theorem will be proved if we show that the eigenprojection P associated with each $\lambda \in \Sigma(T)$, $\lambda \neq 0$, is finite-dimensional.

P is given by (6.19) where Γ is a small circle around λ excluding the origin. $R(\zeta) = R(\zeta, T)$ is in general not compact, but $R(\zeta) + \zeta^{-1} = \zeta^{-1}\, T R(\zeta)$ is compact with T (Theorem 4.8). Since $\int_\Gamma \zeta^{-1}\, d\zeta = 0$, P is equal to the integral along Γ of the compact operator $R(\zeta) + \zeta^{-1}$ and is itself compact (since the integral is the limit *in norm* of finite sums of compact operators). In virtue of Problem 4.5, it follows that P is finite-dimensional.

Remark 6.27. Since every complex number $\lambda \neq 0$ either belongs to $\mathrm{P}(T)$ or is an isolated eigenvalue with finite multiplicity, Remark 6.23 applies to λ. This result is known as the *Riesz-Schauder theorem* and generalizes a classical result of FREDHOLM for integral equations.

Remark 6.28. Let λ_n, $n = 1, 2, \ldots$, be the eigenvalues of a compact operator T, with the associated eigenprojections P_n and eigennilpotents D_n. If we set $Q_n = P_1 + \cdots + P_n$ and $Q_n \mathsf{X} = \mathsf{M}_n$, $\{\mathsf{M}_n\}$ is an increasing sequence of finite-dimensional subspaces of X: $\mathsf{M}_1 \subset \mathsf{M}_2 \subset \mathsf{M}_3 \ldots$. In each M_n, which is invariant under T, we have the spectral representation of T in the form

$$T Q_n = \sum_{h=1}^{n} (\lambda_h P_h + D_h) \tag{6.55}$$

[cf. (6.38)]. This suggests the expression $T = \sum_{h=1}^{\infty} (\lambda_h P_h + D_h)$, but this is not correct without further assumption[1]. In fact T may have no eigen-

[1] It is an interesting but difficult problem to decide when such a spectral decomposition is possible. For this question see DUNFORD and SCHWARTZ [1], Chapter 11. It is also related to the theory of spectral operators due to DUNFORD (see DUNFORD [1]).

values at all (for example, a quasi-nilpotent operator has no nonzero eigenvalue and an integral operator of Volterra type is usually quasi-nilpotent, see Example 3.15). We shall see, however, that the suggested expansion is valid if T is a normal compact operator in a Hilbert space (Theorem V-2.10).

8. Operators with compact resolvent

Another class of operators which have spectra analogous to the spectra of operators in a finite-dimensional space is provided by operators with compact resolvent. Let T be a closed operator in X such that $R(\zeta) = R(\zeta, T)$ exists and is compact at least for some $\zeta = \zeta_0$. According to the result of the preceding paragraph, $\Sigma(R(\zeta_0))$ is a countable set having no accumulation point different from zero. Since $\Sigma(R(\zeta_0))$ is the image of $\tilde{\Sigma}(T)$ under the map $\zeta \to (\zeta - \zeta_0)^{-1}$ (see Problem 6.16), it follows that $\Sigma(T)$ also consists of isolated points alone (with no accumulation point different from ∞). The eigenprojection P for each $\lambda \in \Sigma(T)$ is identical with the eigenprojection for the eigenvalue $\mu = (\lambda - \zeta_0)^{-1}$ of $R(\zeta_0)$, as is seen from (6.18) and (6.19) by transformation of the integration variable. In particular we have $\dim P < \infty$ so that λ is an eigenvalue of T with finite multiplicity. Furthermore, for any $\zeta \in \mathrm{P}(T)$ the relation $R(\zeta) = R(\zeta_0)(1 + (\zeta - \zeta_0) R(\zeta))$ implied by the resolvent equation shows that $R(\zeta)$ is again compact. Thus we have proved

Theorem 6.29. *Let T be a closed operator in X such that the resolvent $R(\zeta)$ exists and is compact for some ζ. Then the spectrum of T consists entirely of isolated eigenvalues*[1] *with finite multiplicities, and $R(\zeta)$ is compact for every $\zeta \in \mathrm{P}(T)$.*

Such an operator T will be called an *operator with compact resolvent* and a spectrum of the kind described will be said to be *discrete*. An operator with compact resolvent has a discrete spectrum. Operators with compact resolvent occur frequently in mathematical physics. It may be said that most differential operators that appear in *classical* boundary problems are of this type.

Problem 6.30. If an operator in X with compact resolvent is bounded, X must be finite-dimensional.

Example 6.31. The differential operators of Examples 2.6-2.7 and of § 2.3 for which the resolvent set is not empty are all operators with compact resolvent, for their resolvents are integral operators with continuous kernels (Example 4.1). This immediately leads to the result that the spectra of these operators consist of isolated eigenvalues with finite multiplicities (cf. Examples 6.19-6.20).

In connection with operators with compact resolvent, the following lemmas and their corollary are useful.

---

[1] See footnote [1] of p. 180.

Lemma 6.32. *Let T_1 $T_2 \in \mathscr{C}(\mathsf{X}, \mathsf{Y})$ have the properties:* i) *T_1 and T_2 are extensions of a common operator T_0, the order of extension for T_1 being finite;* ii) *T_1^{-1} and T_2^{-1} exist and belong to $\mathscr{B}(\mathsf{Y}, \mathsf{X})$. Then $A = T_1^{-1} - T_2^{-1}$ is degenerate, and $\mathsf{N}(A) \supset \mathsf{R}(T_0)$ where* $\operatorname{codim} \mathsf{R}(T_0) < \infty$. *The orders of the extensions T_1 and T_2 of T_0 are equal (so that T_2 is also a finite extension of T_0).*

Proof. Set $\mathsf{D}_1 = \mathsf{D}(T_1)$, $\mathsf{D}_2 = \mathsf{D}(T_2)$, $\mathsf{D}_0 = \mathsf{D}(T_0)$, $\mathsf{R}_0 = \mathsf{R}(T_0)$. Each $v \in \mathsf{R}_0$ is of the form $v = T_0 u = T_1 u$ so that $T_1^{-1} v = u$ and similarly $T_2^{-1} v = u$. Hence $A v = 0$ and $\mathsf{R}_0 \subset \mathsf{N}(A)$. Since the mapping by T_1 as well as by T_2 is one to one, we have $\dim(\mathsf{Y}/\mathsf{R}_0) = \dim(T_1 \mathsf{D}_1/T_1 \mathsf{D}_0) = \dim(\mathsf{D}_1/\mathsf{D}_0)$ and similarly $\dim(\mathsf{Y}/\mathsf{R}_0) = \dim(\mathsf{D}_2/\mathsf{D}_0)$. Hence $\dim (\mathsf{Y}/\mathsf{N}(A)) \leqq \dim(\mathsf{Y}/\mathsf{R}_0) = \dim(\mathsf{D}_2/\mathsf{D}_0) = \dim(\mathsf{D}_1/\mathsf{D}_0) < \infty$ and A is degenerate by Problem 4.11.

Lemma 6.33. *Let T_1, $T_2 \in \mathscr{C}(\mathsf{X}, \mathsf{Y})$ have the properties:* i) *T_1 and T_2 are restrictions of a common operator T, the order of restriction for T_1 being finite;* ii) *T_1^{-1} and T_2^{-1} exist and belong to $\mathscr{B}(\mathsf{Y}, \mathsf{X})$. Then $A = T_1^{-1} - T_2^{-1}$ is degenerate, and $\mathsf{R}(A) \subset \mathsf{N}(T)$ where* $\dim \mathsf{N}(T) < \infty$. *The orders of the restrictions T_1, T_2 of T are equal.*

Proof. For any $v \in \mathsf{Y}$, $T T_1^{-1} v = T_1 T_1^{-1} v = v$ and similarly $T T_2^{-1} v = v$. Hence $T A v = 0$ and $\mathsf{R}(A) \subset \mathsf{N}(T)$. But since T is a finite extension of T_1 and T_1 maps $\mathsf{D}_1 = \mathsf{D}(T_1)$ onto Y, we have $\dim \mathsf{N}(T) = \dim(\mathsf{D}/\mathsf{D}_1)$ where $\mathsf{D} = \mathsf{D}(T)$. Similarly $\dim \mathsf{N}(T) = \dim(\mathsf{D}/\mathsf{D}_2)$. Hence $\dim \mathsf{R}(A) \leqq \dim \mathsf{N}(T) = \dim(\mathsf{D}/\mathsf{D}_2) = \dim(\mathsf{D}/\mathsf{D}_1) < \infty$.

Corollary 6.34. *Let T_1, $T_2 \in \mathscr{C}(\mathsf{X})$ have non-empty resolvent sets. Let T_1, T_2 be either extensions of a common operator T_0 or restrictions of a common operator T, with the order of extension or restriction for T_1 being finite. Then T_1 has compact resolvent if and only if T_2 has compact resolvent.*

Example 6.35. The result of Example 6.31 that the operators considered there have compact resolvents is not accidental. For these operators are finite extensions of a common operator denoted by T_0 and, at the same time, finite restrictions of a common operator denoted by T.

For convenience we add another lemma related to the lemmas proved above.

Lemma 6.36. *Let the assumptions of both* Lemmas 6.32, 6.33 *be satisfied. Then $A = T_1^{-1} - T_2^{-1}$ has the form*

$$A = \sum_{j=1}^{m} (\ , g_j)\, u_j\,, \quad u_j \in \mathsf{N}(T)\,, \quad g_j \in \mathsf{R}(T_0)^{\perp}\,. \tag{6.56}$$

Proof. Since $\mathsf{R}(A) \subset \mathsf{N}(T)$ by Lemma 6.33, we may write $A v = \sum_{j=1}^{m} g_j[v]\, u_j$ where the u_j are linearly independent vectors of $\mathsf{N}(T)$. Obviously the $g_j[v]$ are bounded linear forms on Y and vanish for $v \in \mathsf{R}(T_0)$ by Lemma 6.32. Hence we may write $g_j[v] = (v, g_j)$ with $g_j \in \mathsf{R}(T_0)^{\perp}$.

Chapter Four

Stability theorems

In this chapter we investigate the stability, under small perturbations, of various spectral properties of linear operators acting between Banach spaces. The basic problems to be treated are the stability or instability of the spectrum and the perturbation of the resolvent. The results will be fundamental for the further development of perturbation theory given in the following chapters. Other subjects discussed include the stability of the Fredholm or semi-Fredholm property, of the nullity, deficiency, index, etc. The endeavor is to treat these problems for unbounded operators and for the most general perturbations.

One of the basic problems here is how to define a "small" perturbation for unbounded operators. One rather general definition useful in applications is based on the notion of a relatively bounded perturbation. But it is still too restricted in a general theory. The most natural and general definition of smallness of a perturbation is given in terms of a metric in the space $\mathscr{C}(X, Y)$ of all closed linear operators from one Banach space X to another one Y. Such a metric has long been known, but so far no systematic use of it has been made in perturbation theory. In this chapter we base the main part of the theory on it.

Since the metric is defined in terms of the graphs of operators, which are closed subspaces of the product space X × Y, the technique is equivalent to introducing a metric in the set of all closed subspaces of a Banach space. For this reason, a considerable part of the chapter is devoted to the theory of a metric on subspaces and to related problems. In this way, for example, we are led to define such notions as the Fredholm or semi-Fredholm property and the nullity, deficiency and index for a pair of subspaces. The results obtained for them lead in a natural way to the corresponding results for operators.

§ 1. Stability of closedness and bounded invertibility

1. Stability of closedness under relatively bounded perturbation

Let $T \in \mathscr{C}(X, Y)$, where X, Y are Banach spaces. [$\mathscr{C}(X, Y)$ is the set of all closed operators from X to Y.] We have already noted (Problem III-5.6) that $T + A$ is also closed if $A \in \mathscr{B}(X, Y)$. This expresses the fact that closedness is *stable* under a bounded perturbation A. We now try to extend this *stability theorem* to a not necessarily bounded perturbation.

An immediate extension of this kind can be made to the case of a relatively bounded perturbation. Let T and A be operators with the same domain space X (but not necessarily with the same range space) such that $\mathsf{D}(T) \subset \mathsf{D}(A)$ and

$$\|A u\| \leqq a \|u\| + b \|T u\|, \quad u \in \mathsf{D}(T), \tag{1.1}$$

where a, b are nonnegative constants. Then we shall say that A is *relatively bounded with respect to* T or simply *T-bounded*. The greatest lower bound b_0 of all possible constants b in (1.1) will be called the *relative bound* of A with respect to T or simply the *T-bound* of A. If b is chosen very close to b_0, the other constant a will in general have to be chosen very large; thus it is in general impossible to set $b = b_0$ in (1.1).

Obviously a bounded operator A is T-bounded for any T with $\mathsf{D}(T) \subset \mathsf{D}(A)$, with T-bound equal to zero.

The extension of the stability theorem for closedness mentioned above is now given by

Theorem 1.1. *Let T and A be operators from X to Y, and let A be T-bounded with T-bound smaller than 1. Then $S = T + A$ is closable if and only if T is closable; in this case the closures of T and S have the same domain. In particular S is closed if and only if T is.*

Proof. We have the inequality (1.1) in which we may assume that $b < 1$. Hence

$$-a\|u\| + (1-b)\|Tu\| \leqq \|Su\| \leqq a\|u\| + (1+b)\|Tu\|, \quad u \in \mathsf{D}(T). \tag{1.2}$$

Applying the second inequality of (1.2) to u replaced by $u_n - u_m$, we see that a T-convergent sequence $\{u_n\}$ (that is, a convergent sequence $\{u_n\}$ for which Tu_n is also convergent, see III-§ 5.2) is also S-convergent. Similarly we see from the first inequality that an S-convergent sequence $\{u_n\}$ is T-convergent. If $\{u_n\}$ is S-convergent to 0, it is T-convergent to 0 so that $Tu_n \to 0$ if T is closable [see III-(5.6)]; then it follows from the second inequality of (1.2) that $Su_n \to 0$, which shows that S is closable. Similarly, T is closable if S is.

Let $\tilde{T}$, $\tilde{S}$ be the closures of T, S, respectively. For any $u \in \mathsf{D}(\tilde{S})$, there is a sequence $\{u_n\}$ S-convergent to u (see III-§ 5.3). Since this $\{u_n\}$ is also T-convergent to u as remarked above, we have $u \in \mathsf{D}(\tilde{T})$ so that $\mathsf{D}(\tilde{S}) \subset \mathsf{D}(\tilde{T})$. The opposite inclusion is proved similarly.

Problem 1.2. (1.1) with $b < 1$ implies

$$\|A u\| \leqq a\|u\| + b\|Tu\| \leqq (1-b)^{-1}(a\|u\| + b\|Su\|). \tag{1.3}$$

In particular A is S-bounded with S-bound not exceeding $b(1-b)^{-1}$. More generally, any operator that is T-bounded with T-bound β is also S-bounded with S-bound $\leqq \beta(1-b)^{-1}$.

The assumptions of Theorem 1.1 are not symmetric with respect to T and S, although the assertions are symmetric. In this connection, the following symmetrized generalization of Theorem 1.1 is of some interest:

Theorem 1.3. *Let T, S be operators from X to Y such that*

$$\|Su - Tu\| \leqq a\|u\| + b'\|Tu\| + b''\|Su\|, \quad u \in \mathsf{D}(T) = \mathsf{D}(S), \tag{1.4}$$

where a, b', b'' are nonnegative constants and $b' < 1$, $b'' < 1$. Then the conclusions of Theorem 1.1 *are true.*

Proof. Set $A = S - T$, $T(\varkappa) = T + \varkappa A$, $0 \leqq \varkappa \leqq 1$. $T(\varkappa)$ has constant domain $\mathsf{D}(T)$, and $T(0) = T$, $T(1) = S$. Since $Tu = T(\varkappa)u - \varkappa Au$ and $Su = T(\varkappa)u + (1-\varkappa)Au$, (1.4) gives $\|Au\| \leqq a\|u\| + (b' + b'')\|T(\varkappa)u\| + b\|Au\|$, where $b = \max(b', b'')$. Hence

$$\|Au\| \leqq \frac{1}{1-b}(a\|u\| + (b' + b'')\|T(\varkappa)u\|). \tag{1.5}$$

This shows that A is $T(\varkappa)$-bounded with $T(\varkappa)$-bound not exceeding $\beta = (1-b)^{-1}(b' + b'')$. Hence $(\varkappa' - \varkappa)A$ is $T(\varkappa)$-bounded with $T(\varkappa)$-bound less than 1 provided $|\varkappa' - \varkappa| < 1/\beta$, so that by Theorem 1.1, $T(\varkappa')$ is closable if and only if $T(\varkappa)$ is. This observation leads immediately to the proof of the theorem; for example, the closability of $T(\varkappa)$ propagates from $\varkappa = 0$ to $\varkappa = 1$ in a finite number of steps if T is closable.

Remark 1.4. Let $T \in \mathscr{C}(\mathsf{X}, \mathsf{Y})$. Set

$$|||u||| = \|u\| + \|Tu\|, \quad u \in \mathsf{D}(T). \tag{1.6}$$

It is easily seen that $\mathsf{D}(T)$ becomes a Banach space $\hat{\mathsf{X}}$ if $|||\;|||$ is chosen as the norm; the completeness of $\hat{\mathsf{X}}$ is a direct consequence of the closedness of T. If A is an operator from X to Y' with $\mathsf{D}(A) \supset \mathsf{D}(T)$, the restriction of A to $\mathsf{D}(T)$ can be regarded as an operator $\hat{A}$ on $\hat{\mathsf{X}}$ to Y'. It is easily seen that A is T-bounded if and only if $\hat{A}$ is bounded.

Remark 1.5. If T is closed and A is closable, the inclusion $\mathsf{D}(T) \subset \mathsf{D}(A)$ already implies that A is T-bounded. To see this define $\hat{\mathsf{X}}$ and $\hat{A}$ as in the preceding remark. Then $\hat{A}$ is closable, for an $\hat{A}$-convergent sequence in $\hat{\mathsf{X}}$ is an A-convergent sequence in X. Since $\hat{A}$ is defined on the whole of $\hat{\mathsf{X}}$, $\hat{A}$ is closed and therefore bounded by Theorem III-5.20. Thus A is T-bounded by Remark 1.4.

2. Examples of relative boundedness

Since the notion of relative boundedness is important in perturbation theory, let us consider several examples[1].

Example 1.6. Let $\mathsf{X} = \mathsf{C}[a, b]$ or $\mathsf{X} = \mathsf{L}^p(a, b)$ for a finite interval (a, b) and let T and A be the maximal operators defined by $Tu = -u''$ and $Au = u'$ (see

[1] The inequalities deduced below are special cases of the Sobolev inequalities; see Sobolev [1], Goldberg [1].

Examples 2.6-2.7 and § 2.3 of Chapter III). We shall show that A is T-bounded with T-bound 0. To this end we use the identity

(1.7) $$u' = Gu'' + Hu$$

where G and H are integral operators with the kernels $g(y, x)$ and $h(y, x)$, respectively, given by

(1.8) $$\begin{aligned} g(y,x) &= \frac{(x-a)^{n+1}}{(b-a)(y-a)^n}, \quad h(y,x) = -\frac{n(n+1)(x-a)^{n-1}}{(b-a)(y-a)^n}, \quad a \leqq x < y \leqq b, \\ g(y,x) &= \frac{-(b-x)^{n+1}}{(b-a)(b-y)^n}, \quad h(y,x) = \frac{n(n+1)(b-x)^{n-1}}{(b-a)(b-y)^n}, \quad a \leqq y < x \leqq b, \end{aligned}$$

where n is any positive number. (1.7) can be verified easily by carrying out appropriate integrations by parts. The operators G and H are bounded, for

(1.9) $$\int_a^b |g(y,x)|\, dx \leqq \frac{b-a}{n+2}, \quad \int_a^b |g(y,x)|\, dy \leqq \frac{b-a}{n-1}.$$

(1.10) $$\int_a^b |h(y,x)|\, dx \leqq \frac{2(n+1)}{b-a}, \quad \int_a^b |h(y,x)|\, dy \leqq \frac{2n(n+1)}{(n-1)(b-a)},$$

where we assumed $n > 1$ for simplicity. It follows from III-(2.9) that

(1.11) $$\|G\| \leqq \frac{b-a}{n-1}, \quad \|H\| \leqq \frac{2n(n+1)}{(n-1)(b-a)},$$

Hence (1.7) gives

(1.12) $$\|u'\|_p \leqq \frac{b-a}{n-1}\|u''\|_p + \frac{2n(n+1)}{(n-1)(b-a)}\|u\|_p, \quad n > 1.$$

Since the factor of $\|u''\|_p$ can be made arbitrarily small by taking n large, this gives the desired result.

Incidentally we note that only the first inequality in each of (1.9) and (1.10) is needed in the case of $\mathsf{X} = \mathsf{L}^\infty$ or $\mathsf{X} = \mathsf{C}$. In this case (1.12) may be replaced by

(1.13) $$\|u'\|_\infty \leqq \frac{b-a}{n+2}\|u''\|_\infty + \frac{2(n+1)}{b-a}\|u\|_\infty, \quad n \geqq 0.$$

Note that in (1.12) or (1.13) no boundary conditions are imposed on $u(x)$.

Suppose now that both u and u'' belong to $\mathsf{L}^p(0, \infty)$. Then (1.12) is true with $a = 0$ and any $b > 0$, and the $\|u\|_p$ and $\|u''\|_p$ may be replaced by the corresponding norms taken on the interval $(0, \infty)$. Now choose $n = b/k$ for a fixed $k > 0$ and let $b \to \infty$. We see then that $\|u'\|_p$ taken on $(0, b)$ is bounded for $b \to \infty$ so that $u' \in \mathsf{L}^p(0, \infty)$ also, with

(1.14) $$\|u'\|_p \leqq k\|u''\|_p + \frac{2}{k}\|u\|_p.$$

The same inequality holds for the interval $(-\infty, \infty)$. Since $k > 0$ is arbitrary, $A = d/dx$ is T-bounded with T-bound zero also in the case of an infinite interval.

Problem 1.7. From (1.14) deduce

(1.15) $$\|u'\|_p \leqq 2\sqrt{2}(\|u\|_p\|u''\|_p)^{1/2} \quad \text{(an infinite interval)}.$$

Example 1.8. Let $\mathsf{X} = \mathsf{L}^p(a, b)$ with finite (a, b) and let $Tu = u'$, $Au = u(c)$, where $c \in [a, b]$. A is a linear *form* and is unbounded if $p < \infty$. We shall show that A is T-bounded with T-bound 0 if $p > 1$ and with a positive T-bound if $p = 1$. We start from the identity

(1.16) $$u(c) = (u', g) + (u, h)$$

where

$$(1.17)\qquad \begin{aligned} g(x) &= \frac{(x-a)^{n+1}}{(b-a)(c-a)^n}, & h(x) &= \frac{(n+1)(x-a)^n}{(b-a)(c-a)^n}, & a \leqq x \leqq c, \\ g(x) &= \frac{-(b-x)^{n+1}}{(b-a)(b-c)^n}, & h(x) &= \frac{(n+1)(b-x)^n}{(b-a)(b-c)^n}, & c < x \leqq b, \end{aligned}$$

where n is any positive number. (1.16) can be verified by integration by parts. A straightforward calculation gives the inequalities

$$(1.18)\qquad \|g\|_q \leqq \left(\frac{b-a}{nq+q+1}\right)^{1/q}, \quad \|h\|_q \leqq \frac{n+1}{(b-a)^{1-1/q}(nq+1)^{1/q}}.$$

for any $q \geqq 1$. Hence we obtain from (1.16) and the Hölder inequalities

$$(1.19)\qquad \begin{aligned} |u(c)| &\leqq \|g\|_q \|u'\|_p + \|h\|_q \|u\|_p \leqq \\ &\leqq \left(\frac{b-a}{nq+q+1}\right)^{1/q} \|u'\|_p + \frac{n+1}{(b-a)^{1/p}(nq+1)^{1/q}} \|u\|_p, \quad p^{-1}+q^{-1}=1. \end{aligned}$$

If $p > 1$, then $q < \infty$ and the coefficients of $\|u'\|_p$ on the right of (1.19) can be made arbitrarily small by taking n large; thus A is T-bounded with T-bound 0. If $p = 1$, then $q = \infty$ and (1.19) gives for $n \to 0$

$$(1.20)\qquad |u(c)| \leqq \|u'\|_1 + \|u\|_1/(b-a).$$

This shows that A is T-bounded with T-bound not larger than 1. For $c = a$ and $c = b$, the T-bound is exactly equal to 1; this follows from the existence of a sequence $\{u_k\}$ such that $u_k(b) = 1$, $\|u_k'\|_1 = 1$ and $\|u_k\|_1 \to 0$ for $k \to \infty$. An example of such a sequence is given by $u_k(x) = (x-a)^k/(b-a)^k$. For $a < c < b$, the T-bound of A is equal to 1/2. In fact we have

$$(1.21)\qquad |u(c)| \leqq \frac{1}{2}\|u'\|_1 + \frac{1}{2}\max\left(\frac{1}{c-a}, \frac{1}{b-c}\right)\|u\|_1;$$

this can be proved by noting that (1.16) is true also for g, h given by

$$(1.22)\qquad g(x) = \frac{x-a}{2(c-a)}, \quad h(x) = \frac{1}{2(c-a)}, \quad \text{for} \quad a \leqq x < c,$$

and similarly for $c < x \leqq b$, for which $\|g\|_\infty = 1/2$ and $\|h\|_\infty = \max((c-a)^{-1}, (b-c)^{-1})/2$. That the T-bound of A in this case cannot be smaller than 1/2 is seen from the example $u_k(x) = (x-a)^k/(c-a)^k$ for $a \leqq x \leqq c$ and $= (b-x)^k/(b-c)^k$ for $c \leqq x \leqq b$ so that $u_k(c) = 1$, $\|u_k'\|_1 = 2$ and $\|u_k\|_1 \to 0$ for $k \to \infty$.

Example 1.9. Let $\mathsf{X} = \mathsf{L}^p(a, b)$ for finite (a, b), let $Tu = u'$ (regarded as an operator in X as above) and let $Au = u$ be an operator from X to $\mathsf{Y} = \mathsf{C}[a, b]$. Since every $u \in \mathsf{D}(T)$ is a continuous function, we have $\mathsf{D}(T) \subset \mathsf{D}(A)$. Furthermore, since (1.19) is true for every $c \in [a, b]$ while the right member of this inequality is independent of c, $\|Au\| = \|u\|_\infty$ satisfies the same inequality. This shows that A is T-bounded, with T-bound 0 if $p > 1$ and T-bound 1 if $p = 1$. Note that (1.21) is useless here since the right member is not bounded for varying c. But this inequality is useful if A is regarded as an operator from $\mathsf{L}^1(a, b)$ to $\mathsf{Y}' = \mathsf{C}[a', b']$ for $a < a' < b' < b$; then (1.21) shows that the T-bound of A is 1/2.

Example 1.10. Consider the maximal differential operator T of III-§ 2.3 constructed from the formal operator (2.13). Consider another formal differential operator obtained from (2.13) by replacing the coefficients $p_0(x)$, $p_1(x)$, $p_2(x)$ by other functions $q_0(x)$, $q_1(x)$, $q_2(x)$, respectively, and construct the corresponding operator S. S and T have the same domain, consisting of all $u \in \mathsf{X}$ such that u', $u'' \in \mathsf{X}$. We shall show that S is T-bounded and estimate its T-bound. For any $u \in \mathsf{D}(T)$ we have

$$(1.23)\qquad \|Su\| \leqq N_0\|u''\| + N_1\|u'\| + N_2\|u\|, \quad N_j = \max_{a \leqq x \leqq b} |q_j(x)|, \quad j = 0, 1, 2.$$

But we have the inequality (1.12), which may be written

$$\|u'\| \leqq \varepsilon \|u''\| + C_\varepsilon \|u\| \tag{1.24}$$

where $\varepsilon > 0$ is arbitrary if C_ε is chosen appropriately. Hence

$$\|S u\| \leqq (N_0 + \varepsilon N_1) \|u''\| + (C_\varepsilon N_1 + N_2) \|u\| . \tag{1.25}$$

On the other hand, setting $m_0 = \min |p_0(x)|$ and $M_j = \max |p_j(x)|$, $j = 1, 2$, we have

$$\|T u\| \geqq m_0 \|u''\| - M_1 \|u'\| - M_2 \|u\| \geqq (m_0 - \varepsilon M_1) \|u''\| - - (C_\varepsilon M_1 + M_2) \|u\| . \tag{1.26}$$

If ε is chosen so small that $m_0 > \varepsilon M_1$, it follows that

$$\|S u\| \leqq \frac{N_0 + \varepsilon N_1}{m_0 - \varepsilon M_1} \|T u\| + \left[\frac{(C_\varepsilon M_1 + M_2)(N_0 + \varepsilon N_1)}{m_0 - \varepsilon M_1} + C_\varepsilon N_1 + N_2\right] \|u\| . \tag{1.27}$$

Letting $\varepsilon \to 0$, we see that the T-bound of S is not larger than N_0/m_0. It should be noted that the coefficients of $\|T u\|$ and $\|u\|$ in (1.27) become arbitrarily small if $N = \max(N_0, N_1, N_2)$ is taken sufficiently small. We note also that $q_0(x)$ need not be positive in the above arguments.

If we consider the restrictions T_1, T_2, etc. of T (see loc. cit.) and define S_1, S_2, etc. from S by the same boundary conditions, it follows from the above result that S_n is T_n-bounded.

3. Relative compactness and a stability theorem

A notion analogous to relative boundedness is that of relative compactness. Again let T and A be operators with the same domain space X (but not necessarily with the same range space). Assume that $D(T) \subset D(A)$ and, for any sequence $u_n \in D(T)$ with both $\{u_n\}$ and $\{T u_n\}$ bounded, $\{A u_n\}$ contains a convergent subsequence. Then A is said to be *relatively compact with respect to T* or simply *T-compact*[1].

If A is T-compact, A is T-bounded. For if A is not T-bounded, there is a sequence $u_n \in D(T)$ such that $\|u_n\| + \|T u_n\| = 1$ but $\|A u_n\| \geqq n$, $n = 1, 2, 3, \ldots$. It is obvious that $\{A u_n\}$ has no convergent subsequence.

Theorem 1.11. *Let T, A be operators from X to Y and let A be T-compact*[2]. *If T is closable, $S = T + A$ is also closable, the closures of T and S have the same domain and A is S-compact. In particular S is closed if T is closed*[3].

Proof. First we prove that A is S-compact if T is closable. Assume that $\{u_n\}$ and $\{S u_n\}$ are bounded sequences; we have to show that $\{A u_n\}$ contains a convergent subsequence. Since A is T-compact, it suffices to show that $\{T u_n\}$ contains a bounded subsequence. Suppose this is not

[1] For examples of relatively compact operators for ordinary differential operators see BALSLEV [1].

[2] Here we make no assumption on the "size" of A, in contrast to Theorem 1.1.

[3] The assertions of this theorem are not symmetric in T and S (unlike Theorem 1.1). It is possible that T is not even closable while S is closed. A simple example is given by choosing $T = -A = f$ as an unbounded linear form on X, Y being the one-dimensional space C (see footnote [3] of III-p. 133). T is not closable (see Problem III-5.18) and A is T-compact, but $S = 0$ is closed.

true so that $\|T u_n\| \to \infty$. Set $u_n' = u_n/\|T u_n\|$. Then $u_n' \to 0$, $S u_n' \to 0$ and $\{T u_n'\}$ is bounded. Hence $\{A u_n'\}$ contains a convergent subsequence. Replacing u_n by a suitable subsequence, we may assume that $A u_n' \to w$. Then $T u_n' = S u_n' - A u_n' \to -w$. Since $u_n' \to 0$ and T is closable, we must have $w = 0$. But this contradicts the fact that $-w$ is the limit of $T u_n'$ where $\|T u_n'\| = 1$.

Next we prove that S is closable. Let $u_n \to 0$ and $S u_n \to v$; we have to show that $v = 0$. Since A is S-compact, $\{A u_n\}$ contains a convergent subsequence. Again we may assume that $A u_n \to w$. Then $T u_n = S u_n - A u_n \to v - w$. Since $u_n \to 0$ and T is closable, we have $T u_n \to v - w = 0$. Since A is T-bounded, we have $A u_n \to 0$. Hence $v = w = 0$.

Let $\tilde{T}$, $\tilde{S}$ be the closures of T, S, respectively. If $u \in \mathsf{D}(\tilde{T})$, there is a sequence $\{u_n\}$ which is T-convergent to u. Since S as well as A is T-bounded, $\{u_n\}$ is also S-convergent to u so that $u \in \mathsf{D}(\tilde{S})$. Suppose, conversely, that $u \in \mathsf{D}(\tilde{S})$. Then there is a sequence $\{u_n\}$ which is S-convergent to u. Then $\{T u_n\}$ is bounded; this can be proved exactly in the same way as in the first part of this proof. Hence we may assume, as before, that $A u_n \to w$ and $T u_n = S u_n - A u_n \to v - w$. Thus u_n is T-convergent to u and $u \in \mathsf{D}(\tilde{T})$. This proves that $\mathsf{D}(\tilde{T}) = \mathsf{D}(\tilde{S})$.

Remark 1.12. Let T be closed and $\mathsf{D}(A) \supset \mathsf{D}(T)$. Define $\hat{\mathsf{X}}$ and $\hat{A}$ as in Remark 1.4. Then A is T-compact if and only if $\hat{A}$ is compact.

Remark 1.13. We can also define *relatively degenerate operators* as special cases of relatively compact operators. A is *T-degenerate* if A is T-bounded and $\mathsf{R}(A)$ is finite-dimensional. It is easy to see that a T-degenerate operator is T-compact.

We shall give an explicit form of a T-degenerate operator. Let T be from X to Y and consider the product space $\mathsf{X} \times \mathsf{Y}$ (see III-§ 5.2). Since the T-boundedness of A implies that $\|A u\| \leq \text{const.}\, \|\{u, T u\}\|$, A may be regarded as a bounded operator from $\mathsf{X} \times \mathsf{Y}$ to Y' (the range space of A), with domain $\mathsf{G}(T)$ (the graph of T). Since $\mathsf{R}(A)$ is finite-dimensional, $A u$ can be written in the form

$$A u = \sum_{j=1}^{m} g_j'[u]\, y_j', \quad y_j' \in \mathsf{Y}'. \tag{1.28}$$

where the g_j' are linear forms defined on $\mathsf{G}(T)$. Since $|g_j'[u]| \leq \text{const.} \cdot \|A u\|$ by I-(1.21), it follows that the g_j' are bounded forms in $\mathsf{X} \times \mathsf{Y}$ with domain $\mathsf{G}(T)$. According to the Hahn-Banach theorem, the g_j' can be extended to bounded forms on $\mathsf{X} \times \mathsf{Y}$. Since $(\mathsf{X} \times \mathsf{Y})^* = \mathsf{X}^* \times \mathsf{Y}^*$, the extended forms can be written in the form $\{u, v\} \to (u, f_j) + (v, g_j)$ with $f_j \in \mathsf{X}^*$ and $g_j \in \mathsf{Y}^*$. Restricting $\{u, v\}$ to $\mathsf{G}(T)$, we have $g_j'[u] = (u, f_j) + (T u, g_j)$. In this way we arrive at the following expression for a T-degenerate operator from X to Y':

$$A u = \sum_{j=1}^{m} [(u, f_j) + (T u, g_j)]\, y_j', \quad f_j \in \mathsf{X}^*, \quad g_j \in \mathsf{Y}^*, \quad y_j' \in \mathsf{Y}'. \tag{1.29}$$

Problem 1.14. Assume that there is a densely defined operator S from Y^* to X^* adjoint to T [see III-(5.7)]. Then a T-degenerate operator A has T-bound zero. [hint: Approximate the g_j of (1.29) by elements of $D(S)$.]

Example 1.15. In Example 1.6, A is T-compact if (a, b) is a finite interval. In fact if $\{u_n\}$ is a sequence such that $T u_n = -u_n''$ is bounded, then $A u_n = u_n'$ is equicontinuous. If in addition $\{u_n\}$ is bounded, $\{u_n'(x)\}$ is uniformly bounded in x and n. Thus it follows from the Ascoli theorem that $\{u_n'\}$ contains a uniformly convergent subsequence, which is a Cauchy sequence in X. A is not T-compact for the interval $(0, \infty)$ or $(-\infty, \infty)$, although A is T-bounded with T-bound zero [see (1.14)]. In Example 1.8, A is not only T-compact but T-degenerate. This is obvious since A is T-bounded and the range space of A is one-dimensional.

4. Stability of bounded invertibility

Let $T \in \mathscr{C}(X, Y)$. We shall now show that the property $T^{-1} \in \mathscr{B}(Y, X)$ is stable under a small perturbation.

Theorem 1.16.[1] *Let T and A be operators from X to Y. Let T^{-1} exist and belong to $\mathscr{B}(Y, X)$ (so that T is closed). Let A be T-bounded, with the constants a, b in (1.1) satisfying the inequality*

$$(1.30) \qquad a\|T^{-1}\| + b < 1 .$$

Then $S = T + A$ is closed and invertible, with $S^{-1} \in \mathscr{B}(Y, X)$ and

$$(1.31) \quad \|S^{-1}\| \leqq \frac{\|T^{-1}\|}{1 - a\|T^{-1}\| - b}, \quad \|S^{-1} - T^{-1}\| \leqq \frac{\|T^{-1}\|(a\|T^{-1}\| + b)}{1 - a\|T^{-1}\| - b} .$$

If in addition T^{-1} is compact, so is S^{-1}.

Proof. Since (1.30) implies $b < 1$, S is closed by Theorem 1.1. The proof of other assertions of the theorem is not essentially different from the finite-dimensional case (see the end of I-§ 4.4): first we note that

$$(1.32) \qquad S = T + A = (1 + A T^{-1}) T, \quad A T^{-1} \in \mathscr{B}(Y) ,$$

for $A T^{-1}$ is an operator on Y to Y and is bounded by $\|A T^{-1} v\| \leqq a\|T^{-1} v\| + b\|v\| \leqq (a\|T^{-1}\| + b)\|v\|$. Thus

$$(1.33) \qquad \|A T^{-1}\| \leqq a\|T^{-1}\| + b < 1$$

and $1 + A T^{-1}$ maps Y onto Y one to one, and the argument of I-§ 4.4 is applicable without essential change. The only change required is the use of (1.33) instead of the simple estimate $\|A T^{-1}\| \leqq \|A\| \|T^{-1}\|$ used there.

If T^{-1} is compact, $S^{-1} = T^{-1}(1 + A T^{-1})^{-1}$ is compact by Theorem III-4.8.

Remark 1.17. If A is bounded in Theorem 1.16, we can take $a = \|A\|$, $b = 0$ and (1.31) reduces to I-(4.24). If, furthermore, A is assumed to commute with T ($Y = X$ being assumed), then we can prove the existence of S^{-1} under a weaker assumption, namely,

[1] This theorem will be generalized in the following section (see Theorem 2.21).

Theorem 1.18. *Let T be an operator in X with $T^{-1} \in \mathscr{B}(\mathsf{X})$.* Let $A \in \mathscr{B}(\mathsf{X})$ *commute with* T. *If* $\operatorname{spr} A < 1/\operatorname{spr} T^{-1}$, *then* $(T + A)^{-1}$ *exists and belongs to* $\mathscr{B}(\mathsf{X})$.

Proof. The commutativity of A with T is equivalent to $A T^{-1} = T^{-1} A$ (see Problem III-5.37). Thus the proof is essentially the same as the proof of a similar result in the finite-dimensional case (Remark I-4.7).

§ 2. Generalized convergence of closed operators

1. The gap between subspaces

When we consider various perturbation problems related to closed operators, it is necessary to make precise what is meant by a "small" perturbation. In the previous section we considered a special kind, namely a relatively bounded perturbation. But this notion is still restricted, and we want to have a more general definition of the smallness of a perturbation for closed operators.

This can be done in a most natural way by introducing a metric in the set $\mathscr{C}(\mathsf{X}, \mathsf{Y})$ of all closed operators from X to Y. If $T, S \in \mathscr{C}(\mathsf{X}, \mathsf{Y})$, their graphs $\mathsf{G}(T)$, $\mathsf{G}(S)$ are closed linear manifolds in the product space $\mathsf{X} \times \mathsf{Y}$. Thus the "distance" between T and S can be measured by the "aperture" or "gap" between the closed linear manifolds $\mathsf{G}(T)$, $\mathsf{G}(S)$. In this way we are led to consider how to measure the gap of two closed linear manifolds of a Banach space.

In this paragraph we shall consider closed linear manifolds $\mathsf{M}, \mathsf{N}, \ldots$ of a Banach space Z.

We denote by S_M the unit sphere of M (the set of all $u \in \mathsf{M}$ with $\|u\| = 1$). For any two closed linear manifolds M, N of Z, we set

$$\delta(\mathsf{M}, \mathsf{N}) = \sup_{u \in \mathsf{S}_\mathsf{M}} \operatorname{dist}(u, \mathsf{N}) \,, \tag{2.1}$$

$$\hat{\delta}(\mathsf{M}, \mathsf{N}) = \max[\delta(\mathsf{M}, \mathsf{N}), \delta(\mathsf{N}, \mathsf{M})]^1 \,. \tag{2.2}$$

(2.1) has no meaning if $\mathsf{M} = 0$; in this case we define $\delta(0, \mathsf{N}) = 0$ for any N. On the other hand $\delta(\mathsf{M}, 0) = 1$ if $\mathsf{M} \neq 0$, as is seen from the definition.

$\delta(\mathsf{M}, \mathsf{N})$ can also be characterized as the smallest number δ such that

$$\operatorname{dist}(u, \mathsf{N}) \leqq \delta \|u\| \quad \text{for all} \quad u \in \mathsf{M} \,. \tag{2.3}$$

$\hat{\delta}(\mathsf{M}, \mathsf{N})$ will be called the *gap* between M, N.

The following relations follow directly from the definition.

$$\delta(\mathsf{M}, \mathsf{N}) = 0 \quad \text{if and only if} \quad \mathsf{M} \subset \mathsf{N} \,. \tag{2.4}$$

$$\hat{\delta}(\mathsf{M}, \mathsf{N}) = 0 \quad \text{if and only if} \quad \mathsf{M} = \mathsf{N} \,. \tag{2.5}$$

$$\hat{\delta}(\mathsf{M}, \mathsf{N}) = \hat{\delta}(\mathsf{N}, \mathsf{M}) \,. \tag{2.6}$$

$$0 \leqq \delta(\mathsf{M}, \mathsf{N}) \leqq 1 \,, \quad 0 \leqq \hat{\delta}(\mathsf{M}, \mathsf{N}) \leqq 1 \,. \tag{2.7}$$

---

[1] See GOHBERG and KREIN [1], T. KATO [12], CORDES and LABROUSSE [1].

(2.5) and (2.6) suggest that $\hat{\delta}(M, N)$ could be used to define a *distance* between M and N. But this is not possible, since the function $\hat{\delta}$ does not in general satisfy the triangle inequality required of a distance function[1].

This incovenience may be removed by slightly modifying the definition (2.1–2.2). Set

$$d(M, N) = \sup_{u \in S_M} \operatorname{dist}(u, S_N), \tag{2.8}$$

$$\hat{d}(M, N) = \max[d(M, N), d(N, M)]^{2}. \tag{2.9}$$

(2.8) does not make sense if either M or N is 0. In such cases we set

$$d(0, N) = 0 \quad \text{for any } N; \quad d(M, 0) = 2 \quad \text{for} \quad M \neq 0. \tag{2.10}$$

Then all the relations (2.4–2.7) are again satisfied by δ, $\hat{\delta}$ replaced by d, $\hat{d}$, respectively if 1 is replaced by 2 in (2.7). Furthermore, d and $\hat{d}$ satisfy the triangle inequalities:

$$d(L, N) \leqq d(L, M) + d(M, N), \quad \hat{d}(L, N) \leqq \hat{d}(L, M) + \hat{d}(M, N). \tag{2.11}$$

The second inequality of (2.11) follows from the first, which in turn follows easily from the definition. The proof will be left to the reader. [The case when some of L, M, N are 0 should be considered separately; note (2.10).]

The set of all closed linear manifolds of Z becomes a *metric space* if the distance between M, N is defined by $\hat{d}(M, N)$. A sequence $\{M_n\}$ of closed linear manifolds *converges* to M if $\hat{d}(M_n, M) \to 0$ for $n \to \infty$. Then we write $M_n \to M$ or $\lim M_n = M$.

Although the gap $\hat{\delta}$ is not a proper distance function, it is more convenient than the proper distance function $\hat{d}$ for applications since its definition is slightly simpler. Furthermore, when we consider the *topology* of the set of all closed linear manifolds, the two functions give the same result. This is due to the following inequalities[3]:

$$\begin{aligned} \delta(M, N) &\leqq d(M, N) \leqq 2\delta(M, N), \\ \hat{\delta}(M, N) &\leqq \hat{d}(M, N) \leqq 2\hat{\delta}(M, N). \end{aligned} \tag{2.12}$$

The second set of inequalities follows directly from the first. Also the first inequality of the first set is trivial. To prove the second inequality $d(M, N) \leqq 2\delta(M, N)$, it suffices to assume $N \neq 0$ and show that

$$\operatorname{dist}(u, S_N) \leqq 2 \operatorname{dist}(u, N) \quad \text{for any} \quad u \in Z \quad \text{with} \quad \|u\| = 1. \tag{2.13}$$

[1] $\hat{\delta}$ does satisfy the triangle inequality if Z is a Hilbert space. This follows from the relation $\hat{\delta}(M, N) = \|P - Q\|$, where P, Q are the orthogonal projections on M, N, respectively, which follows from Theorem I-6.34. In this case $\hat{\delta}$ is a more convenient metric than the $\hat{d}$ to be introduced below.

[2] See Gohberg and Markus [1]. A different but equivalent metric is introduced by Newburgh [2]. $\hat{d}$ is the *Hausdorff distance* defined on the set of all S_M (except for $M = 0$); see Hausdorff [1], p. 145. $\hat{d}(M, N)$ is equal to $\varrho(S_M, S_N)$ in the notation of Hausdorff. For the discussion of various metrics see Berkson [1].

[3] Cf. Gohberg and Markus [1].

For any $\varepsilon > 0$, there is a $v \in \mathsf{N}$ such that $\|u - v\| < \operatorname{dist}(u, \mathsf{N}) + \varepsilon$. We may assume that $v \neq 0$, for otherwise we can change v slightly without affecting the inequality. Then $v_0 = v/\|v\| \in \mathsf{S_N}$ and $\operatorname{dist}(u, \mathsf{S_N}) \leqq \|u - v_0\| \leqq \|u - v\| + \|v - v_0\|$. But $\|v - v_0\| = |\|v\| - 1| = |\|v\| - \|u\|| \leqq \|v - u\|$, so that $\operatorname{dist}(u, \mathsf{S_N}) \leqq 2\|u - v\| < 2\operatorname{dist}(u, \mathsf{N}) + 2\varepsilon$. Since $\varepsilon > 0$ is arbitrary, this proves (2.13).

(2.12) shows that $\hat{d}(\mathsf{M}_n, \mathsf{M}) \to 0$ is equivalent to $\hat{\delta}(\mathsf{M}_n, \mathsf{M}) \to 0$. Thus the convergence $\mathsf{M}_n \to \mathsf{M}$ can be defined by $\hat{\delta}(\mathsf{M}_n, \mathsf{M}) \to 0$ without reference to the $\hat{d}$ function. In what follows we shall use almost exclusively the gap $\hat{\delta}$ rather than the distance $\hat{d}$.

Remark 2.1. The metric space of all closed linear manifolds of Z defined above is *complete*; if $\{\mathsf{M}_n\}$ is a Cauchy sequence ($\hat{d}(\mathsf{M}_n, \mathsf{M}_m) \to 0$ for $n, m \to \infty$), then there is a closed linear manifold M such that $\hat{d}(\mathsf{M}_n, \mathsf{M}) \to 0$. Since we do not need this theorem, we shall not prove it here.

The following lemma will be needed later.

Lemma 2.2. *For any closed linear manifolds* M, N *of* Z *and any* $u \in \mathsf{Z}$, *we have*

$$(1 + \delta(\mathsf{M}, \mathsf{N}))\operatorname{dist}(u, \mathsf{M}) \geqq \operatorname{dist}(u, \mathsf{N}) - \|u\|\,\delta(\mathsf{M}, \mathsf{N})\,. \tag{2.14}$$

Proof. For any $\varepsilon > 0$ there is a $v \in \mathsf{M}$ such that $\|u - v\| < \operatorname{dist}(u, \mathsf{M}) + \varepsilon$, and for this v there is a $w \in \mathsf{N}$ such that $\|v - w\| < \operatorname{dist}(v, \mathsf{N}) + \varepsilon$. Hence $\operatorname{dist}(u, \mathsf{N}) \leqq \|u - w\| \leqq \operatorname{dist}(u, \mathsf{M}) + \operatorname{dist}(v, \mathsf{N}) + 2\varepsilon \leqq \operatorname{dist}(u, \mathsf{M}) + \|v\|\,\delta(\mathsf{M}, \mathsf{N}) + 2\varepsilon$. But $\|v\| \leqq \|u\| + \|u - v\| \leqq \|u\| + \operatorname{dist}(u, \mathsf{M}) + \varepsilon$. Hence $\operatorname{dist}(u, \mathsf{N}) \leqq (1 + \delta(\mathsf{M}, \mathsf{N}))\operatorname{dist}(u, \mathsf{M}) + \|u\|\,\delta(\mathsf{M}, \mathsf{N}) + 2\varepsilon + \varepsilon\,\delta(\mathsf{M}, \mathsf{N})$. On letting $\varepsilon \to 0$ we obtain (2.14).

2. The gap and the dimension

The following lemma is basic in the study of the gaps between closed linear manifolds.

Lemma 2.3.[1] *Let* M, N *be linear manifolds in a Banach space* Z. *If* $\dim \mathsf{M} > \dim \mathsf{N}$, *there exists a* $u \in \mathsf{M}$ *such that*

$$\operatorname{dist}(u, \mathsf{N}) = \|u\| > 0\,. \tag{2.15}$$

Remark 2.4. If the quotient space $\tilde{\mathsf{Z}} = \mathsf{Z}/\mathsf{N}$ is introduced (see III-§ 1.8), (2.15) can be written

$$\|\tilde{u}\| = \|u\| > 0\,. \tag{2.16}$$

Note that N is closed since $\dim \mathsf{N} < \infty$ by hypothesis.

[1] See KREIN, KRASNOSEL'SKII and MIL'MAN [1], GOHBERG and KREIN [1], T. KATO [12].

Proof of Lemma 2.3. We may assume that both M and N are finite-dimensional, for $\dim N < \infty$ and we may replace M by any of its finite-dimensional subspaces with dimension equal to $\dim N + 1$. Hence Z itself may also be assumed to be finite-dimensional, for it suffices to consider the problem in the subspace $M + N$.

For the moment assume that Z is *strictly convex*, by which we mean that $\|u + v\| < \|u\| + \|v\|$ whenever u, v are linearly independent. Then it is easily seen that each $u \in Z$ has a unique *nearest point* $v = Au$ in N and that the map $u \to Au$ is continuous[1]. The operator A is in general nonlinear, but it has the property that $A(-u) = -Au$. According to a theorem by Borsuk[2], there exists a $u \in M$ such that $\|u\| = 1$ and $Au = 0$. This u satisfies the requirements of the lemma.

In the general case we regard Z as a real Banach space and choose a basis $f_1, \ldots, f_m$ of the real adjoint space of Z. Then

$$\|u\|_n = \left\{\|u\|^2 + \frac{1}{n}\left[(u, f_1)^2 + \cdots + (u, f_m)^2\right]\right\}^{1/2}$$

defines a new norm in Z and converts Z into a strictly convex space. For each $n = 1, 2, \ldots$, there exists a $u_n \in M$ such that $\mathrm{dist}_n(u_n, N) = \|u_n\|_n = 1$, where dist_n denotes the distance in the sense of the norm $\| \ \|_n$. Since $\|u_n\| \leq \|u_n\|_n = 1$, the sequence $\{u_n\}$ contains a convergent subsequence. The limit u of this subsequence is easily seen to satisfy the requirements of the lemma.

Remark 2.5. The nonlinearity of the operator A used above gives the lemma a non-elementary character. If Z is a unitary space, A is simply the orthogonal projection on N, and the proof of the lemma is quite elementary.

Corollary 2.6. *Let* M, N *be closed linear manifolds.* $\delta(M, N) < 1$ *implies* $\dim M \leq \dim N$. $\hat{\delta}(M, N) < 1$ *implies* $\dim M = \dim N$.

Remark 2.7. The last corollary shows that the space of closed linear manifolds of Z is the union of disjoint open sets, each of which consists of closed linear manifolds with a fixed dimension.

3. Duality

There is a simple relationship between the gap function in a Banach space Z and that in the adjoint space Z^*. For any closed linear manifold $M \subset Z$, $M^\perp$ denotes the annihilator of M; $M^\perp$ is the closed linear manifold of Z^* consisting of all $f \in Z^*$ such that $f \perp M$ (see III-§ 1.4).

---

[1] The existence of a nearest point v in N to u follows from the local compactness of N. The uniqueness of v and its continuous dependence follows from the strict convexity of Z.

[2] See Alexandroff and Hopf [1], p. 483.

Lemma 2.8. *Let* M *be a closed linear manifold of* Z, $0 \neq M \neq Z$. *Then*

$$\text{dist}(f, M^{\perp}) = \sup_{u \in S_M} |(u, f)| = \|f_M\|, \quad f \in Z^*, \tag{2.17}$$

$$\text{dist}(u, M) = \sup_{f \in S_{M^{\perp}}} |(u, f)|, \quad u \in Z, \tag{2.18}$$

where f_M *is the restriction of* f *to* M.

Proof. Let $f \in Z^*$. By the Hahn-Banach theorem, there is a $g \in Z^*$ which is an extension of f_M with $\|g\| = \|f_M\|$. Then $h = f - g \in M^{\perp}$ since $(u, f) = (u, g)$ for $u \in M$. Thus $\text{dist}(f, M^{\perp}) \leq \|f - h\| = \|g\| = \|f_M\| = \sup_{u \in S_M} |(u, f)|$.

On the other hand, for any $h \in M^{\perp}$ we have $|(u, f)| = |(u, f - h)| \leq \leq \|f - h\|$ if $u \in S_M$. Hence $|(u, f)| \leq \text{dist}(f, M^{\perp})$, which gives the opposite inequality to the above and completes the proof of (2.17).

Let $u \in Z$. For each $f \in S_{M^{\perp}}$ we have $|(u, f)| = |(u - v, f)| \leq \|u - v\|$ for any $v \in M$. Hence $|(u, f)| \leq \text{dist}(u, M)$ and so $\sup_{f \in S_{M^{\perp}}} |(u, f)| \leq \text{dist}(u, M)$. The opposite inequality follows from the fact that there exists an $f \in S_{M^{\perp}}$ such that $|(u, f)| = \text{dist}(u, M)$ (which is a direct consequence of Theorem III-1.22).

Theorem 2.9. *For closed linear manifolds* M, N *of* Z, *we have*

$$\delta(M, N) = \delta(N^{\perp}, M^{\perp}), \quad \hat{\delta}(M, N) = \hat{\delta}(M^{\perp}, N^{\perp}). \tag{2.19}$$

Proof. The second equality follows from the first, which in turn follows from Lemma 2.8, for

$$\begin{aligned} \delta(M, N) &= \sup_{u \in S_M} \text{dist}(u, N) = \sup_{u \in S_M} \sup_{g \in S_{N^{\perp}}} |(u, g)| \\ &= \sup_{g \in S_{N^{\perp}}} \sup_{u \in S_M} |(u, g)| = \sup_{g \in S_{N^{\perp}}} \text{dist}(g, M^{\perp}) = \delta(N^{\perp}, M^{\perp}). \end{aligned}$$

[The above proof applies to the case where $M \neq 0$ and $N \neq Z$. If $M = 0$, then $M^{\perp} = Z^*$ so that $\delta(M, N) = 0 = \delta(N^{\perp}, M^{\perp})$. If $N = Z$, then $N^{\perp} = 0$ so that $\delta(M, N) = 0 = \delta(N^{\perp}, M^{\perp})$.]

4. The gap between closed operators

Let us consider the set $\mathscr{C}(X, Y)$ of all closed operators from X to Y. If $T, S \in \mathscr{C}(X, Y)$, their graphs $G(T)$, $G(S)$ are closed linear manifolds of the product space $X \times Y$. We set

$$\begin{aligned} \delta(T, S) = \delta(G(T), G(S)), \quad \hat{\delta}(T, S) &= \hat{\delta}(G(T), G(S)) \\ &= \max[\delta(T, S), \delta(S, T)]. \end{aligned} \tag{2.20}$$

$\hat{\delta}(T, S)$ will be called the *gap* between T and S[1].

[1] A similar notion is introduced, and some of the theorems given below are proved, in NEWBURGH [2]. In the special case where X, Y are Hilbert spaces, most of the following results are simplified and strengthened; see CORDES and LABROUSSE [1].

Similarly we can define the *distance* $\hat{d}(T, S)$ between T and S as equal to $\hat{d}(\mathsf{G}(T), \mathsf{G}(S))$. Under this distance function $\mathscr{C}(\mathsf{X}, \mathsf{Y})$ becomes a metric space. In this space the convergence of a sequence $T_n \in \mathscr{C}(\mathsf{X}, \mathsf{Y})$ to a $T \in \mathscr{C}(\mathsf{X}, \mathsf{Y})$ is defined by $\hat{d}(T_n, T) \to 0$. But since $\hat{\delta}(T, S) \leq \hat{d}(T, S) \leq 2\hat{\delta}(T, S)$ in virtue of (2.12), this is true if and only if $\hat{\delta}(T_n, T) \to 0$. In this case we shall also say that the operator T_n converges to T (or $T_n \to T$) *in the generalized sense.*

It should be remarked that earlier we defined the convergence of operators only for operators of the class $\mathscr{B}(\mathsf{X}, \mathsf{Y})$. Actually we introduced several different notions of convergence: convergence in norm, strong and weak convergence. We shall show in a moment that the notion of generalized convergence introduced above for closed operators is a generalization of convergence *in norm* for operators of $\mathscr{B}(\mathsf{X}, \mathsf{Y})$.

Remark 2.10. When T varies over $\mathscr{C}(\mathsf{X},\mathsf{Y})$, $\mathsf{G}(T)$ varies over a proper subset of the set of all closed linear manifolds of $\mathsf{X} \times \mathsf{Y}$. This subset is not closed and, consequently, $\mathscr{C}(\mathsf{X}, \mathsf{Y})$ is not a complete metric space (assuming, of course, that $\dim \mathsf{X} \geqq 1$, $\dim \mathsf{Y} \geqq 1$). It is not trivial to see this in general, but it is easily seen if $\mathsf{Y} = \mathsf{X}$. Consider the sequence $\{nI\}$ where I is the identity operator in X. $\mathsf{G}(nI)$ is the subset of $\mathsf{X} \times \mathsf{X}$ consisting of all elements $\{n^{-1} u, u\}$, $u \in \mathsf{X}$, and it is readily seen that $\lim \mathsf{G}(nI)$ exists and is equal to the set of all elements of the form $\{0, u\}$, $u \in \mathsf{X}$. But this set is not a graph. Thus $\{nI\}$ is a Cauchy sequence in $\mathscr{C}(\mathsf{X}) = \mathscr{C}(\mathsf{X}, \mathsf{X})$ without a limit.

Lemma 2.11. *Let* $T \in \mathscr{B}(\mathsf{X}, \mathsf{Y})$. *If* $S \in \mathscr{C}(\mathsf{X}, \mathsf{Y})$ *and* $\delta(S, T) < (1 + \|T\|^2)^{-1/2}$, *then S is bounded (so that $\mathsf{D}(S)$ is closed).*

Proof. Let φ be an element of the unit sphere of $\mathsf{G}(S)$: $\varphi = \{u, Su\} \in \mathsf{G}(S)$, $u \in \mathsf{D}(S)$ and

$$\|u\|^2 + \|Su\|^2 = \|\varphi\|^2 = 1^1 . \tag{2.21}$$

Let δ' be any number such that $\delta(S, T) < \delta' < (1 + \|T\|^2)^{-1/2}$. Then φ has distance smaller than δ' from $\mathsf{G}(T)$, so that there exists a $\psi = \{v, Tv\} \in \mathsf{G}(T)$ such that $\|\varphi - \psi\| < \delta'$:

$$\|u - v\|^2 + \|Su - Tv\|^2 = \|\varphi - \psi\|^2 < \delta'^2 . \tag{2.22}$$

Set $A = S - T$; we have $\|Au\|^2 = \|Su - Tv - T(u - v)\|^2 \leq (\|Su - Tv\| + \|T\| \|u - v\|)^2 \leq \delta'^2(1 + \|T\|^2)$ by the Schwarz inequality and (2.22). Since

$$1 = \|u\|^2 + \|Tu + Au\|^2 \leq (1 + \|T\|^2) \|u\|^2 + 2\|T\| \|u\| \|Au\| + \|Au\|^2,$$

by (2.21), we have

$$\|Au\|^2 \leq \delta'^2(1 + \|T\|^2) [(1 + \|T\|^2) \|u\|^2 + 2 \|T\| \|u\| \|Au\| + \|Au\|^2].$$

---

[1] It should be recalled that we defined the norm in $\mathsf{X} \times \mathsf{Y}$ by $\|\{u, v\}\| = (\|u\|^2 + \|v\|^2)^{1/2}$.

Solving this inequality for $\|Au\|$, we obtain

$$\|Au\| \leqq \frac{\delta'(1+\|T\|^2)\,[(1-\delta'^2)^{1/2}+\delta'\|T\|]}{1-\delta'^2(1+\|T\|^2)}\,\|u\| \leqq \tag{2.23}$$
$$\leqq \frac{\delta'(1+\|T\|^2)}{1-\delta'(1+\|T\|^2)^{1/2}}\,\|u\|\,;$$

note that the denominators are positive.

Since (2.23) is homogeneous in u, it is true for every $u \in \mathsf{D}(S)$ without any normalization. Thus A is bounded and so is $S = T + A$.

Lemma 2.12. *Let* $T \in \mathscr{B}(\mathsf{X}, \mathsf{Y})$. *If* $S \in \mathscr{C}(\mathsf{X}, \mathsf{Y})$ *and* $\hat{\delta}(T, S) < (1 + \|T\|^2)^{-1/2}$, *then* S *is densely defined*[1].

Proof. Let v be any vector of X so normalized that $\psi = \{v, Tv\}$ has norm 1:

$$\|v\|^2 + \|Tv\|^2 = \|\psi\|^2 = 1\,. \tag{2.24}$$

Let δ' be such that $\hat{\delta}(T, S) < \delta' < (1 + \|T\|^2)^{-1/2}$. Then there is a $\varphi = \{u, Su\}$ satisfying (2.22) [but not necessarily (2.21)]. Hence $\|v - u\| < \delta'$ and so $\mathrm{dist}(v, \mathsf{M}) < \delta'$ where M is the closure of $\mathsf{D}(S)$. But since $1 \leqq (1 + \|T\|^2)\,\|v\|^2$ by (2.24), $\mathrm{dist}(v, \mathsf{M}) < \delta'(1 + \|T\|^2)^{1/2}\,\|v\|$. The last inequality is homogeneous in v and therefore true for every $v \in \mathsf{X}$. Since $\delta'(1 + \|T\|^2)^{1/2} < 1$, it follows that $\mathsf{M} = \mathsf{X}$; otherwise there would exist a $v \neq 0$ such that $\mathrm{dist}(v, \mathsf{M}) > \delta'(1 + \|T\|^2)^{1/2}\,\|v\|$, see Lemma III-1.12. Thus $\mathsf{D}(S)$ is dense in X.

Theorem 2.13. *Let* $T \in \mathscr{B}(\mathsf{X}, \mathsf{Y})$. *If* $S \in \mathscr{C}(\mathsf{X}, \mathsf{Y})$ *is so close to* T *that* $\hat{\delta}(S, T) < (1 + \|T\|^2)^{-1/2}$, *then* $S \in \mathscr{B}(\mathsf{X}, \mathsf{Y})$ *and*[2]

$$\|S - T\| \leqq \frac{(1 + \|T\|^2)\,\hat{\delta}(S, T)}{1 - (1 + \|T\|^2)^{1/2}\,\hat{\delta}(S, T)}\,. \tag{2.25}$$

Proof. It follows from Lemmas 2.11 and 2.12 that S is bounded, $\mathsf{D}(S)$ is closed and dense in X. Hence $\mathsf{D}(S) = \mathsf{X}$ and $S \in \mathscr{B}(\mathsf{X}, \mathsf{Y})$. Then (2.25) follows from (2.23) since δ' can be chosen arbitrarily close to $\hat{\delta}(S, T)$.

Theorem 2.14. *Let* $T \in \mathscr{C}(\mathsf{X}, \mathsf{Y})$ *and let* A *be* T*-bounded with relative bound less than* 1, *so that we have the inequality* (1.1) *with* $b < 1$. *Then* $S = T + A \in \mathscr{C}(\mathsf{X}, \mathsf{Y})$ *and*

$$\hat{\delta}(S, T) \leqq (1 - b)^{-1}(a^2 + b^2)^{1/2}\,. \tag{2.26}$$

In particular if $A \in \mathscr{B}(\mathsf{X}, \mathsf{Y})$, *then*

$$\hat{\delta}(T + A, T) \leqq \|A\|\,. \tag{2.27}$$

Proof. $S \in \mathscr{C}(\mathsf{X}, \mathsf{Y})$ was proved in Theorem 1.1. To prove (2.26), let $\varphi = \{u, Su\} \in \mathsf{G}(S)$ with $\|\varphi\| = 1$, so that we have (2.21). Setting

[1] Actually we have a stronger result that $S \in \mathscr{B}(\mathsf{X}, \mathsf{Y})$; see Problem 5.21.

[2] In (2.25) $\hat{\delta}(S, T)$ may be replaced by $\hat{\delta}(T, S)$; see Problem 5.21.

$\psi = \{u, Tu\} \in \mathsf{G}(T)$, we have $\|\varphi - \psi\| = \|(S - T)\, u\| = \|A u\| \leq$ $\leq (1 - b)^{-1}(a\,\|u\| + b\,\|S u\|)$ by (1.3). It follows by the Schwarz inequality and (2.21) that $\|\varphi - \psi\| \leq (1 - b)^{-1}(a^2 + b^2)^{1/2}$. Hence $\operatorname{dist}(\varphi, \mathsf{G}(T)) \leq (1 - b)^{-1}(a^2 + b^2)^{1/2}$ and, since φ is an arbitrary element of the unit sphere of $\mathsf{G}(S)$, $\delta(S, T) = \delta(\mathsf{G}(S), \mathsf{G}(T)) \leq (1 - b)^{-1}(a^2 + b^2)^{1/2}$.

$\delta(T, S)$ can be estimated similarly, using (1.1) rather than (1.3); the result is $\delta(T, S) \leq (a^2 + b^2)^{1/2}$. Thus we obtain the estimate (2.26) for $\hat{\delta}(S, T) = \max[\delta(S, T), \delta(T, S)]$.

Problem 2.15. If we assume (1.4) with $b = \max(b', b'') < 1$, then

$$\hat{\delta}(S, T) \leq (1 - b)^{-1}[a^2 + (b' + b'')^2]^{1/2}. \tag{2.28}$$

Remark 2.16. Theorem 2.13 shows that $\mathscr{B}(\mathsf{X}, \mathsf{Y})$ is an open subset of $\mathscr{C}(\mathsf{X}, \mathsf{Y})$. (2.25) and (2.27) show that within this open subset $\mathscr{B}(\mathsf{X}, \mathsf{Y})$, the topology defined by the distance function $\hat{d}$ (or, equivalently, by the gap function $\hat{\delta}$) is identical with the norm topology.

Theorem 2.17. *Let $T, S \in \mathscr{C}(\mathsf{X}, \mathsf{Y})$ and $A \in \mathscr{B}(\mathsf{X}, \mathsf{Y})$. Then*

$$\hat{\delta}(S + A, T + A) \leq 2(1 + \|A\|^2)\, \hat{\delta}(S, T). \tag{2.29}$$

Proof. $T \in \mathscr{C}(\mathsf{X}, \mathsf{Y})$ implies that $T + A \in \mathscr{C}(\mathsf{X}, \mathsf{Y})$ with $\mathsf{D}(T + A) = \mathsf{D}(T)$. Similarly $S + A \in \mathscr{C}(\mathsf{X}, \mathsf{Y})$.

Let $\varphi \in \mathsf{G}(S + A)$ with $\|\varphi\| = 1$. Then there is a $u \in \mathsf{D}(S)$ such that $\varphi = \{u, (S + A)\, u\}$ and

$$\|u\|^2 + \|(S + A)\, u\|^2 = \|\varphi\|^2 = 1. \tag{2.30}$$

Set $\|u\|^2 + \|S u\|^2 = r^2$, $r > 0$. $r^{-1}\{u, Su\}$ is an element of the unit sphere of $\mathsf{G}(S)$. For any $\delta' > \hat{\delta}(S, T) = \hat{\delta}(\mathsf{G}(S), \mathsf{G}(T))$, therefore, $r^{-1}\{u, Su\}$ has distance $< \delta'$ from $\mathsf{G}(T)$. Hence there is a $v \in \mathsf{D}(T)$ such that $\|u - v\|^2 + \|Su - Tv\|^2 < r^2 \delta'^2$. Then, setting $\psi = \{v, (T + A)\, v\}$, we have

$$\begin{aligned} \|\varphi - \psi\|^2 &= \|u - v\|^2 + \|(S + A)\, u - (T + A)\, v\|^2 \leq \\ &\leq \|u - v\|^2 + 2\|Su - Tv\|^2 + 2\|A\|^2\, \|u - v\|^2 \leq \\ &\leq 2(1 + \|A\|^2)\, r^2 \delta'^2. \end{aligned} \tag{2.31}$$

On the other hand $r^2 = \|u\|^2 + \|S u\|^2 = \|u\|^2 + \|(S + A)\, u - A u\|^2 \leq$ $\leq \|u\|^2 + 2\|(S + A)\, u\|^2 + 2\|A\|^2\, \|u\|^2 \leq 2 + 2\|A\|^2\, \|u\|^2 \leq 2 + 2\|A\|^2$ by (2.30). Hence $\|\varphi - \psi\|^2 \leq 4(1 + \|A\|^2)^2\, \delta'^2$. Since $\psi \in \mathsf{G}(T + A)$, this implies that $\operatorname{dist}(\varphi, \mathsf{G}(T + A)) \leq 2(1 + \|A\|^2)\, \delta'$ and, since φ is an arbitrary element of the unit sphere of $\mathsf{G}(S + A)$, that $\delta(S + A, T + A) = \delta(\mathsf{G}(S + A), \mathsf{G}(T + A)) \leq 2(1 + \|A\|^2)\, \delta'$. Since S and T may be exchanged in the above argument and since δ' may be arbitrarily close to $\hat{\delta}(S, T)$, we obtain (2.29).

Theorem 2.18. *Let $T, S \in \mathscr{C}(\mathsf{X}, \mathsf{Y})$ be densely defined. Then $\delta(T, S) = \delta(S^*, T^*)$ and $\hat{\delta}(T, S) = \hat{\delta}(T^*, S^*)$.*

Proof. $\delta(S^*, T^*) = \delta(G(S^*), G(T^*)) = \delta(G'(S^*), G'(T^*)) = \delta(G(-S)^\perp, G(-T)^\perp) = \delta(G(-T), G(-S)) = \delta(G(T), G(S)) = \delta(T, S)$, where $G(T) \subset X \times Y$ is the graph of T and $G'(T^*) \subset X^* \times Y^*$ is the inverse graph of $T^* \in \mathscr{C}(Y^*, X^*)$; note that $G'(T^*) = G(-T)^\perp$ by III-(5.9) and that $\delta(N^\perp, M^\perp) = \delta(M, N)$ by (2.19). $\delta(G(S^*), G(T^*)) = \delta(G'(S^*), G'(T^*))$ is due to the special choice of the norm in the product space (see par. 5 below).

Problem 2.19. Let $T \in \mathscr{C}(X, Y)$. T is bounded if and only if $\delta(T, 0) < 1$. $T \in \mathscr{B}(X, Y)$ if and only if $\hat{\delta}(T, 0) < 1$.

5. Further results on the stability of bounded invertibility

The graph $G(R)$ of an $R \in \mathscr{C}(Y, X)$ is a closed linear manifold of $Y \times X$, and the inverse graph $G'(R)$ of R is a closed linear manifold of $X \times Y$ obtained as the image of $G(R)$ under the map $\{y, x\} \to \{x, y\}$ (see III-§ 5.2). Since this map preserves the norm and hence the gap between two closed linear manifolds, $\delta(R_1, R_2) = \delta(G(R_1), G(R_2)) = \delta(G'(R_1), G'(R_2))$ and the same is true with δ replaced by $\hat{\delta}$, d or $\hat{d}$. Thus in the discussion of the gap or distance between operators of $\mathscr{C}(Y, X)$, we can replace their graphs by their inverse graphs.

If $T \in \mathscr{C}(X, Y)$ is invertible, $T^{-1} \in \mathscr{C}(Y, X)$ and $G'(T^{-1}) = G(T)$ [see III-(5.5)]. The following theorem is an immediate consequence of these observations.

Theorem 2.20. *If $T, S \in \mathscr{C}(X, Y)$ are invertible, then*

$$\delta(S^{-1}, T^{-1}) = \delta(S, T), \quad \hat{\delta}(S^{-1}, T^{-1}) = \hat{\delta}(S, T) . \tag{2.32}$$

If we denote by $\mathscr{C}_i(X, Y)$ the subset of $\mathscr{C}(X, Y)$ consisting of all invertible operators, Theorem 2.20 means that $T \to T^{-1}$ is an *isometric* mapping of $\mathscr{C}_i(X, Y)$ onto $\mathscr{C}_i(Y, X)$. In general the structure of $\mathscr{C}_i(X, Y)$ in $\mathscr{C}(X, Y)$ would be quite complicated. We shall show, however, that the set of $T \in \mathscr{C}_i(X, Y)$ such that $T^{-1} \in \mathscr{B}(Y, X)$ is open in $\mathscr{C}(X, Y)$. This is the *principle of stability of bounded invertibility* in its most general form.

Theorem 2.21. *Let $T \in \mathscr{C}(X, Y)$ be invertible with $T^{-1} \in \mathscr{B}(Y, X)$. If $S \in \mathscr{C}(X, Y)$ with $\hat{\delta}(S, T) < (1 + \|T^{-1}\|^2)^{-1/2}$, then S is invertible and $S^{-1} \in \mathscr{B}(Y, X)$.*

Proof. If S is known to be invertible, then $\hat{\delta}(S^{-1}, T^{-1}) = \hat{\delta}(S, T) < (1 + \|T^{-1}\|^2)^{-1/2}$ so that $S^{-1} \in \mathscr{B}(Y, X)$ by Theorem 2.13 (applied to the pair S^{-1}, T^{-1}). Thus it suffices to show that S is invertible.

Suppose that $Su = 0$, $\|u\| = 1$. Then $\{u, 0\}$ is on the unit sphere of $G(S)$, so that there is a $\{v, Tv\} \in G(T)$ such that $\|u - v\|^2 + \|Tv\|^2 < \delta'^2$, where δ' is a number such that $\hat{\delta}(S, T) < \delta' < (1 + \|T^{-1}\|^2)^{-1/2}$. Then

$1 = \|u\|^2 \leqq (\|u - v\| + \|v\|)^2 \leqq (\|u - v\| + \|T^{-1}\| \, \|Tv\|)^2 \leqq (1 + \|T^{-1}\|^2)\, \delta'^2 <$ < 1, a contradiction.

Remark 2.22. The theorems proved in this and the preceding paragraphs are not very strong from the quantitative point of view, for many crude estimates have been used in their proof. For example, Theorems 2.14 and 2.21 give the result that $(T + A)^{-1} \in \mathscr{B}(\mathsf{Y}, \mathsf{X})$ exists if $\|A\| < (1 + \|T^{-1}\|^2)^{-1/2}$. But this condition is unnecessarily strong, for we know that it is sufficient to assume $\|A\| < \|T^{-1}\|^{-1}$ (a special case of Theorem 1.16).

It is easy to improve this unsatisfactory result by an auxiliary argument. If we apply the result just obtained to the pair αT, αA with $\alpha > 0$, we see that $\alpha(T + A)$ has an inverse in $\mathscr{B}(\mathsf{Y}, \mathsf{X})$ if $\alpha\|A\| < (1 + \alpha^{-2}\|T^{-1}\|^2)^{-1/2}$, that is, if $\|A\| < (\alpha^2 + \|T^{-1}\|^2)^{-1/2}$. Since α can be chosen arbitrarily small, it follows that $(T + A)^{-1} \in \mathscr{B}(\mathsf{Y}, \mathsf{X})$ if $\|A\| < \|T^{-1}\|^{-1}$. In this way we regain the specific result from general theorems. In many cases this kind of auxiliary argument can be used to improve the numerical results.

6. Generalized convergence

We recall that T_n converges to T ($T_n \to T$) in the generalized sense if $\hat{\delta}(T_n, T) \to 0$. The following theorem is a direct consequence of Remark 2.16, Theorems 2.17, 2.18 and 2.20.

Theorem 2.23. *Let $T, T_n \in \mathscr{C}(\mathsf{X}, \mathsf{Y})$, $n = 1, 2, \ldots$.*

a) *If $T \in \mathscr{B}(\mathsf{X}, \mathsf{Y})$, $T_n \to T$ in the generalized sense if and only if $T_n \in \mathscr{B}(\mathsf{X}, \mathsf{Y})$ for sufficiently large n and $\|T_n - T\| \to 0$.*

b) *If T^{-1} exists and belongs to $\mathscr{B}(\mathsf{Y}, \mathsf{X})$, $T_n \to T$ in the generalized sense if and only if T_n^{-1} exists and belongs to $\mathscr{B}(\mathsf{Y}, \mathsf{X})$ for sufficiently large n and $\|T_n^{-1} - T^{-1}\| \to 0$.*

c) *If $T_n \to T$ in the generalized sense and if $A \in \mathscr{B}(\mathsf{X}, \mathsf{Y})$, then $T_n + A \to T + A$ in the generalized sense.*

d) *If T_n, T are densely defined, $T_n \to T$ in the generalized sense if and only if $T_n^* \to T^*$ in the generalized sense.*

Another sufficient condition for generalized convergence is obtained from Theorem 2.14.

Theorem 2.24. *Let $T \in \mathscr{C}(\mathsf{X}, \mathsf{Y})$. Let A_n, $n = 1, 2, \ldots$, be T-bounded so that $\|A_n u\| \leqq a_n\|u\| + b_n\|Tu\|$ for $u \in \mathsf{D}(T) \subset \mathsf{D}(A_n)$. If $a_n \to 0$ and $b_n \to 0$, then $T_n = T + A_n \in \mathscr{C}(\mathsf{X}, \mathsf{Y})$ for sufficiently large n and $T_n \to T$ in the generalized sense.*

The conditions b), c) of Theorem 2.23 lead to a very convenient criterion for generalized convergence in case $\mathsf{Y} = \mathsf{X}$.

Theorem 2.25. *Let $T \in \mathscr{C}(\mathsf{X})$ have a non-empty resolvent set $\mathrm{P}(T)$. In order that a sequence $T_n \in \mathscr{C}(\mathsf{X})$ converge to T in the generalized sense, it is* *necessary that each $\zeta \in P(T)$ belong to $P(T_n)$ for sufficiently large n and*

$$\|R(\zeta, T_n) - R(\zeta, T)\| \to 0\,, \tag{2.33}$$

while it is sufficient that this be true for some $\zeta \in P(T)$.

This theorem is useful since most closed operators that appear in applications have non-empty resolvent sets. It follows also from this theorem that if (2.33) is true for some $\zeta \in P(T)$, then it is true for every $\zeta \in P(T)$. We shall come back to this question in later sections.

Theorem 2.26. *Let $T_n, T \in \mathscr{C}(X)$ and let $T_n \to T$ in the generalized sense. If all the T_n have compact resolvents and if T has non-empty resolvent set, then T has compact resolvent.*

Proof. Let $\zeta \in P(T)$. Then we have (2.33) where $R(\zeta, T_n)$ is compact. Hence $R(\zeta, T)$ is compact (see Theorem III-4.7).

Remark 2.27. The converse of Theorem 2.26 is not true: T_n need not have compact resolvent even if T has. A simple counter-example: let $X = l^p$ and let $S \in \mathscr{B}(X)$ be defined by a diagonal matrix with diagonal elements $1/k$, $k = 1, 2, \ldots$, and let S_n be a similar operator with diagonal elements $(n + k)/n\,k$, $k = 1, 2, \ldots$. Let $T = S^{-1}$, $T_n = S_n^{-1}$, which exist and belong to $\mathscr{C}(X)$. It is easily verified that 0 belongs to the resolvent sets of all the T_n and of T, $R(0, T_n) = S_n \xrightarrow[u]{} R(0, T) = S$ and that T has compact resolvent (because S is compact). But the resolvent of T_n is not compact for any $n = 1, 2, \ldots$ (since S_n is not compact). It should be noted, however, that the converse of Theorem 2.26 is true if $T_n - T$ is T-bounded; for a more precise statement see Theorem 3.17. See also Theorem VI-3.6.

Problem 2.28. Let $T \in \mathscr{C}(X, Y)$. Under what conditions does $n^{-1} T$ [resp. $(1 + n^{-1}) T$] converge to 0 [resp. T] in the generalized sense?

Finally we give another sufficient condition for generalized convergence[1].

Theorem 2.29. *Let $T_n, T \in \mathscr{C}(X, Y)$. Let there be a third Banach space Z and operators U_n, $U \in \mathscr{B}(Z, X)$ and V_n, $V \in \mathscr{B}(Z, Y)$ such that U_n, U map Z onto $D(T_n)$, $D(T)$, respectively, one-to-one and $T_n U_n = V_n$, $TU = V$. If $\|U_n - U\| \to 0$ and $\|V_n - V\| \to 0$, $n \to \infty$, then $T_n \to T$ in the generalized sense.*

Proof. The mapping $z \to \varphi = \{Uz, Vz\} = \{Uz, TUz\}$ is a one-to-one, bounded linear operator on Z onto $G(T)$. Since $G(T)$ is closed, this operator has a bounded inverse:

$$\|z\|^2 \leq c^2 \|\varphi\|^2 = c^2(\|Uz\|^2 + \|Vz\|^2)\,. \tag{2.34}$$

Let $\varphi = \{Uz, Vz\}$ be an arbitrary element of $G(T)$. Then $\varphi_n = \{U_n z, V_n z\} \in G(T_n)$ and

$$\|\varphi - \varphi_n\|^2 \leq (\|U - U_n\|^2 + \|V - V_n\|^2)\,\|z\|^2 \leq c^2 \delta_n^2 \|\varphi\|^2 \tag{2.35}$$

where $\delta_n^2 = \|U - U_n\|^2 + \|V - V_n\|^2$. This implies that $\mathrm{dist}\,(\varphi, G(T_n)) \leq$ $\leq c\,\delta_n \|\varphi\|$ and hence $\delta(T, T_n) = \delta(G(T), G(T_n)) \leq c\,\delta_n \to 0$, $n \to \infty$.

[1] This is a discrete version of the definition, due to RELLICH [3], of the analytic dependence of a family $T(\varkappa)$ of operators on the parameter $\varkappa$.

Similarly we have $\delta(T_n, T) \leqq c_n \delta_n$ where c_n is the c of (2.34) for U, V replaced by U_n, V_n. But $\{c_n\}$ is bounded; in fact (2.35) gives $\|\varphi\| \leqq \|\varphi_n\| + \|\varphi - \varphi_n\| \leqq \|\varphi_n\| + c\delta_n\|\varphi\|$, hence $\|\varphi\| \leqq (1 - c\delta_n)^{-1}\|\varphi_n\|$ and $\|z\| \leqq c\|\varphi\| \leqq c(1 - c\delta_n)^{-1}\|\varphi_n\|$ by (2.34), which means that we can take $c_n = c(1 - c\delta_n)^{-1}$. It follows that $\delta(T_n, T) \to 0$ and hence $\hat{\delta}(T_n, T) \to 0$.

§ 3. Perturbation of the spectrum

1. Upper semicontinuity of the spectrum

In this section we consider the change of the spectrum $\Sigma(T)$ and the resolvent $R(\zeta) = R(\zeta, T)$ of an operator $T \in \mathscr{C}(X)$ under "small" perturbations[1].

Theorem 3.1. *Let $T \in \mathscr{C}(X)$ and let Γ be a compact subset of the resolvent set $P(T)$. Then there is a $\delta > 0$ such that $\Gamma \subset P(S)$ for any $S \in \mathscr{C}(X)$ with $\hat{\delta}(S, T) < \delta$.*

Proof. Let $\zeta \in \Gamma \subset P(T)$. Since $(T - \zeta)^{-1} = R(\zeta) \in \mathscr{B}(X)$, it follows from the generalized stability theorem for bounded invertibility (Theorem 2.21) that $(S - \zeta)^{-1} \in \mathscr{B}(X)$ or $\zeta \in P(S)$ if $\hat{\delta}(S - \zeta, T - \zeta) < (1 + \|R(\zeta)\|^2)^{-1/2}$. According to Theorem 2.17, this is true if

$$2(1 + |\zeta|^2)\, \hat{\delta}(S, T) < (1 + \|R(\zeta)\|^2)^{-1/2} . \tag{3.1}$$

Since $\|R(\zeta)\|$ is continuous in ζ, we have

$$\min_{\zeta \in \Gamma} \frac{1}{2} (1 + |\zeta|^2)^{-1} (1 + \|R(\zeta)\|^2)^{-1/2} = \delta > 0 . \tag{3.2}$$

It follows that $\Gamma \subset P(S)$ if $\hat{\delta}(S, T) < \delta$.

Remark 3.2. If $S = T + A$ where $A \in \mathscr{B}(X)$, we have the sharper result that

$$\Gamma \subset P(S) \quad \text{if} \quad \|A\| < \min_{\zeta \in \Gamma} \|R(\zeta)\|^{-1} \tag{3.3}$$

in virtue of Theorem 1.16. If in addition $T \in \mathscr{B}(X)$, Γ may be any closed (not necessarily bounded) subset of $P(T)$; then $\|R(\zeta)\|^{-1}$ still has positive minimum for $\zeta \in \Gamma$ since $\|R(\zeta)\| \to 0$ for $\zeta \to \infty$.

Remark 3.3. Theorem 3.1 and Remark 3.2 show that $\Sigma(T)$ is an *upper semicontinuous* function of $T \in \mathscr{B}(X)$. In other words, for any $T \in \mathscr{B}(X)$ and $\varepsilon > 0$ there exists a $\delta > 0$ such that[2] $\varrho(\Sigma(S), \Sigma(T)) =$

---

[1] Perturbation of the spectra of operators (and of elements of a Banach algebra) is discussed in detail by NEWBURGH [1].

[2] For ϱ see footnote [2] of p. 198. The Hausdorff distance is obtained by symmetrizing ϱ. It is a distance between two *sets* of points and is different, when applied to the spectra of operators in a finite-dimensional space, from the distance between two N-tuples consisting of *repeated eigenvalues*, which was defined in II-§ 5.2; the multiplicities of eigenvalues are not taken into account here.

$\sup_{\lambda \in \Sigma(S)} \text{dist}(\lambda, \Sigma(T)) < \varepsilon$ if $\|S - T\| < \delta$. This is seen by choosing the Γ of Theorem 3.1 as the set of points with distance not smaller than ε from $\Sigma(T)$. Even for $T \in \mathscr{C}(X)$, Theorem 3.1 may be interpreted to imply that $\Sigma(T)$ is upper semicontinuous in a slightly weaker sense.

Problem 3.4. For $T, T_n \in \mathscr{C}(X)$, it is possible that $\hat{\delta}(T_n, T) \to 0$ and yet the distance between $\Sigma(T_n)$ and $\Sigma(T)$ is infinite for each n. Verify this for the following example: $X = l^p$, T is given by a diagonal matrix with diagonal elements k, $k = 1, 2, \ldots$, and $T_n = (1 + i\, n^{-1})\, T$. [hint: $T_n - T$ is T-bounded.]

Remark 3.5. In Theorem 3.1 it is in general not easy to give a simple expression for δ, although it is expected that δ will be large if the distance of Γ from $\Sigma(T)$ is large. Under some commutativity assumptions, this distance alone determines the necessary size of the perturbation. We have namely

Theorem 3.6. *Let $T \in \mathscr{C}(X)$ and let $A \in \mathscr{B}(X)$ commute with T. Then the distance between $\Sigma(T)$ and $\Sigma(T + A)$ does not exceed* $\text{spr} A$ *and, a fortiori,* $\|A\|$.

Proof. $R(\zeta)$ commutes with A (see Theorem III-6.5). Hence $(T + A - \zeta)^{-1} \in \mathscr{B}(X)$ if $\text{spr} A < 1/\text{spr} R(\zeta)$, by virtue of Theorem 1.18. Since $\text{spr} R(\zeta) = 1/\text{dist}(\zeta, \Sigma(T))$ by Problem III-6.16, it follows that $\zeta \in P(T + A)$ if $\text{spr} A < \text{dist}(\zeta, \Sigma(T))$. In other words,

$$\text{dist}(\zeta, \Sigma(T)) \leq \text{spr} A \quad \text{if} \quad \zeta \in \Sigma(T + A)\,. \tag{3.4}$$

Since A commutes with $T + A$ too, we can apply (3.4) to the pair T, A replaced by the pair $T + A, -A$. Then we see that $\text{dist}(\zeta, \Sigma(T + A)) \leq \text{spr} A$ if $\zeta \in \Sigma(T)$. This proves the theorem.

Remark 3.7. Theorem 3.6 shows that $\Sigma(T)$ changes continuously when T is subjected to a small change which commutes with T.

2. Lower semi-discontinuity of the spectrum

In a finite-dimensional space, the eigenvalues of an operator T depend on T continuously (see II-§ 5.8). Even in a general Banach space X, the spectrum $\Sigma(T)$ changes continuously with $T \in \mathscr{B}(X)$ if the perturbation commutes with T (see Remark 3.7)[1]. But this is not true for more general perturbation; only upper semicontinuity for the spectrum can be proved in general.

Roughly speaking, the upper semicontinuity proved in the preceding paragraph says that $\Sigma(T)$ does not *expand* suddenly when T is changed continuously. But it may well *shrink* suddenly, as is seen from the following examples.

[1] There are other kinds of restricted continuity of the spectrum. For example, $\Sigma(T)$ changes continuously with T if T varies over the set of selfadjoint operators in a Hilbert space. For more precise formulation, see VIII-§ 1.2.

Example 3.8. Let $X = l^p(-\infty, \infty)$, where each $u \in X$ is a bilateral sequence $u = (\xi_j)$, $j = \cdots, -1, 0, 1, 2, \ldots$, with $\|u\| = (\sum |\xi_j|^p)^{1/p}$. Let $\{x_n\}$ be the canonical basis: $x_n = (\delta_{nj})$ and let $T \in \mathscr{B}(X)$ be such that $T x_0 = 0$, $T x_n = x_{n-1}$, $n \neq 0$. T is a left shift operator (cf. Example III-3.16). Since III-(3.9) is true in this case also, we have $\operatorname{spr} T = 1$ so that $\Sigma(T)$ is a subset of the closed unit disk. Actually $\Sigma(T)$ coincides with this disk; in fact, for any ζ with $|\zeta| < 1$ the vector $u = \sum_{n=0}^{\infty} \zeta^n x_n$ is an eigenvector of $T - \zeta$ with the eigenvalue ζ so that $\zeta \in \Sigma(T)$.

Let $A \in \mathscr{B}(X)$ be such that $A x_0 = x_{-1}$, $A x_n = 0$, $n \neq 0$. Let $T(\varkappa) = T + \varkappa A$. $T(\varkappa)$ is again a left shift operator with $T(\varkappa) x_0 = \varkappa x_{-1}$, $T(\varkappa) x_n = x_{n-1}$, $n \neq 0$, so that $\operatorname{spr} T(\varkappa) = 1$ as above and $\Sigma(T(\varkappa))$ is a subset of the unit disk. If $\varkappa \neq 0$, however, the interior of this disk is contained in $P(T(\varkappa))$. This is seen by observing that $T(\varkappa)^{-1}$ exists and is a right shift operator such that $T(\varkappa)^{-1} x_{-1} = x_0/\varkappa$, $T(\varkappa)^{-1} x_{n-1} = x_n$, $n \neq 0$; hence $\operatorname{spr} T(\varkappa)^{-1} = 1$ as above and the exterior of the unit disk belongs to $P(T(\varkappa)^{-1})$, which means that the interior of this disk belongs to $P(T(\varkappa))$ (see Theorem III-6.15).

In this example the perturbation $\varkappa A$ is small not only in the sense that $\|\varkappa A\| = |\varkappa| \to 0$ for $\varkappa \to 0$ but also in the sense that A is a degenerate operator of rank one. $\varkappa A$ is "small" in any stronger topology which might reasonably be put on a subset of $\mathscr{B}(X)$. Thus the lower semicontinuity of $\Sigma(T)$ cannot be established by strengthening the topology for the operators.

Example 3.9. In the above example $\Sigma(T)$ shrinks from a full disk to its circumference by a small perturbation. There is an example in which the shrinkage is more drastic, being from a full disk to its center[1]. Of course such a change cannot be caused by a degenerate perturbation as in the preceding example[2].

Problem 3.10. In Example 3.8 let $R(\zeta, \varkappa) = R(\zeta, T + \varkappa A)$. Then

$$\|R(\zeta, \varkappa)\| \leqq (|\zeta| - 1)^{-1} \quad \text{for} \quad |\zeta| > 1, |\varkappa| \leqq 1,$$
$$\|R(\zeta, \varkappa)\| \leqq |\varkappa|^{-1}(1 - |\zeta|)^{-1} \quad \text{for} \quad |\zeta| < 1, 0 < |\varkappa| \leqq 1.$$

3. Continuity and analyticity of the resolvent

We shall now show that the resolvent $R(\zeta, T)$ is not only continuous but analytic in T in a certain sense.

Theorem 3.11. *Let $T_0 \in \mathscr{C}(X)$ be fixed. Then $R(\zeta, T)$ is piecewise holomorphic in $T \in T_0 + \mathscr{B}(X)$ and $\zeta \in P(T)$ jointly.*

Here $T_0 + \mathscr{B}(X)$ is the set of all operators $T_0 + B$ where B varies over $\mathscr{B}(X)$, in which we introduce the metric (distance function) $\|T - S\|$. Now Theorem 3.11 means the following. First, the set of all pairs ζ, T such that $\zeta \in P(T)$ is open in the product space $C \times [T_0 + \mathscr{B}(X)]$ (C is the complex plane); in other words, for any $T \in T_0 + \mathscr{B}(X)$ and $\zeta_0 \in P(T)$, $\zeta \in P(S)$ is true if $|\zeta - \zeta_0|$ and $\|S - T\|$ are sufficiently small. Second, $R(\zeta, S)$ can be expressed as a convergent double power series in $\zeta - \zeta_0$

---

[1] See Rickart [1], p. 282. In this example there is a sequence T_n of *nilpotent* operators such that $\|T_n - T\| \to 0$, where T is not quasi-nilpotent and $\Sigma(T)$ is a disk of positive radius with center 0.

[2] A compact perturbation preserves the essential spectrum, which is the unit circle for the T in question; see Theorem 5.35 and Example 5.36.

and A, where $A \in \mathscr{B}(X)$, $S = T + A$. In particular for $T_0 = 0$, $R(\zeta, T)$ is piecewise holomorphic in $T \in \mathscr{B}(X)$ and $\zeta \in P(T)$.

The proof of Theorem 3.11 is essentially the same as the proof of Theorem II-1.5; it suffices to replace the $T(\varkappa)$ there by S and $A(\varkappa)$ by A.

It is difficult to extend Theorem 3.11 and assert that $R(\zeta, T)$ is piecewise holomorphic in $T \in \mathscr{C}(X)$, for $\mathscr{C}(X)$ is not an algebra (not even a linear space) like $\mathscr{B}(X)$. Nevertheless, we have a generalization in a different form.

Theorem 3.12. *When T varies over $\mathscr{C}(X)$, $R(\zeta, T)$ is piecewise holomorphic in ζ and $R(\zeta_0, T)$, where ζ_0 is any fixed complex number, in the following sense. There exists a function $\Phi(\eta, B)$, defined in an open subset of $C \times \mathscr{B}(X)$ and taking values in $\mathscr{B}(X)$, with the following properties:*

i) *$\Phi(\eta, B)$ is piecewise holomorphic in η and B jointly (in the sense stated after* Theorem 3.11, *with $T_0 = 0$).*

ii) *Let $T \in \mathscr{C}(X)$ and $\zeta_0 \in P(T)$. Then $\zeta \in P(T)$ if and only if $\Phi(\zeta - \zeta_0, R(\zeta_0, T))$ is defined. In this case we have*

$$R(\zeta, T) = \Phi(\zeta - \zeta_0, R(\zeta_0, T)) . \tag{3.5}$$

Proof. We have the identity

$$R(\zeta, T) = -(\zeta - \zeta_0)^{-1} - (\zeta - \zeta_0)^{-2} R((\zeta - \zeta_0)^{-1}, R(\zeta_0, T)); \tag{3.6}$$

this follows from III-(6.18) and is valid if $\zeta, \zeta_0 \in P(T)$ and $\zeta \neq \zeta_0$. Define $\Phi(\eta, B)$ by

$$\Phi(0, B) = B,\ \Phi(\eta, B) = -\eta^{-1} - \eta^{-2} R(\eta^{-1}, B) \quad \text{if} \quad \eta \neq 0,\ \eta^{-1} \in P(B) . \tag{3.7}$$

Then (3.5) is satisfied whenever $\zeta, \zeta_0 \in P(T)$.

The domain of definition of $\Phi(\eta, B)$ is the set of all pairs $\eta, B \in C \times \mathscr{B}(X)$ such that either $\eta = 0$ or $\eta^{-1} \in P(B)$. This domain is open in $C \times \mathscr{B}(X)$ by Theorem 3.1 and Remark 3.2. It is obvious from Theorem 3.11 that $\Phi(\eta, B)$ is holomorphic in η and B as long as $\eta \neq 0$. On the other hand, the identity

$$\begin{aligned} \Phi(\eta, B) &= -\eta^{-1} - \eta^{-2}(B - \eta^{-1})^{-1} \\ &= -\eta^{-1} + \eta^{-1}(1 - \eta B)^{-1} = B(1 - \eta B)^{-1} \end{aligned} \tag{3.8}$$

together with $\Phi(0, B) = B$ shows that it is holomorphic also for $\eta = 0$.

It remains to prove that when $\zeta_0 \in P(T)$, $\zeta \in P(T)$ if and only if $\Phi(\zeta - \zeta_0, R(\zeta_0, T))$ is defined. For $\zeta = \zeta_0$ this is obvious. Otherwise this follows from the fact that (3.6) is true whenever one of the two members exists.

Remark 3.13. Theorem 3.12 shows explicitly that if $\|R(\zeta, S) - R(\zeta, T)\|$ is small for some ζ, then it is small for every ζ. More precisely, for any $T \in \mathscr{C}(X)$ and $\zeta, \zeta_0 \in P(T)$, there is a constant M such that

$$\|R(\zeta, S) - R(\zeta, T)\| \le M \|R(\zeta_0, S) - R(\zeta_0, T)\| \tag{3.9}$$

for any $S \in \mathscr{C}(X)$ for which $\zeta_0 \in P(S)$ and $\|R(\zeta_0, S) - R(\zeta_0, T)\|$ is sufficiently small (then $\zeta \in P(S)$ is a consequence). This is another proof of a remark given after Theorem 2.25.

Problem 3.14. A more explicit formula than (3.9) is

$$\|R(\zeta, S) - R(\zeta, T)\| \leq \frac{\left\|\frac{T-\zeta_0}{T-\zeta}\right\|^2 \|R(\zeta_0, S) - R(\zeta_0, T)\|}{1 - |\zeta - \zeta_0| \left\|\frac{T-\zeta_0}{T-\zeta}\right\| \|R(\zeta_0, S) - R(\zeta_0, T)\|}, \tag{3.10}$$

which is valid if $\|R(\zeta_0, S) - R(\zeta_0, T)\|$ is so small that the denominator on the right is positive. Here $\frac{T-\zeta_0}{T-\zeta}$ is a convenient expression for $(T - \zeta_0)(T - \zeta)^{-1} = (T - \zeta_0) R(\zeta, T) = 1 + (\zeta - \zeta_0) R(\zeta, T)$.

Theorem 3.15. *$R(\zeta, T)$ is continuous in ζ and $T \in \mathscr{C}(X)$ jointly in the following sense. For any $T \in \mathscr{C}(X)$, $\zeta_0 \in P(T)$ and $\varepsilon > 0$, there exists a $\delta > 0$ such that $\zeta \in P(S)$ and $\|R(\zeta, S) - R(\zeta_0, T)\| < \varepsilon$ if $|\zeta - \zeta_0| < \delta$ and $\hat{\delta}(S, T) < \delta$.*

Proof. By Theorem 2.25 $R(\zeta_0, S)$ exists and $\|R(\zeta_0, S) - R(\zeta_0, T)\|$ is arbitrarily small if $\hat{\delta}(S, T)$ is sufficiently small. Then the result follows from Theorem 3.12 since $R(\zeta, S)$ is a double power series in $\zeta - \zeta_0$ and $R(\zeta_0, S) - R(\zeta_0, T)$.

4. Semicontinuity of separated parts of the spectrum

We have proved above the upper semicontinuity of $\Sigma(T)$ in $T \in \mathscr{C}(X)$. We now prove a somewhat finer result that *each separated part of $\Sigma(T)$ is upper semicontinuous*. The separation of the spectrum and the related decomposition of the space X and of the operator T were discussed in III-§ 6.4. For simplicity we state the result for the case in which the spectrum is separated into two parts, but the generalization of the result to the case of more than two parts is obvious.

Theorem 3.16. *Let $T \in \mathscr{C}(X)$ and let $\Sigma(T)$ be separated into two parts $\Sigma'(T)$, $\Sigma''(T)$ by a closed curve Γ as in* III-§ 6.4. *Let $X = M'(T) \oplus M''(T)$ be the associated decomposition of X. Then there exists a $\delta > 0$, depending on T and Γ, with the following properties. Any $S \in \mathscr{C}(X)$ with $\hat{\delta}(S, T) < \delta$ has spectrum $\Sigma(S)$ likewise separated by Γ into two parts $\Sigma'(S)$, $\Sigma''(S)$ (Γ itself running in $P(S)$). In the associated decomposition $X = M'(S) \oplus M''(S)$, $M'(S)$ and $M''(S)$ are respectively isomorphic with $M'(T)$ and $M''(T)$. In particular* $\dim M'(S) = \dim M'(T)$, $\dim M''(S) = \dim M''(T)$ *and both $\Sigma'(S)$ and $\Sigma''(S)$ are nonempty if this is true for T. The decomposition $X = M'(S) \oplus M''(S)$ is continuous in S in the sense that the projection $P[S]$ of X onto $M'(S)$ along $M''(S)$ tends to $P[T]$ in norm as $\hat{\delta}(S, T) \to 0$.*

Proof. It follows from Theorem 3.1 and its proof that $\Gamma \subset \mathrm{P}(S)$ if $\hat{\delta}(S, T) < \delta = \min_{\zeta \in \Gamma} 2^{-1}(1+|\zeta|^2)^{-1}(1+\|R(\zeta, T)\|^2)^{-1/2}$. Hence $\Sigma(S)$ is separated by Γ into two parts $\Sigma'(S)$, $\Sigma''(S)$ and we have the associated decomposition of X as stated. The projection $P[S]$ of X onto $\mathsf{M}'(S)$ along $\mathsf{M}''(S)$ is given by III-(6.19):

$$P[S] = -\frac{1}{2\pi i}\int_{\Gamma} R(\zeta, S)\, d\zeta . \tag{3.11}$$

Since $R(\zeta, T)$ is continuous in ζ and T as shown in Theorem 3.15 and since Γ is compact, $\|R(\zeta, S) - R(\zeta, T)\|$ is small uniformly for $\zeta \in \Gamma$ if $\hat{\delta}(S, T)$ is sufficiently small. Thus we see from (3.11) that, for any $\varepsilon > 0$, $\|P[S] - P[T]\| < \varepsilon$ if $\hat{\delta}(S, T)$ is sufficiently small. The isomorphism of $\mathsf{M}'(S)$ with $\mathsf{M}'(T)$ etc. then follows from a result of I-§ 4.6 (which is valid in a Banach space too).

5. Continuity of a finite system of eigenvalues

We have seen above that the spectrum $\Sigma(T)$ of a $T \in \mathscr{C}(\mathsf{X})$ does not in general depend on T continuously, even when T is restricted to operators of $\mathscr{B}(\mathsf{X})$. This is a remarkable contrast to the situation in finite-dimensional spaces. We shall show, however, that a part of $\Sigma(T)$ consisting of a *finite system of eigenvalues* (see III-§ 6.5) changes with T continuously, just as in the finite-dimensional case.

Let $\Sigma'(T)$ be such a finite system of eigenvalues. $\Sigma'(T)$ is separated from the rest $\Sigma''(T)$ of $\Sigma(T)$ by a closed curve Γ in the manner described in III-§ 6.4. The corresponding decomposition $\mathsf{X} = \mathsf{M}'(T) \oplus \mathsf{M}''(T)$ of the space X has the property that $\dim \mathsf{M}'(T) = m < \infty$, m being the total multiplicity of the system of eigenvalues under consideration (see loc. cit).

Suppose now that $\{T_n\}$ converges to T in the generalized sense. Then $\Sigma(T_n)$ is likewise separated into two parts $\Sigma'(T_n)$, $\Sigma''(T_n)$ with the corresponding decomposition $\mathsf{X} = \mathsf{M}'(T_n) \oplus \mathsf{M}''(T_n)$ in such a way that $\mathsf{M}'(T_n)$, $\mathsf{M}''(T_n)$ are respectively isomorphic with $\mathsf{M}'(T)$, $\mathsf{M}''(T)$, provided n is sufficiently large (Theorem 3.16). In particular $\dim \mathsf{M}'(T_n) = m$ so that the part of $\Sigma(T_n)$ inside Γ consists of a finite system of eigenvalues with total multiplicity m.

The same result holds when $\Sigma'(T)$ is replaced by any one of the eigenvalues in $\Sigma'(T)$. Thus we conclude that the change of a finite system $\Sigma'(T)$ of eigenvalues of a closed operator T is small (in the sense of II-§ 5.1) when T is subjected to a small perturbation in the sense of generalized convergence.

Not only is the system $\Sigma'(T)$ continuous in T in this sense, but the total projection $P[T]$ of X on $\mathsf{M}'(T)$ along $\mathsf{M}''(T)$ changes with T continuously, as is seen easily from Theorem 3.16.

Summing up, the behavior of a finite system of eigenvalues of a closed operator under a small perturbation is much the same as in the finite-dimensional case. We may pursue this analogy further by introducing the unordered m-tuple $\mathfrak{S}'(T)$ of repeated eigenvalues to represent the finite system $\Sigma'(T)$, as described in II-§5.2. Then the distance between $\mathfrak{S}'(T_n)$ and $\mathfrak{S}'(T)$ converges to zero if T_n converges to T in the generalized sense.

6. Change of the spectrum under relatively bounded perturbation

The results on the upper semicontinuity of the spectrum $\Sigma(T)$ and analyticity of the resolvent $R(\zeta, T)$ as functions of ζ and $R(\zeta_0, T)$ given in par. 3 are rather general but not very convenient for application. Here we shall give a theorem which is less general but more directly useful.

Theorem 3.17. *Let T be a closed operator in X and let A be an operator in X which is T-bounded, so that $\mathsf{D}(A) \supset \mathsf{D}(T)$ and the inequality* (1.1) *holds. If there is a point ζ of $\mathrm{P}(T)$ such that*

$$a\|R(\zeta, T)\| + b\|TR(\zeta, T)\| < 1 , \tag{3.12}$$

then $S = T + A$ is closed and $\zeta \in \mathrm{P}(S)$, with

$$\|R(\zeta, S)\| \leqq \|R(\zeta, T)\| \, (1 - a\|R(\zeta, T)\| - b\|TR(\zeta, T)\|)^{-1} . \tag{3.13}$$

If in particular T has compact resolvent, S has compact resolvent.

This theorem follows easily from Theorem 1.16. It is convenient since (3.12) gives an explicit condition for ζ to belong to $\mathrm{P}(S)$. For example, it can be used to deduce

Theorem 3.18. *Let T, S and A be as in the preceding theorem. Let $\Sigma(T)$ be separated into two parts by a closed curve Γ as in* Theorem 3.16. *If*

$$\sup_{\zeta \in \Gamma} (a\|R(\zeta, T)\| + b\|TR(\zeta, T)\|) < 1 , \tag{3.14}$$

then $\Sigma(S)$ is likewise separated by Γ and the results of Theorem 3.16 *hold.*

Proof. The proof is similar to that of Theorem 3.16; only the following points should be noted. We see from (3.11) and II-(1.11) that $\|P[S] - P[T]\|$ is arbitrarily small if $\|AR(\zeta, T)\|$ is sufficiently small for all $\zeta \in \Gamma$, which is the case if a, b are sufficiently small. In this case the results follow as in the proof of Theorem 3.16. Actually a, b need not be too small but the condition (3.14) suffices. To see this, it is convenient to introduce a parameter $\varkappa$ and set $T(\varkappa) = T + \varkappa A$, $0 \leqq \varkappa \leqq 1$.

Then it follows from II-(1.11) that $R(\zeta, T + \varkappa A)$ is a continuous (even holomorphic) function of ζ and $\varkappa$ for $\zeta \in \Gamma$ and $0 \leqq \varkappa \leqq 1$, so that $P(\varkappa) = -(2\pi i)^{-1} \int_\Gamma R(\zeta, T + \varkappa A)\, d\zeta$ is continuous for $0 \leqq \varkappa \leqq 1$.

Example 3.19. Let T, S be respectively the T_1, S_1 of Example 1.10. $A = S - T$ satisfies (3.14) if the differences $q_i(x) - p_i(x)$, $i = 0, 1, 2$, are sufficiently small.

7. Simultaneous consideration of an infinite number of eigenvalues

If an infinite number of eigenvalues of T are considered simultaneously, their change need not be uniformly small under small change of T. For instance let T have a discrete spectrum consisting of an unbounded set $\{\lambda_n\}$ of eigenvalues, and let $T(\varkappa) = T + \varkappa T$. The perturbation $\varkappa T$ is T-bounded, but the eigenvalues of $T(\varkappa)$ are $(1 + \varkappa)\lambda_n$ and the change $\varkappa \lambda_n$ may be arbitrarily large for large n, no matter how small $|\varkappa|$ is.

There are, however, some cases in which the change of the spectrum is uniform. Instead of dealing with such a case in general, we shall content ourselves by considering an example.

Example 3.20. Consider the ordinary differential operator T of Example III-6.21 [$T u = -u''$ with boundary condition $u(0) = u(\pi) = 0$]. The eigenvalues of T are $\lambda_n = n^2$, $n = 1, 2, 3, \ldots$, and we have the estimates III-(6.47–6.48) for the resolvent $R(\zeta)$. For any given $\delta > 0$, consider the equation

$$\frac{\pi}{|\alpha \sin \pi \alpha|} = \delta\,. \tag{3.15}$$

If N is a positive integer such that $N > \pi/\delta$, there are sequences $\{\alpha_n'\}$, $\{\alpha_n''\}$ consisting of roots of (3.15) such that

$$N < \alpha_N'' < N + \frac{1}{2}\,, \quad n - \frac{1}{2} < \alpha_n' < n < \alpha_n'' < n + \frac{1}{2}\,, \quad n > N, \tag{3.16}$$

$$\alpha_n' = n - \frac{1}{n\delta} + O\left(\frac{1}{n^2}\right), \quad \alpha_n'' = n + \frac{1}{n\delta} + O\left(\frac{1}{n^2}\right), \quad n \to \infty\,. \tag{3.17}$$

According to the remark at the end of Example III-6.21, we have

$$\|R(\zeta)\| \leqq \delta \tag{3.18}$$

along each of the parabolas $\xi = \alpha^2 - \dfrac{\eta^2}{4\alpha^2}$ with $\alpha = \alpha_n'$ and α_n'' in the ζ-plane. Also we see from III-(6.48) that (3.18) holds along the horizontal lines $\eta = \pm 2/\delta$. Let us denote by Γ_n, $n > N$, the curvilinear rectangle consisting of the two of the above parabolas corresponding to $\alpha = \alpha_n', \alpha_n''$ and the straight lines $\eta = \pm 2/\delta$. Then each Γ_n encloses exactly one eigenvalue $\lambda_n = n^2$ and (3.18) holds for $\zeta \in \Gamma_n$. The remaining eigenvalues n^2 for $n \leqq N$ can be enclosed in the curve Γ_0 consisting of the above parabola for $\alpha = \alpha_N''$, the two horizontal lines mentioned above and a vertical line sufficiently far to the left of the imaginary axis; we have again (3.18) for $\zeta \in \Gamma_0$. Let us denote by Γ the aggregate of the Γ_n, $n > N$, and Γ_0. Note that for large n, Γ_n is approximately a square with center n^2 and side length $4/\delta$.

Consider now a perturbed operator $S = T + A$ where $A \in \mathscr{B}(X)$. Choose a δ such that $\|A\| = a < 1/\delta$. Then (3.14) is satisfied in virtue of (3.18), which is true for $\zeta \in \Gamma$ (set $b = 0$). Thus Theorem 3.18 shows that each Γ_n, $n > N$, encloses exactly one eigenvalue of S and Γ_0 encloses exactly N (repeated) eigenvalues of S. These eigenvalues exhaust $\Sigma(S)$, since it is easily seen that any complex number lying outside all the Γ_n and Γ_0 satisfies (3.18) and hence belongs to $\mathrm{P}(S)$. Since the size of Γ_n is bounded for $n \to \infty$ as noted above, we see that the change of the eigenvalues of T by any bounded perturbation A is *uniform*: each eigenvalue λ_n for large n stays within the approximate square of side length $4\|A\|$ with center at the unperturbed eigenvalue.

If we denote by $P_n[T]$ and $P_n[S]$, $n > N$, the eigenprojections for the eigenvalue of T and S inside Γ_n, $P_n[S] - P_n[T]$ is also uniformly bounded for all n. In fact,

$$P_n[S] - P_n[T] = -\frac{1}{2\pi i} \int_{\Gamma_n} [R(\zeta, S) - R(\zeta, T)]\, d\zeta$$

$$= \frac{1}{2\pi i} \int_{\Gamma_n} R(\zeta, S)\, A\, R(\zeta, T)\, d\zeta$$

and

$$\|P_n[S] - P_n[T]\| \leqq \frac{|\Gamma_n|}{2\pi} \frac{\|A\|}{\delta(\delta - \|A\|)} \leqq C$$

where $|\Gamma_n|$, the length of Γ_n, is bounded in n.

Remark 3.21. As we shall see later, we have a better estimate than III-(6.47) if we consider the same operator T in $\mathsf{X} = \mathsf{L}^2(0, \pi)$ and, consequently, an apparently sharper result will be obtained for the perturbation of eigenvalues of T. But it should be remarked that there are operators A which are bounded when considered in $\mathsf{C}[0, \pi]$ but not in L^2. Thus the result of the above example retains its own interest.

8. An application to Banach algebras. Wiener's theorem

The notions of spectrum and resolvent can be defined not only for linear operators but, in a more abstract way, for elements of a *Banach algebra*. We have mentioned earlier that $\mathscr{B}(\mathsf{X})$ is a special case of Banach algebra (I-§ 4.1 and III-§ 3.2). Here we shall give a brief account of Banach algebras and show that many of the results obtained above remain true in this more general setting[1].

A Banach algebra $\mathscr{B}$ is by definition a Banach space in which the product TS of any two elements T, S is defined and belongs to $\mathscr{B}$, with $\|TS\| \leqq \|T\|\,\|S\|$. A *unit (element)* 1 of $\mathscr{B}$ is an element such that $1T = T1 = T$ for all $T \in \mathscr{B}$. $\mathscr{B}$ need not have a unit, but a unit is unique if it exists. We shall consider only Banach algebras with a unit. For any scalar α, $\alpha 1$ will be denoted simply by α.

$T \in \mathscr{B}$ is *invertible* if there is an $S \in \mathscr{B}$ such that $TS = ST = 1$; S is uniquely determined by T and is denoted by T^{-1} (the *inverse* of T).

The *resolvent set* $\mathrm{P}(T)$ of $T \in \mathscr{B}$ is the set of all scalars ζ such that $T - \zeta$ is invertible. The function $R(\zeta) = (T - \zeta)^{-1}$ defined for $\zeta \in \mathrm{P}(T)$

[1] For more details on Banach algebras see e. g. Hille and Phillips [1], Rickart [1].

is the *resolvent* of T. The complementary set $\Sigma(T)$ (in the complex plane) of $P(T)$ is the *spectrum* of T.

Many other notions introduced for $\mathscr{B}(X)$ can be taken over without change to the case of an abstract Banach algebra $\mathscr{B}$ and many theorems for $\mathscr{B}(X)$ remain true for $\mathscr{B}$. As a rule, this is the case if such notions or theorems can be defined or proved without explicit reference to the underlying space X. For example: the spectral radius spr T can be defined for $T \in \mathscr{B}$ exactly as in I-§ 4.2. The Neumann series I-(4.22) is valid for $T \in \mathscr{B}$ and $1 - T$ is invertible if spr $T < 1$. In particular we have the first Neumann series I-(5.6) for the resolvent, from which it follows that $P(T)$ is open and $\Sigma(T)$ is closed, etc.

Most of the results on the perturbation of spectra are again valid in $\mathscr{B}$, but we have to restrict ourselves to those results which are related to operators belonging to $\mathscr{B}(X)$, for an unbounded operator has no analogue in $\mathscr{B}$. For example, $\Sigma(T)$ is an upper semicontinuous function of $T \in \mathscr{B}$, but it is in general not lower semicontinuous. It is continuous, however, for a restricted change of T in a sense similar to Remark 3.7.

Here we have something new and interesting which did not exist in $\mathscr{B}(X)$. It is possible that a Banach algebra $\mathscr{B}$ is *commutative*, which means that all elements of $\mathscr{B}$ commute with one another, whereas this is not true for $\mathscr{B}(X)$ if $\dim X \geqq 2$. If $\mathscr{B}$ is commutative, Theorem 3.6 and Remark 3.7 are valid without any restriction. Thus $\Sigma(T)$ *is continuous in* $T \in \mathscr{B}$ *if* $\mathscr{B}$ *is commutative.*

Example 3.22. A typical example of a Banach algebra is the Banach space l (the set of all sequences $T = (\tau_k)$ with $\|T\| = \sum_k |\tau_k| < \infty$, see Example III-1.1) in which the product TS of two elements $T = (\tau_k)$, $S = (\sigma_k)$ is defined by *convolution:*

$$TS = (\varrho_k), \quad \varrho_k = \sum_j \tau_j \sigma_{k-j}. \tag{3.19}$$

It is easy to verify that $\|TS\| \leqq \|T\| \|S\|$. Also it is easily seen that l is commutative. In the definition of l given in Example III-1.1 the index k runs from 1 to ∞; in this case l has no unit element. If we modify this slightly and assume that k runs either from 0 to ∞ or from $-\infty$ to ∞, then l has the unit $1 = (\delta_{k0})$. In what follows we take the latter case and set $\mathscr{B} = l$.

With each $T \in \mathscr{B}$ we can associate a complex-valued function

$$T(e^{i\theta}) = \sum_{k=-\infty}^{\infty} \tau_k e^{ik\theta} \tag{3.20}$$

defined on the unit circle (for real θ); note that the Fourier series on the right of (3.20) is absolutely convergent. It follows directly from (3.19) that

$$TS(e^{i\theta}) = T(e^{i\theta}) S(e^{i\theta}). \tag{3.21}$$

We shall now show that

$$\Sigma(T) \quad \textit{is exactly the range of} \quad T(e^{i\theta}). \tag{3.22}$$

This implies, in particular, that T is invertible if $T(e^{i\theta}) \neq 0$. If we set $S = T^{-1}$, then $S(e^{i\theta}) = 1/T(e^{i\theta})$. But $S(e^{i\theta})$ has an absolutely convergent Fourier series by definition. Hence the stated result implies the following theorem by Wiener: *If a complex-valued function $T(e^{i\theta})$ with an absolutely convergent Fourier series does not vanish anywhere, $1/T(e^{i\theta})$ has also an absolutely convergent Fourier series*[1].

To prove the proposition (3.22), we first note that it is true if $T(e^{i\theta})$ has an analytic continuation $T(z)$ in a neighborhood of the unit circle K. In fact, let ζ be a complex number not in the range of $T(e^{i\theta})$. Then[2] $(T-\zeta)(z) = T(z) - \zeta$ does not vanish anywhere in some neighborhood of K, so that $R(z) = (T(z) - \zeta)^{-1}$ is again analytic in a neighborhood of K. It follows from the Cauchy inequalities for the Laurent coefficients that $R(e^{i\theta})$ has an absolutely convergent Fourier series. If the sequence of the Fourier coefficients of $R(e^{i\theta})$ is denoted by R, we have $R \in \mathscr{B}$ and $R(T-\zeta) = (T-\zeta)R = 1$, so that $T - \zeta$ is invertible. On the other hand, it is easily seen from (3.21) that $T - \zeta$ is not invertible if ζ lies in the range of $T(e^{i\theta})$.

Now (3.22) is a simple consequence of the theorem stated above that $\Sigma(T)$ is continuous in T (note that $\mathscr{B}$ is commutative). It suffices to note that for any $T = (\tau_k) \in \mathscr{B}$, there is a sequence $\{T_n\}$ such that $\|T_n - T\| \to 0$ and each $T_n(e^{i\theta})$ has an analytic continuation in a neighborhood of K. Such a sequence is given, for example, by setting $T_n(e^{i\theta}) = \sum_{k=-n}^{n} \tau_k e^{ik\theta}$. Since $\Sigma(T_n)$ is the range of $T_n(e^{i\theta})$, $\Sigma(T) = \lim \Sigma(T_n)$ must be the range of $T(e^{i\theta}) = \lim T_n(e^{i\theta})$.

§ 4. Pairs of closed linear manifolds

1. Definitions

This section is intended to be preliminary to the following one, where perturbation theory for the nullity, deficiency and index of an operator is developed. But the subject will probably have its own interest.

The problems to be considered are related to pairs M, N of closed linear manifolds in a Banach space Z. The results obtained will be applied in the following section to operators of the class $\mathscr{C}$(X, Y), with Z = X × Y the product space of X, Y. Some of the basic results in this direction have been obtained in § 2, but we need more results for later applications.

Let Z be a Banach space and let M, N be closed linear manifolds of Z. Then M ∩ N is also a closed linear manifold. We shall define the *nullity* of the pair M, N, in symbol nul (M, N), by

$$\text{nul}(\mathsf{M}, \mathsf{N}) = \dim(\mathsf{M} \cap \mathsf{N}) . \tag{4.1}$$

M + N is a linear manifold (not necessarily closed); we define the *deficiency* of the pair M, N, in symbol def (M, N), by

$$\text{def}(\mathsf{M}, \mathsf{N}) = \text{codim}(\mathsf{M} + \mathsf{N}) = \dim \mathsf{Z}/(\mathsf{M} + \mathsf{N})\,^{3} . \tag{4.2}$$

---

[1] Cf. Riesz and Sz.-Nagy [1], p. 434.

[2] Note that $1(e^{i\theta}) = 1$ for the unit element 1.

[3] A different definition of def (M, N) is to use the closure of M + N instead of M + N in (4.2). But the two definitions coincide for semi-Fredholm pairs.

The *index* of the pair M, N, in symbol ind(M, N), is defined by

$$\text{ind}(M, N) = \text{nul}(M, N) - \text{def}(M, N) \tag{4.3}$$

if at least one of nul(M, N) and def(M, N) is finite.

The pair M, N will be said to be *Fredholm* [*semi-Fredholm*] if $M + N$ is a closed linear manifold and both of [at least one of] nul(M, N) and def(M, N) are [is] finite.

Finally we define the quantity

$$\gamma(M, N) = \inf_{u \in M,\, u \notin N} \frac{\text{dist}(u, N)}{\text{dist}(u, M \cap N)} (\leq 1) . \tag{4.4}$$

By this formula $\gamma(M, N)$ is defined only when $M \not\subset N$. If $M \subset N$ we set $\gamma(M, N) = 1$. Obviously $\gamma(M, N) = 1$ if $M \supset N$. $\gamma(M, N)$ is not symmetric with respect to M, N. We set

$$\hat{\gamma}(M, N) = \min[\gamma(M, N), \gamma(N, M)] \tag{4.5}$$

and call it the *minimum gap* between M and N.

Problem 4.1. $\gamma(M, N) \leq \delta(M, N)$ except for $M \subset N$. [hint: $\text{dist}(u - z, N) = \text{dist}(u, N)$ for $u \in M$, $z \in M \cap N$].

Although $\gamma(M, N)$ and $\gamma(N, M)$ are not in general equal, they are not completely independent. We have namely

$$\gamma(N, M) \geq \frac{\gamma(M, N)}{1 + \gamma(M, N)} . \tag{4.6}$$

The proof of (4.6) will be given in a moment.

Theorem 4.2. *In order that* $M + N$ *be closed, it is necessary and sufficient that* $\gamma(M, N) > 0$.

Proof. First consider the special case $M \cap N = 0$. If $M + N = Z_0$ is closed, Z_0 is a Banach space and each $u \in Z_0$ has a unique expression $u = v + w$, $v \in M$, $w \in N$. Thus $Pu = v$ defines the projection P of Z_0 onto M along N. P is bounded by a remark after Theorem III-5.20. But $\|P\| = \sup_{u \in Z_0} \|Pu\|/\|u\| = \sup_{v \in M,\, w \in N} \|v\|/\|v + w\| = \sup_{v \in M} \|v\|/\text{dist}(v, N) = 1/\gamma(M, N)$ since $\text{dist}(v, M \cap N) = \|v\|$ by $M \cap N = 0$. Thus

$$\gamma(M, N) = 1/\|P\| > 0 . \tag{4.7}$$

The above argument is not correct if $M = 0$. In this case $\gamma(M, N) = 1 > 0$ by definition, but it is not equal to $\|P\|^{-1} = \infty$.

Suppose, conversely, that $\gamma(M, N) > 0$; we shall show that $M + N$ is closed. We may assume that $M \neq 0$. Let $v_n + w_n \to u$, $v_n \in M$, $w_n \in N$. Then $\|v_n - v_m\| \leq \text{dist}(v_n - v_m, N)/\gamma(M, N) \leq \|v_n - v_m + w_n - w_m\|/\gamma(M, N) \to 0$. Hence $\lim v_n = v$ exists and $w_n = (v_n + w_n) - v_n \to u - v$. Since M, N are closed, $v \in M$, $u - v \in N$ and hence $u \in M + N$.

In this case (4.6) follows from (4.7) and the corresponding formula $\gamma(\mathsf{N}, \mathsf{M}) = 1/\|1 - P\|$, noting that $\|1 - P\| \leqq 1 + \|P\|$. The proof is not valid in the exceptional case $\mathsf{M} = 0$ or $\mathsf{N} = 0$, but it is clear that (4.6) is true in these cases too.

In the general case $\mathsf{M} \cap \mathsf{N} \neq 0$ set $\mathsf{L} = \mathsf{M} \cap \mathsf{N}$ and introduce the quotient space $\tilde{\mathsf{Z}} = \mathsf{Z}/\mathsf{L}$. Since L is closed, $\tilde{\mathsf{Z}}$ is a Banach space (see III-§ 1.8). Let $\tilde{\mathsf{M}}$ be the set of all $\tilde{u} \in \tilde{\mathsf{Z}}$ such that $\tilde{u} \subset \mathsf{M}$; note that the whole coset $\tilde{u}$ is contained in M if there is one element u of $\tilde{u}$ contained in M. Similarly we define $\tilde{\mathsf{N}}$. It is easily seen that $\tilde{\mathsf{M}}$, $\tilde{\mathsf{N}}$ are closed linear manifolds of $\tilde{\mathsf{Z}}$ with $\tilde{\mathsf{M}} \cap \tilde{\mathsf{N}} = 0$. Furthermore, $\tilde{\mathsf{M}} + \tilde{\mathsf{N}}$ is closed in $\tilde{\mathsf{Z}}$ if and only if $\mathsf{M} + \mathsf{N}$ is closed in Z.

Thus the proof of the theorem in the general case is reduced to the special case considered above if we show that

$$\gamma(\tilde{\mathsf{M}}, \tilde{\mathsf{N}}) = \gamma(\mathsf{M}, \mathsf{N}) . \tag{4.8}$$

But this follows from the identity

$$\operatorname{dist}(\tilde{u}, \tilde{\mathsf{M}}) = \operatorname{dist}(u, \mathsf{M}) \tag{4.9}$$

and similar identities with M replaced by N and L. To show (4.9) it suffices to note that $\operatorname{dist}(\tilde{u}, \tilde{\mathsf{M}}) = \inf_{\tilde{v} \in \tilde{\mathsf{M}}} \|\tilde{u} - \tilde{v}\| = \inf_{v \in \mathsf{M}} \inf_{z \in \mathsf{L}} \|u - v - z\| = \inf_{v \in \mathsf{M}} \|u - v\| = \operatorname{dist}(u, \mathsf{M})$. Again the exceptional case $\mathsf{M} \subset \mathsf{N}$ in (4.8) should be treated independently, but the proof is trivial.

Remark 4.3. In view of (4.8), (4.7) is true even in the general case if $\tilde{\mathsf{M}} \neq 0$ and P is the projection of $\tilde{\mathsf{M}} + \tilde{\mathsf{N}}$ onto $\tilde{\mathsf{M}}$ along $\tilde{\mathsf{N}}$. Also we note that (4.6) is true in the general case if $\mathsf{M} + \mathsf{N}$ is closed. It is true even when $\mathsf{M} + \mathsf{N}$ is not closed, for then we have $\gamma(\mathsf{M}, \mathsf{N}) = \gamma(\mathsf{N}, \mathsf{M}) = 0$ by Theorem 4.2.

Lemma 4.4. *Let $\mathsf{M} + \mathsf{N}$ be closed. Then we have for any $u \in \mathsf{Z}$*

$$\operatorname{dist}(u, \mathsf{M}) + \operatorname{dist}(u, \mathsf{N}) \geqq \frac{1}{2} \gamma(\mathsf{M}, \mathsf{N}) \operatorname{dist}(u, \mathsf{M} \cap \mathsf{N}) . \tag{4.10}$$

Proof. In view of (4.8) and (4.9), it suffices to prove (4.10) in which u, M, N are replaced by $\tilde{u}$, $\tilde{\mathsf{M}}$, $\tilde{\mathsf{N}}$, respectively. Changing the notation, we may thus prove (4.10) assuming that $\mathsf{M} \cap \mathsf{N} = 0$ (so that $\operatorname{dist}(u, \mathsf{M} \cap \mathsf{N}) = \|u\|$).

For any $\varepsilon > 0$, there exists a $v \in \mathsf{M}$ and $w \in \mathsf{N}$ such that $\operatorname{dist}(u, \mathsf{M}) > \|u - v\| - \varepsilon$, $\operatorname{dist}(u, \mathsf{N}) > \|u - w\| - \varepsilon$.

If $\|v\| \leqq \|u\|/2$, we have $\operatorname{dist}(u, \mathsf{M}) > \|u\| - \|v\| - \varepsilon \geqq \|u\|/2 - \varepsilon$. If $\|v\| \geqq \|u\|/2$, then $\operatorname{dist}(u, \mathsf{M}) + \operatorname{dist}(u, \mathsf{N}) \geqq \|u - v\| + \|u - w\| - 2\varepsilon \geqq \|v - w\| - 2\varepsilon \geqq \operatorname{dist}(v, \mathsf{N}) - 2\varepsilon \geqq \|v\| \gamma(\mathsf{M}, \mathsf{N}) - 2\varepsilon \geqq \frac{1}{2} \|u\| \gamma(\mathsf{M}, \mathsf{N}) - 2\varepsilon$.

In either case the left member of (4.10) is not smaller than $\frac{1}{2}\|u\|\,\gamma(M, N) - 2\varepsilon$. Since $\varepsilon > 0$ is arbitrary, this proves (4.10).

Problem 4.5. $\operatorname{ind}(M, 0) = -\operatorname{codim} M$, $\operatorname{ind}(M, Z) = \dim M$.

Problem 4.6. Let $M' \supset M$ with $\dim M'/M = m < \infty$. Then $\operatorname{ind}(M', N) = \operatorname{ind}(M, N) + m$.

Problem 4.7. If $\operatorname{def}(M, N) < \infty$, then $M + N$ is closed. [hint: There is an M' such that $M' \supset M$, $M' \cap N = M \cap N$, $M' + N = Z$, $\dim M'/M < \infty$. Then $0 < \gamma(M', N) \leqq \gamma(M, N)$.]

2. Duality

For any subset S of Z, the annihilator $S^\perp$ is a closed linear manifold of Z^* consisting of all $f \in Z^*$ such that $f \perp S$. For any closed linear manifolds M, N of Z, it is easily seen that

$$(M + N)^\perp = M^\perp \cap N^\perp . \tag{4.11}$$

The dual relation $M^\perp + N^\perp = (M \cap N)^\perp$ is not always true, for the simple reason that $(M \cap N)^\perp$ is always closed but $M^\perp + N^\perp$ need not be closed. We shall show, however, that it is true if and only if $M + N$ is closed.

Theorem 4.8. *Let* M, N *be closed linear manifolds of* Z. *Then* $M + N$ *is closed in* Z *if and only if* $M^\perp + N^\perp$ *is closed in* Z^*. *In this case we have, in addition to* (4.11),

$$M^\perp + N^\perp = (M \cap N)^\perp , \tag{4.12}$$

$$\operatorname{nul}(M^\perp, N^\perp) = \operatorname{def}(M, N), \quad \operatorname{def}(M^\perp, N^\perp) = \operatorname{nul}(M, N) , \tag{4.13}$$

$$\gamma(M^\perp, N^\perp) = \gamma(N, M), \quad \hat\gamma(M^\perp, N^\perp) = \hat\gamma(M, N) . \tag{4.14}$$

[(4.14) is true even if $M + N$ is not closed.]

The proof of this theorem will be given in several steps.

Lemma 4.9. *If* $M + N$ *is closed, then* (4.12) *is true. In particular* $M^\perp + N^\perp$ *is closed.*

Proof. It is easily seen that $M^\perp + N^\perp \subset (M \cap N)^\perp$. Thus it suffices to prove the opposite inclusion.

Let $f \in (M \cap N)^\perp$ and consider (f, u) for $u \in M + N$. u has a form $u = v + w$, $v \in M$, $w \in N$, but such an expression may not be unique. If $u = v' + w'$ is another expression of this kind, we have $v - v' = w' - w \in M \cap N$ so that $(f, v - v') = (f, w - w') = 0$. Thus $(f, v) = (f, v')$ and (f, v) is determined uniquely by u. The functional $g[u] = (f, v)$ thus defined for $u \in M + N$ is obviously semilinear. Similarly we define $h[u] = (f, w)$. We note that

$$g[u] = 0 \quad \text{for} \quad u \in N \quad \text{and} \quad h[u] = 0 \quad \text{for} \quad u \in M . \tag{4.15}$$

g and h are bounded. In fact we have $|g[u]| = |(f, v)| \leqq \|f\|\,\|v\|$, where v may be replaced by $v - z$ for any $z \in L = M \cap N$. Hence $|g[u]| \leqq$

$\|f\|$ dist (v, L). Since $\|u\| = \|v + w\| \geqq \text{dist}(v, \mathsf{N}) \geqq \gamma(\mathsf{M}, \mathsf{N})$ dist (v, L), it follows that $|g[u]| \leqq \|f\| \|u\|/\gamma(\mathsf{M}, \mathsf{N})$. In other words, g is bounded with

$$\|g\| \leqq \|f\|/\gamma(\mathsf{M}, \mathsf{N}) . \tag{4.16}$$

g can be extended to a bounded semilinear form on Z by the Hahn-Banach theorem, preserving the bound (4.16); we shall denote this extension by the same symbol g. Similarly h can be extended to an element of Z^*, denoted again by h. Then (4.15) shows that

$$g \in \mathsf{N}^\perp, \quad h \in \mathsf{M}^\perp . \tag{4.17}$$

Since $(f, u) = (f, v) + (f, w) = g[u] + h[u] = (g, u) + (h, u)$ for $u \in \mathsf{M} + \mathsf{N}$, the forms f and $g + h$ coincide on $\mathsf{M} + \mathsf{N}$. Thus $f - g - h = k \in (\mathsf{M}+\mathsf{N})^\perp \subset \mathsf{M}^\perp$. Hence $h + k \in \mathsf{M}^\perp$ and $f = g + (h+k) \in \mathsf{M}^\perp + \mathsf{N}^\perp$. This proves the lemma.

Lemma 4.10. *If* $\mathsf{M} \dot{+} \mathsf{N}$ *is closed, then* $\gamma(\mathsf{M}, \mathsf{N}) \leqq \gamma(\mathsf{N}^\perp, \mathsf{M}^\perp)$.

Proof. For any $g_0 \in \mathsf{N}^\perp$ and $h_0 \in \mathsf{M}^\perp$ set $f = g_0 + h_0$. Then $f \in \mathsf{M}^\perp + \mathsf{N}^\perp = (\mathsf{M} \cap \mathsf{N})^\perp$ by Lemma 4.9. According to the proof of Lemma 4.9, f can be written as $f = g + h$, $g \in \mathsf{N}^\perp$, $h \in \mathsf{M}^\perp$, in such a way that (4.16) is true. But since $g - g_0 = h_0 - h \in \mathsf{M}^\perp \cap \mathsf{N}^\perp$, we have $\|g\| = \|g_0 + g - g_0\| \geqq$ $\geqq \text{dist}(g_0, \mathsf{M}^\perp \cap \mathsf{N}^\perp)$. Thus (4.16) implies $\text{dist}(g_0, \mathsf{M}^\perp \cap \mathsf{N}^\perp) \leqq$ $\leqq \|g_0 + h_0\|/\gamma(\mathsf{M}, \mathsf{N})$. Since this is true for any $h_0 \in \mathsf{M}^\perp$, we have $\text{dist}(g_0, \mathsf{M}^\perp \cap \mathsf{N}^\perp) \leqq \text{dist}(g_0, \mathsf{M}^\perp)/\gamma(\mathsf{M}, \mathsf{N})$. Since this is true for any $g_0 \in \mathsf{N}^\perp$, we obtain the desired result. (Again the exceptional case $\mathsf{M} \subset \mathsf{N}$ needs separate discussion but it is trivial.)

Lemma 4.11. *If* $\mathsf{M} + \mathsf{N}$ *is closed, then* $\gamma(\mathsf{N}^\perp, \mathsf{M}^\perp) \leqq \gamma(\mathsf{M}, \mathsf{N})$.

Proof. Again the exceptional case is trivial and we assume that $\mathsf{M} \not\subset \mathsf{N}$. For simplicity we write $\gamma(\mathsf{M}, \mathsf{N}) = \gamma$ and $\mathsf{M} \cap \mathsf{N} = \mathsf{L}$.

By definition, for any $\varepsilon > 0$ there exists a $v \in \mathsf{M}$ such that

$$0 < \text{dist}(v, \mathsf{N}) < (\gamma + \varepsilon) \, \text{dist}(v, \mathsf{L}) . \tag{4.18}$$

For this v, there exists a $f \in \mathsf{Z}^*$ such that $f \in \mathsf{L}^\perp$ and $0 < (f, v) = \|f\| \, \text{dist}(v, \mathsf{L})$ (see Theorem III-1.22). Since $\mathsf{L}^\perp = \mathsf{M}^\perp + \mathsf{N}^\perp$ by Lemma 4.9, f can be written as $f = g + h$ with $g \in \mathsf{N}^\perp$, $h \in \mathsf{M}^\perp$. Thus $\|f\| \, \text{dist}(v, \mathsf{L}) = (f, v)$ $= (g + h, v) = (g, v) = (g, v - w) = (g - k, v - w) \leqq \|g - k\| \, \|v - w\|$, where $w \in \mathsf{N}$ and $k \in \mathsf{M}^\perp \cap \mathsf{N}^\perp$ are arbitrary. Hence $0 < \|f\| \, \text{dist}(v, \mathsf{L}) \leqq$ $\leqq \text{dist}(g, \mathsf{M}^\perp \cap \mathsf{N}^\perp) \, \text{dist}(v, \mathsf{N})$. But since $\text{dist}(g, \mathsf{M}^\perp) \leqq \|g + h\| = \|f\|$, we obtain

$$\text{dist}(g, \mathsf{M}^\perp) \, \text{dist}(v, \mathsf{L}) \leqq \text{dist}(v, \mathsf{N}) \, \text{dist}(g, \mathsf{M}^\perp \cap \mathsf{N}^\perp) . \tag{4.19}$$

It follows from (4.18) and (4.19) that $\text{dist}(g, \mathsf{M}^\perp) \leqq (\gamma + \varepsilon) \cdot$ $\cdot \, \text{dist}(g, \mathsf{M}^\perp \cap \mathsf{N}^\perp)$. Since $g \in \mathsf{N}^\perp$, this shows that $\gamma(\mathsf{N}^\perp, \mathsf{M}^\perp) \leqq \gamma + \varepsilon$. Since $\varepsilon > 0$ is arbitrary, this proves the Lemma.

Lemma 4.12. *If* $\mathsf{M}^\perp + \mathsf{N}^\perp$ *is closed, then* $\mathsf{M} + \mathsf{N}$ *is closed.*

Proof. Let Z_0 be the closure of $M + N$. Let B_M, B_N be the unit balls of M, N, respectively. We shall first show that the closure $\bar{S}$ of the set $S = B_M + B_N$ contains a ball of Z_0.

Let $u_0 \in Z_0$ be outside $\bar{S}$. Since $\bar{S}$ is a closed convex set, there exists a closed hyperplane of Z_0 separating u_0 from $\bar{S}$. In other words, there exists an $f_0 \in Z_0^*$ such that[1]

$$\text{(4.20)} \qquad \operatorname{Re}(f_0, v + w) < \operatorname{Re}(f_0, u_0) \quad \text{for all} \quad v \in B_M,\ w \in B_N .$$

f_0 can be extended to an element of Z^*, denoted again by f_0, preserving the bound.

Since in (4.20) v and w may be multiplied by arbitrary phase factors (complex numbers with absolute value one), the left member can be replaced by $|(f_0, v)| + |(f_0, w)|$. But $\sup_{v \in B_M} |(f_0, v)| = \operatorname{dist}(f_0, M^\perp)$ by (2.17) and similarly for $|(f_0, w)|$. Hence

$$\text{(4.21)} \qquad \operatorname{dist}(f_0, M^\perp) + \operatorname{dist}(f_0, N^\perp) \leqq \operatorname{Re}(f_0, u_0) \leqq \|u_0\| \, \|f_0\| .$$

But the left member of (4.21) is not smaller than $\gamma' \operatorname{dist}(f_0, M^\perp \cap N^\perp)$ by Lemma 4.4, where $\gamma' = \gamma(M^\perp, N^\perp)/2 > 0$. Furthermore, we have $\operatorname{dist}(f_0, M^\perp \cap N^\perp) = \operatorname{dist}(f_0, (M + N)^\perp) = \operatorname{dist}(f_0, Z_0^\perp) =$ (bound of f_0 restricted to Z_0) $= \|f_0\|$ by the definition of f_0. Hence we obtain $\|u_0\| \geqq \gamma'$.

This means that any $u \in Z_0$ with $\|u\| < \gamma'$ belongs to $\bar{S}$. In other words, $\bar{S}$ contains the ball of Z_0 with center 0 and radius γ'. Then an argument similar to the one used in the proof of the closed graph theorem (Theorem III-5.20) can be applied to show that S itself contains a ball of Z_0. Since $M + N$ is a linear set containing S, it must be identical with Z_0 and therefore closed.

Lemmas 4.9 to 4.12 put together prove Theorem 4.8. Note that (4.13) follows immediately from (4.11) and (4.12) (see Lemma III-1.40).

Corollary 4.13. *A pair* M, N *of closed linear manifolds is Fredholm [semi-Fredholm] if and only if the pair* $M^\perp$, $N^\perp$ *is Fredholm [semi-Fredholm]. In this case we have*

$$\text{(4.22)} \qquad \operatorname{ind}(M, N) = -\operatorname{ind}(M^\perp, N^\perp) .$$

3. Regular pairs of closed linear manifolds

We have remarked above that in general $\gamma(M, N) \neq \gamma(N, M)$, although these two quantities are both zero or both non-zero (see (4.6)). If $\gamma(M, N) = \gamma(N, M)$, we shall say that the pair M, N is *regular*.

[1] Let $d = \operatorname{dist}(u_0, S) > 0$ and let S' be the set of all $u \in Z_0$ such that $\operatorname{dist}(u, S) < d/2$. S' is an open convex set not containing u_0. Thus the existence of f_0 follows from the Hahn-Banach theorem (see footnote [1] of p. 135).

It is known that any pair M, N is regular if Z is a Hilbert space[1]. Another example of a regular pair is given by the pair X, Y in the product space $Z = X \times Y$. Here X is identified with the closed linear manifold of Z consisting of all elements of the form $\{u, 0\}$ with u varying over X, and similarly for Y. It is easy to see that $\gamma(X, Y) = \gamma(Y, X) = 1$.

It is interesting, however, to notice that there are other nontrivial regular pairs in the product space $Z = X \times Y$. In fact, there are closed linear manifolds N of Z such that M, N is a regular pair for any $M \subset Z$. Such an N will be called *a distinguished closed linear manifold*.

Theorem 4.14. *In* $Z = X \times Y$, X *is a distinguished closed linear manifold. Similarly for* Y.

Proof. According to our definition, the norm in Z is given by

$$\|\{x, y\}\|^2 = \|x\|^2 + \|y\|^2, \quad x \in X, y \in Y. \tag{4.23}$$

Let M be a closed linear manifold of Z. We have to show that $\gamma(M, X) = \gamma(X, M)$. We set $L = M \cap X$.

First let us compute $\gamma(M, X)$. Let $u = \{x, y\} \in M$. Then $\operatorname{dist}(u, X)^2 = \inf_{x' \in X} (\|x - x'\|^2 + \|y\|^2) = \|y\|^2$, and $\operatorname{dist}(u, L)^2 = \inf_{x' \in L} (\|x - x'\|^2 + \|y\|^2) = \|\tilde{x}\|^2 + \|y\|^2$ where $\tilde{x} \in \tilde{X} = X/L$. Hence

$$\gamma(M, X) = \inf_{\{x, y\} \in M} \frac{\|y\|}{(\|\tilde{x}\|^2 + \|y\|^2)^{1/2}} = \frac{\gamma}{(1 + \gamma^2)^{1/2}}, \tag{4.24}$$

where

$$\gamma = \inf_{\{x, y\} \in M} \|y\| / \|\tilde{x}\|. \tag{4.25}$$

On the other hand

$$\gamma(X, M) = \inf_{x \in X} \frac{\operatorname{dist}(x, M)}{\operatorname{dist}(x, L)} = \inf_{x \in X} \frac{\inf_{\{x', y'\} \in M} (\|x - x'\|^2 + \|y'\|^2)^{1/2}}{\|\tilde{x}\|}$$
$$= \inf_{\substack{\{x' y'\} \in M \\ x'' \in X}} \frac{(\|x''\|^2 + \|y'\|^2)^{1/2}}{\|\tilde{x}' + \tilde{x}''\|} = \inf_{\substack{\{x', y'\} \in M \\ \tilde{x}'' \in \tilde{X}}} \frac{(\|\tilde{x}''\|^2 + \|y'\|^2)^{1/2}}{\|\tilde{x}' + \tilde{x}''\|}.$$

For a given $\{x', y'\} \in M$, there is an $\tilde{x}'' \in \tilde{X}$ with $\|\tilde{x}' + \tilde{x}''\| = \|\tilde{x}'\| + \|\tilde{x}''\|$ and with an arbitrary $\|\tilde{x}''\|$. Hence

$$\gamma(X, M) = \inf_{\substack{\{x', y'\} \in M \\ \tilde{x}'' \in \tilde{X}}} \frac{(\|\tilde{x}''\|^2 + \|y'\|^2)^{1/2}}{\|\tilde{x}'\| + \|\tilde{x}''\|} \tag{4.26}$$
$$= \inf_{\{x', y'\} \in M} \left(1 + \frac{\|\tilde{x}'\|^2}{\|y'\|^2}\right)^{-1/2} = \gamma / (1 + \gamma^2)^{1/2},$$

where the Schwarz inequality was used. (4.24) and (4.26) show that $\gamma(M, X) = \gamma(X, M)$. [The above computations do not make sense if $M \subset X$ or $X \subset M$. But in these exceptional cases $\gamma(M, X) = \gamma(X, M) = 1$.]

[1] This is due to the identity $\|1 - P\| = \|P\|$ which is valid for any (oblique) projection P, $0 \neq P \neq 1$, in a Hilbert space; see Problem I-6.31.

4. The approximate nullity and deficiency

Let M, N be closed linear manifolds of a Banach space Z. We define the *approximate nullity* of the pair M, N, in symbol nul′(M, N), as the least upper bound (actually the greatest number, as will be shown below) of the set of integers m ($m = \infty$ being permitted) with the property that, for any $\varepsilon > 0$, there is an m-dimensional closed linear manifold $M_\varepsilon \subset M$ with $\delta(M_\varepsilon, N) < \varepsilon$.

Problem 4.15. nul′(M, N) $\geqq$ nul(M, N).

We define the *approximate deficiency* of the pair M, N, denoted by def′(M, N), by

$$\text{def}'(M, N) = \text{nul}'(M^\perp, N^\perp)\,. \tag{4.27}$$

It should be remarked that nul(M, N) and def(M, N) have been defined in a purely algebraic fashion, without referring to any topology. The definition of nul′(M, N) and def′(M, N), on the other hand, depends essentially on the topology of the underlying space.

As mentioned above, nul′(M, N) is not only the least upper bound but the greatest number m with the properties stated above. This is obvious if nul′(M, N) is finite. When nul′(M, N) $= \infty$, this remark is equivalent to the following lemma.

Lemma 4.16. *Suppose that for any $\varepsilon > 0$ and any finite m, there exists an m-dimensional $M_\varepsilon \subset M$ with $\delta(M_\varepsilon, N) < \varepsilon$. Then for any $\varepsilon > 0$ there exists an ∞-dimensional $M_\varepsilon \subset M$ with $\delta(M_\varepsilon, N) < \varepsilon$.*

Proof. For any closed linear manifold $M' \subset M$ with $\dim M/M' < \infty$, there is an $M_\varepsilon \subset M$ with $\dim M_\varepsilon > \dim M/M'$ and $\delta(M_\varepsilon, N) < \varepsilon$. Then $\dim(M' \cap M_\varepsilon) > 0$ (see Problem III-1.42), so that there is a $u \neq 0$ in M′ such that $\text{dist}(u, N) < \varepsilon \|u\|$. Thus Lemma 4.16 is a consequence of the following lemma.

Lemma 4.17. *Assume that for any $\varepsilon > 0$ and any closed linear manifold $M' \subset M$ with* $\dim M/M' < \infty$, *there exists a $u \neq 0$ in M′ such that* $\text{dist}(u, N) < \varepsilon \|u\|$. *Then there is, for any $\varepsilon > 0$, an ∞-dimensional $M_\varepsilon \subset M$ such that* $\delta(M_\varepsilon, N) < \varepsilon$. *In particular* nul′(M, N) $= \infty$.

Proof. We construct two sequences u_n, f_n with the following properties.

$$\begin{gathered} u_n \in M\,, \quad f_n \in Z^*\,, \quad \|u_n\| = 1\,, \quad \|f_n\| = 1\,, \\ (u_n, f_n) = 1\,, \quad (u_n, f_k) = 0 \quad \text{for} \quad k < n\,, \\ \text{dist}(u_n, N) \leqq 3^{-n}\varepsilon\,. \end{gathered} \tag{4.28}$$

Supposing that u_k, f_k have been constructed for $k = 1, 2, \ldots, n-1$, u_n and f_n can be found in the following way. Let M′ be the set of all $u \in M$ such that $(u, f_k) = 0$ for $k = 1, \ldots, n-1$. Since M′ is a closed linear manifold with $\dim M/M' \leqq n-1$, there is a $u_n \in M'$ such that

$\|u_n\| = 1$ and $\operatorname{dist}(u_n, N) \leqq 3^{-n}\varepsilon$. For this u_n there is an $f_n \in Z^*$ such that $\|f_n\| = 1$ and $(u_n, f_n) = 1$ (see Corollary III-1.24).

It follows from (4.28) that the u_n are linearly independent, so that their span M'_ε is infinite-dimensional. Each $u \in M'_\varepsilon$ has the form

$$u = \xi_1 u_1 + \cdots + \xi_n u_n \tag{4.29}$$

with some integer n. We shall show that the coefficients ξ_k satisfy the inequalities

$$|\xi_k| \leqq 2^{k-1} \|u\|, \quad k = 1, 2, \ldots, n. \tag{4.30}$$

To prove this we note that by (4.29) and (4.28)

$$(u, f_j) = \xi_1 (u_1, f_j) + \cdots + \xi_{j-1} (u_{j-1}, f_j) + \xi_j . \tag{4.31}$$

If we assume that (4.30) has been proved for $k < j$, (4.31) gives

$$\begin{aligned} |\xi_j| &\leqq |(u, f_j)| + |\xi_1| \, |(u_1, f_j)| + \cdots + |\xi_{j-1}| \, |(u_{j-1}, f_j)| \\ &\leqq \|u\| + \|u\| + \cdots + 2^{j-2} \|u\| = 2^{j-1} \|u\| , \end{aligned}$$

completing the proof of (4.30) by induction.

From (4.28–4.30) we obtain

$$\begin{aligned} \operatorname{dist}(u, N) &\leqq |\xi_1| \operatorname{dist}(u_1, N) + \cdots + |\xi_n| \operatorname{dist}(u_n, N) \leqq \\ &\leqq (3^{-1} + 2 . 3^{-2} + \cdots + 2^{n-1} 3^{-n}) \, \varepsilon \|u\| \leqq \varepsilon \|u\| . \end{aligned}$$

The same inequality is true for all u in the closure $M_\varepsilon \subset M$ of M'_ε. Hence $\delta(M_\varepsilon, N) \leqq \varepsilon$.

Theorem 4.18. *If* $M + N$ *is closed, then*

$$\operatorname{nul}'(M, N) = \operatorname{nul}(M, N), \quad \operatorname{def}'(M, N) = \operatorname{def}(M, N) . \tag{4.32}$$

Proof. $M^\perp + N^\perp$ is closed if and only if $M + N$ is closed, by Theorem 4.8. In view of (4.13) and (4.27), therefore, it suffices to prove the first equality of (4.32).

Suppose that there is an $M_\varepsilon \subset M$ such that $\dim M_\varepsilon > \operatorname{nul}(M, N) = \dim(M \cap N)$ and $\delta(M_\varepsilon, N) < \varepsilon$. Then there is a $u \in M_\varepsilon$ such that $\operatorname{dist}(u, M \cap N) = \|u\| = 1$ (see Lemma 2.3). Then $\operatorname{dist}(u, N) \geqq \gamma \operatorname{dist}(u, M \cap N) = \gamma$, where $\gamma = \gamma(M, N) > 0$ by Theorem 4.2. On the other hand, $\operatorname{dist}(u, N) \leqq \|u\| \, \delta(M_\varepsilon, N) < \varepsilon$. Hence ε cannot be smaller than γ. This shows that $\operatorname{nul}'(M, N) \leqq \operatorname{nul}(M, N)$. Since the opposite inequality is also true (Problem 4.15), we have the required equality.

Theorem 4.19. *If* $M + N$ *is not closed, then*

$$\operatorname{nul}'(M, N) = \operatorname{def}'(M, N) = \infty . \tag{4.33}$$

Proof. Again it suffices to prove $\operatorname{nul}'(M, N) = \infty$. For any $M' \subset M$ with $\dim M/M' < \infty$, $M' + N$ is not closed (otherwise $M + N$ would be closed by Lemma III-1.9). Thus $\gamma(\mathsf{M}', \mathsf{N}) = 0$ by Theorem 4.2. Hence for any $\varepsilon > 0$, there is a nonzero $u \in \mathsf{M}'$ with $\operatorname{dist}(u, \mathsf{N}) \leqq \varepsilon \operatorname{dist}(u, \mathsf{M}' \cap \mathsf{N}) \leqq \varepsilon \|u\|$. Thus the assumptions of Lemma 4.17 are satisfied.

Problem 4.20. $\operatorname{nul}'(\mathsf{M}, \mathsf{N})$, $\operatorname{def}'(\mathsf{M}, \mathsf{N})$ are symmetric in M, N.

Problem 4.21. $\operatorname{def}'(\mathsf{M}, \mathsf{N}) \geqq \operatorname{def}(\mathsf{M}, \mathsf{N})$.

Problem 4.22. $\operatorname{nul}'(\mathsf{M}, \mathsf{N}) = \operatorname{def}'(\mathsf{M}^\perp, \mathsf{N}^\perp)$.

Finally we add a simple criterion for $\operatorname{nul}'(\mathsf{M}, \mathsf{N}) = \infty$.

Theorem 4.23. *We have* $\operatorname{nul}'(\mathsf{M}, \mathsf{N}) = \infty$ *if and only if there is a sequence* $u_n \in \mathsf{M}$ *with* $\|u_n\| = 1$ *and* $\operatorname{dist}(u_n, \mathsf{N}) \to 0$ *which contains no convergent subsequence.*

Proof. Suppose that $\operatorname{nul}'(\mathsf{M}, \mathsf{N}) = \infty$. We shall construct a sequence $u_n \in \mathsf{M}$ such that $\|u_n\| = 1$, $\operatorname{dist}(u_n, \mathsf{N}) \leqq 1/n$ and $\|u_n - u_m\| \geqq 1$ for $n \neq m$. Assume that $u_1, \ldots, u_n$ have already been constructed, and let M_n be their span. Since $\operatorname{nul}'(\mathsf{M}, \mathsf{N}) = \infty$, there is an $(n + 1)$-dimensional linear manifold $\mathsf{M}' \subset \mathsf{M}$ such that $\delta(\mathsf{M}', \mathsf{N}) \leqq 1/(n + 1)$. Since $\dim \mathsf{M}' > \dim \mathsf{M}_n$, there is a $u \in \mathsf{M}'$ such that $\operatorname{dist}(u, \mathsf{M}_n) = \|u\| = 1$ (Lemma 2.3). $u = u_{n+1}$ satisfies all the requirements for u_{n+1}.

Suppose, conversely, that $\operatorname{nul}'(\mathsf{M}, \mathsf{N}) < \infty$ and $u_n \in \mathsf{M}$ is a sequence with $\|u_n\| = 1$ and $\operatorname{dist}(u_n, \mathsf{N}) \to 0$. We shall show that $\{u_n\}$ contains a convergent subsequence. Since $\mathsf{M} + \mathsf{N}$ is closed by Theorem 4.19, $\gamma(\mathsf{M}, \mathsf{N}) = \gamma > 0$ by Theorem 4.2 and $\operatorname{nul}(\mathsf{M}, \mathsf{N}) < \infty$ by Theorem 4.18. Thus $\operatorname{dist}(u_n, \mathsf{M} \cap \mathsf{N}) \leqq \gamma^{-1} \operatorname{dist}(u_n, \mathsf{N}) \to 0$. This means that there is a sequence $z_n \in \mathsf{M} \cap \mathsf{N}$ such that $u_n - z_n \to 0$. Since $\{z_n\}$ is thus bounded and $\dim(\mathsf{M} \cap \mathsf{N}) = \operatorname{nul}(\mathsf{M}, \mathsf{N}) < \infty$, $\{z_n\}$ contains a convergent subsequence. The same is true of $\{u_n\}$ by $u_n - z_n \to 0$.

5. Stability theorems

Let us now show that $\operatorname{nul}(\mathsf{M}, \mathsf{N})$, $\operatorname{def}(\mathsf{M}, \mathsf{N})$, $\operatorname{ind}(\mathsf{M}, \mathsf{N})$ and the closedness of $\mathsf{M} + \mathsf{N}$ have a certain stability when M is subjected to a small perturbation.

Theorem 4.24. *Let* $\mathsf{M}, \mathsf{N}, \mathsf{M}'$ *be closed linear manifolds of* Z *and let* $\mathsf{M} + \mathsf{N}$ *be closed. Then* $\delta(\mathsf{M}', \mathsf{M}) < \gamma(\mathsf{N}, \mathsf{M})$ *implies* $\operatorname{nul}'(\mathsf{M}', \mathsf{N}) \leqq \operatorname{nul}(\mathsf{M}, \mathsf{N})$, *and* $\delta(\mathsf{M}, \mathsf{M}') < \gamma(\mathsf{M}, \mathsf{N})$ *implies* $\operatorname{def}'(\mathsf{M}', \mathsf{N}) \leqq \operatorname{def}(\mathsf{M}, \mathsf{N})$. [*Note that* $\delta(\mathsf{M}, \mathsf{N})$ *and* $\gamma(\mathsf{M}, \mathsf{N})$ *are in general not symmetric in* M, N.]

Proof. Assume that $\delta(\mathsf{M}', \mathsf{M}) < \gamma(\mathsf{N}, \mathsf{M})$. Suppose that there is a closed linear manifold $\mathsf{N}_\varepsilon \subset \mathsf{N}$ such that $\dim \mathsf{N}_\varepsilon > \operatorname{nul}(\mathsf{M}, \mathsf{N}) = \dim(\mathsf{M} \cap \mathsf{N})$; we shall show that $\delta(\mathsf{N}_\varepsilon, \mathsf{M}')$ cannot be too small. Then we shall have $\operatorname{nul}'(\mathsf{M}', \mathsf{N}) = \operatorname{nul}'(\mathsf{N}, \mathsf{M}') \leqq \operatorname{nul}(\mathsf{M}, \mathsf{N})$ (see Problem 4.20).

$\dim \mathsf{N}_\varepsilon > \dim(\mathsf{M} \cap \mathsf{N})$ implies that there exists a $u \in \mathsf{N}_\varepsilon \subset \mathsf{N}$ such that $\operatorname{dist}(u, \mathsf{M} \cap \mathsf{N}) = \|u\| = 1$ (Lemma 2.3). Hence $\operatorname{dist}(u, \mathsf{M}) \geqq \gamma(\mathsf{N}, \mathsf{M})$ by (4.4). If we make the substitutions $\mathsf{N} \to \mathsf{M}$, $\mathsf{M} \to \mathsf{M}'$ in (2.14), we thus obtain $\operatorname{dist}(u, M') \geqq [1 + \delta(M', M)]^{-1} [\gamma(N, M) - \delta(M', M)]$. This shows that $\operatorname{dist}(u, M')$ and, a fortiori, $\delta(N_\varepsilon, M')$ cannot be arbitrarily small.

The second assertion of the theorem follows from the first by considering $M^\perp$, $N^\perp$, $M'^\perp$; note (2.19), (4.13), (4.14) and (4.27).

Corollary 4.25. *Let* M, N *be a Fredholm [semi-Fredholm] pair. Then the same is true of the pair* M', N *if* $\hat{\delta}(M', M) < \hat{\gamma}(M, N)$, *and we have* $\operatorname{nul}(M', N) \leqq \operatorname{nul}(M, N)$, $\operatorname{def}(M', N) \leqq \operatorname{def}(M, N)$.

Proof. The assumption implies that both of the two conditions of Theorem 4.24 are satisfied. Hence both of the two conclusions are true. If M, N is a semi-Fredholm pair, at least one of $\operatorname{nul}(M, N)$ and $\operatorname{def}(M, N)$ is finite. Hence at least one of $\operatorname{nul}'(M', N)$ and $\operatorname{def}'(M', N)$ is finite. Then $M' + N$ is closed by Theorem 4.19, and at least one of $\operatorname{nul}(M', N) = \operatorname{nul}'(M', N)$ and $\operatorname{def}(M', N) = \operatorname{def}'(M', N)$ is finite (see Theorem 4.18). Thus M', N is a semi-Fredholm pair. If M, N is a Fredholm pair, both $\operatorname{nul}(M, N)$ and $\operatorname{def}(M, N)$ are finite and hence both $\operatorname{nul}(M', N)$ and $\operatorname{def}(M', N)$ are finite.

Remark 4.26. In Theorem 4.24, $M' + N$ need not be closed if $\operatorname{nul}(M, N) = \operatorname{def}(M, N) = \infty$. There is an example of a pair M, N for which there is an M' with arbitrarily small $\hat{\delta}(M', M)$ such that $M' + N$ is not closed.

Remark 4.27. In Corollary 4.25, $\hat{\gamma}(M', N) > 0$ since M', N is a semi-Fredholm pair. But it is in general difficult to estimate $\hat{\gamma}(M', N)$ from below in terms of $\hat{\gamma}(M, N)$ and $\hat{\delta}(M', M)$. In other words, $\hat{\gamma}(M, N)$ can change discontinuously when M is changed slightly. This discontinuity is due to the discontinuity of $M \cap N$ that appears in the definition of $\gamma(M, N)$ [see (4.4)]. But there is lower semicontinuity of $\gamma(M, N)$ if $M \cap N = 0$ or $M + N = Z$. In such a case we have more detailed results given by the following lemmas.

Lemma 4.28. *Let* $M + N$ *be closed with* $\operatorname{nul}(M, N) = 0$. *If* $\hat{\delta}(M', M) < \gamma(N, M)/[2 + \gamma(N, M)]$, *then* $M' + N$ *is closed*, $\operatorname{nul}(M', N) = 0$ *and* $\operatorname{def}(M', N) = \operatorname{def}(M, N)$.

Proof. In view of (4.6), we have $\hat{\delta}(M', M) < \min[\gamma(M, N), \gamma(N, M)] = \hat{\gamma}(M, N)$. Hence $M' + N$ is closed and $\operatorname{nul}(M', N) = 0$, $\operatorname{def}(M', N) \leqq \operatorname{def}(M, N)$ by Corollary 4.25. It remains to be shown that $\operatorname{def}(M, N) \leqq \operatorname{def}(M', N)$.

To this end it suffices to show that $\delta(M', M) < \gamma(M', N)$, for then we can apply the second part of Theorem 4.24 with M and M' exchanged.

Let $u \in N$. Then $\operatorname{dist}(u, M) \geqq \gamma(N, M) \operatorname{dist}(u, M \cap N) = \gamma(N, M) \|u\|$ since $M \cap N = 0$. It follows from (2.14) (with the substitution $N \to M$, $M \to M'$) that $\operatorname{dist}(u, M') \geqq [1 + \delta(M', M)]^{-1} [\gamma(N, M) - \delta(M', M)] \|u\|$. Since this is true for any $u \in N$, we have

$$\gamma(N, M') \geqq \frac{\gamma(N, M) - \delta(M', M)}{1 + \delta(M', M)} . \tag{4.34}$$

Application of the inequality (4.6) then gives the desired inequality:

$$(4.35)\quad \gamma(\mathsf{M}',\mathsf{N}) \geqq \frac{\gamma(\mathsf{N},\mathsf{M}')}{1+\gamma(\mathsf{N},\mathsf{M}')} = \frac{\gamma(\mathsf{N},\mathsf{M}) - \delta(\mathsf{M}',\mathsf{M})}{1+\gamma(\mathsf{N},\mathsf{M})} > \delta(\mathsf{M}',\mathsf{M})$$

since $\delta(\mathsf{M}',\mathsf{M}) < \gamma(\mathsf{N},\mathsf{M})/[2+\gamma(\mathsf{N},\mathsf{M})]$ by hypothesis.

Lemma 4.29. *Let* $\mathsf{M}+\mathsf{N}=\mathsf{Z}$ *(so that* $\operatorname{def}(\mathsf{M},\mathsf{N})=0$*). If* $\hat\delta(\mathsf{M}',\mathsf{M}) < \gamma(\mathsf{M},\mathsf{N})/[2+\gamma(\mathsf{M},\mathsf{N})]$*, then* $\mathsf{M}'+\mathsf{N}=\mathsf{Z}$ *(so that* $\operatorname{def}(\mathsf{M}',\mathsf{N})=0$*) and* $\operatorname{nul}(\mathsf{M}',\mathsf{N}) = \operatorname{nul}(\mathsf{M},\mathsf{N})$.

Proof. It suffices to apply Lemma 4.28 to M, N, M′ replaced by their annihilators. Note Theorems 4.8 and 2.9.

Theorem 4.30. *Let* M, N *be a Fredholm [semi-Fredholm] pair. Then there is a* $\delta > 0$ *such that* $\hat\delta(\mathsf{M}',\mathsf{M}) < \delta$ *implies that* M′, N *is a Fredholm [semi-Fredholm] pair and* $\operatorname{ind}(\mathsf{M}',\mathsf{N}) = \operatorname{ind}(\mathsf{M},\mathsf{N})$.

Proof. It suffices to consider the semi-Fredholm case, for then $\operatorname{ind}(\mathsf{M}',\mathsf{N}) = \operatorname{ind}(\mathsf{M},\mathsf{N})$ implies that the pair M′, N is Fredholm if and only if the pair M, N is. Furthermore, we may assume that $\operatorname{def}(\mathsf{M},\mathsf{N}) < \infty$; the case $\operatorname{nul}(\mathsf{M},\mathsf{N}) < \infty$ can be reduced to this case by considering the annihilators.

If $\operatorname{def}(\mathsf{M},\mathsf{N}) = m < \infty$, we can find an $\mathsf{N}_0 \supset \mathsf{N}$ such that $\dim \mathsf{N}_0/\mathsf{N} = m$, $\mathsf{N}_0 \cap \mathsf{M} = \mathsf{N} \cap \mathsf{M}$ and $\mathsf{M}+\mathsf{N}_0 = \mathsf{Z}$. According to Lemma 4.29, there is a $\delta > 0$ such that $\hat\delta(\mathsf{M}',\mathsf{M}) < \delta$ implies that $\operatorname{def}(\mathsf{M}',\mathsf{N}_0) = 0$ and $\operatorname{nul}(\mathsf{M}',\mathsf{N}_0) = \operatorname{nul}(\mathsf{M},\mathsf{N}_0) = \operatorname{nul}(\mathsf{M},\mathsf{N})$. Hence $\operatorname{ind}(\mathsf{M}',\mathsf{N}_0) = \operatorname{nul}(\mathsf{M},\mathsf{N})$, and it follows from Problem 4.6 that $\operatorname{ind}(\mathsf{M}',\mathsf{N}) = \operatorname{ind}(\mathsf{M}',\mathsf{N}_0) - m = \operatorname{nul}(\mathsf{M},\mathsf{N}) - \operatorname{def}(\mathsf{M},\mathsf{N}) = \operatorname{ind}(\mathsf{M},\mathsf{N})$.

Remark 4.31. Theorem 4.30 shows that the index of a semi-Fredholm pair M, N is stable under a small perturbation of M. It could be shown that the same is true for simultaneous perturbation of M and N, but the proof would be more complicated. Also it is not easy to give a simple estimate of δ in Theorem 4.30. In particular, it is in general not known[1] whether ind(M′, N) is constant for all M′ considered in Corollary 4.25, except that this is true if Z is a Hilbert space[2].

§ 5. Stability theorems for semi-Fredholm operators

1. The nullity, deficiency and index of an operator

In this section we define the nullity, deficiency, index, etc. of a linear operator $T \in \mathscr{C}(\mathsf{X},\mathsf{Y})$ and establish several stability theorems for these quantities[3]. The general results follow from the corresponding

[1] Cf. NEUBAUER [1].

[2] Cf. T. KATO [9].

[3] General references for this section are ATKINSON [1], [2], BROWDER [1], [2], [3], CORDES and LABROUSSE [1], DIEUDONNÉ [1], GOHBERG and KREIN [1], KAASHOEK [1], KANIEL and SCHECHTER [1], T. KATO [12], KREIN and KRASNOSEL'SKII [1], SZ.-NAGY [3], NEUBAUER [1], [2], YOOD [1].

results for pairs of closed linear manifolds studied in the preceding section[1]. But these will be supplemented by more specific results pertaining to operators.

As in the finite-dimensional case (I-§ 3.1), the *nullity*, $\operatorname{nul} T$, of an operator T from X to Y is defined as the dimension of $\mathsf{N}(T)$. Since $\mathsf{N}(T)$ is the *geometric* eigenspace of T for the eigenvalue zero, $\operatorname{nul} T$ is the geometric multiplicity of this eigenvalue. The *deficiency*, $\operatorname{def} T$, of T is the codimension in Y of $\mathsf{R}(T)$: $\operatorname{def} T = \dim \mathsf{Y}/\mathsf{R}(T)$[2]. Each of $\operatorname{nul} T$ and $\operatorname{def} T$ takes values $0, 1, 2, \ldots$ or ∞. The *index* of T is defined by[3]

$$\operatorname{ind} T = \operatorname{nul} T - \operatorname{def} T \tag{5.1}$$

if at least one of $\operatorname{nul} T$ and $\operatorname{def} T$ is finite.

The notions of nullity and deficiency involve certain arbitrariness. An operator T from X to Y could as well be regarded as an operator from X to a space Y' containing Y as a subspace; then $\operatorname{def} T$ would be increased by $\dim \mathsf{Y}'/\mathsf{Y}$. Again, T could be regarded as an operator from a space X' to Y, where X' is the direct sum $\mathsf{X} \oplus \mathsf{X}_0$ of X and another space X_0, with the stipulation that $Tu = 0$ for $u \in \mathsf{X}_0$; then $\operatorname{nul} T$ would be increased by $\dim \mathsf{X}_0$. Once the domain and range spaces have been fixed, however, $\operatorname{nul} T$ and $\operatorname{def} T$ are well-defined quantities with important properties to be discussed below.

In what follows we assume that X, Y are two fixed Banach spaces. The stability of bounded invertibility (Theorem 2.21) is the stability of the property of $T \in \mathscr{C}(\mathsf{X}, \mathsf{Y})$ that $\operatorname{nul} T = \operatorname{def} T = 0$. We are now going to generalize this stability theorem to other values of $\operatorname{def} T$ and $\operatorname{nul} T$ under some additional conditions.

One of these additional conditions will be that $\mathsf{R}(T)$ be closed. This condition is automatically satisfied if $\operatorname{def} T < \infty$, but should be assumed independently in the general case. An operator $T \in \mathscr{C}(\mathsf{X}, \mathsf{Y})$ is said to be *Fredholm*[4] if $\mathsf{R}(T)$ is closed and both $\operatorname{nul} T$ and $\operatorname{def} T$ are finite. T is said to be *semi-Fredholm* if $\mathsf{R}(T)$ is closed and at least one of $\operatorname{nul} T$ and $\operatorname{def} T$ is finite. The index (5.1) is well-defined for a semi-Fredholm operator T. The main result to be proved in this section is that the property of being Fredholm [semi-Fredholm] is stable under small perturbations.

---

[1] This method of deducing the results for operators from those for pairs of subspaces seems to be new. It has the advantage of being able to make full use of duality theorems for subspaces. The corresponding theorems for operators are restricted to densely defined operators for which the adjoints are defined.

[2] Sometimes $\operatorname{def} T$ is defined as equal to $\dim \mathsf{Y}/\overline{\mathsf{R}(T)}$, when $\overline{\mathsf{R}(T)}$ is the closure of $\mathsf{R}(T)$. The two definitions coincide for semi-Fredholm operators.

[3] Some authors choose the opposite sign for the index.

[4] It should be noted that there is quite a different use of the term "Fredholm"; see e. g. GROTHENDIECK [1].

To this end it is necessary to study the properties of closed operators with closed range.

For any $T \in \mathscr{C}(X, Y)$, the null space $N = N(T)$ is a closed linear manifold of X. Therefore, the quotient space $\tilde{X} = X/N$ is a Banach space with the norm defined by (see III-§1.8)

$$\|\tilde{u}\| = \inf_{u \in \tilde{u}} \|u\| = \inf_{z \in N} \|u - z\| = \operatorname{dist}(u, N), \quad u \in \tilde{u}. \tag{5.2}$$

If $u \in D(T)$, all $u' \in \tilde{u}$ belong to $D(T)$ by $u' - u \in N \subset D(T)$. Moreover, we have $Tu = Tu'$ since N is the null space of T. Thus we can define an operator $\tilde{T}$ from $\tilde{X}$ to Y by

$$\tilde{T}\tilde{u} = Tu. \tag{5.3}$$

The domain $D(\tilde{T})$ of $\tilde{T}$ is the set of all $\tilde{u} \in \tilde{X}$ such that every $u \in \tilde{u}$ belongs to $D(T)$.

It is obvious that $\tilde{T}$ is linear. Furthermore $\tilde{T}$ is closed. To see this let $\tilde{u}_n$ be a $\tilde{T}$-convergent sequence in $\tilde{X}$: $\tilde{u}_n \in D(\tilde{T})$, $\tilde{u}_n \to \tilde{u} \in \tilde{X}$, $\tilde{T}\tilde{u}_n \to v \in Y$. Let $u_n \in \tilde{u}_n$, $u \in \tilde{u}$. $\tilde{u}_n \to \tilde{u}$ implies that $\operatorname{dist}(u_n - u, N) \to 0$. Hence there are $z_n \in N$ such that $u_n - u - z_n \to 0$. Since $T(u_n - z_n) = Tu_n = \tilde{T}\tilde{u}_n \to v$, the $u_n - z_n$ form a sequence T-convergent to u. It follows from the closedness of T that $u \in D(T)$ and $Tu = v$. Hence $\tilde{u} \in D(\tilde{T})$ and $\tilde{T}\tilde{u} = Tu = v$. This proves that $\tilde{T}$ is closed.

$\tilde{T}$ is invertible. In fact $\tilde{T}\tilde{u} = 0$ implies $Tu = 0$, hence $u \in N$ and $\tilde{u} = N$; but N is the zero element of $\tilde{X}$.

We now define a number $\gamma(T)$ by $\gamma(T) = 1/\|\tilde{T}^{-1}\|$; it is to be understood that $\gamma(T) = 0$ if $\tilde{T}^{-1}$ is unbounded and $\gamma(T) = \infty$ if $\tilde{T}^{-1} = 0$. It follows from (5.3) that $\gamma(T)$ is the greatest number γ such that

$$\|Tu\| \geqq \gamma \|\tilde{u}\| = \gamma \operatorname{dist}(u, N) \quad \text{for all} \quad u \in D(T). \tag{5.4}$$

It should be noted that $\gamma(T) = \infty$ occurs if and only if $\tilde{T}$ is a trivial operator with both $D(\tilde{T})$ and $R(\tilde{T})$ 0-dimensional, and this is the case if and only if $T \subset 0$ [$Tu = 0$ for all $u \in D(T)$]. To make the statement (5.4) correct even for this case, one should stipulate that $\infty \times 0 = 0$.

We shall call $\gamma(T)$ the *reduced minimum modulus* of T. If $N(T) = 0$, $\gamma(T)$ is equal to the *minimum modulus*[1] of T, which is defined as $\inf \|Tu\|/\|u\|$ for $0 \neq u \in D(T)$.

Problem 5.1. $\gamma(\tilde{T}) = \gamma(T)$.

Theorem 5.2. $T \in \mathscr{C}(X, Y)$ *has closed range if and only if* $\gamma(T) > 0$.

Proof. By definition $\gamma(T) > 0$ if and only if $\tilde{T}^{-1}$ is bounded, and this is true if and only if $D(\tilde{T}^{-1}) = R(\tilde{T}) = R(T)$ is closed (see III-§ 5.4).

Example 5.3.[2] Let $X = l^p$ and let $\{x_j\}$ be the canonical basis of X. Let $T \in \mathscr{B}(X)$ be such that $Tx_1 = 0$, $Tx_2 = x_1$, $Tx_3 = x_2$, ... T is a shift operator (see Example

[1] See GINDLER and TAYLOR [1].

[2] $\gamma(T)$ for differential operators T is discussed in detail in GOLDBERG [1].

III-3.16). As is easily verified, $N(T)$ is the one-dimensional subspace spanned by x_1, and $R(T) = X$. Thus nul $T = 1$, def $T = 0$, ind $T = 1$. For any $u = (\xi_j)$, we have $\|\tilde{u}\| = \operatorname{dist}(u, N(T)) = \left(\sum_{j=2}^{\infty} |\xi_j|^p\right)^{1/p} = \|Tu\|$. Hence $\|Tu\|/\|\tilde{u}\| = 1$ for every $u \in X$ so that $\gamma(T) = 1$. T is a Fredholm operator.

Example 5.4. Let X and x_j be as above and let $T \in \mathscr{B}(X)$ be such that $Tx_1 = x_2$, $Tx_2 = x_3, \ldots$. This T is the adjoint of the T of Example 5.3 for an appropriate p. It is easily verified that $N(T) = 0$, and that $R(T)$ is the subspace spanned by $x_2, x_3, \ldots$. Hence nul $T = 0$, def $T = 1$, ind $T = -1$. In this case $\tilde{X} = X$, $\tilde{T} = T$, $\|Tu\| = \|u\|$ so that $\gamma(T) = 1$. T is Fredholm.

Example 5.5. Let X be as above but write each $u \in X$ in the form of a bilateral sequence $u = (\ldots, \xi_{-1}, \xi_0, \xi_1, \ldots)$. Let $\{x_j\}$ be the canonical basis of X and let $T \in \mathscr{B}(X)$ be such that $Tx_0 = 0$, $Tx_j = x_{j-1}$ $(j = \pm 1, \pm 2, \ldots)$. $N(T)$ is the one-dimensional subspace spanned by x_0 and $R(T)$ is the subspace spanned by all the x_j except x_{-1}. Hence nul $T = 1$, def $T = 1$, ind $T = 0$. A consideration similar to that in Example 5.3 shows that $\gamma(T) = 1$. T is Fredholm (cf. Example 3.8).

Problem 5.6. Let $T \in \mathscr{C}(X, Y)$. Then $\zeta \in P(T)$ if and only if $\operatorname{nul}(T - \zeta) = \operatorname{def}(T - \zeta) = 0$.

2. The general stability theorem

The nullity, deficiency and the closed range property can be expressed conveniently in terms of the graph of the operator. Such expressions will be important in perturbation theory since a "small" change of a closed operator is expressed by means of its graph.

Consider the product space $Z = X \times Y$. For convenience we shall identify each $u \in X$ with $\{u, 0\} \in Z$ and each $v \in Y$ with $\{0, v\} \in Z$. Then X is identified with the subspace $X \times 0$ of Z and similarly for Y. Z is identical with the direct sum $X \oplus Y$. Similarly, any subset of X or of Y is identified with a subset of Z.

Let $T \in \mathscr{C}(X, Y)$. The graph $G(T)$ of T is the closed linear manifold of Z consisting of all elements $\{u, Tu\}$ where $u \in D(T)$. $u \in X$ belongs to $N(T)$ if and only if $\{u, 0\} \in G(T)$. According to the identification just introduced, this means that

$$N(T) = G(T) \cap X . \tag{5.5}$$

Also we have

$$R(T) + X = G(T) + X . \tag{5.6}$$

In fact $R(T) + X$ is the set of all $\{v, Tu\}$ where $u \in D(T)$ and $v \in X$, while $G(T) + X$ is the set of all $\{u + v, Tu\}$ where $u \in D(T)$ and $v \in X$. Obviously these two sets are identical.

It follows from (5.5) and (5.6) that [see (4.1) and (4.2)]

$$\begin{aligned} \operatorname{nul} T &= \dim(G(T) \cap X) = \operatorname{nul}(G(T), X) , \\ \operatorname{def} T &= \operatorname{codim}(G(T) + X) = \operatorname{def}(G(T), X) . \end{aligned} \tag{5.7}$$

Thus the nullity and deficiency of T are exactly equal to the corresponding quantities for the pair $\mathsf{G}(T)$, X of closed linear manifolds of Z.

Furthermore, it is easily seen that $\mathsf{R}(T)$ is closed in Y if and only if $\mathsf{X} + \mathsf{R}(T)$ is closed in Z. According to (5.6), therefore, *$\mathsf{R}(T)$ is closed if and only if $\mathsf{G}(T) + \mathsf{X}$ is closed.*

Even the reduced minimum modulus $\gamma(T)$ of T is related in a simple way to the minimum gap $\gamma(\mathsf{G}(T), \mathsf{X})$ of the pair $\mathsf{G}(T)$, X. By (5.4), $\gamma(T)$ is equal to $\inf_{u \in \mathsf{D}(T)} \|Tu\|/\|\tilde{u}\| = \inf_{\{u,v\} \in \mathsf{G}(T)} \|v\|/\|\tilde{u}\|$ where $\tilde{u} \in \tilde{\mathsf{X}} = \mathsf{X}/\mathsf{N}(T)$. In view of (5.5), this is exactly equal to the γ of (4.25) for $\mathsf{M} = \mathsf{G}(T)$. Hence, by (4.24) and (4.26), we have

$$\gamma(\mathsf{G}(T), \mathsf{X}) = \gamma(\mathsf{X}, \mathsf{G}(T)) = \frac{\gamma(T)}{[1 + \gamma(T)^2]^{1/2}} \,. \tag{5.8}$$

In this way we see that all the important quantities introduced for T can be expressed in terms of the pair $\mathsf{G}(T)$, X of closed linear manifolds of Z.

To complete this correspondence, we shall define the *approximate nullity and approximate deficiency* of T by

$$\operatorname{nul}' T = \operatorname{nul}'(\mathsf{G}(T), \mathsf{X}) \,, \quad \operatorname{def}' T = \operatorname{def}'(\mathsf{G}(T), \mathsf{X}) \,. \tag{5.9}$$

The following results are direct consequences of the corresponding ones for pairs of closed linear manifolds proved in the preceding paragraph.

Problem 5.7. Let $T \in \mathscr{C}(\mathsf{X}, \mathsf{Y})$. Then $\operatorname{def} T < \infty$ implies $\gamma(T) > 0$ (see Problem 4.7).

Problem 5.8. If T_1 is an extension of T of order $m < \infty$, $\operatorname{ind} T_1 = \operatorname{ind} T + m$ (see Problem 4.6).

Theorem 5.9. $\operatorname{nul}' T$ *is the greatest number $m \leqq \infty$ with the following property: for any $\varepsilon > 0$ there exists an m-dimensional closed linear manifold $\mathsf{N}_\varepsilon \subset \mathsf{D}(T)$ such that $\|Tu\| \leqq \varepsilon\|u\|$ for every $u \in \mathsf{N}_\varepsilon$ (see the definition in* § 4.4*).*

Theorem 5.10. $\operatorname{nul}' T \geqq \operatorname{nul} T$, $\operatorname{def}' T \geqq \operatorname{def} T$ *for any* $T \in \mathscr{C}(\mathsf{X}, \mathsf{Y})$. *The equalities hold if $\mathsf{R}(T)$ is closed. If $\mathsf{R}(T)$ is not closed, then* $\operatorname{nul}' T = \operatorname{def}' T = \infty$ *(see* Problem 4.15, Theorems 4.18 *and* 4.19*).*

Theorem 5.11.[1] $\operatorname{nul}' T = \infty$ *if and only if there is a sequence $u_n \in \mathsf{D}(T)$ with $\|u_n\| = 1$ and $Tu_n \to 0$ which contains no convergent subsequence (see* Theorem 4.23*).*

Problem 5.12. $\operatorname{nul}'(\alpha T) = \operatorname{nul}' T$, $\operatorname{def}'(\alpha T) = \operatorname{def}' T$ for $\alpha \neq 0$. [hint: Theorem 5.10].

Let us now assume that T is densely defined so that the adjoint operator T^* exists and belongs to $\mathscr{C}(\mathsf{Y}^*, \mathsf{X}^*)$. For convenience we

[1] This theorem is due to Wolf [4].

consider the inverse graph $\mathsf{G}'(T^*)$ of T^* rather than the graph $\mathsf{G}(T^*)$; $\mathsf{G}'(T^*)$ is the closed linear manifold of $\mathsf{X}^* \times \mathsf{Y}^* = \mathsf{Z}^*$ consisting of all elements $\{T^* g, g\}$ where g varies over $\mathsf{D}(T^*) \subset \mathsf{Y}^*$. As was shown in III-§ 5.5, we have the basic relation

$$\mathsf{G}'(-T^*) = \mathsf{G}(T)^\perp . \tag{5.10}$$

Since $\mathsf{G}'(T^*)$ is simply the image of $\mathsf{G}(T^*)$ under the map $\{g, f\} \to \{f, g\}$ of $\mathsf{Y}^* \times \mathsf{X}^*$ onto $\mathsf{X}^* \times \mathsf{Y}^*$, the properties of $\mathsf{G}(T^*)$ correspond to those of $\mathsf{G}'(T^*)$ in an obvious manner. Thus we have, corresponding to (5.7) to (5.9),

$$\mathsf{N}(T^*) = \mathsf{N}(-T^*) = \mathsf{G}'(-T^*) \cap \mathsf{Y}^* = \mathsf{G}(T)^\perp \cap \mathsf{X}^\perp , \tag{5.11}$$

$$\mathsf{R}(T^*) + \mathsf{Y}^* = \mathsf{R}(-T^*) + \mathsf{Y}^* = \mathsf{G}'(-T^*) + \mathsf{Y}^* = \mathsf{G}(T)^\perp + \mathsf{X}^\perp , \tag{5.12}$$

$$\begin{cases} \operatorname{nul} T^* = \dim(\mathsf{G}(T)^\perp \cap \mathsf{X}^\perp) = \operatorname{nul}(\mathsf{G}(T)^\perp, \mathsf{X}^\perp) , \\ \operatorname{def} T^* = \operatorname{codim}(\mathsf{G}(T)^\perp + \mathsf{X}^\perp) = \operatorname{def}(\mathsf{G}(T)^\perp, \mathsf{X}^\perp) , \end{cases} \tag{5.13}$$

$$\gamma(\mathsf{G}(T)^\perp, \mathsf{X}^\perp) = \gamma(\mathsf{X}^\perp, \mathsf{G}(T)^\perp) = \frac{\gamma(T^*)}{[1 + \gamma(T^*)^2]^{1/2}} , \tag{5.14}$$

$$\operatorname{nul}' T^* = \operatorname{nul}'(\mathsf{G}(T)^\perp, \mathsf{X}^\perp) , \quad \operatorname{def}' T^* = \operatorname{def}'(\mathsf{G}(T)^\perp, \mathsf{X}^\perp) . \tag{5.15}$$

Here $\mathsf{G}(T)^\perp$ and $\mathsf{X}^\perp = \mathsf{Y}^*$ are considered as linear manifolds of $\mathsf{Z}^* = \mathsf{X}^* \times \mathsf{Y}^*$ [for (5.15) note Problem 5.12].

Thus Theorem 4.8 gives

Theorem 5.13.[1] *Assume that T^* exists. $\mathsf{R}(T)$ is closed if and only if $\mathsf{R}(T^*)$ is closed. In this case we have*[2]

$$\mathsf{R}(T)^\perp = \mathsf{N}(T^*) , \quad \mathsf{N}(T)^\perp = \mathsf{R}(T^*) , \tag{5.16}$$

$$\operatorname{nul} T^* = \operatorname{def} T , \quad \operatorname{def} T^* = \operatorname{nul} T , \tag{5.17}$$

$$\gamma(T^*) = \gamma(T) . \tag{5.18}$$

(5.18) *is true even when $\mathsf{R}(T)$ is not closed.*

Corollary 5.14. *Assume that T^* exists. T is Fredholm [semi-Fredholm] if and only if T^* is. In this case we have*

$$\operatorname{ind} T^* = -\operatorname{ind} T . \tag{5.19}$$

Problem 5.15. $\operatorname{nul}' T^* = \operatorname{def}' T$, $\operatorname{def}' T^* = \operatorname{nul}' T$.

---

[1] This theorem is given in Banach [1] for bounded T. For unbounded T it is proved by several authors; see Browder [1], [3], Joichi [1], T. Kato [12], Rota [1]. In view of (5.16), a densely defined $T \in \mathscr{C}(\mathsf{X}, \mathsf{Y})$ with closed range is also said to be *normally solvable*.

[2] Here $\mathsf{N}(T)$, $\mathsf{R}(T)$, $\mathsf{N}(T^*)$, $\mathsf{R}(T^*)$ are regarded as subsets of X, Y, Y^*, X^*, respectively. To prove (5.16), for example, we note that $(\mathsf{R}(T) + \mathsf{X})^\perp = (\mathsf{G}(T) + \mathsf{X})^\perp = \mathsf{G}(T)^\perp \cap \mathsf{X}^\perp = \mathsf{G}(-T^*) \cap \mathsf{Y}^* = \mathsf{N}(T^*)$ by (4.11), where all members are regarded as subsets of $\mathsf{Z}^* = \mathsf{X}^* \times \mathsf{Y}^*$. It follows that $\mathsf{R}(T)^\perp = \mathsf{N}(T^*)$ where both members are regarded as subsets of Y^*. Similarly, the second relation of (5.16) follows from (4.12) applied to the pair $\mathsf{G}(T)$, X.

Finally, the stability theorems for closed linear manifolds proved in par. 4 immediately give stability theorems for operators (see Theorems 4.24, 4.30 and Corollary 4.25).

Theorem 5.16. *Let* $T, S \in \mathscr{C}(X, Y)$ *and let* $R(T)$ *be closed (so that* $\gamma(T) = \gamma > 0$*). Then* $\delta(S, T) < \gamma(1+\gamma^2)^{-1/2}$ *implies* $\operatorname{nul}' S \leq \operatorname{nul} T$, *and* $\delta(T, S) < \gamma(1+\gamma^2)^{-1/2}$ *implies* $\operatorname{def}' S \leq \operatorname{def} T$.

Theorem 5.17.[1] *Let* $T, S \in \mathscr{C}(X, Y)$ *and let* T *be Fredholm [semi-Fredholm]. If* $\hat{\delta}(S, T) < \gamma(1+\gamma^2)^{-1/2}$ *where* $\gamma = \gamma(T)$, *then* S *is Fredholm [semi-Fredholm] and* $\operatorname{nul} S \leq \operatorname{nul} T$, $\operatorname{def} S \leq \operatorname{def} T$. *Furthermore, there is a* $\delta > 0$[2] *such that* $\hat{\delta}(S, T) < \delta$ *implies* $\operatorname{ind} S = \operatorname{ind} T$[3].

The above results have been deduced by considering the pair of subspaces $G(T)$ and X of the product space $X \times Y$ and the pair of their annihilators. It is interesting to see what is obtained by considering the pair $G(T)$ and Y. It turns out that the results are related to *bounded* operators rather than to Fredholm operators. We shall state some of the results in the form of problems, with hints for the proofs.

Problem 5.18. Let $T \in \mathscr{C}(X, Y)$. Then $\gamma(G(T), Y) > 0$ $[\gamma(Y, G(T)) > 0]$ if and only if T is bounded. If $T \in \mathscr{B}(X, Y)$, then $\gamma(G(T), Y) = \gamma(Y, G(T)) = (1 + \|T\|^2)^{-1/2}$ (cf. Theorem 4.14).

Problem 5.19. Let $T \in \mathscr{C}(X, Y)$. Then $\operatorname{nul}(G(T), Y) = 0$. T is bounded if and only if $G(T) + Y$ is closed. $T \in \mathscr{B}(X, Y)$ if and only if $\operatorname{def}(G(T), Y) = 0$. [hint: $G(T) + Y = D(T) + Y$.]

Problem 5.20. Let $T \in \mathscr{C}(X, Y)$ be densely defined. Then $T \in \mathscr{B}(X, Y)$ if and only if T^* is bounded[4]. [hint: Apply the preceding problem to T^*.]

---

[1] It appears that this stability theorem for $\operatorname{nul} T$, $\operatorname{def} T$ and $\operatorname{ind} T$ is new in the general form given here. For $T, S \in \mathscr{B}(X, Y)$ and small $\|S - T\|$, this theorem has long been known (Atkinson [1], Dieudonné [1], Krein and Krasnosel'skii [1]). Sz.-Nagy [3] extended the results to unbounded T and relatively bounded perturbations $S - T$. Gohberg and Krein [1] and T. Kato [12] have similar results, with quantitative refinements. Cordes and Labrousse [1] consider a more general kind of perturbation like the one in Theorem 5.17 but assume $X = Y$ to be a Hilbert space. Recently, Neubauer [1] proved a theorem similar to Theorem 5.17.

[2] We can choose $\delta = \gamma(1+\gamma^2)^{-1/2}$ if X, Y are Hilbert spaces. In general it is difficult to give a simple estimate for δ (but see Neubauer [1]).

[3] This stability theorem for the index is important in many respects. In particular it is one of the most powerful methods for proving the existence of solutions of functional equations. As an example we refer to the proof of Lemma X-5.14. Here we have two families of operators $W(\varkappa), Z(\varkappa) \in \mathscr{B}(X)$, depending on a parameter $\varkappa$ holomorphically, such that $Z(\varkappa)\, W(\varkappa) = 1$ and $W(0) = Z(0) = 1$. Application of the stability theorem shows that $W(\varkappa)$ maps X onto X [i. e., $W(\varkappa)\, u = v$ has a solution u for any $v \in X$].

[4] This is a nontrivial result and aroused some interest recently; see Brown [1], Neubauer [1], Goldberg [1]. There is a close relationship between this proposition, the closed range theorem (Theorem 5.13) and Theorem 4.8. We have deduced the first two from the last. Brown [1] deduces the second from the first. Browder gave a deduction of the last from the second (oral communication).

Problem 5.21. Let $T \in \mathscr{B}(\mathsf{X}, \mathsf{Y})$. If $S \in \mathscr{C}(\mathsf{X}, \mathsf{Y})$ and $\delta(T, S) < (1 + \|T\|^2)^{-1/2}$, then $S \in \mathscr{B}(\mathsf{X}, \mathsf{Y})$ and (2.25) holds with $\delta(S, T)$ replaced by $\delta(T, S)$[1]. [hint: See Theorem 4.24 and Problem 5.19.]

3. Other stability theorems

Theorem 5.17 may be called the *general stability theorem* since the only assumption involved is that $\hat{\delta}(S, T)$ be small; there is no assumption on the relationship between the domains of S and T. If we add such an assumption, however, we can deduce a somewhat stronger result.

Theorem 5.22. *Let $T \in \mathscr{C}(\mathsf{X}, \mathsf{Y})$ be semi-Fredholm (so that $\gamma = \gamma(T) > 0$). Let A be a T-bounded operator from X to Y so that we have the inequality* (1.1), *where*

$$a < (1 - b)\,\gamma\,. \tag{5.20}$$

Then $S = T + A$ belongs to $\mathscr{C}(\mathsf{X}, \mathsf{Y})$, S is semi-Fredholm and

$$\operatorname{nul} S \leqq \operatorname{nul} T\,, \quad \operatorname{def} S \leqq \operatorname{def} T\,, \quad \operatorname{ind} S = \operatorname{ind} T\,. \tag{5.21}$$

Proof. (5.20) implies $b < 1$, so that $S \in \mathscr{C}(\mathsf{X}, \mathsf{Y})$ by Theorem 1.1. First we shall show that the problem can be reduced to the case in which T and A are bounded.

Let us introduce a new norm in $\mathsf{D}(T)$ by

$$|||u||| = (a + \varepsilon)\,\|u\| + (b + \varepsilon)\,\|T u\| \geqq \varepsilon\,\|u\|\,, \tag{5.22}$$

where $\varepsilon > 0$ is fixed. Under this norm $\mathsf{D}(T)$ becomes a Banach space, which we shall denote by $\hat{\mathsf{X}}$ (see Remark 1.4), and we can regard T and A [or rather the restriction of A to $\mathsf{D}(T)$] as operators from $\hat{\mathsf{X}}$ to Y; in this sense they will be denoted by $\hat{T}$ and $\hat{A}$ respectively (Remark 1.5). These belong to $\mathscr{B}(\hat{\mathsf{X}}, \mathsf{Y})$ and

$$\|\hat{T}\| \leqq (b + \varepsilon)^{-1}\,, \quad \|\hat{A}\| \leqq 1 \tag{5.23}$$

in virtue of (1.1) and (5.22).

Since $\mathsf{R}(\hat{T}) = \mathsf{R}(T)$, $\hat{T}$ has closed range. Furthermore, we have the obvious relations

$$\begin{aligned} &\operatorname{nul} \hat{T} = \operatorname{nul} T\,, \quad \operatorname{def} \hat{T} = \operatorname{def} T\,, \quad \operatorname{nul} \hat{S} = \operatorname{nul} S\,, \\ &\operatorname{def} \hat{S} = \operatorname{def} S\,, \quad \mathsf{R}(\hat{S}) = \mathsf{R}(S) \quad \text{where} \quad \hat{S} = \hat{T} + \hat{A}\,. \end{aligned} \tag{5.24}$$

---

[1] This partly strengthens Theorem 2.13. To prove that (2.25) is true with $\hat{\delta}(S, T)$ replaced by $\delta(T, S)$, we note that for any $v \in \mathsf{X}$ there is a $u \in \mathsf{X}$ satisfying (2.22), where δ' is any number such that $\delta(T, S) < \delta' < (1 + \|T\|^2)^{-1/2}$ (see the proof of Lemma 2.12). Then we have $\|A u\| \leqq \delta'(1 + \|T\|^2)^{1/2}$ as in the proof of Lemma 2.11. Hence $\|A v\| \leqq \|A u\| + \|A\|\,\|u - v\| < \delta'(1 + \|T\|^2)^{1/2} + \|A\|\,\delta'$, since $\|u - v\| < \delta'$ by (2.22). Since $\|v\|^2(1 + \|T\|^2) \geqq 1$ by (2.24), it follows that $\|A\| \leqq \delta'(1 + \|T\|^2) + \|A\|\,\delta'(1 + \|T\|^2)^{1/2}$ or $\|A\| \leqq \delta'(1 + \|T\|^2)\,[1 - \delta'(1 + \|T\|^2)^{1/2}]^{-1}$, which gives the required result when $\delta' \to \delta(T, S)$.

Hence $\hat{T}$ is semi-Fredholm, and it suffices to show that $\hat{S}$ is semi-Fredholm and (5.21) holds with S, T replaced by $\hat{S}$, $\hat{T}$.

Let us express $\gamma(\hat{T})$ in terms of $\gamma = \gamma(T)$. By definition $\gamma(\hat{T}) = \inf\|\hat{T}u\|/|||\tilde{u}||| = \inf\|Tu\|/|||\tilde{u}|||$ where $\tilde{u} \in \hat{\mathsf{X}}/\mathsf{N}$, $\mathsf{N} = \mathsf{N}(\hat{T}) = \mathsf{N}(T)$ (N is closed both in X and in $\hat{\mathsf{X}}$). But

$$(5.25)\qquad |||\tilde{u}||| = \inf_{z\in\mathsf{N}} |||u - z||| = \inf_{z\in\mathsf{N}} [(a+\varepsilon)\|u - z\| + (b+\varepsilon)\|T(u-z)\|] = (a+\varepsilon)\|\tilde{u}\| + (b+\varepsilon)\|Tu\|$$

by $Tz = 0$. Hence

$$(5.26)\qquad \gamma(\hat{T}) = \inf_{u\in\mathsf{D}(T)} \frac{\|Tu\|}{(a+\varepsilon)\|\tilde{u}\| + (b+\varepsilon)\|Tu\|} = \frac{\gamma}{(a+\varepsilon)+(b+\varepsilon)\gamma}$$

by the definition $\gamma = \inf\|Tu\|/\|\tilde{u}\|$.

It follows from (5.20) and (5.26) that we can make $\gamma(\hat{T}) > 1$ by choosing ε sufficiently small. Since $\|\hat{A}\| \leqq 1$ by (5.23), we have $\|\hat{A}\| < \gamma(\hat{T})$. Rewriting $\hat{T}$, $\hat{A}$, $\hat{S}$ as T, A, S, respectively, we see that it suffices to prove the theorem in the special case in which all T, A, S belong to $\mathscr{B}(\mathsf{X}, \mathsf{Y})$ and $\|A\| < \gamma(T) = \gamma$.

In this special case, however, the theorem follows immediately from the general stability Theorem 5.17. In fact let $\alpha > 0$ be chosen sufficiently small so that $\|A\| < \gamma/(1+\alpha^2\gamma^2)^{1/2}$. Then $\|\alpha A\| = \alpha\|A\| < \gamma(\alpha T)/(1+\gamma(\alpha T)^2)^{1/2}$ since $\alpha\gamma(T) = \gamma(\alpha T)$. Since, on the other hand, $\hat{\delta}(\alpha S, \alpha T) < \|\alpha A\|$ by (2.27), we see that the assumptions of Theorem 5.17 are satisfied for the pair S, T replaced by αS, αT. It follows that αS, and hence S, is semi-Fredholm and $\operatorname{nul} S = \operatorname{nul}\alpha S \leqq \operatorname{nul}\alpha T = \operatorname{nul} T$ and similarly for $\operatorname{def} S$.

It remains to show that $\operatorname{ind} S = \operatorname{ind} T$. It follows from Theorem 5.17 that this is true at least if $\|A\|$ is sufficiently small. Now we can connect S and T by a continuous path; it suffices to consider the family $T(\varkappa) = T + \varkappa A$, $0 \leqq \varkappa \leqq 1$. Since $\|\varkappa A\| \leqq \|A\| < \gamma$, $T(\varkappa)$ is semi-Fredholm for each $\varkappa$. It follows from Theorem 5.17 that $\operatorname{ind} T(\varkappa)$ is continuous in $\varkappa$. Since it is an integer (including $\pm\infty$), it must be constant for $0 \leqq \varkappa \leqq 1$, showing that $\operatorname{ind} S = \operatorname{ind} T$.

Remark 5.23. Let T and A be operators in a finite-dimensional space X to itself. Then $\operatorname{ind}(T + A)$ is always zero so that the third formula of (5.21) is true without restriction [see I-(3.2), (3.3)]. Suppose that $\operatorname{nul} T = \operatorname{def} T > 0$ and that $A = -\zeta$. Then $\operatorname{nul}(T - \zeta) = \operatorname{def}(T - \zeta) = 0$ for sufficiently small $|\zeta| \neq 0$, for the eigenvalues of T are isolated. This shows that in the first two inequalities of (5.21), the equalities cannot be expected in general.

Example 5.24. Consider the T of Example 5.3 and set $A = -\zeta$. Let us compute $\operatorname{nul}(T + A) = \operatorname{nul}(T - \zeta)$. Let $u = (\xi_k) \in \mathsf{N}(T - \zeta)$. Then $\xi_{k+1} = \zeta\xi_k$ so that $\xi_k = \zeta^{k-1}\xi_1$. If $|\zeta| < 1$, this determines u for any ξ_1. If $|\zeta| \geqq 1$, we have to set

$\xi_1 = 0$ in order that $u \in X$ (we assume $1 < p < \infty$ for simplicity). Thus $\text{nul}(T - \zeta) = 1$ for $|\zeta| < 1$ and $\text{nul}(T - \zeta) = 0$ for $|\zeta| \geqq 1$.

A similar consideration applied to T^* (which is the T of Example 5.4) shows that $\text{nul}(T^* - \zeta) = 0$ for any ζ. On the other hand $T - \zeta$ has closed range for $|\zeta| \neq 1$; this follows from the general result of Theorem 5.22 since $\gamma(T) = \gamma(1) = 1$. Thus $T - \zeta$ is Fredholm for $|\zeta| \neq 1$ and $\text{nul}(T - \zeta) = 1$, $\text{def}(T - \zeta) = 0$, $\text{ind}(T - \zeta) = 1$ for $|\zeta| < 1$ while $\text{nul}(T - \zeta) = \text{def}(T - \zeta) = \text{ind}(T - \zeta) = 0$ for $|\zeta| > 1$.

It should be noted that $T - \zeta$ is semi-Fredholm for no ζ with $|\zeta| = 1$. In fact, if $T - \zeta_0$ were semi-Fredholm for some ζ_0 with $|\zeta_0| = 1$, $\text{ind}(T - \zeta)$ would have to be constant in some neighborhood of ζ_0, contrary to the above result. This shows that the assumption (5.20) of Theorem 5.22 cannot in general be replaced by a weaker one.

Problem 5.25. Let T be as in Example 5.5. If $|\zeta| < 1$, $T - \zeta$ is Fredholm with $\text{nul}(T - \zeta) = \text{def}(T - \zeta) = 1$, $\text{ind}(T - \zeta) = 0$. If $|\zeta| > 1$, $T - \zeta$ is Fredholm with $\text{nul}(T - \zeta) = \text{def}(T - \zeta) = \text{ind}(T - \zeta) = 0$. (Check these results directly whenever possible.) $T - \zeta$ is not semi-Fredholm if $|\zeta| = 1$.

Theorem 5.22 (or the more general Theorem 5.17) is sometimes called the *first stability theorem*. The *second stability theorem* is concerned with a perturbation which need not be restricted in "size" [as in (5.20)] but which is assumed to be relatively compact[1].

Theorem 5.26. *Let $T \in \mathscr{C}(X, Y)$ be semi-Fredholm. If A is a T-compact operator from X to Y, then $S = T + A \in \mathscr{C}(X, Y)$ is also semi-Fredholm with* $\text{ind}\, S = \text{ind}\, T$.

Proof. $S \in \mathscr{C}(X, Y)$ by Theorem 1.11. Again the theorem can be reduced to the special case where $T, A \in \mathscr{B}(X, Y)$ and A is compact. To this end it suffices to introduce the space $\hat{X}$ and the operators $\hat{T}$, $\hat{A}$, $\hat{S}$ as in the proof of Theorem 5.22 [here we have nothing to do with the constants a, b, so we set $a = b = 1$ and $\varepsilon = 0$ in (5.22)]. Then the T-compactness of A implies that $\hat{A}$ is compact (see Remark 1.12).

Assume, therefore, that $T, A \in \mathscr{B}(X, Y)$ and A is compact.

First we assume that $\text{nul}\, T < \infty$ and prove that $\text{nul}'\, S < \infty$; then it follows from Theorem 5.10 that S is semi-Fredholm. To this end it is convenient to apply the result of Theorem 5.11. Suppose that there is a sequence $u_n \in X$ such that $\|u_n\| = 1$ and $S u_n \to 0$; we have to show that $\{u_n\}$ has a convergent subsequence. Since A is compact, there is a subsequence $\{v_n\}$ of $\{u_n\}$ such that $A v_n \to w \in Y$. Then $T v_n = (S - A)\, v_n \to -w$. Since $R(T)$ is closed, $-w \in R(T)$. Thus there is a $u \in X$ such that $-w = Tu$. Then $T(v_n - u) \to 0$. Since $\text{nul}'\, T < \infty$, Theorem 5.11 shows that $\{v_n - u\}$ contains a convergent subsequence. Hence the same is true of $\{u_n\}$, as required.

[1] For the second stability theorem, see ATKINSON [1], FRIEDMAN [1], GOHBERG and KREIN [1], SZ.-NAGY [3], YOOD [1]. This theorem has also been generalized to perturbations by not necessarily relatively compact but "strictly singular" operators; see T. KATO [12], GOHBERG, MARKUS and FEL'DMAN [1], GOLDBERG [1], and other papers cited by GOLDBERG.

Next suppose that $\operatorname{def} T < \infty$. Then $\operatorname{nul} T^* < \infty$. Since T^* is semi-Fredholm and A^* is compact, it follows from what was just proved that $S^* = T^* + A^*$, and therefore S too, is semi-Fredholm.

Once it is known that S is semi-Fredholm, it is easy to prove that $\operatorname{ind} S = \operatorname{ind} T$. Again introduce the family $T(\varkappa) = T + \varkappa A$, $0 \leqq \varkappa \leqq 1$. Since $\varkappa A$ is compact with A, $T(\varkappa)$ is semi-Fredholm for every $\varkappa$. Hence $\operatorname{ind} T(\varkappa)$, being continuous by the first stability theorem, must be constant.

Example 5.27. Let T be as in Example 5.5, namely $T x_0 = 0$, $T x_j = x_{j-1}$, $j = \pm 1, \pm 2, \ldots$. We have seen (Problem 5.25) that $T - \zeta$ is Fredholm for $|\zeta| \neq 1$ and

$$\begin{aligned} &\operatorname{nul}(T - \zeta) = \operatorname{def}(T - \zeta) = 1, \quad |\zeta| < 1, \\ &\operatorname{nul}(T - \zeta) = \operatorname{def}(T - \zeta) = 0, \quad |\zeta| > 1. \end{aligned} \tag{5.27}$$

Let $A \in \mathscr{B}(X)$ be such that $A x_0 = x_1$, $A x_j = 0$, $j \neq 0$. A is compact; it is even degenerate with rank one. Thus it follows from Theorem 5.26 that $T(\varkappa) - \zeta$ is Fredholm for every complex $\varkappa$ and $|\zeta| \neq 1$, where $T(\varkappa) = T + \varkappa A$. But we know that $\Sigma(T(\varkappa))$ is the unit circle $|\zeta| = 1$ for $\varkappa \neq 0$ (see Example 3.8). Hence

$$\operatorname{nul}(T(\varkappa) - \zeta) = \operatorname{def}(T(\varkappa) - \zeta) = 0, \quad |\zeta| \neq 1, \varkappa \neq 0. \tag{5.28}$$

4. Isolated eigenvalues

Let $T \in \mathscr{C}(X)$ and let λ be an isolated point of the spectrum $\Sigma(T)$. Let P be the projection associated with λ (see III-§ 6.5).

Theorem 5.28. *Let T, λ, P be as above. If* $\dim P < \infty$, *$T - \lambda$ is Fredholm and we have* $\operatorname{nul}'(T - \lambda) = \operatorname{nul}(T - \lambda) \leqq \dim P$, $\operatorname{def}'(T - \lambda) = \operatorname{def}(T - \lambda) \leqq \dim P$. *If* $\dim P = \infty$, *we have* $\operatorname{nul}'(T - \lambda) = \operatorname{def}'(T - \lambda) = \infty$.

Proof. T is decomposed according to the decomposition $X = M' \oplus M''$, $M' = PX$, $M'' = (1 - P)X$ (see III-§ 6.4–6.5), and $\lambda \in P(T_{M''})$ so that $\operatorname{nul}'(T_{M''} - \lambda) = \operatorname{def}'(T_{M''} - \lambda) = 0$. On the other hand, $T_{M'} - \lambda$ is bounded and quasi-nilpotent. If $\dim P < \infty$, M' is finite-dimensional and the approximate nullity and deficiency of $T_{M'} - \lambda$ cannot exceed $\dim P$. If $\dim P = \infty$, we have $\operatorname{nul}'(T_{M'} - \lambda) = \operatorname{def}'(T_{M'} - \lambda) = \infty$ by Theorem 5.30 to be proved below. Theorem 5.28 follows easily from these properties of the parts $T_{M'}$ and $T_{M''}$.

Lemma 5.29. *Let $T \in \mathscr{C}(X, Y)$ have closed range with* $\operatorname{nul} T < \infty$. *Then TM is closed for any closed linear manifold M of X.*

Proof. Define $\tilde{X} = X/N$, $N = N(T)$, and $\tilde{T}$ as before (par. 1). Then $TM = \tilde{T}\tilde{M}$, where $\tilde{M}$ is the set of all $\tilde{u} \in \tilde{X}$ such that the coset $\tilde{u}$ contains at least one element of M. Since $\tilde{T}$ has bounded inverse (Theorem 5.2), it suffices to show that $\tilde{M}$ is closed in $\tilde{X}$.

Suppose that $\tilde{u}_n \in \tilde{M}$, $\tilde{u}_n \to \tilde{u} \in \tilde{X}$. This implies that $\operatorname{dist}(u_n - u, N) \to 0$, and there are $z_n \in N$ with $u_n - u - z_n \to 0$. Since we may assume that

$u_n \in \mathsf{M}$ and $\mathsf{M} + \mathsf{N}$ is closed by $\dim \mathsf{N} < \infty$ (Lemma III-1.9), it follows that $u \in \mathsf{M} + \mathsf{N}$ or $\tilde{u} \in \tilde{\mathsf{M}}$. This proves that $\tilde{\mathsf{M}}$ is closed.

Theorem 5.30. *Let $T \in \mathscr{B}(\mathsf{X})$ be quasi-nilpotent. If* $\dim \mathsf{X} = \infty$, *then* $\text{nul}' T = \text{def}' T = \infty$.

Proof. In view of Theorem 5.10, it suffices to show that we have necessarily $\dim \mathsf{X} < \infty$ if $\mathsf{R}(T)$ is closed and $\text{nul}\, T < \infty$ or if $\mathsf{R}(T)$ is closed and $\text{def}\, T < \infty$. Since the latter case can be reduced to the former by considering T^* instead of T (T^* is quasi-nilpotent with T), it suffices to prove that $\dim \mathsf{X} < \infty$ if $\mathsf{R}(T)$ is closed and $\text{nul}\, T < \infty$.

Since $\mathsf{R} = \mathsf{R}(T) = T\mathsf{X}$ is closed, it follows from Lemma 5.29 successively that all $T^n \mathsf{X}$ are closed, $n = 1, 2, \ldots$.

Since $\mathsf{X} \supset T\mathsf{X} \supset T^2\mathsf{X} \supset \cdots$, the intersections of the $T^n\mathsf{X}$ with $\mathsf{N} = \mathsf{N}(T)$ also form a descending sequence. Since the latter are subspaces of the finite-dimensional space N, $\mathsf{N}_0 = \mathsf{N} \cap T^n\mathsf{X}$ must be independent of n for sufficiently large n, say for $n \geqq m$. Set $\mathsf{X}_0 = T^m\mathsf{X}$. Since X_0 is a closed linear manifold of X and $T\mathsf{X}_0 \subset \mathsf{X}_0$, the part T_0 of T in X_0 is defined. We have $\mathsf{N}(T_0) = \mathsf{N} \cap \mathsf{X}_0 = \mathsf{N}_0$. Furthermore, $\mathsf{N}_0 \subset T^{m+n}\mathsf{X} = T^n\mathsf{X}_0 = T_0^n\mathsf{X}_0$ for all $n = 1, 2, \ldots$. Thus $\mathsf{N}(T_0)$ is contained in all $\mathsf{R}(T_0^n)$.

This in turn implies that $\mathsf{N}(T_0^n) \subset \mathsf{R}(T_0)$[1] for all n. Let us prove this by induction. Since this is known for $n = 1$, assume that it has been proved for n and let $T_0^{n+1} u = 0$; we have to show that $u \in \mathsf{R}(T_0)$. $T_0^{n+1} u = 0$ implies $T_0^n u \in \mathsf{N}_0 \subset \mathsf{R}(T_0^{n+1})$ or $T_0^n u = T_0^{n+1} v$ for some $v \in \mathsf{X}_0$. Then $u - T_0 v \in \mathsf{N}(T_0^n) \subset \mathsf{R}(T_0)$, which implies $u \in \mathsf{R}(T_0)$.

Since $\dim \mathsf{N}_0 < \infty$, there is a complementary subspace M_0 of N_0 in $\mathsf{X}_0: \mathsf{X}_0 = \mathsf{N}_0 \oplus \mathsf{M}_0$. T_0 maps M_0 onto $\mathsf{R}(T_0) = \mathsf{R}_0$ one to one. Let S_0 be the inverse map of R_0 onto M_0; S_0 is bounded by Problem III-5.21. (Note that $\mathsf{R}_0 = T_0\mathsf{X}_0 = T\mathsf{X}_0 = T^{m+1}\mathsf{X}$ is closed.)

Let $u_0 \in \mathsf{N}_0$. Since $\mathsf{N}_0 \subset \mathsf{R}_0$, $u_1 = S_0 u_0$ is defined. Since $T_0^2 u_1 = T_0 u_0 = 0$, $u_1 \in \mathsf{N}(T_0^2) \subset \mathsf{R}_0$. Hence $u_2 = S_0 u_1$ is defined. Since $T_0^3 u_2 = T_0^2 u_1 = 0$, $u_2 \in \mathsf{N}(T_0^3) \subset \mathsf{R}_0$ and $u_3 = S_0 u_2$ is defined. In this way we see that there is a sequence $u_n \in \mathsf{R}_0$ such that $u_n = S_0 u_{n-1}$, $n = 1, 2, \ldots$. Thus $u_n = S_0^n u_0$ and $\|u_n\| \leqq \|S_0\|^n \|u_0\|$. On the other hand, we have $u_0 = T_0^n u_n = T^n u_n$ and so $\|u_0\| \leqq \|T^n\| \|S_0\|^n \|u_0\|$. But as $\lim \|T^n\|^{1/n} = 0$ by hypothesis, we have $\|T^n\| \|S_0\|^n < 1$ for sufficiently large n. This gives that $u_0 = 0$. Since $u_0 \in \mathsf{N}_0$ was arbitrary, we conclude that $\mathsf{N}_0 = 0$, $\mathsf{M}_0 = \mathsf{X}_0$.

This implies that T_0^{-1} exists and is equal to S_0. Hence $\|u\| \leqq \|S_0\| \|T_0 u\|$ for any $u \in \mathsf{X}_0$ and so $\|u\| \leqq \|S_0\|^n \|T_0^n u\|$. Since $\|T_0^n u\| \leqq \|T^n\| \|u\|$ as above, the same argument as above leads to the result that $u = 0$, $\mathsf{X}_0 = 0$, that is, $T^m\mathsf{X} = 0$.

[1] The sequences of subspaces $\mathsf{N}(T^n)$ and $\mathsf{R}(T^n)$ have been studied in various problems. See DUNFORD and SCHWARTZ [1], p. 556, HUKUHARA [1], KANIEL and SCHECHTER [1].

Then $\mathsf{X} \subset \mathsf{N}(T^m)$ and X must be finite-dimensional, for it is easily seen that $\dim \mathsf{N}(T^m) \leqq m \dim \mathsf{N}$. This completes the proof of Theorem 5.30.

5. Another form of the stability theorem

According to Theorem 5.17, the nullity and deficiency of a semi-Fredholm operator do not increase under a small perturbation. It is in general not easy to see when these quantities are conserved. But a sharper result is obtained if the perturbation is restricted to have the form $\varkappa A$ with a fixed A.

Theorem 5.31. *Let $T \in \mathscr{C}(\mathsf{X}, \mathsf{Y})$ be semi-Fredholm and let A be a T-bounded operator from X to Y. Then $T + \varkappa A$ is semi-Fredholm and* $\mathrm{nul}(T + \varkappa A)$, $\mathrm{def}(T + \varkappa A)$ *are constant for sufficiently small* $|\varkappa| > 0$.

Proof. It suffices to consider the case in which both T and A belong to $\mathscr{B}(\mathsf{X}, \mathsf{Y})$; the general case can be reduced to this as in the proof of Theorem 5.22.

I. First we assume $\mathrm{nul}\, T < \infty$. We define sequences $\mathsf{M}_n \subset \mathsf{X}$ and $\mathsf{R}_n \subset \mathsf{Y}$ successively by[1]

$$(5.29) \quad \mathsf{M}_0 = \mathsf{X},\ \mathsf{R}_0 = \mathsf{Y},\ \mathsf{M}_n = A^{-1}\, \mathsf{R}_n,\ \mathsf{R}_{n+1} = T\mathsf{M}_n,\ n = 0, 1, 2, \ldots;$$

here $A^{-1}\, \mathsf{R}$ denotes the inverse image under A of $\mathsf{R} \subset \mathsf{Y}$. We have

$$(5.30) \qquad \mathsf{M}_0 \supset \mathsf{M}_1 \supset \mathsf{M}_2 \supset \cdots, \quad \mathsf{R}_0 \supset \mathsf{R}_1 \supset \mathsf{R}_2 \supset \cdots,$$

as is easily seen by induction. All the M_n and R_n are closed linear manifolds. This can also be proved by induction; if R_n is closed, M_n is closed as the inverse image under a continuous map A of a closed set R_n, and then $\mathsf{R}_{n+1} = T\mathsf{M}_n$ is closed by Lemma 5.29.

Let $\mathsf{X}' = \bigcap_n \mathsf{M}_n$ and $\mathsf{Y}' = \bigcap_n \mathsf{R}_n$; X' and Y' are closed. Let T' be the restriction of T with domain X'. If $u' \in \mathsf{X}'$, then $u' \in \mathsf{M}_n$ and $T'u' = Tu' \in T\mathsf{M}_n = \mathsf{R}_{n+1}$ for all n so that $\mathsf{R}(T') \subset \mathsf{Y}'$. We shall show that $\mathsf{R}(T') = \mathsf{Y}'$.

Let $v' \in \mathsf{Y}'$. Since $v' \in \mathsf{R}_{n+1} = T\mathsf{M}_n$ for each n, the inverse image $T^{-1}\{v'\}$ has a common element with M_n. But $T^{-1}\{v'\}$ is an inhomogeneous linear manifold of the form $u + \mathsf{N}(T)$. Since $\dim \mathsf{N}(T) = \mathrm{nul}\, T < \infty$, the $T^{-1}\{v'\} \cap \mathsf{M}_n$ form a descending sequence of finite-dimensional inhomogeneous linear manifolds which are not empty. Hence $T^{-1}\{v'\} \cap \mathsf{M}_n$ does not depend on n for sufficiently large n and, therefore, must coincide with $T^{-1}\{v'\} \cap \mathsf{X}'$, which is thus not empty. Let u' be one of its elements. Then $u' \in \mathsf{X}'$ and $T'u' = Tu' = v'$. This shows that $\mathsf{R}(T') = \mathsf{Y}'$.

T' may be regarded as an operator on X' to Y': $T' \in \mathscr{B}(\mathsf{X}', \mathsf{Y}')$. Let A' be the restriction of A with domain X'. Since $u' \in \mathsf{X}'$ implies $u' \in \mathsf{M}_n = A^{-1}\, \mathsf{R}_n$ for all n, $Au' \in \mathsf{R}_n$ and hence $Au' \in \mathsf{Y}'$. Thus A' too

[1] When $\mathsf{Y} = \mathsf{X}$ and $A = 1$, $\mathsf{M}_n = \mathsf{R}_n$ is equal to $\mathsf{R}(T^n)$.

can be regarded as an operator on X' to Y': $A' \in \mathscr{B}(\mathsf{X}', \mathsf{Y}')$[1]. We can now apply Theorem 5.17 to the pair T', $\varkappa A'$. It follows that $\operatorname{def}(T' + \varkappa A') = \operatorname{def} T' = 0$, $\operatorname{nul}(T' + \varkappa A') = \operatorname{ind}(T' + \varkappa A') = \operatorname{ind} T' = \operatorname{nul} T'$ for sufficiently small $|\varkappa|$, for $\operatorname{def} T' = 0$. Thus both $\operatorname{nul}(T' + \varkappa A')$ and $\operatorname{def}(T' + \varkappa A')$ are constant for small $|\varkappa|$.

On the other hand, we have

$$\mathsf{N}(T + \varkappa A) = \mathsf{N}(T' + \varkappa A') \quad \text{for} \quad \varkappa \neq 0 . \tag{5.31}$$

In fact, let $u \in \mathsf{N}(T + \varkappa A)$. Then $Tu = -\varkappa A u$, and it follows by induction that $u \in \mathsf{M}_n$ for all n and so $u \in \mathsf{X}'$.

(5.31) implies that $\operatorname{nul}(T + \varkappa A) = \operatorname{nul}(T' + \varkappa A')$ is constant for small $|\varkappa| > 0$. Since $\operatorname{ind}(T + \varkappa A)$ is constant, it follows that $\operatorname{def}(T + \varkappa A)$ is also constant.

II. The case $\operatorname{def} T < \infty$ can be reduced to the case I by considering the adjoints T^*, A^*; see Theorem 5.13.

Problem 5.32. Let $\operatorname{nul} T < \infty$. In order that in Theorem 5.31 $\operatorname{nul}(T + \varkappa A)$ and $\operatorname{def}(T + \varkappa A)$ be constant including $\varkappa = 0$, it is necessary and sufficient that $\mathsf{N}(T) \subset \mathsf{M}_n$ for all n.[2]

6. Structure of the spectrum of a closed operator

Let $T \in \mathscr{C}(\mathsf{X})$. A complex number ζ belongs to the resolvent set $\mathrm{P}(T)$ of T if and only if $\operatorname{nul}(T - \zeta) = \operatorname{def}(T - \zeta) = 0$. This suggests introducing the functions

$$\begin{aligned} \nu(\zeta) &= \operatorname{nul}(T - \zeta) , \quad & \mu(\zeta) &= \operatorname{def}(T - \zeta) , \\ \nu'(\zeta) &= \operatorname{nul}'(T - \zeta) , \quad & \mu'(\zeta) &= \operatorname{def}'(T - \zeta) , \end{aligned} \tag{5.32}$$

and classifying the complex numbers ζ according to the values of these functions[3].

Let Δ be the set of all complex numbers ζ such that $T - \zeta$ is semi-Fredholm[4]. Let Γ be the complementary set of Δ. It follows from Theorem 5.17 that Δ is open and Γ is closed. Theorem 5.10 shows that

$$\nu'(\zeta) = \mu'(\zeta) = \infty \quad \text{if and only if} \quad \zeta \in \Gamma . \tag{5.33}$$

[1] This means that X', Y' constitute an "invariant pair" of subspaces for both T and A. If in particular $\mathsf{Y} = \mathsf{X}$ and $A = 1$, we have $\mathsf{Y}' = \mathsf{X}'$ because $\mathsf{M}_n = \mathsf{R}_n$ for all n, and X' is invariant under T. A similar but more detailed decomposition of X, Y is considered in T. KATO [12] and GAMELIN [1].

[2] When this condition is not satisfied, there is a decomposition of the spaces X, Y into "invariant pairs" of subspaces; see the preceding footnote.

[3] The results of this section have many applications in spectral theory. An application to the theory of integral equations of Wiener-Hopf type is given in GOHBERG and KREIN [1].

[4] Δ is called the semi-Fredholm domain for T. We can define in the same way the Fredholm domain Δ_F as the subset of Δ consisting of all ζ such that $\nu(\zeta) < \infty$ and $\mu(\zeta) < \infty$.

In general Δ is the union of a countable number of components (connected open sets) Δ_n. By Theorem 5.17, $\operatorname{ind}(T-\zeta)=\nu(\zeta)-\mu(\zeta)$ is constant in each Δ_n. According to Theorem 5.31 (applied with $A=1$, $T\to T-\zeta_0$, $\varkappa=\zeta_0-\zeta$), both $\nu(\zeta)$ and $\mu(\zeta)$ are constant in each Δ_n except for an isolated set of values of ζ. Denoting by ν_n, μ_n these constant values and by λ_{nj} these exceptional points in Δ_n, we have

$$\begin{aligned}(5.34)\qquad &\nu(\zeta)=\nu_n,\ \mu(\zeta)=\mu_n,\ \zeta\in\Delta_n,\ \zeta\neq\lambda_{nj},\\ &\nu(\lambda_{nj})=\nu_n+r_{nj},\ \mu(\lambda_{nj})=\mu_n+r_{nj},\ 0<r_{nj}<\infty.\end{aligned}$$

If $\nu_n=\mu_n=0$, Δ_n is a subset of $\mathrm{P}(T)$ except for the λ_{nj}, which are isolated points of $\Sigma(T)$; the λ_{nj} are isolated eigenvalues of T with finite *(algebraic)* multiplicities, as is seen from Theorem 5.28, and the r_{nj} are their *geometric* multiplicities. In the general case (in which one or both of ν_n, μ_n are positive), the λ_{nj} are also eigenvalues of T and behave like "isolated eigenvalues" (although they are in reality not isolated eigenvalues if $\nu_n>0$), in the sense that their geometric multiplicities are larger by r_{nj} than other eigenvalues in their immediate neighborhood.

We shall call Γ the *essential spectrum* of T and denote it by $\Sigma_e(T)$[1]. It is a subset of $\Sigma(T)$ and consists of all ζ such that either $\mathsf{R}(T-\zeta)$ is not closed or $\mathsf{R}(T-\zeta)$ is closed but $\nu(\zeta)=\mu(\zeta)=\infty$. A simple characterization of $\Sigma_e(T)$ is given by (5.33).

The boundary of Δ and the boundaries of the Δ_n are subsets of $\Sigma_e(T)$. If the number of the components Δ_n of Δ is larger than one (that is, if Δ is not connected), then $\Sigma_e(T)$ necessarily contains an uncountable number of points. In other words, Δ consists of a single component and hence $\nu(\zeta)$, $\mu(\zeta)$ are constant except for an isolated set of points $\zeta=\lambda_j$, provided that $\Sigma_e(T)$ is at most countable. Both of these constant values must be zero if $T\in\mathscr{B}(X)$, for $\mathrm{P}(T)$ is then not empty and hence must coincide with Δ. Thus we have

Theorem 5.33. *An operator $T\in\mathscr{B}(X)$ with at most countable essential spectrum $\Sigma_e(T)$ has at most countable spectrum $\Sigma(T)$, and any point of $\Sigma(T)$ not belonging to $\Sigma_e(T)$ is an isolated eigenvalue with finite (algebraic) multiplicity*[2].

Remark 5.34. A compact operator is a special case of an operator of Theorem 5.33, with at most one point 0 in $\Sigma_e(T)$.

[1] There is considerable divergence in the literature concerning the definition of $\Sigma_e(T)$. $\Sigma_e(T)$ by our definition is fairly small. Another definition (see Wolf [3]) of $\Sigma_e(T)$ is "the complementary set of the Fredholm domain Δ_F (see preceding footnote)"; it is larger than our $\Sigma_e(T)$ by those components Δ_n of Δ for which one of ν_n and μ_n is infinite. One could add to it all other components of Δ except the components of the resolvent set, obtaining another possible definition of $\Sigma_e(T)$ (see Browder [2]). See also Schechter [1].

[2] It follows that $\Sigma_e(T)$ is not empty if $T\in\mathscr{B}(X)$ and $\dim X=\infty$.

Theorem 5.35. *The essential spectrum is conserved under a relatively compact perturbation. More precisely, let $T \in \mathscr{C}(X)$ and let A be T-compact. Then T and $T + A$ have the same essential spectrum*[1].

Proof. It suffices to note that $S = T + A \in \mathscr{C}(X)$ and A is S-compact by Theorem 1.11. By virtue of Theorem 5.26, then, $T - \zeta$ is semi-Fredholm if and only if $S - \zeta$ is.

Example 5.36.[2] In Example 5.24, we have $\nu(\zeta) = 1$, $\mu(\zeta) = 0$ for $|\zeta| < 1$ and $\nu(\zeta) = \mu(\zeta) = 0$ for $|\zeta| > 1$, $R(T)$ being closed for $|\zeta| \neq 1$. Hence there are exactly two components of Δ: the interior and the exterior of the unit circle, and the unit circle $|\zeta| = 1$ is exactly the essential spectrum. A similar result holds for Example 5.25.

Problem 5.37. Any boundary point of $P(T)$ belongs to $\Sigma_e(T)$ unless it is an isolated point of $\Sigma(T)$.

Problem 5.38. Two operators $T, S \in \mathscr{C}(X)$ have the same essential spectrum if there is a $\zeta \in P(T) \cap P(S)$ such that $(S - \zeta)^{-1} - (T - \zeta)^{-1}$ is compact[3].

§ 6. Degenerate perturbations

1. The Weinstein-Aronszajn determinants

In the case of degenerate (or, more generally, relatively degenerate) perturbations, there are certain explicit formulas, deduced by Weinstein and Aronszajn[4], by which the change of the eigenvalues is related to the zeros and poles of a certain meromorphic function in the form of a determinant. This is a quantitative refinement of some of the results of the stability theorems. We now turn to the derivation of these formulas.

Here we have usually two types of problems. One is concerned with an operator of the form $T + A$ in which T is the unperturbed operator and A is the perturbation assumed to be *relatively degenerate with respect to T*. The other is concerned with an operator of the form PTP, where T is the unperturbed operator and P is a projection with a finite deficiency (or, equivalently, with a finite nullity). But the second problem can be reduced to the first, as we shall show below.

---

[1] This theorem remains valid if we adopt the second definition of $\Sigma_e(T)$ (as the complementary set of the Fredholm domain Δ_F), for $T - \zeta$ is Fredholm if and only if $S - \zeta$ is Fredholm. The theorem was proved by Weyl [1] for selfadjoint operators; see also Hartman [1].

[2] For the determination of the essential spectra of differential operators, see Balslev and Gamelin [1], Kreith and Wolf [1], Rota [1], Wolf [3], [4].

[3] Then $(T - \zeta)^{-1}$ and $(S - \zeta)^{-1}$ have the same essential spectrum by Theorem 5.35. But $\Sigma_e((T - \zeta)^{-1})$ is obtained from $\Sigma_e(T)$ by the transformation $\lambda \to (\lambda - \zeta)^{-1}$ (just as for the spectrum). For the application of Problem 5.38 to differential operators, see Birman [5].

[4] See Weinstein [1], [2], [3], Aronszajn [1], [2], Aronszajn and Weinstein [1], [2]. For degenerate perturbations see also Foguel [2], Kleinecke [1], [2], I. M. Lifšic [1], [2], [3], Wolf [2].

Let $T \in \mathscr{C}(\mathsf{X})$, where X is a Banach space, and let A be an operator in X relatively degenerate with respect to T (Remark 1.13). This means that A is T-bounded and $\mathsf{R}(A)$ is finite-dimensional. For any $\zeta \in \mathsf{P}(T)$, $A(T-\zeta)^{-1}$ is then a degenerate operator belonging to $\mathscr{B}(\mathsf{X})$ and

$$(6.1)\qquad \omega(\zeta) = \omega(\zeta; T, A) = \det(1 + A(T-\zeta)^{-1}) = \det[(T + A - \zeta)(T-\zeta)^{-1}]$$

is defined (see III-§ 4.3). $\omega(\zeta; T, A)$ is called the *W-A determinant* (of the first kind) associated with T and A.

If A is not only relatively but absolutely degenerate, Au can be expressed in the form

$$(6.2)\qquad A u = \sum_{j=1}^{m} (u, e_j)\, x_j\,, \quad x_j \in \mathsf{X}\,, \quad e_j \in \mathsf{X}^*$$

[see III-(4.6)]. Then $(R(\zeta) = (T-\zeta)^{-1})$

$$(6.3)\qquad A(T-\zeta)^{-1} u = \sum (R(\zeta)\, u, e_j)\, x_j = \sum (u, R(\zeta)^* e_j)\, x_j$$

and [see III-(4.13)]

$$(6.4)\qquad \omega(\zeta; T, A) = \det\big(\delta_{jk} + (x_j, R(\zeta)^* e_k)\big) = \det\big(\delta_{jk} + (R(\zeta)\, x_j, e_k)\big)\,.$$

If A is only relatively degenerate, (6.2) has to be replaced by a modified expression of the form

$$(6.5)\qquad A u = \sum_{j=1}^{m} ((T-\zeta_0)\, u, f_j)\, x_j\,, \quad u \in \mathsf{D}(T)\,, \quad f_j \in \mathsf{X}^*\,,$$

where $\zeta_0 \in \mathsf{P}(T)$ is fixed. (6.5) is obtained by applying (6.2) to A replaced by the degenerate operator $A(T-\zeta_0)^{-1}$. As above, this leads to the expression

$$(6.6)\qquad \omega(\zeta; T, A) = \det\big(\delta_{jk} + ((T-\zeta_0)(T-\zeta)^{-1} x_j, f_k)\big) = \det\big(\delta_{jk} + (x_j, f_k) + (\zeta-\zeta_0)((T-\zeta)^{-1} x_j, f_k)\big)\,.$$

Thus $\omega(\zeta; T, A)$ is a meromorphic function of ζ in any domain of the complex plane consisting of points of $\mathsf{P}(T)$ and of isolated eigenvalues of T with finite (algebraic) multiplicities.

The *W-A determinant of the second kind*, on the other hand, is associated with a $T \in \mathscr{C}(\mathsf{X})$ and a projection P in X such that $\mathsf{N} = \mathsf{N}(P) = (1-P)\,\mathsf{X}$ is finite-dimensional and contained in $\mathsf{D}(T)$. Let $\{x_1, \ldots, x_m\}$ be a basis of N and let $e_1, \ldots, e_m \in \mathsf{N}^* = (1-P^*)\,\mathsf{X}^*$ form a biorthogonal set with $\{x_j\}$:

$$(6.7)\qquad (x_j, e_k) = \delta_{jk}\,, \quad j, k = 1, \ldots, m\,.$$

The W-A determinant associated with T and P is now defined by

$$(6.8)\qquad \omega(\zeta) = \omega_P(\zeta; T) = \det((T-\zeta)^{-1} x_j, e_k)\,.$$

$\omega_P(\zeta; T)$ is thus defined in terms of a basis of N, but it is independent of the choice of the basis. We shall prove this by showing that it is, apart from a scalar factor, a special case of the W-A determinant of the first kind associated with T and A, where A is given by

$$A = PTP - T = -PT(1-P) - (1-P)T . \tag{6.9}$$

A is relatively degenerate with respect to T, for $(1-P)T$ is T-bounded, $PT(1-P) \in \mathscr{B}(\mathsf{X})$ (see Problem III-5.22) and these operators have finite rank.

The relation to be proved is

$$(-\zeta)^m \omega_P(\zeta; T) = \omega(\zeta; T, A) . \tag{6.10}$$

To prove this, we start from the identity

$$PTP - \zeta = (1 + \zeta^{-1} PT(1-P))(PT - \zeta), \quad \zeta \neq 0 . \tag{6.11}$$

Since $T + A = PTP$, we see from (6.1) and (6.11) that

$$\begin{aligned} \omega(\zeta; T, A) &= \det((PTP - \zeta)(T-\zeta)^{-1}) \\ &= \det((1 + \zeta^{-1} PT(1-P))(PT-\zeta)(T-\zeta)^{-1}) \\ &= \det(1 + \zeta^{-1} PT(1-P)) \det((PT-\zeta)(T-\zeta)^{-1}) \end{aligned} \tag{6.12}$$

[note III-(4.14) and (6.13) below]. Here the first determinant on the right is equal to one, for $PT(1-P)$ is nilpotent (its square is zero) (see problem III-4.19). To calculate the second determinant, we note that

$$(PT - \zeta)(T-\zeta)^{-1} = 1 - (1-P)\, T(T-\zeta)^{-1} \tag{6.13}$$

where the second term on the right is degenerate and $(1-P)\, T(T-\zeta)^{-1} u = \sum (T(T-\zeta)^{-1} u, e_j) x_j$. Hence the determinant of (6.13) is equal to [note the biorthogonality (6.7) and that $1 - P = \sum (\ , e_j) x_j$]

$$\begin{aligned} \det(\delta_{jk} - (T(T-\zeta)^{-1} x_j, e_k)) &= \det(-\zeta((T-\zeta)^{-1} x_j, e_k)) \\ &= (-\zeta)^m \det(((T-\zeta)^{-1} x_j, e_k)) = (-\zeta)^m \omega_P(\zeta; T) . \end{aligned}$$

This proves (6.10).

We have assumed above that $\zeta \neq 0$. But since both $\omega(\zeta; T, A)$ and $\omega_P(\zeta; T)$ are meromorphic, (6.10) is true even for $\zeta = 0$ if they are defined there.

2. The W-A formulas

To state the W-A formulas we have to introduce certain functions. Let $\phi(\zeta)$ be a numerical meromorphic function defined in a domain Δ of the complex plane. We define the *multiplicity function* $\nu(\zeta; \phi)$ of ϕ by

$$\nu(\zeta; \phi) = \begin{cases} k, & \text{if } \zeta \text{ is a zero of } \phi \text{ of order } k, \\ -k, & \text{if } \zeta \text{ is a pole of } \phi \text{ of order } k, \\ 0 & \text{for other } \zeta \in \Delta . \end{cases} \tag{6.14}$$

Thus $\nu(\zeta;\phi)$ takes the values $0, \pm 1, \pm 2, \ldots$ or $+\infty$, but we have the alternatives: either $\nu(\zeta;\phi)$ is finite for all $\zeta \in \Delta$ or $\nu(\zeta;\phi) = +\infty$ identically. (The latter case occurs if and only if $\phi \equiv 0$).

Also we define the multiplicity function $\tilde{\nu}(\zeta; T)$ for a $T \in \mathscr{C}(X)$ by

$$\tilde{\nu}(\zeta; T) = \begin{cases} 0 & \text{if } \zeta \in \mathrm{P}(T), \\ \dim P & \text{if } \zeta \text{ is an isolated point of } \Sigma(T), \\ +\infty & \text{in all other cases,} \end{cases} \tag{6.15}$$

where P is the projection associated with the isolated point of $\Sigma(T)$ (see III-§ 6.5). Thus $\tilde{\nu}(\zeta; T)$ is defined for all complex numbers ζ and takes on the values $0, 1, 2, \ldots$ or $+\infty$.

Problem 6.1. $\tilde{\nu}(\zeta; T) \geqq \mathrm{nul}(T - \zeta)$.

We now state the main theorems of this section, the proof of which will be given in par. 3.

Theorem 6.2. *Let $T \in \mathscr{C}(X)$, let A be a T-degenerate operator*[1] *in X and let $\omega(\zeta) = \omega(\zeta; T, A)$ be the associated W-A determinant of the first kind. If Δ is a domain of the complex plane consisting of points of $\mathrm{P}(T)$ and of isolated eigenvalues of T with finite multiplicities, $\omega(\zeta)$ is meromorphic in Δ and we have for $S = T + A$*

$$\tilde{\nu}(\zeta; S) = \tilde{\nu}(\zeta; T) + \nu(\zeta; \omega), \quad \zeta \in \Delta. \tag{6.16}$$

(6.16) will be called the *first W-A formula*. Several remarks are in order concerning this formula. In the first place, $\tilde{\nu}(\zeta; T)$ is finite for all $\zeta \in \Delta$ by hypothesis. Since $\nu(\zeta;\omega)$ is either finite for all $\zeta \in \Delta$ or identically equal to $+\infty$, we have the alternatives: either $\tilde{\nu}(\zeta; S)$ is finite for all $\zeta \in \Delta$ or $\tilde{\nu}(\zeta; S) = +\infty$ identically. In the first case Δ is contained in $\mathrm{P}(S)$ with the exception of at most countably many eigenvalues with finite multiplicities, whereas in the second case Δ is wholly contained in $\Sigma(S)$. That the second alternative cannot in general be excluded will be shown below by an example. We shall also give some criteria for the absence of such a singular case.

Suppose now that the first alternative occurs. (6.16) implies that each zero λ of ω of order k is an eigenvalue of S with multiplicity exceeding by k the multiplicity of λ as an eigenvalue of T (if λ is not an eigenvalue of T it is considered an eigenvalue of multiplicity zero). Similarly, each pole λ of ω of order k reduces by k the multiplicity of λ as an eigenvalue. Such a pole cannot occur unless λ is an eigenvalue of T with

---

[1] For a generalization of this theorem to a not necessarily T-degenerate operator A (but in a Hilbert space), see KURODA [5].

multiplicity $\geqq k$. In this way the change of isolated eigenvalues of T by the addition of A is completely determined by the W-A determinant $\omega(\zeta)$.

Theorem 6.3. *Let* $T \in \mathscr{C}(\mathsf{X})$ *and let* P *be a projection in* X *such that* $(1-P)\mathsf{X}$ *is finite-dimensional. If* $\omega(\zeta) = \omega_P(\zeta; T)$ *is the associated W-A determinant of the second kind, the relation* (6.16) *holds for* $S = PTP$ *with the exception of* $\zeta = 0$.

In this case (6.16) will be called the *second W-A formula*. Since we know that $\omega_P(\zeta; T)$ is a special case of $\omega(\zeta; T, A)$, where $S = T + A$, except for a scalar factor $(-\zeta)^m$ [see (6.10)], Theorem 6.3 is a direct consequence of Theorem 6.2.

3. Proof of the W-A formulas

Since Theorem 6.3 is a consequence of Theorem 6.2 as remarked above, it suffices to prove the latter.[1] There are two cases to be considered separately.

Suppose first that $\omega(\zeta) = 0$ identically. By (6.1) this implies that the operator $(T + A - \zeta)(T - \zeta)^{-1}$ has an eigenvalue zero for all $\zeta \in \Delta$. Hence ζ is an eigenvalue of $S = T + A$ for all $\zeta \in \Delta \cap \mathrm{P}(T)$. By definition this implies that $\tilde{\nu}(\zeta; S) = +\infty$ identically. Since $\nu(\zeta; \omega) = +\infty$ identically in this case, (6.16) is true.

We next consider the case where $\omega(\zeta)$ is not identically zero. We shall show that Δ belongs to $\mathrm{P}(S)$ except for countably many points and that the following identity holds:[2]

$$\frac{1}{\omega(\zeta)} \frac{d}{d\zeta} \omega(\zeta) = \operatorname{tr}((T-\zeta)^{-1} - (S-\zeta)^{-1}) . \tag{6.17}$$

The required result (6.16) is obtained by integrating (6.17) along a small circle Γ enclosing a given point λ of Δ; recall that

$$(2\pi i)^{-1} \int_\Gamma \omega'(\zeta)\, \omega(\zeta)^{-1}\, d\zeta = \nu(\lambda; \omega) ,$$ [3]

that

$$-\frac{1}{2\pi i} \int_\Gamma (T-\zeta)^{-1}\, d\zeta = P \tag{6.18}$$

is the eigenprojection for the eigenvalue λ of T, and that $\tilde{\nu}(\zeta; T) = \dim P = \operatorname{tr} P$. Note further that $(T-\zeta)^{-1} - (S-\zeta)^{-1} = (S-\zeta)^{-1} \times A(T-\zeta)^{-1}$ is degenerate.

---

[1] The proof given here is due to KURODA [5]. For other proofs see ARONSZAJN [1], GOULD [1].

[2] This identity was proved by KREIN [6] and KURODA [5].

[3] See KNOPP [1], p. 131.

To prove (6.17) we consider a $\zeta \in \mathrm{P}(T)$ such that $\omega(\zeta) \neq 0$. We have [see I-(5.60)[1]]

$$(6.19)\quad \omega(\zeta) = \det(1 + B(\zeta)) = \exp(\operatorname{tr} \log(1 + B(\zeta))),\ B(\zeta) = A(T-\zeta)^{-1}.$$

That $\omega(\zeta) \neq 0$ implies that $B(\zeta)$ has no eigenvalue -1, and the logarithmic function in (6.19) is defined by

$$(6.20)\qquad \log(1 + B(\zeta)) = \frac{1}{2\pi i} \int_C \log(1+z)\,(z - B(\zeta))^{-1}\,dz\,,$$

where C is a closed curve enclosing all the eigenvalues of the degenerate operator $B(\zeta)$ but excluding the point -1 [see I-(5.57)].

It follows from (6.19) and (6.20) that ($' = d/d\zeta$)

$$(6.21)\quad \frac{\omega'(\zeta)}{\omega(\zeta)} = \frac{d}{d\zeta} \operatorname{tr} \log(1 + B(\zeta))$$

$$= \frac{1}{2\pi i} \int_C \operatorname{tr}\,[\log(1+z)\,(z - B(\zeta))^{-1} B'(\zeta)\,(z - B(\zeta))^{-1}]\,dz \qquad \text{[see I-(4.28)]}$$

$$= \frac{1}{2\pi i} \int_C \operatorname{tr}\,[\log(1+z)\,(z - B(\zeta))^{-2} B'(\zeta)]\,dz$$

$$= \frac{1}{2\pi i} \operatorname{tr} \int_C (1+z)^{-1} (z - B(\zeta))^{-1} B'(\zeta)\,dz$$

$$= \operatorname{tr}\,[(1 + B(\zeta))^{-1} B'(\zeta)] \qquad \text{[see I-(5.53)]}\,.$$

Here we have used the formula $\operatorname{tr} AB = \operatorname{tr} BA$, done an integration by parts, and changed the order of the integral and trace signs when possible. Substituting $B(\zeta) = A(T-\zeta)^{-1}$ and $B'(\zeta) = A(T-\zeta)^{-2}$, the last member of (6.21) becomes

$$\operatorname{tr}\,[(T-\zeta)^{-1}(1+B(\zeta))^{-1} A(T-\zeta)^{-1}] = \operatorname{tr}\,[(T+A-\zeta)^{-1} A(T-\zeta)^{-1}]$$
$$= \operatorname{tr}\,[(T-\zeta)^{-1} - (S-\zeta)^{-1}]\,.$$

This completes the proof of (6.17). Note that $(S-\zeta)^{-1} = (T-\zeta)^{-1} \times (1 + B(\zeta))^{-1} \in \mathscr{B}(X)$ so that $\zeta \in \mathrm{P}(S)$.

4. Conditions excluding the singular case

As mentioned before, the possibility is in general not excluded that $\omega(\zeta) = 0$ identically in (6.16) so that $\tilde{\nu}(\zeta; S) = +\infty$ identically [this is equivalent to $\Delta \in \Sigma(S)$].

---

[1] The results of Chapter I can be applied to $\omega(\zeta)$ since $A(T-\zeta)^{-1}$ is degenerate; see III-§ 4.3.

Example 6.4. Consider the operators discussed in Example 3.8. If we change the notation by $T \to S$, $A \to -A$, $T + A \to S - A = T$, the open unit disk Δ belongs to $P(T)$ so that $\tilde{\nu}(\zeta; T) \equiv 0$ in Δ. But since $\Delta \in \Sigma(S)$, we have $\tilde{\nu}(\zeta; S) \equiv \infty$ in Δ. In view of (6.16) this implies $\nu(\zeta, \omega) \equiv \infty$, hence $\omega(\zeta) \equiv 0$. This can also be verified directly as follows. A is of rank one ($m = 1$) and the x_1, e_1 of (6.2) should be replaced by $-x_{-1}$, e_0, respectively (e_0 is the vector x_0 regarded as an element of X^*). Hence $\omega(\zeta) = \omega(\zeta; T, A) = 1 - ((T - \zeta)^{-1} x_{-1}, e_0)$. But since it is easily verified that $(T - \zeta)^{-1} x_{-1} = \sum_{j=0}^{\infty} \zeta^j x_j$ for $\zeta \in \Delta$, we have indeed $\omega(\zeta) = 0$ identically in Δ.

Theorem 6.5. *In order that the $\omega(\zeta)$ of* Theorem 6.2 *not be identically zero for $\zeta \in \Delta$, it is necessary and sufficient that at least one ζ of Δ belong to* $P(S)$.

This is obvious from Theorem 6.2 and the remarks thereafter.

We shall now give several sufficient conditions for $\omega(\zeta)$ not to be identically zero.

a) $\omega(\zeta) \not\equiv 0$ if $\|A(T - \zeta)^{-1}\| < 1$ for *some* $\zeta \in \Delta$. For then we have $\zeta \in P(S)$ because $(S - \zeta)^{-1}$ can be constructed by the Neumann series as in II-(1.13).

b) $\omega(\zeta) \not\equiv 0$ if T^* exists and is densely defined in X^* and if there exists a sequence $\zeta_n \in \Delta$ such that $|\zeta_n| \to \infty$ and $\{|\zeta_n| \, \|(T - \zeta_n)^{-1}\|\}$ is bounded. To see this it suffices by a) to show that $\|A(T - \zeta_n)^{-1}\| \to 0$. But we have by (6.5)

$$A(T - \zeta_n)^{-1} u = \sum ((T - \zeta_0)(T - \zeta_n)^{-1} u, f_j) x_j = \\ = \sum (u, (T^* - \bar{\zeta}_0)(T^* - \bar{\zeta}_n)^{-1} f_j) x_j$$

so that

$$\|A(T - \zeta_n)^{-1}\| \leqq \sum \|(T^* - \bar{\zeta}_0)(T^* - \bar{\zeta}_n)^{-1} f_j\| \, \|x_j\| . \tag{6.22}$$

Now the sequence of operators $B_n^* = (T^* - \bar{\zeta}_0)(T^* - \bar{\zeta}_n)^{-1}$ acting in X^* is uniformly bounded since the $B_n = 1 + (\zeta_n - \zeta_0)(T - \zeta_n)^{-1}$ are uniformly bounded by hypothesis. Furthermore, $B_n^* f = (T^* - \bar{\zeta}_n)^{-1} \times \times (T^* - \bar{\zeta}_0) f \to 0$ if $f \in D(T^*)$ since $\|(T^* - \bar{\zeta}_n)^{-1}\| = \|(T - \zeta_n)^{-1}\| \to 0$. It follows (see Lemma III-3.5) that B_n^* converges strongly to zero, and the right member of (6.22) tends to zero for $n \to \infty$, as we wished to show.

c) $\omega(\zeta) \not\equiv 0$ if $T \in \mathscr{B}(X)$ and there is a $\zeta \in \Delta$ such that $|\zeta| \geqq \|T\|$. In this case Δ can be extended so as to include the whole exterior of the circle $|\zeta| = \|T\|$. Then it is easy to see that b) is satisfied in the extended domain Δ'. Hence $\omega(\zeta) \not\equiv 0$ in Δ' and, a fortiori, in Δ.

Remark 6.6. The condition b) is satisfied if X is a Hilbert space and T is selfadjoint (see Chapter V).

Chapter Five

Operators in Hilbert spaces

Hilbert space is a special case of Banach space, but it deserves separate consideration because of its importance in applications. In Hilbert spaces the general results deduced in previous chapters are strengthened and, at the same time, new problems arise.

The salient feature of Hilbert space is the existence of an inner product and the ensuing notion of orthogonality. For linear operators this leads to the notions of symmetric, selfadjoint and normal operators. Also there is a more general class of accretive (or dissipative) operators, the importance of which has been recognized in application to differential operators. In perturbation theory, these new notions give rise to such problems as perturbation of orthonormal families, perturbation of selfadjoint operators (with applications to the Schrödinger and Dirac operators), etc., which will be discussed in this chapter.

This chapter is again partly preliminary, beginning with an elementary exposition of the specific results in the operator theory of Hilbert spaces which are not covered in the general theory of Banach spaces. Accretive and sectorial operators are given special attention in view of later applications to analytic and asymptotic perturbation theory.

§ 1. Hilbert space

1. Basic notions

Hilbert space is a generalization of the unitary space considered in I-§ 6 to the infinite-dimensional case[1].

A Hilbert space H is a Banach space in which the norm is defined in terms of an inner product (u, v) defined for all pairs u, v of vectors and satisfying the conditions stated in I-§ 6.1. The norm $\|u\|$ is defined by I-(6.3). H is assumed to be *complete* with respect to this norm[2].

Most of the results of I-§ 6 can be taken over to the Hilbert space H, and we shall do this freely without comment whenever it is obvious. But there are also results that need modification or at least some comments. These will be given in this and the following paragraphs. It should be remarked at this point that the Schwarz inequality I-(6.4) is valid and that (u, v) is continuous in u and v jointly, exactly as before.

---

[1] General references to Hilbert space theory are AKHIEZER and GLAZMAN [1], DUNFORD and SCHWARTZ [1], [2], HALMOS [1], SZ.-NAGY [1], RIESZ and SZ.-NAGY [1], STONE [1], YOSIDA [1], ZAANEN [1].

[2] A vector space H with an inner product, which is not necessarily complete, is called a *pre-Hilbert space*. Such a space can be completed to a Hilbert space $\tilde{H}$. Again $\tilde{H}$ can be defined as the set of equivalence classes of Cauchy sequences $\{u_n\}$ in H (see footnote[2] of p. 129). The inner product of two elements of $\tilde{H}$ represented by the Cauchy sequences $\{u_n\}$, $\{v_n\}$ is then defined by $\lim (u_n, v_n)$.

Example 1.1. l^2 is a Hilbert space (see Example III-1.1) if the inner product is defined by

$$(1.1)\qquad (u, v) = \sum_k \xi_k \bar{\eta}_k \quad \text{for} \quad u = (\xi_k) \quad \text{and} \quad v = (\eta_k) .$$

$L^2(E)$ is a Hilbert space (see Example III-1.3) if the inner product is defined by

$$(1.2)\qquad (u, v) = \int_E u(x) \overline{v(x)}\, dx .$$

If H is a Hilbert space, the adjoint space H^* can be identified with H. In particular $H^{**} = H^* = H$ and H is reflexive. The proof of this fact given in I-§ 6.2, depending essentially on the finite dimensionality, cannot be applied to the general case.

For the proof we start with the projection theorem: *every subspace (closed linear manifold)* M *of* H *has an orthogonal (perpendicular) projection*. In other words, we have the decomposition of H

$$(1.3)\qquad H = M \oplus M^\perp ,$$

where $M^\perp$ is the orthogonal complement of M (the set of all $u \in H$ such that $(u, v) = 0$ for all $v \in M$). Note that $M^\perp$ is also a closed subspace of H; this is a direct consequence of the continuity of the inner product.

For any $u \in H$, set $d = \operatorname{dist}(u, M)$ and take a sequence $u_n \in M$ such that $\|u_n - u\| \to d$. We have

$$(1.4)\qquad \left\|\frac{1}{2}(u_n + u_m) - u\right\|^2 + \left\|\frac{1}{2}(u_n - u_m)\right\|^2 = \frac{1}{2}\|u_n - u\|^2 + \frac{1}{2}\|u_m - u\|^2$$

by I-(6.7). For $n, m \to \infty$ the right member tends to d^2, whereas the first term on the left is not smaller than d^2 because $(u_n + u_m)/2 \in M$. Thus $\|u_n - u_m\| \to 0$ and there exists a $u' \in H$ such that $u_n \to u'$(completeness of H). The closedness of M then implies that $u' \in M$. Moreover, we have $\|u - u'\| = \lim \|u - u_n\| = d$.

We shall now show that $u'' = u - u' \in M^\perp$. For this it suffices to show that $(u'', x) = 0$ for any $x \in M$ with $\|x\| = 1$. But this follows from

$$d^2 = \|u''\|^2 = \|u'' - (u'', x)\, x\|^2 + |(u'', x)|^2 \geq d^2 + |(u'', x)|^2$$

[set $u = u''$, $v = x$ in I-(6.5)].

The possibility of the decomposition $u = u' + u''$ proves (1.3), for the uniqueness of this decomposition can easily be proved.

(1.3) implies that

$$(1.5)\qquad M^{\perp\perp} = M$$

as in the finite-dimensional case. Also (1.3) defines the *orthogonal projection* $P = P_M$ on M by $Pu = u'$ as before.

Now the possibility of the identification of H^* with H can be proved. Let $f \in H^*$; f is a bounded semilinear form on H. Let $N = N(f)$, the null space of f. N is a closed linear manifold of H. If $f \neq 0$, $N \neq H$ so that $N^\perp \neq 0$. Let $v \in N^\perp$, $v \neq 0$. We may assume $f[v] = 1$. For any $w \in H$,

$w' = w - \overline{f[w]}\, v \in \mathsf{N}$ since $f[w'] = 0$. Hence $0 = (v, w') = (v, w) - f[w]\, \|v\|^2$ or $f[w] = (u, w)$ with $u = v/\|v\|^2$. *Thus any $f \in \mathsf{H}^*$ can be identified with a $u \in \mathsf{H}$* (Theorem of RIESZ). Other arguments necessary to complete the proof are quite the same as in I-§ 6.2.

As in general Banach spaces, we can define in H the notion of weak convergence (see III-§ 1.6) besides the ordinary (strong) convergence. In virtue of the identification $\mathsf{H}^* = \mathsf{H}$, $u_n \underset{w}{\to} u$ is equivalent to $(u_n, v) \to (u, v)$ for every $v \in \mathsf{H}$.

Lemma 1.2. *If $u_n \underset{w}{\to} u$ and $\limsup \|u_n\| \leq \|u\|$, then $u_n \underset{s}{\to} u$.*

Proof. $\|u_n - u\|^2 = \|u_n\|^2 - 2\,\mathrm{Re}\,(u_n, u) + \|u\|^2$. But $(u_n, u) \to (u, u) = \|u\|^2$ and $\limsup \|u_n\|^2 \leq \|u\|^2$. Hence $\lim \|u_n - u\| = 0$.

Lemma 1.3. *H is weakly complete: if $\{u_n\}$ is weakly convergent, then $u_n \underset{w}{\to} u$ for some $u \in \mathsf{H}$.*

Proof. Since $\{u_n\}$ is bounded (see III-§ 1.6), $\lim (u_n, v) = f[v]$ defines a bounded semilinear form f. Hence $f[v] = (u, v)$ for some $u \in \mathsf{H}$ and $u_n \underset{w}{\to} u$.

Lemma 1.4. *H is sequentially weakly compact. In other words: if $u_n \in \mathsf{H}$ is a bounded sequence, then there is a subsequence $\{v_n\}$ of $\{u_n\}$ such that $v_n \underset{w}{\to} v$ for some $v \in \mathsf{H}$.*

Proof. Since $|(u_n, u_1)| \leq \|u_n\|\, \|u_1\|$ is bounded in n, there is a subsequence $\{u_n^1\}$ of $\{u_n\}$ such that (u_n^1, u_1) is convergent. Since (u_n^1, u_2) is similarly bounded in n, there is a subsequence $\{u_n^2\}$ of $\{u_n^1\}$ such that (u_n^2, u_2) is convergent. Proceeding in the same way, we can construct a sequence of subsequences $\{u_n^m\}$ of $\{u_n\}$ such that $\{u_n^{m+1}\}_{n=1,2,\ldots}$ is a subsequence of $\{u_n^m\}_{n=1,2,\ldots}$ and $\lim\limits_n\, (u_n^m, u_m)$ exists. Then the diagonal sequence $\{v_n\}$, $v_n = u_n^n$, has the property that $\lim\limits_n\, (v_n, u_m)$ exists for all m. Since $\|v_n\|$ is bounded in n, it follows (cf. Lemma III-1.31) that $\lim (v_n, u)$ exists for any u in the closed span M of the set $\{u_n\}$.

On the other hand $\lim (v_n, u)$ exists for any $u \in \mathsf{M}^\perp$ since $(v_n, u) = 0$. Thus $\{v_n\}$ is weakly convergent, hence $v_n \underset{w}{\to} v$ for some v (Lemma 1.3).

A sesquilinear form $\mathfrak{t}[u, u']$ on $\mathsf{H} \times \mathsf{H}'$, where H and H' are Hilbert spaces, can be defined as in I-§ 6.2, and we have again the formula I-(6.11). $\mathfrak{t}$ is said to be *bounded* if there is a constant M such that

$$|\mathfrak{t}[u, u']| \leq M \|u\|\, \|u'\|\,. \tag{1.6}$$

We denote by $\|\mathfrak{t}\|$ the smallest M with this property. *A bounded sesquilinear form $\mathfrak{t}[u, u']$ is continuous in u and u' jointly*; the proof is the same as that of the continuity of the inner product. The converse is also true. *A sesquilinear form $\mathfrak{t}[u, u']$ is bounded if it is continuous in u for each fixed u' and also in u' for each fixed u.* This is a consequence of the principle of uniform boundedness (III-§ 1.5).

Problem 1.5. Prove the last statement.

Problem 1.6. Let $\{t_n\}$ be a sequence of bounded sesquilinear forms on $H \times H'$ and let $\{t_n[u, u']\}$ be bounded for $n = 1, 2, \ldots$ for each fixed $u \in H$, $u' \in H'$. Then the sequence $\{\|t_n\|\}$ is bounded. [hint: Another application of the principle of uniform boundedness.]

2. Complete orthonormal families

We did not define the notion of a basis in a general Banach space. In a Hilbert space, however, a *complete orthonormal family* plays the role of a basis.

A family $\{x_\mu\}$ of vectors of H, with the index μ running over a certain set, is *orthonormal* if

$$(x_\mu, x_\nu) = \delta_{\mu\nu} = \begin{cases} 1, & \mu = \nu, \\ 0, & \mu \neq \nu. \end{cases} \tag{1.7}$$

For any $u \in H$, the scalars

$$\xi_\mu = (u, x_\mu) \tag{1.8}$$

are the *coefficients* of u with respect to $\{x_\mu\}$. Since any finite subfamily $\{x_{\mu_1}, \ldots, x_{\mu_n}\}$ of $\{x_\mu\}$ is also orthonormal, it follows from I-(6.15) that $\sum_{k=1}^{n} |\xi_{\mu_k}|^2 \leq \|u\|^2$. Since this is true for any choice of $\mu_1, \ldots, \mu_n$, we have the *Bessel inequality*

$$\sum_\mu |\xi_\mu|^2 = \sum_\mu |(u, x_\mu)|^2 \leq \|u\|^2 . \tag{1.9}$$

This implies, in particular, that for each u, only a countable number of the coefficients ξ_μ are different from zero.

(1.9) implies also that

$$u' = \sum_\mu \xi_\mu x_\mu = \sum_\mu (u, x_\mu) x_\mu \tag{1.10}$$

exists. In fact, the right member of (1.10) contains only a countable number of non-vanishing coefficients. Let these be denoted by ξ_{μ_k}, $k = 1, 2, \ldots$, in an arbitrary order. Then the $u'_n = \sum_{k=1}^{n} \xi_{\mu_k} x_{\mu_k}$ form a Cauchy sequence because

$$\|u'_n - u'_m\|^2 = \sum_{k=m+1}^{n} |\xi_{\mu_k}|^2 \to 0, \quad m, n \to \infty ,$$

in virtue of (1.9). It is easy to see that the limit u' of the sequence $\{u'_n\}$ does not depend on the way in which the sequence $\{\mu_k\}$ is numbered.

u' is equal to the orthogonal projection Pu of u on the closed span M of $\{x_\mu\}$. For it is obvious that $u' \in M$, whereas $u'' = u - u' \in M^\perp$ because $(u'', x_\mu) = (u - u', x_\mu) = \xi_\mu - (\Sigma\, \xi_\nu x_\nu, x_\mu) = 0$ for every μ. In particular we have

$$\|u'\|^2 = \sum_\mu |\xi_\mu|^2 = \sum_\mu |(u, x_\mu)|^2 , \quad \|u''\|^2 = \|u\|^2 - \|u'\|^2 . \tag{1.11}$$

The orthonormal family $\{x_\nu\}$ is said to be *complete* if $M = H$. A necessary and sufficient condition for completeness is that $u'' = 0$ for all $u \in H$, and this is equivalent to the *Parseval equality*

$$\sum_\mu |\xi_\mu|^2 = \sum_\mu |(u, x_\mu)|^2 = \|u\|^2 \tag{1.12}$$

for every $u \in H$. In this case we have also

$$u = \sum_\mu \xi_\mu x_\mu = \sum_\mu (u, x_\mu) x_\mu , \tag{1.13}$$

which is the *expansion* of u in terms of the complete orthonormal family $\{x_\mu\}$.

If H is *separable* (see III-§ 1.2), any orthonormal family $\{x_\mu\}$ of H consists of at most a countable number of elements. To see this, let $\{u_n\}$ be a countable subset of H which is dense in H. For each μ there is a u_n such that $\|x_\mu - u_n\| < 1/2$. Since $\|x_\mu - x_\nu\| = \sqrt{2}$ for $\mu \neq \nu$, the index n is different for different μ. Thus $\{x_\mu\}$ can be put in correspondence with a subset of the set of positive numbers $\{n\}$.

Conversely, H is separable if there is a *complete* orthonormal family $\{x_n\}$ consisting of a countable number of elements. For the set of all (finite) linear combinations of the x_n with rational coefficients is countable and dense in H.

If H is separable, a complete orthonormal family can be constructed by the Schmidt orthogonalization procedure (see I-§ 6.3) applied to any sequence which is dense in H. A similar construction could be used even in a non-separable space by using the axiom of choice, but we shall not go into details since we are mostly concerned with separable spaces.

Problem 1.7. An orthonormal family $\{x_n\}$ is complete if and only if there is no nonzero vector orthogonal to all x_n.

Example 1.8. A complete orthonormal family of l^2 is given by the canonical basis consisting of $x_n = (\delta_{nk})$.

Example 1.9. The set of trigonometric functions $(2\pi)^{-1/2} e^{inx}$, $n = 0, \pm 1, \pm 2, \ldots$, forms a complete orthonormal family in $L^2(a, a + 2\pi)$, where a is any real number[1].

Example 1.10. Let $\varphi_k(x)$, $\psi_j(y)$ be two sequences of functions on E and F, constituting complete orthonormal families in $L^2(E)$ and $L^2(F)$, respectively. Then the functions $\chi_{kj}(x, y) = \varphi_k(x)\,\psi_j(y)$ form a complete orthonormal family in $L^2(E \times F)$, the space of all measurable functions $w(x, y)$ defined for $x \in E$, $y \in F$ such that $\|w\|^2 = \int_{E\times F} |w(x, y)|^2\, dx\, dy$ is finite. The orthonormality of χ_{kj} is obvious from $(\chi_{kj}, \chi_{il}) = (\varphi_k, \varphi_l)\,(\psi_j, \psi_i) = \delta_{kl}\delta_{ji}$. To prove the completeness, it suffices to show that $(w, \chi_{kj}) = 0$ for all k, j implies $w = 0$ (Problem 1.7). Set

$$w_k(y) = \int_E w(x, y)\,\overline{\varphi_k(x)}\, dx; \tag{1.14}$$

then we have $w_k \in L^2(F)$ since $|w_k(y)|^2 \leqq \int_E |w(x, y)|^2\, dx$ by the Schwarz inequality and $\int_F |w_k(y)|^2\, dy \leqq \int_{E\times F} |w(x, y)|^2\, dx\, dy = \|w\|^2$. This enables one to write (w, χ_{kj})

[1] See any textbook on Fourier series, or on Hilbert space, e. g. STONE [1].

$= (w, \varphi_k \psi_j) = (w_k, \psi_j)$. Thus $(w, \chi_{kj}) = 0$ for all k, j implies that $(w_k, \psi_j) = 0$ and hence $w_k = 0$ by the completeness of $\{\psi_j\}$. On the other hand, $w_k(y)$ is defined for almost every $y \in F$ and this value must be zero for every k and almost every y. In virtue of the completeness of φ_k, (1.14) shows that $w(x, y) = 0$ for almost every x for almost every y. Thus $w = 0$ as an element of $L^2(E \times F)$.

§ 2. Bounded operators in Hilbert spaces

1. Bounded operators and their adjoints

A (linear) operator T from a Hilbert space H to another one H' is defined as in general Banach spaces. There are, however, some properties peculiar to operators in (or between) Hilbert spaces. We begin with bounded operators $T \in \mathscr{B}(H, H')$.

First we note that

$$T^{**} = T; \tag{2.1}$$

this is true since H is reflexive (see III-§ 3.3).

An operator $T \in \mathscr{B}(H, H')$ is closely related to a bounded sesquilinear form $\mathfrak{t}$ on $H \times H'$ by

$$\mathfrak{t}[u, u'] = (Tu, u') = (u, T^* u'); \tag{2.2}$$

note that $T^* \in \mathscr{B}(H', H)$ because $H^* = H$, $H'^* = H'$. The relationship (2.2) gives a one-to-one correspondence between the set of all bounded sesquilinear forms $\mathfrak{t}$ on $H \times H'$, the set of all $T \in \mathscr{B}(H, H')$ and the set of all $T^* \in \mathscr{B}(H', H)$. This can be proved in the same way as in the finite-dimensional case (I-§ 6.4). That (Tu, u') is a bounded form is obvious by $|(Tu, u')| \leq \|Tu\| \|u'\| \leq \|T\| \|u\| \|u'\|$. Conversely, a bounded sesquilinear form $\mathfrak{t}[u, u']$ can be written in the form (v', u'), $v' \in H'$, since it is a bounded semilinear form in u' for fixed u. Setting $v' = Tu$ defines a linear operator T on H to H', and the boundedness of T follows from $\|T\| = \sup |(Tu, u')|/\|u\| \|u'\| = \sup |\mathfrak{t}[u, u']|/\|u\| \|u'\| = \|\mathfrak{t}\|$.

Problem 2.1. Let $T_n \in \mathscr{B}(H, H')$ be such that $(T_n u, u')$ is bounded for each fixed pair $u \in H$, $u' \in H'$. Then $\{T_n\}$ is uniformly bounded (that is, $\{\|T_n\|\}$ is bounded). [hint: The principle of uniform boundedness.]

For an operator $T \in \mathscr{B}(H, H')$, a matrix representation is possible in the same fashion as in the finite-dimensional case (I-§ 6.4). For simplicity assume that H and H' are separable and let $\{x_k\}$, $\{x'_j\}$ be complete orthonormal families in H, H', respectively. The matrix of T is then defined exactly by I-(6.27) and the expansion I-(6.28) holds true (where we have in general an infinite series on the right).

Remark 2.2. For an *unbounded* operator T from H to H', (Tu, u') is also a sesquilinear form, which is, however, not necessarily defined for all $u \in H$. The relationship between an operator and a sesquilinear form in this general case is rather complicated. Later we shall consider it in detail in the special case of a so-called sectorial operator.

A *symmetric* sesquilinear form $\mathfrak{t}[u, v]$ in a Hilbert space H is defined by $\mathfrak{t}[v, u] = \overline{\mathfrak{t}[u, v]}$ as in I-§ 6.5. If in addition $\mathfrak{t}$ is assumed to be bounded, all the results stated there are valid. The operator $T \in \mathscr{B}(\mathsf{H})$ associated by (2.2) with a bounded symmetric sesquilinear form $\mathfrak{t}$ has the property that

(2.3) $$T^* = T;$$

T is said to be *symmetric*. The notion of a positive (or non-negative) symmetric operator or form and the order relation $S \leqq T$ for symmetric $S, T \in \mathscr{B}(\mathsf{H})$ are defined as before (see loc. cit.).

The results of I-§ 6.7 on projections are also valid without modification; here we are concerned with a $P \in \mathscr{B}(\mathsf{H})$ such that $P^2 = P$. Compare also the results of III-§ 3.4 on projections in Banach spaces. We add the following lemma.

Lemma 2.3. *Let $\{P_n\}$ be a sequence of orthogonal projections in H such that $P_m P_n = \delta_{mn} P_n$. Then $\sum_{n=1}^{\infty} P_n = P$ exists in the strong sense and P is an orthogonal projection. The range of P is the closed span of the union of the ranges of all the P_n.*

Proof. For any $u \in \mathsf{H}$ we have $\sum_{k=1}^{n} \|P_k u\|^2 = \|\sum_{k=1}^{n} P_k u\|^2 \leqq \|u\|^2$ since $\sum_{k=1}^{n} P_k$ is an orthogonal projection [see I-(6.49)]. Hence $\sum_{k=1}^{\infty} \|P_k u\|^2$ is convergent and $\|\sum_{k=n}^{n+p} P_k u\|^2 = \sum_{k=n}^{n+p} \|P_k u\|^2 \to 0$, $n \to \infty$. Thus $\sum_{n=1}^{\infty} P_n = P$ is strongly convergent. It is easy to see that the P thus defined has the properties stated.

We shall say that the $\{P_n\}$ of Lemma 2.3 is *complete* if $\Sigma P_n = 1$.

A normal operator $T \in \mathscr{B}(\mathsf{H})$ is defined by the property $T^* T = T T^*$, and the results of I-§ 6.6 remain valid. For example

(2.4) $$\operatorname{spr} T = \|T\| \quad \text{if } T \text{ is normal}.$$

Problem 2.4. A quasi-nilpotent normal operator is zero.

Example 2.5. Consider the integral operator T of Example III-2.4, with the kernel $t(x, y)$. If $\mathsf{E} = \mathsf{F}$ and $t(x, y) = \overline{t(y, x)}$ (Hermitian symmetric kernel), T is a symmetric operator in $\mathsf{H} = \mathsf{L}^2(\mathsf{E})$ (see Example III-3.17). The same is true of the operator T of Example III-2.3, defined from a given matrix (τ_{jk}), if considered in $\mathsf{H} = l^2$ and if (τ_{jk}) is Hermitian symmetric $(\tau_{kj} = \overline{\tau}_{jk})$, see Example III-3.18.

Unbounded symmetric and normal operators will be dealt with in detail in later sections.

2. Unitary and isometric operators

An operator $T \in \mathscr{B}(\mathsf{H}, \mathsf{H}')$ is said to be *isometric* if it preserves the norm:

(2.5) $$\|T u\| = \|u\| \quad \text{for every} \quad u \in \mathsf{H}.$$

As in the finite-dimensional case (I-§ 6.6), this implies that $T^* T = 1_{\mathsf{H}}$ (the identity operator in H) and $(Tu, Tv) = (u, v)$ for all $u, v \in \mathsf{H}$.

An isometric operator T is invertible, for $Tu = 0$ implies $u = 0$. But T^{-1} need not belong to $\mathscr{B}(\mathsf{H}', \mathsf{H})$ since its domain may not be the whole space H'. Note that this may happen even when $\mathsf{H}' = \mathsf{H}$, in contradistinction to the finite-dimensional case.

An isometric operator T is said to be *unitary* if T^{-1} has domain H', that is, if T has range H'. Then T^{-1} belongs to $\mathscr{B}(\mathsf{H}', \mathsf{H})$ and is itself unitary. Thus a $T \in \mathscr{B}(\mathsf{H}, \mathsf{H}')$ is unitary if and only if

$$T^* T = 1_{\mathsf{H}} \quad \text{and} \quad T T^* = 1_{\mathsf{H}'} , \tag{2.6}$$

which is in turn equivalent to

$$T^{-1} = T^* . \tag{2.7}$$

The existence of a unitary operator T on H to H' implies that H and H' have identical structure as Hilbert spaces, for T preserves the linear relations and inner products. An operator A' in H' is said to be *unitarily equivalent* to an operator A in H if there is a unitary operator T from H to H' such that

$$A' T = T A \quad \text{or} \quad A' = T A T^{-1} = T A T^* . \tag{2.8}$$

This means that $\mathsf{D}(A')$ is exactly the image of $\mathsf{D}(A)$ under the unitary map T and $A' T u = T A u$ holds for every $u \in \mathsf{D}(A)$. A' and A have the same internal structure, for the correspondence $u \leftrightarrow u' = T u$ is unchanged under the operators A and A' in virtue of the relation $A u \leftrightarrow T A u = A' T u = A' u'$.

Let $T \in \mathscr{B}(\mathsf{H}, \mathsf{H}')$ be an isometric operator and let $\mathsf{M}' = \mathsf{R}(T)$. The isometry implies that M' is a closed linear manifold of H'. The operator $T^* \in \mathscr{B}(\mathsf{H}', \mathsf{H})$ is an extension of T^{-1} as is readily seen from $T^* T = 1_{\mathsf{H}}$, whereas $T^* u' = 0$ if and only if $u' \in \mathsf{M}'^{\perp}$ [see III-(5.10)]. Thus T^* is not isometric unless T is unitary; it is partially isometric.

An operator $W \in \mathscr{B}(\mathsf{H}, \mathsf{H}')$ is *partially isometric* if there is a closed linear manifold M of H such that $\|Wu\| = \|u\|$ for $u \in \mathsf{M}$ while $Wu = 0$ for $u \in \mathsf{M}^{\perp}$. M is called the *initial set* and $\mathsf{M}' = W\mathsf{M}$ the *final set* of W; M' is a closed linear manifold. An equivalent definition is

$$\|Wu\| = \|Pu\| \quad \text{for every} \quad u \in \mathsf{H} , \tag{2.9}$$

where P is the orthogonal projection of H on M. (2.9) is in turn equivalent to

$$W^* W = P . \tag{2.10}$$

For any $u \in \mathsf{H}$ we have $W(1 - P) u = 0$ since $(1 - P) u \in \mathsf{M}^{\perp}$. Hence

$$W = W P . \tag{2.11}$$

We have $W^* u' = 0$ for $u' \in M'^{\perp}$, as in the case of an isometric operator. Furthermore, $\|W^* W u\| = \|P u\| = \|W u\|$ shows that $\|W^* u'\| = \|u'\|$ for $u' \in M'$. Hence W^* is likewise partially isometric with the initial set M'. Application of this result to W replaced by W^* shows that the final set of W^* is identical with the initial set M of $W^{**} = W$. Thus there is complete symmetry between W and W^*. The following relations are direct consequences of these considerations (P' is the orthogonal projection of H' on M').

$$\begin{aligned} &W^* W = P\,, \quad W P = P' W = W\,, \quad W W^* W = W\,, \\ &W W^* = P'\,, \quad W^* P' = P W^* = W^*\,, \quad W^* W W^* = W^*\,. \end{aligned} \tag{2.12}$$

Problem 2.6. $W \in \mathscr{B}(H, H')$ is partially isometric if and only if $W = W W^* W$. [hint: $W = W W^* W$ implies $W^* W = W^* W W^* W$. Thus $P = W^* W$ is an orthogonal projection of H on some subspace M.]

Example 2.7. Let $H = L^2(E)$ and $\hat{H} = L^2(E')$, where both E and E' are the real n-dimensional euclidean space (so that they may be identified), the points of which will be denoted respectively by $x = (x_1, \ldots, x_n)$ and $k = (k_1, \ldots, k_n)$. The *Fourier transformation* $\hat{u} = T u$ given by

$$\hat{u}(k) = (2\pi)^{-n/2} \int_E e^{-i k \cdot x} u(x)\, dx\,, \tag{2.13}$$

$$k \cdot x = k_1 x_1 + \cdots + k_n x_n\,,$$

defines a unitary operator T on H to $\hat{H}$. The inverse operator $T^{-1} \hat{u} = u$ is given by

$$u(x) = (2\pi)^{-n/2} \int_{E'} e^{i k \cdot x} \hat{u}(k)\, dk\,. \tag{2.14}$$

These results are known as the Fourier-Plancherel theorem. Actually, these integrals should be taken in the sense of *limit in mean*: the integral (2.13) is first to be taken over a bounded subset K of E, the resulting function converging in the norm $\|\ \|$ of $\hat{H}$ to the required transform $\hat{u}$ when K expands to the whole space E[1]. Thus T or T^{-1} is not an integral operator in the sense of Example III-2.4.

Example 2.8. Let $H = L^2(-\pi, \pi)$. For any $u \in H$ the Fourier coefficients

$$\xi_n = (2\pi)^{-1/2} \int_{-\pi}^{\pi} u(x)\, e^{-i n x}\, dx\,, \quad n = 0, \pm 1, \pm 2, \ldots, \tag{2.15}$$

are defined. The operator T that sends u into the vector $v = (\xi_n)$ is a unitary operator on H to $H' = l^2$. The inverse T^{-1} assigns to each $v = (\xi_n) \in H'$ the function $u(x)$ defined by the Fourier series

$$u(x) = (2\pi)^{-1/2} \sum_{n=-\infty}^{\infty} \xi_n e^{i n x} \tag{2.16}$$

which is convergent in norm $\|\ \|$. These results are another expression of the completeness of the family of trigonometric functions $e^{i n x}$. Similar results hold when it is replaced by any complete orthonormal family of $H = L^2(-\pi, \pi)$ or of any abstract Hilbert space.

Example 2.9. An example of an isometric operator that is not unitary is given by a shift operator (cf. Example III-3.16). Let $\{x_n\}_{n=1,2,\ldots}$ be a complete orthonormal family in a Hilbert space H and let $T \in \mathscr{B}(H)$ be such that $T x_n = x_{n+1}$. T is isometric and T^* is partially isometric, with $T^* x_{n+1} = x_n$ and $T^* x_1 = 0$. The initial set of T is H and the final set is the subspace of H spanned by $x_2, x_3, \ldots$.

[1] For details see e. g. STONE [1].

3. Compact operators

Let $T \in \mathscr{B}(\mathsf{H})$ be a compact operator. We know that the spectrum $\Sigma(T)$ consists of at most a countable number of eigenvalues with finite multiplicities, possibly excepting zero (Theorem III-6.26). Let $\lambda_1, \lambda_2, \ldots$ be these nonzero eigenvalues arranged, say, in decreasing order of magnitude, and let $P_1, P_2, \ldots$ be the associated eigenprojections. We note that $|\lambda_1|$ is equal to the spectral radius of T:

$$|\lambda_1| = \operatorname{spr} T = \lim_{n\to\infty} \|T^n\|^{1/n} \tag{2.17}$$

(see III-§ 6.2; read 0 for $|\lambda_1|$ if there is no nonzero eigenvalue).

Suppose, in addition, that T is normal. Then the P_h are orthogonal projections as in the finite-dimensional case (see I-§ 6.9), and the associated eigennilpotents are zero. On setting $Q_n = P_1 + P_2 + \cdots + P_n$, we see from III-(6.55) that $TQ_n = \lambda_1 P_1 + \cdots + \lambda_n P_n$. Since Q_n commutes with T, $T(1 - Q_n)$ is normal and has the eigenvalues λ_{n+1}, $\lambda_{n+2}, \ldots$ and possibly 0. For a normal operator, however, the spectral radius coincides with the norm [see (2.4)]. In view of (2.17), we have therefore $\|T(1 - Q_n)\| = \operatorname{spr} T(1 - Q_n) = |\lambda_{n+1}| \to 0$, $n \to \infty$. This gives[1]

$$\|T - (\lambda_1 P_1 + \lambda_2 P_2 + \cdots + \lambda_n P_n)\| \to 0\,, \quad n \to \infty\,. \tag{2.18}$$

We have thus proved the spectral theorem, analogous to the one obtained in the finite-dimensional case [I-(6.65)].

Theorem 2.10. *If $T \in \mathscr{B}(\mathsf{H})$ is normal and compact, we have the spectral representation*

$$T = \sum_h \lambda_h P_h\,, \quad P_h^* = P_h\,, \quad \dim P_h < \infty\,, \tag{2.19}$$

in the sense of convergence in norm. The projections P_h form a complete orthogonal family together with the orthogonal projection P_0 on the null space $\mathsf{N}(T)$.

Proof. Only the last statement remains to be proved. Let Q be the orthogonal projection on the subspace of H spanned by all the $P_h \mathsf{H}$; we have $Q = \text{s-lim}\, Q_n$ (see Lemma 2.3). Q commutes with T since all the Q_n do. Thus $Q\mathsf{H}$ and $(1 - Q)\mathsf{H}$ are invariant under T. But the part T_0 of T in $(1 - Q)\mathsf{H}$ has no nonzero eigenvalue, for $T_0 u = \lambda u$ with $u \in (1 - Q)\mathsf{H}$, $\lambda \neq 0$, would imply $Tu = \lambda u$, hence $\lambda = \lambda_h$ for some h and $u \in P_h \mathsf{H}$, so that $u = P_h u = P_h Q u = 0$ by $P_h = P_h Q$ and $Qu = 0$. Since T_0 is normal too, it follows from (2.19) applied to T_0 that $T_0 = 0$. This means that $(1 - Q)\mathsf{H} \subset \mathsf{N}(T)$. Since, on the other hand, T has no zeros in $Q\mathsf{H}$, we obtain $(1 - Q)\mathsf{H} = \mathsf{N}(T)$.

---

[1] It is possible that T has only a finite number n of eigenvalues; then (2.18) should be interpreted to mean $T = \sum_{h=1}^{n} \lambda_h P_h$. Similar modification will be needed in some of the following formulas.

Sometimes it is convenient to consider the eigenvectors of T rather than the eigenprojections. From each subspace $P_h\mathsf{H}$ choose an arbitrary orthonormal basis consisting of $m_h = \dim P_h$ vectors. These vectors together form an orthonormal basis $\{\varphi_k\}$ of the subspace $Q\mathsf{H}$, and (2.19) may be written

$$T = \sum_{k=1}^{\infty} \mu_k(\ ,\varphi_k)\,\varphi_k\,, \tag{2.20}$$

where the μ_k are the repeated eigenvalues of T with the eigenvectors φ_k (eigenvalues λ_j each counted repeatedly according to its multiplicity). [For the notation (2.20) cf. III-(4.7).]

Let us now consider an arbitrary compact operator $T \in \mathscr{B}_0(\mathsf{H}, \mathsf{H}')$ where H' is another Hilbert space. Since T^*T is a nonnegative symmetric, compact operator in H, we have the spectral expression (note that the eigenvalues of T^*T are nonnegative)

$$T^*T = \sum_{k=1}^{\infty} \alpha_k^2(\ ,\varphi_k)\,\varphi_k\,, \quad (\varphi_j, \varphi_k) = \delta_{jk}\,, \tag{2.21}$$

where $T^*T\varphi_k = \alpha_k^2\varphi_k$, $k = 1, 2, \ldots$ and $\alpha_1 \geqq \alpha_2 \geqq \cdots > 0$. Set

$$\varphi_k' = \alpha_k^{-1} T\varphi_k \in \mathsf{H}'\,, \quad k = 1, 2, \ldots. \tag{2.22}$$

The φ_k' form an orthonormal family in H', for $(\varphi_j', \varphi_k') = (\alpha_j\alpha_k)^{-1}(T\varphi_j, T\varphi_k) = (\alpha_j\alpha_k)^{-1}(T^*T\varphi_j, \varphi_k) = \alpha_j\alpha_k^{-1}(\varphi_j, \varphi_k) = \delta_{jk}$. We now assert that

$$T = \sum_{k=1}^{\infty} \alpha_k(\ ,\varphi_k)\,\varphi_k'\,. \tag{2.23}$$

Since $\{\varphi_k\}$, $\{\varphi_k'\}$ are orthonormal and $\alpha_k \to 0$, the series on the right of (2.23) converges in norm. In fact,

$$\left\|\sum_{k=n}^{n+p} \alpha_k(u, \varphi_k)\,\varphi_k'\right\|^2 = \sum_{k=n}^{n+p} \alpha_k^2|(u, \varphi_k)|^2 \leqq \alpha_n^2 \sum_{k=n}^{n+p} |(u, \varphi_k)|^2 \leqq \alpha_n^2\|u\|^2$$

so that $\left\|\sum_{k=n}^{n+p} \alpha_k(\ ,\varphi_k)\,\varphi_k'\right\| \leqq \alpha_n \to 0$ for $n \to \infty$. Therefore, it suffices to show that (2.23) is true when applied to vectors u of a dense subset of H. For this it suffices, in turn, to consider $u = \varphi_n$ and $u = \psi_n$, where $\{\psi_n\}$ is an orthonormal basis of $\mathsf{N}(T) = \mathsf{N}(T^*T)$ (eigenvectors of T^*T for the eigenvalue zero). But this is obvious by (2.22) and $\varphi_n \perp \psi_m$.

(2.23) is called the *canonical expansion* of T. Strictly speaking, it is not unique since the choice of $\{\varphi_k\}$ is not unique when there are degenerate eigenvalues α_k^2 of T^*T. $\alpha_1, \alpha_2, \ldots$ are called the (repeated) nonzero *singular values*[1] of T.

---

[1] In an obvious way we can define the set of non-repeated singular values of T and their multiplicities. 0 should be included in this set if T^*T has the eigenvalue 0.

It follows from (2.23) that

$$T^* = \sum_{k=1}^{\infty} \alpha_k (\;, \varphi'_k) \varphi_k \tag{2.24}$$

is the canonical expansion of T^*. Also we have

$$T T^* = \sum \alpha_k^2 (\;, \varphi'_k) \varphi'_k . \tag{2.25}$$

Hence the repeated nonzero eigenvalues of $T T^*$ are exactly the α_k^2. In other words, *T and T^* have the same nonzero singular values.*

Also we define

$$|T| = \sum \alpha_k (\;, \varphi_k) \varphi_k , \quad |T^*| = \sum \alpha_k (\;, \varphi'_k) \varphi'_k . \tag{2.26}$$

Note that $|T|$ is determined by T, independently of the particular choice of the φ_k.

Problem 2.11. Any compact operator of $\mathscr{B}(\mathsf{H}, \mathsf{H}')$ is the limit in norm of a sequence of degenerate operators.

Problem 2.12. If T is compact and symmetric, the singular values of T are the absolute values of the eigenvalues of T.

Problem 2.13. For any unitary operators U, V, the two operators T and $U T V$ have the same singular values.

4. The Schmidt class

One of the most important classes of compact operators of $\mathscr{B}(\mathsf{H}, \mathsf{H}')$ is the *Schmidt class*. In this paragraph we assume for simplicity that both H and H' are separable, though the results are valid in the general case.

Let $T \in \mathscr{B}(\mathsf{H}, \mathsf{H}')$ and define

$$\|T\|_2 = \Big(\sum_{k=1}^{\infty} \|T \varphi_k\|^2 \Big)^{1/2} , \tag{2.27}$$

where $\{\varphi_k\}$ is a complete orthonormal family in H. If the series on the right of (2.27) does not converge, we set $\|T\|_2 = \infty$. $\|T\|_2$ is called the *Schmidt norm* of T.

The Schmidt norm is independent of the choice of the family $\{\varphi_k\}$ employed in the definition. To see this we note that

$$\begin{aligned} \sum_k \|T \varphi_k\|^2 &= \sum_k \sum_j |(T \varphi_k, \varphi'_j)|^2 = \sum_j \sum_k |(\varphi_k, T^* \varphi'_j)|^2 \\ &= \sum_j \|T^* \varphi'_j\|^2 \end{aligned} \tag{2.28}$$

where $\{\varphi'_j\}$ is a complete orthonormal family in H'; the change of order of summation in (2.28) is permitted because all terms involved are non-negative. The left member of (2.28) is independent of the particular family $\{\varphi'_j\}$, while the right member is independent of the particular family $\{\varphi_k\}$. Hence both members are independent of the choice of $\{\varphi_k\}$ or $\{\varphi'_j\}$.

Incidentally, we have proved that

$$\|T\|_2 = \|T^*\|_2 . \tag{2.29}$$

The subset of $\mathscr{B}(\mathsf{H}, \mathsf{H}')$ consisting of all T with $\|T\|_2 < \infty$ is called the Schmidt class; it will be denoted by $\mathscr{B}_2(\mathsf{H}, \mathsf{H}')$.

We have the inequalities

$$\|S T\|_2 \leq \|S\| \, \|T\|_2, \quad \|T S\|_2 \leq \|T\|_2 \|S\| . \tag{2.30}$$

These should be read in the following sense: if $T \in \mathscr{B}_2(\mathsf{H}, \mathsf{H}')$ and $S \in \mathscr{B}(\mathsf{H}', \mathsf{H}'')$, then $S T \in \mathscr{B}_2(\mathsf{H}, \mathsf{H}'')$ and the first inequality holds; and similarly for the second.

Problem 2.14. $\mathscr{B}_2(\mathsf{H}, \mathsf{H}')$ **is a vector space, that is,** $\alpha T + \beta S$ **belongs to** $\mathscr{B}_2(\mathsf{H}, \mathsf{H}')$ whenever S, T do.

Problem 2.15. $\|T\| \leq \|T\|_2$.

We can introduce an inner product (S, T) for S, $T \in \mathscr{B}_2(\mathsf{H}, \mathsf{H}')$ so that $\mathscr{B}_2(\mathsf{H}, \mathsf{H}')$ becomes a Hilbert space with the Schmidt norm as its norm. We set

$$(S, T) = \sum_k (S \varphi_k, T \varphi_k) . \tag{2.31}$$

The series converges absolutely since $2|(S \varphi_k, T \varphi_k)| \leq \|S \varphi_k\|^2 + \|T \varphi_k\|^2$ and $\|S\|_2$, $\|T\|_2$ are finite. In virtue of the polarization principle I-(6.8), we have

$$(S, T) = \frac{1}{4} (\|S + T\|_2^2 - \|S - T\|_2^2 + i\|S + i T\|_2^2 - i\|S - i T\|_2^2) , \tag{2.32}$$

which shows that (S, T) is independent of the choice of $\{\varphi_k\}$ used in the definition (2.31). It is easy to see that (S, T) satisfies all the requirements of an inner product and that

$$\|T\|_2^2 = (T, T) . \tag{2.33}$$

This in turn shows that $\| \;\|_2$ has indeed the properties of a norm.

To see that $\mathscr{B}_2(\mathsf{H}, \mathsf{H}')$ is a complete space, let $\{T_n\}$ be a Cauchy sequence in $\mathscr{B}_2(\mathsf{H}, \mathsf{H}')$: $\|T_m - T_n\|_2 \to 0$, $m, n, \to \infty$. Then $\|T_m - T_n\| \to 0$ by Problem 2.15, so that there is a $T \in \mathscr{B}(\mathsf{H}, \mathsf{H}')$ such that $\|T_n - T\| \to 0$. Since

$$\sum_{k=1}^{s} \|(T_n - T_m) \varphi_k\|^2 \leq \|T_n - T_m\|_2^2 < \varepsilon^2$$

for sufficiently large m, n, and any s, we have for $m \to \infty$

$$\sum_{k=1}^{s} \|(T_n - T) \varphi_k\|^2 \leq \varepsilon^2$$

for sufficiently large n and any s. Hence $\|T_n - T\|_2 \leq \varepsilon$ for sufficiently large n. This implies that $T_n - T \in \mathscr{B}_2(\mathsf{H}, \mathsf{H}')$, hence $T \in \mathscr{B}_2(\mathsf{H}, \mathsf{H}')$, and $\|T_n - T\|_2 \to 0$.

We note that $\mathscr{B}_2(\mathsf{H}, \mathsf{H}') \subset \mathscr{B}_0(\mathsf{H}, \mathsf{H}')$: each operator of $\mathscr{B}_2(\mathsf{H}, \mathsf{H}')$ is compact. To see this let n be so large that $\sum_{k=n+1}^{\infty} \|T\varphi_k\|^2 < \varepsilon^2$. Set $T_n \varphi_k = T\varphi_k$ for $k \leqq n$ and $T_n \varphi_k = 0$ for $k > n$. Clearly T_n can be extended to a degenerate linear operator, and $\|T_n - T\| \leqq \|T_n - T\|_2 < \varepsilon$. Hence T, as the limit in norm of $\{T_n\}$, is compact (see Theorem III-4.7).

Problem 2.16. If α_k are the repeated singular values of T,

$$\|T\|_2 = \Big(\sum_k \alpha_k^2\Big)^{1/2}. \tag{2.34}$$

Problem 2.17. If $T \in \mathscr{B}_2(\mathsf{H}, \mathsf{H}')$, the canonical expansion (2.23) of T converges in the norm $\|\ \|_2$.

Problem 2.18. $T \in \mathscr{B}_2(\mathsf{H}, \mathsf{H}')$ if and only if $T^* \in \mathscr{B}_2(\mathsf{H}', \mathsf{H})$ and

$$(S, T) = (T^*, S^*)\,. \tag{2.35}$$

Example 2.19. *Integral operators of Schmidt type.* Let $t(y, x)$ be a kernel defined for $x \in \mathsf{E}$, $y \in \mathsf{F}$, where E, F are measurable sets in euclidean spaces, and let

$$\|t\|^2 = \iint_{\mathsf{E}\times\mathsf{F}} |t(y, x)|^2\, dx\, dy < \infty\,. \tag{2.36}$$

Then the integral operator with kernel $t(y, x)$ defines an operator $T \in \mathscr{B}_2(\mathsf{H}, \mathsf{H}')$ where $\mathsf{H} = \mathsf{L}^2(\mathsf{E})$ and $\mathsf{H}' = \mathsf{L}^2(\mathsf{F})$.

To see this we first note that the formal expression III-(2.7) for Tu defines an operator $T \in \mathscr{B}(\mathsf{H}, \mathsf{H}')$ (see Problem III-2.5). To show that $T \in \mathscr{B}_2(\mathsf{H}, \mathsf{H}')$, let $\{\varphi_k(x)\}$ and $\{\varphi'_j(y)\}$ be complete orthonormal families in H and H', respectively. We have by (2.27) and (2.28)

$$\begin{aligned}\|T\|_2^2 &= \sum_{j,k} |(T\varphi_k, \varphi'_j)|^2 = \sum_{j,k} \Big|\iint_{\mathsf{E}\times\mathsf{F}} t(y, x)\, \varphi_k(x)\, \overline{\varphi'_j(y)}\, dx\, dy\Big|^2 \\ &= \iint_{\mathsf{E}\times\mathsf{F}} |t(y, x)|^2\, dx\, dy = \|t\|^2\,,\end{aligned} \tag{2.37}$$

where we have used the fact that the functions $\overline{\varphi_k(x)}\, \varphi'_j(y)$ form a complete orthonormal family in the Hilbert space $\mathsf{L}^2(\mathsf{E} \times \mathsf{F})$ (see Example 1.10).

Let $s(y, x)$, $t(y, x) \in \mathsf{L}^2(\mathsf{E} \times \mathsf{F})$ and define the associated integral operators $S, T \in \mathscr{B}_2(\mathsf{H}, \mathsf{H}')$ as above. Then

$$(S, T) = \iint_{\mathsf{E}\times\mathsf{F}} s(y, x)\, \overline{t(y, x)}\, dx\, dy\,; \tag{2.38}$$

this follows from (2.37) by polarization. (2.38) shows that $t \to T$ is an isometric transformation from $\mathsf{L}^2(\mathsf{E} \times \mathsf{F})$ to $\mathscr{B}_2(\mathsf{H}, \mathsf{H}')$. Actually this transformation is unitary: every $T \in \mathscr{B}_2(\mathsf{H}, \mathsf{H}')$ is obtained in this way from a kernel $t \in \mathsf{L}^2(\mathsf{E} \times \mathsf{F})$. To see this it suffices to recall the canonical expansion (2.23) of T, which converges in the norm $\|\ \|_2$ (Problem 2.17). Since the partial sum of this expansion is an integral operator with the kernel

$$t_n(y, x) = \sum_{k=1}^{n} \alpha_k\, \varphi'_k(y)\, \overline{\varphi_k(x)} \quad \text{with} \quad \|t_n\|^2 = \sum_{k=1}^{n} \alpha_k^2, \tag{2.39}$$

it follows that T is the integral operator with the kernel $t(y, x)$ which is the limit in $\mathsf{L}^2(\mathsf{E} \times \mathsf{F})$ of $t_n(y, x)$.

5. Perturbation of orthonormal families

As an application of the theory of the Schmidt class, let us consider perturbation of complete orthonormal families in a separable Hilbert space H. Let $\{\varphi_j\}$, $j = 1, 2, \ldots$, be a complete orthonormal family in H. Let $\{\psi_j\}$ be a sequence of vectors of H, not necessarily orthonormal, such that the differences $\psi_j - \varphi_j$ are small in some sense or other. The question is under what conditions $\{\psi_j\}$ is fundamental (complete) or is a basis of H. We recall that $\{\psi_j\}$ is fundamental if the set of all linear combinations of the ψ_j is dense in H. $\{\psi_j\}$ is called a *basis* of H if every vector u of H can be expressed in the form

$$u = \sum_{j=1}^{\infty} \eta_j \psi_j = \text{s-lim}_{n\to\infty} \sum_{j=1}^{n} \eta_j \psi_j \tag{2.40}$$

by means of a *unique* sequence $\{\eta_j\}$ of complex numbers. For a sequence $\{\psi_j\}$, being a basis is a stronger property than being fundamental.

A convenient measure of the smallness of the $\psi_j - \varphi_j$ is given by

$$r^2 = \sum_{j=1}^{\infty} \|\psi_j - \varphi_j\|^2 . \tag{2.41}$$

Theorem 2.20.[1] *Let $\{\varphi_j\}$ be a complete orthonormal family in* H *and let $\{\psi_j\}$ be a sequence such that $r^2 < \infty$. Then $\{\psi_j\}$ is a basis of* H *if* (2.40) *with $u = 0$ implies that all $\eta_j = 0$*[2].

Proof. We define a linear operator T in H by

$$Tu = \sum_{j=1}^{\infty} \xi_j \psi_j \quad \text{for} \quad u = \sum_{j=1}^{\infty} \xi_j \varphi_j . \tag{2.42}$$

To see that this definition is possible, it suffices to note that any $u \in$ H has a unique series expansion as in (2.42) with $\sum |\xi_j|^2 = \|u\|^2$ and that the series $Tu - u = \sum \xi_j (\psi_j - \varphi_j)$ converges absolutely by the Schwarz inequality:

$$\|Tu - u\|^2 \leq (\sum |\xi_j|^2)(\sum \|\psi_j - \varphi_j\|^2) = r^2 \|u\|^2 . \tag{2.43}$$

On setting $\xi_j = \delta_{jk}$ in (2.42), we see that $T\varphi_k = \psi_k$, $k = 1, 2, \ldots$. Hence

$$\|T - 1\|_2^2 = \sum \|(T - 1)\varphi_j\|^2 = \sum \|\psi_j - \varphi_j\|^2 = r^2 < \infty , \tag{2.44}$$

so that $A = T - 1 \in \mathscr{B}_2(\mathsf{H})$ with $\|A\|_2 \leq r$. In particular A is compact and so $T = 1 + A$ has an inverse $T^{-1} \in \mathscr{B}(\mathsf{H})$, for 0 is not an eigenvalue of T (see Theorem III-6.26). In fact $Tu = 0$ implies $\Sigma\, \xi_j \psi_j = 0$, and hence all $\xi_j = 0$ by hypothesis, so that $u = 0$.

Thus the range of T is the whole space H, and (2.42) shows that any element $v = Tu$ of H has an expansion $\Sigma\, \xi_j \psi_j$. The uniqueness of this expansion follows from the last assumption of the theorem.

It is easily seen that this assumption is satisfied if r is sufficiently small; for example, $r < 1$ is sufficient, as is seen from the above proof by noting that $T = 1 + A$ has a bounded inverse because $\|A\| \leq \|A\|_2 \leq$

[1] See BARY [1], KREIN [4].

[2] The last condition is expressed by saying that the ψ_j are ω-linearly independent.

$\leq r < 1$ (Neumann series for T^{-1}). A somewhat weaker condition is given by

Theorem 2.21. *Let $\{\varphi_j\}$ be a complete orthonormal family in H. Then a sequence $\{\psi_j\}$ of nonzero vectors of H is a basis if*

$$\sum_{j=1}^{\infty} \left(\|\psi_j - \varphi_j\|^2 - \frac{|(\psi_j - \varphi_j, \psi_j)|^2}{\|\psi_j\|^2}\right) < 1 . \tag{2.45}$$

Proof. As is readily seen from the definition, $\{\psi_j\}$ is a basis if and only if $\{\varrho_j \psi_j\}$ is a basis, where the ϱ_j are any nonzero complex numbers. Therefore, $\{\psi_j\}$ is a basis if we can choose the ϱ_j in such a way that $\sum\|\varrho_j \psi_j - \varphi_j\|^2 < 1$ (see the above remark). Naturally the best choice of the ϱ_j is obtained by minimizing each $\|\varrho_j \psi_j - \varphi_j\|$. If P_j is the orthogonal projection on the one-dimensional subspace spanned by ψ_j, this minimum r_j' is given by $r_j'^2 = \|(1 - P_j)\varphi_j\|^2 = \|(1 - P_j)(\varphi_j - \psi_j)\|^2 = \|\varphi_j - \psi_j\|^2 - \|P_j(\varphi_j - \psi_j)\|^2 = \|\varphi_j - \psi_j\|^2 - |(\varphi_j - \psi_j, \psi_j)|^2/\|\psi_j\|^2$. (2.45) is exactly the condition that $\sum r_j'^2 < 1$. Note that, under (2.45), the minimizing ϱ_j are not zero because $r_j' < 1$.

Corollary 2.22.[1] *Each of the following conditions is sufficient for $\{\psi_j\}$ to be a basis, where we write $\|\psi_j - \varphi_j\| = r_j$.*

i) $\sum r_j^2 < 1$.

ii) $\|\psi_j\| = 1$ *for all j and* $\sum r_j^2 \left(1 - \frac{1}{4} r_j^2\right) < 1$.

iii) $(\psi_j, \varphi_j) = 1$ *for all j and* $\sum r_j^2/(1 + r_j^2) < 1$.

Proof. It is trivial that i) implies (2.45). To deal with the cases ii) and iii), we set $\chi_j = \psi_j - \varphi_j$; then, if $\|\psi_j\| = 1$, $1 = \|\varphi_j\|^2 = \|\psi_j - \chi_j\|^2 = 1 - 2\,\mathrm{Re}(\chi_j, \psi_j) + \|\chi_j\|^2$, hence $|(\chi_j, \psi_j)| \geq \mathrm{Re}(\chi_j, \psi_j) = \|\chi_j\|^2/2 = r_j^2/2$ and ii) implies (2.45). If $(\psi_j, \varphi_j) = 1$, then $(\chi_j, \varphi_j) = 0$ so that $(\psi_j - \varphi_j, \psi_j) = (\chi_j, \varphi_j + \chi_j) = \|\chi_j\|^2 = r_j^2$ and $\|\psi_j\|^2 = \|\varphi_j + \chi_j\|^2 = 1 + \|\chi_j\|^2 = 1 + r_j^2$. Thus iii) implies (2.45).

Corollary 2.23.[2] *Let $\{\varphi_j\}$ be a complete orthonormal family and let $\{\psi_j\}$ be an orthonormal family. Then $\{\psi_j\}$ is complete if $\sum r_j^2 \left(1 - \frac{1}{4} r_j^2\right) < \infty$, where $r_j = \|\psi_j - \varphi_j\|$.*

Proof. As in the proof of Corollary 2.22, the assumption implies that $\sum \|\varrho_j \psi_j - \varphi_j\|^2$ converges for a suitable sequence ϱ_j of complex numbers. It may happen that some of the ϱ_j are zero, but there are only a finite number of such ϱ_j since $\|\varphi_j\| = 1$. Hence we can replace these vanishing ϱ_j by $\varrho_j = 1$, say, with the series $\sum\|\varrho_j \psi_j - \varphi_j\|^2$ still converging with all $\varrho_j \neq 0$. Furthermore, $\sum \eta_j \varrho_j \psi_j = 0$ implies $\eta_j = 0$ since the ψ_j are orthogonal and nonvanishing. Hence the $\varrho_j \psi_j$ form a basis of H by Theorem 2.20 and the same is true of the ψ_j.

---

[1] See HILDING [1].

[2] See ISEKI [1], BIRKHOFF and ROTA [1].

§ 3. Unbounded operators in Hilbert spaces

1. General remarks

Let H, H′ be Hilbert spaces and let T be an operator from H to H′. If T is densely defined, the adjoint T^* is defined and is an operator from $H' = H'^*$ to $H = H^*$. According to III-(5.9), the inverse graph $G'(-T^*)$ of $-T^*$ is the annihilator of the graph $G(T)$ of T. But now the product space $H \times H'$ in which $G(T)$ and $G(-T^*)$ lie is a Hilbert space, the inner product of two elements $\{u, u'\}$ and $\{v, v'\}$ being defined as equal to $(u, v) + (u', v')$ [which is compatible with our definition III-(5.4) of the norm in $H \times H'$]. If T is closed, $G(T)$ and $G'(-T^*)$ are thus orthogonal complements to each other in $H \times H'$.

In particular we have Theorem III-5.29: if T is closable, then T^* is closed, densely defined and $T^{**} = \tilde{T}$ (the closure of T). If in particular $T \in \mathscr{C}(H, H')$, then $T^{**} = T$. Conversely T is closable if T^* is densely defined, for $T^{**} \supset T$.

Let $T \in \mathscr{C}(H, H')$ be densely defined. Since the relationship between T and T^* is symmetric, we have (see Problem III-5.27)

$$N(T^*) = R(T)^{\perp}, \quad N(T) = R(T^*)^{\perp}. \tag{3.1}$$

We note further that $R(T^*)$ is closed if and only if $R(T)$ is (Theorem IV-5.13); in this case we have

$$H = N(T) \oplus R(T^*), \quad H' = N(T^*) \oplus R(T). \tag{3.2}$$

2. The numerical range

For operators in a Hilbert space H, the notion of *numerical range* (or *field of values*) is important in various applications.

Let T be an operator in H. The numerical range $\Theta(T)$ of T is the set of all complex numbers (Tu, u) where u changes over all $u \in D(T)$ with $\|u\| = 1$. (We assume $\dim H > 0$).

In general $\Theta(T)$ is neither open nor closed, even when T is a closed operator. The following theorem of HAUSDORFF is important, but we omit the proof[1].

Theorem 3.1. *$\Theta(T)$ is a convex set.*

Let us denote by Γ the closure of $\Theta(T)$; Γ is a closed convex set. Let Δ be the complement of Γ in the complex plane. In view of the convexity of Γ, a little geometric consideration leads to the following result: Δ is a *connected* open set except in the special case in which Γ is a strip bounded by two parallel straight lines (the limiting case is included in which the two lines coincide). In this exceptional case, Δ consists of two components Δ_1, Δ_2 which are half-planes.

[1] For the proof see STONE [1], p. 131.

Theorem 3.2. *Let $T \in \mathscr{C}(\mathsf{H})$, and let Γ, Δ, Δ_1, Δ_2 be as above. For any $\zeta \in \Delta$, $T - \zeta$ has closed range,* $\operatorname{nul}(T - \zeta) = 0$, *and* $\operatorname{def}(T - \zeta)$ *is constant for $\zeta \in \Delta$, except in the special case mentioned above, in which* $\operatorname{def}(T - \zeta)$ *is constant in each of Δ_1 and Δ_2. [This constant value (or pair of values) is called the deficiency index of T.] If* $\operatorname{def}(T - \zeta) = 0$ *for $\zeta \in \Delta$ ($\zeta \in \Delta_1$ or $\zeta \in \Delta_2$) then Δ (Δ_1 or Δ_2) is a subset of* $\mathrm{P}(T)$ *and* $\|R(\zeta, T)\| \leqq 1/\operatorname{dist}(\zeta, \Gamma)$.

Proof. This theorem is a consequence of the first stability theorem for nullity and deficiency (see Theorem IV-5.17). First we note that $|(Tu, u) - \zeta| = |((T - \zeta) u, u)| \leqq \|(T - \zeta) u\|$ for any $u \in \mathsf{D}(T)$ with $\|u\| = 1$ and any complex number ζ. If $\zeta \in \Delta$ so that $\operatorname{dist}(\zeta, \Gamma) = \delta > 0$, it follows that $\|(T - \zeta) u\| \geqq \delta$ for $\|u\| = 1$ or

$$\|(T - \zeta) u\| \geqq \delta \|u\| \quad \text{for every} \quad u \in \mathsf{D}(T) . \tag{3.3}$$

This implies that $\operatorname{nul}(T - \zeta) = 0$ and $\gamma(T - \zeta) \geqq \delta$ (see IV-§ 5.1 for the definition of γ), and hence that $\mathsf{R}(T - \zeta)$ is closed (Theorem IV-5.2). Now it follows from Theorem IV-5.17 that $\operatorname{def}(T - \zeta) = -\operatorname{ind}(T - \zeta)$ is constant in Δ if Δ is connected and in each Δ_k otherwise.

Corollary 3.3. *If $T \in \mathscr{B}(\mathsf{H})$, then $\Sigma(T)$ is a subset of the closure of $\Theta(T)$.*

Proof. $\Theta(T)$ and its closure Γ are bounded since $|(Tu, u)| \leqq \|T\|$ for $\|u\| = 1$. Hence Δ is a connected open set containing the exterior of the circle $|\zeta| = \|T\|$. But this exterior belongs to $\mathrm{P}(T)$ so that $\operatorname{def}(T - \zeta) = 0$ there. Hence the same must be true for all $\zeta \in \Delta$ by Theorem 3.2. Since we have also $\operatorname{nul}(T - \zeta) = 0$ for $\zeta \in \Delta$, it follows that $\Delta \subset \mathrm{P}(T)$, which is equivalent to $\Gamma \supset \Sigma(T)$.

Theorem 3.4. *If T is densely defined and $\Theta(T)$ is not the whole complex plane, then T is closable (hence T^* is also densely defined).*

Proof. Since $\Theta(T)$ is a convex set not identical with the whole plane, $\Theta(T)$ is contained in a half-plane. Replacing T by $\alpha T + \beta$ with some complex numbers α, β, we may thus assume that $\Theta(T)$ is contained in the right half-plane. This implies that

$$\operatorname{Re}(Tu, u) \geqq 0 \quad \text{for} \quad u \in \mathsf{D}(T) . \tag{3.4}$$

Suppose now that we have a sequence $u_n \in \mathsf{D}(T)$ such that $u_n \to 0$, $Tu_n \to v$; it suffices to show that $v = 0$.

For any $w \in \mathsf{D}(T)$, we have

$$\begin{aligned} 0 &\leqq \operatorname{Re}(T(u_n + w), u_n + w) \\ &= \operatorname{Re}[(Tu_n, u_n) + (Tu_n, w) + (Tw, u_n) + (Tw, w)] . \end{aligned}$$

This gives for $n \to \infty$

$$0 \leqq \operatorname{Re}(v, w) + \operatorname{Re}(Tw, w) .$$

If we replace w by αw, $\alpha > 0$, divide the resulting inequality by α and let $\alpha \to 0$, we obtain $0 \leqq \operatorname{Re}(v, w)$. Since $w \in \mathsf{D}(T)$ is arbitrary and $\mathsf{D}(T)$ is dense in H, v must be zero.

Problem 3.5. What is the numerical range of the operator T defined by the matrix $\begin{pmatrix} \alpha & \beta \\ \gamma & \delta \end{pmatrix}$ in C^2 (two dimensional unitary space)?

Problem 3.6. If $T \in \mathscr{C}(H)$, the essential spectrum of T (see IV-§ 5.6) is a subset of the closure of $\Theta(T)$.

Problem 3.7. If T is closable, then $\Theta(T)$ is dense in $\Theta(\tilde{T})$, and $\Theta(T)$ and $\Theta(\tilde{T})$ have the same closure.

3. Symmetric operators

An operator T in a Hilbert space H is said to be *symmetric* if T is densely defined and

$$T^* \supset T . \tag{3.5}$$

T is said to be *selfadjoint* if in addition

$$T^* = T . \tag{3.6}$$

If $D(T) = H$, (3.5) implies $T^* = T$. Thus the definition of a symmetric operator given here is in accordance with that given in § 2.1 for bounded operators.

Problem 3.8. T is symmetric if and only if it is densely defined and

$$(Tu, v) = (u, Tv) \quad \text{for every} \quad u, v \in D(T) . \tag{3.7}$$

Problem 3.9. A densely defined operator T is symmetric if and only if the numerical range $\Theta(T)$ is a subset of the real line.

(3.7) shows that (Tu, u) is real. If $(Tu, u) \geqq 0$, the symmetric operator T is said to be nonnegative (in symbol $T \geqq 0$). (3.5) shows that *a symmetric operator is closable,* for T^* is closed. Since (3.5) implies $T^{***} \supset T^{**}$ and since T^{**} is the closure of T, it follows that *the closure of a symmetric operator is symmetric.* A symmetric operator T is said to be *essentially selfadjoint* if its closure T^{**} is selfadjoint.

Problem 3.10. If T is symmetric, the following propositions are equivalent:
a) T is essentially selfadjoint.
b) T^* is symmetric.
c) T^* is selfadjoint.
d) T^{**} is selfadjoint.

Problem 3.11. If T is symmetric and invertible with dense range, then T^{-1} is symmetric.

Problem 3.12. A closed symmetric operator is bounded if and only if it has domain H. [hint: The closed graph theorem.]

Example 3.13. Let (τ_{jk}) be a (Hermitian) symmetric matrix:

$$\overline{\tau_{jk}} = \tau_{kj} , \quad j, k = 1, 2, 3, \ldots , \tag{3.8}$$

and assume that

$$\sum_k |\tau_{jk}|^2 < \infty \quad \text{for each } j . \tag{3.9}$$

We shall show that we can associate with (τ_{jk}) a symmetric operator T_0 in $H = l^2$. Let D be the linear manifold of H spanned by the canonical basis $x_n = (\delta_{kn})$. D is a subset of the domain of the maximal operator T defined as in Example III-2.3, for $T x_n = (\tau_{kn})$ belongs to H by (3.8) and (3.9). We define T_0 as the restriction of T with domain D. T_0 is symmetric since D is dense in H and (3.7) is satisfied for T replaced by T_0 in virtue of (3.8).

In general T_0 is neither closed nor essentially selfadjoint, but it can be shown[1] that $T_0^* = T$. Thus T is symmetric only if T_0 is essentially selfadjoint, in which case T is even selfadjoint (see Problem 3.10). It is in general a rather complicated task to decide whether or not this is true for a given matrix (τ_{jk}). A very special case in which this is true is that of a diagonal matrix: $\tau_{jk} = \lambda_j \delta_{jk}$ with real diagonal elements λ_j. The proof will be given later, but a direct proof is also simple.

Example 3.14. Let $\mathsf{H} = \mathsf{L}^2(a, b)$ for a finite or infinite interval (a, b), and consider the differential operator d/dx. Let T and $\dot{T}$ be the associated maximal and minimal operators in H (see Example III-2.7). It follows from Example III-5.31 that $-\dot{T}^* = T \supset \dot{T}$, which implies that $i\dot{T}$ is symmetric. Similarly, it can be shown that $i T_0$ in the same example is symmetric [for finite (a, b)].

iT is not symmetric for a finite interval (a, b), for $-T^* = T_0$ is a proper restriction of T. If $(a, b) = (-\infty, +\infty)$, on the other hand, $iT = (i\dot{T})^*$ is not only symmetric but selfadjoint. To see this it suffices to show that iT is symmetric (Problem 3.10). Let $u, v \in \mathsf{D}(T)$. Since $Tu = u'$ we have $(Tu)\bar{v} + u\overline{Tv} = (u\bar{v})'$ and

$$\lim_{b' \to \infty} u(b')\overline{v(b')} = u(0)\overline{v(0)} + \lim_{b' \to \infty} \int_0^{b'} ((Tu)\bar{v} + u\overline{Tv})\, dx \tag{3.10}$$

exists; note that u, v, Tu, Tv all belong to $\mathsf{H} = \mathsf{L}^2$. The limit (3.10) must be zero, for otherwise $u\bar{v}$ would not be integrable in spite of the fact that $u, v \in \mathsf{L}^2$. Similarly $u(x)\overline{v(x)} \to 0$ for $x \to -\infty$, so that

$$(Tu, v) + (u, Tv) = \lim_{b' \to \infty} u(b')\overline{v(b')} - \lim_{a' \to -\infty} u(a')\overline{v(a')} = 0\,. \tag{3.11}$$

This shows that iT is symmetric.

For a finite (a, b), we have defined the operators T_1, T_2, T_3 which are restrictions of T and extensions of T_0 (see Example III-2.7). Neither of iT_1 and iT_2 is symmetric, as is easily verified. iT_3 is symmetric if and only if the constant k that appears in the boundary condition $u(b) = k\,u(a)$ is of absolute value one ($k = e^{i\theta}$ with real θ).

4. The spectra of symmetric operators

Let T be a closed symmetric operator in H; since every symmetric operator is closable, there is no essential restriction in assuming T to be closed. An important property of T is the equality

$$\|(T - \zeta)\,u\|^2 = \|(T - \operatorname{Re}\zeta)\,u\|^2 + (\operatorname{Im}\zeta)^2\|u\|^2\,, \quad u \in \mathsf{D}(T)\,. \tag{3.12}$$

This is easily verified by noting that $T - \operatorname{Re}\zeta$ is symmetric as well as T. (3.12) gives the inequality

$$\|(T - \zeta)\,u\| \geqq |\operatorname{Im}\zeta|\,\|u\|\,, \quad u \in \mathsf{D}(T)\,. \tag{3.13}$$

This implies that $T - \zeta$ has a bounded inverse with bound not exceeding $|\operatorname{Im}\zeta|^{-1}$. Hence $\mathsf{R}(T - \zeta)$ is closed for nonreal ζ. It follows from the first stability theorem of nullity and deficiency (Theorem IV-5.17) that $\operatorname{def}(T - \zeta)$ is constant in each of the half-planes $\operatorname{Im}\zeta \gtrless 0$. This is also a

[1] See STONE [1], p. 90.

direct consequence of Theorem 3.2; note that the numerical range $\Theta(T)$ of T is a subset of the real axis and therefore Δ of Theorem 3.2 either is a connected set or consists of the two half-planes mentioned. The pair (m', m'') of the constant values of $\operatorname{def}(T - \zeta)$ for $\operatorname{Im}\zeta \gtrless 0$ is called the *deficiency index* of the symmetric operator T. If T is not closed, the deficiency index of T is defined as equal to that of $\tilde{T}$, the closure of T.

According to (3.1) (applied to $T - \zeta$), $T^* - \zeta = (T - \bar{\zeta})^*$ has range H and nullity m'' or m' according as $\operatorname{Im}\zeta \gtrless 0$.

If $m' = 0$, $\mathsf{R}(T - \zeta)$ for $\operatorname{Im}\zeta > 0$ is H and $R(\zeta, T) = (T - \zeta)^{-1} \in \mathscr{B}(\mathsf{H})$ exists. In other words, the upper half-plane belongs to the resolvent set $\mathrm{P}(T)$. If $m' > 0$, no ζ in the upper half-plane belongs to $\mathrm{P}(T)$. Similar results hold for m'' and ζ with $\operatorname{Im}\zeta < 0$. Thus there are the following alternatives:

i) $m' = m'' = 0$. All nonreal numbers belong to $\mathrm{P}(T)$; the spectrum $\Sigma(T)$ is a subset of the real axis.

ii') $m' = 0, m'' > 0$. $\mathrm{P}(T)$ is the open upper half-plane $\operatorname{Im}\zeta > 0$ and $\Sigma(T)$ is the closed lower half-plane $\operatorname{Im}\zeta \leqq 0$.

ii'') $m' > 0, m'' = 0$. "upper" and "lower" exchanged in ii').

iii) $m' > 0, m'' > 0$. $\mathrm{P}(T)$ is empty and $\Sigma(T)$ is the whole plane.

Problem 3.15. The case i) occurs whenever $\mathrm{P}(T)$ contains at least one real number.

A closed symmetric operator is said to be *maximal* if at least one of m', m'' is zero [the case i), ii') or ii'')]. *A maximal symmetric operator has no proper symmetric extension.* To show this, we may assume that $m' = 0$. Let T_1 be a symmetric extension of T; T_1 may be assumed to be closed (otherwise take the closure). For any $u \in \mathsf{D}(T_1)$, there is a $v \in \mathsf{D}(T)$ such that $(T - i)\, v = (T_1 - i)\, u$ because $\mathsf{R}(T - i) = \mathsf{H}$. Since $T \subset T_1$, we may write $(T_1 - i)\,(u - v) = 0$. Since T_1 is symmetric, this gives $u - v = 0$. This implies that $\mathsf{D}(T_1) \subset \mathsf{D}(T)$ and hence $T_1 = T$.

If both m' and m'' are zero [case i)], T is selfadjoint. We have namely

Theorem 3.16. *A closed symmetric operator T has deficiency index* (0, 0) *if and only if T is selfadjoint. In this case the resolvent* $R(\zeta, T) = (T - \zeta)^{-1}$ *exists for* $\operatorname{Im}\zeta \neq 0$ *and*

$$\|R(\zeta, T)\| \leqq |\operatorname{Im}\zeta|^{-1}, \quad \|(T - \operatorname{Re}\zeta)\, R(\zeta, T)\| \leqq 1\,. \tag{3.14}$$

Proof. Assume that T has deficiency index (0, 0). Then both $T \pm i$ have range H, so that for each $u \in \mathsf{D}(T^*)$ there is a $v \in \mathsf{D}(T)$ such that $(T - i)\, v = (T^* - i)\, u$. Since $T \subset T^*$, this may be written $(T^* - i) \cdot (u - v) = 0$. But $T^* - i$ has nullity zero by $m'' = 0$, so that $u - v = 0$. This shows that $\mathsf{D}(T^*) \subset \mathsf{D}(T)$ and hence $T^* = T$.

Conversely assume that $T^* = T$. Then T^* is also symmetric and the nullities of $T^* \pm i$ must be zero. Since these nullities are the deficiencies of $T \mp i$, the deficiency index of T is (0, 0).

(3.14) follows directly from (3.12).

Problem 3.17. In order that a symmetric operator T be essentially selfadjoint, each of the following conditions is necessary and sufficient (cf. also Problem 3.10):

a) T^* has no nonreal eigenvalue.

b) $\mathsf{R}(T-\zeta)$ is dense in H for every nonreal ζ.

c) $\mathsf{R}(T-\zeta)$ is dense in H for some $\zeta=\zeta'$ with $\operatorname{Im}\zeta'>0$ and also for some $\zeta=\zeta''$ with $\operatorname{Im}\zeta''<0$.

Problem 3.18. If T is selfadjoint and invertible, T^{-1} is also selfadjoint. [hint: Use, for example, Problem 3.11 and Theorem III-6.15. Note that T^{-1} is densely defined since $u\in\mathsf{R}(T)^\perp$ implies that $Tu=T^*u=0$, $u=0$.]

Problem 3.19. If T is a closed symmetric operator with finite deficiency index, then T^* is a finite extension of T (see III-§ 2.1).

Problem 3.20. Let T, S be closed symmetric operators and let S be a finite extension of T of order r. If T has deficiency index (m', m''), then S has deficiency index $(m'-r, m''-r)$.

Example 3.21. Let us determine the deficiency index of the closed symmetric operator $i\,T_0$ of Example 3.14. Since $(i\,T_0)^*=i\,T$, it suffices to solve the equations $i\,Tu=\pm i\,u$, that is, $u'=\pm u$. The solutions of these differential equations are $u(x)=c\,e^{\pm x}$ (c is a constant). If (a,b) is finite, these solutions belong to $\mathsf{D}(T)$ for any c and satisfy $i\,Tu=\pm i\,u$. Thus both $i\,T\mp i$ have nullity one; the deficiency index of $i\,T_0$ is $(1,1)$.

If $(a,b)=(-\infty,+\infty)$, the above solutions do not belong to H unless $c=0$. Thus the nullity of $i\,T\pm i$ is zero and the deficiency index of $i\,\dot{T}$ is $(0,0)$[1]. This is in conformity with the result stated in Example 3.14 that $i\,T=(i\,\dot{T})^*$ is in this case selfadjoint.

Next consider the case of a semi-infinite interval (a,b), say $a=0$, $b=\infty$. Then the function e^{-x} belongs to $\mathsf{D}(T)$ but e^x does not. Thus the nullity of $i\,T-i$ is zero whereas that of $i\,T+i$ is one, and the deficiency index of $i\,\dot{T}$ is $(1,0)$. The closure of $i\,\dot{T}$ is maximal symmetric but not selfadjoint and has no selfadjoint extension. (Actually $\tilde{\dot{T}}=T_1$.)

Finally consider the symmetric operator $i\,T_3\subset i\,T$ for a finite (a,b), with the boundary condition $u(b)=e^{i\theta}u(a)$. Since $i\,T_3$ is symmetric and a direct extension of $i\,T_0$ and since $i\,T_0$ has deficiency index $(1,1)$, $i\,T_3$ has deficiency index $(0,0)$ and is therefore selfadjoint.

Problem 3.22. Let $\mathsf{H}=\mathsf{L}^2(E)$ and let T be the maximal multiplication operator by a real-valued measurable function $f(x)$ (see Example III-2.2). T is selfadjoint.

5. The resolvents and spectra of selfadjoint operators

Let T be a selfadjoint operator in H. Its spectrum $\Sigma(T)$ is a subset of the real axis, and the resolvent $R(\zeta)=R(\zeta,T)=(T-\zeta)^{-1}$ is defined at least for all nonreal ζ. We have from III-(6.50)

$$R(\zeta)^*=R(\bar{\zeta})\,. \tag{3.15}$$

It follows that $R(\zeta)$ is normal, for the $R(\zeta)$ for different ζ commute with each other.

In particular, we have $\|R(\zeta)\|=\operatorname{spr}R(\zeta)$ by (2.4) and hence by Problem III-6.16

$$\|R(\zeta)\|=1/\operatorname{dist}(\zeta,\Sigma(T))\leq|\operatorname{Im}\zeta|^{-1}\,. \tag{3.16}$$

[1] Note that $i\,\dot{T}$ is not closed.

In the same way we have

$$\|TR(\zeta)\| = \sup_{\lambda\in\Sigma(T)} |\lambda|\,|\lambda-\zeta|^{-1}, \tag{3.17}$$

for $TR(\zeta) = 1 + \zeta R(\zeta)$ so that $\|TR(\zeta)\| = \operatorname{spr}(1+\zeta R(\zeta)) = \sup|1 + \zeta(\lambda-\zeta)^{-1}| = \sup|\lambda(\lambda-\zeta)^{-1}|$. [Note that the spectrum of $1+\zeta R(\zeta)$ is the image of $\Sigma(T)$ under the map $\lambda \to 1 + \zeta(\lambda-\zeta)^{-1}$.]

If a real number α belongs to the resolvent set $P(T)$, $P(T)$ contains a neighborhood of α; we say then that $\Sigma(T)$ has a *gap* at α. Note that $P(T)$ is a connected set if $\Sigma(T)$ has at least one gap.

Suppose that $\Sigma(T)$ has gaps at α and β, $\alpha<\beta$. Let Γ be a closed curve passing through α, β and enclosing the part of $\Sigma(T)$ between α and β. Further we assume that Γ is symmetric with respect to the real axis [for example we may take as Γ a circle with the segment (α,β) as a diameter]. Then $\Sigma(T)$ is separated by Γ into two parts, one Σ' being in the interval (α,β) and the other Σ'' outside this interval. Let

$$\mathsf{H} = \mathsf{M}' \oplus \mathsf{M}'' \tag{3.18}$$

be the associated decomposition of H (see III-§ 6.4). Here M', M'' are orthogonal complements to each other, for the projection P on M' along M'' is orthogonal:

$$P^* = P, \tag{3.19}$$

as is seen from the expressions for P and P^* given by III-(6.52), which coincide in virtue of $T^* = T$ and the symmetry of Γ with respect to the real axis.

If $\Sigma(T)$ has an isolated point λ (which is necessarily real), we have the Laurent expansion III-(6.32) for $R(\zeta) = R(\zeta, T)$. Here not only P but D and S are also symmetric (and bounded). $D^* = D$ follows from III-(6.28) and $S^* = S$ from III-(6.31). But D as a symmetric quasinilpotent operator must be zero (see Problem 2.4). Hence $(T-\lambda)P = 0$ and λ is an *eigenvalue* of T with the associated eigenspace $\mathsf{M}' = P\mathsf{H}$. Note that, in this case, M' is the geometric as well as the algebraic eigenspace. We have further

$$\|S\| = 1/d \quad \text{with} \quad d = \operatorname{dist}(\lambda, \Sigma'') \tag{3.20}$$

where Σ'' is the spectrum of T with the single point λ excluded. We shall call d the *isolation distance* of the eigenvalue λ. (3.20) can be proved as in (3.16), noting that S may be regarded as the value at λ of the resolvent of the part $T_{\mathsf{M}''}$ of T, of which the spectrum is exactly Σ'' [see III-(6.31)] and that $\mathsf{M}' \perp \mathsf{M}''$.

Similar results hold when there are several isolated points $\lambda_1, \ldots, \lambda_s$ of $\Sigma(T)$. We have [see III-(6.35)]

$$R(\zeta) = -\sum_{h=1}^{s} \frac{P_h}{\zeta-\lambda_h} + R_0(\zeta), \tag{3.21}$$

where the P_h have the properties

(3.22) $$P_h^* = P_h, \quad P_h P_k = \delta_{hk} P_h,$$

and where $R_0(\zeta)$ is holomorphic at $\zeta = \lambda_h$, $h = 1, \ldots, s$. Again the λ_h are eigenvalues of T with the associated eigenspaces $\mathsf{M}_h = P_h \mathsf{H}$, which are orthogonal to one another.

6. Second-order ordinary differential operators

Consider the formal differential operator

(3.23) $$Lu = -\frac{d}{dx} p(x) \frac{du}{dx} + q(x) u, \quad a < x < b,$$

where $p(x)$ is positive and continuously differentiable and $q(x)$ is real and continuous, both in the open interval (a, b). L is *formally selfadjoint*: $M = L$ where M is the formal adjoint to L defined by III-(2.27). Thus we have

(3.24) $$\int_{a'}^{b'} ((Lu) v - u L v)\, dx = [p u v' - p u' v]_{a'}^{b'}, \quad a < a' < b' < b.$$

Let us consider the linear operators T, $\dot{T}$, etc. constructed from L in III-§ 2.3, choosing now $\mathsf{X} = \mathsf{H} = \mathsf{L}^2(a, b)$ (see also Example III-5.32). We have $\dot{T}^* = T \supset \dot{T}$ by III-(5.14), so that *the minimal operator $\dot{T}$ is symmetric. $\dot{T}$ is essentially selfadjoint if and only if the maximal operator T* is symmetric (see Problem 3.10). Whether or not this is true depends on the properties of the coefficients $p(x)$, $q(x)$.

It is certainly not true if (a, b) is finite and $p(x)$, $q(x)$ are continuous on the closed interval $[a, b]$ with $p > 0$ (the *regular* case). In this case T is not symmetric; the closure of $\dot{T}$ is exactly the operator T_0 with the boundary condition $u(a) = u'(a) = u(b) = u'(b) = 0$ and T_0 has deficiency index (2, 2). This is obvious since $T_0^* = T$ and the equation $(T \pm i) u = 0$ has two linearly independent solutions (belonging to H). There are infinitely many selfadjoint operators H such that $T_0 \subset H \subset T$. Without trying to determine all such H[1], we note only that the T_1, T_2 and T_3 of III-§ 2.3 are examples of H; this follows immediately from the last results of Example III-5.32 in view of $S = T$.

As a typical case of a *singular* differential operator, let us now consider the case with $(a, b) = (-\infty, +\infty)$. We shall show that *$\dot{T}$ is essentially selfadjoint (and T is selfadjoint) if $p(x) > 0$ is bounded and $q(x)$ is bounded from below on $(-\infty, +\infty)$.* To this end we may assume without loss of generality that $q(x) \geqq 1$, for the addition of a real scalar to T does not affect its selfadjointness. Then

(3.25) $$(\dot{T} u, u) = \int_{-\infty}^{\infty} (p |u'|^2 + q |u|^2)\, dx \geqq \|u\|^2, \quad u \in \mathsf{D}(\dot{T}),$$

so that the numerical range $\Theta(\dot{T})$ of $\dot{T}$ lies to the right of $\zeta = 1$, and the same is true of $\Theta(\tilde{T})$ (see Problem 3.7). It follows that the exterior of $\Theta(\tilde{T})$ is a connected open set containing the origin. To show that $\tilde{T}$ is selfadjoint, it is therefore sufficient to show that $\mathsf{R}(\tilde{T})$ has deficiency 0 or, equivalently, that $T = \dot{T}^*$ has nullity zero (see par. 3 and 4).

Suppose now that $Tu = 0$. Take any smooth real-valued function $w(x)$ that vanishes outside of a finite interval. Then $v = w u$ belongs to $\mathsf{D}(\tilde{T})$ (see Problem III-5.33), and a direct calculation gives

$$\tilde{T} v = L v = w L u - (p w')' u - 2 p w' u'.$$

[1] For a complete determination of the H see STONE [1].

Since $Lu = Tu = 0$, it follows that

$$(3.26)\qquad (\tilde{T}v, v) = -\int_{-\infty}^{\infty} (p\,w')'\,w|u|^2\,dx - 2\int_{-\infty}^{\infty} p\,w'\,w\,u'\,\bar{u}\,dx$$

$$= \int_{-\infty}^{\infty} p\,w'^2|u|^2\,dx + \int_{-\infty}^{\infty} p\,w'\,w\,(u\,\bar{u}' - u'\bar{u})\,dx$$

(integration by parts). Since $(\tilde{T}v, v)$ is real while the first term on the right of (3.26) is real and the second pure imaginary, this last term must vanish. Since $(\tilde{T}v, v) \geqq \geqq \|v\|^2$, we thus obtain

$$(3.27)\qquad \int_{-\infty}^{\infty} w^2|u|^2\,dx = \|v\|^2 \leqq \int_{-\infty}^{\infty} p\,w'^2|u|^2\,dx \leqq c\int_{-\infty}^{\infty} w'^2|u|^2\,dx$$

where $p(x) \leqq c$ by assumption.

It is easy to deduce from (3.27) that $u = 0$. To this end let $w(x)$ be equal to one for $|x| \leqq r$ with $|w'| \leqq s$ everywhere. Then (3.27) gives

$$(3.28)\qquad \int_{-r}^{r} |u|^2\,dx \leqq \int_{-\infty}^{\infty} w^2|u|^2\,dx \leqq c\int_{-\infty}^{\infty} w'^2|u|^2\,dx \leqq c\,s^2 \int_{|x|\geqq r} |u|^2\,dx\,.$$

It is possible to choose r as large as we please while s is fixed. On letting $r \to \infty$ for a fixed s, we obtain $\int_{-\infty}^{\infty} |u|^2\,dx = 0$ since it is known that $u \in \mathsf{L}^2$. This gives the desired result $u = 0$.

Problem 3.23. The assumption made above that $p(x)$ is bounded can be replaced by the weaker one that the integrals of $p(x)^{-1/2}$ on the intervals $(-\infty, 0)$ and $(0, \infty)$ both diverge.

Finally let us consider the case $(a, b) = (0, \infty)$ and assume that $p(x)$ is positive and bounded and $q(x)$ is bounded from below on $[0, \infty)$[1]. An argument similar to that given above shows that $\tilde{T}$ has deficiency index (1, 1). A typical selfadjoint extension T_1 of $\tilde{T}$ is obtained by restricting T with the boundary condition $u(0) = 0$. Again the assumption that $p(x)$ be bounded may be weakened to $\int_0^{\infty} p(x)^{-1/2}\,dx = \infty$. Details may be left to the reader.

7. The operators $T^* T$

The following theorem due to VON NEUMANN is of fundamental importance.

Theorem 3.24. *Let H, H' be Hilbert spaces. Let $T \in \mathscr{C}(\mathsf{H}, \mathsf{H}')$ be densely defined. Then $T^* T$ is a selfadjoint operator in H, and $\mathsf{D}(T^* T)$ is a core of T.*

Proof. As noted in par. 1, the graphs $\mathsf{G}(T)$ and $\mathsf{G}'(-T^*)$ are complementary subspaces of the Hilbert space $\mathsf{H} \times \mathsf{H}'$. This implies that any vector $\{u, u'\}$ of $\mathsf{H} \times \mathsf{H}'$ can be expressed in the form $\{v, Tv\} + \{-T^*v', v'\}$ with some $v \in \mathsf{D}(T)$ and $v' \in \mathsf{D}(T^*)$. If in particular $u' = 0$, we have $u = v - T^*v'$ and $0 = Tv + v'$. Hence $Tv = -v' \in \mathsf{D}(T^*)$ and $u = (1 + T^*T)\,v$. Since $u \in \mathsf{H}$ was arbitrary, it follows that $S = 1 + T^*T$ has range H. But it is easily seen that S^{-1} is symmetric and $\|S^{-1}\| \leqq 1$.

[1] $p(x)$ may tend to 0 and $q(x)$ may go to $+\infty$ for $x \to \infty$.

Thus S^{-1} is symmetric and belongs to $\mathscr{B}(\mathsf{H})$ so that it is selfadjoint. Hence S and T^*T are also selfadjoint (see Problem 3.18). Note that this implies that T^*T is densely defined, which is by no means obvious.

To prove that $\mathsf{D} = \mathsf{D}(T^*T)$ is a core of T, it suffices to show that the set of all elements $\{v, Tv\}$ with $v \in \mathsf{D}$ is dense in $\mathsf{G}(T)$ (see III-§ 5.3). Thus we need only to show that an element $\{u, Tu\}$ with $u \in \mathsf{D}(T)$ orthogonal to all $\{v, Tv\}$ with $v \in \mathsf{D}$ must be zero. But this orthogonality implies that $0 = (u, v) + (Tu, Tv) = (u, (1 + T^*T)v) = (u, Sv)$. Since Sv fills the whole space H when v varies over D, this gives $u = 0$ as required.

Example 3.25. Consider the operators T, T_0, etc. of Example 3.14 for a finite interval (a, b). We know that $T_0^* = -T$, $T_1^* = -T_2$, $T_2^* = -T_1$, $T^* = -T_0$ (see also Example III-5.31). Thus we have $T^*T = -T_0T$; it is the differential operator $-d^2/dx^2$ with the boundary condition $Tu \in \mathsf{D}(T_0)$, that is, $u'(a) = u'(b) = 0$. (This is identical with the T_2 of III-§ 2.3 in the special case $p_0 = -1$, $p_1 = p_2 = 0$, $h_a = h_b = 0$.) Next we have $T_0^*T_0 = -TT_0$; this is the same differential operator $-d^2/dx^2$ with the boundary condition $u(a) = u(b) = 0$ (and coincides with the T_1 of the same example). Again, $T_1^*T_1 = -T_2T_1$ is the same differential operator with the boundary condition $u(a) = u'(b) = 0$, and $T_2^*T_2 = -T_1T_2$ is obtained from it by exchanging a and b. That these differential operators are all selfadjoint is a direct consequence of Theorem 3.24.

Example 3.26. A construction similar to the above one applied to the second-order differential operators of III-§ 2.3 for $\mathsf{X} = \mathsf{H} = \mathsf{L}^2$ will give rise to various differential operators of fourth order which are selfadjoint.

8. Normal operators

Selfadjoint operators are special cases of *normal operators*. The definition of a not necessarily bounded normal operator T in a Hilbert space H is formally the same as that of a normal operator of $\mathscr{B}(\mathsf{H})$: T is normal if T is closed, densely defined and

$$T^*T = TT^*; \tag{3.29}$$

note that both T^*T and TT^* are selfadjoint (preceding paragraph). The implication of (3.29) is rather complicated, however, on account of the domain relations involved.

(3.29) implies that $\|Tu\| = \|T^*u\|$ for $u \in \mathsf{D} = \mathsf{D}(T^*T) = \mathsf{D}(TT^*)$. Since D is a core for both T and T^*, it follows that $\mathsf{D}(T) = \mathsf{D}(T^*) = \mathsf{D}_1 (\supset \mathsf{D})$ and $\|Tu\| = \|T^*u\|$ for any $u \in \mathsf{D}_1$.

For any complex numbers ζ, ζ', we have then $\mathsf{D}[(T^* - \zeta')(T - \zeta)] = \mathsf{D}$. In fact, $u \in \mathsf{D}[(T^* - \zeta')(T - \zeta)]$ implies $u \in \mathsf{D}_1$ and $Tu - \zeta u \in \mathsf{D}_1$, hence $Tu \in \mathsf{D}_1$, hence $u \in \mathsf{D}$. The opposite implication is obvious. Since $\mathsf{D}[(T - \zeta)(T^* - \zeta')] = \mathsf{D}$ similarly, we see from (3.29) that $(T^* - \zeta')(T - \zeta) = (T - \zeta)(T^* - \zeta')$. If in particular $\zeta \in \mathrm{P}(T)$ and $\zeta' \in \mathrm{P}(T^*)$, it follows that [we write $R(\zeta) = R(\zeta, T)$, $R^*(\zeta) = R(\zeta, T^*)$]

$$R^*(\zeta')R(\zeta) = R(\zeta)R^*(\zeta'). \tag{3.30}$$

(3.30) means simply that the resolvents $R(\zeta)$ and $R^*(\zeta')$ commute. Since $R^*(\zeta') = R(\bar{\zeta}')^*$ by III-(6.50), it follows that $R(\zeta)$ is also normal. As in (3.16) and (3.17), we have

$$\|R(\zeta, T)\| = 1/\operatorname{dist}(\zeta, \Sigma(T)) = \sup_{\lambda \in \Sigma(T)} |\zeta - \lambda|^{-1} \tag{3.31}$$
$$\|TR(\zeta, T)\| = \sup_{\lambda \in \Sigma(T)} |\lambda(\zeta - \lambda)^{-1}| .$$

Suppose now that the spectrum $\Sigma(T)$ of a normal operator T is separated into two parts Σ', Σ''. The projection P associated with this decomposition and its adjoint P^* are again given by III-(6.52). It follows from (3.30) that P and P^* commute so that P is normal. As in the finite-dimensional case, this implies that P is an orthogonal projection [see Problem I-6.30]. If Σ' consists of a single point, the associated quasi-nilpotent operator D is similarly seen to be normal, which implies that $D = 0$ (see Problem 2.4). Thus we are led to the same expression (3.21) for $R(\zeta)$ as for a selfadjoint operator T whenever $\Sigma(T)$ contains isolated eigenvalues $\lambda_1, \ldots, \lambda_s$, with the only difference that the eigenvalues λ_h need not be real.

Problem 3.27. If T is normal, T and T^* have the same null space.

Let T be a normal operator with compact resolvent (see III-§ 6.8). Application of (2.19) to the resolvent $R(\zeta) = R(\zeta, T)$ leads to the formula $R(\zeta) = \sum_{h=1}^{\infty} \mu_h P_h$, where the μ_h are the eigenvalues of $R(\zeta)$ and $\mu_h \to 0$, $h \to \infty$, and where the P_h are the associated eigenprojections. The orthogonal family $\{P_h\}$ is complete since $R(\zeta)$ has the null space 0 (see Theorem 2.10). But $R(\zeta)$ and T have the same set of eigenprojections, and the eigenvalues λ_h of T are related to the μ_h by $\mu_h = (\lambda_h - \zeta)^{-1}$. Thus we have

$$R(\zeta) = \sum_{h=1}^{\infty} \frac{1}{\lambda_h - \zeta} P_h , \quad \lambda_h \to \infty , \quad h \to \infty , \tag{3.32}$$

$$\sum_{h=1}^{\infty} P_h = 1 \quad \text{(strong convergence)} . \tag{3.33}$$

Example 3.28. (3.32) is true for *all* selfadjoint restrictions of the iT of Example 3.21. The same is true of all selfadjoint restrictions of the T of par. 6 in the *regular* case (see Example III-6.31). All these operators have discrete spectra consisting of *real* eigenvalues with finite multiplicities (which do not exceed the order m of the differential operator in question, since a differential equation of order m has at most m linearly independent solutions).

9. Reduction of symmetric operators

Suppose that a symmetric operator T is decomposed according to the decomposition $\mathsf{H} = \mathsf{M} \oplus \mathsf{M}^\perp$ of H into the direct sum of mutually orthogonal subspaces M and $\mathsf{M}^\perp$. According to III-(5.24), this is the case if and only if T commutes with the orthogonal projection P on M:

$$PT \subset TP. \tag{3.34}$$

In this case we say simply that T is *reduced* by M. Since $1 - P$ is the projection on $\mathsf{M}^\perp$, T is reduced by M if and only if it is reduced by $\mathsf{M}^\perp$.

A symmetric operator T is reduced by M if and only if $u \in \mathsf{D}(T)$ implies $Pu \in \mathsf{D}(T)$ and $TPu \in \mathsf{M}$. The "only if" part of this theorem is obvious from (3.34), so it suffices to prove the "if" part. The assumption implies that $TPu = PTPu$ for every $u \in \mathsf{D}(T)$. Hence $(u, PTPv) = (PTPu, v) = (TPu, v) = (u, PTv)$ for all $u, v \in \mathsf{D}(T)$, which gives $PTPv = PTv$. Thus $TPu = PTPu = PTu$ for $u \in \mathsf{D}(T)$, which is equivalent to (3.34).

Problem 3.29. Let $\mathsf{H} = \mathsf{M}_1 \oplus \mathsf{M}_2 \oplus \cdots \oplus \mathsf{M}_s$ be a decomposition of H into the direct sum of mutually orthogonal subspaces. If a symmetric operator T is reduced by each M_j, then T is decomposed according to the above decomposition of H in the sense of III-§ 5.6.

10. Semibounded and accretive operators

A symmetric operator T is said to be bounded from below if its numerical range (which is a subset of the real axis) is bounded from below, that is, if

$$(Tu, u) \geqq \gamma(u, u), \quad u \in \mathsf{D}(T). \tag{3.35}$$

In this case we simply write $T \geqq \gamma$. The largest number γ with this property is the *lower bound* of T. Similarly one defines the notions of boundedness from above and the upper bound. A symmetric operator bounded either from below or from above is said to be *semibounded.*

If T is bounded both from below and from above, then T is bounded, with the bound equal to the larger one of the absolute values of the lower and upper bounds. The proof is the same as in the finite-dimensional case [I-(6.33)]. In this case $T \in \mathscr{B}(\mathsf{H})$ if T is closed.

If T is selfadjoint, T is bounded from below (with lower bound γ_T) if and only if $\Sigma(T)$ is bounded from below (with lower bound γ_Σ). In fact, suppose that T is bounded from below. Then the open set Δ complementary to the closure of $\Theta(T)$ is connected, including real numbers ζ such that $\zeta < \gamma_T$, and is contained in the resolvent set $\mathrm{P}(T)$ (see Theorem 3.2). Thus $\Sigma(T)$ is bounded below with lower bound $\gamma_\Sigma \geqq \gamma_T$. Conversely suppose that $\Sigma(T)$ is bounded below with lower bound γ_Σ. Set $T' = T - \gamma_\Sigma$. T' has its spectrum on the nonnegative real axis, and for any $\alpha > 0$ we have $\|(T' + \alpha)^{-1}\| = \alpha^{-1}$ by (3.16). For any $u \in \mathsf{D}(T)$, we have therefore $\|u\| \leqq \alpha^{-1}\|(T' + \alpha)\, u\|$ and so

$$\|u\|^2 \leqq \alpha^{-2}\|T'u\|^2 + 2\alpha^{-1}(T'u, u) + \|u\|^2 \tag{3.36}$$

or $0 \leqq \alpha^{-1}\|T'u\|^2 + 2(T'u, u)$. On letting $\alpha \to \infty$ we have $(T'u, u) \geqq 0$, that is, $T' \geqq 0$, $T \geqq \gamma_\Sigma$. Thus T is bounded from below with $\gamma_T \geqq \gamma_\Sigma$.

An operator T in H is said to be *accretive*[1] if the numerical range $\Theta(T)$ is a subset of the right half-plane, that is, if

$$\text{(3.37)}\qquad \operatorname{Re}(Tu, u) \geqq 0 \quad \text{for all} \quad u \in \mathsf{D}(T).$$

If T is closed, it follows from par. 2 that $\operatorname{def}(T - \zeta) = \mu =$ constant for $\operatorname{Re}\zeta < 0$. If $\mu = 0$, the left open half-plane is contained in the resolvent set $\mathsf{P}(T)$ with

$$\text{(3.38)}\qquad \left.\begin{array}{l}(T+\lambda)^{-1} \in \mathscr{B}(\mathsf{H}),\\ \|(T+\lambda)^{-1}\| \leqq (\operatorname{Re}\lambda)^{-1}\end{array}\right\} \text{for } \operatorname{Re}\lambda > 0\,.$$

An operator T satisfying (3.38) will be said to be *m-accretive*[2].

An m-accretive operator T is maximal accretive, in the sense that T is accretive and has no proper accretive extension. In fact, the argument used above gives (3.36) with T' replaced by T and $(T'u, u)$ by $\operatorname{Re}(Tu, u)$ and leads to the result $\operatorname{Re}(Tu, u) \geqq 0$. Thus T is accretive. Suppose that T_1 is an accretive extension of T. Then $(T_1 + \lambda)^{-1}$ exists and is an extension of $(T + \lambda)^{-1}$ for $\operatorname{Re}\lambda > 0$. But since the latter has domain H, the two must coincide and so $T_1 = T$.

An m-accretive operator T is necessarily densely defined. Since $\mathsf{D}(T)$ is the range of $(T+\lambda)^{-1}$, $\operatorname{Re}\lambda > 0$, it suffices to show that $((T+\lambda)^{-1} u, v) = 0$ for all $u \in \mathsf{H}$ implies $v = 0$. On setting $u = v$ and $(T + \lambda)^{-1} v = w$, we have $0 = \operatorname{Re}((T + \lambda)^{-1} v, v) = \operatorname{Re}(w, (T + \lambda) w) \geqq \operatorname{Re}\lambda \|w\|^2$ and hence $w = 0$, $v = 0$.

We shall say that T is *quasi-accretive* if $T + \alpha$ is accretive for some scalar α. This is equivalent to the condition that $\Theta(T)$ is contained in a half-plane of the form $\operatorname{Re}\zeta \geqq$ const. In the same way we say that T is *quasi-m-accretive* if $T + \alpha$ is m-accretive for some α.

Like an m-accretive operator, a quasi-m-accretive operator is maximal quasi-accretive and densely defined.

Problem 3.30. If T is accretive and invertible, T^{-1} is accretive.

Problem 3.31. If T is m-accretive, then T^* and $(T + \lambda)^{-1}$ are also m-accretive for $\operatorname{Re}\lambda > 0$; $T(T + \lambda)^{-1}$ is m-accretive for $\lambda > 0$, with $\|T(T + \lambda)^{-1}\| \leqq 1$. If T is m-accretive and invertible, then T^{-1} is m-accretive.

Problem 3.32 If T is symmetric, T is m-accretive if and only if T is selfadjoint and nonnegative. If T is selfadjoint and nonnegative, then the same is true of $(T + \lambda)^{-1}$ and $T(T + \lambda)^{-1}$ for $\lambda > 0$. Furthermore, we have

$$\text{(3.39)}\qquad 0 \leqq (T(1 + \alpha T)^{-1} u, u) \leqq (Tu, u)\,, \quad u \in \mathsf{D}(T)\,, \quad \alpha \geqq 0\,.$$

Problem 3.33. If T is m-accretive, then $(1 + n^{-1} T)^{-1} \to 1$ strongly. [hint: $\|(1 + n^{-1} T)^{-1}\| \leqq 1$ and $\|(1 + n^{-1} T)^{-1} u - u\| = n^{-1}\|(1 + n^{-1} T)^{-1} Tu\| \leqq \leqq n^{-1}\|Tu\| \to 0$ if $u \in \mathsf{D}(T)$.]

[1] In this case $-T$ is said to be *dissipative*. Dissipative operators have been studied by FRIEDRICHS [6] (where the term "accretive" appears), PHILLIPS [2], [3], [4]. See also DOLPH [1], [2], DOLPH and PENZLIN [1], M. S. LIVŠIC [1], BRODSKII and LIVŠIC [1]. Some authors define a dissipative operator T by $\operatorname{Im}(Tu, u) \geqq 0$. LUMER and PHILLIPS [1] defines dissipative operators in Banach spaces.

[2] An m-accretive operator is equivalent to a closed, maximal accretive operator, though we do not prove this equivalence completely; see PHILLIPS [3].

For some quasi-accretive operators T, the numerical range $\Theta(T)$ is not only a subset of the half-plane $\operatorname{Re}\zeta \geqq$ const. but a subset of a sector $|\arg(\zeta - \gamma)| \leqq \theta < \pi/2$. In such a case T is said to be *sectorially-valued* or simply *sectorial;* γ and θ will be called a *vertex* and a *semi-angle* of the sectorial operator T (these are not uniquely determined). T is said to be *m-sectorial* if it is sectorial and quasi-m-accretive.

If T is m-sectorial with a vertex γ and a semi-angle θ, $\Sigma(T)$ is a subset of the sector $|\arg(\zeta - \gamma)| \leqq \theta$; in other words $\mathrm{P}(T)$ covers the exterior of this sector. This is clear from Theorem 3.2 since the exterior Δ of $\Theta(T)$ is a connected set.

Example 3.34. Consider the formal regular differential operator $Lu = p_0(x)\,u'' + p_1(x)\,u' + p_2(x)\,u$ on a finite interval $[a, b]$ (see III-§ 2.3), where the $p_k(x)$ are real and $p_0(x) < 0$. Let T_1 be the operator defined from L in the Hilbert space $\mathsf{H} = \mathsf{L}^2(a, b)$ by the boundary condition $u(a) = u(b) = 0$ (see loc. cit.). We shall show that T_1 is m-sectorial.

For $u \in \mathsf{D}(T_1)$ we have

$$\begin{aligned}(T_1 u, u) &= \int (p_0 u'' + p_1 u' + p_2 u)\,\bar{u}\,dx\\ &= -\int_a^b p_0 |u'|^2\,dx + \int_a^b [(p_1 - p_0')\,u' + p_2 u]\,\bar{u}\,dx\,.\end{aligned}$$

Since $-p_0(x) \geqq m_0 > 0$, $|p_1(x) - p_0'(x)| \leqq M_1$, $|p_2(x)| \leqq M_2$ for some positive constants m_0, M_1, M_2, we have

$$\begin{aligned}&\operatorname{Re}(T_1 u, u) \geqq m_0 \int |u'|^2\,dx - M_1 \int |u'|\,|u|\,dx - M_2 \int |u|^2\,dx\,,\\ &|\operatorname{Im}(T_1 u, u)| = \left|\operatorname{Im} \int (p_1 - p_0')\,u'\,\bar{u}\,dx\right| \leqq M_1 \int |u'|\,|u|\,dx\,.\end{aligned}$$

Hence, for any given $k > 0$,

$$\begin{aligned}&\operatorname{Re}(T_1 u, u) - k|\operatorname{Im}(T_1 u, u)| \geqq\\ &\geqq m_0 \int |u'|^2\,dx - (1 + k)\,M_1 \int |u'|\,|u|\,dx - M_2 \int |u|^2\,dx \geqq\\ &\geqq [m_0 - \varepsilon(1 + k)\,M_1] \int |u'|^2\,dx - \left(\frac{(1 + k)\,M_1}{4\varepsilon} + M_2\right) \int |u|^2\,dx\,,\end{aligned}$$

where $\varepsilon > 0$ is arbitrary. If ε is chosen in such a way that $m_0 - \varepsilon(1 + k)\,M_1 \geqq 0$, we have $\operatorname{Re}(T_1 u, u) - k|\operatorname{Im}(T_1 u, u)| \geqq \gamma(u, u)$ for some negative number γ. In other words

$$|\operatorname{Im}(T_1 u, u)| \leqq \frac{1}{k} \operatorname{Re}((T_1 - \gamma)\,u, u)\,.$$

This means that $\Theta(T_1)$ is contained in a sector with vertex γ and semi-angle $\theta = \arctan(1/k)$. Thus T_1 is sectorial with an arbitrarily small semi-angle.

To show that T_1 is quasi-m-accretive, it suffices to note that $(T_1 + \lambda)^* = S_1 + \lambda$, where λ is real and S_1 is an operator in H defined from the formal adjoint M to L (see Example III-5.32). S_1 is sectorial as well as T_1, and $S_1 + \lambda$ has nullity 0 if λ is sufficiently large. Thus $T_1 + \lambda$ has deficiency zero, showing that T_1 is quasi-m-accretive.

It is not difficult to prove that a similar result holds for second-order partial differential operators of elliptic type.

11. The square root of an m-accretive operator

The purpose of this paragraph is to prove the following theorem[1].

Theorem 3.35. *Let T be m-accretive. Then there is a unique m-accretive square root $T^{1/2}$ of T such that $(T^{1/2})^2 = T$. $T^{1/2}$ has the following additional properties.*

i) *$T^{1/2}$ is m-sectorial with the numerical range contained in the sector $|\arg\zeta| \leqq \pi/4$.*

ii) *$\mathsf{D}(T)$ is a core of $T^{1/2}$.*

iii) *$T^{1/2}$ commutes with any $B \in \mathscr{B}(\mathsf{H})$ that commutes with T.*

iv) *If T is selfadjoint and nonnegative, $T^{1/2}$ has the same properties.*

For the proof we start with the special case in which T is *strictly accretive*, that is, $\mathrm{Re}(Tu, u) \geqq \delta\|u\|^2$, $u \in \mathsf{D}(T)$, with a constant $\delta > 0$. This implies that $T - \delta$ is m-accretive, so that

(3.40) $$\|(T + \zeta)^{-1}\| \leqq (\mathrm{Re}\,\zeta + \delta)^{-1}, \quad \mathrm{Re}\,\zeta \geqq 0 .$$

We now define an operator A by the Dunford-Taylor integral

(3.41) $$A = \frac{1}{2\pi i} \int_\Gamma \zeta^{-1/2} (T - \zeta)^{-1} d\zeta ;$$

here the integration path Γ runs from $-\infty$ to $-\infty$ in the resolvent set of T, making a turn around the origin in the positive direction [this is possible since $\mathsf{P}(T)$ contains the half-plane $\mathrm{Re}\,\zeta < \delta$]. The values of $\zeta^{-1/2}$ should be chosen in such a way that $\zeta^{-1/2} > 0$ at the point where Γ meets the positive real axis. The integral of (3.41) is absolutely convergent in virtue of (3.40). Thus A is well defined and belongs to $\mathscr{B}(\mathsf{H})$.

Lemma 3.36. $A^2 = T^{-1}$.

Proof. We take another expression for A in which the integration path Γ is replaced by a slightly shifted one Γ' not intersecting Γ, and multiply out the two expressions. Since the two integrals are absolutely convergent, the order of the integrals is irrelevant. Making use of the resolvent equation for $(T - \zeta)^{-1}(T - \zeta')^{-1}$, we arrive at the result

(3.42) $$A^2 = \frac{1}{2\pi i} \int_\Gamma \zeta^{-1} (T - \zeta)^{-1} d\zeta = T^{-1} .$$

This procedure is exactly the same as the proof of I-(5.17).

$A^2 = T^{-1}$ implies that A is invertible, for $Au = 0$ implies $T^{-1}u = A^2u = 0$, $u = 0$. We now define $T^{1/2}$ as equal to A^{-1}, so that $T^{-1/2} = (T^{1/2})^{-1} = A$. We note that $T^{-1/2}$, and so $T^{1/2}$ too, commutes with the resolvent $(T - \zeta)^{-1}$, as is obvious from (3.41).

A more convenient expression for $T^{-1/2}$ is obtained by reducing the path Γ to the union of the upper and lower edges of the negative real

[1] Cf. Langer [1].

axis. Writing $\zeta = -\lambda$ and noting that $\zeta^{-1/2} = \mp i\,\lambda^{-1/2}$, we obtain

$$T^{-1/2} = \frac{1}{\pi}\int_0^\infty \lambda^{-1/2}(T+\lambda)^{-1}\,d\lambda\,. \tag{3.43}$$

This shows, in particular, that $T^{-1/2}$ is nonnegative selfadjoint if T is selfadjoint, for $(T+\lambda)^{-1}$ has the same properties. In the general case $T^{-1/2}$ is accretive since $(T+\lambda)^{-1}$ is accretive, and this implies that $T^{1/2}$ is m-accretive (see Problem 3.31). Furthermore, we have

$$\|T^{-1/2}\| \leqq \frac{1}{\pi}\int_0^\infty \lambda^{-1/2}\|(T+\lambda)^{-1}\|\,d\lambda \leqq \frac{1}{\pi}\int_0^\infty \lambda^{-1/2}(\lambda+\delta)^{-1}\,d\lambda = \delta^{-1/2}. \tag{3.44}$$

Lemma 3.37. $(T^{1/2})^2 = T$.

Proof. Let $u \in \mathsf{D}(T)$. Since $A^2 T u = u$, $u \in \mathsf{R}(A) = \mathsf{D}(T^{1/2})$ and $T^{1/2}u = A^{-1}u = A\,Tu$. Hence $T^{1/2}u \in \mathsf{D}(T^{1/2})$ and $T^{1/2}(T^{1/2}u) = A^{-1}\cdot A\,Tu = Tu$. Conversely, let $u \in \mathsf{D}((T^{1/2})^2)$ and set $v = (T^{1/2})^2 u$. Then $T^{-1}v = A^2(T^{1/2})^2 u = A(T^{1/2}u) = u$, so that $u \in \mathsf{D}(T)$ and $Tu = v = (T^{1/2})^2 u$. This proves the lemma.

Lemma 3.38. $\mathsf{D}(T)$ *is a core of* $T^{1/2}$.

Proof. Let $u \in \mathsf{D}(T^{1/2})$; we have to show that there is a sequence $u_n \in \mathsf{D}(T)$ such that $u_n \to u$ and $T^{1/2}(u_n - u) \to 0$. Such a sequence is given, for example, by $u_n = (1 + n^{-1}T)^{-1}u = n(T+n)^{-1}u$. In fact, it is obvious that $u_n \in \mathsf{D}(T)$. Set $v = T^{1/2}u$; then $u = Av$ and $u_n = n(T+n)^{-1}Av = nA(T+n)^{-1}v$ (see the remark above) and so $T^{1/2}u_n = n(T+n)^{-1}v \to v = T^{1/2}u$ (see Problem 3.33).

Incidentally, we note the formula

$$T^{1/2}u = \frac{1}{\pi}\int_0^\infty \lambda^{-1/2}(T+\lambda)^{-1}\,Tu\,d\lambda\,,\quad u \in \mathsf{D}(T)\,, \tag{3.45}$$

which is obtained by applying (3.43) to u replaced by Tu and noting that $T^{-1/2}\,Tu = T^{1/2}u$ by Lemma 3.37.

Problem 3.39. $\mathrm{Re}(T^{1/2}u, u) \geqq \delta^{1/2}\|u\|^2$, $u \in \mathsf{D}(T^{1/2})$. [hint: $\mathrm{Re}((T+\lambda)^{-1}\,Tu, u) \geqq \delta(\delta+\lambda)^{-1}\|u\|^2$.]

Lemma 3.40. *Proposition* i) *of* Theorem 3.35 *holds when* T *is strictly accretive.*

Proof. We deduce from (3.41) another expression for A by reducing Γ to the union of the upper and lower edges of the ray $\zeta = -\lambda\,e^{i\theta}$, $0 < \lambda < \infty$, where θ is a fixed angle with $|\theta| < \pi/2$. A calculation similar to that used in deducing (3.43) gives

$$T^{-1/2} = \frac{1}{\pi}\,e^{i\theta/2}\int_0^\infty \lambda^{-1/2}(T+\lambda\,e^{i\theta})^{-1}\,d\lambda\,; \tag{3.46}$$

note that the integral is absolutely convergent since $\|(T+\lambda\,e^{i\theta})^{-1}\| \leqq (\delta + \lambda\cos\theta)^{-1}$ by (3.40). Now $T + \lambda\,e^{i\theta}$ is accretive with T because

$\mathrm{Re}\,\lambda e^{i\theta} = \lambda\cos\theta > 0$, so that $(T + \lambda e^{i\theta})^{-1}$ is also accretive. Hence $e^{-i\theta/2}(T^{-1/2}u, u)$ has nonnegative real part by (3.46). Since this is true for $|\theta| < \pi/2$, it follows that $(T^{-1/2}u, u)$ lies in the sector $|\arg\zeta| \leqq \pi/4$. This result can be written

$$|\mathrm{Im}(T^{1/2}u, u)| \leqq \mathrm{Re}(T^{1/2}u, u)\,, \quad u \in \mathsf{D}(T^{1/2})\,. \tag{3.47}$$

Proposition iii) follows from (3.41), which shows that $A = T^{-1/2}$ commutes with B. Summing up the results obtained so far, we have proved the theorem in the special case of a strictly accretive T except for the uniqueness of $T^{1/2}$. We now prove this uniqueness under some restrictions.

Lemma 3.41. *Assuming that T is strictly accretive, suppose there is an m-accretive operator S such that $S^{-1} = B \in \mathscr{B}(\mathsf{H})$ and $S^2 = T$. Then $S = T^{1/2}$.*

Proof. $B = S^{-1}$ commutes with $B^2 = T^{-1}$, hence with T and also with the resolvent $(T - \zeta)^{-1}$ (see Theorem III-6.5). It follows from (3.41) that $BA = AB$. Since $A^2 = T^{-1} = B^2$, we have $(A + B)(A - B) = A^2 - B^2 = 0$. For any $u \in \mathsf{H}$, we have therefore $(A + B)v = 0$ for $v = (A - B)u$. Then $\mathrm{Re}(Av, v) + \mathrm{Re}(Bv, v) = \mathrm{Re}((A + B)v, v) = 0$. But both (Av, v) and (Bv, v) have nonnegative real parts since A, B are accretive. Thus $\mathrm{Re}(Av, v) = 0$ or $\mathrm{Re}(w, T^{1/2}w) = 0$ for $w = Av$. Since $\mathrm{Re}(w, T^{1/2}w) \geqq \delta^{1/2}\|w\|^2$ by Problem 3.39, we conclude that $w = 0$, $(A - B)u = v = T^{1/2}w = 0$. Since u was arbitrary, this proves $B = A$[1] and hence $S = T^{1/2}$.

We shall now remove the assumption made above that T is strictly accretive. Let T be m-accretive. For any $\varepsilon > 0$, $T_\varepsilon = T + \varepsilon$ is strictly accretive, so that $T_\varepsilon^{1/2} = S_\varepsilon$ can be constructed as above. We shall show that $\lim\limits_{\varepsilon\to 0} S_\varepsilon = S$ exists in a certain sense.

For any $u \in \mathsf{D}(T_\varepsilon) = \mathsf{D}(T)$, we have the expression (3.45) for $S_\varepsilon u$ where T is to be replaced by T_ε. Since $(T_\varepsilon + \lambda)^{-1}T_\varepsilon u = u - \lambda(T_\varepsilon + \lambda)^{-1}u = u - \lambda(T + \varepsilon + \lambda)^{-1}u$, we have $\frac{d}{d\varepsilon}(T_\varepsilon + \lambda)^{-1}T_\varepsilon u = \lambda(T + \varepsilon + \lambda)^{-2}u = \lambda(T_\varepsilon + \lambda)^{-2}u = -\lambda\frac{d}{d\lambda}(T_\varepsilon + \lambda)^{-1}u$. Hence

$$\begin{aligned} \frac{d}{d\varepsilon}S_\varepsilon u &= -\frac{1}{\pi}\int_0^\infty \lambda^{1/2}\frac{d}{d\lambda}(T_\varepsilon + \lambda)^{-1}u\,d\lambda \\ &= \frac{1}{2\pi}\int_0^\infty \lambda^{-1/2}(T_\varepsilon + \lambda)^{-1}u\,d\lambda = \frac{1}{2}T_\varepsilon^{-1/2}u\,, \end{aligned} \tag{3.48}$$

where we made an integration by parts and used (3.43) for $T_\varepsilon^{-1/2}$. Integrating (3.48) between η and ε, where $0 < \eta < \varepsilon$, we obtain $S_\varepsilon u - S_\eta u$

[1] The argument used here is a generalization of that of Riesz and Sz.-Nagy [1], p. 264, where symmetric operators $T \geqq 0$ are considered.

$= \frac{1}{2} \int_\eta^\varepsilon T_\varepsilon^{-1/2} u \, d\varepsilon$. Making use of the estimate $\|T_\varepsilon^{-1/2}\| \leqq \varepsilon^{-1/2}$, which follows from (3.44), we have then

$$(3.49) \qquad \|S_\varepsilon u - S_\eta u\| \leqq \frac{1}{2} \int_\eta^\varepsilon \|T_\varepsilon^{-1/2} u\| \, d\varepsilon \leqq \frac{1}{2} \int_\eta^\varepsilon \varepsilon^{-1/2} \|u\| \, d\varepsilon = (\varepsilon^{1/2} - \eta^{1/2}) \|u\| .$$

This shows that $\lim_{\varepsilon \to 0} S_\varepsilon u = S' u$ exists for $u \in \mathsf{D}(T)$ and that

$$(3.50) \qquad \|S_\varepsilon u - S' u\| \leqq \varepsilon^{1/2} \|u\| , \quad u \in \mathsf{D}(T) .$$

Thus the operator $S' - S_\varepsilon$ [with domain $\mathsf{D}(T)$] is bounded with bound $\leqq \varepsilon^{1/2}$. Since $\mathsf{D}(T) = \mathsf{D}(T_\varepsilon)$ is a core of S_ε (see Lemma 3.38), it follows from the stability theorem of closedness (Theorem IV-1.1) that S' is closable and the closure $S = \tilde{S}'$ has the same domain as S_ε [which implies, in turn, that $\mathsf{D}(S_\varepsilon)$ is independent of ε]. At the same time (3.50) is extended to all $u \in \mathsf{D}(S)$ with the S' on the left replaced by S. Thus we may write

$$(3.51) \qquad S_\varepsilon = S + B_\varepsilon , \quad B_\varepsilon \in \mathscr{B}(\mathsf{H}) , \quad \|B_\varepsilon\| \leqq \varepsilon^{1/2} .$$

Since $(S_\varepsilon u, u) \to (S u, u)$, $\varepsilon \to 0$, $u \in \mathsf{D}(S) = \mathsf{D}(S_\varepsilon)$, S is sectorial with $\Theta(S)$ contained in the sector $|\arg \zeta| \leqq \pi/4$, for this is true for S_ε. Furthermore, any ζ outside of this sector belongs to the resolvent set of S, as is seen by constructing $R(\zeta, S)$ by the Neumann series from $S = S_\varepsilon - B_\varepsilon$, in which ε is chosen so small that $\varepsilon^{1/2}$ is smaller than the distance of ζ from that sector. This shows that S is m-accretive.

Lemma 3.42. $S^2 = T$.

Proof. Let $u \in \mathsf{D}(T)$ and set $v_\varepsilon = S_\varepsilon u$. We have $v_\varepsilon \to S u$, $\varepsilon \to 0$. Since we know that $S_\varepsilon^2 = T_\varepsilon$, $v_\varepsilon \in \mathsf{D}(S_\varepsilon) = \mathsf{D}(S)$ and $S_\varepsilon v_\varepsilon = T_\varepsilon u$. Hence $S v_\varepsilon = (S_\varepsilon - B_\varepsilon) v_\varepsilon = T_\varepsilon u - B_\varepsilon v_\varepsilon \to T u$, $\varepsilon \to 0$. It follows from the closedness of S that $S u \in \mathsf{D}(S)$ and $S S u = T u$. Thus we have proved that $T \subset S^2$. This implies that $T + 1 \subset S^2 + 1 = (S + i)(S - i)$ and so $(T + 1)^{-1} \subset (S - i)^{-1} (S + i)^{-1}$. But $(T + 1)^{-1}$ and $(S \pm i)^{-1}$ belong to $\mathscr{B}(\mathsf{H})$. Hence we must have equality instead of the last inclusion, and this in turn implies that $T = S^2$.

On writing $S = T^{1/2}$, we have thus proved that i), ii) of Theorem 3.35 are satisfied [ii) is implied by $S = \tilde{S}'$]. iv) is also clear, for S_ε and S are symmetric and hence selfadjoint if T is selfadjoint.

Lemma 3.43. (3.45) *is true in the general case.*

Proof. The integral in (3.45) is absolutely convergent, for $\|\lambda^{-1/2} \cdot (T + \lambda)^{-1}\| \leqq \lambda^{-3/2}$ for $\lambda \to \infty$ and $\|\lambda^{-1/2} (T + \lambda)^{-1} T u\| \leqq 2 \lambda^{-1/2} \|u\|$ for $\lambda \to 0$. The corresponding integral expression for $T_\varepsilon^{1/2} u$ has the same properties, the estimate being independent of ε. Since $(T_\varepsilon + \lambda)^{-1} T_\varepsilon u - (T + \lambda)^{-1} T u = -\lambda [(T_\varepsilon + \lambda)^{-1} u - (T + \lambda)^{-1} u] \to 0$ for $\varepsilon \to 0$ for each $\lambda > 0$, it follows by the principle of dominated convergence that $T_\varepsilon^{1/2} u$ converges to the right member of (3.45). Since $T_\varepsilon^{1/2} u \to T^{1/2} u$, $T^{1/2} u$ must be equal to the right member of (3.45).

Lemma 3.44. iii) *of Theorem* 3.35 *is satisfied.*

Proof. B commutes with $T_\varepsilon = T + \varepsilon$, hence with $S_\varepsilon = T_\varepsilon^{1/2}$ by what was already proved. Since it is easily seen that $(S_\varepsilon + 1)^{-1} \underset{s}{\to} (S + 1)^{-1}$, $\varepsilon \to 0$, B commutes with $(S + 1)^{-1}$, hence with $S = T^{1/2}$ (see Theorem III-6.5).

Lemma 3.45. *Let R be any m-accretive square root of T: $R^2 = T$. Then R commutes with the resolvent of T and $\mathsf{D}(T)$ is a core of R.*

Proof. Let $\lambda \neq 0$ be real. Then $(R - i\lambda)(R + i\lambda) = R^2 + \lambda^2 = T + \lambda^2$ has range H and so does $R - i\lambda$. On the other hand $R - i\lambda$ has nullity zero, for $Ru = i\lambda u$ implies $Tu = R^2 u = i\lambda Ru = -\lambda^2 u$, $(T + \lambda^2) u = 0$, $u = 0$. Thus $\pm i\lambda \in \mathrm{P}(R)$ and $(T + \lambda^2)^{-1} = (R - i\lambda)^{-1}(R + i\lambda)^{-1}$. Hence R commutes with $(T + \lambda^2)^{-1}$. It follows also that $\mathsf{D}(R) \supset \mathsf{D}(T)$ and $(R - i\lambda)\,\mathsf{D}(T) = \mathsf{R}[(R - i\lambda)(T + \lambda^2)^{-1}] = \mathsf{R}[(R + i\lambda)^{-1}] = \mathsf{D}(R)$ is dense in H, so that $\mathsf{D}(T)$ is a core of R.

Lemma 3.46. *An m-accretive square root of T is unique.*

Proof. Let R be any m-accretive square root of T. Then $R + \varepsilon$ is strictly m-accretive for any $\varepsilon > 0$, and $(R + \varepsilon)^2 = R^2 + 2\varepsilon R + \varepsilon^2 = T + 2\varepsilon R + \varepsilon^2 \equiv Q_\varepsilon$ is also strictly accretive. In fact it is also m-accretive; to see this it suffices to note that R is relatively bounded with respect to T because $\mathsf{D}(R) \supset \mathsf{D}(T)$ (see Remark IV-1.5), so that $(Q_\varepsilon + 1)^{-1} \in \mathscr{B}(\mathsf{H})$ for sufficiently small ε (see Theorem IV-1.16). According to Lemma 3.41, $R + \varepsilon$ must be identical with the unique square root $Q_\varepsilon^{1/2}$.

Let $u \in \mathsf{D}(T)$. Then $u \in \mathsf{D}(R)$ and $(R + \varepsilon) u \to Ru$, $\varepsilon \to 0$. On the other hand, we have $Q_\varepsilon^{1/2} u \to T^{1/2} u$ as we shall prove in a moment. Thus we obtain $Ru = T^{1/2} u$ for $u \in \mathsf{D}(T)$. Since $\mathsf{D}(T)$ is a core for both R and $T^{1/2}$ (see Lemma 3.45), it follows that $R = T^{1/2}$, as we wished to show.

To prove that $Q_\varepsilon^{1/2} u \to T^{1/2} u$, $\varepsilon \to 0$, for $u \in \mathsf{D}(T)$, we make use of (3.45) and a similar expression for $Q_\varepsilon^{1/2} u$ (see Lemma 3.43). Since $(T + \lambda)^{-1} T u = u - \lambda (T + \lambda)^{-1} u$, we have

$$(3.52)\qquad Q_\varepsilon^{1/2} u - T^{1/2} u = -\frac{1}{\pi}\int_0^\infty \lambda^{1/2}[(Q_\varepsilon + \lambda)^{-1} u - (T + \lambda)^{-1} u]\, d\lambda$$

$$= \frac{\varepsilon}{\pi}\int_0^\infty \lambda^{1/2}(Q_\varepsilon + \lambda)^{-1}(2R + \varepsilon)\,(T + \lambda)^{-1} u\, d\lambda\,.$$

To estimate this integral, it is convenient to split it into two parts $\int_0^\varepsilon$ and $\int_\varepsilon^\infty$ and to use the integrand in the form of the middle member of

(3.52) in the first part and in the form of the last member in the second part. Thus, noting the commutativity of R with $(T+\lambda)^{-1}$ (see Lemma 3.45), we obtain

$$\pi\|Q_\varepsilon^{1/2} u - T^{1/2} u\| \leq \int_0^\varepsilon \lambda^{1/2} [\|(Q_\varepsilon + \lambda)^{-1}\| + \|(T+\lambda)^{-1}\|] \, \|u\| \, d\lambda +$$

$$+ \varepsilon \int_\varepsilon^\infty \lambda^{1/2} \|(Q_\varepsilon + \lambda)^{-1}\| \, \|(T+\lambda)^{-1}\| \, \|(2R+\varepsilon) u\| \, d\lambda \leq$$

$$\leq \left(\int_0^\varepsilon 2\lambda^{-1/2} d\lambda\right) \|u\| + \varepsilon \left(\int_\varepsilon^\infty \lambda^{-3/2} d\lambda\right) \|(2R+\varepsilon) u\|$$

$$= 4\varepsilon^{1/2} \|u\| + 2\varepsilon^{1/2} \|(2R+\varepsilon) u\| \to 0 \,, \quad \varepsilon \to 0 \,.$$

Problem 3.47. If $T \in \mathscr{B}(H)$ is symmetric and nonnegative, then $\|T^{1/2}\| = \|T\|^{1/2}$.

Problem 3.48. If T is m-accretive, the following conditions on u are equivalent: a) $Tu = 0$, b) $T^{1/2} u = 0$, c) $(\lambda + T)^{-1} u = \lambda^{-1} u$, $\lambda > 0$.

Theorem 3.49. *Let T be m-accretive. T has compact resolvent if and only if $T^{1/2}$ has.*

Proof. Set $S = T^{1/2}$. The formula $(T+1)^{-1} = (S+i)^{-1}(S-i)^{-1}$ proved above (in the proof of Lemma 3.42) shows that T has compact resolvent if S has. To prove the converse statement, define $T_\varepsilon = T + \varepsilon$ and $S_\varepsilon = T_\varepsilon^{1/2}$ as above. S_ε^{-1} is given by (3.43) with T replaced by T_ε on the right. Since the integral converges in norm and since $(T_\varepsilon + \lambda)^{-1}$ is compact for all $\lambda > 0$, it follows that S_ε^{-1} is compact. Thus S_ε has compact resolvent, so that $(1 + S_\varepsilon)^{-1}$ is compact. Now $\|(1+S_\varepsilon)^{-1} - (1+S)^{-1}\| = \|(1+S_\varepsilon)^{-1}(1+S)^{-1}(S - S_\varepsilon)\| \leq \|B_\varepsilon\| \leq \varepsilon^{1/2} \to 0$ for $\varepsilon \to 0$ by (3.51). Hence $(1+S)^{-1}$ is compact and S has compact resolvent.

Remark 3.50. More generally we can define the fractional powers T^α, $0 < \alpha < 1$, for an m-accretive operator T by a construction similar to that given above for $T^{1/2}$. If T is strictly m-accretive, we have simply to replace $\zeta^{-1/2}$ by $\zeta^{-\alpha}$ in (3.41) to obtain $T^{-\alpha}$. The formula corresponding to (3.43) is

$$T^{-\alpha} = \frac{\sin \pi\alpha}{\pi} \int_0^\infty \lambda^{-\alpha} (T+\lambda)^{-1} d\lambda \,. \tag{3.53}$$

The general case can be dealt with by the method of this paragraph. It will be noted that T^α can be defined for a more general class of operators acting in a Banach space.

Problem 3.51. Let T_n, T be m-accretive operators such that $T_n \to T$ in the generalized sense. Then $(T_n + \varepsilon)^{-1/2} \to (T+\varepsilon)^{-1/2}$ in norm for any $\varepsilon > 0$.

Problem 3.52. Let T_n, T be m-accretive and $(T_n + \lambda)^{-1} \xrightarrow[s]{} (T+\lambda)^{-1}$ for all $\lambda > 0$. Then $(T_n + \varepsilon)^{-1/2} \xrightarrow[s]{} (T+\varepsilon)^{-1/2}$ for any $\varepsilon > 0$.

§ 4. Perturbation of selfadjoint operators

1. Stability of selfadjointness

Since the selfadjoint operators form the most important class of operators that appear in applications, the perturbation of selfadjoint operators and the stability of selfadjointness is one of our principal problems[1].

The first question is under what conditions the selfadjointness is preserved under a "small" perturbation. A fundamental result in this direction is given by

Theorem 4.1. *Let T be a selfadjoint operator. Then there is a $\delta > 0$ such that any closed symmetric operator S with $\hat{\delta}(S, T) < \delta$ is selfadjoint, where $\hat{\delta}(S, T)$ denotes the gap between S and T.*

Proof. Since $\pm i$ belong to $\mathrm{P}(T)$, it follows from Theorem IV-3.15 that there is a $\delta > 0$ such that $\hat{\delta}(S, T) < \delta$ implies $\pm i \in \mathrm{P}(S)$. Then S is selfadjoint by Theorem 3.16.

Corollary 4.2. *Let T, T_n be closed symmetric operators and let $\{T_n\}$ converge to T in the generalized sense (see* IV-§ 2.4*). If T is selfadjoint, then T_n is selfadjoint for sufficiently large n.*

Although this theorem is rather general, it is not very convenient for application since the definition of the gap $\hat{\delta}(S, T)$ is complicated. A less general but more convenient criterion is furnished by relatively bounded perturbations.

We recall that an operator A is relatively bounded with respect to T (or T-bounded) if $\mathrm{D}(A) \supset \mathrm{D}(T)$ and

$$\|A u\| \leq a\|u\| + b\|T u\|, \quad u \in \mathrm{D}(T); \tag{4.1}$$

see IV-(1.1). An equivalent condition is

$$\|A u\|^2 \leq a'^2\|u\|^2 + b'^2\|T u\|^2, \quad u \in \mathrm{D}(T), \tag{4.2}$$

where the constants a', b' are, of course, in general different from a, b. As is easily seen, (4.2) implies (4.1) with $a = a'$, $b = b'$, whereas (4.1) implies (4.2) with $a'^2 = (1 + \varepsilon^{-1})\, a^2$ and $b'^2 = (1 + \varepsilon)\, b^2$ with an arbitrary $\varepsilon > 0$. Thus the T-bound of A (defined as the greatest lower bound of the possible values of b) may as well be defined as the greatest lower bound of the possible values of b'.

Theorem 4.3.[2] *Let T be selfadjoint. If A is symmetric and T-bounded with T-bound smaller than 1, then $T + A$ is also selfadjoint. In particular $T + A$ is selfadjoint if A is bounded and symmetric with $\mathrm{D}(A) \supset \mathrm{D}(T)$.*

[1] In this section all operators are assumed to act in a Hilbert space H, unless otherwise stated.

[2] Theorems 4.3 and 4.4 are due to Rellich [3]. See also T. Kato [3], [4]. These theorems have been found to be convenient for establishing the selfadjointness or essential selfadjointness of various operators that appear in applications. The applications to the Schrödinger and Dirac operators will be discussed in § 5. For the applications to the quantum theory of fields, see Y. Kato [1], Y. Kato and Mugibayashi [1]; Cf. also Cook [2].

Proof. Obviously $T + A$ has domain $\mathsf{D}(T)$ and is symmetric. We may assume without loss of generality that (4.2) holds with constants a', b' such that $a' > 0$, $0 < b' < 1$. In view of the identity (3.12), (4.2) can be written

$$\|A u\| \leqq \|(b' T \mp i a') u\| , \quad u \in \mathsf{D}(T) . \tag{4.3}$$

With $(T \mp i c') u = v$, this gives

$$\|A R(\pm i c') v\| \leqq b' \|v\| , \quad c' = a'/b' , \tag{4.4}$$

where $R(\zeta) = R(\zeta, T)$ is the resolvent of T. Since T is selfadjoint, v changes over the whole space H when u changes over $\mathsf{D}(T)$, so that we have

$$B_\pm \equiv - A R(\pm i c') \in \mathscr{B}(\mathsf{H}) , \quad \|B_\pm\| \leqq b' . \tag{4.5}$$

Since $b' < 1$, $(1 - B_\pm)^{-1}$ exists and belongs to $\mathscr{B}(\mathsf{H})$ (the Neumann series), so that $1 - B_\pm$ maps H onto itself one to one. But $T + A \mp i c' = (1 - B_\pm)(T \mp i c')$ and $T \mp i c'$ has range H because T is selfadjoint. Hence $T + A \mp i c'$ also has range H, which proves that $T + A$ is selfadjoint. Note that the proof given here is essentially the same as the proof of Theorem IV-3.17.

Theorem 4.4. *Let T be essentially selfadjoint. If A is symmetric and T-bounded with T-bound smaller than* 1, *then $T + A$ is essentially selfadjoint and its closure $(T + A)^\sim$ is equal to $\tilde{T} + \tilde{A}$. In particular this is true if A is symmetric and bounded with $\mathsf{D}(A) \supset \mathsf{D}(T)$.*

Proof. We first prove that $\tilde{A}$ is $\tilde{T}$-bounded, that is,

$$\mathsf{D}(\tilde{A}) \supset \mathsf{D}(\tilde{T}) \text{ and } \|\tilde{A} u\|^2 \leqq a'^2 \|u\|^2 + b'^2 \|\tilde{T} u\|^2 , \; u \in \mathsf{D}(\tilde{T}), \tag{4.6}$$

if (4.2) is assumed. For any $u \in \mathsf{D}(\tilde{T})$, there is a sequence $\{u_n\}$ which is T-convergent to u (that is, $u_n \to u$ and $T u_n \to \tilde{T} u$). (4.2) shows that $\{u_n\}$ is also A-convergent so that $u \in \mathsf{D}(\tilde{A})$ and $A u_n \to \tilde{A} u$. (4.6) is now obtained as the limit of (4.2) with u replaced by u_n. Incidentally we have also $(T + A) u_n \to (\tilde{T} + \tilde{A}) u$, so that $u \in \mathsf{D}((T + A)^\sim)$ and $(T + A)^\sim u = (\tilde{T} + \tilde{A}) u$. This shows that

$$(T + A)^\sim \supset \tilde{T} + \tilde{A} . \tag{4.7}$$

Note that we have so far not used the fact that $b' < 1$, which may be assumed as in the proof of Theorem 4.3.

On the other hand, it follows from Theorem 4.3 applied to the pair $\tilde{T}$, $\tilde{A}$ that $\tilde{T} + \tilde{A}$ is selfadjoint (here we use the assumption $b' < 1$). Thus $\tilde{T} + \tilde{A}$ is a closed extension of $T + A$ and therefore also of $(T + A)^\sim$. Combined with (4.7), this gives $\tilde{T} + \tilde{A} = (T + A)^\sim$ and completes the proof.

Theorems 4.3 and 4.4 are not symmetric with respect to T and $S = T + A$. The following symmetrized version generalizes these theorems. The proof is similar to that of a corresponding theorem for the stability of closedness (Theorem IV-1.3) and may be omitted.

Theorem 4.5. *Let T, S be two symmetric operators such that* $\mathsf{D}(T) = \mathsf{D}(S) = \mathsf{D}$ and

$$(4.8)\qquad \|(S - T)\, u\| \leq a\|u\| + b(\|T u\| + \|S u\|)\,, \quad u \in \mathsf{D}\,,$$

where a, b are nonnegative constants with $b < 1$. Then S is essentially selfadjoint if and only if T is; in this case $\tilde{S}$ and $\tilde{T}$ have the same domain. In particular, S is selfadjoint if and only if T is.

2. The case of relative bound 1

In the theorems proved in the preceding paragraph, the assumption that the relative bound be smaller than 1 cannot be dropped in general (at least in the "selfadjoint" case). This is seen from the simple example in which T is unbounded and selfadjoint and $A = -T$; here A is T-bounded with T-bound 1, but $T + A$ is a proper restriction of the operator 0 and is not selfadjoint.

In this connection the following theorem is of interest, which is concerned with the case of relative bound 1 (but which does *not* cover all such cases).

Theorem 4.6. *Let T be essentially selfadjoint and let A be symmetric. If A is T-bounded and* (4.2) *holds with $b' = 1$, then $T + A$ is essentially selfadjoint.*

Proof. First we assume that T is selfadjoint and define the operators $B_\pm$ as in the proof of Theorem 4.3. We have then $\|B_\pm\| \leq 1$ by (4.5), and the ranges of $1 - B_\pm$ need not be the whole space H. Nevertheless we can show that these ranges are dense in H; then the argument given before will show that the ranges of $T + A \mp i c'$ are dense in H and therefore $T + A$ is essentially selfadjoint [see Problem 3.17, c)].

To see that the range of $1 - B_+$ is dense in H ($1 - B_-$ can be dealt with in the same way), it suffices to show that a $v \in \mathsf{H}$ orthogonal to this range must be zero. Now such a v satisfies $B_+^* v = v$. According to Lemma 4.7 proved below, this implies also that $B_+ v = v$, that is, $A R(i a') v + v = 0$ (note that $c' = a'$ by $b' = 1$). Setting $u = R(i a') v \in \mathsf{D}(T)$, we have $(T + A - i a') u = 0$. But since $T + A$ is symmetric and $a' > 0$ (as assumed before), this gives $u = 0$ and hence $v = 0$. This proves the theorem under the strengthened assumption that T is selfadjoint.

We now consider the general case in which T is assumed only essentially selfadjoint. In this case we have the inclusion (4.7); recall that this has been proved without the assumption $b' < 1$. Now $\tilde{T}$ is selfadjoint and (4.6) is satisfied with $b' = 1$. Applying what was proved above to the pair $\tilde{T}$, $\tilde{A}$, we see that $\tilde{T} + \tilde{A}$ is essentially selfadjoint. (4.7) thus shows that the closed symmetric operator $(T + A)^\sim$ is an extension of an essentially selfadjoint operator. Thus $(T + A)^{\sim}$ is selfadjoint, that is, $T + A$ is essentially selfadjoint.

Lemma 4.7. *Let $B \in \mathscr{B}(\mathsf{H})$ and $\|B\| \leqq 1$. Then $Bu = u$ is equivalent to $B^* u = u$. (Such an operator B is called a contraction.)*

Proof. Since $B^{**} = B$ and $\|B^*\| = \|B\| \leqq 1$, it suffices to show that $Bu = u$ implies $B^* u = u$. But this is obvious from

$$\|B^* u - u\|^2 = \|B^* u\|^2 + \|u\|^2 - 2\operatorname{Re}(B^* u, u) \leqq 2\|u\|^2 - 2\operatorname{Re}(u, Bu) = 0.$$

3. Perturbation of the spectrum

The results of IV-§ 3 on the perturbation of the spectra of closed linear operators apply to selfadjoint operators, often with considerable simplifications. As an example, let us estimate the change of isolated eigenvalues.

Let T be selfadjoint and let A be T-bounded, with T-bound smaller than 1; thus we have the inequality (4.1) with $b < 1$. In general we need not assume that A is symmetric; thus S is closed by Theorem IV-1.1 but need not be symmetric. Let us ask when a given complex number ζ belongs to $\mathrm{P}(S)$.

A sufficient condition for this is given by IV-(3.12). Since T is selfadjoint, however, we have [see (3.16)–(3.17)]

$$\|R(\zeta, T)\| = \sup_{\lambda' \in \Sigma(T)} |\lambda' - \zeta|^{-1}, \quad \|T R(\zeta, T)\| = \sup_{\lambda' \in \Sigma(T)} |\lambda'|\, |\lambda' - \zeta|^{-1}. \tag{4.9}$$

Hence $S \in \mathrm{P}(S)$ if

$$a \sup_{\lambda' \in \Sigma(T)} |\lambda' - \zeta|^{-1} + b \sup_{\lambda' \in \Sigma(T)} |\lambda'|\, |\lambda' - \zeta|^{-1} < 1\,. \tag{4.10}$$

In particular suppose that T has an isolated eigenvalue λ with multiplicity $m < \infty$ and with isolation distance d (see § 3.5). Let Γ be the circle with center λ and radius $d/2$. If $\zeta \in \Gamma$, we have then $|\lambda' - \zeta|^{-1} \leqq 2/d$ and $|\lambda'(\lambda' - \zeta)^{-1}| \leqq 1 + (|\zeta - \lambda| + |\lambda|)\, |\lambda' - \zeta|^{-1} \leqq 2 + 2|\lambda|/d$. Hence (4.10) is satisfied if

$$a + b(|\lambda| + d) < d/2\,. \tag{4.11}$$

It follows from Theorem IV-3.18 that Γ *encloses exactly m (repeated) eigenvalues* of $S = T + A$ and no other points of $\Sigma(S)$. If, in addition, A is symmetric, then S is selfadjoint and any eigenvalue of S must be real. Thus S has exactly m repeated eigenvalues [and no other points of $\Sigma(S)$] in the interval $\left(\lambda - \frac{d}{2}, \lambda + \frac{d}{2}\right)$ provided (4.11) is satisfied.

The condition (4.11) is not always very general; a weaker sufficient condition may be deduced by a more direct application of Theorem IV-3.18.

Problem 4.8. Let T be normal and $A \in \mathscr{B}(\mathsf{H})$. Let $d(\zeta) = \operatorname{dist}(\zeta, \Sigma(T))$. Then $d(\zeta) > \|A\|$ implies $\zeta \in \mathrm{P}(T + A)$ and $\|R(\zeta, T + A)\| \leqq 1/(d(\zeta) - \|A\|)$.

Remark 4.9. We have seen in IV-§ 3.1–3.2 that the spectrum of a closed operator T is upper semicontinuous in T but not necessarily lower semicontinuous. If T is restricted to vary over the set of selfadjoint operators, however, it can be shown that $\Sigma(T)$ is also lower semicontinuous so that it is *continuous* in T. Here the lower semicontinuity of $\Sigma(T)$ means that any open set of the complex plane containing a point of $\Sigma(T)$ contains also a point of $\Sigma(T_n)$ for sufficiently large n, if T and T_n are all selfadjoint and $\{T_n\}$ converges to T in the generalized sense. This lower semicontinuity of the spectrum will be proved later under weaker conditions. Here we content ourselves with proving the following weaker theorem, which however expresses the stated continuity of the spectrum very clearly.

Theorem 4.10. *Let T be selfadjoint and $A \in \mathscr{B}(\mathsf{H})$ symmetric. Then $S = T + A$ is selfadjoint and* $\operatorname{dist}(\Sigma(S), \Sigma(T)) \leqq \|A\|$, *that is,*

$$\sup_{\zeta \in \Sigma(S)} \operatorname{dist}(\zeta, \Sigma(T)) \leqq \|A\|, \quad \sup_{\zeta \in \Sigma(T)} \operatorname{dist}(\zeta, \Sigma(S)) \leqq \|A\|. \tag{4.12}$$

Proof. Since S is selfadjoint as well as T, it suffices by symmetry to show the first inequality. For this it suffices in turn to show that any ζ such that $\operatorname{dist}(\zeta, \Sigma(T)) > \|A\|$ belongs to $\mathrm{P}(S)$. But this is obvious since $\|R(\zeta, T)\| < \|A\|^{-1}$ by (4.9) and the second Neumann series for $R(\zeta, T + A)$ converges [see II-(1.13)].

4. Semibounded operators

An important property of a relatively bounded symmetric perturbation is that it preserves semiboundedness. More precisely, we have

Theorem 4.11. *Let T be selfadjoint and bounded from below. Let A be symmetric and relatively bounded with respect to T with relative bound smaller than* 1. *Then $S = T + A$ is selfadjoint and bounded from below. If the inequality* (4.1) *holds with $b < 1$, the following inequality holds for the lower bounds γ_T, γ_S of T and S, respectively*:

$$\gamma_S \geqq \gamma_T - \max\left(\frac{a}{1-b}, \ a + b\,|\gamma_T|\right). \tag{4.13}$$

Proof. Since the selfadjointness of S is known by Theorem 4.3, it suffices to show that any real number ζ smaller than the right member of (4.13) (which we denote by γ) belongs to the resolvent set $\mathrm{P}(S)$ (see § 3.10). Considering the second Neumann series for $R(\zeta, S) = R(\zeta, T + A)$, we have only to show that $\|A\,R(\zeta)\| < 1$ for $\zeta < \gamma$, where $R(\zeta) = R(\zeta, T)$.

Now we have by (4.1) and (3.16), (3.17)

$$\begin{aligned}(4.14)\qquad \|A R(\zeta)\| &\leqq a\|R(\zeta)\| + b\|T R(\zeta)\| \leqq \\ &\leqq a(\gamma_T - \zeta)^{-1} + b \sup_{\lambda\in\Sigma(T)} |\lambda|\,(\lambda-\zeta)^{-1} \\ &\leqq a(\gamma_T - \zeta)^{-1} + b\max(1, |\gamma_T|\,(\gamma_T-\zeta)^{-1}).\end{aligned}$$

The last member of (4.14) is smaller than 1 if $\zeta < \gamma$.

Theorem 4.11 leads to

Theorem 4.12.[1] *Let T be selfadjoint and nonnegative. Let A be symmetric,* $\mathsf{D}(A) \supset \mathsf{D}(T)$ *and* $\|Au\| \leqq \|Tu\|$ *for* $u \in \mathsf{D}(T)$. *Then*

$$(4.15)\qquad |(Au, u)| \leqq (Tu, u), \quad u \in \mathsf{D}(T).$$

Proof. For any real $\varkappa$ such that $-1 < \varkappa < 1$, Theorem 4.11 can be applied to the pair T, $\varkappa A$ with $a = 0$, $b = |\varkappa|$ and $\gamma_T = 0$. The result is that $T + \varkappa A$ is bounded from below with lower bound $\geqq 0$. Hence $-\varkappa(Au, u) \leqq (Tu, u)$, and (4.15) follows on letting $\varkappa \to \pm 1$.

Remark 4.13. Theorem 4.11 is not necessarily true if the perturbation is not relatively bounded. When a sequence of selfadjoint operators T_n converges in the generalized sense to a selfadjoint operator T bounded from below, the T_n need not be bounded from below and, even if each T_n is bounded from below, its lower bound may go to $-\infty$ for $n \to \infty$. This is seen from the following example[2].

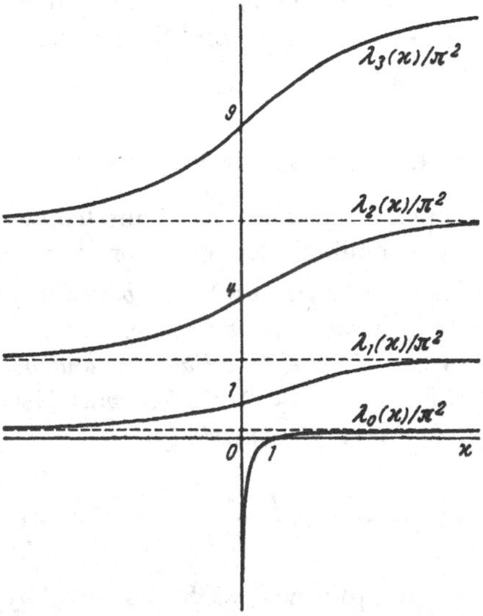

Fig. 1. The eigenvalues of $-u'' = \lambda u$ on [0, 1] with boundary conditions $u(0) = 0$, $\varkappa u'(1) = u(1)$

[1] This is a special case of a more general inequality due to Löwner [1] and Heinz [1]. See also Rellich [7], T. Kato [5], [14], Cordes [1].

[2] This example is due to Rellich [5], [6].

Example 4.14. Let $H = L^2(0, 1)$ and let $T(\varkappa)$ be the differential operator $-d^2/dx^2$ with the boundary condition $u(0) = 0$, $\varkappa u'(1) - u(1) = 0$. It is easily seen that $T(\varkappa)$ is selfadjoint for any real $\varkappa$ [cf. § 3.6; note that the boundary condition at $x = 0$ is of type III-(2.15) while that at $x = 1$ is of type III-(2.16)]. The resolvent $R(\zeta, \varkappa) = R(\zeta, T(\varkappa))$ can be represented by an integral operator with the kernel (Green function)

$$g(y, x; \zeta, \varkappa) = \frac{\sin \zeta^{1/2} y}{\sin \zeta^{1/2} - \varkappa \zeta^{1/2} \cos \zeta^{1/2}} [\zeta^{-1/2} \sin \zeta^{1/2}(1 - x) - \varkappa \cos \zeta^{1/2}(1 - x)] \tag{4.16}$$

for $0 \leq y \leq x \leq 1$ (and x, y exchanged for $x \leq y$). This resolvent exists except for those ζ for which the denominator vanishes; the exceptional values of ζ are exactly the eigenvalues of $T(\varkappa)$. As is easily seen, there is one and only one negative eigenvalue $\lambda_0(\varkappa)$ of $T(\varkappa)$ for $0 < \varkappa < 1$, and this $\lambda_0(\varkappa)$, which is at the same time the lower bound of $T(\varkappa)$, tends to $-\infty$ for $\varkappa \searrow 0$, whereas $T(0)$ is nonnegative; see Fig. 1. Nevertheless, $T(\varkappa)$ converges for $\varkappa \to 0$ to $T(0)$ in the generalized sense; this is seen from the fact that $g(y, x; \zeta, \varkappa) \to g(y, x; \zeta, 0)$, $\varkappa \to 0$, uniformly for a fixed nonreal ζ, which implies that $R(\zeta, \varkappa) \to R(\zeta, 0)$ *in norm*. (The verification of these results is left to the reader. Cf. also Example VII-1.11.)

5. Completeness of the eigenprojections of slightly non-selfadjoint operators

In IV-§ 3.7 we mentioned the difficulties encountered when all the eigenvalues or the eigenprojections of a perturbed operator $S = T + A$ are considered simultaneously, and then gave an example of a perturbed operator in a Banach space in which uniform estimates can be given to all perturbed eigenvalues and eigenprojections. Here we shall consider the *completeness* of the eigenprojections of a non-selfadjoint operator S in a Hilbert space as a perturbation problem for a selfadjoint operator T. Actually we shall go a little further and deduce a sufficient condition for S to be a *spectral operator*.

Theorem 4.15a. *Let T be a selfadjoint operator with compact resolvent, with simple eigenvalues $\lambda_1 < \lambda_2 < \cdots$. Let P_h, $h = 1, 2, \ldots$, be the associated eigenprojections (so that $P_h P_k = \delta_{hk} P_h$, $P_h^* = P_h$,* dim $P_h = 1$). *Assume further that $\lambda_h - \lambda_{h-1} \to \infty$* as $h \to \infty$. *Let $A \in \mathscr{B}(H)$ (not necessarily symmetric). Then $S = T + A$ is closed with compact resolvent, and the eigenvalues and eigenprojections of S can be indexed as $\{\mu_{0k}, \mu_h\}$* and *$\{Q_{0k}, Q_h\}$, respectively, where $k = 1, \ldots, m < \infty$ and $h = n + 1, n + 2, \ldots$ with $n \geq 0$, in such a way that the following results hold.* i) $|\mu_h - \lambda_h|$ *is bounded as $h \to \infty$.* ii) *There is a $W \in \mathscr{B}(H)$ with $W^{-1} \in \mathscr{B}(H)$ such that*

$$\sum_{k=1}^{m} Q_{0k} = W^{-1} \Big(\sum_{h \leq n} P_h \Big) W, \quad Q_h = W^{-1} P_h W \quad \textit{for} \quad h > n.$$

Remark 4.16a. (a) ii) implies that S is a spectral operator[1]. In particular, $\{Q_{0k}, Q_h\}$ is a *complete family* in the sense that

$$\sum_{k=1}^{m} Q_{0k} + \sum_{h>n} Q_h = 1, \tag{4.29}$$

where the series converges strongly and *unconditionally* (with an arbitrary order of summation). This is easily seen by noting that the system $\left\{\sum_k Q_{0k}, Q_{n+1}, Q_{n+2}, \ldots\right\}$ is *similar* to $\left\{\sum_{h\le n} P_h, P_{n+1}, P_{n+2}, \ldots\right\}$, the latter being complete because T is selfadjoint.

(b) The assumption that T be bounded from below is not essential. The proof given below can be modified in an obvious way to deal with the case in which T has simple eigenvalues $\cdots < \lambda_{-1} < \lambda_0 < \lambda_1 < \cdots$ such that $\lambda_h - \lambda_{h-1} \to \infty$ as $h \to \pm\infty$.

(c) The theorem can be generalized to the case in which A is not bounded but relatively bounded with respect to T or some fractional power of T, with a corresponding assumption on the growth rate of $\lambda_h - \lambda_{h-1}$ as $h \to \infty$[2].

The proof of Theorem 4.15a is based on the following lemma[3].

Lemma 4.17a. *Let $\{P_j\}_{j=0,1,\ldots}$ be a complete family of orthogonal projections, and let $\{Q_j\}_{j=0,1,\ldots}$ be a family of (not necessarily orthogonal) projections such that $Q_j Q_k = \delta_{jk} Q_j$. Assume that*

$$\dim P_0 = \dim Q_0 = m < \infty\,, \tag{4.30}$$

$$\sum_{j=1}^{\infty} \|P_j (Q_j - P_j) u\|^2 \le c\, \|u\|^2 \quad \textit{for every } u \in \mathsf{H}, \tag{4.31}$$

where c is a constant smaller than 1. Then there is a $W \in \mathscr{B}(\mathsf{H})$ with $W^{-1} \in \mathscr{B}(\mathsf{H})$ such that $Q_j = W^{-1} P_j W$ for $j = 0,1,2,\ldots$.

Proof. First we shall show that

$$W = \sum_{j=0}^{\infty} P_j Q_j \in \mathscr{B}(\mathsf{H}) \tag{4.32}$$

exists in the strong sense. Since $\sum P_j = 1$ by the completeness of $\{P_j\}$ [see (3.33)], it suffices to show that $\sum (P_j - P_j Q_j) = \sum P_j (P_j - Q_j)$ converges strongly. But this is true since

$$\left\| \sum_{j=k}^{k+p} P_j (P_j - Q_j) u \right\|^2 = \sum_{j=k}^{k+p} \|P_j (P_j - Q_j) u\|^2 \to 0\,,\ k \to \infty\,,$$

[1] For spectral operators, see Dunford [1], Dunford-Schwartz [2].
[2] See Clark [1], H. Kramer [1], Schwartz [1], Turner [1, 2, 3].
[3] See T. Kato [20].

by (4.31). Also (4.31) shows that $\|B\| \leq c^{1/2} < 1$ if

$$B = \sum_{j=1}^{\infty} P_j(P_j - Q_j) = 1 - P_0 - \sum_{j=1}^{\infty} P_j Q_j . \tag{4.33}$$

Now (4.32) implies that $WQ_j = P_j Q_j = P_j W$, $j = 0,1,2,\ldots$. Thus the lemma is proved if we show that $W^{-1} \in \mathscr{B}(\mathsf{H})$ exists.

To this end we define

$$W_1 = \sum_{j=1}^{\infty} P_j Q_j = W - P_0 Q_0 = 1 - P_0 - B . \tag{4.34}$$

Since P_0 is an orthogonal projection with $\dim P_0 = m < \infty$, $1 - P_0$ is a Fredholm operator with $\operatorname{nul}(1 - P_0) = m$, $\operatorname{ind}(1 - P_0) = 0$, and $\gamma(1 - P_0) = 1$ (γ is the reduced minimum modulus, see IV-§ 5.1). Since $\|B\| < 1 = \gamma(1 - P_0)$, it follows that W_1 is also Fredholm, with

$$\operatorname{nul} W_1 \leq \operatorname{nul}(1 - P_0) = m , \quad \operatorname{ind} W_1 = \operatorname{ind}(1 - P_0) = 0 \tag{4.35}$$

(see IV-Theorem 5.22). Since $W = P_0 Q_0 + W_1$, where $P_0 Q_0$ is compact by $\dim P_0 < \infty$, W is also Fredholm with $\operatorname{ind} W = \operatorname{ind} W_1 = 0$ (see IV-Theorem 5.26). To show that $W^{-1} \in \mathscr{B}(\mathsf{H})$ exists, therefore, it suffices to show that $\operatorname{nul} W = 0$.

To this end we first prove that

$$\mathsf{N}(W_1) = Q_0 \mathsf{H} . \tag{4.36}$$

Indeed, we have $W_1 Q_0 = 0$ by (4.34) so that $\mathsf{N}(W_1) \supset Q_0 \mathsf{H}$. But since $\dim Q_0 = m \geq \operatorname{nul} W_1$, we must have (4.36).

Suppose now that $Wu = 0$. Then $0 = P_0 W u = P_0 Q_0 u$ and $W_1 u = Wu - P_0 Q_0 u = 0$. Hence $u = Q_0 u$ by (4.36) and so $P_0 u = P_0 Q_0 u = 0$. Thus $(1 - B) u = W_1 u + P_0 u = 0$. Since $\|B\| < 1$, we obtain $u = 0$. This shows that $\operatorname{nul} W = 0$ and completes the proof.

Proof of Theorem 4.15a. Let $d_h = \max\{\lambda_h - \lambda_{h-1}, \lambda_{h+1} - \lambda_h\}$ be the isolation distance of the eigenvalue λ_h; then $d_h \to \infty$ as $h \to \infty$. Hence there is an integer N and $\delta > 0$ such that $\|A\| < \delta \leq d_h/2$ for $h > N$. For these h let Γ_h be the circle with center λ_h and radius δ, and let Γ_0 be the circle with a diameter $[\lambda_1 - \delta, \lambda_N + \delta]$. Then each Γ_h with $h > N$ encloses exactly one eigenvalue μ_h of S, Γ_0 encloses exactly N repeated eigenvalues of S, and these eigenvalues exhaust the spectrum of S. This is seen by the same argument as was used in par. 3; note that all the circles considered are outside one another and $\|R(\zeta, T)\| \leq 1/\delta < \|A\|^{-1}$ if ζ is on one of the circles or outside all of them, so that $R(\zeta, S)$ exists for such ζ.

Let Q_h, $h > N$, be the one-dimensional eigenprojection for the eigenvalue μ_h of S. Then

$$(4.37)\qquad Q_h - P_h = -\frac{1}{2\pi i}\int_{\Gamma_h} (R(\zeta, S) - R(\zeta, T))\,d\zeta$$

$$= \frac{1}{2\pi i}\int_{\Gamma_h} R(\zeta, S) A R(\zeta, T)\,d\zeta.$$

If ζ is inside Γ_h, $h > N$, we have (see III-(6.32))

$$R(\zeta, T) = -(\zeta - \lambda_h)^{-1} P_h + Z_h + (\zeta - \lambda_h) Z_h^2 + \cdots,$$

where Z_h is the reduced resolvent of T at λ_h. Similarly

$$R(\zeta, S) = -(\zeta - \mu_h)^{-1} Q_h + Z'_h + (\zeta - \mu_h) Z'^2_h + \cdots;$$

note that $R(\zeta, S)$ has a simple pole at $\zeta = \mu_h$ because $\dim Q_h = 1$. Substitution of these expansions into (4.37) gives

$$(4.38)\qquad Q_h - P_h = -Q_h A Z_h - Z'_h A P_h, \quad h > N.$$

Now we use the following lemma to be proved later.

Lemma 4.17b. *We have* (a) $\|Q_h\| \leqq M = \text{const}$, $h > N$; (b) $\|Z'_h\| \to 0$ *as* $h \to \infty$; *and* (c) $\sum_{h>n} \|Z_h u\|^2 \leqq c_n \|u\|^2$ *for* $u \in \mathsf{H}$, *where* $c_n \to 0$ *as* $n \to \infty$.

It follows from this lemma that

$$(4.39)\qquad \sum_{h>n} \|Q_h A Z_h u\|^2 \leqq M^2 \|A\|^2 c_n \|u\|^2, \quad u \in \mathsf{H}, \quad n > N;$$

$$(4.40)\qquad \sum_{h>n} \|Z'_h A P_h u\|^2 \leqq c'_n \|A\|^2 \sum_{h>n} \|P_h u\|^2 \leqq c'_n \|A\|^2 \|u\|^2, \quad n > N,$$

where $c'_n = \sup_{h>n} \|Z'_h\|^2 \to 0$ as $n \to \infty$. (4.38) to (4.40) give

$$(4.41)\qquad \sum_{h>n} \|(Q_h - P_h) u\|^2 \leqq c''_n \|u\|^2, \quad c''_n = 2(M^2 c_n + c'_n) \|A\|^2,$$

where $c''_n \to 0$ as $n \to \infty$. If we choose n so large that $c''_n < 1$, we can apply Lemma 4.17a to deduce the assertion of Theorem 4.15a. Indeed, it suffices to apply the lemma to the two systems $\left\{\sum_{h \leqq n} P_h, P_{n+1}, P_{n+2}, \ldots\right\}$ and $\left\{Q_0 + \sum_{N<h\leqq n} Q_h, Q_{n+1}, Q_{n+2}, \ldots\right\}$, where Q_0 is the total projection for the eigenvalues of S inside Γ_0.

Proof of Lemma 4.17b. (a) follows from

$$Q_h = -\frac{1}{2\pi i}\int_{\Gamma'_h} R(\zeta, S)\,d\zeta\,,$$

where Γ'_h is the circle with center λ_h and radius $d_h/2$; note that $\|R(\zeta, S)\| \leqq ((d_h/2) - \|A\|)^{-1}$ for $\zeta \in \Gamma'_h$ (see Problem 4.8). Similarly (b) follows from

$$Z'_h = \frac{1}{2\pi i}\int_{\Gamma'_h} (\zeta - \mu_h)^{-1} R(\zeta, S)\,d\zeta\,, \quad |\mu_h - \lambda_h| < \delta\,.$$

To prove (c), we note that [see I-(5.32)]

$$\sum_{h>n} \|Z_h u\|^2 = \sum_{h>n}\ \sum_{j \neq h} |\lambda_j - \lambda_h|^{-2} \|P_j u\|^2\,.$$

But since $\sum \|P_j u\|^2 \leqq \|u\|^2$, it suffices to show that

$$\text{(4.42)} \qquad \sup_{j \geqq 1} \sum_{\substack{h>n \\ h \neq j}} |\lambda_j - \lambda_h|^{-2} \to 0 \quad \text{as} \quad n \to \infty\,.$$

(4.42) follows easily from the assumption that $\lambda_h - \lambda_{h-1} \to 0$ as $h \to \infty$. The detail may be left to the reader.

Example 4.18. Consider the differential operator $S = -d^2/dx^2 + q(x)$ on $0 \leqq x \leqq 1$ with the boundary condition $u(0) = u(1) = 0$, where q is a bounded, complex-valued function. If we denote by A the operator of multiplication by $q(x)$, we have $A \in \mathscr{B}(\mathsf{H})$ wehre $\mathsf{H} = \mathsf{L}^2(0,1)$. $T = -d^2/dx^2$ with the same boundary condition is selfadjoint, with the eigenvalues $\lambda_n = \pi^2 n^2$, $n = 1, 2, \ldots$. Thus Theorem 4.15a applies to this case, showing that S is a spectral operator. In this example A can be any bounded operator in H, not necessarily an operator of multiplication. A may even be unbounded to some degree, since we have $\lambda_n - \lambda_{n-1} = O(n)$, which is stronger than necessary in Theorem 4.15a.

§ 5. The Schrödinger and Dirac operators

1. Partial differential operators

We shall now apply the foregoing results to some partial differential operators, in particular to the Schrödinger and Dirac operators, which appear in quantum mechanics.

First we consider the Schrödinger operator

$$L = -\Delta + q(x) \tag{5.1}$$

in a region E of the 3-dimensional euclidean space R^3 with the coordinates $x = (x_1, x_2, x_3)$. Here Δ denotes the Laplacian

$$\Delta = \frac{\partial^2}{\partial x_1^2} + \frac{\partial^2}{\partial x_2^2} + \frac{\partial^2}{\partial x_3^2} \tag{5.2}$$

and $q(x) = q(x_1, x_2, x_3)$ is a real-valued function defined on E. Starting from the *formal* differential operator L, various linear operators in different function spaces can be constructed (just as was the case with ordinary differential operators, see III-§ 2.3). In what follows we shall exclusively consider linear operators in the Hilbert space $\mathsf{H} = \mathsf{L}^2(\mathrm{E})$, which is most important in applications.

It is not at all obvious that (5.1) defines an operator in H. A function $u = u(x)$ must be rather smooth if Δu is to be meaningful, but $q(x)\, u(x)$ may not belong to $\mathsf{H} = \mathsf{L}^2$ for such functions if $q(x)$ is too singular. Therefore, we shall once for all assume that $q(x)$ is *locally square integrable*; this means that $q \in \mathsf{L}^2(\mathrm{K})$ in each compact subset K of E. Then $q(x)\, u(x)$ belongs to H if u is a smooth function with a *compact support* in E (that is, u vanishes outside a compact subset of E, which depends on u). Let us denote by $\mathsf{C}_0^\infty = \mathsf{C}_0^\infty(\mathrm{E})$ the set of all infinitely differentiable functions with compact supports in E, and let $\dot{S}$ be the restriction of L with $\mathsf{D}(\dot{S}) = \mathsf{C}_0^\infty$. $\dot{S}$ is a densely defined linear operator in H. $\dot{S}$ will be called the ***minimal operator*** constructed from the formal differential operator L.

The domain of $\dot{S}$ is narrower than necessary; in the above definition we could replace C_0^∞ by C_0^2 (the set of twice continuously differentiable functions with compact supports) because we need only second-order derivatives in constructing Δu, thereby obtaining an extension of $\dot{S}$. Also we can define a larger extension S of $\dot{S}$ by admitting in its domain all functions $u \in \mathsf{H}$ such that $u \in \mathsf{C}^2(\mathrm{E})$ (twice continuously differentiable in E) and $Lu = -\Delta u + q u \in \mathsf{H}$ (here u need not have compact support). In a certain sense S is the largest operator in H constructed from L. Since $S \supset \dot{S}$, S is densely defined. [S could formally be defined without assuming that $q(x)$ be locally square integrable, but it is then not easy to see whether S is densely defined.] One of the basic problems for differential operators consists in investigating the relationship between the operators $\dot{S}$, S and their adjoints.

L is *formally selfadjoint*; this means that the Green identity holds:

$$\int_{\mathrm{E}_0} ((Lu)\, v - u\, Lv)\, dx = \int_{\partial \mathrm{E}_0} \left(\frac{\partial u}{\partial n} v - u \frac{\partial v}{\partial n}\right) dS\,, \tag{5.3}$$

where E_0 is any subdomain of E with compact closure in E and with smooth boundary ∂E_0 and where dS denotes the surface element and $\partial/\partial n$ the inward normal derivative on ∂E_0. If $u \in D(\dot{S})$ and $v \in D(S)$, we obtain from (5.3) with v replaced by $\bar{v}$

$$(\dot{S}u, v) = (u, Sv) . \tag{5.4}$$

This implies that $\dot{S}$ and S are adjoint; hence

$$\dot{S} \subset S \subset \dot{S}^*, \quad \dot{S} \subset \tilde{\dot{S}} = \dot{S}^{**} \subset S^* \subset \dot{S}^* . \tag{5.5}$$

In particular $\dot{S}$ is symmetric and so is its closure $\tilde{\dot{S}} = \dot{S}^{**}$. The question arises whether the latter is selfadjoint, that is, whether $\dot{S}$ is essentially selfadjoint. In general the answer is no, as is inferred from the case of an ordinary differential operator (see examples in § 3.6). But we note that *$\dot{S}$ is essentially selfadjoint only if S is symmetric*. In fact, the selfadjointness of $\dot{S}^{**}$ implies $\dot{S}^{**} = \dot{S}^{***} = \dot{S}^*$ so that (5.5) gives $S^* = \dot{S}^* \supset S$. Conversely, it can be shown that $\dot{S}$ is essentially selfadjoint if S is symmetric, under a mild additional assumption on the local property of q.

In general there are many selfadjoint operators between $\dot{S}$ and $\dot{S}^*$ (in the sense of the extension-restriction relation); these are obtained usually by restricting $\dot{S}^*$ (which is seen to be a differential operator in a generalized sense) by appropriate *boundary conditions* (just as in the case of ordinary differential operators, see § 3.6). For example suppose that $q = 0$. Then the boundary condition $u = 0$ on ∂E gives an operator corresponding to the *Dirichlet problem*, and similarly the boundary condition $\partial u/\partial n = 0$ is related to the *Neumann problem*. We shall call $\dot{S}^*$ the *maximal operator* constructed from L.

2. The Laplacian in the whole space

In this paragraph we consider the special case in which $E = R^3$ (the whole space) and $q(x) = 0$. We shall show that the minimal operator $\dot{T}$ constructed from the formal Laplace operator $L = -\Delta$ is essentially selfadjoint and, furthermore, give a complete description of the selfadjoint operator $H_0 = \dot{T}^{**} = \dot{T}^*$.

These results are easily deduced by means of the Fourier transformation. According to the Fourier-Plancherel theorem (see Example 2.7), to each $u(x) \in L^2$ corresponds the Fourier transform $\hat{u}(k) \in L^2$ given by (2.13). For convenience we regard $u(x)$ and $\hat{u}(k)$ as belonging to different Hilbert spaces $H = L^2(x)$ and $\hat{H} = L^2(k)$. The map $u \to \hat{u} = U u$ defines a unitary operator U on H to $\hat{H}$.

If $u \in C_0^\infty$, it is easily seen that $\dot{T}u = -\Delta u$ has the Fourier transform $|k|^2 \hat{u}(k)$, where $|k|^2 = k_1^2 + k_2^2 + k_3^2$. Now let K^2 be the maximal multiplication operator by $|k|^2$ in the Hilbert space $\hat{H}$ (see Example III-2.2). K^2 is selfadjoint (see Problem 3.22). We denote by H_0 the operator K^2 transformed by U^{-1}:

$$(5.6) \qquad H_0 = U^{-1} K^2 U .$$

H_0 is a selfadjoint operator in H, being a unitary image of the selfadjoint operator K^2 in $\hat{H}$, with $D(H_0) = U^{-1} D(K^2)$. In other words, $D(H_0)$ *is the set of all* $u \in L^2(x)$ *whose Fourier transforms belong to* $L^2(k)$ *after multiplication by* $|k|^2$. It follows from the remark given above that $H_0 \supset \dot{T}$. That $\dot{T}$ is essentially selfadjoint is therefore equivalent to the assertion that $\dot{T}$ has closure H_0, that is, $D(\dot{T}) = C_0^\infty$ *is a core of* H_0 (see III-§ 5.3).

Before giving the proof of this proposition, we note that there are other subsets of $D(H_0)$ that can be easily seen to be cores of H_0. For example, the set S of all functions of the form

$$(5.7) \qquad e^{-|x|^2/2} P(x) \quad [P(x) \text{ are polynomials in } x_1, x_2, x_3]$$

is a core of H_0. To see this, it suffices to show that the set $\hat{S}$ of the Fourier transforms of the functions (5.7) forms a core of K^2. Since K^2 is nonnegative, this is true if $K^2 + 1$ maps $\hat{S}$ onto a set dense in $\hat{H}$ (see Problem III-6.3). But the Fourier transforms of the functions (5.7) have a similar form with x replaced by k. Noting that $(|k|^2 + 1) e^{-|k|^2/4} = f(k)$ is bounded, we see that $(K^2 + 1) \hat{S}$ is the image under F [the maximal multiplication operator by $f(k)$] of the set $\hat{S}'$ of all functions of the form

$$(5.8) \qquad e^{-|k|^2/4} P(k) .$$

But the set of functions of the form (5.8) is dense in $L^2(k)$ since it includes the whole set of Hermite functions[1]. Since F is a bounded symmetric operator with nullity zero and, consequently, $R(F)$ is dense in $\hat{H}$, $(K^2 + 1) \hat{S}$ is dense in $\hat{H}$ (see Problem III-2.9).

We can now prove that $\tilde{T} = H_0$. Let T_1 be the restriction of H_0 with the set (5.7) as the domain. Since we have shown above that $\tilde{T}_1 = H_0$, it suffices to show that $\tilde{T} \supset T_1$ (which implies $\tilde{T} \supset \tilde{T}_1 = H_0$). To this end we shall construct, for each u of the form (5.7), a sequence $u_n \in C_0^\infty$ such that $u_n \underset{s}{\to} u$ and $\dot{T} u_n = -\Delta u_n \underset{s}{\to} -\Delta u = T_1 u$. Such a sequence is given, for example, by

$$(5.9) \qquad u_n(x) = w\left(\frac{x}{n}\right) u(x)$$

[1] For the completeness of the orthonormal family of Hermite functions, see COURANT and HILBERT [1], 95. The three-dimensional Hermite functions considered here are simply products of one-dimensional ones.

where $w(x)$ is a real-valued function of C_0^∞ such that $0 \leq w(x) \leq 1$ everywhere and $w(x) = 1$ for $|x| \leq 1$. It is obvious that $u_n \in C_0^\infty$ and $u_n \underset{s}{\longrightarrow} u$. Also

$$\Delta u_n(x) = w\left(\frac{x}{n}\right) \Delta u(x) + \frac{2}{n} (\operatorname{grad} w) \left(\frac{x}{n}\right) \cdot \operatorname{grad} u(x) + \\ + \frac{1}{n^2} (\Delta w) \left(\frac{x}{n}\right) u(x) \tag{5.10}$$

shows that $\Delta u_n \underset{s}{\longrightarrow} \Delta u$.

Apart from a constant factor and a unit of length, H_0 is *the Hamiltonian operator for a free particle* in non-relativistic quantum mechanics and K^2 is its *representation in momentum space.*

Problem 5.1. Let T be the operator S of par. 1 in the special case $L = -\Delta$. T is symmetric and essentially selfadjoint with $\tilde{T} = H_0$. Also prove *directly* that (Tu, u) is real for each $u \in D(T)$.

Remark 5.2. The selfadjoint operator H_0 is not a differential operator in the proper sense, for a function $u(x)$ of $D(H_0)$ need not be differentiable everywhere. But $u(x)$ has derivatives in the generalized sense up to the second order, all belonging to L^2. The generalized derivative $\partial u/\partial x_j$ is the inverse Fourier transform of $i k_j \hat{u}(k)$, as a natural generalization from the case of a "smooth" u, and similarly the generalized $\partial^2 u/\partial x_j \partial x_l$ is the inverse Fourier transform of $-k_j k_l \hat{u}(k)$. Since $u \in D(H_0)$ implies $(1 + |k|^2) \hat{u}(k) \in L^2$, all these generalized derivatives belong to L^2[1].

In virtue of this generalized differentiability, functions of $D(H_0)$ are more "regular" than general functions of L^2. In fact *every $u \in D(H_0)$ is (equivalent to) a bounded, uniformly continuous function.* To see this we note that

$$\left(\int |\hat{u}(k)|\, dk\right)^2 \leq \int \frac{dk}{(|k|^2 + \alpha^2)^2} \int (|k|^2 + \alpha^2)^2 |\hat{u}(k)|^2\, dk \\ = \frac{\pi^2}{\alpha} \|(H_0 + \alpha^2) u\|^2 < \infty \tag{5.11}$$

where $\alpha > 0$ is arbitrary. But it is a well known (and easily verifiable) fact that a function $u(x)$ whose Fourier transform $\hat{u}(k)$ is integrable is bounded and continuous, with

$$|u(x)| \leq (2\pi)^{-3/2} \int |\hat{u}(k)|\, dk \leq c\, \alpha^{-1/2} \|(H_0 + \alpha^2) u\| \leq \\ \leq c\, (\alpha^{-1/2} \|H_0 u\| + \alpha^{3/2} \|u\|)\,, \tag{5.12}$$

where c is a numerical constant. In particular the value $u(x)$ of u at any fixed point x is relatively bounded with respect to H_0 with relative bound zero.

---

[1] These generalized derivatives are special cases of derivatives in the theory of *generalized functions* (or *distributions*). For generalized functions see e. g. YOSIDA [1].

A similar calculation gives

$$(5.13)\quad |u(x) - u(y)| \leqq C_\gamma |x-y|^\gamma \left(\alpha^{-\left(\frac{1}{2}-\gamma\right)} \|H_0 u\| + \alpha^{\frac{3}{2}+\gamma} \|u\|\right)$$

where γ is any positive number smaller than 1/2, C_γ is a numerical constant depending only on γ and α is any positive constant. In deducing (5.13) we note the following facts. $|e^{ik\cdot x} - e^{ik\cdot y}| = |e^{ik\cdot(x-y)} - 1|$ does not exceed 2 or $|k|\,|x-y|$, hence it does not exceed $2^{1-\gamma}|k|^\gamma |x-y|^\gamma$. And $\int |k|^\gamma |\hat{u}(k)|\, dk$ is finite if $0 < \gamma < 1/2$, with an estimate similar to (5.11). (5.13) means that $u(x)$ is *Hölder continuous* with any exponent smaller than 1/2.

(5.12) and (5.13) are special cases of the *Sobolev inequalities*[1].

Remark 5.3. The above results can be extended to the case of n variables $x_1, \ldots, x_n$ instead of 3, with the exception of (5.12) and (5.13) which depend essentially on the dimension 3 of the space.

3. The Schrödinger operator with a static potential

We now consider the Schrödinger operator (5.1) in the whole space $E = R^3$. Here we assume that $q(x)$ is not only locally square integrable but can be expressed as

$$(5.14)\qquad q = q_0 + q_1 \quad \text{where} \quad q_0 \in L^\infty(R^3)\,, \quad q_1 \in L^2(R^3)\,.$$

As before we consider the minimal operator $\dot{S}$ and the "larger" operator S constructed from $L = -\Delta + q$. Let $\dot{T}$ be the minimal operator for $q = 0$ investigated in the preceding paragraph. Then $\dot{S}$ can be written

$$(5.15)\qquad \dot{S} = \dot{T} + Q = \dot{T} + Q_0 + Q_1\,,$$

where Q, Q_0 and Q_1 denote the maximal multiplication operators by $q(x)$, $q_0(x)$ and $q_1(x)$ respectively; note that $\dot{S}$ and $\dot{T}$ have the same domain $C_0^\infty(R^3)$. S cannot be put into a similar form, however, for there is no simple relationship between the domains of S and T (the corresponding operator for $q = 0$).

We shall now prove that $\dot{S}$ *is essentially selfadjoint*. To this end it suffices to apply Theorem 4.4, regarding $\dot{S}$ as obtained from $\dot{T}$ by adding the perturbing term Q and showing that Q *is relatively bounded with respect to* $\dot{T}$ *with relative bound* 0. Denoting by H and H_0 the selfadjoint closures of $\dot{S}$ and $\dot{T}$ respectively, we shall further show that

$$(5.16)\qquad H = H_0 + Q\,, \quad D(H) = D(H_0) \subset D(Q)\,.$$

Each $u(x) \in D(H_0)$ is a bounded function in virtue of (5.12). Therefore $Q_1 u$ belongs to L^2, with $\|Q_1 u\| \leqq \|q_1\|_2 \|u\|_\infty \leqq c\|q_1\|_2 (\alpha^{-1/2}\|H_0 u\| + \alpha^{3/2}\|u\|)$.

---

[1] See SOBOLEV [1].

Also $Q_0 u \in \mathsf{L}^2$ and $\|Q_0 u\| \leq \|q_0\|_\infty \|u\|$. Hence

$$(5.17) \qquad \mathsf{D}(Q) \supset \mathsf{D}(H_0) \supset \mathsf{D}(\dot T), \quad \|Q u\| \leq a\|u\| + b\|H_0 u\|$$

$$\text{with} \quad b = c\,\alpha^{-1/2}\|q_1\|_2, \quad a = c\,\alpha^{3/2}\|q_1\|_2 + \|q_0\|_\infty .$$

Since α may be chosen as large as one pleases, (5.17) shows that Q is H_0-bounded (and $\dot T$-bounded a fortiori) with relative bound 0. Thus we see from Theorem 4.3 that $H_0 + Q$ is selfadjoint and from Theorem 4.4 that $\dot S = \dot T + Q$ is essentially selfadjoint. Since $H_0 + Q \supset \dot S$ is obvious, $H_0 + Q$ coincides with the closure of $\dot S$. Thus *the perturbed operator H has the same domain as the unperturbed operator H_0*. Finally we note that H is bounded from below by virtue of Theorem 4.11. Summing up, we have proved

Theorem 5.4[1]. *If (5.14) is satisfied, $\dot S$ is essentially selfadjoint. The selfadjoint extension H of $\dot S$ is equal to $H_0 + Q$ with $\mathsf{D}(H) = \mathsf{D}(H_0)$ and is bounded from below.*

Problem 5.5. *The Coulomb potential* $q(x) = e|x|^{-1}$ satisfies (5.14). For what values of β is (5.14) satisfied for $q(x) = e|x|^{-\beta}$?

Remark 5.6. The results obtained above can be extended to the Schrödinger operator for a system of s particles interacting with each other by Coulomb forces. In this case the formal Schrödinger operator has the form (5.1) in which Δ is the $3s$-dimensional Laplacian and the potential $q(x)$ has the form

$$(5.18) \qquad q(x) = \sum_{j=1}^{s} \frac{e_j}{r_j} + \sum_{j<k} \frac{e_{jk}}{r_{jk}},$$

where e_j and e_{jk} are constants and

$$r_j = (x_{3j-2}^2 + x_{3j-1}^2 + x_{3j}^2)^{1/2},$$

$$r_{jk} = [(x_{3j-2} - x_{3k-2})^2 + (x_{3j-1} - x_{3k-1})^2 + (x_{3j} - x_{3k})^2]^{1/2}.$$

It can be proved that the minimal operator $\dot T$ constructed from the formal operator $-\Delta$ is essentially selfadjoint with the selfadjoint closure H_0 (Remark 5.3) and that Q is relatively bounded with respect to H_0 as well as to $\dot T$ with relative bound 0. The proof given above for $s = 1$ is not directly applicable, for (5.12) is no longer true, but it can be modified

[1] The proof given here is due to T. Kato [4]. This result is, however, not very strong, although the proof is quite simple. Many stronger results have since been obtained (see Stummel [1], Wienholtz [1], Brownell [1], Ikebe and Kato [1], Rohde [1], G. Hellwig [1], B. Hellwig [1], Jörgens [1]), but most of these works take into account special properties of differential operators. Babbitt [1] and Nelson [1] deduce interesting results in this direction using Wiener integrals in function spaces.

by taking into account the fact that $q(x)$ consists of terms each of which depends essentially only on 3 variables[1].

Regarding the spectrum of H, we have a more detailed result if we assume slightly more on q.

Theorem 5.7. *In addition to* (5.14) *assume that* $q_0(x) \to 0$ for $|x| \to \infty$. *Then the essential spectrum of H is exactly the nonnegative real axis. (Thus the spectrum of H on the negative real axis consists only of isolated eigenvalues with finite multiplicities.)*[2]

Proof. It is easy to see that the spectrum of H_0 is exactly the nonnegative real axis, which is at the same time the essential spectrum of H_0. In view of Theorem IV-5.35, therefore, it suffices to prove the following lemma.

Lemma 5.8. *Under the assumptions of* Theorem 5.7, *Q is relatively compact with respect to H_0.*

Proof. First we consider the special case $q_0 = 0$. Let $\{u_n\}$ be a bounded sequence in H such that $\{H_0 u_n\}$ is also bounded; we have to show that $\{Q_1 u_n\}$ contains a convergent subsequence. According to (5.12) and (5.13), the $u_n(x)$ are uniformly bounded in x and n and are equicontinuous. According to the Ascoli theorem, $\{u_n\}$ contains a subsequence $\{v_n\}$ that converges uniformly on any bounded region of R^3. Let v be the limit function; v is bounded, continuous and belongs to H. Hence we have

$$Q_1 v_n = q_1 v_n \to q_1 v \quad \text{in} \quad \mathsf{H}$$

by the dominated convergence theorem.

The general case $q_0 \neq 0$ can be reduced to the above case by approximating q_0 uniformly by a sequence $\{q_0^n\}$ of bounded functions with compact support; for example set $q_0^n(x) = q_0(x)$ for $|x| \leqq n$ and $= 0$ for $|x| > n$. If Q_0^n denotes the operator of multiplication by q_0^n, we obtain $\|Q_0^n - Q_0\| \to 0$. But $Q_1 + Q_0^n$ is H_0-compact by what was proved above, since $q_1 + q_0^n \in \mathsf{L}^2(\mathsf{R}^3)$. Hence $(Q_1 + Q_0^n)(H_0 + 1)^{-1}$ is compact. Since $\|(Q_0^n - Q_0)(H_0 + 1)^{-1}\| \leqq \|Q_0^n - Q_0\| \to 0$, it follows from Theorem III-4.7 that $(Q_1 + Q_0)(H_0 + 1)^{-1}$ is compact. This implies that $Q = Q_1 + Q_0$ is H_0-compact.

Remark 5.9. In the theorems proved above we assumed that $q(x)$ is real-valued, but this is not essential in a certain sense. Of course $\dot{S}$ is not symmetric if q is not real-valued, and the essential selfadjointness of $\dot{S}$ is out of the question. Nevertheless we can consider the spectrum

---

[1] For details see T. Kato [4].

[2] For the spectrum of the Schrödinger operators, see T. Kato [4], [4a], Povzner [1], Birman [5], Brownell [2], Žislin [1], Ikebe [1].

of $\tilde{S}$ and, for example, Theorem 5.7 is true if H is replaced by $\tilde{S}$. Even the proof is not changed; it is only essential that Q_0 and Q_1 are relatively compact with respect to H_0, which is true whether or not q_0 and q_1 are real-valued.

4. The Dirac operator

It is of some interest to consider the Dirac operator in conjunction with the Schrödinger operator[1]. For a particle moving freely in space, this operator has the form (apart from a numerical factor and unit of length)[2]

$$L = i^{-1}\,\alpha\cdot \mathrm{grad} + \beta\,. \tag{5.19}$$

L is again a formal differential operator, but it acts on a vector-valued (or, rather, *spinor*-valued) function $u(x) = (u_1(x), \ldots, u_4(x))$, with 4 components, of the space variable $x = (x_1, x_2, x_3)$. We denote by C^4 the 4-dimensional complex vector space in which the values of $u(x)$ lie. α is a 3-component vector $\alpha = (\alpha_1, \alpha_2, \alpha_3)$ with components α_k which are operators in C^4 and may be identified with their representations by 4×4 matrices. Similarly β is a 4×4 matrix. Thus $Lu = v = (v_1, \ldots, v_4)$ has components

$$v_j(x) = i^{-1}\sum_{l=1}^{3}\sum_{h=1}^{4}(\alpha_l)_{jh}\frac{\partial u_h}{\partial x_l} + \sum_{h=1}^{4}\beta_{jh}\,u_h(x)\,. \tag{5.20}$$

The matrices α_h and β are Hermitian symmetric and satisfy the commutation relations

$$\alpha_j\,\alpha_h + \alpha_h\,\alpha_j = 2\delta_{jh}\,1\,,\quad j, h = 1, 2, 3, 4\,, \tag{5.21}$$

with the convention of writing $\alpha_4 = \beta$ (1 is the unit 4×4 matrix). It is known that such a system of matrices α_j, β exists; it is not necessary for our purpose to give their explicit form.

Since L is a formal differential operator, we can construct from L various operators in the basic Hilbert space $\mathsf{H} = (\mathsf{L}^2(\mathsf{R}^3))^4$ consisting of all C^4-valued functions such that

$$\|u\|^2 = \int |u(x)|^2\,dx < \infty\,,\quad |u(x)|^2 = \sum_{j=1}^{4}|u_j(x)|^2\,. \tag{5.22}$$

The associated inner product is

$$(u, v) = \int u(x)\cdot\overline{v(x)}\,dx\,,\quad u(x)\cdot\overline{v(x)} = \sum_{j=1}^{4}u_j(x)\,\overline{v_j(x)}\,. \tag{5.23}$$

[1] For the selfadjointness of the Dirac operator, see T. Kato [4], Prosser [1].
[2] See e. g. Schiff [1].

In particular we denote by $\dot{T}$ the minimal operator $\dot{T}u = Lu$ with domain $\mathsf{D}(\dot{T}) = (\mathsf{C}_0^\infty)^4$ consisting of all $u(x)$ with the components $u_j(x)$ lying in $\mathsf{C}_0^\infty(\mathsf{R}^3)$.

$\dot{T}$ *is essentially selfadjoint.* The proof is again easily carried out by using the Fourier transformation. Let $\hat{u}(k)$ be the Fourier transform of $u(x)$; this means that $\hat{u}(k)$ has 4 components $\hat{u}_j(k)$ which are the Fourier transforms of the $u_j(x)$. The mapping $u \to \hat{u} = Uu$ defines again a unitary operator on H to $\hat{\mathsf{H}} = (\mathsf{L}^2(k))^4$. It is now easily verified that for $u \in \mathsf{D}(\dot{T})$, $\dot{T}u = Lu = v$ has the Fourier transform

(5.24) $$\hat{v}(k) = (k \cdot \alpha + \beta)\, \hat{u}(k) = (k_1\alpha_1 + k_2\alpha_2 + k_3\alpha_3 + \beta)\, \hat{u}(k)\,.$$

Let us now define an operator K in $\hat{\mathsf{H}}$ as the maximal multiplication operator by $k \cdot \alpha + \beta$; here we use the term "multiplication operator" in a generalized sense, for $k \cdot \alpha + \beta$ is not a scalar but a 4×4 matrix for each fixed k. Since this matrix is of finite order, it can easily be shown that K is selfadjoint (just as for ordinary multiplication operators). Let H_0 be the transform of K by the unitary operator U^{-1}:

(5.25) $$H_0 = U^{-1} K U\,.$$

H_0 is a selfadjoint operator in H with $\mathsf{D}(H_0) = U^{-1}\,\mathsf{D}(K)$, and $\mathsf{D}(K)$ is by definition the set of all $\hat{u} \in (\mathsf{L}^2)^4$ such that $(k \cdot \alpha + \beta)\, \hat{u}(k) \in (\mathsf{L}^2)^4$. Since the matrix $k \cdot \alpha + \beta$ is Hermitian symmetric and

(5.26) $$(k \cdot \alpha + \beta)^2 = (k^2 + 1)\, 1$$

in virtue of (5.21), $\hat{u} \in \mathsf{D}(K)$ is equivalent to

(5.27) $$\int (k^2 + 1)\, |\hat{u}(k)|^2\, dk = \|K\hat{u}\|^2 < \infty\,.$$

It follows from the remark given above that $\dot{T} \subset H_0$. Thus $\dot{T}$ is essentially selfadjoint if and only if $\tilde{\dot{T}} = H_0$.

To prove the essential selfadjointness of $\dot{T}$, it is convenient to show that K is a direct sum of 4 operators, two of which are isomorphic to the multiplication operator by $(1 + k^2)^{1/2}$ and two by $-(1 + k^2)^{1/2}$. To see this we need only to introduce, for each k, a new orthonormal basis of C^4 consisting of eigenvectors of the Hermitian matrix $k \cdot \alpha + \beta$. In view of (5.26), the only eigenvalues of this matrix are $\pm(k^2 + 1)^{1/2}$, and it is known that there are indeed two eigenvectors for each sign forming an orthonormal set of 4 eigenvectors. With the introduction of such a basis (which is by no means unique owing to the degeneracy of the eigenvalues) for each k, the Hilbert space $\hat{\mathsf{H}}$ is seen to be decomposed into the direct sum of 4 subspaces in each of which K acts simply as a multiplication by $(k^2 + 1)^{1/2}$ or $-(k^2 + 1)^{1/2}$. The proof of the essential selfadjointness of $\dot{T}$ can be effected, then, as in the case of the Schrödinger operator for a free particle. Details may be left to the reader[1].

[1] Cf. Prosser [1].

The Dirac operator for a particle in a static field with potential $q(x)$ is given by

$$L = i^{-1}\,\alpha\cdot\mathrm{grad} + \beta + Q \tag{5.28}$$

where Q is the multiplication operator by $q(x)\,1$. If $q(x)$ is real and locally square integrable, the minimal operator $\dot{S}$ with domain $(C_0^\infty)^4$ constructed from (5.28) is densely defined and symmetric (cf. par. 1). To ensure that $\dot{S}$ is essentially selfadjoint, however, we need a further assumption. Without aiming at generality, we shall content ourselves by showing that the Coulomb potential

$$q(x) = e/|x|\,, \quad |x| = (x_1^2 + x_2^2 + x_3^2)^{1/2}\,, \tag{5.29}$$

is sufficient for the essential selfadjointness of $\dot{S}$ provided $|e|$ is not too large. To see this it suffices to verify that Q is relatively bounded with respect to H_0 with relative bound smaller than 1 (see a similar argument in the preceding paragraph). But it is well known that [see VI-(4.24)]

$$\int |x|^{-2}|u(x)|^2\,dx \leqq 4\int |\mathrm{grad}\,u(x)|^2\,dx = 4\int k^2|\hat{u}(k)|^2\,dk \tag{5.30}$$

at least for a scalar-valued function $u \in C_0^\infty$; the same inequality for a vector-valued function $u \in (C_0^\infty)^4$ follows by simply adding the results for the components $u_j(x)$. In virtue of (5.29) and (5.27), this gives

$$\|Q u\| \leqq 2|e|\,\|K\hat{u}\| = 2|e|\,\|H_0 u\| \tag{5.31}$$

for $u \in \mathsf{D}(\dot{T})$, and it can be extended to all $u \in \mathsf{D}(H_0)$ since H_0 is the closure of $\dot{T}$ as proved above. Thus Q is H_0-bounded with relative bound $\leqq 2|e|$. In this way we have proved (note Theorems 4.4 and 4.6)

Theorem 5.10. *If $q(x) = e/|x|$ with a real e such that $|e| \leqq 1/2$, then the minimal Dirac operator $\dot{S}$ is essentially selfadjoint. If $|e| < 1/2$ the closure of $\dot{S}$ has the same domain as H_0.*

Problem 5.11. A similar result is also valid if

$$q(x) = \sum_j e_j/|x - a_j| + q_0(x) \tag{5.32}$$

where a_j, $j = 1, 2, \ldots$, are points of $\mathbf{R}^3$, $q_0(x)$ is a bounded function and $\sum |e_j| \leqq 1/2$.

Remark 5.12. Theorem 5.10 is unsatisfactory inasmuch as the condition $|e| \leqq 1/2$ is too strong to cover all interesting cases of the hydrogen-like atoms. Returning to the ordinary system of units, $|e| \leqq 1/2$ corresponds to the condition $|Z| \leqq 137/2 = 68.5$, where Z is the atomic number and $1/137$ is the fine structure constant. In this connection it is of some interest to note that the condition $|e| \leqq 1/2$ can be slightly improved if we use Theorem VI-3.11 (to be proved in Chapter VI). In fact, it is easily seen that $|H_0| = (H_0^2)^{1/2}$ is just the multiplication operator by $(k^2+1)^{1/2}$ in $\hat{\mathsf{H}}$. But we have

$$\int |x|^{-1}|u(x)|^2\,dx \leqq \frac{\pi}{2}\int |k|\,|\hat{u}(k)|^2\,dk \leqq \frac{\pi}{2}\,(|H_0|\,u, u)\,; \tag{5.33}$$

this can be proved by using the transform by U of the operator $|x|^{-1}$, which is an integral operator with kernel $1/2\,\pi^2|k-k'|^2$. It follows from the cited theorem that *if*

$$|e| < 2/\pi = 1/1.57\,, \tag{5.34}$$

$\dot{S}$ *has a selfadjoint extension with domain contained in* $\mathsf{D}(|H_0|^{1/2})$. The condition (5.34) corresponds to $|Z| \leq 137/1.57 = 87$.

It should be further remarked that, although Theorem VI-3.11 does not assert the uniqueness of the selfadjoint extension of $\dot{S}$ for each fixed e, the extension considered is "analytic" in e and it is a ***unique*** extension with the analytic property. This remarkable result follows from Theorem VII-5.1 to be proved in Chapter VII.

Chapter Six

Sesquilinear forms in Hilbert spaces and associated operators

In a finite-dimensional unitary space, the notion of a sesquilinear form and that of a linear operator are equivalent, symmetric forms corresponding to symmetric operators. This is true even in an infinite-dimensional Hilbert space as long as one is concerned with bounded forms and bounded operators. When we have to consider unbounded forms and operators, however, there is no such obvious relationship. Nevertheless there exists a closed theory on a relationship between semibounded symmetric forms and semibounded selfadjoint operators[1]. This theory can be extended to non-symmetric forms and operators within certain restrictions. Since the results are essential in applications to perturbation theory, a detailed exposition of them will be given in this chapter[2]. Some of the immediate applications are included here, and further results will be found in Chapters VII and VIII.

The final section of this chapter deals with the spectral theorem and perturbation of spectral families. The subjects are not related to sesquilinear forms in any essential manner, though the proof of the spectral theorem given there depends on some results in the theory of forms. In view of the fact that the spectral theorem is of a rather special character, we have avoided its use whenever possible. But since there are problems in perturbation theory which require it in their very formulation, the theorem cannot be dispensed with altogether; and the end of our preliminary treatment of Hilbert space seems an appropriate place to discuss it.

§ 1. Sesquilinear and quadratic forms

1. Definitions

In V-§ 2.1 we considered bounded sesquilinear forms defined on the product space $\mathsf{H} \times \mathsf{H}'$. We shall now be concerned with *unbounded* forms $\mathfrak{t}[u, v]$ but restrict ourselves to forms defined for u, v both belonging to a

[1] This theory is due to FRIEDRICHS [1]. See also T. KATO [8].
[2] For other approaches to this problem see ARONSZAJN [4], LIONS [1].

linear manifold D of a Hilbert space H. Thus $\mathfrak{t}[u, v]$ is complex-valued and linear in $u \in D$ for each fixed $v \in D$ and semilinear in $v \in D$ for each fixed $u \in D$. D will be called the *domain* of $\mathfrak{t}$ and is denoted by $D(\mathfrak{t})$. $\mathfrak{t}$ is *densely defined* if $D(\mathfrak{t})$ is dense in H.

$\mathfrak{t}[u] = \mathfrak{t}[u, u]$ will be called the *quadratic form* associated with $\mathfrak{t}[u, v]$. $\mathfrak{t}[u]$ determines $\mathfrak{t}[u, v]$ uniquely according to I-(6.11), namely

$$\mathfrak{t}[u, v] = \frac{1}{4}(\mathfrak{t}[u+v] - \mathfrak{t}[u-v] + i\,\mathfrak{t}[u+iv] - i\,\mathfrak{t}[u-iv]) . \tag{1.1}$$

We repeat the remark that (1.1) is true only in a complex Hilbert space (Remark I-6.10). We shall call $\mathfrak{t}[u, v]$ or $\mathfrak{t}[u]$ simply a *form* when there is no possibility of confusion.

Two forms $\mathfrak{t}$ and $\mathfrak{t}'$ are equal, $\mathfrak{t} = \mathfrak{t}'$, if and only if they have the same domain D and $\mathfrak{t}[u, v] = \mathfrak{t}'[u, v]$ for all pairs u, v in D. *Extensions* and *restrictions* of forms ($\mathfrak{t}' \supset \mathfrak{t}$ or $\mathfrak{t} \subset \mathfrak{t}'$ in symbol) are defined in an obvious way as in the case of operators.

The situation that the domain of a form need not be the whole space makes the operations with forms rather complicated, just as for operators. The sum $\mathfrak{t} = \mathfrak{t}_1 + \mathfrak{t}_2$ of two forms $\mathfrak{t}_1, \mathfrak{t}_2$ will be defined by

$$\mathfrak{t}[u, v] = \mathfrak{t}_1[u, v] + \mathfrak{t}_2[u, v] , \quad D(\mathfrak{t}) = D(\mathfrak{t}_1) \cap D(\mathfrak{t}_2) . \tag{1.2}$$

The product $\alpha\,\mathfrak{t}$ of a form $\mathfrak{t}$ by a scalar α is given by

$$(\alpha\,\mathfrak{t})[u, v] = \alpha\,\mathfrak{t}[u, v] , \quad D(\alpha\,\mathfrak{t}) = D(\mathfrak{t}) . \tag{1.3}$$

The *unit form* $1[u, v]$ is by definition equal to the inner product (u, v) and the *zero form* $0[u, v]$ takes the value zero for all u, v, both forms having the domain H. Thus we have $0\,\mathfrak{t} \subset 0$ for any form $\mathfrak{t}$, and $\mathfrak{t} + \alpha = \mathfrak{t} + \alpha\,1$ is defined for any form $\mathfrak{t}$ by

$$(\mathfrak{t} + \alpha)[u, v] = \mathfrak{t}[u, v] + \alpha(u, v) , \quad D(\mathfrak{t} + \alpha) = D(\mathfrak{t}) . \tag{1.4}$$

A form $\mathfrak{t}$ is said to be *symmetric* if

$$\mathfrak{t}[u, v] = \overline{\mathfrak{t}[v, u]} , \quad u, v \in D(\mathfrak{t}) . \tag{1.5}$$

As is seen from (1.1), $\mathfrak{t}$ is symmetric if and only if $\mathfrak{t}[u]$ is real-valued.

With each form $\mathfrak{t}$ is associated another form $\mathfrak{t}^*$ defined by

$$\mathfrak{t}^*[u, v] = \overline{\mathfrak{t}[v, u]} , \quad D(\mathfrak{t}^*) = D(\mathfrak{t}) . \tag{1.6}$$

$\mathfrak{t}^*$ is called the *adjoint form* of $\mathfrak{t}$. $\mathfrak{t}$ is symmetric if and only if $\mathfrak{t}^* = \mathfrak{t}$. We have the identity

$$(\alpha_1\,\mathfrak{t}_1 + \alpha_2\,\mathfrak{t}_2)^* = \bar{\alpha}_1\,\mathfrak{t}_1^* + \bar{\alpha}_2\,\mathfrak{t}_2^* . \tag{1.7}$$

For any form $\mathfrak{t}$, the two forms $\mathfrak{h}, \mathfrak{k}$ defined by

$$\mathfrak{h} = \frac{1}{2}(\mathfrak{t} + \mathfrak{t}^*) , \quad \mathfrak{k} = \frac{1}{2i}(\mathfrak{t} - \mathfrak{t}^*) , \tag{1.8}$$

are symmetric and

$$\mathfrak{t} = \mathfrak{h} + i\mathfrak{k} . \tag{1.9}$$

$\mathfrak{h}$ and $\mathfrak{k}$ will be called the *real* and *imaginary parts* of $\mathfrak{t}$, respectively, and be denoted by $\mathfrak{h} = \operatorname{Re}\mathfrak{t}$, $\mathfrak{k} = \operatorname{Im}\mathfrak{t}$. This notation is justified by

$$\mathfrak{h}[u] = \operatorname{Re}\mathfrak{t}[u] , \quad \mathfrak{k}[u] = \operatorname{Im}\mathfrak{t}[u] , \tag{1.10}$$

although $\mathfrak{h}[u, v]$ and $\mathfrak{k}[u, v]$ are not real-valued and have nothing to do with $\operatorname{Re}(\mathfrak{t}[u, v])$ and $\operatorname{Im}(\mathfrak{t}[u, v])$.

2. Semiboundedness

A symmetric form $\mathfrak{h}$ is said to be bounded from below if the set of (real) values $\mathfrak{h}[u]$ for $\|u\| = 1$ is bounded from below or, equivalently,

$$\mathfrak{h}[u] \geqq \gamma \|u\|^2 , \quad u \in \mathsf{D}(\mathfrak{h}) . \tag{1.11}$$

This will be written simply

$$\mathfrak{h} \geqq \gamma . \tag{1.12}$$

The largest number γ with this property is called the *lower bound* of $\mathfrak{h}$ and will be denoted by $\gamma_{\mathfrak{h}}$. If $\mathfrak{h} \geqq 0$, $\mathfrak{h}$ is said to be nonnegative. Similarly we define the notions of boundedness from above, upper bound, non-positiveness, etc. Since we are mostly concerned with forms bounded from below, the symbol $\gamma_{\mathfrak{h}}$ will be used exclusively for the lower bound.

If $\mathfrak{h}$ is a nonnegative symmetric form, we have the inequalities I-(6.32). If a symmetric form $\mathfrak{h}$ is bounded from below as well as from above, then $\mathfrak{h}$ is bounded with bound equal to the larger one of the absolute values of the upper and lower bounds. In other words, $|\mathfrak{h}[u]| \leqq M\|u\|^2$ for all $u \in \mathsf{D}(\mathfrak{h})$ implies $|\mathfrak{h}[u, v]| \leqq M\|u\| \|v\|$ for all $u, v \in \mathsf{D}(\mathfrak{h})$. The proof is the same as in the finite-dimensional case [see I-(6.33)]. It should be remarked that, more generally, $|\mathfrak{k}[u]| \leqq M\mathfrak{h}[u]$ for all $u \in \mathsf{D} = \mathsf{D}(\mathfrak{h}) = \mathsf{D}(\mathfrak{k})$ implies $|\mathfrak{k}[u, v]| \leqq M\mathfrak{h}[u]^{1/2}\mathfrak{h}[v]^{1/2}$ for all $u, v \in \mathsf{D}$, provided both $\mathfrak{h}$ and $\mathfrak{k}$ are symmetric and $\mathfrak{h}$ is nonnegative.

Let us now consider a nonsymmetric form $\mathfrak{t}$. The set of values of $\mathfrak{t}[u]$ for $u \in \mathsf{D}(\mathfrak{t})$ with $\|u\| = 1$ is called the *numerical range* of $\mathfrak{t}$ and will be denoted by $\Theta(\mathfrak{t})$. As in the case of operators, $\Theta(\mathfrak{t})$ is a convex set in the complex plane (cf. Theorem V-3.1). A symmetric form $\mathfrak{h}$ is bounded from below if and only if $\Theta(\mathfrak{h})$ is a finite or semi-infinite interval of the real axis bounded from the left. Generalizing this, we shall say that a form $\mathfrak{t}$ is *bounded from the left* if $\Theta(\mathfrak{t})$ is a subset of a half-plane of the form $\operatorname{Re}\zeta \geqq \gamma$. In particular, $\mathfrak{t}$ will be said to be *sectorially bounded from the left* (or simply *sectorial*) if $\Theta(\mathfrak{t})$ is a subset of a sector of the form

$$|\arg(\zeta - \gamma)| \leqq \theta , \quad 0 \leqq \theta < \frac{\pi}{2} , \quad \gamma \text{ real} . \tag{1.13}$$

This means that

$$(1.14)\qquad \mathfrak{h} \geqq \gamma \quad\text{and}\quad |\mathfrak{k}[u]| \leqq (\tan\theta)\,(\mathfrak{h}-\gamma)\,[u]\,,\quad u\in \mathsf{D}(\mathfrak{t})\,,$$

where $\mathfrak{h}=\operatorname{Re}\mathfrak{t}$, $\mathfrak{k}=\operatorname{Im}\mathfrak{t}$. The numbers γ and θ are not uniquely determined by $\mathfrak{t}$; we shall call γ a *vertex* and θ a corresponding *semi-angle* of the form $\mathfrak{t}$. It follows from a remark given above that

$$(1.15)\qquad \begin{aligned} &|(\mathfrak{h}-\gamma)[u,v]| \leqq (\mathfrak{h}-\gamma)[u]^{1/2}(\mathfrak{h}-\gamma)[v]^{1/2}\,,\\ &|\mathfrak{k}[u,v]| \leqq (\tan\theta)\,(\mathfrak{h}-\gamma)[u]^{1/2}(\mathfrak{h}-\gamma)[v]^{1/2}\,,\\ &|(\mathfrak{t}-\gamma)[u,v]| \leqq (1+\tan\theta)\,(\mathfrak{h}-\gamma)[u]^{1/2}(\mathfrak{h}-\gamma)[v]^{1/2}\,. \end{aligned}$$

In the following sections we shall be concerned mostly with sectorial forms.

Problem 1.1. It follows further from (1.15) that

$$(1.16)\qquad \begin{aligned} &(\mathfrak{h}-\gamma)[u] \leqq |(\mathfrak{t}-\gamma)[u]| \leqq (\sec\theta)\,(\mathfrak{h}-\gamma)[u]\,,\\ &|(\mathfrak{t}-\gamma)[u+v]|^{1/2} \leqq (\sec\theta)^{1/2}\{|(\mathfrak{t}-\gamma)[u]|^{1/2}+|(\mathfrak{t}-\gamma)[v]|^{1/2}\}\,,\\ &|(\mathfrak{t}-\gamma)[u+v]| \leqq 2(\sec\theta)\,\{|(\mathfrak{t}-\gamma)[u]|+|(\mathfrak{t}-\gamma)[v]|\}\,. \end{aligned}$$

Example 1.2. Let T be an operator in H and set

$$(1.17)\qquad \mathfrak{t}[u,v]=(Tu,v)\,,\quad \mathsf{D}(\mathfrak{t})=\mathsf{D}(T)\,.$$

T and $\mathfrak{t}$ have the same numerical range. In particular, $\mathfrak{t}$ is symmetric if T is symmetric, and $\mathfrak{t}$ is bounded from below if T is bounded from below; $\mathfrak{t}$ is bounded from the left if T is quasi-accretive; $\mathfrak{t}$ is sectorial if T is sectorial (see V-§ 3.10).

Example 1.3. Let S be an operator from H to another Hilbert space H' (which may be identical with H) and set

$$(1.18)\qquad \mathfrak{h}[u,v]=(Su,Sv)\,,\quad \mathsf{D}(\mathfrak{h})=\mathsf{D}(S)\,,$$

where the inner product is, of course, that of H'. $\mathfrak{h}$ is a nonnegative symmetric form.

Example 1.4. Let $\mathsf{H}=l^2$ and

$$(1.19)\qquad \mathfrak{t}[u,v]=\sum_{j=1}^{\infty}\alpha_j\,\xi_j\,\bar{\eta}_j \text{ for } u=(\xi_j) \text{ and } v=(\eta_j)\,,$$

where $\{\alpha_j\}$ is a sequence of complex numbers. $\mathsf{D}(\mathfrak{t})$ is by definition the set of all $u=(\xi_j)\in\mathsf{H}$ such that $\sum|\alpha_j|\,|\xi_j|^2<\infty$. Let $\dot{\mathfrak{t}}$ be the restriction of $\mathfrak{t}$ with $\mathsf{D}(\dot{\mathfrak{t}})$ consisting of all $u\in\mathsf{H}$ with a finite number of nonvanishing components. $\dot{\mathfrak{t}}$ is densely defined and so is $\mathfrak{t}$ a fortiori. $\mathfrak{t}$ is symmetric if and only if all the α_j are real. $\mathfrak{t}$ is sectorial with a vertex γ and a semi-angle θ if and only if all the α_j lie in the sector (1.13).

A further restriction $\dot{\mathfrak{t}}_1$ of $\dot{\mathfrak{t}}$ is obtained by requiring each $u\in\mathsf{D}(\dot{\mathfrak{t}}_1)$ to satisfy the condition [in addition to $u\in\mathsf{D}(\dot{\mathfrak{t}})$]

$$(1.20)\qquad \sum_{j=1}^{\infty}\bar{\beta}_j\,\xi_j=0\,,$$

where $\{\beta_j\}$ is a given sequence of complex numbers not all equal to zero. $\dot{\mathfrak{t}}_1$ *is densely defined if and only if* $\sum|\beta_j|^2=\infty$. To see this, let $w=(\zeta_j)\in\mathsf{H}$ be orthogonal to $\mathsf{D}(\dot{\mathfrak{t}}_1)$. Since the vector $(\bar{\beta}_k, 0, \ldots, -\bar{\beta}_1, 0, \ldots)$ with $-\bar{\beta}_1$ in the k-th component belongs to $\mathsf{D}(\dot{\mathfrak{t}}_1)$, we have $\beta_k\,\zeta_1-\beta_1\,\zeta_k=0$. Since this is true for all k, we have

$\zeta_1 : \zeta_2 : \ldots = \beta_1 : \beta_2 : \ldots$. If $\Sigma|\beta_j|^2 = \infty$, we must have all $\zeta_j = 0$ so that $w = 0$. This shows that $\mathsf{D}(\dot{\mathfrak{t}}_1)$ is dense in H. If $\Sigma|\beta_j|^2 < \infty$, $(\beta_j) = b$ belongs to H and (1.20) implies that $\mathsf{D}(\dot{\mathfrak{t}}_1)$ is orthogonal to b. Hence $\mathsf{D}(\dot{\mathfrak{t}}_1)$ is not dense in H.

Example 1.5. Let $\mathsf{H} = \mathsf{L}^2(E)$ and

$$\mathfrak{t}[u, v] = \int_E f(x)\, u(x)\, \overline{v(x)}\, dx\,, \tag{1.21}$$

where $f(x)$ is a complex-valued measurable function on E. $\mathsf{D}(\mathfrak{t})$ is by definition the set of all $u \in \mathsf{H}$ such that $\int |f(x)|\, |u(x)|^2\, dx < \infty$. $\mathfrak{t}$ is densely defined (as in the case of a maximal multiplication operator, see Example III-2.2). $\mathfrak{t}$ is symmetric if $f(x)$ is real-valued. $\mathfrak{t}$ is sectorial with a vertex γ and a semi-angle θ if the values of $f(x)$ lie in the sector (1.13).

Suppose that E is a bounded open set. Then the restriction $\dot{\mathfrak{t}}$ of $\mathfrak{t}$ with domain $\mathsf{D}(\dot{\mathfrak{t}}) = \mathsf{C}_0^\infty(E)$ is also densely defined. Let $\dot{\mathfrak{t}}_1$ be a restriction of $\dot{\mathfrak{t}}$ with $\mathsf{D}(\mathfrak{t}_1)$ consisting of all $u \in \mathsf{D}(\dot{\mathfrak{t}})$ such that

$$\int_E \overline{g(x)}\, u(x)\, dx = 0 \tag{1.22}$$

where $g(x)$ is a locally integrable function. Then $\dot{\mathfrak{t}}_1$ is densely defined if $g \notin \mathsf{L}^2(E)$ and not densely defined if $0 \neq g \in \mathsf{L}^2(E)$. The proof is similar to that in the preceding example.

Example 1.6. Let $\mathsf{H} = \mathsf{L}^2(0, 1)$ and

$$\mathfrak{h}[u, v] = u(0)\, \overline{v(0)}\,. \tag{1.23}$$

$\mathsf{D}(\mathfrak{h})$ is by definition the set of all $u(x)$ continuous (more exactly, equivalent to a continuous function) on the closed interval $[0, 1]$; then $u(0)$ is well defined for $u \in \mathsf{D}(\mathfrak{h})$. $\mathfrak{h}$ is densely defined, symmetric and nonnegative. Actually $\mathfrak{h}$ is a special case of (1.18) in which $Su = u(0)$, S being an operator from H into a one-dimensional space C.

Example 1.7. Let $\mathsf{H} = \mathsf{L}^2(a, b)$ where (a, b) is a finite interval and set

$$\begin{aligned}\mathfrak{t}[u, v] = \int_a^b \{&p(x)\, u'(x)\, \overline{v'(x)} + q(x)\, u(x)\, \overline{v(x)} + \\ &+ r(x)\, u'(x)\, \overline{v(x)} + s(x)\, u(x)\, \overline{v'(x)}\}\, dx + \\ &+ h_a\, u(a)\, \overline{v(a)} + h_b\, u(b)\, \overline{v(b)}\,,\end{aligned} \tag{1.24}$$

where $p(x)$, $q(x)$, $r(x)$, $s(x)$ are given complex-valued functions and h_a, h_b are constants. Suppose that p, q, r, s are continuous on the closed interval $[a, b]$ (regular case), and let $\mathsf{D}(\mathfrak{t})$ be the set of all $u \in \mathsf{H}$ such that $u(x)$ is absolutely continuous on $[a, b]$ with $u'(x) \in \mathsf{H}$ [so that the right member of (1.24) is meaningful for $u, v \in \mathsf{D}(\mathfrak{t})$]. $\mathfrak{t}$ is densely defined. $\mathfrak{t}$ is symmetric if p, q are real-valued, $\overline{r(x)} = s(x)$ and h_a, h_b are real. If we restrict $\mathsf{D}(\mathfrak{t})$ to the set of all $u(x)$ with $u(a) = u(b) = 0$, we obtain a restriction $\mathfrak{t}_0$ of $\mathfrak{t}$. $\mathfrak{t}_0$ is densely defined. We define a further restriction $\dot{\mathfrak{t}}$ of $\mathfrak{t}_0$ by $\mathsf{D}(\dot{\mathfrak{t}}) = \mathsf{C}_0^\infty(a, b)$.

$\mathfrak{t}$ is sectorial if $p(x) > 0$ on $[a, b]$. To see this we note that there are positive constants δ and M such that $p(x) \geqq \delta$, $|q(x)| \leqq M$, $|r(x)| \leqq M$, $|s(x)| \leqq M$. Thus

$$\mathrm{Re}\,\mathfrak{t}[u] \geqq \int \{\delta|u'|^2 - M(|u|^2 + 2|u\,u'|)\}\, dx - |h_a|\, |u(a)|^2 - |h_b|\, |u(b)|^2\,.$$

Since $2|u\,u'| \leqq \varepsilon|u'|^2 + \varepsilon^{-1}|u|^2$, we obtain, making use of IV-(1.19) with $p = 2$, the inequality

$$\mathfrak{h}[u] = \mathrm{Re}\,\mathfrak{t}[u] \geqq \delta\|u'\|^2 - M\|u\|^2 \tag{1.25}$$

in which the positive constants δ and M need not be the same as above. Similarly we have, noting that $p(x)$ is real,

$$|\mathfrak{k}[u]| = |\mathrm{Im}\,\mathfrak{t}[u]| \leqq \varepsilon\|u'\|^2 + M_\varepsilon\|u\|^2 \tag{1.26}$$

where $\varepsilon > 0$ may be chosen arbitrarily small. Thus it is easy to see that we have an inequality of the form (1.14) with appropriate constants γ and θ; θ can even be chosen arbitrarily small if $-\gamma$ is taken sufficiently large.

Problem 1.8. The $\mathfrak{t}$ of the above example is sectorial even if $p(x)$ is not real-valued, provided $\mathrm{Re}\,p(x) > 0$ [the continuity of $p(x)$ being assumed as above].

Example 1.9. Let $\mathsf{H} = \mathsf{L}^2(\mathsf{R}^3)$ and

$$\mathfrak{t}[u, v] = \int [\mathrm{grad}\, u(x) \cdot \overline{\mathrm{grad}\, v(x)} + q(x)\, u(x)\, \overline{v(x)}]\, dx\,, \tag{1.27}$$

where $q(x)$ is a given function assumed to be locally integrable. As $\mathsf{D}(\mathfrak{t})$ we may take $\mathsf{C}_0^1(\mathsf{R}^3)$ (the set of continuously differentiable functions with compact supports). $\mathfrak{t}$ is densely defined, and symmetric if $q(x)$ is real-valued. We may restrict $\mathsf{D}(\mathfrak{t})$ by requiring more differentiability of $u(x)$, $v(x)$ (for example, $\mathsf{D}(\mathfrak{t}) = \mathsf{C}_0^\infty(\mathsf{R}^3)$) without losing these properties of $\mathfrak{t}$; or we could extend $\mathsf{D}(\mathfrak{t})$ somewhat so as to include functions with non-compact supports.

3. Closed forms

Let $\mathfrak{t}$ be a sectorial form. A sequence $\{u_n\}$ of vectors will be said to be *$\mathfrak{t}$-convergent (to $u \in \mathsf{H}$)*, in symbol

$$u_n \underset{\mathfrak{t}}{\to} u\,,\quad n \to \infty\,, \tag{1.28}$$

if $u_n \in \mathsf{D}(\mathfrak{t})$, $u_n \to u$ and $\mathfrak{t}[u_n - u_m] \to 0$ for $n, m \to \infty$. Note that u may or may not belong to $\mathsf{D}(\mathfrak{t})$. Compare also the notion of T-convergence where T is an operator (see III-§ 5.2).

It follows immediately from the definition that *$\mathfrak{t}$-convergence is equivalent to $\mathfrak{t} + \alpha$-convergence for any scalar α.* Also, *$\mathfrak{t}$-convergence is equivalent to $\mathfrak{h}$-convergence where $\mathfrak{h} = \mathrm{Re}\,\mathfrak{t}$*; this follows from (1.16) by which $(\mathfrak{h} - \gamma)[u_n - u_m] \to 0$ if and only if $(\mathfrak{t} - \gamma)[u_n - u_m] \to 0$. Again, $u_n \underset{\mathfrak{t}}{\to} u$ and $v_n \underset{\mathfrak{t}}{\to} v$ imply $\alpha u_n + \beta v_n \underset{\mathfrak{t}}{\to} \alpha u + \beta v$. For the proof it suffices to note that $\mathfrak{t}$ may be assumed to be symmetric and nonnegative by the above remark; then the result follows from I-(6.32).

A sectorial form $\mathfrak{t}$ is said to be *closed* if $u_n \underset{\mathfrak{t}}{\to} u$ implies that $u \in \mathsf{D}(\mathfrak{t})$ and $\mathfrak{t}[u_n - u] \to 0$. It follows from the above remark that *$\mathfrak{t}$ is closed if and only if $\mathrm{Re}\,\mathfrak{t}$ is closed. $\mathfrak{t}$ is closed if and only if $\mathfrak{t} + \alpha$ is closed.*

Problem 1.10. A bounded form is closed if and only if its domain is a closed linear manifold.

Let $\mathfrak{h}$ be a symmetric nonnegative form and set

$$(u, v)_{\mathfrak{h}} = (\mathfrak{h} + 1)[u, v] = \mathfrak{h}[u, v] + (u, v)\,,\quad u, v \in \mathsf{D}(\mathfrak{h})\,. \tag{1.29}$$

$(u, v)_{\mathfrak{h}}$ may be regarded as an inner product in $\mathsf{D}(\mathfrak{h})$, since it is symmetric and strictly positive:

$$\|u\|_{\mathfrak{h}}^2 = (u, u)_{\mathfrak{h}} = (\mathfrak{h} + 1)[u] = \mathfrak{h}[u] + \|u\|^2 \geqq \|u\|^2 . \tag{1.30}$$

$\mathsf{D}(\mathfrak{h})$ will be denoted by $\mathsf{H}_{\mathfrak{h}}$ when it is regarded as a pre-Hilbert space with the inner product $(\, , \,)_{\mathfrak{h}}$.

When a sectorial form $\mathfrak{t}$ is given, we define a pre-Hilbert space $\mathsf{H}_{\mathfrak{t}}$ as equal to $\mathsf{H}_{\mathfrak{h}'}$ with $\mathfrak{h}' = \mathrm{Re}\,\mathfrak{t} - \gamma \geqq 0$ where γ is a vertex of $\mathfrak{t}$. $\mathsf{H}_{\mathfrak{t}}$ will be called a pre-Hilbert space associated with $\mathfrak{t}$. $\mathsf{H}_{\mathfrak{t}}$ is identical with $\mathsf{D}(\mathfrak{t})$ as a vector space, but the inner product $(\, , \,)_{\mathfrak{t}} = (\, , \,)_{\mathfrak{h}'}$ of $\mathsf{H}_{\mathfrak{t}}$ depends on the choice of γ so that $\mathsf{H}_{\mathfrak{t}}$ is not uniquely determined by $\mathfrak{t}$. In any case, the following are immediate consequences of (1.30). *A sequence $u_n \in \mathsf{D}(\mathfrak{t})$ is $\mathfrak{t}$-convergent if and only if it is a Cauchy sequence in $\mathsf{H}_{\mathfrak{t}}$. When $u \in \mathsf{D}(\mathfrak{t})$, $u_n \underset{\mathfrak{t}}{\to} u$ is equivalent to $\|u_n - u\|_{\mathfrak{t}} \to 0$.*

From (1.15) and (1.30) we have

$$\begin{aligned} |\mathfrak{t}[u, v]| &\leqq |\gamma| \, |(u, v)| + (1 + \tan\theta) \, \mathfrak{h}'[u]^{1/2} \, \mathfrak{h}'[v]^{1/2} \leqq \\ &\leqq |\gamma| \, \|u\| \, \|v\| + (1 + \tan\theta) \, \|u\|_{\mathfrak{t}} \, \|v\|_{\mathfrak{t}} \leqq (|\gamma| + 1 + \tan\theta) \, \|u\|_{\mathfrak{t}} \, \|v\|_{\mathfrak{t}} . \end{aligned} \tag{1.31}$$

This shows that $\mathfrak{t}[u, v]$ *is a bounded sesquilinear form on* $\mathsf{H}_{\mathfrak{t}}$.

Theorem 1.11. *A sectorial form $\mathfrak{t}$ in H is closed if and only if the pre-Hilbert space $\mathsf{H}_{\mathfrak{t}}$ is complete.*

Proof. Since $\mathfrak{t}$ is closed if and only if $\mathfrak{h}' = \mathrm{Re}\,\mathfrak{t} - \gamma$ is closed, we may assume that $\mathfrak{t} = \mathfrak{h}$ is symmetric and nonnegative and the inner product of $\mathsf{H}_{\mathfrak{t}} = \mathsf{H}_{\mathfrak{h}}$ is given by (1.29). Suppose that $\mathfrak{h}$ is closed, and let $\{u_n\}$ be a Cauchy sequence in $\mathsf{H}_{\mathfrak{h}}$: $\|u_n - u_m\|_{\mathfrak{h}} \to 0$. Then $\{u_n\}$ is also Cauchy in H by (1.30), so that there is a $u \in \mathsf{H}$ such that $u_n \to u$. Since $\mathfrak{h}[u_n - u_m] \to 0$ by (1.30), $\{u_n\}$ is $\mathfrak{h}$-convergent and the closedness of $\mathfrak{h}$ implies that $u \in \mathsf{D}(\mathfrak{h}) = \mathsf{H}_{\mathfrak{h}}$ and $\mathfrak{h}[u_n - u] \to 0$. Hence $\|u_n - u\|_{\mathfrak{h}} \to 0$ by (1.30), which shows that $\mathsf{H}_{\mathfrak{h}}$ is complete.

Suppose, conversely, that $\mathsf{H}_{\mathfrak{h}}$ is complete. Let $u_n \underset{\mathfrak{h}}{\to} u$. This implies that $\mathfrak{h}[u_n - u_m] \to 0$ and $\|u_n - u_m\| \to 0$ so that $\|u_n - u_m\|_{\mathfrak{h}} \to 0$ by (1.30). By the completeness of $\mathsf{H}_{\mathfrak{h}}$ there is a $u_0 \in \mathsf{H}_{\mathfrak{h}} = \mathsf{D}(\mathfrak{h})$ such that $\|u_n - u_0\|_{\mathfrak{h}} \to 0$. It follows from (1.30) that $\mathfrak{h}[u_n - u_0] \to 0$ and $\|u_n - u_0\| \to 0$. Hence we must have $u = u_0 \in \mathsf{D}(\mathfrak{h})$ and $\mathfrak{h}[u_n - u] \to 0$. This proves that $\mathfrak{h}$ is closed.

Theorem 1.12. *Let $\mathfrak{t}$ be a sectorial form. If $u_n \underset{\mathfrak{t}}{\to} u$ and $v_n \underset{\mathfrak{t}}{\to} v$, then* $\lim \mathfrak{t}[u_n, v_n]$ *exists. If $\mathfrak{t}$ is closed, this limit is equal to* $\mathfrak{t}[u, v]$.

Proof. We have $|\mathfrak{t}[u_n, v_n] - \mathfrak{t}[u_m, v_m]| \leqq |\mathfrak{t}[u_n - u_m, v_n]| + |\mathfrak{t}[u_m, v_n - v_m]| \leqq (1 + \tan\theta) \, \{\mathfrak{h}[u_n - u_m]^{1/2} \, \mathfrak{h}[v_n]^{1/2} + \mathfrak{h}[u_m]^{1/2} \, \mathfrak{h}[v_n - v_m]^{1/2}\}$ where we have assumed $\gamma = 0$ [see (1.15)]. Since $u_n \underset{\mathfrak{t}}{\to} u$ implies that $\mathfrak{h}[u_n - u_m] \to 0$ and $\mathfrak{h}[u_n]$ is bounded and similarly for v_n, $\lim \mathfrak{t}[u_n, v_n]$

exists. The second part of the theorem can be proved in the same way by considering $\mathfrak{t}[u_n, v_n] - \mathfrak{t}[u, v]$.

Example 1.13. Consider the symmetric nonnegative form $\mathfrak{h}[u, v] = (Su, Sv)$ of Example 1.3. $\mathfrak{h}$ *is closed if and only if* S *is a closed operator.* To see this, it suffices to note that $u_n \underset{\mathfrak{h}}{\to} u$ is equivalent to $u_n \underset{S}{\to} u$ and $\mathfrak{h}[u_n - u] \to 0$ is equivalent to $Su_n \to Su$ (see III-§ 5.2).

Example 1.14. Consider the form $\mathfrak{t}[u, v] = \Sigma\, \alpha_j\, \xi_j\, \bar{\eta}_j$ defined in Example 1.4 and assume that all the α_j lie in the sector (1.13) so that $\mathfrak{t}$ is sectorial. Then $\mathfrak{t}$ is closed. In fact we have $\|u\|_{\mathfrak{t}}^2 = \Sigma(1 + \mathrm{Re}\,\alpha_j - \gamma)\, |\xi_j|^2$ for $u = (\xi_j)$. But $\mathsf{H}_{\mathfrak{t}} = \mathsf{D}(\mathfrak{t})$ is the set of all $u \in \mathsf{H}$ such that $\Sigma\, |\alpha_j|\, |\xi_j|^2 < \infty$, and it is easily seen that, under the assumptions made, this condition is equivalent to $\|u\|_{\mathfrak{t}}^2 < \infty$. Thus $\mathsf{H}_{\mathfrak{t}}$ comprises all $u = (\xi_j)$ with $\|u\|_{\mathfrak{t}} < \infty$ and therefore it is complete (just as l^2 is).

Example 1.15. Consider the form $\mathfrak{t}[u, v] = \int f(x)\, u(x)\, \overline{v(x)}\, dx$ defined in Example 1.5. If the values of $f(x)$ are assumed to lie in the sector (1.13) so that $\mathfrak{t}$ is sectorial, $\mathfrak{t}$ is closed; the proof is similar to that in the preceding example.

Before closing this paragraph, we state a theorem which is important in deciding whether a given $u \in \mathsf{H}$ belongs to the domain of a closed form $\mathfrak{t}$. We know from the definition of closedness that $u \in \mathsf{D}(\mathfrak{t})$ if there is a sequence $u_n \in \mathsf{D}(\mathfrak{t})$ such that $u_n \to u$ and $\mathfrak{t}[u_n - u_m] \to 0$. Actually, however, a weaker condition is sufficient, namely,

Theorem 1.16. *Let* $\mathfrak{t}$ *be a closed sectorial form. Let* $u_n \in \mathsf{D}(\mathfrak{t})$, $u_n \to u$ *and let* $\{\mathfrak{t}[u_n]\}$ *be bounded. Then* $u \in \mathsf{D}(\mathfrak{t})$ *and* $\mathrm{Re}\,\mathfrak{t}[u] \leqq \liminf \mathrm{Re}\,\mathfrak{t}[u_n]$.

The proof will be given later, after the representation theorem has been proved (see § 2.2).

4. Closable forms

A sectorial form is said to be *closable* if it has a closed extension.

Theorem 1.17. *A sectorial form* $\mathfrak{t}$ *is closable if and only if* $u_n \underset{\mathfrak{t}}{\to} 0$ *implies* $\mathfrak{t}[u_n] \to 0$. *When this condition is satisfied,* $\mathfrak{t}$ *has the closure (the smallest closed extension)* $\tilde{\mathfrak{t}}$ *defined in the following way.* $\mathsf{D}(\tilde{\mathfrak{t}})$ *is the set of all* $u \in \mathsf{H}$ *such that there exists a sequence* $\{u_n\}$ *with* $u_n \underset{\mathfrak{t}}{\to} u$, *and*

$$(1.32)\qquad \tilde{\mathfrak{t}}[u, v] = \lim \mathfrak{t}[u_n, v_n] \quad \textit{for any} \quad u_n \underset{\mathfrak{t}}{\to} u,\ v_n \underset{\mathfrak{t}}{\to} v\,.$$

Any closed extension of $\mathfrak{t}$ *is also an extension of* $\tilde{\mathfrak{t}}$.

Proof. Let $\mathfrak{t}_1$ be a closed extension of $\mathfrak{t}$. $u_n \underset{\mathfrak{t}}{\to} 0$ implies $u_n \underset{\mathfrak{t}_1}{\to} 0$ so that $\mathfrak{t}[u_n] = \mathfrak{t}_1[u_n] = \mathfrak{t}_1[u_n - 0] \to 0$ by the closedness of $\mathfrak{t}_1$. This proves the necessity part.

To prove the sufficiency part, let $\mathsf{D}(\tilde{\mathfrak{t}})$ be defined as in the theorem. Then the limit on the right of (1.32) exists by Theorem 1.12. We shall show that this limit depends only on u, v and not on the particular sequences $\{u_n\}$, $\{v_n\}$. Let $\{u_n'\}$, $\{v_n'\}$ be other sequences such that $u_n' \underset{\mathfrak{t}}{\to} u$, $v_n' \underset{\mathfrak{t}}{\to} v$. Then $u_n' - u_n \underset{\mathfrak{t}}{\to} 0$, $v_n' - v_n \underset{\mathfrak{t}}{\to} 0$ (see par. 3) so that $\mathfrak{t}[u_n' - u_n]$

$\to 0$, $\mathfrak{t}[v'_n - v_n] \to 0$ by the assumption. Hence $\mathfrak{t}[u'_n, v'_n] - \mathfrak{t}[u_n, v_n] = \mathfrak{t}[u'_n - u_n, v'_n] + \mathfrak{t}[u_n, v'_n - v_n] \to 0$ as in the proof of Theorem 1.12.

We note that

$$\tilde{\mathfrak{t}}[u_n - u] \to 0 \quad \text{if} \quad u_n \underset{\mathfrak{t}}{\to} u. \tag{1.33}$$

In fact, $\lim_{n\to\infty} \tilde{\mathfrak{t}}[u_n - u] = \lim_{n\to\infty} \lim_{m\to\infty} \mathfrak{t}[u_n - u_m]$ by (1.32) and this is zero by $u_n \underset{\mathfrak{t}}{\to} u$.

We have defined a form $\tilde{\mathfrak{t}}$ with domain $\mathsf{D}(\tilde{\mathfrak{t}})$. Obviously $\tilde{\mathfrak{t}}$ is sectorial and $\tilde{\mathfrak{t}} \supset \mathfrak{t}$. Let us show that $\tilde{\mathfrak{t}}$ is closed. Consider the pre-Hilbert space $\mathsf{H}_{\tilde{\mathfrak{t}}}$ associated with $\tilde{\mathfrak{t}}$; it suffices to show that $\mathsf{H}_{\tilde{\mathfrak{t}}}$ is complete (see Theorem 1.11).

Since $\tilde{\mathfrak{t}} \supset \mathfrak{t}$, $\mathsf{H}_{\mathfrak{t}}$ is a linear manifold of $\mathsf{H}_{\tilde{\mathfrak{t}}}$. The construction of $\mathfrak{t}$ given above shows that $\mathsf{H}_{\mathfrak{t}}$ is dense in $\mathsf{H}_{\tilde{\mathfrak{t}}}$ and that any Cauchy sequence in $\mathsf{H}_{\mathfrak{t}}$ has a limit in $\mathsf{H}_{\tilde{\mathfrak{t}}}$ [see (1.33)]. As is well known, this implies that $\mathsf{H}_{\tilde{\mathfrak{t}}}$ is complete[1].

That $\tilde{\mathfrak{t}}$ is the smallest closed extension of $\mathfrak{t}$ follows also from the construction, for if $\mathfrak{t}_1 \supset \mathfrak{t}$ is closed and $u_n \underset{\mathfrak{t}}{\to} u$, then $u_n \underset{\mathfrak{t}_1}{\to} u$ and so $u \in \mathsf{D}(\mathfrak{t}_1)$. Hence $\mathsf{D}(\tilde{\mathfrak{t}}) \subset \mathsf{D}(\mathfrak{t}_1)$ and $\tilde{\mathfrak{t}}[u, v] = \mathfrak{t}_1[u, v]$ by (1.32) and Theorem 1.12.

Theorem 1.18. *Let $\mathfrak{t}$, $\tilde{\mathfrak{t}}$ be as above. The numerical range $\Theta(\mathfrak{t})$ of $\mathfrak{t}$ is a dense subset of the numerical range $\Theta(\tilde{\mathfrak{t}})$ of $\tilde{\mathfrak{t}}$.*

Proof. For any $u \in \mathsf{D}(\tilde{\mathfrak{t}})$ with $\|u\| = 1$, there is a sequence $u_n \in \mathsf{D}(\mathfrak{t})$ such that $u_n \underset{\mathfrak{t}}{\to} u$. We may assume that $\|u_n\| = 1$; otherwise we need only to replace u_n by $u_n/\|u_n\|$. Since $\mathfrak{t}[u_n] \to \tilde{\mathfrak{t}}[u]$, the result follows immediately.

Corollary 1.19. *A vertex γ and a semi-angle θ for $\tilde{\mathfrak{t}}$ can be chosen equal to the corresponding quantities for $\mathfrak{t}$. If $\mathfrak{h}$ is a closable symmetric form bounded from below, $\mathfrak{h}$ and $\tilde{\mathfrak{h}}$ have the same lower bound.*

Theorem 1.20. *A closable sectorial form $\mathfrak{t}$ with domain H is bounded.*

Proof. $\mathfrak{t}$ is necessarily closed, for $\mathfrak{t}$ and $\tilde{\mathfrak{t}}$ have the same domain H and hence $\mathfrak{t} = \tilde{\mathfrak{t}}$. We may again assume that $\mathfrak{t}$ has a vertex zero. The complete Hilbert space $\mathsf{H}_{\mathfrak{t}}$ associated with $\mathfrak{t}$ coincides with H as a vector space, with the norm $\|u\|_{\mathfrak{t}} \geqq \|u\|$. According to the closed graph theorem, these two norms are equivalent, that is, there is a constant M such that $\|u\|_{\mathfrak{t}} \leqq M\|u\|$. (For the proof consider the operator T on $\mathsf{H}_{\mathfrak{t}}$ to H defined by $Tu = u$; T is bounded with $\|T\| \leqq 1$ and maps $\mathsf{H}_{\mathfrak{t}}$ onto H one to one, hence T^{-1} is bounded by Problem III-5.21.) Hence $0 \leqq \mathfrak{h}[u] \leqq M^2\|u\|^2$ and $|\mathfrak{t}[u, v]| \leqq (1 + \tan\theta)\, M^2\|u\|\,\|v\|$ by (1.15).

---

[1] Let $\{u_n\}$ be a Cauchy sequence in $\mathsf{H}_{\tilde{\mathfrak{t}}}$. Since $\mathsf{H}_{\mathfrak{t}}$ is dense in $\mathsf{H}_{\tilde{\mathfrak{t}}}$, there is a sequence $v_n \in \mathsf{H}_{\mathfrak{t}}$ such that $\|v_n - u_n\|_{\tilde{\mathfrak{t}}} \leqq 1/n$. Then $\{v_n\}$ is Cauchy, so that $v_n \to v$ for some $v \in \mathsf{H}_{\tilde{\mathfrak{t}}}$. Then $u_n \to v$, and $\mathsf{H}_{\tilde{\mathfrak{t}}}$ is complete.

When $\mathfrak{t}$ is a closed sectorial form, a linear submanifold D' of $\mathsf{D}(\mathfrak{t})$ is called a *core* of $\mathfrak{t}$ if the restriction $\mathfrak{t}'$ of $\mathfrak{t}$ with domain D' has the closure $\mathfrak{t}: \tilde{\mathfrak{t}}' = \mathfrak{t}$. (Cf. the corresponding notion for an operator, III-§ 5.3.) Obviously D' is a core of $\mathfrak{t}$ if and only if it is a core of $\mathfrak{t} + \alpha$ for any scalar α. If $\mathfrak{t}$ is bounded, D' is a core of $\mathfrak{t}$ if and only if D' is dense in $\mathsf{D}(\mathfrak{t})$ [note that $\mathsf{D}(\mathfrak{t})$ is then a closed linear manifold].

The next two theorems follow directly from the proof of Theorem 1.17.

Theorem 1.21. *Let* $\mathfrak{t}$ *be a closed sectorial form. A linear submanifold* D' *of* $\mathsf{D}(\mathfrak{t})$ *is a core of* $\mathfrak{t}$ *if and only if* D' *is dense in the Hilbert space* $\mathsf{H}_{\mathfrak{t}}$ *associated with* $\mathfrak{t}$.

Theorem 1.22. *Let* $\mathfrak{t}', \mathfrak{t}''$ *be two sectorial forms such that* $\mathfrak{t}' \subset \mathfrak{t}''$. *Let* $\mathsf{H}_{\mathfrak{t}'}$ *and* $\mathsf{H}_{\mathfrak{t}''}$ *be the associated pre-Hilbert spaces*[1]. *If* $\mathfrak{t}''$ *is closed,* $\mathfrak{t}'$ *is closed if and only if* $\mathsf{H}_{\mathfrak{t}'}$ *is a closed subspace of* $\mathsf{H}_{\mathfrak{t}''}$. *When* $\mathfrak{t}''$ *is closable,* $\tilde{\mathfrak{t}}' = \tilde{\mathfrak{t}}''$ *if and only if* $\mathsf{H}_{\mathfrak{t}'}$ *is dense in* $\mathsf{H}_{\mathfrak{t}''}$.

Example 1.23. Consider the form $\mathfrak{h}[u, v] = (Su, Sv)$ of Examples 1.3 and 1.13. *$\mathfrak{h}$ is closable if and only if S is closable.* In fact, $u_n \underset{\mathfrak{h}}{\to} 0$ and $\mathfrak{h}[u_n] \to 0$ are equivalent to $u_n \underset{S}{\to} 0$ and $Su_n \to 0$, respectively. When S is closable, we have

$$\tilde{\mathfrak{h}}[u, v] = (\tilde{S}u, \tilde{S}v) \quad \text{with} \quad \mathsf{D}(\tilde{\mathfrak{h}}) = \mathsf{D}(\tilde{S}) . \tag{1.34}$$

We know also that $\mathfrak{h}$ is closed if and only if S is closed; in this case a subset D' of $\mathsf{D}(\mathfrak{h}) = \mathsf{D}(S)$ is a core of $\mathfrak{h}$ if and only if it is a core of S.

Example 1.24. Consider the form $\mathfrak{t}[u, v] = \Sigma\, \alpha_j\, \xi_j\, \bar{\eta}_j$ of Examples 1.4 and 1.14. Assume that all the α_j lie in the sector (1.13) so that $\mathfrak{t}$ is sectorial and closed. $\dot{\mathfrak{t}}$ is closable with $\tilde{\dot{\mathfrak{t}}} = \mathfrak{t}$; this follows from the fact that $\mathsf{D}(\dot{\mathfrak{t}})$ is dense in $\mathsf{H}_{\mathfrak{t}}$, which can be easily verified. Next let us consider the closure of $\dot{\mathfrak{t}}_1$. The additional condition (1.20) can be written in the form $\Sigma(1 + \operatorname{Re}\alpha_j - \gamma)\, \overline{\beta_j'}\, \xi_j = 0$ where $\beta_j' = \beta_j(1 + \operatorname{Re}\alpha_j - \gamma)^{-1}$. Noting the expression $\|u\|_{\mathfrak{t}}^2 = \Sigma(1 + \operatorname{Re}\alpha_j - \gamma)\, |\xi_j|^2$ given in Example 1.14, we see that $\mathsf{D}(\dot{\mathfrak{t}}_1)$ is dense in $\mathsf{H}_{\mathfrak{t}}$ if and only if $\Sigma(1 + \operatorname{Re}\alpha_j - \gamma)\, |\beta_j'|^2 = \infty$ (see a similar argument in Example 1.4). As is easily seen, this condition is equivalent to

$$\Sigma(|\alpha_j| + 1)^{-1}\, |\beta_j|^2 = \infty . \tag{1.35}$$

Thus we have $\tilde{\dot{\mathfrak{t}}}_1 = \mathfrak{t}$ if and only if (1.35) is satisfied. If the left member of (1.35) is convergent, on the other hand, $\mathsf{D}(\dot{\mathfrak{t}}_1)$ is a dense subset of the orthogonal complement M in $\mathsf{H}_{\mathfrak{t}}$ of the one-dimensional subspace spanned by the vector (β_j'), which belongs to $\mathsf{H}_{\mathfrak{t}}$, and $\mathsf{D}(\tilde{\dot{\mathfrak{t}}}_1)$ is identical with M. Thus $u \in \mathsf{D}(\tilde{\dot{\mathfrak{t}}}_1)$ is characterized by $\Sigma(|\alpha_j| + 1)\, |\xi_j|^2 < \infty$ and $\Sigma\, \bar{\beta}_j\, \xi_j = 0$ [note that $\Sigma\, \bar{\beta}_j\, \xi_j$ is absolutely convergent for $u \in \mathsf{D}(\mathfrak{t})$, as is seen by applying the Schwarz inequality].

Example 1.25. Consider the form $\mathfrak{t}[u, v] = \int f(x)\, u(x)\, \overline{v(x)}\, dx$ of Examples 1.5 and 1.15 and assume that the values of $f(x)$ lie in the sector (1.13) so that $\mathfrak{t}$ is sectorial and closed. By an argument similar to that of the preceding example, it can be shown that $\tilde{\dot{\mathfrak{t}}} = \mathfrak{t}$, and that $\tilde{\dot{\mathfrak{t}}}_1 = \mathfrak{t}$ if and only if

$$\int_E (|f(x)| + 1)^{-1} |g(x)|^2\, dx = \infty . \tag{1.36}$$

---

[1] Here we assume that a common vertex γ is chosen for $\mathfrak{t}', \mathfrak{t}''$ and the norms in $\mathsf{H}_{\mathfrak{t}'}$, $\mathsf{H}_{\mathfrak{t}''}$ are respectively given by $(\operatorname{Re}\mathfrak{t}' - \gamma)\,[u] + \|u\|^2$, $(\operatorname{Re}\mathfrak{t}'' - \gamma)\,[u] + \|u\|^2$.

Example 1.26. An example of a densely defined, sectorial (even symmetric and nonnegative) form which is not closable is given by the form $\mathfrak{h}[u, v] = u(0)\,\overline{v(0)}$ of Example 1.6. To see this, it suffices to note that $u_n \underset{\mathfrak{h}}{\to} 0$ implies that $\|u_n\| \to 0$ and $\lim u_n(0) = \alpha$ exists (where $\{u_n\}$ is a sequence of continuous functions); but this does *not* imply $\alpha = 0$.

5. Forms constructed from sectorial operators

Let us recall that an operator T in H is said to be sectorial if its numerical range is a subset of a sector of the form (1.13) (see V-§ 3.10). Let T be sectorial and set (see Example 1.2)

$$\mathfrak{t}[u, v] = (Tu, v) \quad \text{with} \quad \mathsf{D}(\mathfrak{t}) = \mathsf{D}(T) . \tag{1.37}$$

The form $\mathfrak{t}$ is obviously sectorial with a vertex γ and semi-angle θ.

Theorem 1.27.[1] *A sectorial operator T is form-closable*[2]*, that is, the form $\mathfrak{t}$ defined above is closable.*

Proof. We may assume without loss of generality that T has a vertex 0 so that $\mathfrak{h} = \operatorname{Re} \mathfrak{t} \geq 0$. Let $u_n \in \mathsf{D}(\mathfrak{t}) = \mathsf{D}(T)$ be $\mathfrak{t}$-convergent to 0: $u_n \to 0$ and $\mathfrak{t}[u_n - u_m] \to 0$; we have to show that $\mathfrak{t}[u_n] \to 0$.

We have by (1.15)

$$\begin{aligned} |\mathfrak{t}[u_n]| &\leq |\mathfrak{t}[u_n, u_n - u_m]| + |\mathfrak{t}[u_n, u_m]| \leq \\ &\leq (1 + \tan\theta)\, \mathfrak{h}[u_n]^{1/2}\, \mathfrak{h}[u_n - u_m]^{1/2} + |(T u_n, u_m)| . \end{aligned} \tag{1.38}$$

For any $\varepsilon > 0$ there is an N such that $\mathfrak{h}[u_n - u_m] \leq \varepsilon^2$ for $n, m \geq N$, for $\mathfrak{h}[u_n - u_m] \to 0$. Since $\mathfrak{h}[u_n]$ is bounded in n for the same reason, it follows from (1.38) that

$$|\mathfrak{t}[u_n]| \leq M\,\varepsilon + |(T u_n, u_m)| , \quad n, m \geq N , \tag{1.39}$$

where M is a constant. On letting $m \to \infty$, we obtain $|\mathfrak{t}[u_n]| \leq M\,\varepsilon$ for $n \geq N$, which proves the required result $\mathfrak{t}[u_n] \to 0$.

Corollary 1.28 *A symmetric operator bounded from below is form-closable.*

Problem 1.29. In Theorem 1.27 we have $\tilde{\mathfrak{t}}[u, v] = (Tu, v)$ for every $u \in \mathsf{D}(T)$ and $v \in \mathsf{D}(\tilde{\mathfrak{t}})$.

Example 1.30. Consider the $\dot{T}$ of III-§ 2.3 for $\mathsf{X} = \mathsf{H} = \mathsf{L}^2(a, b)$, which is the minimal operator in H constructed from the formal differential operator $Lu = p_0 u'' + p_1 u' + p_2 u$. An integration by parts gives

$$(\dot{T}u, v) = \int_a^b \{-p_0 u' \bar{v}' + (p_1 - p_0') u' \bar{v} + p_2 u \bar{v}\}\, dx . \tag{1.40}$$

This is of the same form as the form $\mathfrak{t}$ considered in Example 1.7, and defines a sectorial form if $p_0(x) < 0$ and p_0, p_0', p_1, and p_2 are continuous on the closed finite interval $[a, b]$. Hence $\dot{T}$ is form-closable.

---

[1] This theorem is due to M. Schechter [3]. The author originally proved it only when $\theta < \pi/4$; cf. T. Kato [15].

[2] Note that a sectorial operator is closable as an operator if it is densely defined (see Theorem V-3.4).

6. Sums of forms

Theorem 1.31 *Let* $\mathfrak{t}_1, \ldots, \mathfrak{t}_s$ *be sectorial forms in* H *and let* $\mathfrak{t} = \mathfrak{t}_1 + \cdots + \mathfrak{t}_s$ [*with* $\mathsf{D}(\mathfrak{t}) = \mathsf{D}(\mathfrak{t}_1) \cap \cdots \cap \mathsf{D}(\mathfrak{t}_s)$]. *Then* $\mathfrak{t}$ *is sectorial. If all the* $\mathfrak{t}_j$ *are closed, so is* $\mathfrak{t}$. *If all the* $\mathfrak{t}_j$ *are closable, so is* $\mathfrak{t}$ *and*

$$\tilde{\mathfrak{t}} \subset \tilde{\mathfrak{t}}_1 + \cdots + \tilde{\mathfrak{t}}_s . \tag{1.41}$$

Proof. Without loss of generality all the $\mathfrak{t}_j$ may be assumed to have a vertex zero. Let θ_j be a semi-angle of $\mathfrak{t}_j$ for vertex zero. Since the numerical range of $\mathfrak{t}_j$ is a subset of the sector $|\arg\zeta| \leq \theta_j < \pi/2$, the numerical range of $\mathfrak{t}$ is a subset of the sector $|\arg\zeta| \leq \theta = \max\theta_j < \pi/2$. Thus $\mathfrak{t}$ is sectorial with a vertex zero and semi-angle θ.

Suppose that all the $\mathfrak{t}_j$ are closed and let $u_n \underset{\mathfrak{t}}{\to} u$. Then $u_n \underset{\mathfrak{h}}{\to} u$ for $\mathfrak{h} = \operatorname{Re}\mathfrak{t} = \mathfrak{h}_1 + \cdots + \mathfrak{h}_s$, $\mathfrak{h}_j = \operatorname{Re}\mathfrak{t}_j$, which implies that $\mathfrak{h}[u_n - u_m] \to 0$, $n, m \to \infty$. Since $\mathfrak{h}_j \geq 0$, it follows that $\mathfrak{h}_j[u_n - u_m] \to 0$ for each j. Thus $u_n \underset{\mathfrak{h}_j}{\to} u$ and hence $u_n \underset{\mathfrak{t}_j}{\to} u$, so that $u \in \mathsf{D}(\mathfrak{t}_j)$ and $\mathfrak{t}_j[u_n - u] \to 0$ for each j. Hence $u \in \mathsf{D}(\mathfrak{t})$ and $\mathfrak{t}[u_n - u] \to 0$. This proves that $\mathfrak{t}$ is closed.

Next suppose that all the $\mathfrak{t}_j$ are closable. Then $\tilde{\mathfrak{t}}_1 + \cdots + \tilde{\mathfrak{t}}_s$ is closed by what was just proved and is an extension of $\mathfrak{t}$. Therefore $\mathfrak{t}$ is closable and (1.41) holds.

Remark 1.32. The inclusion $\subset$ in (1.41) cannot in general be replaced by equality, even when all the $\mathfrak{t}_j$ are symmetric[1]. Theorem 1.31 can be extended to the case in which there are an infinite number of forms to be summed, with certain assumptions to ensure the convergence and the sectorial boundedness of the sum.

Let $\mathfrak{t}$ be a sectorial form in H. A form $\mathfrak{t}'$ in H, which need not be sectorial, is said to be *relatively bounded with respect to* $\mathfrak{t}$, or simply $\mathfrak{t}$-*bounded*, if $\mathsf{D}(\mathfrak{t}') \supset \mathsf{D}(\mathfrak{t})$ and

$$|\mathfrak{t}'[u]| \leq a\|u\|^2 + b|\mathfrak{t}[u]| , \quad u \in \mathsf{D}(\mathfrak{t}) , \tag{1.42}$$

where a, b are nonnegative constants. The greatest lower bound for all possible values of b will be called the $\mathfrak{t}$-*bound* of $\mathfrak{t}'$.

Obviously $\mathfrak{t}$-boundedness is equivalent to $(\mathfrak{t} + \alpha)$-boundedness for any scalar α. Also $\mathfrak{t}$-boundedness is equivalent to $\mathfrak{h}$-boundedness for $\mathfrak{h} = \operatorname{Re}\mathfrak{t}$, for (1.42) is equivalent to [see (1.16)]

$$|\mathfrak{t}'[u]| \leq a\|u\|^2 + b\,\mathfrak{h}[u] , \quad u \in \mathsf{D}(\mathfrak{t}) = \mathsf{D}(\mathfrak{h}) , \tag{1.43}$$

with a, b not necessarily equal to the a, b of (1.42).

If both $\mathfrak{t}$ and $\mathfrak{t}'$ are sectorial and closable, (1.42) is extended to all $u \in \mathsf{D}(\tilde{\mathfrak{t}})$ with the same constants a, b and with $\mathfrak{t}$, $\mathfrak{t}'$ replaced by $\tilde{\mathfrak{t}}$, $\tilde{\mathfrak{t}}'$, respectively [$\mathsf{D}(\tilde{\mathfrak{t}}') \supset \mathsf{D}(\tilde{\mathfrak{t}})$ being a consequence]. To see this it suffices

[1] For a counter-example see T. Kato [8].

to consider, for each $u \in \mathsf{D}(\tilde{\mathfrak{t}})$, a sequence $\{u_n\}$ such that $u_n \underset{\mathfrak{t}}{\to} u$ and use an appropriate limiting argument.

For symmetric forms we can define relative semiboundedness. Let $\mathfrak{h}$ be a symmetric form bounded from below. A symmetric form $\mathfrak{h}'$, which need not be semibounded, is said to be *relatively semibounded from below with respect to* $\mathfrak{h}$, or simply $\mathfrak{h}$*-semibounded from below*, if $\mathsf{D}(\mathfrak{h}') \supset \mathsf{D}(\mathfrak{h})$ and

$$\mathfrak{h}'[u] \geqq -a'\|u\|^2 - b'\,\mathfrak{h}[u]\,,\quad u \in \mathsf{D}(\mathfrak{h})\,, \tag{1.44}$$

where a', b' are nonnegative constants. Similarly, $\mathfrak{h}'$ is $\mathfrak{h}$*-semibounded from above* if $\mathsf{D}(\mathfrak{h}') \supset \mathsf{D}(\mathfrak{h})$ and

$$\mathfrak{h}'[u] \leqq a''\|u\|^2 + b''\,\mathfrak{h}[u]\,,\quad u \in \mathsf{D}(\mathfrak{h})\,. \tag{1.45}$$

If $\mathfrak{h}'$ is $\mathfrak{h}$-semibounded both from below and from above, $\mathfrak{h}'$ is $\mathfrak{h}$-bounded. Note that $\mathfrak{h}$-semiboundedness from above and from below are not quite parallel since $\mathfrak{h}$ is assumed to be semibounded from below.

Theorem 1.33 *Let* $\mathfrak{t}$ *be a sectorial form and let* $\mathfrak{t}'$ *be* $\mathfrak{t}$*-bounded with* $b < 1$ *in* (1.43). *Then* $\mathfrak{t} + \mathfrak{t}'$ *is sectorial.* $\mathfrak{t} + \mathfrak{t}'$ *is closed if and only if* $\mathfrak{t}$ *is closed.* $\mathfrak{t} + \mathfrak{t}'$ *is closable if and only if* $\mathfrak{t}$ *is closable; in this case* $\mathsf{D}(\mathfrak{t} + \mathfrak{t}')^{\sim} = \mathsf{D}(\tilde{\mathfrak{t}})$.

Proof. We have $\mathsf{D}(\mathfrak{t} + \mathfrak{t}') = \mathsf{D}(\mathfrak{t})$ since $\mathsf{D}(\mathfrak{t}') \supset \mathsf{D}(\mathfrak{t})$. We may assume without loss of generality that $\mathfrak{t}$ has a vertex zero so that $\mathfrak{h} = \operatorname{Re}\mathfrak{t} \geqq 0$; note that replacement of $\mathfrak{h}$ by $\mathfrak{h} - \gamma$ does not change the b of (1.43) though it may change a. Denoting the semi-angle of $\mathfrak{t}$ by θ and setting $\mathfrak{k} = \operatorname{Im}\mathfrak{t}$, $\mathfrak{h}' = \operatorname{Re}\mathfrak{t}'$, $\mathfrak{k}' = \operatorname{Im}\mathfrak{t}'$, we have

$$\begin{aligned} |(\mathfrak{k} + \mathfrak{k}')\,[u]| &\leqq |\mathfrak{k}[u]| + |\mathfrak{k}'[u]| \leqq |\mathfrak{k}[u]| + |\mathfrak{t}'[u]| \leqq \\ &\leqq (\tan\theta + b)\,\mathfrak{h}[u] + a\|u\|^2\,, \end{aligned} \tag{1.46}$$

$$(\mathfrak{h} + \mathfrak{h}')\,[u] \geqq \mathfrak{h}[u] - |\mathfrak{h}'[u]| \geqq (1 - b)\,\mathfrak{h}[u] - a\|u\|^2\,. \tag{1.47}$$

Hence

$$|(\mathfrak{k} + \mathfrak{k}')\,[u]| \leqq (1 - b)^{-1}(\tan\theta + b)\,((\mathfrak{h} + \mathfrak{h}')\,[u] + a) + a \tag{1.48}$$

for $\|u\| = 1$. This shows that the numerical range of $\mathfrak{t} + \mathfrak{t}'$ is a subset of the sector $|\arg(\zeta + R)| \leqq \theta'$ with sufficiently large R and $\theta' < \pi/2$: $\mathfrak{t} + \mathfrak{t}'$ is sectorial.

It follows from (1.42) that $u_n \underset{\mathfrak{t}}{\to} u$ implies $u_n \underset{\mathfrak{t}'}{\to} u$ and hence $u_n \underset{\mathfrak{t}+\mathfrak{t}'}{\longrightarrow} u$. Conversely, it follows from (1.47) that $u_n \underset{\mathfrak{h}+\mathfrak{h}'}{\longrightarrow} u$ (which is equivalent to $u_n \underset{\mathfrak{t}+\mathfrak{t}'}{\longrightarrow} u$) implies $u_n \underset{\mathfrak{h}}{\to} u$ (which is equivalent to $u_n \underset{\mathfrak{t}}{\to} u$). Similarly we see that $\mathfrak{t}[u_n - u] \to 0$ and $(\mathfrak{t} + \mathfrak{t}')\,[u_n - u] \to 0$ are equivalent. The remaining statements of the theorem are direct consequences of these equivalences. We note, incidentally, that all the above inequalities are extended to $u \in \mathsf{D}(\tilde{\mathfrak{t}})$ with $\mathfrak{t}, \mathfrak{t} + \mathfrak{t}'$, etc. replaced by $\tilde{\mathfrak{t}}$, $(\mathfrak{t} + \mathfrak{t}')^{\sim}$, etc. when $\mathfrak{t}$ is closable.

Remark 1.34. If $\mathfrak{t}'$ is also sectorial and closable in Theorem 1.33, then $\mathsf{D}(\tilde{\mathfrak{t}}') \supset \mathsf{D}(\tilde{\mathfrak{t}})$ and $(\mathfrak{t}+\mathfrak{t}')^{\sim} = \tilde{\mathfrak{t}} + \tilde{\mathfrak{t}}'$. In fact, we have $(\mathfrak{t}+\mathfrak{t}')^{\sim} \subset \tilde{\mathfrak{t}} + \tilde{\mathfrak{t}}'$ by Theorem 1.31 but also $\mathsf{D}((\mathfrak{t}+\mathfrak{t}')^{\sim}) = \mathsf{D}(\tilde{\mathfrak{t}})$ by Theorem 1.33.

Remark 1.35. If $\mathfrak{t}$, $\mathfrak{t}'$ are symmetric in Theorem 1.33, it suffices to assume that $\mathfrak{t}'$ is $\mathfrak{t}$-semibounded from below and from above with $b' < 1$ in (1.44).

Example 1.36. Consider the form $\mathfrak{t}$ of Example 1.7. $\mathfrak{t}$ can be expressed as $\mathfrak{t} = \mathfrak{t}_1 + \mathfrak{t}_2 + \mathfrak{t}_3$ with

$$\begin{aligned} &\mathfrak{t}_1[u, v] = \int p(x)\, u' \bar{v}'\, dx\,, \\ (1.49)\qquad &\mathfrak{t}_2[u, v] = \int \{q(x)\, u\bar{v} + r(x)\, u'\bar{v} + s(x)\, u\bar{v}'\}\, dx\,, \\ &\mathfrak{t}_3[u, v] = h_a\, u(a)\, \overline{v(a)} + h_b\, u(b)\, \overline{v(b)}\,, \end{aligned}$$

and $\mathsf{D}(\mathfrak{t}_1) = \mathsf{D}(\mathfrak{t}_2) = \mathsf{D}(\mathfrak{t}_3) = \mathsf{D}(\mathfrak{t})$. If we assume that $p(x) > 0$ as before, $\mathfrak{t}_1$ may be written as $\mathfrak{t}_1[u, v] = (Su, Sv)$ where S is a linear operator defined by $Su(x) = p(x)^{1/2} u'(x)$ with $\mathsf{D}(S) = \mathsf{D}(\mathfrak{t})$. With the $\mathsf{D}(\mathfrak{t})$ described in Example 1.7, S is a closed operator in H [see Example III-5.14 and Problem III-5.7; note that $p(x)^{-1}$ is bounded]. Hence $\mathfrak{t}_1$ is nonnegative and closed by Example 1.13.

On the other hand $\mathfrak{t}_2$ and $\mathfrak{t}_3$ are $\mathfrak{t}_1$-bounded with $\mathfrak{t}_1$-bound 0; in effect this was proved in Example 1.7. Thus we see from Theorem 1.33 that $\mathfrak{t}$ is closed. Similarly, the restriction $\mathfrak{t}_0$ of $\mathfrak{t}$ mentioned in Example 1.7 is closed; this is due to the fact that the corresponding restriction S_0 of S is closed (see Example III-5.14).

Problem 1.37. Let $\mathfrak{t}$, $\mathfrak{t}'$ be sectorial, $\mathfrak{t}$ closed and $\mathfrak{t}'$ closable. If $\mathsf{D}(\mathfrak{t}') \supset \mathsf{D}(\mathfrak{t})$, then $\mathfrak{t}'$ is $\mathfrak{t}$-bounded [hint: Theorem 1.20].

7. Relative boundedness for forms and operators

We have introduced the notions of relative boundedness, which are important in perturbation theory, for *operators* and for *quadratic forms* (the latter only in Hilbert spaces). Both notions can be applied to sectorial operators S and S': on the one hand S' may be bounded relative to S and, on the other, the sectorial form $(S'u, u)$ may be bounded relative to the form (Su, u). Under appropriate assumptions on closedness and the smallness of the relative bound, $S - S'$ and S have the same domain in the first case, and the corresponding closed forms have the same domain in the second case.

In general it is not clear whether there is any relationship between these two kinds of relative boundedness. If we restrict ourselves to considering only symmetric operators S, S', however, form-relative boundedness is weaker than operator-relative boundedness. More precisely, we have

Theorem 1.38. *Let T be selfadjoint and bounded from below, let A be symmetric and T-bounded with T-bound b. Then the form (Au, u) is relatively bounded with respect to the form (Tu, u) with relative bound $\leqq b$, and the same is true of their closures*[1].

[1] This theorem is essentially the same as Theorem V-4.11.

Proof. We have the inequality $\|Au\| \leqq a'\|u\| + b'\|Tu\|$, $u \in \mathsf{D}(T)$, where b' may be chosen arbitrarily close to b. It follows from Theorem V-4.11 that $T(\varkappa) = T + \varkappa A$ is selfadjoint and bounded from below for real $\varkappa$ with $|\varkappa| < b'^{-1}$, the lower bound $\gamma(\varkappa)$ of $T(\varkappa)$ satisfying

$$\gamma(\varkappa) \geqq \gamma_T - |\varkappa| \max\left(\frac{a'}{1 - b'|\varkappa|}, a' + b'|\gamma_T|\right). \tag{1.50}$$

That $\gamma(\varkappa)$ is the lower bound of $T(\varkappa)$ implies that

$$-\varkappa(Au, u) \leqq -\gamma(\varkappa)(u, u) + (Tu, u), \quad u \in \mathsf{D}(T). \tag{1.51}$$

Since $\pm\varkappa$ can be taken arbitrarily near b'^{-1}, and hence to b^{-1}, (1.51) shows that the form (Au, u) is relatively bounded with respect to (Tu, u) with relative bound $\leqq b$. The last assertion of the theorem follows from a remark in par. 6.

§ 2. The representation theorems

1. The first representation theorem

If $\mathfrak{t}[u, v]$ is a bounded form defined everywhere on H, there is a bounded operator $T \in \mathscr{B}(\mathsf{H})$ such that $\mathfrak{t}[u, v] = (Tu, v)$ (see V-§ 2.1). We can now generalize this theorem to an unbounded form $\mathfrak{t}$, assuming that $\mathfrak{t}$ is densely defined, sectorial and closed. The operator T that appears will be sectorial as is expected from the sectorial boundedness of $\mathfrak{t}$. Actually T turns out to be m-sectorial (see V-§ 3.10), which implies that T is closed and the resolvent set $\mathsf{P}(T)$ covers the exterior of $\Theta(T)$. In particular T is selfadjoint and bounded from below if $\mathfrak{t}$ is symmetric. The precise result is given by

Theorem 2.1. *(The first representation theorem)*[1]. *Let $\mathfrak{t}[u, v]$ be a densely defined, closed, sectorial sesquilinear form in H. There exists an m-sectorial operator T such that*

i) $\mathsf{D}(T) \subset \mathsf{D}(\mathfrak{t})$ *and*

$$\mathfrak{t}[u, v] = (Tu, v) \tag{2.1}$$

for every $u \in \mathsf{D}(T)$ and $v \in \mathsf{D}(\mathfrak{t})$;

ii) *$\mathsf{D}(T)$ is a core of $\mathfrak{t}$;*

iii) *if $u \in \mathsf{D}(\mathfrak{t})$, $w \in \mathsf{H}$ and*

$$\mathfrak{t}[u, v] = (w, v) \tag{2.2}$$

holds for every v belonging to a core of $\mathfrak{t}$, then $u \in \mathsf{D}(T)$ and $Tu = w$. The m-sectorial operator T is uniquely determined by the condition i).

---

[1] When $\mathfrak{t}$ is symmetric, this theorem is due to FRIEDRICHS [1]. The generalization to nonsymmetric $\mathfrak{t}$ appears to have been given by many authors, at least implicitly; a systematic treatment may be found in LIONS [1], where the theorem is given in a somewhat different form.

Corollary 2.2. *If a form* $\mathfrak{t}_0$ *is defined from the* T *of* Theorem 2.1 *by* $\mathfrak{t}_0[u, v] = (Tu, v)$ *with* $\mathsf{D}(\mathfrak{t}_0) = \mathsf{D}(T)$, *then* $\mathfrak{t} = \tilde{\mathfrak{t}}_0$.

Corollary 2.3. *The numerical range* $\Theta(T)$ *of* T *is a dense subset of the numerical range* $\Theta(\mathfrak{t})$ *of* $\mathfrak{t}$.

Corollary 2.4. *If* S *is an operator such that* $\mathsf{D}(S) \subset \mathsf{D}(\mathfrak{t})$ *and* $\mathfrak{t}[u, v] = (Su, v)$ *for every* $u \in \mathsf{D}(S)$ *and every* v *belonging to a core of* $\mathfrak{t}$, *then* $S \subset T$.

T will be called *the m-sectorial operator* (or simply *the operator*) *associated with* $\mathfrak{t}$. We shall often write $T = T_{\mathfrak{t}}$ to express this correspondence.

Theorem 2.5. *If* $T = T_{\mathfrak{t}}$, *then* $T^* = T_{\mathfrak{t}^*}$. *In other words, if* T *is the operator associated with a densely defined, closed sectorial form* $\mathfrak{t}$, *then* T^* *is associated in the same way with* $\mathfrak{t}^*$, *the adjoint form of* $\mathfrak{t}$ *(which is also densely defined, sectorial and closed).*

Theorem 2.6. *If* $\mathfrak{h}$ *is a densely defined, symmetric closed form bounded from below, the operator* $T = T_{\mathfrak{h}}$ *associated with* $\mathfrak{h}$ *is selfadjoint and bounded from below.* T *and* $\mathfrak{h}$ *have the same lower bound.*

Theorem 2.7. $\mathfrak{t} \to T = T_{\mathfrak{t}}$ *is a one-to-one correspondence between the set of all densely defined, closed sectorial forms and the set of all m-sectorial operators.* $\mathfrak{t}$ *is bounded if and only if* T *is bounded.* $\mathfrak{t}$ *is symmetric if and only if* T *is selfadjoint.*

Remark 2.8. These results will show that the closed sectorial forms are convenient means for constructing m-sectorial operators (in particular selfadjoint operators bounded from below), for such forms are easy to construct owing to the fact that there are no "maximal" sectorial forms.

2. Proof of the first representation theorem

In the proof of Theorems 2.1 and 2.5, we may assume without loss of generality that $\mathfrak{t}$ has a vertex zero so that $\mathfrak{h} = \operatorname{Re} \mathfrak{t} \geq 0$. Let $\mathsf{H}_{\mathfrak{t}}$ be the associated (complete) Hilbert space into which $\mathsf{D}(\mathfrak{t})$ is converted by introducing the inner product $(u, v)_{\mathfrak{t}} = (u, v)_{\mathfrak{h}}$ given by (1.29).

Consider the form $\mathfrak{t}_1 = \mathfrak{t} + 1$. $\mathfrak{t}_1$ as well as $\mathfrak{t}$ is a bounded form on $\mathsf{H}_{\mathfrak{t}}$. Hence there is an operator $B \in \mathscr{B}(\mathsf{H}_{\mathfrak{t}})$ such that

$$\mathfrak{t}_1[u, v] = (Bu, v)_{\mathfrak{t}}, \quad u, v \in \mathsf{H}_{\mathfrak{t}} = \mathsf{D}(\mathfrak{t}) . \tag{2.3}$$

Since $\|u\|_{\mathfrak{t}}^2 = (\mathfrak{h} + 1)[u] = \operatorname{Re} \mathfrak{t}_1[u] = \operatorname{Re}(Bu, u)_{\mathfrak{t}} \leq \|Bu\|_{\mathfrak{t}} \|u\|_{\mathfrak{t}}$, we have $\|u\|_{\mathfrak{t}} \leq \|Bu\|_{\mathfrak{t}}$. Hence B has a bounded inverse B^{-1} with closed domain in $\mathsf{H}_{\mathfrak{t}}$. This domain is the whole of $\mathsf{H}_{\mathfrak{t}}$ so that $B^{-1} \in \mathscr{B}(\mathsf{H}_{\mathfrak{t}})$ with $\|B^{-1}\|_{\mathfrak{t}} \leq 1$[1]. To see this, it suffices to show that a $u \in \mathsf{H}_{\mathfrak{t}}$ orthogonal in $\mathsf{H}_{\mathfrak{t}}$ to $\mathsf{D}(B^{-1}) = \mathsf{R}(B)$ is zero. This is obvious from $\|u\|_{\mathfrak{t}}^2 = \operatorname{Re}(Bu, u)_{\mathfrak{t}} = 0$.

[1] This result is known as the Lax-Milgram theorem, see LAX-MILGRAM [1].

For any fixed $u \in \mathsf{H}$, consider the semilinear form $v \to l_u[v] = (u,v)$ defined for $v \in \mathsf{H}_\mathfrak{t}$. l_u is a bounded form on $\mathsf{H}_\mathfrak{t}$ with bound $\leqq \|u\|$, since $|l_u[v]| \leqq \|u\| \|v\| \leqq \|u\| \|v\|_\mathfrak{t}$. By the Riesz theorem (see V-§ 1.1), there is a unique $u' \in \mathsf{H}_\mathfrak{t}$ such that $(u, v) = l_u[v] = (u', v)_\mathfrak{t}$, $\|u'\|_\mathfrak{t} \leqq \|u\|$. We now define an operator A by $A u = B^{-1} u'$. A is a linear operator with domain H and range in $\mathsf{H}_\mathfrak{t}$. Regarded as an operator in H, A belongs to $\mathscr{B}(\mathsf{H})$ with $\|A\| \leqq 1$, for $\|A u\| = \|B^{-1} u'\| \leqq \|B^{-1} u'\|_\mathfrak{t} \leqq \|u'\|_\mathfrak{t} \leqq \|u\|$. It follows from the definition of A that

$$(u, v) = (u', v)_\mathfrak{t} = (B A u, v)_\mathfrak{t} = \mathfrak{t}_1 [A u, v] = (\mathfrak{t} + 1)\, [A u, v]\,. \tag{2.4}$$

Hence

$$\mathfrak{t}[A u, v] = (u - A u, v)\,, \quad u \in \mathsf{H}\,, \quad v \in \mathsf{H}_\mathfrak{t} = \mathsf{D}(\mathfrak{t})\,. \tag{2.5}$$

A is invertible, for $A u = 0$ implies by (2.4) that $(u, v) = 0$ for all $v \in \mathsf{D}(\mathfrak{t})$ and $\mathsf{D}(\mathfrak{t})$ is dense in H. On writing $w = A u$, $u = A^{-1} w$ in (2.5), we get $\mathfrak{t}[w, v] = ((A^{-1} - 1)\, w, v) = (T w, v)$, where $T = A^{-1} - 1$, for every $w \in \mathsf{D}(T) = \mathsf{R}(A) \subset \mathsf{D}(\mathfrak{t})$ and $v \in \mathsf{D}(\mathfrak{t})$. This proves i) of Theorem 2.1.

T is a closed operator in H since $A \in \mathscr{B}(\mathsf{H})$. T is sectorial, for $\Theta(T) \subset \subset \Theta(\mathfrak{t})$ because $(T u, u) = \mathfrak{t}[u]$ by (2.1). T is m-sectorial, for $\mathsf{R}(T + 1) = \mathsf{R}(A^{-1}) = \mathsf{D}(A) = \mathsf{H}$ (see V-§ 3.10).

To prove ii) of Theorem 2.1, it suffices to show that $\mathsf{D}(T) = \mathsf{R}(A)$ is dense in $\mathsf{H}_\mathfrak{t}$ (see Theorem 1.21). Since B maps $\mathsf{H}_\mathfrak{t}$ onto itself bicontinuously, it suffices to show that $B\mathsf{R}(A) = \mathsf{R}(BA)$ is dense in $\mathsf{H}_\mathfrak{t}$. Let $v \in \mathsf{H}_\mathfrak{t}$ be orthogonal in $\mathsf{H}_\mathfrak{t}$ to $\mathsf{R}(BA)$. Then (2.4) shows that $(u, v) = 0$ for all $u \in \mathsf{H}$ and so $v = 0$. Hence $\mathsf{R}(BA)$ is dense in $\mathsf{H}_\mathfrak{t}$.

Corollary 2.2 is another expression of ii) just proved, and Corollary 2.3 follows from it by Theorem 1.18.

To prove further results of the theorems, it is convenient at this point to consider $\mathfrak{t}^*$, the adjoint form of $\mathfrak{t}$. Since $\mathfrak{t}^*$ is also densely defined, sectorial with a vertex zero, and closed, we can construct an associated m-sectorial operator T' in the same way as we constructed T from $\mathfrak{t}$. For any $u \in \mathsf{D}(\mathfrak{t}^*) = \mathsf{D}(\mathfrak{t})$ and $v \in \mathsf{D}(T')$, we have then

$$\mathfrak{t}^*[v, u] = (T' v, u) \quad \text{or} \quad \mathfrak{t}[u, v] = (u, T' v)\,. \tag{2.6}$$

In particular let $u \in \mathsf{D}(T) \subset \mathsf{D}(\mathfrak{t})$ and $v \in \mathsf{D}(T') \subset \mathsf{D}(\mathfrak{t})$. (2.1) and (2.6) give $(T u, v) = (u, T' v)$. This implies that $T' \subset T^*$. But since T^* and T' are both m-sectorial (which implies that they are maximal accretive, see V-§ 3.10), we must have $T' = T^*$ and hence $T'^* = T$ too.

This leads to a simple proof of iii) of Theorem 2.1. If (2.2) holds for all v of a core of $\mathfrak{t}$, it can be extended to all $v \in \mathsf{D}(\mathfrak{t})$ by continuity. Specializing v to elements of $\mathsf{D}(T')$, we have then $(u, T' v) = \mathfrak{t}[u, v] = (w, v)$. Hence $u \in \mathsf{D}(T'^*) = \mathsf{D}(T)$ and $w = T'^* u = T u$ by the definition of T'^*. Corollary 2.4 is a direct consequence of iii), and the uniqueness of an m-sectorial operator T satisfying i) follows from this Corollary. This completes the proof of Theorem 2.1.

Theorem 2.5 follows immediately from $T' = T^*$ proved above. Theorem 2.6 results from it since $\mathfrak{t} = \mathfrak{t}^*$ implies $T = T^*$; the assertion on the lower bounds of $\mathfrak{t}$ and T follows from Corollary 2.3.

Proof of Theorem 2.7. We first note that the mapping $\mathfrak{t} \to T = T_\mathfrak{t}$ is one to one; this is a direct consequence of Theorem 2.1 and Corollary 2.2. It remains to show that any m-sectorial operator T is associated in this way with a densely defined, closed sectorial form $\mathfrak{t}$. As is suggested by Corollary 2.2, such a $\mathfrak{t}$ is obtained as the closure of the form

$$\mathfrak{t}_0[u, v] = (T u, v)\,, \quad \mathsf{D}(\mathfrak{t}_0) = \mathsf{D}(T)\,. \tag{2.7}$$

$\mathfrak{t}_0$ is densely defined and sectorial; it is closable by Theorem 1.27. Let $\mathfrak{t} = \tilde{\mathfrak{t}}_0$ and define the associated operator $T_\mathfrak{t}$; then $T_\mathfrak{t} \supset T$ by Corollary 2.4. But since both T and $T_\mathfrak{t}$ are m-sectorial, we must have $T = T_\mathfrak{t}$.

Proof of Theorem 1.16. We can now give the promised proof of Theorem 1.16. By hypotheses the sequences $\{u_n\}$ and $\{(\operatorname{Re}\mathfrak{t} - \gamma)[u_n]\}$ are bounded. Since $\mathfrak{h}' = \operatorname{Re}\mathfrak{t} - \gamma$ is closed with $\mathfrak{t}$ and since $u_n \underset{\mathfrak{t}}{\to} u$ is equivalent to $u_n \underset{\mathfrak{h}'}{\to} u$ (see § 1.3), we may assume without loss of generality that $\mathfrak{t} = \mathfrak{h}$ is symmetric and nonnegative.

Consider the Hilbert space $\mathsf{H}_\mathfrak{h}$ defined before. Since $\{\|u_n\|\}$ and $\{\mathfrak{h}[u_n]\}$ are bounded, $\{u_n\}$ forms a bounded sequence in $\mathsf{H}_\mathfrak{h}$. Consequently, there exists a subsequence $\{v_n\}$ of $\{u_n\}$ which is weakly convergent in $\mathsf{H}_\mathfrak{h}$ (see Lemma V-1.4). Let $v \in \mathsf{H}_\mathfrak{h}$ be the weak limit of $\{v_n\}$, and let $H = T_\mathfrak{h}$. For any $w \in \mathsf{D}(H)$, we have

$$(v_n, (H + 1)\, w) = (\mathfrak{h} + 1)\,[v_n, w] = (v_n, w)_\mathfrak{h} \to$$
$$\to (v, w)_\mathfrak{h} = (\mathfrak{h} + 1)\,[v, w] = (v, (H + 1)\, w)\,.$$

Since we have $(v_n, (H + 1)\, w) \to (u, (H + 1)\, w)$ on the other hand, we have $(u - v, (H + 1)\, w) = 0$. But $(H + 1)\, w$ varies over the whole of H when w varies over $\mathsf{D}(H)$. This gives $u = v$ so that $u \in \mathsf{H}_\mathfrak{h} = \mathsf{D}(\mathfrak{h})$.

We have also $\|u\|_\mathfrak{h} = \|v\|_\mathfrak{h} \leq \liminf \|v_n\|_\mathfrak{h}$, which is, in view of (1.30), equivalent to $\mathfrak{h}[u] \leq \liminf \mathfrak{h}[v_n]$. Since we could have replaced $\{u_n\}$ by any subsequence, this gives $\mathfrak{h}[u] \leq \liminf \mathfrak{h}[u_n]$.

3. The Friedrichs extension

In this paragraph we denote by S a densely defined, sectorial operator. Define the form $\mathfrak{s}$ by $\mathfrak{s}[u, v] = (Su, v)$ with $\mathsf{D}(\mathfrak{s}) = \mathsf{D}(S)$. Then $\mathfrak{s}$ is closable by Theorem 1.27. Let $\mathfrak{t} = \tilde{\mathfrak{s}}$ and let $T = T_\mathfrak{t}$ be the associated m-sectorial operator. It follows from Corollary 2.4 that $T \supset S$, for $\mathsf{D}(S) = \mathsf{D}(\mathfrak{s})$ is a core of $\mathfrak{t}$. T will be called the *Friedrichs extension* of S.

Originally the Friedrichs extension was defined for a semibounded *symmetric* operator S, in which case T is selfadjoint by Theorem 2.6[1].

Theorem 2.9. *If S is m-sectorial, the Friedrichs extension of S is S itself. In particular, the Friedrichs extension of the Friedrichs extension T of a densely defined, sectorial operator is T itself.*

This is clear since an m-sectorial operator has no proper sectorial extension (see V-§ 3.10).

The following two theorems characterize the Friedrichs extension. Here S, $\mathfrak{s}$, $\mathfrak{t}$, T are as above.

Theorem 2.10. *Among all m-sectorial extensions T' of S, the Friedrichs extension T has the smallest form-domain (that is, the domain of the associated form $\mathfrak{t}$ is contained in the domain of the form associated with any other T').*

Proof. Define the form $\mathfrak{t}'$ by $\mathfrak{t}'[u, v] = (T' u, v)$ with $\mathsf{D}(\mathfrak{t}') = \mathsf{D}(T')$. Then T' is associated with the form $\tilde{\mathfrak{t}}'$ (see Theorem 2.7). But since $T' \supset S$, we have $\mathfrak{t}' \supset \mathfrak{s}$ and so $\tilde{\mathfrak{t}}' \supset \tilde{\mathfrak{s}} = \mathfrak{t}$. This implies $\mathsf{D}(\tilde{\mathfrak{t}}') \supset \mathsf{D}(\mathfrak{t})$.

Theorem 2.11. *The Friedrichs extension of S is the only m-sectorial extension of S with domain contained in $\mathsf{D}(\mathfrak{t})$*[2].

Proof. Let T' be any m-sectorial extension of S with $\mathsf{D}(T') \subset \mathsf{D}(\mathfrak{t})$. Let $\mathfrak{t}'$ be as above. For any $u \in \mathsf{D}(T')$ and $v \in \mathsf{D}(\mathfrak{t})$, we have $(T' u, v) = \tilde{\mathfrak{t}}'[u, v] = \mathfrak{t}[u, v]$ since $T' = T_{\tilde{\mathfrak{t}}'}$, $\tilde{\mathfrak{t}}' \supset \mathfrak{t}$, and $u \in \mathsf{D}(T') \subset \mathsf{D}(\mathfrak{t})$. It follows from Corollary 2.4 that $T' \subset T$. But since T' and T are both m-sectorial, we must have $T' = T$.

Remark 2.12. The importance of the Friedrichs extension lies in the fact that it assigns a special m-sectorial extension to each densely defined, sectorial operator S, even when the closure $\tilde{S}$ of S is not m-sectorial.

4. Other examples for the representation theorem

Example 2.13. Consider the form $\mathfrak{h}[u, v] = (Su, Sv)$ studied in Examples 1.3, 1.13 and 1.23. Assume that S is densely defined and closed so that $\mathfrak{h}$ has the same properties. Let $T = T_{\mathfrak{h}}$. Since $(Su, Sv) = (Tu, v)$ for all $u \in \mathsf{D}(T)$ and $v \in \mathsf{D}(\mathfrak{h}) = \mathsf{D}(S)$, it follows that $T \subset S^* S$. Since $S^* S$ is obviously symmetric and T is selfadjoint, we must have $T = S^* S$. Also $\mathsf{D}(T)$ is a core of $\mathfrak{t}$ and hence of S by Theorem 2.1 and Example 1.23. This gives another proof of the facts that *$S^* S$ is a selfadjoint operator in H whenever S is a densely defined, closed operator from H to H' and that $\mathsf{D}(S^* S)$ is a core of S* (see Theorem V-3.24).

---

[1] See FRIEDRICHS [1], FREUDENTHAL [1]. For the generalization of the Friedrichs extension to operators from a Banach space X to its adjoint space X^* see BIRMAN [2].

[2] Cf. FREUDENTHAL [1]. We do not consider the problem of determining all m-sectorial extensions of S. For semibounded selfadjoint extensions of a semibounded symmetric S, this problem is solved by KREIN [1], [2]. See also BIRMAN [1], VIŠIK [1].

Example 2.14. Consider the form $\mathfrak{t}[u, v] = \Sigma \alpha_j \xi_j \bar{\eta}_j$ of Examples 1.4, 1.14 and 1.24. Let $T = T_{\mathfrak{t}}$, $u = (\xi_j) \in \mathsf{D}(T)$, $Tu = w = (\zeta_j)$ and $v = (\eta_j) \in \mathsf{D}(\mathfrak{t})$. We have $\Sigma \zeta_j \bar{\eta}_j = (w, v) = (Tu, v) = \mathfrak{t}[u, v] = \Sigma \alpha_j \xi_j \bar{\eta}_j$. In particular set $\eta_j = \delta_{jk}$; then we obtain $\zeta_k = \alpha_k \xi_k$. Since $w \in l^2$, it is necessary that $\Sigma |\alpha_j|^2 |\xi_j|^2 < \infty$. This condition is also sufficient for a $u \in l^2$ to belong to $\mathsf{D}(T)$, for then $\mathfrak{t}[u, v] = \Sigma \alpha_j \xi_j \bar{\eta}_j = \Sigma \zeta_j \bar{\eta}_j = (w, v)$ so that Tu exists and is equal to $w = (\alpha_j \xi_j)$ by Theorem 2.1, iii).

Next consider the restriction $\dot{\mathfrak{t}}_1$ of $\mathfrak{t}$ with the additional condition $\Sigma \bar{\beta}_j \xi_j = 0$ for $u \in \mathsf{D}(\dot{\mathfrak{t}}_1)$, where we assume

$$\Sigma |\beta_j|^2 = \infty, \quad \Sigma (|\alpha_j| + 1)^{-1} |\beta_j|^2 < \infty. \tag{2.8}$$

Then $\dot{\mathfrak{t}}_1$ is densely defined but $\tilde{\dot{\mathfrak{t}}}_1$ is a proper restriction of $\mathfrak{t}$ (see loc. cit.). Let $T_1 = T_{\tilde{\dot{\mathfrak{t}}}_1}$, $u = (\xi_j) \in \mathsf{D}(T_1)$, $T_1 u = w = (\zeta_j)$ and $v \in \mathsf{D}(\tilde{\dot{\mathfrak{t}}}_1)$. Then we have as above $\Sigma (\zeta_j - \alpha_j \xi_j) \bar{\eta}_j = 0$. In particular set $\eta_1 = \bar{\beta}_k$, $\eta_k = -\bar{\beta}_1$, other η_j being zero [then $v \in \mathsf{D}(\dot{\mathfrak{t}}_1)$]. This gives $\zeta_1 - \alpha_1 \xi_1 : \zeta_k - \alpha_k \xi_k = \beta_1 : \beta_k$ and, since this is true for all k, we have $\zeta_j - \alpha_j \xi_j = \varrho \beta_j$ with a number ϱ independent of j. $w \in l^2$ requires that ϱ should be such that $\Sigma |\alpha_j \xi_j + \varrho \beta_j|^2 < \infty$. In view of (2.8), there is at most one ϱ with this property. Each $u \in \mathsf{D}(T_1)$ therefore satisfies the conditions

$$\Sigma (|\alpha_j| + 1) |\xi_j|^2 < \infty, \quad \Sigma \bar{\beta}_j \xi_j = 0 \quad \text{and} \quad \Sigma |\alpha_j \xi_j + \varrho \beta_j|^2 < \infty \quad \text{for some } \varrho. \tag{2.9}$$

These conditions are also sufficient for $u \in l^2$ to belong to $\mathsf{D}(T_1)$, for then we have $u \in \mathsf{D}(\tilde{\dot{\mathfrak{t}}}_1)$ and $\tilde{\dot{\mathfrak{t}}}_1[u, v] = \Sigma \alpha_j \xi_j \bar{\eta}_j = \Sigma (\alpha_j \xi_j + \varrho \beta_j) \bar{\eta}_j = (w, v)$ with $w = (\alpha_j \xi_j + \varrho \beta_j) \in l^2$, for every $v = (\eta_j) \in \mathsf{D}(\dot{\mathfrak{t}}_1)$. Thus $T_1 u$ exists and equals w by Theorem 2.1, iii).

Example 2.15. The form $\mathfrak{t}[u, v] = \int f u \bar{v}\, dx$ of Examples 1.5, 1.15 and 1.25 can be dealt with as the preceding example. The result is that $T = T_{\mathfrak{t}}$ is the maximal multiplication operator by $f(x)$ (see Example III-2.2). If $g(x)$ is such that $\int |g(x)|^2 dx = \infty$ but the integral on the left of (1.36) is convergent, then $T_1 = T_{\tilde{\dot{\mathfrak{t}}}_1}$ is given by $T_1 u(x) = f(x) u(x) + \varrho g(x)$ where ϱ is determined by the condition $T_1 u \in \mathsf{L}^2$.

Example 2.16. Consider the form $\mathfrak{t}$ of Example 1.7 and 1.36 under the assumptions stated there. Let $T = T_{\mathfrak{t}}$ and $u \in \mathsf{D}(T)$, $Tu = w$. The relation $(w, v) = (Tu, v) = \mathfrak{t}[u, v]$, $v \in \mathsf{D}(\mathfrak{t})$, means

$$\int_a^b w \bar{v}\, dx = \int_a^b \{p u' \bar{v}' + q u \bar{v} + r u' \bar{v} + s u \bar{v}'\} dx + h_a u(a) \overline{v(a)} + h_b u(b) \overline{v(b)}. \tag{2.10}$$

Let z be an indefinite integral of $w - qu - ru'$ (which is integrable):

$$z' = w - qu - ru'. \tag{2.11}$$

Then $\int_a^b (w - qu - ru') \bar{v}\, dx = \int_a^b z' \bar{v}\, dx = z(b) \overline{v(b)} - z(a) \overline{v(a)} - \int_a^b z \bar{v}'\, dx$, and (2.10) gives

$$\int_a^b (p u' + z + s u) \bar{v}'\, dx + [h_a u(a) + z(a)] \overline{v(a)} + [h_b u(b) - z(b)] \overline{v(b)} = 0. \tag{2.12}$$

(2.12) is true for every $v \in \mathsf{D}(\mathfrak{t})$, that is, every v such that $v(x)$ is absolutely continuous and $v' \in \mathsf{L}^2(a, b)$. For any $v' \in \mathsf{L}^2$ such that $\int_a^b v'\, dx = 0$, $v(x) = \int_a^x v'(x)\, dx$ satisfies the conditions $v \in \mathsf{D}(\mathfrak{t})$ and $v(a) = v(b) = 0$, so that $p u' + z + s u$ is orthogonal to v' by (2.12). Thus $p u' + z + s u$ must be equal to a constant c, being orthogonal to all functions orthogonal to 1. Substituting this into (2.12), we obtain

$$[-c + h_a u(a) + z(a)]\, \overline{v(a)} + [c + h_b u(b) - z(b)]\, \overline{v(b)} = 0 . \tag{2.13}$$

Since $v(a)$ and $v(b)$ vary over all complex numbers when v varies over $\mathsf{D}(\mathfrak{t})$, their coefficients in (2.13) must vanish. Noting that $c = p(a)\, u'(a) + z(a) + s(a)\, u(a) = p(b)\, u'(b) + z(b) + s(b)\, u(b)$, we thus obtain

$$p(a)\, u'(a) + (s(a) - h_a)\, u(a) = 0, \quad p(b)\, u'(b) + (s(b) + h_b)\, u(b) = 0 . \tag{2.14}$$

From $p u' + z + s u = c$ it follows that $p u'$ is absolutely continuous and $(p u')' = -z' - (s u)' = -w + q u + r u' - (s u)'$ or $w = -(p u')' + q u + r u' - (su)'$. In this way we have proved that each $u \in \mathsf{D}(T)$ has the following properties:

i) $u(x)$ and $u'(x)$ are absolutely continuous and $u'' \in \mathsf{L}^2(a, b)$;

ii) $u(x)$ satisfies the boundary condition (2.14).

Conversely, any $u \in \mathsf{L}^2$ satisfying i) and ii) belongs to $\mathsf{D}(T)$ and

$$T u = w = -(p u')' + q u + r u' - (s u)' . \tag{2.15}$$

In fact, an integration by parts shows that $\mathfrak{t}[u, v] = (w, v)$ for every $v \in \mathsf{D}(\mathfrak{t})$ and the assertion follows immediately from Theorem 2.1, iii). Thus we have the characterization: *$T = T_{\mathfrak{t}}$ is the second-order differential operator* (2.15) *with the boundary condition* (2.14). T is an operator analogous to the T_3 of III-§ 2.3.

This gives a proof that such a differential operator is m-sectorial, in particular selfadjoint if q is real-valued, $\overline{r(x)} = s(x)$ and h_a, h_b are real.

Example 2.17. Consider the restriction $\mathfrak{t}_0$ of the form $\mathfrak{t}$ of the preceding example defined by restricting the domain to a subset of $\mathsf{D}(\mathfrak{t})$ consisting of all u such that $u(a) = u(b) = 0$. $\mathfrak{t}_0$ can be shown to be closed in the same way as $\mathfrak{t}$. Let us determine $T_0 = T_{\mathfrak{t}_0}$. If $u \in \mathsf{D}(T_0)$ and $T_0 u = w$, we have again (2.10) where $v(a) = v(b) = 0$. The same argument as above thus leads to the result that w is given by (2.15). Therefore, each $u \in \mathsf{D}(T_0)$ has the property i) of the preceding example and

ii') $u(x)$ satisfies the boundary condition $u(a) = u(b) = 0$.

These conditions are also sufficient for a $u \in \mathsf{L}^2$ to belong to $\mathsf{D}(T_0)$; in this case $T_0 u$ is given by the right member of (2.15). The proof again follows from the fact that $\mathfrak{t}_0[u, v] = (w, v)$ for every $v \in \mathsf{D}(\mathfrak{t}_0)$, as is seen by an integration by parts. *Thus T_0 is the differential operator* (2.15) *with the boundary condition* $u(a) = u(b) = 0$.

Note that this boundary condition was already required in the definition of $\mathfrak{t}_0$. On the other hand, the boundary condition (2.14) for T in the preceding example is not imposed on $\mathfrak{t}$ but only on T. In this sense (2.14) is called a ***natural boundary condition***[1].

Problem 2.18. Let $\dot{T}$ be the minimal operator in $\mathsf{L}^2(a, b)$ constructed from the *formal* differential operator (2.15). All the operators T of Example 2.16 with different constants h_a, h_b and the T_0 of Example 2.17 are m-sectorial extensions of $\dot{T}$. Which of them is the Friedrichs extension of $\dot{T}$ (cf. Example 1.30) ?

5. Supplementary remarks

Selfadjointness is an important property, but a rather delicate one and hence not easy to establish. The correspondence $\mathfrak{h} \to H = T_{\mathfrak{h}}$ furnishes a convenient means of producing selfadjoint operators, inas-

[1] By the analogy of a similar notion in the calculus of variations.

much as it is relatively easy to construct closed forms. The examples given in par. 4 show indeed that various kinds of selfadjoint operators can be constructed in this way. The only defect of this method is that not all but only semibounded selfadjoint operators can be obtained. If we consider nonsymmetric forms, however, all m-sectorial operators can be constructed by means of sesquilinear forms.

The convenience of this way of constructing selfadjoint or m-sectorial operators is further illustrated by the following considerations. If $\mathfrak{t}_1$ and $\mathfrak{t}_2$ are closed sectorial forms, the same is true of their sum $\mathfrak{t} = \mathfrak{t}_1 + \mathfrak{t}_2$ by Theorem 1.31. If $\mathfrak{t}$ is densely defined, the associated m-sectorial operators T, T_1, T_2 are defined. T may be regarded as the *sum* of T_1, T_2 in a generalized sense, and we may express this by writing

$$T = T_1 \dotplus T_2 . \tag{2.16}$$

For any two selfadjoint operators T_1, T_2 bounded from below, the associated forms $\mathfrak{t}_1, \mathfrak{t}_2$ exist and the generalized sum of T_1, T_2 can be defined as equal to $T_\mathfrak{t}$ if $\mathfrak{t} = \mathfrak{t}_1 + \mathfrak{t}_2$ is densely defined. The stated condition is weaker than the requirement that the ordinary sum $S = T_1 + T_2$ be densely defined, and S need not be selfadjoint or essentially selfadjoint even when densely defined. In any case T is an extension of $T_1 + T_2$ and so its unique selfadjoint extension if $T_1 + T_2$ is essentially selfadjoint, as is easily seen by applying Theorem 2.1, iii).

This result can be extended to the case in which T_1 and T_2 are m-sectorial; then the associated closed forms $\mathfrak{t}_1, \mathfrak{t}_2$ exist and the generalized sum (2.16) is defined if $\mathsf{D}(\mathfrak{t}_1) \cap \mathsf{D}(\mathfrak{t}_2)$ is dense (see Remark 2.8).

If T_1, T_2 are selfadjoint and bounded from below and if $T_1 + T_2$ is densely defined, the Friedrichs extension T_F of $T_1 + T_2$ is defined (see par. 3). In general, however, T_F differs from (2.16). This is illustrated by the following example.

Example 2.19. Let $\mathfrak{t}_1, \mathfrak{t}_2$ be the forms (1.24) for different choices of the pair of constants h_a, h_b which we assume to be real. For simplicity we further assume that $p(x) = 1$ and $q = r = s = 0$. Then $\mathfrak{t} = \frac{1}{2}(\mathfrak{t}_1 + \mathfrak{t}_2)$ is again of the same form. Thus $\mathfrak{t}_1, \mathfrak{t}_2$ and $\mathfrak{t}$ are all symmetric, and the associated operators T, T_1, T_2 are selfadjoint operators formally given by $-d^2/dx^2$, with boundary conditions of the form (2.14) with $p = 1$, $s = 0$ and with different pairs of constants h_a, h_b. Thus $S = \frac{1}{2}(T_1 + T_2)$ has domain $\mathsf{D}(S) = \mathsf{D}(T_1) \cap \mathsf{D}(T_2)$ consisting of all u with $u'' \in \mathsf{L}^2$ that satisfy the boundary conditions $u(a) = u(b) = u'(a) = u'(b) = 0$. Now the closure of the form (Su, v) defined with domain $\mathsf{D}(S)$ is the $\mathfrak{t}_0$ of Example 2.17 (see also Example 1.30) so that T_F is the differential operator $-d^2/dx^2$ with the boundary condition $u(a) = u(b) = 0$ and differs from T.

Another convenience in considering symmetric forms rather than symmetric operators is the ease of extending such a form: except for bounded forms with domain H, any closed symmetric form bounded from below admits a proper closed symmetric extension bounded from below; there is no such thing as a *maximal* symmetric form, whereas maximal symmetric operators do exist (selfadjoint operators are maximal symmetric). Similar remarks can be made for more general, sectorial forms on the one hand and m-sectorial operators on the other.

A question arises: what is the relationship between the m-sectorial operators H_1, H_2 associated with two forms $\mathfrak{h}_1$, $\mathfrak{h}_2$ such that $\mathfrak{h}_1 \subset \mathfrak{h}_2$? The answer is not straightforward; there is no simple relationship between the domains of H_1 and H_2. Later we shall give a partial answer to this question.

Another question concerns the relationship between the selfadjoint operators H_1, H_2 associated with two symmetric forms $\mathfrak{h}_1$, $\mathfrak{h}_2$ such that $\mathfrak{h}_1 \geqq \mathfrak{h}_2$. We find it convenient to define the order relation $\mathfrak{h}_1 \geqq \mathfrak{h}_2$ for any two symmetric forms $\mathfrak{h}_1$, $\mathfrak{h}_2$ bounded from below by

$$\text{(2.17)} \qquad \mathsf{D}(\mathfrak{h}_1) \subset \mathsf{D}(\mathfrak{h}_2) \quad \text{and} \quad \mathfrak{h}_1[u] \geqq \mathfrak{h}_2[u] \quad \text{for} \quad u \in \mathsf{D}(\mathfrak{h}_1) .$$

Note that according to this definition, if $\mathfrak{h}_1$, $\mathfrak{h}_2$ are symmetric and bounded from below, then $\mathfrak{h}_1 \subset \mathfrak{h}_2$ implies $\mathfrak{h}_1 \geqq \mathfrak{h}_2$.

Problem 2.20. $\mathfrak{h}_1 \geqq \mathfrak{h}_2$ implies $\tilde{\mathfrak{h}}_1 \geqq \tilde{\mathfrak{h}}_2$.

Let H_1, H_2 be the selfadjoint operators bounded from below associated respectively with closed symmetric forms $\mathfrak{h}_1$, $\mathfrak{h}_2$ bounded from below. We write $H_1 \geqq H_2$ if $\mathfrak{h}_1 \geqq \mathfrak{h}_2$ in the sense defined above. This notation is in accordance with the usual one if H_1, H_2 are symmetric operators belonging to $\mathscr{B}(\mathsf{H})$.

Theorem 2.21. *Let H_1, H_2 be selfadjoint operators bounded from below with lower bounds γ_1, γ_2, respectively. In order that $H_1 \geqq H_2$, it is necessary that $\gamma_1 \geqq \gamma_2$ and $R(\zeta, H_1) \leqq R(\zeta, H_2)$ for every real $\zeta < \gamma_2$, and it is sufficient that $R(\zeta, H_1) \leqq R(\zeta, H_2)$ for some $\zeta < \min(\gamma_1, \gamma_2)$ (R denotes the resolvent).*

Proof. Necessity. Let $\mathfrak{h}_1$, $\mathfrak{h}_2$ be the associated symmetric forms bounded from below. $H_1 \geqq H_2$ is equivalent to $\mathfrak{h}_1 \geqq \mathfrak{h}_2$ by definition, which implies $\gamma_{\mathfrak{h}_1} \geqq \gamma_{\mathfrak{h}_2}$. Since $\gamma_1 = \gamma_{\mathfrak{h}_1}$, $\gamma_2 = \gamma_{\mathfrak{h}_2}$ by Theorem 2.6, it follows that $\gamma_1 \geqq \gamma_2$. Thus the resolvents $R(\zeta, H_1)$ and $R(\zeta, H_2)$ exist for $\zeta < \gamma_2$. Replacing $\mathfrak{h}_1 - \zeta$, $\mathfrak{h}_2 - \zeta$, $H_1 - \zeta$, $H_2 - \zeta$ by $\mathfrak{h}_1$, $\mathfrak{h}_2$, H_1, H_2 respectively, it suffices to show that $H_1 \geqq H_2 \geqq \delta > 0$ implies $H_1^{-1} \leqq H_2^{-1}$ [where $H_1^{-1}, H_2^{-1} \in \mathscr{B}(\mathsf{H})$].

For any $u \in \mathsf{H}$, set $v_1 = H_1^{-1} u$, $v_2 = H_2^{-1} u$. Then we have

$$\begin{aligned}(H_1^{-1} u, u)^2 = (v_1, H_2 v_2)^2 = \mathfrak{h}_2[v_1, v_2]^2 \leqq \mathfrak{h}_2[v_1]\, \mathfrak{h}_2[v_2] \leqq \\ \leqq \mathfrak{h}_1[v_1]\, \mathfrak{h}_2[v_2] = (H_1 v_1, v_1)\, (H_2 v_2, v_2) = (u, H_1^{-1} u)\, (u, H_2^{-1} u) ,\end{aligned}$$

which gives the desired result $(H_1^{-1} u, u) \leqq (H_2^{-1} u, u)$. Note that $v_1 \in \mathsf{D}(H_1) \subset \mathsf{D}(\mathfrak{h}_1) \subset \mathsf{D}(\mathfrak{h}_2)$, and that H_1, H_2, H_1^{-1}, H_2^{-1} are all symmetric and nonnegative.

Sufficiency. Again replacing $H_1 - \zeta$, etc. by H_1, etc., it suffices to show that, if both H_1 and H_2 have positive lower bounds and $H_1^{-1} \leqq H_2^{-1}$, then $H_1 \geqq H_2$, that is, $\mathfrak{h}_1 \geqq \mathfrak{h}_2$. To this end we first apply the foregoing result of the necessity part to the pair $S_1 = H_1^{-1}$, $S_2 = H_2^{-1}$ of nonnegative *bounded* operators, obtaining $(S_1 + \alpha)^{-1} \geqq (S_2 + \alpha)^{-1}$, $\alpha > 0$, from the assumption $S_1 \leqq S_2$. Denoting by $\mathfrak{h}_{1n}$, $\mathfrak{h}_{2n}$ the forms associated with the bounded symmetric operators $(S_1 + n^{-1})^{-1} = H_1(1 + n^{-1} H_1)^{-1}$ and $H_2(1 + n^{-1} H_2)^{-1}$, we thus have $\mathfrak{h}_{1n} \geqq \mathfrak{h}_{2n}$.

On the other hand we have $\mathfrak{h}_{1n} \leqq \mathfrak{h}_1$, $\mathfrak{h}_{2n} \leqq \mathfrak{h}_2$. In fact, let $u \in \mathsf{D}(H_1)$; then $\mathfrak{h}_{1n}[u] = (H_1(1 + n^{-1} H_1)^{-1} u, u) \leqq (H_1 u, u) = \mathfrak{h}_1[u]$ by Problem V-3.32. The result is generalized to all $u \in \mathsf{D}(\mathfrak{h}_1)$ because $\mathsf{D}(H_1)$ is a core of $\mathfrak{h}_1$.

Let $u \in \mathsf{D}(\mathfrak{h}_1)$. Then we have by what is proved above $\mathfrak{h}_{2n}[u] \leqq \mathfrak{h}_{1n}[u] \leqq \mathfrak{h}_1[u]$ so that $\mathfrak{h}_{2n}[u]$ is bounded. Set $u_n = (1 + n^{-1} H_2)^{-1} u \in \mathsf{D}(H_2) \subset \subset \mathsf{D}(\mathfrak{h}_2)$. Then $\mathfrak{h}_{2n}[u] = (H_2(1 + n^{-1} H_2)^{-1} u, u) = (H_2 u_n, (1 + n^{-1} H_2) u_n) = (H_2 u_n, u_n) + n^{-1}\|H_2 u_n\|^2 \geqq (H_2 u_n, u_n)$ so that $(H_2 u_n, u_n) = \mathfrak{h}_2[u_n]$ is bounded from above by $\mathfrak{h}_1[u]$.

Since $u_n \to u$ by Problem V-3.33, we conclude by Theorem 1.16 that $u \in \mathsf{D}(\mathfrak{h}_2)$ and that $\mathfrak{h}_2[u] \leqq \mathfrak{h}_1[u]$. This proves the required result that $\mathfrak{h}_1 \geqq \mathfrak{h}_2$[1].

Problem 2.22. Let K be a symmetric operator bounded from below and let H be its Friedrichs extension. Then $H \geqq H'$ for any selfadjoint extension H' of K which is bounded from below.

6. The second representation theorem

Let $\mathfrak{h}$ be a densely defined, closed symmetric form bounded from below, and let $H = T_\mathfrak{h}$ be the associated selfadjoint operator. The relationship $\mathfrak{h}[u, v] = (Hu, v)$ connecting $\mathfrak{h}$ with H is unsatisfactory in that it is not valid for all $u, v \in \mathsf{D}(\mathfrak{h})$ because $\mathsf{D}(H)$ is in general a proper subset of $\mathsf{D}(\mathfrak{h})$. A more complete representation of $\mathfrak{h}$ is furnished by the following theorem.

Theorem 2.23. *(Second representation theorem). Let $\mathfrak{h}$ be a densely defined, closed symmetric form, $\mathfrak{h} \geqq 0$, and let $H = T_\mathfrak{h}$ be the associated selfadjoint operator. Then we have $\mathsf{D}(H^{1/2}) = \mathsf{D}(\mathfrak{h})$ and*

$$\mathfrak{h}[u, v] = (H^{1/2} u, H^{1/2} v), \quad u, v \in \mathsf{D}(\mathfrak{h}). \tag{2.18}$$

A subset D' of $\mathsf{D}(\mathfrak{h})$ is a core of $\mathfrak{h}$ if and only if it is a core of $H^{1/2}$.

Remark 2.24. We recall that $H^{1/2}$ was defined in V-§ 3.11, for a nonnegative selfadjoint operator H is m-accretive. What is essential in

[1] The proof of Theorem 2.21 given above is not simple. A somewhat simpler proof is given in par. 6 by using the second representation theorem.

Theorem 2.23 is that $H^{1/2}$ is selfadjoint, nonnegative, $(H^{1/2})^2 = H$ and that $\mathsf{D}(H)$ is a core of $H^{1/2}$ (see Theorem V-3.35).

Proof of Theorem 2.23. Let us define the symmetric form $\mathfrak{h}'[u, v] = (H^{1/2} u, H^{1/2} v)$ with $\mathsf{D}(\mathfrak{h}') = \mathsf{D}(H^{1/2})$. Since $H^{1/2}$ is densely defined and closed (because it is selfadjoint), the same is true of $\mathfrak{h}'$ (see Example 1.13). Since $\mathsf{D}(H)$ is a core of $H^{1/2}$, it is also a core of $\mathfrak{h}'$ (see Example 1.23). On the other hand, $\mathsf{D}(H)$ is a core of $\mathfrak{h}$ by Theorem 2.1. But the two forms $\mathfrak{h}$ and $\mathfrak{h}'$ coincide on $\mathsf{D}(H)$, for we have

$$\mathfrak{h}[u, v] = (H u, v) = (H^{1/2} u, H^{1/2} v), \quad u, v \in \mathsf{D}(H). \tag{2.19}$$

Thus $\mathfrak{h}$ and $\mathfrak{h}'$ must be identical, being the closure of one and the same form – the restriction of $\mathfrak{h}$ to $\mathsf{D}(H)$. This proves (2.18). The last statement of the theorem follows from $\mathfrak{h} = \mathfrak{h}'$ and Example 1.23.

Problem 2.25. Let $\mathfrak{h}$ be a densely defined, symmetric, closed form bounded from below with lower bound γ and let $H = T_{\mathfrak{h}}$. For any $\xi \leqq \gamma$, we have then $\mathsf{D}(\mathfrak{h}) = \mathsf{D}((H - \xi)^{1/2})$, $\mathfrak{h}[u, v] = ((H - \xi)^{1/2} u, (H - \xi)^{1/2} v) + \xi (u, v)$.

Theorem 2.26. *Let $\mathfrak{h}$, H be as in* Theorem 2.23 *and, in addition, let $\mathfrak{h}$ have a positive lower bound. Then a subset D' of $\mathsf{D}(\mathfrak{h})$ is a core of $\mathfrak{h}$ if and only if $H^{1/2}\mathsf{D}'$ is dense in H.*

Proof. This follows from Theorem 2.23 and Problem III-5.19 [note that $H^{1/2}$ has positive lower bound, as is seen from Theorem 2.23, so that its inverse belongs to $\mathscr{B}(\mathsf{H})$].

Corollary 2.27. *Let $\mathfrak{h}$ be a densely defined, symmetric form, $\mathfrak{h} \geqq 0$, and let H be the selfadjoint operator associated with its closure $\tilde{\mathfrak{h}}$. Then $\mathsf{D}(\mathfrak{h})$ is a core of $H^{1/2}$. If $\mathfrak{h}$ has a positive lower bound, $H^{1/2}\mathsf{D}(\mathfrak{h})$ is dense in H.*

Remark 2.28. In Corollary 2.27, $\mathsf{D}(\mathfrak{h})$ need not be a core of H even when it is a subset of $\mathsf{D}(H)$.

Remark 2.29. The second representation theorem was proved only for symmetric forms. A corresponding theorem for nonsymmetric forms is not known. A natural generalization of that theorem to a nonsymmetric form $\mathfrak{t}$ would be that $\mathfrak{t}[u, v] = (T^{1/2} u, T^{*1/2} v)$ with $\mathsf{D}(\mathfrak{t}) = \mathsf{D}(T^{1/2}) = \mathsf{D}(T^{*1/2})$ where $T = T_{\mathfrak{t}}$; but the question is open whether this is true for a general closed sectorial form $\mathfrak{t}$ (with a vertex $\geqq 0$), although the operators $T^{1/2}$ and $T^{*1/2}$ are well-defined (V-§ 3.11)[1].

As an application of the second representation theorem, we shall give another formulation of the order relation $H_1 \geqq H_2$ for two selfadjoint operators bounded from below. In the preceding paragraph this was defined as equivalent to $\mathfrak{h}_1 \geqq \mathfrak{h}_2$ for the associated closed forms $\mathfrak{h}_1$, $\mathfrak{h}_2$, which means that $\mathsf{D}(\mathfrak{h}_1) \subset \mathsf{D}(\mathfrak{h}_2)$ and $\mathfrak{h}_1[u] \geqq \mathfrak{h}_2[u]$ for all $u \in \mathsf{D}(\mathfrak{h}_1)$. According to Theorem 2.23, this is in turn equivalent to

$$\mathsf{D}(H_1^{1/2}) \subset \mathsf{D}(H_2^{1/2}) \text{ and } \|H_1^{1/2} u\| \geqq \|H_2^{1/2} u\| \text{ for } u \in \mathsf{D}(H_1^{1/2}), \tag{2.20}$$

[1] For this question see Lions [1], T. Kato [16].

if we assume that both H_1 and H_2 are nonnegative (which does not restrict the generality, for $H_1 \geqq H_2$ is equivalent to $H_1 + \alpha \geqq H_2 + \alpha$). If we further assume that H_1 has a positive lower bound so that $H_1^{-1/2} = (H_1^{1/2})^{-1} \in \mathscr{B}(\mathsf{H})$, then (2.20) is equivalent to

$$H_2^{1/2} H_1^{-1/2} \in \mathscr{B}(\mathsf{H}) , \quad \|H_2^{1/2} H_1^{-1/2}\| \leqq 1 . \tag{2.21}$$

We could have chosen (2.21) as the definition of $H_1 \geqq H_2$ in the case under consideration.

We can now give a simpler proof of Theorem 2.21. As noted in the proof of this theorem given before, the essential point is that $H_1 \geqq H_2$ is equivalent to $H_1^{-1} \leqq H_2^{-1}$ when both H_1 and H_2 have positive lower bounds. In view of the equivalence of $H_1 \geqq H_2$ with (2.21), what is to be proved is contained in the following lemma.

Lemma 2.30. *Let S, T be densely defined closed operators in H such that their adjoints S^*, T^* are invertible. If $\mathsf{D}(S) \supset \mathsf{D}(T)$ and $\|Su\| \leqq \|Tu\|$ for all $u \in \mathsf{D}(T)$, then $\mathsf{D}(T^{*-1}) \supset \mathsf{D}(S^{*-1})$ and $\|T^{*-1} u\| \leqq \|S^{*-1} u\|$ for all $u \in \mathsf{D}(S^{*-1})$.*

Proof. Let $u \in \mathsf{D}(S^{*-1}) = \mathsf{R}(S^*)$; $u = S^* g$ for some $g \in \mathsf{D}(S^*)$. For any $v \in \mathsf{D}(T) \subset \mathsf{D}(S)$, we have then $(u, v) = (S^* g, v) = (g, Sv)$ and so $|(u, v)| \leqq \|g\| \|Sv\| \leqq \|g\| \|Tv\|$. Thus $Tv = 0$ implies $(u, v) = 0$ and $Tv_1 = Tv_2$ implies $(u, v_1) = (u, v_2)$. Therefore, (u, v) may be regarded as a function of $w = Tv$ and as such it is semilinear in w. Since $\mathsf{R}(T)^\perp = \mathsf{N}(T^*) = 0$, w varies over a dense linear manifold of H when v varies over $\mathsf{D}(T)$. Since this semilinear form is bounded with bound $\leqq \|g\|$ as is seen from the inequality given above, it can be extended by continuity to all $w \in \mathsf{H}$. Then it can be represented in the form (f, w) with a uniquely determined $f \in \mathsf{H}$ such that $\|f\| \leqq \|g\|$. Thus we obtain $(u, v) = (f, w) = (f, Tv)$ for all $v \in \mathsf{D}(T)$, which implies that $f \in \mathsf{D}(T^*)$ and $u = T^* f$. Thus $u \in \mathsf{R}(T^*) = \mathsf{D}(T^{*-1})$ and $\|T^{*-1} u\| = \|f\| \leqq \|g\| = \|S^{*-1} u\|$, as we wished to show.

Example 2.31. Consider the form $\mathfrak{t}[u, v]$ of Examples 1.4, 1.14, 1.24 and 2.14. Assume that all the α_j are real and nonnegative, so that $\mathfrak{t}$ is symmetric and nonnegative. Let $T = T_\mathfrak{t}$; T has been characterized in Example 2.14. It is now easy to see that $T^{1/2}$ is the maximal diagonal-matrix operator $(\alpha_j^{1/2})$. Similarly, for the form $\mathfrak{t}[u, v] = \int_E f\, u\, \bar{v}\, dx$ of Examples 1.5, 1.15, 1.25, and 2.15, $T^{1/2}$ is the maximal multiplication operator by $f(x)^{1/2}$, provided we assume that $f(x) \geqq 0$ so that $\mathfrak{t}$ is nonnegative symmetric.

Remark 2.32. Except in simple examples such as those given above, the operator $T^{1/2}$ is hard to describe in elementary terms, even when T can be easily described. In particular this is the case for differential operators T as in Example 2.16. In this sense, the expression (2.18) is of theoretical rather than of practical interest. As we shall see later, however, there are some results of practical significance that can be deduced more easily from (2.18) than by other methods.

7. The polar decomposition of a closed operator

Let T be a densely defined, closed operator from a Hilbert space H to another one H'. Consider the symmetric form $\mathfrak{h}[u, v] = (Tu, Tv)$. As we have seen in Example 2.13, $\mathfrak{h}$ is nonnegative and closed, with the associated selfadjoint operator $T_{\mathfrak{h}} = H = T^* T$. Let $G = H^{1/2}$. The second representation theorem gives

$$(2.22)\quad (Tu, Tv) = (Gu, Gv)\,, \quad \|Tu\| = \|Gu\|\,, \quad u, v \in \mathsf{D}(T) = \mathsf{D}(G)\,.$$

This implies that $Gu \to Tu$ defines an isometric mapping U of $\mathsf{R}(G) \subset \mathsf{H}$ onto $\mathsf{R}(T) \subset \mathsf{H}'$: $Tu = UGu$. By continuity U can be extended to an isometric operator on $\overline{\mathsf{R}(G)}$ [the closure of $\mathsf{R}(G)$] onto $\overline{\mathsf{R}(T)}$. Furthermore, U can be extended to an operator of $\mathscr{B}(\mathsf{H}, \mathsf{H}')$, which we shall denote again by U, by setting $Uu = 0$ for $u \in \mathsf{R}(G)^\perp = \mathsf{N}(G)$. Thus defined, U is partially isometric with the initial set $\overline{\mathsf{R}(G)}$ and the final set $\overline{\mathsf{R}(T)}$ (see V-§ 2.2), and we have

$$(2.23)\quad T = UG\,, \quad \mathsf{D}(T) = \mathsf{D}(G)\,.$$

(2.23) is called the *polar decomposition* of T; here G is nonnegative selfadjoint and U is partially isometric. If U is required as above to have initial set $\overline{\mathsf{R}(G)}$, the decomposition (2.23) is unique. In fact it is easily seen from (2.23) that

$$(2.24)\quad T^* = GU^*$$

[which implies that $\mathsf{D}(T^*)$ is the inverse image of $\mathsf{D}(G)$ under U^*], and so $T^* T = GU^* UG = G^2$ because $U^* Uu = u$ for u in the initial set of U. Thus G must be a nonnegative square root of $T^* T$, which determines G uniquely (see V-§ 3.11). Then (2.23) determines U on $\mathsf{R}(G)$ and hence completely since $U = 0$ on $\mathsf{R}(G)^\perp$ by hypothesis.

By analogy with the case of complex numbers, G is called the ***absolute value*** of T and is denoted by $|T|$. Thus $|T|$ is defined for any densely defined closed operator from H to H' and is a nonnegative selfadjoint operator in H; of course it should be distinguished from the scalar $\|T\|$.

Similarly, $|T^*|$ is a nonnegative selfadjoint operator in H'; it is related to $|T|$ by

$$(2.25)\quad |T^*| = U |T|\, U^*\,.$$

To see this set $G' = U|T|\, U^* = UGU^*$. Then $G'u = 0$ for $u \in \mathsf{R}(T)^\perp$ since $U^* u = 0$, and $\mathsf{R}(G') \subset \mathsf{R}(U) = \overline{\mathsf{R}(T)}$. Thus G' is zero on $\mathsf{R}(T)^\perp$, and on $\overline{\mathsf{R}(T)}$ it is unitarily equivalent to the part in $\overline{\mathsf{R}(G)}$ of G. Hence G' is selfadjoint and nonnegative. But $G'^2 = UGU^* UGU^* = UGGU^* = TT^*$. Hence G' must be equal to $|T^*|$ by the uniqueness of the square root.

From (2.23), (2.24), (2.25) we obtain the following relations:

$$(2.26)\quad T = U|T| = |T^*|U = UT^*U, \quad T^* = U^*|T^*| = |T|U^* = U^*TU^*,$$
$$|T| = U^*T = T^*U = U^*|T^*|U, \quad |T^*| = UT^* = TU^* = U|T|U^*.$$

In particular $T^* = U^*|T^*|$ is the polar decomposition of T^*.

Problem 2.33. $\mathsf{N}(T) = \mathsf{N}(|T|)$, $\mathsf{R}(T) = \mathsf{R}(|T^*|)$.

Example 2.34. The canonical expansion of a compact operator $T \in \mathscr{B}(\mathsf{H}, \mathsf{H}')$ discussed in V-§ 2.3 is a special case of the polar decomposition. Let $U\varphi_k = \varphi_k'$, $k = 1, 2, \ldots$, in V-(2.23). If $\{\varphi_k\}$ is not complete, set $Uu = 0$ for $u \perp \varphi_k$, $k = 1, 2, \ldots$. Then $|T| = \Sigma\, \alpha_k(\ ,\varphi_k)\,\varphi_k$ [see V-(2.26)] and V-(2.23) is exactly identical with $T = U|T|$.

Let us now consider the special case in which $T = H$ is selfadjoint: $H^* = H$. Let $H = U|H|$ be its polar decomposition. Since $H = H^* = U^*|H^*| = U^*|H|$ is the polar decomposition of the same operator, we must have $U^* = U$ by the uniqueness proved above. Furthermore, the initial set $\overline{\mathsf{R}(|H|)}$ and the final set $\overline{\mathsf{R}(H)}$ of U coincide by Problem 2.33; we denote by R this common subspace of H. We have $U^2u = U^*Uu = u$ for $u \in \mathsf{R}$ and $Uu = 0$ for $u \in \mathsf{R}^\perp$. Any $u \in \mathsf{R}$ can be written as $u = u_+ + u_-$ where $Uu_+ = u_+$ and $Uu_- = -u_-$; it suffices to set $u_\pm = (1 \pm U)\,u/2$. Moreover, it is easy to see that such a decomposition is unique. Let $\mathsf{M}_\pm$ be the subspaces of R consisting of all u such that $Uu = \pm u$. We have thus the decomposition

$$(2.27)\qquad \mathsf{H} = \mathsf{M}_+ \oplus \mathsf{M}_- \oplus \mathsf{M}_0\,, \quad \mathsf{M}_0 = \mathsf{R}^\perp\,,$$

the three subspaces being orthogonal to one another.

It follows also from (2.26) that $UH = U^*H = |H| = |H^*| = HU^* = HU$; hence H commutes with U. Similarly $|H|$ commutes with U. From the definition of R, $\mathsf{M}_\pm$, it then follows that H as well as $|H|$ is decomposed according to the decomposition (2.27). For $u \in \mathsf{M}_0$, we have $Hu = |H|\,u = 0$, for $\mathsf{M}_0 = \mathsf{R}(|H|)^\perp = \mathsf{N}(|H|) = \mathsf{N}(H)$ (see Problem 2.33). For $u \in \mathsf{M}_+$ we have $Hu = HUu = |H|\,u$, and similarly $Hu = -|H|\,u$ for $u \in \mathsf{M}_-$. Since $|H|$ is positive in R, it follows that the part of H in M_+ is positive and that in M_- is negative. Thus (2.27) gives a decomposition of H into positive, negative and zero parts and the associated decomposition of $|H|$.

In particular it follows that H and $|H|$ commute, because this is obviously true in each of the three subspaces. [By this we mean that the resolvents $R(\zeta, H)$ and $R(\zeta', |H|)$ commute, for the commutativity of two unbounded operators has not been defined.]

Problem 2.35. If H is selfadjoint, $\mathsf{D}(H) = \mathsf{D}(|H|)$ and for $u \in \mathsf{D}(H)$

$$(2.28)\qquad \begin{aligned} &\|Hu\| = \|\,|H|\,u\|\,, \quad |(Hu, u)| \le (|H|\,u, u)\,, \\ &\|(H+\alpha)\,u\| \le \|(|H| + |\alpha|)\,u\|\,, \quad (|H+\alpha|\,u, u) \le ((|H| + |\alpha|)\,u, u)\,. \end{aligned}$$

[hint for the last inequality: V-(4.15).]

Problem 2.36. The orthogonal projections onto the three subspaces M_+, M_- and M_0 are respectively given by

(2.29) $$P_+ = (U^2 + U)/2, \quad P_- = (U^2 - U)/2, \quad P_0 = 1 - U^2.$$

Lemma 2.37. *If $A \in \mathscr{B}(H)$ commutes with H, then A commutes with $|H|$ and U.*

Proof. A commutes with the resolvent of H, hence also with the resolvent of H^2 in virtue of $(H^2 - \zeta)^{-1} = (H - \zeta^{1/2})^{-1}(H + \zeta^{1/2})^{-1}$. Hence A commutes with $|H| = (H^2)^{1/2}$ by Theorem V-3.35. To show that A commutes with U, we note that $AU|H| = AH \subset HA$ and $UA|H| \subset U|H|A = HA$. Hence $AUu = UAu$ for $u \in R$. If $u \in R^\perp = M_0 = N(H)$, on the other hand, we have $AUu = 0 = UAu$ since $Hu = 0$ implies $HAu = AHu = 0$. Hence $AU = UA$.

Lemma 2.38. *The decomposition (2.27) of H into positive, negative and zero parts for H is unique in the following sense. Suppose that H is reduced by a subspace M' and $(Hu, u) \geqq 0$ $[(Hu, u) \leqq 0]$ for every $u \in M' \cap D(H)$. Then M' must be a subspace of $M_+ \oplus M_0 = M_-^\perp$ $(M_- \oplus M_0 = M_+^\perp)$.*

Proof. Let P' be the orthogonal projection on M'. Since M' reduces H, P' commutes with H. It follows from Lemma 2.37 and (2.29) that P' commutes with U and $P_\pm$, P_0. Thus $P'P_- = P_-P'$ is the projection on $M' \cap M_-$. But this subspace is 0, for $u \in D' \equiv D(H) \cap M' \cap M_-$ implies $u = 0$ [otherwise we have the contradiction that $(Hu, u) \geqq 0$ and $(Hu, u) < 0$] and D' is dense in $M' \cap M_-$ because the latter reduces H. This proves that $P'P_- = P_-P' = 0$ and hence that $M' \subset M_-^\perp$.

§ 3. Perturbation of sesquilinear forms and the associated operators

1. The real part of an m-sectorial operator

In this section we shall consider the perturbation of an m-sectorial operator T when the associated form $\mathfrak{t}$ undergoes a small perturbation.

Lemma 3.1. *Let $\mathfrak{h}$ be a densely defined, symmetric, nonnegative closed form with the associated nonnegative selfadjoint operator $H = T_\mathfrak{h}$. Let $\mathfrak{a}$ be a form relatively bounded with respect to $\mathfrak{h}$, such that*

(3.1) $$|\mathfrak{a}[u]| \leqq b\,\mathfrak{h}[u], \quad u \in D(\mathfrak{h}).$$

Then there is an operator $C \in \mathscr{B}(H)$ with $\|C\| \leqq \varepsilon b$ (where $\varepsilon = 1$ or 2 according as $\mathfrak{a}$ is symmetric or not) such that

(3.2) $$\mathfrak{a}[u, v] = (CGu, Gv), \quad G = H^{1/2}, \quad u, v \in D(\mathfrak{h}) = D(G).$$

Proof. (3.1) implies that

(3.3) $$|\mathfrak{a}[u, v]| \leqq \varepsilon\, b\,\mathfrak{h}[u]^{1/2}\,\mathfrak{h}[v]^{1/2} = \varepsilon\, b\|Gu\|\,\|Gv\|,$$

as is seen by applying (1.15) to Re $\mathfrak{a}$ and Im $\mathfrak{a}$ and noting that $\mathfrak{h}[u] = \|Gu\|^2$ by the second representation theorem. (3.3) implies that the value of $\mathfrak{a}[u, v]$ is determined by Gu, Gv, for $Gu = Gu'$, $Gv = Gv'$ implies that $\mathfrak{a}[u', v'] - \mathfrak{a}[u, v] = \mathfrak{a}[u' - u, v'] + \mathfrak{a}[u, v' - v] = 0$. Thus $\mathfrak{a}[u, v]$ may be regarded as a bounded sesquilinear form in $x = Gu$, $y = Gv$, and it can be extended to all x, y in the closure M of the range of G. It follows that there exists a bounded operator C on M to itself such that $\|C\| \leq \varepsilon b$ and $\mathfrak{a}[u, v] = (Cx, y)$. For convenience C may be extended, without increasing the bound, to an operator of $\mathscr{B}(\mathsf{H})$ by setting $Cx = 0$ for $x \in \mathsf{M}^\perp$.

As an application of this lemma, we shall deduce an expression connecting an m-sectorial operator T with its *real part* H. Let T be associated with a closed, densely defined sectorial form $\mathfrak{t}$ by Theorem 2.7. Let $H = T_\mathfrak{h}$ where the symmetric form $\mathfrak{h} = \operatorname{Re}\mathfrak{t}$ is also closed. H is by definition the real part of T, $H = \operatorname{Re} T$ in symbol. It is obvious that $H = \frac{1}{2}(T + T^*)$ if T is bounded, but this is in general not true. It follows directly from Theorem 2.5 that $\operatorname{Re} T^* = \operatorname{Re} T$.

Theorem 3.2. *Let T be an m-sectorial operator with a vertex 0 and semi-angle θ. Then $H = \operatorname{Re} T$ is nonnegative, and there is a symmetric operator $B \in \mathscr{B}(\mathsf{H})$ such that $\|B\| \leq \tan\theta$ and*

$$T = G(1 + iB)\,G\,, \quad G = H^{1/2}\,. \tag{3.4}$$

Proof. Let $T = T_\mathfrak{t}$ and $\mathfrak{h} = \operatorname{Re}\mathfrak{t}$, $\mathfrak{k} = \operatorname{Im}\mathfrak{t}$. Since $|\mathfrak{k}[u]| \leq (\tan\theta)\,\mathfrak{h}[u]$ by hypothesis, we have $\mathfrak{k}[u, v] = (BGu, Gv)$ by Lemma 3.1, where B is symmetric and $\|B\| \leq \tan\theta$. Hence

$$\mathfrak{t}[u, v] = (\mathfrak{h} + i\mathfrak{k})\,[u, v] = ((1 + iB)\,Gu, Gv)\,. \tag{3.5}$$

Now let $u \in \mathsf{D}(T)$. We have $\mathfrak{t}[u, v] = (Tu, v)$ for all $v \in \mathsf{D}(\mathfrak{t}) = \mathsf{D}(\mathfrak{h}) = \mathsf{D}(G)$ by the definition of T. Comparing this with (3.5) and noting that G is selfadjoint, we see that $G(1 + iB)\,Gu$ exists and is equal to Tu. This shows that $T \subset G(1 + iB)\,G$. But it is easy to see that $G(1 + iB)\,G$ is accretive. In view of the fact that T is m-accretive, we must have equality instead of inclusion. This proves (3.4).

Theorem 3.3. *Let T be m-sectorial with $H = \operatorname{Re} T$. T has compact resolvent if and only if H has.*

Proof. We may assume without loss of generality that T has a positive vertex, so that $\zeta = 0$ belongs to the resolvent sets of T and of H [that is, T^{-1} and H^{-1} belong to $\mathscr{B}(\mathsf{H})$]. Suppose that H has compact resolvent. Then the same is true of $G = H^{1/2}$ (see Theorem V-3.49). Since $T^{-1} = G^{-1}(1 + iB)^{-1}\,G^{-1}$ by (3.4), it follows that T^{-1} is compact, which implies that T has compact resolvent. The proof of the converse proposition is somewhat complicated. It is known that $GT^{-\alpha} \in \mathscr{B}(\mathsf{H})$ for $1/2 < \alpha < 1$[1] and that $T^{-(1-\alpha)}$ is compact if T has compact resolvent[2]. Hence $GT^{-1} = GT^{-\alpha}\, T^{-(1-\alpha)}$ is compact and so is $G^{-1} = (1 + iB)\, GT^{-1}$. Thus $H^{-1} = G^{-2}$ is compact, and H has compact resolvent.

2. Perturbation of an m-sectorial operator and its resolvent

Let $\mathfrak{t}$ be a densely defined, closed sectorial form and let $T = T_{\mathfrak{t}}$ be the associated m-sectorial operator. Let us ask how T is changed when $\mathfrak{t}$ is subjected to a "small" perturbation. We shall consider this problem in the case in which the perturbation of $\mathfrak{t}$ is *relatively bounded.*

We recall that $\mathfrak{t} + \mathfrak{a} = \mathfrak{s}$ is also sectorial and closed if $\mathfrak{a}$ is relatively bounded with respect to $\mathfrak{t}$ with relative bound smaller than 1 (Theorem 1.33). It is not easy to compare the associated operator $S = T_{\mathfrak{s}}$ directly with the unperturbed operator T, for S and T need not have the same domain. But we can compare the resolvents $R(\zeta, S)$ and $R(\zeta, T)$ and estimate their difference in terms of the relative bound of $\mathfrak{a}$ with respect to $\mathfrak{t}$.

Theorem 3.4. *Let $\mathfrak{t}$ be a densely defined, closed sectorial form with $\mathfrak{h} = \mathrm{Re}\,\mathfrak{t} \geqq 0$ and let $T = T_{\mathfrak{t}}$ be the associated m-sectorial operator. Let $\mathfrak{a}$ be a form relatively bounded with respect to $\mathfrak{t}$, with*

$$(3.6)\qquad |\mathfrak{a}[u]| \leqq a\|u\|^2 + b\,\mathfrak{h}[u]\,, \quad u \in \mathsf{D}(\mathfrak{h}) = \mathsf{D}(\mathfrak{t}) \subset \mathsf{D}(\mathfrak{a})\,,$$

where a, b are nonnegative constants and $b < 1$. Then $\mathfrak{s} = \mathfrak{t} + \mathfrak{a}$ is also sectorial and closed. Let $S = T_{\mathfrak{s}}$ be the associated m-sectorial operator. If $b < 1/2$, the resolvents $R(\zeta, T)$ and $R(\zeta, S)$ exist, with

$$(3.7)\qquad \|R(\zeta, S) - R(\zeta, T)\| \leqq$$

$$\leqq \begin{cases} \dfrac{2a}{(-\mathrm{Re}\,\zeta - 2a)\,(-\mathrm{Re}\,\zeta)} & \text{for } -\dfrac{a}{b} < \mathrm{Re}\,\zeta < -2a\,, \\[2ex] \dfrac{2b}{(1-2b)\,(-\mathrm{Re}\,\zeta)} & \text{for } \mathrm{Re}\,\zeta \leqq -\dfrac{a}{b}\,, \end{cases}$$

(if $b = 0$ read ∞ for a/b). If T has compact resolvent, the same is true of S.

Proof. The closedness of $\mathfrak{s}$ was proved in Theorem 1.33. Let $\varrho > 0$ (to be determined later) and $\mathfrak{t}' = \mathfrak{t} + \varrho$, $\mathfrak{h}' = \mathrm{Re}\,\mathfrak{t}' = \mathfrak{h} + \varrho$, and let $T' = T + \varrho$, $H' = H + \varrho \geqq \varrho > 0$ be the associated operators. By (3.5) we have $\mathfrak{t}'[u, v] = ((1 + iB')\, G'u, G'v)$ with $G' = H'^{1/2}$ and $B'^* = B' \in \mathscr{B}(\mathsf{H})$. On the other hand, (3.6) can be written

$$(3.8)\qquad |\mathfrak{a}[u]| \leqq k\,\mathfrak{h}'[u] \quad \text{with} \quad k = \max(b, a/\varrho)\,.$$

[1] See T. Kato [15], [16].

[2] See Theorem V-3.49 and Remark V-3.50.

Hence $\mathfrak{a}[u, v] = (C G' u, G' v)$, $\|C\| \leqq 2k$, by Lemma 3.1. For the form $\mathfrak{s}' = \mathfrak{s} + \varrho$, we have the expression $\mathfrak{s}'[u, v] = (\mathfrak{t}' + \mathfrak{a})[u, v] = ((1 + i B' + C) G' u, G' v)$. As in the proof of (3.4), this leads to

$$S + \varrho = S' = G'(1 + iB' + C) G', \quad \|C\| \leqq 2k. \tag{3.9}$$

We now have $S - \zeta = S' - \zeta' = G'(1 - \zeta' H'^{-1} + iB' + C) G'$ where $\zeta' = \zeta + \varrho$, so that

$$R(\zeta, S) = R(\zeta', S') = G'^{-1}(1 - \zeta' H'^{-1} + iB' + C)^{-1} G'^{-1}, \tag{3.10}$$

provided the middle factor on the right exists and belongs to $\mathscr{B}(\mathsf{H})$. But $(1 - \zeta' H'^{-1} + iB')^{-1} \in \mathscr{B}(\mathsf{H})$ exists and has norm $\leqq 1$ if ϱ is chosen so that $\operatorname{Re} \zeta' = \operatorname{Re} \zeta + \varrho \leqq 0$, for then $1 - \zeta' H'^{-1} + iB' \in \mathscr{B}(\mathsf{H})$ has numerical range in the half-plane $\operatorname{Re} z \geqq 1$. It follows [see I-(4.24)] that the factor in question exists and

$$\|(1 - \zeta' H'^{-1} + iB' + C)^{-1} - (1 - \zeta' H'^{-1} + iB)^{-1}\| \leqq \leqq 2k(1 - 2k)^{-1} \tag{3.11}$$

if $2k < 1$. Then we see from (3.10) and a similar expression for $R(\zeta, T)$ obtained by setting $C = 0$ in (3.10) that

$$\|R(\zeta, S) - R(\zeta, T)\| \leqq 2k(1 - 2k)^{-1} \varrho^{-1}, \tag{3.12}$$

where we have also used the fact that $\|G'^{-1}\| \leqq \varrho^{-1/2}$. If we set $\varrho = -\operatorname{Re} \zeta$, (3.12) gives the desired result (3.7), by the definition (3.8) of k.

If T has compact resolvent, the same is true of H, H' and G' by Theorems 3.3 and V-3.49. Hence G'^{-1} is compact and so is $R(\zeta, S)$ by (3.10).

Remark 3.5. Theorem 3.4 shows that $\|R(\zeta, S) - R(\zeta, T)\|$ is small if a, b are sufficiently small. (3.7) shows this explicitly only for ζ with $\operatorname{Re} \zeta < 0$. According to Remark IV-3.13, however, it is true for any $\zeta \in \mathrm{P}(T)$; an explicit formula estimating $R(\zeta, S) - R(\zeta, T)$ could be obtained by using IV-(3.10), though we shall not write it down. Such a formula will be given in the next paragraph in the special case of a *symmetric* T, since it is then particularly simple.

Theorem 3.6. *Let* $\mathfrak{t}$ *be a densely defined, closable sectorial form, and let* $\{\mathfrak{t}_n\}$ *be a sequence of forms with* $\mathsf{D}(\mathfrak{t}_n) = \mathsf{D}(\mathfrak{t})$ *such that*

$$|(\mathfrak{t} - \mathfrak{t}_n)[u]| \leqq a_n \|u\|^2 + b_n \mathfrak{h}[u], \quad u \in \mathsf{D}(\mathfrak{t}), \tag{3.13}$$

where $\mathfrak{h} = \operatorname{Re} \mathfrak{t}$ *and the constants* a_n, $b_n > 0$ *tend to zero as* $n \to \infty$. *Then the* $\mathfrak{t}_n$ *are also sectorial and closable for sufficiently large* n. *Let* $\tilde{\mathfrak{t}}$, $\tilde{\mathfrak{t}}_n$ *be respectively the closures of* $\mathfrak{t}$, $\mathfrak{t}_n$ *and let* T, T_n *be the associated m-sectorial operators. Then* $\{T_n\}$ *converges to* T *in the generalized sense (of* IV-§ 2.4*). If* T *has compact resolvent, the same is true of* T_n *for sufficiently large* n.

Proof. That $\mathfrak{t}_n$ is sectorial and closable follows from Theorem 1.33. We note also that (3.13) holds with $\mathfrak{t}, \mathfrak{t}_n$ replaced by $\tilde{\mathfrak{t}}, \tilde{\mathfrak{t}}_n$ respectively.

We may assume without loss of generality that $\mathfrak{h} \geqq 0$; otherwise we need only to add a common constant to $\mathfrak{t}$ and $\mathfrak{t}_n$. We may further assume that $a_n = b_n$; otherwise it suffices to replace a_n and b_n by the common constant $\max(a_n, b_n)$. Then it follows from (3.7) that $\|R(-1,T_n) - R(-1, T)\| \leqq 2a_n(1 - 2a_n)^{-1} \to 0$, which proves that T_n converges to T in the generalized sense. The last assertion of the theorem also follows from Theorem 3.4.

Remark 3.7. Theorem 3.6 gives a convenient criterion for a sequence of operators T_n to converge to T in the generalized sense. Compare a similar criterion, related to a relatively bounded perturbation of an operator, given by Theorem IV-2.24.

Remark 3.8. All the results pertaining to a sequence of operators converging in the generalized sense can be applied to the T_n of Theorem 3.6. For instance, we note that the spectrum of T does not expand suddenly and, in particular, any finite system of eigenvalues is stable when T is changed to T_n (see IV-§ 3.5).

3. Symmetric unperturbed operators

Theorem 3.9. *Let $\mathfrak{h}$ be a densely defined, closed symmetric form bounded from below and let $\mathfrak{a}$ be a (not necessarily symmetric) form relatively bounded with respect to $\mathfrak{h}$, so that $\mathsf{D}(\mathfrak{a}) \supset \mathsf{D}(\mathfrak{h})$ and*

$$|\mathfrak{a}[u]| \leqq a\|u\|^2 + b\,\mathfrak{h}[u]\,, \tag{3.14}$$

where $0 \leqq b < 1$ but a may be positive, negative or zero. Then $\mathfrak{s} = \mathfrak{h} + \mathfrak{a}$ is sectorial and closed. Let H, S be the operators associated with $\mathfrak{h}$, $\mathfrak{s}$, respectively. If $\zeta \in \mathrm{P}(H)$ and

$$\varepsilon\,\|(a + b\,H)\,R(\zeta, H)\| < 1\,, \tag{3.15}$$

then $\zeta \in \mathrm{P}(S)$ and

$$\|R(\zeta, S) - R(\zeta, H)\| \leqq \frac{4\varepsilon\,\|(a + b\,H)\,R(\zeta, H)\|}{(1 - \varepsilon\,\|(a + b\,H)\,R(\zeta, H)\|)^2}\,\|R(\zeta, H)\|\,. \tag{3.16}$$

Here $\varepsilon = 1$ or 2 according as $\mathfrak{a}$ is symmetric or not.

Proof. First we assume that $b > 0$ and set $\mathfrak{h}' = \mathfrak{h} + a\,b^{-1} + \delta$, $\mathfrak{s}' = \mathfrak{s} + a\,b^{-1} + \delta$ with a $\delta > 0$ to be determined later. $\mathfrak{s}'$ is closed as well as $\mathfrak{s}$ by Theorem 1.33. The operators associated with $\mathfrak{h}'$ and $\mathfrak{s}'$ are respectively $H' = H + a\,b^{-1} + \delta$ and $S' = S + a\,b^{-1} + \delta$.

We have $\mathfrak{h} + a\,b^{-1} \geqq 0$ by (3.14) so that $\mathfrak{h}' \geqq \delta$. (3.14) implies also $|\mathfrak{a}[u]| \leqq b\,\mathfrak{h}'[u]$. We can now apply the same argument that was used in the proof of Theorem 3.4; we note that $\mathfrak{a}[u, v]$ may be written in the form $(CG'u, G'v)$ with $\|C\| \leqq \varepsilon b$, where $\varepsilon = 1$ or 2 according as $\mathfrak{a}$ is symmetric or not (see Lemma 3.1), and that $G' \geqq \delta^{1/2}$ or $\|G'^{-1}\| \leqq \delta^{-1/2}$. In this way we obtain

$$\|R(\zeta, S) - R(\zeta, H)\| \leqq \varepsilon b M^2 (1 - \varepsilon b M)^{-1} \delta^{-1} \quad \text{if} \quad \varepsilon b M < 1, \tag{3.17}$$

where $M = \|(1 - \zeta' H'^{-1})^{-1}\|$ for $\zeta' = \zeta + a b^{-1} + \delta$ (note that here $B' = 0$ because $T - H$ is symmetric). Thus

$$\begin{aligned} M = \|(H'(H' - \zeta')^{-1}\| &= \|(H + a b^{-1} + \delta)(H - \zeta)^{-1}\| \leqq \\ &\leqq b^{-1}\|(a + bH) R(\zeta, H)\| + \delta \|R(\zeta, H)\| . \end{aligned} \tag{3.18}$$

The desired inequality (3.16) follows from (3.17) by substituting for M from (3.18) and setting $\delta = \alpha(1 - \alpha)(1 + \alpha)^{-1}\beta^{-1}$, where $\alpha = \varepsilon\|(a + bH) R(\zeta, H)\|$ and $\beta = \varepsilon b \|R(\zeta, H)\|$; note that $1 - \varepsilon b M = (1 - \alpha) \cdot (1 + \alpha)^{-1} > 0$ if $\alpha < 1$.

The case $b = 0$ can be dealt with by going to the limit $b \to 0$.

Remark 3.10. The condition (3.15) for $\zeta \in \mathrm{P}(S)$ is fairly satisfactory, but the estimate (3.16) is not very sharp, as is seen by considering the special case $b = 0$; in this case one has a sharper estimate with the right member $\varepsilon a \|R\|^2 (1 - \varepsilon a \|R\|)^{-1}$ by I-(4.24) ($R = R(\zeta, H)$), for $|\mathfrak{a}[u]| = |(\mathfrak{s} - \mathfrak{h})[u]| \leqq a\|u\|^2$ implies $|\mathfrak{a}[u, v]| \leqq \varepsilon a \|u\| \|v\|$ so that $\mathfrak{a}[u, v] = (Cu, v)$, $S = H + C$ with $\|C\| \leqq \varepsilon a$. But the merit of Theorem 3.9 lies rather in the fact that a may be negative. A more satisfactory result could be obtained by using the lower bound $\gamma \geqq 0$ of $a + bH$ explicitly in the estimate.

4. Pseudo-Friedrichs extensions

The Friedrichs extension was defined for a densely defined, sectorial operator; it is thus inherently connected with sectorial boundedness. We shall now introduce a new kind of extension, similar to the Friedrichs extension, which can be applied to a not necessarily sectorial operator and which produces a selfadjoint operator when applied to a symmetric operator[1].

Theorem 3.11. *Let H be a selfadjoint operator and let A be an operator such that $\mathsf{D} = \mathsf{D}(A) \subset \mathsf{D}(H)$ and*

$$|(Au, u)| \leqq a\|u\|^2 + b(|H| u, u), \quad u \in \mathsf{D}(A), \tag{3.19}$$

where $0 \leqq b < 1$ or $0 \leqq b < 1/2$ according as A is symmetric or not. If $\mathsf{D}(A)$ is a core of $|H|^{1/2}$, there is a unique closed extension T of $H + A$

[1] The results of this paragraph are not related to sesquilinear forms in any essential way. We consider them here simply because the techniques used in the proofs are similar to those of the preceding paragraphs.

such that $\mathsf{D}(T) \subset \mathsf{D}(|H|^{1/2})$, $\mathsf{D}(T^*) \subset \mathsf{D}(|H|^{1/2})$ *and* $i\eta \in \mathrm{P}(T)$ *for all real* η *with sufficiently large* $|\eta|$. T *is selfadjoint if* A *is symmetric. (*T *will be called the pseudo-Friedrichs extension of* $H + A$.)

Proof. Recall that $|H| = (H^2)^{1/2}$ is selfadjoint with $\mathsf{D}(|H|) = \mathsf{D}(H)$ and commutes with H in the sense stated in § 2.7[1].

We may assume $b > 0$ without loss of generality. Set $H' = |H| + a b^{-1} + \delta$ with a $\delta > 0$. (3.19) implies $|(Au, u)| \leq b(H'u, u) = b\|G'u\|^2$, where $G' = H'^{1/2}$. Since $\mathsf{D}(A)$ is a core of $|H|^{1/2}$, $\mathsf{D}(A)$ is also a core of G' and the last inequality implies that (Au, v) can be extended to a form $\mathfrak{a}[u, v]$ with $\mathsf{D}(\mathfrak{a}) = \mathsf{D}(G')$ such that $|\mathfrak{a}[u, v]| \leq \varepsilon b \|G'u\| \|G'v\|$, where $\varepsilon = 1$ or 2 according as $\mathfrak{a}$ is symmetric or not (see the preceding paragraph). This gives (see par. 1)

$$\mathfrak{a}[u, v] = (CG'u, G'v), \quad C \in \mathscr{B}(\mathsf{H}), \quad \|C\| \leq \varepsilon b. \tag{3.20}$$

We now claim that the operator T defined by

$$T = G'(HH'^{-1} + C)\, G' \tag{3.21}$$

has the properties stated in the theorem. It is clear that $\mathsf{D}(T) \subset \mathsf{D}(G') = \mathsf{D}(|H|^{1/2})$. Let $u \in \mathsf{D}(A)$. Then $G'HH'^{-1}G'u$ exists and equals Hu since $H' = G'^2$ and G'^{-1} and H commute. Since, furthermore, $(Au, v) = \mathfrak{a}[u, v] = (CG'u, G'v)$ for all $v \in \mathsf{D}(A)$ where $\mathsf{D}(A)$ is a core of G', $G'CG'u$ exists and equals Au. Thus Tu exists and equals $(H + A)\,u$, that is, $T \supset H + A$.

Let $\zeta \in \mathrm{P}(H)$. Then $T - \zeta = G'[(H - \zeta)\, H'^{-1} + C]\, G'$ since $1 \supset G'H'^{-1}G'$, so that

$$(T - \zeta)^{-1} = G'^{-1}[(H - \zeta)\, H'^{-1} + C]^{-1}\, G'^{-1} \tag{3.22}$$

provided $(H - \zeta)\, H'^{-1} + C$ has an inverse in $\mathscr{B}(\mathsf{H})$. This is true if $C[(H - \zeta)\, H'^{-1}]^{-1} = CH'(H - \zeta)^{-1}$ has norm less than 1 (the Neumann series), or if

$$\|C\|\, \|(|H| + a b^{-1} + \delta)\, (H - \zeta)^{-1}\| < 1\,. \tag{3.23}$$

Recalling that $\|C\| \leq \varepsilon b$, we see that this condition is satisfied if

$$\varepsilon \|(a + b|H| + b\,\delta)\, (H - \zeta)^{-1}\| < 1\,. \tag{3.24}$$

Since $|a + b\,\delta|\, \|(H - i\,\eta)^{-1}\|$ can be made arbitrarily small and $\||H|\,(H - i\,\eta)^{-1}\| = \|H(H - i\,\eta)^{-1}\| \leq 1$ by choosing η real and very large, $(T - i\,\eta)^{-1} \in \mathscr{B}(\mathsf{H})$ exists if $\varepsilon b < 1$ and $|\eta|$ is sufficiently large. Thus T is closed with non-empty $\mathrm{P}(T)$.

If A is symmetric, then C is symmetric and so is T (note that HH'^{-1} is symmetric). Hence T must be selfadjoint.

---

[1] This commutativity is essential in the proof; this is the reason why H must be assumed selfadjoint.

It follows from (3.22) that $(T^* - \bar{\zeta})^{-1} = G'^{-1}[(H - \bar{\zeta}) H'^{-1} + C^*]^{-1} G'^{-1}$; note that $[(H - \bar{\zeta}) H'^{-1} + C^*]^{-1}$ exists as above in virtue of $\|C^*\| = \|C\|$. It follows that $\mathsf{D}(T^*) = \mathsf{R}((T^* - \bar{\zeta})^{-1}) \subset \mathsf{R}(G'^{-1}) = \mathsf{D}(G')$ and

$$T^* = G'(HH'^{-1} + C^*) G' . \tag{3.25}$$

Finally we prove the uniqueness of T. Suppose T_1 is a closed extension of $H + A$ with the properties stated in the theorem. Let $u \in \mathsf{D}(T_1^*)$ and $v \in \mathsf{D}(A)$. Then $u \in \mathsf{D}(G')$ and $(T_1^* u, v) = (u, T_1 v) = (u, (H + A) v) = (u, Tv) = (u, G'(HH'^{-1} + C) G' v) = ((HH'^{-1} + C^*) G' u, G' v)$. Since this is true for all $v \in \mathsf{D}(A)$ and since $\mathsf{D}(A)$ is a core of G', $T^* u = G'(HH'^{-1} + C^*) G' u$ exists and is equal to $T_1^* u$. This shows that $T_1^* \subset T^*$, hence $T_1 \supset T$ and $T_1 - \zeta \supset T - \zeta$. But since $\zeta = i\eta$ belongs to both $\mathrm{P}(T)$ and $\mathrm{P}(T_1)$ for sufficiently large $|\eta|$, we must have $T_1 - \zeta = T - \zeta$ or $T_1 = T$.

Corollary 3.12. *Let H be selfadjoint and let A_n, $n = 1, 2, \ldots$, satisfy (3.19) for A with constants a, b replaced by a_n, b_n such that $a_n \to 0$, $b_n \to 0$, $n \to \infty$, where it is assumed that $\mathsf{D}(A_n) \subset \mathsf{D}(H)$ is a core of $|H|^{1/2}$ for each n. Then the pseudo-Friedrichs extension T_n of $H + A_n$ is defined for sufficiently large n and $T_n \to H$ in the generalized sense.*

Proof. First we note that we may assume $a_n = b_n > 0$. We can apply the proof of Theorem 3.11 with $H' = |H| + 1$ (set $\delta = 0$). (3.24) is true for any $\zeta \in \mathrm{P}(H)$ if a, b are replaced by a_n, b_n and n is sufficiently large. For such n the resolvent $(T_n - \zeta)^{-1}$ exists and converges to $(H - \zeta)^{-1}$ in norm, for

$$[(H - \zeta) H'^{-1} + C_n]^{-1} \to [(H - \zeta) H'^{-1}]^{-1}$$

by $\|C_n\| \to 0$ (with the obvious definition of C_n). Thus $T_n \to H$ in the generalized sense (see Theorem IV-2.25).

Remark 3.13. Since $\mathsf{D}(|H|) = \mathsf{D}(H)$ and $|(Hu, u)| \leq (|H| u, u)$ by (2.28), (3.19) is satisfied if

$$|(Au, u)| \leq a\|u\|^2 + b|(Hu, u)| , \quad u \in \mathsf{D}(A) . \tag{3.26}$$

Problem 3.14. Let H be selfadjoint and A symmetric with $\mathsf{D}(A) = \mathsf{D}(H)$. If A is H-bounded with $\|Au\|^2 \leq a^2\|u\|^2 + b^2\|Hu\|^2$, then A satisfies (3.19). [hint: Note that $\|Au\| \leq \|(a + b|H|) u\|$ and use Theorem V-4.12.]

§ 4. Quadratic forms and the Schrödinger operators

1. Ordinary differential operators

Since the simple types of *regular* differential operators have been dealt with in several examples above (see Examples 2.16, 2.17), we consider here a *singular* differential operator of the form[1]

$$L = -d^2/dx^2 + q(x)\,, \quad 0 < x < \infty\,. \tag{4.1}$$

For simplicity we assume that $q(x)$ is real-valued, though this is not necessary. For the moment we further assume that $q(x) \geqq 0$ and that $q(x)$ is *locally integrable* on the open interval $(0, \infty)$.

As in V-§ 5, we are mainly interested in constructing from the formal differential operator L a selfadjoint operator acting in $\mathsf{H} = \mathsf{L}^2(0, \infty)$. It should be remarked at this point that, under the rather weak assumption made on $q(x)$, the "minimal" operator such as the $\dot{T}$ of III-§ 2.3 or V-§ 5.1 cannot be defined, for Lu need not belong to L^2 for $u \in \mathsf{C}_0^\infty$ [because $q(x)$ is not assumed to be locally L^2].

Instead we consider the sesquilinear form

$$\mathfrak{h} = \mathfrak{h}_0 + \mathfrak{h}'\,, \tag{4.2}$$

where

$$\mathfrak{h}_0[u, v] = \int\limits_0^\infty u'\,\bar{v}'\,dx = (u', v')\,, \quad u(0) = v(0) = 0\,, \tag{4.3}$$

$$\mathfrak{h}'[u, v] = \int\limits_0^\infty q(x)\,u\,\bar{v}\,dx\,. \tag{4.4}$$

$\mathsf{D}(\mathfrak{h}_0)$ is by definition the set of all $u \in \mathsf{H}$ such that $u(x)$ is absolutely continuous, $u' \in \mathsf{H}$ and $u(0) = 0$. As before (cf. Example 1.36) $\mathfrak{h}_0$ is symmetric, nonnegative and closed. $\mathsf{D}(\mathfrak{h}')$ is the set of all $u \in \mathsf{H}$ such that $\int q(x)\,|u|^2\,dx < \infty$; $\mathfrak{h}'$ is also symmetric, nonnegative and closed (Example 1.15). Thus $\mathfrak{h} = \mathfrak{h}_0 + \mathfrak{h}'$ is symmetric, nonnegative and closed by Theorem 1.31. Moreover, $\mathfrak{h}$ is densely defined since $\mathsf{C}_0^\infty(0, \infty) \subset \mathsf{D}(\mathfrak{h}) = \mathsf{D}(\mathfrak{h}_0) \cap \mathsf{D}(\mathfrak{h}')$.

According to the representation theorem, there is a nonnegative selfadjoint operator $H = T_{\mathfrak{h}}$ associated with the form $\mathfrak{h}$. As in Example 2.16, H can be described as follows: *$u \in \mathsf{D}(H)$ is characterized by the conditions that* i) *u and u' are absolutely continuous on $(0, \infty)$ and belong to* H, ii) $u(0) = 0$, iii) $\int\limits_0^\infty q\,|u|^2\,dx < \infty$, *and* iv) $Lu = -u'' + q\,u \in \mathsf{H}$; for such u we have $Hu = Lu$. In deducing this result, one should note the following points. In the proof of the necessity of the conditions i) to iv), we consider an identity similar to (2.12) in which $v \in \mathsf{C}_0^\infty$ and the boundary terms are absent and in which an indefinite integral z of $w - q\,u$ is used (set $r = s = 0$); note that $w - q\,u$ is locally integrable because $w = Hu \in \mathsf{L}^2$, $u \in \mathsf{D}(H) \subset \mathsf{D}(\mathfrak{h})$ is continuous and q is locally integrable. In the

[1] We take the semi-infinite interval $(0, \infty)$ because it is more important for applications than the whole interval $(-\infty, \infty)$. The latter case can be dealt with in a similar way.

proof of the sufficiency part, we note that if u satisfies the stated conditions i) to iv), then $u \in \mathsf{D}(\mathfrak{h})$ and for any $v \in \mathsf{D}(\mathfrak{h})$,

$$(4.5)\quad \mathfrak{h}[u, v] = \lim_{a\to 0,\, b\to\infty} \int_a^b u' \overline{v'}\, dx + \int_a^b q\, u\, \bar{v}\, dx$$

$$= - \lim_{a\to 0} u'(a)\, \overline{v(a)} + \lim_{b\to\infty} u'(b)\, \overline{v(b)} + \int_0^\infty (-u'' + q\, u)\, \bar{v}\, dx\,.$$

If the first two terms on the right are zero, we have $\mathfrak{h}[u, v] = (-u'' + q\, u, v)$ and we obtain $u \in \mathsf{D}(H)$, $H u = -u'' + q\, u$ by Theorem 2.1, iii). Now (4.5) implies the *existence* of the limits $\lim u'(x)\, \overline{v(x)}$ for $x \to 0$ and $x \to \infty$. Then $\lim_{x\to\infty} u'(x)\, \overline{v(x)}$ must be zero since $u'\bar{v}$ is integrable on $(0, \infty)$, both u' and v being in H. $\lim_{x\to 0} u'\bar{v} = 0$ follows in the same way from the fact that v/x is square-integrable near $x = 0$ while $1/x$ is not integrable. In fact, we have the inequality[1]

$$(4.6)\quad \int_0^\infty x^{-2} |u(x)|^2\, dx \leqq 4 \int_0^\infty |u'(x)|^2\, dx = 4\, \mathfrak{h}_0[u]\,, \quad u \in \mathsf{D}(\mathfrak{h}_0)\,.$$

Remark 4.1. $H u = -u'' + q\, u$ belongs to $\mathsf{H} = \mathsf{L}^2$ for $u \in \mathsf{D}(H)$, but u'' and $q\, u$ need not belong separately to L^2, *even locally*. But they do belong separately to L^1 locally[2].

We shall now weaken the assumption $q(x) \geqq 0$ made above. Suppose that

$$(4.7)\quad q = q_1 + q_2 + q_3\,, \quad q_1(x) \geqq 0\,,$$

where $q_1(x)$ is *locally integrable* while $q_2(x)$ is *locally uniformly integrable* (though not necessarily nonnegative) and $q_3(x)$ satisfies the inequality

$$(4.8)\quad |q_3(x)| \leqq \alpha/4x^2\,, \quad \alpha < 1\,.$$

That $q_2(x)$ is locally uniformly integrable means that

$$(4.9)\quad \int_a^{a+1} |q_2(x)|\, dx \leqq M < \infty\,, \quad a \geqq 0\,,$$

with M independent of a.

[1] If $u \in \mathsf{C}_0^\infty$, (4.6) can be proved by integrating the identity $d(x^{-1}|u|^2)/dx = -x^{-2}|u|^2 + 2x^{-1} \operatorname{Re} u'\bar{u}$ and noting that

$$\int x^{-2}|u|^2\, dx = 2 \operatorname{Re} \int x^{-1} u' \bar{u}\, dx \leqq 2 \left(\int x^{-2}|u|^2\, dx\right)^{1/2} \left(\int |u'|^2\, dx\right)^{1/2}.$$

For general $u \in \mathsf{D}(\mathfrak{h}_0)$, (4.6) follows by going to the limit, for C_0^∞ is a core of $\mathfrak{h}_0$.

[2] H is of course densely defined; although this is by no means trivial, it is not difficult to prove directly by making use of special properties of the differential operator L.

We now define the forms $\mathfrak{h}_j'$, $j = 1, 2, 3$, by (4.4) with q replaced by q_j. *$\mathfrak{h}_2'$ is $\mathfrak{h}_0$-bounded with relative bound* 0. To see this we use the inequality

$$|u(x)|^2 \leq \varepsilon \int_x^{x+1} |u'(y)|^2 \, dy + \delta \int_x^{x+1} |u(y)|^2 \, dy \,, \tag{4.10}$$

which follows from IV-(1.19) with $p = 2$; here $\varepsilon > 0$ may be taken arbitrarily small if δ is chosen large enough. (4.9) and (4.10) give [set $q_2(x) = 0$ for $x < 0$]

$$\begin{aligned}\int_0^\infty |q_2(x)|\, |u(x)|^2 \, dx &\leq \int_0^\infty (\varepsilon |u'(y)|^2 + \delta |u(y)|^2) \, dy \int_{y-1}^{y} |q_2(x)| \, dx \leq \\ &\leq M(\varepsilon\, \mathfrak{h}_0[u] + \delta \|u\|^2) \,.\end{aligned} \tag{4.11}$$

On the other hand, $\mathfrak{h}_3'$ is $\mathfrak{h}_0$-bounded with relative bound $\alpha < 1$, as is seen from (4.6) and (4.8). Hence $\mathfrak{h}_3'$ is also $(\mathfrak{h}_0 + \mathfrak{h}_2')$-bounded with relative bound α (cf. Problem IV-1.2, a similar proposition holds for forms in place of operators). Since $\mathfrak{h}_0$ is closed, it follows that $\mathfrak{h}_0 + \mathfrak{h}_2'$ and then $\mathfrak{h}_0 + \mathfrak{h}_2' + \mathfrak{h}_3'$ are bounded from below and closed (see Theorem 1.33).

Since $\mathfrak{h}_1'$ is nonnegative and closed as before, the sum $\mathfrak{h} = (\mathfrak{h}_0 + \mathfrak{h}_2' + \mathfrak{h}_3') + \mathfrak{h}_1' = \mathfrak{h}_0 + \mathfrak{h}'$ is closed by Theorem 1.31. The characterization of the associated operator $H = T_{\mathfrak{h}}$ can be obtained exactly in the same way as above. In this way we have proved

Theorem 4.2. *Let $q(x) = q_1 + q_2 + q_3$, where all the q_j are real, q_1 is nonnegative and locally integrable, q_2 is uniformly locally integrable, and q_3 satisfies the inequality* (4.8). *Let the operator H be defined by $Hu = -u'' + q(x)\, u$, with $\mathsf{D}(H)$ consisting of all $u \in \mathsf{H} = \mathsf{L}^2(0, \infty)$ such that* i) *u and u' are absolutely continuous on* $(0, \infty)$ *and* $u' \in \mathsf{H}$, ii) $u(0) = 0$, iii) $q_1^{1/2} u \in \mathsf{H}$, *and* iv) $-u'' + qu \in \mathsf{H}$. *Then H is selfadjoint in H and bounded from below.*

Remark 4.3. Note that Theorem 4.2 is stated without any reference to the theory of forms. It could be proved without using the theory of forms, but the proof would not be very simple.

Remark 4.4. It can be shown that C_0^∞ is a core of $\mathfrak{h}$. If one knows this, one can dispense with the consideration of the integrated terms in (4.5), for v may be restricted to functions in C_0^∞ and no such terms appear.

2. The Dirichlet form and the Laplace operator

We now consider the *Dirichlet form*

$$\begin{aligned}\mathfrak{h}[u, v] &= (\operatorname{grad} u, \operatorname{grad} v) \\ &= \int_{\mathsf{R}^3} \left(\frac{\partial u}{\partial x_1} \frac{\partial \bar{v}}{\partial x_1} + \frac{\partial u}{\partial x_2} \frac{\partial \bar{v}}{\partial x_2} + \frac{\partial u}{\partial x_3} \frac{\partial \bar{v}}{\partial x_3} \right) dx\end{aligned} \tag{4.12}$$

in the 3-dimensional space R^3. The dimension 3 is not essential; most of the following results are valid in the m-dimensional space. We regard $\mathfrak{h}$ as a sesquilinear form defined in the basic Hilbert space $\mathsf{H} = \mathsf{L}^2(R^3)$. For the moment the domain $\mathsf{D}(\mathfrak{h})$ is assumed to consist of all functions $u = u(x)$ with continuous first derivatives such that $\mathfrak{h}[u]$ is finite. Obviously $\mathfrak{h}$ is densely defined, symmetric and nonnegative.

$\mathfrak{h}$ is closable. This can be proved in various ways. For instance, (4.12) may be written as $\mathfrak{h}[u, v] = (Tu, Tv)$ where $Tu = \operatorname{grad} u$ is a linear operator from H to $\mathsf{H}' = (\mathsf{L}^2(R^3))^3$ consisting of all vector-valued functions with 3 components each belonging to $\mathsf{L}^2(R^3)$. [The notation $(\operatorname{grad} u, \operatorname{grad} v)$ in (4.12) corresponds exactly to this interpretation.] The adjoint T^* of T exists and is formally given by $T^* u' = -\operatorname{div} u'$. In fact we have $(\operatorname{grad} u, u') = -(u, \operatorname{div} u')$ at least if the vector-valued function $u'(x)$ is continuously differentiable and has compact support. Since such functions form a dense set in H', T^* is densely defined and hence T is closable (see V-§ 3.1). Hence $\mathfrak{h}$ is closable by Example 1.23.

Another way to deal with the form $\mathfrak{h}$ is to introduce the Fourier transforms $\hat{u}(k)$, $\hat{v}(k)$ of $u(x)$, $v(x)$ (see Example V-2.7). Then (4.12) can be written

$$\mathfrak{h}[u, v] = \int |k|^2 \hat{u}(k) \overline{\hat{v}(k)}\, dk\,, \quad |k|^2 = k_1^2 + k_2^2 + k_3^2\,. \tag{4.13}$$

(4.13) defines a closed form $\mathfrak{h}$ if its domain consists of all $u \in \mathsf{H}$ such that $\int |k|^2 |\hat{u}(k)|^2 dk$ is finite (see Example 1.15), for the map $u \to \hat{u}$ is unitary. Actually our definition of $\mathsf{D}(\mathfrak{h})$ as originally given is not as wide as that, but this shows at least that $\mathfrak{h}$ is closable. Furthermore, the closed form just mentioned is equal to the closure of the original $\mathfrak{h}$; this can be proved as in V-§ 5.2 where we proved the essential selfadjointness of the Laplace operator. In fact we could have restricted $\mathsf{D}(\mathfrak{h})$ to $C_0^\infty(R^3)$ without affecting the result.

An immediate consequence is that the operator $H = T_{\tilde{\mathfrak{h}}}$ associated with the closure $\tilde{\mathfrak{h}}$ of $\mathfrak{h}$ is given by (see Example 2.15)

$$(Hu)^\wedge(k) = |k|^2 \hat{u}(k)\,, \tag{4.14}$$

with domain $\mathsf{D}(H)$ consisting of all $u \in \mathsf{H}$ such that $|k|^2 \hat{u}(k) \in \mathsf{L}^2(R^3)$. Thus H is identical with the closure of the negative Laplacian $-\Delta$ considered before (V-§ 5.2).

In particular we have

$$\mathfrak{h}[u, v] = (-\Delta u, v) \tag{4.15}$$

for any $v \in \mathsf{D}(\mathfrak{h})$ if, for instance, $u(x)$ has continuous second derivatives and $u, \Delta u \in \mathsf{H}$. (4.15) is even true for all $u \in \mathsf{D}(H)$ and $v \in \mathsf{D}(\tilde{\mathfrak{h}})$ if $\mathfrak{h}$ is replaced by $\tilde{\mathfrak{h}}$ and the differentiation in Δu is interpreted in the generalized sense (cf. Remark V-5.2).

3. The Schrödinger operators in R^3

Let us consider the Schrödinger operator

$$L = -\Delta + q(x) \tag{4.16}$$

in the whole space R^3, where $q(x)$ is assumed to be real-valued. Here we do not assume that q is square-integrable even locally. Thus the "minimal" operator such as $\dot{S}$ of V-§ 5.1 does not in general exist as an operator in $H = L^2(R^3)$, since Lu need not belong to H for $u \in C_0^\infty(R^3)$.

But we shall assume that q is *locally integrable* on R^3. Then we can construct the form $\mathfrak{h} = \mathfrak{h}_0 + \mathfrak{h}'$, where $\mathfrak{h}_0[u] = \|\text{grad}\, u\|^2$ is the closed *Dirichlet form* (the closure of the $\mathfrak{h}$ considered in the preceding par.) and where

$$\mathfrak{h}'[u, v] = \int_{R^3} q(x) u(x) \overline{v(x)}\, dx\,. \tag{4.17}$$

Since $\mathfrak{h}'$ need not be semibounded, it is convenient to write

$$q = q_1 + q_2 \quad \text{where} \quad q_1 \geqq 0\,, \quad q_2 \leqq 0\,, \tag{4.17a}$$

q_1, q_2 being locally integrable, and set

$$\mathfrak{h}' = \mathfrak{h}_1' + \mathfrak{h}_2'\,, \quad \mathfrak{h}_j'[u, v] = \int q_j u \bar{v}\, dx\,, \quad j = 1,2\,, \tag{4.17b}$$

$D(\mathfrak{h}_j')$ being by definition the set of all $u \in H$ with $\int |q_j|\, |u|^2 dx < \infty$.

$\mathfrak{h}_1'$ is a nonnegative, symmetric closed form, so that $\mathfrak{h}_0 + \mathfrak{h}_1'$ is (densely defined and) nonnegative and closed by Theorem 1.31. $\mathfrak{h}_2'$ is not necessarily bounded from below, but we shall *assume* that

$$\mathfrak{h}_2' \text{ is } \mathfrak{h}_0\text{-bounded with } \mathfrak{h}_0\text{-bound smaller than } 1\,. \tag{4.17c}$$

Later we shall give several sufficient conditions for (4.17c) to be true.

Under condition (4.17c), $\mathfrak{h}_2'$ is a fortiori bounded relative to $\mathfrak{h}_0 + \mathfrak{h}_1'$ with relative bound smaller than 1. It follows from Theorem 1.33 that $\mathfrak{h} = (\mathfrak{h}_0 + \mathfrak{h}_1') + \mathfrak{h}_2'$ is (densely defined and) bounded from below and closed. Let H be the selfadjoint operator associated with $\mathfrak{h}$ by Theorem 2.1. We shall show that H is a restriction of L in a generalized sense. Suppose $u \in D(H)$ and $v \in C_0^\infty(R^3)$. Then

$$(Hu, v) = \mathfrak{h}[u, v] = \mathfrak{h}_0[u, v] + \mathfrak{h}'[u, v] = (u, -\Delta v) + \int q u \bar{v}\, dx \tag{4.18}$$

by (2.1) and (4.15). Hence

$$(u, -\Delta v) = \int (Hu - qu) \bar{v}\, dx\,. \tag{4.19}$$

Here $Hu \in H = L^2(R^3)$ and qu is locally integrable, since $2|qu| \leqq |q| + |q|\, |u|^2$ and both $|q|$ and $|q|\, |u|^2$ are locally integrable (note that

$u \in D(\mathfrak{h}_1') \cap D(\mathfrak{h}_2'))$. Hence $Hu - qu$ is locally integrable, and (4.19) implies

$$-\Delta u = Hu - qu \quad \text{or} \quad Hu = -\Delta u + qu = Lu, \tag{4.20}$$

the differentiation Δu being interpreted in the generalized sense[1].

This shows that H is indeed a restriction of L. But we want to go one step further and obtain a *characterization* of $D(H)$. The result is given by the following theorem.

Theorem 4.6a. *Let* $q = q_1 + q_2$ *be as above. Let* D *be the set of all* $u \in H$ *such that* i) *the generalized derivative* grad u *belongs to* $H' = (L^2(R^3))^3$, ii) $\int q_1(x) |u|^2 dx < \infty$, *and* iii) *the generalized derivative* Δu *exists and* $Lu = -\Delta u + qu$ *belongs to* H. *Define the operator* H *by* $Hu = Lu$ *with* $D(H) = D$. *Then* H *is selfadjoint in* H *and is bounded from below. (Note that* ii) *implies that* $qu \in L^1_{\text{loc}}(R^3)$ *as shown above.)*

The operator H we have constructed above was already shown to satisfy $D(H) \subset D$. Thus it only remains to show that $D \subset D(H)$. To this end let $u \in D$. Then $u \in D(\mathfrak{h}_0) \subset D(\mathfrak{h}_2')$ by i) and $u \in D(\mathfrak{h}_1')$ by ii). Hence $u \in D(\mathfrak{h})$ and we have for each $v \in C_0^\infty(R^3)$

$$\begin{aligned} \mathfrak{h}[u, v] &= \mathfrak{h}_0[u, v] + \mathfrak{h}'[u, v] \\ &= (u, -\Delta v) + \int q u \bar{v}\, dx = (Lu, v) \end{aligned} \tag{4.22}$$

since Lu exists and belongs to H. According to Theorem 2.1, (4.22) implies that $u \in D(H)$ with $Hu = Lu$ *provided we know that* $C_0^\infty(R^3)$ *is a core of* $\mathfrak{h}$. Thus the proof is reduced to

Lemma 4.6b. $C_0^\infty(R^3)$ *is a core of* $\mathfrak{h}$.

Proof. Since $\mathfrak{h} = (\mathfrak{h}_0 + \mathfrak{h}_1') + \mathfrak{h}_2'$ in which the second term is bounded relative to the first term, a simple consideration reduces Lemma 4.6b to

Lemma 4.6c $C_0^\infty(R^3)$ *is a core of* $\mathfrak{h}_0 + \mathfrak{h}_1'$.

Proof. $\mathfrak{h}_0 + \mathfrak{h}_1'$ is nonnegative and closed as shown above. Hence it suffices to show that C_0^∞ is dense in the Hilbert space $D(\mathfrak{h}_0 + \mathfrak{h}_1') = D(\mathfrak{h}_0) \cap D(\mathfrak{h}_1)$ equipped with the inner product $(\mathfrak{h}_0 + \mathfrak{h}_1' + 1)[u, v]$ (see Theorem 1.21). Suppose that u is orthogonal to C_0^∞ in this space:

$$(\mathfrak{h}_0 + \mathfrak{h}_1' + 1)[u, v] = 0 \quad \text{for all} \quad v \in C_0^\infty(R^3); \tag{4.22a}$$

we have to show that $u = 0$.

An argument similar to the one used to deduce (4.20) from (4.18) shows that (4.22a) implies

$$\Delta u = (q_1 + 1) u \tag{4.22b}$$

---

[1] This is exactly the definition of Δ in the generalized sense; see e. g. Yosida [1].

in the generalized sense, where $q_1 u$ is locally integrable. Now it can be shown that the only function $u \in \mathsf{H}$ that satisfies (4.22b) is $u = 0$. This is a nontrivial result in partial differential equations, and we shall not prove it here[1].

Remark 4.7a. Although we have characterized $\mathsf{D}(H)$, the question remains whether or not there are other restrictions of L that are selfadjoint. It is known that if q_2 is bounded, then there are no selfadjoint restrictions of L other than H. To see this we may assume $q_2 = 0$ (by changing L into $L +$ const. if necessary). It suffices to show that if $u \in \mathsf{H}$ and if Lu exists (by which is implied that qu is locally integrable) and belongs to H, then $u \in \mathsf{D}(H)$. Let $v = (H + 1)^{-1} (L + 1) u \in \mathsf{D}(H)$. Since $H \subset L$, it follows that $(L + 1)(u - v) = 0$. Thus $u - v \in \mathsf{H}$ satisfies (4.22b) and hence must vanish as above. This shows that $u = v \in \mathsf{D}(H)$, as required. It can be shown that the same result holds if q_2 is locally square-integrable with local square-integrals not growing too rapidly at infinity.

We shall now give several sufficient conditions for (4.17c) to be met. A simple potential satisfying (4.17c) is given by

$$q_2(x) = \sum_j e_j |x - a_j|^{-2} \quad \text{with} \quad \sum e_j > -1/4\,, \tag{4.23a}$$

where the a_j are fixed points in R^3. (Here it suffices to consider only negative e_j, for the terms with positive e_j can be absorbed into q_1.) (4.23a) implies (4.17c) by the well-known inequality

$$\int |x - a|^{-2} |u(x)|^2 dx \leqq 4 \int |\operatorname{grad} u|^2 dx \leqq 4\, \mathfrak{h}_0[u]\,, \tag{4.24}$$

which can be proved in the same way as (4.6) (use the polar coordinates with origin at $x = a$).

Another sufficient condition for (4.17c) is given in terms of the function

$$M_{q_2}(x; r) = \int_{|x-y|<r} |x - y|^{-1} |q_2(y)|\, dy\,. \tag{4.25a}$$

Lemma 4.8a. *Assume that* $M_{q_2}(x; r) \to 0$ *as* $r \to 0$, *uniformly in* $x \in \mathrm{R}^3$. *Then* (4.17c) *is true.*

Proof. We have to show that $\| |q_2|^{1/2} u\|^2 = |\mathfrak{h}_2'[u]| \leqq b(\mathfrak{h}_0[u] + k^2\|u\|^2) = b\|(H_0 + k^2)^{1/2} u\|^2$ for an arbitrarily small $b > 0$ if k^2 is chosen appropriately. Thus it suffices to show that

$$\|T_k\| \to 0 \quad \text{as} \quad k \to \infty\,, \quad \text{where} \quad T_k = |q_2|^{1/2} (H_0 + k^2)^{-1/2}. \tag{4.25b}$$

[1] For the proof see T. Kato [25].

Since $\|T_k\| = \|T_k T_k^*\|^{1/2}$, it suffices in turn to show that $\|T_k T_k^*\| \to 0$ as $k \to \infty$. But $T_k T_k^* \supset |q_2|^{1/2}(H_0 + k^2)^{-1}|q_2|^{1/2}$, where $(H_0 + k^2)^{-1}$ is an integral operator with the kernel

(4.25c) $$g(y, x; k) = e^{-k|y-x|}/4\pi|y-x|$$

[see IX-(1.67)]. Let us estimate $\|T_k T_k^* v\|$. If $v \in \mathrm{D}(|q_2|^{1/2})$ and $w = (H_0 + k^2)^{-1}|q_2|^{1/2}v$, we have

$$w(y) = \int g(y, x; k)\, |q_2(x)|^{1/2} v(x)\, dx .$$

By the Schwarz inequality we obtain

$$|w(y)|^2 \leqq c_k \int g(y, x; k)\, |v(x)|^2 dx ,$$

where

(4.25d) $$c_k = \sup_{y \in \mathrm{R}^3} \int g(y, x; k)\, |q_2(x)|\, dx .$$

Since $T_k T_k^* v = |q_2|^{1/2} w$, we have (anticipating that it is in H)

$$\begin{aligned} \|T_k T_k^* v\|^2 = \| |q_2|^{1/2} w\|^2 &\leqq c_k \int\int |q_2(y)|\, g(y, x; k)\, |v(x)|^2 dx\, dy \\ &\leqq c_k^2 \int |v(x)|^2 dx = c_k^2 \|v\|^2 . \end{aligned}$$

Hence $\|T_k T_k^*\| \leqq c_k$. Thus the lemma will be proved if we show that $c_k \to 0$ as $k \to \infty$.

To this end, let $\varepsilon > 0$ and choose $r > 0$ so small that the integral in (4.25d) taken on $|x - y| < r$ is smaller than ε; this is possible by the assumption that $M_{q_2}(x; r) \to 0$ as $r \to 0$ uniformly in x. The remaining part of the integral can be made smaller than ε by choosing k sufficiently large, uniformly in y. We may leave the detail to the reader.

Lemma 4.8b. (4.17c) *is true if* $q_2 \in \mathrm{L}^{3/2}(\mathrm{R}^3)$.

Proof. We may assume that the $\mathrm{L}^{3/2}$-norm of q_2 is as small as we please, since q_2 may be approximated in $\mathrm{L}^{3/2}$-norm by a bounded function, which contributes to $\mathfrak{h}_2'$ only a bounded form. Then the desired result follows from the Sobolev embedding theorem[1], which shows that $u \in \mathrm{D}(\mathfrak{h}_0)$ implies $u \in \mathrm{L}^6(\mathrm{R}^3)$ with $\|u\|_{\mathrm{L}^6}^2 \leqq \text{const.}\ \mathfrak{h}_0[u]$, combined with the Hölder inequality

$$\int |q_2|\, |u|^2 dx \leqq \text{const.}\ \|q_2\|_{\mathrm{L}^{3/2}} \|u\|_{\mathrm{L}^6}^2 .$$

Remark 4.9a. Theorem 4.6a and the lemmas given above can be extended without essential change to the case of Schrödinger operators in R^m with $m > 3$. The only changes required are to replace: $-1/4$ in

---

[1] See Nirenberg [1].

(4.23a) by $-(m-2)^2/4$, 4 in (4.24) by $4/(m-2)^2$, $|x-y|^{-1}$ in (4.25a) by $|x-y|^{m-2}$, and $L^{3/2}(R^3)$ in Lemma 4.8b by $L^{m/2}(R^m)$. Lemma 4.8a is still true for $m=2$ if in the definition of $M_{q_2}(x; r)$, $|x-y|^{-1}$ is replaced by $-\log|x-y|$.

4. Bounded regions

If the region E of R^3 or R^m in which the differential operator (4.16) is to be considered is bounded, there are complications arising from the possibility of imposing various boundary conditions. To simplify the matter, let us assume that E is open and has a smooth boundary ∂E and that $q(x)$ is a smooth function in the closed region $\bar{E}$. Then the minimum operator $\dot{S}$ in $H = L^2(E)$ can be defined as in V-§ 5.1, and it would seem that one need not invoke the theory of forms in constructing selfadjoint operators from L. Nevertheless considering a form $\mathfrak{h} = \mathfrak{h}_0 + \mathfrak{h}'$, such as the one discussed in the preceding paragraph (the integrations should now be taken over E), is a convenient way to obtain various selfadjoint extensions of $\dot{S}$.

First let us assume that $D(\mathfrak{h}) = C_0^\infty(E)$, the set of infinitely differentiable functions with compact supports in E; as before it can be shown that $\mathfrak{h}$ is densely defined, symmetric, bounded from below and closable. We denote this form by $\mathfrak{h}_1$ and set $H_1 = T_{\tilde{\mathfrak{h}}_1}$. Next we extend the domain of $\mathfrak{h}$ and include in $D(\mathfrak{h})$ all functions continuously differentiable in the closed region $\bar{E}$; the resulting form, which we denote by $\mathfrak{h}_2$, is densely defined, symmetric, bounded from below and closable. We set $H_2 = T_{\tilde{\mathfrak{h}}_2}$. Obviously $\mathfrak{h}_1 \subset \mathfrak{h}_2$ and so $\tilde{\mathfrak{h}}_1 \subset \tilde{\mathfrak{h}}_2$. According to the definition of the order relation for selfadjoint operators given in § 2.5, this implies that $\tilde{\mathfrak{h}}_1 \geqq \tilde{\mathfrak{h}}_2$ and $H_1 \geqq H_2$.

We define a third form $\mathfrak{h}_3$ by

$$\mathfrak{h}_3[u, v] = \mathfrak{h}_2[u, v] + \int_{\partial E} \sigma u \bar{v} \, dS, \quad D(\mathfrak{h}_3) = D(\mathfrak{h}_2), \tag{4.35}$$

where the integral is taken on the boundary ∂E and σ is a given smooth function defined on ∂E. It can be shown that the additional term in (4.35) is *relatively bounded* with respect to $\mathfrak{h}_2$ with relative bound 0. The proof is omitted, but the assertion corresponds to the fact, in Example 1.36, that t_2 is t_1-bounded.

We set $H_3 = T_{\tilde{\mathfrak{h}}_3}$. Note that $\mathfrak{h}_3 \supset \mathfrak{h}_1$ and so $\tilde{\mathfrak{h}}_3 \supset \tilde{\mathfrak{h}}_1$, for the boundary term in (4.35) vanishes for $u \in D(\mathfrak{h}_1)$. Again this implies that $\tilde{\mathfrak{h}}_3 \leqq \tilde{\mathfrak{h}}_1$ and $H_3 \leqq H_1$.

It can now be shown that H_1, H_2, H_3 are all selfadjoint extensions of the minimal operator $\dot{S}$ and restrictions of the maximal operator $\dot{S}^*$. Roughly speaking, these are the formal differential operator $L = -\Delta + q(x)$ with the following *boundary conditions* on ∂E:

$$\begin{aligned} u &= 0 \quad \text{for} \quad H_1, \\ \partial u/\partial n &= 0 \quad \text{for} \quad H_2, \\ \partial u/\partial n - \sigma u &= 0 \quad \text{for} \quad H_3, \end{aligned} \tag{4.36}$$

where $\partial/\partial n$ denotes the inward normal derivative, as is suggested by comparison with ordinary differential operators (see Example 2.16). But this statement needs some comments. A function u in the domain of any of these operators need not be differentiable in the ordinary sense, and the operator Lu must be interpreted in a generalized sense; also the boundary conditions (4.36) must be interpreted accordingly. In any case it is easily seen that a function u with continuous second derivatives in $\bar{E}$ and satisfying the boundary condition $\partial u/\partial n = 0$ belongs to $D(H_2)$

and $H_2 u = Lu$. For this we need only to verify that $\mathfrak{h}_2[u, v] = (Lu, v)$ for all $v \in \mathsf{D}(\mathfrak{h}_2)$ (see Theorem 2.1); here we are dealing only with smooth functions and there is no difficulty.

These results can be generalized to the form

$$(4.37) \qquad \mathfrak{h}[u, v] = \int_{\mathsf{E}} \sum_{j,k=1}^{3} \left[p_{jk}(x) \frac{\partial u}{\partial x_j} \frac{\partial \bar{v}}{\partial x_k} + q(x)\, u \bar{v} \right] dx + \int_{\partial \mathsf{E}} \sigma u \bar{v}\, dS,$$

where $(p_{jk}(x))$ is a symmetric 3×3 matrix whose elements are real-valued smooth functions of x, and it is assumed that the matrix is *uniformly positive definite*; this means that there exist constants $\alpha, \beta > 0$ such that

$$(4.38) \qquad \alpha \sum |\xi_j|^2 \leq \sum_{j,k} p_{jk}(x)\, \xi_j \bar{\xi}_k \leq \beta \sum |\xi_j|^2$$

for all complex vectors (ξ_1, ξ_2, ξ_3). (4.38) is equivalent to the condition that the eigenvalues of the symmetric matrix $(p_{jk}(x))$ lie between α and β. Incidentally, these eigenvalues are continuous functions of x; this is a consequence of the perturbation theory for symmetric operators (see II-§ 5.5).

(4.38) implies that the $\mathfrak{h}$ of (4.37) is *comparable* with the old $\mathfrak{h}$ in the sense that the two are relatively bounded with respect to each other, so that the two have the same domain as do their closures. (As before we have to distinguish three different forms $\mathfrak{h}_n$, $n = 1, 2, 3$.) Thus we are led to the selfadjoint operators H_n, $n = 1, 2, 3$, which are formally the same differential operator

$$(4.39) \qquad Lu = -\sum_{j,k} \frac{\partial}{\partial x_j} p_{jk}(x) \frac{\partial u}{\partial x_k} + q(x)\, u$$

but with distinct boundary conditions. These are similar to (4.36) but differ in that $\partial/\partial n$ means differentiation in the direction of the *conormal*, which is determined by $(p_{jk}(x))$.

The fact stated above — that the domain of the closure of $\mathfrak{h}_n$ is identical with that of the special case $p_{jk}(x) = \delta_{jk}$ and so independent of the coefficients $p_{jk}(x)$, $q(x)$ and $\sigma(x)$ as long as $(p_{jk}(x))$ is uniformly positive definite and $q(x)$, $\sigma(x)$ are smooth — is one of the merits in considering the forms $\mathfrak{h}_n$. There are in general no simple relationships between the domains of the H_n for different choices of coefficients.

A deeper theory of differential operators[1] shows, however, that the dependence of the domain of H_n on $p_{jk}(x)$ etc. is due exclusively to the change of the *boundary condition*. The *interior properties* of functions u of $\mathsf{D}(H_n)$ are the same for all n and independent of $p_{jk}(x)$; the main point is that u should have generalized derivatives of second order that belong to L^2. Among the various boundary conditions considered above, the Dirichlet condition ($u = 0$ on $\partial\mathsf{E}$) is independent of $p_{jk}(x)$ and $q(x)$. Hence it follows that $\mathsf{D}(H_1)$ *is independent of the coefficients* $p_{jk}(x)$ *and* $q(x)$.

§ 5. The spectral theorem and perturbation of spectral families

1. Spectral families

Let H be a Hilbert space, and suppose there is a nondecreasing family $\{\mathsf{M}(\lambda)\}$ of closed subspaces of H depending on a real parameter λ, $-\infty < \lambda < +\infty$, such that the intersection of all the $\mathsf{M}(\lambda)$ is 0 and their

[1] See e. g. LIONS [1].

union is dense in H. By "nondecreasing" we mean that $\mathsf{M}(\lambda') \subset \mathsf{M}(\lambda'')$ for $\lambda' < \lambda''$.

For any fixed λ, then, the intersection $\mathsf{M}(\lambda + 0)$ of all $\mathsf{M}(\lambda')$ with $\lambda' > \lambda$ contains $\mathsf{M}(\lambda)$. Similarly we have $\mathsf{M}(\lambda) \supset \mathsf{M}(\lambda - 0)$, where $\mathsf{M}(\lambda - 0)$ is the *closure* of the union of all $\mathsf{M}(\lambda')$ with $\lambda' < \lambda$. We shall say that the family $\{\mathsf{M}(\lambda)\}$ is *right continuous* at λ if $\mathsf{M}(\lambda + 0) = \mathsf{M}(\lambda)$, *left continuous* if $\mathsf{M}(\lambda - 0) = \mathsf{M}(\lambda)$ and *continuous* if it is right as well as left continuous. As is easily seen, $\{\mathsf{M}(\lambda + 0)\}$ has the same properties as those required of $\{\mathsf{M}(\lambda)\}$ above and, moreover, it is everywhere right continuous.

These properties can be translated into properties of the associated family $\{E(\lambda)\}$ of orthogonal projections on $\mathsf{M}(\lambda)$. We have:

(5.1) $E(\lambda)$ is nondecreasing: $E(\lambda') \leqq E(\lambda'')$ for $\lambda' < \lambda''$.

$$\text{s-}\lim_{\lambda \to -\infty} E(\lambda) = 0\,, \quad \text{s-}\lim_{\lambda \to +\infty} E(\lambda) = 1\,. \tag{5.2}$$

(5.1) is equivalent to

$$E(\lambda)\, E(\mu) = E(\mu)\, E(\lambda) = E(\min(\lambda, \mu))\,. \tag{5.3}$$

A family $\{E(\lambda)\}$ of orthogonal projections with the properties (5.1), (5.2) is called a *spectral family* or a *resolution of the identity*.

The projections $E(\lambda \pm 0)$ on $\mathsf{M}(\lambda \pm 0)$ are given by

$$E(\lambda \pm 0) = \text{s-}\lim_{\varepsilon \to +0} E(\lambda \pm \varepsilon)\,. \tag{5.4}$$

Thus $\{\mathsf{M}(\lambda)\}$ is right (left) continuous if and only if $\{E(\lambda)\}$ is strongly right(left) continuous. Usually a spectral family is assumed to be right continuous everywhere:

$$E(\lambda + 0) = E(\lambda)\,, \quad -\infty < \lambda < +\infty\,, \tag{5.5}$$

and we shall follow this convention. In any case the family $\{E(\lambda + 0)\}$ is right continuous everywhere.

$\{E(\lambda)\}$ will be said to be *bounded from below* if $E(\mu) = 0$ for some finite μ [then $E(\lambda) = 0$ for $\lambda < \mu$ a fortiori]; the least upper bound of such μ is the *lower bound* of $\{E(\lambda)\}$. Similarly, $\{E(\lambda)\}$ is *bounded from above* if $E(\mu) = 1$ for some finite μ and the *upper bound* is defined accordingly. Note that $E(\lambda)$ need not be zero when λ is equal to the lower bound, while $E(\lambda) = 1$ if λ is equal to the upper bound; this is due to the convention (5.5).

For any semiclosed interval $\mathrm{I} = (\lambda', \lambda'']$ of the real line we set

$$E(\mathrm{I}) = E(\lambda'') - E(\lambda')\,; \tag{5.6}$$

$E(\mathrm{I})$ is the projection on the subspace $\mathsf{M}(\mathrm{I}) = \mathsf{M}(\lambda'') \ominus \mathsf{M}(\lambda')$[1]. If two such intervals I_1, I_2 have no point in common, $\mathsf{M}(\mathrm{I}_1)$ and $\mathsf{M}(\mathrm{I}_2)$ are

[1] $\mathsf{M} \ominus \mathsf{N} = \mathsf{M} \cap \mathsf{N}^\perp$ is the orthogonal complement of N in M (see I-§ 6.3).

orthogonal; for, if I_1 is to the left of I_2, $M(I_2) = M(\lambda_2'') \ominus M(\lambda_2') \perp \perp M(\lambda_2') \supset M(\lambda_1'') \supset M(I_1)$. The corresponding relation for the projections is

$$E(I_1)\, E(I_2) = E(I_2)\, E(I_1) = 0 \quad \text{for disjoint} \quad I_1, I_2, \tag{5.7}$$

which can also be verified directly by using (5.3)

The projection on $M(\lambda) \ominus M(\lambda - 0)$ is

$$P(\lambda) = E(\lambda) - E(\lambda - 0). \tag{5.8}$$

As above we have

$$P(\lambda)\, P(\mu) = P(\mu)\, P(\lambda) = 0 \quad \text{for} \quad \lambda \neq \mu. \tag{5.9}$$

$P(\lambda) \neq 0$ if and only if $E(\lambda)$ is discontinuous at λ. If H is separable, an orthogonal set of nonzero projections is at most countable. Hence there are at most countably many points of discontinuity of $E(\lambda)$ in a separable space.

If S is the union of a finite number of intervals (open, closed or semi-closed) on the real line, S can be expressed as the union of disjoint sets of the form I or $\{\lambda\}$ stated above. If we define $E(S)$ as the sum of the corresponding $E(I)$ or $P(\lambda)$, it is easily seen that $E(S)$ has the property that $E(S')\, E(S'') = E(S' \cap S'')$. $E(S)$ is called a ***spectral measure*** on the class of all sets S of the kind described. This measure $E(S)$ can then be extended to the class of all Borel sets S of the real line by a standard measure-theoretic construction.

For any $u \in H$, $(E(\lambda)\, u, u)$ is a nonnegative, nondecreasing function of λ and tends to zero for $\lambda \to -\infty$ and to $\|u\|^2$ for $\lambda \to +\infty$. For any $u, v \in H$, the polar form $(E(\lambda)\, u, v)$ can be written as a linear combination of functions of the form $(E(\lambda)\, w, w)$ by I-(6.26). Hence the complex-valued function $(E(\lambda)\, u, v)$ of λ is ***of bounded variation***. This can be seen more directly as follows. For any $I = (\lambda', \lambda'']$ we have

$$\begin{aligned} |(E(\lambda'')\, u, v) - (E(\lambda')\, u, v)| &= |(E(I)\, u, v)| \\ &= |(E(I)\, u, E(I)\, v)| \leq \|E(I)\, u\|\, \|E(I)\, v\|. \end{aligned} \tag{5.10}$$

If $I_1, \ldots, I_n$ is a set of disjoint intervals of the above form, we have

$$\begin{aligned} \sum_j |(E(I_j)\, u, v)| &\leq \Sigma \|E(I_j)\, u\|\, \|E(I_j)\, v\| \leq \\ &\leq (\Sigma \|E(I_j)\, u\|^2)^{1/2} (\Sigma \|E(I_j)\, v\|^2)^{1/2} \\ &= [\Sigma (E(I_j)\, u, u)]^{1/2} [\Sigma (E(I_j)\, v, v)]^{1/2} \\ &= (\Sigma\, E(I_j)\, u, u)^{1/2} (\Sigma\, E(I_j)\, v, v)^{1/2} \leq \|u\|\, \|v\|. \end{aligned} \tag{5.11}$$

Thus the total variation of $(E(\lambda)\, u, v)$ does not exceed $\|u\|\, \|v\|$.

$\lambda = \alpha$ is called a ***point of constancy*** with respect to $\{E(\lambda)\}$ if $E(\lambda)$ is constant in a neighborhood of α: $E(\alpha + \varepsilon) = E(\alpha - \varepsilon)$, $\varepsilon > 0$. α is then an internal point of a maximal interval in which $E(\lambda)$ is constant. The set of all points which are not points of constancy is called the *support* of $\{E(\lambda)\}$ [or of the spectral measure $E(\mathsf{S})$]. $\{E(\lambda)\}$ is bounded from below (above) if and only if the support of $E(\lambda)$ is.

2. The selfadjoint operator associated with a spectral family

To any spectral family $\{E(\lambda)\}$, there is associated a selfadjoint operator H expressed by

$$H = \int_{-\infty}^{+\infty} \lambda \, dE(\lambda) . \tag{5.12}$$

$\mathsf{D}(H)$ is the set of all $u \in \mathsf{H}$ such that[1]

$$\int_{-\infty}^{+\infty} \lambda^2 \, d(E(\lambda) u, u) < \infty . \tag{5.13}$$

For such u, (Hu, v) is given by

$$(Hu, v) = \int_{-\infty}^{+\infty} \lambda \, d(E(\lambda) u, v) , \quad v \in \mathsf{H} . \tag{5.14}$$

The convergence of the integral in (5.14) follows from the estimate (5.10), which may be written $|d(E(\lambda) u, v)| \leqq [d(E(\lambda) u, u) \, d(E(\lambda) v, v)]^{1/2}$, and from (5.13) by means of the Schwarz inequality. That H is symmetric is obvious. The selfadjointness of H will be proved below. We note that

$$\begin{aligned} \|Hu\|^2 = (Hu, Hu) &= \int_{-\infty}^{+\infty} \lambda \, d(E(\lambda) u, Hu) \\ &= \int_{-\infty}^{+\infty} \lambda \, d_\lambda \int_{-\infty}^{+\infty} \mu \, d_\mu (E(\lambda) u, E(\mu) u) \\ &= \int_{-\infty}^{+\infty} \lambda \, d_\lambda \int_{-\infty}^{\lambda} \mu \, d(E(\mu) u, u) = \int_{-\infty}^{+\infty} \lambda^2 \, d(E(\lambda) u, u) \end{aligned} \tag{5.15}$$

for $u \in \mathsf{D}(H)$, where (5.3) has been used.

More generally, we can define the operators

$$\phi(H) = \int_{-\infty}^{+\infty} \phi(\lambda) \, dE(\lambda) \tag{5.16}$$

in a similar way. (5.14) corresponds to the special case $\phi(\lambda) = \lambda$. $\phi(\lambda)$ may be any complex-valued, continuous function[2]. If $\phi(\lambda)$ is bounded on the support Σ of $\{E(\lambda)\}$, the condition corresponding to (5.13) is

---

[1] The integrals in (5.13) and (5.14) are *Stieltjes integrals*. These are quite elementary since the integrands λ and λ^2 are continuous functions.

[2] More general functions ϕ can be allowed, but then the integral $(\phi(H) u, v) = \int \phi(\lambda) \, d(E(\lambda) u, v)$ must be taken in the sense of the Radon-Stieltjes integral. For details see e. g. STONE [1].

always fulfilled, so that $\mathsf{D}(\phi(H)) = \mathsf{H}$ and $\phi(H)$ is bounded [cf. (5.15)]:

$$\|\phi(H)\| \leqq \sup_{\lambda \in \Sigma} |\phi(\lambda)| . \tag{5.17}$$

Thus $\phi(H)$ belongs to $\mathscr{B}(\mathsf{H})$. Furthermore it is *normal*. This follows from the general relations

$$\phi(H)^* = \bar{\phi}(H) \quad \text{if} \quad \bar{\phi}(\lambda) = \overline{\phi(\lambda)} , \tag{5.18}$$

$$\phi(H) = \phi_1(H)\, \phi_2(H) \quad \text{if} \quad \phi(\lambda) = \phi_1(\lambda)\, \phi_2(\lambda) , \tag{5.19}$$

$$\phi(H) = \alpha_1\, \phi_1(H) + \alpha_2\, \phi_2(H) \quad \text{if} \quad \phi(\lambda) = \alpha_1\, \phi_1(\lambda) + \alpha_2\, \phi_2(\lambda) , \tag{5.20}$$

where ϕ, ϕ_1 and ϕ_2 are assumed to be bounded on Σ. The proofs of (5.18) and (5.20) are simple. (5.19) can be proved by a calculation similar to that given in (5.15). These relations justify the notation $\phi(H)$ for the operator (5.16).

An important special case is $\phi(\lambda) = (\lambda - \zeta)^{-1}$ for a nonreal constant ζ. The corresponding $\phi(H)$ is seen to be the resolvent $(H - \zeta)^{-1}$, which shows that H is selfadjoint. To prove this, we first note that $\phi(H)(H - \zeta)u = u$ for all $u \in \mathsf{D}(H)$; again this is easily proved by a calculation similar to (5.15). This implies that $\phi(H) \supset (H - \zeta)^{-1}$. On the other hand, any $u \in \mathsf{R}(\phi(H))$ satisfies the condition (5.13); this is due to the fact that $\lambda(\lambda - \zeta)^{-1}$ is a bounded function [again a calculation similar to (5.15) is to be used here]. It follows that $(H - \zeta)^{-1} = \phi(H)$ as we wished to show.

$(H - \zeta)^{-1} = \int (\lambda - \zeta)^{-1}\, dE(\lambda)$ is defined and belongs to $\mathscr{B}(\mathsf{H})$ not only for nonreal ζ but also for any real ζ which is a point of constancy for $E(\lambda)$, for $(\lambda - \zeta)^{-1}$ is bounded for $\lambda \in \Sigma$. Hence such a ζ belongs to the resolvent set $\mathrm{P}(H)$. This implies that $\Sigma(H) \subset \Sigma$. Actually we have $\Sigma(H) = \Sigma$. To show this we note that

$$\|(H - \mu)\, u\|^2 = \int_{-\infty}^{+\infty} (\lambda - \mu)^2\, d(E(\lambda)\, u, u) , \quad u \in \mathsf{D}(H) , \tag{5.21}$$

as is seen by applying (5.15) to $H - \mu$. If $\mu \in \Sigma$, $E' = E(\mu + \varepsilon) - E(\mu - \varepsilon) \neq 0$ for any $\varepsilon > 0$. Hence there is a $u \neq 0$ such that $E' u = u$, which implies $E(\mu - \varepsilon)\, u = 0$ and $E(\mu + \varepsilon)\, u = u$. Then it follows from (5.21) that $\|(H - \mu)\, u\|^2 = \int_{\mu-\varepsilon}^{\mu+\varepsilon} \leqq \varepsilon^2 \|u\|^2$. Since such a u exists for each $\varepsilon > 0$, μ does not belong to $\mathrm{P}(H)$. This implies $\Sigma \subset \Sigma(H)$ and hence $\Sigma = \Sigma(H)$.

It follows also from (5.21) that $(H - \mu)\, u = 0$ if and only if $(E(\lambda) u, u)$ is constant except for a discontinuity at $\lambda = \mu$ so that $E(\mu)\, u = u$, $E(\mu - 0)\, u = 0$. Hence μ is an eigenvalue of H if and only if $P(\mu) \neq 0$, and u is an associated eigenvector if and only if $P(\mu)\, u = u$.

An operational calculus can be developed for the operators $\phi(H)$. $\phi(H)$ is in general unbounded if $\phi(\lambda)$ is unbounded; $\mathsf{D}(\phi(H))$ is the set of all $u \in \mathsf{H}$ such that $\int |\phi(\lambda)|^2 d(E(\lambda)\, u, u) < \infty$. When ϕ or ϕ_1, ϕ_2 are unbounded, it can be shown that (5.18) is still true but (5.19), (5.20) must be replaced in general by

$$\phi(H) \supset \phi_1(H)\, \phi_2(H)\,, \quad \phi(H) \supset \alpha_1\, \phi_1(H) + \alpha_2\, \phi_2(H)\,, \tag{5.22}$$

respectively. But (5.19) remains true if both ϕ_2 and $\phi_1\, \phi_2 = \phi$ are bounded on Σ.

Problem 5.1. All the $E(\lambda)$ and $E(\lambda - 0)$ commute with H. Equivalently, all the $\mathsf{M}(\lambda)$ and $\mathsf{M}(\lambda - 0)$ reduce H.

Problem 5.2. The union of the ranges $\mathsf{M}(\mu) \ominus \mathsf{M}(\lambda)$ of $E(\mu) - E(\lambda)$ for all λ, μ with $-\infty < \lambda < \mu < \infty$ is a core of H.

Problem 5.3. $H \geqq \gamma$ if and only if $E(\gamma - 0) = 0$.

Problem 5.4. If $H \geqq 0$, then $H^{1/2} = \int\limits_0^\infty \lambda^{1/2}\, dE(\lambda)$.

Problem 5.5. If $H_1 \geqq H_2$ in the sense of § 2.5, then $\dim E_1(\lambda) \leqq \dim E_2(\lambda)$ (a generalization of Theorem I-6.44)[1].

We have shown that any spectral family $\{E(\lambda)\}$ determines a self-adjoint operator H by (5.12). We shall now show that different spectral families lead to different selfadjoint operators. To this end, it suffices to give an explicit formula for determining $E(\lambda)$ from H.

Define the operator

$$|H| = \int\limits_{-\infty}^{+\infty} |\lambda|\, dE(\lambda) = \int\limits_{0}^{+\infty} \lambda\, dE(\lambda) - \int\limits_{-\infty}^{0} \lambda\, dE(\lambda)\,. \tag{5.23}$$

$\mathsf{D}(|H|)$ is the same as $\mathsf{D}(H)$. It is easily seen that $|H|$ is selfadjoint and nonnegative, and that $\mathsf{N}(|H|) = \mathsf{N}(H) = \mathsf{R}(P(0))$. Also it is easy to see that $|H|\, u = H u$ if $E(0)\, u = 0$ and $|H|\, u = -H u$ if $u = E(-0)\, u$. Hence

$$H = [1 - E(0) - E(-0)]\, |H|\,. \tag{5.24}$$

But $U = 1 - E(0) - E(-0)$ has the properties that $U u = 0$ for $u \in \mathsf{N}(H) = \mathsf{R}(P(0))$ and $\|U u\| = \|u\|$ if $P(0)\, u = 0$. Thus (5.24) is exactly the polar decomposition of H (see § 2.7), which determines $|H|$ and U uniquely from H.

Applying the same argument to H replaced by $H - \lambda$, we see that

$$U(\lambda) \equiv 1 - E(\lambda) - E(\lambda - 0) \tag{5.25}$$

is the unique partially isometric operator that appears in the polar decomposition $H - \lambda = U(\lambda)\, |H - \lambda|$ of $H - \lambda$. Since (5.25) implies

$$E(\lambda) = 1 - \frac{1}{2}\, [U(\lambda) + U(\lambda)^2]\,, \tag{5.26}$$

$E(\lambda)$ is determined by H.

---

[1] If H is finite-dimensional, $E(\lambda)$ is purely discontinuous in λ; the points of discontinuity are exactly the eigenvalues of H.

The determination of $U(\lambda)$ given above is not explicit, however. An explicit formula is given by

Lemma 5.6. *We have*

$$\text{s-}\lim_{\substack{\delta\to 0\\ \varrho\to\infty}} U_{\delta,\varrho}(\lambda) = U(\lambda) = 1 - E(\lambda) - E(\lambda - 0)\,, \tag{5.27}$$

where

$$U_{\delta,\varrho}(\lambda) = \frac{1}{\pi}\left[\int_{-\varrho}^{-\delta} + \int_{\delta}^{\varrho}\right](H - \lambda - i\,\eta)^{-1}\,d\eta \tag{5.28}$$

$$= \frac{2}{\pi}\int_{\delta}^{\varrho}(H-\lambda)\,[(H-\lambda)^2 + \eta^2]^{-1}\,d\eta\,.$$

Proof. (5.27) is obviously true if applied to any $u \in \mathsf{R}(P(\lambda)) = \mathsf{N}(H-\lambda)$, for then $U_{\delta,\varrho}(\lambda)\,u = 0$ identically and $(1 - E(\lambda) - E(\lambda-0))\,u = 0$. Thus it suffices to prove that (5.27) is true when applied to any $u \in \mathsf{M}(\lambda-0)$ or $u \in \mathsf{M}(\lambda)^\perp$. Since the two cases can be treated similarly, we may assume $u \in \mathsf{M}(\lambda)^\perp$, that is, $E(\lambda)\,u = 0$, and prove

$$U_{\delta,\varrho}(\lambda)\,u \to u = (1 - E(\lambda) - E(\lambda - 0))\,u\,. \tag{5.29}$$

Now it is easy to see that

$$U_{\delta,\varrho}(\lambda) = \phi_{\delta,\varrho,\lambda}(H)\,, \tag{5.30}$$

where

$$\phi_{\delta,\varrho,\lambda}(\mu) = \frac{2}{\pi}\int_{\delta}^{\varrho}\frac{\mu-\lambda}{(\mu-\lambda)^2+\eta^2}\,d\eta \tag{5.31}$$

$$= \frac{2}{\pi}\left[\arctan\frac{\eta}{\mu-\lambda}\right]_{\eta=\delta}^{\eta=\varrho} \quad \text{for} \quad \mu \neq \lambda\,.$$

Thus $|\phi_{\delta,\varrho,\lambda}(\mu)| \leqq 1$, which implies $\|U_{\delta,\varrho}(\lambda)\| \leqq 1$. To prove (5.29) for $E(\lambda)\,u = 0$, it is therefore sufficient to prove it for any u such that $E(\lambda+\varepsilon)\,u = 0$ for some $\varepsilon > 0$ [such u form a dense subset of $\mathsf{M}(\lambda)^\perp$]. We have

$$(U_{\delta,\varrho}(\lambda)\,u, u) = \int_{\lambda+\varepsilon}^{\infty}\phi_{\delta,\varrho,\lambda}(\mu)\,d(E(\mu)\,u, u) \to$$

$$\to \int_{\lambda+\varepsilon}^{\infty} d(E(\mu)\,u, u) = (u, u)\,, \quad \delta \to 0\,, \quad \varrho \to \infty\,,$$

since $\phi_{\delta,\varrho,\lambda}(\mu) \to 1$ uniformly for $\mu \geqq \lambda + \varepsilon$. Hence $\|U_{\delta,\varrho}(\lambda)\,u - u\|^2 = \|U_{\delta,\varrho}(\lambda)\,u\|^2 + \|u\|^2 - 2\,\mathrm{Re}(U_{\delta,\varrho}(\lambda)\,u, u) \leqq 2\|u\|^2 - 2\,\mathrm{Re}(U_{\delta,\varrho}(\lambda)\,u, u) \to$ $\to 0$, which proves (5.29).

Problem 5.7. For any real λ and μ such that $\lambda < \mu$, we have

$$\frac{1}{2}\,[E(\mu) + E(\mu - 0)] - \frac{1}{2}\,[E(\lambda) + E(\lambda - 0)] = \text{s-}\lim_{\varepsilon\to+0}\frac{-1}{2\pi i}\int_{\Gamma_\varepsilon}(H-\zeta)^{-1}\,d\zeta\,, \tag{5.32}$$

where Γ_ε is the union of the two rectifiable curves Γ_ε^+ and Γ_ε^- defined in the following way: Γ_ε^+ is any curve in the upper half-plane beginning at $\mu + i\,\varepsilon$ and ending at $\lambda + i\,\varepsilon$, and Γ_ε^- is any curve in the lower half-plane beginning at $\lambda - i\,\varepsilon$ and ending at $\mu - i\,\varepsilon$.

Problem 5.8. $A \in \mathscr{B}(\mathsf{H})$ commutes with H if and only if A commutes with $E(\lambda)$ for all λ. In particular a subspace M of H reduces H if and only if the orthogonal projection P on M commutes with all $E(\lambda)$. This is in turn true if and only if M is invariant under all $E(\lambda)$.

Problem 5.9. Let $H = \int \lambda\, dE(\lambda)$. $\lambda \in \mathrm{P}(H)$ if and only if $E(\lambda + \varepsilon) - E(\lambda - \varepsilon) = 0$ for some $\varepsilon > 0$, and $\lambda \in \Sigma(H)$ if and only if $E(\lambda + \varepsilon) - E(\lambda - \varepsilon) \neq 0$ for any $\varepsilon > 0$.

3. The spectral theorem

We have shown above that every spectral family $\{E(\lambda)\}$ determines a selfadjoint operator by (5.12). The spectral theorem asserts that every selfadjoint operator H admits an expression (5.12) by means of a spectral family $\{E(\lambda)\}$ which is uniquely determined by H.

A natural proof of the spectral theorem is obtained by reversing the order of the argument given in the preceding paragraph. We *define* $E(\lambda)$ by (5.26), where $U(\lambda)$ is the partially isometric operator that appears in the polar decomposition $H - \lambda = U(\lambda)\,|H - \lambda|$ of $H - \lambda$. We have to show that the $E(\lambda)$ form a spectral family and that the H determined by (5.12) coincides with the given H.

First we note that $E(\lambda) = 1 - P_+(\lambda) = P_-(\lambda) + P_0(\lambda)$, where $P_\pm(\lambda)$, $P_0(\lambda)$ are the $P_\pm$, P_0 of (2.29) corresponding to H replaced by $H - \lambda$. Hence $((H - \lambda)\,u, u) \leqq 0$ for $u \in \mathsf{D}(H) \cap \mathsf{M}(\lambda)$ where $\mathsf{M}(\lambda) = \mathsf{R}(E(\lambda))$. If $\mu > \lambda$, $((H - \mu)\,u, u) \leqq 0$ a fortiori for such a u. Since $\mathsf{M}(\lambda)$ reduces H (see § 2.7), it follows from Lemma 2.38 that $\mathsf{M}(\lambda) \subset \mathsf{M}(\mu)$. This is equivalent to (5.1).

Since $\{E(\lambda)\}$ is thus a monotonic family of projections, the strong limits $E(\pm\infty) = \operatorname*{s-lim}_{\lambda\to\pm\infty} E(\lambda)$ exist [cf. (5.4)]. Since all the $\mathsf{M}(\lambda)$ reduce H, $\mathsf{M}(\pm\infty) = \mathsf{R}(E(\pm\infty))$ also reduce H and $\mathsf{D}(H) \cap \mathsf{M}(\pm\infty)$ is dense in $\mathsf{M}(\pm\infty)$. Let $u \in \mathsf{D}(H) \cap \mathsf{M}(-\infty)$. Then $u \in \mathsf{M}(\lambda)$ for all λ and so $((H - \lambda)\,u, u) \leqq 0$ for all $\lambda < 0$. Hence $(u, u) \leqq \lambda^{-1}(Hu, u)$ for all $\lambda < 0$ and u must be 0. This proves $\mathsf{M}(-\infty) = 0$ or $E(-\infty) = 0$. Similarly we can prove $E(\infty) = 1$.

Thus $\{E(\lambda)\}$ has been shown to form a spectral family. It remains to show that $E(\lambda)$ is right continuous. $E(\lambda + 0)$ is the projection on $\mathsf{M}(\lambda + 0)$, which is the intersection of all the $\mathsf{M}(\mu)$ for $\mu > \lambda$. Since the $\mathsf{M}(\mu)$ reduce H, the same is true of $\mathsf{M}(\lambda + 0)$. For every $u \in \mathsf{D}(H) \cap \mathsf{M}(\lambda + 0)$, we have $(Hu, u) \leqq \mu(u, u)$ for all $\mu > \lambda$ so that $(Hu, u) \leqq \lambda(u, u)$. It follows from Lemma 2.38 that $\mathsf{M}(\lambda + 0) \subset \mathsf{M}(\lambda)$. This shows that $E(\lambda + 0) \leqq E(\lambda)$. Since the opposite inequality is true, it follows that $E(\lambda + 0) = E(\lambda)$.

Finally we have to show that the selfadjoint operator $H' = \int \lambda\, dE(\lambda)$ coincides with H. Since both H and H' are selfadjoint and since the union of the ranges $\mathsf{M}(\lambda, \mu) = \mathsf{M}(\mu) \ominus \mathsf{M}(\lambda)$ of $E(\mu) - E(\lambda)$ is a core of H' (see Problem 5.2), it suffices to prove that $\mathsf{M}(\lambda, \mu) \subset \mathsf{D}(H)$ and $Hu = H'u$ for $u \in \mathsf{M}(\lambda, \mu)$.

To this end we first note that $\mathsf{M}(\lambda, \mu)$ reduces H and that $\lambda(u, u) \leqq (Hu, u) \leqq \mu(u, u)$ for u in $\mathsf{D}' = \mathsf{M}(\lambda, \mu) \cap \mathsf{D}(H)$. Thus H is bounded in the latter subset and, since D' is dense in $\mathsf{M}(\lambda, \mu)$, D' must coincide with $\mathsf{M}(\lambda, \mu)$ by the closedness of H. In other words, we have $\mathsf{M}(\lambda, \mu) \subset \mathsf{D}(H)$. Since H has upper bound μ and lower bound λ in $\mathsf{M}(\lambda, \mu)$, $H - (\lambda + \mu)/2$ has bound $(\mu - \lambda)/2$.

Now divide the interval $\mathrm{I} = (\lambda, \mu]$ into n equal subintervals $\mathrm{I}_1, \ldots, \mathrm{I}_n$ and let λ_k be the middle point of I_k, $k = 1, \ldots, n$. Define $E(\mathrm{I}_k)$ as in (5.6) and set $u_k = E(\mathrm{I}_k)\, u$. We have $u = u_1 + \cdots + u_n$ and the u_k are mutually orthogonal. Since each $E(\mathrm{I}_k)\,\mathsf{H}$ reduces H, Hu_k belongs to $E(\mathrm{I}_k)\,\mathsf{H}$ and so the vectors $(H - \lambda_k)\, u_k$ are also mutually orthogonal. Furthermore, we have $\|(H - \lambda_k)\, u_k\| \leqq (\mu - \lambda)\, \|u_k\|/2n$ by the remark given above. Thus

$$(5.33)\quad \|Hu - \sum_k \lambda_k u_k\|^2 = \|\sum_k (H - \lambda_k)\, u_k\|^2 = \sum_k \|(H - \lambda_k)\, u_k\|^2 \leqq \frac{(\mu - \lambda)^2}{4n^2} \sum_k \|u_k\|^2 = \frac{(\mu - \lambda)^2}{4n^2} \|u\|^2 .$$

On letting $n \to \infty$ we obtain $\lim \sum_k \lambda_k E(\mathrm{I}_k)\, u = Hu$. This implies that $(H'u, v) = \int \lambda (dE(\lambda)\, u, v) = \lim \sum_k \lambda_k (E(\mathrm{I}_k)\, u, v) = (Hu, v)$ for every $v \in \mathsf{H}$, hence $H'u = Hu$ as we wished to show.

4. Stability theorems for the spectral family

Since the spectral family $\{E(\lambda)\}$ associated with a given selfadjoint operator H is a very important quantity, a question suggests itself whether $E(\lambda)$ depends on H continuously in some sense or other[1]. It is clear that in general the answer is no; for even when the underlying space H is finite-dimensional, $E_\varkappa(\lambda)$ for the operator $H_\varkappa = H + \varkappa$ will be discontinuous at a value of $\varkappa$ for which $H + \varkappa$ has an eigenvalue λ.

But this kind of discontinuity of $E(\lambda)$ as a function of H is inherently related to the discontinuity of $E(\lambda)$ as a function of λ. It is reasonable to expect that $E(\lambda)$ will change continuously with H if λ is a point of continuity of $E(\lambda)$. It turns out that this may or may not be true according to the meaning of "continuously"[2].

[1] In what follows $\{E(\lambda)\}$, $\{E'(\lambda)\}$, $\{E_n(\lambda)\}$ etc. denote respectively the spectral families for the selfadjoint operators denoted by H, H', H_n, etc.

[2] In this paragraph we are concerned with the continuity of $E(\lambda)$ *in norm*. The continuity of $E(\lambda)$ in the strong sense will be considered in Chapter VIII.

The simplest result of this kind is obtained when we consider the case in which $\Sigma(H)$ has gaps at $\lambda = \alpha$ and $\lambda = \beta$, $\alpha < \beta$ (see V-§ 3.5). Then α and β are points of constancy with respect to $\{E(\lambda)\}$. It follows from (5.32) that

$$E(\beta) - E(\alpha) = -\frac{1}{2\pi i}\int_{\Gamma} (H-\zeta)^{-1}\, d\zeta\,, \tag{5.34}$$

where Γ is, say, a closed, positively oriented circle with a diameter connecting α, β. (5.34) is exactly the orthogonal projection P on the subspace M' corresponding to the part Σ' of $\Sigma(H)$ enclosed in Γ [see V-(3.18)]. We know that such a part Σ' of the spectrum is upper-semicontinuous and that P changes continuously if H changes continuously in the sense of generalized convergence (see Theorem IV-3.16). Thus we have

Theorem 5.10. *Let H be selfadjoint and let $\Sigma(H)$ have gaps at α, $\beta\,(\alpha < \beta)$. Then there is a $\delta > 0$ such that for any selfadjoint operator H' with $\hat\delta(H', H) < \delta$, $\Sigma(H')$ has gaps at α, β and*

$$\|[E'(\beta) - E'(\alpha)] - [E(\beta) - E(\alpha)]\| \to 0 \tag{5.35}$$

if $\hat\delta(H', H) \to 0$.

Remark 5.11. $\hat\delta(H', H)$ will be small if $H' = H + A$ where A is symmetric and H-bounded with small coefficients a, b in V-(4.1) (see Theorem IV-2.14). In this case, furthermore, H' is bounded from below if H is (see Theorem V-4.11) and the lower bound of H' tends to that of H when $a, b \to 0$. If α is chosen sufficiently small (algebraically), therefore, we have $E'(\alpha) = E(\alpha) = 0$ in (5.35). Thus $\|E'(\beta) - E(\beta)\| \to 0$, $a, b \to 0$, if $\Sigma(H)$ has a gap at β. It is a remarkable fact, however, that this is true even if H is not semibounded. We have namely

Theorem 5.12.[1] *Let H be selfadjoint and let A_n be symmetric and H-bounded: $\|A_n u\| \leqq a_n\|u\| + b_n\|Hu\|$, $u \in \mathsf{D}(H) \subset \mathsf{D}(A_n)$, where $a_n \to 0$, $b_n \to 0$ as $n \to \infty$. Then $H_n = H + A_n$ is selfadjoint for sufficiently large n and $H_n \to H$ in the generalized sense. If $\Sigma(H)$ has a gap at β, then $\Sigma(H_n)$ has a gap at β for sufficiently large n and $\|E_n(\beta) - E(\beta)\| \to 0$, $n \to \infty$.*

This theorem can further be generalized to[2]

Theorem 5.13. *Let H be selfadjoint, and let A_n be symmetric with $\mathsf{D}(A_n) \subset \mathsf{D}(H)$. Assume that*

$$|(A_n u, u)| \leqq a_n(u, u) + b_n(|H|\, u, u)\,, \quad u \in \mathsf{D}(A_n)\,, \tag{5.36}$$

where $a_n \to 0$, $b_n \to 0$. If $\mathsf{D}(A_n)$ is a core of $|H|^{1/2}$ for each n, the pseudo-Friedrichs extension H_n of $H + A_n$ is defined for sufficiently large n and

---

[1] This theorem was proved by HEINZ [1]. The proof of the generalized Theorem 5.13 given below is essentially due to HEINZ.

[2] Note that (5.36) is satisfied if A_n is H-bounded as in Theorem 5.12 [set $\mathsf{D}(A_n) = \mathsf{D}(H)$ and use Problem 3.14].

$H_n \to H$ in the generalized sense. If $\Sigma(H)$ has a gap at β, then $\Sigma(H_n)$ has a gap at β for sufficiently large n and $\|E_n(\beta) - E(\beta)\| \to 0$, $n \to \infty$.

Proof. It was proved in Corollary 3.12 that the H_n exist and $H_n \to H$ in the generalized sense.

To simplify the proof, we note that the $|H|$ on the right of (5.36) may be replaced by $|H - \beta|$, with a possible change of a_n, b_n, for $(|H|\, u, u) \leqq |\beta|\, (u, u) + (|H - \beta|\, u, u)$ by (2.28). This means that we may assume $\beta = 0$ without loss of generality. Then $|H|$ has a positive lower bound, so that we may further assume that $a_n = 0$, with a change of b_n if necessary. Then we can set $\delta = 0$ in the construction of the pseudo-Friedrichs extension given in the proof of Theorem 3.11, for δ was introduced to make H' have a positive lower bound. With these simplifications, we shall use the notations of the proof of Theorem 3.11; in particular we have $H' = |H|$.

Now (3.22) gives

$$(5.37) \qquad \begin{aligned} R_n(\zeta) &= G'^{-1} H' R(\zeta)\, [1 + C_n H' R(\zeta)]^{-1} G'^{-1} \\ &= R(\zeta) + G'^{-1} K(\zeta) \sum_{p=1}^{\infty} [-C_n K(\zeta)]^p G'^{-1}, \end{aligned}$$

where

$$(5.38) \qquad K(\zeta) = H' R(\zeta) = |H|\, (H - \zeta)^{-1}.$$

The generalized convergence $H_n \to H$ already implies that $\Sigma(H_n)$ has a gap at β (see Theorem IV-2.25). Thus we have by Lemma 5.6

$$1 - 2E_n(0) = U_n(0) = \operatorname*{s-lim}_{\varrho \to \infty} \frac{1}{\pi} \int_{-\varrho}^{\varrho} R_n(i\,\eta)\, d\eta$$

and a similar expression for $1 - 2E(0)$. In virtue of (5.37), we have for any $u, v \in \mathsf{H}$

$$(5.39) \qquad \begin{aligned} -2\pi([E_n(0) - E(0)]\, u, v) &= \lim_{\varrho \to \infty} \int_{-\varrho}^{\varrho} ([R_n(i\,\eta) - R(i\,\eta)]\, u, v)\, d\eta \\ &= \lim_{\varrho \to \infty} \int_{-\varrho}^{\varrho} (G'^{-1} K(i\,\eta) \sum_{p=1}^{\infty} [-C_n K(i\,\eta)]^p G'^{-1} u, v)\, d\eta\,. \end{aligned}$$

Let us estimate the last integral. We have $\|K(i\,\eta)\| = \||H|\,(H - i\,\eta)^{-1}\| \leqq 1$. Since $\|C_n\| \leqq b_n$ by (3.20), the integrand on the right of (5.39) is majorized by the series

$$\begin{aligned} &\sum_{p=1}^{\infty} b_n^p \|K(i\,\eta)\, G'^{-1} u\|\, \|K(i\,\eta)^* G'^{-1} v\| \\ &\qquad = \frac{b_n}{1 - b_n} \|G' R(i\,\eta)\, u\|\, \|G' R(i\,\eta)^* v\| \end{aligned}$$

(note that $|H| = H' = G'^2$).

Hence the integral in (5.39) is majorized by

$$\frac{b_n}{1-b_n}\left(\int_{-\infty}^{\infty}\|G' R(i\eta)\, u\|^2\, d\eta\right)^{1/2}\left(\int_{-\infty}^{\infty}\|G' R(-i\eta)\, v\|^2\, d\eta\right)^{1/2}.$$

But

$$\|G' R(i\eta)\, u\|^2 = (G'^2 R(i\eta)\, R(-i\eta)\, u, u) = (|H|\, (H^2+\eta^2)^{-1} u, u)\,,$$

$$\int_{-\infty}^{\infty}\|G' R(i\eta)\, u\|^2\, d\eta = \int_0^{\infty}((H^2+\lambda)^{-1} u, |H|\, u)\, \lambda^{-1/2}\, d\lambda$$

$$= \pi((H^2)^{-1/2} u, |H|\, u) = \pi(|H|^{-1} u, |H|\, u) = \pi\|u\|^2$$

[see V-(3.43) and note that $|H| = (H^2)^{1/2}$ by definition]. It follows that

$$2\pi\, |([E_n(0) - E(0)]\, u, v)| \leqq \frac{b_n}{1-b_n}\, \pi\|u\|\, \|v\|\,, \quad \text{or}$$

$$\|E_n(\beta) - E(\beta)\| \leqq \frac{b_n}{2(1-b_n)} \to 0\,, \quad n\to\infty\,.$$

Remark 5.14. $\|E_n(\beta) - E(\beta)\| \to 0$ is not necessarily true when $H_n \to H$ in the generalized sense. A counter-example is given by Example V-4.14, where $T(\varkappa)$ is selfadjoint for real $\varkappa$ and $\dim E_\varkappa(0) = 1$ for $0 < \varkappa < 1$ [because $T(\varkappa)$ has one negative eigenvalue] but $E_0(0) = 0$ [because $T(0)$ is positive].

Chapter Seven

Analytic perturbation theory

The theory of analytic perturbation is historically the first subject discussed in perturbation theory. It is mainly concerned with the behavior of isolated eigenvalues and eigenvectors (or eigenprojections) of an operator depending on a parameter holomorphically.

We have already studied this problem in the special case in which the basic space is finite-dimensional. Formally there is very little that is new in the general case discussed in this chapter. Once the notion of holomorphic dependence of an (in general unbounded) operator on a parameter is introduced, it is rather straightforward to show that the results obtained in the finite-dimensional case can be extended, at least for isolated eigenvalues, without essential modification.

Thus the main problems in this chapter are the definition of holomorphic families of operators and various sufficient conditions for a given family to be holomorphic. A general definition can be obtained in a natural way from the theory of generalized convergence discussed in Chapter IV. There are several useful criteria for a given family to be holomorphic. We thus consider different types of holomorphic families: (A), (B), (B_0), and (C). Type (A) is defined by the relative boundedness of the perturbation with respect to the unperturbed operator. Type (B) is defined in a Hilbert space in terms of a holomorphic family of sesquilinear forms, where the perturbing form is relatively bounded with respect to the unperturbed form. Type (B_0) is a special case of (B) and is related to the notion of Friedrichs extension. Type (C) is formally similar to (B_0) but differs from it in many respects. It is a matter of course that each of these special types has its own special properties.

§ 1. Analytic families of operators

1. Analyticity of vector- and operator-valued functions

In what follows we are concerned with a family $u(\varkappa)$ of vectors in a Banach space X or a family $T(\varkappa)$ of operators from X to another Banach space Y, and we are particularly interested in the case in which $u(\varkappa)$ or $T(\varkappa)$ is holomorphic in $\varkappa$ in a domain D of the complex plane. We have already defined $u(\varkappa)$ to be holomorphic if it is differentiable at each $\varkappa \in D$; it is immaterial whether the differentiation is taken in the strong or weak sense (see III-§ 1.6). Thus $u(\varkappa)$ is holomorphic if and only if $(u(\varkappa), f)$ is holomorphic in $\varkappa$ for every $f \in X^*$. For applications, the following criterion is often convenient: *$u(\varkappa)$ is holomorphic in $\varkappa$ if and only if each $\varkappa \in D$ has a neighborhood in which $\|u(\varkappa)\|$ is bounded and $(u(\varkappa), f)$ is holomorphic for all f in a fundamental subset of* X^* (see Remark III-1.38).

Sometimes we consider a family $u(\varkappa)$ depending on a real parameter $\varkappa$, $a < \varkappa < b$. $u(\varkappa)$ is *real-holomorphic* if $u(\varkappa)$ admits a Taylor expansion at each $\varkappa$. In such a case $u(\varkappa)$ can be extended by the Taylor series to complex values of $\varkappa$ in some neighborhood D of the real axis. The extended $u(\varkappa)$ is a holomorphic family for $\varkappa \in D$.

Example 1.1. Let $X = L^p(0, \infty)$, $1 \leq p < \infty$, and let $u(\varkappa) = u(x; \varkappa) = e^{-\varkappa x}$. In order that $u(\varkappa) \in X$, $\varkappa$ must be restricted to the half-plane $\operatorname{Re} \varkappa > 0$. Since $du(\varkappa)/d\varkappa = -x\, u(\varkappa) \in X$, $u(\varkappa)$ is holomorphic for $\operatorname{Re} \varkappa > 0$. If we take $X = L^p(a, b)$ for a finite interval (a, b), $u(\varkappa) = e^{-\varkappa x}$ is holomorphic for *all* $\varkappa$. Next set $u(\varkappa) = u(x; \varkappa) = (x - \varkappa)^{-1}$ and $X = L^p(a, b)$, (a, b) being finite or infinite. $u(\varkappa)$ is holomorphic in the domain obtained from the whole plane by removing the points $a \leq \varkappa \leq b$ on the real axis.

In considering an operator-valued holomorphic function $T(\varkappa)$, we first restrict ourselves to the case $T(\varkappa) \in \mathscr{B}(X, Y)$. $T(\varkappa)$ is holomorphic if it is differentiable in norm for all $\varkappa$ in a complex domain. Again we have the criterion: *$T(\varkappa) \in \mathscr{B}(X, Y)$ is holomorphic if and only if each $\varkappa$ has a neighborhood in which $T(\varkappa)$ is bounded and $(T(\varkappa)\, u, g)$ is holomorphic for every u in a fundamental subset of X and every g in a fundamental subset of* Y^* (see III-§ 3.1).

We note also that if $T(\varkappa) \in \mathscr{B}(X, Y)$ is holomorphic and $T(\varkappa_0)^{-1} \in \mathscr{B}(Y, X)$ exists, then $T(\varkappa)^{-1}$ exists, belongs to $\mathscr{B}(Y, X)$ and is holomorphic for sufficiently small $|\varkappa - \varkappa_0|$. This is a consequence of the stability theorem for bounded invertibility (see the Neumann series expansion of the inverse in Theorem IV-1.16); we have $(d/d\varkappa)\, T(\varkappa)^{-1} = -T(\varkappa)^{-1}(dT(\varkappa)/d\varkappa)\, T(\varkappa)^{-1}$ by I-(4.28).

A real-holomorphic family $T(\varkappa)$ can be defined as in the case of a vector-valued, real-holomorphic family $u(\varkappa)$; $T(\varkappa)$ can be extended to a holomorphic family on some domain D of the complex plane.

2. Analyticity of a family of unbounded operators

For applications it is not sufficient to consider bounded operators only, and the notion of analytic dependence of a family of operators on a parameter needs to be generalized to unbounded operators. In order to distinguish the current sense of holomorphy from the extended sense to be introduced in a moment, we shall say $T(\varkappa)$ is *bounded-holomorphic* if $T(\varkappa) \in \mathscr{B}(X, Y)$ and $T(\varkappa)$ is holomorphic in the sense so far considered.

The extension to unbounded operators is suggested by the notion of generalized convergence introduced in IV-§ 2.4. A family of operators $T(\varkappa) \in \mathscr{C}(X, Y)$ defined in a neighborhood of $\varkappa = 0$ is said to be *holomorphic* at $\varkappa = 0$ (in the generalized sense) if there is a third Banach space Z and two families of operators $U(\varkappa) \in \mathscr{B}(Z, X)$ and $V(\varkappa) \in \mathscr{B}(Z, Y)$ which are bounded-holomorphic at $\varkappa = 0$ such that $U(\varkappa)$ maps Z *onto* $D(T(\varkappa))$ one to one and

$$T(\varkappa)\, U(\varkappa) = V(\varkappa)\,. \tag{1.1}$$[1]

$T(\varkappa)$ is holomorphic in a domain D of the complex plane if it is holomorphic at every $\varkappa$ of D. If $T(\varkappa)$ is holomorphic, then it is continuous in $\varkappa$ in the generalized sense: $T(\varkappa) \to T(\varkappa_0)$ in the generalized sense if $\varkappa \to \varkappa_0$, as is seen from Theorem IV-2.29.

The new notion is a generalization of the old one: $T(\varkappa) \in \mathscr{B}(X, Y)$ is holomorphic in the new sense if and only if it is bounded-holomorphic. To prove the "only if" part, it suffices to note that the $U(\varkappa)$ of (1.1) maps Z onto X so that $U(\varkappa)^{-1} \in \mathscr{B}(Y, X)$ (see Problem III-5.21). Then $U(\varkappa)^{-1}$ is bounded-holomorphic (see par. 1), and the same is true of $T(\varkappa) = V(\varkappa)\, U(\varkappa)^{-1}$.

We note also that if $T(\varkappa)$ is holomorphic and $T(\varkappa_0) \in \mathscr{B}(X, Y)$, then $T(\varkappa) \in \mathscr{B}(X, Y)$ for sufficiently small $|\varkappa - \varkappa_0|$ [so that $T(\varkappa)$ is bounded-holomorphic for such $\varkappa$]. This is a consequence of Theorem IV-2.23, for $T(\varkappa)$ is continuous in $\varkappa$ in the sense of generalized convergence.

The notion of a real-holomorphic family $T(\varkappa)$ could be defined in the same way by requiring only that $U(\varkappa)$, $V(\varkappa)$ be real-holomorphic.

---

[1] This definition corresponds to Theorem IV-2.29, which gives a sufficient condition for generalized convergence. It is a formal generalization of a definition given by RELLICH [3], in which Z = X = Y is assumed.

But it is in general impossible to extend a real-holomorphic family to a complex-holomorphic family. It is true that $U(\varkappa)$ and $V(\varkappa)$ can be extended to complex $\varkappa$, but $U(\varkappa)$ may not be a one-to-one mapping for nonreal $\varkappa$ [1].

Problem 1.2. If $T(\varkappa)$ is holomorphic and $A(\varkappa)$ is bounded-holomorphic, $T(\varkappa) + A(\varkappa)$ is holomorphic.

Theorem 1.3.[2] *Let $T(\varkappa) \in \mathscr{C}(X)$ be defined in a neighborhood of $\varkappa = 0$ and let $\zeta \in P(T(0))$. Then $T(\varkappa)$ is holomorphic at $\varkappa = 0$ if and only if $\zeta \in P(T(\varkappa))$ and the resolvent $R(\zeta, \varkappa) = (T(\varkappa) - \zeta)^{-1}$ is bounded-holomorphic for sufficiently small $|\varkappa|$. $R(\zeta, \varkappa)$ is even bounded-holomorphic in the two variables on the set of all $\zeta, \varkappa$ such that $\zeta \in P(T(0))$ and $|\varkappa|$ is sufficiently small (depending on ζ).*

Proof. The proof is similar to the corresponding one for generalized convergence (see IV-§ 2.6). Suppose $T(\varkappa)$ is holomorphic and let $U(\varkappa)$, $V(\varkappa)$ be as in (1.1). We have $(T(\varkappa) - \zeta)\, U(\varkappa) = V(\varkappa) - \zeta U(\varkappa)$, and $V(0) - \zeta U(0)$ maps Z onto X one to one. Therefore its inverse belongs to $\mathscr{B}(X, Z)$ and $[V(\varkappa) - \zeta U(\varkappa)]^{-1}$ is bounded-holomorphic near $\varkappa = 0$ (see a similar argument above). Thus $(T(\varkappa) - \zeta)^{-1} = U(\varkappa)\, [V(\varkappa) - \zeta U(\varkappa)]^{-1}$ is bounded-holomorphic. Conversely, suppose that $R(\zeta, \varkappa)$ is bounded-holomorphic in $\varkappa$. We need only to set $Z = X$, $U(\varkappa) = R(\zeta, \varkappa)$, $V(\varkappa) = 1 + \zeta U(\varkappa)$ to satisfy (1.1). That $R(\zeta, \varkappa)$ is holomorphic in the two variables follows from Theorem IV-3.12, according to which $R(\zeta, \varkappa)$ is holomorphic in ζ and $R(\zeta_0, \varkappa)$.

The holomorphy of $T(\varkappa)$ as defined here is peculiar in that a $T(\varkappa)$ may be holomorphic for *all* complex $\varkappa$, including $\varkappa = \infty$ [3], without being constant. An example is given by

Example 1.4. Let $T(\varkappa)$ be the ordinary differential operator on the interval $(0, 1)$ given by $T(\varkappa)\, u = -u''$ with the boundary condition $u(0) = 0$, $u'(1) + \varkappa u(1) = 0$. $T(\varkappa)$ depends on $\varkappa$ only through the boundary condition at $x = 1$. It is defined even for $\varkappa = \infty$; then the boundary condition is $u(1) = 0$ by definition. The resolvent $R(\zeta, \varkappa)$ is an integral operator with kernel (the Green function) $g(y, x; \zeta, \varkappa)$ given by

$$g(y, x; \zeta, \varkappa) = \frac{\sin \zeta^{\frac{1}{2}} y\, [\cos \zeta^{\frac{1}{2}} (1 - x) + \varkappa \zeta^{-\frac{1}{2}} \sin \zeta^{\frac{1}{2}} (1 - x)]}{\zeta^{\frac{1}{2}} \cos \zeta^{\frac{1}{2}} + \varkappa \sin \zeta^{\frac{1}{2}}} \tag{1.2}$$

---

[1] For example let $T(\varkappa) = (U - \varkappa)^{-1}$ for real $\varkappa$, where $U \in \mathscr{B}(X)$. Suppose U has a set of nonreal eigenvalues dense in a complex neighborhood of 0. Then $T(\varkappa)$ is real-holomorphic [as is seen by taking $U(\varkappa) = U - \varkappa$, $V(\varkappa) = 1$], but it cannot be extended to complex $\varkappa$. In fact $0 \in P(T(0))$ since $T(0)^{-1} = U \in \mathscr{B}(X)$; hence $0 \in P(T(\varkappa))$ and $T(\varkappa)^{-1}$ is bounded-holomorphic for sufficiently small $|\varkappa|$ if such an extension exists (Theorem 1.3). Then we must have $T(\varkappa)^{-1} = U - \varkappa$ since this is true for real $\varkappa$. But this is a contradiction since $U - \varkappa$ is not invertible if $\varkappa$ is an eigenvalue.

[2] This theorem is due to RELLICH [3] when $T(\varkappa)$ is a selfadjoint family (see § 3).

[3] As usual, $T(\varkappa)$ is holomorphic at $\varkappa = \infty$ if $T(1/\varkappa)$ is holomorphic at $\varkappa = 0$ with a suitable definition of $T(\infty)$.

for $0 \leqq y \leqq x \leqq 1$, with x, y exchanged if $x \leqq y$. For a fixed ζ, (1.2) is holomorphic in $\varkappa$ with the single exception of a pole at $\varkappa = -\zeta^{1/2} \cot \zeta^{1/2}$. For any given $\varkappa_0$, a ζ can be found such that $\varkappa$ is not a pole if it is near enough to $\varkappa_0$. According to Theorem 1.3, therefore, $T(\varkappa)$ is holomorphic in the whole complex plane including $\varkappa = \infty$.

Example 1.5. Similarly, let $T(\varkappa)$ be the operator $T(\varkappa) u = -u''$ with the boundary condition $u(0) = 0$, $\varkappa u'(1) - u(1) = 0$. $T(\varkappa)$ is identical with the $T(-\varkappa^{-1})$ of Example 1.4 and is holomorphic for all $\varkappa$. $R(\zeta, \varkappa)$ is given by the Green function V-(4.16).

Remark 1.6. Does the holomorphic function $T(\varkappa)$ defined above possess the *unique continuation property*? In other words, let $T_1(\varkappa)$ and $T_2(\varkappa)$ be holomorphic in a (connected) domain D and let $T_1(\varkappa_n) = T_2(\varkappa_n)$ for a sequence $\{\varkappa_n\}$ converging to a point $\varkappa_0 \in$ D different from any of the $\varkappa_n$. Then is it true that $T_1(\varkappa) = T_2(\varkappa)$ for all $\varkappa \in$ D? It is not clear whether this is true in general. But it is if Y = X and $T_1(\varkappa)$, $T_2(\varkappa)$ have nonempty resolvent sets for every $\varkappa \in$ D.

To see this, we first note that $T_1(\varkappa_n) = T_2(\varkappa_n) = T_n$ implies $T_1(\varkappa_0) = T_2(\varkappa_0) \equiv T_0$ in virtue of the uniqueness of the generalized limit. Let $\zeta_0 \in \mathrm{P}(T_0)$; then $R(\zeta_0, T_1(\varkappa))$ and $R(\zeta_0, T_2(\varkappa))$ exist and are bounded-holomorphic in a neighborhood of $\varkappa_0$ by Theorem 1.3. Since the unique continuation theorem holds for bounded-holomorphic functions just as for numerical functions [it is sufficient to consider the function $(R(\zeta_0, T_1(\varkappa)) u, f)$ for fixed $u \in$ X and $f \in$ X*], we conclude that $R(\zeta_0, T_1(\varkappa)) = R(\zeta_0, T_2(\varkappa))$, and therefore $T_1(\varkappa) = T_2(\varkappa)$, in this neighborhood.

It is now easy to show that $T_1(\varkappa) = T_2(\varkappa)$ holds for all $\varkappa \in$ D. This extension of the domain of validity of $T_1(\varkappa) = T_2(\varkappa)$ from a neighborhood to the whole D follows the standard pattern[1] and may be omitted.

3. Separation of the spectrum and finite systems of eigenvalues

Theorem 1.7. *If a family $T(\varkappa) \in \mathscr{C}(\mathrm{X})$ depending on $\varkappa$ holomorphically has a spectrum consisting of two separated parts, the subspaces of* X *corresponding to the separated parts also depend on $\varkappa$ holomorphically.*

Comment and proof. To be precise, this theorem means the following. Let $T(\varkappa)$ be holomorphic in $\varkappa$ near $\varkappa = 0$ and let $\Sigma(0) = \Sigma(T(0))$ be separated into two parts $\Sigma'(0)$, $\Sigma''(0)$ by a closed curve Γ in the manner described in III-§ 6.4. Since $T(\varkappa)$ converges to $T(0)$ as $\varkappa \to 0$ in the generalized sense, it follows from Theorem IV-3.16 that $\Gamma \subset \mathrm{P}(\varkappa) = \mathrm{P}(T(\varkappa))$ for sufficiently small $|\varkappa|$ and $\Sigma(\varkappa) = \Sigma(T(\varkappa))$ is likewise separated by Γ into two parts $\Sigma'(\varkappa)$, $\Sigma''(\varkappa)$ with the associated decomposition $\mathrm{X} = \mathrm{M}'(\varkappa) \oplus \mathrm{M}''(\varkappa)$ of the space. The projection on $\mathrm{M}'(\varkappa)$ along

---

[1] See Knopp [1], p. 87.

$M''(\varkappa)$ is given by

$$P(\varkappa) = -\frac{1}{2\pi i}\int_\Gamma R(\zeta, \varkappa)\, d\zeta \tag{1.3}$$

by IV-(3.11). Since $R(\zeta, \varkappa)$ is bounded-holomorphic in ζ and $\varkappa$ jointly by Theorem 1.3, it follows from (1.3) that $P(\varkappa)$ is bounded-holomorphic near $\varkappa = 0$. This is what is meant by saying that the subspaces $M'(\varkappa)$, $M''(\varkappa)$ depend on $\varkappa$ holomorphically.

When $\Sigma(\varkappa)$ is separated as in Theorem 1.7, the eigenvalue problem for $T(\varkappa)$ is reduced to the eigenvalue problems for the parts of $T(\varkappa)$ in the two subspaces $M'(\varkappa)$, $M''(\varkappa)$. But it is somewhat inconvenient that these subspaces depend on $\varkappa$. This inconvenience can be avoided by the following device.

Since the projection $P(\varkappa)$ on $M'(\varkappa)$ is holomorphic in $\varkappa$, there exists a transformation function $U(\varkappa) \in \mathscr{B}(X)$ which is bounded-holomorphic together with its inverse $U(\varkappa)^{-1} \in \mathscr{B}(X)$ and which transforms $P(0)$ into $P(\varkappa)$:

$$P(\varkappa) = U(\varkappa)\, P(0)\, U(\varkappa)^{-1}; \tag{1.4}$$

see II-§ 4.2, the results of which are valid in a Banach space X as well. Since $T(\varkappa)$ commutes with $P(\varkappa)$ (see III-§ 6.4), it follows that the operator

$$\check{T}(\varkappa) = U(\varkappa)^{-1}\, T(\varkappa)\, U(\varkappa) \tag{1.5}$$

commutes with $P(0)$. Thus the pair $M'(0) = P(0)X$ and $M''(0) = (1 - P(0))X$ decomposes $\check{T}(\varkappa)$ for all $\varkappa$. But the eigenvalue problem for the part of $T(\varkappa)$ in $M'(\varkappa)$ is equivalent to that for the part of $\check{T}(\varkappa)$ in $M'(0)$, for the eigenvalues (if any) are the same for the two parts and the eigenprojections and eigennilpotents for them are related with each other through transformation by $U(\varkappa)$.

Let us describe this relationship in more detail under the assumption that $\Sigma'(0)$ consists of a finite system of eigenvalues (III-§ 6.5). Then $M'(0)$ is finite-dimensional and the same is true of $M'(\varkappa)$ by (1.4). Let

$$\check{T}(\varkappa)\, \check{P}_h(\varkappa) = \lambda_h(\varkappa)\, \check{P}_h(\varkappa) + \check{D}_h(\varkappa), \quad h = 1, \ldots, s, \tag{1.6}$$

be the solution of the eigenvalue problem for the part of $\check{T}(\varkappa)$ in $M'(0)$, where

$$\begin{aligned} \check{P}_h(\varkappa)\, P(0) &= P(0)\, \check{P}_h(\varkappa) = \check{P}_h(\varkappa), \\ \check{D}_h(\varkappa)\, P(0) &= P(0)\, \check{D}_h(\varkappa) = \check{D}_h(\varkappa) \end{aligned} \tag{1.7}$$

[see I-(5.35)]. Then the solution for $T(\varkappa)$ in $M'(\varkappa)$ is given by

$$T(\varkappa)\, P_h(\varkappa) = \lambda_h(\varkappa)\, P_h(\varkappa) + D_h(\varkappa), \quad h = 1, \ldots, s, \tag{1.8}$$

with

(1.9) $$P_h(\varkappa) = U(\varkappa)\, \check{P}_h(\varkappa)\, U(\varkappa)^{-1}, \quad D_h(\varkappa) = U(\varkappa)\, \check{D}_h(\varkappa)\, U(\varkappa)^{-1}.$$

Now all the results of Chapter II on the analytic perturbation theory in a finite-dimensional space can be applied to the eigenvalue problem for $\check{T}(\varkappa)$ in the fixed subspace $\mathsf{M}'(0)$. In this way the study of a finite system of eigenvalues for $T(\varkappa)$ has been reduced to that of a problem in a finite-dimensional space. Note that the part of $\check{T}(\varkappa)$ in $\mathsf{M}'(0)$ is equal to the part in $\mathsf{M}'(0)$ of the *bounded-holomorphic* function $\check{T}(\varkappa)\, P(0) = U(\varkappa)^{-1}\, T(\varkappa)\, P(\varkappa)\, U(\varkappa)$; $T(\varkappa)\, P(\varkappa)$ is bounded-holomorphic as is seen from III-(6.24).

Recalling the results of II-§ 1.8, we have thus proved

Theorem 1.8.[1] *If $T(\varkappa)$ is holomorphic in $\varkappa$ near $\varkappa = 0$, any finite system of eigenvalues $\lambda_h(\varkappa)$ of $T(\varkappa)$ consists of branches of one or several analytic functions which have at most algebraic singularities near $\varkappa = 0$. The same is true of the corresponding eigenprojections $P_h(\varkappa)$ and eigennilpotents $D_h(\varkappa)$.*

More detailed properties of these functions can be read from the results of II-§ 1, all of which are valid as long as we are concerned with a *finite system* of eigenvalues.

As an application, let us prove some theorems on "nonlinear eigenvalue problems"[2].

Theorem 1.9. *Let $T(\varkappa)$ be a family of compact operators in X holomorphic for $\varkappa \in \mathrm{D}_0$. Call $\varkappa$ a singular point if 1 is*[3] *an eigenvalue of $T(\varkappa)$. Then either*[4] *all $\varkappa \in \mathrm{D}_0$ are singular points or there are only a finite number of singular points in each compact subset of D_0.*

Proof. Let $\varkappa_0 \in \mathrm{D}_0$. If $\varkappa_0$ is not singular, $1 \in \mathrm{P}(T(\varkappa_0))$ since $T(\varkappa_0)$ is compact. Then $1 \in \mathrm{P}(T(\varkappa))$ for sufficiently small $|\varkappa - \varkappa_0|$, by Theorem 1.3. If $\varkappa_0$ is singular, 1 is an isolated eigenvalue of $T(\varkappa_0)$ with finite multiplicity (see III-§ 6.7). It follows from Theorem 1.8 that $T(\varkappa)$ has only a finite system of eigenvalues in some neighborhood of 1 if $|\varkappa - \varkappa_0|$ is small and that these eigenvalues have at most an algebraic singularity at $\varkappa = \varkappa_0$. If some of these eigenvalues are identically equal to 1, all the $\varkappa$ near $\varkappa_0$ are singular. Otherwise there is no singular point $\varkappa \neq \varkappa_0$ if $|\varkappa - \varkappa_0|$ is sufficiently small.

[1] This theorem was proved by Sz.-Nagy [2], Wolf [1], T. Kato [6]. See also Baumgärtel [1] (which contains the most complete description of the analytic functions $\lambda_h(\varkappa)$ etc.), Schäfke [1]—[5].

[2] See footnote [1] of p. 35. If $T(\varkappa) = \varkappa T$ where T is compact, a singular point $\varkappa$ in Theorem 1.9 is the reciprocal of an eigenvalue of T. This is the reason why we speak of a "nonlinear eigenvalue problem". Theorem 1.9 is due to Atkinson [2] (see also Sz.-Nagy [4]) where $T(\varkappa)$ is assumed to be a polynomial in $\varkappa$.

[3] The value 1 has no particular meaning; it can be replaced by any $\alpha \neq 0$.

[4] The first alternative is excluded if $\|T(\varkappa_0)\| < 1$ for some $\varkappa_0 \in \mathrm{D}_0$.

Summing up, for each $\varkappa_0 \in D_0$ there is a neighborhood D such that either every $\varkappa \in D$ is singular or no $\varkappa \in D$ is singular except $\varkappa_0$. The result follows immediately.

Theorem 1.10. *Let $T(\varkappa) \in \mathscr{C}(X)$ be a holomorphic family for $\varkappa \in D_0$ and let $T(\varkappa)$ have compact resolvent for each $\varkappa$. Call $\varkappa$ a singular point if* 0 *is*[1] *an eigenvalue of $T(\varkappa)$. Then the same result holds as in* Theorem 1.9.

Proof. For each $\varkappa_0$, there is a nonzero $\zeta_0 \in P(T(\varkappa_0))$ and $\zeta_0 R(\zeta_0, \varkappa)$ is compact and holomorphic in $\varkappa$ near $\varkappa_0$. A singular point $\varkappa$ is a singular point for the compact family $-\zeta_0 R(\zeta_0, \varkappa)$ in the sense of Theorem 1.9. Hence the result follows at least in some neighborhood of $\varkappa_0$. But since $\varkappa_0 \in D_0$ is arbitrary, the same is true in the whole of D_0.

4. Remarks on infinite systems of eigenvalues

When an infinite number of eigenvalues of $T(\varkappa)$ are considered simultaneously, various complications arise[2]. To make the situation simpler, suppose that $T(\varkappa)$ has compact resolvent for each $\varkappa \in D_0$. For convenience assume that $\varkappa = 0$ belongs to D_0. Then each eigenvalue $\lambda(0)$ of $T(0)$ is isolated with finite multiplicity and can be continued to one or several eigenvalues $\lambda(\varkappa)$ of $T(\varkappa)$, each of which is analytic in a certain domain $D' \subset D_0$. D' may be large for some $\lambda(\varkappa)$ and small for others. Furthermore, these $\lambda(\varkappa)$ need not exhaust all the eigenvalues of $T(\varkappa)$, no matter how small $|\varkappa|$ is. In an extreme case $T(0)$ may have no eigenvalue at all (empty spectrum) while $T(\varkappa)$ for $\varkappa \neq 0$ has an infinite number of eigenvalues. Thus it is in general impossible to speak of "the set of all eigenvalues of $T(\varkappa)$ as analytic functions of $\varkappa$."

Example 1.11. Consider the family $T(\varkappa)$ of Example 1.5. Here $T(\varkappa)$ is holomorphic in the *whole extended complex plane* and has compact resolvent. For a fixed $\varkappa$, the eigenvalues of $T(\varkappa)$ are given as the zeros of the entire function

$$\zeta^{-1/2} \sin \zeta^{1/2} - \varkappa \cos \zeta^{1/2}$$

that appears in V-(4.16) as the denominator of the Green function. The eigenvalues for $\varkappa = 0$ are $\lambda_n = n^2 \pi^2$, $n = 1, 2, \ldots$ and the associated eigenfunctions $\sqrt{2} \sin n \pi x$ form a complete orthonormal family in $H = L^2(0, 1)$. Since each of these eigenvalues is simple, the corresponding eigenvalue $\lambda_n(\varkappa)$ of $T(\varkappa)$ is holomorphic in a neighborhood of $\varkappa = 0$. For real $\varkappa$, it is evident from Fig. 1 (see p. 292) that $\lambda_n(\varkappa) - \lambda_n$ grows with n for fixed $\varkappa$, suggesting that the convergence of the power series of $\lambda_n(\varkappa)$ becomes worse with increasing n.

There is another peculiarity in this example. It is easily verified that the $\lambda_n(\varkappa)$ are the only eigenvalues of $T(\varkappa)$ for $\varkappa \leq 0$, but another eigenvalue $\lambda_0(\varkappa)$ exists for $\varkappa > 0$. If $0 < \varkappa < 1$, $\lambda_0(\varkappa)$ is given by $\lambda_0(\varkappa) = -\varrho(\varkappa)^2$ where $\varrho(\varkappa)$ is determined as the unique positive root of the equation $\tanh \varrho = \varkappa \varrho$. $\lambda_0(\varkappa)$ is holomorphic on and

[1] Again the value 0 may be replaced by any complex number.

[2] See Rellich [5], which contains Example 1.11 below and other examples.

near the positive real axis, and $\lambda_0(\varkappa) \to -\infty$ when $\varkappa \searrow 0$; it cannot be continued to $\varkappa < 0$ as long as $\varkappa$ is confined to a *real* neighborhood of $\varkappa = 0$[1].

If complex values of $\varkappa$ are considered, it turns out that all the $\lambda_n(\varkappa)$, including $n = 0$, have analytic continuations and constitute a single analytic function $\lambda(\varkappa)$, which is exactly the inverse function of the meromorphic function $\zeta^{-1/2}\tan\zeta^{1/2}$. This analytic function $\lambda(\varkappa)$ behaves in a very complicated way near $\varkappa = 0$; there are branch points which accumulate at $\varkappa = 0$. This explains the absence of the graph of $\lambda_0(\varkappa)$ for $\varkappa < 0$; within any neighborhood of $\varkappa = 0$, it is possible to go from the branch $\lambda_0(\varkappa)$ to any $\lambda_n(\varkappa)$ with sufficiently large n by circling around an appropriate branch point.

The convergence radii of the series expansion of the $\lambda_n(\varkappa)$ could be determined if these branch points were computed.

Example 1.12. As a simpler example than the preceding one, consider the first order differential operator $T(\varkappa) = -i\, d/dx$, $0 \leqq x \leqq 1$, with the boundary condition $(1 + i\varkappa)\, u(0) = (1 - i\varkappa)\, u(1)$. As in the preceding example, it is easily seen that $T(\varkappa)$ is holomorphic for all complex $\varkappa$ including $\varkappa = \infty$. The eigenvalues of $T(\varkappa)$ are

$$\lambda_n(\varkappa) = 2\arctan\varkappa + 2n\pi,\ n = 0, \pm 1, \pm 2, \ldots . \tag{1.10}$$

$T(\varkappa)$ is selfadjoint for real $\varkappa$ if considered in the Hilbert space $\mathsf{L}^2(0, 1)$, and the eigenvalues $\lambda_n(\varkappa)$ are real for real $\varkappa$. It is seen from (1.10) that the Taylor series for each $\lambda_n(\varkappa)$ has the convergence radius 1. If $\varkappa$ is unrestricted, all the $\lambda_n(\varkappa)$ form a single analytic function $2\arctan\varkappa$, which has logarithmic singularities at $\varkappa = \pm i$.

It is instructive to inquire what these singularities mean with respect to $T(\varkappa)$. The answer is that $T(\pm i)$ have no eigenvalues at all: their spectra are empty. If $T(i)$ is regarded as the unperturbed operator with the perturbation parameter $\varkappa' = \varkappa - i$, there is *no* perturbation series for $T(\varkappa) = T(i + \varkappa')$ to be considered (hence there is no contradiction to the general result that the perturbation series for simple eigenvalues are convergent Taylor series).

Remark 1.13. In the above examples, $T(\varkappa)$ is holomorphic but $\mathsf{D}(T(\varkappa))$ varies with $\varkappa$. Better behavior may be expected of the eigenvalues in the case where $T(\varkappa)$ has a constant domain. We shall discuss this question later (see §§ 3.5, 4.7).

5. Perturbation series

The formal series in powers of $\varkappa$ for the eigenvalues etc. of $T(\varkappa)$ in a finite system, as considered in par. 3, can be obtained by the method of II-§§ 2, 3 applied to the operator

$$T_r(\varkappa) = T(\varkappa)\, P(\varkappa) = T_r + \varkappa\, T_r^{(1)} + \varkappa^2\, T_r^{(2)} + \cdots, \tag{1.11}$$

which is bounded-holomorphic in $\varkappa$. The following remarks should be added.

---

[1] For real $\varkappa$, $T(\varkappa)$ is selfadjoint and its eigenvectors $\varphi_n(\varkappa)$ form a complete orthonormal family. If $\varkappa > 0$, this family contains the eigenvector $\varphi_0(\varkappa)$ for the eigenvalue $\lambda_0(\varkappa)$. Thus if one starts from the *complete* family $\{\varphi_n(0)\}$, $n \geqq 1$, for $T(0)$ and considers the perturbed family $\{\varphi_n(\varkappa)\}$, $n \geqq 1$, one sees that the latter is not complete for $\varkappa > 0$.

a) The transformation function $U(\varkappa)$ used in par. 3 need not be introduced explicitly; it suffices to apply the results of II-§§ 2, 3 directly to (1.11).

b) If one is interested only in the λ-group eigenvalues (see II-§ 1.2), where λ is an isolated eigenvalue of $T = T(0)$, it is convenient to choose as Γ a circle enclosing the single eigenvalue λ of T; then $P(\varkappa)$ coincides with the $P(\varkappa)$ of II-§§ 2, 3. It should be noted that the series II-(2.1) is now meaningless because $T(\varkappa)$ need not be bounded-holomorphic. Thus the coefficients $T^{(n)}$ should be replaced by the coefficients of (1.11). For the operator S [see II-(2.11)], the S given by III-(6.31) should be used. The definition of $P_j^{(1)}$, $S_j^{(1)}$ etc. is exactly as before.

c) The traces of various operators, for instance $\operatorname{tr} T(\varkappa) P(\varkappa)$ used in II-(2.5), are also defined in the present case and obey the rules of calculation with the trace. This is due to the fact that at least one factor in each expression under the trace sign is a *degenerate operator* in the present situation, whereas the remaining factors belong to $\mathscr{B}(\mathsf{X})$ (see the theory of trace given in III-§ 4.3). For example, the operators $P(\varkappa)$, P, D, $P_j^{(1)}$, $S_j^{(1)}$ are all degenerate with rank $\leqq m$, m being the multiplicity of λ. Moreover, the coefficients $T_r^{(n)}$ of (1.11), which should replace the $T^{(n)}$ of the finite-dimensional case according to the remark above, are also degenerate; this follows from Lemma II-2.12, for $T_r(\varkappa) = T_r(\varkappa) P(\varkappa)$ where $T_r(\varkappa)$ is bounded-holomorphic and $P = P(0)$ is degenerate.

6. A holomorphic family related to a degenerate perturbation

Let $T(\varkappa) \in \mathscr{C}(\mathsf{X})$ be defined for $\varkappa$ near $\varkappa = 0$ and suppose that there are two operators T_0, T^0 in $\mathscr{C}(\mathsf{X})$ such that

$$T_0 \subset T(\varkappa) \subset T^0, \; [T(\varkappa)/T_0] = [T^0/T(\varkappa)] = m < \infty. \tag{1.12}$$

In other words, $T(\varkappa)$ is an extension of finite order m of a fixed "minimal operator" T_0 and at the same time a restriction of order m of a fixed "maximal operator" T^0. (Such a situation occurs very often with ordinary differential operators, as will be seen from Example 1.15 below.) Furthermore, we assume that the resolvents $R(\zeta, \varkappa) = (T(\varkappa) - \zeta)^{-1}$ exist for some ζ for all $\varkappa$.

We now ask *under what conditions $T(\varkappa)$ is holomorphic in $\varkappa$*. To answer this question, we first note that

$$A(\zeta, \varkappa) = R(\zeta, \varkappa) - R(\zeta, 0) \tag{1.13}$$

is degenerate and can be expressed as

$$A(\zeta, \varkappa)\, u = \sum_{j,k=1}^{m} a_{jk}(\zeta, \varkappa)\, (u, f_j)\, w_k \tag{1.14}$$

where w_k, $k = 1, \ldots, m$, form a fixed basis of $\mathsf{N}(T^0 - \zeta)$ and f_j, $j = 1, \ldots, m$, form a fixed basis of $\mathsf{R}(T_0 - \zeta)^\perp$ (see Lemma III-6.36).

Let u_j, $j = 1, \ldots, m$, be m vectors of X linearly independent modulo $\mathsf{R}(T_0 - \zeta)$ and constituting a biorthogonal set with $\{f_j\}$. On setting $u = u_j$ in (1.14) and substituting (1.13), we obtain

$$R(\zeta, \varkappa)\, u_j - R(\zeta, 0)\, u_j = \sum_{k=1}^{m} a_{jk}(\zeta, \varkappa)\, w_k . \tag{1.15}$$

Let e_j, $j = 1, \ldots, m$, be any m linearly independent *linear* forms defined on $\mathsf{D}(T^0)$ such that $\det(e_i[w_k]) \neq 0$. Application of e_i to (1.15) gives

$$e_i[R(\zeta, \varkappa)\, u_j] - e_i[R(\zeta, 0)\, u_j] = \sum_{k=1}^{m} a_{jk}(\zeta, \varkappa)\, e_i[w_k] , \quad i, j = 1, \ldots, m . \tag{1.16}$$

(1.16) determines the coefficients $a_{jk}(\zeta, \varkappa)$, so that these coefficients are holomorphic in $\varkappa$ if

$$e_i[R(\zeta, \varkappa)\, u_j] , \quad i, j = 1, \ldots, m , \tag{1.17}$$

are holomorphic in $\varkappa$. Returning to (1.14) and then to (1.13), we see that $R(\zeta, \varkappa)$ is then holomorphic in $\varkappa$. Thus we have proved

Theorem 1.14. *Under the above assumptions, $T(\varkappa)$ is holomorphic if the m^2 functions* (1.17) *are holomorphic in $\varkappa$.*

Example 1.15. Consider the ordinary differential operator

$$L u = p_0(x)\, u'' + p_1(x)\, u' + p_2(x)\, u \tag{1.18}$$

on a finite interval (a, b), with the boundary condition

$$u'(a) - h_a\, u(a) = 0, \quad u'(b) + h_b\, u(b) = 0 . \tag{1.19}$$

Suppose that h_a and h_b are holomorphic functions of $\varkappa$. Then the operators $T(\varkappa)$ defined from L in $\mathsf{X} = \mathsf{L}^p(a, b)$ or $\mathsf{X} = \mathsf{C}[a, b]$ with the boundary condition (1.19) (as the T_3 of III-§ 2.3) form a holomorphic family. To see this we can apply Theorem 1.14 with the "minimal" operator T_0 and the "maximal" operator $T^0 = T$ (see III-§ 2.3). Here $m = 2$, and $w_k \in \mathsf{N}(T - \zeta)$ and $f_j \in \mathsf{R}(T_0 - \zeta)^\perp = \mathsf{N}(T_0^* - \zeta)$ may be any two linearly independent solutions of the equation $(L - \zeta)\, w = 0$ and the formal adjoint equation $(M - \bar{\zeta})\, f = 0$, respectively. As the linear forms e_i we may choose, for example, $e_i[w_k] = w_k(c_i)$, where c_1, c_2 are any two points of (a, b) such that $\det(w_k(c_i)) \neq 0$. Then the functions (1.17) are holomorphic and Theorem 1.14 is applicable. In fact, for any $u \in \mathsf{X}$, $v = R(\zeta, \varkappa)\, u$ is the unique solution of $(L - \zeta)\, v = u$ under the boundary condition (1.19), and it is easily seen that $v(x)$ for any fixed x is a holomorphic function of $\varkappa$ if h_a, h_b are holomorphic in $\varkappa$, unless ζ happens to be one of the eigenvalues of $T(\varkappa)$.

In this way we see that *a boundary condition holomorphic in the parameter gives rise to a holomorphic family of operators.* It should be remarked that under "holomorphic" boundary conditions, we may even allow conditions of the form

$$l_a\, u'(a) - h_a\, u(a) = 0 , \quad l_b\, u'(b) + h_b\, u(b) = 0 \tag{1.20}$$

where l_a, h_a, l_b, h_b are holomorphic in $\varkappa$ with nonvanishing determinant. This implies that, in the original form (1.19) of the boundary condition, $h_a(\varkappa)$ and $h_b(\varkappa)$ may take the value ∞. This is the case if, for example, h_a, h_b are linear in $\varkappa$. Thus it is possible that $T(\varkappa)$ is holomorphic for all $\varkappa$ of the complex plane including $\varkappa = \infty$ (cf. Examples 1.4, 1.5).

§ 2. Holomorphic families of type (A)

1. Definition

An important special case of a holomorphic family $T(\varkappa)$ of operators is furnished by what we shall call a *holomorphic family of type* (A)[1]. A family $T(\varkappa) \in \mathscr{C}(\mathsf{X}, \mathsf{Y})$, defined for $\varkappa$ in a domain D_0 of the complex plane, is said to be holomorphic of type (A) if i) $\mathsf{D}(T(\varkappa)) = \mathsf{D}$ is independent of $\varkappa$ and ii) $T(\varkappa)\, u$ is holomorphic for $\varkappa \in \mathrm{D}_0$ for every $u \in \mathsf{D}$.

In this case $T(\varkappa)\, u$ has a Taylor expansion at each $\varkappa \in \mathrm{D}_0$. If, for example, $\varkappa = 0$ belongs to D_0, we can write

$$(2.1) \qquad T(\varkappa)\, u = T u + \varkappa T^{(1)} u + \varkappa^2 T^{(2)} u + \cdots, \quad u \in \mathsf{D},$$

which converges in a disk $|\varkappa| < r$ independent of u. $T = T(0)$ and the $T^{(n)}$ are linear operators from X to Y with domain D [their linearity follows from the uniqueness of the expansion (2.1) and the linearity of $T(\varkappa)$].

That a holomorphic family $T(\varkappa)$ of type (A) is actually holomorphic in the sense of § 1.2 can be seen as follows. Fix an arbitrary point of D_0, which may be assumed to be $\varkappa = 0$ without loss of generality. Since $T = T(0)$ is closed, $\mathsf{D} = \mathsf{D}(T)$ is converted into a Banach space Z by introducing the new norm $|||u||| = \|u\| + \|T u\|$ (see Remark IV-1.4). Denote by U the operator that sends each $u \in \mathsf{Z}$ to $u \in \mathsf{X}$; U is bounded since $\|u\| \leqq |||u|||$. Each $T(\varkappa)$ may be regarded as an operator from Z into Y; when thus interpreted it will be denoted by $V(\varkappa)$. The closedness of $T(\varkappa)$ implies the same of $V(\varkappa)$ in virtue of $\|u\| \leqq |||u|||$. But since $V(\varkappa)$ is defined everywhere on Z, $V(\varkappa)$ is bounded and belongs to $\mathscr{B}(\mathsf{Z}, \mathsf{Y})$ (Theorem III-5.20). Since $V(\varkappa)\, u = T(\varkappa)\, u$ is holomorphic for each $u \in \mathsf{Z}$, $V(\varkappa)$ is bounded-holomorphic (see Theorem III-3.12). Since U maps Z onto D one to one and $T(\varkappa)\, U = V(\varkappa)$, this proves that $T(\varkappa)$ is holomorphic.

Incidentally, that $V(\varkappa)$ is bounded-holomorphic implies that $(\|u\| + \|V(\varkappa_1)\, u\|)/(\|u\| + \|V(\varkappa_2)\, u\|)$ is bounded when $\varkappa_1$ and $\varkappa_2$ vary over a

---

[1] This type has been studied extensively, including the special case of bounded-holomorphic families, starting with Rellich [1] and [3] (which are restricted to selfadjoint families). See Heinz [1], Hölder [1], T. Kato [1], [3], [6], Sz.-Nagy [1], [2], Porath [1], [2], Rellich [6], [7], [8], Rosenbloom [1], Schäfke [3], [4], [5], Schröder [1], [2], [3], Šmul'yan [1], Wolf [1].

compact subset D of D_0 and u varies over Z. Since $V(\varkappa)\,u = T(\varkappa)\,u$, we have

$$(2.2)\qquad (\|u\| + \|T(\varkappa_1)\,u\|)/(\|u\| + \|T(\varkappa_2)\,u\|) \leq M,\quad \varkappa_1, \varkappa_2 \in \mathrm{D},\quad u \in \mathsf{D}\,.$$

Furthermore we have $T'(\varkappa)\,u = V'(\varkappa)\,u$, where the operator $T'(\varkappa)$ with domain D is defined by $T'(\varkappa)\,u = d\,T(\varkappa)\,u/d\varkappa$. But since $\|V'(\varkappa)\,u\|/(\|u\| + \|V(\varkappa)\,u\|)$ is bounded for $\varkappa \in \mathrm{D}$ and $u \in \mathsf{Z}$, we have an inequality of the form

$$(2.3)\qquad \|T'(\varkappa)\,u\| \leq a'\|u\| + b'\|T(\varkappa)\,u\|\,,\quad \varkappa \in \mathrm{D}\,,\quad u \in \mathsf{D}\,.$$

Note, however, that $T'(\varkappa)$ (with domain D) may not be closed; hence it need not be a holomorphic family.

Similarly it can be shown that, for any $\varepsilon > 0$, there is a $\delta > 0$ such that

$$(2.4)\qquad \|T(\varkappa_1)\,u - T(\varkappa_2)\,u\| \leq \varepsilon(\|u\| + \|T(\varkappa)\,u\|)\,,\quad |\varkappa_1 - \varkappa_2| < \delta\,,$$

as long as $\varkappa_1$, $\varkappa_2$ and $\varkappa$ belong to D. These results may be expressed roughly by saying that any $T(\varkappa)$ is relatively bounded with respect to any other and $T(\varkappa)$ is relatively continuous.

Example 2.1. In what follows we shall encounter many holomorphic families of type (A). Here we give an almost trivial example. Let $T \in \mathscr{C}(\mathsf{X})$ and let $T(\varkappa) = \varkappa\,T$. $T(\varkappa)$ is holomorphic of type (A) for $\varkappa \neq 0$. In general $\varkappa = 0$ must be excluded, for the operator 0 with domain $\mathsf{D}(T)$ is not closed if T is unbounded. [The exclusion of $\varkappa = 0$ might have been unnecessary if we did not require the closedness of $T(\varkappa)$ in the definition of a holomorphic family, but it is convenient for many purposes to admit only closed $T(\varkappa)$.]

Furthermore, $T(\varkappa) = \varkappa\,T$ *cannot* be made holomorphic [to say nothing of type (A)] including $\varkappa = 0$ by *defining* $T(0) = 0$ (with domain X). $T(0)$ is then indeed closed, but it is easy to see that the resolvent $R(\zeta, \varkappa) = (T(\varkappa) - \zeta)^{-1}$ is not bounded-holomorphic at $\varkappa = 0$ for any $\zeta \neq 0$ [even when there exists a $\zeta \neq 0$ belonging to $\mathrm{P}(T(\varkappa))$ for all $\varkappa$ near $\varkappa = 0$]. In view of Theorem 1.3, $T(\varkappa)$ cannot be holomorphic.

Remark 2.2. The question is open whether or not a holomorphic family $T(\varkappa)$ with constant domain is necessarily of type (A)[1]. But this is true under a slight additional condition; namely we have

Theorem 2.3. *Let $T(\varkappa) \in \mathscr{C}(\mathsf{X})$ be holomorphic near $\varkappa = 0$ and have constant domain $\mathsf{D} = \mathsf{D}(T(\varkappa))$. Furthermore, assume that $T(0)$ has nonempty resolvent set, that $T(0)$ and $T(0)^*$ are densely defined (in X and X^*, respectively) and that $T(0)\,R(\zeta, \varkappa)$ is uniformly bounded near $\varkappa = 0$ for some fixed $\zeta \in \mathrm{P}(T(0))$. Then $T(\varkappa)$ is of type* (A) *near $\varkappa = 0$.*

---

[1] In this connection it is interesting to note that RELLICH [5] gives an example of a *real*-holomorphic family $T(\varkappa)$, with $\mathsf{D} = \mathsf{D}(T(\varkappa))$ independent of $\varkappa$, for which $T(\varkappa)\,u$ is not real-holomorphic for all $u \in \mathsf{D}$. This $T(\varkappa)$ has a (complex-)holomorphic extension $T_1(\varkappa)$, for which, however, $\mathsf{D}(T_1(\varkappa))$ is no longer constant.

Proof. Before giving the proof, we note that $\mathsf{D}(T(0)) = \mathsf{D}(T(\varkappa))$ already implies that $T(0)R(\zeta,\varkappa)$ and $T(\varkappa)R(\zeta,0)$ belong to $\mathscr{B}(\mathsf{X})$ (see Problem III-5.22). The essential assumption is that $T(0)R(\zeta,\varkappa)$ be *uniformly* bounded near $\varkappa = 0$.

For the proof we may assume $\zeta = 0$ for simplicity. Then $R(0,\varkappa)$ is bounded-holomorphic by Theorem 1.3. Hence $\phi(\varkappa; f) = (T(0)R(0,\varkappa)u, f) = (R(0,\varkappa)u, T(0)^* f)$ is holomorphic near $\varkappa = 0$ for each $u \in \mathsf{X}$ and $f \in \mathsf{D}(T(0)^*) = \mathsf{D}^* \in \mathsf{X}^*$. Since D^* is dense and $\|T(0)R(0,\varkappa)\|$ is bounded in $\varkappa$ by hypothesis, it follows from Theorem III-3.12 and a remark following it that $T(0)R(0,\varkappa)$ is bounded-holomorphic. Hence the same is true of the inverse $(T(0)R(0,\varkappa))^{-1} = T(\varkappa)R(0,0)$ (see the remark at the end of § 1.1). This implies that $T(\varkappa)u$ is holomorphic for every $u \in \mathsf{D}(T(0))$.

A holomorphic family of type (A) has many properties not shared by general holomorphic families, as we shall see in the course of the following sections. Here we note

Theorem 2.4. *Let $T(\varkappa) \in \mathscr{C}(\mathsf{X})$ be holomorphic of type* (A) *and have nonempty resolvent set for each $\varkappa$. Then $T(\varkappa)$ has compact resolvent either for all $\varkappa$ or for no $\varkappa$.*

Proof. Let D′ be the set of all $\varkappa$ such that $T(\varkappa)$ has compact resolvent; we have to show that D′ is open and closed relative to D_0, the domain of definition of the family $T(\varkappa)$. That D′ is open follows from Theorem IV-3.17, for $T(\varkappa_2)$ is relatively bounded with respect to $T(\varkappa_1)$ if $|\varkappa_2 - \varkappa_1|$ is sufficiently small [see (2.4)]. On the other hand, Theorem IV-2.26 shows that D′ is closed relative to D_0 [here $T(\varkappa)$ need not be of type (A)].

Example 2.5. An example opposite, in a certain sense, to Example 2.1 is furnished by a family $T(\varkappa) = (U + \varkappa)^{-1}$, where $U \in \mathscr{B}(\mathsf{X})$ is *quasi-nilpotent* but invertible. Then $\mathsf{D}(T(\varkappa)) = \mathsf{X}$ for $\varkappa \neq 0$ while $\mathsf{D}(T(0)) = \mathsf{R}(U)$ is a proper subset of X. $T(\varkappa)$ is holomorphic of type (A) for $\varkappa \neq 0$; in fact it is even bounded-holomorphic. It is interesting to note that $T(\varkappa)$ is holomorphic even when $\varkappa = 0$ is included, though it is no longer of type (A). This is verified by writing $T(\varkappa)U(\varkappa) = 1$, $U(\varkappa) = U + \varkappa$.

Suppose now that U is in addition *compact*. Then $T(0) = U^{-1}$ has compact resolvent, for $T(0)^{-1} = U$ is compact. But $T(\varkappa)$ does not possess this property if $\varkappa \neq 0$. This remark shows that Theorem 2.4 is not true for a general holomorphic family $T(\varkappa)$.

2. A criterion for type (A)

Theorem 2.6.[1] *Let T be a closable operator from X to Y with $\mathsf{D}(T) = \mathsf{D}$. Let $T^{(n)}$, $n = 1, 2, \ldots$, be operators from X to Y with domains containing D, and let there be constants a, b, $c \geq 0$ such that*

$$\|T^{(n)}u\| \leq c^{n-1}(a\|u\| + b\|Tu\|)\,, \quad u \in \mathsf{D},\ n = 1, 2, \ldots. \tag{2.5}$$

[1] This theorem is essentially due to Rellich [3] (where it is restricted to selfadjoint families).

Then the series (2.1) *defines an operator* $T(\varkappa)$ *with domain* D *for* $|\varkappa| < 1/c$. *If* $|\varkappa| < (b+c)^{-1}$, $T(\varkappa)$ *is closable and the closures* $\tilde{T}(\varkappa)$ *for such* $\varkappa$ *form a holomorphic family of type* (A).

Remark 2.7. The form of the condition (2.5) is so chosen as to be particularly convenient when $T^{(2)} = T^{(3)} = \cdots = 0$. In this case we can choose $c = 0$ if

$$\|T^{(1)} u\| \leqq a\|u\| + b\|Tu\|\,, \quad u \in \mathsf{D}\,. \tag{2.6}$$

Note also that the required constants a, b, c exist if the series (2.1) is finite and the $T^{(n)}$ are T-bounded (see IV-§ 1.1).

Proof of Theorem 2.6. $T(\varkappa)$ is well defined by (2.1) if $|\varkappa| < c^{-1}$, for

$$\begin{aligned}\|\varkappa^n T^{(n)} u + \cdots + \varkappa^{n+p} T^{(n+p)} u\| &\leqq \\ &\leqq (c^{n-1}|\varkappa|^n + \cdots + c^{n+p-1}|\varkappa|^{n+p})\,(a\|u\| + b\|Tu\|) \leqq \\ &\leqq c^{n-1}|\varkappa|^n (1 - c|\varkappa|)^{-1}(a\|u\| + b\|Tu\|) \to 0\,, \quad n \to \infty\,,\end{aligned}$$

and the series on the right of (2.1) converges.

If we write (2.1) in the form $T(\varkappa)\,u = Tu + A(\varkappa)\,u$, it follows from the above inequality that

$$\|A(\varkappa)\,u\| \leqq |\varkappa|\,(1 - c|\varkappa|)^{-1}(a\|u\| + b\|Tu\|)\,, \quad u \in \mathsf{D}\,. \tag{2.7}$$

Thus $A(\varkappa)$ is T-bounded with T-bound not exceeding $b|\varkappa|\,(1 - c|\varkappa|)^{-1}$, which is smaller than 1 if $|\varkappa| < (b+c)^{-1}$. It follows from Theorem IV-1.1 that $T(\varkappa)$ is closable for such $\varkappa$. Incidentally we note the following inequality, which follows from IV-(1.3):

$$a\|u\| + b\|Tu\| \leqq (1 - c|\varkappa|)\,(1 - (b+c)\,|\varkappa|)^{-1}(a\|u\| + b\|T(\varkappa)u\|). \tag{2.8}$$

Let $\tilde{T}$ be the closure of T, with domain $\tilde{\mathsf{D}} \supset \mathsf{D}$. $\tilde{\mathsf{D}}$ is converted into a Banach space Z by introducing the norm $|||u||| = (a + \varepsilon)\,\|u\| + b\|Tu\|$, where $\varepsilon > 0$. That $\tilde{T}$ is the closure of T implies that D is dense in Z. Let $V^{(n)}$ be the operator $T^{(n)}$ regarded as an operator from Z to Y; $V^{(n)}$ is densely defined and $\|V^{(n)} u\| \leqq c^{n-1}\,|||u|||$. Hence $V^{(n)}$ has closure $\tilde{V}^{(n)} \in \mathscr{B}(\mathsf{Z}, \mathsf{Y})$ with $\|\tilde{V}^{(n)}\| \leqq c^{n-1}$. If we denote by $V(\varkappa)$ and V the $T(\varkappa)$ and T, respectively, as operators from Z to Y, we have

$$V(\varkappa) = V + \varkappa V^{(1)} + \varkappa^2 V^{(2)} + \cdots\,, \tag{2.9}$$

and $V(\varkappa)$ obviously has the closure $\tilde{V}(\varkappa)$ obtained by replacing V and the $V^{(n)}$ by $\tilde{V}$ and the $\tilde{V}^{(n)}$ on the right of (2.9). Thus $\tilde{T}(\varkappa)\,u = \tilde{V}(\varkappa)\,u$ is holomorphic for each $u \in \tilde{\mathsf{D}}$ if $|\varkappa| < (b+c)^{-1}$.

Remark 2.8. The inequalities (2.5) are also *necessary* for $T(\varkappa)$ to be holomorphic of type (A), in the following sense. If $T(\varkappa)$ is holomorphic of type (A) in a neighborhood of $\varkappa = 0$, we have the expansion (2.1) with the coefficients $T^{(n)}$ satisfying (2.5). This follows from the fact that the $V(\varkappa)$ used above is bounded-holomorphic near $\varkappa = 0$ and, consequently, developable into a power series of the form (2.9). Then we have $\|V^{(n)}\| \leqq \leqq M r^{-n}$ where r is any positive number smaller than the convergence radius of this series (Cauchy's inequality). Hence $\|T^{(n)} u\| = \|V^{(n)} u\| \leqq \leqq M r^{-n} |||u|||$, which is equivalent to (2.5).

3. Remarks on holomorphic families of type (A)

The results concerning a holomorphic family $T(\varkappa)$ of operators are simplified and become more explicit when $T(\varkappa)$ is of type (A). Here we collect such results in a series of remarks.

Remark 2.9. Suppose the spectrum $\Sigma = \Sigma(T)$ of T is separated into two parts by a closed curve Γ. Then $\Sigma(\varkappa) = \Sigma(T(\varkappa))$ is likewise separated by Γ for sufficiently small $\varkappa$ by Theorem 1.7. Here we can estimate how small $|\varkappa|$ should be. According to Theorem IV-3.18, it suffices that IV-(3.14) be satisfied for all $\zeta \in \Gamma$, with the constants a, b replaced by $a|\varkappa| (1 - c|\varkappa|)^{-1}$, $b|\varkappa| (1 - c|\varkappa|)^{-1}$ in view of (2.7). This means that separation of the spectrum occurs at least if

$$|\varkappa| < r_0 = \min_{\zeta \in \Gamma} (a\|R(\zeta, T)\| + b\|T R(\zeta, T)\| + c)^{-1} . \tag{2.10}$$

Then the projection $P(\varkappa)$ given by (1.3) is holomorphic in $\varkappa$. If the part of Σ inside Γ is a finite system of eigenvalues, it follows that the eigenvalues, eigenprojections and eigennilpotents of $T(\varkappa)$ in question are analytic for (2.10) with only algebraic singularities (see § 1.3).

Remark 2.10. The existence of the formal power series for $T(\varkappa)$, which did not occur in the case of a general holomorphic family, makes it possible to take over directly the results of II-§§ 1, 2 without replacing $T^{(n)}$ by $T_r^{(n)}$ (see § 1.5). This is due to the fact that II-(1.13) is valid in spite of the unboundedness of the $T^{(n)}$; in fact, these operators appear in the formulas there only in such combinations as $T^{(n)} R(\zeta)$, $T^{(n)} S$, $T^{(n)} P$, $T^{(n)} D = T^{(n)} PD$ *which all belong to* $\mathscr{B}(X)$ (by Problem III-5.22). In this way we see that all formulas of II-§§ 1, 2 are valid in the present case.

There is one important point to be noticed, however. The use of the *trace* in a formula such as II-(2.28) is not justified here, for the operators that appear under the trace sign are in general not degenerate. Nevertheless the final result II-(2.29) (or similar formulas that follow) is true, in which all operators under the trace sign are degenerate because they contain at least one factor P. This is easily seen by calculating the integral explicitly in the fashion of II-(2.18); then it becomes a polynomial in operators of $\mathscr{B}(X)$ each term of which contains, as just mentioned, at least one degenerate factor. When one takes the trace of such a monomial, the order of the factors can be changed cyclically. As is easily verified, this amounts to the validity of II-(2.30).

Remark 2.11. For applications it is often convenient to generalize (2.5) to[1]

$$(2.11)\qquad \|T^{(n)} u\| \leqq c^{n-1} \chi(\|u\|, \|Tu\|), \quad u \in D, \quad n = 1, 2, \ldots .$$

Here $\chi(s, t)$ is a nonnegative function of two variables $s, t \geqq 0$ with the following properties: i) $\chi(s, t)$ is positive-homogeneous of order 1: $\chi(k s, k t) = k \chi(s, t)$ for $k > 0$; ii) $\chi(s, t)$ is monotonically increasing in s, t: $\chi(s, t) \leqq \chi(s', t')$ for $0 \leqq s \leqq s'$, $0 \leqq t \leqq t'$. We can now generalize Theorem 2.6 by replacing (2.5) by (2.11), the conclusion being that $\tilde{T}(\varkappa)$ *is holomorphic of type* (A) *for* $|\varkappa| < (\chi(+0, 1) + c)^{-1}$. Also (2.10) should be replaced by

$$(2.12)\qquad r_0 = \min_{\zeta \in \Gamma} (\chi(\| R(\zeta, T)\|, \|T R(\zeta, T)\|) + c)^{-1} .$$

Examples of the function $\chi(s, t)$ are:

$$(2.13)\qquad a s + b t, \quad a (s t)^{1/2}, \quad (a s^2 + b t^2)^{1/2} .$$

It should be noticed that (2.11) actually implies (2.5) with appropriate constants a, b. The point is that by strengthening the assumption, the conclusions are also strengthened.

Example 2.12. Consider the formal differential operator

$$(2.14)\qquad L(\varkappa) u = p_0(x, \varkappa) u'' + p_1(x, \varkappa) u' + p_2(x, \varkappa) u$$

on a finite interval $[a, b]$, in which the coefficients $p_j(x, \varkappa)$ are assumed to be holomorphic in $\varkappa$ near $\varkappa = 0$ (uniformly with respect to x) and real-valued for real $\varkappa$, with appropriate continuity properties with respect to x. Furthermore, let $-p_0(x, \varkappa) \geqq \delta > 0$ for real $\varkappa$. Let $T(\varkappa)$ be the operator in $X = C[a, b]$ or $X = L^p(a, b)$ constructed from $L(\varkappa)$ as the T_1 of III-§ 2.3 [boundary condition $u(a) = u(b) = 0$]. $T(\varkappa)$ satisfies the assumptions of Theorem 2.6. In fact $T(\varkappa)$ can be written in the form (2.1) with $T^{(n)} u = L^{(n)} u$ ($T = T^{(0)}$) with the same boundary condition as above, where

$$(2.15)\qquad L^{(n)} u = p_0^{(n)}(x) u'' + p_1^{(n)}(x) u' + p_2^{(n)}(x) u$$

and the $p_j^{(n)}(x)$ are the coefficients of $\varkappa^n$ in the Taylor expansion of $p_j(x, \varkappa)$. Since these coefficients satisfy inequalities of the form $|p_j^{(n)}(x)| \leqq K N^n$, the inequalities (2.5) follow from IV-(1.27). Recalling that the spectrum of $T = T(0)$ consists of isolated eigenvalues of multiplicity one (see Example III-6.20), we see from Theorem 2.6 that these eigenvalues are holomorphic in $\varkappa$ near $\varkappa = 0$ together with the associated eigenprojections.

Example 2.13. Consider the Schrödinger operator

$$(2.16)\qquad L(\varkappa) u = -\Delta u + q(x) u + \varkappa q^{(1)}(x) u$$

in R^3. Assume that the potential $q(x)$ is real-valued and can be written in the form $q = q_0 + q_1$ where $q_1 \in L^2(R^3)$ and q_0 is bounded. Then there is a selfadjoint restriction H of $L(0)$ [acting in $H = L^2(R^3)$] which is defined as the closure of the

---

[1] See SCHRÖDER [1], SCHÄFKE [4], PORATH [1].

minimal operator (see V-§ 5.3). Actually $\mathsf{D}(H)$ is identical with $\mathsf{D}(H_0)$, where H_0 is the special case of H for $q = 0$, the multiplication operator $Q = q(x)$ being H_0-bounded with relative bound zero. It follows that $Q^{(1)} = q^{(1)}(x)$ is also H_0-bounded and hence H-bounded with relative bound zero if $q^{(1)}$ is likewise the sum of two real-valued functions of L^2 and of L^∞. Hence the operators $T(\varkappa)$ defined as the closure of the minimal operators for the $L(\varkappa)$ form a selfadjoint family of type (A) defined for all complex $\varkappa$. It follows that the isolated eigenvalues and eigenprojections of $T(\varkappa)$ are holomorphic in $\varkappa$ at least in a neighborhood of the real axis[1].

4. Convergence radii and error estimates

The estimates of the convergence radii and the errors for the perturbation series given in the finite-dimensional case (see II-§ 3) can be taken over to the case of type (A) without essential modification. The projection $P(\varkappa)$ as given by (1.3) is holomorphic for $|\varkappa| < r_0$, where r_0 is given by (2.10) or (2.12) as the case may be. In particular this is true for the total projection $P(\varkappa)$ for the λ-group eigenvalues of $T(\varkappa)$ when λ is an isolated eigenvalue of T with finite multiplicity m; then Γ should be chosen as a closed curve around $\zeta = \lambda$ in such a way that r_0 becomes as large as possible. The weighted mean $\hat{\lambda}(\varkappa)$ of these eigenvalues is also holomorphic for $|\varkappa| < r_0$, so that the power series for $P(\varkappa)$ and $\hat{\lambda}(\varkappa)$ have convergence radii not smaller than r_0. If the λ-group consists of a single eigenvalue $\lambda(\varkappa)$ (no splitting), which is certainly the case if $m = 1$, then $\lambda(\varkappa)$ itself is a power series convergent for $|\varkappa| < r_0$.

The results of II-§ 3.2–3 on majorizing series hold true without modification; note that the operators in II-(3.10) are bounded in the case under consideration (see Remark 2.10).

Other results of II-§ 3 are also useful in the present case. Naturally we have to set $N = \infty$ in general in the formula II-(3.45) and the following ones in which the demension N of the underlying space X appears explicitly. These estimates may in general not be very sharp, but most of them are still the best possible of their kind, since this was true even in the finite-dimensional case (see II-§ 3.5, in particular).

The estimate II-(3.32) for $(T - \lambda)\,\varphi(\varkappa)$, which is also valid here, deserves special attention. Since $T - \lambda$ is in general an unbounded operator, this estimate is rather strong. Combined with the estimate II-(3.30) for $\varphi(\varkappa)$ itself, it makes possible an error estimate of $\varphi(\varkappa)$ with respect to a norm such as

$$|||u||| = \alpha\|u\| + \beta\|(T - \lambda)\,u\| \tag{2.17}$$

which is stronger than the original norm $\|u\|$. In this way one is able to estimate, for example, the error of the eigenfunction in L^∞-norm in a

[1] Here we used some results on selfadjoint families (see § 3).

theory within an L^2-space, or estimate even the error of the derivatives of the eigenfunction.

Example 2.14. Let $X = C[0, \pi]$ [1]. Consider the family $T(\varkappa) = T + \varkappa T^{(1)}$ where T is the differential operator $Tu = -u''$ on the interval $0 \leqq x \leqq \pi$ with the boundary condition $u(0) = u(\pi) = 0$ (see Examples III-6.21 and IV-3.20) and where $T^{(1)}$ is a *bounded* operator. Each eigenvalue $\lambda_n = n^2$ of T is isolated and simple, so that the corresponding eigenvalue $\lambda_n(\varkappa)$ of $T(\varkappa)$ is holomorphic at least for small $|\varkappa|$. Let us estimate the convergence radius for the Taylor series of $\lambda_n(\varkappa)$.

To this end we consider a closed curve Γ'_n similar to the Γ_n used in Example IV-3.20 but larger in size: Γ'_n consists of the two parabolas $\xi = \alpha^2 - \eta^2/4\alpha^2$ [see III-(6.49)] corresponding to $\alpha = n \pm \frac{1}{2}$ and the two horizontal lines $\eta = \pm(2n-1)/\pi$. It follows then from III-(6.48) that $\|R(\zeta)\| \leqq \pi\left(n - \frac{1}{2}\right)^{-1}$ for $\zeta \in \Gamma'_n$. Since Γ'_n encloses exactly one eigenvalue λ_n of T, a lower bound for the convergence radius of $\lambda_n(\varkappa)$ is given by (2.10) with $b = c = 0$ and $a = \|T^{(1)}\|$. In this way we obtain

$$r_0 \geqq \left(n - \frac{1}{2}\right)\Big/\pi a \quad \text{where} \quad a = \|T^{(1)}\|. \tag{2.18}$$

The coefficients $\lambda_n^{(\nu)}$ in the Taylor series of $\lambda_n(\varkappa)$ can be estimated, for example, by II-(3.5). For this we need the maximum distance ϱ of $\lambda_n = n^2$ from Γ'_n. A straightforward calculation gives the inequality $\varrho \leqq (1 + 4\pi^{-2})^{1/2} n \leqq 1.2n$, so that

$$|\lambda_n^{(\nu)}| \leqq 1.2n \left(\frac{\pi a}{n - 1/2}\right)^{\nu}, \quad n, \nu = 1, 2, \ldots. \tag{2.19}$$

This estimate is not very sharp, at least for small ν. A better estimate can be obtained from the method of majorizing series. For this we need the quantities p, q, s of II-(3.15) for $\lambda_n = n^2$. Now the operator $S = S_n$ is an integral operator with the kernel $s(y, x)$ given by III-(6.46), so that $\|S\|$ is estimated by $\|S\| \leqq \sup \int_0^\pi |s(y, x)|\, dx$ (see Example III-2.11; note that $X = C[0, \pi]$ here and the estimate for $p = \infty$ is applicable). Thus we have [2] $\|S\| < 1.7/n$ and $\|S - P/4n^2\| \leqq \sqrt{2}/n$. Since we know that $\|P\| = 4/\pi < 1.3$ (see Example III-3.23), p, q, s are estimated by II-(3.15) as

$$p < 1.3\, a, \quad q < 1.7\, a/n, \quad s < 1.5/n. \tag{2.20}$$

Substitution of (2.20) into II-(3.21) gives a lower bound for the convergence radius in question:

$$r > 0.136\, n/a, \tag{2.21}$$

which is not so sharp as (2.18) obtained above. But II-(3.22) gives

$$|\lambda_n^{(1)}| < 1.3\, a, \quad |\lambda_n^{(2)}| < 2.3\, a^2/n, \quad |\lambda_n^{(3)}| < 8.1\, a^3/n^2, \tag{2.22}$$

which are much sharper than (2.19). Similarly, the remainder for $\lambda_n(\varkappa)$ after the first order term is given by II-(3.18):

$$|\lambda_n(\varkappa) - n^2 - \varkappa \lambda_n^{(1)}| \leqq 8.8\, |\varkappa|^2 a^2/n \quad \text{for} \quad |\varkappa| \leqq 0.136\, n/a. \tag{2.23}$$

---

[1] We shall later consider the same problem in $X = L^2(0, \pi)$; see Example 2.17.

[2] See ROSENBLOOM [1], where the same example is treated by a different method. Note that $S - P/4n^2$ has the kernel represented by the first two terms in the kernel of S.

Note that the coefficient 8.8 on the right of (2.23) depends on the range of $\varkappa$ considered; it can be reduced if this range is narrowed. If, for example, we restrict $|\varkappa|$ to the range $0.12\, n/a$, II-(3.18) gives

$$|\lambda_n(\varkappa) - n^2 - \varkappa \lambda_n^{(1)}| \leq 4.9\, |\varkappa|^2 a^2/n \quad \text{for} \quad |\varkappa| \leq 0.12\, n/a\,. \tag{2.24}$$

The eigenfunction normalized according to II-(3.26) is estimated by II-(3.38) as

$$\|\varphi_n(\varkappa) - \sin n\, x\| \leq 3.7\, |\varkappa|\, a/n \quad \text{for} \quad |\varkappa| \leq 0.12\, n/a\,, \tag{2.25}$$

where $\sin n\, x$ is the unperturbed eigenfunction with norm 1. Note that, in II-(3.26), we have $\psi(x) = 2\pi^{-1} \sin n\, x$ so that the normalization implies that $\varphi_n(\varkappa) - \sin n\, x$ should be orthogonal to $\sin n\, x$.

The estimates for $(T - \lambda_n)\, \varphi_n(\varkappa) = -\varphi_n''(\varkappa) - n^2 \varphi_n(\varkappa)$ are given by II-(3.40). Since its right member is the same as that of II-(3.38) with one factor s_0 omitted, we have (for the same range of $\varkappa$ as above)

$$\|\varphi_n''(\varkappa) + n^2 \varphi_n(\varkappa)\| \leq 2.2\, |\varkappa|\, a\,. \tag{2.26}$$

(2.25) and (2.26) give

$$\|d^2/d x^2 (\varphi_n(\varkappa) - \sin n\, x)\| \leq (2.2 + 3.7\, n)\, |\varkappa|\, a \tag{2.27}$$

and, on integration,

$$\|d/d x (\varphi_n(\varkappa) - \sin n\, x)\| \leq \pi (2.2 + 3.7\, n)\, |\varkappa|\, a\,; \tag{2.28}$$

note that the function $\varphi_n(\varkappa) - \sin n\, x$ vanishes at $x = 0$ and π and so its derivative vanishes somewhere in $(0, \pi)$.

Remark 2.15. The estimate (2.28) is rather crude. A better result can be obtained by using the formula

$$\varphi_n(\varkappa) - \sin n\, x = S_n (T - \lambda_n)\, (\varphi_n(\varkappa) - \sin n\, x)\,; \tag{2.29}$$

note that $S_n(T - \lambda_n) \subset 1 - P_n$ and $P_n(\varphi_n(\varkappa) - \sin n\, x) = 0$ by the normalization of $\varphi_n(\varkappa)$ employed. Since S_n is the integral operator with the kernel $s = s(y, x)$ mentioned above, we have by differentiating (2.29)

$$\frac{d}{dx} [\varphi_n(\varkappa) - \sin n\, x] = S_n'(T - \lambda_n)\, (\varphi_n(\varkappa) - \sin n\, x)\,, \tag{2.30}$$

where S_n' is the integral operator with the kernel $\frac{\partial}{\partial y} s(y, x)$, which is piecewise continuous in x, y. Hence

$$\left\|\frac{d}{dx} [\varphi_n(\varkappa) - \sin n\, x]\right\| \leq 2.2 |\varkappa|\, a \|S_n'\|\,. \tag{2.31}$$

Here $\|S_n'\|$ is majorized by $\sup_y \int_0^\pi \left|\frac{\partial}{\partial y} s(y, x)\right| dx$, which can be shown to be bounded for $n \to \infty$. Thus (2.31) would lead to a sharper result than (2.28).

Problem 2.16. In Example 2.14, estimate by II-(3.19) and (3.41) the remainders after the second-order term of $\lambda_n(\varkappa)$ and after the first-order term of $\varphi_n(\varkappa)$.

5. Normal unperturbed operators

If $\mathsf{X} = \mathsf{H}$ is a Hilbert space and $T = T(0)$ is a normal operator, the foregoing results are considerably simplified. The situation is similar to the finite-dimensional case (see II-§ 3.5), and few modifications are needed.

In virtue of the inequalities V-(4.9), (2.12) becomes

$$r_0 = \min_{\zeta\in\Gamma} [\chi(\sup_{\lambda'\in\Sigma(T)} |\lambda' - \zeta|^{-1}, \ \sup |\lambda'|\,|\lambda' - \zeta|^{-1}) + c]^{-1} \tag{2.32}$$

if (2.11) is assumed. If we consider an isolated eigenvalue λ of T with isolation distance d, it is convenient to take Γ as the circle $|\zeta - \lambda| = d/2$. Then $|\lambda' - \zeta|^{-1} \leqq 2/d$ and $|\lambda'|\,|\lambda' - \zeta|^{-1} \leqq 1 + |\zeta|\,|\lambda' - \zeta|^{-1} \leqq 1 + (|\lambda| + 2^{-1} d)\, 2d^{-1} = 2 + 2|\lambda|\, d^{-1}$ for $\lambda' \in \Sigma(T)$ and $\zeta \in \Gamma$. Hence

$$r_0 \geqq 1\Big/\Big[\chi\Big(\frac{2}{d}, \ 2 + \frac{2|\lambda|}{d}\Big) + c\Big]. \tag{2.33}$$

If (2.5) is assumed, we have

$$r_0 \geqq 1\Big/\Big[\frac{2(a + b|\lambda|)}{d} + 2b + c\Big] \tag{2.34}$$

corresponding to II-(3.51).

For majorizing series discussed in the preceding paragraph, we can take over the results of II-§ 3.2–3 by setting

$$p = \|T^{(1)} P\|, \quad q = \|T^{(1)} S\|, \quad s = \|S\| = 1/d \tag{2.35}$$

[see II-(3.15) and V-(3.20)]. If $T^{(1)}$ is bounded, we can use

$$p = \|T^{(1)}\|, \quad q = \|T^{(1)}\|/d, \quad s = 1/d, \tag{2.36}$$

since $\|P\| = 1$.

Example 2.17. Let us consider the problem of Example 2.14 in the Hilbert space $\mathsf{H} = \mathsf{L}^2(0, \pi)$. Now the operator T is selfadjoint; it has the eigenvalues $\lambda_n = n^2$ as before, but the normalized eigenfunctions $\varphi_n = (2/\pi)^{1/2} \sin n\,x$ have different factors due to the different norm employed. The resolvent $R(\zeta)$ of T is again given by the integral operator with the kernel $s(y, x)$, but $R(\zeta)$ has a different norm from the former case. In particular, the norm of the reduced resolvent S for the eigenvalue $\lambda_n = n^2$ is now given by $\|S\| = 1/d$ [see (2.35)], where the isolation distance $d = d_n$ for λ_n is

$$d_n = 2n - 1 \quad \text{for} \quad n \geqq 2 \quad \text{and} \quad d_1 = 3. \tag{2.37}$$

Consider now a perturbed operator $T(\varkappa) = T + \varkappa T^{(1)}$ where $T^{(1)}$ is T-bounded as in (2.6). The convergence radius for the n-th eigenvalue $\lambda_n(\varkappa)$ of $T(\varkappa)$ is estimated by (2.34) with $c = 0$. If, in particular, $T^{(1)}$ is bounded, we can set $a = \|T^{(1)}\|$, $b = 0$ and obtain

$$r_0 \geqq d/2a = (n - 1/2)/a \quad \text{(replace } n - 1/2 \text{ by } 3/2 \text{ for } n = 1) \tag{2.38}$$

as a lower bound for the convergence radius. This is apparently sharper by a factor π than the corresponding estimate (2.18) obtained in the case $\mathsf{X} = \mathsf{C}[0, \pi]$, but it must be noticed that an operator $T^{(1)}$ bounded in C need not be bounded in L^2, and need not have the same bound even if bounded. If $T^{(1)}$ is a multiplication operator by a function $q(x)$, however, $\|T^{(1)}\| = \sup|q(x)|$ is the same whether considered in C or in L^2; in such a case the L^2-theory certainly gives a sharper estimate for r_0.

In this connection it is instructive to calculate various estimates furnished by the method of majorizing series. Substituting the value of $d = d_n$ from (2.37) into

(2.36) and using II-(3.21), we get $r = (n - 1/2)/2a$ as a lower bound for the convergence radius, which is less sharp by a factor 2 than (2.38). On the other hand, the following estimates are obtained in the same way as in Example 2.14:

$$(2.39)\quad \lambda_n(\varkappa) - \lambda \prec \Psi(\varkappa) \equiv a\varkappa + \frac{\dfrac{a^2\varkappa^2}{n-1/2}}{1 - \dfrac{a\varkappa}{n-1/2} + \left(1 - \dfrac{2a\varkappa}{n-1/2}\right)^{1/2}}$$
$$= \left(n - \frac{1}{2}\right)\left[1 - \left(1 - \frac{2a\varkappa}{n-1/2}\right)^{1/2}\right] = \frac{2a\varkappa}{1 + \left(1 - \dfrac{2a\varkappa}{n-1/2}\right)^{1/2}},$$

$$(2.40)\quad |\lambda_n^{(1)}| \leq a,\ |\lambda_n^{(2)}| \leq a^2/2\,(n - 1/2),\ |\lambda_n^{(3)}| \leq a^3/2\,(n - 1/2)^2, \ldots,$$

$$(2.41)\quad |\lambda_n(\varkappa) - n^2 - \varkappa\,\lambda_n^{(1)}| \leq 2\,|\varkappa|^2\,a^2/(n - 1/2) \quad \text{for} \quad |\varkappa| \leq \frac{n - 1/2}{2a},$$

$$(2.42)\quad \left\|\varphi_n(\varkappa) - \sqrt{\frac{2}{\pi}}\,\sin n\,x\right\| \leq \frac{2\,|\varkappa|\,a}{n - 1/2}, \quad |\varkappa| \leq \frac{n - 1/2}{2a}.$$

In all these formulas $n - 1/2$ should be replaced by $3/2$ for $n = 1$. The results should be compared with the corresponding results of Example 2.14. It must be borne in mind that the norm $\|\ \|$ used here is the L^2-norm whereas that in the former example is the maximum norm, but the eigenvalues are the same in both cases and eigenfunctions differ only in the normalization factor. It will be seen that the L^2-theory as a rule gives sharper results than the C-theory, at least if $\|T^{(1)}\|$ is the same in both cases (as for a multiplication operator).

In estimating the eigenvectors the C-theory has its own interest, for the C-norm is stronger than the L^2-norm. However, we can even use the result of the L^2-theory in the C-estimate of the eigenfunctions. In fact, in deducing the majorizing function (in C-norm) II-(3.30) for the eigenfunction, we have made use of the majorizing function $\Psi(\varkappa)$ of the eigenvalue $\lambda(\varkappa) - \lambda$, but this $\Psi(\varkappa)$ can be taken from the result of L^2-theory if convenient (because the eigenvalue is the same for the two theories).

Finally it should be noted that estimates for the eigenfunctions in C-norm can be deduced in the framework of pure L^2-theory. To this end we first estimate $(T - \lambda_n)\left(\varphi_n(\varkappa) - \sqrt{\frac{2}{\pi}}\,\sin n\,x\right)$ by II-(3.32) and then use (2.29) $\Big($in which a factor $\sqrt{\frac{2}{\pi}}$ should be attached to $\sin n\,x$ owing to the different normalization$\Big)$. If we calculate the bound of S_n *as an operator from* L^2 *to* C, we can estimate the C-norm of $\varphi_n(\varkappa) - \sqrt{\frac{2}{\pi}}\,\sin n\,x$. Such a bound for S_n can be estimated by $\sup_y \|s(y,\cdot)\|_2$ in terms of the kernel $s(y, x)$ of S_n [$\|s(y,\cdot)\|_2$ means the L^2-norm of $s(y, x)$ as a function of x for a fixed y].

§ 3. Selfadjoint holomorphic families

1. General remarks

When we consider a holomorphic family $T(\varkappa)$ of operators in a Hilbert space H, the most important case is the one in which $T(\varkappa)$ is selfadjoint for real $\varkappa$. More specifically, suppose that $T(\varkappa) \in \mathscr{C}(\mathsf{H})$ is holomorphic for $\varkappa$ in a domain D_0 of the complex plane symmetric with respect to the real axis, $T(\varkappa)$ is densely defined for each $\varkappa$ and that $T(\varkappa)^* = T(\bar{\varkappa})$. Then we shall call $T(\varkappa)$ a *selfadjoint holomorphic family*. It is clear that $T(\varkappa)$ is selfadjoint for each real $\varkappa \in D_0$.

The selfadjointness of $T(\varkappa)$ permits much simplification in the general results on a holomorphic family given in the preceding sections. For example, suppose that for some real $\varkappa_0 \in D_0$ the spectrum $\Sigma(\varkappa_0) = \Sigma(T(\varkappa_0))$ has gaps at α and β. Let Γ be a closed curve passing through α, β, such as considered in V-§ 3.5, with the resulting decomposition of the space $\mathsf{H} = \mathsf{M}'(\varkappa_0) \oplus \mathsf{M}''(\varkappa_0)$. According to the general results of § 1.3, $\Sigma(\varkappa) = \Sigma(T(\varkappa))$ is also separated by Γ into two parts $\Sigma'(\varkappa)$, $\Sigma''(\varkappa)$ with the associated decomposition $\mathsf{H} = \mathsf{M}'(\varkappa) \oplus \mathsf{M}''(\varkappa)$ if $\varkappa$ is sufficiently close to $\varkappa_0$. In particular the spectrum $\Sigma(\varkappa)$ of the selfadjoint operator $T(\varkappa)$ for real $\varkappa$ close to $\varkappa_0$ has gaps at α and β. The projections $P(\varkappa)$ on $\mathsf{M}'(\varkappa)$ along $\mathsf{M}''(\varkappa)$ form a selfadjoint family:

$$P(\varkappa)^* = P(\bar{\varkappa}) . \tag{3.1}$$

To see this, it suffices to note that (3.1) is true for real $\varkappa$ because $T(\varkappa)$ is then selfadjoint and so $P(\varkappa)$ is an orthogonal projection. Then (3.1) is extended to all $\varkappa$ by the unique continuation theorem [which is here trivial since $P(\varkappa)$ is bounded-holomorphic].

The part of $T(\varkappa)$ in the invariant subspace $\mathsf{M}'(\varkappa)$ may be identified with $T_r(\varkappa) = T(\varkappa)\,P(\varkappa) = P(\varkappa)\,T(\varkappa)\,P(\varkappa)$, which is bounded-holomorphic and selfadjoint. To avoid the inconvenience that $\mathsf{M}'(\varkappa)$ depends on $\varkappa$, we could introduce the transformation function $U(\varkappa)$ as in § 1.3. *This operator $U(\varkappa)$ is now unitary for real $\varkappa$*, as in the finite-dimensional case (II-§ 6.2). Thus the $\check{T}(\varkappa)$ of (1.5) form a selfadjoint family, and it may be considered a selfadjoint bounded-holomorphic family in the (fixed) Hilbert space $\mathsf{M}'(\varkappa_0)$ [which is invariant under $\check{T}(\varkappa)$]. In particular if $\mathsf{M}'(\varkappa_0)$ is of finite dimension m, the same is true of $\mathsf{M}'(\varkappa)$, and the results of II-§ 6 can then be applied directly to $\check{T}(\varkappa)$.

These considerations lead to the following results. If $T(\varkappa_0)$ has a finite system of eigenvalues for a real $\varkappa_0$, these eigenvalues change with $\varkappa$ *holomorphically* in a neighborhood of $\varkappa_0$. There may be splitting of eigenvalues, but they have *no singularities*. The associated eigenprojections are also holomorphic in $\varkappa$ and the eigennilpotents are identically zero. In short, there is nothing different from the situation in the finite-dimensional case as long as we are concerned with a limited part of the spectrum $\Sigma(\varkappa)$ for $\varkappa$ close to $\varkappa_0$.

Concerning the general properties of the perturbation series for a selfadjoint holomorphic family $T(\varkappa)$, we refer to the results of § 1.5 and II-§ 2. The selfadjointness insures that the reduction process, starting from an isolated eigenvalue λ of T with multiplicity $m < \infty$, can be continued indefinitely (as in the finite-dimensional case, see II-§ 6.1), until a stage is finally reached after which there is no more splitting and therefore the weighted mean of the group of eigenvalues at that stage is the eigenvalue itself. This indefinite reducibility is due to the fact that, at each stage, the operator $T^{(n)}(0)$ is a (finite-dimensional) selfadjoint operator. We need not repeat the details, for there is nothing new compared with the results of II-§ 6.1 once we start from the bounded-holomorphic family $T_r(\varkappa) = T(\varkappa)\,P(\varkappa)$, which is essentially the part of $T(\varkappa)$ in the m-dimensional subspace $\mathsf{M}(\varkappa) = P(\varkappa)\,\mathsf{H}$.

Problem 3.1. A necessary condition for no splitting of $\lambda(\varkappa)$ is that $\tilde{T}^{(1)}(0)$ be a scalar operator. [When $T(\varkappa)$ is defined by (2.1), $\tilde{T}^{(1)}(0)$ is equal to $P\,T^{(1)}\,P$.]

If $T(\varkappa)$ is a selfadjoint holomorphic family in a neighborhood of $\varkappa = 0$ and, in addition, is of type (A), $T(\varkappa)$ has the expression (2.1), in which T is selfadjoint and all the $T^{(n)}$ are *symmetric* [because $(T^{(n)}u, u)$ must be real]. Conversely, the family $T(\varkappa)$ defined by (2.1) is selfadjoint and of type (A) if T is selfadjoint, the $T^{(n)}$ are symmetric and satisfy the conditions (2.5) or (2.11); this is a direct consequence of Theorems 2.6 and V-4.4.

2. Continuation of the eigenvalues

If $T(\varkappa)$ is a selfadjoint holomorphic family, every isolated eigenvalue λ of $T = T(0)$ with finite multiplicity splits into one or several eigenvalues of $T(\varkappa)$ which are holomorphic at $\varkappa = 0$ [assuming that $\varkappa = 0$ is in the domain of definition D_0 of the family $T(\varkappa)$]. Each one $\lambda(\varkappa)$ of these holomorphic functions, together with the associated eigenprojection $P(\varkappa)$, can be continued analytically along the real axis, and the resulting pair represents an eigenvalue and an eigenprojection of $T(\varkappa)$. This is true even when the graph of $\lambda(\varkappa)$ crosses the graph of another such eigenvalue, as long as the eigenvalue is isolated and has finite multiplicity. In this way there is determined a *maximal interval* I of the real axis in which $\lambda(\varkappa)$ and $P(\varkappa)$ are holomorphic and represent an eigenvalue and an eigenprojection of $T(\varkappa)$[1].

In general this maximal interval I differs from one $\lambda(\varkappa)$ to another. At one or the other end of I, $\lambda(\varkappa)$ can behave in various ways: it may tend to infinity or be absorbed into the continuous spectrum[2]. We shall illustrate this by some examples.

Example 3.2. An example in which $\lambda(\varkappa) \to -\infty$ at the end of the maximal interval is furnished by the $\lambda_0(\varkappa)$ of Example 1.11. Here the maximal interval is $(0, \infty)$ and $\lambda(\varkappa) \to -\infty$ as $\varkappa \to +0$ (see Fig. 1 on p. 292).

[1] Then the transformation function $U(\varkappa)$ exists and is unitary for $\varkappa \in \mathrm{I}$, so that an orthonormal basis $\{\varphi_j(\varkappa)\}$ of $P(\varkappa)\,\mathsf{H}$ exists such that each $\varphi_j(\varkappa)$ is holomorphic for $\varkappa \in \mathrm{I}$; see § 1.3 and II-§ 6.2.

[2] Exact definition of the continuous spectrum is given in X-§ 1.

Example 3.3. An example of absorption of the eigenvalue by the continuous spectrum is given by the Schrödinger operator with the so-called square-well potential. Consider the differential operator $T(\varkappa) = -d^2/dx^2 + \varkappa\, q(x)$ in $\mathsf{H} = \mathsf{L}^2(0, \infty)$ with the boundary condition $u(0) = 0$, where $q(x)$ is assumed to have the form

$$q(x) = -1\,, \quad 0 < x < b\,; \quad q(x) = 0\,, \quad x \geqq b\,. \tag{3.2}$$

As is seen from the remark at the end of par. 1, $T(\varkappa)$ is a selfadjoint holomorphic family of type (A); here we have even a bounded perturbation since the multiplication operator by $q(x)$ is bounded. Suppose that $T(\varkappa)$ has a negative eigenvalue $\lambda = \lambda(\varkappa)$, with the eigenfunction $\varphi(x) = \varphi(x\,;\varkappa)$. φ is a solution of the differential equation

$$\varphi'' + (\lambda + \varkappa)\,\varphi = 0 \quad \text{for} \quad 0 < x < b \quad \text{and} \quad \varphi'' + \lambda\,\varphi = 0 \quad \text{for} \quad x \geqq b\,. \tag{3.3}$$

Hence $\varphi = \alpha \sin(\lambda + \varkappa)^{1/2} x$ for $x < b$ and $\varphi = \beta\, e^{-(-\lambda)^{1/2} x}$ for $x > b$. The constants λ, α, β should be determined in such a way that $\varphi(x)$ and $\varphi'(x)$ are continuous at $x = b$, for any function in $\mathsf{D}(T(\varkappa)) = \mathsf{D}$ must satisfy these conditions. This leads to the following equation to be satisfied by $\mu = -\lambda(\varkappa) > 0$:

$$\sqrt{\varkappa - \mu}\, \cot\left(\sqrt{\varkappa - \mu}\; b\right) = -\sqrt{\mu}\,. \tag{3.4}$$

It can be shown without difficulty that (3.4) has exactly N positive roots μ if

$$\varkappa_N < \varkappa < \varkappa_{N+1}\,, \quad \text{where} \quad \varkappa_N = \left(N - \frac{1}{2}\right)^2 \pi^2 b^{-2}\,, \tag{3.5}$$

and that each of these roots is an increasing function of $\varkappa$, starting with 0 at the value of $\varkappa$ where it appears (see Fig. 2).

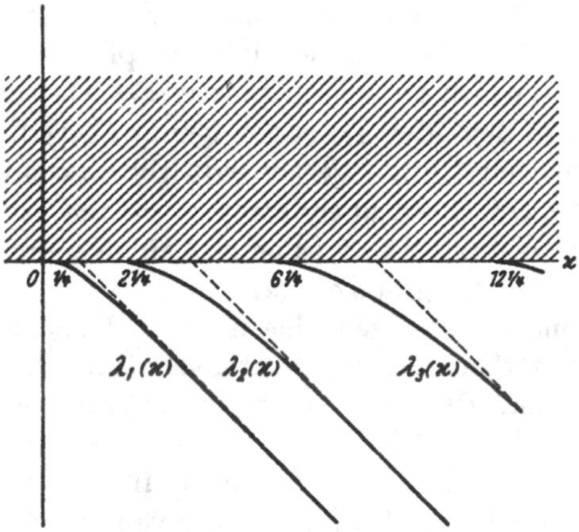

Fig. 2. The spectrum of $-u'' + \varkappa q(x) u = \lambda u$ on $(0, \infty)$, with boundary condition $u(0) = 0$, for a "square-well" potential $\varkappa\, q(x)$ with width π and depth $\varkappa$

On the other hand it is known that $T(\varkappa)$ has no nonnegative eigenvalue, the spectrum of $T(\varkappa)$ consisting of negative isolated eigenvalues and a continuous spectrum covering exactly the positive real axis[1].

Suppose now that $\varkappa$ is increased continuously from $\varkappa = -\infty$. There is no eigenvalue of $T(\varkappa)$ if $\varkappa \leqq \varkappa_1$, the first eigenvalue $\lambda_1(\varkappa)$ appears at $\varkappa = \varkappa_1 + 0$ and

[1] A rigorous proof of this fact will be given in Chapter X; see footnote [2] of p. 546.

decreases from zero, the second eigenvalue $\lambda_2(\varkappa)$ appears at $\varkappa = \varkappa_2 + 0$ and decreases from zero, and so on. The maximal interval for $\lambda_n(\varkappa)$ is $(\varkappa_n, \infty)$. At the left end of this interval $\lambda_n(\varkappa)$ merges into the continuous spectrum. A simple calculation shows that $\lambda_n(\varkappa)$ behaves at the ends of $(\varkappa_n, \infty)$ like

$$\lambda_n(\varkappa) = -\frac{b^2}{4}(\varkappa - \varkappa_n)^2 + \cdots, \quad \varkappa \searrow \varkappa_n . \tag{3.6}$$
$$\lambda_n(\varkappa) = -\varkappa + \frac{N^2\pi^2}{b^2} - \cdots, \quad \varkappa \nearrow \infty .$$

3. The Mathieu, Schrödinger, and Dirac equations

We shall give other examples of selfadjoint holomorphic families of type (A).

Example 3.4. The *Mathieu equation* is

$$u'' + (\lambda + 2\varkappa \cos 2x)\, u = 0, \quad -\pi \leq x \leq \pi . \tag{3.7}$$

Let us consider the solutions of this equation (Mathieu functions) which are *periodic* in x with period 2π. These are the eigenfunctions of the operator

$$T(\varkappa) = T + \varkappa T^{(1)}, \quad T = -d^2/dx^2, \quad T^{(1)} = -2\cos 2x , \tag{3.8}$$

where T is associated with the *periodic boundary condition*

$$u(-\pi) = u(\pi), \quad u'(-\pi) = u'(\pi) . \tag{3.9}$$

It is convenient to regard the operators (3.8) as acting in the Hilbert space $\mathsf{H} = \mathsf{L}^2(-\pi, \pi)$. Then T is selfadjoint with a discrete spectrum. The eigenvalues and the normalized eigenfunctions of T are

$$\begin{aligned} &\lambda_0 = 0, && \varphi_0(x) = (2\pi)^{-1/2}, \\ &\lambda_n^+ = n^2, && \varphi_n^+(x) = \pi^{-1/2}\cos n x, \\ &\lambda_n^- = n^2, && \varphi_n^-(x) = \pi^{-1/2}\sin n x, \quad n = 1, 2, 3, \ldots, \end{aligned} \tag{3.10}$$

each positive eigenvalue n^2 having multiplicity 2. The multiplication operator $T^{(1)}$ by $-2\cos 2x$ is symmetric and bounded with $\|T^{(1)}\| = \max|2\cos 2x| = 2$. Obviously $T(\varkappa)$ is a selfadjoint holomorphic family of type (A). The degeneracy of the eigenvalues of T does not complicate the eigenvalue problem for $T(\varkappa)$, for both T and $T^{(1)}$ are decomposed according to

$$\mathsf{H} = \mathsf{M}_0^+ \oplus \mathsf{M}_1^+ \oplus \mathsf{M}_0^- \oplus \mathsf{M}_1^- \tag{3.11}$$

where the superscripts $\pm$ denote respectively the sets of functions symmetric and antisymmetric with respect to $x = 0$ (even and odd functions) and where the subscripts 0 and 1 denote respectively the sets of functions symmetric and antisymmetric with respect to $x = \pi/2$ (here we suppose all functions continued on the whole real line periodically with period 2π). Thus, for example, M_0^+ is the set of functions symmetric with respect to $x = 0$ as well as with respect to $x = \pi/2$. It is obvious that the four subspaces of (3.11) are mutually orthogonal and span the whole space, and that T and $T^{(1)}$ are decomposed according to (3.11), the parts of these operators in each subspace being again selfadjoint. But each part of T has only simple eigenvalues as given by the following schema:

$$\begin{aligned} &\mathsf{M}_0^+ : \lambda_0, \varphi_0 \quad \text{and} \quad \lambda_n^+, \varphi_n^+, \quad n = 2, 4, 6, \ldots, \\ &\mathsf{M}_1^+ : \lambda_n^+, \varphi_n^+, \quad n = 1, 3, 5, \ldots, \\ &\mathsf{M}_0^- : \lambda_n^-, \varphi_n^-, \quad n = 1, 3, 5, \ldots, \\ &\mathsf{M}_1^- : \lambda_n^-, \varphi_n^-, \quad n = 2, 4, 6, \ldots . \end{aligned}$$

In considering the perturbed operator $T(\varkappa)$, we may therefore regard each eigenvalue λ_0 or $\lambda_n^\pm$ as simple, and its isolation distance d_0 or $d_n^\pm$ may be evaluated in the appropriate subspace, with the results that

$$d_0 = 4,\ d_1^\pm = 8,\ d_2^+ = 4,\ d_2^- = 12, \qquad (3.12)$$
$$d_n^\pm = n^2 - (n-2)^2 = 4(n-1) \quad \text{for} \quad n \geqq 3.$$

Since $\|T^{(1)}\| = 2$, lower bounds for the convergence radii $r_n^\pm$ for the series of the corresponding eigenvalues and eigenfunctions of $T(\varkappa)$ are given by $d_n^\pm/2\|T^{(1)}\|$ [see (2.34), put $b = c = 0$], that is,

$$r_0 \geqq 1,\ r_1^\pm \geqq 2,\ r_2^+ \geqq 1,\ r_2^- \geqq 3,\ r_n^\pm \geqq n-1 \quad \text{for} \quad n \geqq 3\,[1]. \qquad (3.13)$$

Also the majorizing series for each eigenvalue and eigenfunction can be written down according to II-§ 3 with the aid of (2.36).

It should be noted that a similar treatment can be given also in the space $C[-\pi, \pi]$ (or rather its subspaces consisting of periodic functions) instead of $L^2(-\pi, \pi)$. Although the results may be less sharp as regards the eigenvalues, those for eigenfunctions may be useful, in particular if they are improved by using the estimates of eigenvalues obtained by the L^2-theory (cf. Example 2.17).

Example 3.5. Consider the Schrödinger operator

$$T(\varkappa) = -\Delta + Q + \varkappa Q' \qquad (3.14)$$

in R^3, where Q and Q' are multiplication operators by real-valued functions $q(x)$ and $q'(x)$, respectively. If we assume that each of q and q' is expressible as the sum of a function in $L^2(R^3)$ and a function in $L^\infty(R^3)$ as in V-§ 5.3, $q + \varkappa q'$ has the same property for each $\varkappa$ and so $T(\varkappa)$ is selfadjoint for real $\varkappa$ (if the Laplacian Δ is interpreted in the generalized sense, see loc. cit.).

Furthermore, since Q as well as Q' is relatively bounded with respect to $H = -\Delta$ with relative bound 0, Q' is relatively bounded with respect to $T(0) = H + Q$ with relative bound 0 (see Problem IV-1.2). Therefore $T(\varkappa)$ *forms a selfadjoint holomorphic family of type* (A). That the eigenvalues and eigenvectors of (3.14) are holomorphic functions of $\varkappa$ is an immediate consequence of this, and estimates for the convergence radii and error estimates for their series could be given as in preceding examples (see also Example 4.23).

We can treat the *Dirac operator*

$$T(\varkappa) = i^{-1}\alpha\cdot\text{grad} + \beta + Q + \varkappa Q' \qquad (3.15)$$

in the same way with the same result if the potentials Q and Q' are of Coulomb type and Q is not too strong (see V-§ 5.4, in particular Remark 5.12).

4. Growth rate of the eigenvalues

It is of some interest to know how rapidly the eigenvalue $\lambda(\varkappa)$ of $T(\varkappa)$ is able to grow with $\varkappa$. For simplicity we shall restrict ourselves to a selfadjoint holomorphic family $T(\varkappa)$ and to real values of $\varkappa$, so that $T(\varkappa)$ is selfadjoint and $\lambda(\varkappa)$ is real.

In general it is rather hard to estimate the growth rate. In Example 1.11 there is one eigenvalue $\lambda_0(\varkappa)$ which exists for $\varkappa > 0$ and tends to

---

[1] See also SCHÄFKE [4]. r_0 has been studied also by WATSON [1] (who showed that $r_0 \geqq \sqrt{2}$) and BOUWKAMP [1] ($r_0 = 1.468\ldots$) by more direct but complicated methods.

$-\infty$ as $\varkappa \to +0$. However, it will now be shown that such a rapid growth of eigenvalues does not occur in the case of a holomorphic family of type (A).

For this purpose it is convenient to consider not only the holomorphic function formed by an individual $\lambda(\varkappa)$, but also any *piecewise holomorphic*, continuous function $\mu(\varkappa)$ formed by connecting some of the $\lambda(\varkappa)$. Such a $\mu(\varkappa)$ is obtained by moving arbitrarily from one $\lambda(\varkappa)$ to another at a crossing point where their graphs cross. The eigenvalues $\mu_j(\varkappa)$ considered in II-§ 6.4 are of this type. Such a $\mu(\varkappa)$ may be defined in a larger interval than any one of the holomorphic functions $\lambda(\varkappa)$ which is a "part" of it.

Suppose now that we have a selfadjoint holomorphic family $T(\varkappa)$ of type (A) defined for $\varkappa \in D_0$. We have then the inequality (2.3), which asserts that $T'(\varkappa)$ is relatively bounded with respect to $T(\varkappa)$:

$$\|T'(\varkappa)u\| \leq a'\|u\| + b'\|T(\varkappa)u\|\,, \quad u \in \mathsf{D}\,, \quad \varkappa \in \mathrm{I}\,, \tag{3.16}$$

where $\mathsf{D} = \mathsf{D}(T(\varkappa))$ and $\mathrm{I} \subset D_0$ is a compact interval of the real axis. In general the constants a', b' may depend on the choice of I. For convenience I is assumed to contain $\varkappa = 0$.

Theorem 3.6. *Let $T(\varkappa)$, I, a', b' be as above and let $\mu(\varkappa)$ be any continuous, piecewise holomorphic eigenvalue of $T(\varkappa)$ of the kind described above. Then*

$$|\mu(\varkappa) - \mu(0)| \leq \frac{1}{b'}(a' + b'|\mu(0)|)(e^{b'|\varkappa|} - 1) \tag{3.17}$$

as long as $\varkappa \in \mathrm{I}$ and $\mu(\varkappa)$ is defined.

Proof. For any $\varkappa$ at which $\mu(\varkappa)$ is holomorphic, we have

$$\mu'(\varkappa) = (T'(\varkappa)\,\varphi(\varkappa), \varphi(\varkappa)), \tag{3.18}$$

where $\varphi(\varkappa)$ is a normalized eigenvector associated with the eigenvalue $\mu(\varkappa)$[1]; (3.18) can be proved in the same way as in II-(6.10), with only the inessential difference that $T'(\varkappa)$ is not constant here. It follows from (3.18) and (3.16) that

$$|\mu'(\varkappa)| \leq \|T'(\varkappa)\,\varphi(\varkappa)\| \leq a' + b'\|T(\varkappa)\,\varphi(\varkappa)\| = a' + b'|\mu(\varkappa)|\,. \tag{3.19}$$

Since $|\mu(\varkappa)|$ is piecewise holomorphic as well as $\mu(\varkappa)$ itself, it is easy to solve this differential inequality to arrive at (3.17).

Remark 3.7. (3.17) shows that $\mu(\varkappa)$ cannot grow faster than an exponential function. Thus an eigenvalue $\lambda(\varkappa)$ can never go to infinity at a real $\varkappa \in D_0$. The same is true of any $\mu(\varkappa)$ made up by joining several $\lambda(\varkappa)$ in the manner stated above.

[1] The existence of a normalized eigenvector $\varphi(\varkappa)$ which is piecewise holomorphic for $\varkappa \in \mathrm{I}$ follows from footnote [1] of p. 387.

Problem 3.8. In Theorem 3.6 suppose that $T(0)$ is bounded from below. Then the same is true of all $T(\varkappa)$ for $\varkappa \in \mathrm{I}$ and the lower bound $\gamma(\varkappa)$ of $T(\varkappa)$ satisfies the same inequality (3.17) as for $\mu(\varkappa)$. This is true even if $T(\varkappa)$ has a continuous spectrum. [hint: Use V-(4.13) for $T = T(\varkappa)$, $S = T(\varkappa + d\varkappa)$ to deduce the same differential inequality (3.19) for $\gamma(\varkappa)$.]

5. Total eigenvalues considered simultaneously

The result of the preceding paragraph is important in the study of the behavior of the set of eigenvalues of $T(\varkappa)$ as a whole, in the case of selfadjoint holomorphic family of type (A). Since there is an essential difficulty if a continuous spectrum is involved, we assume that $T(\varkappa)$ has *compact resolvent*. It suffices to assume this for a single value of $\varkappa$, for then the same must be true for all $\varkappa$ in virtue of Theorem 2.4. In this case, any eigenvalue $\lambda(\varkappa)$ starting from an eigenvalue $\lambda = \lambda(0)$ of $T = T(0)$ can be continued holomorphically to all real $\varkappa \in \mathrm{D}_0$ (again we assume that $\varkappa = 0$ belongs to D_0); in other words, the maximal interval for $\lambda(\varkappa)$ coincides with the whole interval I_0 of real $\varkappa$ belonging to D_0 (for simplicity we assume that I_0 is connected).

To see this, let I be the maximal interval of $\lambda(\varkappa)$ considered and suppose that the right end $\varkappa_1$, say, of I is interior to I_0. Since $\lambda(\varkappa)$ is bounded for $\varkappa \nearrow \varkappa_1$ by Theorem 3.6, it is impossible for $|\lambda(\varkappa)|$ to tend to ∞. We shall show that $\lambda(\varkappa)$ tends as $\varkappa \nearrow \varkappa_1$ to an eigenvalue μ of $T(\varkappa_1)$. Since $T(\varkappa_1)$ has compact resolvent, there are only a finite number of isolated eigenvalues $\mu_1, \ldots, \mu_N$ of $T(\varkappa_1)$ in the interval $|\lambda| < M$, where M is a constant such that $|\lambda(\varkappa)| < M$ for $\varkappa \nearrow \varkappa_1$, all other points of this interval belonging to $\mathrm{P}(T(\varkappa_1))$. It follows from the upper semicontinuity of the spectrum (see IV-§ 3.1) that the part of $\Sigma(T(\varkappa))$ within this interval is confined to an arbitrarily small neighborhood of these N points μ_j, provided $|\varkappa - \varkappa_1|$ is sufficiently small. Hence $\lambda(\varkappa)$, being an eigenvalue of $T(\varkappa)$, must converge to some $\mu = \mu_j$ as $\varkappa \nearrow \varkappa_1$. Then $\lambda(\varkappa)$ must coincide with one of the holomorphic functions representing the eigenvalues of $T(\varkappa)$ that arise from the splitting of the eigenvalue μ of $T(\varkappa_1)$. This implies that $\lambda(\varkappa)$ admits an analytic continuation beyond $\varkappa_1$, contradicting the supposition that $\varkappa_1$ is the right end of the maximal interval for $\lambda(\varkappa)$.

Theorem 3.9.[1] *Let $T(\varkappa)$ be a selfadjoint holomorphic family of type* (A) *defined for $\varkappa$ in a neighborhood of an interval I_0 of the real axis. Furthermore, let $T(\varkappa)$ have compact resolvent. Then all eigenvalues of $T(\varkappa)$ can be represented by functions which are holomorphic on I_0. More precisely, there is a sequence of scalar-valued functions $\mu_n(\varkappa)$ and a sequence of vector-valued functions $\varphi_n(\varkappa)$, all holomorphic on I_0, such that for $\varkappa \in \mathrm{I}_0$,*

[1] This theorem was proved by RELLICH [5] by a different method.

the $\mu_n(\varkappa)$ represent all the repeated eigenvalues of $T(\varkappa)$ and the $\varphi_n(\varkappa)$ form a complete orthonormal family of the associated eigenvectors of $T(\varkappa)$.

Proof. The first half of the theorem has already been proved above[1]. It remains to show that the eigenvectors form a complete set. But this is another expression of the fact that all the eigenprojections of a selfadjoint operator with compact resolvent form a complete set (see V-§ 3.8).

Remark 3.10. Each $\lambda_n(\varkappa)$ is holomorphic in a complex neighborhood of I_0, but this neighborhood will depend on n, so that it will be in general impossible to find a complex neighborhood in which all the $\lambda_n(\varkappa)$ exist.

Remark 3.11. The assumption in Theorem 3.9 that $T(\varkappa)$ has compact resolvent is rather crucial. A similar result might be expected for a selfadjoint holomorphic family $T(\varkappa)$ of *compact* operators, but the existence of the limit point 0 of the eigenvalues of a compact operator causes some difficulty. Every eigenvalue $\lambda(\varkappa)$ of $T(\varkappa)$ can indeed be continued analytically as long as it does not reach the value 0. Once $\lambda(\varkappa) = 0$ for some $\varkappa$, $\lambda(\varkappa)$ may have no further analytic continuation. There is a sufficient condition that $\lambda(\varkappa)$ never equals 0 unless it is identically zero; then the result of Theorem 3.9 remains true if the system $\varphi_n(\varkappa)$ is supplemented with the eigenvectors belonging to the eigenvalue $\lambda_0(\varkappa) \equiv 0$ (if any) of $T(\varkappa)$. An example of such a sufficient condition[2] is that there exist positive constants m, M such that for real $\varkappa$

$$m\|T(0)\,u\| \leq \|T(\varkappa)\,u\| \leq M\|T(0)\,u\|\,, \quad u \in \mathsf{H}. \tag{3.20}$$

In fact, (3.20) implies that the null space N of $T(\varkappa)$ is independent of $\varkappa$ for any real $\varkappa$. Thus any orthonormal family of N can be a part of $\{\varphi_n(\varkappa)\}$ with $\lambda_n(\varkappa) \equiv 0$. Since $T(\varkappa)$ is selfadjoint for real $\varkappa$, it is decomposed according to $\mathsf{H} = \mathsf{N} \oplus \mathsf{N}^\perp$. In the subspace $\mathsf{N}^\perp$, $T(\varkappa)$ has no eigenvalue 0 so that the result of Theorem 3.9 is true.

§ 4. Holomorphic families of type (B)

1. Bounded-holomorphic families of sesquilinear forms

Let $\{\mathfrak{t}(\varkappa)\}$ be a family of *sesquilinear forms* in a Hilbert space H. Suppose that $\mathfrak{t}(\varkappa)$ is a *bounded* sesquilinear form with domain H for each $\varkappa \in D_0$, where D_0 is a domain in the complex plane, and that $\mathfrak{t}(\varkappa)\,[u]$ is holomorphic in D_0 for each fixed $u \in \mathsf{H}$. Then the family $\{\mathfrak{t}(\varkappa)\}$ is said to be *bounded-holomorphic*.

---

[1] For the existence of an orthonormal family $\{\varphi_n(\varkappa)\}$ of eigenprojections see footnote [1] on p. 387.

[2] Another sufficient condition is $0 < m\, T(0) \leq T(\varkappa) \leq M\, T(0)$; the proof is the same as for (3.20). These conditions are essentially given by RELLICH [5].

It follows by polarization that $\mathfrak{t}(\varkappa)[u, v]$ is holomorphic in D_0 for each fixed $u, v \in H$ [see VI-(1.1)]. The family of operators $T(\varkappa) \in \mathscr{B}(H)$ defined by $(T(\varkappa) u, v) = \mathfrak{t}(\varkappa)[u, v]$ is a bounded-holomorphic family of operators; this follows from Theorem III-3.12. In particular it follows that $\mathfrak{t}(\varkappa)$ is uniformly bounded in each compact subset of D_0.

A similar result holds for a family $\mathfrak{t}(\varkappa)$ originally defined for real $\varkappa$. Suppose $\mathfrak{t}(\varkappa)$ is defined for $-r < \varkappa < r$ and $\mathfrak{t}(\varkappa)[u]$ has a power series expansion convergent for $-r < \varkappa < r$. Then the family $T(\varkappa)$ of the associated operators can be extended to all complex $\varkappa$ with $|\varkappa| < r$ so as to form a bounded-holomorphic family.

To see this, it suffices to show that $\mathfrak{t}(\varkappa)$ can be continued analytically to all $\varkappa$ with $|\varkappa| < r$, preserving its property of being a *bounded* form. Let

$$\mathfrak{t}(\varkappa)[u, v] = \mathfrak{t}[u, v] + \varkappa\, \mathfrak{t}^{(1)}[u, v] + \varkappa^2\, \mathfrak{t}^{(2)}[u, v] + \cdots \tag{4.1}$$

be the Taylor expansion of the numerical function $\mathfrak{t}(\varkappa)[u, v]$ of the real variable $\varkappa$, $-r < \varkappa < r$, obtained by polarization from the expansion of the $\mathfrak{t}(\varkappa)[u]$. $\mathfrak{t} = \mathfrak{t}(0)$ is obviously a bounded sesquilinear form. We shall show that all the $\mathfrak{t}^{(n)}$ are also bounded sesquilinear forms. The sesquilinearity of $\mathfrak{t}^{(n)}$ is a simple consequence of that of $\mathfrak{t}(\varkappa)$ and the uniqueness of the Taylor expansion (4.1).

(4.1) implies that $\mathfrak{t}^{(1)}[u, v] = \lim_{\varkappa \to 0} \varkappa^{-1}(\mathfrak{t}(\varkappa) - \mathfrak{t})[u, v]$. Since this limit exists for every $u, v \in H$ and since $\mathfrak{t}(\varkappa) - \mathfrak{t}$ is a bounded form for every real $\varkappa$, it follows from the principle of uniform boundedness that $\mathfrak{t}^{(1)}$ is bounded (Problem III-1.30). Then we have from (4.1) that $\mathfrak{t}^{(2)}[u, v] = \lim_{\varkappa \to 0} \varkappa^{-2}(\mathfrak{t}(\varkappa) - \mathfrak{t} - \varkappa\, \mathfrak{t}^{(1)})[u, v]$, and a similar argument shows that $\mathfrak{t}^{(2)}$ is bounded, and so on.

Thus there are operators $T^{(n)} \in \mathscr{B}(H)$ such that $\mathfrak{t}^{(n)}[u, v] = (T^{(n)} u, v)$, and (4.1) can be written

$$(T(\varkappa) u, v) = (T u, v) + \varkappa (T^{(1)} u, v) + \varkappa^2 (T^{(2)} u, v) + \cdots. \tag{4.2}$$

This series converges for $|\varkappa| < r$ for all $u, v \in H$. Hence $r'^n (T^{(n)} u, v) \to 0$, $n \to \infty$, for all u, v and any $r' < r$. Again the principle of uniform boundedness (see Problem III-3.13) shows that $\{r'^n \|T^{(n)}\|\}$ is bounded as $n \to \infty$. Hence

$$T(\varkappa) = T + \varkappa\, T^{(1)} + \varkappa^2\, T^{(2)} + \cdots \tag{4.3}$$

is absolutely convergent (in norm) for $|\varkappa| < r$ and defines $T(\varkappa) \in \mathscr{B}(H)$. Thus $\mathfrak{t}(\varkappa)[u, v] = (T(\varkappa) u, v)$, valid for real $\varkappa$, has been extended to all $\varkappa$ with $|\varkappa| < r$, $T(\varkappa)$ forming a bounded-holomorphic family.

Remark 4.1. Let $\mathfrak{t}(\varkappa)$ be a family of sesquilinear forms with domain H and let $\mathfrak{t}(\varkappa)[u]$ be holomorphic for $\varkappa \in D_0$ for every $u \in H$. Then $\mathfrak{t}(\varkappa)$ is a bounded-holomorphic family of forms if there is a sequence $\varkappa_n \in D_0$, converging to a $\varkappa_0 \in D_0$ different from any $\varkappa_n$, such that $\mathfrak{t}(\varkappa_n)$ is a bounded form for every $n = 1, 2, \ldots$.

The point is that the assumptions imply that $\mathfrak{t}(\varkappa)$ is *bounded* for every $\varkappa \in D_0$. This is seen again by invoking the principle of uniform boundedness. We may assume $\varkappa_0 = 0$, $\varkappa_n \neq 0$. The existence of $\lim_{n\to\infty} (T(\varkappa_n) u, v) = \lim \mathfrak{t}(\varkappa_n) [u, v] = \mathfrak{t}[u, v]$ implies that $T(\varkappa_n)$ is uniformly bounded and hence $\mathfrak{t}$ is bounded with the associated T. Then $\mathfrak{t}^{(1)}[u, v] = \lim \varkappa_n^{-1}(\mathfrak{t}(\varkappa_n) - \mathfrak{t}) [u, v]$ shows that $\mathfrak{t}^{(1)}$ is bounded, and so on as above. Thus we see that $\mathfrak{t}(\varkappa)$ is bounded near $\varkappa = 0$, and this property propagates to all $\varkappa \in D_0$.

2. Holomorphic families of forms of type (a) and holomorphic families of operators of type (B)

We now consider a family $\mathfrak{t}(\varkappa)$ of (not necessarily bounded) sesquilinear forms defined for $\varkappa \in D_0$. $\mathfrak{t}(\varkappa)$ will be called a *holomorphic family of type* (a) if i) each $\mathfrak{t}(\varkappa)$ is sectorial and closed with $\mathsf{D}(\mathfrak{t}(\varkappa)) = \mathsf{D}$ independent of $\varkappa$ and dense in H, and ii) $\mathfrak{t}(\varkappa) [u]$ is holomorphic for $\varkappa \in D_0$ for each fixed $u \in \mathsf{D}$. Note that ii) implies, by polarization, that $\mathfrak{t}(\varkappa) [u, v]$ is holomorphic in $\varkappa$ for each fixed pair $u, v \in \mathsf{D}$. (For sectorial forms see Chapter VI.)

Theorem 4.2. *Let* $\mathfrak{t}(\varkappa)$ *be a holomorphic family of forms of type* (a). *For each* $\varkappa$ *let* $T(\varkappa) = T_{\mathfrak{t}(\varkappa)}$ *be the associated m-sectorial operator* (see Theorem VI-2.7). *Then the* $T(\varkappa)$ *form a holomorphic family of operators, and the* $T(\varkappa)$ *are locally uniformly sectorial.*

A holomorphic family of m-sectorial operators associated with a holomorphic family of forms of type (a) according to Theorem 4.2 will be called a *holomorphic family of type* (B)[1].

Proof of Theorem 4.2. We may assume that $\varkappa = 0$ belongs to D_0 and $\mathfrak{h} = \operatorname{Re} \mathfrak{t} \geq 1$, $\mathfrak{t} = \mathfrak{t}(0)$; otherwise this situation can be attained by shifting the origin of the $\varkappa$-plane and adding a suitable constant to $\mathfrak{t}(\varkappa)$.

Let $H = T_{\mathfrak{h}} \geq 1$ be the selfadjoint operator associated with the closed form $\mathfrak{h} \geq 1$ [H is by definition the real part of $T = T(0)$, see VI-§ 3.1], and consider the forms $\mathfrak{t}_0(\varkappa) [u, v] = \mathfrak{t}(\varkappa) [G^{-1} u, G^{-1} v]$ where $G = H^{1/2}$. Since $G^{-1} u \in \mathsf{D}(G) = \mathsf{D}(\mathfrak{h}) = \mathsf{D}$ by the second representation theorem (Theorem VI-2.23), $\mathfrak{t}_0(\varkappa)$ is a sectorial form defined everywhere on H. Since $\mathfrak{t}_0(\varkappa)$ is closable as is easily verified, it follows from Theorem VI-1.20 that $\mathfrak{t}_0(\varkappa)$ is bounded. Since $\mathfrak{t}_0(\varkappa) [u, v]$ is obviously holomorphic for $\varkappa \in D_0$ for each fixed $u, v \in \mathsf{H}$, $\mathfrak{t}_0(\varkappa)$ forms a bounded-holomorphic family of forms. Thus we have the expression $\mathfrak{t}_0(\varkappa) [u, v] = (T_0(\varkappa) u, v)$, where $\{T_0(\varkappa)\}$ is a bounded-holomorphic family of operators (see the

[1] A special case (selfadjoint) of this type was introduced in T. Kato [8].

preceding paragraph). Replacing u, v respectively by Gu, Gv, we thus obtain the expression

$$\mathfrak{t}(\varkappa)[u, v] = (T_0(\varkappa) Gu, Gv), \quad u, v \in \mathsf{D}, \quad G = H^{1/2}. \tag{4.4}$$

An argument similar to that used in deducing VI-(3.4) then gives

$$T(\varkappa) = G T_0(\varkappa) G, \tag{4.5}$$

$$T(\varkappa)^{-1} = G^{-1} T_0(\varkappa)^{-1} G^{-1}. \tag{4.6}$$

Here $G^{-1} \in \mathscr{B}(\mathsf{H})$ and $T_0(\varkappa)^{-1}$ is bounded-holomorphic near $\varkappa = 0$, for $T_0(\varkappa)$ is bounded-holomorphic and $T_0(0)^{-1} \in \mathscr{B}(\mathsf{H})$ exists because $T_0(0)$ is of the form $1 + iB$, $B^* = B$ [see VI-(3.4)]. Therefore $T(\varkappa)$ is holomorphic near $\varkappa = 0$ by Theorem 1.3. Since $\varkappa = 0$ is not distinguished from other points, $T(\varkappa)$ is holomorphic for $\varkappa \in \mathsf{D}_0$. Finally it is clear from (4.4) that the $\mathfrak{t}(\varkappa)$ are uniformly sectorial near $\varkappa = 0$, for $T_0(0) = 1 + iB$ and the $T_0(\varkappa)$ are uniformly bounded near $\varkappa = 0$. Hence the same is true of the $T(\varkappa)$.

Theorem 4.3. *Let $T(\varkappa)$ be a holomorphic family of operators of type* (B). *$T(\varkappa)$ has compact resolvent either for all $\varkappa$ or for no $\varkappa$.*

The proof is similar to that of Theorem 2.4; we need only replace Theorem IV-3.17 used there by Theorem VI-3.4. (Note that an m-sectorial operator has a non-empty resolvent set.)

Remark 4.4. In the definition of a holomorphic family $\mathfrak{t}(\varkappa)$ of forms of type (a), it suffices to require in i) the closedness of $\mathfrak{t}(\varkappa)$ only for a sequence $\varkappa = \varkappa_n$ converging to a $\varkappa_0 \in \mathsf{D}_0$ different from any of the $\varkappa_n$. To see this we note that, in the proof of Theorem 4.2, $\mathfrak{t}_0(\varkappa)$ is first seen to be bounded for $\varkappa = \varkappa_n$. Then we see by Remark 4.1 that $\mathfrak{t}_0(\varkappa)$ is bounded for all $\varkappa = \mathsf{D}_0$ and the proof of Theorem 4.2 is valid. Then it follows from (4.4) that $\mathfrak{t}(\varkappa)$ is closed for sufficiently small $|\varkappa|$, and then we can generalize this to all $\varkappa \in \mathsf{D}_0$.

This remark is useful since we need to verify that $\mathfrak{t}(\varkappa)$ is closed, say, only for real $\varkappa$ when D_0 meets the real axis.

Remark 4.5. For a holomorphic family $\mathfrak{t}(\varkappa)$ of type (a) we have the following inequalities, which correspond to similar inequalities (2.2)–(2.4) for a family of operators of type (A). (Here we assume that $\mathfrak{h} \geqq 1$, as in the proof of Theorem 4.2.)

$$\begin{aligned} &|\mathfrak{t}(\varkappa_1)[u]| \leqq b|\mathfrak{t}(\varkappa)[u]|, \\ &|\mathfrak{t}'(\varkappa_1)[u, v]| \leqq b'|\mathfrak{t}(\varkappa)[u]|^{1/2}| |\mathfrak{t}(\varkappa)[v]|^{1/2}, \\ &|(\mathfrak{t}(\varkappa_1) - \mathfrak{t}(\varkappa_2))[u, v]| \leqq \varepsilon |\mathfrak{t}(\varkappa)[u]|^{1/2} |\mathfrak{t}(\varkappa)[v]|^{1/2}. \end{aligned} \tag{4.7}$$

Here $u \in \mathsf{D}$ and a, b, a', b' are constants as long as $\varkappa$ and $\varkappa_1$ remain in a compact subset of D_0; ε may be made arbitrarily small if $|\varkappa_1 - \varkappa_2|$

is sufficiently small. (4.7) can be easily proved by making use of (4.4). Also we note that the $|\mathfrak{t}(\varkappa)[u]|$ on the right of these inequalities may be replaced by $\mathfrak{h}(\varkappa)[u] = \mathrm{Re}\,\mathfrak{t}(\varkappa)[u]$ with possible change of the constants a, b, etc. [see VI-(1.42)–(1.43)]. (4.7) expresses that, roughly speaking, the $\mathfrak{t}(\varkappa)$ are relatively bounded with respect to one another and the change of $\mathfrak{t}(\varkappa)$ is relatively continuous with respect to $\mathfrak{t}(\varkappa)$.

Remark 4.6. There are several useful identities for the resolvent $R(\zeta, \varkappa)$ of $T(\varkappa)$. Assuming $\mathfrak{h} \geqq 1$ as in the proof of Theorem 4.2, we have

$$(4.8)\quad ((T(\varkappa_1)^{-1} - T(\varkappa_2)^{-1})u, v) = -(\mathfrak{t}(\varkappa_1) - \mathfrak{t}(\varkappa_2))[T(\varkappa_1)^{-1}u, T(\varkappa_2)^{*-1}v].$$

This follows from $\mathfrak{t}(\varkappa_1)[T(\varkappa_1)^{-1}u, g] = (u, g)$ and $\mathfrak{t}(\varkappa_2)[f, T(\varkappa_2)^{*-1}v] = \overline{\mathfrak{t}(\varkappa_2)^*[T(\varkappa_2)^{*-1}v, f]} = \overline{(v, f)} = (f, v)$, where $f = T(\varkappa_1)^{-1}u \in \mathsf{D}$, $g \in T(\varkappa_2)^{*-1}v \in \mathsf{D}$.

Letting $\varkappa_1 \to \varkappa_2$ in (4.8), we obtain

$$(4.9)\qquad \left(\frac{d}{d\varkappa} T(\varkappa)^{-1}u, v\right) = -\mathfrak{t}'(\varkappa)[T(\varkappa)^{-1}u, T(\varkappa)^{*-1}v],$$

since $T(\varkappa)^{-1}u$ is continuous (even holomorphic) in the norm $\|w\|_{\mathfrak{t}} = \|Gw\|$ in virtue of (4.6).

If we replace $\mathfrak{t}(\varkappa)$ and $T(\varkappa)$ by $\mathfrak{t}(\varkappa) - \zeta$ and $T(\varkappa) - \zeta$, respectively, (4.8) and (4.9) give

$$(4.10)\quad ((R(\zeta,\varkappa_1) - R(\zeta,\varkappa_2))u, v) = -(\mathfrak{t}(\varkappa_1) - \mathfrak{t}(\varkappa_2))[R(\zeta,\varkappa_1)u, R(\zeta,\varkappa_2)^*v],$$

$$(4.11)\qquad \left(\frac{d}{d\varkappa} R(\zeta,\varkappa)u, v\right) = -\mathfrak{t}'(\varkappa)[R(\zeta,\varkappa)u, R(\zeta,\varkappa)^*v].$$

In view of (4.7), we obtain from (4.11)

$$\left|\left(\frac{d}{d\varkappa} R(\zeta,\varkappa)u, v\right)\right| \leqq b'|\mathfrak{t}(\varkappa)[R(\zeta,\varkappa)u]|^{1/2}\,|\mathfrak{t}(\varkappa)^*[R(\zeta,\varkappa)^*v]|^{1/2}$$
$$= b'|(T(\varkappa)R(\zeta,\varkappa)u, R(\zeta,\varkappa)u)|^{1/2}\,|(T^*(\varkappa)R(\zeta,\varkappa)^*v, R(\zeta,\varkappa)^*v)|^{1/2}.$$

But $\|T(\varkappa)R(\zeta,\varkappa)\| \leqq 1$ if $\zeta < 0$ (Problem V-3.31) and $\|T(\varkappa)^* R(\zeta,\varkappa)^*\| \leqq 1$ similarly. Hence

$$(4.12)\qquad \left\|\frac{d}{d\varkappa} R(\zeta,\varkappa)\right\| \leqq b'\|R(\zeta,\varkappa)\|^{1/2}\|R(\zeta,\varkappa)^*\|^{1/2} = b'\|R(\zeta,\varkappa)\| \leqq \frac{b'}{1-\zeta},\quad \zeta < 0.$$

Remark 4.7. If $\mathfrak{t}(\varkappa)$ is a holomorphic family of forms of type (a) and if $\mathfrak{t}(\varkappa)[u]$ is real for real $\varkappa$ (assuming that D_0 intersects the real axis), then it is easily seen that $\mathfrak{t}(\varkappa)^* = \mathfrak{t}(\bar{\varkappa})$; in this sense $\mathfrak{t}(\varkappa)$ is a *selfadjoint family*. Then $T(\varkappa)^* = T(\bar{\varkappa})$, and $T(\varkappa)$ forms a selfadjoint family of operators (see § 3.1); in particular $T(\varkappa)$ is a selfadjoint operator for real $\varkappa$.

In this case $(T(\varkappa) + \lambda)^{1/2}$ has constant domain D for real $\varkappa$ and large λ, by the second representation theorem. But it is not clear whether the same is true for complex $\varkappa$, though $(T(\varkappa)+\lambda)^{1/2}$ is well defined by V-§ 3.11. Thus we do not know whether $(T(\varkappa)+\lambda)^{1/2}$ forms a holomorphic family of type (A) when $T(\varkappa)$ is of type (B), even when it is a selfadjoint family.

In any case, however, $(T(\varkappa)+\lambda)^{1/2}$ is a holomorphic family. To see this we note that

$$(4.13)\qquad (T(\varkappa)+\lambda)^{-1/2} = \frac{1}{\pi}\int\limits_0^\infty \mu^{-1/2}(T(\varkappa)+\lambda+\mu)^{-1}\,d\mu$$

by V-(3.43). (4.13) can be differentiated under the integral sign, for we have the estimate $\|d(T(\varkappa)+\lambda+\mu)^{-1}/d\varkappa\| \leqq b'(\lambda+\mu)^{-1}$ by (4.12), so that the resulting integral converges absolutely. This shows that $(T(\varkappa)+\lambda)^{-1/2}$ is bounded-holomorphic in $\varkappa$. Thus $(T(\varkappa)+\lambda)^{1/2}$ is holomorphic by Theorem 1.3.

3. A criterion for type (B)

Theorem 4.8. *Let* $\mathfrak{t}^{(n)}$, $n = 0, 1, 2, \ldots$, *be a sequence of sesquilinear forms in* H. *Let* $\mathfrak{t} = \mathfrak{t}^{(0)}$ *be densely defined, with* $\mathsf{D}(\mathfrak{t}) = \mathsf{D}$, *sectorial and closable. Let* $\mathfrak{t}^{(n)}$, $n \geqq 1$, *be relatively bounded with respect to* $\mathfrak{t}$ *so that* $\mathsf{D}(\mathfrak{t}^{(n)}) \supset \mathsf{D}$ *and*

$$(4.14)\qquad |\mathfrak{t}^{(n)}[u]| \leqq c^{n-1}(a\|u\|^2 + b\,\mathfrak{h}[u])\,,\quad u\in\mathsf{D}\,,\quad n\geqq 1\,,$$

where $\mathfrak{h} = \mathrm{Re}\,\mathfrak{t}$ *and* $a, b \geqq 0$. *Then the form*

$$(4.15)\qquad \mathfrak{t}(\varkappa)\,[u] = \sum_{n=0}^{\infty} \varkappa^n\,\mathfrak{t}^{(n)}[u]\,,\quad \mathsf{D}(\mathfrak{t}(\varkappa)) = \mathsf{D}\,,$$

and the associated polar form $\mathfrak{t}(\varkappa)\,[u, v]$ *are defined for* $|\varkappa| < 1/c$. $\mathfrak{t}(\varkappa)$ *is sectorial and closable for* $|\varkappa| < 1/(b+c)$. *Let* $\tilde{\mathfrak{t}}(\varkappa)$ *be its closure.* $\{\tilde{\mathfrak{t}}(\varkappa)\}$ *is a holomorphic family of forms of type* (a) *for* $|\varkappa| < 1/(b+c)$. *The operator* $T(\varkappa)$ *associated with* $\tilde{\mathfrak{t}}(\varkappa)$ *forms a holomorphic family of type* (B).

Proof. The right member of (4.15) converges for $|\varkappa| < 1/c$ and defines a quadratic form. By polarization the sesquilinear form $\mathfrak{t}(\varkappa)\,[u, v]$ is defined also for $|\varkappa| < 1/c$. We have

$$(4.16)\qquad |(\mathfrak{t}(\varkappa)-\mathfrak{t})\,[u]| \leqq \frac{|\varkappa|}{1-c|\varkappa|}\,(a\|u\|^2 + b\,\mathfrak{h}[u])\,,\quad |\varkappa| < 1/c.$$

By Theorem VI-1.33, $\mathfrak{t}(\varkappa)$ is sectorial and closable if $b|\varkappa|/(1-c|\varkappa|) < 1$, that is, if $|\varkappa| < 1/(b+c)$. It follows also from the same theorem that the closure $\tilde{\mathfrak{t}}(\varkappa)$ has the same domain $\tilde{\mathsf{D}} = \mathsf{D}(\tilde{\mathfrak{t}})$ as $\tilde{\mathfrak{t}}$. What remains is to show that $\tilde{\mathfrak{t}}(\varkappa)\,[u]$ has a power series similar to (4.15) for every $u\in\tilde{\mathsf{D}}$:

$$(4.17)\qquad \tilde{\mathfrak{t}}(\varkappa)\,[u] = \sum_{n=0}^{\infty} \varkappa^n\,\tilde{\mathfrak{t}}^{(n)}[u]\,,\quad u\in\mathsf{D}\,;$$

here $\mathfrak{t}^{(0)} = \mathfrak{t}$ and $\mathfrak{t}^{(n)}$ is an extension of $\mathfrak{t}^{(n)}$ with domain $\tilde{\mathsf{D}}$ (not necessarily the closure of $\mathfrak{t}^{(n)}$), for which the same inequality (4.14) holds with $\mathfrak{h}$ replaced by $\tilde{\mathfrak{h}} = \operatorname{Re}\tilde{\mathfrak{t}}$. $\tilde{\mathfrak{t}}^{(n)}[u]$ is defined for $u \in \tilde{\mathsf{D}}$ by $\tilde{\mathfrak{t}}^{(n)}[u] = \lim_{k\to\infty} \mathfrak{t}^{(n)}[u_k]$ where $\{u_k\}$ is a sequence such that $u_k \underset{\mathfrak{t}}{\to} u$ and the existence of this limit is insured by (4.14). (4.17) follows from (4.15) with u replaced by u_k on going to the limit $k \to \infty$.

Since $T(\varkappa)$ is holomorphic in $\varkappa$ in Theorem 4.8, the resolvent $R(\zeta, \varkappa) = R(\zeta, T(\varkappa))$ is holomorphic in ζ and $\varkappa$ jointly in an appropriate region of the variables. Let us estimate this region and the convergence radius and errors for the Taylor series. Such estimates can be obtained from the proof of Theorem VI-3.4. For simplicity assume $\mathfrak{h} = \operatorname{Re}\mathfrak{t} \geqq 0$ and $a \geqq 0$. It suffices to replace the S of VI-(3.10) by $T(\varkappa)$; then the C on the right of this formula should be replaced by the operator $C(\varkappa)$ defined by $(\tilde{\mathfrak{t}}(\varkappa) - \mathfrak{t})[u, v] = (C(\varkappa) G' u, G' v)$. Recall that $G' = H'^{1/2}$, $H' = H + \varrho \geqq 0$. In view of (4.17), $C(\varkappa)$ has the form

$$(4.18)\quad C(\varkappa) = \sum_{n=1}^{\infty} \varkappa^n C^{(n)}, \quad C^{(n)} \in \mathscr{B}(\mathsf{H}), \quad (C^{(n)} G' u, G' v) = \tilde{\mathfrak{t}}^{(n)}[u, v].$$

We have

$$(4.19)\qquad \|C^{(n)}\| \leqq 2c^{n-1} k, \quad k = \max(b, a/\varrho)$$

[see VI-(3.8)–(3.9)]. Thus $R(\zeta, S) = R(\zeta, T(\varkappa)) = R(\zeta, \varkappa)$ becomes

$$(4.20)\quad R(\zeta, \varkappa) = G'^{-1}(1 - \zeta' H'^{-1} + i B' + C(\varkappa))^{-1} G'^{-1}$$
$$= R(\zeta) + G'^{-1} \sum_{p=1}^{\infty} J(\zeta)\,(-C(\varkappa)\, J(\zeta))^p\, G'^{-1},$$

where

$$(4.21)\quad R(\zeta) = R(\zeta, 0) = R(\zeta, T), \; J(\zeta) = (1 - \zeta' H'^{-1} + i B')^{-1}, \; \zeta' = \zeta + \varrho,$$

if $|\varkappa|$ is so small that

$$(4.22)\qquad \|C(\varkappa)\|\, \|J(\zeta)\| \leqq \|J(\zeta)\| \sum_{n=1}^{\infty} |\varkappa|^n \|C^{(n)}\| < 1.$$

In view of (4.19), (4.22) is satisfied if

$$(4.23)\qquad |\varkappa| < (2k \|J(\zeta)\| + c)^{-1}.$$

Here it has been tacitly assumed that $J(\zeta) \in \mathscr{B}(\mathsf{H})$; in general it is not easy to see when this is satisfied, but it is certainly the case if $\operatorname{Re}\zeta' \leqq 0$, for we have then $\|J(\zeta)\| \leqq 1$ (see loc. cit.) and (4.23) is satisfied if

$$(4.24)\qquad |\varkappa| < (2k + c)^{-1}, \quad \operatorname{Re}\zeta' = \operatorname{Re}\zeta + \varrho \leqq 0.$$

In particular if $-\operatorname{Re}\zeta > a/b$, we can choose $\varrho > a/b$ without violating the condition $\operatorname{Re}\zeta' \leqq 0$; then we have $k = b$ and (4.24) becomes $|\varkappa| < (2b + c)^{-1}$. In other words *$R(\zeta, \varkappa)$ is a convergent power series in $\varkappa$ for $|\varkappa| < (2b + c)^{-1}$ if $-\operatorname{Re}\zeta > 0$ is sufficiently large.* This gives another proof that $T(\varkappa)$ is holomorphic at $\varkappa = 0$.

These results admit some refinements if the unperturbed form $\mathfrak{t} = \mathfrak{h}$ is symmetric so that the associated operator $T = H$ is selfadjoint. Then we need not assume $\mathfrak{h} \geqq 0$, the constant a may be negative and $J(\zeta) \in \mathscr{B}(\mathsf{H})$ is true for any $\zeta \in \mathrm{P}(H)$. To see this we need only make the following modifications. We set $\varrho = a\,b^{-1} + \delta$, $\delta > 0$; then $H' = H + \varrho \geqq \delta > 0$ by (4.14), and (4.19) is true with $k = b$ and $R(\zeta, \varkappa)$ converges as a power series in $\varkappa$ for $|\varkappa| < (2b\|J(\zeta)\| + c)^{-1}$ [see (4.23)]. Since we have $B' = 0$ now, $J(\zeta) = (1 - \zeta' H'^{-1})^{-1} = H'(H' - \zeta')^{-1} = (H + a\,b^{-1} + \delta)\,(H - \zeta)^{-1} = b^{-1}(a + bH + b\,\delta)\,R(\zeta)$. Since $\delta > 0$ is arbitrary, we conclude that $R(\zeta,\varkappa)$ converges as a power series in $\varkappa$ if $|\varkappa| < (2\|(a + bH)\,R(\zeta)\| + c)^{-1}$. Here even the factor 2 can be dropped if all the $\mathfrak{t}^{(n)}$ are symmetric, for then $|\mathfrak{t}^{(n)}[u]| \leqq b\,\mathfrak{h}'[u]$ implies $|\mathfrak{t}^{(n)}[u, v]| \leqq b\|G' u\|\,\|G' v\|$. The case $b = 0$ can be dealt with by going to the limit. These modifications yield

Theorem 4.9. *In* Theorem 4.8 *let $\mathfrak{t} = \mathfrak{h}$ be symmetric (with the associated operator $T = H$ selfadjoint) and let a in* (4.14) *be not necessarily nonnegative. For any $\zeta \in \mathrm{P}(H)$, the resolvent $R(\zeta, \varkappa)$ of $T(\varkappa)$ exists and is a convergent power series in $\varkappa$ for*

$$|\varkappa| < (\varepsilon\|(a + bH)\,R(\zeta, H)\| + c)^{-1}, \tag{4.25}$$

where $\varepsilon = 1$ or 2 according as all the $\mathfrak{t}^{(n)}$ are symmetric or not.

Remark 4.10. Not only the convergence radius but $R(\zeta, \varkappa)$ itself can be estimated as in VI-(3.16), in which a and b must now be replaced by appropriate functions of $\varkappa$. We omit the details.

Example 4.11. Consider the form

$$\mathfrak{t}(\varkappa)\,[u, v] = \int_a^b p(x)\,u'\,\bar{v}'\,dx + \int_a^b [q(x) + \varkappa\,q^{(1)}(x)]\,u\,\bar{v}\,dx + (h_a + \varkappa\,h_a^{(1)})\,u(a)\,\overline{v(a)} + (h_b + \varkappa\,h_b^{(1)})\,u(b)\,\overline{v(b)}\,, \tag{4.26}$$

where $p(x)$, $q(x)$, $q^{(1)}(x)$ are continuous on $a \leqq x \leqq b$ and $p(x) < 0$ [see Example VI-1.7; here we have set $r(x) = s(x) = 0$ for simplicity, but we could have retained them without affecting the following results]. As we have seen in Example VI-1.36, $\mathfrak{t}(\varkappa)$ satisfies the assumptions of Theorem 4.8 with an arbitrarily small b because all forms on the right of (4.26) except the first are relatively bounded with respect to the first term, with relative bound 0. The operator $T(\varkappa)$ associated with $\mathfrak{t}(\varkappa)$ is given by (see Example VI-2.16)

$$T(\varkappa)\,u = -(p\,u')' + (q + \varkappa\,q^{(1)})\,u \tag{4.27}$$

with the boundary condition

(4.28) $p(a)\,u'(a) - (h_a + \varkappa h_a^{(1)})\,u(a) = 0,\ p(b)\,u'(b) + (h_b + \varkappa h_b^{(1)})\,u(b) = 0\,.$

Thus the $T(\varkappa)$ form a holomorphic family of type (B) defined for all complex $\varkappa$.

If we start from the restriction $\mathfrak{t}_0(\varkappa)$ of $\mathfrak{t}(\varkappa)$ with domain restricted by the boundary condition

(4.29) $$u(a) = 0\,,\quad u(b) = 0\,,$$

we get another holomorphic family $T_0(\varkappa)$, which is given formally by (4.27) with boundary condition (4.29) (see Example VI-2.17). But this is rather trivial, for $T_0(\varkappa)$ has a constant domain and the perturbing term $\varkappa q^{(1)}$ in (4.26) is a bounded operator, so that $T_0(\varkappa)$ is actually of type (A).

By restricting the domain of $\mathfrak{t}(\varkappa)$ by the condition $u(a) = 0$ only, we arrive at a third family of operators (4.27) with the boundary condition: (4.28) at $x = b$ and (4.29) at $x = a$, which is again holomorphic of type (B).

4. Holomorphic families of type (B_0)

In Theorem 4.8, the unperturbed form $\mathfrak{t}$ may not be closed. If $\mathfrak{t}$ is closed, the perturbed form $\mathfrak{t}(\varkappa)$ is closed too, for we have then $\mathsf{D}(\tilde{\mathfrak{t}}(\varkappa)) = \mathsf{D}$. But it is convenient to have the theorem in its generality without assuming $\mathfrak{t}$ to be closed.

As an important application, suppose we are given a family of operators $S(\varkappa)$ of the form

(4.30) $$S(\varkappa) = S + \varkappa S^{(1)} + \varkappa^2 S^{(2)} + \cdots,$$

where S is densely defined and sectorial, $\mathsf{D}(S^{(n)}) \supset \mathsf{D}(S)$ and

(4.31) $$|(S^{(n)} u, u)| \leqq c^{n-1}(a\|u\|^2 + b\,\mathrm{Re}(Su, u))\,,\quad u \in \mathsf{D}(S)\,,$$

with constants a, b such that $a \geqq 0, 0 \leqq b < 1$.

Then the assumptions of Theorem 4.8 are satisfied by setting $\mathfrak{t}[u, v] = (Su, v)$ and $\mathfrak{t}^{(n)}[u, v] = (S^{(n)} u, v)$, for the form $\mathfrak{t}$ is closable (see Theorem VI-1.27). The resulting operator $T(\varkappa)$ is the Friedrichs extension of $S(\varkappa)$ (see VI-§ 2.3). Thus we have

Theorem 4.12. *Let* $S(\varkappa)$ *be as above. The Friedrichs extension* $T(\varkappa)$ *of* $S(\varkappa)$ *exists and forms a holomorphic family of type* (B) *for* $|\varkappa| < (b + c)^{-1}$.

A family $T(\varkappa)$ as in Theorem 4.12 will be called a holomorphic family of type (B_0)[1]. It is a special case of type (B), and the domains of the $T(\varkappa)$ for different $\varkappa$ have a dense common part $\mathsf{D}(S)$, which is not true for a general family of type (B).

$T(\varkappa)$ is a closed extension of $S(\varkappa)$ but it may or may not be identical with the closure of $S(\varkappa)$. If not, it is possible that there is another family $T_1(\varkappa)$ consisting of m-sectorial extensions of $S(\varkappa)$. In this connection, the following theorem, essentially due to RELLICH [3], is remarkable.

[1] A special case (selfadjoint) of type (B_0) was introduced by RELLICH [3].

Theorem 4.13. *If the closure* $\tilde{S}$ *of* S *is m-sectorial, the family* $T(\varkappa)$ *of* Theorem 4.12 *is the unique holomorphic family*[1] *defined near* $\varkappa = 0$ *consisting of extensions of* $S(\varkappa)$ *such that* $T(0) = \tilde{S}$.

Proof. We have $T = T(0) = \tilde{S}$ since T is an m-sectorial extension of S. Suppose that $T_1(\varkappa)$ is a holomorphic family such that $T_1(\varkappa) \supset S(\varkappa)$ near $\varkappa = 0$ and $T_1 = T_1(0) = \tilde{S} = T$. Then $P(T_1)$ is not empty. If $\zeta \in P(T_1)$,

$$R(\zeta, T_1(\varkappa)) = R + \varkappa R_1 + \varkappa^2 R_2 + \cdots \tag{4.32}$$

is bounded-holomorphic near $\varkappa = 0$ by Theorem 1.3. For any $u \in D(S)$, we have $R(\zeta, T_1(\varkappa))(S(\varkappa) - \zeta)u = u$ since $T_1(\varkappa) \supset S(\varkappa)$. Hence

$$(R + \varkappa R_1 + \varkappa^2 R_2 + \cdots)(S - \zeta + \varkappa S^{(1)} + \varkappa^2 S^{(2)} + \cdots)u = u, \tag{4.33}$$

and comparison of coefficients gives

$$\begin{aligned} &R(S - \zeta)u = u, \\ &R_1(S - \zeta)u = -RS^{(1)}u, \\ &R_2(S - \zeta)u = -RS^{(2)}u - R_1 S^{(1)}u, \\ &\cdots\cdots\cdots\cdots\cdots\cdots \end{aligned} \tag{4.34}$$

for any $u \in D(S)$. These equations determine the operators R, R_1, $R_2, \ldots \in \mathscr{B}(H)$ *uniquely*, for $(S - \zeta)D(S) = (T - \zeta)D(S)$ is dense in H because $T = \tilde{S}$ (see Problem III-6.3).

Remark 4.14. The essential point in Theorem 4.13 is that the closure of $S(\varkappa)$ need not coincide with $T(\varkappa)$ for $\varkappa \neq 0$ and yet $T(\varkappa)$ is the unique family with the stated property. In many problems in applications $T(\varkappa)$ is actually the closure of $S(\varkappa)$ or even $T(\varkappa) = S(\varkappa)$. Even in such a case, application of Theorem 4.8, in particular the estimate (4.25) when $T = H$ is selfadjoint, often leads to an improvement of results obtained otherwise.

Example 4.15. Consider the formal differential operator

$$L(\varkappa)u = -u'' + \varkappa x^{-2}u, \quad 0 < x < \infty. \tag{4.35}$$

Let $\dot{T}(\varkappa)$ be the minimal operator in $H = L^2(0, \infty)$ constructed from (4.35), with $D(\dot{T}(\varkappa)) = C_0^\infty(0, \infty)$. $\dot{T}(\varkappa)$ is symmetric for real $\varkappa$; it is sectorial for every complex $\varkappa$ with $\operatorname{Re}\varkappa > -1/4$ in virtue of the inequality VI-(4.6) and $-(u'', u) = \int_0^\infty |u'|^2 dx$. Thus the family $\dot{T}(\varkappa)$ satisfies the assumptions for $S(\varkappa)$ in Theorem 4.12, and the Friedrichs extension $T(\varkappa)$ of $\dot{T}(\varkappa)$ forms a holomorphic family of type (B_0).

Is $T(\varkappa)$ the unique holomorphic family consisting of m-sectorial extensions of $\dot{T}(\varkappa)$? The answer depends on the region D_0 of $\varkappa$ considered. It is yes if D_0 contains

[1] As is seen from the proof, the uniqueness holds even if $T(\varkappa)$ is assumed to be only real-holomorphic.

$\varkappa = 3/4$. In fact, it is known that $\dot{T}(3/4)$ is essentially selfadjoint[1]; thus we have the situation of Theorem 4.13 if $\varkappa = 0$ is replaced by $\varkappa = 3/4$ (see also Remark 1.6). It is interesting to note that for $\varkappa < 3/4$, $\dot{T}(\varkappa)$ is not essentially selfadjoint and, consequently, $T(\varkappa)$ is not the only selfadjoint extension of $\dot{T}(\varkappa)$. Nevertheless it is unique as a *holomorphic* extension of the family $\dot{T}(\varkappa)$.

This unique family $T(\varkappa)$ is characterized by the condition $\mathsf{D}(T(\varkappa)) \subset \mathsf{D}(\bar{\dot{\mathfrak{t}}})$, where $\bar{\dot{\mathfrak{t}}}$ is the closure of the form $\dot{\mathfrak{t}} = \dot{\mathfrak{t}}(3/4)$ associated with $\dot{T}(3/4)$, namely, $\dot{\mathfrak{t}}[u] = (u', u') + (3/4) \int\limits_0^\infty x^2 |u|^2 dx$, $u \in \mathsf{C}_0^\infty$. In view of VI-(4.6), it is easy to see that $\mathsf{D}(\bar{\dot{\mathfrak{t}}})$ is the set of all $u \in \mathsf{H}$ such that $u' \in \mathsf{H}$ and $u(0) = 0$ (cf. Problem VI-2.18). Therefore, $T(\varkappa)$ must be the restriction of $L(\varkappa)$ such that $u \in \mathsf{D}(T(\varkappa))$ implies $u' \in \mathsf{H}$ and $u(0) = 0$.

5. The relationship between holomorphic families of types (A) and (B)

It is natural to ask whether there is any relationship between a holomorphic family of type (A) and one of type (B). In general there is no such relation, for the former is defined in any Banach space while the latter is defined only in Hilbert spaces. If we restrict ourselves to selfadjoint families of sectorial operators, however, the families of type (B) form a wider class than those of type (A). More precisely, we have

Theorem 4.16. *A selfadjoint holomorphic family $T(\varkappa)$ of type* (A) *is also of type* (B_0), *at least in a neighborhood of the real axis, if $T(\varkappa)$ is bounded from below for some real $\varkappa$ (whereupon $T(\varkappa)$ is bounded from below for all real $\varkappa$).*

Proof. We may assume that $\varkappa = 0$ belongs to the domain D_0 in which the family is defined and that $T = T(0)$ is selfadjoint and nonnegative. Then we have the expression (2.1) for $T(\varkappa)u$, $u \in \mathsf{D} = \mathsf{D}(T(\varkappa))$, where the coefficients $T^{(n)}$ are symmetric and satisfy inequalities of the form (2.5) (see Remark 2.8). Noting that $a\|u\| + b\|Tu\| \leq \sqrt{2}\,\|(a + bT)u\|$, we see from Theorem V-4.12 that

$$|(T^{(n)} u, u)| \leq \sqrt{2}\, c^{n-1}((a + bT)u, u), \quad n = 1, 2, \ldots. \tag{4.36}$$

Let $T_F(\varkappa)$ be the Friedrichs extension of $T(\varkappa)$. In view of (4.36), we see from Theorem 4.12 that $T_F(\varkappa)$ exists in a neighborhood of $\varkappa = 0$ and forms a holomorphic family of type (B). But we have $T_F(\varkappa) = T(\varkappa)$ for real $\varkappa$, since $T(\varkappa)$ is selfadjoint and $T_F(\varkappa)$ is a selfadjoint extension of $T(\varkappa)$. Since both $T(\varkappa)$ and $T_F(\varkappa)$ are holomorphic families, it follows

[1] At $x = 0$, the differential equation (4.35) is in the *limit point* case if $\varkappa \geq 3/4$ and in the *limit circle* case if $\varkappa < 3/4$ (cf. Coddington and Levinson [1], p. 225). In fact, $L(\varkappa)u = 0$ has two linearly independent solutions $u_\pm = x^{\alpha_\pm}$, where $2\alpha_\pm = 1 \pm (1 + 4\varkappa)^{1/2}$; both $u_\pm$ belong to $\mathsf{L}^2(0, 1)$ if $4\varkappa < 3$ but u_- does not belong to $\mathsf{L}^2(0, 1)$ if $4\varkappa \geq 3$. It follows that for real $\varkappa$, $\dot{T}(\varkappa)$ is essentially selfadjoint if (and only if) $\varkappa \geq 3/4$ (at $x = \infty$ the equation is in the limit point case for any real $\varkappa$).

from the unique continuation property (Remark 1.6) that $T(\varkappa) = T_F(\varkappa)$ for all $\varkappa$ in the common region of definition of $T(\varkappa)$ and $T_F(\varkappa)$. This proves that $T(\varkappa)$ is of type (B) in a neighborhood of $\varkappa = 0$; note that this implies that $T(\varkappa)$ is sectorial and, in particular, bounded from below for real $\varkappa$.

To complete the proof of the theorem by showing that the same result holds in a neighborhood of any real $\varkappa$, it suffices to prove that $T(\varkappa)$ is bounded from below for every real $\varkappa \in \mathrm{D}_0$ where $T(\varkappa)$ is defined. Suppose $T(\varkappa_0)$ is bounded from below for a real $\varkappa_0$. Since by (2.4) $T(\varkappa) - T(\varkappa_0)$ is relatively bounded with respect to $T(\varkappa_0)$ with relative bound < 1 for sufficiently small $|\varkappa - \varkappa_0|$, it follows from Theorem V-4.11 that $T(\varkappa)$ is bounded from below for real $\varkappa$ near $\varkappa_0$. Since the required smallness of $|\varkappa - \varkappa_0|$ is uniform in any compact subset of D_0, the semiboundedness of $T(\varkappa)$ propagates from $\varkappa = 0$ to all real $\varkappa \in \mathrm{D}_0$.

6. Perturbation series for eigenvalues and eigenprojections

The general results of § 1.3 on finite systems of eigenvalues of $T(\varkappa)$ apply to a holomorphic family $T(\varkappa)$ of type (B). We collect here some results that are peculiar to this type. For simplicity we restrict ourselves to dealing with the case of Theorem 4.9.

Consider an isolated eigenvalue λ of H with the eigenprojection P, $\dim P = m < \infty$. The total projection $P(\varkappa) = P + \sum_{n=1}^{\infty} \varkappa^n P^{(n)}$ for the λ-group eigenvalues of $T(\varkappa)$ is holomorphic in $\varkappa$, being given by (1.3). Substitution of (4.20) gives [cf. II-(2.8)]

$$P^{(n)} = - \sum_{p=1}^{n} (-1)^p \sum_{\nu_1 + \cdots + \nu_p = n} \frac{1}{2\pi i} \int_\Gamma G'^{-1} J(\zeta)\, C^{(\nu_1)} \cdot \\ \cdot J(\zeta) \ldots J(\zeta)\, C^{(\nu_p)}\, J(\zeta)\, G'^{-1}\, d\zeta\,. \tag{4.37}$$

Here $G' = H'^{1/2}$, $H' = H + \varrho = H + a\, b^{-1} + \delta$ and $J(\zeta)$ is given by (4.21) where $B' = 0$ since we assumed $\mathfrak{t}$ to be symmetric:

$$J(\zeta) = H'(H' - \zeta')^{-1} = H' R(\zeta)\,, \quad R(\zeta) = (H - \zeta)^{-1}\,. \tag{4.38}$$

The $C^{(n)}$ are given by (4.18). The power series for the weighted mean of the eigenvalues under consideration can be obtained as in II-§ 2.2. It should be remarked, however, that the expression II-(2.23) cannot be used now, for the operators $T^{(n)}$ are not defined. Instead we have

$$\hat{\lambda}(\varkappa) - \lambda = - \frac{1}{2\pi i m} \operatorname{tr} \int_\Gamma \log \left(1 + \left(\sum_{n=1}^{\infty} \varkappa^n C^n\right) J(\zeta)\right) d\zeta\,. \tag{4.39}$$

This can be proved by using the property of the trace $\operatorname{tr} A B = \operatorname{tr} B A$ and by integration by parts, starting from the expression II-(2.25) where we substitute $R(\zeta, \varkappa)$ from (4.20). The integration by parts makes use of the fact that $J(\zeta)\, G'^{-2}\, J(\zeta) = H' R(\zeta)^2 = (d/d\zeta)\, H' R(\zeta)$. The use of the formal equality $\operatorname{tr} A B = \operatorname{tr} B A$ under the integration sign is not directly justified, but it can be justified by the consideration given in § 2.3 in the case of a family of type (A).

(4.39) gives the following expressions for the coefficients of the series $\hat\lambda(\varkappa) = \lambda + \Sigma\, \varkappa^n \hat\lambda^{(n)}$ [cf. II-(2.30)]:

(4.40)
$$\hat\lambda^{(n)} = \frac{1}{2\pi i m} \operatorname{tr} \sum_{p=1}^{\infty} \frac{(-1)^p}{p} \sum_{\nu_1+\cdots+\nu_p = n} \int_\Gamma C^{(\nu_1)} J(\zeta)\, C^{(\nu_2)} \ldots J(\zeta)\, C^{(\nu_p)}\, J(\zeta)\, d\zeta .$$

If the reduced resolvent S of H for the eigenvalue λ is introduced according to $R(\zeta) = -(\zeta - \lambda)^{-1} P + S + (\zeta - \lambda) S^2 + \cdots$ [see I-(5.18) and III-(6.32)], (4.37) and (4.40) reduce to [cf. II-(2.12, 31)]

$$(4.41)\quad P^{(n)} = -\sum_p (-1)^p \sum_{\substack{\nu_1+\cdots+\nu_p = n\\ k_1+\cdots+k_{p+1} = p}} G'^{-1} J^{(k_1)} C^{(\nu_1)} \ldots J^{(k_p)} C^{(\nu_p)} J^{(k_{p+1})} G'^{-1},$$

$$(4.42)\quad \hat\lambda^{(n)} = \frac{1}{m} \sum_p \frac{(-1)^p}{p} \sum_{\substack{\nu_1+\cdots+\nu_p = n\\ k_1+\cdots+k_p = p-1}} \operatorname{tr} C^{(\nu_1)} J^{(k_1)} \ldots C^{(\nu_p)} J^{(k_p)} ,$$

where

$$(4.43)\quad J^{(0)} = -H' P = -(\lambda + a\, b^{-1} + \delta)\, P , \quad J^{(k)} = H' S^k , \quad k \geqq 1 .$$

It is desirable to express these results in terms of the given forms $\mathfrak{t}^{(n)}$ alone, without using such auxiliary quantities as H', G', $C^{(n)}$, etc. This is not simple in general, but can be done without difficulty at least for the first few coefficients. For example,

$$(4.44)\quad \hat\lambda^{(1)} = -\frac{1}{m} \operatorname{tr} C^{(1)} J^{(0)} = \frac{1}{m} \operatorname{tr} C^{(1)} H' P = \frac{1}{m} \sum_{j=1}^{m} (C^{(1)} H' \varphi_j, \varphi_j)$$
$$= \frac{1}{m} \sum_{j=1}^{m} \mathfrak{t}^{(1)} [G' \varphi_j, G'^{-1} \varphi_j] = \frac{1}{m} \sum_{j=1}^{m} \mathfrak{t}^{(1)} [\varphi_j] ,$$

where φ_j, $j = 1, \ldots, m$, form an orthonormal basis of the eigenspace $P\mathsf{H}$ [note that $G'^{\pm} \varphi_j = (\lambda + a\, b^{-1} + \delta)^{\pm 1/2} \varphi_j$].

Problem 4.17. (4.37) and (4.40) are *formally* identical with II-(2.8) and II-(2.30), respectively, after substitution of (4.38) for $J(\zeta)$ and of $C^{(n)} = G'^{-1}\, T^{(n)}\, G'^{-1}$.

Let us now estimate the convergence radius of the series for $P(\varkappa)$. According to (4.25), a lower bound for this convergence radius is given by

$$(4.45)\quad r_0 = \inf_{\zeta \in \Gamma} (\varepsilon \| (a + b H)\, R(\zeta) \| + c)^{-1} = \inf_{\substack{\zeta \in \Gamma\\ \mu \in \Sigma(H)}} \left(\varepsilon \left| \frac{a + b\mu}{\mu - \zeta} \right| + c \right)^{-1} ,$$

where $\varepsilon = 1$ or 2 according as all the $\mathfrak{t}^{(n)}$ are symmetric or not. As before, the knowledge of r_0 will lead to estimates for the convergence radius, the coefficients and the remainders for the series of $\hat\lambda(\varkappa)$.

If Γ is chosen to be the circle $|\zeta - \lambda| = d/2$ where d is the isolation distance of the eigenvalue λ of H, we have $\|R(\zeta)\| \leqq 2/d$ and $\|(H - \lambda) R(\zeta)\| \leqq 2$ for $\zeta \in \Gamma$ (see § 2.5). Hence

$$(4.46)\quad \|(a + bH) R(\zeta)\| \leqq (a + b\lambda) \|R(\zeta)\| + b\|(H - \lambda) R(\zeta)\| \leqq \frac{2}{d}(a + b\lambda) + 2b$$

(note that, although a need not be nonnegative, $a + b\lambda \geqq 0$ since $a + bH \geqq 0$ and λ is an eigenvalue of H). Thus (4.45) gives

$$(4.47)\quad r_0 \geqq 1\Big/\Big[\frac{2\varepsilon(a + b\lambda)}{d} + 2\varepsilon b + c\Big].$$

It is interesting to note the similarity and the difference between (4.47) and the estimate (2.34) for the case of type (A).

Of course a more careful choice of Γ will lead to an improvement of (4.47) (see the examples below).

Remark 4.18. The reduction process used in determining the series for the eigenvalues and eigenprojections should start from the operator $T_r(\varkappa)$ given by (1.11). The process is rather complicated since $T(\varkappa)$ is not given directly in the form of a power series [as in the case of type (A)].

This inconvenience can be avoided to some extent by considering the eigenvalues and eigenprojections of the operator $R(\zeta, \varkappa) = (T(\varkappa) - \zeta)^{-1}$ for a fixed ζ. Since $R(\zeta, \varkappa)$ is bounded-holomorphic in $\varkappa$ with the Taylor expansion (4.20), its eigenvalues and eigenprojections can be calculated by the reduction process described in the finite-dimensional case. Since these quantities are in a simple relationship with those of $T(\varkappa)$, we obtain in this way the required perturbation series for $T(\varkappa)$[1].

For example, the repeated eigenvalues of $R(\zeta, \varkappa)$ corresponding to the eigenvalue $(\lambda - \zeta)^{-1}$ of the "unperturbed" operator $R(0, \zeta) = R(\zeta)$ have the form

$$(4.48)\quad (\lambda - \zeta)^{-1} - \varkappa \nu'_j + \cdots, \quad j = 1, \ldots, m,$$

where the ν_j are the eigenvalues of the operator

$$(4.49)\quad P G'^{-1} J(\zeta) C^{(1)} J(\zeta) G'^{-1} P = \frac{\lambda + a b^{-1} + \delta}{(\lambda - \zeta)^2} P C^{(1)} P$$

regarded as an operator in the finite-dimensional space $P\mathsf{H}$. But $PC^{(1)}P$ is the bounded operator associated with the bounded form $(PC^{(1)}Pu, v) = (C^{(1)}Pu, Pv) = \mathfrak{t}^{(1)}[G'^{-1}Pu, G'^{-1}Pv] = (\lambda + a b^{-1} + \delta)^{-1} \mathfrak{t}^{(1)}[Pu, Pv]$. Hence we have $\nu'_j = (\lambda - \zeta)^{-2} \mu'_j$ where the μ'_j are the m repeated eigenvalues of the m-dimensional operator associated with the form $\mathfrak{t}^{(1)}[u, v]$ restricted to the m-dimensional space $P\mathsf{H}$. It follows from (4.48),

---

[1] The use of the family $T(\varkappa)^{-1}$ instead of $T(\varkappa)$ itself was made by T. KATO [3], [7] and V. KRAMER [1], [2] (mainly for asymptotic perturbation theory).

then, that the eigenvalues of $T(\varkappa)$ must have the form

$$(4.50)\qquad \lambda + \varkappa\,\mu_j' + \cdots.$$

Note that this refines (4.44), which gives only the mean of the m eigenvalues μ_j'. Higher order coefficients of the series for the eigenvalues can be computed in a similar fashion.

Problem 4.19. A necessary condition that there be no splitting at the first order is that $\tilde{t}^{(1)}[u, v] = \lambda'(u, v)$ for $u, v \in P\mathsf{H}$ with a constant λ'. This constant is equal to the common value $\mu_1' = \cdots = \mu_m'$.

Example 4.20. Consider the differential eigenvalue problem

$$(4.51)\qquad -u'' + \varkappa\,q(x)\,u = \lambda(\varkappa)\,u\,,\quad u(0) = u(\pi) = 0\,,\quad 0 \leqq x \leqq \pi\,.$$

Assume, for simplicity, that $q(x)$ is continuous on $[0, \pi]$. This problem is a special case of Example 2.17 in which $T^{(1)}$ is the multiplication operator by $q(x)$, so that the results of that example are applicable with $\|T^{(1)}\| = \|q\|_\infty = \max|q(x)|$. But the estimates for the convergence radii or other quantities may be quite crude if $\max|q(x)|$ is very large while $|q(x)|$ is not very large on the average. In such a case the results of this paragraph may be useful. Note that here we have a family $T(\varkappa) = T + \varkappa\,T^{(1)}$ of type (A) and (B) simultaneously; this is obvious since $T^{(1)}$ is bounded.

Let $s(x)$ be an indefinite integral of $q(x)$. We have

$$(4.52)\qquad \Big|\int_0^\pi q|u|^2\,dx\Big| = \Big|\int_0^\pi s'|u|^2\,dx\Big| = \Big|\int_0^\pi s(u'\bar{u} + u\bar{u}')\,dx\Big| \leqq$$
$$\leqq (\max|s|)\,2\|u\|\,\|u'\| \leqq a\|u\|^2 + b(Tu, u)$$

where $a, b > 0$ are arbitrary if $ab = (\max|s|)^2$ [note that $\|u'\|^2 = (Tu, u)$]. An estimate for the convergence radius for the unperturbed eigenvalue $\lambda = \lambda_n = n^2$ is then given by (4.47) with $c = 0$, $d = 2n - 1$ ($d = 3$, if $n = 1$) (see loc. cit.). The best choice of a under the restriction stated above is $a = (n^2 + 2n - 1)^{1/2}\cdot(\max|s|)$; then (set $\varepsilon = 1$ if $q(x)$ is real)

$$(4.53)\qquad r_0 \geqq \frac{2n - 1}{4\varepsilon(n^2 + 2n - 1)^{1/2}(\max|s|)}\,.$$

The indefinite additive constant in $s(x) = \int^x q(x)\,dx$ should be chosen so as to minimize $\max|s|$. Since $\max|s|$ may be small while $\max|q|$ is very large (even infinite), the estimate (4.53) is independent of the ones given in Example 2.17 [such as (2.38)] in which $\|T^{(1)}\| = \max|q|$ was used.

The same method could be applied to the Mathieu equation (see Example 3.4). But the coefficient $q(x) = \cos 2x$ is too smooth for (4.53) to give an improvement over the former results.

7. Growth rate of eigenvalues and the total system of eigenvalues

For a *selfadjoint* family $T(\varkappa)$ of type (B), we can estimate the growth rate of the eigenvalues in the same way as in the case of type (A) (see § 3.4). We use the inequality [see (4.7)]

$$(4.54)\qquad |\mathfrak{t}'(\varkappa)[u]| \leqq a'(u, u) + b'\,\mathfrak{t}(\varkappa)[u]\,,\quad u \in \mathsf{D}\,,\quad \varkappa \in \mathrm{I}\,,$$

where $\mathsf{D} = \mathsf{D}(\mathfrak{t}(\varkappa)) =$ const. and I is an interval of the real axis, which is assumed to contain $\varkappa = 0$. Again we consider a piecewise holomorphic function $\mu(\varkappa)$ formed by connecting several isolated eigenvalues of $T(\varkappa)$ in the manner described in § 3.4.

Theorem 4.21. *Let $\mu(\varkappa)$ be as above. Then*

$$|\mu(\varkappa) - \mu(0)| \leqq \frac{1}{b'}(a' + b'\,\mu(0))\,(e^{b'|\varkappa|} - 1)\,. \tag{4.55}$$

Proof. (4.55) is similar to the corresponding estimate (3.17) for type (A), with the slight difference that here $\mu(0)$ appears in place of $|\mu(0)|$. The proof of (4.55) is also similar to that of Theorem 3.6, but some modification is needed since $T'(\varkappa)\,\varphi(\varkappa)$ does not make sense here. Instead we use the result of Problem 4.19, which shows that

$$\mu'(\varkappa) = \mathfrak{t}'(\varkappa)\,[\varphi(\varkappa)] \tag{4.56}$$

for any $\varkappa$ at which no crossing of $\mu(\varkappa)$ with other eigenvalues occurs. Hence

$$\begin{aligned}\mu'(\varkappa) \leqq a' + b'\,\mathfrak{t}(\varkappa)\,[\varphi(\varkappa)] &= a' + b'(T(\varkappa)\,\varphi(\varkappa),\,\varphi(\varkappa))\\ &= a' + b'\,\mu(\varkappa)\,.\end{aligned} \tag{4.57}$$

This differs from (3.19) only in the lack of the absolute value signs. Solution of this differential inequality leads to (4.55).

Remark 4.22. As in § 3.5, it follows from Theorem 4.21 that Theorem 3.9 is true also for a selfadjoint family $T(\varkappa)$ of type (B) with compact resolvent [again $T(\varkappa)$ has compact resolvent for all $\varkappa$ if this is the case for some $\varkappa$, see Theorem 4.3]. Thus there exists a complete family of normalized eigenvectors and the associated eigenvalues of $T(\varkappa)$ which are holomorphic on the whole interval of $\varkappa$ being considered[1].

8. Application to differential operators

The theory of holomorphic families of operators of type (B) has wide application in the perturbation theory of differential operators. Simple examples were given in Examples 4.11 and 4.20 for regular ordinary differential operators. Here we add some further examples related to regular partial differential operators and certain singular differential operators, both ordinary and partial.

Example 4.23. Consider the formal differential operator

$$L(\varkappa)\,u = -\sum_{j,k=1}^{m} \frac{\partial}{\partial x_j}\,p_{jk}(x,\varkappa)\,\frac{\partial u}{\partial x_k} + q(x,\varkappa)\,u \tag{4.58}$$

in a bounded region E of $\mathbf{R}^m$. The coefficients $p_{jk}(x,\varkappa)$ and $q(x,\varkappa)$ are assumed to be sufficiently smooth functions of x in the closed region $\bar{\mathrm{E}}$ and to depend on the real parameter $\varkappa$ holomorphically. Furthermore, it is assumed that $p(x,\varkappa)$ and $q(x,\varkappa)$ are real-valued and the matrix $(p_{jk}(x,\varkappa))$ is symmetric and positive definite *uniformly* with respect to x and $\varkappa$.

As we have seen in VI-§ 4.4, several selfadjoint operators in $\mathsf{H} = \mathsf{L}^2(\mathrm{E})$ can be constructed from the formal differential operator $L(\varkappa)$ by considering the quadratic

[1] Cf. Rellich [5], where this is proved for families of type (B_0).

form

$$\mathfrak{h}(\varkappa)[u] = \int_E \Big[\sum p_{jk}(x,\varkappa)\,\frac{\partial u}{\partial x_k}\,\frac{\partial \bar u}{\partial x_j} + q(x,\varkappa)\,|u|^2\Big]\,dx\,. \tag{4.59}$$

The restriction $\mathfrak{h}_1(\varkappa)$ of this form with $\mathsf{D}(\mathfrak{h}_1)$ consisting of functions u satisfying the zero boundary condition $u = 0$ on ∂E leads to the operator $H_1(\varkappa) = L(\varkappa)$ with the same boundary condition. The form $\mathfrak{h}_2(\varkappa) = \mathfrak{h}(\varkappa)$ without any boundary condition leads to $H_2(\varkappa) = L(\varkappa)$ with the generalized Neumann condition [$\partial u/\partial n = 0$ where n is the conormal to the boundary, which is determined by $p_{jk}(x,\varkappa)$ and so depends on $\varkappa$]. The form $\mathfrak{h}_3(\varkappa)$ obtained by adding to $\mathfrak{h}_2(\varkappa)$ a boundary term such as $\int_{\partial E} \sigma(x,\varkappa)\,|u|^2\,dS$ will lead to $H_3(\varkappa) = L(\varkappa)$ with a generalized boundary condition of the third kind $\partial u/\partial n - \sigma u = 0$.

According to the results of VI-§ 4.4, these forms $\mathfrak{h}_n(\varkappa)$ with different $\varkappa$ (but with fixed $n = 1, 2, 3$) are relatively bounded with respect to one another and the same is true of their closures. It follows easily from Theorem 4.8 that the closed $\mathfrak{h}_n(\varkappa)$ has an analytic continuation $\mathfrak{t}_n(\varkappa)$ which is a holomorphic family of forms of type (a) defined for $\varkappa$ on and near the real axis. Hence $H_n(\varkappa)$, defined above for real $\varkappa$, has an analytic continuation $T_n(\varkappa)$ for $\varkappa$ on and near the real axis. $T_n(\varkappa)$ forms a *selfadjoint holomorphic family of type* (B). In particular it follows that the eigenvalues and eigenprojections of $T_n(\varkappa)$ are holomorphic on and near the real axis of $\varkappa$.

Actually $T_1(\varkappa)$ is not only of type (B) but also of type (A). We shall not prove it here, but it is related to the fact that $\mathsf{D}(H_1(\varkappa))$ is independent of $\varkappa$, as was noted earlier (loc. cit.). It should be noted, however, that the analyticity of the eigenvalues and eigenprojections of $H_1(\varkappa)$ follows already from the fact that it is holomorphic of type (B), which has been proved very simply.

Example 4.24. Consider the singular differential operator

$$L(\varkappa)\,u = -u'' + q(x)\,u + \varkappa\,q^{(1)}(x)\,u\,,\quad 0 < x < \infty\,, \tag{4.60}$$

in which $q(x)$ and $q^{(1)}(x)$ are allowed to have certain singularities. We assume, as in VI-§ 4.1, that q can be written $q = q_1 + q_2 + q_3$, where all the q_j are real-valued, $q_1 \geqq 0$ is locally integrable, q_2 is uniformly locally integrable and q_3 is majorized by $1/4\,x^2$ as in VI-(4.8). Then a closed form $\mathfrak{h}[u] = \int_0^\infty (|u'|^2 + q(x)\,|u|^2)\,dx$ can be constructed and the associated selfadjoint operator $H(0) = L(0)$ in $\mathsf{H} = \mathsf{L}^2(0,\infty)$ with an appropriate domain [including the boundary condition $u(0) = 0$] is defined (see loc. cit.). We further assume that $q^{(1)}$ can likewise be written as the sum of three functions $q_j^{(1)}$ with properties similar to those stated above. In addition we assume that $q_1^{(1)} \leqq \beta\,q_1$ with some constant β. Then it is easily seen that the form $\mathfrak{h}^{(1)}[u] = \int q^{(1)}(x)\,|u|^2\,dx$ is relatively bounded with respect to $\mathfrak{h}$, for $q_2^{(1)}$ and $q_3^{(1)}$ are relatively bounded with respect to $\mathfrak{h}_0[u] = \int |u'|^2\,dx$ (see loc. cit.). It follows from Theorem 4.8 that $\mathfrak{t}(\varkappa) = \mathfrak{h} + \varkappa\,\mathfrak{h}^{(1)}$ is a selfadjoint holomorphic family of forms of type (a) and that the associated operator $T(\varkappa) = L(\varkappa)$ (with an appropriate domain) forms a selfadjoint holomorphic family of type (B). The domain of $T(\varkappa)$ consists of all $u \in \mathsf{H} = \mathsf{L}^2(0,\infty)$ such that i) u and u' are absolutely continuous for $0 < x < \infty$ and $u' \in \mathsf{H}$, ii) $u(0) = 0$ and iii) $q_1^{1/2}\,u$ and $L(\varkappa)\,u$ belong to H (see Theorem VI-4.2). Note that $\mathsf{D}(T(\varkappa))$ depends on $\varkappa$ through iii).

It should be noticed that a perturbing "potential" $q^{(1)}(x)$ of such a high singularity as $1/x^2$ still leads to a holomorphic family $T(\varkappa)$.

Example 4.25. Consider the Schrödinger operator

$$L(\varkappa)\,u = -\Delta u + q(x)\,u + \varkappa\,q^{(1)}(x)\,u \tag{4.61}$$

in the whole 3-dimensional space R^3. We assume (see VI-§ 4.3) that q can be written in the form $q = q_1 + q_2$, where the q_k are real-valued, locally integrable, $q_1 \geqq 0$, $q_2 \leqq 0$, and where q_2 satisfies condition (4.17c). [Several sufficient conditions for (4.17c) to be met were given there.] Then a selfadjoint restriction H of $L(0)$ can be defined by Theorem 4.6a. We assume further that $q^{(1)}$ can also be written as $q^{(1)} = q_1^{(1)} + q_2^{(1)}$ with the $q_k^{(1)}$ satisfying conditions similar to the above ones, except that $q_2^{(1)}$ have $\mathfrak{h}_0$-bound smaller than 1 (which is not necessary since we are considering only small $|\varkappa|$). In addition we assume that $q_1^{(1)} \leqq \beta q_1$ for some positive constant β. Then it can be shown, as in the preceding example, that there is a selfadjoint holomorphic family of operators $T(\varkappa)$ in $\mathsf{H} = \mathsf{L}^2(R^3)$ of type (B) defined for small $|\varkappa|$, such that $T(\varkappa)$ is a restriction of $L(\varkappa)$ with $\mathsf{D}(T(\varkappa))$ defined as follows:

$u \in \mathsf{D}(T(\varkappa))$ if and only if i) grad u belongs to $\mathsf{H}' = (\mathsf{L}^2(R^3))^3$, ii) $\int q_1 |u|^2 dx < \infty$, and iii) the generalized derivative Δu exists and $L(\varkappa) u \in \mathsf{H}$. Here again $\mathsf{D}(T(\varkappa))$ depends in general on $\varkappa$ unless the singularity of $q^{(1)}$ is weak enough (as in Example 2.13, for example).

9. The two-electron problem[1]

As another application, we consider a typical problem in quantum mechanics. Consider an atomic system consisting of a fixed nucleus and two electrons. The Schrödinger operator for such a system is given by[2]

$$H = -\Delta_1 - \Delta_2 - \frac{2}{r_1} - \frac{2}{r_2} + \frac{2}{Z r_{12}} \tag{4.62}$$

in an appropriate system of units. Here the basic space is the 6-dimensional euclidean space R^6, the coordinates of which will be denoted by $x_1, y_1, z_1, x_2, y_2, z_2$; Δ_j is the 3-dimensional Laplacian $\partial^2/\partial x_j^2 + \partial^2/\partial y_j^2 + \partial^2/\partial z_j^2$, $j = 1, 2$; $r_j = (x_j^2 + y_j^2 + z_j^2)^{1/2}$, and $r_{12} = [(x_1 - x_2)^2 + (y_1 - y_2)^2 + (z_1 - z_2)^2]^{1/2}$. Z is the atomic number of the nucleus, $Z = 1, 2, 3, \ldots$.

H is selfadjoint in $\mathsf{H} = \mathsf{L}^2(R^6)$ if the differentiations Δ_1 and Δ_2 are taken in the generalized sense. Or one could define H first on $\mathsf{C}_0^\infty(R^6)$ and afterwards take its closure. In any case the selfadjoint operator H is determined uniquely, with the domain D identical with that of $-\Delta_1 - \Delta_2$. These results follow directly from the results of V-§ 5.3 (see Remark V-5.6).

We now write

$$H = H(\varkappa) = H_0 + \varkappa H^{(1)}, \tag{4.63}$$
$$H_0 = -\Delta_1 - \Delta_2 - 2r_1^{-1} - 2r_2^{-1}, \quad H^{(1)} = 2r_{12}^{-1}, \quad \varkappa = Z^{-1},$$

and regard $H(\varkappa)$ as produced from H_0 by the perturbation $\varkappa H^{(1)}$. We are interested in how the eigenvalues of $H(\varkappa)$ depend on $\varkappa$, starting from those of $H_0 = H(0)$.

First of all it should be remarked that $H^{(1)}$ is relatively bounded with respect to H_0 with relative bound 0. This follows from the results (see loc. cit.) that each of $2r_1^{-1}$, $2r_2^{-1}$ and $2r_{12}^{-1}$ is relatively bounded with respect to H_0 with relative bound 0 (see Problem IV-1.2). Therefore, $H(\varkappa)$ (or rather its extension to complex $\varkappa$) forms a selfadjoint family of type (A) defined for all complex $\varkappa$ (see Theorem 2.6). This implies that the eigenvalues and eigenprojections of $H(\varkappa)$ are holomorphic in $\varkappa$ on and near the real axis of the $\varkappa$-plane, as long as the eigenvalues are isolated and have finite multiplicities.

---

[1] See T. Kato [3].

[2] See Kemble [1], p. 209; we assume that the nucleus has infinite mass.

In what follows we are interested in the convergence radius for the power series representing the lowest eigenvalue $\lambda_1(\varkappa)$ of $H(\varkappa)$. This could be estimated by the formulas given in § 2.5, but we shall rather use the estimates of § 4.6. As we shall see, this is much easier; in fact $H(\varkappa)$ is of type (B_0) as well as of type (A) since $H(\varkappa)$ is bounded from below for any real $\varkappa$ (see Theorem 4.16).

The structure of H_0 is well known. The basic space $\mathsf{H} = \mathsf{L}^2(\mathrm{R}^6)$ can be regarded as the *tensor product*[1] $\mathsf{H}_1 \otimes \mathsf{H}_2$ of two copies H_1, H_2 of $\mathsf{L}^2(\mathrm{R}^3)$, and H_0 has the form $(H_1 \otimes 1) + (1 \otimes H_2)$ where H_1, H_2 are copies of the operator $-\Delta + 2r^{-1}$ acting in $\mathsf{L}^2(\mathrm{R}^3)$. This last one is the Schrödinger operator for the hydrogen atom (in a suitable system of units), and its spectrum is known to consist of isolated negative eigenvalues $-n^{-2}$, $n = 1, 2, 3, \ldots$, with multiplicities n^2 and of a continuous spectrum covering the whole positive real axis[2]. According to the structure of H_0 stated above, the lowest part of the spectrum of H_0 consists of isolated eigenvalues

$$\lambda_1 = -2,\ \lambda_2 = -5/4, \ldots, \lambda_n = -1 - n^{-2}, \ldots \tag{4.64}$$

which converge to -1. λ_1 is simple, but the other λ_n have multiplicities $2n^2$.

Since λ_1 is simple, the convergence radius for the power series representing the corresponding eigenvalue $\lambda_1(\varkappa)$ of $H(\varkappa)$, as well as the series for the associated eigenprojection $P_1(\varkappa)$, has a lower bound r_0 given by (4.45). To calculate r_0 we have to know the constants a, b. To this end we first consider the operator

$$\begin{aligned} H_0 - \beta H^{(1)} = (-\alpha\Delta_1 - 2r_1^{-1}) + (-\alpha\Delta_2 - 2r_2^{-1}) + \\ + [-(1-\alpha)(\Delta_1 + \Delta_2) - 2\beta r_{12}^{-1}]\,, \end{aligned} \tag{4.65}$$

where α, β are constants such that $0 < \alpha < 1$, $\beta > 0$. The first term on the right has the form $H_{1\alpha} \otimes 1$ with $H_{1\alpha}$ a copy of the operator $-\alpha\Delta - 2r^{-1}$ in $\mathsf{L}^2(\mathrm{R}^3)$. In general the operator $-\alpha\Delta - 2\beta r^{-1}$ has the lowest eigenvalue (that is, the lower bound) $-\beta^2/\alpha$; this follows from the special case $\alpha = \beta = 1$ stated above by a simple scale transformation. Hence the first term on the right of (4.65) has lower bound $-1/\alpha$, and the same is true of the second term. The third term can be reduced to a similar form by a linear transformation of the coordinates (corresponding to the separation of the motion of the center of gravity from the relative motion), which shows that this term has the lower bound $-\beta^2/2(1-\alpha)$. Thus we obtain

$$((H_0 - \beta H^{(1)})\,u, u) \geqq -\left(\frac{2}{\alpha} + \frac{\beta^2}{2(1-\alpha)}\right)(u, u)\,. \tag{4.66}$$

This leads to the inequality

$$0 \leqq (H^{(1)} u, u) \leqq a(u, u) + b(H_0 u, u) \tag{4.67}$$

with

$$a = \frac{2}{\alpha\beta} + \frac{\beta}{2(1-\alpha)}\,, \quad b = \frac{1}{\beta}\,; \tag{4.68}$$

here α, β are arbitrary except for the restrictions $0 < \alpha < 1$, $\beta > 0$. The best choice of a for a given b is easily seen to be given by

$$a = 2b + \frac{1}{2b} + 2\,. \tag{4.69}$$

We now apply (4.45) with these constants a, b and with $c = 0$, $\varepsilon = 1$ (note that $H^{(1)}$ is symmetric). The curve Γ must be chosen so as to enclose $\lambda_1 = -2$ but to exclude other eigenvalues of (4.64). In calculating (4.45), the critical value of ζ

---

[1] For the tensor product see Dixmier [1]. We need only elementary results of the theory of tensor products, which are familiar in quantum mechanics.

[2] See e. g. Kemble [1], p. 157.

is the point ζ_0 at which Γ intersects the negative real axis between $\lambda_1 = -2$ and $\lambda_2 = -5/4$; other ζ are not essential if the shape of Γ is chosen suitably, for example as a rectangle with sufficiently large vertical sides. To make r_0 as large as possible, ζ_0 is to be chosen in such a way that

$$\frac{a+b\lambda_1}{\zeta_0-\lambda_1} = \frac{a+b\lambda_2}{\lambda_2-\zeta_0}\left(=\frac{1}{r_0}\right) \tag{4.70}$$

(note that $a + b\lambda_1 > 0$). This gives

$$\zeta_0 = -\frac{13a-20b}{8a-13b}, \quad r_0 = \frac{3}{8a-13b} = \frac{3}{3b+4b^{-1}+16}. \tag{4.71}$$

Since $b > 0$ was arbitrary, the best value of r_0 is attained for $b = 2/3^{1/2}$:

$$r_0 = \frac{3}{16+4\sqrt{3}} = \frac{1}{7.64}. \tag{4.72}$$

Thus we arrive at the conclusion that the series for $\lambda_1(\varkappa)$ and $P_1(\varkappa)$ are convergent if $|\varkappa| < 1/7.7$ or $|Z| > 7.7$.

Remark 4.26. There is no reason why one must decompose H into the particular form (4.63). Some fraction of $-2r_1^{-1} - 2r_2^{-1}$ could as well be included in the perturbation. Let us see whether we can in this way improve the above result.

Set

$$\begin{aligned} H_\gamma(\varkappa) &= H_{0\gamma} + \varkappa H_\gamma^{(1)}, \\ H_{0\gamma} &= -\Delta_1 - \Delta_2 - 2(1-\gamma)(r_1^{-1}+r_2^{-1}), \quad 0<\gamma<1. \\ H_\gamma^{(1)} &= -2\gamma(r_1^{-1}+r_2^{-1}) + 2Z^{-1}r_{12}^{-1}. \end{aligned} \tag{4.73}$$

$H_\gamma(\varkappa)$ differs from $H(\varkappa)$ except for $\varkappa = 1$, though the two are essentially similar. The constant γ will be determined later.

We proceed as above by considering the operator $H_{0\gamma} - \beta H_\gamma^{(1)}$. A simple calculation similar to that used in deducing (4.67) leads to

$$(H_\gamma^{(1)}u, u) \leqq \left[\frac{2(1-\gamma-\beta\gamma)^2}{\alpha\beta} + \frac{\beta}{2(1-\alpha)Z^2}\right](u,u) + \frac{1}{\beta}(H_{0\gamma}u, u). \tag{4.74}$$

Since $H_\gamma^{(1)}$ is not positive-definite, unlike $H^{(1)}$, we must also estimate $-(H_\gamma^{(1)}u, u)$ from above. To this end we note that

$$\begin{aligned} H_{0\gamma} + \beta H_\gamma^{(1)} &\geqq -\Delta_1 - \Delta_2 - 2(1-\gamma+\beta\gamma)(r_1^{-1}+r_2^{-1}) \geqq \\ &\geqq -2(1-\gamma+\beta\gamma)^2, \end{aligned} \tag{4.75}$$

obtaining

$$-(H_\gamma^{(1)}u, u) \leqq \frac{2}{\beta}(1-\gamma+\beta\gamma)^2(u,u) + \frac{1}{\beta}(H_{0\gamma}u, u). \tag{4.76}$$

Again we minimize the coefficient of the first term on the right of (4.74) with respect to varying α. The result is $\frac{2}{\beta}\left(1-\gamma-\beta\gamma+\frac{\beta}{2Z}\right)^2$ if $1-\gamma-\beta\gamma \geqq 0$. It is convenient to choose γ in such a way that this coefficient is equal to the coefficient of the first term on the right of (4.76). This is attained by taking

$$\gamma = 1/4Z. \tag{4.77}$$

Then we have

$$|(H_\gamma^{(1)}u, u)| \leqq a(u,u) + b(H_{0\gamma}u, u), \tag{4.78}$$

where $b = 1/\beta$ and

$$a = 2b\left(1 - \frac{1}{4Z} + \frac{1}{4Zb}\right)^2, \quad b \geqq \frac{\gamma}{1-\gamma} = \frac{1}{4Z-1}. \tag{4.79}$$

In applying (4.45) with these constants, it must be borne in mind that the eigenvalues of $H_{0\gamma}$ are not (4.64) but $-(1-\gamma)^2(1+n^{-2})$, $n = 1, 2, 3, \ldots$. This remark changes (4.71) to

$$(4.80)\quad r_0 = 3\left(\frac{8a}{(1-\gamma)^2} - 13b\right)^{-1} = \frac{3}{b}\left[\frac{16\left(1-\frac{1}{4Z}+\frac{1}{4Zb}\right)^2}{\left(1-\frac{1}{4Z}\right)^2} - 13\right]^{-1}.$$

Maximizing r_0 with respect to varying b gives finally

$$(4.81)\quad r_0 = \frac{3\left(Z-\frac{1}{4}\right)}{8+2\sqrt{3}} = \frac{2Z-\frac{1}{2}}{7.64} \quad \text{for} \quad b = \frac{1}{\sqrt{3}\left(Z-\frac{1}{4}\right)} > \frac{1}{4Z-1}.$$

If $r_0 > 1$, the perturbation series under consideratiom converges for the value $\varkappa = 1$ for the real system. It follows from (4.81) that this is the case if $Z > 4.1$. This is a considerable improvement on the condition $Z > 7.7$ deduced above, but still it is not good enough to cover the most important case of the helium atom ($Z = 2$).

§ 5. Further problems of analytic perturbation theory

1. Holomorphic families of type (C)

In § 4.4 we considered a family $T(\varkappa)$ of operators defined as the Friedrichs extensions of operators of the form

$$(5.1)\quad S(\varkappa) = S + \varkappa S^{(1)} + \varkappa^2 S^{(2)} + \cdots$$

in which S was assumed to be sectorial. This assumption is essential since we have defined the Friedrichs extension only for sectorial operators. If S is symmetric, this necessarily restricts S to be bounded from below.

If $S = H$ is selfadjoint, however, there is a certain case in which we can define a holomorphic family $T(\varkappa)$ similar to the above one even when H is not semibounded. Such a $T(\varkappa)$ is given by the pseudo-Friedrichs extension discussed in VI-§ 3.4.

Theorem 5.1. *Let $S = H$ be selfadjoint, let $\mathsf{D} = \bigcap_{n=1}^{\infty} \mathsf{D}(S^{(n)}) \subset \mathsf{D}(H)$ and*

$$(5.2)\quad |(S^{(n)} u, u)| \leq c^{n-1}(a\|u\|^2 + b(|H|\, u, u))\,, \quad u \in \mathsf{D}\,, \quad n = 1, 2, 3, \ldots$$

where the constants b, $c \geq 0$ but a is arbitrary. If D is a core of $|H|^{1/2}$ (or, equivalently, a core of the form associated with $|H|$), the pseudo-Friedrichs extension $T(\varkappa)$ of the operator $S(\varkappa)$ given by (5.1) is defined for $|\varkappa| < (\varepsilon b + c)^{-1}$ and forms a holomorphic family, where $\varepsilon = 1$ or 2 according as all the $S^{(n)}$ are symmetric or not. [$T(\varkappa)$ will be called a holomorphic family of type (C).] The family $T(\varkappa)$ is selfadjoint if the $S^{(n)}$ are symmetric. $T(\varkappa)$ is the unique holomorphic extension of $S(\varkappa)$ if D is a core of H.

Proof. The proof is essentially the same as in Theorem VI-3.11. With the notations used there, we have $|(S^{(n)} u, u)| \leq b\, c^{n-1}\|G' u\|^2$ so that $(S^{(n)} u, u)$ can be extended to a form $\mathfrak{t}^{(n)}[u]$ with domain $\mathsf{D}(G')$.

$\mathfrak{t}^{(n)}$ can be expressed as $\mathfrak{t}^{(n)}[u, v] = (C^{(n)} G'u, G'v)$ where $C^{(n)} \in \mathscr{B}(\mathsf{H})$, $\|C^{(n)}\| \leqq \varepsilon b c^{n-1}$ [see VI-(3.20)]. If we set

$$T(\varkappa) = G'(HH'^{-1} + C(\varkappa))\,G', \quad C(\varkappa) = \sum_{n=1}^{\infty} \varkappa^n C^{(n)}, \tag{5.3}$$

it is easily seen that $T(\varkappa) \supset S(\varkappa)$ and $T(\varkappa)$ is a holomorphic family in some neighborhood of $\varkappa = 0$ [cf. VI-(3.21)–(3.22)]. Since $\|C(\varkappa)\| \leqq \varepsilon b \sum_{n=1}^{\infty} c^{n-1} |\varkappa|^n = \varepsilon b |\varkappa| (1 - c|\varkappa|)^{-1}$, the permitted range of $|\varkappa|$ is given by $\varepsilon b|\varkappa| (1 - c|\varkappa|)^{-1} < 1$ or $|\varkappa| < (\varepsilon b + c)^{-1}$.

The uniqueness of $T(\varkappa)$ can be proved exactly as in Theorem 4.13.

Remark 5.2. If S is symmetric and bounded from below, the Friedrichs extension $T(\varkappa)$ of $S(\varkappa)$ defined in § 4.4 is a special case of the pseudo-Friedrichs extension considered here; it suffices to set H equal to the Friedrichs extension of S and replace the $S^{(n)}$ by their restrictions to $\mathsf{D} = \mathsf{D}(S)$. As is easily seen, (4.31) then implies (5.2) and D is a core of $|H|^{1/2}$ (cf. Remark VI-3.13). On the other hand, Theorem 2.6 for a family of type (A) is also a special case of the pseudo-Friedrichs extension provided the unperturbed operator $T \geqq 0$ is essentially selfadjoint and the $T^{(n)}$ are all symmetric, for (2.5) implies $|(T^{(n)} u, u)| \leqq \sqrt{2}\, c^{n-1}((a + b|\tilde{T}|)\, u, u)$ by (4.36). Thus the family of type (C) is rather general, including the important cases of holomorphic selfadjoint families of types (A) and (B_0).

Remark 5.3. Most of the results regarding a family of type (B_0) can be taken over to a family of type (C) by obvious modifications. Here we only note that the estimates (4.45) and (4.47) are valid if H and λ are respectively replaced by $|H|$ and $|\lambda|$.

2. Analytic perturbation of the spectral family

In analytic perturbation theory, we are concerned with the analytic dependence of various quantities on the parameter $\varkappa$, assuming that the given family $T(\varkappa)$ of operators is analytic. Among the quantities that have been considered so far, there are the resolvent $R(\zeta, \varkappa) = (T(\varkappa) - \zeta)^{-1}$, the isolated eigenvalues $\lambda_n(\varkappa)$, the associated eigenprojections $P_n(\varkappa)$ and eigennilpotents $D_n(\varkappa)$, and also the projection $P(\varkappa)$ on a subspace corresponding to a separated part of the spectrum $\Sigma(T(\varkappa))$.

There are many other quantities that could be considered in this respect. For instance, for any function $\phi(\zeta)$ of a complex variable, the question can be raised whether or not the "function" $\phi(T(\varkappa))$ of $T(\varkappa)$ depends on $\varkappa$ analytically. In general $\phi(\zeta)$ should be itself holomorphic in a certain domain of the complex plane. If $T(\varkappa)$ is a selfadjoint family, however, $\phi(\zeta)$ could belong to a wider class of functions as long as only real $\varkappa$ are considered, according to the definition of $\phi(H)$ for a selfadjoint operator H given in VI-§ 5.2. One of the most important functions is $\phi(\zeta) = e^{it\zeta}$ with a real parameter t. This problem will be dealt with in Chapter IX.

The spectral theorem for selfadjoint operators furnishes other functions to be considered. One of them is the spectral family $E(\lambda, \varkappa)$ for $T(\varkappa)$, defined for real $\varkappa$, when $T(\varkappa)$ is a selfadjoint family. A simple consideration excludes, however, the possibility that $E(\lambda, \varkappa)$ depends analytically on $\varkappa$ for each fixed λ, even if $T(\varkappa)$ is a holomorphic family (see VI-§ 5.4). Nevertheless, there are certain cases in which $E(\lambda, \varkappa)$ is holomorphic in $\varkappa$.

Suppose H is a selfadjoint operator with the spectral family $\{E(\lambda)\}$. If the spectrum $\Sigma = \Sigma(H)$ of H has gaps at α and β, $\alpha < \beta$, we know that $E(\beta) - E(\alpha)$ changes continuously with H (see Theorem VI-5.10). Suppose now that $H(\varkappa)$ is a holomorphic selfadjoint family. Suppose further that the spectrum of $H = H(0)$ has gaps at α and β. Then we know that the spectrum $\Sigma(H(\varkappa))$ of the selfadjoint operator $H(\varkappa)$ with real $\varkappa$ also has gaps at α, β and that the projection $P(\varkappa)$ corresponding to the separated part of $\Sigma(H(\varkappa))$ lying between α and β is holomorphic for small $|\varkappa|$. Since $P(\varkappa) = E(\beta, \varkappa) - E(\alpha, \varkappa)$, this shows that *$E(\beta, \varkappa) - E(\alpha, \varkappa)$ is holomorphic near $\varkappa = 0$ if $\Sigma(H)$ has gaps at α and β*. It should be noted that, although the $E(\lambda, \varkappa)$ are defined only for real $\varkappa$, $E(\beta, \varkappa) - E(\alpha, \varkappa)$ has an analytic continuation $P(\varkappa)$ which is holomorphic near $\varkappa = 0$.

In general $E(\beta, \varkappa)$ itself is not holomorphic in $\varkappa$ when $\Sigma(H)$ has a gap at β. A counter-example is furnished by Example 1.11; here $\Sigma(H(\varkappa))$ is a subset of the positive real axis for $\varkappa \leqq 0$ but a negative eigenvalue $\lambda_0(\varkappa)$ exists for $0 < \varkappa < 1$ such that $\lambda_0(\varkappa) \to -\infty$ for $\varkappa \searrow 0$. Thus $\dim E(0, \varkappa)$ is 0 for $\varkappa \leqq 0$ and 1 for $0 < \varkappa < 1$, so that $E(0, \varkappa)$ is not holomorphic at $\varkappa = 0$ although $\Sigma(H(\varkappa))$ has a gap at 0 for all real $\varkappa < 1$.

However, $E(\beta, \varkappa)$ can be shown to be holomorphic in $\varkappa$ in the special but important case of a holomorphic selfadjoint family of type (A) or, more generally, of type (C).

Theorem 5.4.[1] *Let $H(\varkappa)$ be a selfadjoint holomorphic family of type* (C) *and let $E(\lambda, \varkappa)$ be the associated spectral family for real $\varkappa$. Let $\Sigma(H(0))$ have a gap at a real β. Then $\Sigma(H(\varkappa))$ has a gap at β for small real $\varkappa$ and $E(\beta, \varkappa)$ is holomorphic near $\varkappa = 0$.*

Proof. It suffices to take over the proof of Theorem VI-5.13 with minor modifications. These are: to replace H_n, $R_n(\zeta)$, $E_n(\lambda)$ by $H(\varkappa)$, $R(\zeta, \varkappa)$, $E(\lambda, \varkappa)$ respectively; to replace C_n by $C(\varkappa)$ used in the proof of Theorem 5.1; to replace a_n, b_n respectively by $a\, c^{n-1}|\varkappa|$, $b\, c^{n-1}|\varkappa|$. It is easy to see that we then arrive at an expression of $E(0, \varkappa) - E(0, 0)$

[1] This theorem was proved by Heinz [2] in the case of type (A). Note that the result is trivial for type (B) (see Problem 5.5).

as a power series in $\varkappa$ which converges for sufficiently small $|\varkappa|$ (again assuming $\beta = 0$ without loss of generality).

Problem 5.5. Let $H(\varkappa)$ be a selfadjoint holomorphic family such that $H(\varkappa)$ has a common finite lower bound γ for real $\varkappa$ near $\varkappa = 0$. Then $E(\beta, \varkappa)$ is holomorphic near $\varkappa = 0$ if $\Sigma(H(0))$ has a gap at β. [hint: $E(\alpha, \varkappa) = 0$ for sufficiently small α.]

3. Analyticity of $|H(\varkappa)|$ and $|H(\varkappa)|^\theta$

We continue to consider the family $H(\varkappa)$ of type (C) discussed in Theorem 5.4, assuming that $\beta = 0$ belongs to $\mathrm{P}(T)$. That $E(0, \varkappa)$ is holomorphic implies the same for $|H(\varkappa)|$ for real $\varkappa$, for we have

$$(5.4) \qquad |H(\varkappa)| = (1 - 2E(0, \varkappa))\, H(\varkappa)\,, \quad |H(\varkappa)|^{-1} = (1 - 2E(0, \varkappa))\, H(\varkappa)^{-1}$$

for real $\varkappa$; note that $|H(\varkappa)| = U(\varkappa)\, H(\varkappa)$ by VI-(2.26) where $U(\varkappa) = 1 - 2E(0, \varkappa)$ by VI-(5.25).

(5.4) implies that $|H(\varkappa)|^{-1}$ is a convergent power series in $\varkappa$ because both $(1 - 2E(0, \varkappa))$ and $H(\varkappa)^{-1}$ are. The family $|H(\varkappa)|$ can even be extended to a holomorphic family $H_1(\varkappa)$ defined near the real axis, although $H_1(\varkappa)$ is not equal to $|H(\varkappa)|$ for nonreal $\varkappa$.

It follows also from (5.4) that if $H(\varkappa)$ is of type (A), then $H_1(\varkappa)$ is also of the same type.

We can generalize these results and assert that the family $|H(\varkappa)|^\theta$ has an analytic continuation which is holomorphic in $\varkappa$, for any θ such that $0 \leqq \theta \leqq 1$[1]. To see this it is convenient to use the formula [see V-(3.53)]

$$(5.5) \qquad |H(\varkappa)|^{-\theta} = \frac{\sin \pi\theta}{\pi} \int\limits_0^\infty \mu^{-\theta} (|H(\varkappa)| + \mu)^{-1}\, d\mu\,, \quad 0 < \theta < 1\,.$$

To prove that $|H(\varkappa)|^{-\theta}$ can be expanded into a convergent power series in $\varkappa$, we can proceed as in the proof of Theorem 5.4; we may omit the details.

Remark 5.6. It can be proved that $|H(\varkappa)|^\theta$ (has an analytic continuation which) is of type (A) if $0 \leqq \theta < 1/2$. If $H(\varkappa)$ is of type (A), $|H(\varkappa)|^\theta$ is also of type (A) for $0 \leqq \theta \leqq 1$.

§ 6. Eigenvalue problems in the generalized form

1. General considerations

So far we have been considering the eigenvalue problems in the form $Tu = \lambda u$ in which T is a linear operator in a Banach space X. In applications one often encounters eigenvalue problems of a more general form

$$(6.1) \qquad Tu = \lambda A u$$

[1] The analyticity of $|H(\varkappa)|^\theta$ was proved by Heinz [1] for a family of type (A). See Remark 5.6.

where T and A are operators in X or, more generally, operators from X to another Banach space Y.

There are several ways of dealing with this general type of eigenvalue problem. For instance, (6.1) may be transformed into

$$A^{-1} T u = \lambda u \tag{6.2}$$

if A^{-1} exists. (6.2) has the standard form so far considered in detail, for $A^{-1} T$ is an operator in X. Or (6.1) may be written

$$T A^{-1} v = \lambda v , \quad v = A u ; \tag{6.3}$$

then we have again a standard form, this time the operator $T A^{-1}$ acting in the space Y.

A transformation into a more symmetric form is given by

$$A^{-1/2} T A^{-1/2} w = \lambda w , \quad w = A^{1/2} u . \tag{6.4}$$

This is particularly convenient when T and A are symmetric operators in a Hilbert space. In any case the construction of $A^{1/2}$ would require that $\mathsf{Y} = \mathsf{X}$.

In each transformation given above, however, it appears that there is something arbitrary involved; there is no reason to choose one over the other. Besides, the relationship between the original eigenvalue problem and the ***spectra*** of the operators $A^{-1} T$ or $T A^{-1}$ is not clear. It is true that any eigenvalue λ of (6.1) is at the same time an eigenvalue of (6.2) or (6.3) and that any eigenvector of (6.1) is also an eigenvector of (6.2), the eigenvector of (6.3) being related to that of (6.1) by the transformation $v = A u$. But it is not clear what is meant by an ***isolated eigenvalue*** of (6.1) or the ***algebraic multiplicity*** of such an eigenvalue; for it might happen that λ is an isolated eigenvalue of (6.2) but not of (6.3) and vice versa.

A more natural generalization of the results obtained for the standard problem $T u = \lambda u$ to the general problem (6.1) would consist in the study of the ***generalized resolvent*** $(T - \zeta A)^{-1}$ and the ***generalized spectrum*** associated with it. Of course several modifications would be required of the former results when one considers such a generalization. For example, the resolvent equation should read

$$(T - \zeta' A)^{-1} - (T - \zeta'' A)^{-1} = (\zeta' - \zeta'') (T - \zeta' A)^{-1} A (T - \zeta' A)^{-1} . \tag{6.5}$$

Furthermore, there is in general no sense in speaking of the commutativity of the resolvents for different ζ, for the resolvent is an operator from Y to X.

In this book, however, we shall not pursue this problem in any generality. Instead we shall content ourselves with considering some special cases in which it suffices to consider the problem in the form (6.2) or (6.3).

Let us assume that $T \in \mathscr{C}(\mathsf{X}, \mathsf{Y})$, $A \in \mathscr{B}(\mathsf{X}, \mathsf{Y})$ and $A^{-1} \in \mathscr{B}(\mathsf{Y}, \mathsf{X})$. Then $A^{-1}T \in \mathscr{C}(\mathsf{X})$ and $TA^{-1} \in \mathscr{C}(\mathsf{Y})$, and (6.1) is equivalent to either of (6.2) and (6.3).

Let λ be an isolated eigenvalue of $A^{-1}T$ with a finite algebraic multiplicity m, with the eigenprojection P and the eigennilpotent D. We have [see III-(6.28)–(6.29)]

$$PA^{-1}T \subset A^{-1}TP = \lambda P + D\,, \quad D = DP = PD\,. \tag{6.6}$$

$Q = APA^{-1}$ is a projection in Y, for $Q \in \mathscr{B}(\mathsf{Y})$ and $Q^2 = Q$. Similarly $G = ADA^{-1}$ is a nilpotent in Y. On multiplying (6.6) by A from the left, we obtain

$$QT \subset TP = \lambda AP + AD = \lambda QA + GA\,, \quad G = GQ = QG\,. \tag{6.7}$$

The consideration of the eigenvalue λ of TA^{-1} leads to the same results; in fact, Q is exactly the eigenprojection associated with the eigenvalue λ of TA^{-1} and G is the associated eigennilpotent.

The resolvent $(A^{-1}T - \zeta)^{-1}$ of $A^{-1}T$ has the expansion

$$\begin{aligned}(A^{-1}T - \zeta)^{-1} \\ = -(\zeta-\lambda)^{-1}P - \sum_{n=1}^{\infty} (\zeta-\lambda)^{-n-1} D^n + \sum_{n=0}^{\infty} (\zeta-\lambda)^n S^{n+1}\end{aligned} \tag{6.8}$$

near $\zeta = \lambda$ [see III-(6.32)]. Here S is the reduced resolvent of $A^{-1}T$ at $\zeta = \lambda$. Multiplication from the right by A^{-1} gives

$$\begin{aligned}(T - \zeta A)^{-1} &= -(\zeta-\lambda)^{-1}PA^{-1} - \sum_{n=1}^{\infty} (\zeta-\lambda)^{-n-1} D^n A^{-1} + \\ &\quad + \sum_{n=0}^{\infty} (\zeta-\lambda)^n S^{n+1} A^{-1} \\ &= -(\zeta-\lambda)^{-1} A^{-1} Q - \sum_{n=1}^{\infty} (\zeta-\lambda)^{-n-1} A^{-1} G^n + \\ &\quad + \sum_{n=0}^{\infty} (\zeta-\lambda)^n A^{-1} U^{n+1}\,,\end{aligned} \tag{6.9}$$

where $U = ASA^{-1}$ is the reduced resolvent of TA^{-1}.

If we consider a system λ_h, $h = 1, \ldots, s$, of isolated eigenvalues of $A^{-1}T$, with the associated eigenprojections P_h and eigennilpotents D_h, we have ($Q_h = AP_hA^{-1}$, $G_h = AD_hA^{-1}$)

$$\begin{aligned}&Q_hT \subset TP_h = \lambda_h A P_h + AD_h = \lambda_h Q_h A + G_h A\,,\\ &D_h = D_hP_h = P_hD_h\,, \quad G_h = G_hQ_h = Q_hG_h\,, \quad P_hP_k = \delta_{hk}P_h\,,\\ &\qquad Q_hQ_k = \delta_{hk}Q_h\,.\end{aligned} \tag{6.10}$$

Suppose, in particular, that both T and A are symmetric operators in a Hilbert space $\mathsf{X} = \mathsf{Y} = \mathsf{H}$. In this case neither $A^{-1}T$ nor TA^{-1} is symmetric in general. But $A^{-1}T$ becomes a symmetric operator if the new inner product

$$(6.11) \qquad ((u, v)) = (Au, v)$$

is introduced, assuming in addition that A is positive. Note that H is also a complete Hilbert space with this new inner product under the stated assumptions on A. Consequently, λ or the λ_h must be real and P or the P_h are orthogonal projections (and therefore have bounds 1) in the new metric, and D or the D_h must be zero. Returning to the old metric, we see that P or the P_h are bounded with bounds not exceeding a fixed number depending only on A. In fact, it can easily be seen that

$$(6.12) \qquad \|P\| \leqq \|A^{-1}\|^{1/2} \|A\|^{1/2} .$$

We also note that (in the symmetric case)

$$(6.13) \qquad Q = P^*, \quad P = Q^* .$$

The proof depends on the observation that, since P is the eigenprojection for λ of $A^{-1}T$, P^* is the eigenprojection for $\bar{\lambda} = \lambda$ of $(A^{-1}T)^* = TA^{-1}$.

An alternative way of dealing with the symmetric case is to use the transformation to (6.4). $A^{-1/2}TA^{-1/2}$ is symmetric and similar to $A^{-1}T$ and to TA^{-1}. It is therefore selfadjoint if it (or $A^{-1}T$ or TA^{-1}) has at least one isolated eigenvalue. The associated eigennilpotent must be zero and the associated orthogonal projection is identical with $A^{1/2}PA^{-1/2}$.

2. Perturbation theory

Here we are mainly interested in the eigenvalue problems of the form

$$(6.14) \qquad T(\varkappa)\, u = \lambda(\varkappa)\, A(\varkappa)\, u ,$$

where $T(\varkappa)$ as well as $A(\varkappa)$ is a holomorphic family of closed operators from X to Y defined near $\varkappa = 0$. We inquire whether or not the eigenvalues $\lambda(\varkappa)$ and the associated eigenvectors can be expressed as holomorphic functions of $\varkappa$.

According to the restrictions introduced in the preceding paragraph, we shall assume that $A(\varkappa)$ is *bounded-holomorphic*, so that it can be expressed as a convergent power series

$$(6.15) \qquad A(\varkappa) = A - \varkappa A^{(1)} + \varkappa^2 A^{(2)} + \cdots, \quad A, A^{(n)} \in \mathscr{B}(\mathsf{X}, \mathsf{Y}) .$$

Furthermore, A will be assumed to have an inverse $A^{-1} \in \mathscr{B}(\mathsf{Y}, \mathsf{X})$, so that $A(\varkappa)^{-1}$ also exists and belongs to $\mathscr{B}(\mathsf{Y}, \mathsf{X})$ for small $|\varkappa|$, with the expansion

$$(6.16)\quad A(\varkappa)^{-1} = A^{-1} - \varkappa A^{-1} A^{(1)} A^{-1} + \varkappa^2 (A^{-1} A^{(1)} A^{-1} A^{(1)} A^{-1} - A^{-1} A^{(2)} A^{-1}) + \cdots .$$

(6.14) is now equivalent to the eigenvalue problem

$$(6.17)\quad T_a(\varkappa)\, u \equiv A(\varkappa)^{-1} T(\varkappa)\, u = \lambda(\varkappa)\, u .$$

Here $T_a(\varkappa)$ is a closed operator acting in X and forming a holomorphic family, as is easily verified from the property of $T(\varkappa)$. Thus the analytic perturbation theory developed in this chapter can be applied to $T_a(\varkappa)$. It follows, for example, that any isolated eigenvalue of the unperturbed operator $T_a = T_a(0) = A^{-1} T(0)$ can be continued as an analytic function $\lambda(\varkappa)$ of $\varkappa$, which is an eigenvalue of $T_a(\varkappa)$ for each $\varkappa$. Similar results hold for the associated eigenprojection $P(\varkappa)$ and the eigennilpotent $D(\varkappa)$ for the operator $T_a(\varkappa)$, and for the eigenprojection $Q(\varkappa) = A(\varkappa) P(\varkappa) A(\varkappa)^{-1}$ and eigennilpotent $G(\varkappa) = A(\varkappa) D(\varkappa) A(\varkappa)^{-1}$ for the operator $T_b(\varkappa) = T(\varkappa) A(\varkappa)^{-1} = A(\varkappa) T_a(\varkappa) A(\varkappa)^{-1}$ (cf. the preceding paragraph).

There is another transformation of (6.14) [which is particularly convenient when $T(\varkappa)$ and $A(\varkappa)$ are selfadjoint]. We write (6.15) in the form

$$(6.18)\quad A(\varkappa) = (1 + \varkappa C_2(\varkappa)) A = (1 + \varkappa C_2(\varkappa))^{1/2} A (1 + \varkappa C_1(\varkappa))^{1/2} ,$$

where

$$(6.19)\quad C_2(\varkappa) = \sum_{n=0}^{\infty} \varkappa^n A^{(n+1)} A^{-1} \in \mathscr{B}(\mathsf{Y}) ,\quad C_1(\varkappa) = \sum_{n=0}^{\infty} \varkappa^n A^{-1} A^{(n+1)} \in \mathscr{B}(\mathsf{X})$$

are bounded-holomorphic. The validity of the first equality of (6.18) is obvious. The second is a consequence of the general formula

$$(6.20)\quad (1 + B A^{-1})^k A = A (1 + A^{-1} B)^k ,\quad A, B \in \mathscr{B}(\mathsf{X}, \mathsf{Y}) ,$$

which holds for any number k and any B with sufficiently small $\|B\|$. To verify (6.20) one need only to develop both sides into binomial series.

(6.18) permits the transformation of (6.14) into

$$(6.21)\quad (1 + \varkappa C_2(\varkappa))^{-1/2} T(\varkappa) (1 + \varkappa C_1(\varkappa))^{-1/2} w = \lambda(\varkappa) A w$$

where

$$(6.22)\quad w = (1 + \varkappa C_1(\varkappa))^{1/2} u .$$

It should be noted that the operator on the left of (6.21) forms a selfadjoint family if both $T(\varkappa)$ and $A(\varkappa)$ are selfadjoint families.

The operator A on the right of (6.21) can be eliminated, if desired, by a further transformation of the form (6.4). This is not necessary, however, since A is independent of $\varkappa$. In the selfadjoint case with $A > 0$, it suffices to introduce the new metric (6.11) so that the operator on the left of (6.21) becomes selfadjoint after multiplication from the left by A^{-1}, thus reducing the problem to the standard form. The procedure described here is particularly convenient inasmuch as no fractional powers (such as $A^{1/2}$) of operators are needed except those which can be calculated by binomial series.

Problem 6.1. Let $T(\varkappa)$ and $A(\varkappa)$ be selfadjoint families with $A > 0$. Then $\lambda(\varkappa)$ is real for real $\varkappa$, $\lambda(\varkappa)$ and $P(\varkappa)$ are holomorphic near the real axis, and $D(\varkappa) \equiv 0$.

3. Holomorphic families of type (A)

Suppose that the $T(\varkappa)$ of (6.14) is holomorphic of type (A) near $\varkappa = 0$, with the expansion (2.1). Then $T_a(\varkappa) = A(\varkappa)^{-1} T(\varkappa)$ has the same property, and we have

$$T_a(\varkappa)\, u = A^{-1} T u + \varkappa (A^{-1} T^{(1)} - A^{-1} A^{(1)} A^{-1} T)\, u + \cdots . \tag{6.23}$$

The series for the eigenvalues $\lambda_j(\varkappa)$ that arise by splitting from an isolated eigenvalue λ of $T_a(0) = A^{-1} T$ and for the associated eigenprojections $P_j(\varkappa)$ can then be calculated by application of the formulas of II-§ 2 to $T_a(\varkappa)$. We shall write only the series for the averaged eigenvalue $\hat{\lambda}(\varkappa)$:

$$\begin{aligned} \hat{\lambda}(\varkappa) &= \lambda + \varkappa \hat{\lambda}^{(1)} + \cdots \\ \hat{\lambda}^{(1)} &= \frac{1}{m} \operatorname{tr}(A^{-1} T^{(1)} - A^{-1} A^{(1)} A^{-1} T)\, P \\ &= \frac{1}{m} \operatorname{tr}(A^{-1} T^{(1)} - \lambda A^{-1} A^{(1)})\, P , \end{aligned} \tag{6.24}$$

where P is the eigenprojection for λ of $A^{-1} T$, $m = \dim P$ and the associated eigennilpotent is assumed to be zero.

If $\{u_k\}$, $k = 1, \ldots, m$, is a basis of the eigenspace $P\mathsf{X}$ and $\{e_k\}$ is the adjoint basis of $P^* \mathsf{X}^*$ (forming a biorthogonal set with $\{u_k\}$), we have

$$\begin{aligned} \hat{\lambda}^{(1)} &= \frac{1}{m} \sum_{k=1}^{m} (A^{-1}(T^{(1)} - \lambda A^{(1)})\, u_k, e_k) \\ &= \frac{1}{m} \sum_{k=1}^{m} ((T^{(1)} - \lambda A^{(1)})\, u_k, f_k) , \end{aligned} \tag{6.25}$$

where

$$\begin{aligned} f_k &= (A^{-1})^* e_k \in A^{*-1} P^* \mathsf{X}^* = (P A^{-1})^* \mathsf{X}^* \\ &= (A^{-1} Q)^* \mathsf{X}^* = Q^* A^{*-1} \mathsf{X}^* = Q^* \mathsf{Y}^* \end{aligned} \tag{6.26}$$

are characterized by the properties that

$$(A u_j, f_k) = \delta_{jk} , \quad (A u, f_k) = 0 \quad \text{for} \quad P u = 0 . \tag{6.27}$$

In the special case in which $\mathsf{X} = \mathsf{Y} = \mathsf{H}$ is a Hilbert space, $T(\varkappa)$ and $A(\varkappa)$ are selfadjoint and A is positive, it is convenient to choose u_k in such a way that $f_k = u_k$. This is equivalent to choosing u_k according to

$$(A\, u_j, u_k) = \delta_{jk}\,. \tag{6.28}$$

If, in particular, $m = 1$, we have

$$\lambda^{(1)} = \frac{((T^{(1)} - \lambda A^{(1)})\, u, u)}{(A\, u, u)}\,, \tag{6.29}$$

where u is an eigenvector for λ of $A^{-1} T$. The coefficients $\lambda^{(n)}$ for $n \geqq 2$ can be calculated by the same method.

Of course one arrives at the same results if one starts from the symmetric form (6.21) of the eigenvalue equation. But here there is some inconvenience arising from the fact that the operator-valued function on the left of (6.21) is not necessarily of type (A), thus preventing the straightforward application of the formulas of II-§ 2.

4. Holomorphic families of type (B)

Suppose that $T(\varkappa)$ of (6.14) is a holomorphic family of type (B) near $\varkappa = 0$. Thus $T(\varkappa)$ is an m-sectorial operator associated with a family $\mathfrak{t}(\varkappa)$ of sectorial forms with a constant domain $\mathsf{D} = \mathsf{D}(\mathfrak{t}(\varkappa))$ which has a Taylor expansion of the form (4.15).

In this case we have again the results of par. 2, but the calculation of the perturbation series is not so straightforward as in the case of type (A) considered above. Here it is convenient to consider $T(\varkappa)^{-1}$ instead of $T(\varkappa)$ itself, assuming for the moment that $\mathfrak{t}(\varkappa)$ has a positive vertex γ independent of $\varkappa$. Then $T(\varkappa)^{-1}$ is bounded-holomorphic. Since the inverse $\lambda(\varkappa)^{-1}$ of the eigenvalue $\lambda(\varkappa)$ of (6.14) is an eigenvalue of the inverted eigenvalue problem

$$T(\varkappa)^{-1} v = \lambda(\varkappa)^{-1} A(\varkappa)^{-1} v\,, \quad v = A(\varkappa)\, u\,, \tag{6.30}$$

$\lambda(\varkappa)^{-1}$ can be calculated by the method of the preceding paragraph, from which the series for $\lambda(\varkappa)$ can be obtained. For example, if we have a simple eigenvalue λ of the unperturbed equation $T u = \lambda A u$ or $T^{-1} v = \lambda^{-1} A^{-1} v$, we obtain in this way

$$\lambda(\varkappa) = \lambda + \frac{\mathfrak{t}^{(1)}[u] - \lambda(A^{(1)} u, u)}{(A\, u, u)}\, \varkappa + \cdots; \tag{6.31}$$

the calculation is similar to that given in § 4.6 and may be omitted (see in particular Remark 4.18).

We have assumed above that $T(\varkappa)$ has a positive vertex, but this assumption can be eliminated if the bounded form $(A(\varkappa)\, u, u)$ has a positive vertex. In this case we have only to consider $T_1(\varkappa) = T(\varkappa) + \alpha A(\varkappa)$ instead of $T(\varkappa)$. If α is a sufficiently large real number, $T_1(\varkappa)$

has a positive vertex, while the eigenvalue $\lambda_1(\varkappa)$ for the problem (6.14) with $T(\varkappa)$ replaced by $T_1(\varkappa)$ is equal to $\lambda(\varkappa) + \alpha$.

Remark 6.2. Eigenvalue problems of the form $Tu = \lambda A u$ arise in a natural way when one considers two quadratic forms $\mathfrak{t}[u]$ and $\mathfrak{a}[u]$. In our treatment of forms $\mathfrak{t}[u]$ in a Hilbert space H, we have been mainly concerned with the relation between $\mathfrak{t}[u]$ and the *unit form* $\|u\|^2$, which is a special quadratic form. When two forms $\mathfrak{t}$ and $\mathfrak{a}$ are to be considered without any reference to a third form such as $\|u\|^2$, there is no longer any point in supposing that $\mathfrak{t}$ and $\mathfrak{a}$ are defined in a given Hilbert space. Instead one could start from a vector space X in which both $\mathfrak{t}$ and $\mathfrak{a}$ are defined and, if desired, make X into a Hilbert space by introducing an inner product suitably related to these forms. If $\mathfrak{a}$ is a positive symmetric form, for example, we may take $\mathfrak{a}[u, v]$ as the inner product. The perturbation theory for an eigenvalue problem (6.14) could also be treated accordingly[1].

We shall not go into this general theory. It will only be remarked that the eigenvalue problem $Tu = \lambda A u$ can be formulated directly in terms of the forms $\mathfrak{t}$ and $\mathfrak{a}$ associated with T and A by

$$(6.32) \qquad \mathfrak{t}[u, v] = \lambda\, \mathfrak{a}[u, v] \quad \text{for all} \quad v \in \mathsf{D}(\mathfrak{t}) \cap \mathsf{D}(\mathfrak{a})\,.$$

It will also be noted that, when both $\mathfrak{t}$ and $\mathfrak{a}$ are *symmetric* and certain conditions are satisfied so that the eigenvalues form a discrete set bounded from below, the eigenvalues of $Tu = \lambda A u$ are characterized by the minimax principle applied to the ratio $\mathfrak{t}[u]/\mathfrak{a}[u]$ in place of I-(6.72).

5. Boundary perturbation

As an application of the preceding results, let us consider the so-called *boundary perturbation*[2]. A simple example is the eigenvalue problem

$$(6.33) \qquad -\Delta\, u(x) = \lambda\, u(x)\,, \quad x \in \mathrm{E} \subset \mathrm{R}^n$$

with the zero boundary condition

$$(6.34) \qquad u(x) = 0\,, \quad x \in \partial \mathrm{E}\,.$$

The main question is how the eigenvalues are changed by a *small deformation* of the region E.

In dealing with this problem, it is convenient to consider a family $\mathrm{E}(\varkappa)$ of bounded regions depending on a small real parameter $\varkappa$ and

---

[1] For a general theory of quadratic forms see ARONSZAJN [2], [4].

[2] See RELLICH [8], COURANT-HILBERT [1], p. 419, SEGEL [1]. For formal theory, see also MORSE-FESHBACH [1], SAITO [1]. For the stability of the essential spectrum under variation of the boundary, see WOLF [3], KREITH and WOLF [1]; for the stability of the absolutely continuous spectrum, see BIRMAN [6].

inquire into the dependence of the associated eigenvalues $\lambda(\varkappa)$ on $\varkappa$. We shall assume that $\mathrm{E}(\varkappa)$ is obtained from $\mathrm{E} = \mathrm{E}(0)$ by a one-to-one transformation

$$x \to y = x + \varkappa\, \phi(x) \tag{6.35}$$

where $\phi(x)$ is a sufficiently smooth function defined in an open set containing the closure $\bar{\mathrm{E}} = \mathrm{E} \cup \partial\mathrm{E}$ of E. $\phi(x)$ may be regarded as a vector-valued function with components $\phi_k(x) = \phi_k(x_1, \ldots, x_n)$, $k = 1, \ldots, n$.

The eigenvalue problem (6.33), (6.34) to be considered in the region $\mathrm{E}(\varkappa)$ is associated with the quadratic form

$$\int_{\mathrm{E}(\varkappa)} |\operatorname{grad} u(y)|^2\, dy \tag{6.36}$$

defined in the Hilbert space $\mathsf{H}(\varkappa) = \mathsf{L}^2(\mathrm{E}(\varkappa))$ with the norm

$$\|u\|^2 = \int_{\mathrm{E}(\varkappa)} |u(y)|^2\, dy\,. \tag{6.37}$$

To avoid the difficulty that the underlying Hilbert space $\mathsf{H}(\varkappa)$ depends on $\varkappa$, we introduce the transformation

$$\hat{u}(x) = u(y) \quad \text{with } x, y \text{ related by (6.35)} \tag{6.38}$$

of functions $u \in \mathsf{H}(\varkappa)$ to functions $\hat{u} \in \mathsf{H} = \mathsf{H}(0)$. Note that $\hat{u}$ satisfies the boundary condition $\hat{u} = 0$ on $\partial\mathrm{E}$ if and only if u satisfies the boundary condition $u = 0$ on $\partial\mathrm{E}(\varkappa)$.

Now (6.36) and (6.37) define two quadratic forms in H depending on $\varkappa$:

$$\mathfrak{t}(\varkappa)\,[\hat{u}] = \int_{\mathrm{E}(\varkappa)} |\operatorname{grad} u(y)|^2\, dy = \int_{\mathrm{E}} \sum_k \Big| \sum_j \frac{\partial x_j}{\partial y_k} \frac{\partial \hat{u}(x)}{\partial x_j} \Big|^2 J(x)\, dx\,, \tag{6.39}$$

$$\mathfrak{a}(\varkappa)\,[\hat{u}] = \int_{\mathrm{E}(\varkappa)} |u(y)|^2\, dy = \int_{\mathrm{E}} |\hat{u}(x)|^2 J(x)\, dx\,, \tag{6.40}$$

where $J(x)$ is the Jacobian of the transformation (6.35):

$$J(x) = J(x, \varkappa) = \det\Big(\delta_{jk} + \varkappa \frac{\partial \phi_j}{\partial x_k}\Big)\,. \tag{6.41}$$

By our assumptions on ϕ, $J(x)$ is a smooth function on $\bar{\mathrm{E}}$ and so $\mathfrak{a}(\varkappa)$ is a bounded symmetric form with positive lower bound. The associated bounded operator $A(\varkappa)$ is simply the multiplication operator by $J(x, \varkappa)$.

Now the eigenvalues $\lambda(\varkappa)$ to be considered are exactly the eigenvalues associated with the two forms $\mathfrak{t}(\varkappa)$ and $\mathfrak{a}(\varkappa)$ in the sense stated in the preceding paragraph, that is,

$$T(\varkappa)\, \hat{u} = \lambda(\varkappa)\, A(\varkappa)\, \hat{u} \tag{6.42}$$

where $T(\varkappa)$ is the operator associated with the form $\mathfrak{t}(\varkappa)$. The forms $\mathfrak{t}(\varkappa)$, $\mathfrak{a}(\varkappa)$ as well as the operators $T(\varkappa)$, $A(\varkappa)$ are dependent on $\varkappa$ but they are defined in the fixed Hilbert space H, so that the preceding results are applicable if $T(\varkappa)$ and $A(\varkappa)$ are shown to be holomorphic in $\varkappa$.

Since the $\phi_k(x)$ are smooth, it follows from (6.41) that $J(x)$ is a polynomial in $\varkappa$ with the constant term 1. Hence $A(\varkappa)$ is bounded-holomorphic in $\varkappa$ and $A(\varkappa) \geqq \delta$ for some $\delta > 0$ if $|\varkappa|$ is sufficiently small. As regards $T(\varkappa)$ or $\mathfrak{t}(\varkappa)$, it should be observed that on the right of (6.39), the $\partial x_j/\partial y_k$ as well as $J(x)$ depend on $\varkappa$. But since the matrix $(\partial x_j/\partial y_k)$ is the inverse of $(\partial y_j/\partial x_k) = (\delta_{jk} + \varkappa\, \partial \phi_j/\partial x_k)$, it is easily seen that $\mathfrak{t}(\varkappa)$ can be expressed in the form

$$\mathfrak{t}(\varkappa)\,[\hat{u}] = \mathfrak{t}[\hat{u}] + \varkappa\, \mathfrak{t}^{(1)}[\hat{u}] + \varkappa^2\, \mathfrak{t}^{(2)}[\hat{u}] + \cdots, \tag{6.43}$$

where

$$\mathfrak{t}[\hat{u}] = \int_E |\mathrm{grad}\, \hat{u}(x)|^2\, dx\,, \qquad \mathfrak{t}^{(r)}[\hat{u}] = \int_E \sum_{j,k} p_{jk}^{(r)}(x)\, \frac{\partial \hat{u}}{\partial x_j}\, \frac{\overline{\partial \hat{u}}}{\partial x_k}\, dx \tag{6.44}$$

and the $p_{jk}^{(r)}(x)$ are smooth functions such that

$$|p_{jk}^{(r)}(x)| \leqq b\, c^{r-1}\,, \quad x \in E\,, \tag{6.45}$$

with positive constants b, c. By making use of the Schwarz inequality, it is then easy to show that the $\mathfrak{t}^{(r)}$ satisfy the inequalities of the form $|\mathfrak{t}^{(r)}[\hat{u}]| \leqq b\, c^{r-1}\, \mathfrak{t}[\hat{u}]$ with b, c not necessarily identical with those in (6.45). This shows that $\mathfrak{t}(\varkappa)$ has an analytic continuation which is holomorphic of type (a), and hence that $T(\varkappa)$ is holomorphic of type (B), according to Theorem 4.8.

It follows from the results of par. 2 that *the eigenvalues $\lambda(\varkappa)$ of the problem* (6.33)–(6.34) *are holomorphic in $\varkappa$ near $\varkappa = 0$ when the underlying region* $E(\varkappa)$ *is obtained from* $E = E(0)$ *by a transformation of the form* (6.35). The eigenfunctions $u(\varkappa) = u(y, \varkappa)$ are also holomorphic in $\varkappa$, but it requires a new definition because the region $E(\varkappa)$ of the variable y depends on $\varkappa$. We shall not go into this question, contenting ourselves with the remark that the transformed eigenfunction $\hat{u}(\varkappa) = \hat{u}(x, \varkappa)$ is holomorphic in $\varkappa$ as a vector-valued function with values in H.

These results can be generalized in several directions. First, the zero boundary condition (6.34) can be replaced by others, for instance the Neumann condition $\partial u/\partial n = 0$. In this case one need only to replace $\mathfrak{t}(\varkappa)$ defined above by its extension $\mathfrak{t}_1(\varkappa)$ whose domain comprises all smooth functions not restricted by boundary conditions. The associated operator $T_1(\varkappa)$ will be a second order differential operator with the Neumann boundary condition, which acts on functions $\hat{u}(x) \in \mathsf{H} = L^2(E)$ as above. The boundary condition of the third kind

$$\frac{\partial u}{\partial n} + \sigma(x)\, u = 0 \quad \text{for} \quad x \in \partial E(\varkappa) \tag{6.46}$$

can be treated in the same way by adding to $\mathfrak{t}_1(\varkappa)\,[\hat{u}]$ a term depending on the boundary value of u determined properly from the transformation (6.35) (cf. VI-§ 4.4).

Second, the differential operator in (6.33) need not be the simple Laplacian $-\Delta$. There is no essential change if it is replaced by a second order differential operator of elliptic type, such as

$$Lu = -\sum_{j,k} \frac{\partial}{\partial x_j} a_{jk}(x) \frac{\partial u}{\partial x_k} + \sum_j b_j(x) \frac{\partial u}{\partial x_j} + c(x)\, u\,. \tag{6.47}$$

Then it suffices to consider, instead of (6.36), the form

$$\int\limits_{E(\varkappa)} \Big[\sum_{j,k} a_{jk}(y) \frac{\partial u}{\partial y_k} \frac{\overline{\partial u}}{\partial y_j} + \sum_j b_j(y) \frac{\partial u}{\partial y_j} \bar{u} + c(y)\, |u|^2\Big]\, dy\,. \tag{6.48}$$

Third, there is no essential difference even if the operator L and the boundary condition (6.46) depend on $\varkappa$ through the coefficients $a_{jk}(x)$, $b_j(x)$, $c(x)$ and $\sigma(x)$ (simultaneous perturbation of the differential operator and the boundary).

Chapter Eight

Asymptotic perturbation theory

In the foregoing chapters we have been concerned almost exclusively with *analytic* or *uniform* perturbation theory, in which the continuity in norm of the resolvent in the parameter plays the fundamental role. We shall now go into a study in which the basic notion is the *strong* continuity of the resolvent. Here the assumptions are weakened to such an extent that the analyticity of the resolvent or of the eigenvalues of the operator as functions of the parameter cannot be concluded, but we shall be able to deduce, under suitable conditions, the possibility of *asymptotic expansions* of these quantities.

As in analytic perturbation, the behavior of the resolvent is basic for the theory. We shall even define the notion of generalized strong convergence for unbounded operators in terms of strong convergence of their resolvents.

An inherent difficulty in this generalized problem is that an isolated eigenvalue λ need not remain isolated under perturbation. Thus it is necessary to distinguish between stable and unstable eigenvalues, with respect to a given perturbation. For stable eigenvalues we can develop a theory of asymptotic expansion; the perturbed eigenvalues and eigenvectors have asymptotic expansions in the parameter up to certain orders, depending on the properties of the unperturbed eigenspace.

If λ is not stable, it may happen that it is absorbed into the "continuous spectrum" as soon as the perturbation is "switched on". Then it does not make sense to talk about the perturbed eigenvalue. But the resulting continuous spectrum is supposed to have a particular concentration near λ — the so-called spectral concentration phenomenon.

The asymptotic theory developed here roughly corresponds to what is called *singular perturbation* in the theory of differential equations[1]. The results given in this chapter are abstract and should be applicable to differential equations. At present, however, the abstract theory is not advanced enough to comprise singular perturbation theory for differential equations.

§ 1. Strong convergence in the generalized sense

1. Strong convergence of the resolvent

Let $\{T_n\}$ be a sequence of closed operators in a Banach space X. In the present section we are concerned with general considerations on strong convergence of the resolvents $R_n(\zeta) = (T_n - \zeta)^{-1}$. Let us recall the fundamental result on the convergence *in norm* of the resolvents: if $R_n(\zeta)$ converges in norm to the resolvent $R(\zeta) = (T - \zeta)^{-1}$ of a closed operator T for some $\zeta \in \mathrm{P}(T)$, then the same is true for every $\zeta \in \mathrm{P}(T)$ (see Theorem IV-2.25, Remark IV-3.13, and Problem IV-3.14).

There is no corresponding theorem for strong convergence of the resolvents. Nevertheless, we can prove several theorems on the set of points ζ where the $R_n(\zeta)$ are strongly convergent or bounded.

Let us define the *region of boundedness*, denoted by Δ_b, for the sequence $\{R_n(\zeta)\}$ as the set of all complex numbers ζ such that $\zeta \in \mathrm{P}(T_n)$ for sufficiently large n and the sequence $\{\|R_n(\zeta)\|\}$ is bounded [for n so large that the $R_n(\zeta)$ are defined][2]. Furthermore, let Δ_s be the set of all ζ such that $\text{s-lim}\, R_n(\zeta) = R'(\zeta)$ exists. Δ_s will be called the *region of strong convergence* for $\{R_n(\zeta)\}$. Similarly we define the region Δ_u of *convergence in norm* for $\{R_n(\zeta)\}$. Obviously we have $\Delta_u \subset \Delta_s \subset \Delta_b$.

Theorem 1.1. *Δ_b is an open set in the complex plane. $\{R_n(\zeta)\}$ is bounded uniformly in n and ζ in any compact subset of Δ_b.*

Proof. Let $\zeta_0 \in \Delta_b$; we have the Neumann series (see III-§ 6.1)

$$R_n(\zeta) = \sum_{k=0}^{\infty} (\zeta - \zeta_0)^k R_n(\zeta_0)^{k+1} \quad \text{for} \quad |\zeta - \zeta_0| < \|R_n(\zeta_0)\|^{-1} . \tag{1.1}$$

If $\|R_n(\zeta_0)\| \leqq M_0$, then $\|R_n(\zeta)\| \leqq M_0(1 - M_0|\zeta - \zeta_0|)^{-1}$ for $|\zeta - \zeta_0| < M_0^{-1}$. The theorem follows immediately.

Theorem 1.1 implies that Δ_b consists of at most a countable number of connected open sets $\Delta_{b1}, \Delta_{b2}, \ldots$ (the *components* of Δ_b).

Theorem 1.2. *Δ_s is relatively open and closed in Δ_b (so that Δ_s is the union of some of the components Δ_{bk} of Δ_b). The strong convergence $R_n(\zeta) \to R'(\zeta)$ is uniform[3] in each compact subset of Δ_s.*

---

[1] For a general discussion see FRIEDRICHS [5].

[2] For convenience we call Δ_b also the region of boundedness for the sequence $\{T_n\}$, if there is no possibility of confusion. Similarly for Δ_s and Δ_u.

[3] The strong convergence $R_n(\zeta) \to R'(\zeta)$ is *uniform* in ζ if $\|R_n(\zeta)\, u - R'(\zeta)\, u\| \to 0$ uniformly in ζ for each fixed $u \in \mathsf{X}$.

Proof. If $\text{s-lim}\, R_n(\zeta_0) = A$ exists, we have $\text{s-lim}\, R_n(\zeta_0)^k = A^k$, $k = 1, 2, \ldots$. Since the series on the right of (1.1) is majorized in norm by the numerical series $\sum M_0^{k+1} |\zeta - \zeta_0|^k$, it follows that $\text{s-lim}\, R_n(\zeta)$ exists and is equal to $A(1 - (\zeta - \zeta_0) A)^{-1}$ for $|\zeta - \zeta_0| < M_0^{-1}$. This shows that Δ_s is open and at the same time proves the last assertion of the theorem.

To prove the relative closedness of Δ_s, let $\zeta \in \Delta_b$ and assume that in each neighborhood of ζ there is a $\zeta_0 \in \Delta_s$. $\zeta \in \Delta_b$ implies that $\|R_n(\zeta)\| \leqq M$ for a constant M. Take a $\zeta_0 \in \Delta_s$ such that $|\zeta - \zeta_0| < 1/2M$. Then $\|R_n(\zeta_0)\| < (1 - 2^{-1})^{-1} M = 2M \equiv M_0$, and $\text{s-lim}\, R_n(\zeta)$ also exists because $|\zeta - \zeta_0| < 1/2M = M_0^{-1}$. Hence $\zeta \in \Delta_s$.

The strong limit $R'(\zeta)$ of $R_n(\zeta)$ for $\zeta \in \Delta_s$ need not be the resolvent of an operator. In any case, however, $R'(\zeta)$ satisfies the resolvent equation

$$R'(\zeta_1) - R'(\zeta_2) = (\zeta_1 - \zeta_2) R'(\zeta_1) R'(\zeta_2), \quad \zeta_1, \zeta_2 \in \Delta_s, \tag{1.2}$$

as the strong limit of operators $R_n(\zeta)$ which satisfy the same equation. For this reason $R'(\zeta)$ is called a *pseudo-resolvent*. Note that $R'(\zeta_1)$ and $R'(\zeta_2)$ commute.

(1.2) implies that the null space $\mathsf{N} = \mathsf{N}(R'(\zeta))$ and the range $\mathsf{R} = \mathsf{R}(R'(\zeta))$ of $R'(\zeta)$ are independent of ζ. In fact, it follows from (1.2) that $R'(\zeta_2) u = 0$ implies $R'(\zeta_1) u = 0$ and that $u = R'(\zeta_2) v$ implies $u = R'(\zeta_1) w$ with $w = v - (\zeta_1 - \zeta_2) u$.

The pseudo-resolvent $R'(\zeta)$ is a resolvent (of a closed operator T) if and only if $\mathsf{N} = 0$. The necessity of this condition is obvious. To prove its sufficiency, we note that every $u \in \mathsf{R}$ can be written as $u = R'(\zeta) v(\zeta)$; here $v(\zeta)$ is uniquely determined if $\mathsf{N} = 0$. Application of (1.2) to u gives

$$R'(\zeta_1) R'(\zeta_2) (v(\zeta_2) - v(\zeta_1)) = R'(\zeta_1) u - R'(\zeta_2) u = (\zeta_1 - \zeta_2) R'(\zeta_1) R'(\zeta_2) u$$

and hence $v(\zeta_2) - v(\zeta_1) = (\zeta_1 - \zeta_2) u$ by $\mathsf{N} = 0$. This implies that $v(\zeta) + \zeta u$ is a vector independent of ζ; we shall denote it by Tu. T is a linear operator in X with $\mathsf{D}(T) = \mathsf{R}$ and $Tu - \zeta u = v(\zeta) = R'(\zeta)^{-1} u$. Hence $R'(\zeta) = (T - \zeta)^{-1}$ is the resolvent of T; the closedness of T follows from the fact that $R'(\zeta) \in \mathscr{B}(\mathsf{X})$. Note that Δ_s is a subset of the resolvent set $\mathrm{P}(T)$.

Theorem 1.3. *Let Δ_s be nonempty. There are the alternatives: either $R'(\zeta)$ is invertible for no $\zeta \in \Delta_s$ or $R'(\zeta)$ is equal to the resolvent $R(\zeta) = (T - \zeta)^{-1}$ of a unique operator $T \in \mathscr{C}(\mathsf{X})$. In the latter case we have $\Delta_s = \mathrm{P}(T) \cap \Delta_b$.*

Proof. Only the last statement remains to be proved. We have $\Delta_s \subset \mathrm{P}(T) \cap \Delta_b$ since we proved above that $\Delta_s \subset \mathrm{P}(T)$. To prove the opposite inclusion, we note the identity

$$(1.3)\quad R_n(\zeta) - R(\zeta) = (1 + (\zeta - \zeta_0) R_n(\zeta))(R_n(\zeta_0) - R(\zeta_0))(1 + (\zeta - \zeta_0) R(\zeta))$$

for $\zeta, \zeta_0 \in \mathrm{P}(T) \cap \Delta_b$; this is a simple consequence of the resolvent equations for $R_n(\zeta)$ and $R(\zeta)$. If $\zeta_0 \in \Delta_s$, we have $\text{s-lim}\, R_n(\zeta_0) = R'(\zeta_0) = R(\zeta_0)$ so that (1.3) gives $\text{s-lim}\, R_n(\zeta) = R(\zeta)$ by the uniform boundedness of $\{R_n(\zeta)\}$. This shows that $\zeta \in \Delta_s$ and completes the proof.

Corollary 1.4. *Let T_n and T be selfadjoint operators in a Hilbert space, with the resolvents $R_n(\zeta)$ and $R(\zeta)$. If $\text{s-lim}\, R_n(\zeta) = R(\zeta)$ for some complex number ζ, then the same is true for every nonreal ζ.*

Proof. Since $\|R_n(\zeta)\| \leq 1/|\mathrm{Im}\,\zeta|$, all nonreal numbers ζ are included in Δ_b as well as in $\mathrm{P}(T)$. Thus $\mathrm{P}(T) \cap \Delta_b$ also includes all nonreal ζ, and the assertion follows from Theorem 1.3.

When the second alternative of Theorem 1.3 is realized, we shall say that *$R_n(\zeta)$ converges strongly to $R(\zeta)$ on Δ_s and that T_n converges strongly to T ($T_n \underset{s}{\rightarrow} T$, in symbol) in the generalized sense*[1].

A criterion for generalized strong convergence is given by

Theorem 1.5. *Let $T_n, T \in \mathscr{C}(\mathsf{X})$ and let there be a core D of T such that each $u \in \mathsf{D}$ belongs to $\mathsf{D}(T_n)$ for sufficiently large n and $T_n u \to T u$. If $\mathrm{P}(T) \cap \Delta_b$ is not empty, T_n converges strongly to T in the generalized sense and $\Delta_s = \mathrm{P}(T) \cap \Delta_b$.*

Proof. For any $\zeta \in \mathrm{P}(T) \cap \Delta_b$, we have $R_n(\zeta)\, u - R(\zeta)\, u = R_n(\zeta) \cdot (T - T_n)\, R(\zeta)\, u \to 0$ if $R(\zeta)\, u \in \mathsf{D}$ (note that the $\|R_n(\zeta)\|$ are bounded). But such u form a dense set in X because D is a core of T (see Problem III-6.3). Since the $\|R_n(\zeta)\|$ are bounded, it follows that $R_n(\zeta) \underset{s}{\rightarrow} R(\zeta)$ (see Lemma III-3.5). Thus $\Delta_s \supset \mathrm{P}(T) \cap \Delta_b$ and the second alternative of Theorem 1.3 must be realized.

Corollary 1.6. *Let T_n, T be selfadjoint operators in a Hilbert space, and let there be a core D of T such that $T_n u \to T u$ for $u \in \mathsf{D}$. Then $R_n(\zeta) \underset{s}{\rightarrow} R(\zeta)$ for every nonreal ζ, and T_n converges to T strongly in the generalized sense.*

Remark 1.7. In general there is no simple relationship between the strong convergence of $\{T_n\}$ in the generalized sense and that of $\{T_n^*\}$ (assuming that the T_n are densely defined so that the T_n^* exist). This is true even when $T_n \in \mathscr{B}(\mathsf{X})$ and $\text{s-lim}\, T_n = T$ exists in the proper sense. But the following theorem should be noted.

Theorem 1.8. *Let $\{T_n\}$ and $\{T_n^*\}$ converge strongly to T and T^*, respectively, in the generalized sense, T_n and T being densely defined. Let Δ_s and Δ_s^* be the regions of strong convergence of the resolvents of T_n and T_n^*, respectively. Then Δ_s^* is identical with the mirror image $\overline{\Delta}_s$ of Δ_s with respect to the real axis.*

---

[1] For a related notion cf. Maslov [2].

Proof. This is a direct consequence of the formula $\Delta_s = \Delta_b \cap P(T)$ and a similar one for Δ_s^*, for we have $\Delta_b^* = \overline{\Delta_b}$ and $P(T^*) = \overline{P(T)}$ (see Theorem III-6.22).

Remark 1.9. We have been considering the strong convergence of a *sequence* $\{R_n(\zeta)\}$ of resolvents, but we can deal in quite the same way with a family of resolvents $R(\zeta, \varkappa) = (T(\varkappa) - \zeta)^{-1}$ depending on a continuous parameter $\varkappa$. In what follows we shall do this without any explicit comments.

Example 1.10. Let H be a selfadjoint operator in a Hilbert space H, and let $T_n = n^{-1} H$. Then $T_n u \to 0$ for every $u \in \mathsf{D}(H)$. Since $\mathsf{D}(H)$ is dense in H, it is a core of the bounded operator $T = 0$. Hence Corollary 1.6 is applicable, with the result that $R_n(\zeta) \to R(\zeta) = -1/\zeta$ strongly for any nonreal ζ. This implies that $(1 - n^{-1} \alpha H)^{-1} \to 1$ strongly for any nonreal α. This result is not quite trivial. If the spectrum $\Sigma(H)$ of H is the whole real axis, Δ_b is the union of the upper and lower half-planes ($\mathrm{Im}\,\zeta \gtrless 0$), and $\Delta_b = \Delta_s = P(T) \cap \Delta_b$ [$P(T)$ is the whole plane with the exception of the origin].

Example 1.11. Let $\mathsf{X} = \mathsf{L}^2(0, \infty)$ and

$$T_n = -d^2/dx^2 + q_n(x), \quad T = -d^2/dx^2 + q(x), \tag{1.4}$$

with the boundary condition $u(0) = 0$, say, where $q(x)$ and the $q_n(x)$ are real-valued functions. Under certain conditions (see V-§ 3.6) T_n and T are selfadjoint and $C_0^\infty(0, \infty)$ is a core of T. Suppose now that

$$\int_a^b |q_n(x) - q(x)|^2 \, dx \to 0 \tag{1.5}$$

for any a, b such that $0 < a < b < \infty$. Then we have $T_n u \to T u$ for $u \in C_0^\infty$. It follows from Corollary 1.6 that $R_n(\zeta) \xrightarrow[s]{} R(\zeta)$ for every nonreal ζ. This means that the solution of the boundary value problem

$$-u'' + q_n(x) u - \zeta u = f(x), \quad u(0) = 0, \quad u \in \mathsf{L}^2(0, \infty), \tag{1.6}$$

converges in $\mathsf{L}^2(0, \infty)$ to the solution of the same problem with q_n replaced by q. See also the examples of par. 4.

Example 1.12. Consider the operator $T(\varkappa) = T + \varkappa A$ in which T and A are the shift operators in $l^p(-\infty, \infty)$ discussed in Example IV-3.8. The spectrum $\Sigma(T(\varkappa))$ for $\varkappa \neq 0$ is the unit circle $|\zeta| = 1$. Let $R(\zeta, \varkappa) = (T(\varkappa) - \zeta)^{-1}$. Since $\|R(\zeta, \varkappa)\| \leqq (|\zeta| - 1)^{-1}$ for $|\zeta| > 1$ and $|\varkappa| \leqq 1$ (see Problem IV-3.10), the exterior of the unit circle belongs to the region of boundedness Δ_b for $R(\zeta, \varkappa)$ for $\varkappa \to 0$. But the interior of this circle does not belong to Δ_b (see loc. cit.). Here we have the relations $P(T) = \Delta_b = \Delta_s =$ the exterior of the unit circle.

Example 1.13. Let T be a closed, maximal symmetric operator in a Hilbert space H and suppose T is not selfadjoint. We may assume that the upper half-plane $\mathrm{Im}\,\zeta > 0$ is the resolvent set $P(T)$, whereas the lower half-plane has no point in $P(T)$. Let $\{P_n\}$ be a sequence of orthogonal projections with the following properties. i) The P_n are nondecreasing ($P_1 \leqq P_2 \leqq \cdots$) with s-lim $P_n = 1$; ii) $P_n \mathsf{X} \subset \mathsf{D}(T)$; iii) $P_n T P_n u \to T u$ if u belongs to the linear manifold $\mathsf{D} = \bigcup_{n=1}^{\infty} P_n \mathsf{X}$; iv) D is a core of T. Such a sequence can be constructed in the following way[1]. Let $u_n \in D(T)$ be a sequence dense in H. Let $\{v_n\}$ be the orthonormal family obtained by applying the Schmidt orthogonalization process to the sequence $\{u_n, (T-i)^{-1} u_n\}$. If P_n is the projection on the n-dimensional subspace spanned by $v_1, \ldots, v_n$, $\{P_n\}$ has the required properties.

Set $T_n = P_n T P_n$. The T_n are obviously bounded and symmetric, hence selfadjoint. Thus the region of boundedness Δ_b for $\{T_n\}$ comprises all nonreal numbers. Furthermore, it follows from iii), iv) and Theorem 1.5 that $\Delta_s = P(T) \cap \Delta_b$ is exactly the upper half-plane. This is an example in which Δ_s is a proper subset of Δ_b.

2. Generalized strong convergence and spectra

We have seen before that the spectrum $\Sigma(T)$ is upper-semicontinuous with respect to the generalized convergence (convergence in gap) of the operator, that is, $\Sigma(T)$ does not expand suddenly when T is changed slightly in the sense of generalized convergence (in gap). The situation is quite different if generalized convergence (in gap) is replaced by generalized strong convergence defined above or even by strong convergence. This is one of the reasons why we had to introduce the region of strong convergence Δ_s in the theorems proved above.

A very simple counter-example is furnished by a sequence $\{E_n\}$ of orthogonal projections in a Hilbert space such that $E_n \xrightarrow[s]{} 0$. The limit operator 0 has the single point $\zeta = 0$ in its spectrum, but $\Sigma(E_n)$ in general contains two points $\zeta = 0, 1$. Since, on the other hand, $\Sigma(T)$ is in general not lower-semicontinuous even in the sense of convergence in norm, we conclude that it is neither upper- nor lower-semicontinuous in the sense of strong convergence.

However, the lower-semicontinuity of the spectrum can be proved under certain restrictions.

Theorem 1.14. *Let H_n, H be selfadjoint operators in a Hilbert space* H *and let H_n converge to H strongly in the generalized sense. Then every open set containing a point of $\Sigma(H)$ contains at least a point of $\Sigma(H_n)$ for sufficiently large n.*

Proof. Let $\lambda \in \Sigma(H)$ and set $\zeta = \lambda + i\varepsilon$, $\varepsilon > 0$. We have $\|R(\zeta)\| = 1/\varepsilon$ since H is selfadjoint, so that there is a $u \in$ H such that $\|R(\zeta) u\| \geqq 1/2\, \varepsilon$, $\|u\| = 1$. But we have $R_n(\zeta) u \to R(\zeta) u$ by hypothesis (see also Corollary 1.4). Hence $\|R_n(\zeta) u\| \geqq 1/3\, \varepsilon$ for sufficiently large n. Since H_n is selfadjoint, this implies that there is a $\lambda_n \in \Sigma(H_n)$ such that $|\lambda_n - \zeta| \leqq 3\varepsilon$ or $|\lambda_n - \lambda| \leqq 4\varepsilon$. Since $\varepsilon > 0$ is arbitrary, this proves the theorem.

Theorem 1.14 can be strengthened in a certain sense to the following one:

[1] See Stone [1], p. 166.

Theorem 1.15. *In* Theorem 1.14 *let* $H_n = \int dE_n(\lambda)$, $H = \int dE(\lambda)$ *be the spectral representations of the operators involved. Then* [1,2]

$$\text{s-}\lim_{n\to\infty} E_n(\lambda) = E(\lambda) \quad \text{if} \quad E(\lambda - 0) = E(\lambda)\,. \tag{1.7}$$

Proof. We may assume without loss of generality that $\lambda = 0$. We recall that $E_n(0) + E_n(-0)$ is given by Lemma VI-5.6 with $\lambda = 0$ and H replaced by H_n. Or we may write

$$(1 - E_n(0) - E_n(-0))\, H_n (H_n^2 + 1)^{-1} = \frac{2}{\pi} \int_0^\infty H_n (H_n^2 + \eta^2)^{-1} H_n (H_n^2 + 1)^{-1}\, d\eta\,; \tag{1.8}$$

in this form we need not write $\lim\limits_{\substack{\varepsilon\to 0\\ \varrho\to\infty}} \int\limits_\varepsilon^\varrho$ on the right, for the integral is absolutely convergent by

$$\|H_n (H_n^2 + \eta^2)^{-1} H_n (H_n^2 + 1)^{-1}\| \leq \min(1, \eta^{-2})\,. \tag{1.9}$$

Furthermore, we note that $(E_n(0) - E_n(-0))\, H_n(H_n^2 + 1)^{-1} = 0$, for $E_n(0) - E_n(-0)$ is the projection on the null space of H_n. Hence the $E_n(-0)$ on the left of (1.8) can be replaced by $E_n(0)$. We have also a formula similar to (1.8) with H_n, E_n replaced by H, E respectively.

The integrand in (1.8) tends to $H(H^2 + \eta^2)^{-1} H (H^2 + 1)^{-1}$ strongly as $n \to \infty$ since $R_n(\zeta) \xrightarrow[s]{} R(\zeta)$ for any nonreal ζ. In view of (1.9), it follows by the principle of dominated convergence that

$$(1 - 2E_n(0))\, H_n (H_n^2 + 1)^{-1} \xrightarrow[s]{} (1 - 2E(0))\, H(H^2 + 1)^{-1}\,. \tag{1.10}$$

On the other hand, $H_n(H_n^2 + 1)^{-1} \xrightarrow[s]{} H(H^2 + 1)^{-1}$ by $R_n(\pm i) \xrightarrow[s]{} R(\pm i)$. Hence

$$(1 - 2E_n(0))\, [H_n (H_n^2 + 1)^{-1} - H(H^2 + 1)^{-1}] \xrightarrow[s]{} 0\,. \tag{1.11}$$

(1.10) and (1.11) together give

$$(E_n(0) - E(0))\, H(H^2 + 1)^{-1} \xrightarrow[s]{} 0\,. \tag{1.12}$$

---

[1] (1.7) can be generalized to $\text{s-}\lim E_n(\lambda_n) = E(\lambda)$ for $\lambda_n \to \lambda$. For the proof it suffices to note that $E_n(\lambda_n) = E_n'(\lambda)$ where E_n' is the spectral family for $H_n' = H_n - (\lambda_n - \lambda)$ and that $H_n' \xrightarrow[s]{} H$ since $\lambda_n - \lambda \to 0$. We remark also that $\text{s-}\lim E_n(\lambda_n - 0) = E(\lambda)$ is true. For the proof note that $E_n(\lambda_n)$ and $E_n(\lambda_n - n^{-1})$ has the same strong limit $E(\lambda)$, so that $E_n(\lambda_n) - E_n(\lambda_n - n^{-1}) \xrightarrow[s]{} 0$; then $0 \leq E_n(\lambda_n) - E_n(\lambda_n - 0) \leq E_n(\lambda_n) - E_n(\lambda_n - n^{-1})$ shows that $E_n(\lambda_n) - E_n(\lambda_n - 0) \xrightarrow[s]{} 0$.

[2] We could also raise the question whether $\phi(H_n) \xrightarrow[s]{} \phi(H)$ is true for a given function ϕ. We shall not consider this problem in the general form. For a related problem for spectral operators see Foguel [1].

This implies that $E_n(0)\,u \to E(0)\,u$ if $u = H(H^2+1)^{-1}v$, $v \in \mathsf{H}$, in other words, if u is in the range of $H(H^2+1)^{-1}$. But this range is dense in H, for $H(H^2+1)^{-1}$ is a selfadjoint operator with nullity zero [see III-(5.10)]. Since $E_n(0)$ is uniformly bounded, it follows that $E_n(0) \underset{s}{\to} E(0)$ (see Lemma III-3.5).

Remark 1.16. The argument used in the above proof may be generalized in the following way. One wants to prove a strong convergence $A_n \to A$, but one first proves $A_n\,\phi(H_n) \underset{s}{\to} A\,\phi(H)$ for a certain function ϕ. One also proves $\phi(H_n) \underset{s}{\to} \phi(H)$. Then $A_n\,\phi(H_n) - A_n\,\phi(H) \underset{s}{\to} 0$ if A_n is known to be uniformly bounded. Hence $(A_n - A)\,\phi(H) \underset{s}{\to} 0$. If ϕ is such that the range of $\phi(H)$ is dense in H, one then concludes that $A_n \underset{s}{\to} A$.

3. Perturbation of eigenvalues and eigenvectors

The lack of upper-semicontinuity of the spectrum under a perturbation in the strong sense is rather embarrassing in the perturbation theory of spectra and, in particular, of *isolated eigenvalues*. Contrary to the case of a small perturbation in the sense of gap, it may well happen that an isolated eigenvalue of an operator T becomes involved in the "continuous spectrum"[1] as soon as T is subjected to a perturbation which is small in the strong sense. Let us show by simple examples how such seemingly singular phenomena sometimes take place and sometimes do not.

Example 1.17. Let $\mathsf{X} = \mathsf{L}^2(-\infty, \infty)$ and let T be an integral operator with the kernel $t(y, x) = -f(y)\,\overline{f(x)}$, where $f(x)$ is a continuous function with $\|f\| = 1$ and $f(x) \neq 0$ everywhere. Let $T^{(1)}$ be the maximal multiplication operator by x: $T^{(1)}u(x) = x\,u(x)$. T is bounded and selfadjoint, and has the eigenvalues $0, -1$; -1 is simple but 0 has multiplicity ∞.

Set $T(\varkappa) = T + \varkappa\,T^{(1)}$ for $\varkappa \neq 0$. $T(\varkappa)$ is selfadjoint for real $\varkappa$. For a complex $\varkappa$ the numerical range $\Theta(T(\varkappa))$ of $T(\varkappa)$ is a subset of the strip $\Pi_\varkappa$ between the two straight lines $\eta = \xi\tan\theta$ and $\eta = -1 + \xi\tan\theta$, where $\zeta = \xi + i\eta$ and $\theta = \arg\varkappa$. Now $(T(\varkappa) - \zeta)^{-1} = \varkappa^{-1}(T^{(1)} + \varkappa^{-1}T - \varkappa^{-1}\zeta)^{-1}$ exists and belongs to $\mathscr{B}(\mathsf{X})$ at least if $\|\varkappa^{-1}T\| = |\varkappa^{-1}| < |\mathrm{Im}(\varkappa^{-1}\zeta)| = \|(T^{(1)} - \varkappa^{-1}\zeta)^{-1}\|^{-1}$, which means that ζ is sufficiently far from the strip $\Pi_\varkappa$. Thus $T(\varkappa)$ has deficiency index $(0, 0)$ and the exterior of that strip belongs to $\mathrm{P}(T(\varkappa))$, with $\|R(\zeta, \varkappa)\| \leq 1/\mathrm{dist}(\zeta, \Pi_\varkappa)$ (see V-§ 3.2).

Suppose now that $\varkappa$ varies over the region D_0: $|\mathrm{Im}\,\varkappa| < M\,|\mathrm{Re}\,\varkappa|$. It follows from the above result that if ζ belongs to one of the two sectors $\eta > \phi(\xi) + \varepsilon$, $\eta < -\phi(\xi) - \varepsilon$, where $\varepsilon > 0$ and

$$\phi(\xi) = \max[-M\,\xi, M(\xi + 1)] > 0, \tag{1.13}$$

then $\|R(\zeta, \varkappa)\| \leq \varepsilon^{-1}(M^2 + 1)^{1/2}$.

[1] We have not defined the notion of the continuous spectrum of an operator. We use this term here in a rather vague sense, implying the set of non-isolated points of the spectrum. We shall give a precise definition later for selfadjoint operators (Chapter X).

Now let $\varkappa \to 0$ with $\varkappa \in D_0$. It follows from the above result that the two sectors

$$\Pi_\pm : \eta > \phi(\xi) \quad \text{and} \quad \eta < -\phi(\xi) \tag{1.14}$$

belong to Δ_b. Furthermore, $T(\varkappa)\, u \to T u$ if $u \in D(T^{(1)})$. Since $D(T^{(1)})$ is a core of T and $\Pi_\pm \in P(T)$, $\Pi_\pm$ also belong to Δ_s by Theorem 1.5. Thus $T(\varkappa) \underset{s}{\to} T$ in the generalized sense.

Similar results hold when $\varkappa$ is restricted to the region $D_1 : |\mathrm{Im}\varkappa| > m|\mathrm{Re}\varkappa|$, $m > 0$. In this case Δ_b and Δ_s contain two sectors $|\eta| < m\,\xi$ and $|\eta| < -m(\xi + 1)$, and $T(\varkappa) \underset{s}{\to} T$ in the generalized sense when $\varkappa \to 0$ within D_1.

Let us consider the spectrum of $T(\varkappa)$. It is clear that $\Sigma(T^{(1)})$ is exactly the real axis, which is at the same time the essential spectrum $\Sigma_e(T^{(1)})$ of $T^{(1)}$ (see IV-§ 5.6). Since T is degenerate with rank 1, $T^{(1)} + \varkappa^{-1} T$ has the same essential spectrum as $T^{(1)}$ (see Theorem IV-5.35). This implies that, for any $\varkappa \neq 0$, the essential spectrum of $T(\varkappa) = \varkappa(T^{(1)} + \varkappa^{-1} T)$ is the straight line passing through 0 and $\varkappa$, so that the rest of $\Sigma(T(\varkappa))$ consists of isolated eigenvalues with finite multiplicities (see loc. cit.).

If, in particular, $\varkappa$ is real, $T(\varkappa)$ has no isolated eigenvalue (since it is selfadjoint and has its spectrum on the real axis). Furthermore, it has no eigenvalue at all. In fact, suppose $T(\varkappa)\, u = \lambda\, u$. This implies that

$$-(u, f)\, f(x) + \varkappa\, x\, u(x) = \lambda\, u(x)\,, \tag{1.15}$$

$$u(x) = \frac{c\, f(x)}{\varkappa\, x - \lambda}\,, \quad c = (u, f)\,. \tag{1.16}$$

But such a u belongs to X only if $u = 0$, for the denominator $\varkappa\, x - \lambda$ has a zero at $x = \lambda/\varkappa$ if λ is real. On the other hand $T(\varkappa)$ has no nonreal eigenvalue as noted above. It can be shown that $T(\varkappa)$ has a pure continuous spectrum over the real axis[1].

If $\varkappa$ is not real, on the other hand, the u of (1.16) can belong to X without being 0. The eigenvalue λ is determined by the equation

$$\int_{-\infty}^{\infty} \frac{|f(x)|^2}{\varkappa\, x - \lambda}\, dx = (u, f)/c = 1\,. \tag{1.17}$$

It can be shown that (1.17) determines λ uniquely as a function of $\varkappa$, at least for small $|\varkappa|$ and $|\lambda + 1|$, and that $\lambda = \lambda(\varkappa)$ tends to $\lambda(0) = -1$ if $\varkappa \to 0$ with $\varkappa \in D_1$.

Thus the isolated eigenvalue -1 of T becomes absorbed into the continuous spectrum of $T(\varkappa)$ when $\varkappa$ is real, though it can be continued as an isolated eigenvalue if $\varkappa \in D_1$. The other eigenvalue 0 of T also diffuses into the essential spectrum in any case; this should be expected from the fact that it has multiplicity ∞.

Example 1.18. Let X and T be as in Example 1.17 and let $T^{(1)} = x^2$. Again $T(\varkappa) = T + \varkappa\, T^{(1)}$ converges strongly to T in the generalized sense, if $\varkappa \to 0$ within D_0 or D_1. In this case, however, $T(\varkappa)$ has just one eigenvalue for real $\varkappa > 0$; for the eigenfunction $u(x)$ corresponding to (1.16) has the denominator $\varkappa\, x^2 - \lambda$, which does not vanish if $\lambda < 0$. It is easy to show that the condition corresponding to (1.17), with $\varkappa\, x - \lambda$ replaced by $\varkappa\, x^2 - \lambda$, determines $\lambda = \lambda(\varkappa)$ as a negative, increasing function of $\varkappa$ for $\varkappa > 0$ and that $\lambda(\varkappa) \to -1$ for $\varkappa \searrow 0$. Furthermore, the essential spectrum of $T(\varkappa)$ coincides with the nonnegative real axis for $\varkappa > 0$. $\Sigma(T(\varkappa))$ consists of one negative eigenvalue $\lambda(\varkappa)$ and a continuous spectrum covering the positive real axis if $\varkappa > 0$. The perturbation of the spectrum looks

[1] This can be proved by observing that $T(\varkappa) = \varkappa(T^{(1)} + \varkappa^{-1} T)$, where $T^{(1)}$ has an absolutely continuous spectrum and T is a degenerate operator (see Theorem X-4.4).

quite normal for $\varkappa > 0$ so far as concerns the isolated eigenvalue -1 of T. Similar behavior of the spectrum of $T(\varkappa)$ may be expected when $\varkappa$ is not real but $|\arg\varkappa| \leq \leq \pi - \varepsilon$, $\varepsilon > 0$.

But the situation is entirely different if $\varkappa$ is real and negative. In this case it is easily seen that $T(\varkappa)$ has no eigenvalue; the isolated eigenvalue -1 of T is absorbed into the continuous spectrum as soon as T is changed into $T(\varkappa)$.

Example 1.19. Let X be any one of the function spaces $\mathsf{L}^p(0, 1)$, $1 \leq p < \infty$, or $\mathsf{C}[0, 1]$. Let

$$T(\varkappa) = 2\alpha \frac{d}{dx} - \varkappa \frac{d^2}{dx^2}, \quad \alpha \neq 0, \tag{1.18}$$

with the boundary condition

$$u(0) = u(1) = 0. \tag{1.19}$$

$T(\varkappa)$ is closed for $\varkappa \neq 0$ with $\mathsf{D} = \mathsf{D}(T(\varkappa))$ independent of $\varkappa$. $T(\varkappa)$ is even selfadjoint if $\mathsf{X} = \mathsf{L}^2$, $\varkappa$ is real, $\varkappa \neq 0$ and α is pure imaginary. A simple calculation shows that if $\varkappa \neq 0$ the spectrum of $T(\varkappa)$ consists of isolated eigenvalues

$$\lambda_n(\varkappa) = n^2 \pi^2 \varkappa + \frac{\alpha^2}{\varkappa}, \quad n = 1, 2, 3, \ldots. \tag{1.20}$$

What is the "limit" of $T(\varkappa)$ as $\varkappa \to 0$? Formally one would have $T(0) = 2\alpha\, d/dx$. If one keeps the boundary condition (1.19) for this first-order differential operator, the spectrum of the closure $T(0)^\sim$ of $T(0)$ is the whole complex plane (Example III-6.8). But $T(0)^\sim$ is not a reasonable limit of $T(\varkappa)$ since it is natural to require not two, but only one boundary condition of a first-order differential operator. There are an infinite number of possible boundary conditions compatible with (1.19), namely

$$u(0) = \theta\, u(1) \tag{1.21}$$

where θ is a fixed complex number (including $\theta = \infty$). The question is the choice of a "correct" boundary condition from among (1.21). This is a typical problem of the *singular perturbation* of differential operators.

The difficulty of this problem lies in that it is not covered by Theorem 1.5. We have indeed $T(\varkappa)\, u \to T(0)^\sim u$, $\varkappa \to 0$, for $u \in \mathsf{D}$, but $\mathrm{P}(T(0)^\sim)$ is empty and it is impossible to find an extension of $T(0)^\sim$ which has a nonempty resolvent set and for which D is a core.

Actually $T(\varkappa)$ converges to a definite closed operator T in the generalized sense, *not only in the strong sense but even in gap*, provided that $\varkappa \to 0$ with $\varkappa/\alpha$ being kept away from the imaginary axis. This is seen most directly by constructing the resolvent $R(\zeta, \varkappa) = (T(\varkappa) - \zeta)^{-1}$ as an integral operator with the kernel $g_\varkappa(y, x; \zeta)$, the Green function. A straightforward calculation gives

$$g_\varkappa(y, x; \zeta) = \frac{e^{\frac{\alpha}{\varkappa}(y - x)}}{\sqrt{\alpha^2 - \varkappa\zeta}\, \sinh \frac{\sqrt{\alpha^2 - \varkappa\zeta}}{\varkappa}} \cdot \begin{cases} \sinh\left(\frac{\sqrt{\alpha^2 - \varkappa\zeta}}{\varkappa}\, y\right) \sinh\left(\frac{\sqrt{\alpha^2 - \varkappa\zeta}}{\varkappa}(1 - x)\right), & y \leq x, \\ \sinh\left(\frac{\sqrt{\alpha^2 - \varkappa\zeta}}{\varkappa}(1 - y)\right) \sinh\left(\frac{\sqrt{\alpha^2 - \varkappa\zeta}}{\varkappa}\, x\right), & y \geq x. \end{cases} \tag{1.22}$$

Now let $\varkappa \to 0$. If this is done under the restriction $|\arg(\varkappa/\alpha)| \leqq \delta < \pi/2$, it can be shown without difficulty that

$$(1.23) \qquad g_\varkappa(y, x; 0) \to g(y, x; 0) = \begin{cases} 0, & y < x, \\ \dfrac{1}{2\alpha}, & y > x, \end{cases}$$

$g_\varkappa(y, x; 0)$ being uniformly bounded in the square $0 \leqq x \leqq 1$, $0 \leqq y \leqq 1$. The limiting kernel $g(y, x; 0)$ is exactly the Green function for the operator $T = 2\alpha\, d/dx$ with the boundary condition $u(0) = 0$, which corresponds to (1.21) with $\theta = 0$. (1.23) implies that

$$(1.24) \qquad \int_0^1 \int_0^1 |g_\varkappa(y, x; 0) - g(y, x; 0)|^2\, dx\, dy \to 0$$

by the principle of bounded convergence. If we consider the case $X = L^2$, $\|R(0, \varkappa) - R(0)\|^2$ does not exceed the left member of (1.24) (see Problem III-2.5), where $R(\zeta) = (T - \zeta)^{-1}$. Hence we have $R(\zeta, \varkappa) \to R(\zeta)$ in norm for $\zeta = 0$ and this implies the same for every $\zeta \in P(T)$ (see Remarks IV-3.13, 14). In other words, $T(\varkappa) \to T$ in the generalized sense (in gap). Thus we conclude that *the correct choice among the boundary conditions* (1.21) *is the one with* $\theta = 0$, *if the limit* $\varkappa \to 0$ *is taken in such a way that* $|\arg(\varkappa/\alpha)| \leqq \delta < \pi/2$. It is interesting to note that the spectrum of $T(\varkappa)$ recedes to infinity as $\varkappa \to 0$; if, for example, α is real positive, $\Sigma(T(\varkappa))$ is contained in the half-plane $\operatorname{Re}\zeta \geqq |\varkappa|^{-1}\cos\delta$. This does not conflict with the upper semicontinuity of the spectrum under perturbation small in gap, for the spectrum of T is empty (see Example III-6.8).

Similarly, the correct boundary condition for the "limit" T of $T(\varkappa)$ is given by $\theta = \infty$ in (1.21) if the limit $\varkappa \to 0$ is taken in such a way that $|\arg(-\varkappa/\alpha)| \leqq \delta < \pi/2$. On the other hand, $T(\varkappa)$ has no generalized limit (even in the strong sense) when $\varkappa \to 0$ with $\varkappa/\alpha$ kept pure imaginary. We shall omit the proof, but this may be expected from the fact that no distinguished θ in (1.21) exists in this case. The spectrum of $T(\varkappa)$ behaves in a complicated way when $\varkappa \to 0$. If, for example, α is pure imaginary and $\varkappa > 0$, the eigenvalues (1.20) are all real. Each $\lambda_n(\varkappa)$ tends to $-\infty$ as $\varkappa \to +0$, whereas the spacing of the eigenvalues becomes more and more dense. Thus the whole real axis may be regarded as the limit of $\Sigma(T(\varkappa))$.

Example 1.20. Let X be as in Example 1.19 and let

$$(1.25) \qquad T(\varkappa) = -\alpha\, d^2/dx^2 + \varkappa\, d^4/dx^4, \quad \alpha > 0,$$

with the boundary condition

$$(1.26) \qquad u(0) = u'(0) = u(1) = u'(1) = 0.$$

For small $\varkappa > 0$, $T(\varkappa)$ represents the operator that governs the motion of a stretched string with small rigidity clamped at both ends. If $\varkappa \to +0$, $T(\varkappa)$ formally tends to $T(0) = -\alpha\, d^2/dx^2$. But the boundary condition (1.26) is too strong for a second order differential operator d^2/dx^2, and we have again a problem of singular perturbation: the basic problem is the determination of the correct boundary condition for $T(0)$. Physical intuition leads immediately to the conjecture that

$$(1.27) \qquad u(0) = u(1) = 0$$

is the required boundary condition. In fact, it can be shown that $T(\varkappa) \to T = -\alpha\, d^2/dx^2$ with the boundary condition (1.27) in the generalized sense, at least for $X = L^2$. This could be seen directly by considering the Green function as in Example 1.19, but we shall deduce this later from a more general point of view (see Example 3.8).

The generalized convergence $T(\varkappa) \to T$ (in gap) implies that each isolated eigenvalue λ_n of T is continued continuously to the corresponding eigenvalue $\lambda_n(\varkappa)$ of $T(\varkappa)$. It is known that[1]

$$\lambda_n(\varkappa) = n^2 \pi^2 \alpha \left[1 + 4\left(\frac{\varkappa}{\alpha}\right)^{1/2} + O(\varkappa)\right], \quad \varkappa > 0 . \tag{1.28}$$

Thus the eigenvalues are continuous at $\varkappa = 0$ but not differentiable.

These examples will suffice to show that the spectrum of an operator can behave in quite a complicated way under a "singular" perturbation. In particular, it should be noticed that the behavior may be essentially different for different directions from which the parameter $\varkappa$ tends to zero. Thus the perturbation is not very singular in Examples 1.19 and 1.20 if $\alpha > 0$ and $\varkappa \searrow 0$, for here we have a generalized convergence in gap of $T(\varkappa)$ to T.

Under such a singular perturbation an isolated eigenvalue does not in general remain isolated; it may be absorbed into the continuous spectrum. If one wants to develop a singular perturbation theory of isolated eigenvalues, therefore, it is necessary to assume explicitly that such an absorption does not take place. But the question of deciding when this assumption is satisfied is complicated, and a satisfactory answer does not seem to exist in general. We shall consider these questions in due course.

If the eigenvalue λ is absorbed by the continuous spectrum, we can no longer speak of the perturbation of the eigenvalue λ. Sometimes, however, the perturbed continuous spectrum has a certain kind of concentration near λ. This phenomenon of *spectral concentration* will be discussed in § 5.

4. Stable eigenvalues

According to the remark given in the preceding paragraph, we consider the perturbation of an isolated eigenvalue under the assumption that it remains isolated in spite of the perturbation. To make this assumption more precise, we introduce the following definition.

Let $T_n \xrightarrow[s]{} T$, $n \to \infty$, in the generalized sense. An isolated eigenvalue λ of T with finite multiplicity is said to be *stable* under this perturbation if the following conditions i), ii) are satisfied.

i) The region of convergence Δ_s for $R_n(\zeta)$ contains a neighborhood of λ with the exception of λ. In other words, there is a $\delta > 0$ such that every ζ with $0 < |\zeta - \lambda| < \delta$ belongs to $\mathrm{P}(T_n)$ for sufficiently large n (depending on ζ) and $R_n(\zeta) \xrightarrow[s]{} R(\zeta)$, $n \to \infty$.

This condition implies that the circle $\Gamma : |\zeta - \lambda| = r$, $0 < r < \delta$, is a subset of Δ_s and that the convergence $R_n(\zeta) \xrightarrow[s]{} R(\zeta)$ is uniform on Γ

[1] See Rayleigh [1], p. 300.

(see Theorem 1.2). Therefore, the projection

$$P_n = -\frac{1}{2\pi i}\int_\Gamma R_n(\zeta)\,d\zeta \tag{1.29}$$

is defined and

$$P_n \underset{s}{\to} P = -\frac{1}{2\pi i}\int_\Gamma R(\zeta)\,d\zeta\,, \tag{1.30}$$

where P is the eigenprojection for the eigenvalue λ of T.

ii) $\dim P_n \leqq \dim P$ for sufficiently large n.

In view of (1.30), ii) actually implies that

(1.31a) $\qquad \dim P_n = \dim P \qquad$ for sufficiently large n,

(1.31b) $\qquad \|(P_n - P)P\| \to 0\,,\quad \|(P_n - P)P_n\| \to 0\,,\quad \|(P_n - P)^2\| \to 0\,,$

(1.31c) $\qquad \hat{\delta}(P_n\mathsf{X}, P\mathsf{X}) \to 0\,,\quad n \to \infty,$

where $\hat{\delta}$ is the gap function introduced in IV, § 2.1. These relations follow from Lemma 1.23 proved below. [We shall express (1.31c) by saying that $P_n \to P$ *in gap*.]

$\dim P_n = \dim P = m < \infty$ implies that the spectrum of T_n inside the circle Γ consists of isolated eigenvalues with total multiplicity m. Since this is true for any choice of the radius r of Γ provided n is sufficiently large, it follows that these isolated eigenvalues of T_n converge to λ. At the same time, (1.31c) implies that the *total projection* associated with these eigenvalues (that is, the projection to the m-dimensional space which is the direct sum of the algebraic eigenspaces for these eigenvalues) tends *in gap* to the eigenprojection P for λ of T. Thus the stability of λ implies already the convergence of the eigenvalues and the eigenspaces involved. It should be remarked, however, that the eigenspace for *each individual* eigenvalue of T_n need not be convergent; such an eigenspace need not even have constant dimension.

Example 1.22. The eigenvalue -1 of T of Example 1.17 is not stable for the perturbation considered if $\varkappa$ is real and $\varkappa \to 0$. But it is stable if $\varkappa \to 0$ along a straight line different from the real axis. In Example 1.18, the eigenvalue -1 of T is stable if $\varkappa \to 0$ along a ray different from the negative real axis. In Example 1.19, T has no eigenvalue at all so that there is no question of stability. In Example 1.20, all the eigenvalues of T are stable if $\varkappa \to 0$ along the positive real axis.

Lemma 1.23. *Let* $\{P_n\}$, $n = 1, 2, \ldots$, *be a sequence of projections in a Banach space* X *such that* $P_n \underset{s}{\to} P \in \mathscr{B}(\mathsf{X})$. *Then* P *is also a projection. Suppose further that* $\dim P_n \leqq \dim P < \infty$ *for all* n. *Then* (1.31a), (1.31b), *and* (1.31c) *are true.*

Proof. $P_n^2 = P_n$ and $P_n \underset{s}{\to} P$ together imply $P^2 = P$, so that P is a projection. Let $\{x_1, \ldots, x_m\}$ be a basis of $P\mathsf{X}$, where $m = \dim P$. Then

$$P = \sum_{k=1}^{m} (\ ,e_k)\, x_k\,, \quad e_j \in P^* \mathsf{X}^*\,, \tag{1.33}$$

by III-(4.6); the fact that $e_j \in P^* \mathsf{X}^*$ can be seen by noting III-(4.9). The condition $P^2 = P$ leads, as is easily seen, to

$$(x_k, e_j) = \delta_{jk}\,, \quad k, j = 1, \ldots, m\,. \tag{1.35}$$

Hence $\{e_1, \ldots, e_m\}$ is exactly the basis of $P^* \mathsf{X}^*$ adjoint to the basis $\{x_1, \ldots, x_m\}$ of $P\mathsf{X}$.

Set

$$\alpha_{jk}^n = (P_n x_k, e_j)\,, \quad j, k = 1, \ldots, m\,; \quad n = 1, 2, \ldots\,. \tag{1.36}$$

Since $P_n x_k \to P x_k = x_k$ as $n \to \infty$, we have

$$\alpha_{jk}^n \to \delta_{jk}\,, \quad n \to \infty\,. \tag{1.37}$$

In particular $\det(\alpha_{jk}^n) \to 1$, which implies that $P_n x_1, \ldots, P_n x_m$ are linearly independent if n is sufficiently large. Since $\dim P_n \leqq m$ by hypothesis, they form a basis of $P_n \mathsf{X}$ with $\dim P_n = m$. This proves (1.31a).

Also we can write

$$P_n = \sum_{k=1}^{m} (\ , f_k^n)\, P_n x_k\,, \tag{1.38}$$

where $\{f_1^n, \ldots, f_m^n\}$ is the basis of $P_n^* \mathsf{X}^*$ adjoint to the basis $\{P_n x_1 \ldots, P_n x_m\}$ of $P\mathsf{X}$ (see the argument for P given above). Thus

$$P_n^* = \sum_{k=1}^{m} (\ , P_n x_k)\, f_k^n$$

(again by III-(4.9)) and, in particular,

$$P_n^* e_j = \sum_{k=1}^{m} (e_j, P_n x_k)\, f_k^n = \sum_{k=1}^{m} \bar{\alpha}_{jk}^n f_k^n$$

by (1.36). Hence

$$f_k^n = \sum_{j=1}^{m} \beta_{kj}^n P_n^* e_j\,, \tag{1.39}$$

where (β_{jk}^n) is the inverse matrix to $(\bar{\alpha}_{jk}^n)$, so that

$$\beta_{jk}^n \to \delta_{jk}\,, \quad n \to \infty\,. \tag{1.40}$$

Since $P_n \underset{s}{\to} P$ implies $P_n^* \underset{w*}{\to} P^*$, we have $P_n^* e_j \underset{w*}{\to} P^* e_j = e_j$. Hence (1.39) and (1.40) give

$$f_k^n \underset{w*}{\to} e_k, \quad n \to \infty. \tag{1.41}$$

Now we can complete the proof of the lemma. The first convergence in (1.31b) follows from $P_n \underset{s}{\to} P$ and $\dim P < \infty$ by virtue of Lemma III-3.7. Indeed, $\|(P_n - P)P\| = \sup\limits_{\|u\| \leq 1} \|(P_n - P)Pu\| \leqq \sup\limits_{\substack{\|v\| \leq \|P\| \\ v \in PX}} \|(P_n - P)v\| \to 0$. To prove the second one, we note that (1.38) implies

$$(P_n - P)P_n = \sum_{k=1}^{m} (\ , f_k^n)(P_n - P)P_n x_k,$$

hence

$$\|(P_n - P)P_n\| \leqq \sum_{k=1}^{m} \|f_k^n\| \, \|(P_n - P)P_n x_k\|.$$

Since $\|f_k^n\| \leqq$ const. by (1.41) (see Problem III-1.30) and since $\|(P_n - P)P_n x_k\| \to 0$ by $P_n \underset{s}{\to} P$, we have $\|(P_n - P)P_n\| \to 0$ as required.

Obviously the last convergence in (1.31b) follows from the preceding two.

To prove (1.31c), it suffices to notice that

$$\delta(PX, P_nX) = \sup_{u \in PX, \|u\| \leq 1} \operatorname{dist}(u, P_nX) \leqq \sup_{u \in PX, \|u\| \leq 1} \|(P_n - P)Pu\|$$
$$\leqq \|(P_n - P)P\| \to 0$$

by (1.31b) and similarly $\delta(P_nX, PX) \leqq \|(P_n - P)P_n\| \to 0$. Then we have $\hat{\delta}(P_nX, PX) = \max\{\delta(P_nX, PX), \delta(PX, P_nX\} \to 0$ as required. This completes the proof of Lemma 1.23.

For later use we add

Lemma 1.24. *In addition to the assumptions of Lemma* 1.21a, *assume that* $P_n^* \underset{s}{\to} P^*$. *Then* $\|P_n - P\| \to 0$.

Proof. In this case we have $P_n^* e_j \to P^* e_j = e_j$. Hence (1.39) and (1.40) give

$$f_k^n \to e_k, \quad n \to \infty. \tag{1.42}$$

Since

$$P_n - P = \sum_{k=1}^{m} [(\ , f_k^n)(P_n x_k - x_k) + (\ , f_k^n - e_k) x_k]$$

by (1.33) and (1.38), we obtain

$$\|P_n - P\| \leqq \sum_{k=1}^{m} [\|f_k^n\| \, \|P_n x_k - x_k\| + \|f_k^n - e_k\| \, \|x_k\|] \to 0.$$

§ 2. Asymptotic expansions

1. Asymptotic expansion of the resolvent

In what follows we shall be concerned with a rather special but practically important case in which $T(\varkappa)$ is given *formally* by $T(\varkappa) = T + \varkappa T^{(1)}$. More precisely, we assume that two operators T, $T^{(1)}$ and a parameter $\varkappa$ are given, such that

i) $T \in \mathscr{C}(\mathsf{X})$;

ii) $\mathsf{D} = \mathsf{D}(T) \cap \mathsf{D}(T^{(1)})$ is a core of T;

iii) $T(\varkappa) \in \mathscr{C}(\mathsf{X})$ and it is an extension of the operator $T + \varkappa T^{(1)}$ defined with domain D[1], where $\varkappa$ is restricted to

$$0 < \varkappa \leqq 1 . \tag{2.1}$$

Let us make some comments on these assumptions. The condition (2.1) on the range of $\varkappa$ is not very restrictive. Even when $\varkappa$ is complex, we have in genèral to restrict ourselves to the case in which $\varkappa$ goes to 0 along a ray, in view of the varying behavior of $T(\varkappa)$ for different directions of approach of $\varkappa$ to 0 (see Examples in the preceding section). In such a case the range of $\varkappa$ can be reduced to (2.1) by writing $\varkappa = |\varkappa|\, e^{i\theta}$ and replacing $|\varkappa|$ by $\varkappa$ and $e^{i\theta}\, T^{(1)}$ by $T^{(1)}$, respectively. If ii) is not satisfied, we can replace T by the closure of its restriction to D; this restores ii) without affecting the definition of $T(\varkappa)$.

We can now define the region of boundedness Δ_b and the region of strong convergence Δ_s for the family of resolvents $R(\zeta, \varkappa) = (T(\varkappa) - \zeta)^{-1}$ when $\varkappa \to 0$ (see Remark 1.9). We have by definition

$$\limsup_{\varkappa \to 0} \|R(\zeta, \varkappa)\| = M(\zeta) < \infty , \quad \zeta \in \Delta_b . \tag{2.2}$$

Since $T(\varkappa)\, u \to T u$, $\varkappa \to 0$, for $u \in \mathsf{D}$ and D is a core of T, it follows from Theorem 1.5 that $T(\varkappa) \underset{s}{\to} T$ in the generalized sense with

$$\Delta_s = \Delta_b \cap \mathrm{P}(T) , \tag{2.3}$$

provided that we assume in addition that

iv) $\Delta_b \cap \mathrm{P}(T)$ is not empty.

Thus we have

$$\operatorname*{s-lim}_{\varkappa \to 0} R(\zeta, \varkappa) = R(\zeta) , \quad \zeta \in \Delta_s . \tag{2.4}$$

---

[1] Thus $T(\varkappa)$ may be quite discontinuous in $\varkappa$ for $\varkappa > 0$, for it need not be uniquely determined by $T + \varkappa T^{(1)}$.

This means that $R(\zeta, \varkappa)\, u \to R(\zeta)\, u$ for every $u \in \mathsf{X}$. We now want to estimate the *rate of this convergence* more precisely. To this end, however, we have to make specific assumptions on each u to be considered. We shall prove a series of theorems on such estimates; in what follows i) to iv) are assumed unless otherwise stated.

Theorem 2.1. *Let $\zeta \in \Delta_s$. If $R(\zeta)\, u \in \mathsf{D}(T^{(1)})$, we have*

$$R(\zeta, \varkappa)\, u = R(\zeta)\, u - \varkappa\, R(\zeta)\, T^{(1)}\, R(\zeta)\, u + o(\varkappa)\,. \tag{2.5}$$

Here $o(\varkappa)$ denotes an element of X with norm of the order $o(\varkappa)$ as $\varkappa \to 0$.

Proof. In view of (2.4), (2.5) follows directly from the identity

$$\begin{aligned} R(\zeta, \varkappa)\, u - R(\zeta)\, u &= -R(\zeta, \varkappa)\,(T(\varkappa) - T)\, R(\zeta)\, u \\ &= -\varkappa\, R(\zeta, \varkappa)\, T^{(1)}\, R(\zeta)\, u\,; \end{aligned} \tag{2.6}$$

note that $R(\zeta)\, u \in \mathsf{D} = \mathsf{D}(T) \cap \mathsf{D}(T^{(1)})$ by assumption.

Theorem 2.2. *Let $\zeta \in \Delta_s$. If $R(\zeta)\, u \in \mathsf{D}(T^{(1)})$ and $R(\zeta)\, T^{(1)}\, R(\zeta)\, u \in$ $\in \mathsf{D}(T^{(1)})$, then*

$$\begin{aligned} R(\zeta, \varkappa)\, u = R(\zeta)\, u &- \varkappa\, R(\zeta)\, T^{(1)}\, R(\zeta)\, u + \\ &+ \varkappa^2\, R(\zeta)\, T^{(1)}\, R(\zeta)\, T^{(1)}\, R(\zeta)\, u + o(\varkappa^2)\,. \end{aligned} \tag{2.7}$$

Proof. Apply (2.5) to the right member of (2.6) with u replaced by $T^{(1)}\, R(\zeta)\, u$.

It is obvious that we can proceed in the same way, obtaining the expansion of $R(\zeta, \varkappa)\, u$ in powers of $\varkappa$ under stronger and stronger assumptions. The series that appear in these expansions are exactly the *second Neumann series* for $R(\zeta, \varkappa)$ [see II-(1.13)]. We know that the second Neumann series is a convergent power series in $\varkappa$ if $T^{(1)}$ is relatively bounded with respect to T (see Theorem IV-3.17). Here we have assumed no such relative boundedness, but the above theorems show that the second Neumann series is valid as an *asymptotic expansion* up to a particular power of $\varkappa$ as long as the relevant terms are significant.

The formulas (2.5) and (2.6) lead to corresponding expansions of the scalar quantity $(R(\zeta, \varkappa)\, u, v)$ for $u \in \mathsf{X}$, $v \in \mathsf{X}^*$, under the stated assumptions on u. Since $(R(\zeta, \varkappa)\, u, v) = (u, R(\zeta, \varkappa)^*\, v)$, a similar expansion is valid if v satisfies conditions similar to those made on u above, under the basic assumptions for T^* and $T^{(1)*}$ similar to the ones for the pair T, $T^{(1)}$. For simplicity, we shall state the results when T and the $T(\varkappa)$ are selfadjoint, assuming X to be a Hilbert space. A remarkable result here is that the expansion of $(R(\zeta, \varkappa)\, u, v)$ can be obtained up to the order $\varkappa^2$ if both u and v satisfy the condition of Theorem 2.1, namely

Theorem 2.3. *Let $\mathsf{X} = \mathsf{H}$ be a Hilbert space and let T and $T(\varkappa)$ be selfadjoint and $T^{(1)}$ symmetric. If both ζ and $\bar\zeta$ belong to Δ_s and if $R(\zeta)\, u \in \mathsf{D}(T^{(1)})$, $R(\bar\zeta)\, v = R(\zeta)^* v \in \mathsf{D}(T^{(1)})$, then*

$$(2.8)\quad (R(\zeta, \varkappa)\, u, v) = (R(\zeta)\, u, v) - \varkappa(R(\zeta)\, T^{(1)} R(\zeta)\, u, v) + \\ + \varkappa^2 (R(\zeta)\, T^{(1)} R(\zeta)\, u, T^{(1)} R(\bar\zeta)\, v) + o(\varkappa^2)\,.$$

Proof. From (2.6) we have

$$(2.9)\quad (R(\zeta, \varkappa)\, u, v) - (R(\zeta)\, u, v) = -\varkappa (T^{(1)} R(\zeta)\, u, R(\bar\zeta, \varkappa)\, v)$$

since $R(\zeta, \varkappa)^* = R(\bar\zeta, \varkappa)$. (2.8) follows from (2.9) by applying (2.5) with ζ and u replaced by $\bar\zeta$ and v, respectively.

In the same way we can obtain the expansion of $(R(\zeta, \varkappa)\, u, v)$ up to the order $\varkappa^4$ if $T^{(1)} R(\zeta)\, T^{(1)} R(\zeta)\, u$ and $T^{(1)} R(\bar\zeta)\, T^{(1)} R(\bar\zeta)\, v$ exist.

Remark 2.4. (2.5) and (2.7) give asymptotic expansions for the solution of $(T(\varkappa) - \zeta)\, v = u$, which is, for example, a boundary value problem when $T(\varkappa)$ is a differential operator. However, the use of these formulas is restricted since the assumptions are rather strong. In §§ 3, 4 we shall develop a more satisfactory theory of asymptotic expansion of $R(\zeta, \varkappa)$ based on the theory of sesquilinear forms in a Hilbert space.

Example 2.5. Consider the $T(\varkappa)$ of Example 1.17. If we write $T(\varkappa) = T + |\varkappa|\, e^{i\theta}\, T^{(1)}$ and replace $|\varkappa|$, $e^{i\theta}\, T^{(1)}$ by $\varkappa$, $T^{(1)}$ respectively, $T(\varkappa)$ satisfies the basic assumptions i) to iv) [for iv) see Example 1.17]. The applicability of Theorems 2.1, 2.2 depends on the properties of u and f. Since $T = -(\ ,f)\, f$, $\|f\| = 1$, it is easy to compute $R(\zeta)\, u$; we have

$$R(\zeta)\, u = -\frac{1}{\zeta}\, u + \frac{(u, f)}{\zeta(\zeta + 1)}\, f\,.$$

If u and f belong to $\mathsf{D}(T^{(1)})$, then $R(\zeta)\, u \in \mathsf{D}(T^{(1)})$ and Theorem 2.1 is applicable. In this case $R(\zeta)\, T^{(1)} R(\zeta)\, u$ is a linear combination of $T^{(1)} u$, $T^{(1)} f$ and f. If $[T^{(1)}]^2 u$ and $[T^{(1)}]^2 f$ exist, then $R(\zeta)\, T^{(1)} R(\zeta)\, u \in \mathsf{D}(T^{(1)})$ and Theorem 2.2 is applicable, and so forth. Similar results can be stated for Example 1.18.

2. Remarks on asymptotic expansions

In general the asymptotic expansions of the resolvent given by (2.5) or (2.7) are valid only under certain specific restrictions on u, as stated in these theorems. The expansions are valid, moreover, only up to a certain order in $\varkappa$ determined by u. Thus the expansions are of a more general kind than what is usually called asymptotic expansions, in which it is required that the expansion be possible to any order of the parameter.

This remark is of particular importance in view of the fact that, in the theory of singular perturbation for ordinary differential operators, the expansion for $R(\zeta, \varkappa)\, u$ is usually given to *any* order in $\varkappa$, with the restriction that one or both ends of the interval considered should be excluded. Consider, for example, the operator $T(\varkappa)$ of Example 1.19 and set $\zeta = 0$, $u(x) = 1$. A simple calculation gives then

$$R(0,\varkappa)\, u(x) = \frac{x}{2\alpha} - \frac{e^{-\frac{2\alpha}{\varkappa}(1-x)} - e^{-\frac{2\alpha}{\varkappa}}}{2\alpha\left(1 - e^{-\frac{2\alpha}{\varkappa}}\right)}, \tag{2.10}$$

in which the first term $x/2\alpha$ is equal to the zeroth approximation $R(0)\, u(x)$. The additional term on the right of (2.10) has the asymptotic expansion 0 (which is valid to any order in $\varkappa$) provided $0 \leqq x < 1$, and this expansion is uniform for $0 \leqq x \leqq b' < 1$.

Thus one might suppose that, in this example, no restriction is necessary for the asymptotic expansion of $R(\zeta, \varkappa)\, u$ and, furthermore, that the expansion is valid to any order, at least if $u(x)$ is a smooth function. But this is not correct. Although the remainder term on the right of (2.10) is smaller than any finite power of $\varkappa$ for a fixed $x < 1$, it is not necessarily small in the whole interval $(0, 1)$. In fact, a simple calculation shows that the L^2-norm of this remainder term is

$$\left(\frac{\varkappa}{16\alpha^3}\right)^{1/2} + \cdots, \tag{2.11}$$

where $\cdots$ denotes a function of $\varkappa$ tending to zero faster than any finite power of $\varkappa$. Therefore, we have to say that *in this example the expansion of $R(0, \varkappa)\, u$ as an element of L^2 is valid only up to the zeroth order of $\varkappa$*, the remainder being of the order $\varkappa^{1/2}$.

It should be noted, furthermore, that the situation cannot be improved by considering the asymptotic expansion of (2.10) in powers of, say, $\varkappa^{1/2}$. A glance at (2.10) is sufficient to exclude such a possibility. In this sense the impossibility of the asymptotic expansion for (2.10) is essential, if one considers the vector $R(0, \varkappa)\, u$ of L^2 and does not confine one's attention on a fixed value of x. Similar remarks apply to all asymptotic expansions encountered in the singular perturbation of differential operators[1].

---

[1] The second term on the right of (2.10), which has the asymptotic expansion 0 for each $x < 1$, is called the *boundary layer term*. In general the asymptotic expansion of $R(0, \varkappa)\, u(x)$ must be supplemented by a boundary layer term, which has the asymptotic expansion 0 for a fixed interior point x but which need not be small globally. For detailed theory of the boundary layer and singular perturbation in general, see e. g. Harris [1], [2], Huet [1]—[7], Kumano-go [1], Ladyženskaja [1], Levinson [1], Morgenstern [1], Moser [2], Nagumo [1], Višik and Lyusternik [2].

3. Asymptotic expansions of isolated eigenvalues and eigenvectors

We now consider the rate of convergence of the eigenvalues and eigenvectors of $T(\varkappa)$ arising from an eigenvalue λ of T (see § 1.4), under the basic assumptions of par. 1 and the additional assumption that λ is *stable*. We assume that λ has multiplicity $m < \infty$. We take over the notations Δ_b, Δ_s, δ, r, Γ, etc. of § 1.4. In particular we note that the region of strong convergence Δ_s contains the set of ζ such that $0 < |\zeta - \lambda| < \delta$.

We recall also that $\Sigma(T(\varkappa))$ inside the circle $\Gamma: |\zeta - \lambda| = r$ consists of isolated eigenvalues with total multiplicity m and that these eigenvalues converge to λ as $\varkappa \to 0$. Let us denote these *repeated* eigenvalues by $\mu_1(\varkappa), \ldots, \mu_m(\varkappa)$:

$$\mu_j(\varkappa) \to \lambda\,, \quad \varkappa \to 0\,, \quad j = 1, \ldots, m\,. \tag{2.12}$$

The total projection

$$P(\varkappa) = -\frac{1}{2\pi i}\int_\Gamma R(\zeta, \varkappa)\, d\zeta \tag{2.13}$$

tends in gap to P, the eigenprojection for λ [see (1.31a, b, c)]:

$$\begin{aligned} &\|(P(\varkappa) - P)P\| \to 0\,, \\ &\|(P(\varkappa) - P)P(\varkappa)\| \to 0\,, \quad \dim P(\varkappa) = \dim P = m\,. \end{aligned} \tag{2.14 a}$$

If $m = 1$, $P(\varkappa)$ is itself the eigenprojection on the one-dimensional eigenspace of $T(\varkappa)$ for the eigenvalue $\lambda(\varkappa) = \mu_1(\varkappa)$, which converges to λ. (2.14) implies that an eigenvector $\varphi(\varkappa)$ of $T(\varkappa)$ for the eigenvalue $\lambda(\varkappa)$ can be chosen in such a way that

$$\varphi(\varkappa) \to \varphi\,, \quad \varkappa \to 0\,, \tag{2.15}$$

where φ is an eigenvector of T for the eigenvalue λ:

$$T\varphi = \lambda\varphi\,, \quad T(\varkappa)\,\varphi(\varkappa) = \lambda(\varkappa)\,\varphi(\varkappa)\,. \tag{2.16}$$

To see this, it suffices to set $\varphi(\varkappa) = P(\varkappa)\,\varphi$.

In general it is impossible to deduce more precise results on the rate of convergence in (2.12) or (2.14) or (2.15). Such a refinement is possible, however, if one makes further assumptions on the eigenspace $P\mathsf{X}$ of T. We have namely[1]

Theorem 2.6. *Let λ be an isolated, semisimple eigenvalue of T with the eigenprojection P, $\dim P = m < \infty$. Let λ be stable and let $P\mathsf{X} \subset \mathsf{D}(T^{(1)})$.*

[1] The results of this and the following paragraphs were proved in T. KATO [1], [3] for selfadjoint operators by making use of the spectral representations. The generalization to operators in Banach spaces given here needs an entirely different proof.

Then the eigenvalues $\mu_j(\varkappa)$ admit asymptotic expansions[1]

$$\mu_j(\varkappa) = \lambda + \varkappa\, \mu_j^{(1)} + o(\varkappa)\,, \quad j = 1, \ldots, m\,, \tag{2.17}$$

where the $\mu_j^{(1)}$ are the repeated eigenvalues of the operator $P T^{(1)} P$ considered in the m-dimensional space $P\mathsf{X}$. The total projection $P(\varkappa)$ for these eigenvalues has the property that

$$P(\varkappa)\, P = P - \varkappa\, S T^{(1)} P + o(\varkappa)_u\,, \tag{2.18}$$

where S is the reduced resolvent for λ of T (see III-§ 6.5) *and $o(\varkappa)_u$ denotes an operator such that $\varkappa^{-1}\|o(\varkappa)_u\| \to 0$ as $\varkappa \to 0$. If in particular $m = 1$, an eigenvector $\varphi(\varkappa)$ for the eigenvalue $\lambda(\varkappa) = \mu_1(\varkappa)$ of $T(\varkappa)$ can be chosen in such a way that*

$$\varphi(\varkappa) = \varphi - \varkappa\, S T^{(1)} \varphi + o(\varkappa)\,, \tag{2.19}$$

where φ is an eigenvector of T for λ.

Proof. I. We start with the remark that $R(\zeta)\, Pv = (\lambda - \zeta)^{-1} Pv \in \mathsf{D}(T^{(1)})$ for every $v \in \mathsf{X}$ so that (2.6) gives

$$\begin{aligned} R(\zeta, \varkappa)\, Pv &= (\lambda - \zeta)^{-1} Pv - \varkappa(\lambda - \zeta)^{-1} R(\zeta, \varkappa)\, T^{(1)} Pv \\ &= (\lambda - \zeta)^{-1} Pv - \varkappa(\lambda - \zeta)^{-1} R(\zeta)\, T^{(1)} Pv + o(\varkappa)\,, \end{aligned}$$

where $o(\varkappa)$ is uniform in ζ for $\zeta \in \Gamma$. Since $\dim P\mathsf{X} = m < \infty$, this implies[2]

$$\begin{aligned} R(\zeta, \varkappa)\, P &= (\lambda - \zeta)^{-1} P - \varkappa(\lambda - \zeta)^{-1} R(\zeta, \varkappa)\, T^{(1)} P \\ &= (\lambda - \zeta)^{-1} P - \varkappa(\lambda - \zeta)^{-1} R(\zeta)\, T^{(1)} P + o(\varkappa)_u\,, \end{aligned} \tag{2.20}$$

where $o(\varkappa)_u$ is uniform in $\zeta \in \Gamma$. Substitution of (2.20) into (2.13) multiplied by P from the right leads immediately to (2.18); recall the expansion $R(\zeta) = (\lambda - \zeta)^{-1} P + S + (\zeta - \lambda)\, S^2 + \cdots$ [see III-(6.32) where $D = 0$ by assumption] and note that

$$T^{(1)} P \in \mathscr{B}(\mathsf{X}) \tag{2.21}$$

by the assumption that $P\mathsf{X} \subset \mathsf{D}(T^{(1)})$. (2.19) follows from (2.18) on setting $\varphi(\varkappa) = P(\varkappa)\, \varphi = P(\varkappa)\, P\, \varphi$ for a $\varphi \in \mathsf{M} = P\mathsf{X}$.

II. To deduce the asymptotic expansion for $\mu_j(\varkappa)$, we introduce the operator [see (2.18) just proved]

$$V(\varkappa) = 1 - P + P(\varkappa)\, P = 1 - \varkappa\, S T^{(1)} P + o(\varkappa)_u\,. \tag{2.22}$$

Since $V(\varkappa)\, P = P(\varkappa)\, P$ and $V(\varkappa)\,(1 - P) = 1 - P$, $V(\varkappa)$ maps the eigenspace $\mathsf{M} = P\mathsf{X}$ onto $\mathsf{M}(\varkappa) = P(\varkappa)\, \mathsf{X}$ and leaves every element of the complementary subspace $(1 - P)\, \mathsf{X}$ unchanged. The map $\mathsf{M} \to V(\varkappa)\, \mathsf{M}$

[1] If T and $T(\varkappa)$ are selfadjoint (in a Hilbert space), we have expansions of the $\mu_j(\varkappa)$ up to the order $\varkappa^2$, see Theorem 2.9 and the attached footnote.

[2] Note that $A_n B \xrightarrow[s]{} 0$ implies $\|A_n B\| \to 0$ if B has finite rank.

$= \mathsf{M}(\varkappa)$ is *onto* since $\| (P(\varkappa) - P)^2 \| \to 0$ by (2.14a) (see Problem I-4.12). It follows that the inverse

(2.23) $$V(\varkappa)^{-1} = 1 + \varkappa S T^{(1)} P + o(\varkappa)_u$$

maps $\mathsf{M}(\varkappa)$ onto M, keeping every element of $(1 - P)\mathsf{X}$ invariant.

We now consider the operator $R_1(\zeta, \varkappa) = V(\varkappa)^{-1} R(\zeta, \varkappa) V(\varkappa) P$. $V(\varkappa) P$ has range $\mathsf{M}(\varkappa)$ by the above remark, $R(\zeta, \varkappa)$ sends $\mathsf{M}(\varkappa)$ into itself because $P(\varkappa)$ commutes with $R(\zeta, \varkappa)$, and $V(\varkappa)^{-1}$ maps $\mathsf{M}(\varkappa)$ onto M; hence $R_1(\zeta, \varkappa)$ has range in M. Thus

$$\begin{aligned} R_1(\zeta, \varkappa) &= P R_1(\zeta, \varkappa) = P V(\varkappa)^{-1} R(\zeta, \varkappa) V(\varkappa) P \\ &= (P + o(\varkappa)_u) R(\zeta, \varkappa) (P - \varkappa S T^{(1)} P + o(\varkappa)_u) \end{aligned}$$

(note that $PS = 0$). Substituting for $R(\zeta, \varkappa) P$ from (2.20) and noting that $R(\zeta, \varkappa) S T^{(1)} P \to R(\zeta) S T^{(1)} P$ in norm, $P R(\zeta) = (\lambda - \zeta)^{-1} P$, and $PS = 0$, we obtain

(2.24) $$\begin{aligned} V(\varkappa)^{-1} & R(\zeta, \varkappa) V(\varkappa) P \\ &= (\lambda - \zeta)^{-1} P - \varkappa (\lambda - \zeta)^{-2} P T^{(1)} P + o(\varkappa)_u , \end{aligned}$$

where $o(\varkappa)_u$ is uniform for $\zeta \in \Gamma$.

Since

(2.25) $$T(\varkappa) P(\varkappa) = -\frac{1}{2\pi i} \int_\Gamma \zeta R(\zeta, \varkappa) \, d\zeta$$

[see III-(6.24)], integration of (2.24) along Γ after multiplication by $-\zeta/2\pi i$ gives[1]

(2.26) $$V(\varkappa)^{-1} T(\varkappa) P(\varkappa) V(\varkappa) P = \lambda P + \varkappa P T^{(1)} P + o(\varkappa)_u .$$

Now the $\mu_j(\varkappa)$ are the repeated eigenvalues of $T(\varkappa)$ considered in the m-dimensional subspace $\mathsf{M}(\varkappa)$, hence equal to the eigenvalues of $T(\varkappa) P(\varkappa)$ in $\mathsf{M}(\varkappa)$ and therefore also to those of $V(\varkappa)^{-1} T(\varkappa) P(\varkappa) V(\varkappa) = T_0(\varkappa)$, which is similar to $T(\varkappa) P(\varkappa)$. Since $V(\varkappa)^{-1} \mathsf{M}(\varkappa) = \mathsf{M}$ as remarked above, these are in turn equal to the eigenvalues of $T_0(\varkappa) P$, that is, of the operator (2.26), considered in M. In view of (2.26), this remark leads immediately to the result (2.17) by Theorem II-5.4 concerning the differentiability of eigenvalues in a finite-dimensional space. This completes the proof of Theorem 2.6.

Remark 2.7. (2.18) shows that $P(\varkappa) P$ admits an asymptotic expansion up to the first order in $\varkappa$. $P(\varkappa)$ itself does not seem to have such an expansion, for the *formal* expansion of $P(\varkappa)$ is $P(\varkappa) = P - \varkappa (S T^{(1)} P + P T^{(1)} S) + \cdots$ [see II-(2.14)], in which the term $P T^{(1)} S$ need not

[1] In view of the unboundedness of $T(\varkappa)$, it would be difficult to deduce (2.26) by simply expanding the left member into a formal power series in $\varkappa$.

make sense under the present assumptions. (This is one reason why the proof of Theorem 2.6 is rather complicated.) Thus the expansion (2.19) of the eigenvector (but not of the eigenprojection) is the best one can obtain even for $m = 1$.

In Theorem 2.6 it is in general impossible to give any statement on the behavior of the eigenvectors or eigenspaces of $T(\varkappa)$ except in the case $m = 1$. If we assume that the $\mu_j^{(1)}$ are distinct, however, we have

Theorem 2.8. *In* Theorem 2.6 *assume that the m eigenvalues $\mu_j^{(1)}$ of $PT^{(1)}P$ are distinct, and let $P_j^{(1)}$ be the associated eigenprojections. Then the $\mu_j(\varkappa)$ are also distinct for sufficiently small $|\varkappa|$. Let $P_j^{(1)}(\varkappa)$ be the one-dimensional eigenprojections of $T(\varkappa)$ for the eigenvalues $\mu_j(\varkappa)$. Then*

$$P_j^{(1)}(\varkappa) \to P_j^{(1)} \quad \text{in gap as} \quad \varkappa \to 0 . \tag{2.27}$$

Proof. $P_j^{(1)}(\varkappa)$ is the eigenprojection of the operator $\tilde{T}^{(1)}(\varkappa) = \varkappa^{-1}(T(\varkappa) - \lambda) P(\varkappa)$ for the eigenvalue $\varkappa^{-1}(\mu_j(\varkappa) - \lambda) = \mu_j^{(1)} + o(1)$. But we have, in virtue of (2.26),

$$V(\varkappa)^{-1} \tilde{T}^{(1)}(\varkappa) V(\varkappa) P = PT^{(1)}P + o(1)_u \tag{2.28}$$

since $V(\varkappa)^{-1} P(\varkappa) V(\varkappa) P = V(\varkappa)^{-1} V(\varkappa) P = P$. Since $PT^{(1)}P$ has m distinct eigenvalues $\mu_j^{(1)}$, the operator (2.28) has the same property. Thus we can choose its m eigenvectors $\psi_j(\varkappa) \in \mathsf{M}$ in such a way that $\psi_j(\varkappa) \to \varphi_j$, $\varphi_j \in \mathsf{M}$ being the eigenvectors of $PT^{(1)}P$ (continuity of eigenvalues and eigenprojections, see II-§ 5.1). Now $\varphi_j(\varkappa) = V(\varkappa)\,\psi_j(\varkappa)$ is an eigenvector of $\tilde{T}^{(1)}(\varkappa)$ and $\varphi_j(\varkappa) \to \varphi_j$ since $V(\varkappa) \to 1$.

The $\varphi_j(\varkappa)$ form a basis of $\mathsf{M}(\varkappa)$, so that for any $w \in \mathsf{X}$ we have

$$P(\varkappa)\, w = \xi_1(\varkappa)\, \varphi_1(\varkappa) + \cdots + \xi_m(\varkappa)\, \varphi_m(\varkappa) .$$

Since $\varphi_j(\varkappa) \to \varphi_j$ and the φ_j form a basis of M and since $P(\varkappa)\, w \to Pw$, it is easily seen that $\lim_{\varkappa \to 0} \xi_j(\varkappa) = \xi_j$ exists and $Pw = \xi_1 \varphi_1 + \cdots + \xi_m \varphi_m$. On the other hand we have $\xi_j(\varkappa)\, \varphi_j(\varkappa) = P_j^{(1)}(\varkappa)\, w$ and $\xi_j\, \varphi_1 = P_j^{(1)} w$. Hence we conclude that $P_j^{(1)}(\varkappa)\, w \to P_j^{(1)} w$, that is, $P_j^{(1)}(\varkappa) \to P_j^{(1)}$ strongly. Since $\dim P_j^{(1)}(\varkappa) = \dim P_j^{(1)} = 1$, the convergence is actually in gap by Lemma 1.23. This completes the proof of Theorem 2.8.

4. Further asymptotic expansions

We shall now show that a higher approximation to the eigenvalues can be obtained if we assume that $PT^{(1)}$ is *bounded*, in addition to the assumptions of Theorem 2.6. Since $PT^{(1)}$ has range in the finite-dimensional subspace $\mathsf{M} = P\mathsf{X}$, it can then be extended to an operator of $\mathscr{B}(\mathsf{X})$ with range in M. Such an extension is not unique unless $T^{(1)}$ is densely defined, but we take any such extension and denote it by $[PT^{(1)}]$;

we note that $P[PT^{(1)}] = [PT^{(1)}]$. If $T^{(1)}$ is densely defined, $PT^{(1)}$ is bounded if and only if $T^{(1)*}P^* \in \mathscr{B}(\mathsf{X}^*)$; in this case $[PT^{(1)}]$ is unique.

To state the theorem that gives the higher approximation, it is convenient to modify the notations of Theorem 2.6 and introduce further ones. Let λ_j, $j = 1, \ldots, s$, be the *distinct* eigenvalues of $PT^{(1)}P$ (that is, the different ones among the μ_j) in the subspace $P\mathsf{X}$ and let $P_j^{(1)}$ be the associated eigenprojections. Then we have

Theorem 2.9.[1] *Assume* D *is a core of* $T(\varkappa)$ *for* $\varkappa > 0$. *In* Theorem 2.6 *suppose that* $PT^{(1)}$ *is bounded and define* $[PT^{(1)}] \in \mathscr{B}(\mathsf{X})$ *as above. Let all the eigenvalues* $\lambda_j^{(1)}$ *of* $PT^{(1)}P$ *be semisimple*[2]. *Then the m eigenvalues* $\mu_j(\varkappa)$ *can be renumbered in the form* $\mu_{jk}(\varkappa)$, $j = 1, \ldots, s$, $k = 1, \ldots, m_j^{(1)}$, *in such a way that they have the asymptotic expansions*

$$\mu_{jk}(\varkappa) = \lambda + \varkappa\,\lambda_j^{(1)} + \varkappa^2\,\mu_{jk}^{(2)} + o(\varkappa^2)\,, \tag{2.29}$$

where the $\mu_{jk}^{(2)}$, $k = 1, \ldots, m_j^{(1)}$, *are the repeated eigenvalues of* $-P_j^{(1)}[PT^{(1)}]\,S\,T^{(1)}P_j^{(1)}$. *The total projection* $P_j^{(1)}(\varkappa)$ *for the* $m_j^{(1)}$ *eigenvalues* $\mu_{jk}(\varkappa)$ *for* $k = 1, \ldots, m_j^{(1)}$ *has the asymptotic expansion*

$$P_j^{(1)}(\varkappa) = P_j^{(1)} + \varkappa\,P_j^{(11)} + o(\varkappa)_s\,, \tag{2.30}$$

where $o(\varkappa)_s$ *denotes an operator such that* $\varkappa^{-1}o(\varkappa)_s \underset{s}{\to} 0$ *and*

$$\begin{aligned} P_j^{(11)} &= -P_j^{(1)}[PT^{(1)}]\,S + P_j^{(1)}[PT^{(1)}]\,S\,T^{(1)}\,S_j^{(1)} - \\ &\quad - S\,T^{(1)}P_j^{(1)} + S_j^{(1)}[PT^{(1)}]\,S\,T^{(1)}P_j^{(1)}\,, \\ S_j^{(1)} &= -\sum_{i \neq j} (\lambda_j^{(1)} - \lambda_i^{(1)})\,P_i^{(1)}\,. \end{aligned} \tag{2.31}$$

The total projection $P(\varkappa)$ *for the m eigenvalues* $\mu_{jk}(\varkappa)$ *has the expansion*

$$P(\varkappa) = P - \varkappa(S\,T^{(1)}P + [PT^{(1)}]\,S) + o(\varkappa)_s\,. \tag{2.32}$$

Remark 2.10. The formulas (2.29) to (2.32) are formally identical with certain formulas of Theorem II-5.11, but they are more complicated than the latter since here we are dealing with unbounded operators. It should be noted, nevertheless, that the operators $T^{(1)}P$, $T^{(1)}P_j^{(1)} = T^{(1)}PP_j^{(1)}$, $T^{(1)}S_j^{(1)} = T^{(1)}PS_j^{(1)}$ belong to $\mathscr{B}(\mathsf{X})$.

---

[1] This theorem gives an expansion of eigenvalues up to the order $\varkappa^2$. No doubt we can generalize this theorem and obtain an expansion to the order $\varkappa^n$ under additional conditions, but there is no existing proof except for the special case in which T and $T(\varkappa)$ are selfadjoint and some assumptions are made on the mode of splitting of eigenvalues (see T. KATO [1], [3]).

[2] All these conditions are satisfied under the assumptions of Theorem 2.6 if X is a Hilbert space, T and $T(\varkappa)$ are selfadjoint and $T^{(1)}$ is symmetric, for $PT^{(1)} \subset (T^{(1)}P)^*$. It should be noticed that in this case we need not assume that D is a core of $T(\varkappa)$. This assumption is used only in the proof of (2.33), but (2.33) follows by taking the adjoint of (2.20) with ζ replaced by $\bar{\zeta}$.

Proof. The proof of this theorem is rather complicated. We shall give it in several steps.

I. We start with the identity

$$PR(\zeta, \varkappa) = (\lambda - \zeta)^{-1} P - \varkappa(\lambda - \zeta)^{-1} [P T^{(1)}] R(\zeta, \varkappa), \tag{2.33}$$

which is in a certain sense dual to (2.20). To prove it, we denote by A the right member; then

$$\begin{aligned} A(T(\varkappa) - \zeta) u &= (\lambda - \zeta)^{-1} P(T - \zeta + \varkappa T^{(1)}) u - \\ &\quad - \varkappa(\lambda - \zeta)^{-1} [P T^{(1)}] u \\ &= Pu + \varkappa(\lambda - \zeta)^{-1} P T^{(1)} u - \varkappa(\lambda - \zeta)^{-1} P T^{(1)} u = Pu \end{aligned} \tag{2.34}$$

for any $u \in \mathsf{D} = \mathsf{D}(T) \cap \mathsf{D}(T^{(1)})$ (note that $PT \subset TP = \lambda P$). Since D is a core of $T(\varkappa)$ by hypothesis, the final result of (2.34) can be extended to all $u \in \mathsf{D}(T(\varkappa))$. This shows that $A = PR(\zeta, \varkappa)$ as required.

Since $R(\zeta, \varkappa) \underset{s}{\to} R(\zeta)$ as $\varkappa \to 0$, (2.33) gives[1]

$$PR(\zeta, \varkappa) = (\lambda - \zeta)^{-1} P - \varkappa(\lambda - \zeta)^{-1} [P T^{(1)}] R(\zeta) + o(\varkappa)_s, \tag{2.35}$$

where $o(\varkappa)_s$ is uniform in $\zeta \in \Gamma$, that is, $\varkappa^{-1} o(\varkappa)_s u \to 0$ uniformly in ζ for each u.

Now we multiply (2.33) from the right by P and substitute (2.20) for the $R(\zeta, \varkappa) P$ that appears on the right, obtaining

$$\begin{aligned} PR(\zeta, \varkappa) P = (\lambda - \zeta)^{-1} P - \varkappa(\lambda - \zeta)^{-2} P T^{(1)} P + \\ + \varkappa^2 (\lambda - \zeta)^{-2} [P T^{(1)}] R(\zeta) T^{(1)} P + o(\varkappa^2)_u \end{aligned} \tag{2.36}$$

[note that $[P T^{(1)}] P = P T^{(1)} P$ since $P X \subset \mathsf{D}(T^{(1)})$].

Integration of (2.35) and (2.36) along Γ gives, as before,

$$PP(\varkappa) = P - \varkappa [P T^{(1)}] S + o(\varkappa)_s. \tag{2.37}$$

$$PP(\varkappa) P = P - \varkappa^2 [P T^{(1)}] S^2 T^{(1)} P + o(\varkappa^2)_u, \tag{2.38}$$

where we used the expansion $R(\zeta) = (\lambda - \zeta)^{-1} P + S + (\zeta - \lambda) S^2 + \cdots$.

II. To prove (2.32), let us write $Q(\varkappa) = P(\varkappa) - P$. Making use of (2.37) and a similar expansion (2.18) for $P(\varkappa) P$ proved above, we obtain

$$\begin{aligned} Q(\varkappa)^2 = P(\varkappa) + P - P(\varkappa) P - PP(\varkappa) = Q(\varkappa) - \varkappa P^{(1)} + o(\varkappa)_s, \\ P^{(1)} = -S T^{(1)} P - [P T^{(1)}] S. \end{aligned} \tag{2.39}$$

Hence

$$(1 - Q(\varkappa))(Q(\varkappa) - \varkappa P^{(1)}) = \varkappa Q(\varkappa) P^{(1)} + o(\varkappa)_s = o(\varkappa)_s \tag{2.40}$$

because $Q(\varkappa) = o(1)_s$. Hence,

$$Q(\varkappa) - \varkappa P^{(1)} = (1 - Q(\varkappa))^{-1} o(\varkappa)_s = o(\varkappa)_s, \tag{2.41}$$

which proves (2.32); note that $\|(1 - Q(\varkappa))^{-1}\| \leq$ const. because $\|Q(\varkappa)^2\| \to 0$ by (2.14a).

---

[1] In (2.35) $o(\varkappa)_s$ cannot be replaced by $o(\varkappa)_u$. Even when B has finite rank, $A_n \underset{s}{\to} 0$ does not imply $\|B A_n\| \to 0$; cf. footnote 2 on p. 444.

III. We now introduce the operator

$$U(\varkappa) = (1 - Q(\varkappa)^2)^{-1/2}[(1 - P(\varkappa))(1 - P) + P(\varkappa) P] \tag{2.42}$$
$$= 1 - P - P(\varkappa) + 2P(\varkappa) P + O(\varkappa^2)_u;$$

note that $Q(\varkappa) = O(\varkappa)_u$ by (2.41), for $o(\varkappa)_s = O(\varkappa)_u$ by the principle of uniform boundedness. By (2.18) and (2.32), we obtain

$$U(\varkappa) = 1 + \varkappa([PT^{(1)}] S - ST^{(1)} P) + o(\varkappa)_s. \tag{2.43}$$

The properties of $U(\varkappa)$ have been studied in detail in I-§ 4.6. It follows from the results proved there that

$$U(\varkappa)^{-1} = (1 - Q(\varkappa)^2)^{-1/2}[(1 - P)(1 - P(\varkappa)) + PP(\varkappa)] \tag{2.44}$$
$$= 1 - P - P(\varkappa) + 2PP(\varkappa) + O(\varkappa^2)_u$$
$$= 1 + \varkappa(ST^{(1)} P - [PT^{(1)}] S) + o(\varkappa)_s,$$

$$P(\varkappa) = U(\varkappa) PU(\varkappa)^{-1}. \tag{2.45}$$

Recall, in particular, that $Q(\varkappa)^2$ *commutes with* P *and* $P(\varkappa)$.

We further note that $PU(\varkappa) P$ can be expanded to the second order:

$$PU(\varkappa) P = P(1 - Q(\varkappa)^2)^{-1/2} P(\varkappa) P \tag{2.46}$$
$$= PP(\varkappa) P + \frac{1}{2} PQ(\varkappa)^2 P + O(\varkappa^3)_u$$
$$= P - \frac{1}{2}\varkappa^2 [PT^{(1)}] S^2 T^{(1)} P + o(\varkappa^2)_u;$$

note (2.38) and that $P(P^{(1)})^2 P = [PT^{(1)}] S^2 T^{(1)} P$ in virtue of $PS = SP = 0$. Here we have also made use of the fact that $o(\varkappa)_s P = o(\varkappa)_u$ because $\dim P < \infty$. In quite the same way it can be shown that

$$PU(\varkappa)^{-1} P = P - \frac{1}{2}\varkappa^2 [PT^{(1)}] S^2 T^{(1)} P + o(\varkappa^2)_u. \tag{2.47}$$

IV. $U(\varkappa)$ will now replace the $V(\varkappa)$ used in the proof of Theorem 2.6. The advantage of $U(\varkappa)$ over $V(\varkappa)$ lies in the transformation property (2.45) that $V(\varkappa)$ does not possess. In this connection, it should be remarked that $U(\varkappa)$ becomes useful only under the present assumptions by which $[PT^{(1)}] S$ makes sense.

We now calculate $PU(\varkappa)^{-1} R(\zeta, \varkappa) U(\varkappa) P$. This can be written in the form

$$PU(\varkappa)^{-1} R(\zeta, \varkappa) U(\varkappa) P = A_1 + A_2 + A_3 + A_4, \tag{2.48}$$
$$A_1 = PU(\varkappa)^{-1} PR(\zeta, \varkappa) PU(\varkappa) P,$$
$$A_2 = PU(\varkappa)^{-1}(1 - P) R(\zeta, \varkappa) PU(\varkappa) P,$$
$$A_3 = PU(\varkappa)^{-1} PR(\zeta, \varkappa)(1 - P) U(\varkappa) P,$$
$$A_4 = PU(\varkappa)^{-1}(1 - P) R(\zeta, \varkappa)(1 - P) U(\varkappa) P.$$

In A_1 we substitute (2.36), (2.46) and (2.47) to obtain (note $P = P^2$)

$$A_1 = (\lambda - \zeta)^{-1} P - \varkappa(\lambda - \zeta)^{-2} P T^{(1)} P - \varkappa^2(\lambda - \zeta)^{-1}[P T^{(1)}] S^2 T^{(1)} P + \\ + \varkappa^2(\lambda - \zeta)^{-2}[P T^{(1)}] R(\zeta) T^{(1)} P + o(\varkappa^2)_u .$$

In A_2 we note that [see (2.44) and (2.20)]

$$PU(\varkappa)^{-1}(1 - P) = -\varkappa[P T^{(1)}] S + o(\varkappa)_s , \tag{2.49}$$

$$(1 - P) R(\zeta, \varkappa) P = -\varkappa(\lambda - \zeta)^{-1}(1 - P) R(\zeta) T^{(1)} P + o(\varkappa)_u , \tag{2.50}$$

obtaining [note $1 - P = (1 - P)^2$ and $SP = 0$]

$$A_2 = \varkappa^2(\lambda - \zeta)^{-1}[P T^{(1)}] S R(\zeta) T^{(1)} P + o(\varkappa^2)_u ,$$

where $o(\varkappa)_s\, o(\varkappa)_u = o(\varkappa^2)_u$ because the factor $o(\varkappa)_u$ involves a factor P to the extreme right. Similarly we have

$$A_3 = \varkappa^2(\lambda - \zeta)^{-1}[P T^{(1)}] R(\zeta) S T^{(1)} P + o(\varkappa^2)_u ,$$

$$A_4 = \varkappa^2[P T^{(1)}] S R(\zeta) S T^{(1)} P + o(\varkappa^2)_u .$$

Collecting these terms, we arrive at [note that $R(\zeta)$ and S commute]

$$\begin{aligned} &PU(\varkappa)^{-1} R(\zeta, \varkappa)\, U(\varkappa) P \\ &= (\lambda - \zeta)^{-1} P - \varkappa(\lambda - \zeta)^{-2} P T^{(1)} P - \varkappa^2(\lambda - \zeta)^{-1}[P T^{(1)}] S^2 T^{(1)} P + \\ &+ \varkappa^2(\lambda - \zeta)^{-2}[P T^{(1)}] R(\zeta) [1 + 2(\lambda - \zeta) S + (\lambda - \zeta)^2 S^2] T^{(1)} P + o(\varkappa^2)_u . \end{aligned} \tag{2.51}$$

Finally we multiply (2.51) by ζ and integrate on Γ. Since all terms $o(\varkappa^2)_u$ are uniform for $\zeta \in \Gamma$, we thus obtain by (2.25)

$$\begin{aligned} &PU(\varkappa)^{-1} T(\varkappa) P(\varkappa) U(\varkappa) P \\ &\qquad = \lambda P + \varkappa P T^{(1)} P - \varkappa^2[P T^{(1)}] S T^{(1)} P + o(\varkappa^2)_u ; \end{aligned} \tag{2.52}$$

note that the coefficient of $\varkappa^2$ in (2.51) simplifies to $[P T^{(1)}]\, [(\lambda - \zeta)^{-3} P + (\lambda - \zeta)^{-2} S]\, T^{(1)} P$.

V. We are now able to complete the proof of Theorem 2.9. As before we consider the operator $\tilde{T}^{(1)}(\varkappa) = \varkappa^{-1}(T(\varkappa) - \lambda) P(\varkappa)$. Since $P(\varkappa)$ commutes with $T(\varkappa)$ and $U(\varkappa)^{-1} P(\varkappa) U(\varkappa) = P$ by (2.45), we have by (2.52)

$$\begin{aligned} \tilde{T}_0^{(1)}(\varkappa) &\equiv U(\varkappa)^{-1} \tilde{T}^{(1)}(\varkappa) U(\varkappa) \\ &= P T^{(1)} P - \varkappa[P T^{(1)}] S T^{(1)} P + o(\varkappa)_u . \end{aligned} \tag{2.53}$$

This can be regarded as an operator in the m-dimensional space $\mathsf{M} = P\mathsf{X}$. The first term on the right of (2.53) has the eigenvalues $\lambda_j^{(1)}$ with the associated eigenprojections $P_j^{(1)}$. Application of Theorem II-5.4 to (2.53) (replace T by $P T^{(1)} P$, λ by $\lambda_j^{(1)}$, P by $P_j^{(1)}$, $T'(0)$ by $-[P T^{(1)}] S T^{(1)} P$)

shows immediately that the $\lambda_j^{(1)}$-group eigenvalues of $\tilde{T}_0^{(1)}(\varkappa)$ have the form

$$\lambda_j^{(1)} + \varkappa\, \mu_{jk}^{(2)} + o(\varkappa)\,, \quad k = 1, \ldots, m_j^{(1)}\,, \tag{2.54}$$

with the $\mu_{jk}^{(2)}$ as given in the theorem. $\tilde{T}^{(1)}(\varkappa)$ has exactly the same eigenvalues, and the eigenvalues of $T(\varkappa)$ are given by (2.29). Also the total projection for the $\lambda_j^{(1)}$-group eigenvalues of $\tilde{T}_0^{(1)}(\varkappa)$ is given by II-(5.9) with P replaced by $P_j^{(1)}$, $T'(0)$ by $-[P T^{(1)}]\, S\, T^{(1)} P$ and S by $S_j^{(1)}$. This gives

$$P_j^{(1)} + \varkappa\{P_j^{(1)}[P T^{(1)}]\, S\, T^{(1)} S_j^{(1)} + S_j^{(1)}[P T^{(1)}]\, S\, T^{(1)} P_j^{(1)}\} + o(\varkappa)_{\mathrm{u}}. \tag{2.55}$$

The corresponding projection $P_j^{(1)}(\varkappa)$ for $\tilde{T}^{(1)}(\varkappa)$, which is the total projection for the $\lambda + \varkappa\,\lambda_j^{(1)}$-group eigenvalues of $T(\varkappa)$, is obtained from (2.55) by multiplication from the left by $U(\varkappa)$ and from the right by $U(\varkappa)^{-1}$. Noting that (2.55) is unchanged by multiplication by P either from the left or from the right and that

$$\begin{aligned} U(\varkappa)\, P &= P - \varkappa\, S\, T^{(1)} P + o(\varkappa)_{\mathrm{u}}\,, \\ P\, U(\varkappa)^{-1} &= 1 - \varkappa[P T^{(1)}]\, S + o(\varkappa)_{\mathrm{s}}\,, \end{aligned} \tag{2.56}$$

we obtain the required result (2.30). This completes the proof of Theorem 2.9.

Example 2.11. The assumptions of Theorems 2.6 and 2.9 are simple and easy to verify[1]. Here we consider the $T(\varkappa)$ of Example 1.17. Rewriting it as $T(\varkappa) = T + \varkappa\, T^{(1)}$ as in Example 2.5, with $\varkappa > 0$ and $T^{(1)} = e^{i\theta} x$, the eigenvalue $\lambda = -1$ of T is stable if $e^{i\theta}$ is not real (see Example 1.22), which will be assumed in the following. Since the eigenvector of T for the simple eigenvalue -1 is f and since $T^{(1)}$ differs from a selfadjoint operator x only by a numerical factor $e^{i\theta}$, Theorem 2.9 is applicable if $f \in \mathsf{D}(T^{(1)})$, that is, if $x\, f(x) \in \mathsf{L}^2$. [Of course this can be verified directly from (1.17).] Similar results can be stated for the $T(\varkappa)$ of Example 1.18; Theorem 2.9 is applicable to the eigenvalue -1 of T if $f \in \mathsf{D}(T^{(1)})$.

§ 3. Generalized strong convergence of sectorial operators

1. Convergence of a sequence of bounded forms

Let $\{\mathfrak{t}_n\}$ be a sequence of bounded sesquilinear forms defined on a Hilbert space H. We say that $\mathfrak{t}_n$ *converges to a form* $\mathfrak{t}$, in symbol $\mathfrak{t}_n \to \mathfrak{t}$, if $\mathfrak{t}$ is also bounded and defined on H and if $\mathfrak{t}_n[u, v] \to \mathfrak{t}[u, v]$ for all $u, v \in \mathsf{H}$. By virtue of the polarization principle VI-(1.1), it suffices to assume that $\mathfrak{t}_n[u] \to \mathfrak{t}[u]$. Let T_n, $T \in \mathscr{B}(\mathsf{H})$ be the operators associated with these

[1] They are satisfied in many problems related to differential operators, except when $T^{(1)}$ is of higher order than T and involves boundary conditions which are not satisfied by elements of $P\,\mathsf{H}$. Some of the examples considered later (§ 4) in connection with the theory of forms can also be regarded as examples of Theorems 2.6, 2.8, and 2.9.

forms. Then $\mathfrak{t}_n \to \mathfrak{t}$ is equivalent to $(T_n u, v) \to (T u, v)$ for all u, v, that is, $T_n \underset{w}{\to} T$. In general, however, it is difficult to draw from the weak convergence $T_n \underset{w}{\to} T$ any interesting conclusion regarding the spectral properties of T_n and T. Thus we have to assume something more on the mode of convergence of the sequence $\{\mathfrak{t}_n\}$ in order to attain results of interest from the point of view of perturbation theory.

A fundamental theorem on the convergence of bounded forms $\mathfrak{t}_n$ is given by

Theorem 3.1. *Let $\{\mathfrak{t}_n\}$ be a sequence of bounded sesquilinear forms defined everywhere on* H. *Let $\{\mathfrak{t}_n\}$ be uniformly sectorial in the sense that*

$$(3.1) \qquad |\mathrm{Im}\,\mathfrak{t}_n[u]| \leqq M\, \mathrm{Re}\,\mathfrak{t}_n[u]\,, \quad u \in \mathsf{H}\,,$$

where $M > 0$ is independent of n. If $\mathfrak{t}_n \to 0$, the operator $T_n \in \mathscr{B}(\mathsf{H})$ associated with $\mathfrak{t}_n$ tends to zero strongly. The same is true of T_n^*.

Proof. (3.1) implies that $\mathfrak{h}_n = \mathrm{Re}\,\mathfrak{t}_n \geqq 0$. It follows by VI-(1.15) that

$$(3.2) \qquad |(T_n u, v)| = |\mathfrak{t}_n[u, v]| \leqq (1 + M)\, \mathfrak{h}_n[u]^{1/2}\, \mathfrak{h}_n[v]^{1/2}\,.$$

Set $v = T_n u$ in (3.2). Since T_n is weakly convergent and thus uniformly bounded ($\|T_n\| \leqq N$), $\mathfrak{h}_n[v] = \mathfrak{h}_n[T_n u] = \mathrm{Re}(T_n^2 u, T_n u)$ is bounded by $N^3\|u\|^2$. Since $\mathfrak{h}_n[u] = \mathrm{Re}\,\mathfrak{t}_n[u] \to 0$ by hypothesis, it follows from (3.2) that $\|T_n u\| \to 0$. Since the adjoint forms $\mathfrak{t}_n^*$ satisfy the same conditions as $\mathfrak{t}_n$, we have also $T_n^* \underset{s}{\to} 0$.

Corollary 3.2. *If a sequence of bounded, nonnegative selfadjoint operators converges weakly to* 0, *it converges strongly to* 0.

A related theorem is

Theorem 3.3. *Let $\{\mathfrak{h}_n\}$ be a nonincreasing sequence of bounded symmetric forms on* H *bounded from below*:

$$(3.3) \qquad \mathfrak{h}_1 \geqq \mathfrak{h}_2 \geqq \cdots \geqq -c\,.$$

Then there is a bounded symmetric form $\mathfrak{h}$ such that $\mathfrak{h}_n \geqq \mathfrak{h}$ and $\mathfrak{h}_n \to \mathfrak{h}$. The associated bounded selfadjoint operators H_n and H have the property that $H_n \to H$ strongly.

Proof. (3.3) implies that, for each $u \in \mathsf{H}$, $\mathfrak{h}_n[u]$ is a nonincreasing sequence bounded below by $-c\|u\|^2$. Hence $\lim \mathfrak{h}_n[u]$ exists, so that $\lim \mathfrak{h}_n[u, v] = \mathfrak{h}[u, v]$ exists for each u, v by polarization. $\mathfrak{h}$ is bounded by the principle of uniform boundedness. We have $\mathfrak{h}_n \geqq \mathfrak{h}$ and $\mathfrak{h}_n \to \mathfrak{h}$, and the assertion on H_n follows from Corollary 3.2.

Remark 3.4. In Corollary 3.2 the sequence need not be monotonic but the existence of the limit is assumed. The situation is in a sense reversed in Theorem 3.3. Obviously there are similar theorems in which the inequalities $\geqq$ are replaced by $\leqq$.

These results will be generalized to unbounded sectorial forms in the following paragraphs. Here we add another theorem.

Theorem 3.5. *In* Theorem 3.3 *let $H_n - H$ be compact for all n. Then* $\|H_n - H\| \to 0$.

Proof. Set $H_n - H = K_n$; we have $K_n \geqq 0$ and $K_n \to 0$ strongly by Theorem 3.3. Since K_1 is compact, we can find, for any $\varepsilon > 0$, a decomposition $\mathsf{H} = \mathsf{M} \oplus \mathsf{N}$ of H into the direct sum of orthogonal subspaces M, N, both invariant under K_1, in such a way that $\dim \mathsf{M} < \infty$ and $\|K_1 u\| \leqq \varepsilon \|u\|$ for $u \in \mathsf{N}$ (see V-§ 2.3). For any $u \in \mathsf{H}$, let $u = u' + u''$, $u' \in \mathsf{M}$, $u'' \in \mathsf{N}$. Since $K_n \geqq 0$, we have by the triangle inequality

$$0 \leqq (K_n u, u) \leqq 2(K_n u', u') + 2(K_n u'', u''). \tag{3.4}$$

Since M is finite-dimensional, the convergence $K_n \to 0$ is locally uniform on M, so that there exists an N such that

$$0 \leqq (K_n u', u') \leqq \varepsilon \|u'\|^2 \quad \text{for} \quad n > N. \tag{3.5}$$

On the other hand, the definition of N gives

$$0 \leqq (K_n u'', u'') \leqq (K_1 u'', u'') \leqq \varepsilon \|u''\|^2 \quad \text{for all} \quad n. \tag{3.6}$$

It follows from (3.4)–(3.6) that

$$0 \leqq (K_n u, u) \leqq 2\varepsilon(\|u'\|^2 + \|u''\|^2) = 2\varepsilon \|u\|^2 \quad \text{for} \quad n > N. \tag{3.7}$$

Hence $\|K_n\| \leqq 2\varepsilon$ for $n > N$ so that $\|K_n\| \to 0$, $n \to \infty$.

2. Convergence of sectorial forms "from above"

Let us now consider the convergence of a sequence of unbounded sesquilinear forms $\mathfrak{t}_n$ in H. The following is a fundamental theorem, which generalizes Theorem 3.1.

Theorem 3.6.[1] *Let* $\mathfrak{t}_n$, $n = 1, 2, \ldots$, *and* $\mathfrak{t}$ *be densely defined, closed sectorial forms in* H *with the following properties*:

i) $\mathsf{D}(\mathfrak{t}_n) \subset \mathsf{D}(\mathfrak{t})$, $n = 1, 2, \ldots$.

ii) $\mathfrak{t}_n' = \mathfrak{t}_n - \mathfrak{t}$ *is uniformly sectorial in the sense that*

$$|\operatorname{Im} \mathfrak{t}_n'[u]| \leqq M \operatorname{Re} \mathfrak{t}_n'[u], \quad u \in \mathsf{D}(\mathfrak{t}_n), \quad n = 1, 2, \ldots, \quad M > 0. \tag{3.8}$$

iii) *There is a core* D *of* $\mathfrak{t}$ *such that* $\mathsf{D} \subset \liminf \mathsf{D}(\mathfrak{t}_n)$ *(that is, each* $u \in \mathsf{D}$ *belongs to* $\mathsf{D}(\mathfrak{t}_n)$ *for sufficiently large n) and*

$$\lim_{n \to \infty} \mathfrak{t}_n[u] = \mathfrak{t}[u] \quad if \quad u \in \mathsf{D}. \tag{3.9}$$

[1] The results of this and the following paragraphs were proved in T. Kato [3], [7], [8], for symmetric forms and the associated selfadjoint operators. For their applications see Killeen [1]. See also Birman [3], [4], Gol'dberg [1]. Similar results for nonsymmetric forms were first given by Huet [1], [2]; these and subsequent papers [3]–[7] by Huet contain applications to partial differential equations.

Let T_n and T be the m-sectorial operators associated with $\mathfrak{t}_n$ *and* $\mathfrak{t}$, *respectively, according to the first representation theorem* (Theorem VI-2.1). *Then T_n and T_n^* converge to T and T^*, respectively, strongly in the generalized sense. More precisely, let Δ_s, Δ_s^* be the regions of strong convergence for $\{T_n\}$, $\{T_n^*\}$, respectively. Then Δ_s^* is the mirror image of Δ_s with respect to the real axis and both Δ_s and Δ_s^* contain the half-plane* $\mathrm{Re}\,\zeta < \gamma$, *where γ is a vertex of* $\mathfrak{t}$, *and we have as $n \to \infty$*

$$(3.10)\quad \begin{aligned} &R_n(\zeta) \underset{s}{\to} R(\zeta)\,, \quad \mathfrak{t}[R_n(\zeta)\,u - R(\zeta)\,u] \to 0\,, \quad \mathfrak{t}_n'[R_n(\zeta)\,u] \to 0\,,\\ &R_n(\zeta)^* \underset{s}{\to} R(\zeta)^*\,, \quad \mathfrak{t}[R_n(\zeta)^*\,u - R(\zeta)^*\,u] \to 0, \quad \mathfrak{t}_n'[R_n(\zeta)^*\,u] \to 0\,,\end{aligned}$$

for $\zeta \in \Delta_s$ and $u \in \mathsf{H}$, where $R_n(\zeta)$, $R(\zeta)$ are the resolvents of T_n, T, respectively. The convergences in (3.10) *are uniform in each compact subset of Δ_s.*

Proof. I. Set $\mathfrak{h}_n = \mathrm{Re}\,\mathfrak{t}_n$, $\mathfrak{h} = \mathrm{Re}\,\mathfrak{t}$, $\mathfrak{h}_n' = \mathrm{Re}\,\mathfrak{t}_n'$. By adding a suitable scalar to $\mathfrak{t}_n$ and to $\mathfrak{t}$ if necessary, we may assume that $\mathfrak{t}$ has a vertex $\gamma = 0$, so that

$$(3.11)\qquad \mathfrak{h}_n[u] \geqq \mathfrak{h}[u] \geqq 0\,, \quad u \in \mathsf{D}(\mathfrak{t}_n) \subset \mathsf{D}(\mathfrak{t})\,,$$

where $\mathfrak{h}_n \geqq \mathfrak{h}$ is a consequence of (3.8). Thus the half-plane $\mathrm{Re}\,\zeta < 0$ belongs to $\mathrm{P}(T_n)$ and $\mathrm{P}(T)$, with [see V-(3.38)]

$$(3.12)\quad \|R_n(\zeta)\| \leqq 1/\mathrm{Re}(-\zeta)\,, \quad \|R(\zeta)\| \leqq 1/\mathrm{Re}(-\zeta)\,, \quad \mathrm{Re}\,\zeta < 0\,.$$

Recalling the definition of T_n and T, we see that, for any $u \in \mathsf{H}$ and $\mathrm{Re}\,\zeta = \xi < 0$,

$$(3.13)\quad \begin{aligned}(\mathfrak{h} - \xi)\,[R_n(\zeta)\,u] &\leqq (\mathfrak{h}_n - \xi)\,[R_n(\zeta)\,u] = \mathrm{Re}(\mathfrak{t}_n - \zeta)\,[R_n(\zeta)\,u]\\ &= \mathrm{Re}((T_n - \zeta)\,R_n(\zeta)\,u,\, R_n(\zeta)\,u) = \mathrm{Re}(u,\, R_n(\zeta)\,u) \leqq\\ &\leqq \|u\|\,\|R_n(\zeta)\,u\| \leqq (-\xi)^{-1/2}\|u\|\,((\mathfrak{h} - \xi)\,[R_n(\zeta)\,u])^{1/2}\,.\end{aligned}$$

It follows from (3.13) first that $(\mathfrak{h} - \xi)\,[R_n(\zeta)\,u] \leqq (-\xi)^{-1}\|u\|^2$ and then, considering the second and the last member, that

$$(3.14)\qquad \left.\begin{aligned}(\mathfrak{h} - \xi)\,[R_n(\zeta)\,u]\\ \mathfrak{h}_n'[R_n(\zeta)\,u]\end{aligned}\right\} \leqq (\mathfrak{h}_n - \xi)\,[R_n(\zeta)\,u] \leqq (-\xi)^{-1}\,\|u\|^2\,.$$

II. Let $v \in \mathsf{D}$ and $u \in \mathsf{H}$. By hypothesis we have $v \in \mathsf{D}(\mathfrak{t})$, $R_n(\zeta)\,u \in \mathsf{D}(\mathfrak{t}_n) \subset \mathsf{D}(\mathfrak{t})$ and

$$(3.15)\quad \begin{aligned}(\mathfrak{t} - \zeta)\,[R_n(\zeta)\,u - R(\zeta)\,u, v] &= (\mathfrak{t}_n - \zeta)\,[R_n(\zeta)\,u, v] -\\ &- \mathfrak{t}_n'[R_n(\zeta)\,u, v] - (\mathfrak{t} - \zeta)\,[R(\zeta)\,u, v] = -\,\mathfrak{t}_n'[R_n(\zeta)\,u, v]\,,\end{aligned}$$

since the first and the third terms in the middle member cancel each other, both being equal to (u, v). It follows by (3.8) and VI-(1.15) that

$$(3.16)\quad |(\mathfrak{t} - \zeta)\,[R_n(\zeta)\,u - R(\zeta)\,u, v]| \leqq (1 + M)\,\mathfrak{h}_n'[R_n(\zeta)\,u]^{1/2}\,\mathfrak{h}_n'[v]^{1/2} \to 0,$$

since the $\mathfrak{h}'_n[R_n(\zeta)\,u]$ are bounded by (3.14) and $\mathfrak{h}'_n[v] = \mathrm{Re}\,\mathfrak{t}'_n[v] \to 0$ by iii).

Let $\mathsf{H}_{\mathfrak{h}}$ be the Hilbert space into which $\mathsf{D}(\mathfrak{t})$ is converted on introducing the inner product $(u, v)_{\mathfrak{h}} = \mathfrak{h}[u, v] + (u, v)$ and the associated norm $\|u\|_{\mathfrak{h}}$ (see VI-§ 1.3). That D is a core of $\mathfrak{t}$ means that D is dense in $\mathsf{H}_{\mathfrak{h}}$ (Theorem VI-1.21). On the other hand, (3.14) shows that the sequence $R_n(\zeta)\,u$ is bounded in $\mathsf{H}_{\mathfrak{h}}$. Therefore, the relation (3.16) is extended to all $v \in \mathsf{H}_{\mathfrak{h}} = \mathsf{D}(\mathfrak{t})$ since $\mathfrak{t}$ is a bounded form on $\mathsf{H}_{\mathfrak{h}}$. In particular we have

$$(3.17)\qquad (\mathfrak{t} - \zeta)\,[R_n(\zeta)\,u - R(\zeta)\,u, R(\zeta)\,u] \to 0\,, \quad n \to \infty\,.$$

III. We have now

$$\begin{aligned}(3.18)\quad (\mathfrak{t} - \zeta)\,[R_n(\zeta)\,u - R(\zeta)\,u] &+ \mathfrak{t}'_n[R_n(\zeta)\,u] \\ &= (\mathfrak{t}_n - \zeta)\,[R_n(\zeta)\,u] + (\mathfrak{t} - \zeta)\,[R(\zeta)\,u] - \\ &\quad - (\mathfrak{t} - \zeta)\,[R_n(\zeta)\,u, R(\zeta)\,u] - (\mathfrak{t} - \zeta)\,[R(\zeta)\,u, R_n(\zeta)\,u] \\ &= (\mathfrak{t} - \zeta)\,[R(\zeta)\,u - R_n(\zeta)\,u, R(\zeta)\,u] \to 0\,,\end{aligned}$$

since the first and the last terms in the middle member cancel each other, both being equal to $(u, R_n(\zeta)\,u)$. Taking the real part of (3.18) and noting that $\mathfrak{h}$, $-\xi$ and $\mathfrak{h}'_n$ are all nonnegative, we see that $R_n(\zeta)\,u \to R(\zeta)\,u$, $\mathfrak{h}[R_n(\zeta)\,u - R(\zeta)\,u] \to 0$ and $\mathfrak{h}'_n[R_n(\zeta)\,u] \to 0$ for $n \to \infty$. In virtue of the sectorial property of the forms $\mathfrak{t}$ and $\mathfrak{t}'_n$, this gives the first three convergences in (3.10) for $\mathrm{Re}\,\zeta < 0$.

IV. So far we have been assuming that $\mathrm{Re}\,\zeta < 0$. We shall now eliminate this restriction. The first formula of (3.10) is true for all $\zeta \in \Delta_s$ by the very definition of Δ_s. To prove the second, it suffices to show that $\mathfrak{h}[R_n(\zeta)\,u - R(\zeta)\,u] \to 0$. Substituting from (1.3), we have

$$\begin{aligned}(3.19)\quad \mathfrak{h}[R_n(\zeta)\,u - R(\zeta)\,u] = \mathfrak{h}[(1 + (\zeta - \zeta_0)\,R_n(\zeta))\,v_n] &\leqq 2\mathfrak{h}[v_n] + \\ &+ 2|\zeta - \zeta_0|^2\,\mathfrak{h}[R_n(\zeta)\,v_n]\,,\end{aligned}$$

where $v_n = (R_n(\zeta_0) - R(\zeta_0))\,(1 + (\zeta - \zeta_0)\,R(\zeta))\,u$ and ζ_0 is any fixed number with $\mathrm{Re}\,\zeta_0 < 0$. Now $\mathfrak{h}[v_n] \to 0$ by what was proved above. Furthermore, $\mathfrak{h}[R_n(\zeta)\,v_n] \leqq \mathfrak{h}_n[R_n(\zeta)\,v_n] = \mathrm{Re}\,(\mathfrak{t}_n - \zeta)\,[R_n(\zeta)\,v_n] + \mathrm{Re}\,\zeta\,\|R_n(\zeta)\,v_n\|^2 = \mathrm{Re}\,(v_n, R_n(\zeta)\,v_n) + \mathrm{Re}\,\zeta\,\|R_n(\zeta)\,v_n\|^2 \to 0$ since $v_n \to 0$ and $R_n(\zeta)$ is uniformly bounded as $n \to \infty$ by $\zeta \in \Delta_s$. Thus the right member of (3.19) goes to zero as $n \to \infty$ as we wished to show. (3.19) also shows that the convergences in question are uniform in ζ on any compact subset of Δ_s; note that $R(\zeta)\,u$ varies over a compact subset of $\mathsf{H}_{\mathfrak{h}}$ and apply Lemma III-3.7.

Once the first two formulas of (3.10) have been extended to every $\zeta \in \Delta_s$, the same is true of the third by (3.18).

Finally we remark that the assumptions of the theorem are also satisfied for the adjoint forms $\mathfrak{t}_n^*$, $\mathfrak{t}^*$. Therefore, the above results hold true when T_n, T are replaced by T_n^*, T^* respectively. This proves (3.10), and $\Delta_s^* = \bar{\Delta}_s$ follows from Theorem 1.8.

Theorem 3.6 is very useful as a criterion for the generalized strong convergence $T_n \underset{s}{\to} T$. In many cases it is more powerful than Theorem 1.5 when $\mathsf{X} = \mathsf{H}$.

Example 3.7. Let $\mathsf{H} = \mathsf{L}^2(0, \infty)$ and

$$\text{(3.20)}\qquad \begin{aligned} \mathfrak{t}[u] &= \int_0^\infty (|u'(x)|^2 + q(x)|u(x)|^2)\,dx\,, \\ \mathfrak{t}_n[u] &= \int_0^\infty (|u'(x)|^2 + q_n(x)\,|u(x)|^2)\,dx\,, \end{aligned}$$

where q and q_n are real-valued functions with the properties stated in VI-§ 4.1, so that $\mathfrak{t}$ and $\mathfrak{t}_n$ are closed symmetric forms bounded from below (with an appropriate definition of their domains). Furthermore, let $q_n(x) \geqq q(x)$ and

$$\text{(3.21)}\qquad \int_a^b (q_n(x) - q(x))\,dx \to 0$$

for any a, b such that $0 < a < b < \infty$. It is easily verified that the assumptions of Theorem 3.6 are satisfied [with $\mathsf{D} = \mathsf{D}(\mathfrak{t})$]. The operators associated with these forms are $T = -d^2/dx^2 + q(x)$ and $T_n = -d^2/dx^2 + q_n(x)$ (see Theorem VI-4.2) with appropriate domains. It follows from Theorem 3.6 that $T_n \underset{s}{\to} T$ in the generalized sense. Note that the condition (3.21) is weaker than the corresponding one (1.5) in Example 1.11. It will also be remarked that q and q_n need not be real-valued; the main condition is that the $q_n - q$ are uniformly sectorial.

Example 3.8. Consider the operator $T(\varkappa)$ of Example 1.20. It is obvious that Theorem 3.6 is applicable to this case with T_n replaced by $T(\varkappa)$. If $\mathrm{Re}\,\alpha > 0$ and $\mathrm{Re}\,\varkappa > 0$, $T(\varkappa)$ is the operator associated with the form

$$\text{(3.22)}\qquad \mathfrak{t}(\varkappa)\,[u, v] = \int_0^1 \big(\alpha\, u'(x)\,\overline{v'(x)} + \varkappa\, u''(x)\,\overline{v''(x)}\big)\,dx$$

with the boundary condition (1.26). It is easy to show that $\mathfrak{t}(\varkappa)$ is sectorial and closed. We have $\mathfrak{t}(\varkappa) \to \mathfrak{t}$, $\varkappa \to 0$, in the sense required in Theorem 3.6, where $\mathfrak{t}[u, v]$ is given by the first term of (3.22) with the boundary condition $u(0) = u(1) = 0$. In fact, we have $\mathfrak{t}(\varkappa)\,[u] \to \mathfrak{t}[u]$ if $u \in \mathsf{D} = \mathsf{D}(\mathfrak{t}(\varkappa))$ (which is independent of $\varkappa$), and D is a core of $\mathfrak{t}$ as is easily seen.

The operator associated with this $\mathfrak{t}$ is exactly the T of Example 1.20. Since, moreover, $\mathfrak{t}(\varkappa) - \mathfrak{t}$ is uniformly sectorial in the sense of (3.8) if $\varkappa$ is restricted to a sector $|\arg\varkappa| \leqq \delta < \pi/2$, Theorem 3.6 shows that $T(\varkappa) \underset{s}{\to} T$ in the generalized sense. If we further assume that α and $\varkappa$ are real positive, $T(\varkappa)$ and T are selfadjoint and have compact resolvents (as is always the case with a regular differential operator on a bounded domain), and the monotonicity of $\mathfrak{t}(\varkappa)[u]$ in $\varkappa$ implies that $T(\varkappa)^{-1}$ is monotone nondecreasing as $\varkappa \to 0$ (see Theorem VI-2.21). It follows from Theorem 3.5 that $T(\varkappa)^{-1} \to T^{-1}$ *in norm*, so that we have also $R(\zeta, \varkappa) \to R(\zeta)$ in norm and $T(\varkappa) \to T$ in the generalized sense (in gap). This gives a complete proof of the statements given in Example 1.20.

Note that Theorem 1.5 is not applicable to this example, for $\mathsf{D}(T(\varkappa))$ is not a core of T. In this connection it should be remarked that even Theorem 3.6 is not applicable to Example 1.19, for the limiting form $\mathfrak{t}[u, v] = 2\alpha \int_0^1 u'(x)\,\overline{v(x)}\,dx$ that would correspond to the limiting operator T is *not* sectorial.

Remark 3.9. Another remark to Theorem 3.6 is that the convergence (3.9) is required only for $u \in \mathsf{D}$, D being some core of $\mathfrak{t}$ such that $\mathsf{D} \subset \subset \liminf \mathsf{D}(\mathfrak{t}_n)$. It is not necessary that (3.9) hold for all $u \in \liminf \mathsf{D}(\mathfrak{t}_n)$. A pertinent example is given by

Example 3.10. Let $\mathsf{H} = L^2(0, 1)$ and define $\mathfrak{t}_n$ by

$$\mathfrak{t}_n[u] = n^{-1} \int_0^1 |u'(x)|^2 \, dx + |u(0)|^2 + |u(1)|^2 \tag{3.23}$$

with no boundary condition. $\mathsf{D}(\mathfrak{t}_n) = \mathsf{D}_0$ is the set of $u \in L^2$ such that $u' \in L^2$; D_0 is independent of n. Theorem 3.6 is applicable if we set $\mathfrak{t} = 0$ [with $\mathsf{D}(\mathfrak{t}) = \mathsf{H}$]. The only condition to be considered is iii) ,the other two conditions being trivially satisfied. iii) is satisfied if D is chosen as the set of $u \in \mathsf{D}_0$ such that $u(0) = u(1) = 0$. Even with these restrictions D is a core of $\mathfrak{t}$, for $\mathfrak{t}$ is bounded and D is dense in H. It follows from Theorem 3.6 that $T_n \underset{s}{\longrightarrow} 0$ in the generalized sense. This result is by no means trivial. T_n is the differential operator $-n^{-1} d^2/dx^2$ with the boundary condition $u'(0) = n\,u(0)$, $u'(1) = -n\,u(1)$.

It should be noted that $\lim \mathfrak{t}_n[u]$ exists for all $u \in \mathsf{D}_0$, but it is not necessarily equal to $\mathfrak{t}[u] = 0$.

3. Nonincreasing sequences of symmetric forms

In Theorem 3.6 it is assumed that the limiting form $\mathfrak{t}$ is given. It would be desirable to have a theorem in which only the sequence $\{\mathfrak{t}_n\}$ is given and the limiting form $\mathfrak{t}$, or at least the limiting operator T, is to be constructed. The circumstance stated in Remark 3.9 and Example 3.10, however, suggests that this is not easy, for in Theorem 3.6 $\lim \mathfrak{t}_n[u]$ need not be equal to $\mathfrak{t}[u]$ for all u for which this limit exists. Thus the attempt to construct $\mathfrak{t}$ by $\mathfrak{t}[u] = \lim \mathfrak{t}_n[u]$ must fail.

At present there seem to exist no theorems of the desired kind for a sequence of nonsymmetric sectorial forms. But we can give at least a theorem on *monotonic sequences of symmetric forms*.

Let us recall that, when $\mathfrak{h}_1$ and $\mathfrak{h}_2$ are symmetric forms bounded from below, $\mathfrak{h}_1 \geqq \mathfrak{h}_2$ means that $\mathsf{D}(\mathfrak{h}_1) \subset \mathsf{D}(\mathfrak{h}_2)$ and $\mathfrak{h}_1[u] \geqq \mathfrak{h}_2[u]$ for $u \in \mathsf{D}(\mathfrak{h}_1)$ (the larger form has the smaller domain!) (see VI-§ 2.5). A sequence $\{\mathfrak{h}_n\}$ of symmetric forms bounded from below is nonincreasing (nondecreasing) if $\mathfrak{h}_n \geqq \mathfrak{h}_{n+1}$ ($\mathfrak{h}_n \leqq \mathfrak{h}_{n+1}$) for all n.

Theorem 3.11. *Let $\{\mathfrak{h}_n\}$ be a nonincreasing sequence of densely defined, closed symmetric forms uniformly bounded from below: $\mathfrak{h}_n \geqq \gamma$, γ being a constant. If H_n is the selfadjoint operator associated with $\mathfrak{h}_n$, H_n converges to a selfadjoint operator $H \geqq \gamma$ strongly in the generalized sense. We have as $n \to \infty$ (writing $R_n(\zeta) = (H_n - \zeta)^{-1}$, $R(\zeta) = (H - \zeta)^{-1}$)*

$$R_n(\zeta) \underset{s}{\longrightarrow} R(\zeta)\,, \quad \operatorname{Re}\zeta < \gamma\,, \tag{3.24}$$

$$(H_n - \xi)^{1/2} u \underset{w}{\longrightarrow} (H - \xi)^{1/2} u \quad \text{for} \quad u \in \bigcup_n \mathsf{D}(\mathfrak{h}_n) \quad \text{and} \quad \xi < \gamma\,. \tag{3.25}$$

If, in particular, the symmetric form $\mathfrak{h}$ *defined by* $\mathfrak{h}[u] = \lim \mathfrak{h}_n[u]$ *with* $D(\mathfrak{h}) = \bigcup_n D(\mathfrak{h}_n)$ *is closable, then* H *is the selfadjoint operator associated with* $\tilde{\mathfrak{h}}$, *the closure of* $\mathfrak{h}$, *and the convergence in* (3.25) *is strong convergence.*

Proof. We may assume without loss of generality that $\gamma = 0$, so that $\mathfrak{h}_n$ and H_n are nonnegative. As we know (see Theorem VI-2.21), the nonincreasing property of $\{\mathfrak{h}_n\}$ implies that the sequence of bounded selfadjoint operators $R_n(\xi)$ with $\xi < 0$ is nondecreasing. Since the $R_n(\xi)$ are uniformly bounded from above by $(-\xi)^{-1}$, it follows from Theorem 3.3 (with the order relation reversed) that s-$\lim R_n(\xi) = R(\xi)$ exists. $R(\xi)$ is invertible, for, since $0 \leqq R_n(\xi) \leqq R(\xi)$, $R(\xi)\,u = 0$ implies $R_n(\xi)\,u = 0$ and so $u = 0$. It follows from Theorem 1.3 that $R(\xi)$ is the resolvent of a closed linear operator H. H is selfadjoint since $R(\xi)$ is. It follows also that $R_n(\zeta) \xrightarrow[s]{} R(\zeta) = (H - \zeta)^{-1}$ for any ζ with $\mathrm{Re}\,\zeta < 0$, for such ζ belong to Δ_b, the region of boundedness.

$0 \leqq R_n(\xi) \leqq R(\xi)$ for $\xi < 0$ implies $H_n - \xi \geqq H - \xi$ by Theorem VI-2.21, which implies that $\mathsf{D}(\mathfrak{h}_n) = \mathsf{D}((H_n - \xi)^{1/2}) \subset \mathsf{D}((H - \xi)^{1/2})$. Similarly we have $\mathsf{D}(\mathfrak{h}_m) \subset \mathsf{D}(\mathfrak{h}_n)$ for $m < n$. For any $u \in \mathsf{D}(\mathfrak{h}_m)$, $v \in \mathsf{H}$ and $n > m$, $\xi < 0$, we have therefore

$$(3.26)\quad ((H_n - \xi)^{1/2}\,u - (H - \xi)^{1/2}\,u,\ (H_1 - \xi)^{-1/2}\,v)$$
$$= (u - (H_n - \xi)^{-1/2}(H - \xi)^{1/2}\,u,\ B_n\,v) \to 0\,, \quad n \to \infty\,,$$

where $B_n = (H_n - \xi)^{1/2}(H_1 - \xi)^{-1/2} \in \mathscr{B}(\mathsf{H})$ because $\mathsf{D}((H_n - \xi)^{1/2}) = \mathsf{D}(\mathfrak{h}_n) \supset \mathsf{D}(\mathfrak{h}_1)$; note that $R_n(\xi) \xrightarrow[s]{} R(\xi)$ implies $R_n(\xi)^{1/2} \xrightarrow[s]{} R(\xi)^{1/2}$ (see Problem V-3.52) and that $\|B_n\| \leqq 1$ because $\mathfrak{h}_n \leqq \mathfrak{h}_1$. Since $\|(H_n - \xi)^{1/2}\,u\| \leqq \|(H_m - \xi)^{1/2}\,u\|$ is bounded for $n \to \infty$ and $(H_1 - \xi)^{-1/2}\,v$ varies over $\mathsf{D}(\mathfrak{h}_1)$, which is dense in H, when v varies over H, it follows that $(H_n - \xi)^{1/2}\,u \xrightarrow[w]{} (H - \xi)^{1/2}\,u$ if $u \in \mathsf{D}(\mathfrak{h}_m)$. Since m is arbitrary, we have proved (3.25).

Suppose now that the limiting form $\mathfrak{h}$ as defined in the theorem is closable. Then we have a situation of Theorem 3.6, with $\mathfrak{t}_n$, $\mathfrak{t}$ replaced respectively by $\mathfrak{h}_n$, $\tilde{\mathfrak{h}}$. Let H_0 be the selfadjoint operator associated with $\tilde{\mathfrak{h}}$. We must have $H_0 = H$, for both $(H - \zeta)^{-1}$ and $(H_0 - \zeta)^{-1}$ are the strong limit of $R_n(\zeta)$. It only remains to show that (3.25) is a strong convergence. Since the weak convergence has been proved, it suffices (see Lemma V-1.2) to show that $\|(H_n - \xi)^{1/2}\,u\|^2 \to \|(H - \xi)^{1/2}\,u\|^2$ for $u \in \bigcup_n \mathsf{D}(\mathfrak{h}_n)$. But this is another expression of the assumption that $\mathfrak{h}_n[u] \to \mathfrak{h}[u] = \tilde{\mathfrak{h}}[u]$.

Remark 3.12. In general the weak convergence in (3.25) cannot be replaced by strong convergence. This is seen from Example 3.10, in which $\mathfrak{h}_n[u] \to \mathfrak{h}[u]$ is not true for all $u \in \mathsf{D}_0$.

4. Convergence from below

Theorem 3.13a. *Let $\mathfrak{h}_1 \leqq \mathfrak{h}_2 \leqq \dots$ be a monotone nondecreasing sequence of closed symmetric forms bounded from below. Set $\mathfrak{h}[u] = \lim_{n\to\infty} \mathfrak{h}_n[u]$ whenever the finite limit exists. $\mathfrak{h}$ is a closed symmetric form bounded from below. Suppose $\mathfrak{h}$, and hence all the $\mathfrak{h}_n$, are densely defined, and let H and H_n be the associated selfadjoint operators, with the resolvents R and R_n. Then we have, as $n\to\infty$,*

$$\mathfrak{h}_n[u, v] \to \mathfrak{h}[u, v], \quad u, v \in \mathsf{D}(\mathfrak{h}), \tag{3.28a}$$

$$R_n(\zeta) \underset{s}{\to} R(\zeta), \quad \operatorname{Re}\zeta < \gamma_1 \quad (\text{the lower bound of } \mathfrak{h}_1), \tag{3.29a}$$

$$(H_n - \xi)^{1/2} u \underset{s}{\to} (H - \xi)^{1/2} u, \quad u \in \mathsf{D}(\mathfrak{h}), \quad \xi < \gamma_1. \tag{3.30a}$$

Before proving this theorem, it is convenient to introduce the *improper extension* $\mathfrak{t}_{\text{ext}}$ of a symmetric form $\mathfrak{t}$ bounded from below. It is defined by $\mathfrak{t}_{\text{ext}}[u] = \mathfrak{t}[u]$ if $u \in \mathsf{D}(\mathfrak{t})$ and $\mathfrak{t}_{\text{ext}}[u] = +\infty$ otherwise.

Lemma 3.14a. *$\mathfrak{t}$ is closed if and only if $\mathfrak{t}_{\text{ext}}$ is lower semicontinuous.*

Proof. Suppose $\mathfrak{t}_{\text{ext}}$ is lower semicontinuous. Let $u_n \in \mathsf{D}(\mathfrak{t})$, $u_n \to u \in \mathsf{H}$, and $\mathfrak{t}[u_n - u_m] \to 0$. Then $\mathfrak{t}_{\text{ext}}[u_n - u] \leqq \lim_m \mathfrak{t}[u_n - u_m]$ by the lower semicontinuity, so that $\limsup_n \mathfrak{t}_{\text{ext}}[u_n - u] \leqq \lim_n \lim_m \mathfrak{t}[u_n - u_m] = 0$. Hence $u_n - u \in \mathsf{D}(\mathfrak{t})$, $u \in \mathsf{D}(\mathfrak{t})$, and $\mathfrak{t}[u_n - u] \to 0$. This proves that $\mathfrak{t}$ is closed.

Suppose, conversely, that $\mathfrak{t}$ is closed. We have to show that $\liminf \mathfrak{t}_{\text{ext}}[u_n] \geqq \mathfrak{t}_{\text{ext}}[u]$ whenever $u_n \to u$. We may assume that the $\liminf = c < \infty$. Going over to a subsequence, we may further assume that $\mathfrak{t}_{\text{ext}}[u_n] = \mathfrak{t}[u_n] \to c$. If we regard $\mathsf{D}(\mathfrak{t})$ as a Hilbert space $\mathsf{H}_\mathfrak{t}$ as in VI-§1.3, the u_n form a bounded set in $\mathsf{H}_\mathfrak{t}$. Hence it contains a subsequence (which we denote again by u_n) converging weakly in $\mathsf{H}_\mathfrak{t}$. The limit must be identical with u. Hence $u \in \mathsf{D}(\mathfrak{t})$, $\mathfrak{t}_{\text{ext}}[u] = \mathfrak{t}[u] \leqq \lim \mathfrak{t}[u_n] = c$ as required [see III-(3.26)].

Proof of Theorem 3.13a. It is easy to see that $\mathsf{D}(\mathfrak{h})$ is a linear manifold, $\mathfrak{h}$ is a quadratic form, and that (3.28a) holds [see VI-(1.1), (1.16)]. To see that $\mathfrak{h}$ is closed, it suffices to note that $\mathfrak{h}_{\text{ext}}$ is lower semicontinuous (Lemma 3.14a). Indeed, the semicontinuity follows from that of the $\mathfrak{h}_{n,\text{ext}}$ because $\mathfrak{h}_{n,\text{ext}}[u] \uparrow \mathfrak{h}_{\text{ext}}[u]$ for each $u \in \mathsf{H}$.

Suppose now that $\mathfrak{h}$ is densely defined, so that the same is true of the $\mathfrak{h}_n$. Then $H_1 \leqq H_2 \leqq \dots \leqq H$ and $R_1(\xi) \geqq R_2(\xi) \geqq \dots \geqq R(\xi)$ for $\xi < \gamma_1$. As in the proof of Theorem 3.11, it follows that $R_0(\xi) = \text{s-}\lim R_n(\xi)$ exists and

$R_0(\xi) \geqq R(\xi)$. Since $R(\xi) \geqq 0$ is invertible, the same is true of $R_0(\xi)$, which is therefore the resolvent of a selfadjoint operator $H_0 \leqq H$. If $\mathfrak{h}_0$ is the associated form, we have $\mathfrak{h}_n \leqq \mathfrak{h}_0 \leqq \mathfrak{h}$. In view of the definition of $\mathfrak{h}$, however, one must have $\mathfrak{h}_0 = \mathfrak{h}$. Hence $H_0 = H$, $R_0 = R$, and we have proved (3.29a) (cf. Proof of Theorem 3.11).

It remains to prove (3.30a). For $u \in \mathsf{D}(\mathfrak{h})$ set $v_n = (H_n - \xi)^{1/2} u$. Since $\|v_n\| \leqq \|(H - \xi)^{1/2} u\|$ by $H_n \leqq H$, we may assume that $v_n \underset{w}{\to} v \in \mathsf{H}$, going over to a subsequence if necessary. For any $f \in \mathsf{H}$, we have then $(u, f) = (R_n(\xi)^{1/2} v_n, f) = (v_n, R_n(\xi)^{1/2} f) \to (v, R(\xi)^{1/2} f)$ because $R_n(\xi)^{1/2} \underset{s}{\to} R(\xi)^{1/2}$ (see Proof of Theorem 3.11). Thus $u = R(\xi)^{1/2} v$, or $v = (H - \xi)^{1/2} u$. Since this implies $\|v_n\| \leqq \|v\|$, we must have $v_n \underset{s}{\to} v$. Since we could have started with any subsequence of v_n with the same limit v, we have proved (3.30a).

5. Spectra of converging operators

Suppose $\{T_n\}$ is the sequence of operators defined in one of Theorems 3.6, 3.11 or 3.13a. $\{T_n\}$ is strongly convergent to T in the generalized sense. Let λ be an isolated eigenvalue of T with finite multiplicity m. If λ is *stable* in the sense defined in § 1.4, there are exactly m repeated eigenvalues of T_n in the neighborhood of λ and these eigenvalues converge to λ. This follows exactly in the same way as in § 1.4.

In the case of Theorem 3.6, λ is stable (with respect to the perturbation $T \to T_n$) if and only if $\bar{\lambda}$ is a stable eigenvalue of T^* (with respect to the perturbation $T^* \to T_n^*$). This follows easily from Theorem 3.6.

In all these cases, (1.31b, c) can be strengthened to

$$\|P_n - P\| \to 0$$

(see Lemma 1.24).

Theorem 3.15. *If the lower part $\lambda < \beta$ of the spectrum of H of* Theorem 3.11 *consists of isolated eigenvalues with finite multiplicities, these eigenvalues are stable under the perturbation considered. The same is true in* Theorem 3.6 *if $\mathfrak{t}_n$ and $\mathfrak{t}$ are symmetric so that $T = H$ and $T_n = H_n$ are selfadjoint. When $n \to \infty$, the eigenvalues $\mu_k^{(n)}$ of H_n tend from above to the corresponding eigenvalues μ_k of H.*

Proof. Since $H_n \geqq H \geqq \gamma$, we have (see Problem VI-5.5)

$$\dim E_n(\lambda) \leqq \dim E(\lambda), \quad -\infty < \lambda < \infty, \tag{3.33}$$

where $E_n(\lambda)$ and $E(\lambda)$ are the spectral families for the selfadjoint operators H_n and H, respectively. If $\lambda < \beta$, we have $\dim E(\lambda) < \infty$ and hence $\dim E_n(\lambda) < \infty$ for $\lambda < \beta$. This implies that H_n also has a spectrum consisting of isolated eigenvalues in the part $\lambda < \beta$. Let us denote the repeated eigenvalues of H and H_n in this part respectively by

$$\mu_1 \leqq \mu_2 \leqq \cdots < \beta\,, \quad \mu_1^{(n)} \leqq \mu_2^{(n)} \leqq \cdots < \beta\,. \tag{3.34}$$

Then (3.33) implies that (cf. Theorem I-6.44)

$$\mu_k^{(n)} \geqq \mu_k\,, \quad n, k = 1, 2, \ldots\,. \tag{3.35}$$

In the case of Theorem 3.11 in which $\mathfrak{h}_n$ and H_n are nonincreasing, each sequence $\mu_k^{(n)}$ for fixed k is nonincreasing in n.

Now we have

$$\|E_n(\lambda) - E(\lambda)\| \to 0\,, \quad n \to \infty\,, \quad \lambda < \beta\,, \quad \lambda \neq \mu_k\,. \tag{3.36}$$

To see this, we first note that $E_n(\lambda) \overset{s}{\to} E(\lambda)$ for every λ such that $E(\lambda - 0) = E(\lambda)$, in particular, for every $\lambda < \beta$ different from any of the μ_k (see Theorem 1.15). In view of (3.33) and $\dim E(\lambda) < \infty$, (3.36) then follows by Lemma 1.23. (3.36) in turn implies that $\dim E_n(\lambda) = \dim E(\lambda)$ for such λ for sufficiently large n. It is easy to see that this implies $\mu_k^{(n)} \to \mu_k$, $n \to \infty$.

Since $\|R_n(\zeta)\| \leqq 1/\mathrm{dist}(\zeta, \Sigma(H_n))$, every real number $\zeta \neq \mu_k$ belongs to the region Δ_b of boundedness and hence to Δ_s by Theorem 1.3. This proves the theorem.

Example 3.16. Applying Theorem 3.15 to Example 3.7, we see that the eigenvalues of $T = -d^2/dx^2 + q(x)$ are stable under the perturbation considered, provided the real-valued potential $q(x)$ is such that T has isolated eigenvalues at the lower part of the spectrum.

Remark 3.17. The above result does not appear to hold in the non-selfadjoint case. At first it might be expected that in Theorem 3.6, in which the forms $\mathfrak{t}_n$ tend to $\mathfrak{t}$ "from above" (in the sense that $\mathrm{Re}\,\mathfrak{t}_n \geqq \mathrm{Re}\,\mathfrak{t}$), the eigenvalues of T_n should also tend to the eigenvalues of T from above, at least in a region of the spectrum of T consisting of isolated eigenvalues. This conjecture is wrong, however. For example, consider Example 1.19 with $\mathsf{X} = \mathsf{L}^2$ and $\alpha > 0$, $\varkappa > 0$. Here the operator $T(\varkappa)$ is associated with the quadratic form $\mathfrak{t}(\varkappa)\,[u] = 2\alpha \int_0^1 [u'(x)\,\overline{u(x)} + \varkappa |u'(x)|^2]\, dx$, which is sectorial and closed for $\varkappa > 0$ and which is decreasing for decreasing $\varkappa$. But the eigenvalues of $T(\varkappa)$ are given by (1.20), each of which is *increasing* for decreasing $\varkappa$ for sufficiently small $\varkappa$.

§ 4. Asymptotic expansions for sectorial operators

1. The problem. The zeroth approximation for the resolvent

We resume the study, begun in § 2, of asymptotic expansions[1] for $\varkappa \to 0$ of the resolvent $R(\zeta, \varkappa)$ and the isolated eigenvalues of an operator $T(\varkappa)$ depending on the parameter $\varkappa$. We shall again be concerned with

[1] The results of this section were given by TITCHMARSH [1], [2], T. KATO [3], [7], [8], V. KRAMER [1], [2] in the symmetric case. Some results in the nonsymmetric case are given by HUET [1], [2]. The proof of the theorems given below can be greatly simplified in the symmetric case; see the papers cited above.

the case in which $T(\varkappa)$ may be written $T + \varkappa T^{(1)}$ *formally*, but now the exact definition of $T(\varkappa)$ will be given in terms of a sectorial form in a Hilbert space H.

Consider a family of sectorial forms $\mathfrak{t}(\varkappa)$ in H given by

$$\mathfrak{t}(\varkappa) = \mathfrak{t} + \varkappa\, \mathfrak{t}^{(1)}, \quad 0 < \varkappa \leqq 1 . \tag{4.1}$$

Throughout the present section, we make the following fundamental assumptions.

i) $\mathfrak{t}$ and $\mathfrak{t}^{(1)}$ are densely defined, closed sectorial forms.

ii) $\mathfrak{t}^{(1)}$ has a vertex zero; $\mathfrak{t}$ has a vertex γ.

iii) $\mathsf{D} = \mathsf{D}(\mathfrak{t}) \cap \mathsf{D}(\mathfrak{t}^{(1)})$ is a core of $\mathfrak{t}$.

Again we make some comments on these assumptions. ii) is not a restrictive assumption, for if $\mathfrak{t}$ and $\mathfrak{t}^{(1)}$ have vertices γ and $\gamma^{(1)}$, we may write $\mathfrak{t}^{(1)} = \gamma^{(1)} + \mathfrak{t}'$, where $\mathfrak{t}'$ is sectorial with a vertex 0, and (4.1) becomes

$$\mathfrak{t}(\varkappa) = \varkappa\, \gamma^{(1)} + \mathfrak{t} + \varkappa\, \mathfrak{t}' . \tag{4.2}$$

Here the first term $\varkappa\, \gamma^{(1)}$ is a small scalar, has no importance in most problems to be considered below and may be omitted, and we may replace $\mathfrak{t}'$ by $\mathfrak{t}^{(1)}$. Again, (4.1) implies that $\mathsf{D}(\mathfrak{t}(\varkappa)) = \mathsf{D}(\mathfrak{t}) \cap \mathsf{D}(\mathfrak{t}^{(1)}) = \mathsf{D}$. If D were not a core of $\mathfrak{t}$, we could replace $\mathfrak{t}$ by the closure of its restriction to D; this does not affect $\mathfrak{t}(\varkappa)$, while D is a core of the new $\mathfrak{t}$. For the same reason we could have assumed that D is also a core of $\mathfrak{t}^{(1)}$, but we need not assume this explicitly.

Since both $\mathfrak{t}$ and $\mathfrak{t}^{(1)}$ are closed sectorial, the same is true of $\mathfrak{t}(\varkappa)$ (see Theorem VI-6.1). $\mathsf{D}(\mathfrak{t}(\varkappa)) = \mathsf{D}$ is dense, since it is a core of the densely defined form $\mathfrak{t}$ by iii). Let $T(\varkappa)$ be the m-sectorial operator associated with $\mathfrak{t}(\varkappa)$ by the first representation theorem. Similarly, we denote by T the m-sectorial operator associated with $\mathfrak{t}$.

Now Theorem 3.6 is applicable to the present problem, with $\mathfrak{t}_n$, $\mathfrak{t}$, T_n replaced by $\mathfrak{t}(\varkappa)$, $\mathfrak{t}$, $T(\varkappa)$, respectively. It follows that $T(\varkappa)$ converges as $\varkappa \to 0$ strongly to T in the generalized sense. More precisely, we have by (3.10)

$$\left.\begin{array}{ll} (4.3) & R(\zeta,\varkappa) \xrightarrow[s]{} R(\zeta), \quad R(\zeta,\varkappa)^* \xrightarrow[s]{} R(\zeta)^* \\ (4.4) & \mathfrak{t}[R(\zeta,\varkappa)u - R(\zeta)u] \to 0, \quad \mathfrak{t}[R(\zeta,\varkappa)^*u - R(\zeta)^*u] \to 0 \\ (4.5) & \varkappa\, \mathfrak{t}^{(1)}[R(\zeta,\varkappa)u] \to 0, \quad \varkappa\, \mathfrak{t}^{(1)}[R(\zeta,\varkappa)^*u] \to 0 \end{array}\right\} \begin{array}{l} \varkappa \to 0, \\ \zeta \in \Delta_s, \\ u \in \mathsf{H}, \end{array}$$

where we write $R(\zeta,\varkappa) = (T(\varkappa) - \zeta)^{-1}$, $R(\zeta) = (T - \zeta)^{-1}$ as usual. The convergences (4.3)–(4.5) are uniform in each compact subset of Δ_s. Δ_s contains the half-plane $\operatorname{Re}\zeta < \gamma$.

In the following paragraphs we shall study under what conditions the expressions $R(\zeta,\varkappa)\, u$ and $(R(\zeta,\varkappa)\, u, v)$ admit asymptotic expansions in powers of $\varkappa$.

We write

(4.6) $\quad \mathfrak{h} = \operatorname{Re}\mathfrak{t},\ \mathfrak{h}^{(1)} = \operatorname{Re}\mathfrak{t}^{(1)},\ \mathfrak{h}(\varkappa) = \operatorname{Re}\mathfrak{t}(\varkappa) = \mathfrak{h} + \varkappa\,\mathfrak{h}^{(1)},\ \xi = \operatorname{Re}\zeta,$

and denote by $T^{(1)}$, H, $H^{(1)}$, $H(\varkappa)$ the operators associated with $\mathfrak{t}^{(1)}$, $\mathfrak{h}$, $\mathfrak{h}^{(1)}$, $\mathfrak{h}(\varkappa)$, respectively. Also we have

(4.7) $\quad |\operatorname{Im}\mathfrak{t}[u]| \leqq M(\mathfrak{h}-\gamma)[u]\,,\quad |\operatorname{Im}\mathfrak{t}^{(1)}[u]| \leqq M'\,\mathfrak{h}^{(1)}[u]\,.$

Where $M = \tan\theta$, $M' = \tan\theta'$ if θ, θ' are the semi-angles for the sectorial forms $\mathfrak{t}, \mathfrak{t}^{(1)}$. Finally we note that there need not be any simple relationship between $\mathsf{D}(T)$ and $\mathsf{D}(T^{(1)})$, so that $T(\varkappa)$ need not be equal to $T + \varkappa\,T^{(1)}$.

Remark 4.1. Since $\mathsf{D}(\mathfrak{t}^*) = \mathsf{D}(\mathfrak{t})$, etc., the assumptions i) ~ iii) imply the same for their adjoint forms.

2. The 1/2-order approximation for the resolvent

Theorem 4.2. *Let* $\zeta \in \Delta_s$, $u \in \mathsf{H}$ *and* $R(\zeta)\,u \in \mathsf{D}(\mathfrak{t}^{(1)})$. *Then we have as* $\varkappa \to 0$

(4.8) $$\|R(\zeta,\varkappa)\,u - R(\zeta)\,u\| = o(\varkappa^{1/2})\,,$$

(4.9) $$\mathfrak{t}\,[R(\zeta,\varkappa) - R(\zeta)\,u] = o(\varkappa)\,,\quad \mathfrak{t}^{(1)}[R(\zeta,\varkappa)\,u - R(\zeta)\,u] = o(1)\,.$$

If $R(\zeta)\,u \in \mathsf{D}(\mathfrak{t}^{(1)})$ *and* $R(\zeta)^*\,v \in \mathsf{D}(\mathfrak{t}^{(1)})$, *then*

(4.10) $$(R(\zeta,\varkappa)\,u, v) = (R(\zeta)\,u, v) - \varkappa\,\mathfrak{t}^{(1)}[R(\zeta)\,u, R(\zeta)^*\,v] + o(\varkappa)\,.$$

Proof. I. Set

(4.11) $$w = w(\varkappa) = R(\zeta,\varkappa)\,u - R(\zeta)\,u\,;$$

note that $w \in \mathsf{D}$ since $R(\zeta)\,u \in \mathsf{D}(T) \subset \mathsf{D}(\mathfrak{t})$ and $R(\zeta)\,u \in \mathsf{D}(\mathfrak{t}^{(1)})$ by assumption and since $R(\zeta,\varkappa)\,u \in \mathsf{D}(T(\varkappa)) \subset \mathsf{D}(\mathfrak{t}(\varkappa)) = \mathsf{D}$. Then we have

(4.12) $$\begin{aligned}(\mathfrak{t}(\varkappa) - \zeta)\,[w] &= (\mathfrak{t}(\varkappa) - \zeta)\,[R(\zeta,\varkappa)\,u - R(\zeta)\,u, w]\\ &= (u, w) - (u, w) - \varkappa\,\mathfrak{t}^{(1)}[R(\zeta)\,u, w] = -\varkappa\,\mathfrak{t}^{(1)}[R(\zeta)\,u, w]\,.\end{aligned}$$

Suppose, for the moment, that $\xi = \operatorname{Re}\zeta < 0$, assuming $\gamma = 0$ for simplicity (which does not affect the generality). Then $\operatorname{Re}(\mathfrak{t}(\varkappa) - \zeta) = \mathfrak{h} - \xi + \varkappa\,\mathfrak{h}^{(1)} \geqq \varkappa\,\mathfrak{h}^{(1)} \geqq 0$, and (4.12) implies

(4.13) $$\varkappa\,\mathfrak{h}^{(1)}[w] \leqq \varkappa\,|\mathfrak{t}^{(1)}[R(\zeta)\,u, w]| \leqq (1 + M')\,\varkappa\,\mathfrak{h}^{(1)}[R(\zeta)\,u]^{1/2}\,\mathfrak{h}^{(1)}[w]^{1/2}$$

[see VI-(1.15)] and hence

(4.14) $$\mathfrak{h}^{(1)}[w] \leqq (1 + M')^2\,\mathfrak{h}^{(1)}[R(\zeta)\,u]\,.$$

This shows that $w = w(\varkappa)$ is bounded as $\varkappa \to 0$ with respect to the norm $\|w\|_{\mathfrak{h}^{(1)}} = (\mathfrak{h}^{(1)}[w] + \|w\|^2)^{1/2}$ by which $\mathsf{D}(\mathfrak{t}^{(1)})$ becomes a Hilbert space $\mathsf{H}_{\mathfrak{t}^{(1)}}$. But we have, for any $v \in \mathsf{H}$, $(\mathfrak{h}^{(1)} + 1)\,[w, (H^{(1)} + 1)^{-1}\,v] = (w, v) \to 0$ because $w \to 0$ by (4.3). Since $(H^{(1)} + 1)^{-1}\,\mathsf{H} = \mathsf{D}(H^{(1)})$ is a core of $\mathfrak{h}^{(1)}$ and hence dense in $\mathsf{H}_{\mathfrak{t}^{(1)}}$, it follows that $w \to 0$ weakly in $\mathsf{H}_{\mathfrak{t}^{(1)}}$. Since $\mathfrak{t}^{(1)}$

is a bounded form on $\mathsf{H}_{\mathfrak{t}^{(1)}}$, this implies that $\mathfrak{t}^{(1)}[R(\zeta)u, w] \to 0$. Thus the right member of (4.12) is $o(\varkappa)$.

Again taking the real part of (4.12) and noting that $\operatorname{Re}(\mathfrak{t}(\varkappa) - \zeta)[w] = \mathfrak{h}[w] + (-\xi)\|w\|^2 + \varkappa\,\mathfrak{h}^{(1)}[w]$, in which all three terms are non-negative, we see that $\mathfrak{h}[w]$, $\|w\|^2$ and $\varkappa\,\mathfrak{h}^{(1)}[w]$ are all $o(\varkappa)$. This proves (4.8) and (4.9).

II. Now (4.8), (4.9) can be extended to the general case in which $\zeta \in \Delta_s$ and $R(\zeta)u \in \mathsf{D}(\mathfrak{t}^{(1)})$. To show this we use a version of (1.3):

$$(4.15)\quad R(\zeta,\varkappa) - R(\zeta) = (1 + (\zeta - \zeta_0)R(\zeta,\varkappa))(R(\zeta_0,\varkappa) - R(\zeta_0))(1 + (\zeta - \zeta_0)R(\zeta)),$$

in which we choose ζ_0 as a fixed number with $\operatorname{Re}\zeta_0 < 0$. Set $x = (1 + (\zeta - \zeta_0)R(\zeta))u$. Then we have $R(\zeta_0)x = R(\zeta)u \in \mathsf{D}(\mathfrak{t}^{(1)})$ by the resolvent equation. Hence $y(\varkappa) \equiv (R(\zeta_0,\varkappa) - R(\zeta_0))x = o(\varkappa^{1/2})$ by what was proved above. Since $R(\zeta,\varkappa)$ is bounded when $\varkappa \to 0$, (4.8) follows from $(R(\zeta,\varkappa) - R(\zeta))u = (1 + (\zeta - \zeta_0)R(\zeta,\varkappa))y(\varkappa)$.

To prove (4.9), we note that

$$(4.16)\quad \mathfrak{h}(\varkappa)[(R(\zeta,\varkappa) - R(\zeta))u] = \mathfrak{h}(\varkappa)[y(\varkappa) + (\zeta - \zeta_0)R(\zeta,\varkappa)y(\varkappa)] \leqq 2\mathfrak{h}(\varkappa)[y(\varkappa)] + 2|\zeta - \zeta_0|^2\,\mathfrak{h}(\varkappa)[R(\zeta,\varkappa)y(\varkappa)].$$

We shall show that the last member of (4.16) is $o(\varkappa)$. This is true for the first term by what has already been proved. In the second term, we may replace $\mathfrak{h}(\varkappa)$ by $\mathfrak{h}(\varkappa) - \xi$ since $\|R(\zeta,\varkappa)y(\varkappa)\|^2 = o(\varkappa)$ as shown above. But

$$(\mathfrak{h}(\varkappa) - \xi)[R(\zeta,\varkappa)y(\varkappa)] = \operatorname{Re}(\mathfrak{t}(\varkappa) - \zeta)[R(\zeta,\varkappa)y(\varkappa)] = \operatorname{Re}(y(\varkappa), R(\zeta,\varkappa)y(\varkappa)) \leqq \|R(\zeta,\varkappa)\|\,\|y(\varkappa)\|^2 = o(\varkappa).$$

Thus (4.16) is $o(\varkappa)$ and hence (4.9) follows immediately.

III. To prove (4.10), we use the identity

$$\begin{aligned}(4.17)\quad (w, v) &= (\mathfrak{t} - \zeta)[w, R(\zeta)^* v] \\ &= (\mathfrak{t} - \zeta)[R(\zeta,\varkappa)u, R(\zeta)^* v] - (\mathfrak{t} - \zeta)[R(\zeta)u, R(\zeta)^* v] \\ &= (u, R(\zeta)^* v) - \varkappa\,\mathfrak{t}^{(1)}[R(\zeta,\varkappa)u, R(\zeta)^* v] - (u, R(\zeta)^* v) \\ &= -\varkappa\,\mathfrak{t}^{(1)}[R(\zeta)u, R(\zeta)^* v] - \varkappa\,\mathfrak{t}^{(1)}[w, R(\zeta)^* v].\end{aligned}$$

This is true since $R(\zeta)^* v \in \mathsf{D}$, which follows from the assumption that $R(\zeta)^* v \in \mathsf{D}(\mathfrak{t}^{(1)})$ as before. (4.10) follows from (4.17) in virtue of (4.9), which implies that $w \to 0$ in $\mathsf{H}_{\mathfrak{t}^{(1)}}$.

3. The first and higher order approximations for the resolvent

Once Theorem 4.2 for the 1/2-order approximation has been established, the further approximations are rather straightforward. For example, we have

Theorem 4.3. *Let $\zeta \in \Delta_s$ and $u \in \mathsf{H}$. If $R(\zeta) u \in \mathsf{D}(T^{(1)})$, we have as $\varkappa \to 0$*

$$(4.18)\qquad R(\zeta,\varkappa)\, u = R(\zeta)\, u - \varkappa\, R(\zeta)\, T^{(1)} R(\zeta) u + o(\varkappa)\,.$$

If $R(\zeta) u \in \mathsf{D}(T^{(1)})$ and $R(\zeta)^* v \in \mathsf{D}(T^{(1)*})$, then

$$(4.19)\qquad (R(\zeta,\varkappa)\, u, v) = (R(\zeta)\, u, v) - \varkappa (R(\zeta)\, T^{(1)} R(\zeta)\, u, v) + \\ + \varkappa^2 (R(\zeta)\, T^{(1)} R(\zeta)\, u, T^{(1)*} R(\zeta)^* v) + o(\varkappa^2)\,.$$

Proof. Using the same notations as in the preceding paragraph, we have for any $v \in \mathsf{H}$

$$(4.20)\qquad (w, v) = (\mathfrak{t}(\varkappa) - \zeta)\, [w, R(\zeta,\varkappa)^* v] \\ = (\mathfrak{t}(\varkappa) - \zeta)\, [R(\zeta,\varkappa)\, u, R(\zeta,\varkappa)^* v] - (\mathfrak{t}(\varkappa) - \zeta)\, [R(\zeta)\, u, R(\zeta,\varkappa)^* v] \\ = (u, R(\zeta,\varkappa)^* v) - (u, R(\zeta,\varkappa)^* v) - \varkappa\, \mathfrak{t}^{(1)} [R(\zeta)\, u, R(\zeta,\varkappa)^* v] \\ = -\varkappa (T^{(1)} R(\zeta)\, u, R(\zeta,\varkappa)^* v) = -\varkappa (R(\zeta,\varkappa)\, T^{(1)} R(\zeta)\, u, v)$$

since $R(\zeta)\, u \in \mathsf{D}(T^{(1)})$. Hence

$$(4.21)\qquad R(\zeta,\varkappa)\, u - R(\zeta)\, u = w = -\varkappa\, R(\zeta,\varkappa)\, T^{(1)} R(\zeta)\, u\,.$$

(4.18) follows from (4.21) in virtue of (4.3).

Suppose now that $R(\zeta)^* v \in \mathsf{D}(T^{(1)*})$. Since T^*, $T(\varkappa)^*$ and $T^{(1)*}$ are the operators associated with the adjoint forms $\mathfrak{t}^*$, $\mathfrak{t}(\varkappa)^*$ and $\mathfrak{t}^{(1)*}$, respectively, application of (4.21) to these adjoint forms with ζ replaced by $\bar{\zeta}$ leads to

$$(4.22)\qquad R(\zeta,\varkappa)^* v - R(\zeta)^* v = -\varkappa\, R(\zeta,\varkappa)^* T^{(1)*} R(\zeta)^* v\,.$$

Taking the inner product of (4.22) with u and using (4.21), we obtain

$$(4.23)\qquad (R(\zeta,\varkappa)\, u, v) - (R(\zeta)\, u, v) = -\varkappa (R(\zeta,\varkappa)\, u, T^{(1)*} R(\zeta)^* v) \\ = -\varkappa (R(\zeta)\, u, T^{(1)*} R(\zeta)^* v) + \varkappa^2 (R(\zeta,\varkappa)\, T^{(1)} R(\zeta)\, u, T^{(1)*} R(\zeta)^* v)\,,$$

from which (4.19) follows in virtue of (4.3).

Theorem 4.4. *If $R(\zeta)\, T^{(1)} R(\zeta)\, u$ exists and belongs to $\mathsf{D}(\mathfrak{t}^{(1)})$,*

$$(4.24)\qquad R(\zeta,\varkappa)\, u = R(\zeta)\, u - \varkappa\, R(\zeta)\, T^{(1)} R(\zeta)\, u + o(\varkappa^{3/2})\,.$$

If, in addition, $R(\zeta)^* T^{(1)*} R(\zeta)^* v$ exists and belongs to $\mathsf{D}(\mathfrak{t}^{(1)})$,

$$(4.25)\qquad (R(\zeta,\varkappa)\, u, v) = (R(\zeta)\, u, v) - \varkappa (R(\zeta)\, T^{(1)} R(\zeta)\, u, v) + \\ + \varkappa^2 (R(\zeta)\, T^{(1)} R(\zeta)\, u, T^{(1)*} R(\zeta)^* v) - \\ - \varkappa^3\, \mathfrak{t}^{(1)} [R(\zeta)\, T^{(1)} R(\zeta)\, u, R(\zeta)^* T^{(1)*} R(\zeta)^* v] + o(\varkappa^3)\,.$$

Proof. (4.24) follows by applying (4.8) to the right member of (4.21). To prove (4.25), it suffices to apply (4.10) to the last term of (4.23) [with u, v replaced by $T^{(1)} R(\zeta)\, u$, $T^{(1)*} R(\zeta)^* v$, respectively].

It is now easy to see how further approximations should proceed. It will not be necessary to write them down explicitly.

Problem 4.5. If $R(\zeta)\, T^{(1)} R(\zeta)\, u$ exists and belongs to $\mathsf{D}(\mathfrak{t}^{(1)})$ and if $R(\zeta)^* v \in \mathsf{D}(\mathfrak{t}^{(1)})$, then

$$(4.26) \qquad (R(\zeta, \varkappa)\, u, v) = (R(\zeta)\, u, v) - \varkappa (R(\zeta)\, T^{(1)} R(\zeta)\, u, v) + \\ + \varkappa^2\, \mathfrak{t}^{(1)} [R(\zeta)\, T^{(1)} R(\zeta)\, u, R(\zeta)^* v] + o(\varkappa^2)\,.$$

Remark 4.6. All the above formulas are, in spite of their different forms, essentially the *second Neumann series* for the resolvent, which may be written *formally*

$$(4.27) \qquad R(\zeta, \varkappa) = R(\zeta) - \varkappa\, R(\zeta)\, T^{(1)} R(\zeta) + \varkappa^2\, R(\zeta)\, T^{(1)} R(\zeta)\, T^{(1)} R(\zeta) + \cdots.$$

For example, the coefficient of $\varkappa^3$ in (4.25) is *formally* equal to

$$(4.28) \qquad -\mathfrak{t}^{(1)} [R(\zeta)\, T^{(1)} R(\zeta)\, u, R(\zeta)^*\, T^{(1)*} R(\zeta)^* v] \\ = -(T^{(1)} R(\zeta)\, T^{(1)} R(\zeta)\, u, R(\zeta)^*\, T^{(1)*} R(\zeta)^* v) \\ = -(R(\zeta)\, T^{(1)} R(\zeta)\, T^{(1)} R(\zeta)\, T^{(1)} R(\zeta)\, u, v)\,,$$

corresponding to the coefficient of $\varkappa^3$ in (4.27). Actually (4.28) is correct only when $T^{(1)} R(\zeta)\, T^{(1)} R(\zeta)\, T^{(1)} R(\zeta)\, u$ makes sense. The left member of (4.28) exists, however, without assuming so much on u if we impose on v the assumptions stated. The formula (4.25) is particularly useful when $\mathfrak{t}$ and $\mathfrak{t}^{(1)}$ are symmetric and ζ is real; then $T^{(1)*} = T^{(1)}$, $R(\zeta)^* = R(\zeta)$ so that it is valid for $v = u$ if we assume only that $R(\zeta)\, T^{(1)} R(\zeta)\, u \in \mathsf{D}(\mathfrak{t}^{(1)})$. A similar remark applies to all the other formulas.

Example 4.7. Consider the differential operator

$$(4.29) \qquad T(\varkappa) = -d^2/dx^2 + q(x) + \varkappa\, q^{(1)}(x)\,, \quad 0 \leqq x < \infty\,,$$

with the boundary condition $u(0) = 0$. $T(\varkappa)$ is associated with the quadratic form

$$(4.30) \qquad \mathfrak{t}(\varkappa)\,[u] = \int_0^\infty [|u'(x)|^2 + q(x)\, |u(x)|^2 + \varkappa\, q^{(1)}(x)\, |u(x)|^2]\, dx\,.$$

This example is a version of Example 3.7 with a continuous parameter $\varkappa$ rather than a discrete one n. If $q(x)$ satisfies the assumptions of Theorem VI-4.2 and if $q^{(1)}(x)$ is nonnegative and locally integrable, $T(\varkappa)$ is selfadjoint and the above results are applicable. The zeroth approximation (4.3) for the resolvent is valid for every $u \in \mathsf{H} = \mathsf{L}^2(0, \infty)$, the 1/2-order approximation (4.8) is true if $R(\zeta)\, u \in \mathsf{D}(\mathfrak{t}^{(1)})$, that is, if

$$(4.31) \qquad \int_0^\infty q^{(1)}(x)\, |R(\zeta)\, u(x)|^2\, dx < \infty\,,$$

and the first order approximations (4.18) and (4.19) with $v = u$ are valid if $R(\zeta)\, u \in \mathsf{D}(T^{(1)})$, that is if

$$(4.32) \qquad \int_0^\infty q^{(1)}(x)^2\, |R(\zeta)\, u(x)|^2\, dx < \infty\,,$$

and so on.

Suppose, in particular, that $q(x) \to 0$ as $x \to \infty$, as is often the case in the Schrödinger operators. $w = R(\zeta)\, u$ is the solution of $(T - \zeta)\, w = u$ or

$$(4.33) \qquad -w'' + q(x)\, w - \zeta\, w = u(x)\,, \quad w(0) = 0\,.$$

If ζ is real and negative, it can be shown that the solution $w(x)$ of (4.33) belonging to L^2 must decrease in magnitude very rapidly as $x \to \infty$, provided this is the case with $u(x)$. In such cases (4.31) and (4.32) are satisfied even when $|q^{(1)}(x)|$ is large as $x \to \infty$.

$q^{(1)}(x)$ may also have a rather strong singularity at $x = 0$. If $q(x)$ is $o(x^{-2})$ as $x \to 0$, $w(x) = R(\zeta)\, u(x)$ is $O(x)$. In order that (4.31) be satisfied, therefore, $q^{(1)}(x)$ need only be of the order $O(x^{-3+\varepsilon})$ with $\varepsilon > 0$. Similarly, a singularity of $O(x^{-1.5+\varepsilon})$ is allowed to $q^{(1)}(x)$ for (4.32) to be satisfied. Thus the above results are quite satisfactory when applied to problems of this kind. A similar remark applies to the perturbation of Schrödinger operators in three or higher dimensional spaces.

Example 4.8. Consider the operator

$$T(\varkappa) = -\alpha\, d^2/dx^2 + \varkappa\, d^4/dx^4\,,\quad 0 \leq x \leq 1\,, \tag{4.34}$$

of Example 1.20. $T(\varkappa)$ is the operator associated with the form $\mathfrak{t}(\varkappa)$ of (3.22). The zeroth order approximation $R(\zeta, \varkappa)\, u \to R(\zeta)\, u$ is again true for any $u \in \mathsf{H} = \mathsf{L}^2(0, 1)$. The assumption of Theorem 4.2 for the 1/2-order approximation requires that $w = R(\zeta)\, u \in \mathsf{D}(\mathfrak{t}^{(1)})$, that is, $w(x)$ be twice differentiable and satisfy the boundary condition $w(0) = w'(0) = w(1) = w'(1) = 0$. The differentiability is automatically satisfied since $w(x)$ is the solution of $(T - \zeta)\, w = u$, which is a second order differential equation. Also the boundary conditions inherent in T imply that $w(0) = w(1) = 0$, but the remaining conditions $w'(0) = w'(1) = 0$ are satisfied only in exceptional cases. A necessary and sufficient condition for this is given by

$$\int_0^1 e^{\pm(-\zeta)^{1/2} x}\, u(x)\, dx = 0\,, \tag{4.35}$$

for the required condition means that u is in the range of the restriction $T_0 - \zeta$ of $T - \zeta$ with the stated four boundary conditions, and this is true only if u is orthogonal to the null space of $(T_0 - \zeta)^*$. But T_0^* is simply the differential operator $-d^2/dx^2$ without any boundary condition, so that the null space in question is spanned by the two functions $e^{\pm(-\bar\zeta)^{1/2} x}$.

Thus we conclude that "in general" the estimates (4.8) – (4.10) are not valid. For example if $u(x) = 1$ and $\zeta = 0$, we have

$$f(x) \equiv R(0, \varkappa)\, u(x) \tag{4.36}$$

$$= \frac{x(1-x)}{2\alpha} - \frac{\sinh \frac{1}{2}\sqrt{\frac{\alpha}{\varkappa}}\left[\cosh \frac{1}{2}\sqrt{\frac{\alpha}{\varkappa}} - \cosh\sqrt{\frac{\alpha}{\varkappa}}\left(x - \frac{1}{2}\right)\right]}{\alpha\sqrt{\frac{\alpha}{\varkappa}}\left(\cosh\sqrt{\frac{\alpha}{\varkappa}} - 1\right)}\,.$$

It is true that this function has the asymptotic expansion

$$f(x) = \frac{x(1-x)}{2\alpha} - \frac{1}{2\alpha}\sqrt{\frac{\varkappa}{\alpha}} + \cdots\,,\qquad \frac{x(1-x)}{2\alpha} = T^{-1} u(x)\,, \tag{4.37}$$

for each *fixed* x such that $0 < x < 1$, the remainder ... being smaller than any finite power of $\varkappa$ as $\varkappa \to 0$. But this remainder is not quite so small in the whole interval (0, 1). In fact, a simple calculation shows that it is exactly of the order $\varkappa^{3/4}$ in norm. Thus

$$R(0, \varkappa)\, u = T^{-1} u - \frac{1}{2\alpha}\sqrt{\frac{\varkappa}{\alpha}} + O(\varkappa^{3/4})\,, \tag{4.38}$$

the order $\varkappa^{3/4}$ being exact. A similar result can be expected for any other $u(x)$ if it is at least smooth enough on [0, 1]. In this way we see that a satisfactory asymptotic expansion is impossible in this example, at least if one is concerned with the *global* behavior of the function $R(\zeta, \varkappa)\, u$.

It is known, on the other hand, that the scalar quantity of the form $(R(\zeta, \varkappa)\, u, v)$ admits an asymptotic expansion in powers of $\varkappa^{1/2}$ *to any order* if $u(x)$, $v(x)$ are sufficiently smooth. Of course our general theorems proved above are not capable of dealing with such asymptotic expansion in *fractional powers*[1] of $\varkappa$.

---

[1] The appearance of fractional powers of $\varkappa$ is quite common in the singular perturbation theory of differential operators; see references on footnote [1] of p. 442.

4. Asymptotic expansions for eigenvalues and eigenvectors

Let us now consider an isolated eigenvalue λ of T with finite multiplicity m and assume that λ is *stable* with respect to the perturbation $T \to T(\varkappa)$ (in the sense of § 1.4). Since $T(\varkappa)$ converges to T strongly in the generalized sense, there are exactly m (repeated) eigenvalues of $T(\varkappa)$ in the neighborhood of λ and these eigenvalues tend as $\varkappa \to 0$ to λ (see § 3.5).

We denote by $\mu_j(\varkappa)$, $j = 1, \ldots, m$, these (repeated) eigenvalues of $T(\varkappa)$, and follow the notations of §§ 2.3–2.4 for other related quantities.

Theorem 4.9. *Let the eigenvalue λ of T be semisimple and let the ranges of P and P^* be contained in $\mathsf{D}(\mathfrak{t}^{(1)})$. Then the eigenvalues $\mu_j(\varkappa)$ admit the asymptotic expansions*

$$\mu_j(\varkappa) = \lambda + \varkappa\, \mu_j^{(1)} + o(\varkappa)\,, \quad j = 1, \ldots, m\,, \tag{4.39}$$

where $\mu_j^{(1)}$, $j = 1, \ldots, m$, are the eigenvalues of the sesquilinear form $\mathfrak{t}^{(1)}[Pu, P^* v]$ considered in the m-dimensional space $P\mathsf{H}$ (see Remark 4.11 *below). The total projection $P(\varkappa)$ for these m eigenvalues $\mu_j(\varkappa)$ has the property that*

$$\|P(\varkappa) - P\| = o(\varkappa^{1/2})\,. \tag{4.40}$$

Remark 4.10. If $\mathfrak{t}$ is symmetric (so that T is selfadjoint), $P^* = P$ and the assumptions reduce to $P\mathsf{H} \subset \mathsf{D}(\mathfrak{t}^{(1)})$.

Remark 4.11. Let $\mathfrak{t}$ be an arbitrary sesquilinear form in H. Let P be a projection such that both $P\mathsf{H}$ and $P^*\mathsf{H}$ are contained in $\mathsf{D}(\mathfrak{t})$. Then $\mathfrak{t}_P[u, v] = \mathfrak{t}[Pu, P^* v]$ defines a form $\mathfrak{t}_P$ with domain H. It is easy to prove that $\mathfrak{t}_P$ is bounded if $\dim P = m < \infty$. Thus there exists an operator $T_P \in \mathscr{B}(\mathsf{H})$ such that $\mathfrak{t}_P[u, v] = (T_P u, v)$ for all $u, v \in \mathsf{H}$. Since $\mathfrak{t}_P[Pu, v] = \mathfrak{t}_P[u, v] = \mathfrak{t}_P[u, P^* v]$ for all u, v, it follows that $T_P P = T_P = P T_P$. Thus the subspace $\mathsf{M} = P\mathsf{H}$ is invariant under T_P. The eigenvalues of T_P considered in M will be called the eigenvalues of the form $\mathfrak{t}_P$ in M.

If $\mathfrak{t}$ is closed sectorial and if $P\mathsf{H} \subset \mathsf{D}(T)$, where $T = T_{\mathfrak{t}}$ is the operator associated with $\mathfrak{t}$, it can be readily seen that $T_P = PTP$. In general, however, PTP need not be defined on a sufficiently wide domain.

Proof of Theorem 4.9. I. First we note that [for notation see (2.18)]

$$R(\zeta, \varkappa)\, P - R(\zeta)\, P = o(\varkappa^{1/2})_{\mathrm{u}}\,, \quad \zeta \in \Delta_s\,, \tag{4.41}$$

$o(\varkappa^{1/2})_{\mathrm{u}}$ being uniform on any compact subset Γ of Δ_s. (4.41) follows from the fact that $u \in P\mathsf{H}$ implies $R(\zeta)\, u = (\lambda - \zeta)^{-1} u \in \mathsf{D}(\mathfrak{t}^{(1)})$ so that $R(\zeta, \varkappa)\, u - R(\zeta)\, u = o(\varkappa^{1/2})$ by (4.8) [cf. the proof of (2.20)]. The uniformity in $\zeta \in \Gamma$ can be proved easily by using (4.15). Taking as Γ the circle $|\zeta - \lambda| = r$ and integrating (4.41) on Γ, we thus obtain

$$P(\varkappa)\, P - P = o(\varkappa^{1/2})_{\mathrm{u}}\,. \tag{4.42}$$

Since $\bar{\lambda}$ is a stable eigenvalue of T^* by the remark given in § 3.5, we have a result similar to (4.42) with P and $P(\varkappa)$ replaced by the adjoints P^* and $P(\varkappa)^*$, respectively. Taking the adjoint of this formula, we have

$$PP(\varkappa) - P = o(\varkappa^{1/2})_{\mathrm{u}} . \tag{4.43}$$

It follows from (4.42) and (4.43), using an argument similar to the one used in the proof of (2.41), that $Q(\varkappa) \equiv P(\varkappa) - P = o(\varkappa^{1/2})_{\mathrm{u}}$. This proves (4.40).

II. We now introduce the operator $U(\varkappa)$ defined by (2.42) and consider the operator $U(\varkappa)^{-1} R(\zeta, \varkappa)\, U(\varkappa)\, P$. Recalling that $U(\varkappa)\, P = P(\varkappa)\, U(\varkappa)$ [see (2.45)], that $Q(\varkappa)^2 = (P(\varkappa) - P)^2 = o(\varkappa)_{\mathrm{u}}$ commutes with both P and $P(\varkappa)$, and that $P(\varkappa)$ commutes with $R(\zeta, \varkappa)$, and noting (2.42) and (2.44), we have

$$\begin{aligned} U(\varkappa)^{-1} R(\zeta, \varkappa)\, U(\varkappa)\, P &= U(\varkappa)^{-1} P(\varkappa)\, R(\zeta, \varkappa)\, P(\varkappa)\, U(\varkappa) \\ &= (1 + o(\varkappa)_{\mathrm{u}})\, P P(\varkappa)\, R(\zeta, \varkappa)\, P(\varkappa)\, P(1 + o(\varkappa)_{\mathrm{u}}) \\ &= P P(\varkappa)\, R(\zeta, \varkappa)\, P + o(\varkappa)_{\mathrm{u}} \\ &= P P(\varkappa)\, P P R(\zeta, \varkappa)\, P + P P(\varkappa)\, (1 - P)\, (1 - P)\, R(\zeta, \varkappa)\, P + o(\varkappa)_{\mathrm{u}} . \end{aligned} \tag{4.44}$$

The second term on the right is $o(\varkappa)_{\mathrm{u}}$; this is seen by noting that $P P(\varkappa) \cdot (1 - P) = P(P(\varkappa) - P)(1 - P) = o(\varkappa^{1/2})_{\mathrm{u}}$ by (4.40) and $(1 - P)\, R(\zeta, \varkappa)\, P = (1 - P)\, (R(\zeta, \varkappa) - R(\zeta))\, P = o(\varkappa^{1/2})_{\mathrm{u}}$ by (4.41). In the first term we note that $P P(\varkappa)\, P = P - P Q(\varkappa)^2 = P + o(\varkappa)_{\mathrm{u}}$. Hence

$$U(\varkappa)^{-1} R(\zeta, \varkappa)\, U(\varkappa)\, P = P R(\zeta, \varkappa)\, P + o(\varkappa)_{\mathrm{u}} . \tag{4.45}$$

The first term on the right of (4.45) can be calculated by using (4.10). Replacing u, v respectively by Pu, $P^* v$ in (4.10), we obtain

$$\begin{aligned} (P R(\zeta, \varkappa)\, Pu, v) &= (R(\zeta, \varkappa)\, Pu, P^* v) \\ &= (\lambda - \zeta)^{-1} (Pu, v) - \varkappa (\lambda - \zeta)^{-2}\, \mathfrak{t}^{(1)} [Pu, P^* v] + o(\varkappa) . \end{aligned} \tag{4.46}$$

Since $\dim P < \infty$, this gives

$$P R(\zeta, \varkappa)\, P = (\lambda - \zeta)^{-1} P - \varkappa (\lambda - \zeta)^{-2}\, T_P^{(1)} + o(\varkappa)_{\mathrm{u}} , \tag{4.47}$$

where $T_P^{(1)}$ is the bounded operator associated with the form $\mathfrak{t}_P^{(1)} [u, v] = \mathfrak{t}^{(1)} [Pu, P^* v]$ (see Remark 4.11).

Substitution of (4.47) into (4.45) gives an expansion similar to (2.24), and the expansion (4.39) for $\mu_j(\varkappa)$ follows as in the proof of Theorem 2.6.

We proceed to the expansion of the eigenvalues $\mu_j(\varkappa)$ up to the second order in $\varkappa$.

Theorem 4.12. *In* Theorem 4.9 *assume further that* $P \mathsf{H} \subset \mathsf{D}(T^{(1)})$ *and* $P^* \mathsf{H} \subset \mathsf{D}(T^{(1)*})$. *If the (distinct) eigenvalues* λ_j, $j = 1, \ldots, s$, *of* $T_P^{(1)}$ *are semisimple, then all the results of* Theorem 2.9 *are true.*

Remark 4.13. The operator $[P T^{(1)}]$ of Theorem 2.9 is here equal to $(T^{(1)*} P^*)^*$; note that $T^{(1)*} P^* \in \mathscr{B}(\mathsf{H})$ by the assumption $P^* \mathsf{H} \subset \mathsf{D}(T^{(1)*})$.

Proof of Theorem 4.12. If $u \in P\mathsf{H}$ we have $R(\zeta) u = (\lambda - \zeta)^{-1} u \in \mathsf{D}(T^{(1)})$, so that we have the expansion (4.18) for $R(\zeta, \varkappa)\, u$. The $o(\varkappa)$ on the right of (4.18) is uniform in ζ on any compact subset of Δ_s, as is seen from (4.21) and Lemma III-3.7. Considering that $\dim P < \infty$, we again obtain the formula (2.20) for $R(\zeta, \varkappa)\, P$.

Considering the adjoints $R(\zeta)^*$ etc., we arrive at the same formula (2.20) with P, $R(\zeta), \ldots$ replaced by P^*, $R(\zeta)^*, \ldots$. Taking the adjoint of this result, we obtain the formula (2.35) for $PR(\zeta, \varkappa)$, with $o(\varkappa)_s$ replaced by $o(\varkappa)_u$.

Once we have these two expressions, the proof of Theorem 2.9 can be applied without change to deduce the required results.

Remark 4.14. In the same way we could proceed to the expansion of the eigenvalues and eigenvectors to higher orders of $\varkappa$. We shall not give the details[1], for the second approximation for the eigenvalues is sufficient for most problems in application. It will only be noted that, in view of the formal expansion of $\lambda(\varkappa)$ in which the third coefficient $\lambda^{(3)}$ contains the term $\operatorname{tr} T^{(1)} S T^{(1)} S T^{(1)} P = \mathfrak{t}^{(1)} [S T^{(1)} \varphi, S^* T^{(1)*} \varphi]$ (assuming $m = 1$), we have to assume that

$$S T^{(1)} P \mathsf{X} \subset \mathsf{D}(\mathfrak{t}^{(1)}), \quad S^* T^{(1)*} P^* \mathsf{X} \subset \mathsf{D}(\mathfrak{t}^{(1)}), \tag{4.48}$$

in order to obtain the expansion of the eigenvalues up to the order $\varkappa^3$. Similarly, the expansion of the eigenvalues up to the order $\varkappa^4$ would require the existence of

$$T^{(1)} S T^{(1)} P, \quad T^{(1)*} S^* T^{(1)*} P^* \in \mathscr{B}(\mathsf{H}). \tag{4.49}$$

Example 4.15. Consider the differential operator $T(\varkappa)$ given by (4.29) with the boundary condition $u(0) = 0$ (see Example 4.7). To be definite, assume that the real-valued unperturbed potential $q(x)$ tends to zero as $x \to \infty$. Then it is well known that the unperturbed operator T has only negative eigenvalues, which are all isolated. Let λ be one of them and let φ be the associated eigenfunction (λ is simple). Since the associated eigenprojection P is the orthogonal projection on the one-dimensional subspace spanned by φ, Theorem 4.9 is applicable to λ if

$$\int_0^\infty q^{(1)}(x)\, |\varphi(x)|^2\, dx < \infty. \tag{4.50}$$

Similarly, Theorem 4.12 is applicable if (note that P and $T^{(1)}$ are selfadjoint here)

$$\int_0^\infty q^{(1)}(x)^2\, |\varphi(x)|^2\, dx < \infty. \tag{4.51}$$

In other words, we have the expansion of $\lambda(\varkappa)$ up to the order $\varkappa$ if (4.50) is satisfied, and up to the order $\varkappa^2$ if (4.51) is satisfied.

It is well known, however, that $\varphi(x)$ behaves like x as $x \to 0$ and like $\exp[-(-\lambda)^{1/2} x]$ as $x \to \infty$ [more precise behavior depends on $q(x)$]. Therefore,

[1] See V. Kramer [1], [2], where a detailed result is given in the special case when $H(\varkappa)$ is a selfadjoint Friedrichs family [that is, $\mathfrak{t}(\varkappa)$ is defined by $\mathfrak{t}(\varkappa)[u] = (K(\varkappa) u, u)$ from a given family $K(\varkappa)$ of symmetric operators].

(4.50) and (4.51) are very weak conditions on $q^{(1)}$. For example, (4.50) is satisfied if $q^{(1)}(x)$ is $O(x^{-3+\varepsilon})$ as $x \to 0$ and $O(x^n)$ as $x \to \infty$ with any n. Also it can be seen that the conditions (4.48) and (4.49) are satisfied under very general conditions on $q^{(1)}$; it suffices to note that $w = S T^{(1)} \varphi$ is the solution of the differential equation $-w'' + q(x) w - \lambda w = \psi(x)$ with the auxiliary condition $(w, \varphi) = 0$, where $\psi = (1 - P) T^{(1)} \varphi$ [that is, $\psi(x) = q^{(1)}(x) \varphi(x) - c \varphi(x)$ with c given by (4.50)].

§ 5. Spectral concentration

1. Unstable eigenvalues

In considering the perturbation of an isolated eigenvalue λ, we have assumed that λ is stable under the perturbation in question (see §§ 1.4, 2.3, 2.4, 3.5, 4.4). Some sufficient conditions for the stability have been given, but these are not always satisfied. It happens very often that a given eigenvalue λ is not stable; then the spectrum of the perturbed operator can behave in various ways and it would be difficult to give a general account of its behavior.

If we restrict ourselves to selfadjoint operators, the most common phenomenon with an unstable eigenvalue λ is the absorption of λ by the continuous spectrum: the perturbed operator has a continuous spectrum that covers an interval containing λ and has no eigenvalue near λ[1]. In such a case it does not make sense to speak of "the perturbation of the eigenvalue λ".

Nevertheless it is often possible to compute the formal series for the "perturbed eigenvalue $\lambda(\varkappa)$" according to the formulas given in Chapter VII, at least up to a certain order in $\varkappa$. The question arises, then, as to what such a series means.

It has been suggested that, although the spectrum of the perturbed operator is continuous, there is a concentration of the spectrum at the points represented by these *pseudo-eigenvalues* computed according to the formal series.

It is the purpose of this section to give a definition of "concentration" and to prove some theorems showing that such concentration does exist exactly where the pseudo-eigenvalues lie[2].

---

[1] There is an opposite phenomenon in which a continuous spectrum is changed into a discrete spectrum by perturbation, but we shall not consider this problem; see MASLOV [1], [3].

[2] Spectral concentration was considered by TITCHMARSH [3], [4], [5] in special cases of ordinary differential operators, where it is shown that the analytic continuation of the Green function, which is the kernel for the resolvent of the perturbed operator, has a pole close to the real axis (as a function of the resolvent parameter ζ), whereas the unperturbed kernel has a pole λ on the real axis. More abstract results were given by FRIEDRICHS and REJTO [1], CONLEY and REJTO [1]. T. KATO [3] contains a somewhat different formulation of the problem. Spectral concentration is closely related to the so-called weak quantization in quantum mechanics; in this connection cf. also BROWNELL [4].

2. Spectral concentration

Let H be a Hilbert space and $\{H_n\}$ a sequence of selfadjoint operators in H, with the associated spectral families $\{E_n(\lambda)\}$. We denote by $E_n(\mathrm{S})$ the spectral measure[1] constructed from $E_n(\lambda)$, where S varies over subsets of the real line R. (In what follows all subsets of R considered are assumed to be Borel sets.)

Let $\mathrm{S}_n \subset \mathrm{R}$, $n = 1, 2, \ldots$. We say that the spectrum of H_n is (asymptotically) concentrated on S_n if

$$E_n(\mathrm{S}_n) \underset{s}{\rightarrow} 1, \quad n \rightarrow \infty, \tag{5.1}$$

or, equivalently,

$$E_n(\mathrm{R} - \mathrm{S}_n) \underset{s}{\rightarrow} 0, \tag{5.2}$$

where $\mathrm{R} - \mathrm{S}_n$ is the complementary set of S_n with respect to R. Here $\mathrm{S}_n = \mathrm{S}$ may be independent of n; then we say that the spectrum of H_n is concentrated on S. It is obvious that if the spectrum of H_n is concentrated on S_n, it is concentrated on any S_n' such that $\mathrm{S}_n \subset \mathrm{S}_n'$.

Thus spectral concentration is an *asymptotic notion* related to a given sequence (or a family) $\{H_n\}$ of operators; it does not make sense to speak of spectral concentration for a single operator H (except in the trivial special case of a sequence $\{H_n\}$ in which all H_n are equal to H).

More generally, we shall say that the part of the spectrum of H_n in a subset T of R is (asymptotically) concentrated on S_n if

$$E_n(\mathrm{T} - \mathrm{S}_n) \underset{s}{\rightarrow} 0, \tag{5.3}$$

where $\mathrm{T} - \mathrm{S}_n = \mathrm{T} \cap (\mathrm{R} - \mathrm{S}_n)$ (the S_n need not be subsets of T). It is obvious that (5.2) implies (5.3) for any T because $E_n(\mathrm{T} - \mathrm{S}_n) \leq E_n(\mathrm{R} - \mathrm{S}_n)$. If, conversely, the parts of the spectrum of H_n in T and in R – T are concentrated on S_n, then the spectrum is concentrated on S_n, for $E_n(\mathrm{R} - \mathrm{S}_n) = E_n(\mathrm{T} - \mathrm{S}_n) + E_n(\mathrm{R} - \mathrm{T} - \mathrm{S}_n)$. For this reason it suffices, for most purposes, to consider concentration with respect to some given set T.

A basic result for spectral concentration is given by

Theorem 5.1. *Let $H_n \underset{s}{\rightarrow} H$ in the generalized sense, where H is selfadjoint. Let* $\mathrm{S} \subset \mathrm{R}$ *be an open set containing $\Sigma(H)$. Then the spectrum of H_n is asymptotically concentrated on* S. *Furthermore, we have $E_n(\mathrm{S} \cap \mathrm{I}) \underset{s}{\rightarrow} E(\mathrm{I})$ for any interval* I *if the end points of* I *are not eigenvalues of H (E denotes the spectral measure for H).*

[1] See VI-§ 5.1 and X-§ 1.2. Here we need only elementary properties of the spectral measure; in most cases it suffices to consider sets S which are unions of intervals.

Proof. The open set S is the union of at most a countable number of disjoint open intervals I_k, all of which are finite except possibly one or two. $\Sigma(H) \subset S$ implies that $\sum_k E(I_k) = E(S) = 1$, in which the series converges in the strong sense. Hence the span of the subspaces $E(I_k)$ H is dense in H. In order to prove (5.1), therefore, it suffices to show that $E_n(S)\, u \to u$ whenever $u \in E(I_k)$ H for some k.

First assume that $I_k = (a_k, b_k)$ is finite. Then a_k, b_k belong to $P(H)$, for they cannot belong to $\Sigma(H)$. Thus there are a'_k, b'_k such that $a_k < a'_k < b'_k < b_k$ and the two intervals $[a_k, a'_k]$, $[b'_k, b_k]$ are subsets of $P(H)$. On setting $I'_k = (a'_k, b'_k)$, we have $E(I_k) = E(I'_k)$ and so $E_n(I'_k)\, u = E_n(b'_k - 0)\, u - E_n(a'_k)\, u \to E(b'_k)\, u - E(a'_k)\, u = E(I'_k)\, u = E(I_k)\, u = u$ (see Theorem 1.15 and the attached footnote). Since $I'_k \subset S$, we have $E_n(I'_k) \leqq E_n(S)$ and so $E_n(S)\, u \to u$, as we wished to show.

When I_k is not a finite interval, we need a slight modification in the above proof. If, for example, $b_k = \infty$, we set $b'_k = \infty$; then $E_n(\infty)\, u = u = E(\infty)\, u$ and the above proof goes through.

The last assertion of the theorem follows from $E_n(S \cap I) = E_n(S)\, E_n(I)$ by noting that $E_n(I) \underset{s}{\to} E(I)$ in virtue of Theorem 1.15.

3. Pseudo-eigenvectors and spectral concentration

Theorem 5.1 shows that if $H_n \underset{s}{\to} H$ in the generalized sense, the spectrum of H_n is asymptotically concentrated on any neighborhood of $\Sigma(H)$. In particular let λ be an isolated eigenvalue of H with isolation distance d, and let $I = (\lambda - d/2, \lambda + d/2)$. Then the part of the spectrum of H_n in I is asymptotically concentrated on an arbitrarily small neighborhood of λ [see (5.3)].

We shall now sharpen this result by localizing the concentration to a smaller interval (or intervals) depending on n, under some additional assumptions.

Theorem 5.2. *Let $H_n \underset{s}{\to} H$ in the generalized sense, and let λ and I be as above. Let P be the eigenprojection of H for λ and let* $\dim P = m < \infty$. *Suppose there exist m sequences of "pseudo-eigenvalues" $\{\lambda_{jn}\}$ and "pseudo-eigenvectors" $\{\varphi_{jn}\}$ of H_n, $j = 1, \ldots, m$, such that*

$$\|(H_n - \lambda_{jn})\, \varphi_{jn}\| = \varepsilon_{jn} \to 0\,, \quad \varphi_{jn} \to \varphi_j\,, \quad n \to \infty\,, \tag{5.4}$$

where $\varphi_1, \ldots, \varphi_m$ form a basis of PH with $\|\varphi_j\| = 1$. Then the part of the spectrum of H_n in I is asymptotically concentrated on the union of m intervals $I_{jn} = (\lambda_{jn} - \alpha_n \varepsilon_{jn}, \lambda_{jn} + \alpha_n \varepsilon_{jn})$, where $\{\alpha_n\}$ is any sequence of positive numbers such that $\alpha_n \to \infty$.

Proof. (5.4) implies that

$$\varepsilon_{jn}^2 = \int_{-\infty}^{\infty} (\mu - \lambda_{jn})^2 \, d(E_n(\mu)\, \varphi_{jn}, \varphi_{jn}) \geqq$$
$$\geqq \alpha_n^2 \, \varepsilon_{jn}^2 \int_{\mu \notin I_{jn}} d(E_n(\mu)\, \varphi_{jn}, \varphi_{jn})$$
$$= \alpha_n^2 \, \varepsilon_{jn}^2 \, \|(1 - E_n(I_{jn}))\, \varphi_{jn}\|^2 .$$

Hence $\|(1 - E_n(I_{jn}))\, \varphi_{jn}\| \leqq 1/\alpha_n \to 0$, $n \to \infty$. Since $\|(1 - E_n(I_{jn})) \cdot (\varphi_{jn} - \varphi_j)\| \leqq \|\varphi_{jn} - \varphi_j\| \to 0$, it follows that

$$(1 - E_n(I_{jn}))\, \varphi_j \to 0 , \quad n \to \infty . \tag{5.5}$$

If we denote by I_n the union of $I_{1n}, \ldots, I_{mn}$, then $1 - E_n(I_n) \leqq 1 - E_n(I_{jn})$ and so (5.5) implies $(1 - E_n(I_n))\, \varphi_j \to 0$. Since the φ_j span $P\mathsf{H}$, it follows that

$$(1 - E_n(I_n))\, P \to 0 \quad \text{in norm} . \tag{5.6}$$

On the other hand we have

$$E_n(I)\,(1 - P) \underset{s}{\to} 0 , \tag{5.7}$$

since $E_n(I) \underset{s}{\to} E(I) = P$ by Theorem 1.15. Multiplying (5.6) by $E_n(I)$, (5.7) by $1 - E_n(I_n)$ and adding the results, we obtain $E_n(I - I_n) = E_n(I)\,(1 - E_n(I_n)) \underset{s}{\to} 0$, as we wished to prove.

Remark 5.3. We can arrange that $|I_{jn}| = 2\alpha_n \varepsilon_{jn} \to 0$ by choosing $\alpha_n \to \infty$ appropriately. Hence the part of the spectrum of H_n in I is asymptotically concentrated on a set I_n with measure $|I_n| \to 0$, and Theorem 5.2 strengthens Theorem 5.1. It should be noted that for any neighborhood $I' \subset I$ of λ, $I_n \subset I'$ is true for sufficiently large n. In fact, let I'' be a proper subinterval of I' containing λ; then $E_n(I'')\, \varphi_j \to P\, \varphi_j = \varphi_j$ so that $E_n(I_{jn} \cap I'')\, \varphi_j = E_n(I_{jn})\, E_n(I'')\, \varphi_j \to \varphi_j$ by (5.5). Since $|I_{jn}| \to 0$, this implies $I_{jn} \subset I'$, and hence $I_n \subset I'$, for sufficiently large n. In particular we must have $\lambda_{jn} \to \lambda$ for every j.

4. Asymptotic expansions

We now consider a family $H(\varkappa)$ of selfadjoint operators formally given by $H + \varkappa H^{(1)}$, where H is selfadjoint and $H^{(1)}$ is symmetric. More precisely, we assume the conditions ii), iii) of § 2.1, namely that $\mathsf{D} = \mathsf{D}(H) \cap \mathsf{D}(H^{(1)})$ is a core of H and that $H + \varkappa H^{(1)}$ (with domain D) has a selfadjoint extension $H(\varkappa)$ for $0 < \varkappa \leqq 1$. Then iv) of § 2.1 is automatically satisfied, for $\Delta_b \cap \mathrm{P}(H)$ contains all nonreal complex numbers.

Under these conditions, it has been shown that $H(\varkappa) \underset{s}{\to} H$, $\varkappa \to 0$, in the generalized sense (see loc. cit.)[1]. It follows from Theorem 5.1 that the spectrum of $H(\varkappa)$ is asymptotically concentrated on any neighborhood of $\Sigma(H)$. If λ is an isolated eigenvalue of H with the isolation distance d and the eigenprojection P, then the part of the spectrum of $H(\varkappa)$ in $\mathrm{I} = (\lambda - d/2, \lambda + d/2)$ is asymptotically concentrated on any small neighborhood of λ, with

$$E(\mathrm{I}, \varkappa) \underset{s}{\to} P, \quad \varkappa \to 0. \tag{5.8}$$

We shall now sharpen this result by making specific assumptions on the unperturbed eigenspace $P\mathsf{H}$.

Theorem 5.4. *Let* $\dim P = m < \infty$ *and* $P\mathsf{H} \subset \mathsf{D}(H^{(1)})$. *Let* $\mu_j^{(1)}, j = 1, \ldots, m$, *be the repeated eigenvalues of the symmetric operator* $PH^{(1)}P$ *in the m-dimensional subspace* $P\mathsf{H}$. *Then the part of the spectrum of* $H(\varkappa)$ *in* I *is asymptotically concentrated on the union of m intervals with the centers* $\lambda + \varkappa\mu_j^{(1)}$ *and with the width* $o(\varkappa)$.

Remark 5.5. The result of Theorem 5.4 may be interpreted to mean that $H(\varkappa)$ has m pseudo-eigenvalues $\lambda + \varkappa\mu_j^{(1)} + o(\varkappa)$ (cf. the corresponding Theorem 2.6 when λ is assumed to be stable). Of course such a statement is not fully justified unless the corresponding pseudo-eigenvectors are constructed, which will be done in the course of the proof.

Proof. Let $\{\varphi_j\}$ be an orthonormal basis of $P\mathsf{H}$ consisting of eigenvectors of $PH^{(1)}P$:

$$PH^{(1)}P\varphi_j = \mu_j^{(1)}\varphi_j, \quad j = 1, \ldots, m. \tag{5.9}$$

We shall construct pseudo-eigenvectors of $H(\varkappa)$ in the form

$$\varphi_j(\varkappa) = \varphi_j + \varkappa\varphi_j^{(1)}(\varkappa), \quad \|\varphi_j^{(1)}(\varkappa)\| \leq M, \tag{5.10}$$

so that as $\varkappa \to 0$

$$\varkappa^{-1}\|(H(\varkappa) - \lambda - \varkappa\mu_j^{(1)})\varphi_j(\varkappa)\| < \varepsilon(\varkappa) \to 0. \tag{5.11}$$

Then an application of Theorem 5.2 will lead to the desired result (with the discrete parameter n replaced by the continuous one $\varkappa$); it suffices to take the interval $\mathrm{I}_{j\varkappa}$ with center $\lambda + \varkappa\mu_j^{(1)}$ and width $2\varkappa\alpha(\varkappa)\varepsilon(\varkappa)$, where $\alpha(\varkappa) \to \infty$ and $\alpha(\varkappa)\varepsilon(\varkappa) \to 0$.

Let S be the reduced resolvent of H for λ (see III-§ 6.5). S has the properties $S \in \mathscr{B}(\mathsf{H})$, $(H - \lambda)S = 1 - P$ [see III-(6.34)]. Let $\varepsilon > 0$. Since $-SH^{(1)}\varphi_j \in \mathsf{D}(H)$ and D is a core of H, there is a $\varphi_j' \in \mathsf{D}$ such that $\|\varphi_j'\| \leq \|SH^{(1)}\varphi_j\| + 1 \leq M$ and $\|(H - \lambda)(\varphi_j' + SH^{(1)}\varphi_j)\| < \varepsilon/2$. Since $(H - \lambda)SH^{(1)}\varphi_j = (1 - P)H^{(1)}\varphi_j = H^{(1)}\varphi_j - \mu_j^{(1)}\varphi_j$, we have [note

[1] The same result holds if we replace the above assumptions by those of § 4.1, with the additional assumption that $\mathfrak{t} = \mathfrak{h}$ and $\mathfrak{t}^{(1)} = \mathfrak{h}^{(1)}$ are symmetric [so that $T = H$ and $T(\varkappa) = H(\varkappa)$ are selfadjoint].

that $(H - \lambda)\,\varphi_j = 0$]

$$\varkappa^{-1}\,\|(H(\varkappa) - \lambda - \varkappa\,\mu_j^{(1)})\,(\varphi_j + \varkappa\,\varphi_j')\| = \varkappa^{-1}\,\|(H - \lambda)\,\varphi_j + \\ + \varkappa(H - \lambda)\,\varphi_j' + \varkappa(H^{(1)} - \mu_j^{(1)})\,\varphi_j + \varkappa^2(H^{(1)} - \mu_j^{(1)})\,\varphi_j'\| \leqq \\ \leqq (\varepsilon/2) + \varkappa\|(H^{(1)} - \mu_j^{(1)})\,\varphi_j'\|\,.$$

Thus there is a $\delta > 0$ such that

$$(5.12)\qquad \varkappa^{-1}\|(H(\varkappa) - \lambda - \varkappa\,\mu_j^{(1)})\,(\varphi_j + \varkappa\,\varphi_j')\| < \varepsilon \quad \text{for} \quad 0 < \varkappa < \delta\,;$$

note that φ_j' also depends on ε.

Now let $\varepsilon_1 > \varepsilon_2 > \cdots$, $\varepsilon_n \to 0$. Let δ_n and φ_{jn}' be the δ and φ_j' of (5.12) corresponding to $\varepsilon = \varepsilon_n$. We may assume that $\delta_1 > \delta_2 > \cdots$, $\delta_n \to 0$. Set

$$(5.13)\qquad \varphi_j(\varkappa) = \varphi_j + \varkappa\,\varphi_{jn}' \quad \text{for} \quad \delta_{n+1} \leqq \varkappa < \delta_n\,.$$

Then (5.11) is satisfied, with

$$\varepsilon(\varkappa) = \varepsilon_n \quad \text{for} \quad \delta_{n+1} \leqq \varkappa < \delta_n\,.$$

This completes the proof.

Remark 5.6. Theorem 5.4 shows that the pseudo-eigenvalues admit asymptotic expansions to the first order in $\varkappa$, with the spectral concentration on the union of m intervals of width $o(\varkappa)$. Similarly, it can be shown that the pseudo-eigenvalues admit asymptotic expansions up to $\varkappa^n$, with the spectral concentration on the union of m intervals of width $o(\varkappa^n)$, if the following condition is satisfied[1]: all possible expressions of the form $X_1 \ldots X_k P$ are defined on the whole of H, where $k \leqq n$ and each X_j is equal to either S or $T^{(1)} S$. It is easily seen that under this condition the formal expansion of the total projection $P(\varkappa)$ given in II-§ 2.1 makes sense up to the order $\varkappa^n$.

Example 5.7. Spectral concentration occurs for the operator $T(\varkappa) = H(\varkappa)$ of Example 1.17 if $\varkappa$ is real, where we now write H, $H^{(1)}$ for T, $T^{(1)}$ (in the notation of Example 2.5, this corresponds to the case $e^{i\theta} = \pm 1$). In fact, we know that the basic assumptions of this paragraph are satisfied (see loc. cit.). If $x\,f(x) \in \mathsf{L}^2$, then the assumption of Theorem 5.4 is satisfied, so that we have a spectral concentration for $H(\varkappa)$ near $\lambda = -1$, with the pseudo-eigenvalue $\lambda + \varkappa \int x|f(x)|^2\,dx$ and with the width $o(\varkappa)$. It can be shown that if $f(x)$ tends to 0 rapidly as $x \to \pm\infty$, there is a concentration of higher order.

Example 5.8. *(Stark effect)* A typical example of spectral concentration is furnished by the operator of the Stark effect. If we restrict ourselves to the simplest case of hydrogen-like atoms, the operator to be considered is defined in $\mathsf{H} = \mathsf{L}^2(\mathsf{R}^3)$ *formally* by

$$(5.14)\qquad -\Delta - \frac{2}{r} + \varkappa\,x_1\,.$$

The unperturbed operator $H = -\Delta - 2/r$ is the Schrödinger operator for the hydrogen atom and has been studied in V-§ 5.3; it is known that H is selfadjoint

[1] This result is proved by Riddell [1,2].

and C_0^∞ is a core of H. It can be shown[1] that (5.14) for real $\varkappa \neq 0$ is also essentially selfadjoint if restricted to have domain C_0^∞; we shall denote by $H(\varkappa)$ the unique selfadjoint extension of this restriction. Then $D = D(H(\varkappa)) \cap D(H)$ contains C_0^∞ and so it is a core of H as well as of $H(\varkappa)$. Thus the fundamental assumptions are satisfied. Now it can be shown that for any eigenvalue λ of H (which is necessarily negative), the condition stated in Remark 5.6 is satisfied for any positive integer n. It follows[2] that the pseudo-eigenvalues have asymptotic expansions up to any order in $\varkappa$ (thus coinciding exactly with the formal perturbation series), with the spectral concentration on the union of a finite number of intervals of width $o(\varkappa^n)$ for any n.

Chapter Nine

Perturbation theory for semigroups of operators

The subject of this chapter originates in the so-called time-dependent perturbation theory in quantum mechanics, in which the main question is the perturbation of the unitary group generated by a given Hamiltonian due to a small change in the latter. This problem is generalized in a natural way to perturbation theory for semigroups, which is no less important in applications.

The chapter begins with a brief account of basic results in the generation theory of semigroups of operators. Only restricted types of semigroups — quasi-bounded semigroups — are considered here. But the importance of the holomorphic semigroups, which are special cases of quasi-bounded semigroups, is emphasized.

In the following section various problems of perturbation theory are considered. It will be seen that holomorphic semigroups are rather well-behaved with respect to perturbation, whereas general quasi-bounded semigroups are not necessarily so. In the last section, an approximation theory of semigroups by discrete semigroups is presented. This theory serves as a basis in the approximation of some differential equations by difference equations.

§ 1. One-parameter semigroups and groups of operators

1. The problem

In this chapter we consider the time-dependent perturbation theory. In applications to quantum mechanics, it is concerned with the solution of the time-dependent Schrödinger equation

$$du/dt = -i\,H u\,, \tag{1.1}$$

where the unknown $u = u(t)$ is a vector in a Hilbert space H and H is a selfadjoint operator in H. If $H = H(\varkappa)$ depends on a small parameter $\varkappa$, the question arises how the solution of (1.1) depends on $\varkappa$.

[1] See STUMMEL [1], IKEBE and KATO [1].
[2] For details see RIDDELL [1].

(1.1) is a particular case of equations of the form

$$du/dt = -Tu \tag{1.2}$$

in a Banach space X. Here T is a linear operator in X and t is usually restricted to the semi-infinite interval $0 < t < \infty$, with the initial condition given at $t = 0$. The solution of (1.2) is formally given by $u = u(t) = e^{-tT} u(0)$. Thus our first problem will be to study how the exponential function e^{-tT} of T can be defined; then we have to investigate how e^{-tT} changes with $\varkappa$ when $T = T(\varkappa)$ depends on a parameter $\varkappa$.

e^{-tT} is a particular "function" of T. So far we have considered perturbation theory for several functions of T. The simplest one among them was the resolvent $(T - \zeta)^{-1}$, which was discussed in detail in preceding chapters. When $T = H$ is a selfadjoint operator in a Hilbert space, we also considered such functions of H as $|H|$, $|H|^{1/2}$, and $E(\lambda)$, the spectral family associated with H. The importance of the function e^{-tT} lies in its relation to the differential equation (1.2), which occurs in various fields of application. In reference to the basic identity $e^{-(s+t)T} = e^{-sT} e^{-tT}$, the family $\{e^{-tT}\}_{t>0}$ is called a *one-parameter semigroup of operators*. If t is allowed to vary over $-\infty < t < \infty$, $\{e^{-tT}\}$ is a *one-parameter group*.

2. Definition of the exponential function

Let X be a Banach space. If $T \in \mathscr{B}(\mathsf{X})$, the operator e^{-tT} can be defined simply by the Taylor series

$$e^{-tT} = \sum_{n=0}^{\infty} \frac{(-1)^n}{n!} t^n T^n , \tag{1.3}$$

which converges absolutely for any complex number t (see Example I-4.4). Thus e^{-tT} also belongs to $\mathscr{B}(\mathsf{X})$ and is a holomorphic function of t in the whole t-plane (entire function). The group property

$$e^{-(s+t)T} = e^{-sT} e^{-tT} \tag{1.4}$$

can be verified directly from (1.3). We have also

$$\frac{d}{dt} e^{-tT} = -T e^{-tT} = -e^{-tT} T , \tag{1.5}$$

where the differentiation may be taken in the sense of the norm. Thus $u(t) = e^{-tT} u_0$ is a solution of the differential equation (1.2) for any $u_0 \in \mathsf{X}$.

If T is unbounded, however, it is difficult to define e^{-tT} by (1.3), for the domain of T^n becomes narrower for increasing n. The formula $e^{-tT} = \lim_{n\to\infty} \left(1 - \frac{t}{n} T\right)^n$, suggested by the numerical exponential function, is not useful for the same reason. But a slight modification of this formula, namely,

(1.6) $$e^{-tT} = \lim_{n\to\infty} \left(1 + \frac{t}{n} T\right)^{-n}$$

is found to be useful. In fact $\left(1 + \frac{t}{n} T\right)^{-1}$ is the resolvent of $-T$, apart from a constant factor, and it can be iterated even when T is unbounded.

The following is a sufficient condition in order that the limit (1.6) exist in an appropriate sense.

i) $T \in \mathscr{C}(\mathsf{X})$ with domain $\mathsf{D}(T)$ dense in X.

ii) The negative real axis belongs to the resolvent set of T, and the resolvent $(T + \xi)^{-1}$ satisfies the inequality

(1.7) $$\|(T + \xi)^{-1}\| \leqq 1/\xi\,, \quad \xi > 0\,.$$

To see this, we note first that (1.7) implies

(1.8) $$\|(1 + \alpha\, T)^{-1}\| \leqq 1\,, \quad \alpha \geqq 0\,.$$

Set

(1.9) $$V_n(t) = \left(1 + \frac{t}{n} T\right)^{-n}, \quad t \geqq 0\,, \quad n = 1, 2, \ldots\,.$$

It follows from (1.8) that $\|V_n(t)\| \leqq 1$, so that the $V_n(t)$ are uniformly bounded. Furthermore, $V_n(t)$ is holomorphic in t for $t > 0$, for $(T + \xi)^{-1}$ is holomorphic in $\xi > 0$; in particular

(1.10) $$V_n'(t) = \frac{d}{dt} V_n(t) = -T\left(1 + \frac{t}{n} T\right)^{-n-1} \in \mathscr{B}(\mathsf{X})\,, \quad t > 0\,.$$

$V_n(t)$ is not necessarily holomorphic at $t = 0$, but it is strongly continuous:

(1.11) $$V_n(t) \underset{s}{\to} V_n(0) = 1 \quad \text{as} \quad t \searrow 0\,;$$

this follows from

(1.12) $$(1 + \alpha\, T)^{-1} \underset{s}{\to} 1\,, \quad \alpha \searrow 0\,,$$

which can be proved in the same way as in Problem V-3.33.

To prove the existence of $\lim V_n(t)$, we estimate $V_n(t)\, u - V_m(t)\, u$. For this we note that, in virtue of (1.11),

(1.13) $$V_n(t)\, u - V_m(t)\, u = \lim_{\varepsilon \searrow 0} \int_{\varepsilon}^{t-\varepsilon} \frac{d}{ds} [V_m(t - s)\, V_n(s)\, u]\, ds$$

$$= \lim_{\varepsilon \searrow 0} \int_{\varepsilon}^{t-\varepsilon} [-V_m'(t - s)\, V_n(s)\, u + V_m(t - s)\, V_n'(s)\, u]\, ds\,.$$

The integrand can be calculated easily by using (1.10); the result is

(1.14) $V_n(t)\,u - V_m(t)\,u$

$$= \lim_{\varepsilon \searrow 0} \int\limits_{\varepsilon}^{t-\varepsilon} \left(\frac{s}{n} - \frac{t-s}{m}\right) T^2 \left(1 + \frac{t-s}{m} T\right)^{-m-1} \left(1 + \frac{s}{n} T\right)^{-n-1} u\, ds.$$

It is not easy to estimate (1.14) for a general u. But it is simple if $u \in \mathsf{D}(T^2)$. Since the resolvent of T commutes with T (in the sense of III-§ 5.6), (1.14) gives then

(1.15) $V_n(t)\,u - V_m(t)\,u$

$$= \int\limits_0^t \left(\frac{s}{n} - \frac{t-s}{m}\right) \left(1 + \frac{t-s}{m} T\right)^{-m-1} \left(1 + \frac{s}{n} T\right)^{-n-1} T^2 u\, ds;$$

note that the integrand is continuous for $0 \leqq s \leqq t$ by (1.11). It follows by (1.8) that

$$(1.16) \quad \|V_n(t)\,u - V_m(t)\,u\| \leqq \|T^2 u\| \int\limits_0^t \left(\frac{s}{n} + \frac{t-s}{m}\right) ds$$

$$= \frac{t^2}{2} \left(\frac{1}{n} + \frac{1}{m}\right) \|T^2 u\|\,.$$

Thus the $V_n(t)\,u$ form a Cauchy sequence and $\lim V_n(t)\,u$ exists, uniformly in t in any finite interval, provided that $u \in \mathsf{D}(T^2)$. But $\mathsf{D}(T^2)$ is dense in X, for $\mathsf{D}(T^2) = (T + \xi)^{-1}\,\mathsf{D}(T)$ for $\xi > 0$ and $(T + \xi)^{-1}$ has range $\mathsf{D}(T)$ which is dense in X (see Problem III-2.9). In view of the uniform boundedness of the $V_n(t)$, it follows that

$$(1.17) \qquad U(t) = \operatorname*{s-lim}_{n\to\infty} V_n(t) = \operatorname*{s-lim}_{n\to\infty} \left(1 + \frac{t}{n} T\right)^{-n}, \quad t \geqq 0,$$

exists (see Lemma III-3.5). In the following paragraph we shall show that $U(t)$ has the properties expected of the exponential function, and we shall *define* e^{-tT} *as equal to* $U(t)$.

Problem 1.1. $\|U(t)\,u - V_n(t)\,u\| \leqq \frac{t^2}{2n} \|T^2 u\|$, $u \in \mathsf{D}(T^2)$.

Problem 1.2. The strong convergence (1.17) is uniform in t in each finite interval [that is, $V_n(t)\,u \to U(t)\,u$ uniformly in t in each finite interval for each fixed $u \in \mathsf{X}$].

Problem 1.3. $\|T(T + \xi)^{-1}\| \leqq 2$ if T satisfies (1.7). Compare Problem V-3.32.

3. Properties of the exponential function

Since $V_n(t)\,u \to U(t)\,u$ uniformly in t in any finite interval (see Problem 1.2) and $V_n(t)\,u$ is continuous in t, $U(t)\,u$ is continuous in t. In other words, $U(t)$ is *strongly continuous* for $t \geqq 0$. Furthermore, we have

$$(1.18) \qquad \|U(t)\| \leqq 1\,, \quad U(0) = 1\,,$$

since $\|V_n(t)\| \leqq 1$ and $V_n(0) = 1$.

Now it follows from (1.10) that

$$(1.19)\quad V_n'(t) = -T\left(1+\frac{t}{n}T\right)^{-1} V_n(t) = -V_n(t)\,T\left(1+\frac{t}{n}T\right)^{-1}$$
$$= -T\,V_n(t)\left(1+\frac{t}{n}T\right)^{-1}.$$

But

$$(1.20)\quad T\left(1+\frac{t}{n}T\right)^{-1} u = \left(1+\frac{t}{n}T\right)^{-1} T u \to T u \quad \text{for} \quad u \in \mathsf{D}(T)$$

by (1.11). Hence the third member of (1.19), when applied to u, tends to[1] $-U(t)\,Tu$. Since the factor after T of the fourth member, when applied to u, tends to $U(t)\,u$ for a similar reason, it follows from the closedness of T that $T\,U(t)\,u$ exists and equals $U(t)\,Tu$. In other words T commutes with $U(t)$:

$$(1.21)\qquad T\,U(t) \supset U(t)\,T\,.$$

It follows also from (1.10) and (1.11) that

$$(1.22)\quad V_n(t)\,u - u = -\int_0^t \left(1+\frac{s}{n}T\right)^{-n-1} Tu\,ds\,,\quad u \in \mathsf{D}(T)\,.$$

Since $\left(1+\frac{t}{n}T\right)^{-n-1} = \left(1+\frac{t}{n}T\right)^{-1} V_n(t) \underset{s}{\to} U(t)$ uniformly in t in each finite interval, we can go to the limit $n \to \infty$ under the integral sign in (1.22), obtaining

$$(1.23)\qquad U(t)\,u - u = -\int_0^t U(s)\,Tu\,ds\,,\quad u \in \mathsf{D}(T)\,.$$

Since $U(s)\,Tu$ is continuous in s, (1.23) shows that $U(t)\,u$ is differentiable in t if $u \in \mathsf{D}(T)$, with

$$(1.24)\quad \frac{d}{dt}\,U(t)\,u = -U(t)\,Tu = -T\,U(t)\,u\,,\quad t \geqq 0\,,\quad u \in \mathsf{D}(T)\,,$$

the second equality being a consequence of (1.21). Thus $u(t) = U(t)\,u_0$ is a solution of the differential equation (1.2) with the initial value $u(0) = u_0$, *provided the initial value u_0 belongs to* $\mathsf{D}(T)$.

This solution is unique. In fact, let $u(t)$ be any solution of (1.2); by this we mean that $u(t)$ is continuous for $t \geqq 0$, the *strong derivative* $du(t)/dt$ exists for[2] $t > 0$, $u(t) \in \mathsf{D}(T)$ for $t > 0$ and (1.2) holds true. Then (see Lemma III-3.11)

$$(1.25)\quad \frac{d}{ds}\,U(t-s)\,u(s) = -U'(t-s)\,u(s) + U(t-s)\,u'(s)$$
$$= U(t-s)\,Tu(s) - U(t-s)\,Tu(s) = 0\,,\quad 0 < s \leqq t\,,$$

---

[1] Here and in what follows we frequently use, without further comments, the lemma that $A_n \underset{s}{\to} A$ and $B_n \underset{s}{\to} B$ imply $A_n B_n \underset{s}{\to} AB$ (Lemma III-3.8).

[2] In the initial value problem it is customary not to require the differentiability at $t = 0$ of the solution.

by (1.24). Thus $U(t-s)\,u(s)$ is constant for $0 \leqq s \leqq t$ (see Lemma III-1.36). On setting $s = t$ and $s = 0$, we have

$$u(t) = U(t-s)\,u(s) = U(t)\,u(0)\,, \quad 0 \leqq s \leqq t\,. \tag{1.26}$$

Applying (1.26) to the solution $u(t) = U(t)\,u_0$, we obtain $U(t)\,u_0 = U(t-s)\,u(s) = U(t-s)\,U(s)\,u_0$. Since this is true for all $u_0 \in \mathsf{D}(T)$, we have $U(t) = U(t-s)\,U(s)$. This may be written

$$U(t+s) = U(t)\,U(s)\,, \quad s, t \geqq 0\,. \tag{1.27}$$

Thus $\{U(t)\}_{t \geqq 0}$ forms a one-parameter semigroup. Since $\|U(t)\| \leqq 1$, $\{U(t)\}$ is called a *contraction semigroup*. $-T$ is called *the infinitesimal generator* (or simply the generator) of this semigroup. We write $U(t) = e^{-tT}$.

Finally we shall show that *different generators generate different semigroups*. For this it suffices to give a formula expressing T in terms of $U(t)$. Such a formula is

$$(T + \zeta)^{-1} = \int\limits_0^\infty e^{-\zeta t}\,U(t)\,dt\,, \quad \mathrm{Re}\,\zeta > 0\,, \tag{1.28}$$

which shows that the resolvent $(T+\zeta)^{-1}$ of T is the *Laplace transform* of the semigroup $U(t)$. Note that the integral on the right of (1.28) is an improper Riemann integral, defined as the limit as $\tau \to \infty$ (which exists in the sense of the norm) of the *strong Riemann integral* $\int\limits_0^\tau$ [which is defined as an operator $A(\tau)$ such that $A(\tau)\,u = \int\limits_0^\tau e^{-\zeta t}\,U(t)\,u\,dt$ for every $u \in \mathsf{X}$, the last integral being defined because the integrand is a continuous function of t]; see III-§ 3.1.

To prove (1.28), let $u \in \mathsf{D}(T)$. Then $dU(t)\,u/dt = -U(t)\,Tu$ and $(d/dt)\,e^{-\zeta t}\,U(t)\,u = -e^{-\zeta t}\,U(t)\,(T+\zeta)\,u$. Integration of this equality gives $u = \int\limits_0^\infty e^{-\zeta t}\,U(t)\,(T+\zeta)\,u\,dt$. With $v = (T+\zeta)\,u$, this becomes $(T+\zeta)^{-1}\,v = \int\limits_0^\infty e^{-\zeta t}\,U(t)\,v\,dt$. If ζ is real and $\zeta > 0$, then $\zeta \in \mathrm{P}(-T)$ and so v varies over X when u varies over $\mathsf{D}(T)$. Hence we obtain (1.28) if $\zeta > 0$. But the right member of (1.28) is holomorphic in ζ for $\mathrm{Re}\,\zeta > 0$, as is easily seen by noting that $\|U(t)\| \leqq 1$. Thus (1.28) must be true for $\mathrm{Re}\,\zeta > 0$ (see Theorem III-6.7).

In particular the half-plane $\mathrm{Re}\,\zeta > 0$ belongs to $\mathrm{P}(-T)$ and

$$\|(T+\zeta)^{-1}\| \leqq \int\limits_0^\infty |e^{-\zeta t}|\,dt = 1/\mathrm{Re}\,\zeta\,, \quad \mathrm{Re}\,\zeta > 0\,, \tag{1.29}$$

although we started from the assumption ii) which means that (1.29) is true for real $\zeta > 0$. Of course (1.29) can be deduced directly from ii).

Remark 1.4. We have constructed a strongly continuous, contraction semigroup $U(t)$ starting from a given generator $-T$. If, conversely, a strongly continuous, contraction semigroup $U(t)$ is given, we can define the generator $-T$ of $U(t)$ by (1.28) and show that T satisfies the conditions i), ii) and that the semigroup generated by $-T$ coincides with $U(t)$. We shall not give a proof of this part of semigroup theory, though it is not difficult[1]. We shall be content with the following remark.

Suppose T satisfies i), ii) and suppose there is an operator-valued function $V(t)$ such that $\|V(t)\|$ is bounded and $\int_0^\infty e^{-\xi t} V(t)\, dt = (T+\xi)^{-1}$ for all $\xi > 0$. Then $V(t) = U(t) = e^{-tT}$.

To prove this, set $W(t) = V(t) - U(t)$. Then $\int_0^\infty e^{-\xi t} W(t)\, dt = 0$ for all $\xi > 0$. Writing $e^{-t} = s$, we see that $\int_0^1 s^{\xi-1} (W(\log s^{-1})\, u, f)\, ds = 0$ for all $\xi > 0$ and for all $u \in \mathsf{X}$, $f \in \mathsf{X}^*$. Thus the continuous function $s(W(\log s^{-1})\, u, f)$ is orthogonal to all s^n, $n = 0, 1, 2, \ldots$, and so must vanish identically. This proves that $W(t) = 0$.

Remark 1.5. We have seen above that $e^{-tT} u$ is strongly differentiable in t if $u \in \mathsf{D}(T)$. The converse is true in the following strengthened form: *if $e^{-tT} u$ is weakly differentiable at $t = 0$ with the weak derivative v, then $u \in \mathsf{D}(T)$ and $-Tu = v$.* To prove this, we note that $h^{-1}(U(t+h)\, u - - U(t)\, u) = U(t)\, h^{-1}(U(h)\, u - u)$ has a weak limit $U(t)\, v$ as $h \searrow 0$. Hence $e^{-\xi t}(U(t)\, u, f)$ has the *right* derivative $e^{-\xi t}(U(t)(v - \xi u), f)$ for every $f \in \mathsf{X}^*$. Since this derivative is continuous in t, we have on integration $(u, f) = -\int_0^\infty e^{-\xi t}(U(t)(v - \xi u), f)\, dt$ and so $u = -\int_0^\infty e^{-\xi t} U(t) \cdot \cdot (v - \xi u)\, dt = -(T+\xi)^{-1}(v - \xi u)$. This shows that $u \in \mathsf{D}(T)$ and that $Tu + \xi u = -v + \xi u$, that is, $Tu = -v$.

Example 1.6. If $\mathsf{X} = \mathsf{H}$ is a Hilbert space and $T = iH$ with H a selfadjoint operator, both T and $-T$ satisfy the conditions i) and ii). Hence $U(t) = e^{-tT}$ is defined for $-\infty < t < +\infty$ and satisfies (1.24). It follows as above that (1.27) is satisfied for all real s, t, positive, negative or zero. In particular $U(t)\, U(-t) = U(0) = 1$, so that $\|U(t)\, u\| = \|u\|$ and $U(t)$ is isometric. Since $U(t)^{-1} = U(-t) \in \in \mathscr{B}(\mathsf{H})$, $U(t)$ is even unitary: $U(t) = e^{-itH}$ forms a *unitary group*.

Example 1.7. Let $\mathsf{X} = L^p(0, \infty)$, $1 \leq p < \infty$, and $T = d/dx$ with the boundary condition $u(0) = 0$. For $\xi > 0$, $(T+\xi)^{-1}$ is the integral operator given by (see Problem III-6.9)

$$(T+\xi)^{-1}\, u(y) = \int_0^y e^{-\xi(y-x)}\, u(x)\, dx\,. \tag{1.30}$$

Since $\int_x^\infty e^{-\xi(y-x)}\, dy = 1/\xi$ and $\int_0^x e^{-\xi(y-x)}\, dx \leq 1/\xi$, we have $\|(T+\xi)^{-1}\| \leq 1/\xi$ (see Example III-2.11). Thus the conditions i), ii) are satisfied and $-T$ generates a

[1] For this result and for more details on semigroup theory, see HILLE and PHILLIPS [1], YOSIDA [1].

semigroup $U(t) = e^{-tT}$. This semigroup is given by $U(t)\,u(x) = u(x - t)$ for $x > t$ and $U(t)\,u(x) = 0$ for $x < t$. This may be seen from the remark above, for

$$\int_0^\infty e^{-\xi t}\, U(t)\, u(x)\, dt = \int_0^x e^{-\xi t}\, u(x - t)\, dt$$

$$= \int_0^x e^{-\xi(x-s)}\, u(s)\, ds = (T + \xi)^{-1}\, u(x)\,.$$

Example 1.8. In the preceding example replace T by another operator T defined by $T = -d/dx$ without any boundary condition. Then (see Problem III-6.9)

$$(1.31) \qquad (T + \xi)^{-1}\, u(y) = \int_y^\infty e^{-\xi(x-y)}\, u(x)\, dx$$

and it can be shown as above that T satisfies i), ii). The semigroup $U(t) = e^{-tT}$ is given by $U(t)\,u(x) = u(x + t)$.

Problem 1.9. Let $\mathsf{X} = \mathsf{L}^p(-\infty, +\infty)$ and $T = d/dx$. Both T and $-T$ satisfy i), ii) so that T generates a group $U(t) = e^{-tT}$, which is given by $U(t)\,u(x) = u(x - t)$ (see Problem III-6.10).

Problem 1.10. In the examples and problems given above, construct e^{-tT} directly according to the formula (1.17).

Problem 1.11. If $Tu = \lambda u$, then $e^{-tT} u = e^{-\lambda t} u$, $t \geqq 0$.

Problem 1.12. In Example 1.6 we have

$$(1.32) \qquad e^{-itH} = \int_{-\infty}^{\infty} e^{-it\lambda}\, dE(\lambda)\,,$$

where $\{E(\lambda)\}$ is the spectral family for H.

4. Bounded and quasi-bounded semigroups

In order that an operator $-T$ generate a semigroup $U(t)$ of bounded linear operators, it is not necessary that T satisfy the conditions i), ii) stated in par. 2. For example, the inequality (1.7) can be replaced by a weaker condition

$$(1.33) \qquad \|(T + \xi)^{-k}\| \leqq M/\xi^k\,, \quad \xi > 0\,, \quad k = 1, 2, 3, \ldots,$$

where M is a constant independent of ξ and k.

In fact, (1.33) implies

$$(1.34) \qquad \|(1 + \alpha T)^{-k}\| \leqq M\,, \quad \alpha \geqq 0\,,$$

so that the operators $V_n(t)$ defined by (1.9) are uniformly bounded by $\|V_n(t)\| \leqq M$. Thus the construction of $U(t) = \text{s-lim}\, V_n(t)$ can be carried out exactly as before; the only modification needed is to put the factor M^2 into the right member of (1.16). The operator $e^{-tT} = U(t)$ is again strongly continuous for $t \geqq 0$ and satisfies

$$(1.35) \qquad \|U(t)\| \leqq M\,, \quad U(0) = 1\,.$$

The semigroup property of $U(t)$ and all the other results of the preceding paragraphs can be proved quite in the same way. We shall call $U(t)$ a *bounded semigroup*.

The condition ii) on T can further be relaxed as follows.

ii') Let the semi-infinite interval $\xi > \beta$ belong to the resolvent set of $-T$ and let

$$(1.36) \quad \|(T+\xi)^{-k}\| \leq M(\xi-\beta)^{-k}, \quad \xi > \beta, \quad k = 1, 2, 3, \ldots .$$

Then $T_1 = T + \beta$ satisfies the assumptions sated above, so that the bounded semigroup $U_1(t) = e^{-tT_1}$ is defined. If we set $U(t) = e^{\beta t}\, U_1(t)$, it is easily verified that $U(t)$ has all the properties stated above, with the following modification of (1.35):

$$(1.37) \qquad \|U(t)\| \leq M e^{\beta t}, \quad U(0) = 1 .$$

We define by $e^{-tT} = U(t)$ the semigroup generated by $-T$. Here $\|U(t)\|$ need not be bounded as $t \to \infty$. We shall call $U(t)$ a *quasi-bounded semigroup*. The set of all operators T satisfying the conditions i) and ii') will be denoted by[1,2] $\mathscr{G}(M, \beta)$. $-T$ is the infinitesimal generator of a contraction semigroup if and only if $T \in \mathscr{G}(1, 0)$.

Problem 1.13. $M \geq 1$ if $\mathscr{G}(M, \beta)$ is not empty[3].

Problem 1.14. $T \in \mathscr{G}(M, \beta)$ implies $T - \alpha \in \mathscr{G}(M, \alpha + \beta)$ and $e^{-t(T-\alpha)} = e^{\alpha t} e^{-tT}$.

Problem 1.15. If $T \in \mathscr{G}(M, \beta)$, we have

$$(1.38) \qquad \int_0^\infty t^k e^{-\zeta t} e^{-tT} dt = k!\,(T+\zeta)^{-k-1}, \quad \mathrm{Re}\,\zeta > \beta .$$

It follows from (1.38) that the half-plane $\mathrm{Re}\,\zeta > \beta$ belongs to the resolvent set $\mathrm{P}(-T)$ and the following generalizations of (1.36) hold [cf. (1.29)]:

$$(1.39) \qquad \|(T+\zeta)^{-k}\| \leq M(\mathrm{Re}\,\zeta - \beta)^{-k}, \quad \mathrm{Re}\,\zeta > \beta, \quad k = 1, 2, \ldots .$$

Problem 1.16. If $T \in \mathscr{G}(M, 0)$, then

$$(1.40) \qquad \|(1+\alpha T)^{-k}(1+\alpha' T)^{-h}\| \leq M, \quad \alpha, \alpha' \geq 0, \quad k, h = 1, 2, \ldots .$$

Also generalize this to the case of more than two factors. [hint: (1.38).]

Problem 1.17. Let $T \in \mathscr{G}(M, 0)$. Use the preceding problem to show that we need only to put the factor M (instead of M^2 as stated above) into the right member of (1.16). Furthermore

$$(1.41) \qquad \|U(t)\,u - V_n(t)\,u\| \leq \frac{M t^2}{2n} \|T^2 u\|, \quad u \in \mathrm{D}(T^2) .$$

Problem 1.18. Let X be a Hilbert space. Then $T \in \mathscr{G}(1, 0)$ if and only if T is m-accretive, and $T \in \mathscr{G}(1, \beta)$ for some β if and only if T is quasi-m-accretive.

[1] There are more general kinds of semigroups (see Hille-Phillips [1]), but we shall be concerned only with quasi-bounded semigroups.

[2] The infimum of the set of β for which $T \in \mathscr{G}(M, \beta)$ is called the *type* of the semigroup $\{e^{-tT}\}$.

[3] Assuming $\dim \mathsf{X} > 0$.

5. Solution of the inhomogeneous differential equation

Let $T \in \mathscr{G}(M, \beta)$. $U(t) = e^{-tT}$ gives the solution of the differential equation $du/dt = -Tu$ in the form $u = u(t) = U(t)\,u(0)$; see (1.26). $U(t)$ can also be used to express the solution of the inhomogeneous differential equation

$$du/dt = -Tu + f(t)\,, \quad t > 0\,, \tag{1.42}$$

where $f(t)$ is a given function with values in X and is assumed to be strongly continuous for $t \geqq 0$.

If $u = u(t)$ is a solution of (1.42), a calculation similar to (1.25) gives $(d/ds)\,U(t-s)\,u(s) = U(t-s)\,f(s)$. Integrating this on $(0, t)$, we obtain [note that $U(t-s)\,f(s)$ is strongly continuous in s]

$$u(t) = U(t)\,u_0 + \int_0^t U(t-s)\,f(s)\,ds\,, \quad u_0 = u(0)\,. \tag{1.43}$$

In particular this implies that the solution of (1.42) is uniquely determined by $u(0)$.

Conversely, we have

Theorem 1.19.[1] *Let $T \in \mathscr{G}(M, \beta)$ and let $f(t)$ be continuously differentiable for $t \geqq 0$. For any $u_0 \in \mathsf{D}(T)$, the $u(t)$ given by* (1.43) *is continuously differentiable for $t \geqq 0$ and is a solution of* (1.42) *with the initial value $u(0) = u_0$.*

Proof. Since we know that the first term $U(t)\,u_0$ on the right of (1.43) satisfies the homogeneous differential equation and the initial condition, it suffices to show that the second term satisfies (1.42) and has initial value 0. Denoting this term by $v(t)$, we have

$$\begin{aligned} v(t) &= \int_0^t U(t-s)\,f(s)\,ds = \int_0^t U(t-s)\left[f(0) + \int_0^s f'(r)\,dr\right]ds \\ &= \left[\int_0^t U(t-s)\,ds\right]f(0) + \int_0^t\left[\int_r^t U(t-s)\,ds\right]f'(r)\,dr\,. \end{aligned} \tag{1.44}$$

But

$$T\int_r^t U(s)\,ds = U(r) - U(t)\,, \quad 0 \leqq r \leqq t\,. \tag{1.45}$$

To see this we note that if $u \in \mathsf{D}(T)$, $TU(s)\,u = -dU(s)\,u/ds$ and hence $\int_r^t TU(s)\,u\,ds = U(r)\,u - U(t)\,u$. Recalling the definition of the integral and the closedness of T, we may write this as $T\int_r^t U(s)\,u\,ds = (U(r) - U(t))\,u$. This can be extended to an arbitrary $u \in \mathsf{X}$ by choosing a sequence $u_n \in \mathsf{D}(T)$ such that $u_n \to u$ and going to the limit, for

[1] See PHILLIPS [1].

$\int_r^t U(s)\, u_n\, ds \to \int_r^t U(s)\, u\, ds$ and $(U(r) - U(t))\, u_n \to (U(r) - U(t))\, u$ (here again the closedness of T is essential). In this way we obtain (1.45).

(1.45) implies that

$$T \int_r^t U(t-s)\, ds = 1 - U(t-r)\,, \quad 0 \leqq r \leqq t\,. \tag{1.46}$$

It follows from (1.44) and (1.46) that $v(t) \in \mathsf{D}(T)$ and

$$\begin{aligned} T v(t) &= (1 - U(t))\, f(0) + \int_0^t (1 - U(t-r))\, f'(r)\, dr \\ &= f(t) - U(t)\, f(0) - \int_0^t U(t-r)\, f'(r)\, dr\,. \end{aligned} \tag{1.47}$$

On the other hand, $v(t) = \int_0^t U(s)\, f(t-s)\, ds$ and so

$$dv(t)/dt = U(t)\, f(0) + \int_0^t U(s)\, f'(t-s)\, ds\,. \tag{1.48}$$

Comparison of (1.47) and (1.48) shows that $dv(t)/dt = -Tv(t) + f(t)$, as we wished to prove. Also it is easy to show that $v(t) \to 0$, $t \to 0$. The continuity of $dv(t)/dt$ can be concluded easily from (1.48) using the continuity of $f'(t)$.

For later use we shall estimate $\|u'(t)\|$ and $\|Tu(t)\|$. We see from (1.43) and (1.48) that

$$\begin{aligned} \|u'(t)\| &\leqq \|U(t)\, T u_0\| + \|v'(t)\| \leqq \\ &\leqq M e^{\beta t} (\|T u_0\| + \|f(0)\|) + M \int_0^t e^{\beta s} \|f'(t-s)\|\, ds\,, \\ \|Tu(t)\| &= \|u'(t) - f(t)\| \leqq \|u'(t)\| + \|f(t)\|\,. \end{aligned} \tag{1.49}$$

6. Holomorphic semigroups

The construction of the semigroup $U(t) = e^{-tT}$ described in the preceding paragraphs is rather complicated. One may raise the question whether a Dunford-Taylor integral of the form

$$U(t) = \frac{1}{2\pi i} \int_\Gamma e^{\zeta t} (T + \zeta)^{-1}\, d\zeta \tag{1.50}$$

cannot be used. Obviously (1.50) is valid if $T \in \mathscr{B}(X)$ and Γ is a positively-oriented, closed curve enclosing the spectrum of $-T$ in its interior. In the more general case of $T \in \mathscr{G}(M, \beta)$, one may try to justify (1.50) by taking as Γ the straight line running from $c - i\infty$ to $c + i\infty$ with $c > \beta$, by analogy with the ordinary inverse Laplace transformation. But it is not easy to prove the convergence of this integral.

(1.50) is useful, however, if we assume slightly more on T. Suppose that T is densely defined and closed, $P(-T)$ contains not only the half-plane $\operatorname{Re}\zeta > 0$ but a sector $|\arg\zeta| < \frac{\pi}{2} + \omega$, $\omega > 0$, and that for any $\varepsilon > 0$,

$$\|(T+\zeta)^{-1}\| \leq M_\varepsilon/|\zeta| \quad \text{for} \quad |\arg\zeta| \leq \frac{\pi}{2} + \omega - \varepsilon \tag{1.51}$$

with M_ε independent of ζ[1]. Then the integral in (1.50) is absolutely convergent for $t > 0$ if Γ is chosen as a curve in $P(-T)$ running, within the sector mentioned, from infinity with $\arg\zeta = -\left(\frac{\pi}{2} + \omega - \varepsilon\right)$ to infinity with $\arg\zeta = \frac{\pi}{2} + \omega - \varepsilon$, where $\varepsilon < \omega$.

The semigroup property of $U(t)$ thus defined can be easily proved by a standard manipulation of Dunford integrals. Suppose $U(s)$ is given by a formula like (1.50) with Γ replaced by another path Γ' similar to Γ but shifted to the right by a small amount. Then

$$\begin{aligned} U(s)\,U(t) &= \left(\frac{1}{2\pi i}\right)^2 \int_{\Gamma'}\int_{\Gamma} e^{\zeta' s + \zeta t}(T+\zeta')^{-1}(T+\zeta)^{-1}\,d\zeta\,d\zeta' \\ &= \left(\frac{1}{2\pi i}\right)^2 \Big[\int_{\Gamma'} e^{\zeta' s}(T+\zeta')^{-1}\,d\zeta' \int_{\Gamma} e^{\zeta t}(\zeta-\zeta')^{-1}\,d\zeta - \\ &\quad - \int_{\Gamma} e^{\zeta t}(T+\zeta)^{-1}\,d\zeta \int_{\Gamma'} e^{\zeta' s}(\zeta-\zeta')^{-1}\,d\zeta'\Big] \\ &= \frac{1}{2\pi i}\int_{\Gamma} e^{\zeta(t+s)}(T+\zeta)^{-1}\,d\zeta = U(t+s)\,, \end{aligned} \tag{1.52}$$

where we have used the resolvent equation $(T+\zeta')^{-1}(T+\zeta)^{-1} = (\zeta-\zeta')^{-1}[(T+\zeta')^{-1} - (T+\zeta)^{-1}]$ and the relations $\int_\Gamma e^{\zeta t}(\zeta-\zeta')^{-1}\,d\zeta = 0$, $\int_{\Gamma'} e^{\zeta' s}(\zeta-\zeta')^{-1}\,d\zeta' = -2\pi i\, e^{\zeta s}$.

(1.50) is defined even for complex t if $|\arg t| < \omega$, for then we can deform Γ to ensure that $|\arg t\zeta| > \frac{\pi}{2}$ for $\zeta \in \Gamma$ and $|\zeta| \to \infty$. Since (1.50) can be differentiated in t under the integral sign, it follows that *$U(t)$ is holomorphic in t in the open sector* $|\arg t| < \omega$. In fact we have

$$U(t)\,T \subset T\,U(t) = -dU(t)/dt \in \mathscr{B}(X)\,, \quad |\arg t| < \omega\,, \tag{1.53}$$

for $dU(t)/dt = \frac{1}{2\pi i}\int_\Gamma e^{\zeta t}\zeta(T+\zeta)^{-1}\,d\zeta = \frac{1}{2\pi i}\int_\Gamma e^{\zeta t}(1 - T(T+\zeta)^{-1})\,d\zeta$ $= \frac{-1}{2\pi i}T\int_\Gamma e^{\zeta t}(T+\zeta)^{-1}\,d\zeta = T\,U(t)$; here the closedness of T is used to justify taking the factor T out of the integral sign.

---

[1] The main difference of (1.51) from (1.39) for $T \in \mathscr{G}(M, \beta)$ is the appearance of $|\zeta|$ instead of $\operatorname{Re}\zeta$.

On changing the integration variable to $\zeta' = \zeta t$, we obtain from (1.50)

$$U(t) = \frac{1}{2\pi i t} \int_{\Gamma'} e^{\zeta'} \left(T + \frac{\zeta'}{t}\right)^{-1} d\zeta' , \tag{1.54}$$

where Γ' may be taken independent of t. Since $\left\|\left(T + \frac{\zeta'}{t}\right)^{-1}\right\| \leqq$ const. $\cdot |t/\zeta'|$ for $|\arg t| \leqq \omega - \varepsilon$, we have $\|U(t)\| \leqq$ const. $\int_{\Gamma'} |e^{\zeta'}| \, |\zeta'|^{-1} \, |d\zeta'|$ = const. Thus $U(t)$ is uniformly bounded:

$$\|U(t)\| \leqq M'_\varepsilon \quad \text{for} \quad |\arg t| \leqq \omega - \varepsilon . \tag{1.55}$$

Similarly we have the estimate

$$\|d U(t)/dt\| = \|T U(t)\| \leqq M''_\varepsilon \, |t|^{-1}, \quad |\arg t| \leqq \omega - \varepsilon . \tag{1.56}$$

Furthermore,

$$\begin{aligned} U(t) - 1 &= \frac{1}{2\pi i} \int_{\Gamma'} e^{\zeta'} ((t T + \zeta')^{-1} - \zeta'^{-1}) \, d\zeta' \\ &= \frac{-t}{2\pi i} \int \frac{e^{\zeta'}}{\zeta'} T (t T + \zeta')^{-1} \, d\zeta', \end{aligned} \tag{1.57}$$

so that $\|U(t) u - u\| \leqq$ const. $|t| \, \|T u\| \int_{\Gamma'} |e^{\zeta'}| \, |\zeta'|^{-2} \, |d\zeta'| \to 0$ for $u \in \mathsf{D}(T)$ when $|t| \to 0$ with $|\arg t| \leqq \omega - \varepsilon$. Since $\mathsf{D}(T)$ is dense and $U(t)$ is uniformly bounded, it follows that

$$\operatorname*{s-lim}_{t \to 0} U(t) = U(0) = 1 \quad (|\arg t| \leqq \omega - \varepsilon) . \tag{1.58}$$

Thus we have proved that $U(t)$ *is holomorphic for* $|\arg t| < \omega$, *uniformly bounded for* $|\arg t| \leqq \omega - \varepsilon$ *and strongly continuous (within this smaller sector) at* $t = 0$ *with* $U(0) = 1$. $U(t) = e^{-tT}$ will be called a *bounded holomorphic semigroup*[1].

Remark 1.20. Since $U(t)$ is holomorphic, it can be differentiated any number of times. A calculation similar to the one used in deducing (1.56) gives the estimates (for simplicity we consider only real $t > 0$)

$$\|d^n U(t)/dt^n\| = \|T^n U(t)\| \leqq M_n t^{-n} , \quad t > 0 . \tag{1.59}$$

Also we note the inequality

$$\|T(U(t) - U(s))\| \leqq M_2 (t - s)/t s , \quad 0 < s \leqq t . \tag{1.60}$$

For the proof, it suffices to note that

$$\begin{aligned} \|T(U(t) - U(s))\| &= \left\| T \int_s^t \frac{dU(r)}{dr} \, dr \right\| \\ &= \left\| \int_s^t T^2 U(r) \, dr \right\| \leqq M_2 \int_s^t r^{-2} \, dr = M_2 (t - s)/st . \end{aligned}$$

[1] This is a special case of holomorphic semigroups discussed in detail in Hille-Phillips [1] and Yosida [1].

We shall denote by $\mathscr{H}(\omega, 0)$ the set of all densely defined, closed operators T with the property (1.51). We denote by $\mathscr{H}(\omega, \beta)$, β real, the set of all operators of the form $T = T_0 - \beta$ with $T_0 \in \mathscr{H}(\omega, 0)$. Obviously $e^{-tT} = e^{\beta t} e^{-tT_0}$ is a semigroup holomorphic for $|\arg t| < \omega$. e^{-tT} need not be uniformly bounded, but it is *quasi-bounded* in the sense that $\|e^{-tT}\| \leqq$ const. $|e^{\beta t}|$ in any sector of the form $|\arg t| \leqq \omega - \varepsilon$.

Remark 1.21. It is not obvious that (1.51) is stronger than the assumption (1.33) for the generator of a bounded semigroup. But this is true, as is seen from Theorem 1.23 below.

Remark 1.22. If $T \in \mathscr{H}(\omega, \beta)$, $u(t) = e^{-tT} u_0$ satisfies $du(t)/dt = -Tu(t)$ for $t > 0$ for *any* $u_0 \in \mathsf{X}$ [see (1.53)]. This is a great difference from the case $T \in \mathscr{G}(M, \beta)$.

Theorem 1.23. *$T \in \mathscr{H}(\omega, 0)$ is equivalent to the existence, for each $\varepsilon > 0$, of a constant M'_ε such that $e^{i\theta} T \in \mathscr{G}(M'_\varepsilon, 0)$ for any real θ with $|\theta| \leqq \omega - \varepsilon$. In particular it implies $T \in \mathscr{G}(M, 0)$ for some M.*

Proof. Let $T \in \mathscr{H}(\omega, 0)$. From (1.53) we can again deduce the inversion formula (1.38). Since $\|U(t)\| \leqq$ const. for real $t > 0$ by (1.55), it follows that $\|(T + \xi)^{-k-1}\| \leqq M' \xi^{-k-1}$, $k = 0, 1, 2, \ldots$, for $\xi > 0$. This shows that $T \in \mathscr{G}(M', 0)$. Furthermore, we may shift the integration path in (1.38) from the positive real axis to the ray $t = r e^{i\theta}$, $r > 0$, provided that $|\theta| \leqq \omega - \varepsilon$. Since $\|U(t)\| \leqq$ const. for such t by (1.55), we obtain the inequality $\|(\xi e^{-i\theta} + T)^{-k-1}\| \leqq M'_\varepsilon \xi^{-k-1}$, $\xi > 0$ (set $\zeta = \xi e^{-i\theta}$). This proves that $e^{i\theta} T \in \mathscr{G}(M'_\varepsilon, 0)$.

Conversely, let $e^{i\theta} T \in \mathscr{G}(M'_\varepsilon, 0)$ for $|\theta| \leqq \omega - \varepsilon$. This implies that the spectrum of $e^{i\theta} T$ is contained in the half-plane $\operatorname{Re}\zeta \geqq 0$ for each θ with $|\theta| < \omega$. Hence the sector $|\arg\zeta| < \frac{\pi}{2} + \omega$ belongs to $\mathrm{P}(-T)$. If $0 \leqq \arg\zeta \leqq \frac{\pi}{2} + \omega - 2\varepsilon$, then $\operatorname{Re} e^{-i(\omega-\varepsilon)}\zeta \geqq |\zeta| \sin\varepsilon > 0$ and so $\|(e^{-i(\omega-\varepsilon)} T + e^{-i(\omega-\varepsilon)}\zeta)^{-1}\| \leqq M'_\varepsilon / \operatorname{Re} e^{-i(\omega-\varepsilon)}\zeta \leqq M'_\varepsilon / |\zeta| \sin\varepsilon$ [see (1.39)]. Since a similar result holds for $-\left(\frac{\pi}{2} + \omega - 2\varepsilon\right) \leqq \arg\zeta \leqq 0$, we have proved (1.51).

Another convenient criterion for $-T$ to be the generator of a holomorphic semigroup is given by

Theorem 1.24. *Let T be an m-sectorial operator in a Hilbert space H with a vertex 0 (so that its numerical range $\Theta(T)$ is a subset of a sector $|\arg\zeta| \leqq \frac{\pi}{2} - \omega$, $0 < \omega \leqq \frac{\pi}{2}$). Then $T \in \mathscr{H}(\omega, 0)$, and e^{-tT} is holomorphic for $|\arg t| < \omega$ and is bounded by $\|e^{-tT}\| \leqq 1$.*

Proof. According to Theorem 1.23, it suffices to show that $e^{i\theta} T \in \mathscr{G}(1, 0)$ for $|\theta| \leqq \omega$. $\Theta(e^{i\theta} T) = e^{i\theta}\Theta(T)$ is a subset of the right half-plane if $|\theta| \leqq \omega$. Thus $e^{i\theta} T$ is m-sectorial and the left half-plane is contained in $\mathrm{P}(e^{i\theta} T)$. If $\operatorname{Re}\zeta > 0$, therefore, $\|(\zeta + e^{i\theta} T)^{-1}\|$ does not exceed the inverse of the distance of ζ from the imaginary axis. In particular it does not exceed ζ^{-1} for real $\zeta > 0$. This shows that $e^{i\theta} T \in \mathscr{G}(1, 0)$.

Example 1.25. If H is a nonnegative selfadjoint operator in a Hilbert space, e^{-tH} is holomorphic for $\mathrm{Re}\, t > 0$ and $\|e^{-tH}\| \leqq 1$.

Example 1.26. Consider a differential operator $Lu = p_0(x) u'' + p_1(x) u' + p_2(x) u$ on $a \leqq x \leqq b$, where $p_0(x) < 0$. It is known that the operator T_1 in $\mathsf{H} = \mathsf{L}^2(a, b)$ defined by L with the boundary condition $u(a) = u(b) = 0$ is m-sectorial (see Example V-3.34). Hence $T_1 \in \mathscr{H}(\omega, \beta)$ for some $\omega > 0$ by Theorem 1.24. A similar result holds for second-order partial differential operators of *elliptic* type.

7. The inhomogeneous differential equation for a holomorphic semigroup

If $-T$ is the generator of a holomorphic semigroup, we have a stronger result than Theorem 1.19 for the solution of the inhomogeneous differential equation (1.42).

Theorem 1.27. *Let $T \in \mathscr{H}(\omega, \beta)$ and let $f(t)$ be Hölder continuous for $t \geqq 0$:*

$$(1.61) \qquad \|f(t) - f(s)\| \leqq L(t - s)^k, \quad 0 \leqq s \leqq t,$$

where L and k are constants, $0 < k \leqq 1$. For any $u_0 \in \mathsf{X}$, the $u(t)$ given by (1.43) is continuous for $t \geqq 0$, continuously differentiable for $t > 0$ and is a solution of (1.42) with $u(0) = u_0$.

Proof. Since $U(t)\, u_0$ satisfies the homogeneous differential equation and the initial condition (see Remark 1.22), it suffices to show that the second term $v(t)$ on the right of (1.43) satisfies (1.42) for $t > 0$. [Again it is easy to show that $v(t) \to 0$, $t \to 0$, since $U(t)$ is bounded as $t \to 0$.]

We have

$$(1.62) \qquad v(t) = \int_0^t U(t - s)\, [f(s) - f(t)]\, ds + \left[\int_0^t U(t - s)\, ds\right] f(t),$$

so that $T v(t)$ exists and is given by

$$(1.63) \qquad T v(t) = T \int_0^t U(t - s)\, [f(s) - f(t)]\, ds + [1 - U(t)]\, f(t)$$

[see (1.46)]; the existence of the first integral on the right follows from the estimate

$$\|T U(t - s)\|\, \|f(s) - f(t)\| \leqq \text{const.}\ (t - s)^{-1} L (t - s)^k = \text{const.}\ (t - s)^{k-1}$$

due to (1.56) and (1.61).

On the other hand, we have from (1.44)

$$(1.64) \qquad v(t + h) = \left[\int_0^t + \int_t^{t+h}\right] U(t + h - s)\, f(s)\, ds$$
$$= U(h)\, v(t) + \int_t^{t+h} U(t + h - s)\, f(s)\, ds,$$

where $t > 0$ and $h > 0$. Computing $\lim_{h \searrow 0} h^{-1}[v(t+h) - v(t)]$, we obtain ($D^+$ denotes the right derivative)

$$D^+ v(t) = -T v(t) + f(t) , \tag{1.65}$$

since it is known that $v(t) \in D(T)$.

Now $T v(t)$ is continuous for $t > 0$ as will be proved in Lemma 1.28 below. Thus $D^+ v(t)$ is also continuous. It follows[1] that $dv(t)/dt$ exists for $t > 0$ and equals $-T v(t) + f(t)$, as we wished to prove.

Lemma 1.28. *$T v(t)$ is continuous for $t \geqq 0$ and Hölder continuous for $t \geqq \varepsilon$ with the exponent k, for any $\varepsilon > 0$.*

Proof. The second term on the right of (1.63) satisfies the assertion of the lemma, for $f(t)$ is Hölder continuous and $U(t)$ is holomorphic for $t > 0$.

Let the first term on the right of (1.63) be denoted by $w(t)$. Then

$$\begin{aligned} w(t+h) - w(t) &= T \int_0^t [U(t+h-s) - U(t-s)]\,[f(s) - f(t)]\,ds \\ &\quad + T \int_0^t U(t+h-s)\,[f(t) - f(t+h)]\,ds + \\ &\quad + T \int_t^{t+h} U(t+h-s)\,[f(s) - f(t+h)]\,ds \\ &= w_1 + w_2 + w_3, \text{ say} . \end{aligned} \tag{1.66}$$

To estimate w_1, we use (1.60) and (1.61):

$$\begin{aligned} \|w_1\| &\leqq \int_0^t \|T[U(t+h-s) - U(t-s)]\|\,\|f(s) - f(t)\|\,ds \leqq \\ &\leqq M_2 L h \int_0^t (t+h-s)^{-1} (t-s)^{-1+k}\,ds \leqq \\ &\leqq M_2 L h \int_0^\infty (s+h)^{-1} s^{-1+k}\,ds \leqq \text{const.}\, h^k . \end{aligned}$$

w_2 is estimated by using (1.45) and (1.61):

$$\|w_2\| = \|[U(h) - U(t+h)]\,[f(t) - f(t+h)]\| \leqq 2 M_0 L h^k .$$

To estimate w_3, we use (1.59) with $n = 1$:

$$\begin{aligned} \|w_3\| &\leqq \int_t^{t+h} \|T U(t+h-s)\|\,\|f(s) - f(t+h)\|\,ds \\ &= M_1 L \int_t^{t+h} (t+h-s)^{-1+k}\,ds \leqq \text{const.}\, h^k . \end{aligned}$$

---

[1] If $D^+ v(t)$ is continuous, then $dv(t)/dt$ exists and equals $D^+ v(t)$. For the proof, set $w(t) = \int_a^t D^+ v(s)\,ds$ where $a > 0$. Then $D^+(w(t) - v(t)) = 0$. Since $w(t) - v(t)$ is continuous, it must be constant (see Lemma III-1.36). Hence $dv(t)/dt = dw(t)\,dt = D^+ v(t)$.

Collecting the above estimates, we see that $w(t)$ is Hölder continuous with the exponent k, *uniformly*[1] for $t \geqq 0$.

8. Applications to the heat and Schrödinger equations

We mentioned above (Example 1.26) that the semigroup e^{-tT} is holomorphic when T is a second-order partial differential operator of elliptic type. Let us consider in some detail the semigroup generated by the Laplacian Δ in R^3.

We have shown before (see V-§ 5.2) that $-\Delta$ defines in a natural way a self-adjoint operator in the Hilbert space $L^2(R^3)$. Here we shall consider $T = -\Delta$ in the Banach space $X = L^p(R^3)$, $1 \leqq p < \infty$, or $X = C(R^3)$. The simplest definition of T is given indirectly by the explicit formula for the resolvent $(T + \zeta)^{-1}$ of T as an integral operator with the kernel

$$(1.67)\qquad g(y, x; \zeta) = \frac{e^{-\sqrt{\zeta}|y-x|}}{4\pi|y-x|}, \quad \operatorname{Re}\sqrt{\zeta} > 0 .$$

It can be seen from Example III-2.11 that this kernel defines an integral operator $G(\zeta) \in \mathscr{B}(X)$, with

$$(1.68)\qquad \|G(\zeta)\| \leqq \int |g(y, x; \zeta)|\, dy = \int |g(y, x; \zeta)|\, dx \leqq (\operatorname{Re}\sqrt{\zeta})^{-2} = 1/|\zeta| \sin^2\frac{\varepsilon}{2}$$

for $|\arg\zeta| \leqq \pi - \varepsilon$. It is a well known fact that $(\zeta - \Delta)\, G(\zeta)\, u = u$ for sufficiently smooth function $u(x)$, say $u \in C_0^\infty(R^3)$. With $-\Delta$ defined on the set of such smooth functions, T should be defined as that extension of $-\Delta$ for which $(T + \zeta)^{-1}$ is exactly equal to $G(\zeta)$.

It follows from (1.68) that $T \in \mathscr{H}\left(\frac{\pi}{2}, 0\right)$ and $-T$ generates a semigroup e^{-tT} holomorphic for $\operatorname{Re} t > 0$. Furthermore, e^{-tT} is a *contraction* semigroup for real $t > 0$:

$$(1.69)\qquad \|e^{-tT}\| \leqq 1 \quad \text{for real} \quad t > 0 ;$$

this is seen from $\|(T + \xi)^{-1}\| = \|G(\xi)\| \leqq 1/\xi$ for $\xi > 0$, which follows from (1.68) for $\varepsilon = \pi$.

For any $u_0 \in X$, $u(t) = e^{-tT} u_0$ is a solution of the differential equation $du(t)/dt = -Tu(t)$, which is an abstract version of the *heat equation*

$$(1.70)\qquad \frac{\partial u}{\partial t} = \Delta u .$$

Such a solution is unique for the given initial value u_0 if it is required that $u(t) \in X$. Thus *any solution of the heat equation belonging to* X *is holomorphic in* t (as an X-valued function). In general this does not imply that the function $u(t, x)$ of the space-time variables is analytic in t for each fixed x. But the latter does follow from the abstract analyticity if we choose $X = C(R^3)$, for convergence in this space means uniform convergence of the functions.

In view of (1.50), e^{-tT} is represented by an integral operator with the kernel

$$(1.71)\qquad h(y, x; t) = \frac{1}{2\pi i} \int_\Gamma e^{\zeta t} g(y, x; \zeta)\, d\zeta = (4\pi t)^{-3/2} e^{-|y-x|^2/4t}, \quad \operatorname{Re} t > 0 ,$$

where Γ is chosen as in par. 6.

---

[1] But $Tv(t)$ is not necessarily Hölder continuous up to $t = 0$, since $U(t)$ need not be so.

(1.71) suggests that e^{-itT} is the integral operator with the kernel

$$h(y, x; i t) = (4\pi i t)^{-3/2} e^{-|y-x|^2/4it}. \tag{1.72}$$

But this is not obvious; it is even not clear in what sense (1.72) is an integral kernel. We know, however, that if $X = L^2(R^3)$, T is selfadjoint and hence $\{e^{-itT}\}$ is a group defined for $-\infty < t < +\infty$. In this case e^{-itT} is in fact the integral operator with the kernel (1.72), in the sense to be stated below.

To simplify the description we shall consider, instead of (1.72), its one-dimensional analogue

$$k(y, x; i t) = (4\pi i t)^{-1/2} e^{-(y-x)^2/4it} \tag{1.73}$$

where x, y, and t vary over the real line $(-\infty, +\infty)$. Then

$$K(t)\, u(y) = \int_{-\infty}^{\infty} k(y, x; i t)\, u(x)\, dx \tag{1.74}$$

is defined at least if $u(x)$ tends to zero sufficiently rapidly as $x \to \pm\infty$. For example, an elementary calculation shows that

$$K(t)\, u(x) = (1 + 4 i b t)^{-1/2} e^{-b(x-a)^2/(1+4ibt)} \tag{1.75}$$
$$\text{for} \quad u(x) = e^{-b(x-a)^2}, \quad b > 0.$$

If we take two functions $u_j(x) = e^{-b_j(x-a_j)^2}$, $j = 1, 2$, of this form and construct $K(t)\, u_j$ by (1.75), we obtain by a straightforward calculation

$$(K(t)\, u_1, K(t)\, u_2) = (u_1, u_2) = \left(\frac{\pi}{b_1 + b_2}\right)^{1/2} e^{-\frac{b_1 b_2}{b_1+b_2}(a_1 - a_2)^2}. \tag{1.76}$$

It follows that any linear combination u of functions of the form (1.75) satisfies the equality $\|K(t)\, u\| = \|u\|$. In other words, $K(t)$ is isometric if defined on the set D of such functions u. But D is dense in L^2. To see this it suffices to verify that the set of Fourier transforms of the functions of D is dense in L^2 (since the Fourier transformation is unitary from L^2 to L^2). The Fourier transform $\hat{u}$ of the u of (1.75) is given by $\hat{u}(p) = (2b)^{1/2} e^{-p^2/4b - iap}$. Suppose a $\hat{v}(p) \in L^2$ is orthogonal to all functions $\hat{u}$ of this form with fixed b and varying a. This means that $\hat{w}(p) = e^{-p^2/4b} \hat{v}(p)$, which belongs to L^1, has the Fourier transform zero; thus $\hat{w}(p) = 0$, $\hat{v}(p) = 0$, showing that the set of u is dense.

Therefore, the $K(t)$ defined on D can be extended uniquely to an isometric operator on L^2 to L^2, which will again be denoted by $K(t)$. $K(t)$ may not be an integral operator in the proper sense, but it is an integral operator in a generalized sense:

$$K(t)\, u(y) = \text{l.i.m.}\ (4\pi i t)^{-1/2} \int_{-\infty}^{\infty} e^{-(y-x)^2/4it} u(x)\, dx, \quad -\infty < t < \infty, \tag{1.77}$$

just as was the case with the Fourier-Plancherel transform.

It remains to show that the generalized integral operator $K(t)$ coincides with e^{-itT}. For this it suffices to show that $K(t)\, u = e^{-itT} u$ for $u \in D$. This is seen to be true by letting t approach a real value from the lower half-plane $\operatorname{Im} t < 0$, where this equality is obvious (see Problem 1.29 below).

In the 3-dimensional case we have only to modify the above arguments by taking as D the set of linear combinations of functions of the form $u(x) = e^{-b|x-a|^2}$, $b > 0$, $a \in R^3$. In this way we see that e^{-itT} is given by the integral operator

$$e^{-itT} u(y) = \text{l.i.m.}\ (4\pi i t)^{-3/2} \int_{R^3} e^{-|y-x|^2/4it} u(x)\, dx, \quad -\infty < t < \infty. \tag{1.78}$$

These results can be further extended to the m-dimensional space, the only modification needed being replacement of 3/2 by $m/2$.

Problem 1.29. When $X = L^2$, e^{-tT} is strongly continuous and $\|e^{-tT}\| \leqq 1$ for $\operatorname{Re} t \geqq 0$.

§ 2. Perturbation of semigroups

1. Analytic perturbation of quasi-bounded semigroups[1]

We now ask, does the infinitesimal generator of a semigroup retain the properties of a generator if subjected to a small perturbation? And if it does, how does the generated semigroup change? First we shall show that the property of being a generator is stable under the addition of a *bounded* operator. More precisely

Theorem 2.1. *Let $T \in \mathscr{G}(M, \beta)$ and $A \in \mathscr{B}(X)$. Then $T + A \in \mathscr{G}(M, \beta + M\|A\|)$ and $e^{-t(T+A)}$ is, for fixed $t \geqq 0$, a holomorphic function of A. In particular, $e^{-t(T+\varkappa A)}$ is an entire function of the complex variable $\varkappa$.*

Proof. If $-(T + A)$ is to be a generator of a quasi-bounded semigroup $V(t) = e^{-t(T+A)}$, we must have the differential equation

$$dv(t)/dt = -(T + A)\, v(t) \tag{2.1}$$

for $v(t) = V(t)\, u_0$, $u_0 \in D(T + A) = D(T)$.

According to § 1.5, the solution of (2.1) must satisfy the integral equation

$$v(t) = U(t)\, u_0 - \int_0^t U(t-s)\, A\, v(s)\, ds\,, \tag{2.2}$$

where $U(t) = e^{-tT}$ is the unperturbed semigroup. Substituting $v(t) = V(t)\, u_0$, this gives

$$V(t) = U(t) - \int_0^t U(t-s)\, A\, V(s)\, ds\,. \tag{2.3}$$

It should be noted that (2.3) has not been proved to be true except when applied to a $u_0 \in D(T)$. But (2.3) is true since the two members belong to $\mathscr{B}(X)$ and $D(T)$ is dense in X. Note further that the integral on the right is a strong integral: $U(t-s)\, A\, V(s)\, u_0$ is continuous in s for every $u_0 \in X$ so that it is integrable.

Let us solve (2.3) by successive approximation:

$$V(t) = \sum_{n=0}^{\infty} U_n(t)\,, \tag{2.4}$$

$$U_{n+1}(t) = -\int_0^t U(t-s)\, A\, U_n(s)\, ds\,, \quad n = 0, 1, 2, \ldots, \tag{2.5}$$

where $U_0(t) = U(t)$. Since $U(t)$ is strongly continuous, it is easily seen by induction that all the $U_n(t)$ are defined [the integral in (2.5) is a strong

[1] For details see HILLE and PHILLIPS [1], Chapter 13; PHILLIPS [1].

integral of a strongly continuous function] and are strongly continuous in t. Furthermore, we have the estimates

$$\|U_n(t)\| \leqq M^{n+1}\|A\|^n e^{\beta t} t^n/n!\,, \quad n = 0, 1, 2, \ldots, \tag{2.6}$$

which can be proved by induction. In fact, (2.6) is true for $n = 0$; assuming it for n, we have from (2.5), noting that $\|U(t-s)\| \leqq M e^{\beta(t-s)}$,

$$\begin{aligned}\|U_{n+1}(t)\| &\leqq M^{n+2}\|A\|^{n+1} n!^{-1} \int_0^t e^{\beta(t-s)} e^{\beta s} s^n\, ds \\ &= M^{n+2}\|A\|^{n+1} e^{\beta t} t^{n+1}/(n+1)!\,.\end{aligned}$$

We see from (2.6) that the series (2.4) is absolutely convergent, the sum $V(t)$ is a solution of the integral equation (2.3) and that

$$\|V(t)\| \leqq \sum_{n=0}^{\infty} \|U_n(t)\| \leqq M e^{(\beta + M\|A\|)t}\,. \tag{2.7}$$

To show that $V(t)$ is in fact the semigroup generated by $-(T + A)$, we multiply (2.3) by $e^{-\zeta t}$ and integrate over $0 \leqq t < \infty$, assuming $\operatorname{Re}\zeta > \beta + M\|A\|$. Then

$$\begin{aligned}R_1(\zeta) &= \int_0^\infty e^{-\zeta t} V(t)\, dt = \int_0^\infty e^{-\zeta t} U(t)\, dt - \\ &\quad - \left[\int_0^\infty e^{-\zeta t} U(t)\, dt\right] A \left[\int_0^\infty e^{-\zeta t} V(t)\, dt\right] \\ &= (T+\zeta)^{-1} - (T+\zeta)^{-1} A\, R_1(\zeta)\,.\end{aligned} \tag{2.8}$$

This shows that $(T+\zeta) R_1(\zeta) = 1 - A R_1(\zeta)$ or $(T + A + \zeta) R_1(\zeta) = 1$. Since $-\zeta \in \mathrm{P}(T + A)$ in virtue of $\|A\| < M^{-1} \operatorname{Re}(\zeta - \beta) \leqq 1/\|(T+\zeta)^{-1}\|$ [see (1.39)], it follows that $R_1(\zeta) = (T + A + \zeta)^{-1}$.

Thus we have for $k = 0, 1, 2, \ldots,$

$$\begin{aligned}\|(T + A + \zeta)^{-k-1}\| &= \frac{1}{k!}\left\|\frac{d^k}{d\zeta^k}(T+A+\zeta)^{-1}\right\| \\ &= \frac{1}{k!}\left\|\frac{d^k}{d\zeta^k} R_1(\zeta)\right\| \leqq \frac{1}{k!}\int_0^\infty t^k |e^{-\zeta t}|\, \|V(t)\|\, dt \\ &= \frac{M}{k!}\int_0^\infty t^k e^{-(\operatorname{Re}\zeta - \beta - M\|A\|)t}\, dt = M(\operatorname{Re}\zeta - \beta - M\|A\|)^{-k-1}\,,\end{aligned} \tag{2.9}$$

which shows that $T + A$ satisfies (1.36) with β replaced by $\beta + M\|A\|$: $T + A \in \mathscr{G}(M, \beta + M\|A\|)$.

Then we see from Remark 1.4 [or rather its generalization to $T \in \mathscr{G}(M, \beta)$] that $V(t) = e^{-t(T+A)}$.

The expression (2.4) for $V(t) = e^{-t(T+A)}$ has the form of an expansion in powers of A. This shows that $V(t)$ is a holomorphic function of A (see Remark II-5.17).

Remark 2.2. The above proof is indirect in so far as it concerns the property of $-(T+A)$ being a generator. It is possible to verify directly the final inequalities of (2.9). This proof is in general not very simple; it is necessary to make some combinatorial computations[1]. But the proof is trivial if $M = 1$. In this case we need only to verify (2.9) for $k = 1$, but this is obvious from the second Neumann series for the resolvent $(T + A + \zeta)^{-1} = \sum_{n=0}^{\infty} (T + \zeta)^{-1} [-A(T+\zeta)^{-1}]^n$ and $\|(T+\zeta)^{-1}\| \leqq$ $\leqq (\operatorname{Re}\zeta - \beta)^{-1}$.

Problem 2.3. Prove by successive approximation that $v_n(t) = U_n(t)\, u_0$ satisfies the differential equation $dv_n/dt = -Tv_n - Av_{n-1}$ (set $v_{-1} = 0$) and $v = \Sigma v_n$ satisfies $dv/dt = -(T+A)\, v$, provided that $u_0 \in \mathsf{D}(T)$. [hint: Theorem 1.19.]

2. Analytic perturbation of holomorphic semigroups

In the preceding paragraph we considered a *bounded* perturbation of the generator $-T$ of a quasi-bounded semigroup. In general it is difficult to add an unbounded operator A to T without destroying the property of $-T$ being a generator. For example, $-(T+A)$ need not be a generator of a quasi-bounded semigroup even if A is relatively bounded with respect to T.

If $-T$ is assumed to be the generator of a *holomorphic* semigroup, however, perturbations of a wider class are permitted.

Theorem 2.4.[2] *For any $T \in \mathscr{H}(\omega, \beta)$ and $\varepsilon > 0$, there exist positive constants γ, δ with the following properties. If A is relatively bounded with respect to T so that*

$$\|Au\| \leqq a\|u\| + b\|Tu\|, \quad u \in \mathsf{D}(T) \subset \mathsf{D}(A), \tag{2.10}$$

with $a < \delta$, $b < \delta$, then $T + A \in \mathscr{H}(\omega - \varepsilon, \gamma)$. If, in particular, $\beta = 0$ and $a = 0$, then $T + A \in \mathscr{H}(\omega - \varepsilon, 0)$.

Proof. As is easily seen, we may assume $\beta = 0$ without loss of generality[3]. (2.10) implies that

$$\|A(T+\zeta)^{-1}\| \leqq a\|(T+\zeta)^{-1}\| + b\|T(T+\zeta)^{-1}\|. \tag{2.11}$$

If $|\arg\zeta| \leqq \frac{\pi}{2} + \omega - \varepsilon$, we have $\|(T+\zeta)^{-1}\| \leqq M_\varepsilon/|\zeta|$ by (1.51) and $\|T(T+\zeta)^{-1}\| = \|1 - \zeta(T+\zeta)^{-1}\| \leqq 1 + M_\varepsilon$. Hence

$$\|A(T+\zeta)^{-1}\| \leqq aM_\varepsilon|\zeta|^{-1} + b(1 + M_\varepsilon), \tag{2.12}$$

[1] See Hille and Phillips [1], p. 389.

[2] See Hille and Phillips [1], p. 418.

[3] If $\beta < 0$, $T \in \mathscr{H}(w, \beta) \subset \mathscr{H}(w, 0)$. If $\beta > 0$, set $T_0 = T + \beta$; then $T_0 \in \mathscr{H}(w, 0)$ and $\|Au\| \leqq (a + b\beta)\|u\| + b\|T_0 u\|$. If γ_0, δ_0 denote the γ, δ of the theorem for $\beta = 0$, then $T_0 + A \in \mathscr{H}(\omega - \varepsilon, \gamma_0)$ for $a + b\beta < \delta_0$, $b < \delta_0$. Hence $T + A \in \mathscr{H}(\omega - \varepsilon, \gamma_0 + \beta)$ if $a < \delta_0/2$, $b < \min(\delta_0, \delta_0/2\beta)$, so that it suffices to take $\gamma = \gamma_0 + \beta$, $\delta = \min(\delta_0/2, \delta_0/2\beta)$.

and the second Neumann series for $(T + \zeta + A)^{-1}$ converges if the right member of (2.12) is smaller than 1, with

(2.13)
$$\|(T+\zeta+A)^{-1}\| \leqq \frac{M_\varepsilon|\zeta|^{-1}}{1 - aM_\varepsilon|\zeta|^{-1} - b(1+M_\varepsilon)} = \frac{M_\varepsilon[1-b(1+M_\varepsilon)]^{-1}}{|\zeta| - aM_\varepsilon[1-b(1+M_\varepsilon)]^{-1}}.$$

If $b < (1 + M_\varepsilon)^{-1}$, (2.13) shows that $\|(T + A + \zeta)^{-1}\| \leqq M'/|\zeta - \gamma|$ for $|\arg(\zeta - \gamma)| \leqq \frac{\pi}{2} + \omega - \varepsilon$, where M' and γ are some positive constants depending on a, b and M_ε. If $a = 0$, we can take $\gamma = 0$.

Corollary 2.5. *If $-T$ is the generator of a quasi-bounded holomorphic semigroup and A is T-bounded with relative bound 0, then $-(T + A)$ is also the generator of a quasi-bounded holomorphic semigroup.*

Proof. It suffices to observe that b can be chosen arbitrarily small in the proof of Theorem 2.4.

Theorem 2.6. *Let $T(\varkappa) \in \mathscr{C}(X)$ be a holomorphic family of type* (A) *defined near $\varkappa = 0$. If $T(0)$ is the generator of a quasi-bounded holomorphic semigroup, the same is true for $T(\varkappa)$ with sufficiently small $|\varkappa|$. In this case $U(t, \varkappa) = e^{-tT(\varkappa)}$ is holomorphic in $\varkappa$ and t when t is in some open sector containing $t > 0$. Moreover, all $\partial^n U(t, \varkappa)/\partial\varkappa^n$ are strongly continuous in t up to $t = 0$*[1]*. If, in particular, $T(\varkappa) = T + \varkappa A$ where T and A satisfy the conditions of* Corollary 2.5, *then $e^{-t(T+\varkappa A)}$ is an entire function of $\varkappa$ for t in such a sector.*

Proof. Since $T(\varkappa) - T(0)$ is relatively bounded with respect to $T(0)$ (see VII-§ 2.1), $T(\varkappa)$ belongs to some $\mathscr{H}(\omega, \beta)$ for sufficiently small $|\varkappa|$. Replacing $T(\varkappa)$ by $T(\varkappa) + \beta$ if necessary, we may assume $T(\varkappa) \in \mathscr{H}(\omega, 0)$.

It follows from (1.50) for $U(t, \varkappa)$ that

$$\frac{\partial^n}{\partial\varkappa^n} U(t,\varkappa) = \frac{1}{2\pi i}\int_\Gamma e^{\zeta t} \frac{\partial^n}{\partial\varkappa^n}(T(\varkappa) + \zeta)^{-1}\, d\zeta. \tag{2.14}$$

But we have

$$\frac{\partial^n}{\partial\varkappa^n}(T(\varkappa) + \zeta)^{-1} = \frac{n!}{2\pi i}\int_C (T(\varkappa') + \zeta)^{-1}(\varkappa' - \varkappa)^{-n-1}\, d\varkappa', \tag{2.15}$$

where C is a small circle in the $\varkappa$-plane and $\varkappa$ is inside C. Since (1.51) is true for $T = T(\varkappa)$ uniformly in $\varkappa$ near $\varkappa = 0$, we have

$$\left\|\frac{\partial^n}{\partial\varkappa^n}(T(\varkappa)+\zeta)^{-1}\right\| \leqq \frac{M_\varepsilon N^n n!}{|\zeta|}, \quad |\arg\zeta| \leqq \frac{\pi}{2} + \omega - \varepsilon, \tag{2.16}$$

[1] As is seen from the proof, the conclusions of this theorem are true for *any* holomorphic family $T(\varkappa)$ provided that $T(\varkappa) \in \mathscr{H}(\omega, \beta)$ for small $|\varkappa|$. The "type (A)" assumption is used to show that $T(0) \in \mathscr{H}(\omega_0, \beta_0)$ implies $T(\varkappa) \in \mathscr{H}(\omega, \beta)$. It is easy to show that "type (A)" can be replaced by "type (B)" (see VII-§ 4); in this case $\mathfrak{t}(\varkappa) - \mathfrak{t}(0)$ is relatively bounded with respect to $\mathfrak{t}(0)$ [$\mathfrak{t}(\varkappa)$ denoting the form associated with $T(\varkappa)$], so that $T(\varkappa) \in \mathscr{H}(\omega, \beta)$ is true with some constants ω, β for all $\varkappa$ in any compact subset of the domain in which $\varkappa$ varies.

with some constant N if $|\varkappa|$ is sufficiently small. Thus the same argument as given in § 1.6 shows that $\partial^n U(t,\varkappa)/\partial \varkappa^n$ is holomorphic in t in some open sector and strongly continuous up to $t=0$. In particular, this implies that $U(t,\varkappa)$ is holomorphic in $\varkappa$ with its Taylor coefficients holomorphic in t.

The last statement of the theorem is true since $T(\varkappa)$ satisfies the above condition for all $\varkappa$ in virtue of Corollary 2.5.

These results show that holomorphic semigroups are rather stable under perturbation.

3. Perturbation of contraction semigroups

We have remarked in the preceding paragraph that when $-T$ is the generator of a bounded semigroup, $-(T+A)$ need not be such a generator even when A is relatively bounded with respect to T. An exception to this statement occurs when both $-T$ and $-A$ are generators of contraction semigroups. We have namely

Theorem 2.7.[1] *Let T and A belong to $\mathscr{G}(1,0)$ and let A be relatively bounded with respect to T with T-bound $<1/2$. Then $T+A\in\mathscr{G}(1,0)$ too.*

Proof. $\mathrm{P}(T)$ covers the half-plane $\operatorname{Re}\zeta<0$ and $\|(T+\xi)^{-1}\|\leq\xi^{-1}$ for $\xi>0$. A satisfies an inequality of the form (2.10), with $b<1/2$, so that $\|A(T+\xi)^{-1}\|\leq a\|(T+\xi)^{-1}\|+b\|T(T+\xi)^{-1}\|\leq a\xi^{-1}+2b<1$ if ξ is sufficiently large. It follows from the second Neumann series that the resolvent $(T+A+\xi)^{-1}$ exists for such ξ.

Let us estimate $(T+A+\xi)^{-1}$. To this end we consider the vector $v(t)=e^{-tA}e^{-t(T+\xi)}u$ where $u\in\mathsf{D}(T)\subset\mathsf{D}(A)$. We have

$$u-v(t)=(u-e^{-tA}u)+e^{-tA}(u-e^{-t(T+\xi)}u). \tag{2.17}$$

Since $e^{-tA}u$ and $e^{-t(T+\xi)}u$ are differentiable in t, we have

$$\lim_{t\searrow 0} t^{-1}(u-v(t))=Au+(T+\xi)u \tag{2.18}$$

by (1.24). On the other hand, we have $\|v(t)\|\leq e^{-t\xi}\|u\|$ since e^{-tA} and e^{-tT} are contraction semigroups. Hence

$$t^{-1}\|u-v(t)\|\geq t^{-1}(\|u\|-e^{-t\xi}\|u\|)\to\xi\|u\|\,,\quad t\to 0\,. \tag{2.19}$$

It follows from (2.18) and (2.19) that

$$\|(T+A+\xi)u\|\geq\xi\|u\|\,, \tag{2.20}$$

which implies that $\|(T+A+\xi)^{-1}\|\leq\xi^{-1}$, at least for ξ sufficiently large that $(T+A+\xi)^{-1}$ exists. Then one sees by means of the first

[1] This theorem and Theorem 2.11 are essentially contained in Trotter [2], in which the semigroup $e^{-t(T+A)}$ is constructed as the limit of $(e^{-(t/n)T}e^{-(t/n)A})^n$ as $n\to\infty$. An interesting application of this method is given by Nelson [1]. Cf. also Babbitt [1].

Neumann series that $(T + A + \xi)^{-1}$ exists with the same inequality for every $\xi > 0$. This shows that $T + A \in \mathscr{G}(1, 0)$, as we wished to show.

Problem 2.8. If $T \in \mathscr{G}(1, \beta)$, $A \in \mathscr{G}(1, \beta')$ and A is T-bounded with T-bound $< 1/2$, then $T + A \in \mathscr{G}(1, \beta + \beta')$.

Problem 2.9. The bound 1/2 for the T-bound of A in Theorem 2.7 and Problem 2.8 can be replaced by 1, if X is a Hilbert space.

Remark 2.10. Theorem 2.7 does not imply that $e^{-t(T+\varkappa A)}$ is holomorphic in $\varkappa$. In fact $\varkappa < 0$ is in general not permissible if $\varkappa A$ is to belong to $\mathscr{G}(1, \beta)$. But it can be shown that $e^{-t(T+\varkappa A)}$ is strongly continuous in $\varkappa$ for $\varkappa \geqq 0$. More generally, $e^{-t(T+A_n)} \underset{s}{\to} e^{-tT}$ if A_n tends to zero in the sense that A_n is T-bounded and $\|A_n u\| \leqq a_n \|u\| + b_n \|T u\|$ with $a_n, b_n \to 0$. This is a consequence of Theorem 2.16 proved below.

The T-boundedness of A in Theorem 2.7 is used essentially only in showing that $-\xi \in \mathrm{P}(T + A)$ for sufficiently large ξ. The proof of (2.20) is valid without any such assumption if $u \in \mathsf{D}(T+A) = \mathsf{D}(T) \cap \mathsf{D}(A)$. If the range of $T + A + \xi$ is dense in X and if $T + A$ is closable, it follows that $-\xi \in \mathrm{P}(S)$ with $\|(S + \xi)^{-1}\| \leqq \xi^{-1}$, where S is the closure of $T + A$. If S is densely defined, it follows as above that $S \in \mathscr{G}(1, 0)$. Thus we have the following theorem, which is symmetric with respect to T and A.

Theorem 2.11. *Let T and A belong to $\mathscr{G}(1, 0)$, let $\mathsf{D}(T) \cap \mathsf{D}(A)$ be dense in X and let $T + A + \xi$ have a dense range for sufficiently large real ξ. If $T + A$ is closable, its closure S belongs to $\mathscr{G}(1, 0)$.*

4. Convergence of quasi-bounded semigroups in a restricted sense

Let us return to the generators of quasi-bounded semigroups in general. When $-T$ is such a generator, it is in general difficult to prove the same for $-S = -(T + A)$ for an unbounded perturbation A. If we *assume* that $-S$ is also a generator, however, we can show that $e^{-tS} - e^{-tT}$ is small relative to T provided A is small relative to T, even if A is unbounded. More precisely, we have

Theorem 2.12. *Let $T, S \in \mathscr{G}(M, \beta)$, $\beta \geqq 0$, and let $S = T + A$ where A is T-bounded so that* (2.10) *is true. Then we have for $\xi > \beta$*

$$\begin{aligned} (2.21) \qquad & \|(e^{-tS} - e^{-tT}) (T + \xi)^{-1}\| \leqq \\ & \leqq M^2 t\, e^{\beta t} [b(M + 1) + (a + b\,\beta)\, M (\xi - \beta)^{-1}] \,. \end{aligned}$$

Proof. Let $u \in \mathsf{D}(T) = \mathsf{D}(S)$. Then we have, as in (2.2),

$$(2.22) \qquad (e^{-tS} - e^{-tT})\, u = -\int_0^t e^{-(t-s) S} A\, e^{-sT} u\, ds\,;$$

here $A e^{-sT} u = A(T+\xi)^{-1} e^{-sT}(T+\xi) u$ is continuous in s because $A(T+\xi)^{-1}$ is bounded. Since $(T+\xi)^{-1} v \in \mathsf{D}(T)$ for any $v \in \mathsf{X}$, (2.22) applied to $u = (T+\xi)^{-1} v$ gives

$$(2.23) \quad (e^{-tS} - e^{-tT})(T+\xi)^{-1} = -\int_0^t e^{-(t-s)S} A (T+\xi)^{-1} e^{-sT} ds .$$

Since $\|e^{-(t-s)S}\| \leq M e^{\beta(t-s)}$, $\|e^{-sT}\| \leq M e^{\beta s}$, and

$$(2.24) \quad \|A(T+\xi)^{-1}\| \leq a\|(T+\xi)^{-1}\| + b\|T(T+\xi)^{-1}\| \leq aM(\xi-\beta)^{-1} + b\|1-\xi(T+\xi)^{-1}\| \leq b(1+M) + M(a+b\beta)(\xi-\beta)^{-1},$$

(2.21) follows from (2.23).

Remark 2.13. Theorem 2.12 shows that $(e^{-tS} - e^{-tT}) u$ tends to zero when $a, b \to 0$, not uniformly for $\|u\| \leq 1$ but uniformly for all $u \in \mathsf{D}(T)$ with $\|(T+\xi) u\| \leq 1$.

In this direction we can go further and obtain a certain kind of estimate for $e^{-tS} - e^{-tT}$ without assuming anything on $S - T$.

Theorem 2.14. *Let $T, S \in \mathscr{G}(M, \beta)$. Then*

$$(2.25) \quad \|(S+\zeta)^{-1}(e^{-tS} - e^{-tT})(T+\zeta)^{-1}\| \leq M^2 t e^{\beta t}\|(S+\zeta)^{-1} - (T+\zeta)^{-1}\|, \quad \mathrm{Re}\,\zeta > \beta .$$

Proof. $e^{-tT}(T+\zeta)^{-1}$ is strongly differentiable in t: $(d/dt)\, e^{-tT}(T+\zeta)^{-1} u = -e^{-tT} T(T+\zeta)^{-1} u = -e^{-tT}[1-\zeta(T+\zeta)^{-1}]$, and similarly for $e^{-tS}(S+\zeta)^{-1}$. Hence

$$(2.26) \quad \frac{d}{ds} e^{-(t-s)S}(S+\zeta)^{-1} e^{-sT}(T+\zeta)^{-1} = e^{-(t-s)S} e^{-sT}(T+\zeta)^{-1} - e^{-(t-s)S}(S+\zeta)^{-1} e^{-sT} = e^{-(t-s)S}[(T+\zeta)^{-1} - (S+\zeta)^{-1}]e^{-sT}.$$

Integrating with respect to s on $(0, t)$, we obtain

$$(2.27) \quad (S+\zeta)^{-1}(e^{-tT} - e^{-tS})(T+\zeta)^{-1} = \int_0^t e^{-(t-s)S}[(T+\zeta)^{-1} - (S+\zeta)^{-1}] e^{-sT} ds ,$$

from which (2.25) follows by $\|e^{-(t-s)S}\| \leq M e^{(t-s)\beta}$ and $\|e^{-sT}\| \leq M e^{\beta s}$.

Example 2.15. Let $T = iH$ where H is a selfadjoint operator in a Hilbert space H. If K is a symmetric operator relatively bounded with respect to H with H-bound < 1, $H + K$ is selfadjoint (see Theorem V-4.3). The unitary operator $e^{-it(H+K)}$ tends to e^{-itH} in the sense of Remark 2.13 when K tends to zero in the sense stated.

5. Strong convergence of quasi-bounded semigroups

If we content ourselves with *strong convergence* of the semigroups generated by a given sequence of generators, we can weaken the assumptions on the perturbation considerably. The following theorem is fundamental in the approximation theory of semigroups.

Theorem 2.16. *Let T and T_n, $n = 1, 2, \ldots$, belong to $\mathscr{G}(M, \beta)$. If*

$$(T_n + \zeta)^{-1} \xrightarrow[s]{} (T + \zeta)^{-1} \tag{2.28}$$

for some ζ with $\mathrm{Re}\,\zeta > \beta$, *then*

$$e^{-tT_n} \xrightarrow[s]{} e^{-tT} \tag{2.29}$$

uniformly in any finite interval of $t \geqq 0$[1]. *Conversely, if (2.29) holds for all t such that $0 \leqq t \leqq b$, $b > 0$, then (2.28) holds for every ζ with* $\mathrm{Re}\,\zeta > \beta$.

Proof. We may assume $\beta = 0$ without loss of generality. We start from the identity (2.27) in which S is replaced by T_n. Since $\|e^{-(t-s)T_n}\| \leqq M$, we have for any $u \in \mathsf{X}$

$$\begin{aligned} \|(T_n + \zeta)^{-1}(e^{-tT_n} - e^{-tT})\,(T + \zeta)^{-1} u\| &\leqq \\ \leqq M \int_0^t \|[(T_n + \zeta)^{-1} - (T + \zeta)^{-1}]\, e^{-sT} u\|\, ds\,. \end{aligned} \tag{2.30}$$

If $(T_n + \zeta)^{-1} \xrightarrow[s]{} (T + \zeta)^{-1}$, the integrand on the right of (2.30) tends as $n \to \infty$ to zero for each fixed s. Furthermore, the integrand is majorized by $2M\,\xi^{-1} M \|u\|$, which is independent of n, where $\xi = \mathrm{Re}\,\zeta > 0$. Therefore, the right member of (2.30) tends to zero by the principle of bounded convergence. Obviously the convergence is uniform in t in any finite interval.

On writing $v = (T + \zeta)^{-1} u$, we have thus shown that

$$(T_n + \zeta)^{-1}(e^{-tT_n} - e^{-tT})\, v \to 0\,, \quad n \to \infty\,, \tag{2.31}$$

uniformly in t. This is true for all $v \in \mathsf{D}(T)$, since v varies over $\mathsf{D}(T)$ when u varies over X. But the operator on the left of (2.31) is uniformly bounded with bound not exceeding $2M^2 \xi^{-1}$. Hence (2.31) is true for all $v \in \mathsf{X}$.

On the other hand we have

$$\begin{aligned} &(T_n + \zeta)^{-1} e^{-tT_n} v - e^{-tT_n} (T + \zeta)^{-1} v \\ &\qquad = e^{-tT_n}[(T_n + \zeta)^{-1} v - (T + \zeta)^{-1} v] \to 0\,, \end{aligned} \tag{2.32}$$

since $\|e^{-tT_n}\| \leqq M$ and $(T_n + \zeta)^{-1} v - (T + \zeta)^{-1} v \to 0$. Similarly,

$$\begin{aligned} &(T_n + \zeta)^{-1} e^{-tT} v - e^{-tT} (T + \zeta)^{-1} v \\ &\qquad = [(T_n + \zeta)^{-1} - (T + \zeta)^{-1}]\, e^{-tT} v \to 0\,. \end{aligned} \tag{2.33}$$

It follows from (2.31)–(2.33) that

$$(e^{-tT_n} - e^{-tT})\,(T + \zeta)^{-1} v \to 0\,. \tag{2.34}$$

[1] Recall that (2.28) means that $T_n \xrightarrow[s]{} T$ in the generalized sense (see VIII-§ 1) Sufficient conditions for (2.28) are extensively studied in VIII-§§ 1 and 3.

This means that $(e^{-tT_n} - e^{-tT})\, w \to 0$ for all $w \in \mathsf{D}(T)$. It follows as above that $e^{-tT_n} - e^{-tT} \underset{s}{\to} 0$.

That this convergence is uniform in t follows from the uniformity of convergence in (2.31)–(2.33). For (2.31) this was remarked above. For (2.32) it is obvious. For (2.33) it follows from the continuity of $e^{-tT} v$ in t, which implies that the set of $e^{-tT} v$ for all t of a finite interval is compact, so that Lemma III-3.7 is applicable.

Conversely, suppose that (2.29) is true for $0 \leqq t \leqq b$. By $e^{-mtT} = (e^{-tT})^m$ etc., it follows that (2.29) is true for any $t \geqq 0$. Then

$$\int\limits_0^\infty e^{-\zeta t}(e^{-tT_n} - e^{-tT})\, dt \underset{s}{\to} 0\,, \quad \mathrm{Re}\,\zeta > 0\,,$$

by dominated convergence. In view of the inversion formula (1.28), this gives (2.28) for every ζ with $\mathrm{Re}\,\zeta > 0$.

In Theorem 2.16 the limit $-T$ of the generators $-T_n$ is *assumed* to be a generator. The question naturally arises whether this is a consequence of the other assumptions. It turns out that the answer is yes if one assumes a certain uniformity on the behavior of the resolvents of the T_n. We have namely

Theorem 2.17.[1] *Let* $T_n \in \mathscr{G}(M, \beta)$, $n = 1, 2, \ldots$, *and let* $\underset{\alpha \searrow 0}{\text{s-lim}}\, (1 + \alpha T_n)^{-1} = 1$ *uniformly in* n. *Let* $\underset{n\to\infty}{\text{s-lim}}\, (T_n + \zeta)^{-1}$ *exist for some* ζ *with* $\mathrm{Re}\,\zeta > \beta$. *Then there is a* $T \in \mathscr{G}(M, \beta)$ *such that the results of* Theorem 2.16 *hold*[2].

Proof. Again we may assume $\beta = 0$. $T_n \in \mathscr{G}(M, 0)$ implies that the $(T_n + \zeta)^{-1}$ are uniformly bounded in n for each fixed complex number ζ with $\mathrm{Re}\,\zeta > 0$. In other words, the right half-plane $\mathrm{Re}\,\zeta > 0$ belongs to the region of boundedness Δ_b for the sequence of the resolvents of $-T_n$ (see VIII-§ 1.1). But it follows from the assumptions that at least one ζ with $\mathrm{Re}\,\zeta > 0$ belongs to the region of strong convergence Δ_s. Since Δ_s is relatively open and closed in Δ_b by Theorem VIII-1.2, Δ_s must contain the half-plane $\mathrm{Re}\,\zeta > 0$.

Thus $\text{s-lim}\,(T_n + \zeta)^{-1} = -R'(\zeta)$ exists for all ζ of this half-plane, and $R'(\zeta)$ is a pseudo-resolvent (see loc. cit.). We shall now show that $R'(\zeta)$ is the resolvent of a densely defined closed operator $-T$.

For any $u \in \mathsf{X}$ we have $u = \underset{\alpha \searrow +0}{\lim}\, (1 + \alpha T_n)^{-1} u = \underset{\xi \to +\infty}{\lim}\, \xi (T_n + \xi)^{-1} u$. Since this convergence is uniform in n by hypothesis, it follows that

$$u = -\lim_{\xi \to +\infty} \xi\, R'(\xi)\, u\,, \quad u \in \mathsf{X}\,. \tag{2.35}$$

Now the $R'(\zeta)$ have a common null space N and a common range D (see loc. cit.). If $u \in \mathsf{N}$, we have $R'(\xi)\, u = 0$ so that $u = 0$ by (2.35).

[1] Cf. Trotter [1].

[2] $\text{s-lim}\,(1 + \alpha T_n)^{-1} = 1$ follows from $T_n \in \mathscr{G}(M, \beta)$; see (1.12). The *uniformity* in n of this convergence is the essential assumption.

Thus $\mathsf{N} = 0$. Also (2.35) implies that D is dense in X, for any $u \in \mathsf{X}$ is the limit of $-\xi R'(\xi) u \in \mathsf{D}$. It follows (see loc. cit.) that $R'(\zeta)$ is the resolvent of a densely defined operator $-T$: $R'(\zeta) = -(T + \zeta)^{-1}$.

The inequalities $\|(T_n + \xi)^{-k}\| \leqq M/\xi^k$ go over for $n \to \infty$ to the same inequalities for $\|(T + \xi)^{-k}\|$. This shows that $T \in \mathscr{G}(M, 0)$, completing the proof.

6. Asymptotic perturbation of semigroups

We have been able to develop a satisfactory *analytic* perturbation theory for quasi-bounded semigroups only for bounded perturbations (see par. 1). In the more restricted class of holomorphic semigroups, it was possible to admit relatively bounded perturbations (see par. 2). In applications, however, we have often to deal with non-holomorphic semigroups, the most important case being the unitary groups in a Hilbert space. Therefore, it is desirable to develop the perturbation theory for non-holomorphic semigroups further by admitting unbounded perturbations. Naturally, then, we have to content ourselves with obtaining weaker results than the analytic dependence of the semigroup on the parameter $\varkappa$. In this way we are led to consider *asymptotic series* in $\varkappa$ which are valid for $\varkappa \to 0$. A simple example of asymptotic behavior of semigroups is obtained if we set $A = \varkappa T^{(1)}$ and let $\varkappa \to 0$ in Theorem 2.12 [assuming that $T + \varkappa T^{(1)} \in \mathscr{G}(M, \beta)$ for all $\varkappa$]. What we are going to do in the sequel is to extend the result to higher orders of the parameter[1] $\varkappa$.

For simplicity we shall consider a family of operators

$$T(\varkappa) = T + \varkappa T^{(1)}, \tag{2.36}$$

though we could as well consider a formal infinite series in $\varkappa$. We assume that $T \in \mathscr{G}(M, \beta)$ and $T^{(1)}$ is T-bounded:

$$\|T^{(1)} u\| \leqq a\|u\| + b\|Tu\|, \quad u \in \mathsf{D}(T) \subset \mathsf{D}(T^{(1)}). \tag{2.37}$$

As remarked before, this does not guarantee that $-T(\varkappa)$ is the generator of a semigroup, though $T(\varkappa)$ is a closed operator with domain $\mathsf{D} = \mathsf{D}(T)$ for $|\varkappa| < 1/b$ by Theorem IV-1.1. Therefore we add the assumption that $T(\varkappa) \in \mathscr{G}(M, \beta)$ for $\varkappa \in \mathrm{D}_0$, where D_0 is a subset of the disk $|\varkappa| < 1/b$. D_0 may be an open set, a subset of the real axis or a subset of the positive real axis. In what follows $\varkappa$ is always assumed to belong to D_0. Also we assume $\beta = 0$, for this does not affect the generality [otherwise consider $T(\varkappa) + \beta$ instead of $T(\varkappa)$].

Under these assumptions, $-T(\varkappa)$ generates a semigroup $U(t, \varkappa) = e^{-tT(\varkappa)}$ for each $\varkappa$. We write $U(t, 0) = U(t)$. The $U(t, \varkappa)$ are uniformly

[1] The following theorems were given by T. Kato [3] in the case in which the $T(\varkappa)$ are selfadjoint.

bounded by

$$\|U(t, \varkappa)\| \leq M, \quad t \geq 0. \tag{2.38}$$

Theorem 2.18. *$U(t, \varkappa) \underset{s}{\rightarrow} U(t)$, $\varkappa \to 0$, uniformly in t in any finite interval.*

Proof. This is a direct consequence of Theorem 2.12.

Theorem 2.19. *Let $u \in \mathsf{D}$. Then*

$$U(t, \varkappa)\, u = U(t)\, u + \varkappa\, u^{(1)}(t) + o(\varkappa), \quad \varkappa \to 0, \tag{2.39}$$

with

$$u^{(1)}(t) = -\int_0^t U(t-s)\, T^{(1)}\, U(s)\, u\, ds, \tag{2.40}$$

where $o(\varkappa)$ denotes a vector with norm $o(\varkappa)$ uniformly in each finite interval of t. The integrand on the right of (2.40) is continuous in s so that the integral is well defined.

Proof. The second term on the right of (2.39) is formally identical with the $v_1(t) = U_1(t)\, u$ of par. 1 if we set $A = \varkappa\, T^{(1)}$ and $u_0 = u$. Although $T^{(1)}$ is not bounded, the integrand of (2.40) is continuous in s because

$$T^{(1)}\, U(s)\, u = B(\zeta)\, U(s)\, (T + \zeta)\, u, \tag{2.41}$$

where [cf. (2.24)]

$$B(\zeta) = T^{(1)} (T + \zeta)^{-1} \in \mathscr{B}(\mathsf{X}), \quad \operatorname{Re}\zeta > 0, \tag{2.42}$$

$$\|B(\zeta)\| \leq a M\, \xi^{-1} + b(1 + M), \quad \xi = \operatorname{Re}\zeta. \tag{2.43}$$

Similarly $T^{(1)}\, U(s, \varkappa)\, u$ is continuous in s. To see this we write

$$T^{(1)}\, U(s, \varkappa)\, u = B(\zeta, \varkappa)\, U(s, \varkappa)\, (T(\varkappa) + \zeta)\, u, \tag{2.44}$$

where

$$B(\zeta, \varkappa) = T^{(1)} (T(\varkappa) + \zeta)^{-1} \in \mathscr{B}(\mathsf{X}) \tag{2.45}$$

because $T^{(1)}$ is $T(\varkappa)$-bounded too (see Problem IV-1.2). $B(\zeta, \varkappa)$ is even holomorphic in ζ and $\varkappa$, for we have from the second Neumann series for $(T(\varkappa) + \zeta)^{-1} = (T + \zeta + \varkappa\, T^{(1)})^{-1}$

$$B(\zeta, \varkappa) = \sum_{k=1}^{\infty} (-\varkappa)^{k-1} B(\zeta)^k. \tag{2.46}$$

Since $u \in \mathsf{D}$ implies that $U(t, \varkappa)\, u \in \mathsf{D}$ and $dU(t, \varkappa)\, u/dt = -T(\varkappa)\, U(t, \varkappa)\, u = -T\, U(t, \varkappa)\, u - \varkappa\, T^{(1)}\, U(t, \varkappa)\, u$, it follows from (1.43) that

$$U(t, \varkappa)\, u = U(t)\, u - \varkappa \int_0^t U(t-s)\, T^{(1)}\, U(s, \varkappa)\, u\, ds. \tag{2.47}$$

(2.39) follows from (2.47) if it is proved that the integral on the right tends as $\varkappa \to 0$ to $-u^{(1)}(t)$ uniformly in t. For this it suffices to show that $\|T^{(1)} U(t, \varkappa) u - T^{(1)} U(t) u\| \to 0$ boundedly in t (principle of bounded convergence). But this is obvious from (2.44) and (2.41) since $B(\zeta, \varkappa)$, $U(t, \varkappa)$, $(T(\varkappa) + \zeta) u = (T + \zeta) u + \varkappa T^{(1)} u$ are all uniformly bounded, $B(\zeta, \varkappa) \underset{u}{\to} B(\zeta)$, and $U(t, \varkappa) \underset{s}{\to} U(t)$ by Theorem 2.18 (we need only a fixed ζ).

Theorem 2.20. *Let $u \in \mathsf{D}(T^2)$. Then*

$$U(t, \varkappa) u = U(t) u + \varkappa u^{(1)}(t) + \varkappa^2 u^{(2)}(t) + o(\varkappa^2), \tag{2.48}$$

uniformly in each finite interval of t, where $u^{(1)}(t)$ is given by (2.40) *and*

$$u^{(2)}(t) = -\int_0^t U(t - s) T^{(1)} u^{(1)}(s)\, ds. \tag{2.49}$$

$T^{(1)} u^{(1)}(t)$ exists and is continuous in t, so that the integrand in (2.49) *is continuous in s.*

Proof. First we show that $T^{(1)} U(t) u$, which exists because $U(t) u \in \mathsf{D}$, is continuously differentiable in t, with

$$\frac{d}{dt} T^{(1)} U(t) u = -T^{(1)} U(t) T u. \tag{2.50}$$

In fact, we have $T^{(1)} U(t) u = B(\zeta) U(t) (T + \zeta) u$ by (2.41) and so $(d/dt) T^{(1)} U(t) u = -B(\zeta) U(t) T(T + \zeta) u = -B(\zeta) U(t) (T + \zeta) T u = -T^{(1)} U(t) T u$ because $T(T + \zeta) u$ exists by hypothesis.

According to Theorem 1.19, it follows that the $u^{(1)}(t)$ given by (2.40) is continuously differentiable and $T u^{(1)}(t)$ is continuous. Then $T^{(1)} u^{(1)}(t) = B(\zeta) (T + \zeta) u^{(1)}(t)$ is also continuous. This proves the last statement of the theorem.

In view of (2.47), (2.41) and (2.49), (2.48) will be proved if it is shown that

$$\int_0^t U(t - s) T^{(1)} \left\{\frac{1}{\varkappa} [U(s, \varkappa) u - U(s) u] - u^{(1)}(s)\right\} ds \to 0 \tag{2.51}$$

uniformly in each finite interval of t. Since $\|U(t - s)\| \leq M$ and $T^{(1)}$ is T-bounded, (2.51) is true if

$$(T + \zeta) \left\{\frac{1}{\varkappa} [U(t, \varkappa) u - U(t) u] - u^{(1)}(t)\right\} \to 0 \tag{2.52}$$

boundedly for a fixed ζ. But we have

$$\begin{aligned}
(T + \zeta) & [U(t, \varkappa) u - U(t) u] \\
&= (T + \zeta) (T(\varkappa) + \zeta)^{-1} U(t, \varkappa) (T(\varkappa) + \zeta) u - U(t) (T + \zeta) u \\
&= (1 + \varkappa B(\zeta))^{-1} U(t, \varkappa) (T + \zeta + \varkappa T^{(1)}) u - U(t) (T + \zeta) u \\
&= -\varkappa B(\zeta) (1 + \varkappa B(\zeta))^{-1} U(t, \varkappa) (T + \zeta + \varkappa T^{(1)}) u + \\
&\qquad + [U(t, \varkappa) - U(t)] (T + \zeta) u + \varkappa U(t, \varkappa) T^{(1)} u.
\end{aligned} \tag{2.53}$$

Hence

$$\frac{1}{\varkappa}(T+\zeta)\,[U(t,\varkappa)\,u - U(t)\,u] \to \\ \to -B(\zeta)\,U(t)\,(T+\zeta)\,u + v^{(1)}(t) + U(t)\,T^{(1)}\,u\,, \tag{2.54}$$

where $v^{(1)}(t)$ is the coefficient of $\varkappa$ in the asymptotic expansion of $U(t,\varkappa)\,(T+\zeta)\,u$, which exists by Theorem 2.19 because $(T+\zeta)\,u \in \mathsf{D}(T)$; in other words,

$$v^{(1)}(t) = -\int_0^t U(t-s)\,T^{(1)}\,U(s)\,(T+\zeta)\,u\,ds\,. \tag{2.55}$$

Since $B(\zeta)\,U(t)\,(T+\zeta)\,u = T^{(1)}\,U(t)\,u$, the right member of (2.54) is equal to

$$-T^{(1)}\,U(t)\,u + U(t)\,T^{(1)}\,u - \int_0^t U(t-s)\,T^{(1)}\,U(s)\,(T+\zeta)\,u\,ds \\ = -(T+\zeta)\int_0^t U(t-s)\,T^{(1)}\,U(s)\,u\,ds\,, \tag{2.56}$$

where we have used (1.47) with $f(t) = T^{(1)}\,U(t)\,u$, noting (2.50). This proves (2.52) and completes the proof of Theorem 2.20.

Remark 2.21. The proof of Theorem 2.20 is somewhat complicated since the assumption $u \in \mathsf{D}(T^2)$ does not imply $u \in \mathsf{D}(T(\varkappa)^2)$. It is not easy to go on in the same way to higher approximations of $U(t,\varkappa)\,u$ without introducing severe restrictions on u. For example, it is not possible to expand $U(t,\varkappa)\,u$ up to the third order in $\varkappa$ under the assumption $u \in \mathsf{D}(T^3)$ alone; it would be necessary to assume that $T^{(1)}\,u \in \mathsf{D}(T)$ in addition. We shall not pursue this problem any further.

Example 2.22. Let $T = iH$ where H is a selfadjoint operator in a Hilbert space H, and $T^{(1)} = iK$ where K is symmetric and H-bounded. Then $H + \varkappa K$ is selfadjoint for sufficiently small real $\varkappa$ and $T(\varkappa) = T + \varkappa T^{(1)}$ generates a unitary group, as does T. Thus the fundamental assumptions are satisfied if we take as D_0 a neighborhood of $\varkappa = 0$ on the real axis, and Theorems 2.18 to 2.20 are applicable to $U(t,\varkappa) = e^{-it(H+\varkappa K)}$.

§ 3. Approximation by discrete semigroups

1. Discrete semigroups

The perturbation theory of semigroups discussed in the preceding section may be regarded as a theory of approximation of a semigroup $U(t)$ by a family of semigroups, which may depend on a discrete parameter n (Theorems 2.16, 2.17) or on a continuous parameter $\varkappa$ (Theorems 2.18–2.20). In this section we consider approximation of a semigroup $U(t)$ by a sequence of discrete semigroups; the result will have applications to the approximation theory of differential equations by difference equations[1].

[1] The following presentation leans heavily on TROTTER [1].

A discrete semigroup is simply a family $\{U^k\}_{k=0,1,2,\ldots}$ consisting of powers of an operator $U \in \mathscr{B}(X)$; it has the semigroup property $U^j U^k = U^{j+k}$. In approximation theory, however, it is necessary to correlate the exponent k with the "time" variable t. To this end we associate a "time unit" $\tau > 0$ with each discrete semigroup $\{U^k\}$ and write $U^k = U(k\tau)$, $k = 0, 1, 2, \ldots$, the function $U(t)$ being thus defined only for a discrete set of values $t = k\tau$. $\{U(k\tau)\}$ will be called a *discrete semigroup with time unit* τ. To distinguish them from discrete semigroups, the semigroups $\{U(t)\}$ considered in preceding sections will be called *continuous semigroups*.

A discrete semigroup $\{U(k\tau)\}$ is said to be bounded if $\|U(k\tau)\| \leq M$, $k = 0, 1, 2, \ldots$, with a constant M, where $M \geqq 1$ since $U(0) = 1$.

For a discrete semigroup $\{U(k\tau)\}$ with time unit τ, set

$$T = \tau^{-1}(1 - U(\tau)) . \tag{3.1}$$

$-T$ is called the *generator* of $\{U(k\tau)\}$. Thus

$$U(k\tau) = (1 - \tau T)^k . \tag{3.2}$$

Since $T \in \mathscr{B}(X)$, $-T$ also generates a continuous semigroup $\{e^{-tT}\}$, which will be called the continuous semigroup associated with $\{U(k\tau)\}$. $\{e^{-tT}\}$ is an approximation to $\{U(k\tau)\}$ in a certain sense, as is seen from

Lemma 3.1. *If $\{U(k\tau)\}$ is bounded by $\|U(k\tau)\| \leq M$, the semigroup $\{e^{-tT}\}$ is also bounded by $\|e^{-tT}\| \leq M$ and*

$$\|U(k\tau)\, u - e^{-k\tau T} u\| \leq \frac{1}{2} M k \tau^2 \|T^2 u\| . \tag{3.3}$$

Proof. e^{-tT} can be defined simply by the Taylor series. Since $-tT = -(t/\tau) + (t/\tau)\, U(\tau)$, we have

$$\begin{aligned} U(k\tau)\, e^{-tT} &= e^{-t/\tau}\, U(k\tau) \sum_{n=0}^{\infty} \frac{(t/\tau)^n}{n!} U(n\tau) \\ &= e^{-t/\tau} \sum_{n=0}^{\infty} \frac{(t/\tau)^n}{n!} U((n+k)\tau) , \end{aligned} \tag{3.4}$$

hence

$$\|U(k\tau)\, e^{-tT}\| \leq M e^{-t/\tau} \sum_{n=0}^{\infty} \frac{(t/\tau)^n}{n!} = M . \tag{3.5}$$

In particular $\|e^{-tT}\| \leq M$.

Since the $U(k\tau)$ and T commute, we have

$$U(k\tau) - e^{-k\tau T} = \left[\sum_{j=0}^{k-1} U((k-j-1)\tau)\, e^{-j\tau T}\right] (U(\tau) - e^{-\tau T}) . \tag{3.6}$$

But

$$U(\tau) - e^{-\tau T} = 1 - \tau T - e^{-\tau T} = -\int_0^\tau (\tau - s)\, e^{-sT} T^2\, ds\,,$$

so that

$$\|U((k-j-1)\tau)\, e^{-j\tau T}(U(\tau) - e^{-\tau T})\, u\| \leqq$$

$$\leqq \int_0^\tau (\tau - s)\, M \|T^2 u\|\, ds = \frac{\tau^2}{2} M \|T^2 u\|$$

by (3.5). Thus (3.3) follows from (3.6).

Remark 3.2. To avoid the inconvenience that a discrete semigroup $U(t)$ is defined only for $t = k\tau$, we set

(3.7) $$U(t) = U\left(\left[\frac{t}{\tau}\right]\tau\right)$$

for any $t \geqq 0$, where $[t/\tau]$ is the greatest integer not exceeding t/τ. Of course $\{U(t)\}$ is then not a continuous semigroup, but we have

(3.8) $$\|U(t)\, u - e^{-tT} u\| \leqq \frac{1}{2} M \tau t \|T^2 u\| + M\tau \|Tu\|\,.$$

In fact, let $k\tau \leqq t < (k+1)\tau$; then $U(t) = U(k\tau)$ and $\|e^{-tT} u - - e^{-k\tau T} u\| \leqq \left\|\int_{k\tau}^t e^{-sT}\, Tu\, ds\right\| \leqq M\tau \|Tu\|$, so that (3.8) follows from (3.3).

2. Approximation of a continuous semigroup by discrete semigroups

Consider a sequence $\{U_n\}$ of discrete semigroups with time units τ_n, $n = 1, 2, \ldots$, where $\tau_n \to 0$, $n \to \infty$. $\{U_n\}$ is said to *approximate* a continuous semigroup $U = \{U(t)\}$ at $t = t_0$ if

(3.9) $$U_n(k_n \tau_n) \underset{s}{\to} U(t_0)\,, \quad n \to \infty\,,$$

for any sequence $\{k_n\}$ of nonnegative integers such that $k_n \tau_n \to t_0$. $\{U_n\}$ is said to approximate U on an interval I of t if (3.9) is true for each $t_0 \in \mathrm{I}$.

Lemma 3.3. *If $\{U_n\}$ approximates U on an interval $0 \leqq t < b$, $\{U_n\}$ approximates U on the whole interval $0 \leqq t < \infty$. (In this case we simply say that $\{U_n\}$ approximates U, and write $U_n \to U$.)*

Proof. Let $t_0 \geqq 0$ and $k_n \tau_n \to t_0$. Let m be an integer such that $t_0 < mb$, and let $k_n = m q_n + r_n$, $0 \leqq r_n < m$ (q_n, r_n are integers). Then $r_n \tau_n \to 0$ and $q_n \tau_n \to t_0/m < b$. Hence $U_n(r_n \tau_n) \underset{s}{\to} U(0) = 1$ and $U_n(q_n \tau_n) \underset{s}{\to} U(t_0/m)$, so that $U_n(k_n \tau_n) = U_n(q_n \tau_n)^m U_n(r_n \tau_n) \underset{s}{\to} U(t_0/m)^m = U(t_0)$.

Lemma 3.4. *If $U_n \to U$, the U_n are uniformly quasi-bounded in the sense that*

$$\|U_n(t)\| \leqq M e^{\beta t}, \quad t \geqq 0, \tag{3.10}$$

where M and β are independent of n and t, and where $U_n(t)$ is defined for all $t \geqq 0$ according to (3.7).

Proof. For each fixed $u \in X$, the values $\|U_n(k\tau_n) u\|$ are bounded for all integers n and k such that $k\tau_n \leqq 1$; otherwise there would exist a sequence $\{k_n\}$ such that $\|U_n(k_n \tau_n) u\| \to \infty$, $k_n \tau_n \to t_0 \leqq 1$, when $n \to \infty$ along a subsequence, contrary to the assumption. It follows from the principle of uniform boundedness that $\|U_n(k\tau_n)\| \leqq M$ for $k\tau_n \leqq 1$ (note that $M \geqq 1$).

If $k > 1/\tau_n$ and $\tau_n < 1$, write $k = q m_n + r$, $0 \leqq r < m_n = [1/\tau_n]$ (q, r are integers). Then

$$\begin{aligned} \|U_n(k\tau_n)\| &= \|U_n(m_n \tau_n)^q U_n(r\tau_n)\| \leqq \\ &\leqq \|U_n(m_n\tau_n)\|^q \|U_n(r\tau_n)\| \leqq M^{q+1}. \end{aligned}$$

But since $m_n + 1 > 1/\tau_n$, we have $q \leqq k/m_n \leqq k\tau_n/(1-\tau_n) \leqq 2k\tau_n$ if $\tau_n \leqq 1/2$. Hence $\|U_n(k\tau_n)\| \leqq M \exp(2k\tau_n \log M)$, which is also true for $k\tau_n \leqq 1$. Since $U_n(t) = U_n(k\tau_n)$ where $k = [t/\tau_n] \leqq t/\tau_n$, we obtain $\|U_n(t)\| \leqq M \exp(2t \log M)$. Since there are only finitely many n for which $\tau_n > 1/2$, this proves the lemma with $\beta = 2\log M$, with some later change in M and β if necessary.

Lemma 3.5. *For $U_n \to U$, it is necessary that $U_n(t) \underset{s}{\to} U(t)$ uniformly in each finite interval of t, and it is sufficient that this be true for some interval* $[0, b]$.

Proof. Let $U_n \to U$ and suppose that for some $u \in X$ and some finite closed interval I, it is not true that $U_n(t) u \to U(t) u$ uniformly for $t \in$ I. Then there is a sequence $t_n \in$ I such that

$$\|U_n(t_n) u - U(t_n) u\| \geqq \varepsilon > 0 \tag{3.11}$$

(replacing $\{U_n\}$ by a subsequence if necessary). Choosing a further subsequence, we may assume that $t_n \to t_0 \in$ I, $n \to \infty$. Set $k_n = [t_n/\tau_n]$. Then $k_n \tau_n \to t_0$ because $\tau_n \to 0$. Since $U_n(t_n) = U_n(k_n \tau_n)$ by (3.7) and since $U(t_n) u \to U(t_0) u$ by the strong continuity of $U(t)$, (3.11) gives $\limsup \|U_n(k_n \tau_n) u - U(t_0) u\| \geqq \varepsilon$, a contradiction to (3.9).

Suppose, conversely, that $U_n(t) u \to U(t) u$ uniformly for $t \in [0, b]$ for each $u \in X$. Then $U_n(k_n \tau_n) u - U(k_n \tau_n) u \to 0$ for $k_n \tau_n \in [0, b]$. In particular it is true if $k_n \tau_n \to t_0 < b$. Since $U(k_n \tau_n) u \to U(t_0) u$ by the strong continuity of $U(t)$, (3.9) follows for $t_0 < b$. Hence $U_n \to U$ by Lemma 3.3.

3. Approximation theorems

In this paragraph we consider a sequence $\{U_n\}$ of discrete semigroups with time units τ_n and generators $-T_n$, where $\tau_n \to 0$. For $\{U_n\}$ to approximate a continuous semigroup, it is necessary that $\{U_n\}$ be uniformly quasi-bounded (see Lemma 3.4). Therefore we consider only such a sequence $\{U_n\}$.

Theorem 3.6. *Let $\{U_n\}$ be uniformly quasi-bounded as in* (3.10), *and let U be a continuous semigroup with the generator $-T$. For $U_n \to U$, it is necessary and sufficient that $T_n \underset{s}{\to} T$ in the generalized sense; in other words, it is necessary that*

$$(T_n + \zeta)^{-1} \underset{s}{\to} (T + \zeta)^{-1} \tag{3.12}$$

for every ζ with $\mathrm{Re}\,\zeta > \beta$, *and it is sufficient that* (3.12) *hold for some ζ with* $\mathrm{Re}\,\zeta > \beta$.

Proof. It is not difficult to see that the general case can be reduced to the case $\beta = 0$ by the transformations $U_n(t) \to e^{-\beta t} U_n(t)$, $U(t) \to e^{-\beta t} U(t)$. Thus we assume that $\{U_n\}$ is uniformly bounded: $\|U_n(t)\| \leqq M$, $t \geqq 0$, where $U_n(t)$ is defined for all $t \geqq 0$ according to (3.7).

Suppose now that $U_n \to U$. Then $U_n(t) \underset{s}{\to} U(t)$ uniformly in each finite interval of t by Lemma 3.5. Hence for $\mathrm{Re}\,\zeta > 0$

$$\begin{aligned}(T_n + \zeta)^{-1} &= [\zeta + \tau_n^{-1}(1 - U_n(\tau_n))]^{-1} \\ &= \tau_n \sum_{k=0}^{\infty} \frac{U_n(k\tau_n)}{(1+\tau_n\zeta)^{k+1}} = \int_0^\infty \frac{U_n(t)\,dt}{(1+\tau_n\zeta)^{[t/\tau_n]+1}}\,.\end{aligned}$$

It follows by the dominated convergence theorem that

$$(T_n + \zeta)^{-1} \underset{s}{\to} \int_0^\infty e^{-\zeta t}\, U(t)\, dt = (T + \zeta)^{-1};$$

here it should be noted that $(1 + \tau_n \zeta)^{-[t/\tau_n]-1} \to e^{-\zeta t}$ *dominatedly*, for

$$\begin{aligned}|(1+\tau_n\zeta)^{-[t/\tau_n]-1}| &\leqq (1+\tau_n \mathrm{Re}\,\zeta)^{-[t/\tau_n]-1} \leqq \\ &\leqq (1+\tau_n \mathrm{Re}\,\zeta)\left(1 + t\,\mathrm{Re}\,\zeta + \frac{1}{2} t^2 (\mathrm{Re}\,\zeta)^2\right)^{-1},\end{aligned}$$

the last member being integrable on $0 \leqq t < \infty$.

Suppose, conversely, that (3.12) is satisfied by some ζ with $\mathrm{Re}\,\zeta > 0$. Since $T_n \in \mathscr{G}(M, 0)$ by Lemma 3.1, it follows from Theorem 2.16 that (3.12) is true for any ζ with $\mathrm{Re}\,\zeta > 0$ and that $e^{-tT_n} \underset{s}{\to} e^{-tT}$ uniformly in each finite interval of t. Thus it remains to show that

$$U_n(t) - e^{-tT_n} \underset{s}{\to} 0 \tag{3.13}$$

uniformly in each finite interval.

To this end we apply (3.8) with U replaced by U_n. Replacing u by $(T_n + 1)^{-2} u$, we see that

$$\|(U_n(t) - e^{-tT_n})(T_n + 1)^{-2}\| \leqq$$
$$\leqq M\tau_n \left(\frac{1}{2} t\|T_n^2(T_n + 1)^{-2}\| + \|T_n(T_n + 1)^{-2}\|\right) \to 0$$

since $\|T_n^2(T_n + 1)^{-2}\| \leqq (1 + M)^2$, $\|T_n(T_n + 1)^{-2}\| \leqq M(1 + M)$. Since $(T_n + 1)^{-2} \underset{s}{\to} (T + 1)^{-2}$ and $U_n(t)$, e^{-tT_n} are uniformly bounded, it follows that

$$(U_n(t) - e^{-tT_n})(T + 1)^{-2} \underset{s}{\to} 0 .$$

Since $\mathsf{R}((T + 1)^{-2})$ is dense in X, (3.13) follows again by using the uniform boundedness of $U_n(t)$ and e^{-tT_n}.

Remark 3.7. (3.12) is satisfied if there is a core D of T such that $T_n u \to Tu$ for $u \in \mathsf{D}$ (see Theorem VIII-1.5).

Example 3.8. Let $T \in \mathscr{G}(M, \beta)$ and $U(t) = e^{-tT}$. Let $U_n(k/n) = (1 + n^{-1} T)^{-k}$. U_n is a discrete semigroup with time unit $\tau_n = 1/n$. The generator $-T_n$ of U_n is given by $T_n = n[1 - (1 + n^{-1} T)^{-1}] = T(1 + n^{-1} T)^{-1}$. Hence $T_n u = (1 + n^{-1} T)^{-1} Tu \to Tu$ if $u \in \mathsf{D}(T)$ [see (1.12)]. Thus $U_n \to U$ by Remark 3.7; this is another formulation of the result of § 1.4, where $U(t)$ was defined exactly in this way.

The following theorem, which does not assume explicitly that the limit in (3.12) is the resolvent of a generator, can be proved exactly as Theorem 2.17.

Theorem 3.9. Let $\{U_n\}$ be uniformly quasi-bounded and let $(1 + \alpha T_n)^{-1} \underset{s}{\to} 1$, $\alpha \searrow 0$, uniformly in n. Let $\operatorname{s-lim}_{n\to\infty} (T_n + \zeta)^{-1}$ exist for some ζ with $\operatorname{Re}\zeta > \beta$. Then there is a continuous semigroup $U(t) = e^{-tT}$ such that $U_n \to U$ and $T_n \underset{s}{\to} T$ in the generalized sense.

4. Variation of the space

The approximation theorems proved in the preceding paragraph are not directly applicable to the approximation of a differential equation by difference equations[1], since the difference operators act in spaces different from the one in which the differential operator acts. To deal with such a problem, it is necessary to consider the approximation of a continuous semigroup U of operators in a Banach space X by a sequence $\{U_n\}$ of discrete semigroups acting in other Banach spaces X_n.

Let X and X_n be Banach spaces. Suppose there exists, for each n, a $P_n \in \mathscr{B}(\mathsf{X}, \mathsf{X}_n)$ such that

i) $\|P_n\| \leqq N$ (N independent of n);

ii) $\|P_n u\| \to \|u\|$, $n \to \infty$, for each $u \in \mathsf{X}$;

[1] Except when only the time variable t is discretized.

iii) there is a constant N' (independent of n) such that each $v \in X_n$ can be expressed as $v = P_n u$ with $\|u\| \leq N' \|v\|$.

Then a sequence $\{u_n\}$, $u_n \in X_n$, is said to converge to $u \in X$, in symbol $u_n \to u$, if[1]

$$\|u_n - P_n u\| \to 0\,, \quad n \to \infty\,. \tag{3.14}$$

A sequence $A_n \in \mathscr{B}(X_n)$ is said to converge strongly to $A \in \mathscr{B}(X)$, in symbol $A_n \underset{s}{\to} A$, if

$$A_n P_n u \to A u\,, \quad \text{that is,} \quad \|A_n P_n u - P_n A u\| \to 0\,. \tag{3.15}$$

With this definition of the strong convergence $A_n \underset{s}{\to} A$, it is easy to show that $A_n \underset{s}{\to} A$ implies that the A_n are uniformly bounded, $A_n \underset{s}{\to} A$ and $B_n \underset{s}{\to} B$ imply $A_n B_n \underset{s}{\to} A B$, etc.

Once the notion of strong convergence is defined with these properties, it is easy to define the approximation $U_n \to U$ of a continuous semigroup U in X by a sequence $\{U_n\}$ of discrete semigroups, where U_n acts in X_n, exactly as before. Then it is easy to verify that Lemmas and Theorems 3.3–3.9 are valid. Details may be left to the reader[2].

Example 3.10. Let X be the subspace of $C[0,1]$ consisting of all functions vanishing at 0 and 1, $X_n = C^{m_n}$ [the set of m_n-dimensional numerical vectors $v = (\xi_1, \ldots, \xi_{m_n})$ with the norm $\|v\| = \max|\xi_j|$]. For each $u \in X$ set $P_n u = v = (\xi_j) \in X_n$ with $\xi_j = u(j h_n)$, $h_n = 1/(m_n + 1)$. $P_n u$ is the approximation of a continuous function $u(x)$ by the set of its values at the m_n mesh points for the mesh width h_n. If $m_n \to \infty$, the conditions i) to iii) are satisfied with $N = N' = 1$. The notion of convergence $u_n \to u$ defined as above is well adapted to such an approximation.

Now consider the operators T and T_n defined as follows. T is an operator in X given by $T u = -d^2 u/d x^2$ with $D(T)$ consisting of all $u \in X$ with $u'' \in X$. T_n is an operator in X_n such that for $v = (\xi_j)$, $T_n v = w = (\eta_j)$ is given by

$$\eta_j = -(\xi_{j+1} - 2\xi_j + \xi_{j-1})/h_n^2 \quad (\text{set } \xi_0 = \xi_{m_n+1} = 0)\,.$$

Then it is easily seen that $T_n P_n u \to T u$ for $u \in D(T)$, in the sense of (3.15). Thus the condition corresponding to Remark 3.7 is satisfied. It follows that the discrete semigroups U_n generated by the $-T_n$ approximate $U(t) = e^{-tT}$ *provided that the U_n are uniformly quasi-bounded*[3]. Whether or not this condition is satisfied depends on the rate of convergence $\tau_n \to 0$. It is known[4] that it is satisfied if $\tau_n/h_n^2 \leq c < 1/2$. If, on the other hand, we set $T'_n = T_n(1 + \tau_n T_n)^{-1}$ and define the discrete semigroup U'_n generated by $-T'_n$ with time unit τ_n, the condition of Theorem 3.6 is satisfied and $U'_n \to U$ is true if only $\tau_n \to 0$ (without any restriction on the ratios τ_n/h_n^2). $\{U_n\}$ and $\{U'_n\}$ correspond respectively to the choice of the forward and backward difference coefficients in time, in the approximation of the heat equation $\partial u/\partial t = \partial^2 u/\partial x^2$ by difference equations.

[1] In what follows we denote by $\| \;\|$ the norms in different spaces X and X_n, but there should be no possibility of confusion.

[2] Cf. Trotter [1].

[3] This is called the *stability condition.*

[4] See e. g. Richtmyer [1].

Chapter Ten

Perturbation of continuous spectra and unitary equivalence

This chapter is concerned with the perturbation theory for continuous spectra. The operators considered are mostly selfadjoint. The stability of the continuous spectrum under a small perturbation has been studied rather extensively, though the results are by no means satisfactory. It is known that the continuous spectrum is rather unstable, even under degenerate perturbations. In this respect it is much worse-behaved than the essential spectrum (which is in general larger than the continuous spectrum). On the other hand, the absolutely continuous spectrum (which is in general smaller than the continuous spectrum) is stable under certain restricted perturbations; furthermore, the absolutely continuous parts of the perturbed and the unperturbed operators are seen to be unitarily equivalent.

These results are closely related to scattering theory in quantum mechanics. They are even proved here by the so-called time-dependent method in scattering theory, for it seems to be the most elementary way to deduce general results. There are other useful methods (stationary methods) in scattering theory. It is impossible to include these methods in full here, partly because they are currently under rapid development. But an account of a special one of them is given in the last section, which has results not covered by the time-dependent method.

§ 1. The continuous spectrum of a selfadjoint operator

1. The point and continuous spectra

Let H be a selfadjoint operator in a Hilbert space H. We have the spectral representation (see VI-§ 5)

$$H = \int_{-\infty}^{\infty} \lambda \, dE(\lambda) , \tag{1.1}$$

where $\{E(\lambda)\}$ is the right-continuous spectral family associated with H. We set

$$P(\lambda) = E(\lambda) - E(\lambda - 0) . \tag{1.2}$$

$P(\lambda) \neq 0$ if and only if λ is an eigenvalue of H; in this case $P(\lambda)$ is the orthogonal projection on the associated eigenspace. The $P(\lambda)$ for different λ are mutually orthogonal: $P(\lambda) P(\mu) = 0$ for $\lambda \neq \mu$.

The set of all eigenvalues of H is called the *point spectrum* of H, $\Sigma_p(H)$ in symbol. It is at most a countable set if H is separable.

Let H_p be the closed linear manifold spanned by all the $P(\lambda)\,\mathsf{H}$. If $\mathsf{H}_p = \mathsf{H}$, H is said to have a ***pure point spectrum***[1], or to be ***spectrally discontinuous***. In general H_p reduces H since each $P(\lambda)$ does. Let H_p be the part of H in H_p. H_p is spectrally discontinuous; in fact $P(\lambda)\,\mathsf{H}$ is, if not 0, exactly the eigenspace of H_p for the eigenvalue λ.

If $\mathsf{H}_p = 0$, H is said to have a ***pure continuous spectrum*** or to be ***spectrally continuous***; in this case $E(\lambda)$ is strongly continuous in λ. In general, the part H_c of H in $\mathsf{H}_c = \mathsf{H}_p^{\perp}$ is spectrally continuous. $\Sigma(H_c)$ is called the *continuous spectrum*[2] of H and is denoted by $\Sigma_c(H)$.

H_p and H_c will be called the (spectrally) *discontinuous* and *continuous parts*, respectively, of H. The subspaces H_p and H_c are called the *subspaces of discontinuity* and *of continuity*, respectively, with respect to H.

Remark 1.1. An eigenvalue of H need not be an isolated point of $\Sigma(H)$, even when H has a pure point spectrum. A point spectrum can be a countable set everywhere dense on the real axis.

Problem 1.2. $u \in \mathsf{H}_c$ if and only if $(E(\lambda)\,u, u)$ is continuous in λ.

A convenient means for separating the point spectrum from the continuous spectrum is furnished by the *mean ergodic theorem*. This theorem gives a formula that expresses $P(\lambda)$ in terms of the group $\{e^{itH}\}$ generated by iH. (For e^{itH} see Example IX-1.6.)

Theorem 1.3. *For any real* λ *we have*

$$P(\lambda) = \operatorname*{s-lim}_{t_2 - t_1 \to \infty} (t_2 - t_1)^{-1} \int_{t_1}^{t_2} e^{itH} e^{-i\lambda t}\, dt\,. \tag{1.3}$$

Proof. We may assume that $\lambda = 0$; this amounts simply to considering $H - \lambda$ instead of H. If $u \in P(0)\,\mathsf{H}$, then $Hu = 0$ and so $e^{itH} u = u$; hence

$$(t_2 - t_1)^{-1} \int_{t_1}^{t_2} e^{itH} u\, dt = u \to u = P(0)\, u\,, \quad t_2 - t_1 \to \infty\,. \tag{1.4}$$

If u is in the range of H so that $u = Hv$, $v \in \mathsf{D}(H)$, then $e^{itH} u = e^{itH} Hv = -i(d/dt)\, e^{itH} v$ by IX-(1.24). Hence the left member of (1.4) is equal to $-i(t_2 - t_1)^{-1}(e^{it_2 H} - e^{it_1 H})\, v \to 0 = P(0)\, u$, for $P(0)\, u = H P(0)\, v = 0$. Thus (1.3) is true when applied to any u belonging to the span of the ranges of $P(0)$ and of H. But this span is dense in H, for any $u \in \mathsf{H}$ orthogonal to the range of H belongs to $P(0)\,\mathsf{H}$, the null space of $H^* = H$

[1] This does not imply $\Sigma(H) = \Sigma_p(H)$, for $\Sigma(H)$ is a closed set but $\Sigma_p(H)$ need not be closed. But it implies that $\Sigma(H)$ is the closure of $\Sigma_p(H)$.

[2] Here we follow Riesz and Sz.-Nagy [1]. There is a different definition of continuous spectrum, which can be applied to any operator $T \in \mathscr{C}(\mathsf{X})$ in a Banach space X. By this definition a complex number λ belongs to the continuous spectrum of T if and only if $T - \lambda$ is invertible with dense range but $(T - \lambda)^{-1}$ is not bounded. It must be admitted that there is a disparity between the definitions of point and continuous spectra.

[see III-(5.10)]. Since the operator on the right of (1.3) is uniformly bounded, it follows that (1.3) is true (see Lemma III-3.5).

Remark 1.4. If λ is an isolated point of the spectrum of H, $P(\lambda)$ can be expressed as a contour integral of the resolvent of H [see VI-(5.34)]. In general this is impossible and (1.3) is a convenient substitute for it.

2. The absolutely continuous and singular spectra

It is convenient to divide the spectrally continuous part H_c of a selfadjoint operator H into two parts.

The spectral family $\{E(\lambda)\}$ determines a spectral measure $E(\mathrm{S})$. If S is the interval $(a, b]$, $E((a, b]) = E(b) - E(a)$. For other types of intervals, we define $E([a, b]) = E(b) - E(a - 0)$, $E([a, b)) = E(b - 0) - E(a - 0)$, $E((a, b)) = E(b - 0) - E(a)$. It is seen that this determines a projection-valued function $E(\mathrm{S})$ defined for all Borel sets S of the real line with the properties[1]

$$(1.5)\qquad E(\mathrm{S} \cap \mathrm{S}') = E(\mathrm{S})\, E(\mathrm{S}')\,,$$

$$(1.6)\qquad E(\mathrm{S} \cup \mathrm{S}') = E(\mathrm{S}) + E(\mathrm{S}') \quad \text{if S and S}' \text{ are disjoint}\,,$$

$$(1.7)\qquad E\Big(\bigcup_{n=1}^{\infty} \mathrm{S}_n\Big) = \sum_{n=1}^{\infty} E(\mathrm{S}_n) \quad \text{if } \mathrm{S}_n, \mathrm{S}_m \text{ are disjoint for } m \neq n\,.$$

For any fixed $u \in \mathsf{H}$, $m_u(\mathrm{S}) = (E(\mathrm{S})\, u, u) = \|E(\mathrm{S})\, u\|^2$ is a nonnegative, countably additive measure[2] defined for Borel sets S. If this measure is absolutely continuous (with respect to the Lebesgue measure $|\mathrm{S}|$), we shall say that *u is absolutely continuous with respect to H*. In other words, u is absolutely continuous with respect to H if and only if $|S| = 0$ implies $E(S)\, u = 0$. If, on the other hand, $m_u(\mathrm{S})$ is singular, we shall say that *u is singular with respect to H*. Thus u is singular if and only if there is a Borel set S_0 with $|\mathrm{S}_0| = 0$ such that $m_u(\mathrm{S}) = m_u(\mathrm{S} \cap \mathrm{S}_0)$. It is equivalent to $(1 - E(\mathrm{S}_0))\, u = 0$.

The set of all $u \in \mathsf{H}$ which are absolutely continuous (singular) with respect to H is denoted by H_{ac} (H_s) and is called the *subspace of absolute continuity (of singularity)* with respect to H (see the following theorem).

Theorem 1.5.[3] *H_{ac} and H_s are closed linear manifolds of H, are orthogonal complements to each other and reduce H.*

Proof. First we show that $\mathsf{H}_{ac} \perp \mathsf{H}_s$. Let $u \in \mathsf{H}_{ac}$ and $v \in \mathsf{H}_s$. There is a Borel set S_0 with $|\mathrm{S}_0| = 0$ such that $(1 - E(\mathrm{S}_0))\, v = 0$. Hence $(u, v) = (u, E(\mathrm{S}_0)\, v) = (E(\mathrm{S}_0)\, u, v) = 0$ since $E(\mathrm{S}_0)\, u = 0$.

---

[1] For the exact definition of the spectral measure see standard books on Hilbert space theory, e. g. HALMOS [1], STONE [1].

[2] For elementary results of measure theory used here and below, see e. g. ROYDEN [1].

[3] This is a special case of a theorem of HALMOS [1], p. 104.

Next we show that $\mathsf{H}_{ac} + \mathsf{H}_s = \mathsf{H}$, that is, any $w \in \mathsf{H}$ can be written as a sum $u + v$ with $u \in \mathsf{H}_{ac}$ and $v \in \mathsf{H}_s$. To this end we decompose the nonnegative measure $m_w(\mathsf{S})$ into the sum of an absolutely continuous measure $m'(\mathsf{S})$ and a singular measure $m''(\mathsf{S})$ (the Lebesgue decomposition). With m'' is associated a Borel set S_0 with $|\mathsf{S}_0| = 0$ such that $m''(\mathsf{S}) = m''(\mathsf{S} \cap \mathsf{S}_0)$. Let $v = E(\mathsf{S}_0) w$ and $u = w - v$; we assert that $u \in \mathsf{H}_{ac}$ and $v \in \mathsf{H}_s$. In fact, $m_v(\mathsf{S}) = \|E(\mathsf{S}) v\|^2 = \|E(\mathsf{S}) E(\mathsf{S}_0) w\|^2 = \|E(\mathsf{S} \cap \mathsf{S}_0) w\|^2 = m_w(\mathsf{S} \cap \mathsf{S}_0) = m''(\mathsf{S})$ since $m'(\mathsf{S} \cap \mathsf{S}_0) = 0$ and $m''(\mathsf{S} \cap \mathsf{S}_0) = m''(\mathsf{S})$, and $m_u(\mathsf{S}) = \|E(\mathsf{S}) u\|^2 = \|E(\mathsf{S})(1 - E(\mathsf{S}_0)) w\|^2 = \|E(\mathsf{S}) w\|^2 - \|E(\mathsf{S} \cap \mathsf{S}_0) w\|^2 = m_w(\mathsf{S}) - m''(\mathsf{S}) = m'(\mathsf{S})$. Thus m_u is absolutely continuous and m_v is singular: $u \in \mathsf{H}_{ac}$ and $v \in \mathsf{H}_s$.

As is easily seen, $\mathsf{H}_{ac} \perp \mathsf{H}_s$ and $\mathsf{H}_{ac} + \mathsf{H}_s = \mathsf{H}$ proved above imply that H_{ac} and H_s are closed linear manifolds of H and are orthogonal complements to each other.

If u is absolutely continuous, the same is true of all $E(\lambda) u$, for $E(\mathsf{S}) E(\lambda) u = E(\lambda) E(\mathsf{S}) u = 0$ for $|\mathsf{S}| = 0$. Thus H_{ac} reduces H, and so does $\mathsf{H}_s = \mathsf{H}_{ac}^{\perp}$. This completes the proof of Theorem 1.5.

Theorem 1.6. *Let M be a subspace of H that reduces H. Then the orthogonal projections E on M and P on H_{ac} commute. In other words, $u \in \mathsf{M}$ implies $Pu \in \mathsf{M}$ and $u \in \mathsf{H}_{ac}$ implies $Eu \in \mathsf{H}_{ac}$.*

Proof. $Q = 1 - P$ is the projection on H_s. For any $w \in \mathsf{H}$, we have $Qw = v = E(\mathsf{S}_0) w$ with an S_0 such that $|\mathsf{S}_0| = 0$, where S_0 may depend on w (see the proof of Theorem 1.5). Since M reduces H, we have $E E(\mathsf{S}_0) = E(\mathsf{S}_0) E$ so that $(1 - E(\mathsf{S}_0)) EQw = 0$ and EQw is singular. Thus $PEQw = 0$ for all $w \in \mathsf{H}$ and so $PEQ = 0$, $EQ = QEQ$. Taking the adjoint gives $QE = QEQ$ and so $QE = EQ$, $PE = EP$.

It is obvious that $\mathsf{H}_{ac} \subset \mathsf{H}_c$ (see Problem 1.2). Hence $\mathsf{H}_s \supset \mathsf{H}_p$. We write $\mathsf{H}_c \ominus \mathsf{H}_{ac} = \mathsf{H}_{sc}$. Thus

$$(1.8) \qquad \mathsf{H} = \mathsf{H}_{ac} \oplus \mathsf{H}_{sc} \oplus \mathsf{H}_p = \mathsf{H}_{ac} \oplus \mathsf{H}_s = \mathsf{H}_c \oplus \mathsf{H}_p .$$

If $\mathsf{H}_{ac} = \mathsf{H}$ (so that $\mathsf{H}_s = 0$), H is said to be *(spectrally) absolutely continuous*. If $\mathsf{H}_s = \mathsf{H}$, H is *(spectrally) singular*; if $\mathsf{H}_{sc} = \mathsf{H}$, H is *(spectrally) singularly continuous*. In general, the part H_{ac} (H_s, H_{sc}) of H in the reducing subspace H_{ac} (H_s, H_{sc}) is called the *spectrally absolutely continuous (singular, singularly continuous) part* of H. $\Sigma(H_{ac})$ ($\Sigma(H_s)$, $\Sigma(H_{sc})$) is the *absolutely continuous (singular, singularly continuous)* spectrum of H and is denoted by $\Sigma_{ac}(H)$ ($\Sigma_s(H)$, $\Sigma_{sc}(H)$).

Theorem 1.7. *If $u \in \mathsf{H}_{ac}$ and $f \in \mathsf{H}$, then $(E(\lambda) u, f)$ is absolutely continuous in λ and*

$$(1.9) \qquad \left|\frac{d}{d\lambda}(E(\lambda) u, f)\right|^2 \leq \frac{d}{d\lambda}(E(\lambda) u, u) \frac{d}{d\lambda}(E(\lambda) f_0, f_0)$$

almost everywhere, where f_0 is the projection of f on H_{ac}.

Proof. Note that the right member of (1.9) is nonnegative and finite almost everywhere, for $(E(\lambda)u, u)$ and $(E(\lambda)f_0, f_0)$ are absolutely continuous and nondecreasing in λ.

Now $(E(\lambda)u, f) = (E(\lambda)u, f_0)$ since $E(\lambda)u \in \mathsf{H}_{ac}$ by Theorem 1.6. But $(E(\lambda)u, f_0)$ is absolutely continuous as is easily seen by the polarization principle I-(6.26). (1.9) follows immediately from the estimate

$$|(E(\mathrm{I})u, f_0)|^2 \leq \|E(\mathrm{I})u\|^2 \|E(\mathrm{I})f_0\|^2 = (E(\mathrm{I})u, u)(E(\mathrm{I})f_0, f_0),$$

where $E(\mathrm{I}) = E(\lambda'') - E(\lambda')$, $\lambda' \leq \lambda''$.

Remark 1.8. For most selfadjoint operators that appear in applications, the (spectrally) continuous part is absolutely continuous so that there is no singularly continuous part. But there are second order ordinary differential operators which are (spectrally) singularly continuous[1].

Example 1.9. Consider the selfadjoint multiplication operator H defined by $Hu(x) = f(x)u(x)$ in $\mathsf{H} = L^2(\mathrm{E})$, where $f(x)$ is a real-valued measurable function on a region E of $\mathbf{R}^n$, say. Then the range of $E(\lambda)$ is the set of $u \in \mathsf{H}$ such that $u(x) = 0$ if $f(x) > \lambda$. Hence $\|E(\mathrm{S})u\|^2 = \int_{f^{-1}(\mathrm{S})} |u(x)|^2 dx$. If $f(x)$ is continuously differentiable and $\operatorname{grad} f(x) \neq 0$ almost everywhere in E, then it follows that H is spectrally absolutely continuous.

Example 1.10. Let H be the selfadjoint operator defined from $-\Delta$ in $\mathsf{H} = L^2(\mathbf{R}^n)$ (see V-§ 5.2). H is unitarily equivalent to the multiplication operator by $|x|^2$, hence H is spectrally absolutely continuous.

Remark 1.11. There is a third way of decomposing the spectrum of a selfadjoint operator H into two parts. Remove from the spectrum all isolated points which are eigenvalues with finite multiplicities; the remaining set $\Sigma_e(H)$ is the *essential spectrum* of H (see IV-§ 5.6)[2].

A point λ of $\Sigma_e(H)$ is characterized by

$$\dim[E(\lambda + \varepsilon) - E(\lambda - \varepsilon)] = \infty \quad \text{for any} \quad \varepsilon > 0. \tag{1.10}$$

Another characterization is that $\operatorname{nul}'(H - \lambda) = \operatorname{def}'(H - \lambda) = \infty$ [see IV-(5.33)]. The equivalence of the latter condition with (1.10) is easily verified if one recalls the definition of nul′ and def′ (see Theorem IV-5.9). Set $\mathsf{M}_\varepsilon = [E(\lambda + \varepsilon) - E(\lambda - \varepsilon)]\mathsf{H}$. If (1.10) is true, $\|(H - \lambda)u\| \leq \varepsilon\|u\|$ for all $u \in \mathsf{M}_\varepsilon$ where $\dim \mathsf{M}_\varepsilon = \infty$. Since $\varepsilon > 0$ is arbitrary, it follows that $\operatorname{nul}'(H - \lambda) = \infty$; note that $\operatorname{def}'(H - \lambda) = \operatorname{nul}'(H - \lambda)$ since H is selfadjoint. If, on the other hand, $\dim \mathsf{M}_\varepsilon = m < \infty$ for some $\varepsilon > 0$, then $\operatorname{nul}'(H - \lambda) < \infty$. To see this, let N_ε be a subspace such that $\|(H - \lambda)u\| \leq \varepsilon\|u\|/2$ for $u \in \mathsf{N}_\varepsilon$. If $\dim \mathsf{N}_\varepsilon > m$, there would exist a

[1] See ARONSZAJN [3].

[2] It was remarked (see loc. cit.) that there are many different definitions of $\Sigma_e(T)$. But all these definitions coincide when T is selfadjoint.

$u \in \mathsf{N}_\varepsilon$, $u \neq 0$, such that $u \perp \mathsf{M}_\varepsilon$. Then $\|(H - \lambda) u\| \geqq \varepsilon \|u\|$, contradicting the above inequality. Thus $\dim \mathsf{N}_\varepsilon \leqq m$ so that $\operatorname{nul}'(H - \lambda) \leqq m < \infty$.

Problem 1.12. $\Sigma_e(H) \supset \Sigma_c(H)$.

3. The trace class

In what follows we are concerned with certain classes of compact operators in separable Hilbert spaces. We have already considered the Schmidt class $\mathscr{B}_2(\mathsf{H}, \mathsf{H}')$ in some detail (V-§ 2.4). We now consider the so-called *trace class*, denoted by $\mathscr{B}_1(\mathsf{H}, \mathsf{H}')$ (or by $\mathscr{B}_1(\mathsf{H})$ if $\mathsf{H}' = \mathsf{H}$).

Let $T \in \mathscr{B}_0(\mathsf{H}, \mathsf{H}')$ (the class of compact operators on H to H') and let α_n be the singular values of T (see V-§ 2.3). We define the *trace norm* of T by

$$\|T\|_1 = \sum_{k=1}^{\infty} \alpha_k . \tag{1.11}$$

We set $\|T\|_1 = \infty$ if the series on the right diverges. The set of all T with $\|T\|_1 < \infty$ is the trace class $\mathscr{B}_1(\mathsf{H}, \mathsf{H}')$. That $\| \ \|_1$ has the properties of a norm can be proved easily except for the triangle inequality, which will be proved later.

Since the nonzero singular values of T, T^* and $|T|$ are identical, we have [see V-(2.34)]

$$\|T\|_1 = \|T^*\|_1 = \||T|\|_1 = \||T|^{1/2}\|_2^2 . \tag{1.12}$$

It is obvious that

$$\|T\|_2 \leqq \|T\|_1 , \quad \mathscr{B}_1(\mathsf{H}, \mathsf{H}') \subset \mathscr{B}_2(\mathsf{H}, \mathsf{H}') . \tag{1.13}$$

Furthermore, we have

$$\|S T\|_1 \leqq \|S\|_2 \|T\|_2 ; \tag{1.14}$$

this should be read in the following sense: if $T \in \mathscr{B}_2(\mathsf{H}, \mathsf{H}')$ and $S \in \mathscr{B}_2(\mathsf{H}', \mathsf{H}'')$, then $S T \in \mathscr{B}_1(\mathsf{H}, \mathsf{H}'')$ and (1.14) is true. To prove it, let

$$S T = \Sigma \beta_k (\ , \psi_k) \psi_k'' , \quad \beta_k > 0 , \tag{1.15}$$

be the canonical expansion of $S T \in \mathscr{B}_0(\mathsf{H}, \mathsf{H}'')$ [see V-(2.23)]. We have by definition

$$\begin{aligned} \|S T\|_1 &= \Sigma \beta_k = \Sigma (S T \psi_k, \psi_k'') = \Sigma (T \psi_k, S^* \psi_k'') \leqq \\ &\leqq \Sigma \|T \psi_k\| \|S^* \psi_k''\| \leqq (\Sigma \|T \psi_k\|^2)^{1/2} (\Sigma \|S^* \psi_k''\|^2)^{1/2} \leqq \\ &\leqq \|T\|_2 \|S^*\|_2 = \|T\|_2 \|S\|_2 . \end{aligned} \tag{1.16}$$

Conversely, any operator $T \in \mathscr{B}_1(\mathsf{H}, \mathsf{H}')$ can be written as the product of two operators of the Schmidt class, and the latter may be taken from $\mathscr{B}_2(\mathsf{H})$ if $T \in \mathscr{B}_1(\mathsf{H})$. For example, we may write $T = U|T| = U|T|^{1/2}|T|^{1/2}$

by VI-(2.26), where $|T|^{1/2}$ belongs to $\mathscr{B}_2(\mathsf{H})$ by (1.12) and $U|T|^{1/2}$ belongs to $\mathscr{B}_2(\mathsf{H}, \mathsf{H}')$ since $U \in \mathscr{B}(\mathsf{H}, \mathsf{H}')$.

We have the further inequalities

$$\|S\,T\|_1 \leqq \|S\|\,\|T\|_1\,, \quad \|T\,S\|_1 \leqq \|T\|_1\,\|S\|\,, \tag{1.17}$$

which should be read in a sense similar to (1.14). To prove (1.17), let $T = U|T|$ as above. We have $S\,T = S\,U|T| = S\,U|T|^{1/2}|T|^{1/2}$, so that

$$\|S\,T\|_1 \leqq \|S\,U|T|^{1/2}\|_2\,\||T|^{1/2}\|_2 \leqq \|S\,U\|\,\||T|^{1/2}\|_2^2 \leqq \|S\|\,\|T\|_1 \tag{1.18}$$

by (1.14), V-(2.30) and (1.12). The second inequality of (1.17) can be proved in the same way, or may be deduced from the first by $\|T\,S\|_1 = \|S^*\,T^*\|_1$.

Finally we shall prove the triangle inequality

$$\|T + S\|_1 \leqq \|T\|_1 + \|S\|_1\,. \tag{1.19}$$

Let $T = U|T|$, $S = V|S|$, $T + S = W|T + S|$ be the polar decompositions of T, S, $T + S$, respectively. Denoting by χ_k the eigenvectors of $|T + S|$ and setting $\chi_k' = W\,\chi_k$, we have, as in (1.16),

$$\begin{aligned} \|T + S\|_1 &= \sum_k ((T + S)\,\chi_k,\,\chi_k') = \sum_k [(T\,\chi_k,\,\chi_k') + (S\,\chi_k,\,\chi_k')] \\ &= \sum_k [(U|T|\,\chi_k,\,\chi_k') + (V|S|\,\chi_k,\,\chi_k')] \\ &= \sum_k [(|T|^{1/2}\,\chi_k,\,|T|^{1/2}\,U^*\,\chi_k') + (|S|^{1/2}\,\chi_k,\,|S|^{1/2}\,V^*\,\chi_k')] \\ &\leqq \||T|^{1/2}\|_2\,\||T|^{1/2}\,U^*\|_2 + \||S|^{1/2}\|_2\,\||S|^{1/2}\,V^*\|_2 \leqq \\ &\leqq \||T|^{1/2}\|_2^2 + \||S|^{1/2}\|_2^2 = \|T\|_1 + \|S\|_1\,. \end{aligned} \tag{1.20}$$

$\mathscr{B}_1(\mathsf{H}, \mathsf{H}')$ is a normed vector space with the norm $\|\;\|_1$. It is a complete normed space, but the proof will be omitted[1].

Problem 1.13. If $T \in \mathscr{B}_1(\mathsf{H}, \mathsf{H}')$, the canonical expansion V-(2.23) of T converges in the norm $\|\;\|_1$.

Problem 1.14. If $T \in \mathscr{B}_1(\mathsf{H})$ is symmetric, $\|T\|_1$ is equal to the sum of the absolute values of the repeated eigenvalues of T.

Remark 1.15. There are other classes of compact operators than the Schmidt and trace classes with more or less analogous properties. Each of these classes can be defined in terms of a certain norm. An example of such a norm is the *p-norm*

$$\|T\|_p = \Big(\sum_k \alpha_k^p\Big)^{1/p}, \quad 1 \leqq p \leqq \infty\,. \tag{1.21}$$

The set of all compact T with $\|T\|_p < \infty$ forms a normed vector space with the norm $\|\;\|_p$. Again this space is complete and is a two-sided ideal of $\mathscr{B}(\mathsf{H})$ if $\mathsf{H}' = \mathsf{H}$.

---

[1] See SCHATTEN [1].

More generally, a similar norm $|||T|||$ can be defined as a certain function of the singular values α_k of T, which is symmetric in the α_k and nondecreasing in each α_k. We shall not state an exact condition for this function to define a true norm. In any case $|||T|||$ is *unitarily invariant* in the sense that

$$|||U\,T\,V||| = |||T||| \tag{1.22}$$

for any unitary operators U, V; this is a simple consequence of Problem V-2.13. The set of all T with $|||T||| < \infty$ forms a Banach space. The norm $||| \; |||$ is usually normalized in such a way that $|||T||| = 1$ if $|T|$ is a one-dimensional orthogonal projection. These norms are called *cross norms*[1].

The p-norm for $p = \infty$ is identical with the ordinary norm: $\| \; \|_\infty = \| \; \|$, as is seen from the fact that $\alpha_1 = \| |T| \| = \|T\|$. The ordinary norm and the trace norm are the smallest and the largest ones among the cross norms; we have namely

$$\|T\| \leqq |||T||| \leqq \|T\|_1 \tag{1.23}$$

for any $T \in \mathscr{B}_0(\mathsf{H}, \mathsf{H}')$.

Problem 1.16. If $T \in \mathscr{B}(\mathsf{H}, \mathsf{H}')$ is of finite rank m, $\|T\|_p \leqq m^{1/p} \|T\|$.

Problem 1.17. If $\|T_n - T\| \to 0$ and $\|T_n\|_p \leqq M$, then $\|T\|_p \leqq M$. [hint: $\|T_n - T\| \to 0$ implies $\| \, |T_n|^2 - |T|^2 \| \to 0$ so that $\alpha_{kn} \to \alpha_k$ for each k, where the α_{kn}, $k = 1, 2, \ldots$, are the singular values of $T^{(n)}$ (continuity of the spectrum for selfadjoint operators). Thus $\sum_{k=1}^{m} \alpha_k^p \leqq \limsup \sum_{k=1}^{m} \alpha_{kn}^p \leqq M^p$ for any m.]

4. The trace and determinant

We have considered the trace $\operatorname{tr} T$ of operators in earlier chapters (III-§ 4.3, IV-§ 6), but it was restricted to degenerate operators T. We shall now show that $\operatorname{tr} T$ can be defined for operators T of the trace class $\mathscr{B}_1(\mathsf{H})$. As we have seen above, $T \in \mathscr{B}_1(\mathsf{H})$ can be written as the product $T = A\,B$ of two operators A, B of the Schmidt class $\mathscr{B}_2(\mathsf{H})$. We define

$$\operatorname{tr} T = (A, B^*) = (B, A^*) \tag{1.24}$$

in the notations of V-(2.31) [see also V-(2.35)]. (1.24) is independent of the particular decomposition $T = A\,B$, for

$$\begin{aligned} \operatorname{tr} T = (B, A^*) &= \sum_k (B\,\varphi_k, A^*\,\varphi_k) = \sum_k (A\,B\,\varphi_k, \varphi_k) \\ &= \sum_k (T\,\varphi_k, \varphi_k) \,. \end{aligned} \tag{1.25}$$

---

[1] For details see SCHATTEN [1].

Thus $\operatorname{tr} T$ is the *diagonal sum* of the matrix of T in any matrix representation of T in terms of a complete orthonormal family. (1.25) implies that the diagonal sum is always absolutely convergent.

It follows from (1.24) that

$$\operatorname{tr} AB = \operatorname{tr} BA \tag{1.26}$$

for $A, B \in \mathscr{B}_2(\mathsf{H})$. It is true more generally for $A \in \mathscr{B}_2(\mathsf{H}, \mathsf{H}')$ and $B \in \mathscr{B}_2(\mathsf{H}', \mathsf{H})$; in this case we have $AB \in \mathscr{B}_1(\mathsf{H}')$ and $BA \in \mathscr{B}_1(\mathsf{H})$, and the proof follows from V-(2.35). (1.26) is true also for $A \in \mathscr{B}_1(\mathsf{H}, \mathsf{H}')$ and $B \in \mathscr{B}(\mathsf{H}', \mathsf{H})$; then we have again $AB \in \mathscr{B}_1(\mathsf{H}')$, $BA \in \mathscr{B}_1(\mathsf{H})$. For the proof let $A = CD$ with $C \in \mathscr{B}_2(\mathsf{H}, \mathsf{H}')$ and $D \in \mathscr{B}_2(\mathsf{H})$. Then $\operatorname{tr} AB = \operatorname{tr} CDB = \operatorname{tr} DBC = \operatorname{tr} BCD = \operatorname{tr} BA$, where we have applied (1.26) first to the pair C, DB (both of the Schmidt class) and then to the pair D, BC.

The determinant has been defined so far only for operators of the form $1 + T$ with a degenerate T. Now we can extend it to the case when $T \in \mathscr{B}_1(\mathsf{H})$. We set

$$\det(1 + T) = e^{\operatorname{tr} \log(1 + T)}. \tag{1.27}$$

This definition is not complete inasmuch as $\log(1 + T)$ is not defined unambiguously for all $T \in \mathscr{B}_1(\mathsf{H})$. If $\|T\| < 1$ (or, more generally, if $\operatorname{spr} T < 1$), however, $\log(1 + T)$ can be defined by the Taylor series $\log(1 + T) = T(1 - T/2 + T^2/3 - \cdots)$ and belongs to $\mathscr{B}_1(\mathsf{H})$ with T. Hence (1.27) defines $\det(1 + T)$ for $\|T\| < 1$. Even in the general case, (1.27) is useful if the operator $\log(1 + T)$ is defined more carefully. Another way to define $\det(1 + T)$ is by the analytic continuation of the function $\det(1 + zT)$, which is defined as above by the Taylor series at least for $|z| < \|T\|^{-1}$. We shall not go into details since we shall not have occasion to use $\det(1 + T)$ for general $T \in \mathscr{B}_1(\mathsf{H})$ [1].

Example 1.18. Let $\mathsf{H} = \mathsf{L}^2(a, b)$ and $T \in \mathscr{B}_1(\mathsf{H})$. T can be written as $T = RS$ with $R, S \in \mathscr{B}_2(\mathsf{H})$. R, S can be regarded as integral operators with kernels $r(y, x)$, $s(y, x)$ of Schmidt type (Example V-2.19). Thus T is an integral operator with the kernel

$$t(y, x) = \int_a^b r(y, z)\, s(z, x)\, dz. \tag{1.28}$$

Then $\operatorname{tr} T = (R, S^*) = \iint r(x, z)\, \overline{s^*(x, z)}\, dx\, dz = \iint r(x, z)\, s(z, x)\, dz\, dx$ or

$$\operatorname{tr} T = \int_a^b t(x, x)\, dx. \tag{1.29}$$

It should be remarked, however, that (1.29) is correct only if one uses the particular kernel given by (1.28). Note that $t(x, x)$ as given by (1.28) is defined for almost all x. If, on the other hand, one is given simply a kernel $t(y, x)$ of T as an integral operator in some sense or other, the right member of (1.29) need not

[1] For details on the trace and determinant, see e. g. Dunford and Schwartz [1].

make sense, for one could change the values of $t(x, x)$ arbitrarily without changing the operator T.

It can be shown that if $T \in \mathscr{B}_1(\mathsf{H})$ is represented by a kernel $t(y, x)$ *continuous* in x, y, then the trace formula (1.29) is correct. But it must be noticed that not all integral operators with a continuous kernel belong to the trace class. In such a case one could define tr T by (1.29), but it might not have the properties of the trace deduced above.

§ 2. Perturbation of continuous spectra

1. A theorem of WEYL-VON NEUMANN

One of the important results on the perturbation of continuous spectra was given by H. WEYL and later generalized by VON NEUMANN. This theorem asserts that any selfadjoint operator H in a *separable* Hilbert space H can be changed into a selfadjoint operator $H + A$ with a pure point spectrum by the addition of a suitable "small" operator[1] A. The smallness of A can be expressed by the condition that A can be chosen from the Schmidt class $\mathscr{B}_2(\mathsf{H})$ with arbitrarily small Schmidt norm $\|A\|_2$. More precisely, we have

Theorem 2.1.[2] *Let H be a selfadjoint operator in a separable Hilbert space H. For any $\varepsilon > 0$, there exists a selfadjoint operator $A \in \mathscr{B}_2(\mathsf{H})$ with $\|A\|_2 < \varepsilon$ such that $H + A$ has a pure point spectrum.*

To prove this theorem, we need a lemma.

Lemma 2.2. *For any $f \in \mathsf{H}$ and $\eta > 0$, there is a finite-dimensional orthogonal projection P and a selfadjoint operator $Y \in \mathscr{B}_2(\mathsf{H})$ such that $\|(1 - P) f\| < \eta$, $\|Y\|_2 < \eta$ and $H + Y$ is reduced by $P\mathsf{H}$.*

Proof. Let $H = \int_{-\infty}^{\infty} \lambda\, dE(\lambda)$ be the spectral representation of H. Let $a > 0$ be chosen in such a way that

$$\|[1 - (E(a) - E(-a))] f\| < \eta , \tag{2.1}$$

which is possible because the left member tends to 0 as $a \to \infty$. Let n be a positive integer and set

$$E_k = E\left(\frac{2k - n}{n} a\right) - E\left(\frac{2k - n - 2}{n} a\right), \quad k = 1, 2, \ldots, n . \tag{2.2}$$

The E_k form an orthogonal family of orthogonal projections: $E_j E_k = \delta_{jk} E_k$. Set

$$f_k = E_k f , \quad g_k = \|f_k\|^{-1} f_k , \quad k = 1, 2, \ldots, n , \tag{2.3}$$

with the agreement that $g_k = 0$ whenever $f_k = 0$. Since $f_k, g_k \in E_k \mathsf{H}$, the g_k form a normalized orthogonal family (if those g_k equal to 0 are

[1] This is a rather negative result as regards the perturbation of continuous spectra.

[2] The proof given below is due to VON NEUMANN [1]. This theorem will be generalized in the following paragraph; see Theorem 2.3.

omitted). We have

$$\sum_{k=1}^{n} \|f_k\| g_k = \sum_{k=1}^{n} f_k = ((E(a) - E(-a)) f . \tag{2.4}$$

The g_k are approximate eigenvectors of H in the sense that

$$\|(H - \lambda_k) g_k\| \leq a/n , \quad \lambda_k = \frac{2k - n - 1}{n} a , \tag{2.5}$$

as is seen from VI-(5.21). Let P be the orthogonal projection on the subspace spanned by $g_1, \ldots, g_n$, so that $\dim P \leq n$. Since $(1 - P) g_k = 0$, we have

$$\|(1 - P) H g_k\| = \|(1 - P) (H - \lambda_k) g_k\| \leq \|(H - \lambda_k) g_k\| \leq a/n . \tag{2.6}$$

Furthermore, we have

$$((1 - P) H g_j, (1 - P) H g_k) = 0 , \quad j \neq k . \tag{2.7}$$

To see this it suffices to show that $(1 - P) H g_k \in E_k \mathsf{H}$. But $H g_k \in E_k \mathsf{H}$ since $g_k \in E_k \mathsf{H}$ and $E_k \mathsf{H}$ reduces H. Hence $H g_k \perp g_j, j \neq k$, and therefore $P H g_k = \sum_j (H g_k, g_j) g_j = (H g_k, g_k) g_k \in E_k \mathsf{H}$ and so $(1 - P) H g_k \in$ $\in E_k \mathsf{H}$.

For any $u \in \mathsf{H}$, we now have by (2.7), (2.6) and the Bessel inequality

$$\begin{aligned} \|(1 - P) H P u\|^2 &= \Big\|\sum_k (u, g_k) (1 - P) H g_k\Big\|^2 \\ &= \sum_k |(u, g_k)|^2 \|(1 - P) H g_k\|^2 \leq a^2 n^{-2} \|u\|^2 , \end{aligned} \tag{2.8}$$

that is,

$$\|(1 - P) H P\| \leq a/n . \tag{2.9}$$

The operator $(1 - P) H P$ is degenerate with rank $\leq n$. Hence (see Problem 1.16)

$$\|(1 - P) H P\|_2 \leq n^{1/2} \|(1 - P) H P\| \leq a/n^{1/2} . \tag{2.10}$$

Now we have

$$H = P H P + (1 - P) H (1 - P) + (1 - P) H P + [(1 - P) H P]^* . \tag{2.11}$$

The first two terms on the right are reduced by $P\mathsf{H}$. Each of the last two terms has Schmidt norm not exceeding $a/n^{1/2}$ by (2.10), which can be made smaller than $\eta/2$ by choosing n large enough. On the other hand, (2.4) implies that $(1 - P) ((E(a) - E(-a)) f = 0$ so that

$$\|(1 - P) f\| = \|(1 - P) [f - ((E(a) - E(-a)) f]\| < \eta \tag{2.12}$$

by (2.1). Thus the lemma is proved by setting $-Y = (1 - P) H P +$ $+ [(1 - P) H P]^*$.

Proof of Theorem 2.1. Let $\{u_k\}$, $k = 1, 2, \ldots$, be a dense subset of H. We apply Lemma 2.2 for H, $f = u_1$ and $\eta = \varepsilon/2$; let the resulting P and Y be denoted by P_1 and Y_1, respectively. Then we apply the same lemma to the part of $H + Y_1$ in the subspace $(1 - P_1)\,\mathsf{H}$ with $f = (1 - P_1)\,u_2$ and $\eta = \varepsilon/2^2$; let the resulting P and Y be denoted by P_2 and Y_2. We extend P_2 and Y_2 to the whole space H by simply setting $P_2 v = 0$, $Y_2 v = 0$ for $v \in P_1\mathsf{H}$; then $H + Y_1 + Y_2$ is reduced by $P_1\mathsf{H}$ and $P_2\mathsf{H}$. We then apply the lemma to the part of $H + Y_1 + Y_2$ in $(1 - P_1 - P_2)\,\mathsf{H}$ with $f = (1 - P_1 - P_2)\,u_3$ and $\eta = \varepsilon/2^3$, denoting the resulting P and Y by P_3 and Y_3 and extending these operators to the whole H by setting them equal to zero in $(P_1 + P_2)\,\mathsf{H}$. Proceeding in the same way, we get series of projections $P_1, P_2, \ldots$ and selfadjoint operators $Y_1, Y_2, \ldots$ such that $\|(1 - P_1 - \cdots - P_k)\,u_k\| \leqq \varepsilon/2^k$ and $\|Y_k\|_2 \leqq \varepsilon/2^k$.

Set $A = Y_1 + Y_2 + \cdots$; this series converges by $\|Y_k\|_2 \leqq \varepsilon/2^k$ in the Schmidt norm and so defines a selfadjoint operator $A \in \mathscr{B}_2(\mathsf{H})$ with $\|A\|_2 \leqq \varepsilon$. We shall show that A has the other properties required by Theorem 2.1.

The P_k form an orthogonal family of projections by construction. This family is complete: $\sum_{k=1}^{\infty} P_k = 1$. To see this, let any $u \in \mathsf{H}$ and $\eta > 0$ be given. Then there is an n such that $\|u_n - u\| < \eta$ and $\varepsilon/2^n < \eta$. Since $\|(1 - P_1 - \cdots - P_n)\,u_n\| \leqq \varepsilon/2^n < \eta$ we have $\|(1 - P_1 - \cdots - P_n)\,u\| \leqq 2\eta$. This shows that $\Sigma P_k u = u$ as required (note Lemma V-2.3).

Next we show that each $P_k\,\mathsf{H}$ reduces $H + A$. By construction $P_n\,\mathsf{H}$ is a subspace of $(1 - P_1 - \cdots - P_{n-1})\,\mathsf{H}$ and reduces $H + Y_1 + \cdots + Y_n$. Hence P_n commutes with $H + Y_1 + \cdots + Y_n$. Since $P_n Y_k = Y_k P_n = 0$ for $k > n$, P_n commutes with $H + A$, that is, $P_n\,\mathsf{H}$ reduces $H + A$.

The fact that $P_n\,\mathsf{H}$ reduces H implies that there are a finite number of eigenvectors of $H + A$ that constitute an orthonormal basis of $P_n\,\mathsf{H}$, which is finite-dimensional by construction. By the completeness of $\{P_n\}$, the totality of these eigenvectors for all n forms a complete orthonormal family of H. This shows that $H + A$ has a pure point spectrum, completing the proof of Theorem 2.1.

2. A generalization

The appearance of the Schmidt norm $\|A\|_2$ in Theorem 2.1 is not essential. We have the following generalization due to S. T. KURODA [1].

Theorem 2.3. *In* Theorem 2.1, *A can be chosen in such a way that $|||A||| < \varepsilon$. Here $|||\ |||$ is any (unitarily invariant) cross norm, with the single exception of the trace norm or its equivalent.*

Proof. It suffices to show that $\|Y\|_2$ in Lemma 2.2 can be replaced by $|||Y|||$; then the proof of Theorem 2.1 works by simply replacing $\| \|_2$ by $||| \; |||$. If one recalls the inequality (2.10) used in the proof of the lemma, in which the essential point is that the right member tends to zero as $n \to \infty$, we need only to prove that

$$|||X_n||| \leqq n\, c_n \|X_n\| \tag{2.13}$$

for any degenerate operator X_n of rank n, where c_n is a constant tending to zero as $n \to \infty$.

Since both $||| \; |||$ and $\| \|$ are unitarily invariant, we may assume that X_n is nonnegative symmetric. Let $\alpha_1 \geqq \alpha_2 \geqq \cdots \geqq \alpha_n$ be the positive repeated eigenvalues of X_n, so that $\|X_n\| = \alpha_1$. Since $|||X_n|||$ is nondecreasing in each α_k (see Remark 1.15) it suffices to prove the statement for $\alpha_1 = \cdots = \alpha_n = 1$, that is, that

$$n^{-1}\, |||E_n||| \equiv c_n \to 0 \tag{2.14}$$

for any orthogonal projection E_n with $\dim E_n = n$. Note that $|||E_n|||$ depends only on n by the unitary invariance.

E_n can be written as the sum $P_1 + \cdots + P_n$ where the P_k form an orthogonal family of one-dimensional orthogonal projections. Hence

$$\begin{aligned} n\, |||E_{n+1}||| &= n\, |||P_1 + \cdots + P_{n+1}||| = |||n P_1 + \cdots + n P_{n+1}||| \leqq \\ &\leqq |||P_1 + \cdots + P_n||| + |||P_1 + \cdots + P_{n-1} + P_{n+1}||| + \cdots + \\ &\qquad + |||P_2 + \cdots + P_{n+1}||| \\ &= (n+1)\, |||P_1 + \cdots + P_n||| = (n+1)\, |||E_n||| \end{aligned} \tag{2.15}$$

since $|||P_1 + \cdots + P_n||| = |||P_2 + \cdots + P_{n+1}|||$ etc. by the unitary invariance of $||| \; |||$. (2.15) implies that c_n is a nonincreasing function of n.

Suppose that (2.14) is not true, that is, $c_n \geqq c > 0$. Then $|||E_n||| \geqq n\, c$ and, for $\alpha_k \geqq 0$,

$$\begin{aligned} n\, c(\alpha_1 + \cdots + \alpha_n) &\leqq (\alpha_1 + \cdots + \alpha_n)\, |||E_n||| \\ &= |||(\alpha_1 + \cdots + \alpha_n)(P_1 + \cdots + P_n)||| \leqq \\ &\leqq |||\alpha_1 P_1 + \cdots + \alpha_n P_n||| + |||\alpha_2 P_1 + \cdots + \alpha_1 P_n||| + \cdots + \\ &\quad + |||\alpha_n P_1 + \cdots + \alpha_{n-1} P_n||| = n\, |||\alpha_1 P_1 + \cdots + \alpha_n P_n|||\,, \end{aligned} \tag{2.16}$$

again by the unitary invariance of $||| \; |||$. With $\alpha_1 P_1 + \cdots + \alpha_n P_n = X_n$, (2.16) gives

$$c\|X_n\|_1 \leqq |||X_n|||\,, \tag{2.17}$$

which is valid for any degenerate operator X_n of arbitrary rank. Now (2.17) can be extended to all X with $|||X||| < \infty$ by approximating X

by a sequence of degenerate operators X_n and going to the limit $n \to \infty$. The resulting inequality $|||X||| \geqq c\|X\|_1$ means, however, that $||| \; |||$ is equivalent to the trace norm $\| \; \|_1$, for the opposite inequality is always true [see (1.23)]. This proves the required generalization of Lemma 2.2 and completes the proof of Theorem 2.3.

Remark 2.4. The question remains whether or not Theorem 2.1 is true with $\|A\|_2$ replaced by $\|A\|_1$. The answer is no; we shall see later that a selfadjoint operator H with a nonvanishing H_{ac} (absolutely continuous part) can never be changed into an operator with a pure point spectrum by the addition of an operator of the trace class (Theorem 4.3)[1].

Remark 2.5. Theorems 2.1 and 2.3 state that a pure continuous spectrum can be converted into a pure point spectrum by perturbing the operator by a "small" amount. It should not be imagined that this point spectrum consists of *isolated* eigenvalues. On the contrary, the distribution of the eigenvalues will be everywhere dense in an interval I, if the continuous spectrum of the unperturbed operator covers I. This is a consequence of the stability of the essential spectrum under the addition of a compact operator (see Theorem IV-5.35), for $\Sigma_c(H)$ is a subset of $\Sigma_e(H)$ so that $\mathrm{I} \subset \Sigma_e(H) = \Sigma_e(H + A)$. (This implies that the set of the eigenvalues of $H + A$ must be dense in I, if $H + A$ has a pure point spectrum.)

§ 3. Wave operators and the stability of absolutely continuous spectra

1. Introduction

In what follows we are exclusively concerned with a *separable* Hilbert space H. Let H_1, H_2 be two selfadjoint operators in H. Consider the one-parameter groups e^{-itH_1}, e^{-itH_2} generated by $-iH_1, -iH_2$, respectively, and the one-parameter family of unitary operators

$$W(t) = e^{itH_2} e^{-itH_1}, \quad -\infty < t < \infty . \tag{3.1}$$

In general $W(t)$ does not form a group. We are interested in the asymptotic behavior of $W(t)$ as $t \to \pm\infty$, which is important in physical applications because $W(t)$ is used to describe the motion of a quantum-mechanical system in the so-called *interaction representation*. In particular, the limits $W_\pm$ of $W(t)$ as $t \to \pm\infty$, if they exist, are called the *wave operators* and $S = W_+^* W_-$ the *scattering operator*; these are basic quanti-

[1] For the perturbation of the continuous spectra of ordinary differential operators (Sturm-Liouville problems), see Aronszajn [3], Putnam [1], [2], Butler [2], [3], Moser [1].

ties in *scattering theory*[1]. On the other hand, the question has an independent mathematical interest, for it is closely related to the problem of the *unitary equivalence* of H_1 and H_2, as will be seen from the theorems proved below.

In these studies, the existence of the wave operators is one of the main problems. Naturally, the wave operators exist only under rather strong restrictions. It would be necessary that H_2 differ from H_1 only slightly in some sense or other; thus the problem essentially belongs to perturbation theory. Another important question is the unitarity of S. As the limits of $W(t)$, $W_\pm$ may be expected to be unitary. But this is not necessarily the case. If we assume $W_\pm$ to be the *strong* limits of $W(t)$, they are isometric but need not be unitary. $S = W_+^* W_-$ is unitary if and only if the ranges of $W_\pm$ are the same[2]. In physical applications it is essential that S be unitary[3]. Thus the second of the main mathematical problems will be to see under what conditions the ranges of $W_\pm$ coincide.

The theory centering around the wave operators is peculiar to an infinite-dimensional Hilbert space, without having any analogue in finite-dimensional spaces. In fact, the wave operators can reasonably be expected to exist only when H_1 has a pure continuous spectrum. Suppose that H_1 has an eigenvalue λ: $H_1 u = \lambda u$, $u \neq 0$. Then $W(t) u = e^{it(H_2-\lambda)} u$. If $W_+ = \operatorname{s-lim}_{t\to\infty} W(t)$ should exist, we must have $\|W(t+a) u - W(t) u\| = \|e^{ia(H_2-\lambda)} u - u\| \to 0$, $t \to \infty$, for any real a. This implies that $e^{ia(H_2-\lambda)} u = u$ and so $H_2 u = \lambda u$. Thus any eigenvector of H_1 must also be an eigenvector of H_2 with the same eigenvalue. Except for such special cases, the wave operators exist only when H_1 has a pure continuous spectrum.

Example 3.1. Let $\mathsf{H} = \mathsf{L}^2(-\infty, +\infty)$ and $H_1 = -i d/dx$, $H_2 = -i d/dx + q(x)$, where $q(x)$ is a real-valued function. H_1 is selfadjoint. H_2 is also selfadjoint at least if $q(x)$ is a bounded function (so that it defines a bounded operator). We know that $-iH_1 = -d/dx$ generates a group e^{-itH_1} given by $e^{-itH_1} u(x) = u(x-t)$

---

[1] For the mathematical formulation of scattering theory see FRIEDRICHS [4], JAUCH [1], [2], KURODA [2]. What we consider in this chapter corresponds to a very special case of scattering phenomena (*simple scattering system* according to JAUCH). There are few mathematical results for more general, *multi-channel* scattering, except JAUCH [2], HACK [2], FADDEEV [1], [2].

[2] Sometimes the $W_\pm$ are themselves unitary so that S is automatically unitary. In such a case the $W_\pm$ have their own spectral representations. Here we are not interested in the spectral properties of $W_\pm$. For these and related problems see PUTNAM [4], [6], [8], [9], [10].

[3] As is easily seen, S commutes with H_1. Therefore, S can be expressed as a *direct integral* of unitary operators $S(\lambda)$, $-\infty < \lambda < \infty$, where $S(\lambda)$ acts in a Hilbert space $\mathsf{H}_1(\lambda)$ and H_1 is the direct integral of the scalar operator λ in $\mathsf{H}_1(\lambda)$ (for direct integrals see DIXMIER [1]). $S(\lambda)$ is called the S-matrix.

(see Problem IX-1.9). Let us find what the group e^{-itH_2} is. As is easily verified, H_2 may be written as $H_2 = W_0^{-1} H_1 W_0$, where W_0 is a unitary operator given by $W_0 u(x) = e^{ip(x)} u(x)$ with $p(x) = \int_0^x q(y)\, dy$. Thus $(H_2 - \zeta)^{-1} = W_0^{-1}(H_1 - \zeta)^{-1} W_0$ for any nonreal ζ, and it follows from the construction of the groups that $e^{itH_2} = W_0^{-1} e^{itH_1} W_0$. Hence $e^{itH_2} u(x) = e^{-ip(x)} e^{itH_1} W_0 u(x) = e^{-ip(x)} e^{ip(x+t)} u(x+t)$. It follows that $W(t)$ is simply the multiplication operator by $e^{i(p(x+t)-p(x))}$. Going to the limit $t \to \infty$, we obtain

$$W_+ = \exp\left(i \int_x^\infty q(y)\, dy\right), \tag{3.2}$$

provided the improper integral $\int_0^\infty q(y)\, dy$ exists. It is easy to verify that the limit (3.2) exists as a strong limit. Obviously W_+ is unitary. Similarly W_- exists and is given by $\exp\left(-i \int_{-\infty}^x q(y)\, dy\right)$ provided the relevant improper integral exists. If both these integrals exist, we have $S = W_+^* W_- = \exp\left(-i \int_{-\infty}^\infty q(y)\, dy\right)$, which is simply a multiplication by a scalar with absolute value 1.

2. Generalized wave operators

We denote by $H_{k,\mathrm{ac}}$, $k = 1, 2$, the spectrally absolutely continuous part of H_k, that is, the part of H_k in the space $\mathsf{H}_{k,\mathrm{ac}}$ of absolute continuity for H_k. The orthogonal projection on $\mathsf{H}_{k,\mathrm{ac}}$ will be denoted by P_k, $k = 1, 2$.

We have remarked above that the wave operators will in general not exist unless H_1 has a pure continuous spectrum. As we shall see, it happens frequently that

$$W_\pm = W_\pm(H_2, H_1) = \operatorname*{s-lim}_{t\to\pm\infty} W(t)\, P_1 \tag{3.3}$$

exist even when the proper wave operators do not exist[1]. For this and other reasons, we prefer to consider the limits (3.3) rather than the proper wave operators. $W_\pm$ will be called the *generalized wave operator(s)* associated with the pair H_1, H_2 whenever one or both of them exist. If in particular H_1 is spectrally absolutely continuous so that $P_1 = 1$, $W_\pm$ coincide with the proper wave operators.

The basic properties of the generalized wave operators are given by the following theorems[2].

Theorem 3.2. *If $W_+ = W_+(H_2, H_1)$ exists, it is partially isometric with initial set $\mathsf{H}_{1,\mathrm{ac}}$ and final set M_+ contained in $\mathsf{H}_{2,\mathrm{ac}}$. M_+ reduces H_2.*

---

[1] The choice of strong limit is essential for $W_\pm$ to be partially isometric. Sometimes, however, one defines $W_\pm$ as the weak limits or in some weaker sense. Cf. Cook [2]. The limits as $t \to \pm\infty$ of operators of the form $W(t)\, A\, W(t)^{-1}$ are considered by Y. Kato and Mugibayashi [1].

[2] See T. Kato [10], [11], [17], Kuroda [2].

Thus

$$W_+^* W_+ = P_1 , \quad W_+ W_+^* = E_+ \leq P_2 , \tag{3.4}$$

$$W_+ = W_+ P_1 = E_+ W_+ = P_2 W_+ , \quad W_+^* = P_1 W_+^* = W_+^* E_+ = W_+^* P_2 , \tag{3.5}$$

where E_+ *is the orthogonal projection on* M_+ *and commutes with* H_2. *Furthermore, we have*

$$H_2 W_+ = H_2 P_2 W_+ = W_+ H_1 P_1 \supset W_+ H_1 , \quad H_1 W_+^* \supset W_+^* H_2 . \tag{3.6}$$

In particular (3.6) *implies that* $H_{1,\mathrm{ac}}$ *is unitarily equivalent to the part of* $H_{2,\mathrm{ac}}$ *in* M_+ *and* $\Sigma_{\mathrm{ac}}(H_1) \subset \Sigma_{\mathrm{ac}}(H_2)$. *Similar results hold for* W_+ *replaced by* W_- *whenever the latter exists.*

Proof. $W_+ = \text{s-lim}\, W(t)\, P_1$ implies $\|W_+ u\| = \lim \|W(t)\, P_1 u\| = \|P_1 u\|$ for every $u \in \mathsf{H}$, which is equivalent to $W_+^* W_+ = P_1$. Hence W_+ is partially isometric with initial set $P_1 \mathsf{H} = \mathsf{H}_{1,\mathrm{ac}}$ (see V-§ 2.2). $E_+ = W_+ W_+^*$ is the projection on the final set of W_+, which we denote by M_+. These remarks prove (3.4) and (3.5) so far as P_2 is not concerned (see loc. cit.).

On the other hand[1], we have for any real s

$$e^{isH_2} W_+ = \underset{t \to \infty}{\text{s-lim}}\, W(s + t)\, e^{isH_1} = W_+ e^{isH_1} . \tag{3.7}$$

Multiply both members by $e^{-i\zeta s}$, $\mathrm{Im}\,\zeta < 0$, and integrate on $0 < s < \infty$ (Laplace transformation); then [see IX-(1.28)]

$$(H_2 - \zeta)^{-1} W_+ = W_+ (H_1 - \zeta)^{-1} . \tag{3.8}$$

The same is true for $\mathrm{Im}\,\zeta > 0$ too; we need only to integrate on $-\infty < s < 0$. (3.8) is equivalent to[2]

$$H_2 W_+ \supset W_+ H_1 . \tag{3.9}$$

As is easily seen, (3.9) implies the adjoint relation $H_1 W_+^* \supset W_+^* H_2$.

It follows that

$$E_+ H_2 = W_+ W_+^* H_2 \subset W_+ H_1 W_+^* \subset H_2 W_+ W_+^* = H_2 E_+ . \tag{3.10}$$

This shows that E_+ commutes with H_2 and so M_+ reduces H_2. Also we have $E_+ H_2 E_+ \subset H_2 E_+ E_+ = H_2 E_+$. Here, however, we must have equality instead of inclusion, for $E_+ H_2 E_+$ and $H_2 E_+$ have the same domain. Hence $H_2 E_+ = E_+ H_2 E_+$ and so (3.10) must give all equalities

[1] (3.7), (3.8), (3.9) and (3.12) are true even if W_+ is a weak limit of $W(t)$ instead of a strong limit. But then W_+ need not be partially isometric and might even be zero; then these results will be void.

[2] The proof of the equivalence of (3.8) and (3.9) is similar to the proof that a bounded operator W commutes with H if and only if W commutes with the resolvent $(H - \zeta)^{-1}$; see Theorem III-6.5.

when multiplied by E_+ from the right. In particular $E_+ H_2 E_+ = W_+ H_1 W_+^* E_+ = W_+ H_1 W_+^*$ and hence

$$(3.11)\quad H_2 W_+ = H_2 E_+ W_+ = E_+ H_2 E_+ W_+ = W_+ H_1 W_+^* W_+ = W_+ H_1 P_1 .$$

Let $E_k(\lambda)$, $k = 1, 2$, be the spectral family for H_k. For every λ at which $E_k(\lambda)$ is continuous, $E_k(\lambda)$ can be expressed as an integral of $(H_k - \zeta)^{-1}$ along a certain curve (see Lemma VI-5.6). Since there are at most a countable number of discontinuities of $E_k(\lambda)$, it follows from (3.8) that

$$(3.12)\quad E_2(\lambda)\, W_+ = W_+ E_1(\lambda)\,, \quad -\infty < \lambda < \infty\,,$$

at first except for the discontinuities of either of $E_k(\lambda)$ and then for all λ by the right continuity of $E_k(\lambda)$. Hence $\|E_2(\lambda)\, W_+ u\|^2 = \|W_+ E_1(\lambda)\, u\|^2 = \|P_1 E_1(\lambda)\, u\|^2 = \|E_1(\lambda)\, P_1 u\|^2$ is absolutely continuous in λ by $P_1 u \in \mathsf{H}_{1,\mathrm{ac}}$. Thus $W_+ u \in \mathsf{H}_{2,\mathrm{ac}} = P_2 \mathsf{H}$ for any $u \in \mathsf{H}$. Since the range of W_+ is $\mathsf{M}_+ = E_+ \mathsf{H}$, this proves that $\mathsf{M}_+ \subset \mathsf{H}_{2,\mathrm{ac}}$ or $E_+ \leqq P_2$. In particular $W_+ = P_2 W_+$ and $W_+^* = W_+^* P_2$, and this completes, together with (3.11), the proof of (3.6) and the remaining equalities in (3.4) and (3.5).

In Theorem 3.2, M_+ is a subset of $\mathsf{H}_{2,\mathrm{ac}}$ and in general a proper subset. If $\mathsf{M}_+ = \mathsf{H}_{2,\mathrm{ac}}$ or, equivalently, $E_+ = P_2$, the wave operator W_+ will be said to be *complete*[1]. A similar definition applies to W_-. *If either W_+ or W_- exists and is complete, then $H_{1,\mathrm{ac}}$ is unitarily equivalent to $H_{2,\mathrm{ac}}$.*

Theorem 3.3. *If $W_+ = W_+(H_2, H_1)$ exists, we have the following strong convergence as $t \to \infty$.*

$$(3.13)\quad e^{itH_2} e^{-itH_1} P_1 \underset{s}{\to} W_+\,, \qquad e^{itH_1} e^{-itH_2} E_+ \underset{s}{\to} W_+^*\,,$$

$$(3.14)\quad e^{-itH_2} W_+ - e^{-itH_1} P_1 \underset{s}{\to} 0\,, \qquad e^{itH_1} e^{-itH_2} W_+ \underset{s}{\to} P_1\,,$$

$$(3.15)\quad (W_+ - 1)\, e^{-itH_1} P_1 \underset{s}{\to} 0\,, \quad (W_+^* - 1)\, e^{-itH_1} P_1 \underset{s}{\to} 0\,,$$

$$(3.16)\quad e^{itH_1} W_+ e^{-itH_1} \underset{s}{\to} P_1\,, \qquad e^{itH_1} W_+^* e^{-itH_1} P_1 \underset{s}{\to} P_1,$$

$$(3.17)\quad (1 - E_+)\, e^{-itH_1} P_1 \underset{s}{\to} 0\,, \quad (1 - P_2)\, e^{-itH_1} P_1 \underset{s}{\to} 0\,.$$

Similar relations hold as $t \to -\infty$ with W_+ replaced by W_- if the latter exists.

Proof. The first of (3.13) is the definition of W_+, from which the first of (3.14) follows by multiplication from the left[2] by e^{-itH_2} and the second

---

[1] For an example of non-complete W_+, see T. Kato and Kuroda [1].

[2] Note that any relation of the form $A_t \underset{s}{\to} 0$ may be multiplied *from the left* by a uniformly bounded operator B_t, while multiplication of B_t *from the right* is in general not permitted unless B_t is independent of t.

by a further multiplication by e^{itH_1}. The second of (3.13) follows from the second of (3.14) by multiplication from the right by W_+^*. The first of (3.15) is the same thing as the first of (3.14) in virtue of (3.7) and $W_+ = W_+ P_1$, and the second of (3.15) follows from it by multiplication from the left by $-W_+^*$ (note that $W_+^* W_+ = P_1$ commutes with e^{-itH_1}). The two relations of (3.16) follow from the corresponding ones of (3.15) by multiplication from the left by e^{itH_1}. Finally, the first of (3.17) follows from $\|(1 - E_+) e^{-itH_1} P_1 u\| = \|e^{itH_2}(1 - E_+) e^{-itH_1} P_1 u\| = \|(1 - E_+) e^{itH_2} e^{-itH_1} P_1 u\| \to \|(1 - E_+) W_+ u\| = 0$, and the second follows from it because $1 - P_2 \leq 1 - E_+$.

Theorem 3.4. *(The chain rule) If $W_+(H_2, H_1)$ and $W_+(H_3, H_2)$ exist, then $W_+(H_3, H_1)$ also exists and*

$$W_+(H_3, H_1) = W_+(H_3, H_2)\, W_+(H_2, H_1)\,. \tag{3.18}$$

A similar result holds for W_+ replaced by W_-.

Proof. We have $W_+(H_2, H_1) = \text{s-lim}\, e^{itH_2} e^{-itH_1} P_1$ and $W_+(H_3, H_2) = \text{s-lim}\, e^{itH_3} e^{-itH_2} P_2$. Hence

$$W_+(H_3, H_2)\, W_+(H_2, H_1) = \text{s-lim}\, e^{itH_3} P_2\, e^{-itH_1} P_1 \tag{3.19}$$

since P_2 and e^{itH_2} commute. The proof of the theorem is complete if we show that $\text{s-lim}\, e^{itH_3} (1 - P_2)\, e^{-itH_1} P_1 = 0$. But this is obvious from the second relation of (3.17).

Theorem 3.5. *Let both $W_+(H_2, H_1)$ and $W_+(H_1, H_2)$ exist. Then*

$$W_+(H_1, H_2) = W_+(H_2, H_1)^*\,; \tag{3.20}$$

$W_+(H_2, H_1)$ is partially isometric with initial set $\mathsf{H}_{1,ac}$ and final set $\mathsf{H}_{2,ac}$, while $W_+(H_1, H_2)$ is partially isometric with initial set $\mathsf{H}_{2,ac}$ and final set $\mathsf{H}_{1,ac}$. Both of these two generalized wave operators are complete, and $H_{1,ac}$ and $H_{2,ac}$ are unitarily equivalent. Similar results hold with W_+ replaced by W_-.

Proof. We can set $H_3 = H_1$ in Theorem 3.4; we obtain $W_+(H_1, H_2)\, W_+(H_2, H_1) = P_1$ since it is obvious that $W_+(H_1, H_1) = P_1$. Writing $W_{21} = W_+(H_2, H_1)$ etc. for simplicity, we have thus $W_{12} W_{21} = P_1$ and so $W_{21} W_{12} = P_2$ by symmetry. In virtue of (3.5) and (3.6), it follows that $W_{12} = P_1 W_{12} = (W_{21}^* W_{21}) W_{12} = W_{21}^* (W_{21} W_{12}) = W_{21}^* P_2 = W_{21}^*$. This proves (3.20), and the other assertions follow automatically.

Remark 3.6. In the theorems proved above it is not at all essential that the P_1 appearing in the definition (3.3) of $W_\pm$ is the projection on $\mathsf{H}_{1,ac}$. We could as well have chosen P_1 to be the projection on $\mathsf{H}_{1,c}$, the space of continuity for H_1. The importance of the choice of $\mathsf{H}_{1,ac}$ will be recognized when we consider the conditions for the *existence* of $W_\pm$.

3. A sufficient condition for the existence of the wave operator

The following theorem gives a sufficient condition for the existence of the generalized wave operator, which is not very strong but is convenient for application.

Theorem 3.7.[1] *Let there exist a fundamental subset* D *of* $\mathsf{H}_{1,ac}$ *with the following properties: if* $u \in \mathsf{D}$, *there exists a real* s *such that* $e^{-itH_1} u \in \mathsf{D}(H_1) \cap \mathsf{D}(H_2)$ *for* $s \leq t < \infty$, $(H_2 - H_1) e^{-itH_1} u$ *is continuous in* t, *and* $\|(H_2 - H_1) e^{-itH_1} u\|$ *is integrable on* (s, ∞). *Then* $W_+(H_2, H_1)$ *exists. (A similar theorem holds for the existence of* $W_-(H_2, H_1)$, *with an obvious modification.)*

Proof. If $u \in \mathsf{D}$ we have $(d/dt)\, W(t)\, u = (d/dt)\, e^{itH_2} e^{-itH_1} u = i\, e^{itH_2}(H_2 - H_1)\, e^{-itH_1} u$ (see IX-§ 1.3), and this derivative is continuous by hypothesis. Hence we have by integration

$$W(t'')\, u - W(t')\, u = i \int_{t'}^{t''} e^{itH_2}(H_2 - H_1)\, e^{-itH_1} u\, dt\,. \tag{3.21}$$

Since $\|e^{itH_2}\| = 1$, we have

$$\|W(t'')\, u - W(t')\, u\| \leq \int_{t'}^{t''} \|(H_2 - H_1)\, e^{-itH_1} u\|\, dt\,. \tag{3.22}$$

Since the integrand on the right is integrable on (s, ∞), the right member tends to zero when $t', t'' \to \infty$ and $\operatorname{s-lim}_{t\to\infty} W(t)\, u$ exists.

Since this is true for all u of D, which is fundamental in $\mathsf{H}_{1,ac}$, and since $W(t)$ is uniformly bounded, it follows (see Lemma III-3.5) that the same limit exists for every $u \in \mathsf{H}_{1,ac} = P_1 \mathsf{H}$. This is equivalent to the existence of $W_+(H_2, H_1)$.

In connection with the identity (3.21), we note the following lemma for future use

Lemma 3.8. *Let* $H_2 = H_1 + A$ *where* $A \in \mathscr{B}(\mathsf{H})$. *If* $W_+ = W_+(H_2, H_1)$ *exists, then we have for any* $u \in \mathsf{H}_{1,ac}$

$$\|W_+ u - W(t)\, u\|^2 = -2\, \mathrm{Im} \int_t^\infty (e^{isH_1} W_+^* A\, e^{-isH_1} u, u)\, ds\,. \tag{3.23}$$

Proof. If $u \in \mathsf{D}(H_1)$, we have the identity (3.21). This identity is even true for every $u \in \mathsf{H}$ if we replace $H_2 - H_1$ by A, for the two members of (3.21) are then elements of $\mathscr{B}(\mathsf{H})$. If in particular $u \in \mathsf{H}_{1,ac}$, $\lim_{t''\to\infty} W(t'')\, u = W_+ u$ exists, so that

$$W_+ u - W(t)\, u = i \int_t^\infty e^{isH_2} A\, e^{-isH_1} u\, ds\,, \quad u \in \mathsf{H}_{1,ac}\,. \tag{3.24}$$

---

[1] See Cook [1], Jauch [1], Kuroda [2].

Since $\|W_+ u\| = \|u\|$ because $u \in \mathsf{H}_{1,\mathrm{ac}}$ and since $W(t)$ is unitary, $\|W_+ u - W(t) u\|^2 = 2\|u\|^2 - 2\,\mathrm{Re}(W(t) u, W_+ u) = 2\,\mathrm{Re}(W_+ u - W(t) u, W_+ u)$. Thus substitution from (3.24) gives (3.23) if we note that $W_+^* e^{isH_1} = e^{isH_1} W_+^*$ [a relation adjoint to (3.7)].

4. An application to potential scattering

Let $\mathsf{H} = \mathsf{L}^2(\mathrm{R}^3)$ and set

$$H_1 = -\Delta\,,\quad H_2 = -\Delta + q(x)\,, \tag{3.25}$$

where Δ is the Laplacian and $q(x)$ is a real-valued function. We know that H_1 is selfadjoint if Δ is taken in a generalized sense and that H_2 is also selfadjoint with $\mathsf{D}(H_2) = \mathsf{D}(H_1)$ at least if $q(x)$ is the sum of a function of $\mathsf{L}^2(\mathrm{R}^3)$ and a bounded function (see Theorem V-5.4). We further note that H_1 is spectrally absolutely continuous (see Example 1.10) so that $P_1 = 1$.

We shall show that Theorem 3.7 is applicable to the pair H_1, H_2 under some additional restrictions on $q(x)$. Set

$$u(x) = e^{-|x-a|^2/2}\,,\quad a \in \mathrm{R}^3\,. \tag{3.26}$$

Then we have

$$e^{-itH_1} u(x) = (1 + 2\,i\,t)^{-3/2}\, e^{-|x-a|^2/2(1+2it)}\,, \tag{3.27}$$

as is seen from the results of IX-§ 1.8 [the corresponding one-dimensional formula is IX-(1.75)]. The function (3.27) is bounded in x with bound $(1 + 4t^2)^{-3/4}$. Consequently,

$$\|(H_2 - H_1)\, e^{-itH_1} u\| \le (1 + 4t^2)^{-3/4} \|q\| \tag{3.28}$$

provided $q \in \mathsf{L}^2(\mathrm{R}^3)$, where $\|q\|$ denotes the L^2-norm of q. Obviously (3.28) is integrable on $-\infty < t < \infty$. Since the set of functions of the form (3.26) with different a is fundamental in H (see loc. cit.), it follows from Theorem 3.7 that $W_\pm(H_2, H_1)$ exist, and they are proper wave operators since $P_1 = 1$. In particular, H_1 is unitarily equivalent to a part of H_2. This implies, in particular, that $\Sigma(H_2)$ includes the whole positive real axis. This is not a trivial result and is not easy to deduce by an independent consideration.

The assumption made above that $q \in \mathsf{L}^2$ can be weakened slightly if one notes that the rate of decrease as $t \to \pm\infty$ of the right member of (3.28) is $|t|^{-3/2}$, which is stronger than necessary for the integrability. To obtain a better estimate, we note that

$$\begin{aligned} v_t(x) &\equiv (H_2 - H_1)\, e^{-itH_1} u(x) \\ &= (1 + 2\,i\,t)^{-1-\frac{\varepsilon}{2}} \cdot \frac{q(x)}{|x-a|^{(1-\varepsilon)/2}} \cdot \frac{|x-a|^{(1-\varepsilon)/2}}{(1+2\,i\,t)^{(1-\varepsilon)/2}} \cdot e^{-|x-a|^2/2(1+2it)} \end{aligned} \tag{3.29}$$

for any ε with $0 < \varepsilon < 1$. The product of the last two factors in the last member is bounded in x with bound independent of t [note that $|\exp(-|x-a|^2/2(1+2it))| = \exp(-|x-a|^2/2(1+4t^2))$]. The second factor is majorized by a constant times $|q(x)|/(1+|x|)^{(1-\varepsilon)/2}$ in the region $|x-a| \geqq 1$. Hence the integral of $|v_t(x)|^2$ in this region does not exceed a constant times $(1+4t^2)^{-1-\frac{\varepsilon}{2}}$ if

$$\int\limits_{R^3} (1+|x|)^{-1+\varepsilon} |q(x)|^2 \, dx < \infty \quad \text{for some} \quad \varepsilon > 0\,. \tag{3.30}$$

On the other hand, the integral of $|v_t(x)|^2$ taken in $|x-a| \leqq 1$ is majorized by $(1+4t^2)^{-3/2}$ times the integral of $|q(x)|^2$ in this bounded region [for the same reason as in (3.28)]. Hence $\|v_t\| \leqq \text{const.}\,(1+4t^2)^{-\frac{1}{2}-\frac{\varepsilon}{4}}$ and $\|v_t\|$ is integrable on $(-\infty, \infty)$.

Thus we have proved

Theorem 3.9.[1] *Let H_1, H_2 be as above, where $q(x)$ is the sum of a function of $L^2(R^3)$ and a bounded function. If $q(x)$ satisfies the additional condition* (3.30), *then $W_\pm(H_2, H_1)$ exist and are proper wave operators. In particular, H_1 is unitarily equivalent to a part of H_2, and $\Sigma_{ac}(H_2)$ includes the positive real axis.*

Remark 3.10. (3.30) is satisfied if $q(x) = O(|x|^{-1-\varepsilon})$ as $|x| \to \infty$.

Remark 3.11. According to Theorem V-5.7, H_2 has only isolated eigenvalues on the negative real axis under a condition similar to that of Theorem 3.9; but it showed only that the positive real axis is $\Sigma_e(H_2)$. Thus Theorem 3.9 and Theorem V-5.7 complement each other.

§ 4. Existence and completeness of wave operators

1. Perturbations of rank one (special case)

We are now going to prove several theorems on the existence and completeness of the generalized wave operators $W_\pm(H_2, H_1)$. The proof needs several steps. We begin by considering a very special case.

Let $H = L^2(-\infty, \infty)$ and let H_1 be the maximal multiplication operator defined by $H_1 u(x) = x\,u(x)$. H_1 is selfadjoint and spectrally absolutely continuous, with $D(H_1)$ consisting of all $u \in H$ with $x\,u(x) \in H$. Let A be an operator of rank one given by

$$A u = c(u, f)\, f\,, \tag{4.1}$$

where c is a real constant and f is a given element of H with $\|f\| = 1$. We further assume that $f = f(x)$ is a smooth function and rapidly decreasing as $x \to \pm\infty$.

[1] See KURODA [2]. Cf. also COOK [1], HACK [1], JAUCH and ZINNES [1], BROWNELL [3]. For the wave operators of many-particle systems, see HACK [2].

Set $H_2 = H_1 + A$; H_2 is selfadjoint with $\mathsf{D}(H_2) = \mathsf{D}(H_1)$. We shall show that $W_\pm(H_2, H_1)$ exist. For this it suffices to show that the assumptions of Theorem 3.7 are satisfied. For any $u \in \mathsf{H}$, we have

$$A e^{-itH_1} u = c(e^{-itH_1} u, f) f \tag{4.2}$$

and

$$\|A e^{-itH_1} u\| = |c| \, |(e^{-itH_1} u, f)| = |c| \left| \int_{-\infty}^{\infty} e^{-itx} u(x) \overline{f(x)} \, dx \right| . \tag{4.3}$$

If $u(x)$ is a smooth, rapidly decreasing function, the Fourier transform of $u \bar{f}$ is again smooth and rapidly decreasing, so that (4.3) is integrable in t on $(-\infty, \infty)$. Thus the assumptions of Theorem 3.7 are satisfied, for the set of such functions u is dense in $\mathsf{H} = \mathsf{H}_{1,ac}$.

Let us estimate the rate of convergence $W(t) \to W_+$, $t \to \infty$. Substitution of (4.2) into (3.23) gives

$$\|W_+ u - W(t) u\|^2 = -2c \operatorname{Im} \int_t^\infty (e^{-isH_1} u, f) (e^{isH_1} W_+^* f, u) \, ds \leq \tag{4.4}$$
$$\leq 2|c| \left[\int_t^\infty |(e^{-isH_1} u, f)|^2 ds\right]^{1/2} \left[\int_t^\infty |(e^{isH_1} W_+^* f, u)|^2 ds\right]^{1/2} .$$

The integrals on the right of (4.4) are convergent. In fact $(e^{-isH_1} u, f)$ is the Fourier transform of $u \bar{f}$ as noted above, so that (Parseval's equality)

$$\int_{-\infty}^{\infty} |(e^{-isH_1} u, f)|^2 ds = 2\pi \|u \bar{f}\|^2 \leq 2\pi \|f\|^2 \|u\|_\infty^2 = 2\pi \|u\|_\infty^2 \tag{4.5}$$

and the same inequality holds for the other integral in (4.4). $\|u\|_\infty$ denotes the maximum norm of $u(x)$, which is smooth by assumption. Note that $\|W_+^* f\| \leq \|f\| = 1$ since W_+^* is partially isometric.

Since we do not know what kind of a function $W_+^* f(x)$ is, we shall replace the second integral on the right of (4.4) by its majorant $2\pi \|u\|_\infty^2$, obtaining

$$\|W_+ u - W(t) u\|^2 \leq 2|c| (2\pi)^{1/2} \|u\|_\infty \left[\int_t^\infty |(e^{-isH_1} u, f)|^2 ds\right]^{1/2} . \tag{4.6}$$

This estimates the rate of convergence $W(t) u \to W_+ u$. Taking the square root of (4.6) and subtracting two inequalities for different values of t, we obtain

$$\|W(t'') u - W(t') u\| \leq (8\pi)^{1/4} |c|^{1/2} \|u\|_\infty^{1/2} \cdot \tag{4.7}$$
$$\cdot \left\{\left[\int_{t'}^\infty |(e^{-isH_1} u, f)|^2 ds\right]^{1/4} + \left[\int_{t''}^\infty |(e^{-isH_1} u, f)|^2 ds\right]^{1/4}\right\} .$$

It should be remarked that (4.7) does not contain W_+ or W_+^*. Moreover, the fact that $f(x)$ is smooth does not appear explicitly in this inequality [the integrals involved are finite if only $f \in \mathsf{H}$, see (4.5)]. This suggests that (4.7) *is true for any* $f \in \mathsf{H}$ *and any* $u \in \mathsf{H} \cap \mathsf{L}^\infty$. We shall prove this by going to the limit[1] from smooth f and u.

First let $f(x)$ be smooth and rapidly decreasing and assume $u \in \mathsf{H} \cap \mathsf{L}^\infty$. There exists a sequence $\{u_n\}$ of smooth, rapidly decreasing functions such that $\|u_n - u\| \to 0$ and $\|u_n\|_\infty \to \|u\|_\infty$. Since (4.7) is true for u replaced by u_n, we obtain (4.7) for u itself by going to the limit $n \to \infty$, making use of the inequality

$$\text{(4.8)} \quad \left| \left[\int_t^\infty |(e^{-isH_1} u_n, f)|^2 \, ds \right]^{1/2} - \left[\int_t^\infty |(e^{-isH_1} u, f)|^2 \, ds \right]^{1/2} \right| \leqq$$
$$\leqq \left[\int_t^\infty |(e^{-isH_1} (u_n - u), f)|^2 \, ds \right]^{1/2} \leqq (2\pi)^{1/2} \|u_n - u\| \, \|f\|_\infty \to 0 .$$

The first inequality in (4.8) is the triangle inequality in $\mathsf{L}^2(t, \infty)$ and the second is the same as (4.5) with the roles of u and f exchanged.

Consider now the general case $f \in \mathsf{H}$, $\|f\| = 1$, and $u \in \mathsf{H} \cap \mathsf{L}^\infty$. Let f_n be a sequence of smooth and rapidly decreasing functions such that $\|f_n\| = 1$ and $\|f_n - f\| \to 0$. If we define $H_{2n} = H_1 + c(\ , f_n) f_n$, (4.7) is true when f is replaced by f_n and $W(t)$ by $W_n(t) = e^{itH_{2n}} e^{-itH_1}$. Then (4.7) follows on letting $n \to \infty$; here we make use of the following facts. Since $\|H_{2n} - H_1\| \leqq 2|c| \, \|f_n - f\| \to 0$, $\|W_n(t) - W(t)\| = \|e^{itH_{2n}} - e^{itH_2}\| \to 0$ by Theorem IX-2.1. Again, the integrals on the right of (4.7) with f replaced by f_n converge to the ones with f as $n \to \infty$; the proof is similar to (4.8), with the roles of u and f exchanged.

Since the integrals on the right of (4.7) exist, (4.7) implies that $\lim_{t\to\infty} W(t) u$ exists for any $u \in \mathsf{H} \cap \mathsf{L}^\infty$. From this we conclude as above that $\operatorname{s-lim}_{t\to\infty} W(t) = W_+(H_2, H_1)$ exists. Since the situation for $t \to -\infty$ is the same, we have proved

Theorem 4.1. *Let* $\mathsf{H} = \mathsf{L}^2(-\infty, \infty)$ *and let* H_1 *be the multiplication operator* $H_1 u(x) = x\,u(x)$. *Then* H_1 *is spectrally absolutely continuous, and the wave operators* $W_\pm(H_2, H_1)$ *exist for any selfadjoint operator* H_2 *obtained from* H_1 *by a perturbation of rank one*: $H_2 = H_1 + c(\ , f) f$, $f \in \mathsf{H}$.

Remark 4.2. For future use we shall give here a formal generalization of Theorem 4.1. Suppose that H_1 is a selfadjoint operator in an abstract Hilbert space H, and that $H_{1,ac}$ is unitarily equivalent to the H_1 of Theorem 4.1. Then we maintain that $W_\pm(H_2, H_1)$ exist for any $H_2 = H_1 + c(\ , f) f$, $f \in \mathsf{H}$.

This can be proved by the same arguments as in the proof of Theorem 4.1. First we note that $\mathsf{H}_{1,ac}$ may be identified with $\mathsf{L}^2(-\infty, \infty)$ and $H_{1,ac}$ with the H_1 of Theorem 4.1. Then we set $f = g + h$, $g = P_1 f$,

[1] It is rather embarrassing that no direct proof of (4.7) has been found.

$h = (1 - P_1) f$. g belongs to $\mathsf{H}_{1,ac} = \mathsf{L}^2$ so that it is represented by a function $g(x)$. We first consider the case in which $g(x)$ is a smooth and rapidly decreasing function. Then the existence of $W_\pm$ can be proved in the same way as above. The only change required is to restrict u to elements of $\mathsf{H}_{1,ac}$ [so that $u = u(x) \in \mathsf{L}^2$]; then $u = P_1 u$ and $(e^{-itH_1} u, f) = (e^{-itH_1} u, P_1 f) = (e^{-itH_1} u, g)$ can be expressed by the integral in (4.3) with f replaced by g. Also the estimate (4.4) is true with f replaced by g ($W_\pm^* f$ should not be changed, but it belongs to $\mathsf{H}_{1,ac}$ and so is represented by a function of L^2). Hence (4.6) and (4.7) remain true with f replaced by g.

Then the extension of (4.7) from smooth $u(x)$ to any $u(x) \in \mathsf{L}^2 \cap \mathsf{L}^\infty$ can be done in the same way as above (always with f replaced by g). The further extension to an arbitrary f can also be achieved in the same way; here we define $f_n = g_n - h$ where $g_n = g_n(x) \in \mathsf{H}_{1,ac}$ are smooth functions such that $\|g_n\| = \|g\|$ and $\|g_n - g\| \to 0$ (so that $\|f_n\| = 1$, $\|f_n - f\| \to 0$), and the proof proceeds without modification.

It follows that $\lim W(t)\, u$ exists for any $u \in \mathsf{L}^2 \cap \mathsf{L}^\infty$. Since such u form a dense set in $\mathsf{L}^2 = \mathsf{H}_{1,ac}$, the existence of s-lim $W(t)$ has been proved.

2. Perturbations of rank one (general case)

In Theorem 4.1 proved above, the unperturbed operator H_1 was a very special one: multiplication by x in $\mathsf{H} = \mathsf{L}^2(-\infty, \infty)$. We shall now remove this restriction.

First we take the case in which $\mathsf{H} = \mathsf{L}^2(\mathsf{S})$, where S is an arbitrary Borel set on the real line $(-\infty, \infty)$, and H_1 is the maximal multiplication operator $H_1 u(x) = x\, u(x)$ with $\mathsf{D}(H_1)$ consisting of all $u(x) \in \mathsf{H}$ such that $x\, u(x) \in \mathsf{H}$; H_1 is spectrally absolutely continuous as before. We shall show that the wave operators $W_\pm(H_2, H_1)$ exist for any $H_2 = H_1 + c(\ , f) f$ obtained from H_1 by a perturbation of rank one.

H can be regarded as a subspace of a larger Hilbert space $\mathsf{H}' = \mathsf{L}^2(-\infty, \infty)$, consisting of all $u \in \mathsf{H}'$ such that $u(x) = 0$ for $x \notin \mathsf{S}$. Let H_1' be the maximal multiplication operator by x in H'. Then the subspace H of H' reduces H_1', and the part of H_1' in H is identical with H_1. Let $H_2' = H_1' + c(\ , f) f$, where f is regarded as an element of H' by setting $f(x) = 0$ for $x \notin \mathsf{S}$. Then H_2' is also reduced by H and its part in H is H_2. Thus $e^{-itH_1'}$ and $e^{-itH_2'}$ are also reduced by H and their parts in H are e^{-itH_1} and e^{-itH_2}, respectively.

But we know from Theorem 4.1 that $W'(t) = e^{itH_2'} e^{-itH_1'}$ has strong limits as $t \to \pm\infty$. Since $W'(t)$ is also reduced by H and its part in H is $W(t) = e^{itH_2} e^{-itH_1}$, it follows that $W_\pm = \operatorname*{s-lim}_{t \to \pm\infty} W(t)$ exist.

Again we can extend the result obtained to the more general case in which H_1 need not be absolutely continuous, in the manner described in Remark 4.2. Suppose that $H_{1,ac}$ is unitarily equivalent to the H_1 considered above: multiplication by x in $L^2(S)$. We assert that, for any $H_2 = H_1 + c(\ ,f)f$, $f \in H$, the generalized wave operators $W_\pm(H_2, H_1)$ exist. For the proof it suffices to proceed in the same way as in Remark 4.2. We may identify $H_{1,ac}$ with $L^2(S)$ and $H_{1,ac}$ with the multiplication operator by x, and regard $H_{1,ac} = L^2(S)$ as a subspace of a larger Hilbert space $H'_0 = L^2(-\infty, \infty)$ and, accordingly, H as a subspace of $H' = H'_0 \oplus \oplus H_{1,s}$. ($H_{1,s}$ is the subspace of singularity for H_1.) Then H_1 may be regarded as the part in H of the operator $H'_1 = H'_{1,ac} \oplus H_{1,s}$ where $H'_{1,ac}$ is multiplication by x in $H'_0 = L^2(-\infty, \infty)$. We set $H'_2 = H'_1 + c(\ ,f)f$, where f is regarded as an element of H' by setting $f = g + h$, $g \in H'_0$, $h \in H_{1,s}$, where $g(x) = 0$ for $x \notin S$ [originally $g(x)$ was defined for $x \in S$]. In this way the same argument as given above applies, using Remark 4.2 in place of Theorem 4.1.

Now we shall consider the perturbation of rank one without any assumption on H_1. Let H_1 be an arbitrary selfadjoint operator in a Hilbert space H and set $H_2 = H_1 + c(\ ,f)f$, where $f \in H$ with $\|f\| = 1$ and c is real. Let H_0 be the smallest subspace containing f which reduces H_1, and let P_0 be the projection on H_0. H_0 may be characterized as the closed span of the set of vectors $\{E_1(\lambda) f\}$ for all real λ. H_2 is also reduced by H_0, for $P_0 u \in D(H_1) = D(H_2)$ if $u \in D(H_2) = D(H_1)$ and $H_2 P_0 u = H_1 P_0 u + c(P_0 u, f) f \in H_0$.

Let $H_0^\perp$ be the orthogonal complement of H_0 in H. $H_0^\perp$ reduces both H_1 and H_2 and $H_1 u = H_2 u$ for $u \in H_0^\perp$. Hence $W(t) u = e^{itH_2} \cdot e^{-itH_1} u = u$ for $u \in H_0^\perp$. Since $u \in H_0^\perp$ implies $P_1 u \in H_0^\perp$ by Theorem 1.6, $W(t) P_1 u = P_1 u$ and so s-lim $W(t) P_1 u = P_1 u$ for $u \in H_0^\perp$. In order to prove the existence of $W_\pm(H_2, H_1) =$ s-lim $W(t) P_1$, it is therefore sufficient to prove the existence of s-lim $W(t) P_1 u$ for $u \in H_0$. Since P_1, H_1 and H_2 are all reduced by H_0 and since $H_{k,ac} \cap H_0$, $k = 1, 2$, is exactly the subspace of absolute continuity for the part of H_k in H_0 by Theorem 1.6, we may assume from the outset that $H_0 = H$ (this amounts simply to changing the notation).

Thus we assume that the smallest subspace containing f which reduces H_1 is the whole space[1] H. Let

$$f = g + h\,, \quad g = P_1 f\,, \quad h = (1 - P_1) f\,. \tag{4.9}$$

Since the subspace spanned by the set $\{E(\lambda) f\}$ is $H_0 = H$, the two subspaces spanned by the sets $\{E(\lambda) g\}$ and $\{E(\lambda) h\}$ together span the whole space H. But these two subspaces consist respectively of absolutely continuous vectors and singular vectors with respect to H_1. Hence they must coincide with the subspaces of absolute continuity and of singularity

[1] In other words, H_1 has a *simple spectrum* and f is a *generating vector*.

for H_1. In other words, $\mathsf{H}_{1,ac}$ is spanned by the set $\{E(\lambda)g\}$ and $\mathsf{H}_{1,s}$ is spanned by $\{E(\lambda)h\}$. Thus $\mathsf{H}_{1,ac}$ is the closure of the set of all vectors of the form $\phi(H_1)g = \int_{-\infty}^{\infty}\phi(\lambda)\,dE(\lambda)\,g$, where $\phi(\lambda)$ is any bounded (Borel measurable) function [or one may restrict $\phi(\lambda)$ to smoother functions].

Now we have

$$(4.10)\quad (\phi_1(H_1)g, \phi_2(H_1)g) = \int_{-\infty}^{\infty}\phi_1(\lambda)\overline{\phi_2(\lambda)}\,d(E_1(\lambda)g,g) = \int_{\mathsf{S}}\psi_1(\lambda)\,\psi_2(\lambda)\,d\lambda,$$

where

$$(4.11)\quad \psi_k(\lambda) = \phi_k(\lambda)\,\varrho(\lambda)^{1/2}, \quad k = 1, 2, \quad \varrho(\lambda) = d(E_1(\lambda)g, g)/d\lambda,$$

and S is the set of all λ such that $d(E_1(\lambda)g, g)/d\lambda$ exists and is positive; note that this derivative exists almost everywhere because $g \in \mathsf{H}_{1,ac}$. S is a Borel set.

If $\phi(\lambda)$ varies over all bounded functions, $\psi(\lambda) = \phi(\lambda)\,\varrho(\lambda)^{1/2}$ varies over a dense subset of $\mathsf{L}^2(\mathsf{S})$. Hence we may identify $\mathsf{H}_{1,ac}$ with $\mathsf{L}^2(\mathsf{S})$ by the correspondence $\phi(H_1)g \to \psi$. In this realization of $\mathsf{H}_{1,ac}$, the operator H_1 is represented by multiplication by λ[1]. Therefore, the absolutely continuous part $H_{1,ac}$ of H_1 is of the type considered above, and the existence of the generalized wave operators follows. This proves the following theorem (note Theorem 3.5).

Theorem 4.3.[2] *Let H_1, H_2 be selfadjoint operators in a Hilbert space* H *such that $H_2 = H_1 + c(\ ,f)f$, $f \in \mathsf{H}$, c real. Then the generalized wave operators $W_\pm(H_2, H_1)$ and $W_\pm(H_1, H_2)$ exist and are complete. In particular, the absolutely continuous parts $H_{1,ac}$, $H_{2,ac}$ of these two operators are unitarily equivalent.*

3. Perturbations of the trace class

Theorem 4.3 can further be generalized to

Theorem 4.4.[3] *Let H_1, H_2 be selfadjoint operators in a Hilbert space* H *such that $H_2 = H_1 + A$ where A belongs to the trace class $\mathscr{B}_1(\mathsf{H})$. Then the generalized wave operators $W_\pm(H_2, H_1)$ and $W_\pm(H_1, H_2)$ exist and are complete*[4]*. The absolutely continuous parts of H_1, H_2 are unitarily equivalent.*

---

[1] This is essentially the theorem that any spectrally absolutely continuous, selfadjoint operator with a simple spectrum is unitarily equivalent to the multiplication operator by λ in some space $\mathsf{L}^2(\mathsf{S})$; see e. g. STONE [1].

[2] Cf. T. KATO [10], in which the theorem is proved by a "stationary method".

[3] This theorem was proved by ROSENBLUM [1] when both H_1, H_2 are spectrally absolutely continuous, and by T. KATO [11] in the general case.

[4] The scattering operator $S = W_+^* W_-$ is unitary on $P_1\,\mathsf{H}$ to itself and commutes with H_1. The corresponding S-matrix $S(\lambda)$ (see footnote [3] on p. 528) has been studied by BIRMAN and KREIN [1] using the *trace formula* due to LIFŠIC [4] and KREIN [3], [6].

Proof. Let the expansion of A [see V-(2.20)] be given by

$$(4.12)\qquad A = \sum_{k=1}^{\infty} c_k(\ ,f_k)\, f_k\,,$$

where the f_k form an orthonormal family of eigenvectors and the c_k (real) are the associated (repeated) eigenvalues. $A \in \mathscr{B}_1(\mathsf{H})$ implies that

$$(4.13)\qquad \sum_{k=1}^{\infty} |c_k| = \|A\|_1 < \infty$$

(see Problem 1.14).

Let A_n be the partial sum of the first n terms of the series in (4.12), and set $H^{(n)} = H_1 + A_n$ with the convention that $H^{(0)} = H_1$. Then each $H^{(n)} - H^{(n-1)}$ is of rank one and $W_\pm(H^{(n)}, H^{(n-1)})$ exist by Theorem 4.3. By successive application of the chain rule (Theorem 3.4), it then follows that $W_{n\pm} = W_\pm(H^{(n)}, H_1)$ exist for $n = 1, 2, \ldots$.

We see from Lemma 3.8 that for each $u \in \mathsf{H}_{1,\mathrm{ac}}$, with $W_n(t) = e^{itH^{(n)}}\, e^{-itH_1}$,

$$(4.14)\qquad \|W_{n+}u - W_n(t)\, u\|^2 = -2\,\mathrm{Im}\int_t^{\infty} (e^{isH_1}\, W_{n+}^*\, A_n\, e^{-isH_1}\, u, u)\, ds\,.$$

Let us estimate the right member of (4.14). Since

$$(4.15)\qquad (e^{isH_1}\, W_{n+}^*\, A_n\, e^{-isH_1}\, u, u) = \sum_{k=1}^{n} c_k (e^{-isH_1}\, u, f_k)\, (e^{isH_1}\, W_{n+}^* f_k, u)\,,$$

we have by the Schwarz inequality

$$(4.16)\qquad \|W_{n+}u - W_n(t)\, u\|^2 \le 2\left[\sum_{k=1}^{\infty} |c_k| \int_t^{\infty} |(e^{-isH_1}\, u, f_k)|^2\, ds\right]^{1/2} \cdot \left[\sum_{k=1}^{\infty} |c_k| \int_t^{\infty} |(e^{isH_1}\, g_{kn}, u)|^2\, ds\right]^{1/2},$$

where $g_{kn} = W_{n+}^* f_k$, $\|g_{kn}\| \le 1$.

The right member of (4.16) is finite, not necessarily for all $u \in P_1\mathsf{H}$ but for some restricted u. This is due to the following lemma, the proof of which will be given later.

Lemma 4.5. *Let* $H = \int_{-\infty}^{\infty} \lambda\, dE(\lambda)$ *be a selfadjoint operator in* H. *Let* $u \in \mathsf{H}$ *be absolutely continuous with respect to* H *and let*

$$(4.17)\qquad |||u|||^2 = \operatorname*{ess\text{-}sup}_{\lambda}\, d(E(\lambda)\, u, u)/d\lambda\,.$$

Then we have[1] *for any* $f \in \mathsf{H}$

$$(4.18)\qquad \int_{-\infty}^{\infty} |(e^{-itH}\, u, f)|^2\, dt \le 2\pi\, |||u|||^2\, \|f\|^2\,.$$

[1] This is a generalization of (4.5).

It follows from this lemma that the first factor (except 2) on the right of (4.16) does not exceed $\sqrt{2\pi}\,|||u|||\,(\Sigma|c_k|)^{1/2} = |||u|||\,(2\pi\|A\|_1)^{1/2}$ since $\|f_k\| = 1$. The same is true of the second factor. We use this estimate only for the second factor, obtaining

$$(4.19)\qquad \|W_{n+}u - W_n(t)\,u\| \leqq |||u|||^{1/2}\,(8\pi\|A\|_1)^{1/4}\,\eta(t;u)^{1/4}\,,$$

where

$$(4.20)\qquad \eta(t;u) = \sum_{k=1}^{\infty} |c_k| \int_t^{\infty} |(e^{-isH_1}u, f_k)|^2\,ds \leqq 2\pi\,|||u|||^2\,\|A\|_1\,.$$

From (4.19) we further obtain

$$(4.21)\quad \|W_n(t'')\,u - W_n(t')\,u\| \leqq |||u|||^{1/2}(8\pi\|A\|_1)^{1/4}\,[\eta(t';u)^{1/4} + \eta(t'';u)^{1/4}]\,,$$

which is valid for any $u \in \mathsf{H}_{1,\mathrm{ac}}$.

We can now go to the limit $n\to\infty$ in (4.21). Since $H_2 = H^{(n)} + A - A_n$ and $\|A - A_n\| \to 0$, we have $e^{itH^{(n)}} \to e^{itH_2}$ in norm (see Theorem IX-2.1) and so $W_n(t) \to W(t)$ in norm, for any fixed t. Thus (4.21) gives

$$(4.22)\quad \|W(t'')\,u - W(t')\,u\| \leqq |||u|||^{1/2}(8\pi\|A\|_1)^{1/4}\,[\eta(t';u)^{1/4} + \eta(t'';u)^{1/4}],$$

which shows clearly that $\lim\limits_{t\to\infty} W(t)\,u$ exists if $|||u||| < \infty$, for then $\eta(t;u)\to 0$ as is seen from (4.20).

The set of all $u \in \mathsf{H}_{1,\mathrm{ac}}$ such that $|||u||| < \infty$ is dense in $\mathsf{H}_{1,\mathrm{ac}}$. In fact, let $v \in \mathsf{H}_{1,\mathrm{ac}}$; we have to show that there are $u_n \in \mathsf{H}_{1,\mathrm{ac}}$ such that $|||u_n||| < \infty$ and $u_n \to v$. $(E_1(\lambda)\,v, v)$ is absolutely continuous and $\varrho(\lambda) = d(E_1(\lambda)\,v, v)/d\lambda$ exists almost everywhere and is nonnegative. Let S_n be the set of values λ for which $\varrho(\lambda) > n$; then $\{\mathrm{S}_n\}$ is a nonincreasing sequence with $|\lim \mathrm{S}_n| = 0$. Set $u_n = (1 - E_1(\mathrm{S}_n))\,v$. Then $(E_1(\lambda)\,u_n, u_n) = ((1 - E_1(\mathrm{S}_n))\cdot E_1(\lambda)\,v, v) = \int_{-\infty}^{\lambda} (1 - \chi_n(\lambda'))\,d(E_1(\lambda')\,v, v) = \int_{-\infty}^{\lambda} (1 - \chi_n(\lambda'))\,\varrho(\lambda')\,d\lambda'$, where $\chi_n(\lambda') = 1$ for $\lambda' \in \mathrm{S}_n$, and $= 0$ for $\lambda' \notin \mathrm{S}_n$. Hence $d(E_1(\lambda)\,u_n, u_n)/d\lambda = (1 - \chi_n(\lambda))\,\varrho(\lambda) \leqq n$ almost everywhere and $|||u_n|||^2 \leqq n$. Since $E_1(\mathrm{S}_n)\,v \to 0$ by the absolute continuity of v, it follows that $u_n \to v$.

Since $W(t)$ is uniformly bounded, it follows that $\lim W(t)\,u$ exists for every $u \in \mathsf{H}_{1,\mathrm{ac}} = P_1\mathsf{H}$. In other words, $\operatorname*{s-lim}\limits_{t\to\infty} W(t)\,P_1 = W_+$ exists. Since the same is true of $\operatorname*{s-lim}\limits_{t\to-\infty} W(t)\,P_1$ and since $H_1 = H_2 - A$ with $-A \in \mathscr{B}_1(\mathsf{H})$, the proof of Theorem 4.4 is complete.

Incidentally we note that, on letting $t''\to\infty$ in (4.22) and writing t for t', we get

$$(4.23)\quad \|W_+u - W(t)\,u\| \leqq |||u|||^{1/2}(8\pi\|A\|_1)^{1/4}\eta(t;u)^{1/4} \leqq |||u|||\,(4\pi\|A\|_1)^{1/2}.$$

Proof of Lemma 4.5. Denote by P the projection on the space H_{ac} of absolute continuity with respect to H. Since $u = Pu$, $(E(\lambda)\,u, f)$ is absolutely continuous and its derivative is majorized by $\left[\frac{d}{d\lambda}(E(\lambda)\,u,u)\cdot\right.$ $\left.\cdot\frac{d}{d\lambda}(E(\lambda)\,f_0,f_0)\right]^{1/2} \leq |||u|||\left[\frac{d}{d\lambda}(E(\lambda)\,f_0,f_0)\right]^{1/2}$ where $f_0 = Pf$ (see Theorem 1.7). Hence $\frac{d}{d\lambda}(E(\lambda)\,u,f)$ belongs to $\mathsf{L}^2(-\infty,\infty)$ with L^2-norm not exceeding $|||u|||\,[\int d(E(\lambda)\,f_0,f_0)]^{1/2} = |||u|||\,\|f_0\| \leq |||u|||\,\|f\|$. But $(e^{-itH}u,f)$ $= \int_{-\infty}^{\infty} e^{-it\lambda}\,d(E(\lambda)\,u,f)$ is just the Fourier transform of $\frac{d}{d\lambda}(E(\lambda)\,u,f)$. Hence Lemma 4.5 follows from the Parseval theorem.

4. Wave operators for functions of operators

We shall now show that not only the (generalized) wave operators $W_\pm = W_\pm(H_2, H_1)$ but also the wave operators $W_\pm(\phi(H_2), \phi(H_1))$ exist for certain functions $\phi(\lambda)$ provided $H_2 - H_1$ belongs to the trace class $\mathscr{B}_1(\mathsf{H})$. This will extend the applicability of Theorem 4.4 to a great extent. Moreover, we have the remarkable result that $W_\pm(\phi(H_2), \phi(H_1))$ do not depend on ϕ for a wide class of functions ϕ. This result will be called the *invariance of the wave operators.*

First we prove

Lemma 4.6. *Let $\phi(\lambda)$ be a real-valued function on $(-\infty,\infty)$ with the following properties: the whole interval $(-\infty,\infty)$ can be divided into a finite number of subintervals in such a way that in each open subinterval, $\phi(\lambda)$ is differentiable with $\phi'(\lambda)$ continuous, locally of bounded variation, and positive. Then, for every $w(\lambda) \in \mathsf{L}^2(-\infty,\infty)$ we have*

$$2\pi\|w\|^2 \geq \int_0^\infty dt\left|\text{l.i.m.}\int_{-\infty}^{\infty} e^{-it\lambda - is\phi(\lambda)}\,w(\lambda)\,d\lambda\right|^2 \to 0 \quad \textit{as } s\to+\infty\,. \tag{4.24}$$

Proof. Let H be the selfadjoint operator $Hw(\lambda) = \lambda\,w(\lambda)$ acting in $\mathsf{H} = \mathsf{L}^2(-\infty,\infty)$, and let U be the unitary operator given by the Fourier transformation. Then the inner integral of (4.24) represents the function $(2\pi)^{1/2}(U e^{-is\phi(H)}\,w)\,(t)$, and the middle member of (4.24) is equal to $2\pi\|E\,U e^{-is\phi(H)}\,w\|^2 \leq 2\pi\,\|w\|^2$ where E is the projection of H onto the subspace consisting of all $f(t)$ such that $f(t) = 0$ for $t < 0$. Hence (4.24) is equivalent to $\operatorname*{s-lim}_{s\to+\infty} E\,U e^{-is\phi(H)} = 0$. Since $E\,U e^{-is\phi(H)}$ is uniformly bounded with norm ≤ 1, it suffices to prove (4.24) for all w belonging to a fundamental subset of H. Thus we may restrict ourselves to considering only characteristic functions $w(\lambda)$ of finite intervals $[a,b]$: $w(\lambda) = 1$ for $\lambda \in [a,b]$ and $= 0$ otherwise. Furthermore, we may assume that $[a,b]$ is contained in an open interval in which $\phi(\lambda)$ is continuously differentiable with the properties stated in the lemma.

We have then

$$(4.25)\quad v(t,s) = \int_{-\infty}^{\infty} e^{-it\lambda - is\phi(\lambda)}\, w(\lambda)\, d\lambda = \int_a^b e^{-it\lambda - is\phi(\lambda)}\, d\lambda$$
$$= i \int_a^b (t + s\,\phi'(\lambda))^{-1} \frac{d}{d\lambda} e^{-it\lambda - is\phi(\lambda)}\, d\lambda\,.$$

If $t, s > 0$, $\psi(\lambda) = (t + s\,\phi'(\lambda))^{-1}$ is positive and of bounded variation. An elementary computation shows that the total variation of $\psi(\lambda)$ on $[a, b]$ satisfies

$$\int_a^b |d\psi(\lambda)| \leqq M s/(t + c\,s)^2 \leqq M/c(t + c\,s)\,,$$

where $c > 0$ is the minimum of $\phi'(\lambda)$ and M is the total variation of $\phi'(\lambda)$, both on $[a, b]$. Integrating (4.25) by parts, we thus obtain

$$v(t,s) = i[\psi(\lambda)\, e^{-it\lambda - is\phi(\lambda)}]_a^b - i \int e^{-it\lambda - is\phi(\lambda)}\, d\psi(\lambda)\,,$$

$$|v(t,s)| \leqq \psi(a) + \psi(b) + \int_a^b |d\psi(\lambda)| \leqq (2c + M)/c(t + c\,s)\,.$$

Hence the middle member of (4.24) is

$$\int_0^\infty |v(t,s)|^2\, dt \leqq (2c + M)^2/c^3 s \to 0\,, \quad s \to +\infty\,.$$

We can now prove the invariance of the wave operators in the following form.

Theorem 4.7.[1] *Let H_1, H_2 be selfadjoint operators such that $H_2 = H_1 + A$ where $A \in \mathscr{B}_1(\mathsf{H})$. If $\phi(\lambda)$ is a function with the properties described in* Lemma 4.6, *the generalized wave operators $W_\pm(\phi(H_2), \phi(H_1))$ exist, are complete, and are independent of ϕ; in particular they are respectively equal to $W_\pm(H_2, H_1)$.*

Proof. We start from the estimate (4.23) in which u is replaced by $v = e^{-is\phi(H_1)} u$. Since $(E_1(\lambda)\, v, v) = (E_1(\lambda)\, u, u)$, we have $|||v||| = |||u|||$ by (4.17). On setting $t = 0$, we obtain

$$(4.26)\quad \|(W_+ - 1)\, e^{-is\phi(H_1)} u\| \leqq |||u|||^{1/2} (8\pi \|A\|_1)^{1/4}\, \eta(0; e^{-is\phi(H_1)} u)^{1/4}\,,$$

where

$$(4.27)\quad \eta(0; e^{-is\phi(H_1)} u) = \sum_{k=1}^{\infty} |c_k| \int_0^\infty |(e^{-itH_1 - is\phi(H_1)} u, f_k)|^2\, dt$$

by (4.20).

[1] This theorem can further be generalized by permitting ϕ to be increasing in some subintervals and decreasing in others. Then $W_{\pm,\phi} = W(\phi(H_2), \phi(H_1))$ still exist, though they are no longer equal to $W_\pm$. Instead we have $W_{\pm,\phi}\, u = W_\pm\, u$ or $W_{\pm,\phi}\, u = W_\mp\, u$ for $u \in E_1(I)\,\mathsf{H}$, where I is one of the subintervals in which ϕ is increasing or decreasing. The proof is essentially the same as that of Theorem 4.17. See T. KATO [17], BIRMAN [9], [10], [11].

The integrals on the right of (4.27) have the form (4.24), where $w(\lambda)$ is to be replaced by $d(E_1(\lambda)\,u, f_k)/d\lambda$, which belongs to $\mathsf{L}^2(-\infty, \infty)$ with L^2-norm $\leqq |||u|||$ as noted at the end of par. 3. According to Lemma 4.6 each term on the right of (4.27) tends to 0 as $s \to +\infty$. Since the series is uniformly (in s) majorized by the convergent series $\sum |c_k|\, 2\pi\, |||u|||^2 = 2\pi\, |||u|||^2\, \|A\|_1$, (4.27) itself tends to 0 as $s \to +\infty$. Thus the left member of (4.26) tends to 0 as $s \to +\infty$. Since the set of u with $|||u||| < \infty$ is dense in $P_1 \mathsf{H}$ as remarked in par. 3, it follows that

$$(4.28) \qquad (W_+ - 1)\, e^{-is\phi(H_1)}\, P_1 \xrightarrow[s]{} 0\,, \quad s \to +\infty\,.$$

But $W_+ e^{-is\phi(H_1)} = e^{-is\phi(H_2)}\, W_+$ by (3.12) [note that $e^{-is\phi(H_1)} = \int_{-\infty}^{\infty} e^{-is\phi(\lambda)}\, dE_1(\lambda)$]. On multiplying (4.28) from the left by $e^{is\phi(H_2)}$, we thus obtain

$$(4.29) \qquad \underset{s \to +\infty}{\text{s-lim}}\, e^{is\phi(H_2)}\, e^{-is\phi(H_1)}\, P_1 = W_+ P_1 = W_+\,.$$

This proves the existence of $W_+(\phi(H_2), \phi(H_1))$ and its identity with $W_+ = W_+(H_2, H_1)$, provided we can show that P_1 is also the projection on the subspace of absolute continuity for $\phi(H_1)$, that is, the latter subspace is identical with $\mathsf{H}_{1,\mathrm{ac}}$.

To see this, let $\{F_1(\lambda)\}$ be the spectral family associated with $\phi(H_1)$. Then[1]

$$(4.30) \qquad F_1(\mathrm{S}) = E_1(\phi^{-1}(\mathrm{S}))$$

for any Borel subset S of the real line, where $\phi^{-1}(\mathrm{S})$ denotes the inverse image of S under the map ϕ. If $|\mathrm{S}| = 0$, we have $|\phi^{-1}(\mathrm{S})| = 0$ by the properties of ϕ, so that $F_1(\mathrm{S})\, u = 0$ if $u \in \mathsf{H}_{1,\mathrm{ac}}$. On the other hand, $F_1(\phi(\mathrm{S})) = E_1(\phi^{-1}[\phi(\mathrm{S})]) \geqq E_1(\mathrm{S})$. If $|\mathrm{S}| = 0$, we have $|\phi(\mathrm{S})| = 0$ so that $\|E_1(\mathrm{S})\, u\| \leqq \|F_1(\phi(\mathrm{S}))\, u\| = 0$ if u is absolutely continuous with respect to $\phi(H_1)$. This proves the required result.

By specializing ϕ we get many useful results[2] from Theorem 4.7. For example,

Theorem 4.8.[3] *Let H_1, H_2 be selfadjoint operators with positive lower bounds (so that their inverses belong to $\mathscr{B}(\mathsf{H})$). If $H_2^{-\alpha} - H_1^{-\alpha}$ belongs to the trace class for some $\alpha > 0$, then $W_\pm(H_2, H_1)$ exist, are equal to $W_\mp(H_2^{-\alpha}, H_1^{-\alpha})$ and are complete.*

---

[1] See STONE [1].

[2] If we take $\phi(\lambda) = 2 \operatorname{arc\,cot} \lambda$, then the unitary operator $e^{-i\phi(H_k)} = (H_k - i) \cdot (H_k + i)^{-1} = U_k$ is the so-called Cayley transform of H_k. Since ϕ satisfies the condition of Theorem 4.7, it follows that $\lim_{n \to \pm\infty} U_2^{-n}\, U_1^n\, P_1 = W_\pm(\phi(H_2), \phi(H_1)) = W_\pm(H_2, H_1)$. This limit is a discrete analogue of (3.3). Cf. BIRMAN and KREIN [1].

[3] See BIRMAN [8]. The idea to consider H_1^{-1}, H_2^{-1} in proving the unitary equivalence of H_1, H_2 is found in PUTNAM [2]. See also BIRMAN and KREIN [1].

Proof. The function $\phi(\lambda)$ defined by $\phi(\lambda) = -\lambda^{-1/\alpha}$ for $\lambda \geqq \gamma$ and $\phi(\lambda) = \lambda$ for $\lambda < \gamma$ satisfies the conditions of Lemma 4.6 (where $\gamma > 0$ is the smaller one of the lower bounds of H_1 and H_2). Theorem 4.8 follows from Theorem 4.7 applied to H_1, H_2 replaced by $H_1^{-\alpha}$, $H_2^{-\alpha}$ respectively, with the above $\phi(\lambda)$; here the values of $\phi(\lambda)$ for $\lambda < \gamma$ are immaterial. Note that $W_\pm(H_2, H_1) = W_\mp(-H_2, -H_1)$.

Theorem 4.9.[1] *Let H_1 be selfadjoint and bounded from below. Let V be symmetric and relatively bounded with respect to H_1 with H_1-bound less than 1. Furthermore, let $V = V'' V'$ where $V'(H_1-\gamma)^{-1}$ and $V''^*(H_1-\gamma)^{-1}$ belong to the Schmidt class for some γ smaller than the lower bound of H_1. Then $H_2 = H_1 + V$ is selfadjoint and bounded from below, and $W_\pm(H_2, H_1)$, $W_\pm(H_1, H_2)$ exist and are complete.*

Proof. H_2 is selfadjoint and bounded from below by Theorem V-4.11. Since $W_\pm(H_2-\gamma, H_1-\gamma) = W_\pm(H_2, H_1)$, we may assume that both H_1, H_2 have positive lower bounds and the assumptions of the theorem are satisfied with $\gamma = 0$. [Note that if $V'(H_1-\zeta)^{-1} \in \mathscr{B}_2(\mathsf{H})$ for $\zeta = \gamma$, then the same is true for any $\zeta \in \mathrm{P}(H_1)$ because $V'(H_1-\zeta)^{-1} = V'(H_1-\gamma)^{-1}(H_1-\gamma)(H_1-\zeta)^{-1}$ where $(H_1-\gamma)(H_1-\zeta)^{-1} \in \mathscr{B}(\mathsf{H})$.]

Thus the theorem follows from Theorem 4.8 with $\alpha = 1$ if we can show that $H_2^{-1} - H_1^{-1} \in \mathscr{B}_1(\mathsf{H})$. Now $H_2^{-1} - H_1^{-1} = -H_2^{-1} V H_1^{-1} = -(V''^* H_2^{-1})^* (V' H_1^{-1})$ belongs to $\mathscr{B}_1(\mathsf{H})$ since $V''^* H_2^{-1}$ and $V' H_1^{-1}$ belong to $\mathscr{B}_2(\mathsf{H})$. For $V' H_1^{-1}$ this is the assumption. For $V''^* H_2^{-1} = (V''^* H_1^{-1})(H_1 H_2^{-1})$, it follows from the assumption and the fact that $H_1 H_2^{-1} \in \mathscr{B}(\mathsf{H})$.

Example 4.10.[2] Let us again consider the operators $H_1 = -\Delta$, $H_2 = -\Delta + q(x)$ in $H = \mathsf{L}^2(\mathrm{R}^3)$, see § 3.4. We have shown (Theorem 3.9) that $W_\pm(H_2, H_1)$ exist under certain general conditions on $q(x)$, but we have not ascertained whether or not they are complete.

We shall show that $W_\pm(H_2, H_1)$ *are complete if*[3]

$$q \in \mathsf{L}^1(\mathrm{R}^3) \cap \mathsf{L}^2(\mathrm{R}^3)\,. \tag{4.31}$$

Since we know that $q \in \mathsf{L}^2$ implies that $V = q(x)$ is H_1-bounded with H_1-bound 0, it suffices to show that $|V|^{1/2}(H_1 + c^2)^{-1}$, $c > 0$, belongs to the Schmidt class,

---

[1] See Kuroda [2], [3]. There is an analogous theorem in which H_1, H_2 are respectively associated with symmetric forms $\mathfrak{h}_1$, $\mathfrak{h}_2$ bounded from below, such that $\mathfrak{h}_2 = \mathfrak{h}_1 + \mathfrak{a}$ where $\mathfrak{a}$ is "of relative trace class with respect to $\mathfrak{h}_1$". We shall not state the exact theorem but refer to Kuroda [2], [4]. See also Problem 4.14.

[2] As other examples in which the above theorems have been applied, we mention: the absolute continuity of the Toeplitz's matrices by Putnam [3], [5], Rosenblum [2]; the invariance of the absolutely continuous spectrum of partial differential operators under variation of the boundary and boundary condition by Birman [6], [7]; some problems in neutron scattering by Shizuta [2].

[3] Similar results hold in the one-dimensional problem: $H_1 = -d^2/dx^2$, $H_2 = -d^2/dx^2 + q(x)$ in $\mathsf{H} = \mathsf{L}^2(0, \infty)$ [with boundary condition $u(0) = 0$], if $q \in \mathsf{L}^1 \cap \mathsf{L}^2$. An improved theory (see footnote [1] above) shows that $q \in \mathsf{L}^1$ is enough (in this case H_2 must be defined as in Theorem VI-4.2).

where $|V|^{1/2}$ is the multiplication operator by $|q(x)|^{1/2}$. (Apply Theorem 4.9 with $V = V''V'$, $V' = |V|^{1/2}$, $V'' = U|V|^{1/2}$, where U is the multiplication operator by sign $q(x)$ and is bounded.)

$(H_1 + c^2)^{-1}$ is an integral operator with kernel $e^{-c|y-x|}/4\pi|y - x|$ [see IX-(1.67)]. Hence $|V|^{1/2}(H_1 + c^2)^{-1}$ is an integral operator with kernel $|q(y)|^{1/2} e^{-c|y-x|}/4\pi|y-x|$. This kernel is of Schmidt type since

$$\int\int_{\mathsf{R}^3\times\mathsf{R}^3} \frac{|q(y)|\, e^{-2c|y-x|}}{|y-x|^2}\, dx\, dy \le \int_{\mathsf{R}^3} |q(y)|\, dy \int_{\mathsf{R}^3} \frac{e^{-2c|x|}}{|x|^2}\, dx < \infty$$

if $q \in \mathsf{L}^1$. The same is true for $U|V|^{1/2}(H_1 + c^2)^{-1}$.

It follows, in particular, that $H_{2,ac}$ is unitarily equivalent to H_1 (H_1 is itself absolutely continuous). In general H_2 will have a singular part $H_{2,s}$, including a discontinuous part. It is not known whether or not H_2 can have a continuous singular part[1].

5. Strengthening of the existence theorems

The assumption that $H_2 - H_1$ be in the trace class, made in Theorems 4.4 and 4.7, will now be weakened. First we prove

Lemma 4.11. *Let $R_k(\zeta) = (H_k - \zeta)^{-1}$ be the resolvent of H_k, $k = 1, 2$. If $R_2(\zeta) - R_1(\zeta) \in \mathscr{B}_1(\mathsf{H})$ for some nonreal ζ, then it is true for every nonreal ζ.*

Proof. Suppose $R_2(\zeta_0) - R_1(\zeta_0) \in \mathscr{B}_1(\mathsf{H})$. From the Neumann series for the $R_k(\zeta)$ given by I-(5.6), we have

$$R_2(\zeta) - R_1(\zeta) = \sum_{n=0}^{\infty} (\zeta - \zeta_0)^n [R_2(\zeta_0)^{n+1} - R_1(\zeta_0)^{n+1}]$$

for $|\zeta - \zeta_0| < |\mathrm{Im}\,\zeta_0|$ [note that $\|R_k(\zeta_0)\| \le |\mathrm{Im}\,\zeta_0|^{-1}$]. But

$$\|R_2(\zeta_0)^{n+1} - R_1(\zeta_0)^{n+1}\|_1 = \left\|\sum_{k=0}^{n} R_2(\zeta_0)^{n-k}(R_2(\zeta_0) - R_1(\zeta_0))\, R_1(\zeta_0)^k\right\|_1 \le$$

$$\le \sum_{k=0}^{n} \|R_2(\zeta_0)\|^{n-k}\, \|R_2(\zeta_0) - R_1(\zeta_0)\|_1\, \|R_1(\zeta_0)\|^k \le$$

$$\le (n+1)\, |\mathrm{Im}\,\zeta_0|^{-n}\, \|R_2(\zeta_0) - R_1(\zeta_0)\|_1 .$$

[1] It is known that H_2 has no continuous singular part under somewhat different assumptions on $q(x)$. This was proved by IKEBE [1] by a stationary method, where the $W_\pm$ are constructed explicitly as singular integral operators, whose kernels are *improper eigenfunctions* (not in L^2) of H_2. For these eigenfunctions see also BUSLAEV [1], HUNZIKER [1], IKEBE [2]. For a similar result regarding the operator $-\Delta$ in a domain exterior to a bounded set in R^3, see SHIZUTA [1], IKEBE [3]. For the phase shift formula see GREEN and LANFORD [1], KURODA [6]. There are scattering problems related to the wave equation (even nonlinear) instead of the Schrödinger equation; see BROWDER and STRAUSS [1], LAX and PHILLIPS [1], NIŽNIK [1], STRAUSS [1], [2].

Hence (see Problem 1.17)

$$\|R_2(\zeta) - R_1(\zeta)\|_1 \leqq \|R_2(\zeta_0) - R_1(\zeta_0)\|_1 \sum_{n=0}^{\infty} (n+1)\,|\mathrm{Im}\,\zeta_0|^{-n}\,|\zeta - \zeta_0|^n =$$
$$= \|R_2(\zeta_0) - R_1(\zeta_0)\|_1 (1 - |\mathrm{Im}\,\zeta_0|^{-1}\,|\zeta - \zeta_0|)^{-2},$$

which shows that $R_2(\zeta) - R_1(\zeta) \in \mathscr{B}_1(\mathsf{H})$ for all ζ inside the circle having center ζ_0 and touching the real axis. Repeated application of the same process shows that all nonreal ζ in the half-plane containing ζ_0 have the same property. The same is true of all ζ on the other half-plane in virtue of $R_2(\zeta) - R_1(\zeta) = (R_2(\bar\zeta) - R_1(\bar\zeta))^*$.

Theorem 4.12.[1] *Let $R_2(\zeta) - R_1(\zeta) \in \mathscr{B}_1(\mathsf{H})$ for some nonreal ζ. If ϕ is a function as in* Lemma 4.6, *$W_\pm(\phi(H_2), \phi(H_1))$ exist, are complete, and are independent of ϕ. In particular $W_\pm(H_2, H_1)$ exist and are complete, and the absolutely continuous parts of H_1 and H_2 are unitarily equivalent.*

Proof. I. Let $r > 0$ and define

$$\psi_r(\lambda) = \lambda/(1 + r^{-2}\lambda^2), \quad -\infty < \lambda < \infty, \tag{4.32}$$

$$\varphi_r(\mu) = 2\mu/[1 + (1 - 4r^{-2}\mu^2)^{1/2}], \quad -r/2 \leqq \mu \leqq r/2. \tag{4.33}$$

$\varphi_r(\mu)$ is the inverse function of $\psi_r(\lambda)$ restricted to $-r \leqq \lambda \leqq r$ (in which ψ_r is univalent). For formal convenience we shall extend $\varphi_r(\mu)$ to all real μ by setting $\varphi_r(\mu) = 2\mu$ for $|\mu| > r/2$, so that φ_r satisfies the conditions of Lemma 4.6.

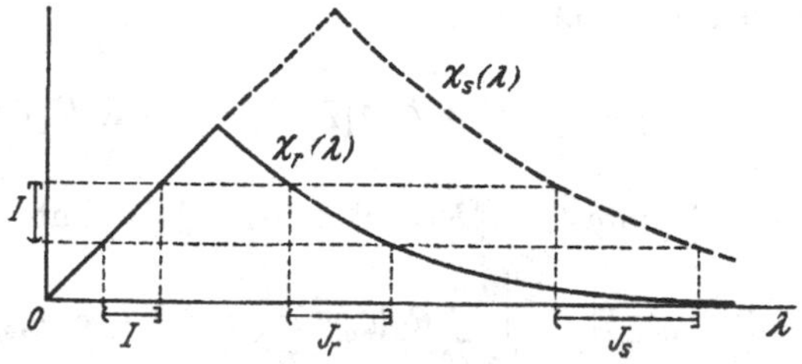

Fig. 3. The functions $\chi_r(\lambda)$, $\chi_s(\lambda)$ and the intervals I, J_r, J_s

Set $\chi_r(\lambda) = \varphi_r(\psi_r(\lambda))$; χ_r is a continuous odd function and

$$\chi_r(\lambda) = \lambda \quad \text{for} \quad -r \leqq \lambda \leqq r. \tag{4.34}$$

For $\lambda \geqq r$, $\chi_r(\lambda)$ decreases monotonically to 0 as $\lambda \to \infty$ (see Fig. 3). Now we set

$$H_{k,r} = \psi_r(H_k), \quad K_{k,r} = \chi_r(H_k), \quad k = 1, 2. \tag{4.35}$$

According to the operational calculus for selfadjoint operators[2], we have

$$K_{k,r} = \varphi_r(\psi_r(H_k)) = \varphi_r(H_{k,r}), \quad k = 1, 2. \tag{4.36}$$

[1] See BIRMAN [9], [10], T. KATO [17].
[2] See STONE [1], Chapter 6.

In virtue of (4.34), however, $K_{k,r}$ and H_k coincide on the subspace $\mathsf{H}_{k,r} \equiv (E_k(r) - E_k(-r))\,\mathsf{H}$, which reduces H_k and $K_{k,r}$. Hence

$$(4.37)\qquad e^{-it\phi(K_{k,r})}\,u = e^{-it\phi(H_k)}\,u \quad \text{for} \quad u \in \mathsf{H}_{k,r}\,.$$

II. Since $H_{k,r} = H_k(1 + r^{-2} H_k^2)^{-1} = (r^2/2)\,[(H_k + i\,r)^{-1} + (H_k - i\,r)^{-1}]$, it follows from Lemma 4.11 that $H_{2,r} - H_{1,r} \in \mathscr{B}_1(\mathsf{H})$. Hence

$$(4.38)\qquad \begin{aligned} W_\pm(\phi(\varphi_r(H_{2,r}), \phi(\varphi_r(H_{1,r})) &= W_\pm(\varphi_r(H_{2,r}), \varphi_r(H_{1,r})) = \\ &= W_\pm(H_{2,r}, H_{1,r}) \equiv W_{\pm,r} \end{aligned}$$

exist by Theorem 4.7 for any ϕ satisfying the conditions of Lemma 4.6, for the composite function $\phi(\varphi_r(\lambda))$ also satisfies the same conditions. In view of (4.36), this implies that

$$\operatorname*{s-lim}_{t\to+\infty} e^{it\phi(K_{2,r})}\,e^{-it\phi(K_{1,r})}\,P_1 = W_{+,r}\,;$$

here P_1 is the projection on $\mathsf{H}_{1,ac}$ but the subspace of absolute continuity for $\phi(K_{1,r})$ coincides with $\mathsf{H}_{1,ac}$ (the proof is the same as in the proof of Theorem 4.7). Using (4.37) for $k = 1$, we have therefore

$$(4.39)\qquad e^{it\phi(K_{2,r})}\,e^{-it\phi(H_1)}\,P_1\,u \to W_{+,r}\,u \quad \text{for} \quad u \in \mathsf{H}_{1,r}\,.$$

III. Let $\mathrm{I} = (a, b]$ where $0 < a < b < r$. The inverse image of I under the map χ_r is the union of I itself and another finite interval J_r lying to the right of r (see Fig. 3). If we denote by $F_{k,r}$ the spectral measure for $K_{k,r} = \chi_r(H_k)$, $k = 1, 2$, we have [see (4.30)]

$$(4.40)\qquad F_{k,r}(\mathrm{I}) = E_k(\chi_r^{-1}(\mathrm{I})) = E_k(\mathrm{I}) + E_k(\mathrm{J}_r) \geqq E_k(\mathrm{I})\,.$$

Let $u \in E_1(\mathrm{I})\,\mathsf{H}$. Then $u \in \mathsf{H}_{1,r}$ and (4.39) holds. Since $W_{+,r} = W_+(K_{2,r}, K_{1,r})$ by (4.38) and $u \in F_{1,r}(\mathrm{I})\,\mathsf{H}$ by (4.40), we have

$$(4.41)\qquad W_{+,r}\,u \in F_{2,r}(\mathrm{I})\,\mathsf{H}\,.$$

From (4.39) and (4.41), we obtain for $u \in E_1(\mathrm{I})\,\mathsf{H}$

$$\begin{aligned} &\|(1 - F_{2,r}(\mathrm{I}))\,e^{-it\phi(H_1)}\,P_1 u\| \\ &\quad = \|(1 - F_{2,r}(\mathrm{I}))\,e^{it\phi(K_{2,r})}\,e^{-it\phi(H_1)}\,P_1 u\| \to \\ &\quad \to \|(1 - F_{2,r}(\mathrm{I}))\,W_{+,r} u\| = 0\,, \quad t \to +\infty\,; \end{aligned}$$

note that $F_{2,r}(\mathrm{I})$ and $e^{it\phi(K_{2,r})}$ commute. Hence

$$(1)_r \qquad e^{-it\phi(H_1)}\,P_1 u \sim F_{2,r}(\mathrm{I})\,e^{-it\phi(H_1)}\,P_1 u\,,$$

where $\sim$ means that the difference of the two members tends to zero as $t \to +\infty$. The same is true if r is replaced by any $s > r$, which result will be called $(1)_s$. If s is sufficiently large, we have $F_{2,r}(\mathrm{I})\,F_{2,s}(\mathrm{I}) = E_2(\mathrm{I})$ as is easily verified. Multiplying $(1)_s$ from the left by $F_{2,r}(\mathrm{I})$ and using

$(1)_r$ we thus obtain

$$e^{-it\phi(H_1)} P_1 u \sim E_2(\mathrm{I})\, e^{-it\phi(H_1)} P_1 u \,,$$

and hence by (4.37) and (4.39)

$$e^{it\phi(H_2)} e^{-it\phi(H_1)} P_1 u \sim e^{it\phi(H_2)} E_2(\mathrm{I})\, e^{-it\phi(H_1)} P_1 u$$

$$= E_2(\mathrm{I})\, e^{it\phi(K_{2,r})} e^{-it\phi(H_1)} P_1 u \to E_2(\mathrm{I})\, W_{+,r} u \,.$$

Thus $\lim e^{it\phi(H_2)} e^{-it\phi(H_1)} P_1 u$ exists and is independent of ϕ.

IV. We have proved that

$$W_\phi(t)\, P_1 u \to W_{+,r} u\,, \quad t \to +\infty\,, \tag{4.43}$$

provided $u \in E_1(\mathrm{I})\, \mathsf{H}$ where $\mathrm{I} = (a, b]$ and $0 < a < b < r$. The same result holds when I is replaced by $(-b, -a]$. The allowable u and their linear combinations form a dense set in $\mathsf{H}_{1,r} = (E_1(r) - E_1(-r))\, \mathsf{H}$, except for the possible eigenspaces of H_1 for the eigenvalues 0 and r. But these eigenspaces are unimportant, being annihilated by P_1.

Since $W_\phi(t)$ is unitary and uniformly bounded, it follows that (4.43) *holds for all* $u \in \mathsf{H}_{1,r}$.

Since s-lim $W_\phi(t)\, P_1 u$ exists for all $u \in \mathsf{H}_{1,r}$ and since the union of all $\mathsf{H}_{1,r}$ for $r > 0$ is dense in H, it follows finally that s-lim $W_\phi(t)\, P_1 = W_+(\phi(H_2), \phi(H_1))$ exists. This limit is independent of ϕ as is seen from (4.43). Obviously the same is true for W_-. Since the assumption of the theorem is symmetric in H_1 and H_2, the same is true when H_1, H_2 are exchanged, so that these wave operators are complete.

Remark 4.13. The assumption $R_2(\zeta) - R_1(\zeta) \in \mathscr{B}_1(\mathsf{H})$ in Theorem 4.12 is not essential. The point is that for any $r > 0$, there is a function $\psi_r(\lambda)$ with the following properties:

1) $\psi_r(\lambda)$ is piecewise monotonic like $\phi(\lambda)$ but may be increasing in some subinterval and decreasing in others;

2) $\psi_r(\lambda)$ is univalent for $-r < \lambda < r$;

3) $\psi_r(H_2) = \psi_r(H_1) + A_r$ where $A_r \in \mathscr{B}_1(\mathsf{H})$.

The proof of Theorem 4.12 can be adapted to this general case with a slight modification[1].

According to this generalization, the results of Theorem 4.12 are true if $R_2(\zeta)^m - R_1(\zeta)^m \in \mathscr{B}_1(\mathsf{H})$ for every pure imaginary ζ (or at least for a sequence $\zeta_n = \pm i\, r_n$ with $r_n \to \infty$). In this case we need only to take $\psi_r(\lambda) = i[(r + i\lambda)^{-m} - (r - i\lambda)^{-m}]$ with $r = r_n$. $\psi_r(\lambda)$ is univalent on a neighborhood of $\lambda = 0$, the size of which is proportional to r.

Problem 4.14. In Theorem 4.9 *remove* the assumptions that H_1 is bounded below and γ is real. The conclusions are still true, except the one that H_2 is bounded from below.

---

[1] This gives very general sufficient conditions for the existence and completeness of the generalized wave operators. There are some conditions not included in this category; see Birman [11], Birman and Entina [1], Stankevič [1].

6. Dependence of $W_\pm(H_2, H_1)$ on H_1 and H_2

Theorem 4.15. *Let H_1, H_2 be selfadjoint operators such that $W_+(H_2, H_1)$ exists. Then $W_+(H_2 + A, H_1)$ and $W_+(H_2, H_1 + A)$ exist for any $A \in \mathscr{B}_1(\mathsf{H})$ and*

$$(4.44)\quad \begin{aligned} W_+(H_2 + A, H_1) &\underset{s}{\to} W_+(H_2, H_1)\,, \\ W_+(H_2, H_1 + A) &\underset{w}{\to} W_+(H_2, H_1) \end{aligned}$$

as $\|A\|_1 \to 0$. Similar results hold with W_+ replaced by W_-.

Proof. $W_+(H_2 + A, H_2)$ exists by Theorem 4.4 and, therefore, $W_+(H_2 + A, H_1)$ exists and equals $W_+(H_2 + A, H_2)\, W_+(H_2, H_1)$ by Theorem 3.4. Similarly, $W_+(H_2, H_1 + A) = W_+(H_2, H_1)\, W_+(H_1, H_1 + A)$ exists. Thus it suffices to prove (4.44) for the special case $H_2 = H_1$.

We see from (4.23), on setting $t = 0$, that $\|W_+(H_1 + A, H_1)\, u - u\| \leq$ $\leq |||u|||\, (4\pi \|A\|_1)^{1/2} \to 0$ as $\|A\|_1 \to 0$. Since the set of u with $|||u||| < \infty$ is dense in $P_1\, \mathsf{H}$, it follows that $W_+(H_1 + A, H_1) = W_+(H_1 + A, H_1)\, P_1 \underset{s}{\to} P_1$ $= W_+(H_1, H_1)$ as $\|A\|_1 \to 0$. Since $W_+(H_1, H_1 + A) = W_+(H_1 + A, H_1)^*$ by Theorem 3.5, it then follows that $W_+(H_1, H_1 + A) \underset{w}{\to} P_1^* = P_1$ $= W_+(H_1, H_1)$.

Remark 4.17. The above results on the continuity of $W_\pm(H_2, H_1)$ as functions of H_1, H_2 are very weak inasmuch as a very strong topology is employed for H_1, H_2. A stronger result is obtained by using Theorem 4.12 instead of Theorem 4.4. Thus, for example, $W_+(H_2 + A, H_1) \underset{s}{\to} W_+(H_2, H_1)$ if A tends to zero in the sense that $\|(H_2 + A - \zeta)^{-1} - (H_2 - \zeta)^{-1}\|_1 \to 0$ for some nonreal ζ; here A need not even be bounded[1]. In general $W_\pm(H_2, H_1)$ are not continuous in H_1, H_2 jointly[2].

Problem 4.18. Discuss the continuity of $W_\pm(H_2, H_1)$ in Example 4.10 when $q(x)$ is changed.

§ 5. A stationary method

1. Introduction

There are other ways to construct the generalized wave operators $W_\pm(H_2, H_1)$. In contrast with the "time-dependent" method developed in preceding sections, these schemes do not make explicit use of the "time variable" t, and in consequence are known as the *stationary*

[1] It is difficult to deduce a stronger continuity of $W_\pm(H_2, H_1)$ by the method used above. But there is a different topology for H_1, H_2 under which $W_\pm(H_2, H_1)$ aquires stronger continuity, for example, continuity in norm. A special case will be considered in the following section. The discontinuity of $W_\pm$ in norm was discussed by Putnam [7].

[2] See T. Kato [11].

methods. There are several varieties among them[1]. In this section we shall give an account of one of them, which seems to complement the time-dependent theory most effectively.

In this method the $W_\pm$ are constructed as solutions of certain operational equations. To deduce these equations, it is convenient to start from the time-dependent formulas (inasmuch as we have defined $W_\pm$ in the time-dependent scheme). For simplicity let us assume that $H_2 = H_1 + A$ where $A \in \mathscr{B}(\mathsf{H})$. Then we have the formula (3.21), from which we obtain the identity

$$W(t'') - W(t') = i \int_{t'}^{t''} e^{itH_2} A e^{-itH_1} dt \tag{5.1}$$

(see the proof of Lemma 3.8). Similarly, we have by exchanging H_1, H_2,

$$W(t'')^{-1} - W(t')^{-1} = -i \int_{t'}^{t''} e^{itH_1} A e^{-itH_2} dt . \tag{5.2}$$

Suppose now that $W_+ = \operatorname*{s-lim}_{t\to\infty} W(t) P_1$ exists. Then $W(t)^{-1} W_+ \underset{s}{\to} P_1$ as $t \to \infty$. Thus (5.2) gives, when multiplied by $-W_+$ from the right, with $t' = 0$ and $t'' \to \infty$

$$W_+ - P_1 = i \int_0^\infty e^{itH_1} A W_+ e^{-itH_1} dt \quad \text{(strong)} , \tag{5.3}$$

where we have also used the relationship $e^{-itH_2} W_+ = W_+ e^{-itH_1}$ [see (3.7)]. The integral on the right of (5.3) exists as the strong limit of $\int_0^{t''}$ as $t'' \to \infty$, which is indicated by the symbol (strong).

At this point it is convenient to introduce the notations

$$\Gamma_1^\pm T = \Gamma_{H_1}^\pm T = i \int_0^{\pm\infty} e^{itH_1} T e^{-itH_1} dt , \tag{5.4}$$

whenever one or both of the integrals on the right exists as the strong limit[2] of $\int_0^{t''}$ as $t'' \to \pm\infty$. Then (5.3) can be written

$$W_+ = P_1 + \Gamma_1^+ (A W_+) . \tag{5.5$_+$}$$

Similarly, if $W_- = W_-(H_2, H_1)$ exists, it must satisfy

$$W_- = P_1 + \Gamma_1^- (A W_-) . \tag{5.5$_-$}$$

Let us now forget all about the assumption on the existence of $W_\pm(H_2, H_1)$. Instead we start from the "integral equations" $(5.5)_\pm$

[1] For stationary methods see BIRMAN and ENTINA [1], DE BRANGES [1], T. KATO [10], KURODA [7], [8] and the papers related to the Friedrichs equation referred to in footnote [1] on page 553.

[2] Sometimes it is convenient to define *weak* $\Gamma_1^\pm$, in which weak convergence is used instead of strong convergence. We do not consider this definition, however.

and try to construct $W_\pm$ as their solutions. These equations are "stationary", for the time t does not appear in them explicitly.

The operations $\Gamma_1^\pm$ and the equations $(5.5)_\pm$ were introduced by FRIEDRICHS, though the original definition is formally different from the above[1]. It should be noted that they are linear operators acting in the space $\mathscr{B}(\mathsf{H})$ of bounded linear operators in H; their domains are not the whole space $\mathscr{B}(\mathsf{H})$ since (5.4) need not exist for all $T \in \mathscr{B}(\mathsf{H})$.

In the following paragraphs we shall study basic properties of $\Gamma_1^\pm$ and, using the results obtained, show that the solutions of $(5.5)_\pm$ are indeed the generalized wave operators $W_\pm(H_2, H_1)$. Then we shall proceed to the problem of solving $(5.5)_\pm$.

2. The Γ operations

The $\Gamma_1^\pm = \Gamma_{H_1}^\pm$ as defined by (5.4) depend on the selfadjoint operator H_1. In this paragraph we write H for H_1 and $\Gamma^\pm$ for $\Gamma_1^\pm$; H may be any selfadjoint operator. The domains and ranges of $\Gamma^\pm$ are denoted by $\mathscr{D}(\Gamma^\pm)$, $\mathscr{R}(\Gamma^\pm)$, respectively; these are linear manifolds of $\mathscr{B}(\mathsf{H})$.

Lemma 5.1.[2] *Let $B \in \mathscr{B}(\mathsf{H})$ commute with H. Then $T \in \mathscr{D}(\Gamma^+)$ implies that BT and TB are in $\mathscr{D}(\Gamma^+)$ and $\Gamma^+(BT) = B(\Gamma^+ T)$, $\Gamma^+(TB) = (\Gamma^+ T)\, B$. Similarly for Γ^-.*

Proof. Obvious since B commutes with $e^{\pm itH}$.

Lemma 5.2. *Let $T \in \mathscr{D}(\Gamma^+)$ and $S = \Gamma^+ T$. Then $S\,\mathsf{D}(H) \subset \mathsf{D}(H)$, $Tu = SHu - HSu$ for every $u \in \mathsf{D}(H)$ and $Se^{-itH} \underset{s}{\to} 0$ as $t \to +\infty$. Similarly for Γ^-, with $+\infty$ replaced by $-\infty$ in the last proposition.*

Proof. We have

$$e^{itH} S e^{-itH} = i \int_t^{+\infty} e^{isH} T e^{-isH}\, ds \quad \text{(strong)}, \tag{5.6}$$

as is seen by multiplying (5.4) by e^{itH} from the left and by e^{-itH} from the right. Denoting by $S(t)$ the right member of (5.6), we have $e^{itH} S = S(t)\, e^{itH}$. Since $dS(t)/dt = -i\, e^{itH} T e^{-itH}$ in the strong sense, we have

$$\frac{d}{dt} e^{itH} Su = \frac{d}{dt} S(t)\, e^{itH} u = -i\, e^{itH} Tu + iS(t)\, e^{itH} Hu$$

for $u \in \mathsf{D}(H)$. Thus $e^{itH} Su$ is strongly differentiable in t, which implies $Su \in \mathsf{D}(H)$ (see Remark IX-1.5) so that $(d/dt)\, e^{itH} Su = i\, e^{itH} HSu$.

[1] Friedrichs defines $\Gamma_1^\pm T$ for integral operators T of special types; see FRIEDRICHS [2], [3], [7]. $(5.5)_\pm$ are called the Friedrichs equations. For these equations see also FADDEEV [3], LADYŽENSKAJA and FADDEEV [1], REJTO [1], [2], SCHWARTZ [2], [3], [4]. A discrete analogue of the Friedrichs equation was considered recently by FREEMAN [1], in which Γ is defined in terms of a discrete semigroup instead of a group e^{-itH}.

[2] The properties of $\Gamma^\pm$ given in the following lemmas are proved in FRIEDRICHS [2], [3]. Here we need different proofs since we employ a formally different definition of $\Gamma^\pm$.

On setting $t = 0$, we thus obtain $HSu = -Tu + SHu$, which proves the first part of the lemma.

The last part follows from (5.6), which implies that $e^{itH} S e^{-itH} \underset{s}{\to} 0$ as $t \to +\infty$; multiplication from the left by the unitary operator e^{-itH} gives $S e^{-itH} \underset{s}{\to} 0$.

Lemma 5.3. *A necessary and sufficient condition for an* $S \in \mathscr{B}(\mathsf{H})$ *to belong to* $\mathscr{R}(\Gamma^+)$ *is that* $S\,\mathsf{D}(H) \subset \mathsf{D}(H)$, $SH - HS$ [defined on $\mathsf{D}(H)$] *be bounded, and* $S e^{-itH} \underset{s}{\to} 0$ *as* $t \to +\infty$. *In this case* $S = \Gamma^+ T$ *if* T *is the closure of* $SH - HS$. *Similarly for* Γ^-, *with* $+\infty$ *replaced by* $-\infty$.

Proof. The necessity part was proved in Lemma 5.2. To prove the sufficiency part, we note that

$$\frac{d}{ds} e^{isH} S e^{-isH} u = i\, e^{isH} (HS - SH)\, e^{-isH} u \tag{5.7}$$

if $u \in \mathsf{D}(H)$, for then $e^{-isH} u \in \mathsf{D}(H)$ and so $S e^{-isH} u \in \mathsf{D}(H)$ by hypothesis. Since $HS - SH$ can be replaced by its closure $-T$, we obtain by integrating (5.7) on $0 \leqq s \leqq t$

$$(e^{itH} S e^{-itH} - S)\, u = -\left(i \int_0^t e^{isH}\, T e^{-isH}\, ds\right) u\,. \tag{5.8}$$

Here the restriction $u \in \mathsf{D}(H)$ can be removed, since the operators involved are bounded and $\mathsf{D}(H)$ is dense in H. Then, going to the limit $t \to +\infty$ and using the assumption $S e^{-itH} \underset{s}{\to} 0$, we see that the right member of (5.8) has the limit $-Su$. This means that $\Gamma^+ T$ exists and equals S.

Lemma 5.4.[1] *Let* $T', T'' \in \mathscr{D}(\Gamma^+)$. *Then* $(\Gamma^+ T')\, T'' + T'(\Gamma^+ T'')$ *belongs to* $\mathscr{D}(\Gamma^+)$ *and*

$$\Gamma^+ [(\Gamma^+ T')\, T'' + T'(\Gamma^+ T'')] = (\Gamma^+ T')\,(\Gamma^+ T'')\,. \tag{5.9}$$

Similarly for Γ^-.

Proof. Set $\Gamma^+ T' = S'$, $\Gamma^+ T'' = S''$. We claim that $S' S'' \in \mathscr{R}(\Gamma^+)$. Since both S' and S'' map $\mathsf{D}(H)$ into itself by Lemma 5.2, $S' S''$ has the same property. If $u \in \mathsf{D}(H)$, we have

$$\begin{aligned} S' S'' Hu - HS' S'' u &= S'(S'' Hu - HS'' u) + (S' H - HS')\, S'' u \\ &= S' T'' u + T' S'' u \end{aligned} \tag{5.10}$$

by Lemma 5.2. Thus $S' S'' H - HS' S''$ has a bounded extension

$$T = S' T'' + T' S'' \in \mathscr{B}(\mathsf{H})\,. \tag{5.11}$$

Finally

$$S' S'' e^{-itH} = S'(S'' e^{-itH}) \underset{s}{\to} 0 \quad \text{as} \quad t \to +\infty \tag{5.12}$$

[1] Similarly one can deduce $\Gamma^\pm [T'(\Gamma^- T'') + (\Gamma^+ T')\, T''] = (\Gamma^+ T')\,(\Gamma^- T'')$, assuming $T', T'' \in \mathscr{D}(\Gamma^+) \cap \mathscr{D}(\Gamma^-)$, and many other formulas of a similar kind.

by Lemma 5.2. It follows from Lemma 5.3 that $T \in \mathscr{D}(\Gamma^+)$ and $\Gamma^+ T = S' S''$, which is exactly (5.9).

Remark 5.5. Lemma 5.3 shows that the operator Γ^+ is in a certain sense inverse to the *commutator operator* $S \to [S, H] = SH - HS$. (5.9) is a relation inverse to the formula

$$[S' S'', H] = S'[S'', H] + [S', H] S'' . \tag{5.13}$$

Lemma 5.6. *If both T and T^* belong to $\mathscr{D}(\Gamma^+)$, then $(\Gamma^+ T)^* = -\Gamma^+(T^*)$ Similarly for Γ^-.*

Proof. Obvious from the definition (5.4). Note, however, that $T \in \mathscr{D}(\Gamma^+)$ need not imply $T^* \in \mathscr{D}(\Gamma^+)$.

Remark 5.7. We have considered above the operations $\Gamma^+ T$ only for $T \in \mathscr{B}(\mathsf{H})$. This restriction is neither necessary nor natural, and can be weakened to some extent. Set

$$(\Gamma_s^+ T)\, u = \int_0^s e^{itH}\, T\, e^{-itH} u\, dt . \tag{5.14}$$

(5.14) has a meaning for $u \in \mathsf{D}(H)$ at least if T is relatively bounded with respect to H. If $(\Gamma_s^+ T)\, u$ has a limit v as $s \to +\infty$, we may write $v = S' u$. If S' is bounded, its closure determines an operator $S \in \mathscr{B}(\mathsf{H})$, which we define as $\Gamma^+ T$. Most of the results stated above remain valid for this generalized definition of Γ^+, but we shall not consider it in detail.

3. Equivalence with the time-dependent theory

We can now prove

Theorem 5.8. *Let H_1 and A be selfadjoint and let $A \in \mathscr{B}(\mathsf{H})$. Suppose that there exists a $W_+ \in \mathscr{B}(\mathsf{H})$ satisfying* (5.5)$_+$. *Then the generalized wave operator $W_+(H_2, H_1)$ exists and coincides with W_+, where $H_2 = H_1 + A$. Similarly for* (5.5)$_-$.

Proof. Since $W_+ - P_1 = \Gamma_1^+(A W_+)$, it follows from Lemma 5.2 that $(W_+ - P_1)\, H_1 u - H_1(W_+ - P_1)\, u = A W_+ u$ for $u \in \mathsf{D}(H_1)$. Since $P_1 H_1 u = H_1 P_1 u$, this implies $H_2 W_+ u = W_+ H_1 u$ or

$$H_2 W_+ \supset W_+ H_1 . \tag{5.15}$$

This further implies $(H_2 - \zeta)^{-1} W_+ = W_+(H_1 - \zeta)^{-1}$ for every nonreal ζ and hence (recall the construction of e^{itH_2}, IX-§ 1.2)

$$e^{itH_2} W_+ = W_+ e^{itH_1} , \quad -\infty < t < +\infty . \tag{5.16}$$

It follows also from Lemma 5.2 that $(W_+ - P_1)\, e^{-itH_1} \underset{s}{\to} 0$ as $t \to +\infty$. In view of (5.16), this implies $e^{-itH_2} W_+ - e^{-itH_1} P_1 \to 0$. On multiplying from the left by e^{itH_2}, we obtain $W_+ - e^{itH_2} e^{-itH_1} P_1 \underset{s}{\to} 0$. This shows that $W_+(H_2, H_1)$ exists and coincides with W_+.

Remark 5.9. The equations $(5.5)_\pm$ make sense, and can have solutions, even if A is not symmetric. The above proof is valid in this general case except for the multiplication by e^{itH_1} used in the final part; note that $iH_2 = i(H_1 + A)$ is still the generator of a quasi-bounded group (see Theorem IX-2.1) but the group may not be bounded. In particular we note that (5.15), (5.16) are valid for a non-symmetric A.

4. The Γ operations on degenerate operators

In this paragraph we consider a selfadjoint operator H with the spectral family $\{E(\lambda)\}$ and the operators $\Gamma^\pm = \Gamma^\pm_H$. We first ask under what conditions an operator

$$T = (\ ,g)\,f \tag{5.17}$$

of rank one belongs to $\mathscr{D}(\Gamma^\pm)$, where f, g are assumed to be absolutely continuous with respect to H.

For any $u \in \mathsf{H}$ we have [for the notation Γ^+_a see (5.14)]

$$(\Gamma^+_a T)\, u = i(2\pi)^{1/2} \int_0^a \phi_{u,g}(t)\, e^{itH} f\, dt\,, \tag{5.18}$$

where

$$\phi_{u,g}(t) = (2\pi)^{-1/2} (e^{-itH} u, g) = (2\pi)^{-1/2} \int_{-\infty}^{\infty} e^{-it\lambda} \varrho_{u,g}(\lambda)\, d\lambda \tag{5.19}$$

with

$$\varrho_{u,g}(\lambda) = \frac{d}{d\lambda} (E(\lambda)\, u, g) \in \mathsf{L}^1(-\infty, \infty)\,; \tag{5.20}$$

note that $(E(\lambda)\, u, g)$ is absolutely continuous if g is (see Theorem 1.7).

Using the spectral formula $e^{itH} = \int e^{it\lambda}\, dE(\lambda)$, we have from (5.18)

$$(\Gamma^+_a T)\, u = i(2\pi)^{1/2} \int_{-\infty}^{\infty} \left(\int_0^a \phi_{u,g}(t)\, e^{it\lambda}\, dt \right) dE(\lambda)\, f = i\, \sigma^+_{a,u,g}(H)\, f\,, \tag{5.21}$$

where

$$\sigma^+_{a,u,g}(\lambda) = (2\pi)^{1/2} \int_0^a \phi_{u,g}(t)\, e^{it\lambda}\, dt\,. \tag{5.22}$$

$(2\pi)^{-1}\, \sigma^+_{a,u,g}$ may be regarded as the inverse Fourier transform of $\chi^+_a(t) \phi_{u,g}(t)$, $\chi^+_a(t)$ being the characteristic function of the interval $(0, a)$.

Suppose now that (we write ϱ_g for $\varrho_{g,g}$)

$$|\!|\!| g |\!|\!|^2 = \|\varrho_g\|_\infty = \sup d(E(\lambda)\, g, g)/d\lambda < \infty\,^1\,. \tag{5.23}$$

Then $\phi_{u,g} \in \mathsf{L}^2(-\infty, \infty)$ with $\|\phi_{u,g}\| \le |\!|\!| g |\!|\!|\, \|u\|$ (see Lemma 4.5). Thus $\chi^+_a \phi_{u,g} \to \chi^+ \phi_{u,g}$ in L^2 as $a \to +\infty$, where χ^+ is the characteristic function

[1] Here and in what follows we write simply sup where we should write ess sup.

of $(0, \infty)$. Going over to the Fourier transforms, we see that $\sigma^+_{a,u,g} \to \sigma^+_{u,g}$ in L^2, where $\sigma^+_{u,g}/2\pi$ is the inverse Fourier transform of $\chi^+ \phi_{u,g}$:

$$(5.24) \qquad \sigma^+_{u,g}(\lambda) = (2\pi)^{1/2} \,\text{l.i.m.} \int_0^\infty \phi_{u,g}(t)\, e^{it\lambda}\, dt\,,$$

so that

$$(5.25) \qquad \|\sigma^+_{u,g}\| \leqq 2\pi \|\phi_{u,g}\| \leqq 2\pi \|u\|\, |||g|||\,.$$

Now it follows from (5.21) that $(\Gamma^+_a T)\,u$ tends as $a \to +\infty$ to

$$(5.26) \qquad (\Gamma^+ T)\, u = i\, \sigma^+_{u,g}(H)\, f$$

provided $|||f||| < \infty$ too. In fact $\|\sigma^+_{a,u,g}(H) f - \sigma^+_{u,g}(H) f\|^2 = \int |\sigma^+_{a,u,g}(\lambda) - \sigma^+_{u,g}(\lambda)|^2 (d/d\lambda)\,(E(\lambda) f, f)\, d\lambda \leqq \|\sigma^+_{a,u,g} - \sigma^+_{u,g}\|^2\, |||f|||^2 \to 0$. This shows that $\Gamma^+ T$ exists and justifies the notation $(\Gamma^+ T)\, u$ used in (5.26). At the same time we have $\|(\Gamma^+ T)\, u\| \leqq \|\sigma^+_{u,g}\|\, |||f||| \leqq 2\pi \|u\|\, |||f|||\, |||g|||$. Hence

$$(5.27) \qquad \|\Gamma^+ T\| \leqq 2\pi\, |||f|||\, |||g|||\,.$$

As was seen above, the Fourier transform of $\sigma^+_{u,g}/2\pi$ is χ^+ times the Fourier transform of $\varrho_{u,g}$. If we set

$$(5.28) \qquad \sigma^+_{u,g} = 2\pi\, G^+ \varrho_{u,g}\,,$$

G^+ is an orthogonal projection in L^2, being the Fourier transform of the multiplication operator by $\chi^+(t)$. G^+ is associated with the so-called *Hilbert transform*.

A convenient expression for G^+ is given by

$$(5.29) \qquad G^+ = \underset{\varepsilon \searrow 0}{\text{s-lim}}\, G^+_\varepsilon\,, \quad G^+_\varepsilon\, \varrho(\lambda) = \frac{1}{2\pi i} \int_{-\infty}^{\infty} \frac{\varrho(\mu)}{\mu - \lambda - i\varepsilon}\, d\mu\,.$$

To prove this, we note that the multiplication operator by $\chi^+(t)$ is the strong limit as $\varepsilon \searrow 0$ of the multiplication operator by $e^{-\varepsilon t} \chi^+(t)$. Going over to the Fourier transform, the latter operator becomes

$$G^+_\varepsilon\, \varrho(\lambda) = (2\pi)^{-1} \int_0^\infty e^{it\lambda}\, e^{-\varepsilon t}\, dt \int_{-\infty}^{\infty} e^{-it\mu}\, \varrho(\mu)\, d\mu$$

which is equal to (5.29).

$\Gamma^- T$ can be dealt with quite in the same way. In the final result, it is only necessary to replace G^+ by G^- defined by replacing ε by $-\varepsilon$ in (5.29). In this way we have proved

Lemma 5.10. *Let $T = (\ ,g)\, f$ be an operator of rank one, where f, g are absolutely continuous with respect to the selfadjoint operator H. If $|||f|||$ and $|||g|||$ are finite, $\Gamma^\pm T$ exist and $(\Gamma^\pm T)\, u = i\, \sigma^\pm_{u,g}(H)\, f$ for any $u \in H$. Here $\sigma^\pm_{u,g} = 2\pi\, G^\pm \varrho_{u,g}$ and $\varrho_{u,g}(\lambda) = d(E(\lambda)\, u, g)/d\lambda \in L^1 \cap L^2$. We have $\|\Gamma^\pm T\| \leqq 2\pi\, |||f|||\, |||g|||$.*

We now consider an operator of finite rank

(5.30) $$T = \sum_{k=1}^{m} (\ ,g_k)\, f_k$$

where all the f_k, g_k are absolutely continuous with respect to H. Since T is the sum of operators of rank one, application of Lemma 5.10 shows that $\Gamma^\pm A$ exist if the $|||f_k|||$, $|||g_k|||$ are finite and $\|\Gamma^\pm A\| \leq 2\pi \sum |||f_k|||\, |||g_k|||$. But we shall deduce a somewhat sharper estimate.

Lemma 5.11. *Let T be as above. Then $\Gamma^\pm T$ exist and $\|\Gamma^\pm T\| \leq$ $\leq 2\pi\alpha\beta$ where $\alpha^2 = \sup \sum_{k=1}^{m} \varrho_{f_k}(\lambda)$ and $\beta^2 = \sup \sum_{k=1}^{m} \varrho_{g_k}(\lambda)$. Here $\varrho_f(\lambda) = \varrho_{f,f}(\lambda) = d(E(\lambda)\, f, f)/d\lambda$.*

Proof. By Lemma 5.10 we have $(\Gamma^\pm T)\, u = i \sum_{k=1}^{m} \sigma^\pm_{u,g_k}(H)\, f_k$. Hence

$$\begin{aligned}
\|(\Gamma^\pm T)\, u\|^2 &= \sum_{j,k=1}^{m} (\sigma^\pm_{u,g_j}(H)\, f_j\,,\ \sigma^\pm_{u,g_k}(H)\, f_k) = \\
&= \sum_{j,k} \int \sigma^\pm_{u,g_j}(\lambda)\, \overline{\sigma^\pm_{u,g_k}(\lambda)}\, d(E(\lambda)\, f_j, f_k) \leq \\
&\leq \int \sum_{j,k} |\sigma^\pm_{u,g_j}(\lambda)|\, |\sigma^\pm_{u,g_k}(\lambda)|\, \varrho_{f_j}(\lambda)^{1/2}\, \varrho_{f_k}(\lambda)^{1/2}\, d\lambda \leq \qquad \text{[by (1.9)]} \\
&\leq \int \Big[\sum_k |\sigma^\pm_{u,g_k}(\lambda)|\, \varrho_{f_k}(\lambda)^{1/2}\Big]^2 d\lambda \leq \\
&\leq \int \Big(\sum_k |\sigma^\pm_{u,g_k}(\lambda)|^2\Big) \Big(\sum_k \varrho_{f_k}(\lambda)\Big)\, d\lambda \leq \\
&\leq \alpha^2 \sum_k \|\sigma^\pm_{u,g_k}\|^2 \leq (2\pi\alpha)^2 \sum_k \|\varrho_{u,g_k}\|^2 = \\
&= (2\pi\alpha)^2 \sum_k \int \Big|\frac{d}{d\lambda} (E(\lambda)\, u, g_k)\Big|^2 d\lambda \leq \\
&\leq (2\pi\alpha)^2 \int \varrho_u(\lambda) \Big(\sum_k \varrho_{g_k}(\lambda)\Big)\, d\lambda \leq \qquad \text{[by (1.9)]} \\
&\leq (2\pi\alpha\beta)^2 \int \varrho_u(\lambda)\, d\lambda = (2\pi\alpha\beta)^2\, \|u\|^2\,.
\end{aligned}$$

5. Solution of the integral equation for rank $A = 1$

We shall now solve the "integral equations" (5.5). For simplicity we assume in the remainder of this section that H_1 is spectrally absolutely continuous so that $P_1 = 1$, and we shall write H for H_1. Introducing for convenience a numerical parameter $\varkappa$, we thus consider the equation

(5.31) $$W = 1 + \varkappa\, \Gamma(A\, W)\,;$$

here Γ and W stand for either of $\Gamma^\pm$ and $W_\pm$, respectively.

A natural idea for solving (5.31) is to use a successive approximation:

(5.32) $$W = W(\varkappa) = \sum_{n=0}^{\infty} \varkappa^n\, W^{(n)}\,, \quad W^{(0)} = 1\,,$$
$$W^{(n+1)} = \Gamma(A\, W^{(n)})\,, \quad n = 0, 1, 2, \ldots\,.$$

But it is not at all clear whether all the $W^{(n)}$ can be constructed, for Γ is not defined everywhere on $\mathscr{B}(\mathsf{H})$.

We shall show, however, that this method works if A is of rank one:

$$A = (\ ,g)\,f \tag{5.33}$$

and if f, g are restricted properly. We shall later extend the results to more general cases. *We do not assume A to be symmetric*; it is an interesting fact that (5.31) can be solved for a non-symmetric A.

Suppose $|||f|||$, $|||g|||$ are finite [see (5.23)]. Then Lemma 5.10 shows that $W^{(1)} = \Gamma A$ exists and $\|W^{(1)}\| \leqq 2\pi\, |||f|||\, |||g|||$. To construct $W^{(2)} = \Gamma(A\,W^{(1)})$, we want to apply Lemma 5.10 with $T = A\,W^{(1)} = (\ ,g^{(1)})\,f$, where

$$g^{(1)} = W^{(1)*} g = (\Gamma A)^* g = -(\Gamma A^*)\, g = -i\,\sigma_{g,f}(H)\, g \tag{5.34}$$

by Lemmas 5.6 and 5.10 [note that $A^* = (\ ,f)\,g$ also satisfies the assumptions of Lemma 5.10]. Hence $(d/d\lambda)\,(E(\lambda)\,g^{(1)}, g^{(1)}) = |\sigma_{g,f}(\lambda)|^2 \cdot (d/d\lambda)\,(E(\lambda)\,g, g)$, so that $|||g^{(1)}||| \leqq M\,|||g||| < \infty$ if we assume that[1]

$$\|\sigma_{g,f}\|_\infty = \sup|\sigma_{g,f}(\lambda)| = M < \infty\,. \tag{5.35}$$

It turns out that we can construct the $W^{(n)}$ for $n = 3, 4, \ldots$ without introducing any further assumptions. In fact we can apply Lemma 5.10 to construct $W^{(n+1)}$ from $W^{(n)}$, where $A\,W^{(n)} = (\ ,g^{(n)})\,f$ is of rank one with

$$g^{(n)} = W^{(n)*} g = (-i)^n\, \sigma_{g,f}(H)^n\, g\,, \tag{5.36}$$

which implies as above

$$|||g^{(n)}||| \leqq M^n\, |||g||| < \infty\,. \tag{5.37}$$

(5.36) can be proved by induction; we have $g^{(n+1)} = W^{(n+1)*}\, g = (\Gamma A\,W^{(n)})^* g = -[\Gamma(A\,W^{(n)})^*]\, g = -i\,\sigma_{g,f}(H)\, g^{(n)}$ since $(A\,W^{(n)})^* = (\ ,f)\, g^{(n)}$.

It follows also that

$$\|W^{(n+1)}\| = \|\Gamma(A\,W^{(n)})\| \leqq 2\pi\, |||g^{(n)}|||\, |||f||| \leqq 2\pi\, M^n\, |||f|||\, |||g|||\,. \tag{5.38}$$

Thus the series (5.32) converges in norm if $|\varkappa| < 1/M$. It is easy to see that the sum W satisfies the equation (5.31): it suffices to show that $\Gamma(A\,W)$ can be calculated term by term by substituting for W the series (5.32). This can be done by noting that $\|\Gamma_a(A\,W^{(n)})\| \leqq 2\pi M^n\, |||f||| \cdot |||g|||$ and that $\Gamma_a(A\,W^{(n)}) \underset{s}{\to} \Gamma(A\,W^{(n)})$ as $a \to \infty$ for each n.

---

[1] $\varrho_{g,f}$ is bounded since $|\varrho_{g,f}(\lambda)|^2 \leqq \varrho_g(\lambda)\,\varrho_f(\lambda)$, but it does not imply that $\sigma_{g,f} = 2\pi\, G\, \varrho_{g,f}\,(G = G\pm)$ is bounded. It is known that $\sigma_{g,f}(\lambda)$ is bounded (and Hölder continuous) if $\varrho_{g,f}(\lambda)$ is Hölder continuous and goes to zero as $|\lambda| \to \infty$ not too slowly. In the problem under consideration, it is convenient to *assume* (5.35).

Thus we have proved

Lemma 5.12. *If* $A = (\ ,g)f$ *where* $|||f|||$, $|||g|||$ *are finite and* $\|\sigma_{g,f}\|_\infty = M < \infty$, (5.31) *has a solution* $W = W(\varkappa)$ *which is holomorphic in* $\varkappa$ *for* $|\varkappa| < 1/M$.

For further study of $W(\varkappa)$, it is convenient to consider an associated equation

$$Z = 1 - \varkappa\, \Gamma(ZA)\,, \tag{5.39}$$

which is dual to (5.31) in a certain sense. (5.39) can be solved quite in the same way as above under the same conditions. We have namely

$$Z = Z(\varkappa) = \sum_{n=0}^{\infty} \varkappa^n Z^{(n)}\,, \quad Z^{(0)} = 1\,, \quad Z^{(n+1)} = -\Gamma(Z^{(n)} A)\,, \tag{5.40}$$

$$Z^{(n)} A = (\ ,g)\, f^{(n)}\,, \quad f^{(n)} = Z^{(n)} f = (-i)^n \sigma_{f,g}(H)^n f\,.$$

Since $\sigma_{f,g}(\lambda) = \overline{\sigma_{g,f}(\lambda)}$, we obtain

Lemma 5.13. *Under the same assumptions as in* Lemma 5.12, (5.39) *has a solution* $Z = Z(\varkappa)$ *holomorphic for* $|\varkappa| < 1/M$.

There is a simple relationship between the solutions of (5.31) and (5.39). We have namely

Lemma 5.14. *For each fixed* $\varkappa$ *with* $|\varkappa| < 1/M$, *the solutions of* (5.31) *and* (5.39) *are unique. These solutions are related to each other by* $Z(\varkappa) = W(\varkappa)^{-1}$, $W(\varkappa) = Z(\varkappa)^{-1}$.

Proof. Let W and Z be any solutions of these equations. Multiplying the two equations, we have

$$ZW = 1 + \varkappa\, \Gamma(AW) - \varkappa\, \Gamma(ZA) - \varkappa^2\, \Gamma(ZA)\, \Gamma(AW)\,.$$

Using (5.9) and the equations (5.31), (5.39) once more, we obtain

$$\begin{aligned} ZW &= 1 + \varkappa\, \Gamma[AW - ZA - \varkappa\, \Gamma(ZA)\, AW - \varkappa\, ZA\, \Gamma(AW)] \\ &= 1 + \varkappa\, \Gamma[ZAW - ZAW] = 1\,. \end{aligned} \tag{5.41}$$

This implies that Z has range H; in other words Z is semi-Fredholm with $\operatorname{def} Z = 0$ (see IV-§ 5.1). In particular it is true of the $Z(\varkappa)$ of Lemma 5.13 for all $\varkappa$. But since $Z(\varkappa)$ is holomorphic in $\varkappa$, it follows from the stability theorem for the index (see Theorem IV-5.17) that $\operatorname{nul} Z(\varkappa)$ is constant. Since $Z(0) = 1$, this constant must be 0. Thus $Z(\varkappa)$ maps H onto itself one to one, and $Z(\varkappa)^{-1}$ has the same property.

Now (5.41) gives $Z(\varkappa)\, W = 1$, hence $W = Z(\varkappa)^{-1}$. Since this is true of *any* solution W of (5.31), the solution of (5.31) is unique. Similarly, using this $W = Z(\varkappa)^{-1}$ in (5.41), we have $ZZ(\varkappa)^{-1} = 1$ or $Z = Z(\varkappa)$. Since this is true of any solution Z of (5.39), the solution of (5.39) is unique[1].

[1] One can argue without using the index theorem. We have $Z(\varkappa)\, W(\varkappa) = 1$ by (5.41). Since $W(\varkappa)$ and $Z(\varkappa)$ are holomorphic in $\varkappa$ with $W(0) = Z(0) = 1$, $W(\varkappa)^{-1}$ exists for sufficiently small $|\varkappa|$ and $Z(\varkappa) = W(\varkappa)^{-1}$, hence $W(\varkappa) Z(\varkappa) = 1$. But since $W(\varkappa) Z(\varkappa)$ is holomorphic for $|\varkappa| < 1/M$, we have $W(\varkappa) Z(\varkappa) = 1$ for $|\varkappa| < 1/M$. Thus $Z(\varkappa) = W(\varkappa)^{-1}$ for $|\varkappa| < 1/M$. The uniqueness then follows from (5.41). For example, any W must satisfy $Z(\varkappa)\, W = 1$, hence $W = Z(\varkappa)^{-1} = W(\varkappa)$.

Summing up the above lemmas, we have

Theorem 5.15. *Let H be selfadjoint and spectrally absolutely continuous, let $A = (\ ,g) f$ where $|||f||| < \infty$, $|||g||| < \infty$ and $\|\sigma^{+}_{f,g}\|_\infty = M < \infty$. Then (5.31) and (5.39) have unique solutions $W_+(\varkappa)$ and $Z_+(\varkappa)$, respectively, for $\Gamma = \Gamma^+_H$ and for $|\varkappa| < 1/M$. These solutions are holomorphic in $\varkappa$ and are inverse to each other. Similar results hold with $\Gamma = \Gamma^-_H$. $H(\varkappa) = H + \varkappa A$ is similar to $H: H(\varkappa) = W_+(\varkappa) H W_+(\varkappa)^{-1}$. If $f = g$ and $\varkappa$ is real, then $H(\varkappa)$ is selfadjoint, the $W_\pm(\varkappa)$ are unitary and coincide with the wave operators $W_\pm(H(\varkappa), H)$, and $H(\varkappa)$ is unitarily equivalent to H. The spectral family $E(\lambda, \varkappa)$ for $H(\varkappa)$ is holomorphic in $\varkappa$ (for real $\varkappa$) for each fixed λ*[1].

The similarity of $H(\varkappa)$ to H follows from Remark 5.9 and the identity of the $W_\pm(\varkappa)$ with the wave operators follows from Theorem 5.8. Note that $\|\sigma^{+}_{f,f}\|_\infty = \|\sigma^{-}_{f,f}\|_\infty$ so that both $W_\pm(\varkappa)$ exist if $g = f$. The assertion on $E(\lambda, \varkappa)$ follows from $E(\lambda, \varkappa) = W_+(\varkappa) E(\lambda) W_+(\varkappa)^{-1}$.

6. Solution of the integral equation for a degenerate A

The results of the preceding paragraph can be generalized to the case in which A is degenerate (of finite rank m)

$$A = \sum_{k=1}^{m} (\ , g_k) f_k, \quad f_k, g_k \in \mathsf{H}, \tag{5.42}$$

in a straightforward fashion. (We keep the assumption that H is spectrally absolutely continuous.) We can use the successive approximation (5.32) to solve (5.31). In fact, it is easy to see as before that, at least formally,

$$A W^{(n)} = \sum_{k=1}^{m} (\ , g_k^{(n)}) f_k, \tag{5.43}$$

$$g_k^{(n)} = W^{(n)*} g_k = -i \sum_{j=1}^{m} \sigma_{k,j}(H) g_j^{(n-1)}, \tag{5.44}$$

where we have written $\sigma_{k,j} = \sigma_{g_k, f_j}$ for brevity. It is easy to see as before that these results are valid if the $|||f_k|||$, $|||g_k|||$ and $\|\sigma_{k,j}\|_\infty$ are all finite.

To estimate $\|W^{(n)}\|$, we first observe that $\varrho_f(\lambda)^{1/2} = [d(E(\lambda) f, f)/d\lambda]^{1/2}$ satisfies the triangle inequality in f for each fixed λ; $\varrho_{f+g}(\lambda)^{1/2} \leqq \varrho_f(\lambda)^{1/2} + \varrho_g(\lambda)^{1/2}$, as is easily seen from (1.9). We apply this inequality to the sum (5.44); on writing $\varrho_k^{(n)}(\lambda)$ for $\varrho_g(\lambda)$ with $g = g_k^{(n)}$ for brevity, we obtain

$$\varrho_k^{(n)}(\lambda)^{1/2} \leqq \sum_{j=1}^{m} |\sigma_{k,j}(\lambda)|\, \varrho_j^{(n-1)}(\lambda)^{1/2} \tag{5.45}$$

[note that $f = \sigma(H) g$ implies $\varrho_f(\lambda) = |\sigma(\lambda)|^2 \varrho_g(\lambda)$].

[1] In the time-dependent theory we did not prove any theorem which asserts the continuity in norm of $W_\pm(H + \varkappa A, H)$ in $\varkappa$ (to say nothing of analyticity). Thus Theorem 5.15 supplements earlier results.

Let $M(\lambda)$ be the norm of the linear operator defined by the matrix $(|\sigma_{k,j}(\lambda)|)$ acting in the m-dimensional unitary space C^m. Then (5.45) gives

$$(\Sigma\, \varrho_k^{(n)}(\lambda))^{1/2} \leqq M(\lambda)\, (\Sigma\, \varrho_k^{(n-1)}(\lambda))^{1/2} .$$

Successive application of this formula gives

$$(\Sigma\, \varrho_k^{(n)}(\lambda))^{1/2} \leqq M(\lambda)^n\, (\Sigma\, \varrho_k(\lambda))^{1/2} . \tag{5.46}$$

Set $M = \sup_\lambda M(\lambda)$; M is finite since we assumed that all the $\|\sigma_{j,k}\|_\infty$ are finite. Applying Lemma 5.11 to $T = A\, W^{(n)}$ [see (5.43)], we thus obtain

$$\begin{aligned} \|W^{(n+1)}\| = \|\Gamma(A\, W^{(n)})\| &\leqq 2\pi\, (\sup \Sigma\, \varrho_k^{(n)}(\lambda))^{1/2} (\sup \Sigma\, \varrho_{f_k}(\lambda))^{1/2} \leqq \\ &\leqq 2\pi\, M^n \|\Sigma\, \varrho_{g_k}\|_\infty^{1/2}\, \|\Sigma\, \varrho_{f_k}\|_\infty^{1/2} . \end{aligned} \tag{5.47}$$

It follows, as in the preceding paragraph, that the series for $W(\varkappa)$ converges for $|\varkappa| < 1/M$ and gives a solution of (5.31). All further results of Theorem 5.15 can now be deduced exactly as before. Thus we have proved

Theorem 5.16. *In* Theorem 5.15 *replace* A *by* $A = \sum_{k=1}^{m} (\ , g_k)\, f_k$ *where* $|||f_k||| < \infty$, $|||g_k||| < \infty$ *and* $\|\sigma_{g_k, f_j}\|_\infty < \infty$, $j, k = 1, \ldots, m$. *Then the assertions remain valid if* $M = \sup_{-\infty < \lambda < \infty} M(\lambda)$, *where* $M(\lambda)$ *is the norm of the* $m \times m$ *matrix* $(|\sigma_{g_k, f_j}(\lambda)|)$ *as an operator in the* m*-dimensional unitary space. (In the last statement of* Theorem 5.15 *replace* $f = g$ *by* $f_k = \pm g_k$, *where* $\pm$ *can be chosen at random for each* k.*)*

Remark 5.17. Let M' be the norm of the $m \times m$ matrix $(\|\sigma_{g_k, f_j}\|_\infty)$; then $M \leqq M'$. This is due to the simple fact that the norm of a matrix with nonnegative elements is not decreased when some of the elements are increased.

Remark 5.18. These results can be further generalized to the case in which A is no longer degenerate and the series in (5.42) is infinite. Looking at (5.47), we can expect the successive approximation to converge if

$$\|\sum_k \varrho_{f_k}\|_\infty < \infty , \quad \|\sum_k \varrho_{g_k}\|_\infty < \infty \quad \text{and} \quad |\varkappa| < 1/M , \tag{5.48}$$

where $M = \sup M(\lambda)$ and $M(\lambda)$ is now the norm of the operator, acting in $C^\infty = l^2$, defined by the infinite matrix $(|\sigma_{g_k, f_j}(\lambda)|)$. It should be remarked, however, that some supplementary assumption such as

$$\Sigma \|f_k\|\, \|g_k\| < \infty \tag{5.49}$$

would be required in order that $A \in \mathscr{B}(H)$, for we have defined ΓX only for bounded X. But (5.49) is not essential and can be replaced by a weaker assumption. Incidentally, (5.49) implies that $A \in \mathscr{B}_1(\mathsf{H})$ (trace class).

Remark 5.19. We can further extend the above results to the case of a continuous analogue of (5.42). Suppose that

$$A = \int (\ ,g_k)\, f_k\, dk \tag{5.50}$$

where f_k and g_k depend on a continuous parameter k. It is natural to expect that the technique of successive approximation can be applied in the same way as above if

$$\|\textstyle\int \varrho_{f_k}\, dk\|_\infty < \infty\,, \quad \|\textstyle\int \varrho_{g_k}\, dk\|_\infty < \infty \quad \text{and} \quad |\varkappa| < 1/M\,, \tag{5.51}$$

where $M = \sup M(\lambda)$ and $M(\lambda)$ is the norm of the integral operator T_λ, acting in L^2, represented by the kernel

$$t(k,j;\lambda) = |\sigma_{g_k, f_j}(\lambda)|\,. \tag{5.52}$$

(5.51) is an analogue of (5.48).

We have tacitly assumed that $f_k, g_k \in \mathsf{H}$. But even this assumption could be omitted. Of course various quantities used above would then lack a rigorous meaning, but they admit natural interpretations in concrete problems. Suppose, for instance, that $\mathsf{H} = \mathsf{L}^2(0,\infty)$ and H is the multiplication operator by the coordinate λ. Suppose further that the $f_k = f_k(\lambda)$ are functions not necessarily belonging to L^2. Then (5.50) should be interpreted as an integral operator with kernel $a(\lambda,\mu) = \int f_k(\lambda)\, \overline{g_k(\mu)}\, dk$, which may well be a bounded operator even if the f_k, g_k do not belong to L^2. Since $d(E(\lambda) f, g)/d\lambda = f(\lambda)\,\overline{g(\lambda)}$ if $f, g \in \mathsf{H}$, we should take $\varrho_{f_j, g_k}(\lambda) = f_j(\lambda)\,\overline{g_k(\lambda)}$ even when the f_j, g_k do not belong to L^2.

Again, a supplementary condition like (5.49) would be required to ensure that A is bounded, if we strictly observe the definition of Γ as given above.

We shall not give a detailed proof of these results. It must be remarked that such a proof could not follow too closely the line of the "discrete" case (5.42), since $\varrho_{f_j, g_k}(\lambda)$ may not have a Fourier transform [1].

7. Application to differential operators

As a simple application of the foregoing results, let us consider the differential operator [2]

$$H = -d^2/dx^2\,, \quad 0 < x < \infty\,, \tag{5.53}$$

[1] A justification of these generalizations is given in T. Kato [18] using a somewhat different method.

[2] For a more complete discussion of this example and its generalizations to higher-dimensional case, see T. Kato [18].

with the boundary condition $u(0) = 0$, and the perturbed operator $H(\varkappa) = H + \varkappa A$ where A is the multiplication operator by a function $q(x)$. We assume for simplicity that $q(x)$ is bounded.

H is selfadjoint in $\mathsf{H} = \mathsf{L}^2(0, \infty)$ (see V-§ 3.6). It can be "diagonalized" by the sine-transformation

$$u(x) \to \hat{u}(k) = (2/\pi)^{1/2} \int_0^\infty \sin k x \, u(x) \, dx \tag{5.54}$$

in the sense that $\|\hat{u}\| = \|u\|$ and

$$(H u)^\wedge(k) = k^2 \hat{u}(k), \quad 0 < k < \infty. \tag{5.55}$$

To transform this multiplication operator k^2 to the standard form, we take $\lambda = k^2$ as a new variable. Noting that

$$\|\hat{u}\|^2 = \int_0^\infty |\hat{u}(k)|^2 \, dk = \frac{1}{2} \int_0^\infty |\hat{u}(\lambda^{1/2})|^2 \lambda^{-1/2} \, d\lambda, \tag{5.56}$$

we set

$$\hat{\hat{u}}(\lambda) = 2^{-1/2} \lambda^{-1/4} \hat{u}(\lambda^{1/2}) = \pi^{-1/2} \lambda^{-1/4} \int_0^\infty \sin(\lambda^{1/2} x) \, u(x) \, dx. \tag{5.57}$$

$u(x) \to \hat{\hat{u}}(\lambda)$ is a unitary transformation of $\mathsf{L}^2(0, \infty)$ to itself and H is thereby transformed into the multiplication operator by λ:

$$(H u)^{\wedge\wedge}(\lambda) = \lambda \hat{\hat{u}}(\lambda). \tag{5.58}$$

As is easily seen, the multiplication operator $q(x)$ in the x-representation is transformed into an integral operator with the kernel

$$\alpha(\mu, \lambda) = \pi^{-1} (\mu \lambda)^{-1/4} \int_0^\infty \sin(\mu^{1/2} x) \sin(\lambda^{1/2} x) \, q(x) \, dx. \tag{5.59}$$

Now (5.59) has the form (5.50) with k replaced by x and with

$$\begin{aligned} f_x(\lambda) &= \pi^{-1/2} \lambda^{-1/4} \sin(\lambda^{1/2} x) \, q_1(x), \\ g_x(\lambda) &= \pi^{-1/2} \lambda^{-1/4} \sin(\lambda^{1/2} x) \, q_2(x), \end{aligned} \tag{5.60}$$

where q_1 and q_2 are such that $q(x) = q_1(x) \overline{q_2(x)}$, and $|q_1(x)| = |q_2(x)| = |q(x)|^{1/2}$. f_x and g_x do not belong to $\mathsf{L}^2(0, \infty)$, but it is not a serious difficulty[1] since A is a bounded operator. In order to apply the results of Remark 5.19, we have to calculate ϱ_{f_x}, ϱ_{g_x} and M.

Since H is multiplication by λ in the λ-representation, we have $\varrho_{f_x}(\lambda) = 0$ for $\lambda < 0$ and (formally)

$$\begin{aligned} \varrho_{f_x}(\lambda) &= \frac{d}{d\lambda} (E(\lambda) f_x, f_x) = |f_x(\lambda)|^2 \\ &= \pi^{-1} \lambda^{-1/2} \sin^2(\lambda^{1/2} x) \, |q(x)| \leq \pi^{-1} x |q(x)| \end{aligned}$$

---

[1] This apparent difficulty can easily be overcome by replacing $f_x(\lambda)$, $g_x(\lambda)$ by $f_x(\lambda) (1 + \varepsilon \lambda)^{-1}$, $g_x(\lambda) (1 + \varepsilon \lambda)^{-1}$ and going to the limit $\varepsilon \searrow 0$.

for $\lambda \geqq 0$ [note that $\sin^2(\lambda^{1/2} x) \leqq |\sin(\lambda^{1/2} x)| \leqq \lambda^{1/2} x$]. Hence the first inequality of (5.51) is satisfied if

(5.61) $$\int_0^\infty x |q(x)|\, dx < \infty .$$

The same is true for the second inequality.

To compute M, we note that

$$\varrho_{gx, fy}(\lambda) = g_x(\lambda)\, \overline{f_y(\lambda)} = \pi^{-1} \lambda^{-1/2} \sin(\lambda^{1/2} x) \sin(\lambda^{1/2} y)\, \overline{q_1(y)}\, q_2(x)$$

for $\lambda \geqq 0$ and $= 0$ for $\lambda < 0$. Hence $\sigma^+_{gx, fy} = 2\pi G^+ \varrho_{gx, fy}$ is given by [see (5.28)]

(5.62) $$\sigma_{gx, fy}(\lambda) = \frac{1}{i\pi} \overline{q_1(y)}\, q_2(x) \lim_{\varepsilon \searrow 0} \int_0^\infty \frac{\mu^{-1/2} \sin(\mu^{1/2} x) \sin(\mu^{1/2} y)}{\mu - \lambda - i\varepsilon}\, d\mu .$$

Assuming that $0 < x \leqq y$, an elementary calculation shows that the limit on the right is equal to

$$- \pi i \lambda^{-1/2} e^{-i y \lambda^{1/2}} \sin(\lambda^{1/2} x) \quad \text{for} \quad \lambda > 0$$

$$- \pi i |\lambda|^{-1/2} e^{-y |\lambda|^{1/2}} \sinh(|\lambda|^{1/2} x) \quad \text{for} \quad \lambda < 0 .$$

As is easily seen, these two expressions are bounded by $\pi x = \pi \min(x, y)$ in absolute value. Hence we have from (5.62)

(5.63) $$|\sigma_{gx, fy}(\lambda)| \leqq \min(x, y)\, |q(x)|^{1/2} |q(y)|^{1/2} .$$

Now the norm $M(\lambda)$ of the integral operator in $L^2(0, \infty)$ with the kernel $|\sigma_{gx, fy}(\lambda)|$ does not exceed its Schmidt norm $N(\lambda)$. But (5.63) shows that

(5.64) $$N(\lambda)^2 = \int_0^\infty \int_0^\infty |\sigma_{gx, fy}(\lambda)|^2\, dx\, dy \leqq \int_0^\infty \int_0^\infty [\min(x, y)]^2 |q(x)|\, |q(y)|\, dx\, dy \equiv N^2 .$$

Therefore, we have $M(\lambda) \leqq N$ for all λ and hence $M = \sup_\lambda M(\lambda) \leqq N$. Thus we conclude from Remark 5.19 that the results of Theorem 5.16 are valid if (5.61) is satisfied and if[1]

(5.65) $$|\varkappa| < 1/N .$$

Remark 5.20. Since $[\min(x, y)]^2 \leqq x y$, we have

(5.66) $$N \leqq N' \equiv \int_0^\infty x |q(x)|\, dx .$$

Hence (5.65) is satisfied if $|\varkappa| < 1/N'$. In other words, $-d^2/dx^2$ and $-d^2/dx^2 + q(x)$ are similar [and unitarily equivalent if $q(x)$ is real] provided $N' < 1$[2]. It is interesting to note that *this condition is the "best possible"*. In fact, it is known (and easily verified) that for any $N' > 1$ there is a real $q(x)$ such that $\int_0^\infty x |q(x)|\, dx = N'$ and yet $-d^2/dx^2 + q(x)$ has a negative eigenvalue so that it cannot be similar to $-d^2/dx^2$ [for example it suffices to set $q(x) = -1/\varepsilon$ for $1 \leqq x \leqq 1 + \varepsilon$ and $q(x) = 0$ otherwise, where ε is to be chosen very small].

---

[1] Actually (5.61) is not necessary if $N < \infty$; see T. KATO [18].

[2] Cf. MOSER [1] where similar results are deduced under a stronger condition. Cf. also SCHWARTZ [2].

Supplementary notes

Chapter I

1. (§ 4.6) Suppose P, Q are orthogonal projections in a unitary space. The fact that $R = (P - Q)^2$ commutes with P, Q leads to a simultaneous spectral decomposition of P and Q. Each eigenspace $\mathsf{M}(\lambda)$ of the selfadjoint operator R $(0 \leqq R \leqq 1)$ for the eigenvalue λ reduces both P and Q. In $\mathsf{M}(\lambda)$, P and Q behave essentially like two one-dimensional projections in a two dimensional space, their ranges having the angle θ if $\lambda = \sin^2\theta$. For details see DAVIS [2].

2. (§ 4.6) There are other invertible operators U that implement the similarity between P and Q than the one given by (4.38). For example, a simple one is given by $U = (1 - R)^{-1/2} (1 - P - Q) = U^{-1}$. But this U does not reduce to 1 when $Q = P$.

3. (§ 6.7) The equality $\|P\| = \|1 - P\|$ in Problem 6.31 was proved earlier by DEL PASQUE [1].

Chapter II

1. Perturbation theory for operators in finite dimensional spaces has been greatly extended and refined by BAUMGÄRTEL [1], [4, 5, 6, 7].

2. (§ 1.3) The following simple but basic theorem should be added.

Theorem 1.5a. *Fix* ζ *in Theorem* 1.5. *Then either* i) $R(\zeta, \varkappa)$ *does not exist for any* $\varkappa \in \mathrm{D}_0$, *or* ii) $R(\zeta, \varkappa)$ *is meromorphic in* $\varkappa \in \mathrm{D}_0$ *(so that it exists for all* $\varkappa$ *except for isolated points).*

Proof. If $\det(T(\varkappa) - \zeta) = 0$ identically in $\varkappa$, we have the case i). Otherwise ii) follows easily from the matrix representation of $R(\zeta, \varkappa)$; note that the matrix elements are polynomials in those of $T(\varkappa) - \zeta$ divided by $\det(T(\varkappa) - \zeta)$, which is holomorphic in $\varkappa$.

3. (§ 5.4–7) The troubles that arose in these paragraphs about the differentiability in $\varkappa$ of the eigenvalues and eigenvectors of $T(\varkappa)$ are solely due to the possibility that the number $s(\varkappa)$ of *distinct* eigenvalues may change discontinuously with $\varkappa$. If $s(\varkappa)$ is assumed to be *constant*, all

the difficulties disappear and eigenvalues and eigenprojections behave as smoothly as the operator $T(\varkappa)$ itself. For example, suppose that $T(\varkappa)$ depends continuously on $\varkappa = (\varkappa_1, \ldots, \varkappa_m) \in D_0 \subset R^m$ and that $s(\varkappa) = s$ is constant. Then the results of par. 1–2 show that the eigenvalues of $T(\varkappa)$ can be numbered as $\lambda_1(\varkappa), \ldots, \lambda_s(\varkappa)$, so that each $\lambda_h(\varkappa)$ is continuous in $\varkappa$, at least in a neighborhood of any given point in D_0, which we may assume to be $\varkappa = 0$. Then the eigenprojection $P_h(\varkappa)$ associated with $\lambda_h(\varkappa)$ is given by the integral (1.16), in which $\Gamma = \Gamma_h$ may be chosen as a small fixed circle about $\lambda_h(0)$ as long as $|\varkappa|$ is sufficiently small. Suppose now that $T(\varkappa)$ is C^k (k-times continuously differentiable). Then it is easy to see that $R(\zeta, \varkappa)$ is also C^k in $\varkappa$ uniformly for $\zeta \in \Gamma_h$. It follows from the integral representation that $P_h(\varkappa)$ is also C^k. Then the formula (2.5) with $m = m_h = \dim P_h(\varkappa) = \text{const}$ shows that $\lambda_h(\varkappa)$ is also C^k. That the eigenvectors (or generalized eigenvectors) can also be chosen as C^k functions is seen by the argument given in § 4.1. Similar results hold when $T(\varkappa)$ is analytic in $\varkappa$ where $\varkappa$ is a set of several real or complex variables $\varkappa_1, \ldots, \varkappa_m$. For another proof of differentiability of a more "real variable" type, see NOMIZU [1].

4. (§ 6.5) Add DAVIS [3], DAVIS and KAHN [1] to footnote on p. 125.

Chapter III

1. The following may be added to the general reference: DIEUDONNÉ [2], DUNFORD and SCHWARTZ [2], GOFFMAN and PEDRICK [1], REED and SIMON [1].

2. (§ 2.1) For integral operators see JÖRGENS [1].

3. (§ 4.2) The proof of Theorem 4.10 given in text is too long. A simple proof is given, for example, in YOSIDA [1], p. 282.

4. (§ 4.3) We have the following theorem on $\det(1 + T)$.

Theorem. *Let T be a degenerate operator in* X. *Then* $(1 + T)^{-1} \in \mathscr{B}(X)$ *exists if and only if* $\det(1 + T) \neq 0$.

Proof. Necessity. Let T_R be defined as in text. If $\det(1 + T) = 0$, then $\det(1_R + T_R) = 0$ and there is a nonzero $u \in R$ such that $u + Tu = 0$. Hence $1 + T$ is not invertible. Sufficiency. First we show that $1 + T$ is invertible if $\det(1 + T) \neq 0$. Suppose $u + Tu = 0$. Then $u = -Tu \in R$ so that $(1_R + T_R)u = 0$. Since $\det(1_R + T_R) \neq 0$, it follows that $u = 0$. Next we show that $1 + T$ is onto X. Let $f \in X$. Then $Tf \in R$. Since $\det(1_R + T_R) \neq 0$ implies that $1_R + T_R$ is onto R and has a bounded inverse, there is $v \in R$ such that $v + Tv = -Tf$ with $\|v\| \leq c\,\|Tf\|$ with c

independent of f. If we write $u = f + v$, we have $u + Tu = f$ with $\|u\| \leq (1 + c\,\|T\|)\,\|f\|$. Hence $1 + T$ is onto X with $(1 + T)^{-1} \in \mathscr{B}(X)$.

Chapter IV

1. (§ 1.2) Inequality (1.15) is not very sharp. The factor $2\sqrt{2}$ can be replaced by 2; this follows from a general theorem for the generator of a contraction semigroup due to KALLMAN and ROTA (1). If $p = 2$, it can even be replaced by $\sqrt{2}$ (see HARDY, LITTLEWOOD, and POLYA [1], p. 187; for an abstract result applicable to this case, see KATO [24]).

2. (§ 2.4) For further results for the gap between subspaces in a Hilbert space, see LABROUSSE [1].

3. (§ 3). For explicit estimates for the eigenvalues of differential operators, see e. g. BAZLEY [1], BAZLEY and FOX [1], GOULD [1], STENGER [1], WEINSTEIN [1–5].

4. (§ 4.1) The minimum gap between two subspaces in a Banach space was studied in detail by DEL PASQUE [1, 2].

5. (§ 5.1) For more results related to nullity, deficiency, and other notions (ascent, descent, etc.) for linear operators, see TAYLOR [1], KAASHOEK [2, 3, 4, 5], KAASHOEK and LAY [1], CARDUS [1], NEUBAUER [3].

6. (§ 5.1) There is an addition theorem for the index: $\operatorname{ind}(ST) = \operatorname{ind}S + \operatorname{ind}T$. Here $S \in \mathscr{C}(Y, Z)$ and $T \in \mathscr{C}(X, Y)$ are assumed to be Fredholm and densely defined. Then $ST \in \mathscr{C}(X, Z)$ and ST is also Fredholm. Moreover, ST depends continuously on S and T in the gap topology. For the proof see NEUBAUER [3].

7. (§ 5.5) A theorem stronger than Theorem 5.31 is found in MARKUS [1]. For more recent results in this direction, see KAASHOEK [1, 2, 3], OLIVER [1], RIBARIČ and VIDAV [1], BART [1], BART, KAASHOEK and LAY [1], BART and LAY [1], FÖRSTER [1], GRAMSCH [1,2].

8. (§ 6.2) For a further generalization of Theorem 6.2, see HOWLAND [6], which also contains application of the $W - A$ formulas to embedded eigenvalues of a selfadjoint operator.

9. (§ 6) Analytic functions with values in the set of degenerate operators have been studied in detail by HOWLAND [7] and some of the authors cited in note 7 above.

Chapter V

1. The following may be added to the general reference: HALMOS [3], HELSON [1], PUTNAM [1], REED and SIMON [2], SZ.-NAGY and FOIAŞ [1].

2. (§ 3.2) The proof of Theorem 3.1 (convexity of the numerical range) is quite simple if one notes that the problem is essentially two-dimensional. Indeed, one wants to show that for each pair $z_1, z_2 \in \Theta(T)$, the segment (z_1, z_2) is contained in $\Theta(T)$. Choose u_j such that $z_j = (Tu_j, u_j)$, $\|u_j\| = 1, j = 1,2$. It suffices to show that if u varies over the unit sphere in the two-dimensional subspace spanned by u_1, u_2, then (Tu, u) covers the segment (z_1, z_2). Thus it suffices to prove the theorem for a two-dimensional space H. In this case, however, it is a direct consequence of Problem 3.5. (The answer to this problem is that the numerical range is an ellipse with the two eigenvalues as foci. For the proof of the theorem, the exact shape of the numerical range of a two-dimensional operator is not necessary, but such a knowledge is useful anyway.)

3. (§ 3.11) Regarding the fractional powers of linear operators in a Banach space [see (3.53)], we refer to a comprehensive study by KOMATSU [1–6] and YOSHIKAWA [1].

4. (§ 4.1) Regarding Theorems 4.3-4, we want to note that it is in general impossible to improve the limit 1 of the T-bound of A, even when both T and A are bounded from below. This is seen from the following example. Let $T = -d^2/dx^2 + (3/2)\,x^{-2}$ in $H = L^2(0, \infty)$. If we choose $D(T) = C_0^\infty(0, \infty)$, it is known that T is essentially selfadjoint (see VII-Example 4.15). Let $A = -d^2/dx^2$ with the same domain as T. Then a simple calculation shows that $\|Au\| \leq \|Tu\|$. But $T + \varkappa A = (1+\varkappa)\,[-d^2/dx^2 + (3/2)\,(1+\varkappa)^{-1}x^{-2}]$ is not essentially selfadjoint if $\varkappa > 1$ because $(3/2)\,(1+\varkappa)^{-1} < 3/4$ (see loc. cit). Note that T and A are nonnegative.

5. (§ 4.1) The problem considered in Theorems 4.3–4 can be generalized to the following. Suppose $T \in \mathscr{C}(X, Y)$ is densely defined, where X, Y are Banach spaces. Let A be a densely defined operator from X to Y such that A is T-bounded and A^* is T^*-bounded, with the relative bounds smaller than 1. Then $(T + A)^* = T^* + A^*$. For this and related problems see HESS and KATO [1].

6. (§ 4.1) Theorem 4.6 was strengthened by WÜST [1] to the case in which the weaker inequality (4.1) [rather than (4.2)] is assumed with $b = 1$. It has further been generalized by OKAZAWA [2] and CHERNOFF [2] to the case in which T and A are accretive (or dissipative) operators in a Banach space, under certain mild restrictions. (T is *accretive* if $\|u + \alpha Tu\| \geq \|u\|$ for all $u \in D(T)$ and $\alpha > 0$; T is *dissipative* if $-T$ is accretive.) The generalized form of Theorem 4.6 also generalizes Theorem IX-2.7, with the limit 1/2 replaced by 1. This was first proved by GUSTAFSON [1, 2]. For further related results see OKAZAWA [1, 3, 4], YOSHIKAWA [3]. It was found that Theorem 4.6 is often useful for the proof of essential

selfadjointness in various delicate problems; see KONRADY [1], SIMON [6], GRÜTTER [1].

7. (§ 4.4) Theorem 4.12 implies that if T, S are nonnegative selfadjoint operators such that $\mathsf{D}(T) \subset \mathsf{D}(S)$, then $\mathsf{D}(T^{1/2}) \subset \mathsf{D}(S^{1/2})$. Indeed, the assumption implies that $\|Su\| \leqq c\,\|(T+1)\,u\|$ for $u \in \mathsf{D}(T)$ (see IV-Remark 1.5). Hence $\|S^{1/2}u\| \leqq c^{1/2}\|(T+1)^{1/2}u\|$ by (4.15). Since $\mathsf{D}(T)$ is a core of $(T+1)^{1/2}$ by Theorem 3.35, it follows that $\mathsf{D}(S^{1/2}) \supset D(T^{1/2})$ with the last inequality extended to all $u \in \mathsf{D}(T^{1/2})$. These results are implicitly contained in VII-§ 4.5. They are special cases of the interpolation theorems for the fractional powers of linear operators (see HEINZ [1], KATO [5, 14], KOMATSU [1–6]).

8. (§ 5.2) Inequality (5.13) is true even for $\gamma = 1/2$. It is a special case of the Sobolev imbedding theorem (see e. g. NIRENBERG [1]). Note that $\mathsf{D}(H_0)$ is exactly the Sobolev space $\mathsf{H}^2(\mathrm{R}^3)$ or $\mathsf{W}^{2,2}(\mathrm{R}^3)$ in the usual notation.

9. (§ 5.3–4) The essential selfadjointness of the Schrödinger and Dirac operators continue to be the object of extensive study. The following is a partial list that may be added to the footnote to Theorem 5.4: FARIS **[1]**, JÖRGENS and WEIDMANN **[1]**, SCHECHTER **[1]**, CARLEMAN [1], CHERNOFF [3], CORDES [2], EVANS [1], FARIS [3], FARIS and LAVINE [1], GRÜTTER [1], HELLWIG [2, 3], JÖRGENS [2, 3], KALF and WALTER [1], KATO [21, 25, 26, 28], REJTO [8], ROHDE [2], SCHMINCKE [1, 2], SIMADER [1], SIMON [6], STETKAER-HANSEN [1], WALTER [1, 2, 3], WEIDMANN [3], WIENHOLTZ [2], J. WOLF [1]. Here we mention a typical result for $L = -\Delta + q(x)$ on R^m. For the minimal operator $\dot{S}$ to be essentially selfadjoint, the following is a sufficient condition for q: $q = q_1 + q_2$, where q_1, q_2 are real-valued and locally square-integrable on R^m; $q_1(x) \geqq -q^*(|x|)$ where $q^*(r) \geqq 0$ is monotone increasing for $0 < r < \infty$ and $q^*(r) = O(r^2)$ as $r \to \infty$; the integral of $|q_2|^2$ on the ball $|x| < r$ is at most of the order $O(r^{2s})$ with some real number s as $r \to \infty$; and

$$\int_{|y|<r} |q_2(x-y)|\,|y|^{m-2} dy \to 0 \quad \text{as} \quad r \to 0 \quad \text{uniformly in} \quad x \in \mathrm{R}^m$$

($|y|^{m-2}$ should be replaced by $-\log|y|$ if $m = 2$ and by 1 if $m = 1$). (In KATO [25] it was assumed that $q^*(r) = o(r^2)$, but it can be replaced by $q^*(r) = O(r^2)$ with a slight modification of the proof, cf. KATO [28]).

10. (§ 5.3–4) For further results on the spectral properties of the Schrödinger and Dirac operators, see ARAI [1, 2], BALSLEV [2–4], HUNZIKER [2], JÖRGENS [2, 3], KATO [21], KONNO and KURODA [1], SCHECHTER **[1]**, [4], SIMON [2], UCHIYAMA [1–3], USHIJIMA [1, 2], WEIDMANN [1, 2], ZISLIN [2, 3].

Chapter VI

1. Symmetric forms that are not semibounded, and their relationship to selfadjoint operators, are considered by Mc Intosh [1, 2]. An extensive study of Hamiltonian operators in quantum mechanics defined in terms of quadratic forms is given by Simon [1], [3].

2. (§ 1.4) Another convenient criterion for closability is given by the following theorem, which is a generalization of Theoren 1.27 but which does not involve an operator in its statement.

Theorem. *A sectorial form* $\mathfrak{t}$ *is closable if and only if* $u_n \in \mathsf{D}(\mathfrak{t})$, $|\mathfrak{t}[u_n]| \leq M$, $n = 1, 2, \ldots$, *and* $u_n \to 0$ *together imply that* $\mathfrak{t}[v, u_n] \to 0$ *for each* $v \in \mathsf{D}(\mathfrak{t})$.

Proof. The sufficiency of the condition can be proved as in Theorem 1.27, with obvious modifications. To prove the necessity, we may assume without loss of generality that $\mathsf{D}(\mathfrak{t})$ is dense in H. Suppose $\mathfrak{t}$ is closable and let T be the m-sectorial operator associated with $\tilde{\mathfrak{t}}$, the closure of $\mathfrak{t}$ (see Theorem 2.1). Then $\tilde{\mathfrak{t}}[v, u_n] = (Tv, u_n) \to 0$ for $v = \mathsf{D}(T)$. If we introduce the Hilbert space $\mathsf{H}_{\mathfrak{h}}$ with $\mathfrak{h} = \operatorname{Re} \tilde{\mathfrak{t}}$ as in par. 3, $\tilde{\mathfrak{t}}$ is a bounded form on $\mathsf{H}_{\mathfrak{h}}$ and $\{u_n\}$ is a bounded sequence in $\mathsf{H}_{\mathfrak{h}}$. Since $\mathsf{D}(T)$ is dense in $\mathsf{H}_{\mathfrak{h}}$ by Theorems 1.21 and 2.1, it follows that $\mathfrak{t}[v, u_n] = \tilde{\mathfrak{t}}[v, u_n] \to 0$ for each $v \in \mathsf{D}(\mathfrak{t})$.

3. (§ 2.3) For the Friedrichs extension of differential operators with a singular potential see Kalf [1].

4. (§ 2.6) The question raised in Remark 2.29 was solved by McIntosh [3], who gave an example of an m-sectorial operator T for which $\mathsf{D}(T^{1/2}) \neq \mathsf{D}(T^{*1/2})$. In this connection it should be noted that $\mathsf{D}(T^{\alpha}) = \mathsf{D}(T^{*\alpha})$ is true for any α with $0 \leq \alpha < 1/2$ and any m-accretive (not necessarily sectorial) operator T (see Kato [16], Sz.-Nagy and Foiaş [1]).

5. (§ 4.3) This paragraph has been completely revised. The new Theorem 4.6a is simpler but stronger than the old Theorem 4.6. The improvement was made possible by the use of a lemma on differential equations given in Kato [25], also cited in note 9, Chapter V. The use of the theory of forms is effective in constructing a selfadjoint restriction H of L, but the *characterization* of $\mathsf{D}(H)$ would require some other means. Construction of a selfadjoint restriction of L under very general assumptions is given by Schechter [5].

Chapter VII

1. (§ 1.2) Sometimes $R(\zeta, \varkappa)$ may be extended to some $\varkappa$ as a pseudo-resolvent (see VIII-§ 1.1), even when $T(\varkappa)$ no longer makes sense. For a related theory and its application to Dirac operators see VESELIĆ [2].

2. (§ 1.3) An interesting application of analytic perturbation theory to the study of Schrödinger operators, in particular for many-particle systems, is contained in the theory of *dilatation analytic potentials* due to COMBES and BALSLEV. See AGUILAR and COMBES [1], BALSLEV and COMBES [1], SIMON [5].

3. (§ 1.3) Theorem 1.9 can be strengthened by adding the following assertion: in the second alternative case, the resolvent $R(1, \varkappa) = (T(\varkappa) - 1)^{-1}$ is meromorphic in $\varkappa \in \mathrm{D}_0$. Similarly for Theorem 1.10, with $R(0, \varkappa) = T(\varkappa)^{-1}$ instead of $R(1, \varkappa)$. These results are direct consequences of the following theorem.

Theorem. *Let $T(\varkappa)$ be a holomorphic family for $\varkappa \in \mathrm{D}_0$. Let λ be an isolated eigenvalue of $T(\varkappa_0)$ with finite multiplicity m (this means $\lambda \in \mathrm{P}(T(\varkappa_0))$ if $m = 0$). Then either* i) *$R(\lambda, \varkappa)$ does not exist for any $\varkappa$ in a neighborhood of $\varkappa_0$, or else* ii) *$R(\lambda, \varkappa)$ is meromorphic in $\varkappa$ in a neighborhood of $\varkappa_0$ (so that $\varkappa_0$ is at most a pole of the meromorphic function $R(\lambda, \varkappa)$).*

Proof. We may assume $\varkappa_0 = 0$. In the proof of Theorem 1.7, choose Γ as a small circle about λ. Let $T'(\varkappa)$, $T''(\varkappa)$be the parts of $T(\varkappa)$ in $\mathsf{M}'(\varkappa)$, $\mathsf{M}''(\varkappa)$, respectively. Then $\lambda \in \mathrm{P}(T''(\varkappa))$ so that $(T''(\varkappa) - \lambda)^{-1} \in \mathscr{B}(\mathsf{X})$ exists and is holomorphic in $\varkappa$ near $\varkappa = 0$. On the other hand,

$$T'(\varkappa) - \lambda = U(\varkappa)\,(\check{T}'(\varkappa) - \lambda)\,U(\varkappa)^{-1},$$

where $\check{T}'(\varkappa)$ is the part in $\mathsf{M}'(0)$ of $\check{T}(\varkappa)$. Since $\dim \mathsf{M}'(0) = m < \infty$, it follows from Theorem 1.5a in note 2 for Chapter II that $(\check{T}'(\varkappa) - \lambda)^{-1}$ either does not exist for any small $\varkappa$ or else is meromorphic. This leads to the required results. (For related results see STEINBERG [1].)

4. (§ 1.4) A detailed function-theoretic study of the behavior of the eigenvalues of the operator $H = -d^2/dx^2 + x^2 + \varkappa x^4$ in $\mathsf{H} = \mathsf{L}^2(-\infty, \infty)$ is given by SIMON [1]. It is analogous to the situation in Example 1.11 but much more complicated, since we have a singular perturbation here.

5. (§ 4.2) In (4.7) we have to assume that $\mathfrak{h}(\varkappa) \geqq 1$ for all $\varkappa$ considered, not only for $\varkappa = 0$. (But this is not an essential restriction as long as $\varkappa$ is confined to a compact subset of D_0.)

Chapter VIII

1. The problem of determining the eigenvalues from *divergent* perturbation series is studied in SIMON [4].

2. (§ 1.1) We add the following theorem.

Theorem. *Let $T_n \underset{s}{\to} T$ in the generalized sense and $A_n \to A$ in the proper sense [so that $A_n, A \in \mathscr{B}(X)$]. Then $T_n + \varkappa A_n \underset{s}{\to} T + \varkappa A$ in the generalized sense for sufficiently small $|\varkappa|$. $\varkappa$ is arbitrary if, for example, T_n and T are selfadjoint operators in a Hilbert space.*

The proof is easy, depending on the Neumann series expansion for $(T_n + \varkappa A_n - \zeta)^{-1}$, etc.

3. (§ 2.5) For approximation of semigoups, see STRANG [1].

4. (§ 3.2) Add HUET [8–16] to footnote to Theorem 3.6.

5. (§3.4) Theorem 3.13a for a monotone nondecreasing sequence of symmetric forms is a strengthening of Theorem 3.13 of the first and second editions. Its proof depends on Lemma 3.14a for the equivalence of closedness and lower semicontinuity of forms. Similar results have been given by B. Simon, Lower semicontinuity of positive quadratic forms, Proc. Roy. Soc. Edinburgh **79**, 267–273 (1977); A canonical decomposition for quadratic forms with applications to monotone convergence theorems, J. Functional Anal. **28**, 377–385 (1978).

6. (§ 4.4) When $T(\varkappa)$ is selfadjoint, the rate of convergence for the resolvent, eigenvalues and eigenprojections was studied in greater detail by GREENLEE [2, 3], based on the theory of forms and interpolation spaces, where the rate is given in the fractional powers of $\varkappa$. (This does not mean that the expansion in powers of a fractional power of $\varkappa$ is given.) See also YOSHIKAWA [2].

7. (§ 5.1) Add CONLEY and REJTO [2], GREENLEE [1], VESELIĆ [1] to the footnote. Spectral concentration naturally occurs when an eigenvalue embedded in the continuous spectrum undergoes a perturbation. Such a problem was already considered by FRIEDRICHS [3]. For a general treatment of this problem and related problems, see HOWLAND [2, 4, 5, 8, 9], SIMON [7], THOMAS [1].

Chapter IX

1. (§ 2.3) It was shown by YOSIDA [1] that if $\mathsf{D}(A) \supset \mathsf{D}(T^\alpha)$ for some α with $0 < \alpha < 1$, then A is T-bounded with an arbitrarily small T-bound, so that the result of Theorem 2.7 holds. Theorem 2.7 itself was improved by GUSTAFSON [1, 2] by replacing 1/2 by 1. For further generalizations of this theorem, see note 6 to Chapter V. For the Trotter product formula, see FARIS [1, 2], CHERNOFF [1, 4], KATO [29].

2. (§ 2.6) Regarding Remark 2.21, a sufficient condition for $U(t, \varkappa)\, u$ to have an asymptotic expansion up to $\varkappa^n$ was obtained by Kai-Nan Chueh.

3. (§ 3.4) The convergence of operators acting in different Banach spaces is studied in greater detail by STUMMEL [2–5].

Chapter X

1. (§ 1) For some interesting problems in perturbation theory see KREIN [7], BAUMGÄRTEL [8].

2. (§ 1.1) Theorem 1.6 can be slightly strengthened by asserting that P commutes with every operator $A \in \mathscr{B}(\mathsf{H})$ that commutes with H. The proof is the same, since A commutes with $E(\mathsf{S}_0)$.

3. (§ 2.1) A problem opposite to Theorem 2.1 in some sense was considered by BAUMGÄRTEL [2], in which the following theorem is proved. Given any selfadjoint operator H with a pure point spectrum, with eigenvalues all simple and dense in an interval $[a, b]$, there exists a selfadjoint operator V of rank one such that $H + \varkappa V$ has pure continuous spectrum on $[a, b]$ for all real $\varkappa$ with sufficiently small $|\varkappa|$.

4. (§ 3.2) Regarding Remark 3.6, we note that a theory with P_1 replaced by a formally different projection is proposed by WILCOX [3].

5. (§ 3.4) For further sufficient conditions for the existence of the wave operators in potential scattering, see LUNDQVIST [1], JÖRGENS and WEIDMANN [1]. Very general (not necessarily differential) operators are considered by VESELIĆ and WEIDMANN [1, 2]. KUPSCH and SANDHAS [1] consider scattering by highly singular potentials and show that only their behavior at infinity matters for the existence of the wave operators. The situation is different, however, for the completeness of the wave operators; see PEARSON [2].

6. (§ 3.4) For the decay rate of wave packets satisfying the Schrödinger equation, see WILCOX [1], HUNZIKER [3].

7. (§ 4.3) For a generalization of the trace formula (see footnote 4 to Theorem 4.3) to non-selfadjoint (or non-unitary) operators, see LANGER [2].

8. The last three sections of Chapter X is concerned with scattering theory, which is currently under rapid progress. Most of the general theory given in the text is still useful, but there has been much improvement in detail. Moreover, many new ideas, methods, and problems have

appeared since the publication of the first edition of the book. In what follows we shall give a brief review of the development. For a general survey, see also NEWTON [1], DOLLARD [3], DOLPH [3], KATO [23, 27].

(a) Abstract theory of stationary methods. The text is inclined towards the time-dependent theory of scattering, only a special kind of stationary method being presented in § 5 (a variant of the Friedrichs method). Recently, however, more emphasis has been laid on stationary rather than time-dependent methods, resulting in greater generality, refinement, and ramification. Basic contributions to the abstract theory of stationary methods are due to REJTO [1, 2, 4–7] (gentle and partly gentle perturbations), BIRMAN [9–13], BIRMAN and ENTINA [1, 2], KURODA [7–12], HOWLAND [1, 3], KATO and KURODA [2, 3], SCHECHTER [7]. In a certain sense stationary methods are the Laplace transform of time-dependent methods. The resolvents $R(\zeta, H_j)$ are the basic tools in the former in place of the unitary groups e^{-itH_j} in the latter, and there are many technical advantages (as well as some disadvantages) in the use of the resolvents. In the stationary theory one constructs the wave operators $W_\pm$ independently of the time-limit formulas (3.3). Then $W_\pm$ are used for the proof of the existence of the time limits $\lim e^{itH_2} e^{-itH_1} P_1$ and their identity with $W_\pm$.

The expression for $W_\pm$ in the stationary methods usually takes the form of "spectral integrals", such as $W_\pm = \int A_\pm(\lambda)\, dE_1(\lambda)\, P_1$ with certain operator-valued functions $A_\pm(\lambda)$, but the interpretation of such integrals require considerable work. Roughly speaking, the difference among various stationary methods lies in the different assumptions and different ways to interpret such integrals. Very often the integrals are interpreted by introducing a topology, in the space of the operators involved, which is different from the usual ones, and this is done by using certain auxiliary subspaces X of H with a topology different from that of H. Sometimes one can dispense with an explicit use of X by employing the so-called *factorization method* (see e.g. KURODA [10], KATO [18], KATO and KURODA [2, 3]). Other attempts have been made to interpret the spectral integrals directly (see BAUMGÄRTEL [3], AMREIN, GEORGESCU and JAUCH [1]). But it seems to the author that the freedom of the choice of X is a great advantage in the stationary theory (see, for example, the application of the general stationary theory to the three-body problem by HOWLAND [11]).

(b) Trace conditions versus smooth perturbations. In most cases the time-dependent methods were explicitly or implicitly tied to a trace condition (which says that a certain operator is in the trace class), at least when the completeness of the wave operators is concerned. A typical example is the result on potential scattering given in Example 4.10. More general second-order elliptic differential operator in R^3 are considered by IKEBE and TAYOSHI [1], also by using a trace condition. Some of the stationary methods also use the trace conditions (e.g. BIRMAN, BIRMAN and ENTINA cited in (a)). On the other hand, some stationary methods are adapted to "smooth perturbations", in which the resolvents are continuous up to the real axis in a certain topology (e.g. HOWLAND). It is possible, however, to construct a general theory that comprises the two types of perturbation and possibly others (e.g. KATO and KURODA).

(c) Potential scattering. The development of the stationary theory has made it possible to improve greatly the results on potential scattering given in Theorem 3.9 and Example 4.10. For example, it has been shown that $W_\pm(H_2, H_1)$ for $H_1 = -\Delta$ and $H_2 = H_1 + q(x)\cdot$ in $\mathsf{H} = \mathsf{L}^2(\mathsf{R}^m)$ exist and are complete if q is real and

$$(1) \qquad |q(x)| \leqq c(1 + |x|)^{-\beta}, \quad \beta > 1,$$

(see KATO [22]). Strictly speaking, (1) is not comparable with (4.31) but practically it is weaker than the latter. More general conditions are given by KURODA [14], REJTO [9, 10], SCHECHTER [6], and others. For related problems see DOLLARD [2, 3], FADDEEV [4], KURODA [15].

Scattering by highly singular potentials was studied by KUPSCH and SANDHAS [1], HUNZIKER [4], AMREIN and GEORGESCU [2], PEARSON [1, 2]. PEARSON shows, among others, that there are spherically symmetric potentials with compact supports for which the wave operators exist but are not complete.

Following the pioneering work of FADDEEV [1, 2], scattering theory for many-particle systems has been developed by many authors: VAN WINTER [1, 2], YAKUBOVSKI [1], HEPP [1], COMBES [1, 2], IORIO and O'CARROL [1], SIGAL [1], DOLLARD [4], THOMAS [4], GINIBRE and MOULIN [1], HOWLAND [11].

For scattering theory for the Dirac equation, see THOMPSON [1], YAMADA [1], ECKARDT [1, 2]. For scattering for crystals, see KURODA [13], THOMAS [2].

(d) Nonexistence of the singular continuous spectrum. The stationary methods are not only able to prove the existence and completeness of the wave operators, but they provide means to answer the question of whether or not the singular continuous spectrum of H_2 exists, a question that appears to be out of reach of the time-dependent theory. Thus it can be shown that under condition (1), H_2 has no singular continuous part. This was first proved by AGMON [1, 2] for a more general elliptic differential operator in R^m, independently of the theory of wave operators. The same result was obtained by KURODA [16, 17] in the framework of abstract scattering theory. For a similar result for uniformly propagative systems [which are related to subjects (g), (h) below], see SUZUKI [1].

Potentials that (roughly) satisfy (1) are called *short range potentials*. For certain *long range potentials* q [which do not satisfy (1)], the nonexistence of the singular continuous part of H_2 was proved by LAVINE [1, 3–5], ARAI [2], IKEBE and SAITO [1]. In such cases, the wave operators are not likely to exist [see (i) below]. But they may exist if q is rapidly oscillating; see BUSLAEV [2], MATVEEV and SKRIGANOV [1].

(e) The invariance principle. The invariance principle, proved in § 4.4 under a trace condition, has been shown to be valid under more general conditions. It was an empirical fact that whenever the existence and completeness of both $W_\pm$ were proved, the invariance principle was found to hold (see KATO and KURODA [2, 3]; for earlier results see also SHENK [2]). On the other hand, it was shown by WOLLENBERG [1] that the existence and completeness of $W_+ (H_2, H_1)$ do not necessarily imply the existence of $W_+(\phi(H_2), \phi(H_1))$. He shows, however, that if $W_+(H_2, H_1)$ exists and is complete and if $W_+(\phi(H_2), \phi(H_1))$ exists, then the invariance principle holds.

In this connection, it is interesting to note that the principle was proved in some cases in which only the existence of W_+ is known; see DONALDSON, GIBSON, and HERSH [1]. The principle was proved also for some non-selfadjoint problems (see GOLDSTEIN [4]) and for the generalized wave operators to be discussed in (i) (see MATVEEV [1], SAHNOVIC [5]).

(f) Eigenfunction expansions. By the stationary methods one can handle the eigenfunction expansions for the continuous spectrum in an abstract setting. Here the eigenfunctions do not belong to the basic Hilbert space, and it is necessary to interpret them as elements of some abstract space [for example the dual space of the auxiliary space X mentioned in (a)]. For these results, see KURODA [11, 12], HOWLAND [3], KATO and KURODA [2, 3].

A variant of the stationary method starts with a concrete construction of eigenfunctions and use them to construct the wave operators. This method is less abstract but gives more detailed results for differential operators. It was first used by POVZNER [1, 2] and IKEBE [1] (cf. also TITCHMARSH [1]) and has been applied to a great variety of problems by ALSHOLM and SCHMIDT [1], ECKARDT [2], GOLDSTEIN [1, 2], IKEBE [4–8], MOCHIZUKI [1, 3, 4], SCHULENBERGER and WILCOX [2], SHENK [1], SHENK and THOE [1, 2], THOMPSON [1], and others. The most complete results are contained in a recent paper by AGMON [2]. Another variant of this approach is to regard the second-order differential operator such as $-\Delta + q(x)\cdot$ in R^m as an ordinary differential operator acting on functions on $(0, \infty)$ with values in $\mathsf{X} = \mathsf{L}^2(\mathsf{S}^{m-1})$, where S^{m-1} is the unit sphere in $\mathbf{R}^m$. This enables one to use techniques in ordinary differential equations and leads to eigenfunction expansions for X-valued functions. It was successfully applied to potential scattering; see JÄGER [1], SAITO [1].

(g) Two-Hilbert space theory. In some problems in scattering theory, in particular those related to classical physics (wave equations, Maxwell equations, etc.), it is necessary to consider selfadjoint operators H_1, H_2 acting in different Hilbert spaces H_1, H_2, respectively. In such a case the wave operators are defined by

$$(2) \qquad W_{\pm} = W_{\pm}(H_2, H_1; J) = \operatorname*{s-lim}_{t \to \pm\infty} e^{itH_2} J e^{-itH_1} P_1 ,$$

where $J \in \mathscr{B}(\mathsf{H}_1, \mathsf{H}_2)$, the *identification operator*, is to be determined from physical considerations. Since J is not necessarily isometric, $W_{\pm}$ need not be either. In many cases, however, they turn out to be isometric. Such wave operators were implicitly considered by WILCOX [2], SCHMIDT [1] and THOE [1]. A general theory and applications of two-space wave operators were given by KATO [19]. See also HUNZIKER [4], BELOPOL'SKII and BIRMAN [1], BIRMAN [13–15], SCHECHTER [7].

Scattering for wave equations and symmetric systems of partial differential equations has been studied by abstract method as well as by eigenfunction expansions; see IKEBE [6, 7], KAKO [1], SCHULENBERGER and WILCOX [1, 2], SUZUKI [1], YAJIMA [1], SCHECHTER [7].

(h) Non-selfadjoint operators. Scattering theory for non-selfadjoint operators (or non-unitary operators) has also been developed. Actually the stationary method in § 5 of the text is of this kind, since A need not be a symmetric operator. But it is a "small perturbation" theory since the parameter $\varkappa$ is assumed to be sufficiently small. The original theory of FRIEDRICHS was even valid in Banach spaces (FRIEDRICHS [2, 3]). A rather general theory of small and smooth perturbations in a Hilbert space was given by KATO [18]. It was partially extended to semigroups (rather than groups) in Banach spaces by LIN [1, 2]. Non-selfadjoint perturbations that are not necessarily small were studied by MOCHIZUKI [2], GOLDSTEIN [3, 4], LJANCE [1–4], and others, under various assumptions, which may or may not be verified easily in applications. For related problems see also SAHANOVIC [1, 2], STANKEVIC [2]. A general (stationary) theory of perturbation of *spectral operators* was given by HUIGE [1].

It should be noted that the scattering theory for non-selfadjoint operators is closely related to the two-Hilbert space theory discussed in (g). Indeed, (2) can be written

$$(3) \qquad J^{-1} W_{\pm}(H_1, H_2; J) = \operatorname*{s-lim}_{t \to \pm\infty} e^{it\hat{H}_2} e^{-itH_1} P_1$$

if we assume that $J^{-1} \in \mathscr{B}(\mathsf{H}_2, \mathsf{H}_1)$ exists and write $\hat{H}_2 = J^{-1}H_2J$. If J is not unitary, $\hat{H}_2$ is not necessarily selfadjoint though it is similar to a selfadjoint operator (and hence spectral). In this sense the two-space theory is essentially a non-selfadjoint theory, although it is a rather special case of the latter. In this connection, it is useful to note that different choices of J often lead to the same $W_\pm$ and that among the possible J's there may be a unitary operator. Since $\hat{H}_2$ with a unitary J is selfadjoint, we have here a means to convert a non-selfadjoint problem into a selfadjoint one (see KATO [19]).

(i) Generalized wave operators of Dollard type. As noted in (c), there now exists a satisfactory theory for potential scattering with a short range potential. It was shown by DOLLARD [1], on the other hand, that the wave operators do not exists if $q(x)$ is a Coulomb potential $c\,|x|^{-1}$ in R^3. Nevertheless, he was able to show that the generalized wave operators

$$W_{D,\pm} = W_{D,\pm}(H_2, H_1) = \operatorname*{s-lim}_{t\to\pm\infty} e^{itH_2}\, e^{-itH_1 - iX_t} \tag{4}$$

exist, have properties of the proper wave operators (isometry, intertwining property, etc.) and lead to a satisfactory scattering operator. Here $X_t = c\,H_1^{-1/2}\log|t|$ in the Coulomb case. This result suggests that the wave operators in the usual sense could not exist for long range potentials that decay like $|x|^{-1}$ or more slowly (though this has not been proved rigorously to the author's knowledge). Dollard's proof also suggests that in a more general case X_t should be chosen as

$$X_t = \int_0^t q(sp)\,ds\,, \quad p = (p_1, p_2, p_3)\,, \quad p_j = i^{-1}\partial/\partial x_j. \tag{5}$$

This conjecture has been verified to be correct if the decay of q is not too slow (roughly like $|x|^{-\beta}$ with $1/2 < \beta \leq 1$); see AMREIN, MARTIN, and MISRA [1], BUSLAEV and MATVEEV [1], ALSHOLM [1], ALSHOLM and KATO [1]. [For a more slowly decaying potential q, (5) must be replaced by another, more complicated expression.] It appears that so far the completeness of the generalized wave operators of Dollard type has not been proved except in the special cases of spherically symmetric potentials. (For related results see SAHANOVIC [3, 4].)

Recently a spectral representation for Schrödinger operators $H_2 = -\Delta + q(x)\cdot$ with certain long range potentials was constructed by IKEBE [9]. His result implies that there exist complete wave operators of a generalized type in the sense of the stationary theory (so that $H_{2,ac}$

is unitarily equivalent to $H_1 = -\Delta$). But the relationship of these wave operators to the time-dependent theory is as yet unclear.

Other kinds of generalized wave operators than those of Dollard type may exist for a given pair H_1, H_2. The operators (2) mentioned above in the two-space theory are examples. Indeed (2) may be useful even when $\mathsf{H}_2 = \mathsf{H}_1$, and some $J \neq 1$ may lead to a pair $W_\pm(H_2, H_1; J)$ with nice properties even when $J = 1$ fails to do so. Although the physical meaning of (2) is vague for an arbitrary choice of J, the resulting $W_\pm$ will give useful information on the structure of H_2 if they do exist. MATVEEV and SKRIGANOV [2] prove the existence and completeness of such generalized wave operators for the radial Schrödinger operator with certain long range potentials.

(j) Other formulations of scattering theory. There are different ways to formulate scattering theory in quantum mechanics. The wave operators are mathematically convenient but not indispensable for physical purposes. What is essential is the existence of the limits

$$\omega_\pm(A) = \lim_{t \to \pm\infty} e^{itH_2} A e^{-itH_2} P_2 \tag{6}$$

for operators A in a certain subset $\mathscr{A}$ of $\mathscr{B}(\mathsf{H})$ (here P_2 should be loosely interpreted as the projection onto $\mathsf{H}_{2,c}$ or $\mathsf{H}_{2,ac}$). In the case of potential scattering in $\mathbf{R}^3$, one may choose $\mathscr{A}$ as the set of all $A = \Phi(p_1, p_2, p_3)$ with $p_j = i^{-1}\partial/\partial x_j$. It can be shown that (6) exist and equal $W_{D,\pm}^* A W_{D,\pm}$ if $W_{D,\pm} = W_{D,\pm}(H_2, H_1)$ exist and are complete (with a choice of X_t that commutes with the p_j). But (6) could exist without $W_{D,\pm}$ existing. A "weak scattering theory" in this sense was developed by LAVINE [2] and applied to potential scattering with a long range potential. On the other hand, it was shown by AMREIN, MARTIN, and MISRA [1] that the existence of (6) for sufficiently large class $\mathscr{A}$ implies the existence of isometric operators $\Omega_\pm$ such that $\omega_\pm(A) = \Omega_\pm^* A \Omega_\pm$ for all $A \in \mathscr{A}$ and, moreover, that $\Omega_\pm$ can be expressed by time-limits of the form (4). (On the other hand, one cannot require that $\mathscr{A} = \mathscr{B}(\mathsf{H})$ unless H_2 is a trivial operator; see HOWLAND [10]). For other formulations of scattering theory, see JAUCH, MISRA, and GIBSON [1], RUELLE [1], WILCOX [3], THOMAS [3], AMREIN and GEORGESCU [1].

(k) Lax-Phillips theory. There is a comprehensive theory of scattering due to Lax and Phillips, which has been developed independently of the theory presented in this book. Naturally there is a close relationship between the two theories, but so far it has not been clarified completely. For the Lax-Phillips theory we refer to LAX and PHILLIPS [1], [1–3], among others.

Bibliography

Articles

ARONSZAJN, N.: [1] The Rayleigh-Ritz and the Weinstein methods for approximation of eigenvalues. I. Operators in a Hilbert space. II. Differential operators. Proc. Nat. Acad. Sci. U.S.A. 34, 474–480, 594–601 (1948). [2] Approximation methods for eigenvalues of completely continuous symmetric operators. Proceedings of the Symposium on Spectral Theory and Differential Problems, Oklahoma A. M. College, 1955, 179–202. [3] On a problem of Weyl in the theory of singular Sturm Liouville equations. Am. J. Math. 79, 597–610 (1957). [4] Quadratic forms on vector spaces. Proceedings of International Symposium on Linear Spaces, Hebrew Univ. Jerusalem, 1960, 29–87.

ARONSZAJN, N., and A. WEINSTEIN: [1] Existence, convergence and equivalence in the unified theory of plates and membranes. Proc. Nat. Acad. Sci. 64, 181–191 (1941). [2] On a unified theory of eigenvalues of plates and membranes. Am. J. Math. 64, 623–645 (1942).

ATKINSON, F. V.: [1] The normal solvability of linear equations in normed spaces. Mat. Sbornik 28, (70), 3–14 (1951) (Russian). [2] A spectral problem for completely continuous operators. Acta Math. Acad. Sci. Hungar. 3, 53–60 (1952). [3] On relatively regular operators. Acta Sci. Math. Szeged. 15, 38–56 (1953).

BABBITT, D.: [1] The Wiener integral and perturbation theory of the Schrödinger operator. Bull. Am. Math. Soc. 70, 254–259 (1964).

BALSLEV, E.: [1] Perturbation of ordinary differential operators. Math. Scand. 11, 131–148 (1962).

BALSLEV, E., and T. W. GAMELIN: [1] The essential spectrum of a class of ordinary differential operators. Pacific. J. Math. 14, 755–776 (1964).

BARY, N.: [1] Sur les systèmes complets de fonctions orthogonales. Mat. Sbornik 14 (56), 51–108 (1944).

BAUMGÄRTEL, H.: [1] Zur Störungstheorie beschränkter linearer Operatoren eines Banachschen Raumes. Math. Nachr. 26, 361–379 (1964).

BAZLEY, N. W.: [1] Lower bounds for eigenvalues. J. Math. Mech. 10, 289–308 (1961).

BAZLEY, N. W., and D. W. FOX: [1] Lower bounds to eigenvalues using operator decompositions of the form B^*B. Arch. Rational Mech. Anal. 10, 352–360 (1962).

BERKSON, E.: [1] Some metrics on the subspaces of a Banach space. Pacific J. Math. 13, 7–22 (1963).

BIRKHOFF, G.: [1] Three observations on linear algebra. Univ. Nac. Tucumán Rev. Ser. A. 5, 147–151 (1946).

BIRMAN, M. SH.: [1] On the theory of selfadjoint extension of positive definite operators. Mat. Sbornik 38 (80), 431–450 (1956) (Russian). [2] On the method of Friedrichs for extending a positive-definite operator to a selfadjoint operator. Leningrad. Gorn. Inst. Zap. 33, 132–136 (1956) (Russian). [3] Method of quadratic forms in the problems of small parameter for the highest derivatives. Vestnik Leningrad. Univ. No. 13, 9–12 (1957) (Russian). [4] On many-dimensional boundary problems with small parameter in the highest derivatives.

Uspehi Mat. Nauk **12**, 6 (78), 212–213 (1957) (Russian). [5] Perturbations of quadratic forms and the spectrum of singular boundary value problems. Dokl. Akad. Nauk SSSR **125**, 471–474 (1959) (Russian). [6] Perturbation of the spectrum of a singular elliptic operator for variation of boundary and boundary conditions. Dokl. Akad. Nauk SSSR **137**, 761–763 (1961) (Russian). [7] Perturbations of the continuous spectrum of a singular elliptic operator for changing boundary and boundary conditions. Vestnik Leningrad. Univ. No. **1**, 22–55 (1962) (Russian). [8] Conditions for the existence of wave operators. Dokl. Akad. Nauk SSSR **143**, 506–509 (1962) (Russian). [9] A test for the existence of wave operators. Dokl. Akad. Nauk SSSR **147**, 1008–1009 (1962) (Russian). [10] On the conditions for the existence of wave operators. Izv. Akad. Nauk SSSR Ser. Mat. **27**, 883–906 (1963) (Russian). [11] A local symptom of the existence of wave operators. Dokl. Akad. Nauk SSSR **159**, 485–488 (1964) (Russian).

Birman, M. Sh., and S. B. Entina: [1] Stationary approach in the abstract theory of scattering. Dokl. Akad. Nauk SSSR **155**, 506–508 (1964) (Russian).

Birman, M. Sh., and M. G. Krein: [1] On the theory of wave operators and scattering operators. Dokl. Akad. Nauk SSSR **144**, 475–478 (1962) (Russian).

Bloch, C.: [1] Sur la théorie des perturbations des états liés. Nuclear Phys. **6**, 329–347 (1958).

Bouwkamp, C. J.: [1] A note on Mathieu functions. Indag. Math. **10**, 319–321 (1948).

Branges, L. de: [1] Perturbations of self-adjoint transformations. Am. J. Math. **84**, 543–560 (1962).

Brodskii, M. S., and M. S. Livšic: [1] Spectral analysis of non-selfadjoint operators and intermediate systems. Uspehi Mat. Nauk **13**, 1 (79), 3–85 (1958) (Russian).

Browder, F. E.: [1] Functional analysis and partial differential equations. I. Math. Ann. **138**, 55–79 (1959). [2] On the spectral theory of elliptic differential operators. I. Math. Ann. **142**, 22–130 (1961). [3] Functional analysis and partial differential equations. II. Math. Ann. **145**, 81–226 (1962).

Browder, F. E., and W. A. Strauss: [1] Scattering for non-linear wave equations. Pacific J. Math. **13**, 23–43 (1963).

Brown, A.: [1] On the adjoint of a closed transformation. Proc. Am. Math. Soc. **15**, 239–240 (1964).

Brownell, F. H.: [1] A note on Kato's uniqueness criterion for Schrödinger operator self-adjoint extensions. Pacific J. Math. **9**, 953–973 (1959). [2] Finite dimensionality of the Schrödinger operator bottom. Arch. Rational Mech. Anal. **8**, 59–67 (1961). [3] A note on Cook's wave-matrix theorem. Pacific J. Math. **12**, 47–52 (1962). [4] Perturbation theory and an atomic transition model. Arch. Rational Mech. Anal. **10**, 149–170 (1962).

Buslaev, V. S.: [1] Trace formulas for Schrödinger's operators in three-dimensional space. Dokl. Akad. Nauk SSSR **143**, 1067–1070 (1962) (Russian).

Butler, J. B. jr.: [1] Perturbation series for eigenvalues of analytic non-symmetric operators. Arch. Math. **10**, 21–27 (1959). [2] Perturbation of the continuous spectrum of even order differential operators. Canad. J. Math. **12**, 309–323 (1960). [3] Perturbation of the continuous spectrum of systems of ordinary differential operators. Canad. J. Math. **14**, 359–379 (1962).

Cloizeaux, J. des: [1] Extension d'une formule de Lagrange à des problèmes de valeurs propres. Nuclear Phys. **20**, 321–346 (1960).

Coester, F.: [1] Scattering theory for relativistic particles. Helv. Phys. Acta **38**, 7–23 (1965).

Conley, C. C., and P. A. Rejto: [1] On spectral concentration. Technical Report IMM-NYU 293, New York Univ., 1962.

COOK, J. M.: [1] Convergence to the Møller wave-matrix. J. Math. Phys. 36, 82–87 (1957). [2] Asymptotic properties of a boson field with given source. J. Mathematical Phys. 2, 33–45 (1961).

CORDES, H. O.: [1] A matrix inequality. Proc. Am. Math. Soc. 11, 206–210 (1960).

CORDES, H. O., and J. P. LABROUSSE: [1] The invariance of the index in the metric space of closed operators. J. Math. Mech. 12, 693–720 (1963).

DAVIS, C.: [1] The rotation of eigenvectors by a perturbation. J. Math. Anal. Appl. 6, 159–173 (1963).

DIEUDONNÉ, J.: [1] Sur les homomorphismes d'espaces normés. Bull. Sci. Math. 67, 72–84 (1943).

DOLPH, C. L.: [1] Recent developments in some non-self-adjoint problems of mathematical physics. Bull. Am. Math. Soc. 67, 1–69 (1961). [2] Positive real resolvents and linear passive Hilbert systems. Ann. Acad. Sci. Fenn. Ser. A. I. No. 336/9, (1963).

DOLPH, C. L., and F. PENZLIN: [1] On the theory of a class of non-self-adjoint operators and its applications to quantum scattering theory. Ann. Acad. Sci. Fenn. Ser. A. I. No. 263 (1959).

DUNFORD, N.: [1] A survey of the theory of spectral operators. Bull. Am. Math. Soc. 64, 217–274 (1958).

FADDEEV, L. D.: [1] The structure of the resolvent of Schrödinger's operator for a system of three particles and the scattering problem. Dokl. Akad. Nauk SSSR 145, 301–304 (1962) (Russian). [2] Mathematical problems of quantum scattering theory for three-particle systems. Trud. Math. Inst. Steklov. 69, 1–122 (1963) (Russian). [3] On the Friedrichs model in the perturbation theory of continuous spectrum. Trud. Math. Inst. Steklov. 73, 292–313 (1964) (Russian).

FOGUEL, S. R.: [1] A perturbation theorem for scalar operators. Comm. Pure Appl. Math. 11, 293–295 (1958). [2] Finite dimensional perturbations in Banach spaces. Am. J. Math. 82, 260–270 (1960).

FREEMAN, J. M.: [1] Perturbations of the shift operator. Trans. Am. Math. Soc. 114, 251–260 (1965).

FREUDENTHAL, H.: [1] Über die Friedrichssche Fortsetzung halbbeschränkter Hermitescher Operatoren. Proc. Acad. Amsterdam 39, 832–833 (1936).

FRIEDMAN, B.: [1] Operators with a closed range. Comm. Pure Appl. Math. 8, 539–550 (1955).

FRIEDRICHS, K. O.: [1] Spektraltheorie halbbeschränkter Operatoren und Anwendung auf die Spektralzerlegung von Differentialoperatoren. I. Math. Ann. 109, 465–487 (1934). [2] Über die Spektralzerlegung eines Integraloperators. Math. Ann. 115, 249–272 (1938). [3] On the perturbation of continuous spectra. Comm. Pure Appl. Math. 1, 361–406 (1948). [4] Zur asymptotischen Beschreibung von Streuprozessen. Nachr. Akad. Wiss. Göttingen Math.-Phys. Kl. IIb, 43–50, 1952. [5] Asymptotic phenomena in mathematical physics. Bull. Am. Math. Soc. 61, 485–504 (1955). [6] Symmetric positive linear differential equations. Comm. Pure Appl. Math. 11, 333–418 (1958).

FRIEDRICHS, K. O., and P. A. REJTO: [1] On a perturbation through which a discrete spectrum becomes continuous. Comm. Pure Appl. Math. 15, 219–235 (1962).

GAMELIN, T. W.: [1] Decomposition theorems for Fredholm operators. Pacific J. Math. 15, 97–106 (1965).

GARRIDO, L. M.: [1] Generalized adiabatic invariance. J. Mathematical Phys. 5, 355–362 (1964).

GARRIDO, L. M., and F. J. SANCHO: [1] Degree of approximate validity of the adiabatic invariance in quantum mechanics. Physica 28, 553–560 (1962).

GINDLER, H. A., and A. E. TAYLOR: [1] The minimum modulus of a linear operator and its use in spectral theory. Studia Math. **22**, 15–41 (1962).

GOHBERG, I. C., and M. G. KREIN: [1] The basic propositions on defect numbers, root numbers, and indices of linear operators. Uspehi Mat. Nauk 12, 2 (74), 43–118 (1957) (Russian).

GOHBERG, I. C., and A. S. MARKUS: [1] Two theorems on the opening of subspaces of Banach space. Uspehi Mat. Nauk **14**, 5 (89) 135–140 (1959)(Russian).

GOHBERG, I. C., A. S. MARKUS, and I. A. FEL'DMAN: [1] On normally solvable operators and related ideals. Izv. Moldav. Filial Akad. Nauk SSSR 10 (76) 51–69 (1960) (Russian).

GOL'DBERG, V. N.: [1] Perturbation of linear operators with a purely discrete spectrum. Dokl. Akad. Nauk SSSR **115**, 643–645 (1957) (Russian).

GREEN, T. A., and O. E. LANFORD, III: [1] Rigorous derivation of the phase shift formula for the Hilbert space scattering operator of a single particle. J. Mathematical Phys. **1**, 139–148 (1960).

GREINER, P. C.: [1] Eigenfunction expansions and scattering theory for perturbed elliptic partial differential operators. Bull. Am. Math. Soc. **70**, 517–521 (1964).

HACK, M. N.: [1] On convergence to the Møller wave operators. Nuovo Cimento **9**, 731–733 (1958). [2] Wave operators in multichannel scattering. Nuovo Cimento **13**, 231–236 (1959).

HARRIS, W. A. JR.: [1] Singular perturbations of two-point boundary problems for systems of ordinary differential equations. Arch. Rational Mech. Anal, **5**, 212–225 (1960). [2] Singular perturbations of eigenvalue problems. Arch. Rational Mech. Anal. **7**, 224–241 (1961).

HARTMAN, P.: [1] On the essential spectra of symmetric operators in Hilbert space. Am. J. Math. **75**, 229–240 (1953).

HEINZ, E.: [1] Beiträge zur Störungstheorie der Spektralzerlegung. Math. Ann. **123**, 415–438 (1951).

HELLWIG, B.: [1] Ein Kriterium für die Selbstadjungiertheit elliptischer Differentialoperatoren im R_n. Math. Z. **86**, 255–262 (1964).

HILDING, S. H.: [1] On the closure of disturbed complete orthonormal sets in Hilbert space. Ark. Mat. Astr. Fys. **32** B, No. 7 (1946).

HOFFMAN, A. J., and H. W. WIELANDT: [1] The variation of the spectrum of a normal matrix. Duke Math. J. **20**, 37–39 (1953).

HÖLDER, F.: [1] Über die Vielfachheiten gestörter Eigenwerte. Math. Ann. **113**, 620–628 (1937).

HUET, D.: [1] Phénomènes de perturbation singulière. C. R. Acad. Sci. Paris **244**, 1438–1440 (1957); **246**, 2096–2098 (1958); **247**, 2273–2276 (1958); **248**, 58–60 (1959). [2] Phénomènes de perturbation singulière dans les problèmes aux limites. Ann. Inst. Fourier Grenoble **10**, 1–96 (1960). [3] Perturbations singulières. C. R. Acad. Sci. Paris **257**, 3264–3266 (1963). [4] Perturbations singulières. C. R. Acad. Sci. Paris **258**, 6320–6322 (1964). [5] Perturbations singulières relatives au problème de Dirichlet dans un demi-espace. Ann. Scuola Norm. Sup. Pisa **18**, 425–448 (1964). [6] Sur quelques problèmes de perturbation singulière dans les espaces L_p, Rev. Fac. Ci Lisboa, Ser. A, Vol. XI, 1965, 137–164. [7] Perturbations singulières. C. R. Acad. Sci. Paris. **259**, 4213–4215 (1964).

HUKUHARA, M.: [1] Théorie des endomorphismes de l'espace vectoriel. J. Fac. Sci. Univ. Tokyo, Sect. I, **7**, 129–192; 305–332 (1954).

HUNZIKER, W.: [1] Regularitätseigenschaften der Streuamplitude im Fall der Potentialstreuung. Helv. Phys. Acta **34**, 593–620 (1961).

IKEBE, T.: [1] Eigenfunction expansions associated with the Schroedinger operators and their applications to scattering theory. Arch. Rational Mech. Anal. **5**, 1–34

(1960). [2] On the phase-shift formula for the scattering operator. Pacific J. Math. 15, 511–523 (1965). [3] Orthogonality of the eigenfunctions for the exterior problem connected with $-\Delta$. Arch. Rational Mech. Anal. 19, 71–73 (1965).

IKEBE, T., and T. KATO: [1] Uniqueness of the self-adjoint extension of singular elliptic differential operators. Arch. Rational Mech. Anal. 9, 77–92 (1962).

ISEKI, K.: [1] On complete orthonormal sets in Hilbert space. Proc. Japan Acad. 33, 450–452 (1957).

JAUCH, J. M.: [1] Theory of the scattering operator. Helv. Phys. Acta 31, 127–158 (1958). [2] Theory of the scattering operator. II. Multichannel scattering. Helv. Phys. Acta 31, 661–684 (1958).

JAUCH, J. M., and I. I. ZINNES: [1] The asymptotic condition for simple scattering systems. Nuovo Cimento 11, 553–567 (1959).

JOICHI, J. T.: [1] On operators with closed range. Proc. Am. Math. Soc. 11, 80–83 (1960).

JÖRGENS, K.: [1] Wesentliche Selbstadjungiertheit singulärer elliptischer Differentialoperatoren zweiter Ordnung in $C_0^\infty(G)$. Math. Scand. 15, 5–17, (1964).

KAASHOEK, M. A.: [1] Closed linear operators on Banach spaces. Thesis, University of Leiden, 1964.

KANIEL, S., and M. SCHECHTER: [1] Spectral theory for Fredholm operators. Comm. Pure Appl. Math. 16, 423–448 (1963).

KATO, T.: [1] On the convergence of the perturbation method, I, II. Progr. Theor. Phys. 4, 514–523 (1949); 5, 95–101; 207–212 (1950). [2] On the adiabatic theorem of quantum mechanics. J. Phys. Soc. Japan 5, 435–439 (1950). [3] On the convergence of the perturbation method. J. Fac. Sci. Univ. Tokyo Sect. I, 6, 145–226 (1951). [4] Fundamental properties of Hamiltonian operators of Schrödinger type. Trans. Am. Math. Soc. 70, 195–211 (1951). [4a] On the existence of solutions of the helium wave equation. Trans. Am. Math. Soc. 70, 212–218 (1951). [5] Notes on some inequalities for linear operators. Math. Ann. 125, 208–212 (1952). [6] On the perturbation theory of closed linear operators. J. Math. Soc. Japan 4, 323–337 (1952). [7] Perturbation theory of semi-bounded operators. Math. Ann. 125, 435–447 (1953). [8] Quadratic forms in Hilbert spaces and asymptotic perturbation series. Technical Report No. 7, Univ. Calif., 1955. [9] Notes on projections and perturbation theory. Technical Report No. 9, Univ. Calif., 1955. [10] On finite-dimensional perturbation of selfadjoint operators. J. Math. Soc. Japan 9, 239–249 (1957). [11] Perturbation of continuous spectra by trace class operators. Proc. Japan Acad. 33, 260–264 (1957). [12] Perturbation theory for nullity, deficiency and other quantities of linear operators. J. Analyse Math. 6, 261–322 (1958). [13] Estimation of iterated matrices, with application to the von Neumann condition. Numer. Math. 2, 22–29 (1960). [14] A generalization of the Heinz Inequality. Proc. Japan Acad. 37, 305–308 (1961). [15] Fractional powers of dissipative operators. J. Math. Soc. Japan 13, 246–274 (1961). [16] Fractional powers of dissipative operators, II. J. Math. Soc. Japan 14, 242–248 (1962). [17] Wave operators and unitary equivalence. Pacific J. Math. 15, 171–180 (1965). [18] Wave operators and similarity for some non-selfadjoint operators. Math. Ann. 162, 258–279 (1966).

KATO, T., and S. T. KURODA: [1] A remark on the unitarity property of the scattering operator. Nuovo Cimento 14, 1102–1107 (1959).

KATO, Y.: [1] Some converging examples of the perturbation series in the quantum field theory. Progr. Theoret. Phys. 26, 99–122 (1961).

KATO, Y., and N. MUGIBAYASHI: [1] Regular perturbation and asymptotic limits of operators in quantum field theory. Progr. Theoret. Phys. 30, 103–133 (1963).

KILLEEN, J.: [1]Asymptotic perturbation of differential equations. Technical Report UCRL-3056, Radiation Lab., Univ. Calif., 1955.

KLEINECKE, D. C.: [1] Degenerate perturbations. Technical Report No. 1, Univ. Calif., 1953. [2] Finite perturbations and the essential spectrum. Technical Report No. 4, Univ. Calif., 1954.

KRAMER, H. P.: [1] Perturbation of differential operators. Pacific J. Math. 7, 1405–1435 (1957).

KRAMER, V. A.: [1] Asymptotic inverse series. Proc. Am. Math. Soc. 7, 429–437 (1956). [2] Asymptotic perturbation series. Trans. Am. Math. Soc. 85, 88–105 (1957).

KREIN, M. G.: [1] On self-adjoint extension of bounded and semibounded Hermitian transformations. Dokl. Akad. Nauk SSSR 48, 303–306 (1945). [2] The theory of self-adjoint extensions of semi-bounded Hermitian transformations and its applications. I. Mat. Sbornik 20 (62) 431–495 (1947) (Russian). [3] On the trace formula in the theory of perturbation. Mat. Sbornik 33, (75) 597–626 (1953) (Russian). [4] On the Bary basis of Hilbert space. Uspehi Mat. Nauk 12, 3 (75), 333–341 (1957). [5] A contribution to the theory of linear non-selfadjoint operators. Dokl. Akad. Nauk SSSR 130, 254–256 (1960) (Russian). [6] Perturbation determinants and trace formula for unitary and selfadjoint operators. Dokl. Akad. Nauk SSSR 144, 268–271 (1962).

KREIN, M. G., and M. A. KRASNOSEL'SKII: [1] Stability of the index of an unbounded operator. Mat. Sbornik 30 (72), 219–224 (1952) (Russian).

KREIN, M. G., M. A. KRASNOSEL'SKII, and D. C. MIL'MAN: [1] On the defect numbers of linear operators in Banach space and on some geometric problems. Sbornik Trud. Inst. Mat. Akad. Nauk Ukr. SSR 11, 97–112 (1948) (Russian).

KREITH, K., and F. WOLF: [1] On the effect on the essential spectrum of the change of the basic region. Indag. Math. 22, 312–315 (1960).

KUMANO-GO, H.: [1] On singular perturbation of linear partial differential equations with constant coefficients, II. Proc. Japan Acad. 35, 541–546 (1959).

KURODA, S. T.: [1] On a theorem of Weyl-von Neumann. Proc. Japan Acad. 34, 11–15 (1958). [2] On the existence and the unitary property of the scattering operator. Nuovo Cimento 12, 431–454 (1959). [3] Perturbation of continuous spectra by unbounded operators, I. J. Math. Soc. Japan 11, 247–262 (1959). [4] Perturbation of continuous spectra by unbounded operators, II. J. Math. Soc. Japan 12, 243–257 (1960). [5] On a generalization of the Weinstein-Aronszajn formula and the infinite determinant. Sci. Papers Coll. Gen. Ed. Univ. Tokyo 11, 1–12 (1961). [6] On a paper of Green and Lanford. J. Mathematical Phys. 3, 933–935 (1962). [7] Finite-dimensional perturbation and a representation of scattering operator. Pacific J. Math. 13, 1305–1318 (1963). [8] On a stationary approach to scattering problem. Bull. Am. Math. Soc. 70, 556–560 (1964).

LADYŽENSKAJA, O. A.: [1] On partial differential equations with a small parameter in the highest derivatives. Vestnik Leningrad Univ. No. 7, 104–120 (1957) (Russian).

LADYŽENSKAJA, O. A., and L. D. FADDEEV: [1] On continuous spectrum perturbation theory. Dokl. Akad. Nauk SSSR 120, 1187–1190 (1958).

LANGER, H.: [1] Über die Wurzeln eines maximalen dissipativen Operators. Acta Math. Acad. Sci. Hungar. 13, 415–424 (1962).

LAX, P. D., and A. N. MILGRAM: [1] Parabolic equations. Contributions to the theory of partial differential equations, Annals of Mathematics Studies, No. 33, 167–190, Princeton, 1954.

LAX, P. D., and R. S. PHILLIPS: [1] Scattering theory. Bull. Am. Math. Soc. 70, 130–142 (1964).

LEVINSON, N.: [1] The first boundary value problem for $\varepsilon\Delta u + Au_x + Bu_y + Cu = D$ for small ε. Ann. Math. 51, 428–445 (1950).

LIDSKII, V. B.: [1] The proper values of the sum and product of symmetric matrices. Dokl. Akad. Nauk SSSR 75, 769–772 (1950) (Russian).

LIFŠIC, I. M.: [1] On the theory of regular perturbations. Dokl. Akad. Nauk SSSR 48, 79–81 (1945). [2] On degenerate regular perturbations I. Discrete spectrum. Ž. Eksper. Teoret. Fiz. 17, 1017–1025 (1947) (Russian). [3] On degenerate regular perturbations. II. Quasicontinuous and continuous spectrum. Ž. Eksper. Teoret. Fiz. 17, 1076–1089 (1947) (Russian). [4] On a problem of perturbation theory related to quantum statistics. Uspehi Mat. Nauk 7, 1 (47) 171–180 (1952) (Russian).

LIONS, J. L.: [1] Espaces d'interpolation et domaines de puissances fractionnaires d'opérateurs. J. Math. Soc. Japan 14, 233–241 (1962).

LIVŠIC, B. L.: [1] Perturbation theory for a simple structure operator. Dokl. Akad. Nauk SSSR 133, 800–803 (1960) (Russian).

LIVŠIC, M. S.: [1] On the spectral resolution of linear non-selfadjoint operators. Mat. Sbornik 34, (76) 145–179 (1954) (Russian).

LÖWNER, K.: [1] Über monotone Matrixfunctionen. Math. Z. 38, 177–216 (1934).

LUMER, G., and R. S. PHILLIPS: [1] Dissipative operators in a Banach space. Pacific J. Math. 11, 679–698 (1961).

MASLOV, V. P.: [1] Perturbation theory for the transition from discrete spectrum to continuous spectrum. Dokl. Akad. Nauk SSSR 109, 267–270 (1956) (Russian). [2] The theory of perturbations of linear operator equations and the problem of the small parameter in differential equations. Dokl. Akad. Nauk SSSR 111, 531–534 (1956) (Russian). [3] The use of the perturbation theory for finding the spectrum of ordinary differential operators involving a small parameter in the term with the highest derivative. Dokl. Akad. Nauk SSSR 111, 977–980 (1956) (Russian).

MORGENSTERN, D.: [1] Singuläre Störungstheorie partieller Differentialgleichungen. J. Rational Mech. Anal. 5, 204–216 (1956).

MOSER, J.: [1] Störungstheorie des kontinuierlichen Spektrums für gewöhnliche Differentialgleichungen zweiter Ordnung. Math. Ann. 125, 366–393 (1953). [2] Singular perturbation of eigenvalue problems for linear differential equations of even order. Comm. Pure Appl. Math. 8, 251–278 (1955).

MOTZKIN, T. S., and O. TAUSSKY: [1] Pairs of matrices with property L. Trans. Am. Math. Soc. 73, 108–114 (1952). [2] Pairs of matrices with property L. II. Trans. Am. Math. Soc. 80, 387–401 (1955).

NAGUMO, M.: [1] On singular perturbation of linear partial differential equations with constant coefficients, I. Proc. Japan Acad. 35, 449–454 (1959).

NAGY, B. Sz. – (See under Sz.-NAGY).

NELSON, E.: [1] Feynman inetgrals and the Schrödinger equation. J. Mathematical Phys. 5, 332–343 (1964).

NEUBAUER, G.: [1] Über den Index abgeschlossener Operatoren in Banachräumen. Math. Ann. 160, 93–130 (1965). [2] Über den Index abgeschlossener Operatoren in Banachräumen II. Math. Ann. 162, 92–119 (1965).

NEUMANN, J. VON: [1] Charakterisierung des Spektrums eines Integraloperators. Actualités Sci. Ind. No. 229, 38–55 (1935).

NEWBURGH, J. D.: [1] The variation of spectra. Duke Math. J. 18, 165–176 (1951). [2] A topology for closed operators. Ann. Math. 53, 250–255 (1951).

NIŽNIK, L. P.: [1] Scattering problem for non-stationary perturbation. Dokl. Akad. Nauk SSSR 132, 40–43 (1960) (Russian).

PHILLIPS, R. S.: [1] Perturbation theory for semi-groups of linear operators. Trans. Am. Math. Soc. 74, 199–221 (1954). [2] Dissipative hyperbolic systems. Trans. Am. Math. Soc. 86, 109–173 (1957). [3] Dissipative operators and hyperbolic systems of partial differential equations. Trans. Am. Math. Soc. 90, 193–254 (1959). [4] Dissipative operators and parabolic partial differential equations. Comm. Pure Appl. Math. 12, 249–276 (1959).

PORATH, G.: [1] Störungstheorie der isolierten Eigenwerte für abgeschlossene lineare Transformationen im Banachschen Raum. Math. Nachr. 20, 175–230 (1959). [2] Störungstheorie für lineare Transformationen im Banachschen Raum. Wiss. Z. Tech. Hochsch. Dresden 9, 1121–1125 (1959/60).

POVZNER, A. YA.: [1] On the expansions of arbitrary functions in terms of the eigenfunctions of the operator $-\Delta u + cu$. Mat. Sbornik 32 (74), 109–156 (1953) (Russian).

PROSSER, R. T.: [1] Relativistic potential scattering. J. Mathematical Phys. 4, 1048–1054 (1963). [2] Convergent perturbation expansions for certain wave operators. J. Mathematical Phys. 5, 708–713 (1964).

PUTNAM, C. R.: [1] On the continuous spectra of singular boundary value problems. Canad. J. Math. 6, 420–426 (1954). [2] Continuous spectra and unitary equivalence. Pacific J. Math. 7, 993–995 (1957). [3] Commutators and absolutely continuous operators. Trans. Am. Math. Soc. 87, 513–525 (1958). [4] On differences of unitarily equivalent self-adjoint operators. Proc. Glasgow Math. Assoc. 4, 103–107 (1960). [5] A note on Toeplits matrices and unitary equivalence. Boll. Un. Mat. Ital. 15, 6–9 (1960). [6] Commutators, perturbations and unitary spectra. Acta Math. 106, 215–232 (1961). [7] On the spectra of unitary half-scattering operators. Quart. Appl. Math. 20, 85–88 (1962/63). [8] Absolute continuity of certain unitary and half-scattering operators. Proc. Am. Math. Soc. 13, 844–846 (1962). [9] Absolutely continuous Hamiltonian operators. J. Math. Anal. Appl. 7, 163–165 (1963). [10] Commutators, absolutely continuous spectra, and singular integral operators. Am. J. Math. 86, 310–316 (1964).

REJTO, P. A.: [1] On gentle perturbations. I. Comm. Pure Appl. Math. 16, 279–303 (1963). [2] On gentle perturbations, II. Comm. Pure Appl. Math. 17, 257–292 (1964).

RELLICH, F.: [1] Störungstheorie der Spektralzerlegung, I. Math. Ann. 113,600–619 (1937). [2] Störungstheorie der Spektralzerlegung, II. Math. Ann. 113, 677–685 (1937). [3] Störungstheorie der Spektralzerlegung, III. Math. Ann. 116, 555–570 (1939). [4] Störungstheorie der Spektralzerlegung, IV. Math. Ann. 117, 356–382 (1940). [5] Störungstheorie der Spektralzerlegung, V. Math. Ann. 118, 462–484 (1942). [6] Störungstheorie der Spektralzerlegung. Proceedings of the International Congress of Mathematicians, 1950, Vol. 1, 606–613. [7] New results in the perturbation theory of eigenvalue problems. Nat. Bur. Standards Appl. Math. Ser. 29, 95–99 (1953). [8] Perturbation theory of eigenvalue problems. Lecture Notes, New York Univ. 1953.

RIDDELL, R. C.: [1] Spectral concentration for self adjoint operators. Thesis, Univ. Calif. 1965.

ROHDE, H.-W.: [1] Über die Symmetrie elliptischer Differentialoperatoren. Math. Z. 86, 21–33 (1964).

ROSENBLOOM, P.: [1] Perturbation of linear operators in Banach spaces. Arch. Math. 6, 89–101 (1955).

ROSENBLUM, M.: [1] Perturbation of the continuous spectrum and unitary equivalence. Pacific J. Math. 7, 997—1010 (1957). [2] The absolute continuity of Toeplitz's matrices. Pacific J. Math. 10, 987—996 (1960).

ROTA, G. C.: [1] Extension theory of differential operators, I. Comm. Pure Appl. Math. 11, 23—65 (1958).

RUSTON, A. F.: [1] On the Fredholm theory of integral equations for operators belonging to the trace class of a general Banach space. Proc. London Math. Soc. 53, 109—124 (1951).

SAITO, T.: [1] The perturbation method due to the small change in the shape of the boundary. J. Phys. Soc. Japan 15, 2069—2080 (1960).

SCHÄFKE, F. W.: [1] Zur Parameterabhängigkeit beim Anfangswertproblem für gewöhnliche lineare Differentialgleichungen. Math. Nachr. 3, 20—39 (1949). [2] Zur Parameterabhängigkeit bei gewöhnlichen linearen Differentialgleichungen mit singulären Stellen der Bestimmtheit. Math. Nachr. 4, 45—50 (1951). [3] Über Eigenwertprobleme mit zwei Parametern. Math. Nachr. 6, 109—124 (1951). [4] Verbesserte Konvergenz- und Fehlerabschätzungen für die Störungsrechnung. Z. angew. Math. Mech. 33, 255—259 (1953). [5] Zur Störungstheorie der Spektralzerlegung. Math. Ann. 133, 219—234 (1957).

SCHECHTER, M.: [1] Invariance of the essential spectrum. Bull. Am. Math. Soc. 71, 365—367 (1965). [2] On the essential spectrum of an arbitrary operator. I. J. Math. Anal. Appl. 13, 205—215 (1966). [3] On the invariance of the essential spectrum of an arbitary operator, II. Ricerche Mat. 16, 3—26 (1967).

SCHRÖDER, J.: [1] Fehlerabschätzungen zur Störungsrechnung bei linearen Eigenwertproblemen mit Operatoren eines Hilbertschen Raumes. Math. Nachr. 10, 113—128 (1953). [2] Fehlerabschätzungen zur Störungsrechnung für lineare Eigenwertprobleme bei gewöhnlichen Differentialgleichungen. Z. angew. Math. Mech. 34, 140—149 (1954). [3] Störungsrechnung bei Eigenwert- und Verzweigungsaufgaben. Arch. Rational Mech. Anal. 1, 436—468 (1958).

SCHRÖDINGER, E.: [1] Quantisierung als Eigenwertproblem. (Dritte Mitteilung: Störungstheorie, mit Anwendung auf den Starkeffekt der Balmerlinien.) Ann. Physik 80, 437—490 (1926).

SCHWARTZ, J.: [1] Perturbations of spectral operators, and applications, I. Bounded perturbations. Pacific J. Math. 4, 415—458 (1954). [2] Some non-selfadjoint operators. Comm. Pure Appl. Math. 13, 609—639 (1960). [3] Some non-selfadjoint operators II. A family of operators yielding to Friedrichs' method. Comm. Pure Appl. Math. 14, 619—626 (1961). [4] Some results on the spectra and spectral resolutions of a class of singular integral operators. Comm. Pure Appl. Math. 15, 75—90 (1962).

SEGEL, L. A.: [1] Application of conformal mapping to boundary perturbation problems for the membrane equation. Arch. Rational Mech. Anal. 8, 228—262 (1961).

SHIZUTA, Y.: [1] Eigenfunction expansion associated with the operator $-\Delta$ in the exterior domain. Proc. Japan Acad. 39, 656—660 (1963). [2] On fundamental equations of spatially independent problems in neutron thermalization theory. Progr. Theoret. Phys. 32, 489—511 (1964).

ŠMUL'YAN, YU. L.: [1] Completely continuous perturbation of operators. Dokl. Akad. Nauk SSSR 101, 35—38 (1955) (Russian).

STANKEVIČ, I. V.: [1] On the theory of perturbation of continuous spectra. Dokl. Akad. Nauk SSSR 144, 279—282 (1962) (Russian).

STRAUSS, W. A.: [1] Scattering for hyperbolic equations. Trans. Am. Math. Soc. 108, 13—37 (1963). [2] Les opérateurs d'onde pour des équations d'onde non linéaires indépendants du temps. C. R. Acad. Sci. Paris. 256, 5045—5046 (1963).

STUMMEL, F.: [1] Singuläre elliptische Differentialoperatoren in Hilbertschen Räumen. Math. Ann. **132**, 150—176 (1956).

SZ.-NAGY, B.: [1] Perturbations des transformations autoadjointes dans l'espace de Hilbert. Comment. Math. Helv. **19**, 347—366 (1946/47). [2] Perturbations des transformations linéaires fermées. Acta Sci. Math. Szeged. **14**, 125—137 (1951). [3] On the stability of the index of unbounded linear transformations. Acta Math. Acad. Sci. Hungar. **3**, 49—51 (1952). [4] On a spectral problem of Atkinson. Acta Math. Acad. Sci. Hungar. **3**, 61—66 (1952).

TITCHMARSH, E. C.: [1] Some theorems on perturbation theory. Proc. Roy. Soc. London Ser. A, **200**, 34—46 (1949). [2] Some theorems on perturbation theory. II. Proc. Roy. Soc. London Ser. A, **201**, 473—479 (1950). [3] Some theorems on perturbation theory. III. Proc. Roy. Soc. London Ser. A, **207**, 321—328 (1951). [4] Some theorems on perturbation theory. IV. Proc. Roy. Soc. London Ser. A. **210**, 30—47 (1951). [5] Some theorems on perturbation theory. V. J. Analyse Math. **4**, 187—208 (1954/56).

TROTTER, H. F.: [1] Approximation of semi-groups of operators. Pacific J. Math. **8**, 887—919 (1958). [2] On the product of semi-groups of operators. Proc. Am. Math. Soc. **10**, 545—551 (1959).

VIŠIK, M. I.: [1] On general boundary problems for elliptic differential equations. Trud. Moskov. Mat. Obšč. **1**, 187—246 (1952) (Russian).

VIŠIK, M. I., and L. A. LYUSTERNIK: [1] Perturbation of eigenvalues and eigenelements for some non-selfadjoint operators. Dokl. Akad. Nauk SSSR **130**, 251—253 (1960) (Russian). [2] Regular degeneration and boundary layer for linear differential equations with small parameter. Uspehi Mat. Nauk **12**, 5 (77) 3—122 (1957).

WATSON, G. N.: [1] The convergence of the series in Mathieu's functions. Proc. Edinburgh Math. Soc. **33**, 25—30 (1915).

WEINSTEIN, A.: [1] Étude des spectres des équations aux dérivées partielles de la théorie des plaques élastique. Memor. Sci. Math. **88**, 1937. [2] The intermediate problems and the maximum-minimum theory of eigenvalues. J. Math. Mech. **12**, 235—246 (1963). [3] Bounds for eigenvalues and the method of intermediate problems. Proceedings of International Conference on Partial Differential Equations and Continuum Mechanics, 39—53. Univ. of Wisconsin Press 1961.

WEYL, H.: [1] Über beschränkte quadratische Formen, deren Differenz vollstetig ist. Rend. Circ. Mat. Palermo **27**, 373—392 (1909).

WIENHOLTZ, E.: [1] Halbbeschränkte partielle Differentialoperatoren zweiter Ordnung vom elliptischen Typus. Math. Ann. **135**, 50—80 (1958).

WILSON, A. H.: [1] Perturbation theory in quantum mechanics, I. Proc. Roy. Soc. London Ser. A, **122**, 589—598 (1929).

WOLF, F.: [1] Analytic perturbation of operators in Banach spaces. Math. Ann. **124**, 317—333 (1952). [2] Perturbation by changes one-dimensional boundary conditions. Indag. Math. **18**, 360—366 (1956). [3] On the invariance of the essential spectrum under a change of boundary conditions of partial differential boundary operators. Indag. Math. **21**, 142—147 (1959). [4] On the essential spectrum of partial differential boundary problems. Comm. Pure Appl. Math. **12**, 211—228 (1959).

YOOD, B.: [1] Properties of linear transformations preserved under addition of a completely continuous transformation. Duke Math. J. **18**, 599—612 (1951).

ŽISLIN, G. M.: [1] Discussion of the Schrödinger operator spectrum. Trud. Moskov. Mat. Obšč. **9**, 82—120 (1960).

Books and monographs

AKHIEZER, N. I., and I. M. GLAZMAN: [1] Theory of linear operators in Hilbert space (English translation), Vol. I and II. New York: Frederick Ungar 1961 and 1963.

ALEXANDROFF. P., and H. HOPF: [1] Topologie, I. Berlin: Springer 1935.

BANACH, S.: [1] Théorie des opérations linéaires. Warsaw 1932.

BAUMGÄRTEL, H.: [1] Endlichdimensionale analytische Störungstheorie. Berlin: Akademie-Verlag 1972.

CODDINGTON, E. A., and N. LEVINSON: [1] Theory of ordinary differential equations. New York-Toronto-London: McGraw-Hill 1955.

COURANT, R., and D. HILBERT: [1] Methods of mathematical physics, I. New York: Interscience 1953.

DIEUDONNÉ, J.: [1] Foundations of modern analysis. New York-London: Academic Press 1960. [2] Treatise on analysis, Vol. II. New York-London: Academic Press 1970.

DIXMIER, J.: [1] Les algèbres d'opérateurs dans l'espace hilbertien. Paris: Gauthiers-Villars 1957.

DUNFORD, N., and J. T. SCHWARTZ: [1] Linear operators, Part I: General theory; Part II: Spectral theory. New York: Interscience 1958, 1963. [2] Linear operators, Part III: Spectral operators. New York-London-Sydney-Toronto: Wiley 1971.

EGGLESTON, H. G.: [1] Convexity. Cambridge: University Press 1963.

FARIS, W. G.: [1] Self-adjoint operators, Lecture Notes in Math. 433, Springer 1975.

FRIEDRICHS, K. O.: [1] Perturbation of spectra in Hilbert space. Providence: Am. Math. Soc. 1965.

GEL'FAND, I. M.: [1] Lectures on linear algebra. (English translation). New York: Interscience 1961.

GOFFMAN, G., and G. PEDRICK: [1] First course in functional analysis. Englewood Cliffs: Prentice Hall 1965.

GOLDBERG, S.: [1] Unbounded linear operators with applications. New York: McGraw-Hill 1966.

GOULD, S. H.: [1] Variational methods for eigenvalue problems. Toronto: University of Toronto Press 1957.

GROTHENDIECK, A.: [1] Produits tensoriels topologiques et espaces nucléaires. Mem. Am. Math. Soc. No. 16, 1955.

HALMOS, P. R.: [1] Introduction to Hilbert space and the theory of spectral multiplicity. New York: Chelsea 1951. [2] Finite-dimensional vector spaces. 2nd Ed. Princeton: D. van Nostrand 1958. [3] A Hilbert space problem book. Princeton-Toronto-London: D. van Nostrand 1967.

HARDY, G. H., J. E. LITTLEWOOD, and G. PÓLYA: [1] Inequalities. 2nd Ed. Cambridge: University Press 1952.

HAUSDORFF, F.: [1] Mengenlehre. 3. Aufl. Berlin-Leipzig: W. de Gruyter 1935.

HELLWIG, G.: [1] Differentialoperatoren der mathematischen Physik. Berlin-Göttingen-Heidelberg: Springer 1964.

HELSON, H.: [1] Lectures on invariant subspaces. New York-London: Academic Press 1964.

HILLE, E., and R. S. PHILLIPS: [1] Functional analysis and semigroups. Revised Ed. Providence: Am. Math. Soc. Colloq. Publ. Vol. 31, 1957.

HOFFMAN, K., and R. KUNZE: [1] Linear algebra. Englewood Cliffs: Prentice-Hall 1961.

JÖRGENS, K.: [1] Lineare Integraloperatoren. Stuttgart: B. G. Teubner 1970.

JÖRGENS, K., and J. WEIDMANN: [1] Spectral properties of Hamiltonian operators. Lecture Notes in Mathematics 313. Berlin-Heidelberg-New York: Springer 1973.

KEMBLE, E. C.: [1] The fundamental principles of quantum mechanics. New York: Dover 1958.

KNOPP, K.: [1], [2] Theory of functions. (English translation) Parts I and II. New York: Dover 1945 and 1947.

LAX, P. D., and R. S. PHILLIPS: [1] Scattering theory. New York-London: Academic Press 1967.

LIONS, J. L.: [1] Équations différentielles opérationnelles et problèmes aux limites. Berlin-Göttingen-Heidelberg: Springer 1961.

LORCH, E. R.: [1] Spectral theory. New York: Oxford University Press 1962.

LYUSTERNIK, L. A., and V. I. SOBOLEV: [1] Elements of functional analysis, (English translation). New York: Frederick Ungar 1955.

MASLOV, V. P.: [1] Perturbation theory and asymptotic methods, Moscow Univ. Press 1965.

MORSE, P. M., and H. FESHBACH: [1] Methods of theoretical physics. 2 vols. New York-Toronto-London: McGraw-Hill 1953.

NAIMARK, M. A.: [1] Linear differential operators. Moskow 1954 (Russian).

NEWTON, R. G.: [1] Scattering theory of waves and particles. McGraw-Hill, New York 1966.

PÓLYA, G., and G. SZEGÖ: [1] Aufgaben und Lehrsätze aus der Analysis, I. 3. Aufl. Berlin-Göttingen-Heidelberg: Springer 1964.

PUTNAM, C. R.: [1] Commutation properties of Hilbert space operators and related topics. Berlin-Heidelberg-New York: Springer 1967.

RAYLEIGH, LORD: [1] The theory of Sound. Vol. I. London: 1927.

REED, M., and B. SIMON: [1] Methods of modern mathematical physics. Vol. I., II. New York-London: Academic Press 1972, 1975.

RICHTMYER, R. D.: [1] Difference methods for initial value problems. New York: Interscience 1957.

RICKART, C. E.: [1] General theory of Banach algebras. Princeton: D. van Nostrand 1960.

RIESZ, F., and B. SZ.-NAGY: [1] Functional analysis (English translation). New York: Frederick Unger 1955.

ROYDEN, H. L.: [1] Real analysis. New York: Macmillan 1963.

SCHATTEN, R.: [1] Norm ideals of completely continuous operators. Ergebnisse der Mathematik und ihrer Grenzgebiete. Berlin-Göttingen-Heidelberg: Springer 1960.

SCHECHTER, M.: [1] Spectra of partial differential operators. North Holland 1971.

SCHIFF, L. I.: [1] Quantum mechanics. New York-Toronto-London: McGraw-Hill 1955.

SCHRÖDINGER, E.: [1] Collected papers on wave mechanics. London and Glasgow: 1928.

SIMON, B.: [1] Quantum mechanics for Hamiltonians defined as quadratic forms. Princeton: Princeton University Press 1971.

SOBOLEV, S. L.: [1] Applications of functional analysis in mathematical physics. Translations of Mathematical Monographs, Vol. 7. Providence: Am. Math. Soc. 1963.

STONE, M. H.: [1] Linear transformations in Hilbert space and their applications to analysis. Providence: Am. Math. Soc. Colloq. Publ. Vol. 15, 1932.

SZ.-NAGY, B.: [1] Spektraldarstellung linearer Transformationen des Hilbertschen Raumes. Ergebnisse der Mathematik und ihrer Grenzgebiete. Berlin: Springer 1942.

SZ.-NAGY, B., and C. FOIAŞ: [1] Analyse harmonique des opérateurs de l'espace de Hilbert. Budapest: Akadémiai Kiadó. 1967.

TAYLOR, A. E.: [1] Introduction to functional analysis. New York: Wiley 1961.
TITCHMARSH, E. C.: [1] Eigenfunction expansions associated with second-order differential equations, Part I. (Second Edition.) Part II. Oxford: Clarendon Press, 1962, 1958.
WILCOX, C. H.: [1] Perturbation theory and its applications in quantum mechanics. New York-London-Sydney: Wiley 1966.
YOSIDA, K.: [1] Functional analysis. Berlin-Göttingen-Heidelberg: Springer 1965.
ZAANEN, A. C.: [1] Linear analysis. New York: Interscience 1953.

Supplementary Bibliography

Articles

AGMON, S.: [1] Spectral properties of Schrödinger operators. Actes, Congrès intern. math., 1970, Tome 2, pp. 679—683. [2] Spectral properties of Schrödinger oerators and scattering theory. Ann. Scuola Norm. Sup. Pisa, Ser. 4, 2, 151—218 (1975).

AGUILAR, J., and J. M. COMBES: [1] A class of analytic perturbations for one-body Schrödinger Hamiltonians. Comm. Math. Phys. 22, 269—279 (1971).

ALSHOLM, P.: [1] Wave operators for long-range scattering. Thesis, Univ. California, 1972.

ALSHOLM, P., and T. KATO: [1] Scattering with long range potentials. Proc. Symp. in Pure Math., Vol. 23, Amer. Math. Soc., Providence 1973, pp. 393—399.

ALSHOLM, P., and G. SCHMIDT: [1] Spectral and scattering theory for Schrödinger operators. Arch. Rational Mech. Anal. 40, 281—311 (1971).

AMREIN, W. O., and V. GEORGESCU: [1] On the characterization of bound states and scattering states in quantum mechanics, Helv. Phys. Acta 46, 635—657 (1973). [2] Strong asymptotic completeness of wave operators for highly singular potentials,. Helv. Phys. Acta 47, 517—533 (1974).

AMREIN, W. O., V. GEORGESCU, and J. M. JAUCH: [1] Stationary state scattering theory. Helv. Phys. Acta 44, 407—434 (1971).

AMREIN, W. O., PH. A. MARTIN, and B. MIȘRA: [1] On the asymptotic condition of scattering theory. Helv. Phys. Acta 43, 313—344 (1970).

ARAI, M.: [1] On the essential spectrum of the many-particle Schrödinger operator with combined Zeeman and Stark effect. Publ. Res. Inst. Math. Sci. 3, 271—287 (1968). [2] Absolute continuity of Hamiltonian operators with repulsive potentials. Publ. Res. Inst. Math. Sci. 7, 621—635 (1972).

BALSLEV, E.: [2] The essential spectrum of elliptic differential operators in $L^p(R_n)$. Trans. Amer. Math. Soc. 116, 193—217 (1965). [3] The singular spectrum of elliptic differential operators in $L^p(R_n)$. Math. Scand. 19, 193—210 (1966). [4] Spectral theory of Schrödinger operators of many-body systems with permutation and rotation symmetries. Ann. Phys. 73, 49—107 (1972).

BALSLEV, E., and J. M. COMBES: [1] Spectral properties of many-body Schrödinger operators with dilatation-analytic interactions. Comm. Math. Phys. 22, 280—294 (1971).

BART, H.: [1] Holomorphic relative inverses of operator valued functions, Math. Ann. 208, 179—194 (1974).

BART, H., M. A. KAASHOEK, and D. C. LAY: [1] Stability properties of finite meromorphic operator functions, Indag. Math. 36, 217—259 (1974).

BART, H., and D. C. LAY: [1] Poles of a generalised resolvent operator, Proc. Royal Irish Acad. 74 A, 147—168 (1974).

BAUMGÄRTEL, H.: [2] Eindimensionale Störung eines selbstadjungierten Operators mit reinem Punktspektrum. Monatsb. Deutsch. Akad. Wiss. Berlin 7, 245—251 (1965). [3] Integraldarstellung der Wellenoperatoren von Streusystemen. Monatsb. Deutsch. Akad. Wiss. Berlin 9, 170—174 (1967). [4] Analytische Störung isolierter Eigenwerte endlicher algebraischer Vielfachheit von nichtselbstadjungierten Operatoren. Monatsb. Deutsch. Akad. Wiss. Berlin 10, 250—257 (1968). [5] Jordansche Normalform holomorpher Matrizen. Monatsb. Deutsch. Akad. Wiss. Berlin 11, 23—24 (1969). [6] Ein Reduktionsprozeß für analytische Störungen nichthalbeinfacher Eigenwerte. Monatsb. Deutsch. Akad. Wiss. Berlin 11, 81—89(1969). [7] Zur Abschätzung der Konvergenzradien von Störungsreihen. Monatsb. Deutsch. Akad. Wiss. Berlin 11, 556—572 (1969). [8] Zu einem Problem von M. G. Krein. Math. Nachr. 58, 279—294 (1973).

Belopol'skii, A. L., and M. Sh. Birman: [1] Existence of wave operators in scattering theory for a pair of spaces. Izv. Akad. Nauk SSSR 32, 1162—1175 (1968).

Birkhoff, G., and G.-C. Rota: [1] On the completeness of Sturm-Liouville expansions. Amer. Math. Monthly 67, 835—841 (1960).

Birman, M. Sh.: [12] A local test for the existence of wave operators, Izv. Akad. Nauk SSSR 32, 914—942 (1968). [13] Some applications of a local criterion for the existence of wave operators. Dokl. Akad. Nauk SSSR 185, 735—738 (1969). [14] Scattering problems for differential operators with constant coefficients. Funkcional. Anal. i Priložen. 3, 1—16 (1969). [15] Scattering problems for differential operators under a perturbation of the space. Izv. Akad. Nauk SSSR 35, 440—455 (1971).

Birman, M. Sh., and S. B. Entina: [2] Stationary approach in abstract scattering theory. Izv. Akad. Nauk SSSR 31, 401—430 (1967).

Buslaev, V. S.: [2] Generalized wave operators. Vestnik Leningrad. Univ. 25 (No. 13), 153—154 (1970).

Buslaev, V. S., and V. B. Matveev: [1] Wave operators for the Schroedinger equation with a slowly decreasing potential. Teoret. Mat. Fiz. 1, 367—376 (1970).

Caradus, S. R.: [1] Operators of Riesz type. Pacific J. Math. 18, 61—71 (1966).

Carleman, T.: [1] Sur la théorie mathématique de l'équation de Schroedinger. Ark. Mat. Astr. Fys. 24 B, No. 11, 1—7 (1934).

Chernoff, P.: [1] Note on product formulas for operator semi-groups. J. Functional Anal. 2, 238—242 (1968). [2] Perturbations of dissipative operators with relative bound one. Proc. Amer. Math. Soc. 33, 72—74 (1972). [3] Essential self-adjointness of powers of generators of hyperbolic equations. J. Functional Anal. 12, 401—414 (1973). [4] Product formulas, nonlinear semigroups, and addition of unbounded operators. Mem. Amer. Math. Soc. No. 140 (1974).

Clark, C.: [1] On relatively bounded perturbations of ordinary differential operators. Pacific J. Math. 25, 59—70 (1968).

Combes, J. M.: [1] Time-dependent approach to nonrelativistic multichannel scattering. Nuovo Cimento 64A, 111—144 (1969). [2] Relatively compact interactions in many particle systems. Comm. Math. Phys. 12, 283—295 (1969).

Conley, C. C., and P. A. Rejto: [2] Spectral concentration II (see Wilcox [1], pp. 129—143).

Cordes, H. O.: [2] Self-adjointness of powers of elliptic operators on non-compact manifolds. Math. Ann. 195, 257—272 (1972).

Davis, C.: [2] Separation of two linear subspaces. Acta Sci. Math. Szeged. 19, 172—189 (1958). [3] The rotation of eigenvectors by a perturbation, II. J. Math. Anal. Appl. 11, 20—27 (1965).

Davis, C., and W. M. Kahan: [1] The rotation of eigenvectors by a perturbation, III. SIAM J. Numer. Anal. 7, 1—46 (1970).

Del Pasque, D.: [1] Su una nozioni di varietà lineari disgiunte di uno spazio di Banach. Rendi. di Mat. 13, 1—17 (1955). [2] Sulle coppie di varietà lineari supplementari di uno spazio di Banach. Rendi, di Mat. 15, 1—11 (1956).

DOLLARD, J. D.: [1] Asymptotic convergence and the Coulomb interaction. J. Mathematical Phys. 5, 729—738 (1964). [2] Scattering into cones, I: potential scattering. Comm. Math. Phys. 12, 193—203 (1969). [3] Quantum-mechanical scattering theory for short-range and Coulomb interactions. Rocky Mountain J. Math. 1, 5—88 (1971); 2, 317 (1972). [4] Scattering into cones, II: n body problems. J. Mathematical Phys. 14, 708—718 (1973).

DOLPH, C. L.: [3] The integral equation method in scattering theory. Problems in Analysis. Princeton University Press, Princeton 1970, pp. 201—227.

DONALDSON, J. A., A. G. GIBSON, and R. HERSH: [1] On the invariance principle of scattering theory. J. Functional Anal. 14, 131—145 (1973).

ECKARDT, K-J.: [1] On the existence of wave operators for Dirac operators. Manuscripta Math. 11, 359—371 (1974). [2] Scattering theory for Dirac operators. Math. Z. 139, 105—131 (1974).

EVANS, W. D.: [1] On the unique self-adjoint extension of the Dirac operator and the existence of the Green matrix. Proc. London Math. Soc. 20, 537—557 (1970).

FADDEEV, L. D.: [4] Factorization of the S-matrix for the multidimensional Schrödinger operator. Dokl. Akad. Nauk SSSR 167, 69—72 (1966).

FARIS, W. G.: [1] The product formula for semigroups defined by Friedrichs extensions. Pacific J. Math. 22, 47—70 (1967). [2] Product formulas for perturbations of linear propagators. J. Functional Anal. 1, 93—108 (1967). [3] Essential self-adjointness of operators in ordered Hilbert space. Technical Report, Battelle, 1972.

FARIS, W. G., and R. B. LAVINE: [1] Commutators and self-adjointness of Hamiltonian operators. Comm. Math. Phys. 35, 39—48 (1974).

FÖRSTER, K-H.: [1] Über lineare, abgeschlossene Operatoren, die analytisch von einem Parameter abhängen. Math. Z. 95, 251—258 (1967).

GINIBRE, J., and M. MOULIN: [1] Hilbert space approach to the quantum mechanical three-body problem. Ann. Inst. Henri Poincaré. Sec. A, 21, 97—145 (1974).

GOLDSTEIN, C. I.: [1] Eigenfunction expansions associated with the Lanlacian for certain domains with infinite boundaries, I. Trans. Amer. Math. Soc. 135, 1—31 (1969). [2] —, II. Applications to scattering theory. Trans. Amer. Math Soc. 135, 33—50 (1969). [3] Perturbation of non-selfadjoint operators, I. Arch. Rational Mech. Anal. 37, 268—296 (1970). [4] —, II. Arch. Rational Mech. Anal. 42, 380—402 (1971).

GRAMSCH, B.: [1] Meromorphie in der Theorie der Fredholmoperatoren mit Anwendungen auf elliptische Differentialoperatoren. Math. Ann. 188, 97—112 (1970). [2] Inversion von Fredholmfunkionen bei stetiger und holomorpher Abhängigkeit von Parametern. Math. Ann. 214, 95—147 (1975).

GREENLEE, W. M.: [1] On spectral concentration for semi-bounded operators. J. Functional Anal. 5, 66—70 (1969). [2] Rate of convergence in singular perturbations. Ann. Inst. Fourier (Grenoble) 18, 135—191 (1968). [3] Singular perturbation of eigenvalues. Arch. Rational Mech. Anal. 34, 143—164 (1969).

GRÜTTER, A.: [1] Wesentliche Selbstadjungiertheit eines Schrödinger-Operators. Math. Z. 135, 289—291 (1974).

GUSTAFSON, K.: [1] A perturbation lemma. Bull. Amer. Math. Soc. 72, 334—338 (1966). [2] Doubling perturbation sizes and preservation of operator indices in normed spaces. Proc. Camb. Phil. Soc. 66, 281—294 (1969).

HELLWIG, B.: [2] Ein Kriterium für die Selbstadjungiertheit singulärer elliptischer Differentialoperatoren im Gebiet G. Math. Z. 89, 333—344 (1965). [3] A criterion for self-adjointness of singular elliptic differential operators. J. Math. Anal. Appl. 26, 279—291 (1969).

HEPP, K.: [1] On the quantum mechanical N-body problem. Helv. Phys. Acta 42, 425—458 (1969).

HESS, P., and T. KATO: [1] Perturbation of closed operators and their adjoints. Comment. Math. Helv. 45, 524—529 (1970).

HOWLAND, J. S.: [1] Banach space techniques in the perturbation theory of self-adjoint operators with continuous spectra. J. Math. Anal. Appl. 20, 22—47 (1967). [2] Perturbation of embedded eigenvalues by operators of finite rank. J. Math. Anal. Appl. 23, 575—584 (1968). [3] A perturbation-theoretic approach to eigenfunction expansions. J. Functional Anal. 2, 1—23 (1968). [4] Spectral concentration and virtual poles. Amer. J. Math. 91, 1106—1126 (1969). [5] Embedded eigenvalues and virtual poles. Pacific J. Math. 29, 565—582 (1969). [6] On the Weinstein-Aronszajn formula. Arch. Rational Mech. Anal. 39, 323—339 (1970). [7] Analyticity of determinants of operators on a Banach space. Proc. Amer. Math. Soc. 28, 177—180 (1971). [8] Spectral concentration and virtual poles, II., Trans. Amer. Math. Soc. 162, 141—156 (1971). [9] Perturbation of embedded eigenvalues. Bull. Amer. Math. Soc. 78, 280—283 (1972). [10] Nonexistence of asymptotic observables. Proc. Amer. Math. Soc. 35, 175—176 (1972). [11] Abstract stationary theory of multichannel scattering. J. Functional Anal., to appear.

HUET, D.: [8, 9, 10] Perturbations singulières. C. R. Acad. Sci. Paris 260, 6800—6801 (1965); 272, 430—432, 789—791 (1971). [11, 12, 13] Perturbations singulières et régularité. C. R. Acad. Sci. Paris 265, 316—318 (1967); 266, 924—926, 1237—1239 (1968). [14] Perturbations singulières d'inégalités variationnelles. C. R. Acad. Sci. Paris 267, 932—934 (1968). [15] Remarque sur un théorème d'Agmon et applications à quelques problèmes de perturbation singulière. Boll. Un. Mat. Ital. 21, 219—227 (1966). [16] Singular perturbations of elliptic problems. Ann. Mat. Pura Appl. 95, 77—114 (1973).

HUIGE, G. E.: [1] Perturbation theory of some spectral operators. Comm. Pure Appl. Math. 24, 741—757 (1971).

HUNZIKER, W.: [2] On the spectra of Schrödinger multiparticle Hamiltonians. Helv. Phys. Acta 39, 451—462 (1966). [3] On the space-time behaviour of Schroedinger wave functions. J. Mathematical Phys. 7, 300—304 (1966). [4] Time-dependent scattering theory for singular potentials. Helv. Phys. Acta 40, 1052—1062 (1967).

IKEBE, T.: [4] On the eigenfunction expansion connected with the exterior problem for the Schrödinger equation. Japan. J. Math. 36, 33—55 (1967). [5] Scattering for the Schrödinger operator in an exterior domain. J. Math. Kyoto Univ. 7, 93—112 (1967). [6] Wave operators for $-\Delta$ in a domain with non-finite boundary. Publ. Res. Inst. Math. Sci. 4, 413—418 (1968). [7] Scattering for uniformly propagative systems. Proc. Intern. Conference on Functional Anal. and Related Topics, Tokyo 1969, pp. 225—230. [8] Remarks on the orthogonality of eigenfunctions for the Schrödinger operator in R^n. J. Fac. Sci. Univ. Tokyo. Sec. I, Vol. 17, 1970, pp. 355—361. [9] Spectral representation for Schrödinger operators with long-range potentials. J. Functional Anal. 20, 158—177 (1975).

IKEBE, T., and Y. SAITO: [1] Limiting absorption method and absolute continuity for the Schrödinger operator. J. Math. Kyoto Univ. 12, 513—542 (1972).

IKEBE, T., and T. TAYOSHI: [1] Wave and scattering operators for second-order elliptic operators in R^3. Publ. Res. Inst. Math. Sci. 4, 483—496 (1968).

IORIO, R. J., and M. O'CARROL: [1] Asymptotic completeness for multi-particle Schroedinger Hamiltonians with weak potentials. Comm. Math. Phys. 27, 137—145 (1972).

JÄGER, W.: [1] Ein gewöhnlicher Differentialoperator zweiter Ordnung für Funktionen mit Werten in einem Hilbertraum. Math. Z. **113**, 68—98 (1970).

JAUCH, J. M., B. MISRA, and A. G. GIBSON: [1] On the asymptotic condition of scattering theory. Helv. Phys. Acta **41**, 513—527 (1968).

JÖRGENS, K.: [2] Zur Spektraltheorie der Schrödinger-Operatoren. Math. Z. **96**, 355—372 (1967). [3] Perturbation of the Dirac operator. Conference on the theory of ordinary and partial differential equations, Dundee 1972. Lecture Notes in Math. **280**, Springer, pp. 87—102.

JÖRGENS, K., and J. WEIDMANN: [1] Zur Existenz der Wellenoperatoren. Math. Z. **131**, 141—151 (1973).

KAASHOEK, M. A.: [2] Closed linear operators on Banach spaces. Indag. Math. **27**, 405—414 (1965). [3] Stability theorems for closed linear operators. Indag. Math. **27**, 452—466 (1965). [4] On the Riesz set of a linear operator. Indag. Math. **30**, 46—53 (1968). [5] Ascent, descent, nullity and defect, a note on a paper by A. E. TAYLOR. Math. Ann. **172**, 105—115 (1967).

KAASHOEK, M. A., and D. C. LAY: [1] On operators whose Fredholm set is the complex plane. Pacific J. Math. **21**, 275—278 (1967).

KAKO, T.: [1] Scattering theory for abstract differential equations of second order. J. Fac. Sci. Univ. Tokyo Sec. I A, Vol. 19, 1972, pp. 377—392.

KALF, H.: [1] On the characterization of the Friedrichs extension of ordinary or elliptic differential operators with a strongly singular potential. J. Functional Anal. **10**, 230—250 (1972).

KALF, H., and J. WALTER: [1] Strongly singular potentials and essential selfadjointness of singular elliptic operators in $C_0^\infty(R^n \setminus \{0\})$. J. Functional Anal. **10**, 114—130 (1972).

KALLMAN, R. R., and G-C. ROTA: [1] On the inequality $\|f'\|^2 \leqq 4\|f\|\,\|f''\|$. Inequalities, Vol. 2, Academic Press, New York and London 1970, pp. 187—192.

KATO, T.: [19] Scattering theory with two Hilbert spaces. J. Functional Anal. **1**, 342—369 (1967). [20] Similarity for sequences of projections. Bull. Amer. Math. Soc. **73**, 904—905 (1967). [21] Some mathematical problems in quantum mechanics. Progr. Theoret. Phys. Suppl. **40**, 3—19 (1967). [22] Some results on potential scattering. Proc. Intern. Conference on Functional Anal. and Related Topics, Tokyo 1969, pp. 206—215. [23] Scattering theory and perturbation of continuous spectra. Actes, Congrès intern. math. 1970, Tome 1, pp. 135—140. [24] On an inequality of Hardy, Littlewood, and Pólya. Advances in Math. **7**, 217—218 (1971). [25] Schrödinger operators with singular potentials. Israel J. Math. **13**, 135—148 (1972). [26] A remark to the preceding paper by CHERNOFF. J. Functional Anal. **12**, 415—417 (1973). [27] Scattering theory, Studies in Mathematics. Math. Assoc. Amer., Vol. 7, 1971, pp. 90—115. [28] A second look at the essential selfadjointness of the Schrödinger operators, Physical Reality and Mathematical Description. D. Reidel Publ. Co., Dordrecht 1974, pp. 193—201. [29] On the Trotter-Lie product formula. Proc. Japan Acad. **50**, 694—698 (1974).

KATO, T., and S. T. KURODA: [2] Theory of simple scattering and eigenfunction expansions. Functional Anal. and Related Fields. Springer 1970, pp. 99—131. [3] The abstract theory of scattering. Rocky Mountain J. Math. **1**, 127—171 (1971).

KOMATSU, H.: [1] Fractional powers of operators. Pacific J. Math. **19**, 285—346 (1966). [2] II. Interpolation spaces. Pacific J. Math. **21**, 89—111 (1967). [3] III. Negative powers. J. Math. Soc. Japan **21**, 205—220 (1969). [4] IV. Potential operators. J. Math. Soc. Japan **21**, 221—228 (1969). [5] V. Dual operators. J. Fac. Sci.

Univ. Tokyo, Sec. I, Vol. 17, 1970, pp. 373—396. [6] VI. Interpolation of non-negative operators and imbedding theorems. J. Fac. Sci. Univ. Tokyo, Sec. I, Vol. 19, 1972, pp. 1—63.

Konno, R., and S. T. Kuroda: [1] On the finiteness of perturbed eigenvalues. J. Fac. Sci. Univ. Tokyo, Sec. I, Vol. 13, 1966, pp. 55—63.

Konrady, J.: [1] Almost positive perturbations of positive selfadjoint operators. Comm. Math. Phys. 22, 295—300 (1971).

Krein, M. G.: [7] Analytic problems and results of the theory of linear operators in Hilbert space. Proc. Intern. Congr. Math., Moscow, 1966, pp. 189—216.

Kupsch, J., and W. Sandhas: [1] Møller operators for scattering on singular potentials. Comm. Math. Phys. 2, 147—154 (1966).

Kuroda, S. T.: [9] Stationary methods in the theory of scattering, (see Wilcox [1], pp. 185—214). [10] An abstract stationary approach to perturbation of continuous spectra and scattering theory. J. d'Analyse Math. 20, 57—117 (1967). [11] Perturbation of eigenfunction expansions. Proc. Nat. Acad. Sci. 57, 1213—1217 (1967). [12] Construction of eigenfunction expansions by the perturbation method and its application to n-dimensional Schrödinger operators. MRC Technical Report No. 744, Univ. Wisconsin, 1967. [13] A stationary method of scattering and some applications. Proc. Intern. Conference on Functional Anal. and Related Topics, Tokyo 1969, pp. 231—239. [14] Some remarks on scattering for Schrödinger operators. J. Fac. Sci. Univ. Tokyo, Sec. I, Vol. 17, 1970, pp. 315—329. [15] Spectral representations and the scattering theory for Schrödinger operators. Actes, Congrès intern. Math. 1970, Tome 2, pp. 441—445. [16] Scattering theory for differential operators, I. Operator theory. J. Math. Soc. Japan 25, 75—104 (1973). [17] II. Self-adjoint elliptic operators. J. Math. Soc. Japan 25, 222—234 (1973).

Labrousse, J. P. [1] On a metric of closed operators on a Hilbert space. Rev. Mat. Fis. Teor. (Tucuman) 16, 45—77 (1966).

Langer, H.: [2] Eine Erweiterung der Spurformel der Störungstheorie. Math. Nachr. 30, 123—135 (1965).

Lavine, R. B.: [1] Absolute continuity of Hamiltonian operators with repulsive potential. Proc. Amer. Math. Soc. 22, 55—60 (1969). [2] Scattering theory for long range potentials. J. Functional Anal. 5, 368—382 (1970). [3] Commutators and scattering theory: I. repulsive interactions. Comm. Math. Phys. 20, 301—323 (1971). [4] II. A class of one body problems. Indiana Univ. Math. J. 21, 643—656 (1972). [5] Absolute continuity of positive spectrum for Schrödinger operators with long-range potentials. J. Functional Anal. 12, 30—54 (1973).

Lax, P. D., and R. S. Phillips: [2] Scattering theory. Rocky Mountain J. Math. 1, 173—223 (1971). [3] Scattering theory for dissipative hyperbolic systems. J. Functional Anal. 14, 172—235 (1973).

Lin, S-C.: [1] Wave operators and similarity for generators of semigroups in Banach spaces. Trans. Amer. Math. Soc. 139, 469—494 (1969). [2] On smoothness of gentle perturbations. Bull. Amer. Math. Soc. 75, 445—449 (1969).

Ljance, V. E.: [1] Nonselfadjoint one-dimensional perturbation of the operator of multiplication by the independent variable. Dokl. Akad. Nauk SSSR 182, 1010—1013 (1968). [2] On the perturbation of a continuous spectrum. Dokl. Akad. Nauk SSSR 187, 514—517 (1969). [3] Completely regular perturbation of a continuous spectrum. Mat. Sb. 82 (124), 126—156 (1970). [4] II. Mat. Sb. 84 (126), 141—158 (1971).

Lundqvist, E.: [1] On the existence of the scattering operator. Ark. Mat. 7, 145—157 (1967).

MARKUS, A. S.: [1] On holomorphic operator-functions. Dokl. Akad. Nauk SSSR **119**, 1099—1102 (1958).

MATVEEV, V. B.: [1] Invariance principle for generalized wave operators. Teoret. Mat. Fiz. **8**, 49—54 (1971).

MATVEEV, V. B., and M. M. SKRIGANOV: [1] Wave operators for the Schrödinger equation with rapidly oscillating potential. Dokl. Akad. Nauk SSSR **202**, 755—758 (1972). [2] Scattering problem for the radial Schrödinger equation with slowly decreasing potential. Teoret. Mat. Fiz. **10**, 238—248 (1972).

McINTOSH, A.: [1] Representation of bilinear forms in Hilbert space by linear operators. Trans. Amer. Math. Soc. **131**, 365—377 (1968). [2] Bilinear forms in Hilbert space. J. Math. Mech. **19**, 1027—1045 (1970). [3] On the comparability of $A^{1/2}$ and $A^{*1/2}$. Proc. Amer. Math. Soc. **32**, 430—434 (1972).

MOCHIZUKI, K.: [1] Eigenfunction expansions associated with the Schrödinger operator with a complex potential and the scattering inverse problem. Proc. Japan Acad. **43**, 638—643 (1967). [2] On the large perturbation by a class of non-selfadjoint operators. J. Math. Soc. Japan **19**, 123—158 (1967). [3] Eigenfunction expansions associated with the Schrödinger operator with a complex potential and the scattering theory. Publ. Res. Inst. Math. Sci. **4**, 419—466 (1968). [4] Spectral and scattering theory for symmetric hyperbolic systems in an exterior domain. Publ. Res. Inst. Math. Sci. **5**, 219—258 (1969).

NEUBAUER, G.: [3] Homotopy properties of semi-Fredholm operators in Banach spaces. Math. Ann. **176**, 273—301 (1968).

NIRENBERG, L.: [1] On elliptic partial differential equations. Ann. Scuola Norm. Sup. Pisa **13**, 115—162 (1959).

NOMIZU, K.: [1] Characteristic roots and vectors of a differentiable family of symmetric matrices. Linear and Multilinear Algebra **1**, 159—162 (1973).

OKAZAWA, N.: [1] Two perturbation theorems for contraction semigroups in a Hilbert space. Proc. Japan Acad. **45**, 850—853 (1969.) [2] A perturbation theorem for linear contraction semigroups on reflexive Banach spaces. Proc. Japan Acad. **47**, 947—949 (1971). [3] Perturbation of linear m-accretive operators. Proc. Amer. Math. Soc. **37**, 169—174 (1973). [4] Remarks on linear m-accretive operators in a Hilbert space. J. Math. Soc. Japan **27**, 160—165 (1975).

OLIVER, R. K.: [1] Note on a duality relation of Kaashoek. Indag. Math. **28**, 364—368 (1966).

PEARSON, D. B.: [1] Time-dependent scattering theory for highly singular potentials, Helv. Phys. Acta **47**, 249—264 (1974). [2] An example in potential scattering illustrating the breakdown of asymptotic completeness. Comm. Math. Phys. **40**, 125—146 (1975).

POVZNER, A. YA.: [2] On expansions in functions which are solutions of a scattering problem. Dokl. Akad. Nauk SSSR **104**, 360—363 (1955).

REJTO, P. A.: [3] On the essential spectrum of the hydrogen energy and related operators. Pacific J. Math. **19**, 109—140 (1966). [4] On gentle perturbations. (See WILCOX **[1]**, pp. 57—95). [5] On partly gentle perturbations, I. J. Math. Anal. Appl. **17**, 453—462 (1967). [6] II. J. Math. Anal. Appl. **20**, 145—187 (1967). [7] III. J. Math. Anal. Appl. **27**, 21—67 (1969). [8] Some essentially self-adjoint one-electron Dirac operators. Israel J. Math. **9**, 144—171 (1971). [9] Some potential perturbations of the Laplacian. Helv. Phys. Acta **44**, 708—736 (1971). [10] On a limiting case of a theorem of Kato and Kuroda. J. Math. Anal. Appl. **39**, 541—557 (1972).

RIBARIČ, M., and I. VIDAV: [1] Analytic properties of the inverse $A^{-1}(z)$ of an analytic linear operator valued function $A(z)$. Arch. Rational Mech. Anal. **32**, 298—310 (1969).

RIDDELL, R. C.: [2] Spectral concentration for self-adjoint operators. Pacific J. Math. 23, 377—401 (1967).

ROHDE, H-W.: [2] Ein Kriterium für das Fehlen von Eigenwerten elliptischer Differentialoperatoren. Math. Z. 112, 375—388 (1969).

RUELLE, D.: [1] A remark on bound states in potential-scattering theory. Nuovo Cimento A 61, 655—662 (1969).

SAHNOVIC, L. A.: [1] Nonunitary operators with absolutely continuous spectrum on the unit circle. Dokl. Akad. Nauk SSSR 181, 558—561 (1968). [2] Dissipative operators with absolutely continuous spectrum. Trudy Moscow. Mat. Obšč. 19, 211—270 (1968). [3] Generalized wave operators and regularization of the series of perturbation theory. Teoret. Mat. Fiz. 2, 80—86 (1970). [4] Generalized wave operators. Mat. Sb. 81, 209—227 (1970). [5] The invariance principle for generalized wave operators. Funkcional. Anal. i Priložen. 5, 61—68 (1971).

SAITO, Y.: [1] Spectral and scattering theory for second-order differential operators with operator-valued coefficients. Osaka J. Math. 9, 463—498 (1972).

SCHECHTER, M.: [4] Essential spectra of elliptic partial differential equations. Bull. Amer. Math. Soc. 73, 567—572 (1967). [5] Hamiltonians for singular potentials. Indiana Univ. Math. J. 22, 483—503 (1972). [6] Scattering theory for elliptic operators of arbitrary order. Comment. Math. Helv. 49, 84—113 (1974). [7] A unified approach to scattering. J. Math. Pures Appl. 53, 373—396 (1974).

SCHMIDT, G.: [1] Scattering theory for Maxwell's equations in an exterior domain. Arch. Rational Mech. Anal. 28, 284—322 (1967/1968).

SCHMINCKE, U-W.: [1] Essential selfadjointness of a Schrödinger operator with strongly singular potential. Math. Z. 124, 47—50 (1972). [2] Essential selfadjointness of Dirac operators with a strongly singular potential. Math. Z. 216, 71—81 (1972).

SCHULENBERGER, J. R., and C. H. WILCOX: [1] Completeness of the wave operators for perturbations of uniformly propagative systems. J. Functional Anal. 7, 447—474 (1971). [2] Eigenfunction expansions and scattering theory for wave propagation problems of classical physics. Arch. Rational Mech. Anal. 46, 280—320 (1972).

SHENK, N.: [1] Eigenfunction expansions and scattering theory for the wave equation in an exterior region. Arch. Rational Mech. Anal. 21, 120—150 (1966). [2] The invariance of wave operators associated with perturbations of $-\Delta$. J. Math. Mech. 17, 1005—1022 (1968).

SHENK, N., and D. THOE: [1] Eigenfunction expansions and scattering theory for perturbations of $-\Delta$. Rocky Mountain J. Math. 1, 89—125 (1971). [2] Eigenfunction expansions and scattering theory for perturbations of $-\Delta$. J. Math. Anal. Appl. 36, 313—351 (1971).

SIGAL, I. M.: [1] Reduction to diagonal form of the Schrödinger operator in Fok space. Teor. Mat. Fiz. 10, 249—258 (1972).

SIMADER, C. G.: [1] Bemerkungen über Schrödinger-Operatoren mit stark singulären Potentialen. Math. Z. 138, 53—70 (1974).

SIMON, B.: [1] Coupling constant analyticity for the anharmonic oscillator. Ann. Phys. 58, 76—136 (1970). [2] On the infinitude or finiteness of the number of bound states of an N-body quantum system, I. Helv. Phys. Acta 43, 607—630 (1970). [3] Hamiltonians defined as quadratic forms. Comm. Math. Phys. 21, 192—210 (1971). [4] Determination of eigenvalues by divergent perturbation series. Advances in Math. 7, 240—253 (1971). [5] Quadratic form techniques and the Balslev-Combes theorem. Comm. Math. Phys. 27, 1—9 (1972). [6] Essential self-adjointness of Schrödinger operators with positive potentials. Math. Ann. 201, 211—220 (1973). [7] Resonances in n-body quantum systems with dilatation analytic potentials and the foundations of time-dependent perturbation theory. Ann. Math. 97, 247—274 (1973).

STANKEVIC, I. V.: [2] Asymptotic behavior for $t \to \infty$ of the solution of Schrödinger's nonstationary equation with a nonselfadjoint Hamiltonian. Dokl. Akad. Nauk SSSR **160**, 1271—1274 (1965).

STEINBERG, S.: [1] Meromorphic families of compact operators. Arch. Rational Mech. Anal. **31**, 372—379 (1968/1969).

STENGER, W.: [1] On perturbations of finite rank. J. Math. Anal. Appl. **28**, 625—635 (1969).

STETKAER-HANSEN, H.: [1] A generalization of a theorem of Wienholtz concerning essential selfadjointness of singular elliptic operators. Math. Scand. **19**, 108—112 (1966).

STRANG, G.: [1] Approximating semigroups and consistency of difference schemes. Proc. Amer. Math. Soc. **20**, 1—7 (1969).

STUMMEL, F.: [2] Diskrete Konvergenz linearer Operatoren. I. Math. Ann. **190**, 45—92 (1970/1971). [3] II. Math. Z. **120**, 231—264 (1971). [4] III. Proc. Oberwolfach Conference on Linear Operators and Approx. 1971, Vol. 20, pp. 196—216. [5] Singular perturbations of elliptic sesquilinear forms. Conference on the Theory of Ordinary and Partial Differential Equztions. Lecture Notes in Math. 280. Springer 1972, pp. 155—180.

SUZUKI, T.: [1] The limiting absorption principle and spectral theory for a certain non-selfadjoint operator and its application. J. Fac. Sci. Univ. Tokyo, Sec. I A, **20**, 401—412 (1973).

TAYLOR, A.: [1] Theorems on ascent, descent, nullity and defect of linear operators. Math. Ann. **163**, 18—49 (1966).

THOE, D.: [1] Spectral theory for the wave equation with a potential term. Arch. Rational Mech. Anal. **22**, 364—406 (1966).

THOMAS, L. E.: [1] On the spectral properties of some one-particle Schrödinger Hamiltonians. Helv. Phys. Acta **45**, 1057—1065 (1972). [2] Time dependent approach to scattering from impurities in a crystal. Comm. Math. Phys. **33**, 335—343 (1973). [3] On the algebraic theory of scattering. J. Functional Anal. **15**, 364—377 (1974). [4] Asymptotic completeness in two- and three-particle quantum mechanical scattering. Ann. Phys. **90**, 127—165 (1975).

THOMPSON, M.: [1] Eigenfunction expansions and the associated scattering theory for potential perturbations of the Dirac equation. Quart. J. Math. Oxford, Ser. II **23**, 17—55 (1972).

TURNER, R. E. L.: [1] Perturbation of compact spectral operators. Comm. Pure Appl. Math. **18**, 519—541 (1965). [2] Perturbation of ordinary differential operators. J. Math. Anal. Appl. **13**, 447—457 (1966). [3] Eigenfunction expansions in Banach spaces. Quart. J. Math. Oxford **19**, 193—211 (1968).

UCHIYAMA, J.: [1] On the discrete eigenvalues of the many-particle system. Publ. Res. Inst. Math. Sci. **2**, 117—132 (1966). [2] On the spectra of integral operators connected with Boltzmann and Schrödinger operators. Publ. Res. Inst. Math. Sci. **3**, 101—107 (1967). [3] Finiteness of the number of discrete eigenvalues of the Schrödinger operator for a three particle system. Publ. Res. Inst. Math. Sci. **5**, 51—63 (1969).

USHIJIMA, T.: [1] Spectral theory of the perturbed homogeneous elliptic operator with real constant coefficients. Sci. Papers College Gen. Ed. Univ. Tokyo **16**, 27—42 (1966). [2] Note on the spectrum of some Schrödinger operators. Publ. Res. Inst. Math. Sci. **4**, 497—509 (1968).

VESELIĆ, K.: [1] On spectral concentration for some classes of selfadjoint operators. Glasnik Mat. **4** (24), 213—229 (1969). [2] Perturbation of pseudoresolvents and analyticity in 1/c in relativistic quantum mechanics. Comm. Math. Phys. **22**, 27—43 (1971).

VESELIĆ, K., and J. WEIDMANN: [1] Existenz der Wellenoperatoren für eine allgemeine Klasse von Operatoren. Math. Z. **134**, 255—274 (1973). [2] Asymptotic estimates of wave functions and the existence of wave operators. J. Functional Anal. **17**, 61—77 (1974).

WALTER, J.: [1] Symmetrie elliptischer Differentialoperatoren. Math. Z. **98**, 401—406 (1967). [2] II. Math. Z. **106**, 149—152 (1968). [3] Note on a paper by Stetkaer-Hansen concerning essential selfadjointness of Schrödinger operators. Math. Skand. **25**, 94—96 (1969).

WEIDMANN, J.: [1] On the continuous spectrum of Schrödinger operators. Comm. Pure Appl. Math. **19**, 107—110 (1966). [2] The virial theorem and its application to the spectral theory of Schrödinger operators. Bull. Amer. Math. Soc. **73**, 452—456 (1967). [3] Spectral theory of partial differential operators. Lecture Notes in Math. **448**. Springer 1975, pp. 71—111.

WEINSTEIN, A.: [4] La teoria di massimo-minimo degli: autovalori ed il metodo dei problemi intermidi. Seminari Ist. Naz. di Alta Mat. 1962/1963, pp. 596—609. [5] Some applications of the new maximum-minimum theory of eigenvalues. J. Math. Anal. Appl. **12**, 58—64 (1965).

WIENHOLTZ, E.: [2] Bemerkungen über elliptische Differentialoperatoren. Arch. Math. **10**, 126—133 (1959).

WILCOX, C. H.: [1] Uniform asymptotic estimates for wave packets in the quantum theory of scattering. J. Mathematical Phys. **6**, 611—620 (1965). [2] Wave operators and asymptotic solutions of wave propagation problems of classical physics. Arch. Rational Mech. Anal. **22**, 37—78 (1966). [3] Scattering states and wave operators in the abstract theory of scattering. J. Functional Anal. **12**, 257—274 (1973).

WINTER, C. VAN: [1] Theory of finite systems of particles, I. The Green function. Mat. Fys. Skr. Dan. Vid. Selsk. **2**, No. 8, 1—60 (1964). [2] II. Scattering theory. Mat. Fys. Skr. Dan. Vid. Selsk. **2**, No. 10, 1—94 (1965).

WOLF, J. A.: [1] Essential self-adjointness for the Dirac operator and its square. Indiana Univ. Math. J. **22**, 611—640 (1973).

WOLLENBERG, M.: [1] The invariance principle for wave operators, Pacific J. Math. **59**, 303—315 (1975).

WÜST, R.: [1] Generalizations of Rellich's theorem on perturbation of (essentially) selfadjoint operators. Math. Z. **119**, 276—280 (1971).

YAJIMA, K.: [1] The limiting absorption principle for uniformly propagative systems. J. Fac. Sci. Univ. Tokyo, Sec. I A, **21**, 119—131 (1974).

YAKUBOVSKI, O. A.: [1] On the integral equations in the theory of N particle scattering. Sov. J. Nucl. Phys. **5**, 937—942 (1967).

YAMADA, O.: [1] On the principle of limiting absorption for the Dirac operator. Pub. Res. Inst. Math. Sci. **8**, 557—577 (1972/1973).

YOSHIKAWA, A.: [1] Fractional powers of operators, interpolation theory and imbedding theorems. J. Fac. Sci. Univ. Tokyo, Sec. I A, **18**, 335—362 (1971). [2] Note on singular perturbation of linear operators. Proc. Japan Acad. **48**, 595—598 (1972). [3] On perturbation of closed operators in a Banach space. J. Fac. Sci. Hokkaido Univ. Ser. I, **22**, 50—61 (1972).

YOSIDA, K.: [1] A perturbation theorem for semi-groups of linear operators. Proc. Japan Acad. **41**, 645—647 (1965).

ZISLIN, G. M.: [2] An investigation of the spectrum of differential operators of many particle quantum mechanical systems in function spaces of given symmetry. Izv. Akad. Nauk SSSR **33**, 590—649 (1969). [3] On the finiteness of discrete spectrum of energy operator of the negative atomic and molecular ions. Theoret. Mat. Fiz. **7**, 332—341 (1971).

Notation index

Author index

Subject index

Springer-Verlag and the Environment

We at Springer-Verlag firmly believe that an international science publisher has a special obligation to the environment, and our corporate policies consistently reflect this conviction.

We also expect our business partners – paper mills, printers, packaging manufacturers, etc. – to commit themselves to using environmentally friendly materials and production processes.

The paper in this book is made from low- or no-chlorine pulp and is acid free, in conformance with international standards for paper permanency.